Lecture Notes in Computer Science 2510
Edited by G. Goos, J. Hartmanis, and J. van Leeuwen

Springer
Berlin
Heidelberg
New York
Barcelona
Hong Kong
London
Milan
Paris
Tokyo

Hassan Shafazand A Min Tjoa (Eds.)

EurAsia-ICT 2002: Information and Communication Technology

First EurAsian Conference
Shiraz, Iran, October 29-31, 2002
Proceedings

Volume Editors

Hassan Shafazand
Shahid Bahonar University, Faculty of Mathematics and Computer Science
Computer Science Department, 22 Bahman Bulvard, Kerman, Iran
E-mail: shafazand@mail.uk.ac.ir
Fraunhofer IPSI, Dolivostr. 15, 64293 Darmstadt, Germany
shafazand@ipsi.fhg.de

A Min Tjoa
Vienna University of Technology, Institute of Software Technology
Favoritenstr. 9/188, 1040 Vienna, Austria
E-mail: tjoa@ifs.tuwien.ac.at

Cataloging-in-Publication Data applied for

Bibliograhpic information published by Die Deutsche Bibliothek
Die Deutsche Bibliothek lists this publication in the Deutsche Nationalbibliografie; detailed bibliographic data is available in the Internet at http://dnb.ddb.de

CR Subject Classification (1998): H, C.2, D, F, I

ISSN 0302-9743
ISBN 3-540-00028-3 Springer-Verlag Berlin Heidelberg New York

Springer-Verlag Berlin Heidelberg New York
a member of BertelsmannSpringer Science+Business Media GmbH

http://www.springer.de

Printed in Germany

Typesetting: Camera-ready by author, data conversion by PTP Berlin, Stefan Sossna e.K.
Printed on acid-free paper SPIN: 10870902 06/3142 5 4 3 2 1 0

Preface

We welcomed participants to the 1st EurAsian Conference on Advances in Information and Communication Technology (EurAsia-ICT 2002) held in Iran. The aim of the conference was to serve as a forum to bring together researchers from academia and commercial developers from industry to discuss the current state of the art in ICT, mainly in Europe and Asia. Inspirations and new ideas were expected to emerge from intensive discussions during formal sessions and social events.

Keynote addresses, research presentation, and discussion during the conference helped to further develop the exchange of ideas among the researchers, developers, and practitioners who attended.

The conference attracted more than 300 submissions and each paper was reviewed by at least three program committee members. The program committee selected 119 papers from authors of 30 different countries for presentation and publication, a task which was not easy due to the high quality of the submitted papers. Eleven workshops were organized in parallel with the EurAsia-ICT conference. The proceedings of these workshops, with more than 100 papers, were published by the Austrian Computer Society.

We would like to express our thanks to our colleagues who helped with putting together the technical program: the program committee members and external reviewers for their timely and rigorous reviews of the papers, and the organizing committee for their help in administrative work and support. We owe special thanks to Thomas Schierer for always being available when his helping hand was needed.

Finally, we would like to thank all the authors who submitted papers, authors who presented papers, and the participants who together made this conference an intellectually stimulating event through their active contributions.

We enjoyed a successful conference and hope that the participants enjoyed the hospitality of Iran.

October 2002

M. Hassan Shafazand
(Shahid Bahonar University of Kerman, Iran)
A Min Tjoa (Vienna University of Technology, Austria)

Program Committee

Honorary Chairpersons

W. Grafendorfer, President of the International Federation of Information Processing, Austria

R. Kneucker, Director – General of the Division of Scientific Research and International Affairs, Austria

M. Shafiee, Deputy Minister of Post, Telegraph and Telephone, Iran

Conference Organizing Committee

E. Neuhold, Fraunhofer-IPSI and Technical University of Darmstadt, Germany

H. Rabiee, AICTC and Sharif University, Iran

Program Chairpersons

M.H. Shafazand, Shahid Bahonar University of Kerman, Iran

A M. Tjoa, Technical University of Vienna, Austria

K. Badie, ITRC, Iran

Program Committee Members

K. Aberer, EPFL-DSC, Switzerland

H. Afsarmanesh, University of Amsterdam, The Netherlands

A. Al-Zobaidie, University of Greenwich, UK

M. Albadavi, Tarbiat Modares University, Iran

M. Arikawa, University of Tokyo, Japan

F. Badran, Conservatoire National des Arts et Metiers, France

T.P. Bagchi, IIT Kanpur, India

J.-P. Bahsoun, Université Paul Sabatier-IRIT, France

K. Barkaoui, CNAM, Paris, France

M. Beikzadeh, ITRC, Iran

B.K. Bhargava, Purdue University, USA

P. Bollerslev, Gyldendal Education, Denmark

A. Bondavalli, University of Florence, Italy

E. Borcoci, University Politehnica Bucharest, Romania

M. Bouzeghoub, University of Versailles, France

C.J. Breiteneder, Vienna University of Technology, Austria

S. Bressan, National University of Singapore, Singapore

B. Buchberger, Johannes Kepler University, Austria

C. Cap, University of Rostock, Germany

W. Cellary, Poznan University of Economics, Poland

N. Chakraborti, Indian Institute of Technology (IIT), India

S. Chaudhury, IIT, India

W.-S.E. Chen, National Chung-Hsing University Taiwan, Taiwan

E. Cloete, University of South Africa, South Africa

P. Cunningham, Trinity College Dublin, Ireland

E. Damiani, Università di Milano, Italy
M. Dastani, Utrecht University, The Netherlands
U. Dayal, Hewlett-Packard Laboratories, USA
V. De Antonellis, University of Brescia, Italy
J. Debenham, University of Technology, Sydney, Australia
J. Diaz, Universitat Politecnica Catalunya, Spain
V. Diekert, University of Stuttgart, Germany
D. Dietrich, Vienna Technical University, Austria
F. Dignum, Utrecht University, The Netherlands
V. Dignum, Achmea and Utrecht University, The Netherlands
A. Djunaidy, Sepuluh Nopember Institute of Technology, Indonesia
M. Dorigo, Université Libre de Bruxelles, Belgium
A. Düsterhöft, University of Wismar, Germany
C. Edwards, Lancaster University, UK
F. Eliassen, Simula Research Laboratory, Norway
A. El-Iraki, National School of Mineral Industry, Morocco
L. Ferreira Pires, University of Twente, The Netherlands
A. Ferko, Comenius University, Bratislava, Slovakia, and TU Graz, Austria
A. Ferscha, University of Linz, Austria
F.G. Filip, Romanian Academy, Romania
E. Flerackers, Limburg University Center, Belgium
M. Gaedke, University of Karlsruhe, Germany
J. Gruska, Masaryk University, Czech Republic
R. Guerraoui, EPFL, Switzerland
R. Hagelauer, Johannes Kepler University, Linz, Austria
A. Hameurlain, IRIT-University Paul Sabatier, Toulouse, France
Z. Hanzalek, Czech Technical University, Czech Republic
S. Harous, University of Sharjah, UAE
S. Hashemi, University of Houston-Downtown, USA
D.L. Hicks, Aalborg University Esbjerg, Denmark
S. Hong, Seoul National University, Korea
C. Huemer, University of Vienna, Austria
I.K. Ibrahim, Utrecht University, The Netherlands
R.E. Indrajit, Perbanas School of Computing, Indonesia
Y.E. Ioannidis, University of Athens, Greece
R. Jalili, Sharif University of Technology, Iran
P. Jalote, I.I.T. Kanpur, India
A. Jaoua, University of Qatar, Qatar
L.A. Kalinichenko, Russian Academy of Sciences, Russia
Y. Kambayashi, Kyoto University, Japan
H. Kawano, Kyoto University, Japan
M. Kitsuregawa, Tokyo University, Japan
W. Klas, University of Vienna, Austria
H. Krawczyk, Technical University of Gdansk, Poland
T.-W. Kuo, National Taiwan University, Taiwan
L. Kutvonen, University of Helsinki, Finland
W. Lamersdorf, University of Hamburg, Germany
J. Lazansky, Czech Technical University, Czech Republic
X. Li, Tsinghua University, China

E.-P. Lim, Nanyang Technological University, Singapore
J. Yi-Bing Lin, National Chiao Tung University, Taiwan
L. Lo Bello, University of Catania, Italy
S.K. Madria, University of Missouri-Rolla, USA
V. Malyshkin, Russian Academy of Sciences, Russia
M. Marchiori, W3C/MIT, University of Venice, Italy
V. Marik, Czech Technical University, Czech Republic
M.J. Matsumoto, University of Tsukuba, Japan
H. Maurer, Graz University of Technology, Austria
J. Misic, Hong Kong University of Science, Hong Kong
B. Mitschang, University of Stuttgart, Germany
R. Moghimi, K.N. Toosi University of Technology, Iran
M. Mohania, IBM India Research Lab, India
F. Moller, University of Wales, Swansea, UK
B. Montazeri, Razi University, Iran
R. Morel, TC3 IFIP, Vice-Chairman, CPTIC, Switzerland
L. Motus, Tallinn Technical University, Estonia
M. Mühlhäuser, Darmstadt Technical University, Germany
F. Naghdy, University of Wollongong, Australia
J. Nawrocki, Poznan University of Technology, Poland
L. Neumann, Technical University of Vienna, Austria
S. Ohsuga, Waseda University, Japan
H. Oinas-Kukkonen, University of Oulu, Finland
J.L. Oliveira, Universidade de Aveiro, Portugal
M. Papazoglou, University of Tilburg, The Netherlands
F.-N. Pavlidou, Aristotle University of Thessaloniki, Greece
G. Pernul, University of Essen, Germany
F. Petit, LaRIA, University of Picardie, France
F. Pichler, Johannes Kepler University of Linz, Austria
G. Pongor, Budapest University of Technology, Hungary
B. Pröll, Johannes Kepler University of Linz, Austria
G. Quirchmayr, University of Vienna, Austria
H. Radha, Michigan State University, USA
H.A. Ramadhan, Sultan Qaboos University, Oman
M. Raynal, IRISA, Université de Rennes, France
S. Reich, Salzburg Research, Austria
W. Reisig, Humboldt University Berlin, Germany
W. Retschitzegger, Johannes Kepler University of Linz, Austria
N. Revell, Middlesex University, UK
L. Rodrigues, University of Lisbon, Portugal
G. Rossi, LIFIA-UNLP, Argentina
M.H. Safar, Kuwait University, Kuwait
D. Saha, Jadavpur University and IIM, India
D. Sanghi, IIT Kanpur, India
E. Schweighofer, University of Vienna, Austria
P. Sénac, ENSICA, France
T.K. Shih, Tamkang University, Taiwan
S.Y. Shin, Korea Advanced Institute of Science and Technology, KAIST, South Korea

M. Sperka, Slovak University of Technology, Slovakia
M. Stal, Siemens AG, Germany
R. Steinmetz, Darmstadt Technical University, Germany
S.Y.W. Su, University of Florida, USA
Y. Suenaga, Nagoya University, Japan
K. Tanaka, Kyoto University, Japan
S. Teufel, University of Fribourg, Switzerland
J. Tiuryn, Warsaw University, Poland
F. Tolba, Ainshams University, Egypt
E. Tovar, University of Porto, Portugal
R. Traunmüller, University of Linz, Austria
J. Veijalainen, University of Jyvaskyla, Finland
R. Weigel, University of Linz, Austria
M. Welzl, University of Innsbruck, Austria
U.K. Wiil, Aalborg University Esbjerg, Denmark
M. Wirsing, University of Munich, Germany
K.Y. Wohn, Korea Advanced Institute of Science and Technology (KAIST), South Korea
V. Wolfengagen, JurInfoR-MSU Institute, Russia
N. Yamanaka, NTT Network Innovation Labs, Japan
J.X. Yu, Chinese University of Hong Kong, Hong Kong, China
A. Zaslavsky, Monash University, Australia
A. Zhou, Fudan University, China

List of External Reviewers

Ana Afonso
Toni Alatalo
Abdullah Al-Mutawa
Osamah Al-Sayegh
Holger Austinat
Tobias Baier
Walter Bartek
Andreas Bartelt
Berta Batista
Hubert Baumeister
Adam Belloum
Ammar Benabdelkader
Lars Braubach
Pavel Burget
Josef Capek
Somchai Chatvichienchai
Silvano Chiaradonna
Kyung-Yong Chwa
Andrea Coccoli
Domenico Cotroneo
Alain Cournier
Zoran Despotovic
João Paulo de Andrade Almeida
S. De Capitani di Vimercati
Remco Dijkman
Felicita Di Giandomenico
Ondrej Dolejs
Fredj Drid
Horst Eidenberger
Mohamed Elaffendi
Ashraf Elnagar
Dietrich Fahrenholtz
Didier Ferment
Luis Lino Ferreira
Cesar Garita
Victor Guevara
Giancarlo Guizzardi
Hyungjin Ha
Rolf Hennicker
Ulrich Hertrampf
Chayoun Hong
Zhisheng Huang
Abdullah Hussein
Mohammed Jaragh
Michal Jasinski
Tudor Jebelean
Ersin Kaletas
Hayat Kara
Mehmet Karaata
Lami Kaya
Zoubida Kedad
Dohan Kim
Taehoon Kim
Piotr Kosiuczenko
Manfred Kufleitner
Temur Kutsia
Inso Kwon
Miroslav Licko
Markus Lohrey
Rui Pedro Lopes
Stephane Lopes
Allooua Maamir
Michael Mackay
Zoubir Mammeri
Jouni Markkula
Philip Meier
M. Melchiori
Stephan Merz
Oliver Meyer
Hugo Miranda
Paulo Monteiro
Jean-Frédéric Myoupo
Miyuki Nakano
Masato Oguchi
Filipe Pacheco
Seppo Pahnila
Chong-Dae Park
Jungkeun Park
Min Je Park
Sangmin Park Holger Petersen
Dimitris Pezaros
Willy Picard
Erhard Plödereder
Alexander Pokahr
Stefano Porcarelli
Iko Pramudiono
Bowo Prasetyo
Torsten Priebe
Dick Quartel
Amar Ramdane-Cherif

Ralf Rantzau
Markus Rosenkranz
Jarogniew Rykowski
Minsoo Ryu
Rachid Samouda
Michael Sampels
Beatriz Sousa Santos
Torsten Schlichting
Holger Schwarz
David Semé
Seunghyup Shin
Mario Silva
Mikko Siponen
Paul Smith
Mikolaj Sobczak
Krzysztof Socha
Sergiusz Strykowski
Katsumi Takahashi
Taro Tezuka
Siegbert Tiga
Masashi Toyoda
Gil Utard
Leon van der Torre
Marten van Sinderen
Vincent Villain
Libor Waszniowski
Nicole Weicker
Harald Weinreich
Marco Wiering
Adam Wojciechowski
Saneyasu Yamaguchi
David Yat Kwan Tang
Haruo Yokota
Christopher Zach
Shuigeng Zhou
Christian Zirpins

Table of Contents

Artificial Intelligence I

Data Mining

Multimedia I

Artificial Intelligence II

Security I

Multimedia II

Neural Network

Security II

Multimedia III

Data and Knowledge Engineering I

XML I

Mobile Communication I

Computer Graphics

Data and Knowledge Engineering II

XML II

Mobile Communication II

Digital Libraries and Natural Language Issues

Mobile Web Information Systems

Wireless Communication Technology I

Web-Based Application

Intelligent Agents I

Real-Time Systems

Wireless Communication Technology II

Software Engineering I

Intelligent Agents II

Algorithm and Computer Theory

Wireless Communication Technology III

Software Engineering II

Intelligent Agents II

Speaker Model and Decision Threshold Updating in Speaker Verification

M. Mehdi Homayounpour

Computer Engineering Department of Amirkabir Univ. of Technology, Tehran, Iran
Institut Dalle Molle d'Intelligence Artificielle Perceptive, Martigny, Switzerland
homayoun@ce.aku.ac.ir

Abstract. This paper deals with the problem of updating reference models and decision thresholds in speaker verification. In the real application of a speaker verification system, reference models of speakers and their decision thresholds should be already determined. Reference model and decision threshold updating can render a speaker verification system more robust against inter-session variability, due to changes in speaker's voice and the diversity in transmission line characteristics. In this paper, different methods for the calculation of decision threshold, the updating of decision threshold and reference model, and BMDR decision criterion are compared and evaluated. In our experiments, Sphericity Measure (a Second Order Statistical Measure) was implemented and used for speaker modeling.

1 Introduction

Speaker verification is a process of automatically verifying the identity of a speaker on a basis of personal information included in speech signal. This technique provides the possibility of access control by voice in various services such as banking transactions, e-commerce, database access services, information services, security control and many other services over telephone or using Internet.

A new client of a speaker verification system should be recorded in a training phase to construct a model for him/her. Some practical limits may not allow the recording of sufficient speech material for an adequate training. When this client starts to use the verification system, more and more speech material is gradually provided. New speech material can be used for a better training of the client's model and probably his/her decision threshold. This procedure called "model updating" guarantees a good verification performance and prevents its degradation due to intra-speaker variation of voices of speakers during time [1]. Speaker Verification is a binary decision. The distance between the speaker's utterance and his/her reference model is compared to a decision threshold which can be computed *a priori* or *a posteriori* and the speaker is accepted or rejected regarding to this comparison. BMDR criterion [2] is another way to decide to accept or reject a speaker. This paper aims to evaluate some methods for updating the speaker's reference model and his/her decision threshold. Second Order Statistical Measures (SOSM) called here as Sphericity measure was used here for speaker modeling. Sphericity measures were used for both speaker Identification [3] and Speaker verification [2]. This technique can be applied for both text-independent and text-dependent speaker verification. It represents a low complexity to calculate a

M.H. Shafazand and A M. Tjoa (Eds.): EurAsia-ICT 2002, LNCS 2510, pp. 1–10, 2002.

model for a client and needs a small memory to memorize the model. The rest of the paper is organized as follows. Speech database used for our experiments is introduced in section 2. In section 3, the Sphericity measure, and in section 4, different methods for decision threshold calculation are described. Experimental results for model and decision threshold updating techniques are given in section 5. Finally section 6 concludes the paper.

2 Speech Database

PolyVar is a French database recorded to provide high speech intra-speaker variability. It comprises 100 speakers uttering each one 100 calls. This database is a good database for text-independent speaker verification experiments. 100 calls from each speaker are not really necessary for text independent speaker verification, in our study. For this reason, a part of PolyVar database was extracted and some preprocessing were done on it to render it more adapted for text-independent speaker verification studies. We call this new database as PolyDat. It includes 56 speakers with 3 calls, 9 speakers with 2 calls, and 61 speakers with 1 call. After extracting these calls from PolyVar, a silence removal process was first done. A reduction of 60% of call duration was achieved. It means that about two third of the calls was the silences that were removed by silence removal process. A hearing of some silence-suppressed calls showed a good removal of silences and a very little suppression of some speech events with small energy such as fricatives. In another processing each call was cut to utterances of 10s duration and the resulted utterances of each speaker were enumerated. For example, for a speaker with 3 calls a total number of about 35 utterances of 10s duration was obtained. Utterances with 10s duration don't prevent verification tests with utterances of smaller durations (through cutting 10s utterances to smaller parts) and/or doing tests with utterances with a longer duration (through the concatenation of two or more 10s utterances). The 56 speakers with 3 calls have a sufficient number of utterances for intra-speaker tests. Other speakers with 1 or two calls may for example be considered as impostors or they may be used as speakers with similar acoustic characteristics to a given client when distance or likelihood normalization [4] or a decision criterion such as BMDR [2] are used. PolyDat includes about 360 KB, 16 bits, 8khz linear form speech data. Utterances from the first 30 speakers with the most number of calls are called the evaluation part of database, and utterances from the other 26 speakers are called the representative part of database.

3 Sphericity Measure

Consider $\{x_t\}$ as a sequence of m dimensional training feature vectors belonging to a client and $\{y_t\}$ as a sequence of test feature vectors. The covariance matrix X of his/her training data represents a reference model for a client $\{x_t\}$:

$$X = \frac{1}{M}\sum_{t=1}^{M} x_t x_t^T \tag{1}$$

If as explained above, X and Y are defined as the covariance matrix of Training and test feature vectors, then the Sphericity measure (distance) between reference and test utterance is defined as follows:

$$\mu_{sphsym}(X,Y) = \rho_{MN}.\log(tr(YX^{-1})) + \rho_{NM}.\log[tr(XY^{-1}))$$

$$-\frac{1}{m}(\rho_{MN} - \rho_{NM}).\log\left[\frac{\det(Y)}{\det(X)}\right] - \log(m) \tag{2}$$

Where: $\rho_{MN} = \frac{M}{N+M}$ $\rho_{NM} = \frac{N}{N+M}$

Here tr(A) means trace of matrix A, and M and N are the number of feature vectors used to obtain X and Y respectively. It can be seen that this measure takes into consideration the difference in total number of parameter vectors used for training and test.

4 Decision Threshold

Here we present three methods for the calculation of decision threshold. These methods are Equal Error Rate method (EER), Linear Regression method (LR) explained in [5], and Best Match Decision Rule (BMDR) method [2].

4.1 Equal Error Rate Method (EER)

EER method for calculation of decision threshold is based on inter- and Intra-speaker verification tests. When these tests are done, False Rejection (FR) and False Acceptance (FA) error rate curves against different distances as decision thresholds are plotted. The intersection point between FR and FA curves is called Equal Error Rate (EER). The distance corresponding to EER can be used as decision threshold and is usually named as EER decision threshold. The use of EER implies a perfect choice for decision threshold, which is not possible in a real application since the threshold would have to be determined *a priori*, while EER decision threshold is inherently *a posteriori*. Although EER is an important performance measure, it is also useful to have a measure of how well a system separates the probability distributions for client speakers and the impostors [6]. In a system-training phase sometimes we may obtain the EER decision thresholds using a different group of speakers, and then use them for a second group of speakers (in test phase). In this case, EER decision thresholds are considered as *a priori* decision thresholds.

4.2 Linear Regression Method (LR)

The second method for calculation of decision method is based on a technique explained in [5] by the following relation:

$$S(k) = C1(\mu - \sigma) + C2 \tag{3}$$

Where μ and σ present the mean and the standard deviation of inter-speaker distances respectively. This relation needs two predetermined constants C1 and C2. To obtain these two constants, we use a database, which is supposed to be representative of future clients of current speaker verification system. The more number of speakers and the amount of speech material yields a better estimation of C1 and C2. C1 and C2 are computed as follows:

For each speaker (client) in the representative database a reference model is constructed. Inter-and Intra-speaker verification tests are done and the decision thresholds corresponding to EER are calculated for all speakers. For each speaker the difference of mean and standard deviation (μ-σ) of inter-speaker distances are also calculated. In this way, each speaker is represented by a point in a plane with (μ-σ) as abscissa and S_{EER} as ordinate. In fact, here, the problem is to fit a straight line to this set of points. This problem is often called linear regression. The slope and the intersection point of this line with ordinate provide the two constants C1 and C2 respectively. To solve this problem the derivatives of a chi-square merit function of C1 and C2 is vanished. Once these two constants are calculated, they are used for any new client (using the relation 3). Usually for many applications of speaker verification systems, a very limited amount of speech material is available from a new client, since it is preferable that the new clients not to be bothered by a long training recording. But it is always possible to possess recordings of a high number of speakers. The above-mentioned technique is attractive for the cases where, as explained above, a limited amount of speech material is available from a new client. To obtain a decision threshold for this new client, inter-speaker verification tests are done using the available speech data of this client and the speech material of a number of other clients or some of speakers in representative database. The mean and the standard deviation of inter-speaker distances and the two constants C1 and C2 are used in relation 3 to obtain the decision threshold for this client.

4.3 Best Match Decision Rule (BMDR)

Another decision criterion is the BMDR criterion [2]. To apply this criterion, we need a general model or population model. This model can be obtained using speech data from a sufficient number of speakers. In each verification test, test utterance is compared once with the client's model, providing distance Di, and once with general model, providing distance Dg. The speaker is accepted if Di< Dg, otherwise he/she is rejected.

5 Experiments

In our experiments, 30 seconds of speech material (corresponding to the first 3 utterances of each speaker of PolyDat) was used to train a reference model for each speaker. 30 speakers were considered as clients. Speech data from the other 26 speakers was used for calculation of LR decision threshold and general model for BMDR decision criterion.

Reference models of speakers were trained using LPCC coefficients. To obtain these coefficients, speech samples were pre emphasized and subsequently multiplied by a Hamming window function. Each analysis frame spanned 30ms and was shifted by 20ms. A vector of 12 LPCCs was retained from each analysis frame and was then normalized by Cepstral Mean Subtraction to compensate telephone channel effects. Several experiments were conducted. These experiments are summarized in table 1. More than 1000 inter and intra-speaker tests were conducted in each experiment. In our experiments, verification performance is presented by verification error rates including FR, FA, and ERR (mean of FR and FA).

Table 1. Results of speaker verification experiments. Training Data Duration=30ms and Test Data Duration=10ms.

EXP NO	Decision Criterion	Threshold Updating	Speaker Model Updating	FR	FA	ERR
1	*A Priori* (LR)	No	No	16.2%	3.3%	9.7%
2	*A Priori* (LR)	No	Yes	7.4%	5.2%	6.3%
3	*A Priori* (LR)	Yes	Yes	8.9%	3.9%	6.4%
4	*A posteriori* (EER)	No	No	–	–	1.0%
5	*A Priori* (EER)	No	No	41.9%	1.0 %	22.5 %
6	*A Priori* (EER)	No	Yes	24.0 %	0.5 %	12.5 %
7	BMDR	–	No	6.4 %	32.6%	19.5%
8	BMDR	–	Yes	1.2%	65.5%	33.2%

In experiment 1, as can be seen in the above table, speaker verification tests are done on 30 speakers of evaluation part of database using the *a priori* LR decision thresholds (relation 3). LR decision thresholds are calculated using the representative part of database. To obtain a LR decision threshold for each client, 10 intra-speaker and 25 inter-speaker tests per speaker are done. In this experiment no speaker model or decision threshold updating is done. A speaker verification error rate of ERR=9.7% is obtained. In experiment 2, when a speaker is accepted during a verification test, his/her model is updated using the test utterance. Table 1 shows a reduction in ERR from 9.7% to 6.3% due to updating the speaker's model. In experiment 3, both reference model and decision threshold updating are done. Decision updating is done after each two updating of a client's model. To do so, inter-speaker tests are done by client's updated model and one utterance from each of other clients in the database. These inter-speaker distances, and C1 and C2 constants are again used in relation 3 to calculate the client's new decision threshold. A comparison of the resulted FR and FA error rates in this experiment and the previous one shows that, updating of the decision thresholds increases FR and reduces the FA. The total error rate of ERR has not changed significantly.

In experiments 4 and 5 EER threshold is used as decision threshold. Here we consider the case where no sufficient speech data is available from each client and we obtain the EER thresholds using the intra- and inter-speaker verification tests using a limited quantity of training data. In experiment 4, we supposed that each client has been recorded in training phase for about 80s. 30s of this data was used for model training and 50s for 5 intra-speaker tests. As can be seen here a small ERR is obtained. It may be explained by two reasons:

- Speech data is obtained by recording the client in only one session (only one call), so there is no inter session variability, and this yields a reduction in FR error rate.
- As said before, EER implies a perfect choice for decision threshold. It provides an upper bound on performance. It is useful to have a measure of how well a system separates the probability distributions for client speakers and the impostors.

In experiment 5, intra- and inter-speaker tests are done on another part of main database that has been used neither for reference model nor for EER threshold calculation. This part of database has been recorded during multiple sessions. The EER threshold obtained previously in experiment 4, are used here as decision thresholds. So these decision thresholds can be considered as a kind of *a priori* decision threshold. Table 1 shows a high increase in the FR error rate for this experiment. This increase is due to inter-session variability, as it has not been taken into consideration for calculation of EER decision thresholds and speaker models. Intra-speaker comparisons between the client model and the test utterances recorded in other sessions result in the distance values, which are higher than EER decision threshold, and this results a high FR error rate.

Experiment 6 was the same as experiment 5, but client's model was updated after each acceptance. The total error rate decreased from 22.5% to 12.5%. This experiment confirms the importance of model updating in reducing verification error rate.

In experiment 7, we used another decision criterion named BMDR. For using this criterion we need a general model or population model. In our experiments we obtained this general model using 1 utterance from all the 26 clients in the representative database. Inter- and intra-speaker verification tests were then done on the evaluation part of PolyDat comprising of 30 speakers. In each verification test, test utterance was compared once with the client's model providing distance Di and once with general model providing distance Dg. Speaker is accepted if Di<Dg, otherwise he/she is rejected. As it can be seen in table 1, an important error rate was obtained specially for FA error rate. An observation of verification distances showed that, in many cases, when an utterance of an impostor is compared to the general model, Dg becomes greater than Di, and this yields a false acceptance.

Experiment 8 was conducted to see the effect of updating of reference models, when the BMDR decision criterion is used. Table 1 shows that FR decreases but FA increases considerably. An observation of the distance values shows that when utterances of impostors are compared with the claimed client's model and with the general model, while Di is less than Dg, based on the BMDR criterion, the impostor is accepted and the client's model is updated using impostor test utterance. This updating results in a more reduction of Di when other tests are done with test utterances of this impostor. In this way after each false acceptance, the client's model loses the client's speaker-dependent characteristics and obtains the acoustic characteristics of impostors; so the difference between the client's model and the general model diminishes gradually. On the other hand comparing experiments 7 and 8 in table 1 shows a reduction of FR error rate from 6.4% to 1.2% due to model updating. It means that client's model updating decreases Di when intra-speaker verification tests are done. Figure 1 and 2 show the distributions of Di and Dg for Intra-speaker and inter-speaker verification tests using BMDR criterion. As can be seen for intra-speaker tests, Di and Dg distances represent the distributions with a small intersection while Dg and Di for inter-speaker verification tests are totally superposed and no separation between them can be observed. Considering the distance distributions in figures 1 and 2, one can conclude that BMDR criterion can well verify the clients of a speaker verifications system, but this criterion has not a good performance for rejection of impostors. Small FR and high FA error rates in table 1 for experiments 7 and 8 confirm this conclusion. To improve the verification results using BMDR, we propose a shift to the left in the distribution of Dg in figure 2. It means that Dg value should be reduced each time an utterance is compared to the general model. In this way when inter-speaker tests are done, Dg may become smaller than Di and the impostor will be rejected. This yields a reduction in FA error rate. It should be noticed that the reduction of Dg,, increases FR. So, the reduction of Dg should be done in a way that the amount of reduction in FA be more than the amount of increase in FR. To achieve this aim, we propose relation 4, where the same constants C1 and C2 in relation 3, obtained previously using the representative database, are used to update Dg. Here Dg_{old} and Dg_{new} are the old and the new values of Dg and C1 is a value less than 1.

$$Dg_{new} = Cl * Dg_{old} + C2 \qquad (4)$$

The results of this experiment with Cl=0.41 and C2= 0.23 and other C1 values are given in table 2. Cl=0.41 and C2=0.23 are the constant values obtained previously in experiment 1 for *a priori* calculation of decision thresholds. The first raw in this table presents the result of experiment 9 where BMDR decision criterion and updating were employed with no modification of Dg. The other experiments present the results obtained by modifying the Dg value (relation 4) using different values for C1. As said before the predetermined value of C1 is 0.41. Several series of verification test are done using different values of C1: starting from 0.2 to a maximum value of 0.5. It is interesting to see that the best total error rate (ERR) was obtained for C1= 0.41. This value of C1 is the same value obtained previously using another part of PolyDat for *a priori* calculation of decision thresholds. Table 2 shows that when C1 is small, the FR is high and FA is small, and when the value of C1 is high, the FR decreases and the FA increases. On the other hand an observation of intra-speaker distances shows that

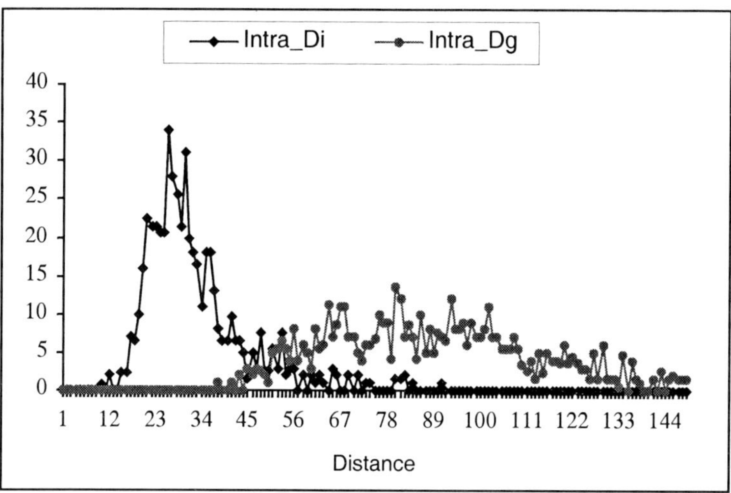

Fig. 1. Distance distributions obtained from comparing clients' test utterances to clients' models and general model.

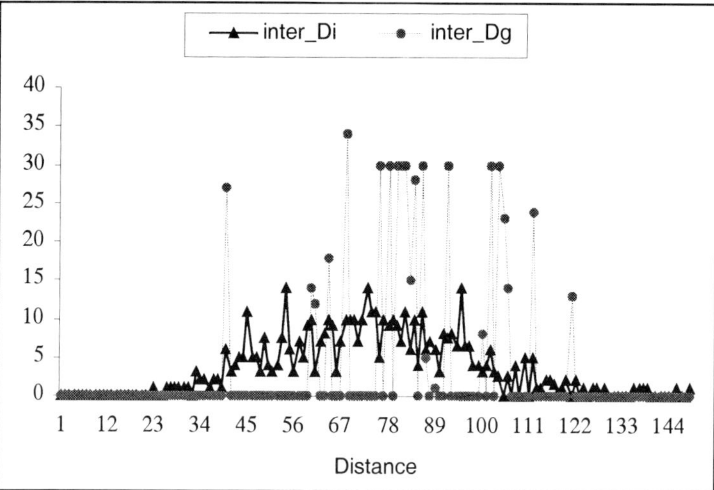

Fig. 2. Distance distributions obtained from comparing impostors' test utterances to clients' models and the general model.

when updating is done the intra-speaker distance values decrease and the inter-speaker distances and Dg values don't change significantly. So, one idea to more improve the verification performance is to start the verification tests with a big value for C1 (big values for Dg) and to decrease it gradually with each updating, using relation 5. Based on table 2, reduction of C1 tends to increase FR, but model updating moderates it and prevents its considerable increase. Reduction of C1, on the other hand, decreases FA considerably. The changes in FR and FA tend to decrease the total error rate. Here we start with the initial value of C1 equal to C_{max}=0.41. C1 will reduce to its minimum value C_{min}, according to relation (5), after a limited number of updating of reference model (here L= 12):

$$C1(i) = C_{min} + \frac{C_{max} - C_{min}}{2}(1 + \sin\pi(0.5 - \frac{i-1}{L})) \tag{5}$$

Where i is the updating number. The last row in the table 2 shows the results obtained for this experiment. As can be seen, the total error rate (ERR) has decreased for about 1.4 % related to the best case obtained previously using relation 4, with C1= 0.41 (experiment 12).

Table 2. Results of updating reference models using BMDR criterion and different C1 values.

Experiment NO	C1	FR	FA	ERR= (FR+ FA) /2
9	-	1.2 %	65.5 %	33.2 %
10	C1= 0.2	33.7 %	4.8 %	19.3 %
11	C1= 0.25	22.5 %	8.3 %	15.4 %
12	C1= 0.41	4.1 %	26.2 %	15.1 %
13	C1= 0.5	1.8 %	39.0 %	20.4 %
14	C1: relation (5)	17.4 %	10.0 %	13.7 %

It should be noticed that in all experiments with PolyDat, two speakers among 30 speakers were in the origin of a great part of total error rates. In this paper we conducted some experiments using PolyDat database. The experiment conditions were not the same as those of real conditions. Usually in real conditions, speakers try to be as cooperative as possible because they don't want to be rejected. In these conditions, the verification system itself influences the speaker's behavior. This influence is ignored when tests are running on pre-recorded speech databases. The environment in which the verification system is installed, details of the user interface, and previous acceptances/rejections can effect the user's interaction with the system. It implies that valid testing in a target environment requires a real implementation of the algorithm and an accurate simulation of the user interface.

6 Conclusion

Inter-session speech variability is an important factor in decreasing the performance of speaker verification systems. As it was shown in this paper, when training and test data belong to the same session, the verification error rate is much less than cases where training and test data belong to different sessions. We observed that when training data are updated, even for the cases where the number of client and impostor attempts is equal, speaker verification performance improves. Usually when reference models are updated, intra-speaker distances have a tendency to decrease, so it becomes necessary to update the decision threshold. One can hope better verification performance in a real application, when the speaker is cooperative and that the interaction between machine and speaker is more realistic, i.e. the speaker does a great effort to be accepted by the system. On the other hand in our experiments the number of impostors' and clients' attempts were equal, i.e. one impostor attempt after each client attempt. This results a mistraining of client's model by the impostor test data whenever the impostor is accepted. In a real application, usually the number of impostor attempts is less than the number of client attempts. So the possibility of mistraining of client models by impostor utterances is much weaker. As it was seen, BMDR and its modified versions (using relations 4 and 5) reduce the necessity for decision threshold calculation and therefore can ease the implementation of speaker verification systems. But BMDR needs yet to be modified to result better speaker verification performance.

7 References

1. Doddington, G. R., "Speaker Recognition, Identifying People from Their Voices: Proc. IEEE, 73(11), pp. 1651–1664, 1985.
2. Homayounpour, M. M., and Chollet, G., "Neural Nets Approaches to Speaker Verification: Comparison with Second Order Statistical Measures", ICASSP 95, pp. 353–356, 1995.
3. Bimbot F., "Second Order Statistical Measures for Text-Independent Speaker Identification, ESCA Workshop on Speaker Recognition, Identification, and Verification, pp. 51–54, 1995.
4. Higgins L., Bahler L., Porter, J., "Speaker Verification Using Randomized Phrase Prompting:, Digital signal processing:, Vol. 1, pp. 89–106, 1991.
5. Furui S., "an Overview of Speaker Recognition Technology", ESCA Workshop on Speaker Recognition, Identification, and Verification, pp. 1–9, 1994.
6. Forsyth, M. E., Bagshaw, P. C., Jack, M. A., "Incorporating Discriminating Observation Probabilities into Semi-Continuous HMM for Speaker Verification", ESCA Workshop on Speaker Recognition, Identification, and Verification, pp. 19–22, 1995.

Application of Constraint Hierarchy to Timetabling Problems

Tatyana Yakhno and Evren Tekin

Dokuz Eylul University, Computer Engineering Department,
Izmir, TURKEY
yakhno@cs.deu.edu.tr

Abstract. Timetable scheduling is a well-known instance of scheduling problems. There are many studies done on this subject and it still attracts many researchers since it is one of the most challenging problems in the domain. The present paper considers the application of the hierarchy of constraints which was used for the University timetabling problem. The hierarchy of constraints allows users to specify their preferences according to which the system is looking for solutions that can satisfy most of the users.

1 Introduction

At the beginning of each school year universities everywhere must undertake boring and time-consuming process of slotting students, teachers and lessons into available classrooms. It is a natural scheduling problem and schedule-makers are looking for simple flexible and effective automatic systems.

Scheduling problems are often NP-complete. There are many approaches to deal with them such as algorithmic approach, operation research approach, etc. Algorithmic (or integer programming) approach is hard for scheduling problems since the domain of the search space is very large. Finding the solution in this search space by traditional search methods is very inefficient. A lot of attention in operation research has been paid to scheduling problems that are based on relatively simple mathematical models. Operation research often aims achieving a high level of efficiency. This approach has some classical models to use when modeling a practical scheduling problem. Main disadvantages of these models are in that they discard many degrees of freedom and side constraints that exist in the practical scheduling situation. Discarding degrees of freedom and side constraints causes optimal solutions to be eliminated, regardless of the solution method used. Discarding side constraints may result in a simplified version of the problem and solving this simplified problem might be easier but the solution found can be impractical for the original problem [2].

While operation research methods are efficiency oriented (i.e., they are specialized algorithms for specialized problems), artificial intelligence research tends to investigate more general scheduling models and tries to solve these problems by using general problem solving techniques. However, in some specific cases, AI algorithms may perform poorly on specific instances compared to operation research

M.H. Shafazand and A M. Tjoa (Eds.): EurAsia-ICT 2002, LNCS 2510, pp. 11–18, 2002.

algorithms. On the other hand, operation research offers us efficient algorithms to solve problems that however can not be suitable in practice, while AI algorithms are more applicable [8].

Naturally, we want the best of both worlds, i.e., we want efficient algorithms that can be applied to a wide range of problems [6].

With the emergence of constraint programming, especially the introduction of finite domain constraints into logic programming, the constraint satisfaction approach has started to attract more and more attention due to its effectiveness in solving real-life planning/scheduling problems. Although constraint programming approach is still a search-based approach, it involves many improvements over the integer and operation research approaches. Generate-and-test nature of problem solving in logic programming is greatly extended with constraint programming, giving it the efficiency of finding optimal solutions in short execution time [7,9].

Timetable scheduling is a well-known instance of scheduling problems. There are many studies done on this subject and it still attracts many researchers, since it is one of the most challenging problems in the domain. As other scheduling problems, the timetable scheduling problem is also NP-complete [4]. In timetable scheduling, resources are instructors, classrooms, groups of students and hours of the weeks. They should be allocated on lessons so that the preferences of a student and lecturer should be maximized (optimal solution) without conflicts on scheduling a room, instructor and student in the timetable [1,10] .

Although many similar systems had been developed so far, the main disadvantage of these systems is the lack of flexibility. In nearly all of these systems, users cannot define their own constraints. Hard coded constraints direct the entire search and may leave the user with some unwanted solutions. Giving a user the ability to define his/her own constraints would give more user satisfaction as well as improve the solution quality.

The present paper considers the application of the hierarchy of constraints which was used for the University timetabling problem. The hierarchy of constraints allows users to specify their preferences according to which the system is looking for solutions that can satisfy most of the users.

2 Over-Constraint Systems and Constraint Hierarchy

Constraint Satisfaction Problem is a triple (V, D, C), where V is a finite set of variables, D is a set of domains of these variables, C is a conjunction of constraints $c_1, c_2, \ldots c_k$. Each constraint describes the relation that should be satisfied. Solution of such a problem is a set of values of all variables which simultaneously satisfy all the constraints [9].

An over-constrained system represents constraint satisfaction problem without a solution. In other words, for the variables of the systems, it is impossible to find valuations that exactly satisfy all of the problem constraints. Generally, real world problems fall into this category. In order to handle with this kind of problems, the constraint hierarchies are used. Instead of exact solutions partial solutions are used that satisfy not all the constraints but different subsets of the given constraints.

In many applications, such as interactive graphics, planning, scheduling, document formatting, and decision support systems, users need to express their preferences as well as strict requirements. In such systems, expressing the preferences in the same way as the strict requirements generally results in insolvability of the problem. In order to overcome this situation, an arbitrary number of levels of preference are used, each successive level being more weakly preferred than the previous one.

Definition 2.1.

*A **labeled constraint** is a constraint labeled with strength. Strength indicates the level of the preference for the constraint.*

Usually in writing labeled constraints symbolic names to the different strengths of constraints are given. These symbolic names can be mapped onto integers 0,…,n where n is the number of levels.

Constraints representing strict requirements are called *Hard Constraints.* Constraints representing preference are called *Soft Constraints.* Most of the systems and constraint programming languages allow users to define an arbitrary number of levels for preference, where each successive level being more weakly preferred than the previous one [5].

Definition 2.2

*A **constraint hierarchy** H is a multiset of labeled constraints. Given a constraint hierarchy H, H_0 denotes the HARD constraints in H. In the same way, the sets $H_1, H_2, \ldots, H_n$ are defined for preference levels 1,2,…n.*

The greater the label is, the weaker the constraint is. A solution of the constraint satisfaction problem embodying constraint hierarchy is not the same as the solution of a constraint satisfaction problem without it. So we need to revise our definition of a solution to the constraint satisfaction problem [3].

Now for each values of index n let us define the proper sets of constraints $H_1,\ldots,H_n$.

H_0 is HARD Constraints,

$H_i = \{c \in H \,/\, \text{strength}(c) = i\}$, $i \leq n$,

$H_k = \emptyset$ if $k > n$.

Definition 2.3.

A solution S to a constraint hierarchy H is asset of valuations for the variables in constraints with the following properties.

- *each valuation in S must be such that after it is applied all the HARD constraints hold;*
- *each valuation in S satisfies the non required constraints as well as possible, respecting their relative strengths.*

To formalize this let us define set S_0 of valuation such that all the H_0 constraints hold

$$S_0 = \{\theta \ / \ \forall c \in H_0, \ c\theta = \text{true}\}.$$

Here $c\theta$ denotes the Boolean result of applying the valuation θ to constraint c.

Then, using S_0, it is possible to define solution set S by eliminating all potential valuations that are worse than some other valuation, using the special predicate *better*

$$S = \{\theta \ / \ \theta \in S_0 \wedge \forall \sigma \in S_0 \ \neg \text{better}(\sigma, \theta, H)\}.$$

There are many different ways to specify the predicate *better* [9]. In our application we will use the so-called *locally-better* predicate, when the system tries to satisfy as much soft constraints as possible taking their strength into account.

3 Problem Description and System Inputs

Computer Engineering Department of Dokuz Eylül University offers undergraduate, Master and PhD degrees to their students. Undergraduate program consists of several courses lectured during 4 years (or 8 semesters). In MS program, each student must take 24 credits for graduation. PhD program is normally a 4 -year program.

Computer Engineering Department has to prepare the timetables at the beginning of the semester. Every student normally enrolls from 5 to 8 courses in one semester.

Master program courses are lectured only two days of the week. So, these courses should be scheduled taking this constraint into account.

Each course consists of theoretical and practical sections. Each course is normally divided into 2 or 3 blocks of 2 hours (sometimes 3 hours) during the week.

The department building has 6 rooms. Only 3 of them are suitable for a large number of students. Elective courses have a relatively small number of students enrolled to, so, while preparing the timetable, small classrooms are assigned to these courses. Some classrooms can also be assigned to specific courses.

Computer Engineering Department currently employs 8 academic staff members and 9 research assistants to help the academic staff. Generally one lecturer conducts 3 lessons per semester. Because of the large number of lecture hours, preferences (preferred hours of the days for delivering lectures, , etc.) of the lecturers are taken into account while preparing the timetable.

The timetable problem discussed here is to generate weekly timetable for Computer Engineering Department, i.e. to schedule all the blocks of all the courses to the available classrooms and hours of the week.

The final result must satisfy the following constraints:

- Blocks of the same semester cannot overlap.
- Two blocks with a common lecturer cannot overlap.
- There cannot be overlapping blocks in the same classroom.
- Lecturers and Classroom of a block must be available during its whole duration.
- Blocks of a course should be scheduled on different days.

All the conditions described above make it very difficult to prepare a timetable for the department manually. Generally, the final timetable for the department is prepared after 4 weeks work. Therefore there is a need to develop an efficient system to solve the problem.

The system mainly contains 4 kinds of objects at the input stage of the program. They are:

- Lecturer Objects
- Classroom Objects
- Lesson Objects
- Constraint Objects

A lecturer object is the entity, where all the information about each academic staff is stored. This information includes: name of the lecturer, available hours of the lecturer. A lecturer may be not available at the department during some periods of the week. So, while entering the information to the system, these hours should be identified to the system in order not to have an invalid schedule, i.e., a schedule that employs a lecturer while he/she is not available in the department. A lecturer may also have some other preferences over days. We will consider them as soft constraints. For example, a lecturer can prefer morning classes. So, while these soft constraints are unimportant when we control the validity of a timetable, they can help us to evaluate the final timetables, i.e., a timetable can be more preferable than the other according to the number of satisfied constraints.

All this availability and preference information is stored in a 5x8 integer array. Each element of the array can take values in the range [–3, 2].

A value –3 in the array means, "the lecturer is not available in the department during the corresponding hour". Values –2 and –1 correspond to unwillingness for lecturing a course during the corresponding hours. The scheduler will try not to assign lessons to hours having a preference of –2 unless there is no other choice.

A value 0 means "neutral" preference for the lecturer. Assignment of a lesson to this hour in the timetable gets no penalty, but it does not also increase the score of the solution found.

Values 1 and 2 in the array correspond to the wish to deliver a course during the marked hours of the timetable. The scheduler will first try to assign a lesson of the lecturer to the hour with the preference equal to 2. By doing this, the solution found will get bonus for making this assignment, thus having a higher score.

The system has got some built in constraints like "there cannot be overlapping lessons in the same classroom". Besides, user can want to define other constraints of different weights. Objects belonging to *Constraint Class* store the user-defined constraints.

All the constraints in the system are binary, so a constraint is a relation between two lesson objects. Since constraint hierarchy is used in the implementation of the system, strength of the constraints must be defined.

4 Search and Constraint Solver

Constraint Solver converts all the inputs of the program into variables which should be evaluated later, specifies the domains of each variable, searches for a solution, evaluates the results and selects the best solution for the user.

All of the functions described above are implemented in a single class named *SchedulerClass*, which contains multiple methods.

SchedulerClass directs the whole search process. Most of the updates over input type objects are also carried out by this class. For example, when we delete a classroom from the system, all of the lesson instances must be checked and the domains containing the deleted classroom should be updated. In order to keep this job simple, all the lists containing pointers to different object types are kept in this class. When a change in one list occurs, it is very easy to go over the references of other lists and carry out the updates

4.1 Building a Constraint Graph

After each variable and their corresponding domain are created, the system builds the constraint graph that will direct the search.

Two constraint graphs are used in the system. One of them contains the hard constraints with no weight and the second one contains the soft constraints with various weights.

Hard constraints are not handled in the constraint graph. Satisfaction of these constraints is guarantied by forward checking and arc-consistency methods [9].

After building the hard constraints defined in the system, the constraint objects, i.e., user-defined constraints, are to be checked. A user can also define his/her hard constraints. For example, although lessons A and B do not belong to the same lecturer, user may want them to be scheduled to different hours because some students might be enrolled to the both courses at the same time.

Hard constraint graph contains the constraints that directly conduct the search. If any inconsistencies are detected in the graph, a *back-jump* occurs [9].

Soft constraints are built using the information from the constraint objects and there can be some other soft user-defined constraints of various weights. Some scoring criteria, such as "blocks of the same lesson should not be scheduled to the days following each other", are used after a valid scheduling is found. Soft constraints do not direct the search process. Soft constraints are only used to score the solutions found. Thus, they are used in finding the optimal solution.

4.2 Ordering of the Variables

Ordering of the problem variables is important when we need to find any solution for the Constraint Satisfaction Problem [9]. In this case finding the first solution as soon as possible is the goal of the search process.

But the timetabling problem is the optimization one and we are looking for the best (optimal or near optimal) solution. To achieve this, the entire search tree should be inspected for solutions. Ordering of the variables is of no importance when the entire tree is to be traversed.

In the system, Branch&Bound technique is used to prune off the branches of the search tree. But in order Branch&Bound technique to be able to work, first a *bound* should be found. This will come from the first solution found to the problem.

In the system, we employed *Fail First Principle* for the ordering of the variables [9]. This is a general heuristic for search. It suggests that the variable which is likely to fail in labeling should be labeled first. By doing this, inconsistencies can be detected earlier. The labeling complexity can be measured differently. For finite domains, the size of a variable domain is important. A variable with a smaller domain is likely to fail earlier when compared to a variable with a large domain. The number of constraints affecting the variable can also make the labeling difficult [4].

We made several tests employing both measures. In some problems, ordering according to the domains was performed well while ordering according to the constraints was very poor in performance. But there were other problems, where ordering according to constraints were very good. The system currently uses domain size measurement for the ordering of the variables.

4.3 Search

The search algorithm is a recursive algorithm and it embodies *back-jumping* and Branch&Bound techniques, which are two of the most efficient search methods in the area of scheduling, together with forward checking algorithm [7].

The algorithm continually tries to go deeper in the search tree if no conflicts arise. First the algorithm checks, whether the lecturer of the currently scheduled block is overloaded for that day or not. To find it out, the algorithm calls a function *Lecturer_Over_Schedule*, which checks the previous schedules of the lecturer. If the lecturer is not overloaded, then forward checking occurs and domains of the other variables are reduced. If any variables domain becomes empty, no further labeling will be done (forward checking). The algorithm rolls back and tries to make another schedule for the current block. If the domains of other variables are reduced successfully, then the partial solutions score is computed. If it is higher than the bound (the best solutions score) then the operations described above work for the next variable. If not, a new labeling is done for the current variable (Branch&Bound).

4.4 Results and Performance

System is developed under Borland Delphi 5.0 programming environment. The first tests were relatively simple problems, which helped us to test the stability of the system and to inspect the behavior of the system under some certain problems.

Real data obtained from the previous semesters were used in performance testing. These tests were also used to determine the quality of solutions found.

During the development, many methods had been tried to improve the search performance and the quality of the solution. We were running the system during 2 months to collect the results and, on this basis, to improve the heuristics that guide the search.

Although Branch&Bound and forward checking helps lot to prune the search tree, the search process still takes several days. But the solution was found on the second day. The rest of the search was just pruning. Since we cannot know whether the last found solution is the optimum result or not, we cannot cut off the rest of the search although it is sometimes useless to continue.

5 Conclusions and Future Work

The implementation of the final version of the system took 4 months. Most of this time was consumed in a search for good heuristic function parameters. Unlike most of the systems developed so far, this system tries to do a more complete scheduling for the university. It tries to allocate not only the blocks to the hours, but also the classrooms. When this is the case, search tree grows wider, in other words, number of candidate solutions increases, which in turn affects the search time in negative way.

It is certain that the hardware development, especially in CPU technology, will help the search process to be carried out more effectively. But in a problem having a huge search space, relying on the power of the hardware is not wise. The developed system works well and can handle lots of different situations.. More efforts, especially on the scoring mechanism should be done.

References

1. Azevedo, F., Barahona, P. Timetabling in Constraint Logic Programming. Proceedings of the 2nd World Congress on Expert Systems, 1994.
2. Baptiste, P., Le Pape, C., Nuijten, W. Incorporating Efficient Operations Research Algorithms in Constraint-Based Scheduling. Proceedings of the First International Joint Workshop on Artificial Intelligence and Operations Research, 1995.
3. Borning, A., Freeman-Benson, B.,& Wilson, M. Constraint Hierarchies. LISP and Symbolic Computation: An International Journal, 5, 1992, pp. 223–270.
4. Cooper, T.B.,& Kingston, J.H. The Complexity of Timetable Construction Problems. Proceedings of the First International Conference on the Practice and Theory of Automated Timetabling (ICPTAT '95), 1995.
5. Freeman-Benson, B.N.,& Borning, A. Integrating Constraints with an Object-Oriented Language. Proceedings of the 1992 European Conference on Object-Oriented Programming, pp. 268–286.
6. Lustig, J.I., Puget, J. Constraint Programming and its Relationship to Mathematical Programming. The Institute for Operations Research and Management Science (INFORMS), Maryland. 2000.
7. Marriott K., Stuckey P. Programming with Constraints. The MIT Press. UK, 1998.
8. Russell, S.J., Norvig, P. Artificial Intelligence – A Modern Approach, Prentice Hall, 1995.
9. Sang, E. Foundations of Constraint Satisfaction. Academic Press, 1993.
10. Yeung, C., et al. Applying Constraint Satisfaction Technique in University Timetable Scheduling. The Practical Application of Prolog: Proceedings of the 3rd International Conference on the Practical Application of Prolog, UK, 1995, pp.683–695.

An Intelligent System for Therapy Control in a Distributed Organization

José Joaquín Cañadas, Isabel María del Águila, Alfonso Bosch, and Samuel Túnez

Department of Languages and Computation. University of Almería. 04120. Almería, Spain.
{jjcanada, imaguila, abosch, stunez}@ual.es
Phone:++34950015988 Fax:++34950015129

Abstract. This paper describes a decision-making system for phytosanitary control advicing. The solution adopted consists of developing a Web-accessible information system based on a multi-agent architecture, integrating knowledge-based techniques and classical information analysis and management techniques. CommonKADS was used for the design of some knowledge-based agents. Internet implementation and integration was done in a knowledge-based system implementation environment, with programs executed on the Web server.

1 Introduction

This work belongs to a project[1] developed over the last three years, the purpose of which is to apply the most recent innovations of information technologies and communications to the agriculture sector, in order to modernize it. This project consists of three subprojects coordinated by the Universities of Almería, Murcia and Granada. Three areas of interest have been defined, which correspond, respectively, to the spheres of action of each university and subproject: a) Optimization of the combined use of irrigation and fertilizers. b) Evaluation of the aptness of soils for cultivation using soil assessment techniques. c) Phytosanitary therapy control for crops produced according to the Integrated Production Quality Standard.

The main objective of this project is the design, development and implementation of a set of interactive web-based tools in three advisory decision-making systems (ADS). These ADS are responsible for advisory services within the scope of each subproject. Figure 1 shows a general view of the system proposed in the project. The agents responsible for specific tasks can be described using the following levels: 1) A tasks and processes level, where the agents that implement the different ADS tools are located. 2) An interface level that allows access by final users. 3) An information access level, holding the mechanisms of query and interaction for information sources located in the databases and in Geographic Information Systems.

This paper focuses on the ADS developed at the University of Almería (highlighted elements in Figure 1). The developed agents cover the advisory requirements in the distributed organization necessary for monitoring the Integrated

[1] "An Intelligent Decision Support System for the South-East Spanish Agricultural Environment" Reference: 1FD97-0255-C03-03 financed by the CICYT and the EC.

M.H. Shafazand and A M. Tjoa (Eds.): EurAsia-ICT 2002, LNCS 2510, pp. 19–26, 2002.

Production quality standard. These agents integrate knowledge-based techniques and classical information analysis and management techniques.

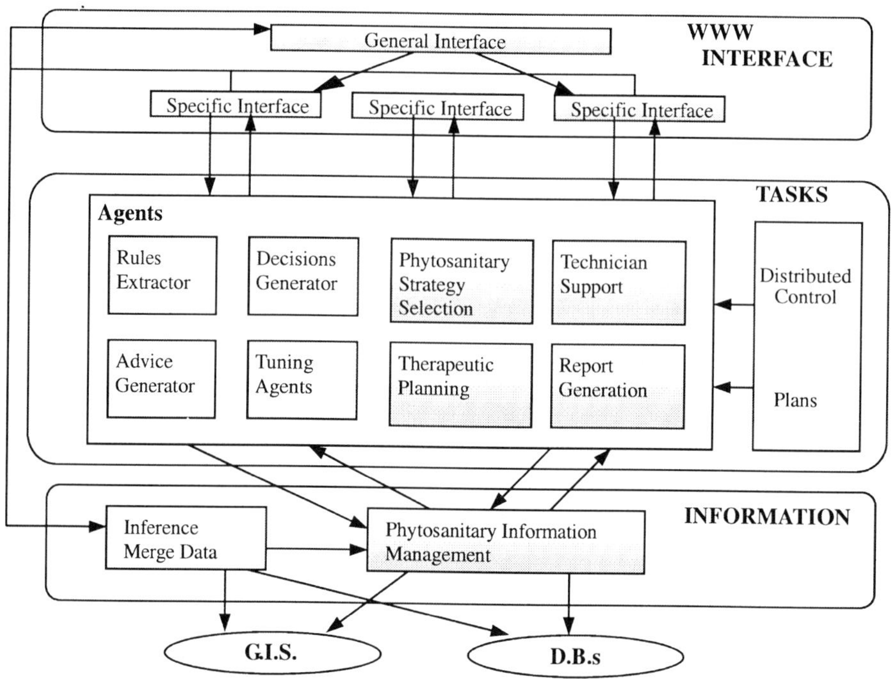

Fig. 1. General architecture of the intelligent decision-making system.

Our group had previously developed the EXPLA knowledge-based system (KBS) [9], for chemical pest control advisory services, using Smart Elements [5] as implementation environment. The implementation uses knowledge-based techniques, the object-oriented paradigm and database access. The combined use of these techniques allows complex applications in very different spheres to be approached, satisfying a set of requirements which are not always easily managed in conventional systems. This combination provides a clear separation between the knowledge base and the facts recorded in a database, facilitating maintenance and portability of the applications, and allows the principles of data abstraction and encapsulation to be used for structuring knowledge. The dynamic object-based design combined databases allows a high degree of independence from the domain, which favors its reusability.

Nowadays there is much interest in incorporating KBS onto the Internet, and several different alternatives have been proposed for it. In the work of Simpson [7], the KBS development and execution environment is JESS (Java Expert System Shell), an inference engine inspired by CLIPS, written in Java and which, therefore, can be combined with any Java program and executed on the Internet. In the works of Grove [2] and Morris [4], the potential offered by Internet and the Web and their associated technologies as an environment for development and deployment of expert systems is analysed. In our case, the solution adopted, with very satisfactory results, is based on the use of the Smart Elements environment already known to the group, together with the Web technology necessary to implement the KBS on the Internet.

This paper is structured as follows. Section 2 analyses the problem context, stating the need for a distributed organization. Section 3 describes the agent architecture used in the system, and the design, communication and Web integration of the most important agents. Section 4 presents the conclusions and future work.

2 Integrated Production Quality Standard

In this section we briefly describe the organization of Integrated Production (IP), the basic reason why a distributed agent architecture becomes necessary. IP is understood as an agricultural production system that uses natural production resources and mechanisms to the maximum. Biological and chemical methods are carefully selected taking into account the requirements of society, profitability and environmental protection. To maintain these agricultural production techniques, a mark of quality called Integrated Production has been created that is carried out according to the Spanish office of patents and trademarks. IP includes extensive regulating standards, as well as monitoring and inspection by the corresponding authorities.

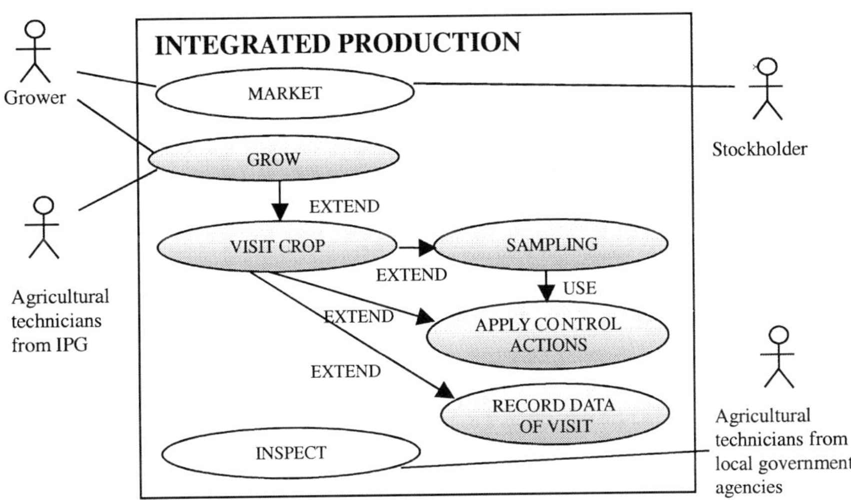

Fig. 2. Integrated production quality marking

Figure 2 shows the IP use case diagram, in which three wide interaction scenarios stand out: *grow*, *market* and *inspect*, along with the growers and technicians involved in the process. When a group of growers (IPG) decides to adopt the IP quality standard, it must submit to discipline in growing implying intervention by technicians, marketing controls and periodical reviews by the certifying companies to see that the standard is being complied with.

In general, the *grow* use case considers the crop as a complex system, made up on one hand by the field/greenhouse, the plants, the parasites (pests and diseases) and the auxiliary fauna (useful fauna). This system is affected by external variables (climate, humidity, market price of the fruit, …) and to maintain balance, control measures may be applied which are especially respectful of the crop, the useful fauna and the environment.

Among interactions during growing, three tasks linked to integrated control of parasites affecting the crop appear. This control is carried out in weekly visits by the agricultural technician, in which he decides what action to take on the crop. The technician must take a sample of the state of the crop, understanding this as the system made up by the plants, the pests and the auxiliary fauna and other factors that enable him to estimate the risk induced by the different harmful agents, and when there is imbalance, advise treatment.

The information tools developed in this work assist the agricultural technicians and the growers in all the monitoring and decision-making tasks indicated above, as shown in the following sections. In future work, the ADS will be extended to the rest of the use cases in the diagram, such as marketing and inspection.

3 Distributed Multi-agent Architecture

To assist in all the aspects of monitoring, advising and management related to phytosanitary control in the scope of the Integrated Production Quality Standard, an information system based on a multi-agent architecture is proposed. As indicated above, some of the agents require knowledge-based techniques; therefore, we have used CommonKADS [6] for the entire KBS analysis and design process. One important result of the organizational analysis carried out was the confirmation that the experts clearly differentiate between two large tasks in the grower advisory process. In the first, a decision is made on whether it is necessary to use chemical control on a crop and, if so, the product to be applied is selected. This gave rise to two subsystems or agents that are dealt with separately: a) Decide whether it is necessary to take action on the crop, and b) decide what kind of action.

Figure 3 shows the ADS agent architecture. Along with the agents that implement the tasks above, there are other agents defined for the distributed organization and that are responsible for IP monitoring. Two agent categories are established: *centralized* and *peripherical*. The first are located on the servers of the Data, Knowledge and Software Engineering Research Laboratory at the University of Almería. They are available at `http://saepi.ual.es/` and enable the study of several pests and diseases of the target crops (tomatoes and grapevines). The second are located in the servers and or work stations of each of the Grower Groups devoted to Integrated Production (IPG). The purpose of each of these agents is described briefly below:

- *Phytosanitary Information Management*: its purpose is to offer the phytosanitary information and tools necessary for agricultural technicians and growers to easily generate tailored reports on plant-care products appropriate to the condition of their crops.
- *Selection of Phytosanitary Strategy*: the purpose is to provide a decision on whether or not it is necessary to take action on a crop. There are two ways of using this agent: the first is interactively on the Web by the technicians or growers (generally inexperienced beginners), for complete analysis of a problem to obtain a decision on the need for action on the crop and a complete report of the

consultation. The second is by communication with the Agent for Agricultural Technician Support to validate a decision made by the technician (usually experienced), that is, compare his own decision, based on data from samples collected during his visit to the crop, with the proposal offered the Phytosanitary Strategy Selection Agent, which strictly follows the IP protocol.

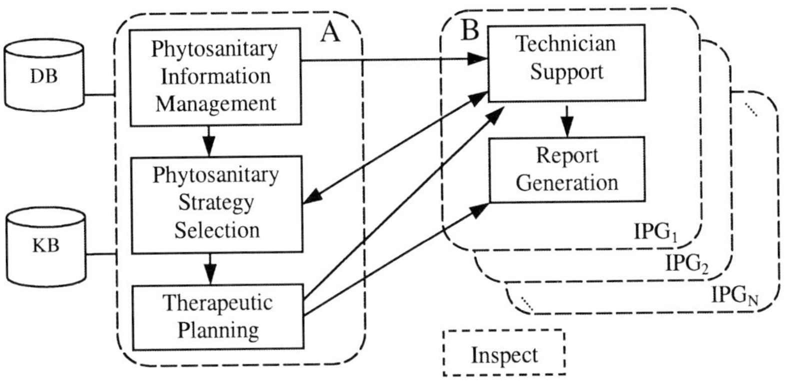

Fig. 3. Agent Architecture: A) centralized, B) peripherical

- *Treatment Planning*: if the Phytosanitary Strategy Selection Agent indicates need to take action on the crop, the type of treatment to be applied must be determined. This agent's purpose is to build up a treatment plan by abductive assembly of therapeutic conjectures, where the best conjectures (phytosanitary products applicable to the problem) are selected, building up the treatment plan necessary to correct a problem.
- *Technician Support*: its objective is to provide a working environment on the Web that facilitates all the tasks that must be performed by the agricultural technician during his IP monitoring visits. It offers tools for collecting the sampling information related to different pests and diseases that may affect the crop and records the type of action recommended by the technician in situ during his visit. One important feature of this agent is that it automatically checks that action, by communicating with the Phytosanitary Strategy Selection Agent.
- *Report Generation*: the purpose is to generate a series of results in the form of reports prescribed by the IP Quality Standard. These reports are necessary to control the greenhouses in so far as visits, types of action recommended, products applied, etc. They also enable periodical summaries to be produced and alarms to be sent out as phytosanitary warnings on possible risk of a certain pest or disease.
- *Inspection*: its purpose is to facilitate some of the IP Quality Standard certifying company tasks, checking on compliance with the standard and the assignment of the corresponding quality certifications.

The following sections deal in more depth with the design of the most important agents in the system: Phytosanitary Information Management, Phytosanitary Strategy Selection and Treatment Planning.

3.1 Phytosanitary Information Management Agent

The management of phytosanitary information is approached as a traditional data management problem. This agent offers the phytosanitary information and the tools necessary for the agricultural technicians and growers to easily generate tailored reports on the most appropriate products for the condition of their crops in general, and on the products permitted under the IP Quality Standard. The phytosanitary information is stored on an extended relational database server with an interface for Web queries using dynamic forms. It enables generation of tailored reports, about which the reliability of the information given and its frequent updating by the Plant Health Services should be mentioned. Its design is shown in the diagram in Figure 4.

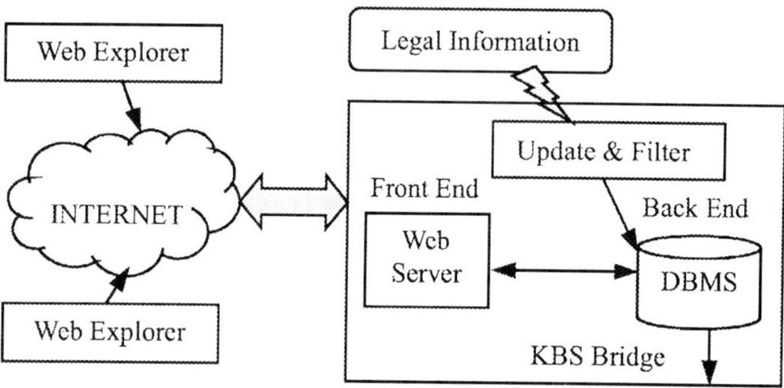

Fig. 4. Management of Phytosanitary Information

3.2 Phytosanitary Strategy Selection Agent

The Phytosanitary Strategy Selection and Therapy Planning agents described below were designed using knowledge-based techniques, forming a KBS devoted to advising growers and agricultural technicians on whether or not to take action on a crop and what type of action to be taken. Two operating modes have been developed for this agent, as indicated above.

The interactive operating mode employs Web forms to provide the KBS with the appropriate information on the growing status of the crop and the physiological condition of the pest or disease necessary to provide a decision. Expert knowledge and the IP standard are represented by sets of rules stored in knowledge bases. Depending on the problem to be studied, the system uses the knowledge bases that contain the knowledge related to that type of problem. Agricultural technicians usually solve the problem in a way analogous to the proposal implemented in TERAP_IA [1], establishing elemental problems (pest or disease present in a crop) and analyzing them to establish what they consider therapeutic objectives.

One result of the system evaluation was that this way of using it is best for growers and beginner agricultural technicians. However, experienced technicians make their decision on action to be taken "in situ" during the visit to the crop, without consulting the helping decision-making system, based on their own skill and experience.

Therefore, communication has been established between the Technician Support agent and the Phytosanitary Strategy Selection agent to enable this agent to offer a report on the action to be taken. This allows the experienced technician to validate and revise his own decision with the one he would have obtained if he had made an interactive query and whether he is considering all the restrictions of the IP.

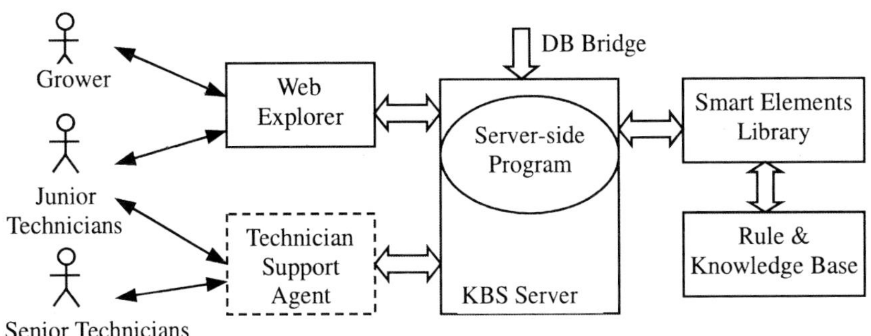

Fig. 5. Phytosanitary Strategy Selection

Figure 5 shows the functional diagram, which was implemented using the following components: 1) The Smart Elements implementation environment and its C function library. 2) Executable programs on the server for execution and integration on the Web. These programs process the input data supplied by the grower or technician on HTML forms and call up the Smart Elements library functions that enable loading the knowledge bases, suggesting the hypotheses, launching the inference process and showing the results.

3.3 Therapy Planning Agent

The construction of the treatment plan is closely related to the phytosanitary information management system and is carried out by an agent of abductive assembly of therapy conjectures [10], which selects the best conjectures by multi-criteria aggregation techniques [3], building up a set of actions that represent the treatment of the overall problem. One key feature is the evaluation of phytosanitary products by the phytosanitary information management agent, in which criteria established by experts and the relative importance of such criteria are applied.

4 Conclusions and Future Work

In this paper we have presented the development of an ADS for phytosanitary control in Integrated Production using a distributed architecture of software agents. This ADS is part of the results of a project that offer set of integrated Web tools for advising and assisting growers and agricultural technicians in making decisions on phytosanitary control, fertirrigation and soil evaluation.

We have described the most important features of the design and implementation of the ADS where some agents were implemented using classic information analysis and management techniques and others with knowledge-based techniques.

CommonKADS was used for the analysis and design of the latter; organizational context analysis established what are the intensive knowledge tasks in the advisory process: definition of therapeutic objectives and generation of a treatment plan.

The Web ADS offers all the phytosanitary information and has the tools necessary for agricultural technicians and growers to analyze several pests in grapevines in Murcia and tomatoes in Almería.

One aspect of implementation to be highlighted is KBS integration in the Internet through use of the Smart Elements environment, along with the necessary Web technology. This solution is easily transferable to other application domains.

In the future, it is planned to extend the distributed agent architecture to aspects of IP Quality Standard inspection and marketing of the fruit, which are not currently included in the ADS. And concerning the ADS development and usage environment, it is also planned to include management of Web users and monitor system queries made and the decisions obtained by it for each farm.

Acknowledgments. We gratefully acknowledge the collaboration of the Provincial Delegation of the Ministry of Agriculture and Fishing of Almería and its Plant Health Services.

References

1. Barrufet, P., Puyol-Gruart, J., Sierra, C.: Terap-IA, a Knowledge-Based System for Pneumonia Treatment. Proceedings of the International ICSC Symposium on Engineering of Intelligent Systems. EIS'98. Vol. 1. (1998) 176–182
2. Grove, R.: Internet-Based Expert Systems. Expert Systems, Vol 17 (2000) 129–135
3. Klein, D. A., Shortliffe, E. H.: Integrating Artificial Intelligence & Decision Theory in Heuristic Process Control Systems. Proceedings Avignon 90 (1991) 165–177
4. Morris, S., Neilson, I., Charlton, C., Little, J.: Interactivity and collaboration on the WWW. Is the 'WWW shell' sufficient?. Interacting with Computers 13 (2001) 717–730
5. Neuron Data: Nexpert Object User Guide. Neuron Data, Inc (1994)
6. Schereiber,G, Akkermans, H. Anjewierden, A. de Hoog, R., Van del Velde, W. Wielinga, B. Knowledge engineering management. The CommonKADS Methodology. MIT Press, Massachusetts (1999)
7. Simpson, J., Kingston, J., Molony, N.: Internet-based decisi6n support for evidence-based medicine. Knowledge Based Systems 12 (1999) 247–255
8. Sycara, K., Pannu, A., Williamson, M., Zeng, D.: Distributed Intelligent Agents. IEEE Expert (1996) 36–45
9. Túnez, S., Aguila, I. M., Bosch, A., Marín, R.: Integrating decision support and knowledge-based system: application to pest control in greenhouses. Proceedings 6th International Congress for Computer Technology in Agriculture. ICCTA'96. Wageningen. (1996) 417–422
10. Túnez, S., Aguila, I. M., Marín, R.: An Expertise Model for Therapy Planning Using Abductive Reasoning. Cybernetics and Systems: An International Journal. 32.(2001) 829–849

Discovering Local Patterns from Multiple Temporal Sequences[1]

Xiaoming Jin, Yuchang Lu, and Chunyi Shi

The National Key Laboratory of Intelligent Technology and System
Computer Science and Technology Dept., Tsinghua University, Beijing, 100084, China
xmjin00@mails.tsinghua.edu.cn lyc@tsinghua.edu.cn
scy@est4.cs.tsinghua.edu.cn

Abstract. In this paper, we address a data-mining problem that is the discovery of local sequential patterns from a set of long sequences. Each local sequential pattern is represented by a pattern *A*→*B* and a time period in which *A*→*B* is frequent. Such patterns are actually very common in practice and are potentially very useful. However it is impractical to use traditional methods on this problem directly. We propose a suffix-tree-like data structure for indexing the instances of the patterns. Based on this index, our mining method can discover all locally frequent patterns after one scan of the sequences. We have analyzed the behavior of the problem and evaluated the performance of our algorithm with both synthetic and real data. The results correspond with the definition of the problem and verify the superiority of our approach.

Keywords. Local sequential pattern, temporal sequence, data mining algorithm.

1 Introduction

In this paper, we consider a data-mining problem that is the discovery of local sequential patterns from a set of long sequences. Temporal sequences are lists of transaction records ordered by transaction time, e.g. sales records, stock prices, weather data, medical data, etc. Each local sequential pattern is represented by a pattern *A*→*B* and a time period in which *A*→*B* is frequent. For example, "In the sale records of a supermarket, customers always buy biscuits followed by soda in summer, whereas biscuits followed by milk in winter."

Such local patterns are actually very common in practice. Efficient discovery of it could benefit the KDD in many real applications. Consider the data-mining problems in stock market where the raw data is a set of long sequences collected daily over years indicting the price movements of each stock, there are many influences on the price movements, e.g. political policies, economic environment, society environment, etc. Apparently, these influences are time varying. Therefore the patterns of price movements related to the influencing factors are also time varying. Thus finding the temporal patterns and corresponding temporal features become equally if not more useful than simply finding the globally frequent patterns, where every record in the

[1] The research has been supported in part of Chinese national key fundamental research program (no, G1998030414) and Chinese national fund of natural science (no. 79990580)

M.H. Shafazand and A M. Tjoa (Eds.): EurAsia-ICT 2002, LNCS 2510, pp. 27–34, 2002.

temporal sequence contributes to the pattern. For example, "In summer, if the stock price of a game producer goes up and stays about level for two days, then it will go up the third day". Then we observe that there may be some correlation of price behavior from July to September so that we can plan buy-sell strategy appropriately in that season, whereas we will not be confused by this pattern when we make a decision for the rest time.

Most of the previous works on frequent pattern discovery mainly considered finding global patterns, e.g. the research work on sequential patterns [3][4][5][8], frequent episodes [6], interval-based events [7], etc. The problem of mining temporal association rules [9][12] and mining of second order knowledge seems similar to the problem we consider, but the formats of both the database and the knowledge are essentially different. It is impractical to use such traditional methods on the problem that we consider directly. A practicable approach is sliding a window through the sequence, and mining global patterns in each window. This method can only find a small portion of all local patterns, which are of the same valid subsequence length. A method that finds all local patterns can be derived by first retrieving all the possible subsequences, and then using previous algorithms to mine each subsequence. However, the time complexity of this method is always extremely poor.

In this paper, we present a suffix-tree-like data structure for indexing the instances of the patterns, which we term the local pattern tree for multi-sequences (MLP-tree). Based on this index, we propose a mining method that can discover all locally frequent patterns after one scan of all the sequences. We have analyzed the behavior of the problem and evaluated the performance of our algorithm with both synthetic and real data. The results correspond with the definition of our problem and verify the superiority of our approach.

2 Problem Description

We consider the database that consists of a set of long temporal sequences collected within the same time period, i.e. $SS=\{S_1\ S_2\ \dots\ S_N\}$. Each sequence $S=A_1, A_2, \dots, A_m$ is a list of records ordered by position number. In actual applications, records are descriptions of interesting events that happened sequentially. Without losing generality, we represent S by a \$-terminated sequence of symbols from an alphabet $\Sigma=\{a_1,\dots,a_k\}$, where each symbol uniquely represents a record at a time point. A subsequence, $S[sp,ep]=A_{sp}, \dots, A_{ep}$, is a continuous part of the original sequence. Given a time period represented by the starting position sp and the ending position ep relative to the sequences, we use the pattern format: B follows A in the period $[sp, ep]$ where A and B are two subsequence. We denoted it as $A\rightarrow B[sp, ep]$. The goal of our problem is to find all such patterns that happen frequently through searching in the sequences.

To evaluate the local patterns, we use *local frequency*, *local support* and *local confidence* in our problem definition and mining process.

Definition 1. The *local frequency* of subsequence A in period $[sp,ep]$ is the occurrences of A in all subsequences $S_n[sp,ep](S_n\in SS)$. We denote it as:

$$\mathrm{Lf}(A,[sp,ep])=|\{<n, i> \mid s=S_n[sp,ep]\wedge s[i,i+|A|-1]=A\}|$$

where $|A|$ is the length of A, i.e. the number of symbols in A.

Definition 2. The *local support* of *A* is defined as:

$$\delta(A) = \min(\{m|A[1,m] = A[|A|-m+1]\} \cup \{|A|\})$$

$$\text{Lsupp}(A, [sp,ep]) = \text{Lf}(A, [sp,ep])\ \delta(A)/((ep-sp+1)\cdot|SS|)$$

where $|SS|$ is the number of items, i.e. sequences, in set *SS*.

The maximal number of the instances of a pattern depends on the length and the repetition of the pattern. As longer pattern is discovered, the maximal number of its instances automatically shrinks with its size. On the other hand, if the prefix and the suffix of a pattern is the same, its maximal instance number automatically expands. Therefore we use $\delta(A)$ to adjust *Local support* comparable for the potential patterns with different length or repetition.

Definition 3. The *local support* of pattern $A \rightarrow B[sp,ep]$ is the support of subsequence *AB* in $[sp,ep]$, i.e.

$$\text{Lsupp}(A \rightarrow B[sp,ep]) = \text{Lsupp}(AB, [sp,ep]).$$

Definition 4. The *local confidence* of the pattern $A \rightarrow B[sp,ep]$ is the ratio of the *local frequency* of *AB* in $[sp,ep]$ over the *local frequency* of *A* in the same period, i.e.

$$\text{Lconf}(A \rightarrow B[sp,ep]) = \text{Lf}(AB, [sp,ep]) / \text{Lf}(A, [sp,ep]).$$

Definition 5. Given a *minimum support* ms and *a minimum confidence* mc, if the *local support* of pattern $A \rightarrow B$ in $[sp,ep]$ is no less than ms and the *local confidence* of that pattern in the same period is no less than mc, we consider $A \rightarrow B\ [sp,ep]$ as a *local sequential pattern* in multiple sequences (M-LSP). Then the problem of M-LSP discovery is: Given a sequence set *SS*, find all $A \rightarrow B\ [sp,ep]$ that are satisfied with:

$$\text{Lsupp}(A \rightarrow B\ [sp,ep]) \geq \text{ms and Lconf}(A \rightarrow B\ [sp,ep]) \geq \text{mc}$$

Example 1. *SS*={"dsdsuddsdsusus$","dsudsddsdusosu$"}, ms=0.5, mc=0.5. we could find d→s [1,10] is a M-LSP, because:

$\delta(\text{"ds"})=2$

Lsupp (d→s[1,10])=Lf("ds",[1,10])δ("ds")/((10-1+1)·|SS|)=7·2/(10·2)≥ms

Lconf (d→s[1,10])=Lf("ds", [1,10])/ Lf ("d", [1,10])=7/10≥mc

But d→s is not a frequent pattern for the whole duration, because the support of d→s is 7·2/(15·2)<0.5. We could find some other M-LSPs, e.g. s→u [11,14] with Lsupp (s→u [10,14]) = 4·2/(5·2) = 0.8, and Lconf (s→u [11,14]) = 4/5 = 0.8.

3 Index Structure

MLP-tree consists of a standard suffix tree, a leaf chain and a set of leaf pointers. For the rest of this paper, we shall use the following notational conventions: *locus*(A) denotes the first node in the suffix tree encountered after *A* is spelled out; *subsequence*(*t*) denotes the subsequence spelled out in the suffix tree by following the path from root to node *t*; *T*(*A*) denotes the sub-tree of *T* of which the root is *Locus*(*A*) and *T*(*t*) is the sub-tree of which the root is node *t*; {*Leaf* (*t*)} denotes the set of all leaf nodes of *T*(*t*).

A suffix tree [1] is a tree for storing strings in leaf nodes. Each internal node corresponds to one common prefix. Any sequence *S* is mapped to a suffix tree *T* whose paths are the suffixes of *S*, and whose leaf nodes correspond uniquely to positions within *S*. Any common subsequence can be spelled out according to the path from the root to a unique internal node, and each internal node except the root has at least two children.

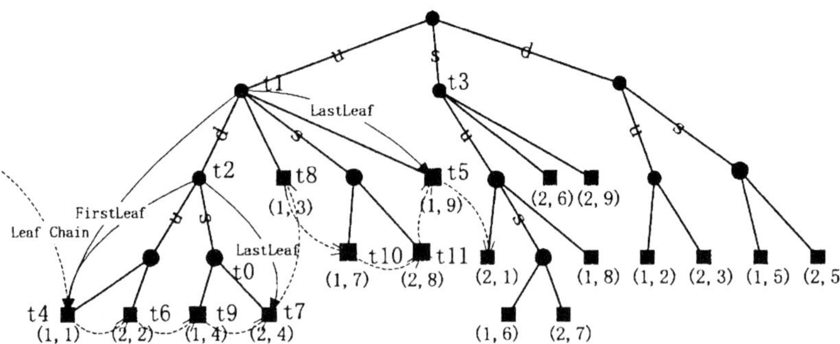

Fig. 1. MLP-tree for dataset *SS*={"uduudsusu$", "sududssus$"}, each leaf node is marked by corresponding (*t.SequenceID*, *t.Position*)

The data structure of MLP-tree is as follows: for any internal node *t*: *t.Child*, *t.Ancestor, t.Next*, , *t.Start*, *t.End*, *t.FirstLeaf*, *t.LastLeaf*, *t.Offset*. *t.Child* stores the pointer to the first child of node *t*, *t.Ancestor* stores the pointer to the ancestor node of *t*, *t.Next* stores the pointer to the next node with the same ancestor node of *t*. *t.SequenceID*, *t.End* and *t.Start* indicate the corresponding common subsequence *s* associated with the node *t*, where $s = S_{t.SequenceID}[t.End, t.Start]$. In order to locate and count the leaf nodes efficiently, all the leaves are linked, forming a leaf chain. For each internal node *t, t.FirstLeaf* and *t.LastLeaf* store the pointer to the first leaf node and the last leaf node of the sub tree *T*(*t*). Finally, $\delta(subsequence(t))$ is stored in *t.Offset* when *t* is created for avoiding repeated calculation of it.

For a leaf nodes *t*, the data structure is: *t.Ancestor*, *t.Next*, *t.SequenceID, t.Position*, *t.NextLeaf* where *t.Ancestor* and *t.Next* have the same meaning as that of an internal node, *t.Position* is the position of correspoding suffix in the sequence of which the ID is *t.SequenceID*, *t.NextLeaf* is the next leaf node in the leaf chain.

The construction method for MLP-tree is similar to that of a suffix tree. A difference with the standard insertion method is that we alternatively insert subsequences of different sequence into one indexing tree. Therefore, the sequence ID is stored for locating common subsequences. In addition, when a node *t* is inserted into a MLP-tree, the relative leaf chain pointers need updating, including the data member *FirstLeaf, LastLeaf* and *NextLeaf*. The updating process is done by a simply scan of the nodes from *t* to the root. For each node, determine whether modification is needed. If no modification is needed for one node, updating is completed.

Suppose *A* and *B* are two subsequences. In a MLP-tree, we can use the positions of all the leaf nodes of the sub tree rooted at *locus*(*A*) to calculate the positions of *A* in a period. Thereby the *local frequency* of *A* in any [*m*, *n*] can be calculated by counting the number of leaf nodes of the sub tree rooted at *locus*(*A*) that satisfies the position restrictions. That is:

$$\mathrm{Lf}\,(A, [m,n]) = |\{ leaf(t_A) | m \le leaf(t_A).Position \le n-|A|+1 \}|$$

where $t_A = locus(A)$. And easily gained:

$$\mathrm{Lf}\,(AB, [m,n]) = |\{ leaf(t_B) | m \le leaf(t_B).Position \le n-|AB|+1 \}|$$

where $t_B = locus\ (AB)$. Benefiting from maintaining the leaf chain, the number of leaf nodes can be easily and efficiently obtained by a simple traverse in the leaf chain between *t.Firstleaf* and *t.Lastleaf*. After *local frequency* has been obtained, the *local support* of a subsequence and the *local confidence* of $A \rightarrow B[m, n]$ are calculated as:

$$\text{Lsupp}(A, [m, n]) = \text{Lf}(A, [m,n]) \cdot t_A.Offset/(n-m+1)$$

$$\text{Lconf}(A \rightarrow B[m, n]) = \text{Lf}(AB, [m, n])/\text{Lf}(A, [m, n])$$

Example 2. The MLP-tree for the sequence set *SS*={"uduudsusu$", "sududssus$"} is shown in Fig. 1. Suppose we are interested in the local pattern d→s[1,5], then the *Lsupp* and *Lconf* can be calculated as follows:

locus ("u")=t1, *locus* ("ud")=t2

leaf(t1)={t4,t5,t6,t7,t8,t9,t10,t11}, leaf(t2)={t4,t6,t7,t9},

Lf (u, [1,5]) = $|\{leaf(t1)|1 \le leaf(t1).\text{Position} \le 5-|\text{"d"}|+1\}| = |\{t4,t6,t7,t8,t9\}| = 5$,

Lf (ds, [1,5]) = $|\{leaf(t2)|1 \le leaf(t2).\text{Position} \le 5-|\text{"ds"}|+1\}| = |\{t4,t6,t7,t9\}| = 4$,

Lsupp(u→d, [1,5]) = Lf ("ud", [1,5])$\cdot\delta$("ud")/(5-1+1)$\cdot|SS|$ = 4·2/5·2=0.8,

Lconf(u→d, [1,5]) = Lf("ud", [1,5])/Lf("u", [1,5])=4/5=0.8.

4 Method and Discussion

The overall mining strategy is as follows: Go through the whole time duration of the dataset *SS* sequentially, and insert each subsequence $S_m[n,\ldots,|S|]$ ($S_m \in SS$) into a MLP-tree. After a leaf node *t* is inserted, only the *local support* and *local confidence* of all the patterns stored in the ancestor of *t* will be changed. So we need only search the nodes in the path of traveling from node *t* to root. For each node t_1 in the travel path, find all the possible periods [*m*,*n*] such that the *local support* of *subsequence*(t_1) [*m*,*n*] is no less than minimal support and the length, i.e. *n*-*m*+1, is large enough. We call these <*subsequence*(t_1), [*m*,*n*]> M-LSP candidates. For each candidate, consider all its ancestor nodes t_2. The pattern *subsequence*(t_2)→*subsequence*(t_1)-*subsequence*(t_2)[*m*,*n*] is outputted as an M-LSP if and only if the *local confidence* of that pattern is no less than minimal confidence.

Let $(p_1,p_2,\ldots p_n)$ denote the starting positions of subsequence *A*, we need only consider the subsequences whose starting positions are p_m instead of scanning all the possible subsequences. It reduces the time complexity dramatically without losing M-LSP.

The detail of our algorithm for M-LSP discovery is given in Algorithm 1. During the mining process, we should consider another special type of pattern individually. For a node *t*, if $|S_{t.SequenceID}[t.Start,t.End]|>1$, $S_{t.SequenceID}[t.Start,t.End]$ can be split into two subsequences, which form a set of patterns whose local confidences all equal to 100. We did not express it in our algorithm, but implemented it in the experiments.

Example 3. Consider the MLP-tree in Fig. 1. When node t7 is inserted, t5, t10, t11 have not been inserted. Then the mining process is as follows:

1) Search nodes t0 for M-LSP candidates in [4,4] of starting positions. Because the duration is too small, no candidate is found.

2) Search nodes t2 for M-LSP candidates in starting duration [1,4](leaf nodes t4,t9) and [2,4](leaf node t6). Results: <"ud",[1,5]>, <"ud",[2,5]>. Then search the ancestor node of t2 for M-LSPs. Results: <u→d,[1,5]>, <u→d,[2,5]>

3) Since the next ancestor node t1 is on the first level of the tree, no M-LSP candidate will be found. Therefore the searching process is terminated.

```
Algorithm 1 M-LSP (SS, ms, mc)
Input: set of sequences SS={S1,S2,…,SM}, minimal support
ms, minimal confidence mc.
Output: M-LSPs in SS.
FOR n=1 TO |S|
  FOR each S∈SS
    insert S[n , |S|] into a MLP-tree T, leaf node p
    update corresponding Leaf Chain
    FOR each ancestor node t2 of p
      AB = subsequence (t2)
      FOR each leaf node t of T(t2)
        D = [t.Position, n+|AB|-1]
        Supp = Lf(AB,D)·t2.offset/|D|
        IF Supp ≥ms THEN
          FOR each ancestor node t1 of t2
          A = subsequence (t1)
          Conf = Lf(AB, D)/Lf(A, D)
          IF Conf ≥mc then output <A→(AB - A) D>
      END FOR
    END FOR
  END FOR
END FOR
```

Similar to that of a suffix tree [2], the storage complexity of a MLP-tree is about O($|SS||S_m|$). The time complexity of the M-LSP algorithm is approximately O($|SS|^2|S|^2$). Compared with traditional mining algorithms in which the core operations are pattern generation and tuple counting, our method costs more storage for maintaining the MLP-tree. The new prospects for the data, which result from our algorithm, may justify the added expense.

The expenses of both time and storage grow with the length of the sequences. This becomes a serious problem when the sequences are extremely huge. On this occasion, we can divide all the sequences into a set of continuers segments with the same length so that *S=S1S2…Sp…*, and apply M-LSP algorithm on each segment. Using this strategy, the MLP-tree is re-initialized in each segment. Therefore the size of MLP-tree is restricted, and then the time for searching through the tree is also reduced to approximately O($|SS|^2|S|$).

5 Experimental Results

We evaluated the behavior of our problem and the performance of our algorithm on both synthetic data and real data. In the experiments, the lowercase letter a..z was used as the symbol set.

Synthetic dataset was generated randomly using the following parameters: *L*: Length of sequences, *N*: Number of sequences, *PN*: The number of potential M-LSPs, *PL1*: Length of the subsequence *A* in potential M-LSP *A→B*, *PL2*: Length of the subsequence *B* in potential M-LSP *A→B*, *PD*: Average duration of potential M-LSPs,

Table 1. Experimental results on synthetic data

Parameters								**Discovered Expected Patterns**
L	N	PN	PL1	PL2	PD	PS	PC	
100	10	1	1	1	20	0.3	0.8	1
100	20	2	1	1	20	0.3	0.8	2
100	20	1	2	1	30	0.3	0.8	1
500	10	2	2	2	30	0.3	0.8	2
500	10	10	1	1	40	0.2	0.5	10
500	20	10	2	2	40	0.2	0.5	10
1K	10	10	1	1	40	0.3	0.8	10
1K	20	20	2	2	40	0.3	0.8	20
10K	20	10	1	2	40	0.3	0.8	10
10K	40	20	2	2	40	0.3	0.8	20

PS: Average support of potential M-LSPs, *PC*: Average confidence of potential M-LSPs.

The parameters for generating datasets and the results are given in Table 1. Our method successfully discovered all the potential M-LSPs. There are some additional patterns that have also been found. Some of them have the form $A \rightarrow C$, where C is any prefix of B for an expected pattern $A \rightarrow B$. Another portion of unexpected patterns are a part of an expected pattern, e.g. A→B of expected pattern AB→C. Furthermore, there are a few unexpected patterns that are not related to the patterns we generated. It comes from the randomness of the method of generating experimental data.

We also implemented the naïve frequency counting algorithm that counting every possible subsequence (introduced in section 1), and carried out experiments using generated sequence with varying length for the purpose of performance comparison. The results verified the superiority of our method to the naïve ones.

The real data used in our experiments is the Standard and Poor 500 index (S&P) historical stock data at http://kumo.swcp.com/stocks/, in which daily price movements of approximately 500 stocks have been collected daily over the time period of one year. We used the time series of closing price, and used a clustering windows method [3] to discretize it into a symbol sequence in which each symbol represents the series behavior at that time point.

Each time, we retrieved the sequences of stocks with the same business focus. From the sequences sets, totally 177 M-LSPs were discovered, such as "during the time [10,43] a period of falls was always followed by surging in sequences set *A*, during [15,48] a period of falls was always followed by slow raising in sequences set *B*", etc. With further explanations from domain experts, such knowledge could benefit the analysis of the stock market.

6 Conclusion

In this paper, we consider a data-mining problem: the discovery of local sequential patterns from a set of long temporal sequences. As we discussed, previous approaches are either inapplicable or have extremely poor time complexity for this problem. We propose using MLP-tree index to support efficient mining of this knowledge. Based on this index, our mining method can discover all M-LSPs after one scan of all the sequences. We also propose a strategy that can keep the growing of the storage expense under a threshold and make the time expense scale linearly. We evaluated our method on both synthetic data and real data. The results correspond with the definition of our problem and verify the superiority of our method to the naïve one.

References

[1] P.Weiner. Linear pattern matching algorithms. Conference Record, the IEEE 14th Annual Symposium on Switching and Automata Theory, 1973.

[2] K. Wang, Discovering patterns from large and dynamic sequential data, Special Issues on Data Mining and Knowledge Discovery, Journal of Intelligent Information Systems, 9(1), 8–33, 1997.

[3] R.Agrawal and R.Srikant. Mining sequential patterns. In Proc. of International Conference On Data Engineering. Taipei, 1995.

[4] R.Srikant, R.Agrawal. Mining sequential patterns: generalizations and performance improvements. The Fifth International Conference on Extending Database Technology. 1996.

[5] K.Wang and J.Tan. Incremental discovery of sequential patterns. ACM SIGMOD Workshop on Research Issues on Data Mining and Knowledge Discovery, Montreal, Canada. 1996.

[6] H.Mannila and H.Toivonen. Discovering generalised episodes using minimal occurences. Second International Conference on Knowledge Discovery and Data Mining (KDD-96). 1996.

[7] P.-s.Kam and A.W.-C.Fu. Discovering temporal patterns for interval-based events. Second International Conference on Data Warehousing and Knowledge Discovery. 2000.

[8] Y.Li, X.S.Wang and S.Jajodia. Discovering temporal patterns in multiple granularities. International Workshop on Temporal, Spatial and Spatio-Temporal Data Mining. Lyon, France. 2000.

[9] A. Tansel, N. Ayan. Discovery of association rules in temporal databases. 4th International Conference on Knowledge Discovery and Data Mining (KDD'98) Distributed Data Mining Workshop, NewYork, USA, August 1998.

[10] G.Das, K.Lin, H.Mannila, G.Renganathan and P.Smyth. Rule discovery from time series. the 4th International Conference on KDD. 1998.

[11] M.Spiliopoulou and J.F.Roddick. Higher order mining: modelling and mining the results of knowledge discovery. Data Mining II – Second International Conference on Data Mining Methods and Databases. 2000.

[12] X. Chen, I. Petrounias. An Integrated query and mining system for temporal association rules. The 2nd International Conference on Data Warehousing and Knowledge Discovery (DaWaK 2000), London, UK. 327–336. 2000.

The Geometric Framework for Exact and Similarity Querying XML Data

Michal Krátký[1], Jaroslav Pokorný[2], Tomáš Skopal[1], and Václav Snášel[1]

[1] Department of Computer Science, VŠB-Technical University of Ostrava, Czech Republic
michal.kratky@vsb.cz, jaroslav.pokorny@ksi.ms.mff.cuni.cz
[2] Department of Software Engineering, Charles University, Prague, Czech Republic
tomas.skopal@vsb.cz, vaclav.snasel@vsb.cz

Abstract. Using the terminology usual in databases, it is possible to view XML as a language for data modeling. To retrieve XML data from XML databases, several query languages have been proposed. The common feature of such languages is the use of regular path expressions. They enable the user to navigate through arbitrary long paths in XML data. If we considered a path content as a vector of path elements, we would be able to model XML paths as points within a multidimensional vector space. This paper introduces a geometric framework for indexing and querying XML data conceived in this way. In consequence, we can use certain data structures for indexing multidimensional points (objects). We use the UB-tree for indexing the vector spaces and the M-tree for indexing the metric spaces. The data structures for indexing the vector spaces lead rather to exact matching queries while the structures for indexing the metric spaces allow us to provide the similarity queries.

1 Introduction

Using the terminology usual in databases, it is possible to view XML as a language for data modelling. The notions like XML database and XML query language logically extend this idea [6,14]. So called native XML databases are implemented in increasing extent. To reach a quality of conventional relational databases, appropriate tools for manipulating have been designed. Among many attempts to query languages over XML data, the language XQuery [15] seems to be the leading approach now. The common feature of such languages is the use of regular path expressions. They enable the user to navigate through arbitrary long paths in XML data. Obviously, in the next step to XML databases some appropriate index structures have to be constructed for their data. Particularly, paths can be objects of indexing. In [9], we consider a path content as a vector of path elements. Then we can model XML paths as points within a multidimensional vector space. To speed-up access to such vectors, either various multidimensional trees (such as the R*-tree [4], X-tree [5] or UB-tree [2]), or metric trees can be used for their indexing (e.g., the M-tree [1] and the mvp-tree [7]). Only few these data structures have been used for indexing XML data. In

M.H. Shafazand and A M. Tjoa (Eds.): EurAsia-ICT 2002, LNCS 2510, pp. 35–46, 2002.

[9], we used UB-trees for indexing path contents for more efficient exact querying XML data. In this work we pursue a different, in some sense complementary, direction that is based on M-trees. Metric trees only require the distance between points to be a metric, thus they can be used even when no vector representation exists. We show how M-trees can be used for indexing XML paths and how similarity querying XML data can be supported. Section 2 introduces to us the geometric framework used in this paper. We shortly describe necessary basics of vector and metric spaces. Section 3 contains the vector model for indexing and querying XML data. The approach is based on the notion of path content. The main contribution of the paper – a similarity indexing XML data with M-tree – is contained in Section 4. We introduce briefly M-trees and propose a cumulated metric based in the Hamming metric for indexing XML paths. The section is completed with experimental evaluation of M-tree index applied on a real XML data set. In conclusions we summarize the approach.

2 Geometric Framework

In our approach to indexing and querying XML data we exploit the properties of two geometric models. Both of these models treat the XML data as objects/points within a space. In the first case within a *vector space* and in the second case within a *metric space*. As we will see, each of the models is suitable for a different purpose. We can say that they are complementary to each other.

There are two initial problems. First, we need to find a technique of transformation (so-called feature transformation) of the XML data into objects within a vector or metric space. Second, we need to find the data structures for storage and effective querying XML data according to the given model.

2.1 Vector Spaces

Vector model treats the XML data as points within multidimensional vector space. This approach allows us to index values and even the structure of XML documents and provides an ability of *exact matching* range queries. High vector space dimension (greater than approx. 20) is unfortunately associated with *curse of dimensionality* which has a negative influence on the range queries efficiency (see [3]). A representative data structure for the vector model is the UB-tree (see [2]). We discuss the vector model for indexing and querying XML data in Section 3.

2.2 Metric Spaces

In a metric space there are generally neither the dimension nor the vectors. However, in this paper we share the same representation of objects for the metric spaces and for the vector spaces – i.e. multidimensional points. An important difference is that each metric space has defined a metric – i.e. function measuring a distance (or similarity) between every two objects. This function d must satisfy following conditions:

$$d(o_i, o_i) = 0 \tag{1}$$
$$d(o_i, o_j) > 0 \qquad (o_i \neq o_j) \tag{2}$$
$$d(o_i, o_j) = d(o_j, o_i) \tag{3}$$
$$d(o_i, o_k) + d(o_k, o_j) \geq d(o_i, o_j) \tag{4}$$

The presence of the metric prompts that the metric model provides an ability of *similarity queries*. A representative data structure for the metric model is the M-tree, see section 4.

3 The Vector Model for Indexing and Querying XML Data

In our approach to indexing XML documents we model the XML data as points within multidimensional vector space and thus we can use certain index structures for multidimensional indexing (for example UB-tree). This approach was introduced in [9]. The data structures for indexing the vector spaces lead rather to *exact queries.*

We distinguish between indexing XML data with and without "mixed content" in [9]. Here we show only the latter case. The example of DTD for documents without "mixed content" and an XML document valid w.r.t. the DTD are in Figure 1a) and 1b), respectively. We will not consider the attributes of elements in our approach.

Example 1 (Querying XML document).
The example of the DTD and the valid XML document is in Figure 1. The path `accounts/account/name` denotes a query for obtaining all account customer names from the document.

```
<!DOCTYPE accounts [
  <!ELEMENT accounts (account*)>
  <!ELEMENT account (id, name)>
  <!ELEMENT id (#PCDATA)>
  <!ELEMENT name (#PCDATA)>
]>
```

a)

```
<?xml version="1.0" ?>
<accounts>
 <account><id>1234-8952</id>
   <name>Thomas Newell</name></account>
 <account><id>1234-4123</id>
  <name>David Moore</name></account>
 <account><id>5842-5321</id>
  <name>David Moore</name></account>
</accounts>
```

b)

Fig. 1. a) An example of DTD for XML documents without "mixed contents". b) An example of valid XML document without "mixed contents".

3.1 Indexing Path Contents and XML Structure

In our approach to indexing XML documents, we consider the *n-dimensional points representing path contents for XML structure indexing* of all paths from the root to all its leafs. The dimension n of the space is equal to the length of the maximal path in XML-tree, i.e. the number of edges from the root to its leaf element. To estimate the number n from DTD, we will consider only the "nonrecursive" DTDs in our approach.

Definition 1 (path content).
Given a path $e = e_1/e_2/\ldots/e_k$, $e \in \mathcal{X}_P$, $\mathcal{X}_P$ *is* ***set of paths****, the* ***path content*** *is defined as a sequence of string values* $s = s_1/s_2/\ldots/s_k$, $s \in \mathcal{X}_{PC}$, $\mathcal{X}_{PC}$ *is* ***set of path contents****. Each* s_i*, except* s_k*, can be empty* (ϵ)*.*

Because string values can have a different length, it is necessary to use a procedure, which maps different strings into binary numbers of the same length. We use the signatures in our approach (e.g. [10]). The main idea of signatures is to reflect the data items into bit patterns and store them in a separate file which acts as a filter to eliminate the non-qualifying data items for an information request. We will denote the function generating signatures by sig(x), where x is a variable of string type.

The XML document is represented by m points within n-dimensional space, where m is the considered number of path contents. All these points are inserted into any index structures for multidimensional indexing. All complete paths contents are stored in other data structures. It is important to create binding between the elements of XML document having the same parent. We can create this binding using the elements unique numbers in the point representing path content for XML structure indexing. Of course, it is possible to index even paths (see Section 3.2).

Example 2 (Transformation of XML data to n-dimensional points).
We will show the transformation of the XML document from 1b) to the points of multidimensional space. We see the space has $n = 3$. We determine the length of the domains as 64b. This signature value is large enough for the signature s_i. But generally, there is not cause for domain cardinalities to be the same. The cardinality of domain for signatures of #PCDATA and for unique numbers of root elements can be different for example. The important role plays here the analysis of DTD.

If we are browsing through document in Figure 1b), then the following path contents are obtained: ϵ/ϵ/`1234-8952`, ϵ/ϵ/`Thomas Newell`, ϵ/ϵ/`1234-4123`, ϵ/ϵ/`David Moore`, ϵ/ϵ/`5842-5321` and ϵ/ϵ/`David Moore`. It is necessary to group these path contents according to the relationship to particular `accounts` and `account` elements. Therefore we nest the unique numbers of `accounts` and `account` elements into 1st (2nd respectively) coordinate of points representing path contents. The points representing path contents will be `(0,0,sig("1234-8952"))`, `(0,0,sig("Thomas Newell"))`, `(0, 1, sig("1234-4123"))`, `(0, 1, sig("David Moore"))`, `(0, 2, sig("5842-5321"))`, and `(0, 2, sig("David`

Moore")). The points in 3-dimensional space are depicted in Figure 3.1. These points are inserted into the indexing structure. If we index paths (see Section 3.2), then we will work with 4-dimensional space.

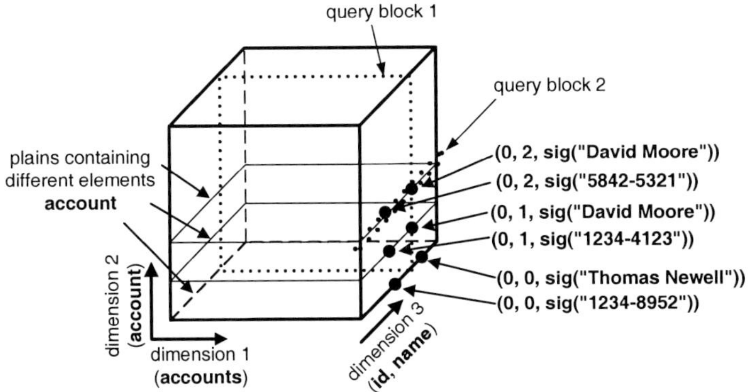

Fig. 2. The 3-dimensional space with indexed XML document without "mixed contents".

Example 3 (Querying XML document).
We show now how it is possible to query the XML document from Figure 1b) transformed to the points within multidimensional space by above mentioned technique. Let as take the query `accounts/account[name='David Moore']`, i.e. we want to get all `account` elements for `David Moore`'s account. First we need to transform this query to the *range query*. It means to find all the points from Figure 3.1, that are contained by *query block 1*. It is necessary to determine the coordinates of two points defining the query block. By means of range query we get the points from the 3-dimensional space which represent the unique numbers of parent elements of `name` element with content `David Moore`. We get the result set and if user will want to obtain the contents of child elements of any `account` element, for example, then the query block like *query block 2* from Figure 3.1 is effected for their retrieval. To distinguish the points representing the path contents for different paths it is necessary to index even the paths (see Section 3.2).

3.2 Indexing Paths

The indexing XML data as it is proposed in Section 3.1 considers only a path content. If the XML document is transformed to points of a space in this way, the element tags are lost. If we consider the XML document from Figure 1 then we will be not able to distinguish the points representing the path contents for paths `accounts/account/id` or `accounts/account/name`.

We consider a binary relation *PPC* [9] between paths and their path contents. All points representing paths will be inserted to other index structure. Besides the point coordinates and pointers to data structures containing the whole paths we insert even the path unique numbers in another dimension of the space which contains the path contents. In fact, the relation PPC is built by adding other dimension to the space which contains path contents, i.e. the dimension of space will be $n+1$. It is hereby possible to index even the documents valid to different DTD in one index structure in this way.

Example 4 (Indexing paths).
We get two different paths `accounts/account/id` and `accounts/account/name` from the XML document in Figure 1b). So we get two points (we get two paths) representing paths in 3-dimensional space (paths contain three elements). These points are inserted into other indexing structure. The point `(sig("accounts"), sig("account"),sig("id"))` representing path `accounts/account/id` is inserted with unique number 0 and point representing path `(sig("accounts"), sig("account"),sig("name"))` with unique number 1. The points representing paths are in Figure 3. The points representing the path contents have last coordinate equal to the unique number of the associated path.

We see the space to have $n = 4$ (one dimension will be for unique numbers of paths). The gained path contents are in Example 2. Let as take the path content ϵ/ϵ/`1234-8952` and point representing the path contents `(0,0,sig("1234-8952"))` for example. The path unique number of path `accounts/account/id` from index structures which contain points representing paths is append as fourth coordinate to the point. We get point `(0,0,sig("1234-8952"),0)` in this manner. The all six points gained by the same way are inserted into index structure containing path contents.

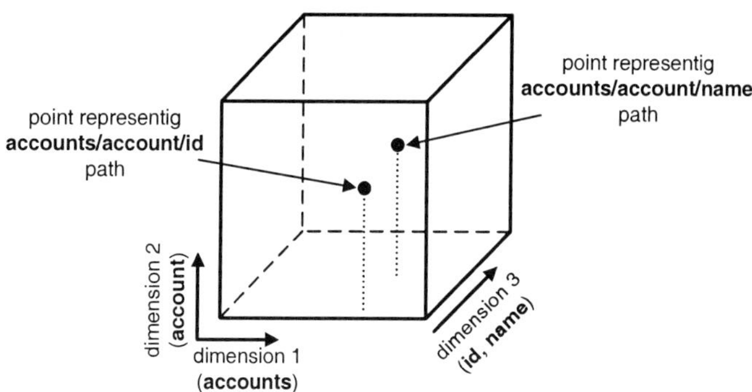

Fig. 3. The 3-dimensional space with points representing paths.

Example 5 (Querying XML document).
It is important to get by a point query the unique number of the desired path from index structure containing paths. After that we get desired points by a range query from the indexing structure containing path contents. It is necessary to

work with four dimensions in the case of defined coordinates of points determining the query block.

4 Similarity Queries

Another aspect of indexing XML data, in addition to the structural indexing, is the *similarity indexing*. In such an XML index we can query for XML objects that are similar to a query object.

Properties of metric spaces, where the metric represents the notion of similarity, are suitable formal basis for indexing similarities inside XML data. Following subsection describes a data structure M-tree which allows to index general objects of metric spaces.

4.1 M-Tree

Data structure M-tree (introduced in [1] and closely discussed in [13]) was developed for indexing and querying objects within metric spaces. Its main characteristic is that M-tree allows to process similarity queries. It is, in fact, dynamic, persistent, paged and balanced tree like e.g. the B-tree. The difference is in the semantics of the nodes. Indexed objects themselves, i.e. *ground objects*, lie in the leaf nodes. The inner nodes contain *routing objects* that represent a hierarchy of specific metric regions.

- The record of a routing object O_r in inner node contains:
  1. a ground object O_r (its significant properties respectively). This ground object determines the center of the metric region.
  2. pointer $ptr(T(O_r))$ to its own subtree $T(O_r)$ – i.e. *covering tree*
  3. value $r(O_r)$ – *covering radius* of the metric region
  4. value $d(O_r, P(O_r))$ – distance to the parent routing object $P(O_r)$

  Notes:

  The ground object in the routing object (inner node) is one of the ground objects remaining in the child leaf nodes of $T(O_r)$. The distance function d is a metric of a metric space.
- The record of a ground object looks similarly, but it also contains $oid(O_j)$ – identifier of the whole object (stored outside of M-tree) – instead of covering tree and covering radius.

Hierarchy of M-tree is based on partition of the metric space onto metric subregions which do not have to be strictly disjunct. This regions are formed by the routing objects O_r where the child routing objects (their regions respectively) and the child ground objects of its covering tree $T(O_r)$ are within the distance $r(O_r)$ to the center of O_r. Formally,

$$\forall O_i \in T(O_r), \qquad d(O_r, O_i) \leq r(O_r)$$

The precalculated distance value $d(O_r, P(O_r))$ to the parent object along with the covering radius $r(O_r)$ allow to eliminate the *untouched regions* from the process of an operation on M-tree (i.e. searching, insertion, deletion). Structure of the M-tree and the routing object relations are depicted in figure 4.

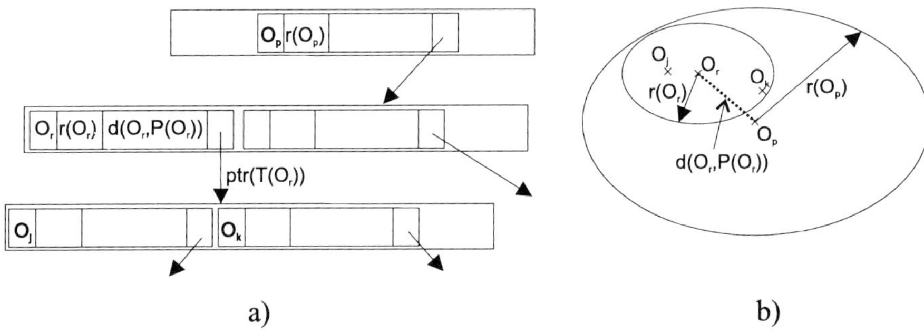

Fig. 4. (a) Nodes of M-tree contain object records. (b) Routing objects – metric regions.

Searching the M-tree. We must take into account two factors of complexity when we make some operation on the M-tree. The first one is the number of *accesses to disk pages* (number of regions being searched respectively) and the second one (specifically to M-tree) is the number of *distance calculations*. The goal is to minimize both these factors.

We can meet two kinds of queries by metric trees. The *range queries* search for all the objects within certain distance to the query object. The *k-nearest neighbours queries* search for the first k nearest objects to the query object. In both cases we can see a tendency to order the metric space – relative to the query object.

Managing the regions. The crucial factor of the M-tree's cost-effectiveness is a "good layout" of the metric regions stored within the M-tree. As we have said earlier, the regions can overlap another ones. This property arises from the M-tree's universality which is caused due to specifying only a metric of the metric space. High "overlap rate" leads, in the worst case, to sequential search – i.e. to linear complexity.

With the design of the M-tree there were also developed some techniques for minimizing this "overlap rate". The first technique is "embedded" into the phase of a tree node(page) splitting and consists of a choice of *split policy* and a mechanism of creating the best routing object – *promoting* phase. This is the dynamic technique. The second technique, more efficient, is the *bulk loading* algorithm. This algorithm takes at the beginning the whole collection of objects and loads all of them into the empty M-tree at once. The loading is based on preliminary clustering where prospective regions of objects are created at once. This is the static method.

Summary. M-tree is balanced, highly parametrizable data structure making possible to index objects of a metric space. The M-tree operations are performed with approximately logarithmic time complexity (if well build) but the M-tree doesn't represent a complete linear order like other trees (B-tree, UB-tree, ...) do. On the other side, M-tree is more general than the *Spatial Access Methods* based on vector spaces.

4.2 Indexing XML Data with M-Tree

If we consider XML paths as simple objects, we can index such objects into a metric space or actually into the M-tree. For example, path `BOOK/AUTHOR/SURNAME` is object to store within M-tree. All paths in given XML document(s) can be transformed in this way into a collection of this simple XML objects. XML objects can also have assigned to every element tag its element content, which will increase the number of unique objects. For example, `BOOK{technical}/AUTHOR{writer}/ SURNAME{Walsh}`, but furthermore, for simplicity, we will ignore the possibility of any content.

XML object o_i (path) can be represented as a variable vector of strings (element tags), $o_i = (o_i^1, o_i^2, \cdots, o_i^{l_i})$.

Choosing metric for paths. Metric chosen for XML indexing must take as arguments two XML objects (paths) and calculate distance between them. We propose as an example *cumulated metric* which is defined as:

$$D(o_i, o_j) = \sum_{k=1}^{max(l_i, l_j)} d(o_i^k, o_j^k)$$

where $d(x, y)$ is an ordinary metric (e.g. Hamming metric) between two strings.

Hamming metric [8] adds up the mismatching pairs of characters where the first character of a pair is located on a position in the first string while the second character is on the same position in the second string. Formally,

$$d_H(x, y) = \sum_{i=1}^{min(|x|,|y|)} sgn(|x[i] - y[i]|) + ||x| - |y||$$

For example, d_H(`AUTHORS`, `AUTOMATON`) $= 0+0+0+1+1+1+1+2 = 6$

Example 6.

Let d be the Hamming metric. Then

D(`BOOK/AUTHOR/SURNAME`, `BOOK/AUTHOR/FIRSTNAME`) $= 0 + 0 + 8 = 8$

Let d be the discrete (yes/no) metric. Then

D(`BOOK/PREFACE/TITLE`, `BOOK/BOOKINFO/TITLE`) $= 0 + 1 + 0 = 1$

Note: In this section, the paths used in examples are generated according to the DocBook DTD, see [12].

Processing queries. We have defined objects of metric space (XML paths) as well as metric (cumulated metric) thus we have accomplished the requirements for indexing with the M-tree.

We can distinguish two types of queries:

1. *similarity queries.* An object o_i in query result is within some distance r (*query radius*) to the query object o_q, i.e. the M-tree is traversed with condition $D(o_q, o_i) \leq r$. This kind of query allows to obtain the similar XML paths.

Example 7 (cumulated Hamming metric).
Query object = `BOOK/PART/CHAPTER/PARA/ACRONYM`, $r = 6$
Query result = {`BOOK/PART/CHAPTER/PARA/ACRONYM` (distance 0)
`BOOK/PART/CHAPTER/PARA/SCREEN` (distance 4)
`BOOK/PART/CHAPTER/TITLE/ACRONYM` (distance 5)
`BOOK/PART/CHAPTER/PARA/FILENAME` (distance 6) }

2. *exact matching queries.* An object o_i in query result must exactly match the query object o_q, i.e. the M-tree is traversed with the condition $D(o_q, o_i) = 0$. This is the special case of similarity query with $r = 0$ – no differences are allowed.

Notes:

- The query object is not expressed by any query language, its structure is the same as the structure of any ground object.
- The syntax of query object can be extended with keyword "*", where using this keyword on the k-th coordinate of object vector brings evaluation of $d(o_q^k, o_i^k)$ always as 0 (match).
  Example: D(`BOOK/AUTHOR/*`, `BOOK/AUTHOR/FIRSTNAME`) $= 0 + 0 + 0 = 0$.
  This extension allows to treat the exact matching queries as range queries.
- The objects in query result give only the information about existence of such paths in XML tree but the objects cannot tell the exact location. This lack of "context" can be removed with additional property of XML object – unique identifier of the last path element pointing into an external data structure (e.g. the source XML tree or UB-tree index). This improvement makes possible the consequential navigation in the external XML tree.

4.3 Testing with M-Tree

We have performed particular tests with M-tree – XML path indexing and XML similarity queries. XML data we have indexed was a XML file containing the documentation to DocBook. The size of this file was about 3MB.

In the first phase, we have transformed the whole file into collection of XML objects (unique paths) – 972 unique paths were extracted. Second, we have inserted all of these objects one-by-one into the M-tree. Page size of the M-tree was 1kB and cumulated Hamming metric was chosen. Each object (path vector) of the M-tree was aligned on size of 256 bytes.

Table 1. M-tree statistics

Level	Pages count	Avg. radius
0 (root)	3	207.33
1	5	183.00
2	10	135.40
3	16	121.75
4	23	102.74
5	34	85.18
6	53	65.79
7	83	54.02
8	141	37.14
9	233	22.95
10	376	13.12
11 (leafs)	590	6.01

After the indexing phase, the M-tree has acquired following statistics: pages(nodes) count: 1568, leafs count: 590. Table 1 shows for each level of the M-tree its pages count and average radius of all routing objects(regions) within the level.

Furthermore, disk access costs test was performed. A series of queries was produced by specifying the query object as:

("BOOK/PART/CHAPTER/SECT1/SECT2/PARA/ACRONYM") and by increasing the query radius from $r = 0$ (exact matching query) to $r = 32$. The results are shown in figure 5.

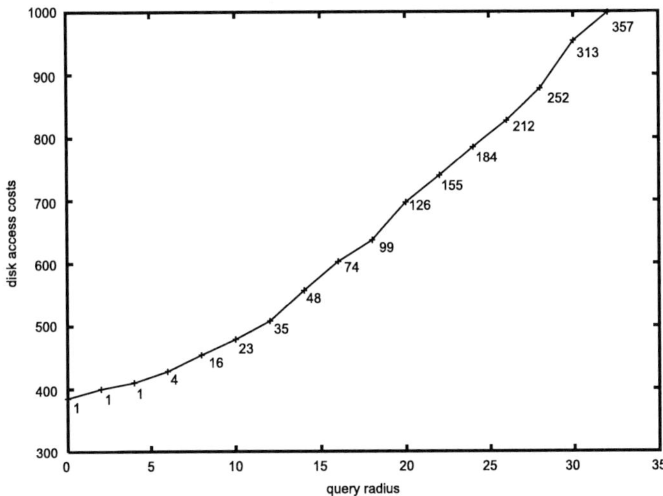

Fig. 5. Results of disk access costs test. The numbers below particular results are the total numbers of objects returned in particular query result (objects similar to the query object within the current radius).

5 Conclusions and Outlook

In this paper we have shown that XML data can be modelled in multidimensional vector spaces and in metric spaces. We use the UB-tree for indexing the vector spaces and the M-tree for indexing metric the metric spaces in our approach of indexing XML data. The data structures for indexing the vector spaces lead rather to the *exact queries* while structures for indexing of the metric spaces allow us to provide *similarity queries.* In the course of writing this paper some interesting questions appeared, e.g. new metric designs or different feature transformations. Their solution will be the topic of our future work. Furthermore, presented data structures are independent and our intention is to integrate them into a single hybrid data structure providing a possibility of XML data storage and also efficient exact and similarity querying.

Acknowledgments. This research was supported in part by GACR grant 201/00/1031.

References

1. Ciaccia P, Pattela M., Zezula P.: M-tree: An Efficient Access Method for Similarity Search in Metric Spaces. *Proc. 23rd Athens Intern. Conf. on VLDB (1997)*, 426–435.
2. Bayer R.: The Universal B-Tree for multidimensional indexing: General Concepts. In: *Proc. Of World-Wide Computing and its Applications 97 (WWCA 97)*. Tsukuba, Japan, 1997.
3. Böhm C., Berchtold S., Keim D.A.: Searching in High-dimensional Spaces – Index Structures for Improving the Performance of Multimedia Databases. *ACM*, 2002
4. Beckmann, N., Kriegel, H.-P., Schneider, R., Seeger, B.: The R*-tree: An efficient and robust access method for points and rectangles. In: *Sigmod'90*, Atlantic City, NY, 1990, pp. 322–331.
5. Brechtold, S., Keim, A., Kriegel, H.-P.: The X-tree: An index structure for high-dimensional data. In: *Proc. of 22nd Intern. Conference on VLDB'96*, Bombay, India, 1996, pp. 28–39.
6. Bourret, R.: *XML and Databases.* `http://www.rpbourret.com/xml/ XMLAndDatabases.htm`. 2001.
7. Bozkaya, T., Özsoyoglu, M.: Distance-based indexing for high-dimensional metric spaces. In: *Sigmod '97*, Tuscon, AZ, 1997, pp. 357–368.
8. Baeza-Yates, R., Ribeiro-Neto, B.: *Modern Information Retrieval.* Addison Wesley, New York, 1999.
9. Krátký M., Pokorný, J., Snášel V.: Indexing XML data with UB-trees. *ADBIS 2002*, Bratislava, Slovakia, accepted.
10. Lee, D.L., Kim, Y.M., Patel, G.: Efficient Signature File Methods for Text Retrieval., *Knowledge and Data Engineering*, Vol. 7, No. 3, 1995, pp. 423–435.
11. Markl, V.: *Mistral: Processing Relational Queries using a Multidimensional Access Technique*, `http://mistral.in.tum.de/results/ publications/Mar99.pdf`, 1999
12. *The DocBook open standard*, Organization for the Advancement of Structured Information Standards (OASIS), 2002, `http://www.oasis-open.org/committees/docbook`
13. M. Patella: *Similarity Search in Multimedia Databases.* Dipartmento di Elettronica Informatica e Sistemistica, Bologna 1999 `http://www-db.deis.unibo.it/ ~patella/MMindex.html`
14. Pokorný, J.: XML: a challenge for databases?, Chap. 13 In: *Contemporary Trends in Systems Development* (Eds.: Maung K. Sein), Kluwer Academic Publishers, Boston, 2001, pp. 147–164.
15. *XQuery 1.0: An XML Query Language.* W3C Working Draft 20 December 2001, `http://www.w3.org/TR/2001/ WD-xquery-20011220/`

A Mobile System for Extracting and Visualizing Protein-Protein Interactions

Kyungsook Han and Hyoungguen Kim

Department of Computer Science and Engineering, Inha University, Inchon 402-751, South Korea
khan@inha.ac.kr

Abstract. Recent improvements in proteomics technology have produced a large volume of protein interaction data. However, understanding protein-protein interactions has not kept pace with the rapidly expanding amount of protein interaction data. In addition to the volume of the data, the difficulties in analyzing protein interaction data come from (1) data is disseminated in many different database, (2) different databases use different HyperText Markup Language (HTML) tags to represent proteins and their interactions, and (3) HTML says nothing about the semantics of a document. This paper presents the development of a new, mobile application for making the task of analyzing protein interaction data easy and customized to individual researchers' needs. To the best of our knowledge, this is the first mobile application for mining and visualizing bioinformatics data. A HTML parser was developed to extract protein interaction data from remote databases and to construct a local database with the extracted data. A program running on a mobile computer dynamically queries the local protein interaction database and visualizes the query results in three-dimensional space, based on our heuristic layout algorithm. The three-dimensional drawing can be explored further by rotating or by zooming in or out of it. Experimental results demonstrate that this system can be used as useful aids in studying protein-protein interactions in a variety of different environments.

1 Introduction

Today's information technology has seen a rapid movement of end-devices from desktop computers to mobile computer systems such as PDAs, handheld PCs, and embedded computers in cellular phones or personal appliances. As mobile computer systems improve in efficiency and multimedia Internet services become widely available on mobile devices, we will undoubtedly see innovations in mobile application programs. Over the past few years, many applications and communication services that were originally designed for use on desktop computers have been extended to mobile computer systems. Despite the widespread use of mobile computer systems in many applications, bioinformatics has received little attention.

The nature of a mobile computer allows it to be used whenever a user needs it and in a variety of different environments. In the next few years, people will be able to use

M.H. Shafazand and A M. Tjoa (Eds.): EurAsia-ICT 2002, LNCS 2510, pp. 47–56, 2002.

mobile computers for bioinformatics applications, whether they are at work or at home. Consider a simple scenario in which a mobile application for bioinformatics is needed. Suppose a researcher in bioinformatics is attending a conference. He has brought a PDA, which is smaller and more lightweight than a laptop computer. In his hotel room, he may wish to be kept informed of new bioinformatics data related to his research as well as to continue his work on analyzing bioinformatics data. With current technology, new bioinformatics data customized to an individual researcher, if obtainable at all, would have to be assembled manually and transmitted to the researcher's mobile computer when it is connected to the network. Collecting such information by manual work is tedious, labor-intensive and error-prone.

The work described in this paper represents our first steps towards realizing such a future. By extending a bioinformatics application to mobile computer systems, the bioinformatics application provides *service mobility* as well as *terminal mobility*. Service mobility refers to access to service from any device such as PDAs, handheld PCs, and desktop computers [1]. Terminal mobility refers to the provision of seamless service during physical movement of the user. A general principle in designing a mobile application program is to always keep in mind the impact the design will have on the power consumption and network traffic load. Mobile computer systems differ from conventional desktop computers in terms of their computing power, memory size, display area, network connection method, and the possibility of data loss, which require a different design of an application. Consequently, the development of an efficient mobile application is the major challenge for the success of the application running on a mobile computer with limited computing power, memory and display area.

In this paper, section 2 introduces the problem we attempt to solve, section 3 describes the method for extracting bioinformatics data from remote databases, section 4 discusses visualization of the data on a mobile computer, section 5 describes our implementation results, and section 6 summarizes the general lessons learned from this project and topics for future work.

2 The Problem

The science of proteomics attempts to understand cellular processes and networks by elucidating interactions between all proteins in a cell or organism [2]. Recent improvements in experimental proteomics techniques such as yeast two-hybrid [3, 10] have produced a rapidly expanding volume of protein-protein interaction data of an unprecedented scale. The interaction data is available either in text files or in databases. However, due to the volume of data (e.g., thousands of interacting proteins), a graphical representation of protein-protein interactions has proven to be much easier to understand than a long list of interacting proteins, prompting visualization studies of protein-protein interaction networks.

Existing drawing programs are of limited use in visualizing protein interactions, either because they produce a cluttered drawing with many edge crossings or a static drawing that is not easy to modify to reflect changes in data, or because they require the input data to be in a specific format rather than taking the data directly from protein-protein interaction databases. As an example of such a drawing program, a Java applet program [5] has been developed for drawing protein interactions based on a relaxation algorithm and tested on the yeast two-hybrid (Y2H) data [10]. This program requires all protein-protein interaction data to be provided as parameters of the applet program in html sources. There is no way to save a visualized graph except by capturing the window. An image captured from the window is a static image and is of a generally low quality. Since it is a static image, it cannot be refined or changed later to reflect an update in data. A user can move a node but cannot select or save a connected component containing a specific protein for later use.

Other visualization works on protein-protein interactions do not have their own algorithms or programs developed for visualization, but use general-purpose drawing tools. PSIMAP [4, 7], for example, displays interactions between protein families by comparing the Y2H data with the DIP data [11] using Structural Classification of Proteins [6]. PSIMAP was drawn by Tom Sawyer software (http://www.tomsawyer.com/) and then refined by significant amount of manual work to remove the edge crossings of the map. From the perspective of graph drawing, PSIMAP is a static image and leaves several things to be improved.

A research group of University of Washington [8, 9] has visualized the Y2H data using another general-purpose drawing tool called AGD (http://www.mpi-sb.mpg.de/AGD/). AGD produces a two-dimensional graph drawing and the visualization result is satisfactory most of the time. Although AGD is powerful, it is a general-purpose drawing tool and does not provide functions that we hold are necessary for studying protein-protein interactions. For example, most protein-protein interaction data, including the Y2H data, generates a disconnected graph consisting of many connected components. The graph is also a nonplanar graph with a large number of edge crossings that cannot be removed in a two-dimensional drawing. One way to analyze such a graph would be to work on individual connected components or subgraphs containing a specified protein. Alternatively, the nonplanar graph could be visualized as a three-dimensional drawing with no edge crossing. However, AGD does not provide these functionalities.

We have developed an integrated modeling system that queries protein-protein interaction databases and dynamically visualizes the query results in three-dimensional space at runtime [15]. Visualized networks can be further refined or used to explore protein-protein interactions. The goal of the work described in this paper is to extend the modeling system to the mobile personal computer and to make the visualization and analysis of updated data easy and customized to individual researchers' needs (see Figure 1 for the system architecture of the entire modeling system). Readers interested in the modeling system are encouraged to contact the authors for more information.

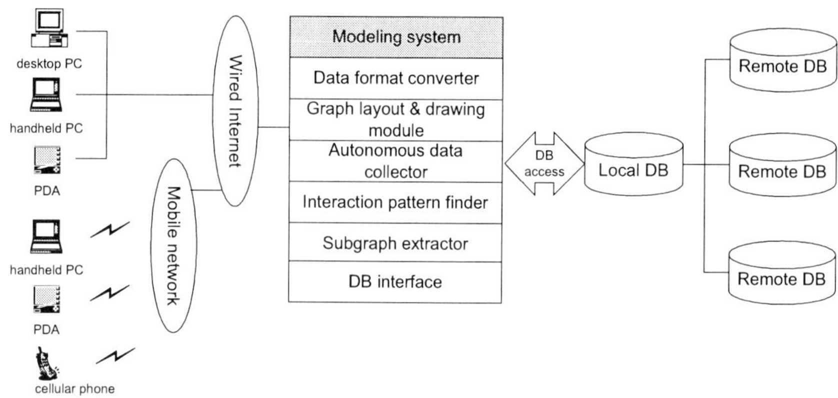

Fig. 1. The architecture of the entire modeling system of protein-protein interactions

3 Extracting Protein Interaction Data

Protein interaction data is available in several databases, which can be accessed using a Web browser. The difficulties in extracting protein interaction data come from (1) data is disseminated in many different database, (2) different databases use different HyperText Markup Language (HTML) tags to represent proteins and their interactions, and (3) HTML says nothing about the semantics of a document.

We have developed a HTML parser to extract protein interaction data from remote databases. The extracted data is maintained in local databases rather than in files because the protein interaction data is of large volume and may change over time. Currently the local database contains protein interaction data from the Biomolecular Interaction Network Database (BIND; http://www.binddb.org/) and both genetic and physical protein interaction data from the MIPS database (http://mips.gsf.de/proj/yeast/tables/interaction/). In the HTML source of the MIPS database, for example, three lines of the following format repeatedly occur.

```
<TD><FONT...><A href="..." >ABP1</A> </FONT></TD>
<TD><FONT...><A href="...">ACT1</A> </FONT></TD>
<TD><FONT...>unlinked ...</FONT></TD>
```

The first two lines contain protein names and the third line describes the interaction between the two proteins. From the entire HTML source code, we extract only the contents enclosed by <TD> and </TD> tags to obtain a filtered HTML source code (see Figure 2A). The <TD> and </TD> tags are then removed from the filtered HTML source code and only protein interaction data (Figure 2B) is stored in the local database. The local database maintains the node table and edge table, as shown in Figure 3. Protein interaction data from the BIND data can be extracted in a similar way. However, protein names are enclosed by <B> and </B> tags, which are in turn enclosed by <DD> and </DD> tags for protein interactions (Figure 4).

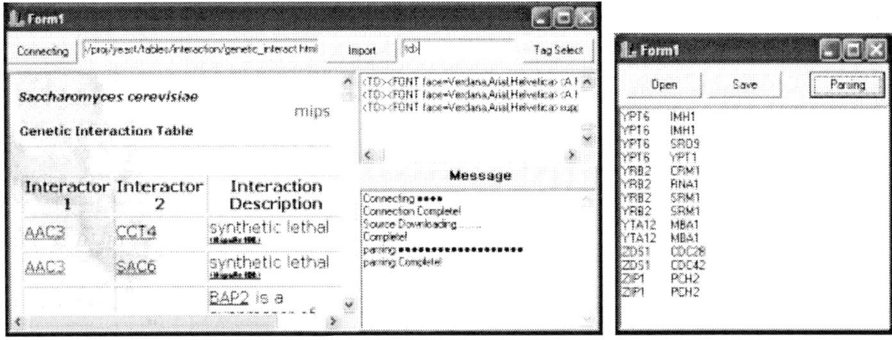

Fig. 2. (A) The HTML parser extracting protein interaction data from the MIPS database. The left windows shows a Web page when connected to the database and the top right window shows a filtered HTML source code of the Web page, which contains only parts enclosed by <TD> and </TD> tags. (B) The final protein interaction data extracted from the MIPS database. Each line contains a pair of interacting proteins.

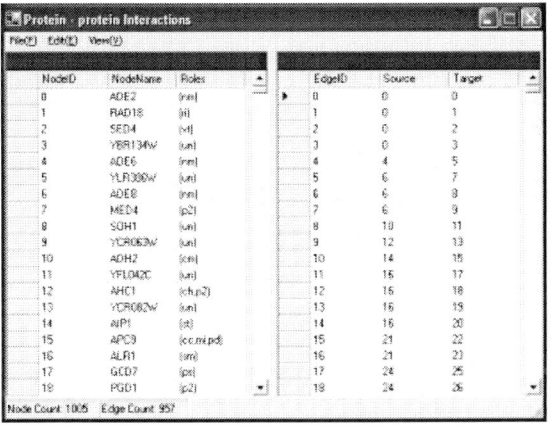

Protein - protein Interactions

File(F) Edit(E) View(V)

NodeID	NodeName	Roles
0	ADE2	(nm)
1	RAD18	(ii)
2	SED4	(vt)
3	YBR134W	(un)
4	ADE6	(nm)
5	YLR380W	(un)
6	ADE8	(nm)
7	MED4	(p2)
8	SOH1	(un)
9	YCR063W	(un)
10	ADH2	(cm)
11	YFL042C	(un)
12	AHC1	(cft,p2)
13	YCR082W	(un)
14	AIP1	(ct)
15	APC9	(cc,mi,pd)
16	ALR1	(im)
17	GCD7	(px)
18	PGD1	(p2)

EdgeID	Source	Target
0	0	0
1	0	1
2	0	2
3	0	3
4	4	5
5	6	7
6	6	8
7	6	9
8	10	11
9	12	13
10	14	15
11	16	17
12	16	18
13	16	19
14	16	20
15	21	22
16	21	23
17	24	25
18	24	26

Node Count 1005 Edge Count 957

Fig. 3. The local database maintains two tables, node table and edge table. NodeID and EdgeID are the primary keys of the two tables.

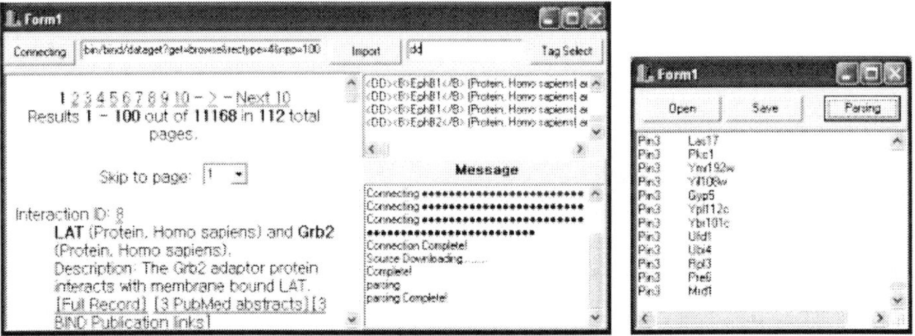

Fig. 4. (A) The HTML parser extracting protein interaction data from the BIND database. (B) Protein interaction data extracted from the BIND database.

4 Visualizing Protein Interaction Networks

Protein interaction data can be visualized as a graph in which nodes represent proteins and edges represent protein-protein interactions. The *degree* of a vertex is the number of its edges. An edge (u, v) with $u=v$ is a *self-loop*. From our experience with protein interaction data, the data can be characterized as follows:

(1) When visualized as a graph, the data yields a disconnected graph with many connected components. The data often contains protein interactions corresponding to self-loops.
(2) The data may contain multiple interactions for an identical pair of proteins.
(3) Each protein has very different number of interacting proteins within a same set of data, so a graph visualizing the data contains nodes of very high degree as well as those of low degree.

A protein interaction network generated by the program can be saved either in an image file, the local database or a text file in what we call a simplified GML format. GML, the graph modeling language, is a popular graph file format [12]. An advantage of saving a graph in GML format is that we can resume visualization of a graph by loading its GML file. The reason we use a simplified GML format instead of a full GML format on a mobile computer is because the simplified format takes much less space. We define the simplified GML format as follows, which is described in BNF notation. In this format, x^* denotes a sequence of zero or more x items. An example of the simplified GML format is shown later in Figure 7A.

```
<SimplifiedGML> ::= <NodeCount><EdgeCount><NodeList><EdgeList>
<NodeCount> ::= <integer>
<EdgeCount> ::= <integer>
<NodeList> ::= (<NodeIndex><NodeName><Annotation><NodePos>)*
<EdgeList> ::= (<NodeIndex><NodeIndex>)*
<NodeIndex> ::= <integer>
<NodeName> ::= <string>
<Annotation> ::= <NULL> | <string>
<NodePos> ::= <x-coord> <y-coord> <z-coord>
<x-coord> ::= <double>
<y-coord> ::= <double>
<z-coord> ::= <double>
```

Most protein interaction data, when visualized, generates nonplanar graph with a large number of edge crossings that cannot be removed in a two-dimensional drawing. Since edge crossings significantly decrease the readability of the graph, our program visualizes a protein interaction network as a three-dimensional drawing using a heuristic layout algorithm. Our heuristic layout algorithm is based on Walshaw's force-directed algorithm [13], which is a variation of Eades' original spring-embedder algorithm [14]. We have made several modifications to the algorithm by Walshaw because of the problems associated with very large-scale graphs [15]. For example, Walshaw's algorithm does not yield a pleasant result when the graph contains a dense subgraph (i.e., the subgraph contains nodes of very high degree). His algorithm stops iterative computation of node positions when the graph size falls below a certain threshold value. But we have found the termination condition makes the algorithm iterate unnecessarily many times; after a certain number of iterations of the algorithm, computation of node positions does not improve the entire layout of the graph any more. Therefore, we make the algorithm terminates after 20 iterations by default. If the user wants to refine a graph further, the algorithm iterates by 5, 10 or 20 times.

Due to the nature and size of protein-protein interaction networks, it should be possible to find subgraphs (such as connected components of a disconnected graph or subgraphs of proteins with a common function) and to work on individual subgraphs. The way we find a connected component containing a certain protein is as follows. When a user enters a protein name *p*, it is compared with the node names in the node table to find the node ID of *p*. The node ID of *p* is then compared with the sources and targets in the edge table to find neighbors of *p*. We repeat this process by degree(*p*) times for each of the neighbors of *p* until we find no new neighbors. The resulting proteins constitute the nodes of a connected component containing the initial protein *p*. If a connected component containing a certain protein is too large to analyze, we can reduce the size of the connected component by restricting the distance level of neighbors. If we apply the procedure of finding a connected component to all nodes in the node table, we obtain a complete list of connected components. The lists show the total number of connected components as well as the size, nodes and edges of each connected component (see Figure 5 for an example).

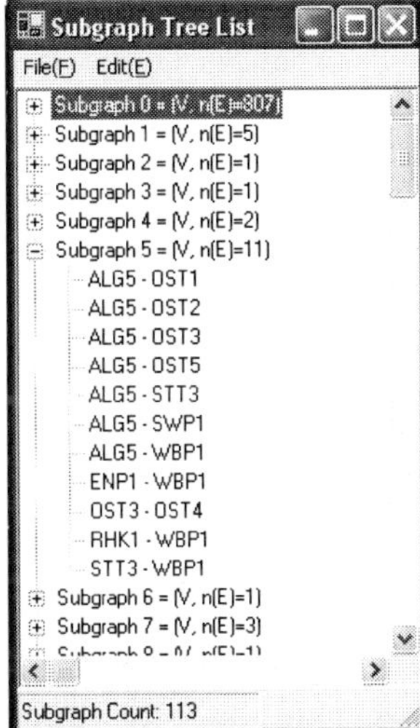

Fig. 5. The list of the connected components of the MIPS genetic interaction data. When the user clicks a connected component, all edges of the connected component are displayed.

5 Implementation and Experimental Results

Microsoft's SQL server 7.0 was used for constructing the local database of protein-protein interactions. The HTML parser was written in Borland C++ Builder. The graph drawing part for a mobile computer was developed using embedded C++. All other parts, including the database interface and the algorithm for finding subgraphs, were implemented in Microsoft's C#, which is a modern, object-oriented programming language. The mobile program runs on any PDA or handheld PC running Windows CE as its operating system.

Figure 6A displays a three-dimensional drawing of a connected component of protein interaction data on a PDA. It appears to have edge crossings, but it actually contains no edge crossing when it is visualized as a three-dimensional drawing on a PDA monitor. The program allows the user to explore a three-dimensional drawing of a protein interaction network by rotating it using the stylus pen of a PDA. Due to the limitation in PDA display area, a large drawing can be examined by zooming in or out of it. When the user clicks a node of the drawing, its protein name is displayed (Figure 6B). Alternatively, all protein names can be displayed by choosing the option of "view node name" in the Edit menu (Figure 6C).

Figure 7A shows an example session when a PDA is connected to the local database. When the user enters a protein name and interaction distance, all proteins interacting with the protein within the specified distance level are found from the database and are displayed in a simplified GML format. A protein interaction network visualized by the program can be saved either in an image file or a file in simplified GML format. The visualized network can be further refined by recomputing the positions of its nodes through 5, 10 or 20 iterations of the algorithm (Figure 7B). An iteration constitutes the computation of the positions of all nodes. Using the memo function of the Edit menu, the user can annotate a node. Annotations written by the user in the dialog box can be saved for later use (Figure 7C).

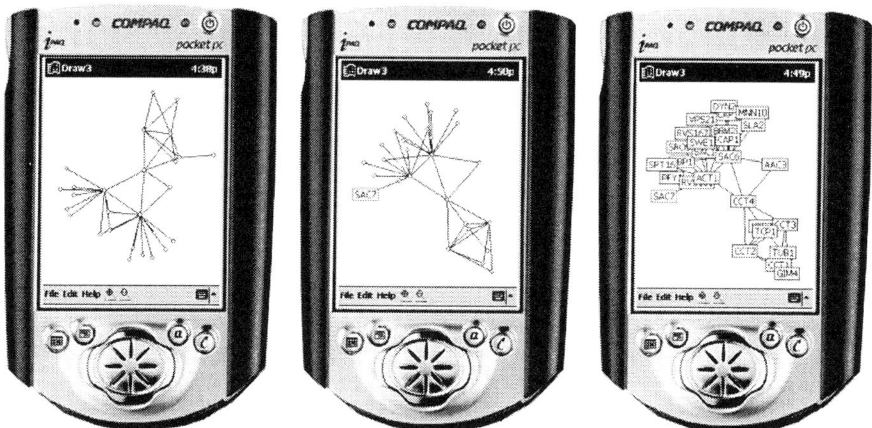

Fig. 6. (A) A three-dimensional drawing of a protein interaction network on a PDA. The three-dimensional drawing can be further explored by rotating or by zooming in or out of it. (B) When the user clicks a node, the name of the node is displayed. (C) All protein names can be displayed using the submenu "view node name" of the Edit menu. The node selected is shown in red color, which is SAC7 in this example.

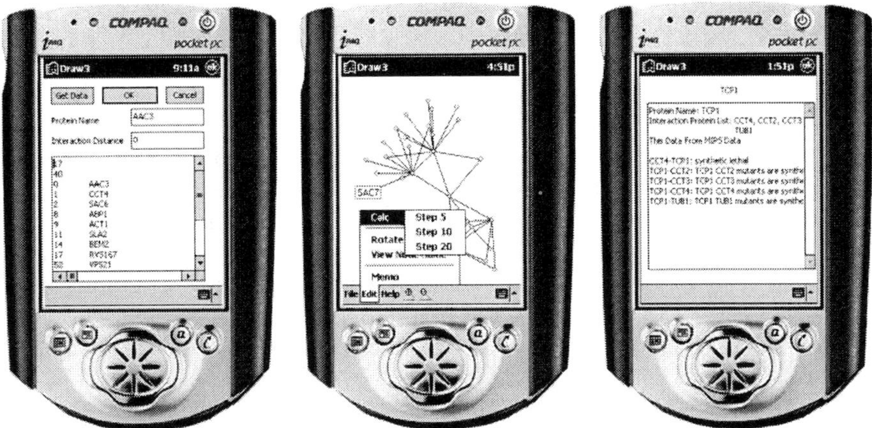

Fig. 7. (A) Interacting proteins with a certain protein within a specified distance level are found from the protein-protein interaction database and are displayed in a simplified GML format. (B) A protein interaction network visualized by the program can be saved either in an image file or a file in simplified GML format. The visualized network can be further refined by iterative computation of the positions of its nodes. (C) The user can annotate a node using the memo function of the Edit menu. Annotations in the dialog box written by the user can be saved for later use.

6 Conclusion and Future Work

This paper has described a new mobile application program for visualizing and analyzing protein interaction networks. Unique features of this program include: (1) this is the first mobile application program for mining and visualizing bioinformatics data, (2) it can be used not only for visualizing protein interactions but also for finding and exploring individual connected components or subgraphs interactively, which we found very useful in studying protein-protein interactions on a large scale, and (3) it provides an integrated framework for dynamically queries protein-protein interaction databases and directly visualizes the query results, making the visualization and analysis of large amounts of updated data easy.

Experimental results showed that the program generates clear and aesthetically pleasing protein interaction networks in three-dimensional space, based on the data provided by the database. The program currently works with a local database, but with changes in the database interface part only it should be possible to work with remote databases. We believe that the program has great potential as useful aids in studying protein-protein interactions whenever a researcher needs it in a variety of different environments.

Acknowledgments

This work was supported by the advanced backbone IT technology development program of the Ministry of Information and Communication (MIC) under grant number IMT2000-C3-4.

References

1. Raman, B., Katz, R., Joseph, A.D.: Universal Inbox: providing extensible personal mobility and service mobility in an integrated communication network. Proceedings of IEEE Workshop on Mobile Computing Systems and Applications (2000) 95-106
2. Bock, J.R., Gough, D.A.: Predicting protein-protein interactions from primary structure. Bioinformatics 17 (2001) 455-460
3. Ito, T., Tashiro, K., Muta, S., Ozawa, R., Chiba, T., Nishizawa, M., Yamamoto, K., Kuhara, S., Sakaki, Y.: Toward a protein-protein interaction map of the budding yeast: A comprehensive system to examine two-hybrid interactions in all possible combinations between the yeast proteins. Proc. Natl. Acad. Sci. USA 97 (2000) 1143-1147
4. Lappe, M., Park, J., Niggemann, O., Holm, L.: Generating protein interaction maps from incomplete data: application to fold assignment. Bioinformatics 17 (2001) s149-s156
5. Mrowka, R.: A Java applet for visualizing protein-protein interaction. Bioinformatics 17 (2001) 669-670
6. Murzin, A.D., Brenner, S.E. et al.: SCOPE: a structural classification of proteins database for the investigation of sequences and structures. J. Mol. Biol. 247 (1995) 536-540
7. Park, J., Lappe, M., Teichmann, S.A.: Mapping protein family interactions: intramolecular and intermolecular protein family interaction repertoires in the PDB and Yeast. J. Mol. Biol. 307 (2001) 929-938
8. Schwikowski, B., Uetz, P., Fields, S.: A network of protein-protein interactions in yeast. Nature Biotechnology 18 (2000) 1257-1261
9. Tucker, C.L., Gera, J.F., Uetz, P.: Towards an understanding of complex protein networks. Trends in Cell Biology 11 (2001) 102-106
10. Uetz, P. et al.: A comprehensive analysis of protein-protein interactions in Saccharomyces cerevisiae. Nature 403 (2000) 623-627
11. Xenarios, I., Fernandez, E., Salwinski, L., Duan, X.J., Thompson, M.J., Marcotte, E.M., Eisenberg, D.: DIP: the database of interacting proteins: 2001 update. Nucleic Acids Res. 29 (2001) 239-241
12. Himsolt, M. GML: Graph Modeling Language. http://www.uni-passau.de/Graphlet/GML (1997)
13. Walshaw, C.: A multilevel Algorithm for Force-Directed Graph Drawing. Lecture Notes in Computer Science 1984 (2000) 171-182
14. Eades, P.: A heuristic for graph drawing. Congresssus Numerantium 42 (1984) 149-160
15. Park, B., Ju, B., Park, J., Han, K.: Dynamic visualization and analysis of protein interactions using a hierarchical force-directed algorithm. Submitted for publication (2002)

Discovering Temporal Relation Rules Mining from Interval Data

Jun Wook Lee[1], Yong Joon Lee[2], Hey Kyu Kim[2], Bu Hun Hwang[3], and Keun Ho Ryu[1]

[1] Database Laboratory, School of Electrical and Computer Engineering, Chungbuk National University, Cheongju 361-763, Republic of Korea
{junux,khryu}@dblab.chungbuk.ac.kr

[2] Postal Technology Development Department,
Electronics and Telecommunications Research Institute, Korea
{yjl,hkkim}@etri.re.kr

[3] Department of Computer Science, Chonnam National University, Korea,
300, Yongbong-dong,Gwangju, Chonnam, Korea
bhhwang@chonnam.chonnam.ac.kr

Abstract. In this paper, we propose a new data mining technique that can address the temporal relation rules of temporal interval data by using Allen's theory. We present two new algorithms for discovering temporal relationships: one is to preprocess an algorithm for the generalization of temporal interval data and to transform timestamp data into temporal interval data; and the other is to use a temporal relation algorithm for mining temporal relation rules and to discover the rules from temporal interval data. This technique can provide more useful knowledge in comparison with other conventional data mining techniques.

1 Introduction

Due to the explosion of data volumes in various application areas, development of new types of data mining algorithms which can discover some useful and unknown knowledge has been focused [21]. Emerging data type of recent applications is from such as e-commerce, bioinformatics, GIS, medical care, and so on. Many works [6,7,8] on data mining have focused on the issues of temporal data mining for discovering temporal knowledge from temporal data, such as sequential patterns[5,9,10,11], similar time sequences [13,14], and temporal association rules [1].

An interesting type of rule is temporal rule from data which encompass temporal elements. Temporal rule is a mining technique that extends conventional data mining techniques. Through its extension, it can discover rules of temporal and causal relationships. This technique includes cyclic associations [15] that discover cyclically repeating association rules, that is, calendric associations [16,17] which mean association rules satisfying the temporal patterns expressed in the form of a calendar. In order to find some meaningful relationships from the temporal data, it is necessary to consider the relationships between data. Allen [2] introduces temporal relationships between intervals and operators for reasoning about relations between intervals. Moreover, the applications of Allen's interval

M.H. Shafazand and A M. Tjoa (Eds.): EurAsia-ICT 2002, LNCS 2510, pp. 57–66, 2002.

operators in geographical information system have been attempted in many works [3,4].

Previous studies have developed their research designs only with the data that were stamped with time points rather than time intervals. Basic studies on mining of useful patterns from interval data have been even partially attempted [18,19] recently. However, a lack of attention on that issue may be attributable to the complicated research process.

In this paper, we present a new data mining technique based on Allen's theory. Despite the scalability problem of the theory, the relationships and operators defined can help effectively discovering temporal relation rules from interval data. In order to reduce the processing time of time point data, algorithm uses the preprocessing method. The core of algorithm is discovering temporal relation rules from temporal interval data using temporal relationship operators introduced by Allen. This algorithm extends the AprioriAll[5], a typical sequential pattern algorithm, for discovering temporal relation rules.

The outline of the paper is as follows. Mining problems of temporal relation rules will be explored in section 2 and algorithms for discovering temporal relation rules will be convincingly explicated through a step by step process in section 3. In section 4, we will prove the effectiveness of algorithm through a series of experiments. Finally, further researches will be discussed.

2 Mining Problem Definition

2.1 Temporal Relation

Definitions. Let $e=(E,t)$ be an event with time point by. E stands for an event type and t means the occurrence time of event e. A sequence is an ordered list of event according to time. We denote an event sequence S by $<e_1, e_2, ..., e_n>$, where $e_i=(E_i,t_i)$ and $t_i \leq t_{i+1}$ for each $i=1,..., n-1$. $[t_1, t_n]$ means the period between the first event e_1 and the last e_n of S.

Also, Let $e'=(E, vs, ve)$ be an event with temporal interval. We call vs the starting point of the temporal interval and ve its end point. They are denoted as $e'.vs$, $e'.ve$. An sequence of this events is expressed as $S'=<e'_1, e'_2, ..., e'_n>$, where $e'_j=(E_j, vs_j, ve_j)$ and $ve_j \leq vs_{j+1}$ for each $j=1, ..., n-1$. Given time granularity U and the criterion time point V, a sequence S can be converted into a sequence S'. The period of S'is defined as [vs1,ven] and can be converted into [1,m], where m≥1.

Let $IE=\{e'_1, e'_2, ..., e'_n\}$ an event set with some temporal interval, where $e'_j=(E_j, vs_j, ve_j)$ for each $j=1, ..., n$. $e'_j.vs$ means the start point of event e'_j with interval and $e'_j.ve$ is the end point. Some event pair (x, y), which is included in event relation $\Omega = \{(x, y) \mid x, y \in IE, x \neq y\}$, has a binary temporal relationship $R(x, y)$. However, x and y are two separate events. A temporal relation is defined as $R(x, y) = \{P(x, y) \mid P \in IO\}$, where $\forall (x, y) \in \Omega$. The set of temporal interval operators is IO={*before, equal, meets, overlaps,* during} and P(x, y) is a binary predicate which expresses temporal relationship between x and y. For Example, event x occurs prior to the event of y in case of before(x,y). If e'_1, e'_2, e'_3 is the three events belongs to IE, two temporal relations $R_1(e_1^{'},e_2^{'})$, $R_2(e_2^{'},e_3^{'})$ can occurs for $e_1^{'},e_2^{'},e_3^{'}$. A new temporal relation $R_1(e_1^{'},e_2^{'})$ can be derived automatically from the two relations. For example, if two relations are $before(e_1^{'},e_2^{'})$ and $meet(e_2^{'},e_3^{'})$, then $before(e_1^{'},e_3^{'})$ can be derived.

Unfortunately, the cost for derivation is too high. Allen's algorithm cannot scale up for large databases. To solve this problem, we present two new algorithms for mining temporal relation rules. One is to preprocess algorithm for the generalization of temporal interval data, summarize timestamp data into temporal interval data so that it can reduce size of input database. The other is to find out the rules through temporal relation algorithm, which satisfy a user-specified minimum support from temporal interval data.

2.2 Generalization of Temporal Interval

As mentioned in [2], there are few advantages to get knowledge when we use the point-based representation. Actually, many applications only hold the states of object represented by time point timestamping simply. However, the useful knowledge is hidden behind the relative relationships between events. Time point data can be converted into interval data with semantic equality.

Let D be a database composed of transactions and each transaction consists of customer-id, transaction-time stamped with time point, a set of event type. The schema of database is expressed as $\{C_{id}, T_{time}, \{E_1, E_2, ..., E_n\}\}$, C_{id} is customer-id, T_{time} is transaction-time and E is an event type. All customers have no more than one transaction with the same transaction-time. The examples of the event type are symptoms of a patient, purchased merchandise, and visited web page.

The transactions of database D are ordered by C_{id} and $T_{time.}$ Let an event sequence of some customer C_{id} be $S(C_{id})=<e_1, e_2, ..., e_n>$ according to the paragraph 2.1. Then we can define $S(C_{id})=<(E_1, t_1), (E_2, t_2), ..., (E_n, t_n)>$ for each i=1, ..., n-1. For some event type, E_i included in $S(C_{id})$, same event set with different time point $\{(E_i,t_1). (E_i, t_2), ..., (E_i, t_m)\}$ may occur such that $t_j \leq t_{j+1}$, j=1, ..., m-1. For example, patient C_1 may undergo event type of symptom E_1 several times.

A series of several identical events with different time points or with different time intervals can be summarized into a single event with temporal intervals. Through this generalization of temporal interval, temporal interval data can be produced from database D and D can be efficiently summarized. The customer set is $C=\{C_1,C_2,\ldots,C_n\}$ in database D and event sequence is $S(C)=\{S(C_1),S(C_2),\ldots,S(C_n)\}$. Every event type included in some sequence $S(C_{id})$ is called candidate event type. For some candidate, event type E_i occurs in $S(C_{id})$, customer C_{id} which causes event type $E_{i,}$ supports it as well. The support of E_i is referred as $Supp(E_i)$ and it indicates the entire number of customers supporting E_i in D. If $Supp(E_i) \geq Supp_{min}$ in regards to the minimum support assigned by user, then E_i is large event type and e_i having E_i is a large event. $Supp_{min}$ is the proportion of the number of supporting customer to the entire number of customers. We can find a set of large event types, LE=$\{E_1,E_2,\ldots,E_m\}$ from D.

As already mentioned in the paragraph 2.1, a sequence S can be converted into a sequence S' with respect to the granularity U and the criterion time point V. Let w be the window size assigned by user and W be some window with the size of w, where w≥vei-vsi. The number of windows in S' is denoted by wn. Then we can get a window sequence of W1,W2,...Wn. If a large event type Ei exists within some window Wj of S', it is said that Ei occurred in Wj. The Freq(Ei) is defined as the number of Ei occurred in window sequence. Given Freqmin, which is the user assigned minimum frequency, if Freq(Ei)≥Freqmin, we say that Ei occurred uniformly within the period [1,m] of S'. Ei is called a uniform event type. So we can discover a set of uniform event types UE={E1,E2,...,Em} from D.

Event sequence for customer C_{id} is defined as $S(C_{id})=<e_1, e_2, ..., e_n>$. Then, $S(C_{id})=<(E_1, t_1), (E_2, t_2), ..., (E_n, t_n)>$, where $t_i \leq t_{i+1}$ for each $i=1, ..., n-1$ and $E_i \in UE$. Also $S(C_{id})$ is partial set of S. For some customer Cid whomever caused uniform event type Ej, whose sequence is S''(Ej), can be both summarized and generalized as a single event. The more detail definitions and proof can be seen in [20].

2.3 Temporal Relation Rule

Using the uniform event type set *UE*, A generalized database *ND* is produced from database *D*. If every customer set of *ND* is $\{C_1, C_2, ..., C_n\}$, then *ND* is composed of event sequence set of $\{S'(C_1), S'(C_2), ..., S'(C_n)\}$.

The event set composing some event sequence $S'(C_{id})$ can be defined as $IE(C_{id}) = \{e_1', e_2', \ldots, e_m'\}$ such that $e_i' = (E_i, vs_i, ve_i)$, $E_i \in UE$ for each $i=1,\ldots,m$. The set of all event pairs is defined as $\Omega = \{(x, y) \mid x, y \in IE(C_{id}), x \neq y\}$ for some event sequence $S'(C_{id})$. Some event pair (x, y) has a binary temporal relation $R(x, y) = \{P(x, y) \mid P \in IO\}$, where $\forall (x, y) \in \Omega$. Then, $R(x, y)$ is termed as candidate temporal relation. Candidate temporal relation set of customer Cid is called as customer temporal relation CR(Cid) of Cid. If CR(Cid) includes a candidate temporal relation R1(x,y), then client Cid support R1(x,y). The support of R1(x,y) is defined as Supp(R1). Given Suppmin which is a is a user assigned minimum support, if Supp(R1)≥Suppmin, then R1(x,y) can be called as a frequent temporal relation. We can discover frequent temporal relation set FR={R1(x,y),R2(x,y),...,Rz(x,y)} from candidate temporal relation set CR.

Let {R1(e'1, e'2), R2(e'2, e'3), R3(e'1, e'3)} be frequent temporal relations which occurred from event set {e'1, e'2, e'3}. A temporal relation rule TR is defined as TR(e'1, e'2, e'3)=R1(e'1, e'2) | Supp(R1)^R2(e'2, e'3) | Supp(R2)^R3(e'1, e'3) | Supp(R3). That is to say, temporal relation rule can be expressed as the connection of frequent temporal relation occurred in some event set. For instance, suppose that frequent temporal relation set is {before(e'1, e'2), overlap(e'2, e'3), before(e'1, e'3)}, then temporal relation rule is TR(e'1, e'2, e'3)=before(e'1, e'2) | Supp(R1)^overlap(e'2, e'3) | Supp(R2)^before(e'1, e'3) | Supp(R3). In summary, we can discover that every set of temporal relation rule is {TR1, TR2, ..., TRn} from database D.

3 Mining Temporal Relation Rules

Based on the problem definition of section 2, the procedure for discovering temporal rules is as follows.

Algorithm of discovering temporal relation rule

Input data. Database D composed of transactions with time point, and the schema of D is $\{C_{id}, T_{time}, \{E_1, E_2, \ldots, E_n\}\}$.

Output result. The set of temporal relation rules, defined as $\{TR_1, TR_2, \ldots, TR_n\}$

Phase 1. Discover a set of large event types, $LE = \{E_1, E_2, \ldots, E_m\}$ which satisfies the $Supp_{\min}$, a user-specified minimum support , from database D.

Phase 2. Suppose that time granularity U and criterion time point V are given, then find event sequence with temporal interval, $S' = \langle (E_1, vs_1, ve_1), (E_2, vs_2, ve_2), \ \ldots, (E_n, vs_n, ve_n) \rangle$, which only has transaction with LE from database D.

Phase 3. Find the set of uniform event types, $UE = \{E_1, E_2, \ldots, E_l\}$ which satisfies the $Freq_{\min}$, a user-specified minimum frequency, from S' with window size w.

Phase 4. Summarize event sequence S' by using UE and then generate the generalized database ND.

Phase 5. Find the set of every candidate temporal relations, $CR = \{R_1(x, y), R_2(x, y), \ldots, R_k(x, y)\}$, from ND.

Phase 6. According to paragraph 3.3, find the set of frequent temporal relations, $FR = \{R_1(x, y), R_2(x, y), \ldots, R_z(x, y)\}$, which satisfy the $Supp_{\min}$ from CR.

Phase 7. Discover temporal relation rules, { $\{TR_1, TR_2, \ldots, TR_n\}$, from FR.

3.1 Preprocessing Algorithm

Sorting Phase

In this phase, input database is sorted by customer-id and transaction-time as Fig. 1. Event types occurred monthly are stored in each transaction of database D.

LE Phase

This phase corresponds to phase 1 of algorithm and finds large event type set, LE, which satisfies user-specified minimum support $Supp_{min}$ from database D. Fig. 2-a is candidate event type set, *CE*={*A, B, C, D. E, F, G*}, produced from *D*. Meanwhile, Fig. 2-b is a large event type set, *LE*={*B, C, D, E*}, which satisfies the $Supp_{min}$ 75% from *CE*.

Generalization Phase

This phase corresponds to phase 2 to 4 of the algorithm and with using *LE* of Fig.

2-b, temporal interval database of Fig. 5 can be produced from database *D* of Fig. 1.

A problem to be solved in changing temporal interval data is how to how to generalize identical events with large event type into a single event with temporal interval. We call the period of [t1,tm] the life span of E1. Time support is defined as the frequency of E1 within time span [t1,tm]. E1, which satisfies a user specified minimum time support, can be summarized into a single event (E1,t1,tm). However, this technique has a problem of summarizing[20]. We, thus, use a following technique in order to solve the problem of time support.

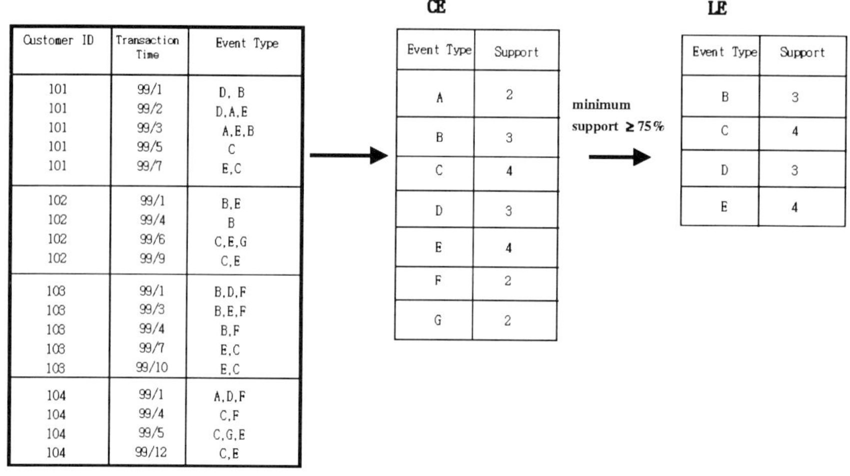

Customer ID	Transaction Time	Event Type
101	99/1	D, B
101	99/2	D,A,E
101	99/3	A,E,B
101	99/5	C
101	99/7	E,C
102	99/1	B,E
102	99/4	B
102	99/6	C,E,G
102	99/9	C,E
103	99/1	B,D,F
103	99/3	B,E,F
103	99/4	B,F
103	99/7	E,C
103	99/10	E,C
104	99/1	A,D,F
104	99/4	C,F
104	99/5	C,G,E
104	99/12	C,E

CE

Event Type	Support
A	2
B	3
C	4
D	3
E	4
F	2
G	2

LE

Event Type	Support
B	3
C	4
D	3
E	4

Fig. 1. Sorted database **Fig. 2-a.** Candidate event type **Fig. 2-b**. Large event type

First, by phase 2 of the algorithm, database *D* in Fig. 1 is converted into event sequence *S'* with temporal interval by a using user-specified time granularity *U* and time point *V*. And *S'* only includes *LE*. The S' can be expressed as in Fig. 3.

Second, as in phase 3 of section 2 algorithm, we find out a uniform event type set, *UE*={*B, C, E*}, which satisfies minimum frequency $Freq_{min}$ from event sequence *S'* in Fig. 3 in regard to a user-specified window size *w*. Fig. 4 is *UE*, which satisfies $Freq_{min}$ 33% in regard to window size 2. *S'* only includes *LE* and only parts of *LE* becomes *UE*. Finally, as in phase 4 of the algorithm, we convert event sequence *S'* of Fig. 3 into identical temporal interval database *ND* like Fig. 5. *ND* is composed of a temporal interval event set having event type *UE*.

3.2 Temporal Relation Algorithm

Temporal relation Algorithm consists of two phases. This algorithm produces temporal relation rule by discovering frequent temporal relation set *FR* from temporal interval database *ND* produced through preprocessing procedure.

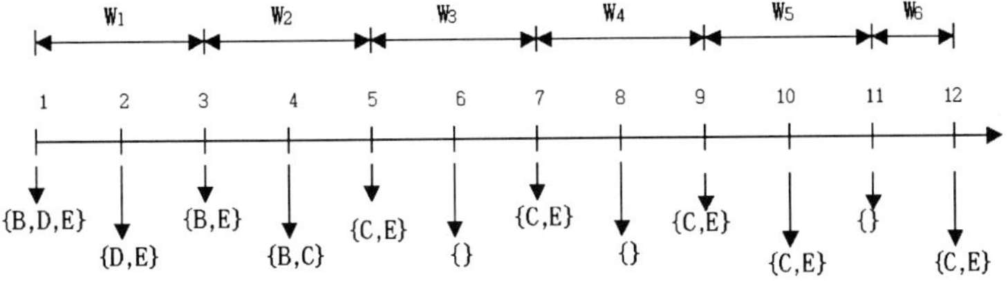

Fig. 3. Event sequence S' with temporal interval

LE

Event Type	Frequency
B	2
C	5
D	1
E	6

minimum frequency ≥ 33% →

UE

Event Type	Frequency
B	2
C	5
E	6

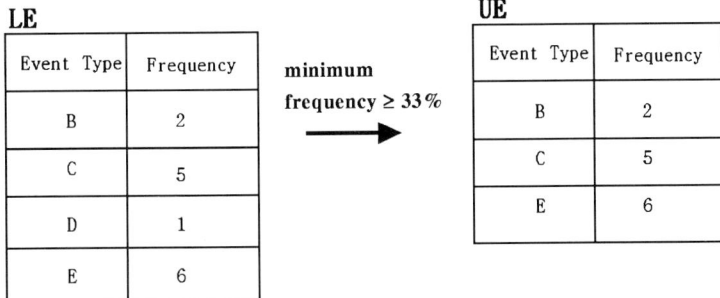

Fig. 4. A set of uniform event types, UE

Temporal Relationship Phase

This phase corresponds to phase 5 of algorithm of section 2 and discovers candidate temporal relation set $CR=\{R_1,(B, C), R_2,(C, E), ..., R_K,(B, E)\}$ from *ND* of Fig. 5. Fig. 6 shows *CR* produced from *ND* and is expressed in form of a tree. A node represents an event type. A connection shows the temporal relation between two events, and leaf node (quadrangle) stands for support of temporal relation. Candidate temporal relations found here are *before*(*B, C*), *during*(*C, E*), *overlap*(*C, E*), *overlap*(*B, E*), and *during*(*B, E*). The support of temporal relation represents the number of temporal relation occurred within database *ND*. The technique of making a tree is as followings: iff temporal relations already exist while scanning the events of *ND*, then increased the supports. However, if new temporal relations exist, then it dynamically produced by adding nodes and connections.

Rule Generation Phase

This phase corresponds to phase 6 to 7 of algorithm. We here attempt to find out a frequent temporal relation set, $FR=\{R_1(x,y), R_2(x,y), \dots, R_z(x,y)\}$, which satisfies the $Supp_{min}$, a user-specified minimum support, from CR tree of Fig. 6 and then produce temporal relation rules $\{TR_1, TR_2, \dots, TR_n\}$ from *FR*.

Until now, we present a technique for mining temporal relation rules that was divided into preprocessing and temporal relation algorithm. This technique has been explained step by step with examples. More complete algorithms are given in [20].

We demonstrate the effectiveness of these algorithms through a series of experiments of following section 4.

Customer ID	vs	ve	Event Type
101	1	3	B
101	5	7	C
101	2	7	E
102	1	4	B
102	6	9	C
102	1	9	E
103	1	4	B
103	7	10	C
103	3	10	E
104	4	12	C
104	5	12	E

Fig. 5. Generalized database, ND

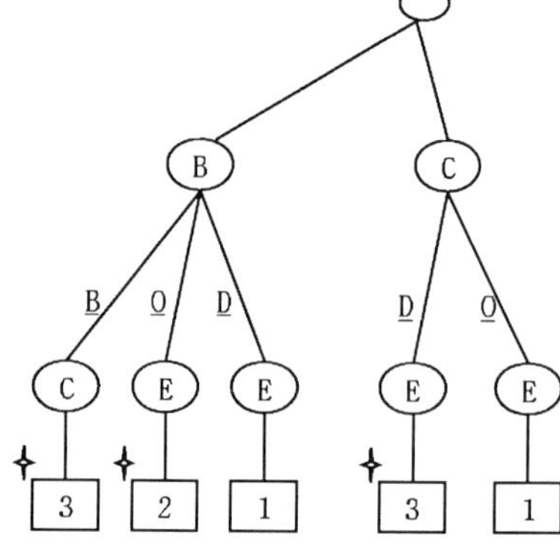

Fig. 6. Tree of candidate temporal relations

4 Experimental Results

To evaluate the performance of the algorithm, several experiments have been done. Due to lack of space, we only summarize the main result in this section. Details of the experiments, including dataset generation and detailed explanations of the results, can be found in [20].

Data Generation. To generate the test datasets, we use the customer database of EC shopping mall as source data, like the database shown in Fig. 1. Datasets are generated from source data by running preprocessing algorithm repeatedly. The preprocessing algorithm generates several datasets by setting parameter values.

Analysis of Performance. We analyzed the performance of temporal relation algorithm by following ways. First, we compared the performance of temporal relation algorithm to the Allen's algorithm. As predicted, execution time of temporal relation algorithm increased linearly in contrast to execution time of Allen's algorithm with time complexity $O(N^2)$ so that it proved the significant performance enhancement compared to the Allen's algorithm. However, this result was applicable to the case of dataset which is almost not summarized. Therefore, source data should be sufficiently summarized by preprocessing. And further researches are necessary in order to improve performance because the execution time is increased continuously by increasing the number of records.

Second, we calculated the time required in each phase of the algorithm, found that the most time was spent in a temporal relationship phase and a generalization phase. Conventional association rule and sequential pattern algorithms required more time-spending in producing a candidate item set. On the other hand, in case of this algorithm, the temporal relationship phase needed the most time in abstracting a candidate temporal relation. However, a rule generation phase, in which attempted to discover the frequent temporal relations and produce the rules from candidate temporal relations, require least cost. It was, therefore, recognized in this phase that time-extended further research should be done.

Thirdly, we analyzed the relationships between the user-specified minimum support and numbers of discovered rules. It shows that the number of rules increase as the minimum support is decreased. So the appropriate minimum support is import to reduce the number of rules.

5 Conclusion

We are given a database composed of transactions with time point, where each transaction can be converted with time interval. Problem of mining temporal rules is how to handle the temporal semantics. We address some limitations of the earlier works considering temporal interval data.

We have presented a new data mining technique in order to efficiently discover useful temporal relation rules from temporal interval data based on Allen's interval operators. It is a complete algorithm in that it guarantees finding temporal relation rules that have a user-specified minimum support and minimum frequency. This technique composes of preprocessing algorithm and temporal relation algorithm.

With comparison to that of existing research, this technique has some significant merits. First, it can help discover fruitful temporal rules from temporal interval data, which cannot be abstracted by the existing algorithm like the sequential patterns. Secondly, it also enables to efficiently find out temporal relation rules from temporal database. That is to say, through the process of generalization, we can not only produce temporal interval data but also reduce the size of the data and save search space and time. Through the several experiments, we showed the effectiveness of the algorithm and some problems remained.

References

1. J. F. Roddick, M. Spiliopoulou: Temporal data mining: survey and issues, Research Report ACRC-99-007, University of South Australia(1999)
2. J. Allen: Maintaining Knowledge about Temporal Intervals, Comm. Of the ACM, Vol.26, No.11, Nov. (1983)
3. J.Y. Lee, K.J. Oh, Keun Ho Ryu: Integration with Spatiotemporal Relationship Operators in SQL, ACM-GIS (1998)
4. Dong Ho Kim, Keun Ho Ryu, Chie Hang Park: Design and implementation of spatiotemporal database query processing system, Journal of systems and software, Dec., (2001)
5. R. Agrawal, R. Srikant: Mining sequential patterns, Int'l. Conf. on Data Engineering, Taipei, Taiwan (1995)
6. X. Chen, I. Petrounias: A framework for temporal data mining, Int'l. Conf. on Database and Expert Systems Applications (1998).
7. C. Rainsford, J. F. Roddick: Temporal data mining in information systems: a model, Australasian Conf. on Information Systems (1996)
8. M. H. Saraee, B. Theodoulidis: Knowledge discovery in temporal databases, IEEE Colloquium on Knowledge Discovery in Databases (1995)

9. R. Srikant, R. Agrawal: Mining sequential patterns: generalisations and performance improvements, In Proc. Int'l. Conf. on Extending Database Technology, Avignon, France, Springer-Verlag(1996)
10. Minos N. Garofalakis, Rajeev Rastogi, Kyuseok Shim: SPIRIT: Sequential Pattern Mining with Regular Expression Constraints, the VLDB Conf., Edinburgh, Scotland, UK (1999)
11. H. Mannila, H. Toivonen: Discovering generalized episodes using minimal occurrences, Int'l Conf. on Knowledge Discovery in Databases and Data Mining (KDD-96), Portland, USA (1996)
12. J. Han, G. Dong, Y. Yin: Efficient Mining of Partial Periodic Patterns in Time Series Database, Int'l. Conf. on Data Engineering, Sydney, Australia (1999)
13. R. Agrawal, King-Ip Lin, Harpreet S. Sawhney, Kyuseok Shim: Fast similarity search in the presence of noise, scaling, and translation in time series databases, the VLDB Conf., Zurich, Switzerland (1995)
14. C. Faloutsos, M. Ranganathan, Y. Manolopoulos: Fast subsequence matching in time-series databases, the ACM SIGMOD Conf. on Management of Data, Minneapolis, USA (1994)
15. B. Ozden, S. Ramaswamy, and A. Silberschatz: Cyclic association rules, Int'l. Conf. on Data Engineering, Orlando, USA (1998)
16. X. Chen, I. Petrounias, H. Heathfield: Discovering temporal association rules in temporal databases, Int'l. Workshop on Issues and Applications of Database Technology (1998)
17. S. Ramaswamy, S. Mahajan, A. Silberschatz: On the discovery of interesting patterns in association rules, the VLDB Conf., New York City, USA (1998)
18. J. M. Ale, G. H. Rossi: An Approach to Discovering Temporal Association Rules, SAC'00, Italy (2000)
19. C. Rainsford: Accommodating Temporal Semantics in Knowledge Discovery and Data Mining, PhD Thesis, University of South Australia (1998)
20. Y. J. Lee: A Data Mining Technique for Discovering Temporal Relation Rules, PhD Thesis, Chungbuk National University (2001)
21. Yun, H., Ha, D., Hwang, B., Ryu, K.: Mining Association Rules on Significant Rare Data using Relative Support. Journal of Systems and Software, 2002 (accepted).

An Abstract Image Representation Based on Edge Pixel Neighborhood Information (EPNI)

Abdolah Chalechale and Alfred Mertins

University of Wollongong
School of Electrical, Computer and Telecommunications Engineering
Wollongong, NSW 2522, Australia
{ac82, mertins}@uow.edu.au

Abstract. In this paper we introduce a new abstract image representation based on Edge Pixel Neighborhood Information (EPNI). It is applied in image retrieval problem when user query is a fast drawn, rough example. The representation consists of two main elements. A neighborhood vector f and a vicinity table v. The former contains the frequencies of edge pixels with similar directions and the latter holds information about neighboring edge directions. An image similarity measure based on EPNI components is also designed and compared with some other measures known from the literature. Experimental results show a good recognition accuracy in a data set containing a wide range of color images.

1 Introduction

Managing the tremendous number of images and video clips in relevant databases and also on the Web needs more efficient and fast algorithms and tools. Image similarity measurement is one of the most important aspects in a large image database for efficient search and retrieval to find the best answer for a user query. Image and video indexing using a content-based approach plays an important role in finding and accessing minimal information. Recently, this area has attracted many new researches. Representative systems are QBIC [1], Photobook [2], FourEyes [3], MetaSEEk and VisualSEEk [4,5].

In most current content-based image retrieval systems the emphasis is on four clues: color, texture, shape and position of objects. The MPEG7 standard suggests some descriptors for color and texture [6], and for visual shape [7]. Although color and texture are significant features for retrieval purposes, there are some situations where they cannot be used efficiently. For instance, when the query is a rough and quick black and white sketched image and a user is asking the system to find the most similar images to his/her query example, then color and texture lose their original importance. In addition, because the sketched query does not contain a well defined object contour, using shape descriptors may yield undesired results.

In sketch based image and video retrieval situations, when the query is a rude, uncolored example image drawn with some primitive tools, the following methods may generate more acceptable results.

M.H. Shafazand and A M. Tjoa (Eds.): EurAsia-ICT 2002, LNCS 2510, pp. 67–74, 2002.

A) A correlation approach was introduced by Hirata and Kato [8]. In this method, the query and target images are resized to 64*64 pixels, then their edges are extracted by a gradient operator and finally, a global correlation factor (C_t) which is used for similarity measuring is calculated. A modified version of this method is used in QBIC [1].

B) Normalized central moments and skew and rotation invariant functions based on them have been used as powerful tools for shape description (see [9]). Mohamad, Sulong and Ipson [10] used this method for trademark matching in an image retrieval context. They conclude that in scanned b&w images, moment values can be taken as standard features for the matching task. The QBIC system takes advantage of digital moments for shape similarity as well [1].

C) The Hausdorff distance measures the similarity of two sets of points. This distance may be applied to determine the extent to which one image resembles another. Huttenlocher, Klanderman and Rucklidge [11] compared the Hausdorff distance with binary correlation on edge maps and conclude the former works better. They also provide algorithms for computing the Hausdorff distance between all possible relative positions of a binary image and a translated model of the image [12].

D) Histograms of edge directions for representing image information is one of the well known methods in the image retrieval field. Recently, C. S. Won et al. [13] showed that the global and semi-global edge histograms have better retrieval performance than the MPEG-7 recommended local edge histogram descriptor. M. Abdel-Mottaleb [14] used the approach by applying the Canny edge operator to find strong edges in an image and then quantized them into 4 directions. Jain and Vailaya [15] also proposed edge directions as an image attribute for shape description.

In this paper we introduce a new method of feature extraction for retrieval purposes based on edge maps using 1) a vector of neighborhood information of edge pixels and 2) a second-order vicinity table. The vector is used for measuring the similarity between two images and the vicinity table is used for reducing the search space and improving the efficiency of the method. The method is scale independent and yields excellent retrieval results in a data set containing a wide range of images.

In the next section the explanation of the method is provided. Experimental results are presented in Section 3. Conclusions and some directions for future work are finally given in Section 4.

2 Edge Pixel Neighborhood Information(EPNI)

The objective of the proposed approach is to transform the image data into a new structure that supports measuring the similarity between a full colored image and a rough sketch given by a user as a query example in a correct, easy and fast way.

At first, the color image is converted to a gray intensity image by eliminating the hue and saturation while retaining the luminance. Applying the Canny edge operator [16] on this gray scale image results in an edge image I, which is the platform for further feature extraction.

The algorithm uses an edge pixel neighbor diagram (see Fig. 1). In this diagram the center is an edge pixel in I. Considering 8-connectivity, each pixel has, in most cases, up to 4 neighbors as it is an edge point. By numbering the directions as indicated in Fig. 1, each pixel neighborhood is coded with a number n, $0 \leq n \leq 240$, by summing up the direction numbers of its neighbors. For instance, $n = 0$ means a singular point (without any neighbor), a pixel point with two horizontal neighbors has n=17 code and $n = 240$ means a point with 4 neighbors in the directions represented by 128, 64, 32 and 16.

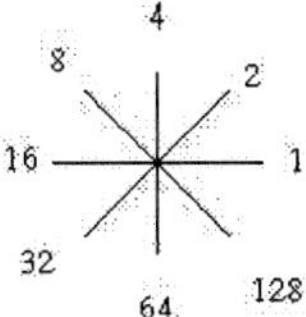

Fig. 1. Edge Pixel Neighbor Diagram.

The frequencies of the neighborhood codes ($n \neq 0$) form a neighborhood vector f with maximally 240 entries. The singular points ($n = 0$) are scarce and not considered. The sum of all edge pixels with the same neighborhood code is stored in the appropriate entry of f. For example, the sum of all edge pixels with two vertical neighbors is stored in entry 68. The numbering scheme simplifies the code finding process since we only set a bit to 1 in the appropriate position in the code byte. The emphasis is on the number of occurrences (frequencies) of each code and not on the code itself. Because for each image, f depends on the size of the image, it is necessary to normalize f to be scale invariant. We found that normalizing f by the size of the image is better than normalizing it by the number of edge pixels as in [15]. Brandt et al. also use this type of normalization in their work [17]. The normalized vectors are defined as:

$$f_i = \frac{\sum_I P_i}{\text{size } I} \tag{1}$$

where $1 \leq i \leq 240$ and P_i is any edge pixel with $n = i$.

The next part of the feature extraction algorithm is a second-order vicinity table v. This table contains information about neighboring edge directions. Most pixel points have only two other neighbors. If P_i is an edge pixel with neighborhood code $n_{1,i}$ and P_j and P_k are its first two neighbors in ascending order with $n_{2,i}$ and $n_{3,i}$ as their neighborhood codes respectively, then there are three neighboring codes for each edge pixel P_i (a triplet). The first code ($n_{1,i}$) is the neighborhood code of the pixel, the second code ($n_{2,i}$) and the third one ($n_{3,i}$) are neighborhood codes of the two neighbors P_j and P_k. See Fig. 2 for an example. If there is only one neighbor point, the third code would be zero. For all

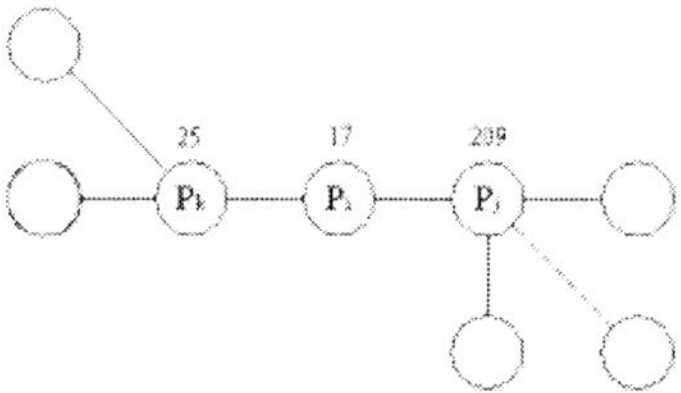

Fig. 2. An example of neighboring edge directions.

triplets $T_i = \{n_{1,i}, n_{2,i}, n_{3,i}\}$ we obtain the frequency of occurrence, denoted as N_i. Sorting with respect to N in descending order results in the vicinity table v whose rows have the structure:

$$v_i = \{n_{1,i}, n_{2,i}, n_{3,i}, N_i\}$$

where
n_1 = neighborhood code,
n_2 = neighborhood code of first neighbor ,
n_3 = neighborhood code of second neighbor and
N = number of triplet $\{n_1, n_2, n_3\}$.

2.1 Similarity Measure

An efficient and effective similarity factor is one essential part in any retrieval approach. Using neighborhood vector f as a feature vector, we can use L_1, L_2 or any other histogram similarity measures to obtain a similarity factor. We designed a new measure to evaluate the similarity between two images. Evaluation method is given in Section 2.3. The goal is to find a similarity factor between a sketched query image q and images in the database (I_d), based on neighborhood vector f. Suppose f_q is the neighborhood vector for query image q and f_d for a database image I_d. To find the similarity between q and I_d we define a measure $\mu(q, I_d)$ as:

$$\mu(q, I_d) = \sum_{i=1}^{240} k_i$$

where

$$k_i = \begin{cases} f_d^i - \theta & \text{if} \quad (f_d^i \geq \theta) \ \& \ (f_q^i < \theta), \\ f_q^i - \theta & \text{if} \quad (f_d^i < \theta) \ \& \ (f_q^i \geq \theta), \\ 0 & \text{else.} \end{cases}$$

θ is a constant value and f^i are the elements of vector f. In Section 3 we will show that this measure is more robust and gives better experimental results than L_1 and L_2.

2.2 Search Space

Comparing q with all images in the database to find $\mu(q, I_d)$, $1 \leq d \leq M$ where M is the number of images in the database, is a time consuming process. Using vicinity tables v and v', representing the images q and I_d respectively, we obtain a correlation factor C as follows:

$$C = \frac{\sum_{r=1}^{j} \sum_{s=1}^{j} (v_{rs} - \bar{v})(v'_{rs} - \bar{v'})}{\sqrt{(\sum_{r=1}^{j} \sum_{s=1}^{j} (v_{rs} - \bar{v})^2))(\sum_{r=1}^{j} \sum_{s=1}^{j} (v'_{rs} - \bar{v'})^2)}}$$

$\bar{v}$ and $\bar{v'}$ are mean values of corresponding vicinity tables. $\bar{v}$ and $\bar{v'}$ are calculated for only j rows and the first 3 columns. j is an arbitrary factor and determines the number of used rows in v. C is in the range of -1 to 1, indicating minimum-to-maximum correlation between the two chosen tables. A lower bound for C limits the number of comparisons. It also improves the overall efficiency as a huge number of non-similar images are disregarded.

2.3 Evaluation Method

The following scoring scheme is defined for similarity measuring evaluation. Let M be the total number of images in the database and $\widehat{q}$ be an image among them which is supposed to be found when providing the query image q. Sorting images by the similarity measure in the descending order will, generally, put $\widehat{q}$ at row k with $0 \le k \le M - 1$. $k = 0$ means that the exact image has been found (score 1), small k's mean good and large k's mean poor findings. A score assigning scheme is defined as :

$$S_q = \frac{M - k}{M}$$

For each query q there is an S_q, and for a set of q's there is a set of S_q's. Therefore, the overall efficiency of the similarity measure should be considered upon a set of S_q's as :

$$\eta = \frac{\sum S_q}{\text{number of } q\text{'s}} * 100$$

A large η indicates a good ability to find the most similar answers to the given sketches. This parameter can serve as an evaluation tool for different similarity measures.

3 Experimental Results

To compare the overall efficiencies of different similarity measures, we created a small (but wide range) data set of color JPEG images of 50 images. Different users were asked to sketch rough, black and white queries (16 queries) that resemble images in the data set (see Fig. 3).

For all following measures, we first converted the JPEG images to single band luminance and then applied the Canny operator [16] to gain edge images I_d. For query images also, we determined the edge image q by the same process. For the methods in Table 1, similarity measure were computed as follows:

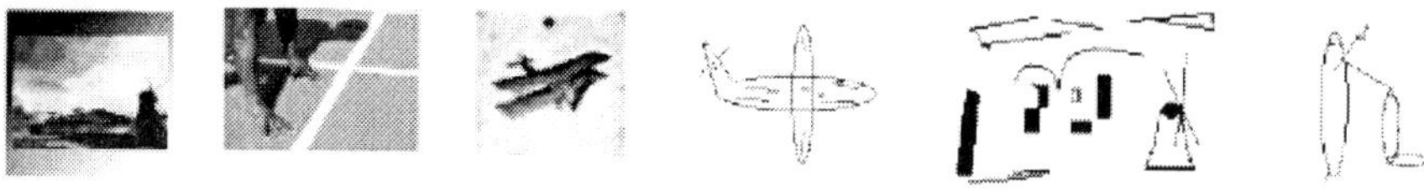

Fig. 3. 3 examples of database images (left) and 3 examples of sketched images (right).

- Correlation. We first resized q and I_d to $64 * 64$ pixels, then divided them to $8 * 8$ blocks. Finally the algorithm given in [8] was applied to calculate C_t as the similarity measure.
- Hausdorff. q and I_d were resized to 64*64 pixels, then they were divided into 4 equal sub images and the Hausdorff distance [12] for corresponding sub images (H_1, H_2, H_3, H_4) was obtained. Finally, the minimum was chosen for the similarity measure.
- $L1$ metric (Manhattan-Cityblock). Let f_q and f_{I_d} be the neighborhood vectors of q and I_d. Instead of using all elements of f_q and f_{I_d} we only apply the $L1$ measure to the t most papular edge directions in f_q. For this, f_q is first sorted in descending order to find the t most popular indices. After storing these indices in a set X, the $L1$ similarity measure was calculated as:

$$L1(q, I_d) = \sum_{i \in X} \left| f_q^i - f_{I_d}^i \right|$$

  We found that the best t in this measure is 13, therefore, X contains only 13 members in our experiments.
- Weighted $L1$. Putting some appropriate weights on the terms of $L1$ summation improves the overall efficiency of the metric. The following set of weights was found to be a good choice for a weighted $L1$ measure $(L1_w)$ as the similarity measure of a length-13 vectors f. The weight set puts more emphasis $(12, 8, 6)$ on more important directions in the 13 sorted ones.

$$w = \{2.5, 12, 8, 6, 4, 3, 2, 3, 4, 6, 6, 8, 2.5\}$$

  The measure then is

$$L1_w(q, I_d) = \sum_{i \in X} w_i \left| f_q^i - f_{I_d}^i \right|$$

- $L2$ metric (Euclidean). Unlike for the $L1$ metric, all indices in f_q and f_{I_d} have to be considered to maximize the efficiency. Euclidean distance between the two neighborhood vectors as a similarity measure is :

$$L2(q, I_d) = \sqrt{\sum_{1}^{240} (f_q^i - f_{I_d}^i)^2}$$

- $\mu(q, I_d)$. As explained in Section 2.1, for each q, the μ measure was computed for all I_d's. The parameter θ was set to 0.0089 in finding the k_i's.
- $\mu_r(q, I_d)$. We obtained the measure μ only for those pairs q and I_d that satisfy the constraint $C \geq -0.5$, as explained in Section 2.2, while the number of used rows of vicinity tables (j) was set to 3.

The overall efficiencies η of the considered techniques are presented in Table 1.

Table 1. Efficiencies of different similarity measures.

Technique	Efficiency%
correlation	76.75
Hausdorff	81.126
$L2$	82.75
$L1$	84
$L1_w$	89.75
μ	91.75
μ_r	92.375

It is worthwhile to mention that, in addition to ascending ability of finding the targeted images by techniques in the table, the correlation method is the most time-consuming and μ_r is the fastest one. The next significant point is that when μ_r is used as the similarity measure, only 667 comparisons, instead of 800 (number of queries * number of images) took place. It means 16.625% reduction of search space.

4 Conclusion

We introduced a new algorithm for image similarity measuring in sketch based context based on edge pixel neighborhood information. It is based on two main elements: 1) a feature vector f which includes the frequencies of edge pixels with similar neighborhood directions and 2) a second-order vicinity table v that contains information about neighboring edge directions. We defined also a measure μ for comparing the similarity between two feature vectors. Using table v in addition to f improves the efficiency and also reduces the search space. The paper presented comparative experimental results that showed a great improvement in finding targeted images when using this algorithm and the similarity measure.

The proposed approach is scale invariant and we intent to expand it to be rotation invariant by grouping rotation-similar entries of f . The μ measure could be extended by a weighting concept to further improve the search capabilities.

Acknowledgment. The first author is financially supported by the Ministry of Science, Research and Technology of I.R. Iran.

References

1. W. Niblack, R. Barber, W. Equitz, M. Flickner, E. Glasman, D. Petkovic, P. Yanker, C. Faloutsos, and G. Taubin, "The QBIC project: querying images by content using color, texture, and shape," in *Proceedings of Spie*, USA, 1993, vol. 1908, pp. 173–187.
2. A. Pentland, R. W. Picard, and S. Sclaroff, "Photobook: content-based manipulation of image databases," *International Journal of Computer Vision*, vol. 18, no. 3, pp. 233–254, June 1996.
3. T. P. Minka and R. W. Picard, "Interactive learning with a "society of models"," *Pattern Recognition*, vol. 30, no. 4, pp. 565–581, Apr. 1997.
4. J. R. Smith and S. Chang, "Visualseek: a fully automated content-based image query system," in *Proceedings ACM Multimedia 96.*, NY, USA, 1996, pp. 87–98.
5. M. Beigi, A. B Benitez, and S. Chang, "Metaseek: a content-based meta-search engine for images," in *Proceedings of Spie*, USA, 1997, vol. 3312, pp. 118–128.
6. B. S. Manjunath, J.-R. Ohm, and V. V. Vasudevan, "Color and texture descriptors," *IEEE Transactions on Circuits and Systems for Video Technology*, vol. 11, no. 6, pp. 703–715, June 2001.
7. M. Bober, "Mpeg-7 visual shape descriptors," *IEEE Transactions on Circuits and Systems for Video Technology*, vol. 11, no. 6, pp. 716–719, June 2001.
8. K. Hirata and T. Kato, "Query by visual example-content based image retrieval," in *Advances in Database Technology - EDBT '92*, Berlin, Germany, 1992, pp. 56–71.
9. A. D. Bimbo, *Visual information retrieval*, Morgan Kaufmann Publishers, 1999.
10. D. Mohamad, G. Sulong, and S. S. Ipson, "Trademark matching using invariant moments," in *Proceedings second Asian Conference on Computer Vision, [ACVV'95].*, Singapore, 1995, vol. 1, pp. 439–444.
11. D. P. Huttenlocher, W. J. Rucklidge, and G. A. Klanderman, "Comparing images using the Hausdorff distance under translation," in *Proceedings 1992 IEEE Computer Society Conference on Computer Vision and Pattern Recognition*, Los Alamitos, CA, USA, 1992, pp. 654–656.
12. D. P. Huttenlocher, G. A. Klanderman, and W. J. Rucklidge, "Comparing images using the Hausdorff distance," *IEEE Transactions on Pattern Analysis and Machine Intelligence*, vol. 15, no. 9, pp. 850–863, Sept. 1993.
13. C. S. Won, D. K. Park, and S Park, "Efficient use of Mpeg-7 edge histogram descriptor," *Etri Journal*, vol. 24, no. 1, pp. 23–30, Feb. 2002.
14. M. Abdel-Mottaleb, "Image retrieval based on edge representation," in *Proceedings 2000 International Conference on Image Processing*, Piscatway, NJ, USA, 2000, vol. 3, pp. 734–737.
15. A. K. Jain and A. Vailaya, "Image retrieval using color and shape," *Pattern Recognition*, vol. 29, no. 8, pp. 1233–1244, Aug. 1996.
16. J. Canny, "A computational approach to edge detection," *IEEE Transactions on Pattern Analysis and Machine Intelligence*, vol. PAMI-8, no. 6, pp. 679–698, Nov. 1986.
17. S. Brandt, J. Laaksonen, and E. Oja, "Statistical shape features in content-based image retrieval," in *Proceedings 15th International Conference on Pattern Recognition. ICPR-2000*, Los Almaitos, CA, USA, 2000, vol. 2, pp. 1062–1065.

Motion Estimation Based on Temporal Correlations

H.S. Yoon[1], G.S. Lee[1], S.H. Kim[1], and J.Y. Chang[2]

[1] Department of Computer Science and the Information & Telecommunication Research Institute, 300 Youngbong-dong, Buk-gu, Kwangju 500-757, Korea
estheryoon@hotmail.com, {shkim,gslee}@chonnam.chonnam.ac.kr
[2] System IC Design Team, Electronics and Telecommunications Research Institute, 161 Gajeong-dong, Yuseong, Daejeon 305-350, Korea
jychang@etri.re.kr

Abstract. To remove temporal redundancy contained in a sequence of images, motion estimation techniques have been developed. However, the high computational complexity of the problem makes such techniques very difficult to be applied to high-resolution applications in a real time environment. If a priori knowledge about the motion of the current block is available before the motion estimation, a better starting point for the search of an optimal motion vector can be selected. In this paper, we present a new motion estimation approach based on temporal correlations of consecutive image frames that defines the search pattern and the location of initial search point adaptively. Experiments show that, comparing with DS(Diamond Search) algorithm, the proposed algorithm is about $0.1 \sim 0.5$(dB) better than DS in terms of PSNR and improves as much as 50% in terms of the average number of search points per motion estimation.

1 Introduction

Recently, great interest has been devoted to the study of different approaches in video compressions. The high correlation between successive frames of a video sequence makes it possible to achieve high coding efficiency by reducing the temporal redundancy. Motion estimation(ME) and motion compensation techniques are an important part of most video encoding, since it could significantly affect the compression ratio and the output quality. The most popular motion estimation and motion compensation method has been the block-based motion estimation, which uses a block matching algorithm (BMA) to find the best matched block from a reference frame. ME based on the block matching are adopted in many existing video coding standards such as H.261/H.263 and MPEG-1/2/4. If the performance in terms of prediction error is the only criterion for BMA, full search block matching algorithm(FS) is the simplest BMA, guaranteeing an exact result. FS can achieve optimal performance by examining all possible points in search area of the reference frame. However, FS is very computationally intensive and it can hardly be applied to any real time applications. Hence, it is inevitable to develop fast motion estimation algorithms for real time video

M.H. Shafazand and A M. Tjoa (Eds.): EurAsia-ICT 2002, LNCS 2510, pp. 75–83, 2002.

coding applications. Many low complexity motion estimation algorithms such as Diamond Search (DS) [1,2], Three Step Search(TSS) [3], New Three Step Search(NTSS) [4], Four Step Search(FSS) [5], Two Step Search(2SS) [6] and Two-dimensional logarithmic search algorithm [7] have been proposed. Regardless of the characteristic of the motion of a block, all these fast block matching algorithms (FBMAs) use fixed search patterns, which results in the use of many checking points to find a good motion vector(MV). Since this class of BMAs do not have any information on the motion of the current block, they use the origin of the search window as the starting point. To improve the accuracy of the existing BMA algorithms, in this paper, the motion correlation between successive frames is used to predict an initial starting point that reflects the current block's motion trend. Because a properly predicted initial starting point makes the global optimum closer to the predicted starting point, it increase the chance of finding the optimum or near-optimum motion vector with less search points.

In this paper, we proposed an adaptive block matching algorithm based on temporal correlations. In this algorithm, the motion vector of the block with the same coordinate in the reference frame is considered to predict the movement of the current block. And then we determine an initial starting point and the search pattern adaptively according to the predictive movement of the current block.

This paper is organized as follows. Section 2 describes the existing motion estimation algorithms. The proposed algorithm is described in Section 3. Section 4 reports the simulation results and conclusions are given in Section 5.

2 Motion Estimation Algorithms

Among many strategies for motion estimation, in particular, we focus on the block-based scheme where the frames are sub-divided into smaller units called blocks. The block-based motion estimation uses BMA to find the best matched block from a reference frame. The displacement between the best-matched block in the reference block and the current block is called a motion vector(MV). Many search algorithms for motion estimation have been developed[1–9]. FS, the simplest algorithm, examines every point in the search area to find the best match. Clearly, it is optimal in terms of finding the best motion vector, but it requires very high computations. This has led to the development of FBMAs such as DS, TSS, NTSS, FSS and 2SS. The TSS is a coarse-to-fine search algorithm. The starting step size for search is large and the center of the search is moved in the direction of the best match at the stage, and the step size is reduced by half. In contrast, FSS starts with a fine step size(usually 2) and the center of the search is moved in the direction of the best match without changing the step size, until the best match at that stage is the center itself. The step size is then halved to 1 to find the best match. In other words, in FSS the search process is performed mostly around the original search point (0,0), or it is more center-biased. Based on the characteristics of a center-biased motion vector distribution, NTSS enhanced TSS by using additional search points, which are around the search origin (0,0) of the first step of TSS. The DS is also a center-biased algorithm by exploiting the shape of the motion vector distribution. Regardless of the characteristic of the motion of a block, DS shows the best performance compared to these FBMA methods in terms of both average number of search points per motion vector and the PSNR (peak signal to noise ratio) of the predicted image. The DS algorithm is summarized as follows.

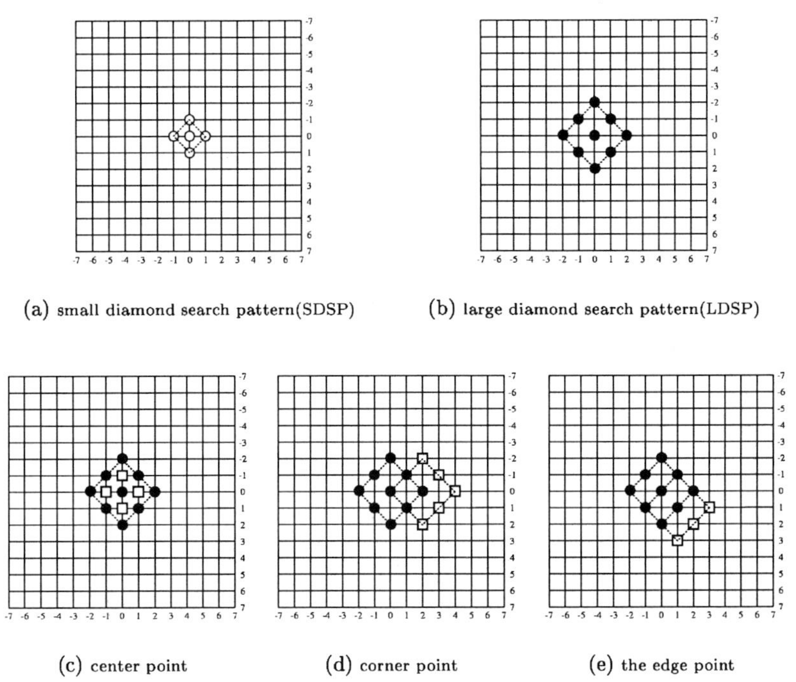

(a) small diamond search pattern(SDSP) (b) large diamond search pattern(LDSP)

(c) center point (d) corner point (e) the edge point

Fig. 1. Diamond Search algorithm (DS)

Step 1. The initial LDSP is centered at the origin of the search window, and the 9 checking points of LDSP are tested. If the minimum block distortion (MBD) point calculated is located at the center position, go to Step 3; otherwise, go to Step 2.

Step 2. The MBD point found in the previous search step is re-positioned as the center point from a new LDSP. If the new MSD point obtained is located at the center position, go to Step 3; otherwise, recursively repeated this step.

Step 3. Switch the search pattern from LDSP to SDSP. The MBD point found in this step is the final motion vector which points to the best matching block.

3 The Proposed Algorithm

If we can use any information on the motion of the current block before the motion estimation, the new starting point for the current can be set so that we can find the motion vector with much less number of search points. Since the time interval between successive frames is very short, there are high temporal correlations between successive frames of a video sequence. In other words, the motion of the current block is very similar to that of the same coordinate block in the reference frame. In this paper, we use the MV of the same coordinate block in the reference frame to determine the starting point of the search and then the search patterns are selected adaptively. In our algorithm, depending on the MV of the same coordinate block in the reference frame, two search patterns as illustrated in Fig. 3 and Fig. 4 are used. If the MV of the same coordinate block in the reference frame is zero vector as in a stationary block, the search pattern called small diamond search pattern (SDSP) [8] as shown in Fig. 3 is used without changing the starting point which is the search origin (0,0). In Fig. 3(a), white circles are the initial search points and in Fig. 3(b), black circles are search points added in the second step. Note that the center of black circles is the position which showed the minimum distortion in the first step. Otherwise, the starting point is moved to the displacement of the MV of the same coordinate block in the reference frame and then the modified diamond search pattern (MDSP) [9] as illustrated in Fig. 4. is used for motion estimation. Based on the fact that about 50%(in large motion case) $\sim$ 98%(in small motion case) of motion vectors are enclosed in a circular support, as shown in Fig. 2, with a radius of 2 pixels [1,2] and centered on the search origin (0,0), the circular support around the starting point becomes the initial search points in MDSP as shown in Fig. 4(a). If one of the positions notated with the '$\oplus$'s in Fig. 4(b) shows the minimum distortion among the search points of the first step of Fig. 4(a), the search procedure terminates. Otherwise, the new search points are set as shown in Fig. 4(c) or Fig. 4(d).

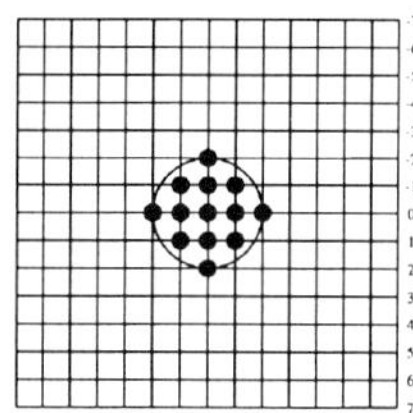

Fig. 2. Motion vector distribution

The block diagram of the proposed algorithm appears in Fig. 5. According to the current block's reference motion vector, the proposed algorithm selects

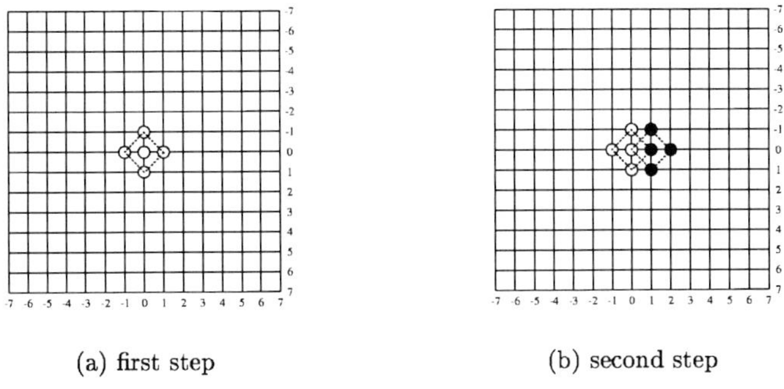

(a) first step (b) second step

Fig. 3. Small Diamond Search Algorithm(SDSP)

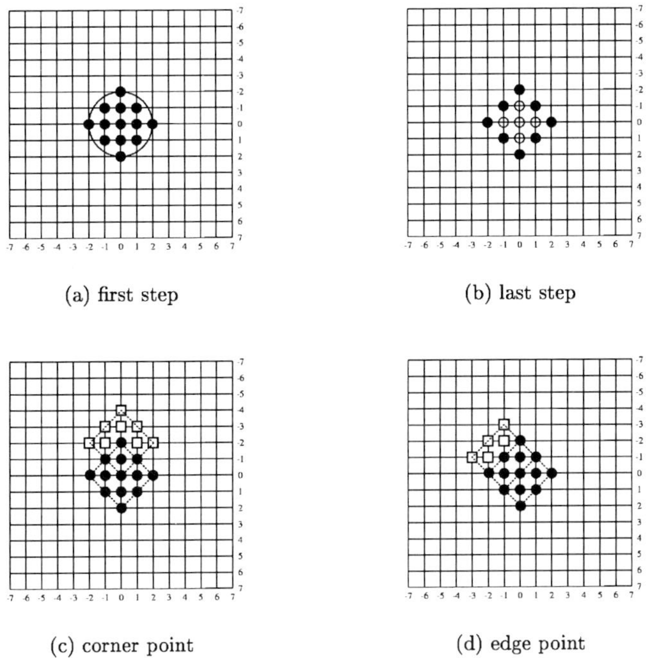

(a) first step (b) last step

(c) corner point (d) edge point

Fig. 4. Modify Diamond Search Algorithm(MDSP)

adaptively the search pattern between SDSP and MDSP. If the current block's reference MV is zero vector, SDSP is selected for the motion estimation. Otherwise, MDSP is chosen. The proposed algorithm is summarized as follows.

Step 1. If the current block's reference motion vector is zero vector, go to Step 2; otherwise, go to Step 3.

Step 2. When the current block's reference motion vector is zero vector, then

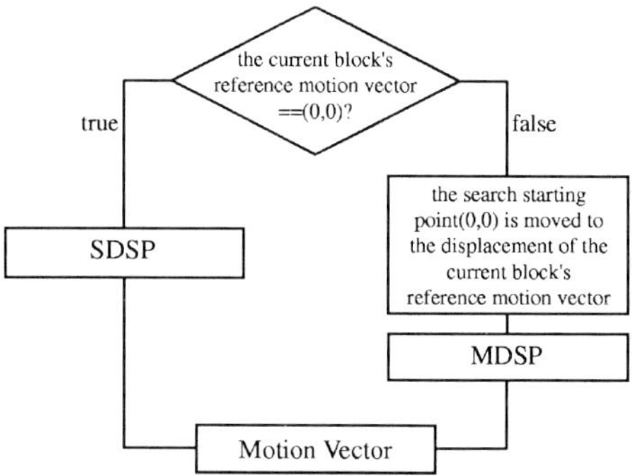

Fig. 5. The block diagram of the proposed algorithm

I. The initial SDSP is centered at the origin of search window, and the 5 checking points of SDSP as seen in Fig. 3(a) are tested. If the minimum block distortion (MBD) point calculated is located at the center position of SDSP, then it is the final solution of the motion vector. otherwise go to II.

II. The MBD point founded in the previous search step is repositioned as the center point to form a new SDSP. If the new MSD point obtained is located at the center position, then it is the final solution of the motion vector. Otherwise, recursively repeated this step.

Step 3. When the current block's reference motion vector is non-zero vector, then

I. The starting point(0,0) is moved to the displacement of the current block's reference motion vector, let's call the moved search starting point the new starting point.

II. The MDSP is disposed at the center of the new search starting point, and the 13 checking points of MDSP as seen in Fig. 4(a) are tested. If the MBD point calculated is located at the center position of MDSP or one of ⊕ points in Fig. 4(b), then it is the final solution of the motion vector. otherwise go to III.

III. If the MBD point is located at the corner of MDSP, eight additional checking points as shown in Fig. 4(c) are used. If the MBD point is located at the edge of MDSP, five additional checking points as shown in Fig. 3(d) are used. And then the MBD point found in the previous search step is repositioned as the center to from a new MDSP. If the minimum block distortion(MBD) point calculated is located at the center position of MDSP or one of ⊕ points in Fig. 4(b), then it is the final solution of the motion vector. Otherwise, recursively repeated this step.

4 Simulation Result

In this section, we show the experiment results for the proposed algorithm. We compared FS, TSS, NTSS, FSS, 2SS and DS to the proposed algorithm in both of image quality and search speed. Eight QCIF test sequences are used for the experiment: Suzie, Foreman, Mother and Daughter, Carphone, Salesman, Stefan, Table, Claire. The mean square error(MSE) distortion function is used as the block distortion measure(BDM). The quality of the predicted image is measured by the peak signal to noise ratio(PSNR), which is defined by

$$MSE = \left(\frac{1}{MN}\right) \sum_{m=1}^{M} \sum_{n=1}^{N} \left[x(m,n) - \hat{x}(m,n)\right]^2 \tag{1}$$

$$PSNR = 10 \; log_{10} \frac{255^2}{MSE} \tag{2}$$

In Eq. (1), $x(m,n)$ denotes the original image and $\hat{x}(m,n)$ denotes the motion compensated prediction image. From Table 1 and 2 we can see that proposed algorithm is better than DS in terms of both the computational complexity (as measured by the average number of search points per motion vector) and PSNR of the predicted image. From Table 3, how much portion of the estimated motion vectors in our approach are exactly same as those result from FS is shown. In other words, the motion vectors of our approach are compared to those found by FS and the numbers in Table 3 show the similarity between those two sets of motion vectors. The proposed algorithm can find the motion vector generated by FS with fewer search points in more than 90% of the stationary sequences and about 40∼50% in motioned sequences. Stationary sequences refer to images involving no movements or very small movements. In terms of PSNR, the proposed algorithm is about 0.1(dB) better than DS in stationary sequences such as Suzie, Salesman, Claire, Mother and Daughter and about 0.5(dB) in motioned sequences such as Stefan, Foreman, Table, Carphone in Table 1. In terms of the average number of search points per motion vector, the proposed algorithm improves as high as 50% compared with DS. The 2SS shows the performance in PSNR very close to our algorithm, but the proposed algorithm requires less computation by up to more than 95% on average as shown in Table 2.

5 Conclusion

Based on the temporal correlation between successive frames, an adaptive block motion algorithm is proposed in this paper. The proposed algorithm chooses the search pattern and the search starting point according to the motion vector of the same coordinate block in the reference frame. Through experiments, compared with DS, the proposed algorithm is about $0.1 \sim 0.5$ (db) better than DS in terms of PSNR and improves as high as 50% compared with DS in terms of average number of search points per motion vector. The proposed algorithm reduces the computational complexity compared with previously developed fast BMAs, while maintaining better quality.

Table 1. Average PSNR of the test image sequence

	FS	TSS	NTSS	FSS	2SS	DS	Proposed
Stefan	23.887	20.117	22.249	22.620	23.858	22.775	23.160
Foreman	29.547	26.735	28.195	28.220	29.243	28.663	29.064
Suzie	32.196	30.395	31.646	31.549	32.162	31.890	32.085
Table	26.509	23.682	25.607	24.811	26.274	25.671	25.702
Carphone	30.889	29.418	30.742	30.150	30.777	30.485	30.709
Salesman	32.703	32.365	32.691	32.538	32.709	32.621	32.695
Claire	35.059	34.618	34.911	34.744	35.014	34.857	34.933
M&D	31.527	31.192	31.478	31.345	31.515	31.429	31.491

Table 2. Average number of search points per motion vector estimation

	FS	TSS	NTSS	FSS	2SS	DS	Proposed
Stefan	961	25	20.06	18.94	255	16.24	8.95
Foreman	961	25	19.39	18.66	255	15.41	7.22
Suzie	961	25	18.65	17.84	255	14.41	7.00
Table	961	25	19.78	18.70	255	15.50	8.22
Carphone	961	25	18.62	17.83	255	14.44	6.89
Salesman	961	25	17.15	17.05	255	13.09	5.17
Claire	961	25	17.24	17.08	255	13.15	5.20
M&D	961	25	17.37	17.15	255	13.27	5.44

Table 3. The similarity of motion vectors between our approach and FS

Radius(pel)	M&D	Claire	Salesman	Carphone	Table	Suzie	Foreman	Stefan
0 (same)	0.918	0.975	0.996	0.494	0.490	0.819	0.417	0.448
1	0.987	0.999	0.999	0.906	0.836	0.978	0.871	0.812
2.5	0.994	0.999	0.999	0.965	0.838	0.993	0.954	0.946

Acknowledgement. This work was supported by grant No. R02-2000-00280 from the Korea Science & Engineering Foundation.

References

1. Tham, J.Y., Ranganath, S., Kassim, A.A.: A Novel Unrestricted Center-Biased Diamond Search Algorithm for Block Motion Estimation. IEEE Transactions on Circuits and Systems for Video Technology. **8(4)** (1998) 369–375
2. Shan, Z., Kai-kuang, M.: A New Diamond Search Algorithm for Fast block Matching Motion Estimation. IEEE Transactions on Image Processing. **9(2)** (2000) 287–290
3. Koga, T., Iinuma, K., Hirano, Y., Iijim, Y., Ishiguro, T.: Motion compensated interframe coding for video conference. In Proc. NTC81, (1981) C9.6.1–9.6.5
4. Renxiang, L., Bing, Z., Liou, M.L.: A New Three Step Search Algorithm for Block Motion Estimation. IEEE Transactions on Circuits and Systems for Video Technology. **4(4)** (1994) 438–442

5. Lai-Man, P., Wing-Chung, M.: A Novel Four-Step Search Algorithm for Fast Block Motion Estimation. IEEE Transactions on Circuits and Systems for Video Technology. **6(3)** (1996) 313–317
6. Yuk-Ying, C., Neil, W.B.: Fast search block-matching motion estimation algorithm using FPGA. Visual Communication and Image Processing 2000. Proc. SPIE. **4067** (2000) 913–922
7. Jain, J., Jain, A.: Dispalcement measurement and its application in interframe image coding. IEEE Transactions on Communications. **COM-29** (1981) 1799–1808
8. Guy. C. , Michael. G. , Faouzi. K.: Efficient Motion Vector Estimation and Coding for H.263-based very low bit rate video compression. ITU-T SG 16, Q15-A-45. (1997) 18
9. Yoon, H.S., Lee. G.S.: A modified Diamond search algorithm for fast block-matching motion estimation. Proceeding of the 2001 korean Signal Processing Conference. **14(1)** (2001) 393–396

A New Boundary Matching Algorithm Based on Edge Detection

NamRye Son[1], YoJin Yang[1], GueeSang Lee[1], and SungJu Park[2]

[1] Department of Computer Science and the Information & Telecommunication Research Institute, 300 Youngbong-dong, Buk-gu, Kwangju 500-757, Korea
{nrson,yojin}@cs.chonnam.ac.kr,gslee@chonnam.ac.kr

[2] Department of Computer Science and Engineering, Hanyang University, 1271 Sa-1 dong, Ansan, Kyunggi-do, 425-791, Korea
parksj@eecs.hanyang.ac.kr

Abstract. In transmitting compressed video bit-stream over Internet, packet loss causes error propagation in both spatial and temporal domain, which in turn leads to severe degradation in image quality. In this paper, a new error concealment algorithm, called EBMA (Edge Detection based Boundary Matching Algorithm), is proposed to repair damaged portions of the video frames in the receiver. Conventional BMA (Boundary Matching Algorithm) assumes that the pixels on the boundary of the missing block and its neighboring blocks are very similar, but has no consideration of edges across the boundary. In our approach, the edges are detected across the boundary of the lost or erroneous block. Once the orientation of each edge is found, only the pixel difference along the expected edges across the boundary is measured instead of the calculation of differences between all adjacent pixels on the boundary. Therefore, the proposed approach needs very few computations and the experiment shows an improvement of the performance over the conventional BMA in terms of both subjective and objective quality of video sequences.

1 Introduction

Hybrid block based MC/DPCM/DCT algorithm has been adopted in several international video coding standards such as the ITU-T H.261, H.263, MPEG-1,2,4,7 [1,2]. Typical applications include video conferencing, video phone and digital TV. For most of these applications, the bitstream will be transmitted over a communication channel where bit error or packet loss sometimes is inevitable. In recognizing the need to provide reliable video communications, error concealment methods have been developed and they play an important role as we find more and more applications of digital video over packet networks and wireless channels.

Due to the coding structure of the hybrid block based MC/DPCM/DCT coding algorithms, Motion Vectors (MVs) is crucial in reconstructing the predicted frames. For example, if one block's variable length coded motion vector is lost by a burst error, the error propagates until a new resynchronization is occurred.

M.H. Shafazand and A M. Tjoa (Eds.): EurAsia-ICT 2002, LNCS 2510, pp. 84–92, 2002.

These prevalent video coding schemes use a row of MBs (macroblocks), called slice(in MPEG-2) or GOB(in H.26x), as the minimal resynchronization unit. Thus the effect of one erroneous block is spread out to the end of underlying slice in spatial domain. In addition, a motion compensation scheme employs the images in the previous frame, an erroneous block has an influence on afterward frames until the next new Intracoded frame (I-picture) appears.

Among many robust video coding schemes, the error concealment technique is considered as one of the most effective ways to give error resiliency to the system, which is used either in sole or together with other robust video coding approaches. In most cases, existing error concealment algorithms fall into two categories. The first one is the technique in which the DCT coefficients, partially lost or erroneously, are recovered [3]. Secondly, lost MBs are compensated by the recovered motion vector based on the boundary matching criterion [4].

In this paper, we propose a new error concealment method which recovers the MV of lost or corrupted MBs in inter-frame. Once a MV is recovered, the image in the previous frame(motion compensated image) is taken to replace the corrupted MB [7,8].

BMA has been proved to be one of the most effective solutions to the problem, in which a fixed number of candidate MVs are examined and the one which results in the minimum pixel variations at the boundaries of the lost block and the neighboring blocks is selected [4]. Since it assumes the adjacent pixels are highly correlated, it provides a good performance when the edges are horizontally or vertically oriented along the boundaries of the erroneous MB. However, it may perform poorly with edges of arbitrary orientations. To deal with such problems, a modified BMA has been proposed[5]. But, it requires very high computations and more than that, it assumes that all edges across a MB boundary has same directions, which may not be true in many cases. We propose algorithms for the detection of edges across the boundaries of the erroneous MB and the decision of edge directions which are used in measuring distortions of the image found by the candidate MV.

In section 2, the ideas in the conventional BMA and the modified BMA are briefly overviewed and the proposed algorithms are given in section 3. Experimental results are shown in section 4 and conclusions follow.

2 Related Works

2.1 Boundary Matching Algorithm (BMA)

Boundary Matching Algorithm aims that the corrupted block is replaced by the most feasible block of the previous frame by means of strong correlations among the neighboring pixels [3]. This concept basically can be explained by the smoothness constraint measure [6]. Let the top-left pixel (p, q) denote the first pixel of $N \times N$ MB X in Fig. 1, and assume that the Variable Length Coded(VLC) data for MB X is confused by a random bit error or burst error. Thus some macro blocks after MB X are lost until the next synchronization codeword is occurred.

To recover each corrupted MB, the missing block is replaced by an image block from the previous frame which is found by a candidate MV and the distortion d_S is computed as shown in equation (1).

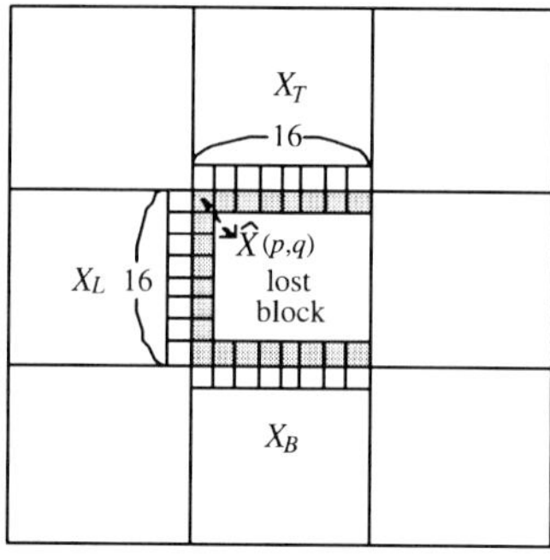

Fig. 1. BMA (Boundary Matching Algorithm)

$$
\begin{aligned}
d_L &= \sum_{i=0}^{N-1} \left[\hat{X}(p,i) - X_L(p-1,i)\right]^2 \\
d_T &= \sum_{i=0}^{N-1} \left[\hat{X}(i,q) - X_T(i,q-1)\right]^2 \\
d_B &= \sum_{i=0}^{N-1} \left[\hat{X}(i,q+N-1) - X_B(i,q+N)\right]^2 \\
d_S &= d_L + d_T + d_B
\end{aligned} \tag{1}
$$

The set of candidate MVs are usually composed of the MV of :

1) The same block in the previous frame
2) Neighboring blocks available (Top, Left and Bottom blocks)
3) Median of the available neighboring blocks
4) Average of the available neighboring blocks
5) The ZERO MV

In this candidate set, a MV which results in the minimum error d_S is finally selected as the MV for the lost MB.

2.2 A Modified BMA

Since the differences of the neighboring pixels are computed along the boundary of the erroneous MB for the distortion measure, BMA works well when there are no specific edges across the boundary or there are edges vertically located across the boundary. However, in most cases diagonal or anti-diagonal edges exist across the boundary and the computation of errors on the boundary in BMA may deviate severely from the exact distortion measure. To solve this problem, a modified BMA (MBMA) is proposed which considers different edge orientations in computing the boundary difference [5]. One of the three edge orientations, diagonal, anti-diagonal and horizontal(vertical), is decided by equation (2), by which pixel differences along the boundary are computed. For example, if E_{Ld} produces the highest value in equation (2), left diagonal edges dominate over other edge orientations on the boundary of the erroneous MB. Now the pixel differences are computed between pixels located left-diagonal across the boundary as shown in Fig. 2.

$$E_{Ld} = \frac{1}{N-1} \sum_{i=0}^{N-2} [X_L(p+i, q-1) - X_L(p+i+1, q-2)]^2$$
$$E_{La} = \frac{1}{N-1} \sum_{i=0}^{N-1} [X_L(p+i, q-1) - X_L(p+i-1, q-2)]^2$$
$$E_{Lh} = \frac{1}{N-1} \sum_{i=0}^{N-1} [X_L(p+i, q-1) - X_L(p+i, q-2)]^2 \quad (2)$$

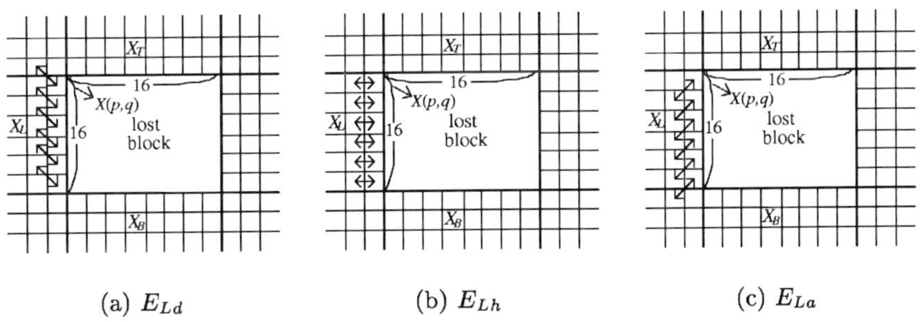

Fig. 2. Decision of edge orientations at the boundary

3 Edge Detection Based BMA (EBMA): Proposed Approach

MBMA requires high computations and more than that, it assumes that all edges across a MB boundary has same directions or at least one of the edge orientation dominates over others, which may not be true in many cases. We propose a simple algorithm for the detection of every edge across the boundaries of the erroneous MB. When the orientation is decided for each edge, the pixel difference between pixels which lie across the boundary along the detected orientation is computed and added to the total distortion of the candidate image block. In this way, each edge is considered separately along the boundary and only the pixel differences along the edges are added to the distortion, leading to very few computations. Details of EBMA for reconstructing the lost block in interframe-coded imges are described as follows.

Step 1) Select a boundary of the erroneous or lost MB. If all the boundaries of the lost MB has been considered, the algorithm terminates.

Step 2) Edge detection : Scanning from the first pixel on the boundary, calculate the difference(*diff*), in equation (3), between adjacent pixels on the boundary of the neighboring block as shown in Fig. 3. If the difference is greater than a threshold th, we assume that an edge is detected.

$$diff = |X_T(p+i, q-1) - X_T(p+i+1, q-1)|,\ i = 0 \sim 15, N = 16 \quad (3)$$

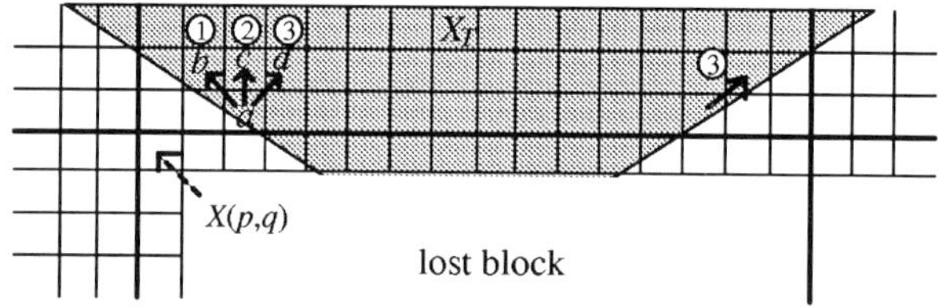

a : pixel where an edge is detected, $X_T(p+i, q-1)$
b : left-top pixel coordinates of the a, $X_T(p+i-1, q-2)$
c : top pixel coordinates of the a, $X_T(p+i, q-2)$
d : right-top pixel coordinates of the a, $X_T(p+i+1, q-2)$

Fig. 3. Detection Method of edge and direction

Step 3) Decision of the edge orientation : Pixel differences between pixel a in Fig. 3 and the three pixels above it are computed as shown in equation (4). The smallest difference decides the edge direction.

$$\begin{aligned} &1)\text{direction} : |X_T(p+i-1, q-2) - X_T(p+i, q-1)| \\ &2)\text{direction} : |X_T(p+i, q-2) - X_T(p+i, q-1)| \\ &3)\text{direction} : |X_T(p+i+1, q-2) - X_T(p+i, q-1)| \end{aligned} \quad (4)$$

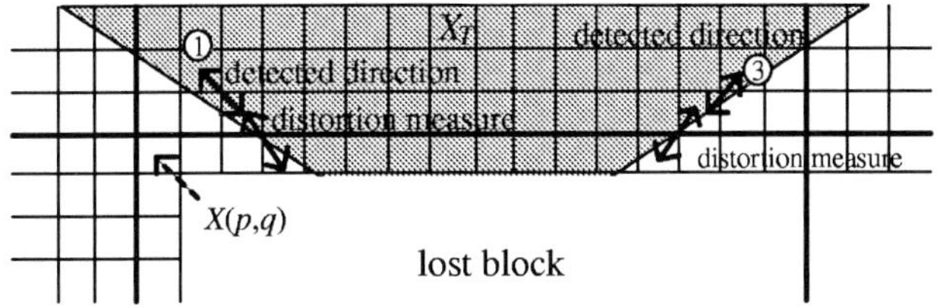

Fig. 4. Measuring distortion by the pixel difference across the boundary

Step 4) Measuring distortion : Assuming that the lost or erroneous MB is replaced by an image block from the previous frame by a candidate MV, the difference between pixel a in Fig. 3 and the pixel located in the opposite place along the edge orientation is computed and added to the total distortion.

Step 5) If the pixel a is not the last pixel on the boundary of the lost MB, goto step 2). Otherwise, goto step 1).

The above procedure is applied to every candidate MV and the one with smallest distortion will be selected for the recovery of the lost MB.

4 Simulation Results

Experiments are carried out with H.263 video coder. Four QCIF test sequences, Suzie, Carphone, Mother&Daughter, and Foreman with a Block Error Rate(BER) of 5% $\sim$ 20% are used for the experiment. To simulate the effect of transmission over practical communication channels, errors are introduced randomly in MBs or GOBs. Note that sometimes GOBs can be lost since the loss of a GOB header implies the loss of the whole data in the GOB. It is assumed that the concealment process is supported by an appropriate transport format which helps to identify lost or damaged blocks at the decoder. Table 1 shows the comparison of the quality of the recovered images using BMA, MBMA and the proposed method. Fig. 5 displays an example image of 47th frame of Suzie sequence recovered from 20% block loss. In Fig. 5(c)$\sim$(h), the distortions incurred by each algorithm are shown. Specially, the phone image is shrunk in Fig. 5(f) and Fig. 5(g) shows very big displacement. EBMA recovers the image mostly close to the original image.

In terms of PSNR, the proposed EBMA is about to 0.1 $\sim$ 1(dB) better than BMA and is not less than MBMA. In terms of processing time of recovery of missing or erroneous MV, we are consider into two categories. first, in decision factor for edge direction, BMA did not used, but MBMA and EBMA made a decision about edge orientations on the boundary of MB. However MBMA is calculated three edge orientations per the boundary of MB, EBMA is computed three edge orientations at edge detected. Therefor EBMA is superior to MBMA by about 18% $\sim$ 25% in decision factor for edge decision. secondly, in calculation

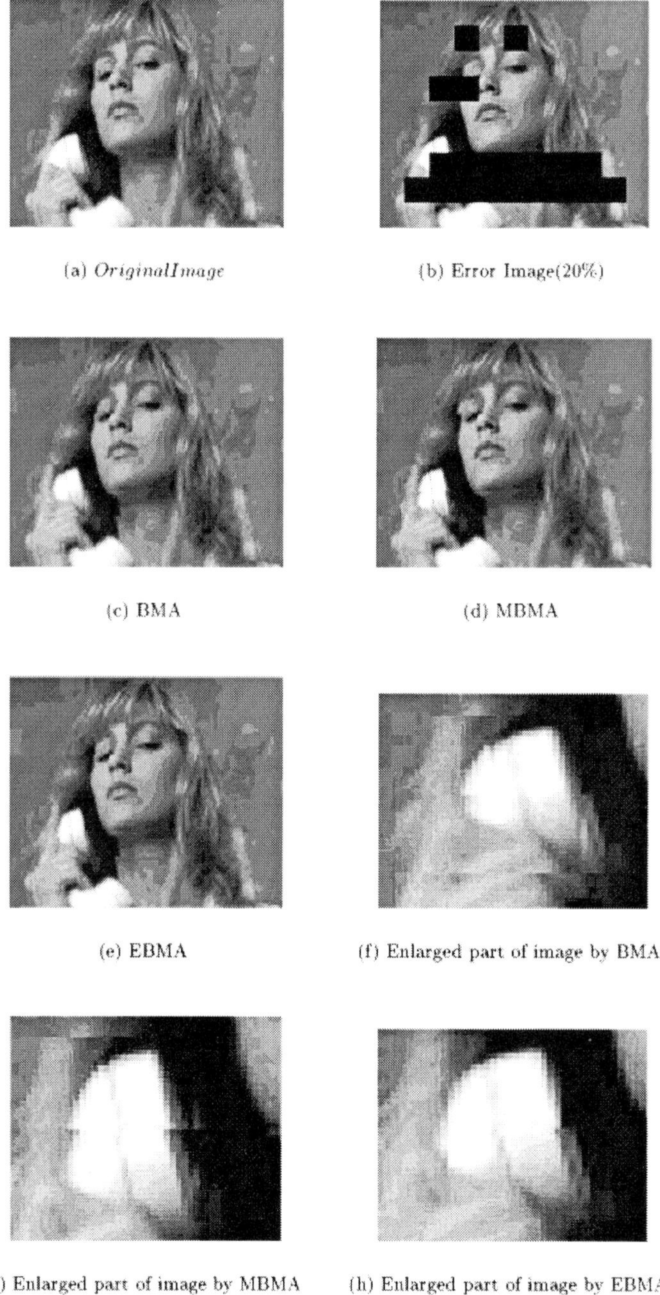

(a) *OriginalImage* (b) Error Image(20%)

(c) BMA (d) MBMA

(e) EBMA (f) Enlarged part of image by BMA

(g) Enlarged part of image by MBMA (h) Enlarged part of image by EBMA

Fig. 5. Reconstructed images using different computation methods

per candidate MVs, both BMA and MBMA are inferior to EBMA by about 20% $\sim$ 84% at the rate of increasing candidate MVs.

Table 1. List of PSNR for the Test Sequences

Image	BER(%)	BMA	MBMA	EBMA
Carphone #20	5	37.01	37.01	37.01
	10	36.99	37.02	37.02
	15	36.97	37.04	37.04
	20	36.97	37.02	37.02
Foreman #3	5	35.70	35.16	35.57
	10	35.61	35.16	35.57
	15	35.61	35.17	35.55
	20	35.47	35.18	35.53
Mother& Daughter #15	5	52.65	53.61	53.61
	10	52.21	52.65	52.75
	15	52.45	52.53	52.65
	20	50.63	50.68	50.69
Suzie #47	5	52.90	53.11	53.20
	10	47.78	48.22	48.66
	15	47.72	47.40	47.90
	20	45.77	47.44	47.86

5 Conclusion

In this paper, a new error concealment algorithm, called EBMA (Edge Detection based Boundary Matching Algorithm), is proposed to repair damaged portions of the video frames in the receiver. Conventional BMA has no consideration for the direction of edges across the boundary. In our approach, the edges are detected across the boundary of the lost or erroneous block and its neighboring blocks. Once the edge direction is decided for each edge detected, only the pixel differences along the expected edges are measured, which results in very few computations, instead of the simple calculation of differences between all adjacent pixels on the boundary of the lost block. The experiments showed that the proposed approach has better performance compared with conventional BMA and MBMA in terms of both subjective and objective quality of video sequences.

Acknowledgement. This work was supported by grant No. R02-2000-00280 from the Korea Science & Engineering Foundation.

References

1. ISO/IEC13818-2.: Information Technology Generic Coding of Moving Pictures and Associated Audio. Draft International Standard. (1994)
2. ITU-T Recommendation H.263 Version 2, ITU-T SG-16.: Video Coding for Low Bitrate Communication. (1997)
3. Jong Wook Park, Jong Won Kim and Sang Uk Lee.: DCT Coefficient Recovery Based Error Concealment Technique and its Application to the MPEG-2 Bit Stream Error. IEEE Transactions on Circuits Systems for Video Technology **7** (1997) 845–854
4. W.-M. Lam, A.R. Reibman, and B.Lin.: Recovery of Lost or Erroneously Received Motion Vectors. In Proc. ICASSP. **5** (1993) 417–420
5. Jian Feng, Kwok-Tung Lo and Hanssna Mehrpour.: Error Concealment for MEPG Video Transmissions. IEEE Transactions on Consumer Electronics. **43(2)** (1997) 183–187
6. Y.Wang, Qin-Fan Shu and Leonard Shaw.: Maximally Smoothness Image Recovery in Transform Coding. IEEE Transactions on Communications. **41** (1993) 1544–1551
7. J.S.Hwang, D.K.Park, C.S.Won, J.C.Jung, and S.Y.Kim.: A Concealment Algorithm Based on the Analysis of Transmission Errors or H.263 Bitstream. In Proc. Korean Signal Processing conference. **10(1)** (1997) 555–558
8. H.C.Shyu and J.J.Leou.: Detection and Concealment of Transmission Errors in MPEG-2 Images – A Genetic Algorithm Approach. IEEE Transactions on Circuits Systems for Video Technology. **9(6)** (1999) 937–948

Lookmark: A 2.5D Web Information Visualization System

Christian Breiteneder, Horst Eidenberger[1], Geert Fiedler, and Markus Raab[2]

Vienna University of Technology, Institute of Software Technology and Interactive Systems, Favoritenstrasse 9–11 – 188/2, A-1040 Vienna, Austria
[1]{breiteneder, eidenberger}@ims.tuwien.ac.at
[2]{a9505913, a9505911}@unet.univie.ac.at

Abstract. Lookmarks are thumbnails of existing web pages that can be arranged within a 2.5-dimensional space, just like documents can be arranged on a normal desk. The Lookmark system offers the user the opportunity of taking individual web pages and structuring and managing them within a 2.5- dimensional space. The paper discusses relevant related work in the field of Information Visualization and Interaction, design issues and the implementation of Lookmark in Java and Java3D.

1 Introduction

Due to the enormous growth of the World Wide Web more and more users have access to a steadily increasing number of documents. All developers of the various web browsers offer web document management solutions based on tree/directory structures. Examples are the Favorites of Microsoft Internet Explorer or the Bookmarks used by Netscape Navigator. The clear advantage of such a solution is the easy integration into browsers and the simple handling (the user just uses the mouse to pull the link over the designated folder). A further criterion is the low demand on hardware. However, the fact that such a system is not always conducive to finding documents that have thus been marked, is an important reason for the development of alternative forms of representation. A tree structure presented in text format can easily seem overloaded and confusing.

A better solution is to address the inherent human ability to think spatially. Just like a user would normally arrange files and folders on his real desk, he should be able to arrange electronic documents and web pages on his electronic desk. In the same context these electronic documents (e.g. an image of a website) should be used physically, and not only its description or links. Documents that are related in content can then be managed in spatial groups, whereby the depth of arrangement displays the hierarchical position of individual documents.

Lookmark, the visualization of bookmarks, is a prototype implemented in the project 'Bhutan – Fortress of the Gods', a virtual exhibition on the religion, history, art and culture of the kingdom of Bhutan [2].

The rest of this paper is organized as follows: Section 2 points out relevant related work, Section 3 describes the idea and design concepts behind Lookmark and Section 4 is dedicated to implementation issues.

M.H. Shafazand and A M. Tjoa (Eds.): EurAsia-ICT 2002, LNCS 2510, pp. 93–101, 2002.

2 Related Work

2.1 Motivation of Information Visualization

The authors of [3] argue that the growth of the internet, the computerization of business and defense and the deployment of data warehouses have created a widespread need and emerging appreciation for Information Visualization (IV) techniques. The media of visual computing and display are quite new, and we do not understand well their advantages and disadvantages. For the authors, the key problem is to discover new visual metaphors and understand what analytical tasks they support [3]. Generally, IV aims at the 2D or 3D arrangement of information that lacks inherit 2D or 3D semantics [5]. The goal of spatial IV methods is to offload part of the burden of conscious information processing to the human perceptual system [6].

The motivation behind IV is that visualization provides an interface between two powerful information processing systems: the human mind and the modern computer [3, 5]. It allows real 'browsing in information'. Documents have content and history that can both be visualized [3]. Apart from that, humans gain more joy from using visual methods. This is relevant, because effectiveness (of retrieval), ease of use and joy are significantly correlated with each other [6].

IV systems can be symbolic (information items are represented by text) or diagrammatic (information items are represented by images or graphics, [8]). The major reasons for representation of information items by images are that images can transport more information and are cognitively direct: seeing is – in comparison to reading – a non-abstract process [3, 8]. On the other hand, text represents information more clearly and images are to some extent culturally dependent [3]. Practical arguments against symbolic representation are that in the average only about 100 data items can be displayed at one time [5]. Visualization offers more capacity, is faster and leads to better results ([5, 8] based on usability evaluations).

2.2 Information Visualization Methods

Spatial IV methods can be split in two groups: 2D and 3D methods. State-of-the-art 2D IV methods include:

- SeeSoft. This system represents the features of source code (e.g. line length, last modification, frequency of use) in diagrammatic form. Each line of code is displayed as one row (line). Colored line segments indicate features (e.g. red lines for recently modified code [3]).
- Tree map visualizes hierarchical information in diagrammatic and symbolic form. A 2D display – similar to a map – allows to fill in available space [3].

Prominent 3D IV approaches are:

- Cone Tree (developed by XEROX PARC). The Cone Tree was one of the first attempts to represent information symbolically in a conic tree structure [3].

– WebBook, Web Forager and DataMountain. These systems allow the diagrammatic organization and manipulation of web pages and other documents. Web Forager embeds WebBook and other objects in a hierarchical workspace. DataMountain is a generalization of the IV concept of WebForager [3, 7].

Subsequently, we point out the major advantages of 3D IV systems over 2D methods. 3D views take advantage of human spatial memory [7], allow displaying more information without incurring additional cognitive load because of pre-attentive processing of perspective views [7] and lead to better retrieval results in user studies (reaction time, number of incorrect retrievals, failed trials, [7]). Additionally, they allow displaying more information items because of scaling and a better global view [8]. Finally, there is experimental evidence that 3D displays enhance subjects' spatial performances [8]. The major open problem of 3D systems is the development of 3D user interaction techniques [7, 5].

3 Lookmark Design

This Section gives an overview on the Lookmark design: we will explain the idea behind Lookmark, the user interface layout and the interaction methods.

3.1 Idea

Traditionally, annotated web-links are stored in symbolic tree structures. In this paper, we use the term 'Bookmark' for this concept. As pointed out above, this method has some serious drawbacks: (1) Tree-structured graphs are limited and easily cluttered, tend to become confusing and often unusable [3]. (2) Symbolic approaches lead to worse retrieval results than diagrammatic representations.

Lookmark is basically a diagrammatic 3D IV approach for websites. The new idea in Lookmark is that we give up one dimension of movement resp. interaction in 3D space to gain advantage in ease of use. Information items are represented as thumbnails and are always displayed in parallel to the screen (image plane). The thumbnails are positioned on the X-Z-plane. Movement and interaction are restricted to the horizontal (left-right) and depth (front-back) dimension. Because this method uses a perspective 3D view and 2D interaction, we call it 2.5 dimensional (2.5D). The limited movement and interaction has two advantages: movement is less confusing than in 3D spaces and can be easily mapped to 2D interaction metaphors (e.g. mouse).

In the existing implementation, Lookmark offers the user the opportunity of taking each of the individual pages of the virtual exhibition and then structuring and managing them within a three-dimensional space, just like documents can be arranged on a normal desk. The name Lookmark should more than hint at the fact that here use is made of a visual filing system. The user decides on the selection of the web pages, how they will be arranged in space and how they are to be grouped together. Grouping describes a content-based linking that can be freely named. The item's hierarchical order is reflected by its designated spatial depth.

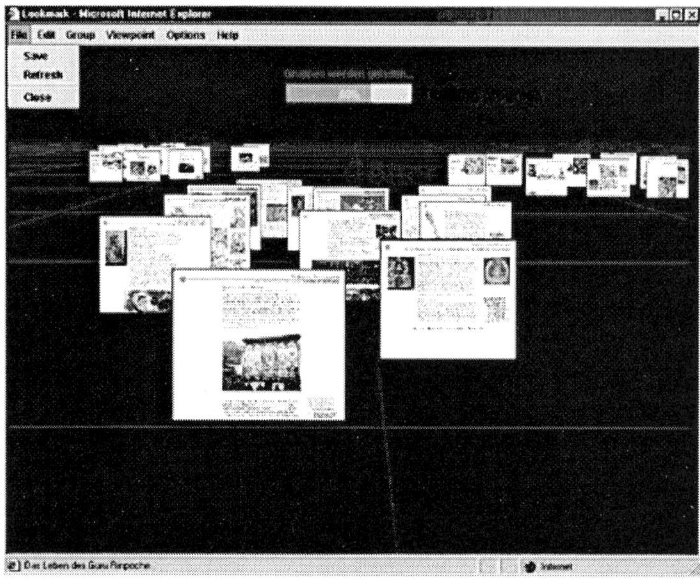

Fig. 1. Lookmark user interface screenshot. The user interface consists of a 2D control panel that contains a menu and a media loading progress bar and a 2.5D visualization and interaction panel.

3.2 User Interface Design

The user interface (see Figure 1 for a screenshot) consists of two panels (the 2.5D manipulation panel and a 2D control panel) and a status bar. The manipulation panel displays the thumbnail representations of web pages and represents the main work area. The chosen design and perspective shrinking of the thumbnails provide a 3D look-and-feel. The control panel consists of a pull down menu and a status bar for the loading progress of the thumbnails. The name of the current selected object is displayed in the status bar. For technical reasons (see Subsection 4.1 for details) the control panel must be extended as far downwards as the longest pull down menu. Because the background colors of both areas are the same, this trick is barely noticed.

After the user interface has been built up, the thumbnails, groups and views are loaded from a database and the progress of the download is displayed in the status bar, which disappears as soon as the download has finished. The download from the database is done in a separate thread, which allows the user to interact with the applet during the download. Groups have associated labels that are shown on the panel as text. The panels are independent from each other and interact through events.

3.3 User Interaction

As pointed out above, one of the major advantages of a 2.5D view is the intuitive handling with 2D interaction devices. In Lookmark both standard devices, mouse and keyboard, can be used. Mouse handling is limited to single click, double click and drag-and-drop. The meaning of mouse events depends on the context of the object that receives the event:

- Thumbnails. With a single mouse click one can add resp. delete the clicked thumbnail to resp. from the current selection. At the same time the name of the corresponding web page is displayed in the status bar of the user interface. By double clicking on a specific thumbnail, one can open the corresponding web page in the associated Web-browser. By using drag-and-drop one can move a thumbnail along the horizontal and depth dimension (X- and Z-axis).
- Groups of thumbnails. Thumbnails in a group are handled like single thumbnails. Groups of thumbnails can be moved simultaneously by dragging the group label.
- Selections. Dragging a thumbnail of a selection moves all selected thumbnails. A selection can be dissolved by clicking somewhere in the manipulation panel.

By using the menus a selection of thumbnails can either be grouped together, deleted or copied. When thumbnails are grouped, the user is required to enter a group name. This group label is then displayed in the manipulation area centered above the selected thumbnails. The color of the selection is replaced by that of the group. By using the menu any grouping can be dissolved. A thumbnail can be member of only one group at a time. However, copies are allowed to participate in other groups. When a thumbnail, selection or group is copied, an exact copy is created, which is displayed as a new object offset behind the selected object.

A fundamental part of Lookmark is the possibility to change and save the camera position. Each time Lookmark is initiated, it uses a default camera position (central view on the center of the given objects). The position of the camera can be changed with the cursor keys on the X- and Z-axis. For the user this movement of the camera causes the impression as if he were flying over the thumbnails. Additionally, the camera can also be rotated along the X- and Y-axis. With the use of the menu the present camera position can be saved and restored.

4 Implementation

Subsequently, we describe the software development environment and the implementation of the visual part (based on scene graphs) of Lookmark.

4.1 Software Development Environment

The Lookmark prototype, although developed for the special application of a virtual museum, should be open and extendible. It should be exclusively based on free software, equipped with a well-documented API and consist of independent components (panels). To satisfy these demands, we selected Java and the JavaSDK as the basis for the prototype. The control panel is based on Java Swing components. For reasons of compatibility and the support of 3D modeling with scene graphs, Java3D was chosen for the 2.5D panel implementation. The Java3D API packet represents a good base to quickly create a virtual world with simple interaction. Many aspects of 3D computer graphics, such as the calculation of transformations, shadings, collision tests, textures etc. are taken over by Java3D. On the other hand, Java3D has the disadvantage that 3D areas are always set to the foreground automatically. Intersecting user interface components (like the entries in the pull down menus and the status bar of the Lookmark control panel) would basically disappear behind the 3D area. This problem was solved with the trick described in Subsection 3.2. To handle the Lookmark thumbnails the development of special Java3D interaction classes was necessary.

In the Bhutan project, web pages consist of page building blocks stored in a database and are generated dynamically. PHP and the macro processor m4 are used for web page generation. The Lookmark thumbnails in the database are constructed with the same method. The database is accessed through JDBC and can be any relational database. We used Oracle 8i for the prototype. To keep the working performance as high as possible, small changes in the data (e.g. new groups, movements, etc.) are not saved immediately, but only when the Lookmark user interface is closed. The user can force saving through the control panel menu.

4.2 Visual Interaction Modeling

The modeling of visual display and user interaction in the 2.5D panel is based on Java3D scene graphs (see [4] for details). The scene graph for the whole application is depicted in Figure 2. The nodes are named after the resources used in the implementation classes.

myUni and *myLocale* are standard objects of the VirtualUniverse and Locale Classes that are elements of each Java3D world. The *myBranch* branch group node forms the root for all objects that are used in the 3D environment. For this node and all its children, both the behaviors *pickzoom* and *clickbehavior* apply. *pickzoom* is responsible for dragging visual objects. The handling of single and double mouse clicks are defined in *clickbehavior*. *myBack* is a background type leaf node, which uses the *backColor* attribute to define the background color of the 3D environment. Another branch group node (grid) is the base node for the panel grid. This allows the grid to be made visible or invisible easily. The form of the grid is defined in the leaf node *landGeom* (a line array node) and the color is defined in the color attribute *gridColor*.

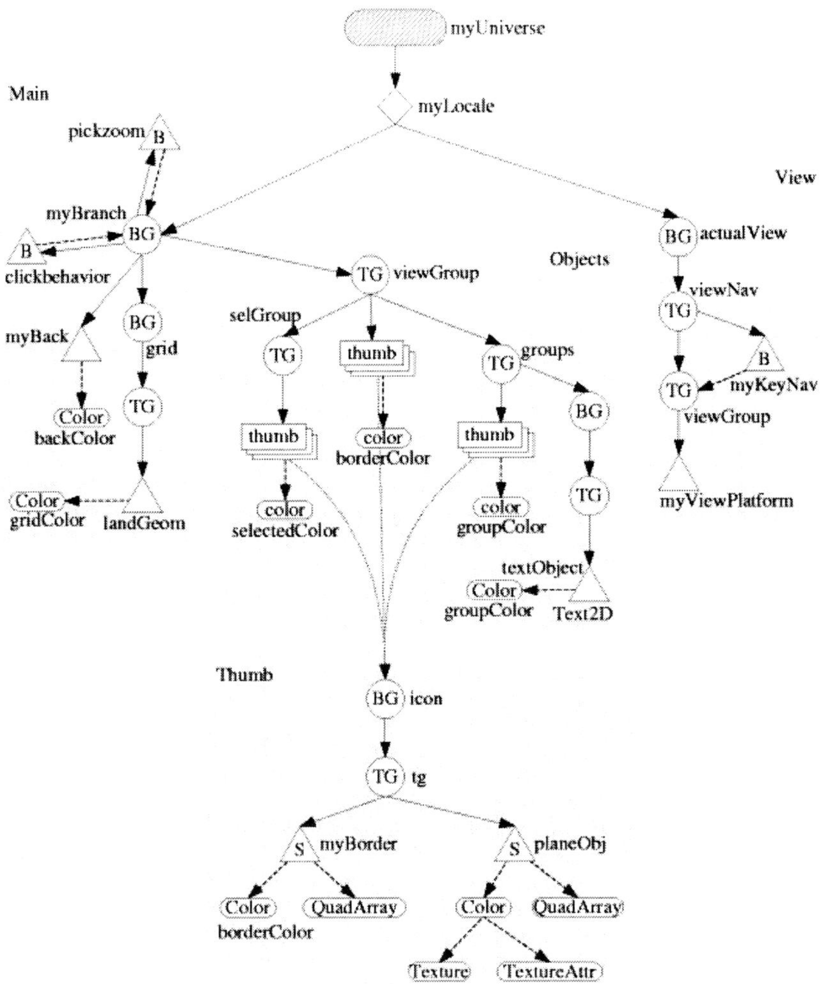

Fig. 2. Lookmark scene graph. Thumb is a self-defined object that is used in the branches of transform group viewGroup.

In the *viewGroup* layer of the scene graph the differences between the three types of visual objects (thumbnails, groups and selections) are visualized. The transform group node *viewGroup* forms the basis for all these different representations. In this node a 0.25rad rotation of the original coordinate system along the X-axis is implemented. This rotation leads to the perspective used in Lookmark. Thumbnails are directly inserted as children of *viewGroup*. The frame color of thumbnails is determined by the attribute *borderColor*. Selections are connected by the transform group *selGroup*. If properties of the selection change (e.g. movement) not all thumbnails in the selection have to be changed since only *selGroup* needs to be transformed. *selectionColor* defines the frame color of the selected thumbnails. The behavior of grouped thumbnails is very similar to selections. Unfortunately, it is necessary to predefine all the transform group nodes of the *groups*, since transform group nodes cannot be added to the scene during runtime. Additionally, each group has a group label (represented as a leaf node *textObject*), which is a Text2D element.

icon is a self-defined object derived from a branch group object, and therefore allows the trouble-free inclusion or deletion of thumbnails in a scene during runtime. If thumbnails are moved, the transformation of the transform group node *tg* is adjusted. An *icon* consists of the leaf node *planeObj*, which is a Shape3D type node and consists of a simple square provided with the texture of the relevant image. The effect of a frame is created by drawing a slightly bigger square, which has the correct frame color (*borderColor*) assigned, behind the *icon* square. Since all thumbnails are descendants of the transform group node *viewGroup*, they have to be rotated by – *0.25rad* along the X-axis to make them appear parallel to the view plane. Finally, the transform group nodes *viewNav* and *viewGroup* and the behavior node *myKeyNav* are used to display the parallel movement of the camera in relation to the grid plane. Java3D-specific rendering constructs were omitted from the scene graph.

5 Conclusion

Lookmark is a prototype for visual information organization originally developed for the virtual exhibition ‘Bhutan – Fortress of the Gods’. Its purpose is to visualize web pages as thumbnails in a 3D environment that following the desktop metaphor and to allow the user quick and comfortable interaction and browsing. Lookmark is a novel approach that integrates many ideas from other spatial Information Visualization systems and goes one step further by using a 2.5D display and reducing interaction to two dimensions. Thus, Lookmark is easy-to-use and less confusing than 3D systems. Users that navigated with Lookmark in the Bhutan virtual exhibition gave the feedback that it allows creative interaction with the topic and is after all great joy.

References

1. Breiteneder, C., Hitz, M., Platzer, H.: Bhutan – Fortress of Gods (2000). http://www.bhutan.at
2. Breiteneder, C., Hitz, M., Platzer, H.: A Reusable Software Framework for Authoring and Managing Web Exhibitions. In: Bearman, D., Trant J. (eds.): Museums an the Web Proceedings. Archives & Museum Informatics (2001)
3. Gershon, N., Eick, S.G., Card, S.: Information visualization. ACM interactions. 5/2 (1998) 9–15
4. Java3D website: http://java.sun.com/products/java-media/3D/index.html
5. Keim, D.A.: Visual exploration of large data sets. ACM Comm. 44/8 (2001) 38–44

6. Modjeska, D., Waterworth, J.: Effects of desktop 3D world design on user navigation and search performance. In: Proceedings IEEE International Conference on Information Visualization. Salt Lake City UT (2000) 215–220
7. Robertson, G., Czerwinski, M., Larson, K.: Data Mountain: Using Spatial Memory for Document Management. In: Proceedings of ACM Symposium on User Interface Software and Technology. San Francisco CA (1997)
8. Tavanti, M., Lind, M.: 2D vs. 3D, implications on spatial memory. In: Proc. IEEE Symp. on Information Visualization. San Diego CA (2001) 139–145

Different Local Search Algorithms in STAGE for Solving Bin Packing Problem

Saeed Bagheri Shouraki[1] and Gholamreza Haffari[2]

[1] Computer Engineering Department,
Sharif University of Technology, Tehran, Iran. Sbagheri@ce.sharif.edu
[2] Gholamreza Haffari, Computer Engineering Department,
Sharif University of Technology, Tehran, Iran. Haffari@ce.sharif.edu

Abstract. Previous researches have shown the success of using Reinforcement Learning in solving combinatorial optimization problems. The main idea of these methods is to learn (near) optimal evaluation functions to improve local searches and find (near) optimal solutions. STAGE algorithm, introduced by Boyan & Moore, is one of the most important algorithms in this area. In this paper, we focus on Bin-Packing problem, an important NP-Complete problem. We analyze cost surface structure of this problem and investigate "big valley" structure for the set of its local minima. The result gives reasons for STAGE's success in solving this problem. Then by comparing the results of experiments on Bin-Packing problem, we analyze the effectiveness of steepest-descent hill climbing, stochastic hill climbing and first-improvement hill climbing as the local search algorithms in STAGE.

1 Introduction

Large scale optimization problems are in great importance in all fields of science, engineering and operation research. The goal of each of these problems is to find the best possible configuration from a large space of possible configurations. Unfortunately, most of these problems are NP-Hard [9], and finding their optimum solution in reasonable amount of time is almost impossible. Thus, there has been a great deal of work on heuristic methods for finding approximate solution in limited amount of time. Some of these methods, called Approximation Algorithms [11], are based on strong theoretical background that guarantees the quality of approximate solution in the specific distance of optima. For example, many approximation algorithms have been proposed and analyzed for Bin-Packing problem in [8]. But these algorithms are special-purpose, i.e., they are specific to particular problems, so general-purpose heuristic methods are emerged.

General-purpose heuristic methods do not guarantee the quality of solution in the way that Approximation Algorithms do, but practically they find good solutions. Frequently, these heuristic search methods, such as Simulated Annealing and Genetic

M.H. Shafazand and A M. Tjoa (Eds.): EurAsia-ICT 2002, LNCS 2510, pp. 102–109, 2002.

Algorithm, are based on iteration and are easy to implement. Algorithms, which use Reinforcement Learning methods in solving combinatorial optimization problems, are in this category.

By adopting the familiar state-space search perspective to a combinatorial optimization problem [5], a good solution is found by starting in some initial state and applying Greedy-Descent policy (usually based on an evaluation function) to eventually reach some final good state. With this viewpoint, Reinforcement Learning methods are used to learn an evaluation function that predicts the outcome of the local search, and to guide search to low-cost solutions using this learned evaluation function. Note that the evaluation function is not limited to the same form as the objective function. In addition to providing a good measure for the features of a state (directly related to the objective function), an evaluation function also gives some hints on which states predictably lead to good states using some local search algorithm [1]. Based on this idea, Zhang and Diettrech applied Reinforcement Learning to the Space Shuttle Payload Processing domain [6], and Boyan & Moore [2] introduced STAGE algorithm that has shown excellent performance on a wide range of optimization problems. By combining aspects of these two works, Reinforcement Learning was used in solving the Dial-A-Ride problem, a complicated variant of TSP [4].

In this paper we focus on Bin-Packing problem and investigate some interesting characteristics of STAGE in more details. First of all, we analyze cost surface structure of the Bin-Packing problem and examine the "big valley" structure for the set of its local minima. The result confirms previous works done for TSP and graph-bisection problem that the cost surfaces exhibit globally convex structure [7]. It gives further insight to the success of STAGE in solving Bin-Packing problem. Then, we investigate some interesting characteristics of STAGE based on the results of experiments on Bin-Packing problem. We compare the effectiveness of steepest-descent hill climbing, stochastic hill climbing, and first-improvement hill climbing as the local search algorithms in STAGE.

The rest of the paper is organized as follows. First, it gives background information and reviews the definition of Bin-Packing problem and STAGE algorithm in section 2. Analyzing cost surface structure of this problem comes in section 3. The effect of changing local search algorithm in STAGE comes in section 4. Finally, we outline conclusions and future works.

2 Background

2.1 Bin-Packing Problem

Bin-Pacing is a classical NP-Complete problem [9]. This problem has many real-world applications, including loading trucks subject to weight limitations, packing commercials into station breaks, and cutting stock materials from standard lengths of cable or lumber [8].

In this problem, we are given a bin capacity C and a list $L = (a_1, a_2, \ldots, a_n)$ of n items, each having a size $s(a_i)>0$. The goal is to pack the items into as few bins as possible, i.e. partition them into a minimum number m of subsets $B_1, B_2, .., B_m$ such that for each $B_j : \sum s(a_i) < C, a_i \in B_j$. To view this problem as a state-space search problem, we need the definition of state and neighborhood structure. A solution state x simply assigns a bin number $b(a_i)$ to each item. Each item is initially placed alone in a bin: $b(a_1) = 1, b(a_2) = 2, \ldots, b(a_n) = n$. Neighboring states can be generated by moving any single item a_i into a random other bin with enough spare capacity to accommodate it [1].

2.2 STAGE Algorithm

The central idea of STAGE is learning to predict which starting state is more promising for some local search algorithm, from sample search trajectories [2]. In addition to the usual objective function, STAGE creates and tries to learn a new evaluation function for predicting how promising a state is as a starting point for some search algorithm. The new evaluation function is approximated with some form of function approximation such as polynomial regression or multi-layer perceptron. Each state is represented as a real-valued feature vector, where the relevant features are handpicked in advance. STAGE repeatedly alternates between two stages of local search: running the original search algorithm on the objective function and running hill climbing on the new evaluation function to find a promising new starting state for the original search algorithm. In each iteration, STAGE tries to learn the new evaluation function from the available search trajectories. The training data can be obtained by Monte-Carlo simulation.

To guarantee convergence, STAGE requires the search algorithm to be proper (terminates with probability one) and behaves as a Markov chain [1]. When the search algorithm satisfies Markov property, all intermediate states on each simulated trajectory can be considered as alternate starting points for that search, thus to obtain many training data from a single search trajectory. When a local minimum for both the original objective function and the new evaluation function is reached, STAGE resets search to a random starting point.

3 Cost Surface Structure of Bin-Packing Problem

Recent analyses of optimization cost surface show that as problems grow large, random local minima are almost surely of "average" quality, and "central limit catastrophe" becomes true for them; Consequently all but an exponentially small number of local minima will have a cost approximately equal to that of the average local minimum [3]. If we exploit a structure for the cost surface of the problem, the best previously found local minima can be used to intelligently suggest next starting points for greedy-descent algorithm, in such a way that lead to lower-cost solutions.

The method, which we use to study cost surface structure for Bin-Packing problem, is similar to that of Boese and others [7] for exploiting cost surface structure for TSP and the graph-bisecting problem. They examined the set of local minima from the perspective of the best local minimum. As we see later in this section, the experimental results confirm the "globally convex" structure for Bin-Packing problem.

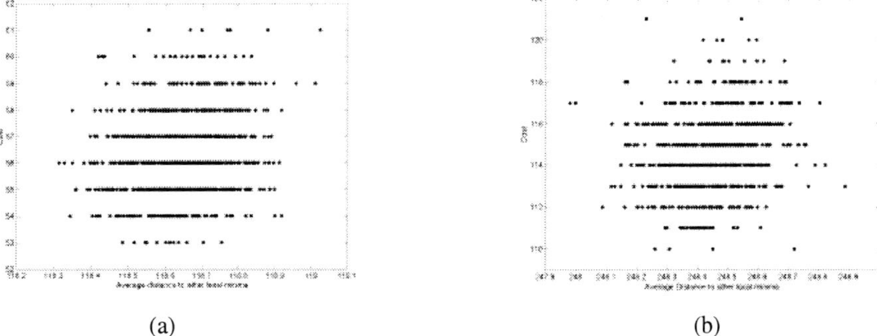

(a) (b)

Fig. 1. Analysis of 4,001 random locally minimum solutions for bin-packing instances The right figure represents data for 4,001 distinct local minima for u120_00 instance problem, and the left figure represents data for 4,001 distinct local minima for u_250_00 instance problem.

For exploiting a structure for the cost surface of Bin-Packing problem, we need to define a neighborhood structure, and equivalently an operator. As we can see from section 2.1, this operator moves one item from a bin to another bin, which has enough free space to accommodate it. We define the distance between two solutions x_1 and x_2 as the minimum number of operators needed to transform x_1 to x_2 and denote it by $d(x_1,x_2)$. Since computation of $d(x_1,x_2)$ is time consuming, we measure the similarity between x_1 and x_2 according to the number of items that are in the same bins in both solutions. We will use the term bond distance, denoted $b(x_1,x_2)$, equal to 2*number_of_bins minus similarity between x_1 and x_2, and use it as the estimation for $d(x_1,x_2)$.

To find how local minima correlate with each other, we obtain 4001 random locally minimum solutions for u120_00 instance of Bin-Packing problem[1], which has 120 bins of capacity 150. Figure 1.a plots the solution cost versus its average bond distance to all (4000) other local minima. A "random" local minimum is found by starting at a random initial solution and executing greedy-descent algorithm. In the figure 1 we can see a clear correlation: the best local minimum appears to be "central" to all other local minima, and indeed a "big valley" structure can be said to govern the set of locally minimum solutions, as illustrated in figure 2. Figure 1.b gives analogous plot for u250_00 instance problem, which has 250 bins of capacity 150.

---

[1] All instances of Bin-Packing problem are from Operation Research Library. For further information see this web page: http://www.ms.ic.ac.uk/info.html.

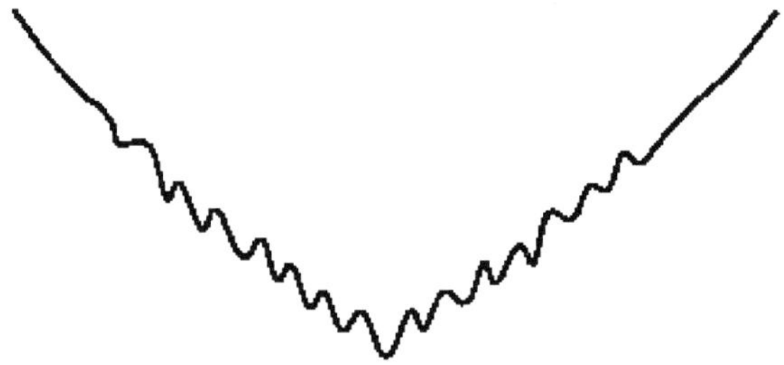

Fig. 2. Intuitive picture of the "big valley" search space structure (Boese 1996).

In Bin-Packing problem, the objective function for STAGE to minimize is the number of bins used. For automatic learning of its secondary evaluation function, Boyan provided STAGE with two state features [1]: the number of bins used, and the variance in bin fullness level. STAGE learned its evaluation function by quadratic regression over these two features. By choosing quadratic regression for evaluation function, STAGE implicitly exploited the cost surface structure for the Bin-Packing problem. To show the relationship between cost surface and evaluation function, figure 3 shows a typical evaluation function learned by STAGE for u250_00 instance problem. The number of neighborhood states grows rapidly as the number of bins increases; so stochastic hill climbing has been used by Boyan as the local search algorithm in STAGE for optimizing evaluation function. In the next section, we want to see the effect of other local search algorithms on STAGE's performance.

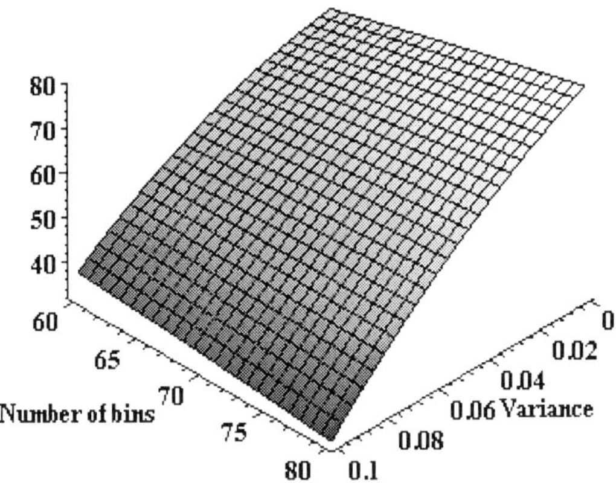

Fig. 3. A typical learned evaluation function for u250_00 instance problem.

4 Different Local Search Algorithms in STAGE

In this section, based on the experimental results from Bin-Packing problem, we compare the effects of different local search algorithms, including stochastic hill climbing, first-improvement hill climbing, and steepest-descent hill climbing, on the STAGE's performance.

Table 1. Summary of STAGE local search algorithms.

ALGORITHM	DESCRIPTION
FIHC	FIRST-IMPROVEMENT HILL CLIMBING
SDHC	STEEPEST-DESCENT HILL CLIMBING
STHC	STOCHASTIC HILL CLIMBING, PATIENCE[2] = 250

Table 2. Experimental results of solving different instances of Bin-Packing by STAGE with different local search algorithms for problem instances with 500 bins of capacity 150. Each line reports the mean, 90% confidence interval[3], best, and worst solutions found by 30 independent runs.

INSTANCE	ALGORITHM	MEAN	BEST	WORST
U500_00	FIHC	208.80 ± 0.472	207	211
	SDHC	214.50 ± 3.021	209	244
	STHC	212.80 ± 0.859	209	216
U500_01	FIHC	211.80 ± 0.505	210	215
	SDHC	219.64 ± 5.402	211	260
	STHC	215.50 ± 0.735	211	218
U500_02	FIHC	212.00 ± 0.439	211	215
	SDHC	223.05 ± 6.475	212	260
	STHC	216.10 ± 0.900	210	219
U500_03	FIHC	214.89 ± 0.439	212	216
	SDHC	225.50 ± 7.371	215	273
	STHC	218.35 ± 0.518	215	220
U500_04	FIHC	216.05 ± 0.670	213	221
	SDHC	231.00 ± 8.251	216	271
	STHC	220.10 ± 0.637	218	224

Steepest-descent hill climbing takes a step from a state to one of its neighbor states that maximally improves objective function. For search problems where number of neighbors of a state is huge, stochastic hill climbing is cheaper to run than steepest-

[2] In stochastic hill climbing, we cut off the search process when Patience consecutive moves produce no improvement.

[3] The confidence interval for the mean is produced by $\mu \pm t_{\frac{\alpha}{2},N-1}\frac{S}{\sqrt{N}}$, where 1-α is confidence factor, N is the number of runs, and t is Student-t distribution.

descent hill climbing. Stochastic hill climbing with no equal-cost move considers limited numbers of neighbors of a state randomly, and takes one of them which enhances the objective function. First-improvement hill climbing systematically examines all of the neighbors and selects the first state, which is better than the current state. If no neighbor improves objective function, the search trajectory terminates. All of these algorithms are strictly monotonic, Markovian, and (if the search space is finite) proper. Table 1 shows the summary of these algorithms.

In our experiments, each instance has 500 bins of capacity 150. STAGE is limited to 500,000 total moves. The results of 30 runs of STAGE with each algorithm for each instance are summarized in Table 2. The effect of each local search algorithm on u500_00 and u250_00 instance problems is displayed in figure 4.

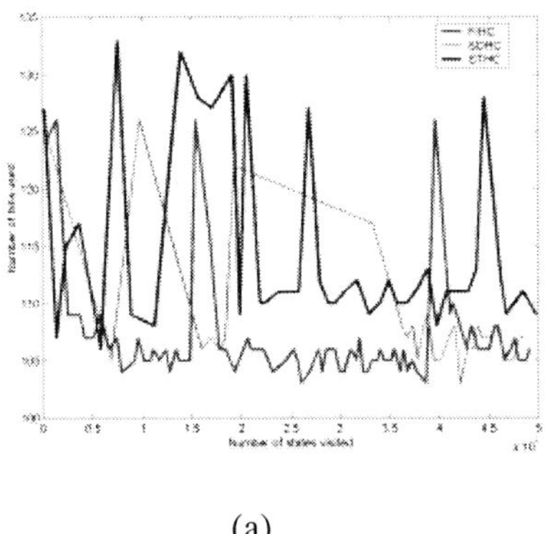

(a)

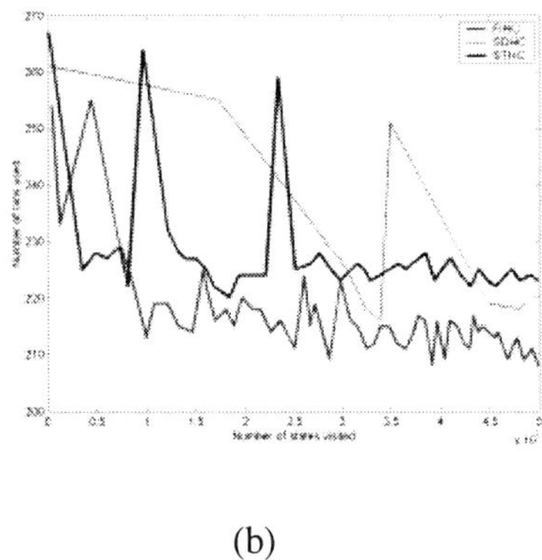

(b)

Fig. 4. Effect of using different local search algorithms on STAGE's performance. Figure 4.a plots the results for u250_00 instance problem, and figure 4.b plots the results for u500_00 instance problem.

As it can be concluded from experimental results of table 2, FIHC outperforms other local search algorithms with respect to the value of mean, best, and worst solutions that it has found. The reason for its good results is that it explores the search space as it could as possible, i.e., it tries paths about which little is yet known, while it exploits a search branch, i.e., it pursues what appears to be the best path given the limited observations made thus far. After FIHC, STHC has shown good effectiveness. This algorithm visits more areas of the search space than FIHC do but uses little information from previously visited states. Finally, SDHC has produced the worst results because it strongly sticks to a search branch to produce a local minimum solution and needs much more time to explore other areas of search space. It also can be seen from figure 5 that STHC and FIHC have similar effectiveness and oscillate much more than SDHC. The reason is that they explore search space far more than SDHC do, so they find many more local minima than SDHC.

5 Conclusions and Future Works

Recent researches have shown the success of using Reinforcement Learning in solving combinatorial optimization problems. STAGE algorithm, one of the most important algorithms based on Reinforcement Learning, was analyzed from the experimental results on Bin-Packing problem. We studied the cost surface structure for the Bin-Packing problem. It was illustrated that "big valley" structure governs its set of local minima which gives us further reasons why STAGE has produced good results for this problem. We also examined different local search algorithms for STAGE and compared their relative effectiveness: first-improvement hill climbing outperforms stochastic hill climbing and steepest descent hill climbing in solving Bin-Packing problem, where the number of neighboring states is nearly large. Further research on STAGE's performance will investigate using other function approximators such as multi-layer perceptron for estimating evaluation function instead of polynomial approximator. It is also interesting to analyze the effects of using previously learned evaluation functions in solving new instances of a problem on STAGE's performance.

References

1. Boyan, J. A.: Learning Evaluation Function for Global Optimization. Doctoral dissertation, Computer Science Department, Carnige Mellon University (1998).
2. Boyan, J. A., & Moore, A. W. : Learning evaluation functions for global optimization and boolean satisfiability. In Proceedings of the Fifteenth National Conference on Artificial Intelligence (AAAI) (1998).
3. Boese, K. D. : Models for Iterative Global Optimization. Doctoral dissertation, Computer Science Department, University of California, Los Angeles (1996).
4. Moll, R., Barto, A. G., Perkins, T. J., & Sutton, R. S. : Learning Instance-Independent Value Functions to Enhance Local Search, Proceeding of NIPS-98. Denver (1998).
5. Bertsekas, D. P., & Tsitsiklis, J. N. : Neuro-Dynamic Programming, Athena Scientific, Belmont, MA (1996).
6. Zhang, W., & Dietterich, T. G. : A reinforcement learning approach to job-shop scheduling. In Proceedings of the International Joint Conference on Artificial Intelligence (IJCAI) (1995), pages 1114–1120.
7. Boese, K.D., Khang, A. B., & Muddu, S. : On the Big Valley and Adaptive Multi-Start for Discrete Global Optimization. Operation Research Letters (1994), 16(2).
8. Coffman, E. G., Garey, M. R., & Johnson, D. S. : Approximation algorithms for bin packing: a survey. In D. Hochbaum, editor, Approximation Algorithms for NP-Hard Problems. PWS Publishing (1996).
9. Garey, M. R., & Johnson, D. S. : Computer and Intractability: A Guide to the Theory of NP-Completeness. W. H. Freeman (1979).
10. Zhang, W. (1996). Reinforcement Learning for Job-Shop Scheduling. Doctoral dissertation, Computer Science Department, Oregon State University (1996).
11. Vazirani, Vijay V. : *Approximation Algorithms*, Springer-Verlag, New York (1999).

A Prototype for Functionality Based Network Management System

V. Neelanarayanan, N. Satyanarayana, N. Subramanian, and E. Usha Rani

National Centre for Software Technology, Bangalore, India
{neela,satya,subbu,usha}@ncb.ernet.in

Abstract. We present the design and implementation of our functionality based network management prototype. The uniqueness of our prototype lies in the three-layered weakly distributed hierarchical architecture; functionalities based network management approach, and Inference Engine. Tasks such as polling, trap handling and log processing are distributed and segregated that reduces the consumption of network bandwidth and processing load. The top-level manager and the middle level managers work cooperatively and provide intelligence, greater autonomy and automation.

1 Introduction

As network installations become larger, more complex and more heterogeneous, the cost increases and the management of the networks becomes more challenging. Problems such as failure, performance inefficiency, security, configuration makes it even more complex. This requires to build an effective network management system to bring out the management possible by segregating the errors under already established functionalities viz., accounting management, configuration management, fault management, and performance management and Security management [6] coupled along with the inbuilt intelligence.

1.1 Segregation Based on Functionalities

Troubleshooting a network starts with best documentation possible. Documentation of the network starts with recording the configurations. MIB-II[1] [2] provides many configuration objects. It is a challenge to segregate them as they are scattered through out various groups. In our prototype, the middle level managers are equipped with the responsibility of functionality-based management. For example the TCP group contains objects that indicate availability attacks, UDP group contains objects that indicate availability and integrity attacks and SNMP group contains objects that indicate authentication failures [1] Configuration, Accounting, Performance and Security objects provide calibration of the network. Faults are the variations in the network. Various groups of MIB-II contain objects indicating faults.

[1] 1 MIB-II Second Version of the Management Information Base.

M.H. Shafazand and A M. Tjoa (Eds.): EurAsia-ICT 2002, LNCS 2510, pp. 110–116, 2002.

1.2 Uniqueness of Our Prototype

The traditional SNMP [3] based network management applications manage the network based on the MIB structure, which are segregated, based on the protocols like TCP, IP, and UDP etc. We differ from this methodology and have implemented the segregation based on the basic network management functionalities. Our prototype also differs in architecture by implementing the weakly distributed hierarchical network management framework where we have introduced the Inference Engine at the middle level and at the top-level managers so that our prototype is an evolving Intelligent NMS system. Provisions have been made to define user-defined traps in our system.

2 Architecture

Our Prototype is a three-layered weakly distributed hierarchical architecture [4] based on SNMP. The first level contains the Manager; the second level contains a number of middle level managers, which performs their tasks according to one or more functionalities assigned to them. The third level contains the nodes that are to be managed i.e., the Agents. We have distributed the five different functionalities of network management amongst various middle level managers decided by the user. This kind of user-defined configuration provides flexibility, and since this is multi-layered the architecture addresses the scalability issue also.

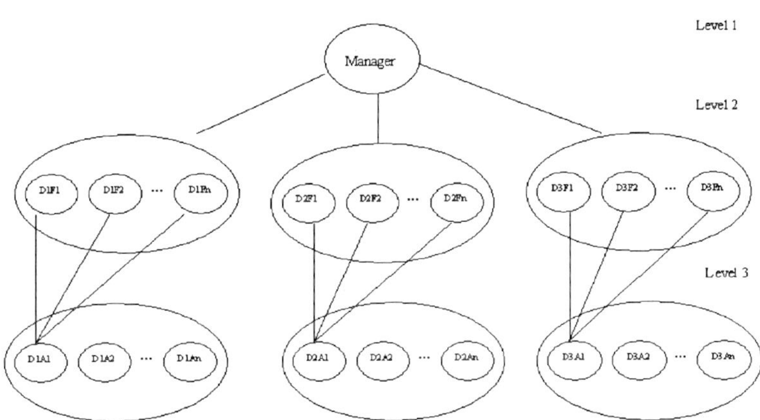

Fig. 1. Architecture of our prototype
(D1F1, D1F2 ...) – Functionality managers at Middle level
(D1A1,D1A2 ...) – Agents at the lower level

3 Design

3.1 Top Level Manager

This manager is the 'Manager of Managers' (MOM [2]) [5]. It sits at the top level and controls the functionality of each middle level manager and maintains the status of whole network. This manager initiates the polling of agents through the respective middle level managers based on functionality. The middle level manager reports to MOM about themselves and about the agents. In case of the failure of any middle level manager, this top-level manager takes the control of the respective agents. This provides the reliability for our prototype. The log information from the middle level managers and the agents are collected and maintained. An Inference Engine is present where by the MOM is made proactive and intelligent.

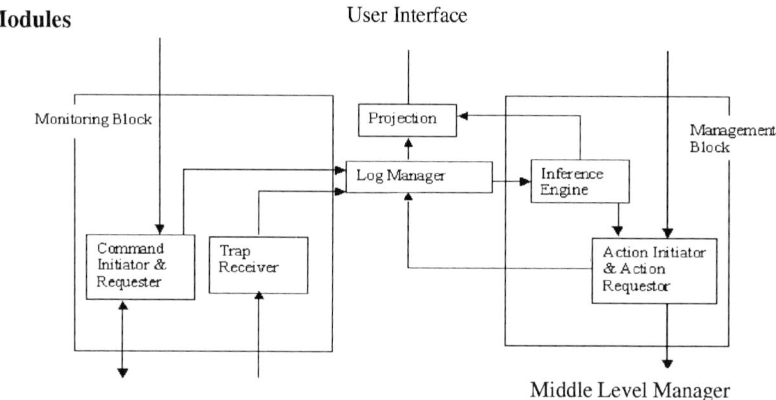

Fig. 2. Modules of the top level manager

Functionality

Monitoring Block: This has sub modules such as Command Initiator and Command Requester through which the polling is initiated periodically as configured by the user and also when ever there is a request from the user. Command Initiator is responsible to poll the respective middle level managers. Command Requestor is responsible to initiate the middle level managers by sending signals to start polling the respective agents. Trap Receiver is another sub module that receives the traps [3] sent from the middle level managers either about themselves or about the agents they are responsible for. All the information collected is sent to the Log Manager.

[2] MOM has the responsibility of initiating an action and retrieving the information through the middle level managers, which are functionality based.

Management Block: This block has sub modules, one Action Initiator that accepts the user command to handle the events reported by the monitoring block and initiate the action over the respective middle level managers, two Action Requestor which sends signals to the middle level managers regarding the action it has to take over its respective agents. Final sub module is the Inference Engine that learns the actions from the user and also analyzes the logs collected in the Log Manager to infer and suggest the user about the current state along with the proposed actions that can be taken.

3.2 Middle Level Manager

Primary task of this unit is to initiate the polling to get the values of respective variables of their agents and themselves, upon receiving a signal from top level manager and report back the information. They are equipped with the Log Manager and the Inference Engine to make them more intelligent and independent.

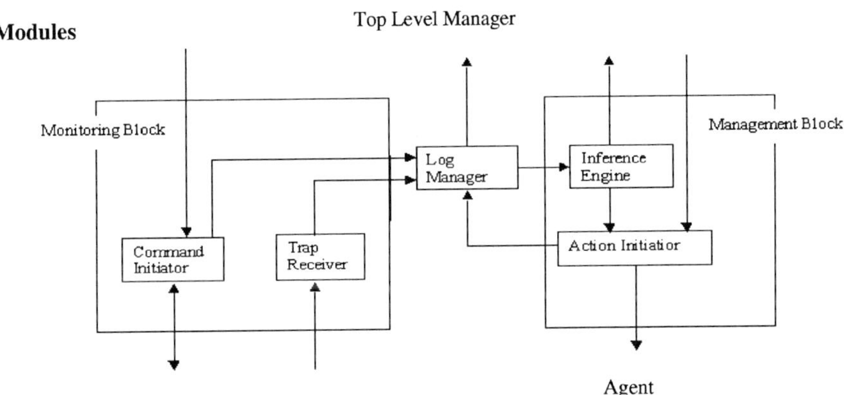

Fig. 3. Modules of the middle level manager

Functionality

Monitoring Block: This block has sub modules such as Command Initiator and Trap Receiver. Command Initiator initiates the polling amongst the agents and Trap Receiver receives any trap message that is being sent by the agents whenever they cross the threshold values. Activities are logged.
Management Block: Action Initiator will take the actions upon receiving the command from the top-level manager. Depending on the intelligence level the Inference Enginc suggests the user, through the top-level manager about possible actions to be taken.

3.3 Agents

Agents are placed at the lower most level and are enabled for user-defined trap[3]. The agents are intelligent enough to report to the respective middle level manager based on the functionality with regard to the trap.

3.4 Common Modules

Irrespective of the Monitoring block and the Management Block modules that are common to both are Projection and the Log Manager. The main job of the Projection module is to notify the event happenings about anything in the network to the user. The main job of the Log Manager is to collect various information and to maintain it as logs for analysis and learning purposes.
Projection: The Projection module provides the Graphical User Interface through which the user can configure and view the system. Users can perform the desired operations such as get, walk, poll, set related to Data Monitoring and Data Management blocks respectively. Statistical Analyzer module is being provided to bring out the analysis results based on user's customized requests.
Log Manager: Log Manager is present in both the top-level manager and in middle level managers. The purpose of this module is to collect information from polling, trap and Inference Engine. This also logs the user actions, which makes our prototype learn for future references.

4 Implementation

We have implemented the segregation of MIB variables based on functionality and hence our SNMP Walk operation will retrieve the values under a particular functionality and similarly the poll operation will poll for particular set of variables that come under the respective functionality periodically as per the user configuration done during the time of installation.
Walk: Implementation of walk is done in such a way, that it retrieves all the MIB values of the chosen functionality unlike the SNMP walk. Get and Set: The get module provides the current information about the polled variable under any functionality of the selected agent or the middle level manager. The set module enables the user respond to a particular event.
Inference Engine: We have implemented the Inference Engine part in the Top Level Manager and currently working to extend the same for the Middle Level Managers. We have an IE Process which gets instantiated based on the polling interval which checks for the values of MIBs and compare it with the Rule Base maintained in a database and acts as per the actions predefined. We have incorporated both the alert mechanisms and also issue of commands for settable[4] MIB variables. The command is issued to the respective Middle Level Manager

[3] User Defined Traps are extensions to basic SNMP traps, which allow the user to define thresholds on MIB variables.
[4] Settable MIB variables are MIB variables having write access

as per the actions defined with parameters such as the IP address of the target and value which gets implemented. The IE has the learning capability based on new user actions.
Statistical Analyzer: We provide interface that brings out the statistics of the happenings in graphical representation. Currently we are working to provide customizable projections.
Fail Safe Mechanisms: The top-level manager is equipped with fail-safe procedures to handle different failures. To avoid single point of failure, a back-up manager is suggested. This will be in synchronization with the top-level manager always in all aspects. Procedures for the back-up manager to keep track of the top-level manager and to take control during failures are mandatory. Similarly procedures are there for the top-level manager to take control of the agents when any middle level manager fails. This makes our prototype more reliable.

5 Deviations from SNMP

Trap: Our prototype implements the trap for user-defined thresholds. In addition our agent is intelligent enough to notify regarding trap to the respective middle level managers based on functionalities and values defined in the rule set. Thus we differ from SNMP trap implementation.
Walk: In SNMP, snmp-walk is implemented using GetBulkRequest [7] PDU, which follows the defined lexicographic order. But in our approach we use the GetRequest PDU and we don't follow the lexicographic order to retrieve variables based on functionalities. Inference Engine: We have incorporated this module in our prototype, which will infer from the log manager and project to the user the result of the analysis. We are currently working to enhance this module to be more intelligent whereby it shall even issue commands without human intervention.

6 Conclusions

Our Prototype has differed from the traditional network management system by addressing scalability through our architecture, better control through the functionality based management and Intelligence through the Inference Engine. We are currently concentrating in enhancing our Inference Engine to make our system more intelligent and to provide greater autonomy to the middle level managers.

References

1. SNMP Network Management – Paul Simoneau – Tata McGraw Hill, 1999 [Pages 266–398].
2. RFC 1213 – MIB II.
3. RFC 1157, RFC 1905 and RFC2571 – RFC 2575 SNMP.

4. A survey of distributed network and systems management paradigms – Jean-Philippe Martin-Flatin, Simon Znaty and Jean-Pierre Huvaux, INSM, Special Issue on Enterprise Network and Systems Management Dec. 1997, pp. 5.
5. Decentralized Approaches for Network Management – Mohsen Kahani, H.W. Peter Beadle, The Institute for Telecommunication Research (TITR) [Page 2], ACM-SIGCOMM Computer Communication Review, vol 27, Number 3, July 1997, pp.36–47.
6. SNMP, SNMPv2, SNMPv3 and RMON 1 and 2 – Third Edition William Stallings – Addision Wesley.
7. A Performance Evaluation of Distributed Paradigms for Network Management, Hyeok Chan Kwon O , Tae Gun Kang, Jae Hoon Nah, Sung Won Sohn, JCCI 2001, Jan 2001.
8. A Hybrid Centralised-Distributed Network Management Architecture, Damianos Gavalas, Dominic Greenwood, Mohammed Ghanbari, Mike O'Mahony, Proc. Fourth IEEE Symposium on Computers and Communication (ISCC'99), July 1999, pp. 434–441.
9. A Scaleable, Platform-Based Architecture for Multiple Domain Network Management, F.Stamatelopoulos, T.Chiotis, B.Maglaris, NETwork Management and Optimal DEsign (NETMODE) Group,
`http://smurfland.cit.buffalo.edu/NetMan/Papers.html/#WhitePapers`, 1995.

Coaching a Soccer Simulation Team in RoboCup Environment

J. Habibi, E. Chiniforooshan, A. HeydarNoori, M. Mirzazadeh, M.A. Safari, and H.R. Younesy

{habibi,chinif,noori,younesy}@ce.sharif.edu
{mirzazad, safari}@math.uwaterloo.ca

Abstract. Constructing soccer robots is an attempt in development of AI researches, done by defining a standard problem and solving it by many researchers all over the world. In this field, every year a formal federation holds international competitions, called RoboCup [1]. The Simulation League is one of the branches of the RoboCup.
We have designed and implemented an online coach for a soccer simulation team, which is able to analyze the simulated match similar to a coach in a real football match, and sends commands to the players to improve their behaviors and get a better result from the match process. This coach is able to exchange the roles between players during the match. Also, it has the capability of recognizing the opponent formation and improving the playing style of the team. This coach got the 1st place of Seattle'2001 RoboCup world championship in the field of online coaches.

1 Introduction

Simulation league is one of the branches of the RoboCup competitions. In this league, A match is carried out in a client/server manner. The central software called SoccerServer simulates the field and the match and controls the players' sensors. Each client takes the role of a player and communicates with the SoccerServer by some specific commands (such as *move*, *turn* or *dash*) in a standard protocol. The server manipulates the status of the field or the world's model, executes the commands of the players and sends information about the place and the direction of visible objects in their view ranges with some purposed noises. So, players have to deal with the inconsistent information about objects (ball, goals and players) to make a model of the world [2].

Although SoccerServer provides a simple environment for soccer simulation, it has many complexities of real soccer. In a real soccer match, there is a coach who observes and analyzes the match and regarding to the states of players, result of the match, opponent team strategy and etc, gives suitable commands to its players to improve their behaviors. Similar to a coach in a real soccer match, each team in the RoboCup simulation league, can have a special client which is called online coach. This client has a global view of the match and receives noise-free information about the positions, directions and velocities of

M.H. Shafazand and A M. Tjoa (Eds.): EurAsia-ICT 2002, LNCS 2510, pp. 117–126, 2002.

ball and all the players from the server [3]. So, online coach can analyze the match with more accuracy and make long-term tactical decisions. Online coach is able to communicate with its players during the match and supply them with the decisions.

We have designed and implemented an online coach for *Sharif-Arvand* soccer simulation team [4], which is able to analyze the game during the simulation matches and to give tactical commands to its players to get better result from the match process. We called this online coach, *"Arvand-Coach"*.

This article is organized as follows. In the second section, we will introduce the coach client and its abilities. In the third section, we will describe the algorithms used in developing Arvand-Coach. Finally, in the fourth section, we evaluate the implemented coach and section five concludes.

2 The Coach Client

Coach is a special client with some abilities used for modeling and analyzing the simulation matches. The coach client has the following abilities:

1. Receives the exact positions, directions and velocities of ball and all the players from the server
2. Communicates with its players

To prevent coaches from micro-controlling every single action of the players, the communication between coach and its players is somewhat limited. In a game with two 3000 simulated cycle halves, online coach can send messages to its players in each 300 simulation cycles using a standard coach language provided by RoboCup federation [5]. Also, when the play-mode is not play-on (such as *offside*, *corner-kick*, *goal*, *etc*), the coach is able to send messages with any desired language. There is no restriction on the format of these messages, but the total message size cannot exceed the predefined character limitation (currently it is 128 characters).

One of the other capabilities of the coach is the ability to substitute 3 players during the match. Also, the coach is able to decide the type of the players. At the beginning of the match, server generates six different types of players with different capabilities and sends their specifications to the coach. Coach is able to choose different player types for different roles.

Because of these abilities of the coach client, coach can be a suitable tool for modeling and analyzing the behavior of the opponent team.

3 Algorithms Used in Arvand-Coach

In this section, we will describe the algorithms used in developing Arvand-Coach.

3.1 Using Heterogeneous Players

As mentioned in section 2, SoccerServer generates six different types of players at the beginning of the match and sends their specifications to the coach. Each player type is identified with 11 different parameters in such a way that it is difficult for the coach to infer directly which player type has the highest velocity or which player type has the highest stamina. So, we used simulation approaches to sort player types according their velocity, stamina and etc. For this purpose, we assumed that each player type runs for some simulation cycles. Then we used the formulas used in the SoccerServer to simulate the behavior of each player type. After this epoch, we sort player types according their stamina, velocity and etc. Now, it is possible for us to find the fastest player type or the most vigorous player type and etc and then we can use the best player type for each role in the team. One of the heuristics which works well, is using faster player types as offensive players and more vigorous player types as midfielders, the same as a real soccer match.

3.2 Recognizing the Skills of Players

As mentioned in section 2, the SoccerServer only sends the exact positions and directions of the players and the ball to the coach. The coach will use this information to recognize the skills of players (such as *"Kicking the ball"* and *"Passing the ball"*). Our coach recognizes the skills of players in different levels. In the low level, it recognizes *"Owning the ball"* and *"Kicking the ball"*. In the high level, it uses that low-level information to recognize other skills (such as *"Passing the ball"* and *"Intercepting the ball"*). For this purpose, we defined some rules in the form of (Condition, Action). If the condition has been satisfied, then that action has occurred.

By recognizing these skills, the coach would be able to analyze the behaviors of players. In the following, we describe how we recognize different skills.

- **Owning the Ball**
  The first thing that the coach should recognize is that which player owns the ball at each simulation cycle. We say player P owns the ball, if the following conditions are satisfied:
  1. It is the nearest player to the ball, and
  2. The distance between that player and ball is less than a defined threshold.
- **Kicking the Ball**
  After recognizing that which player owns the ball, it is possible to recognize that which player has kicked the ball [6]. We say that player P has kicked the ball at time $t-1$, if:
  1. Player P owns the ball at time $t-1$
  2. In consecutive cycles $(t-2, t-1)$ and $(t-1, t)$, the direction or the velocity of ball has been changed.
- **Passing and Intercepting the Ball**
  With the definition of consecutive kicks, it is possible to recognize high-level skills such as passing and intercepting the ball. Assume that player P_1 kicks the ball at time T_1 and player P_2 kicks the ball at time T_2 and there are no other kicks in the interval (T_1, T_2). Now, we say that ball is passed from player P_1 to player P_2, if both P_1 and P_2 belong to the same team, and we say that ball is intercepted by player P_2, if P_1 and P_2 belong to different teams.

3.3 Processing Referee Messages

By default, there is an internal referee module within the SoccerServer that controls the match. This referee is able to recognize different situations of the match (such as *goal*, *offside* and *etc*). We know these situations as play-modes. After these situations, the referee sends a message to the coach and informs it about the situation. The coach should have proper capabilities to reaction against these situations.

After the goal situations, it is possible to change the players' positions without wasting stamina. So, it is one of the best times for the coach to change the formation of players. At these situations, first, our coach considers whether changing the formation would be beneficial for the team or not?. If yes, it chooses a new formation for the team and informs the players. Otherwise, it exchanges the roles between players in the current formation. For exchanging the roles between players, we sort players by their remaining stamina. Then, we exchange the role of the most tired player with the role of the most vigorous player.

When the play-mode is not play-on (such as *offside*, *corner*, *ball-out* and *etc*), we exchange the roles between players again. But, in these situations, it is not possible to change the players' positions directly and if a player wants to change its position to a new location, it should consume some of its stamina. Because of the great number of non-play-on situations, and because of the valuable stamina, we use another approach to exchange roles between players. First, we sort players by their remaining stamina. Then with the probability of %50, we exchange roles between players.

There are some advantages for exchanging roles between players. For example, Some teams use learning methods to learn the behavior of the opponent team and select a suitable behavior against it. Exchanging roles between players, causes a player with a uniform number (e.g. *2*) behaves differently in the field. This causes some problems in the learning algorithms of the opponent team. This results in that the opponent teams need to use more complex algorithms against us.

3.4 Gathering Game Statistics

It is possible for a coach to gather statistical information from the game and use them. Our coach gathers the following statistics during the match:

- Ball's average position and its variance
- Each player's average position and its variance
- Average position of all of team's players taken as a whole
- Number of goal, corner-kick, offside and ball-out situations
- Number of shoots towards the opponent goal
- Number of successful ball passes and ball interceptions
- Number of goals which each player has scored

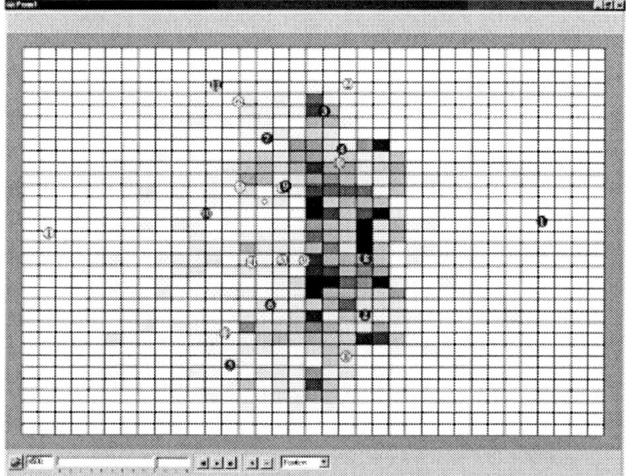

Fig. 1. Presence rate of opponent players in different regions

Recognizing the Opponent Formation by Statistical Methods. There are different methods for recognizing the opponent formation [7, 8, 9, 10]. But, most of these methods use the exact positions of players to construct a predefined model of players. But, as mentioned in section 3.3, our coach exchanges roles between players to complicate the learning algorithm of the opponent team. It is possible that some other teams use this method, too. To solve this problem, we used statistical approaches.

We divide field into rectangular areas. After that, at each simulation cycle, we check that each opponent player is within which area and then, our coach increases the counter of that area by one. By this method, we can determine the aggregation position of opponent players. Fig. 1 shows the result. In this figure, the more dark in color, the more aggregation is in that region. The problem in this method is that it assumes the rectangular areas separate. Though it is not so in real and there is some kind of relationship between close areas. Here is an example. Player P_1 is in region R_1. If it changes its position on the X-axis a little (for example *0.5* meters), then it would be in region R_2. To solve this problem, we used *windowing* method. In this method, the value of each region is the sum of the values of that region and its neighbor regions in a surrounding window. So, we assume some kind of relationship between neighbor regions. The result of this method is shown in Fig. 2.

As shown in Fig. 2, we can't determine the formation of players yet. The reason of this problem is that we used the exact positions of players instead of relative positions of them. Here is an example: Player P_i is a defensive player and its default position is *(30, 10)*. But, now it is in the position *(20, 20)* on the field. The important note we should notice is that, though the players change their positions on the field. But, they retain their relative distances to each other. For example, offensive players usually ambulate in front of defensive players. To

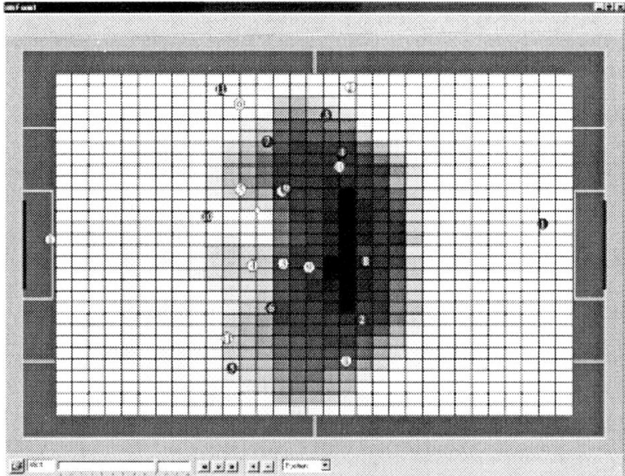

Fig. 2. Presence rate of opponent players in different regions using windowing method

solve the problem, we draw the rectangle around the players in each simulation cycle (Fig. 3) and then, we map the relative positions of players in this rectangle to a point on the field. In the other words, we stretch the rectangle around the players on the field. The result is shown in Fig. 4.

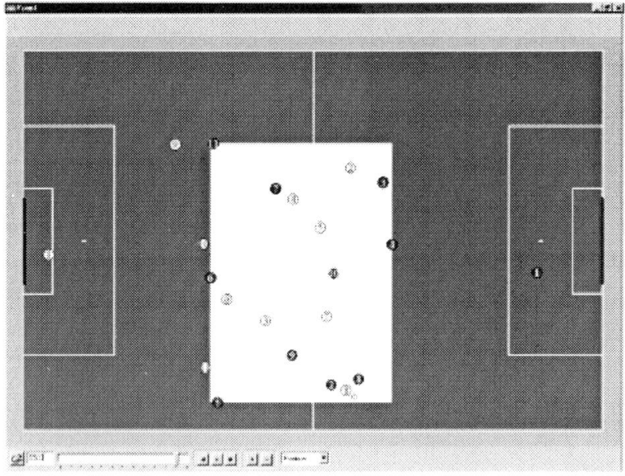

Fig. 3. Rectangle around the opponent players

As shown in Fig. 4, now we have contradistinctive areas and we can detect the formation of players. There are different clustering algorithms to extract the formation of players from this structure. Some of these algorithms are *Self*

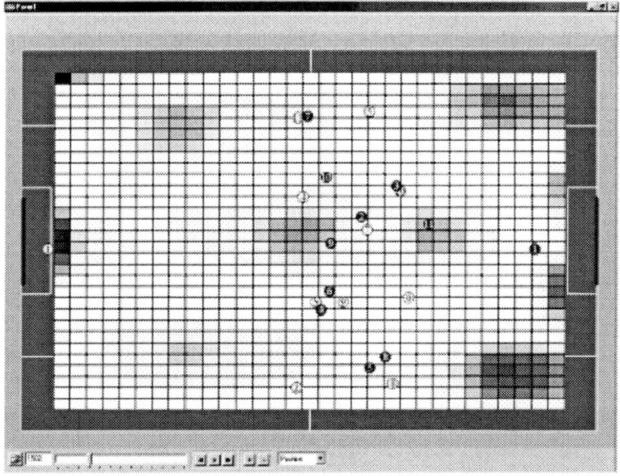

Fig. 4. Aggregation of opponent players in the field

Organizing Map (SOM) neural networks [11] and *labeling* method which is used in Image Processing to detect connected components of a binary image [12]. Because of the simplicity and similarity of labeling method with our purpose, we used this method to determine the formation of players.

3.5 Modifying the Formation of Players

One of the abilities of our coach is modifying the formation of its players regarding to the behavior and formation of opponent team. There are different methods for this purpose. For example, by detecting the opponent formation (section 3.4.1), or designing a number of formations before the match and using them during the match depended on the result of the match [13].

In our coach, we have used a completely new approach to change the formation of our players. For this purpose, our coach saves the successful trajectories of opponent players with ball. We call a trajectory, successful, if it resulted in a goal or ball was moved to the penalty area of opponent team. The algorithm has the following steps:

1. *Determining The Trajectories*
   We save the trajectories of opponent players with ball since they took the ball until they lose the ball.
2. *Optimizing The Trajectories*
   The number of lines in each trajectory is usually great and some of the lines are very short and are not useful. So, we need some algorithms for removing these unimportant lines from trajectories. We used the following two algorithms for this purpose:
   a) For each two consecutive lines AB and AC, if length AB is smaller than a threshold, then we merge these two lines:

if *length(AB)* $<$ *DistThr* **then**
merge(AB, BC)

Example:

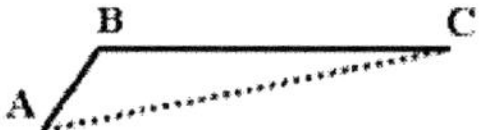

Fig. 5. Merging two consecutive lines regarding their lengths

b) For each two consecutive lines AB and AC, if the angle between AB and AC is smaller than a threshold, then we merge these two lines:

if *Angle(AB, BC)* $<$ *AngleThr* **then**
merge(AB, BC)

Example:

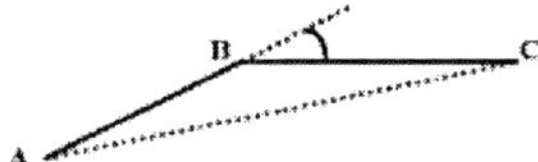

Fig. 6. Merging two consecutive lines regarding their angle

3. *Valuating The Trajectories*
Now, we have trajectories with smaller number of lines and we can valuate them. We define the value of trajectories by the values of its vertices. There are different parameters for valuating the vertices. For example, the closer points to the opponent's goal, will have higher values.
4. *Using the Trajectories*
In this step of the method, for each line, AB, we compute a midpoint C:

$$C = \frac{A \times Value(A) + B \times Value(B)}{Value(A) + Value(B)} \tag{1}$$

The point C is biased to the vertex with higher value. In the next step, for each midpoint C, we find a player with nearest default position to it. Then we add C to the corresponding set of that player. At the end, for each player ($Player_i$) we have a list of points, P_js. The new default position of $Player_i$ is computed by the following formula:

$$C_i = \frac{Player_i \times Value(Player_i) + \sum(P_j \times Value(P_j))}{Value(Player_i) + \sum Value(P_j)} \tag{2}$$

In the above formula, $Value(Player_i)$ is the value of the current default Position of $Player_i$. By computing the default position of players with this formula, their default positions are biased to the successful opponent trajectories with ball. In the other words, by this method we refine the formation of our team to improve the performance of players in the strategic regions.

4 Evaluating the Coach

In this section, we present the results of the evaluation of our coach. We have used *FCPortugal2000* team for this purpose. FCPortugal2000 was the champion of RoboCup'2000 world championship held in Melbourne, Australia [14]. Without a coach, Sharif-Arvand team loses the match with the result of 0.75 against 2.25 on the average. But, with using coach, Sharif-Arvand team wins the match with the result of 2.5 against 1 on the average.

5 Conclusion and Future Work

In this paper, we described the algorithms used in the online coach of *Sharif-Arvand* RoboCup simulation team. Our coach is able to observe and analyze a soccer simulation match in the RoboCup environment and based on this analysis, give tactical commands to its players. Our coach is able to use heterogeneous players for different roles in the team. Other capabilities of our coach are exchanging roles between players, detecting the formation of opponent's players and improving our players' formation, based on the successful trajectories of opponent players. This coach got the 1st place of Seattle'2001 RoboCup world championship in the field of online coaches.

The coach we have developed is a good basis for further work in this field. In the future, we want to use AI and learning algorithms to learn the behavior of the opponent team. Also, we want to use planning algorithms to generate plans for our players during the simulation match.

References

1. "RoboCup official site", *http://www.robocup.org*, visited: May 10, 2002
2. H. Matsubara, I. Noda and K. Hiraki, "Learning of Cooperative Actions in Multi-Agent Systems: A Case Study of Pass Play in Soccer", *papers from the 1996 AAAI spring symposium, Menlo Park, CA*, 1996
3. "Soccer Server Manual", *http://sourceforge.net/projects/sserver/*, visited: May 10, 2002
4. J. Habibi et al, "Sharif-Arvand Simulation Team", *In P. Stone, T. Balch and G. Kraetzschmar, editors, RoboCup 2000: Robot Soccer WorldCup IV, pp. 433–436, Springer*, 2001
5. L. Paulo and N. Lau, "COACH UNILANG - A Standard Language for Coaching a (Robo)Soccer Team", *In Proceedings of the Fifth International Workshop on RoboCup, Seattle, USA*, 2001

6. T. Takahashi, "LogMonitor: From Player's Action Analysis to Collaboration Analysis and Advice on Formation", *In M. Veloso, E. Pagello and H. Kitano, editors, RoboCup-99: Robot Soccer WorldCup III, pp. 103–113, Springer-Verlag, Berlin*, 2000
7. C. Ducker, S. Hubner, E. Schmidt, "Virtual Werder: Using the Online-Coach to Change Team Formation", *In Proceedings of the Fourth International Workshop on RoboCup, pp. 217–223, Melbourne, Australia*, 2000
8. U. Visser et al, "Recognizing Formations in Opponent Teams", *In P. Stone, T. Balch and G. Kraetzschmar, editors, RoboCup 2000: Robot Soccer WorldCup IV, pp. 433–436, Springer*, 2001
9. C. Drucker et al, "As time goes by: using time series based decision tree induction to analyze the behavior of opponent players", *http://www.virtualwerder.de/*, visited: May 10, 2002
10. P. Riley and M. Veloso, "Adaptive Team Coaching Using Opponent Model Selection", *Submitted to the Fifth International Conference on Autonomous Agents*, 2001
11. L. Fausett, "Fundamentals of Neural Networks", *Prentice Hall*, 1994
12. R. Jain, R. Kasturi and B.G. Schunck, "Machine Vision", *McGraw-Hill*, 1995
13. T. Takahashi, "Kasugabito III", *In M. Veloso, E. Pagello and H. Kitano, editors, RoboCup-99: Robot Soccer WorldCup III, pp. 593–595, Springer*, 2000
14. L. Paulo and N. Lau, "FC Portugal Team Description: RoboCup 2000 Simulation League Champion", *In P. Stone, T. Balch and G. Kraetzschmar, editors, RoboCup-2000: Robot Soccer World Cup IV, Springer, Berlin*, 2001

Improving Information Retrieval System Security via an Optimal Maximal Coding Scheme

Dongyang Long[1]

Department of Computer Science, City University of Hong Kong, 83 Tat Chee Avenue
Kowloon, Hong Kong SAR, PRC
dylong@cs.cityu.edu.hk
Department of Computer Science, Zhongshan University, Guangzhou 510275, PRC
dylong25112002@yahoo.com

Abstract. Novel maximal coding compression techniques for the most important file-the text file of any full-text retrieval system are discussed in this paper. As a continuation of our previous work, we show that the optimal maximal coding schemes coincide with the optimal uniquely decodable coding schemes. An efficient algorithm generating an optimal maximal code (or an optimal uniquely decodable code) is also given. Similar to the Huffman codes, from the computational difficulty and the information-theoretic impossibility point of view, the problem of breaking an optimal maximal code is further investigated. Due to the Huffman code being a proper subclass of the optimal maximal code, which is good at applying to a large information retrieval system and consequently improving the system security.

1 Introduction

The Huffman coding [9] has been widely used in data, image, and video compression [2-6,11-15]. The ideal of using data compression schemes for encryption is very old, dating back at least to Roger Bacon in the 13th century [5]. The field of data compression has grown vigorously since Huffman's algorithm that is published in 1952. Rubin [13] and Jones [11] present the ways in which data compression algorithms may be used as encryption techniques. Klein *et al.* [6] have discussed the cryptographic properties of Huffman codes in the context of a large, compressed natural language database on CD-ROM. Based on the same problem, Fraenkel and Klein [4] have proven that, given a natural language cleartext and a ciphertext obtained by Huffman coding, the complexity of guessing the Huffman code is NP-complete. Gillman *et al.* [5] have also considered the problem of deciphering a file that has been Huffman coded but not otherwise encrypted, from the information-theoretic impossibility but not the computational difficulty point of view. They find

[1] This work was partially sponsored by the 2002 Open Project of the State Key Laboratory of Information Security (SKLOIS) (project No. 01-02), the National Natural Science Foundation of China (project No. 60073056) and the Guangdong Provincial Natural Science Foundation (project No. 001174).

M.H. Shafazand and A M. Tjoa (Eds.): EurAsia-ICT 2002, LNCS 2510, pp. 127–134, 2002.

that a Huffman code can be surprisingly difficult to cryptanalyze. The authors [7-8] have introduced novel optimal uniquely decodable, prefix, maximal prefix, and maximal coding schemes. We have shown that all Huffman codes have to be optimal uniquely decodable, prefix, maximal prefix, and maximal codes. Conversely, none of the optimal uniquely decodable, prefix, maximal prefix, and maximal codes is necessarily the Huffman code. To see difference between four types of the optimal codes above and Huffman codes, we first consider the following example.

Example 1.1 Let an information source $I = (S = \{s_1, s_2, s_3, s_4, s_5, s_6\}$, $P = \{0.26, 0.24, 0.14, 0.13, 0.12, 0.11\})$ and input alphabet $\Sigma = \{0,1\}$. The following Table 1.1 shows two Huffman codes and a non-Huffman code. According to the Huffman's algorithm, we know that the codes of source alphabets s_1 and s_2 must start with different bits, but in C_3 they both start with 0. This code C_3 is therefore impossible to generate by any re-labeling of the nodes of the Huffman trees. That is, C_3 cannot be generated by the Huffman method! We easily verify that C_3 is an optimal uniquely decodable, prefix, maximal prefix, and maximal code. And the code C_4 is clearly not a prefix code and consequently it cannot be the optimal prefix or maximal prefix code. But, We easily calculate

$$2^{-l(00)} + 2^{-l(01)} + 2^{-l(100)} + 2^{-l(101)} + 2^{-l(110)} + 2^{-l(111)}$$
$$= 2^{-2} + 2^{-2} + 2^{-3} + 2^{-3} + 2^{-3} + 2^{-3} = 1$$

Table 1.1 Two Huffman and a non-Huffman codes

Source letter	Probability	Huffman Code C_1	Huffman Code C_2	Code C_3	Code C_4
s_1	0.26	01	10	00	00
s_2	0.24	10	01	01	10
s_3	0.14	000	111	100	001
s_4	0.13	001	110	101	101
s_5	0.12	110	001	110	011
s_6	0.11	111	000	111	111

Thus, by Theorem 1.4.2 in [14], C_4 is a maximal code. Since C_4 has the same average code word length as the Huffman code C_1, C_4 is not only an optimal maximal code but also an optimal uniquely decodable code (by Theorem in [7] and Theorem 1 in [8]).

Example 1.1 shows that the class of Huffman codes is a proper subclass of the above four types of optimal codes and that the optimal uniquely decodable code (maximal code) is different from the optimal prefix code (maximal prefix code). Motivated by the same problem as breaking a Huffman code [4-5], the problem of breaking an optimal uniquely decodable code (maximal code) will be presented. Because there is a quite difference between the uniquely decodable code and the prefix code, breaking an optimal prefix code (maximal prefix code) will be investigated in a separate paper. Additionally, although the terms and notions such as *Huffman coding (encoding)*, *Huffman code*, *optimal code*, and *optimal prefix code* are easily found in many literatures [3,12,14], relationships of these concepts has been rather vague and have not detailed yet.

2 Optimal Uniquely Decodable, Optimal Maximal, and Huffman Codes

In general, the class of maximal codes is much less than the class of uniquely decodable codes [1,10]. For optimal uniquely decodable and optimal maximal codes, however, they are strong connected. Further relation between them is given below.

First, we have the following Theorem 2.1.

Theorem 2.1 *Every optimal maximal code has to be an optimal uniquely decodable code.*

Proof: The details of proof of the theorem are omitted here. #

Conversely, the following result is given.

Theorem 2.2 *Every optimal uniquely decodable code has to be maximal.*

Proof: We first show that Theorem 2.2 is true for the alphabet $\{0.1\}$. Suppose that (S,P) is a finite information source. Let $C=\{c_1,c_2,...,c_n\}$ be an optimal uniquely decodable code. And $l(c_1)=l_1, l(c_2)=l_2,...,l(c_n)=l_n$, and $l_1 \le l_2 \le ... \le l_n$. Without loss of generality, suppose that C is an optimal prefix code. In fact, for the uniquely decodable code C there exists a prefix code D such that D has the same sequence of code word lengths as C. By definitions, it is easy to verify that D is optimal. Assume that $P=\{p_1,p_2,...,p_n\}$ with $p_1 \ge p_2 \ge ... \ge p_n$. We will show that C is a maximal prefix code by reduction to absurdity. Suppose that C is not a maximal prefix code. By definitions, there exists at least a code word $c \in \{0,1\}^+ - C$ such that $C \cup \{c\}$ is still a prefix code. When $l(c) < l(c_i) = l_i$ for some i, we easily construct a prefix code $C_1 = (C - \{c_i\}) \cup \{c\} = \{c_1, c_2, ..., c, ..., c_n\}$ such that the average code word length of the prefix code C_1 is less than the one of the optimal prefix code C. Therefore, this is impossible. Thus we have that $l_1 \le l_2 \le ... \le l_n \le l(c)$. According to the choice of the code word *c*, we can take the code word *c* satisfying $l_n = l(c)$. Otherwise, we easily get the code word *c* by replacing *c* with any prefix c', which is of the length l_n, of the word c. Now, let $c = d0$ or $c = d1$, where d is the proper prefix of c and of the length $l_n - 1$. If $c = d0$ and the word $d1$ is not in C, by $C \cup \{c\}$ being a prefix code, then $C \cup \{d\}$ is a prefix code and consequently $C_2 = (c_1, c_2, ..., c_{n-1}, d\}$ is also a prefix code. Clearly, the average code word length of the prefix code C_2 is less than the one of the optimal prefix code C. This is impossible too. Similarly, when $c = d1$ and the word $d0$ is not in C, we will get a contradiction too. Therefore, without loss of generality, assume that $c = d0 \notin C$ and the word $d1 \in C$. Since C is a prefix code, the set $C_3 = (C - \{d1\}) \cup \{d\}$ is also a prefix code. By $l_n = l(c) = l(d0) = l(d1)$, the average code word length of the prefix code $C_3 = (C - \{d1\}) \cup \{d\}$ is $p_1 l_1 + p_2 l_2 + ... + p_{n-1} l_{n-1} + p_n (l_n - 1)$. Clearly,

$$p_1 l_1 + p_2 l_2 + ... + p_{n-1} l_{n-1} + p_n (l_n - 1) < p_1 l_1 + p_2 l_2 + ... + p_{n-1} l_{n-1} + p_n l_n$$

which is the average code word length of the optimal prefix code C. This contradicts with C being an optimal prefix code. Combining the above discussion, we have that C is a maximal prefix code. That is, an optimal prefix code has to be maximal.

Next, consider the number of the alphabet being greater than 2. The details of proof are omitted here.

Combining the above discussion, we have that C is a maximal code. #

Therefore, by Theorems 2.1 and 2.2 we immediately get Corollary 2.1 below.

Corollary 2.1 *Optimal uniquely decodable codes coincide with optimal maximal codes.*

Remark 2.1 It is very interesting that the word 'optimal' concerns the economy of a code. As seen in [10], if *C* is a maximal code then every code word occurs as part of a message, hence no part of all words over the alphabet is 'wasted'. Every optimal uniquely decodable code has to be a maximal code. However, this particular property does not belong to general coding schemes. Note that in all the following sections, optimal code, optimal uniquely decodable code, and optimal maximal code are only different names for the same thing.

Although Huffman codes are a proper subclass of maximal codes, Theorem 2.3, which follows, shows nearly relation between Huffman codes and maximal codes. We will omit the details of proof of Theorem 2.3.

Theorem 2.3 *If* $C \subseteq \Sigma^+$ *is any maximal code, then there exist some suitable information source* $I = (\Sigma_1, P)$ *and a Huffman code* H *for* I *such that* C *has the* *same average code word length as* H . *And consequently* C *is an optimal code for* $I = (\Sigma_1, P)$.

Remark 2.2 By Theorem 1 in [7], we have that a Huffman code has to be a maximal code. Conversely, making use of Theorem 2.3, for any maximal code H , we are able to construct a suitable probability distribution H (i.e., a suitable information source $I = (\Sigma, P)$, because of the alphabet Σ determined by P) such that H is exactly a Huffman code for $I = (\Sigma, P)$. Therefore, when taking out all probability distributions *P*, a maximal code can be considered as a Huffman code. In other words, the maximal coding schemes are very near to the Huffman coding schemes.

In additional, for a special information source with a *dyadic* [6] probability distribution, we easily construct an optimal maximal code, i.e., we have Theorem 2.4 below. Proof of the theorem is also omitted.

Theorem 2.4 *Let* $I = (\Sigma, P)$ *be a finite information source with a dyadic probability distribution* $P = \{2^{-l_1}, 2^{-l_2}, ..., 2^{-l_n}\}$ *with* $l_1 \le l_2 \le ... \le l_n$. *Then any maximal code* $C = \{c_1, c_2, ..., c_n\}$ *satisfying the conditions* $l(c_1) = l_1, l(c_2) = l_2, ...,$ *and* $l(c_n) = l_n$ *is the optimal maximal code for* $I = (\Sigma, P)$.

3 Application to Data Compression

As the simplest example, consider a special file $\mathbf{A^3B^4A^{90}B^3}$ over the alphabet {**A**, **B**}. Regardless of the probabilities, Huffman coding will assign a single bit to each of the letters **A** and **B**, giving no compression, thus the file $\mathbf{0^3 1^4 0^{90} 1^3}$ is 100 bits. But in dictionary methods but not traditional statistical modeling, we take a maximal coding such that $\mathbf{A^3B^4A^{89} \rightarrow 1}$, $\mathbf{AB \rightarrow 01}$, and $\mathbf{BB \rightarrow 00}$, where {1, 01, 00} is clearly a maximal code. And the file **01011** is 5 bits. Therefore, we get a compression ratio of **100/3** ≈ **33.3**.

For example, we will encode the file M: STATUS REPORT ON THE FIRST ROUND OF THE DEVELOPMENT OF THE ADVANCED ENCRYPTION STANDARD. By making use of Table 3.1, in traditional statistical modeling, we easily calculate that the average code word length of the block code C_1 is **5** bits/symbol, and that the average code word length of a Huffman code C_2 is **342/87** bits/symbol. Furthermore, the encoded file by the block code C_1 and the Huffman code C_2 will take up **87×5 = 435** bits and **87×342/87 = 342** bits, respectively. Thus the compression ratio is **435/342 = 1.27:1**.

We consider the code D_1 in Table 3.1. It is easy to verify that the code D_1 is an optimal code without the Huffman code. In fact, since the word 011 is a proper prefix of the word 0111, D_1 is not a Huffman code (by Theorem 5.2.1). Clearly, the code D_1 has the same average code word length as the Huffman code C_2, thus D_1 is an optimal code. By the results of Section 5.5, D_1 has the same compression ratio **435/342 = 1.27:1** as the Huffman code C_2. By Table 3.2, we directly follow that the two codes D_3 and D_4 are not prefix codes. And D_4 is an optimal code without the Huffman code. By directly calculating, we know that D_3 and D_4 have the compression ratio **435/90** ≈ **4.83:1** and **435/74** ≈ **5.87:1**, respectively (comparing with the block code C_1 in Table 3.1). And they have the compression ratio **27×4/90** ≈ **1.2:1** and **27×4/74** ≈ **1.46:1**, respectively (comparing with the block code C_5 in Table 3.2).

Table 3.1 An optimal coding

Source Letter	Probability	Block Code C_1	Optimal Code D_1	Huffman code C_2
(Space)	13/87	00000	110	011
T	10/87	00001	011	110
E	9/87	00010	0000	0000
N	7/87	00011	1000	0001
O	7/87	00100	1100	0011
D	6/87	00101	1010	0101
R	6/87	00110	0001	1000
A	4/87	00100	1001	1001
S	4/87	01000	0111	1110
C	3/87	01001	1111	1111
F	3/87	01010	00100	00100
H	3/87	01011	10100	00101
P	3/87	01100	00010	01000
I	2/87	01101	10010	01001
U	2/87	01110	00101	10100
V	2/87	01100	10101	10101
L	1/87	10001	01101	10110
M	1/87	10011	011101	101110
Y	1/87	10101	111101	101111

Table 3.2 Coding schemes based on source words

Source Word	Probability	Code D_3	Optimal Code D_4	Huffman Code C_4	Block Code C_5
(Space)	13/27	000	0	0	0000
THE	3/27	100	001	100	0100
ON	2/27	001	0101	1010	0010
ENCRYPTION	1/27	011	1011	1101	0110
STANDARD	1/27	110	11111	11111	1111
ADVANCED	1/27	0010	01111	11110	1001
STATUS	1/27	1010	00111	11100	1101
REPORT	1/27	1101	10111	11101	1011
FIRST	1/27	0111	00011	11000	1000
ROUND	1/27	1111	10111	11001	1100
DEVELOPMENT	1/27	00101	11101	10111	1010
ON	1/27	10101	01101	10110	1110

4 Breaking an Optimal Maximal Code

As seen from [4], from the issue of computational difficulty, it easily follows the following theorem 4.1.

Theorem 4.1 *Given an original file and a corresponding encoded file by the optimal coding, the complexity of guessing the optimal code is NP-complete.*

We further have the following Theorem 4.2 ([8], Theorem 4.2).

Theorem 4.2 *Let M' be the encoded file by the uniquely decodable coding (the maximal coding) and the length of M' be m. Then M' is encoded by at most 2^{m-1} uniquely decodable codes (maximal codes).*

Note that Theorem 4.1 is an immediately corollary of Theorem 4.2. Moreover, in the proof of Theorem 4.2 we assume that it is simple to verify that a given set of words is a uniquely decodable code. But, in fact, it is very difficult to decide whether a finite set of words is a uniquely decodable code [1], even if there has been the Sardinas and Patterson algorithm ([1], p.82). Therefore, from the computational complexity point of view, breaking an optimal maximal code is much more difficult than the complexity provided by Theorem 4.2.

Next, from the information-theoretic impossibility point of view [5], we will discuss the problem of breaking an optimal uniquely decodable code.

First, an efficient algorithm generating an optimal maximal code is given below.

Theorem 4.3 *For a given finite information source there exists an efficient algorithm constructing an optimal uniquely decodable code.*

Proof: The details of proof of the theorem are omitted here. #

Note that the optimal codes constructed by the way in the above theorem 4.3 are not the Huffman codes in general. Additionally, as seen [5], we easily verify that: *Breaking the encoded file by an optimal uniquely decodable code can be surprisingly difficult.*

5 Conclusion

As we have seen from [6], the Huffman codes are good at using in a large information retrieval system. Important for a large information retrieval system is the issue of the cryptographic security of storing the text in compressed form, as might be required for copyrighted material. And in the usual approach to full-text retrieval, the processing of queries does not directly involve the original text files (in which key words may be located using some pattern matching technique), but rather the auxiliary dictionary and concordance files. An optimal maximal coding scheme based on the words of the original file is suitable for storing these auxiliary dictionary and concordance files.

On the other hand, although the adaptive Huffman coding [2,15] and the Lempel-Ziv coding [2] are preferred in some real-time applications and for communication, they are not suitable for storing a large body of static text.

Finally, we have known that the Huffman code is a proper subclass of the uniquely decodable code. From Theorem 4.2 and 4.3, it easily follows that breaking an optimal uniquely decodable code is much more difficult than breaking a Huffman code. Therefore, the issue of the cryptographic security of a large information retrieval system will be further improved by an optimal uniquely decodable code compressed.

References

1. Berstel, J., Perrin, D.: Theory of Codes. Academic Press, Orlando (1985)
2. Bell, T.C., Cleary, J.G., Witten, I.H.: Text Compression. Prentice Hall. Englewood Cliffs, NJ (1990)
3. Cover, T, Thomas, J.: Elements of Information Theory. New York, Wiley (1991)
4. Fraenkel, A.S., Klein, S.T.: Complexity Aspects of Guessing Prefix Codes. Algorithmica, Vol.12(1994), 409-419
5. Gillman, David, W., Mohtashemi, M., Rivest, R.L.: On Breaking a Huffman Code. IEEE Trans. Inform. Theory, IT- 42(1996)3, 972-976
6. Klein, S.T., Bookstein, A., Deerwester, S.: Storing Text-Retrieval Systems on CD-ROM: Compression and Encryption Considerations. ACM Trans. Inform. Syst., Vol.7(1989), 230-245
7. Long, D., Jia, W.: Optimal Maximal Encoding Different From Huffman Encoding. Proc. of International Conference on Information Technology: Coding and Computing (ITCC 2001), Las Vegas, IEEE Computer Society (2001) 493-497
8. Long, D., Jia, W.: On the Optimal Coding. Advances in Multimedia Information Processing, Lecture Notes in Computer Science 2195, Springer-Verlag, Berlin (2001) 94-101.
9. Huffman, D.A.: A Method for the Construction of Minimum-Redundancy Codes. Proc. IRE, Vol.40(1952), 1098-1101
10. Jürgensen, H., Konstantinidis, S.: Codes. in: G. Rozenberg, A. Salomaa (editors), Handbook of Formal Languages, Vol.1 Springer-Verlag Berlin Heidelberg (1997) 511-607
11. Jones, D.W.: Applications of Splay Trees to Data Compression. Communication of ACM, Vol.31(1988), 996-1007
12. Linder, T., Tarokh, V., Zeger, K.: Existence of Optimal Prefix Codes for Infinite Source Alphabets. IEEE Trans. Inform. Theory, 43(1997)6, 2026-2028
13. Rubin, F.: Cryptographic Aspects of Data Compression Codes. Cryptologia, Vol.3(1979), 202-205
14. Roman, S.: Introduction to Coding and Information Theory. Springer-Verlag New York (1996)
15. Vitter, J.S.: Design and Analysis of Dynamic Huffman Codes. Journal of the Association for Computing Machinery, 34(1987)4, 825-845

A New Scheme Based on Semiconductor Lasers with Phase-Conjugate Feedback for Cryptographic Communications

A. Iglesias

Department of Applied Mathematics and Computational Sciences, University of Cantabria, Avda. de los Castros s/n E-39005, Santander, Spain

Abstract. In this paper a new scheme for cryptographic communications is introduced. The proposal consists of an optical fiber communication network in which the transmitter and the receiver are both semiconductor lasers subjected to phase-conjugate feedback. The laser parameters are carefully chosen in such a way that they exhibit a chaotic behavior, which is used to mask the message from the transmitter to the receiver. Potential applications of this scheme as well as its advantages with respect to other cryptographic schemes are also discussed.

1 Introduction

In the last few years we have witnessed an extraordinary worldwide growth of electronic data storage and digital communications (Internet, cellular telephones, high-speed industrial networks, high-bandwidth optical fibers, etc.). The users of these new technologies demand an effective protection of their information. To this end, different cryptographic schemes have been applied [7,20]. The basic elements of these schemes are: a *sender* or *transmitter*, a *receiver* and a *message* to be sent from the transmitter to the receiver. It is assumed that any communication between sender and receiver may be read or intercepted by a hostile person, the *attacker*. The primary objective of cryptography is to encode the message in such a way that the attacker cannot understand it. Furthermore, the most recent cryptographic models incorporate additional methods for many other tasks, such as access control, authentication, confidentiality, integrity, non-repudiation, availability, etc. Of course, some of the previous features can be combined. For example, user authentication is often used for access control purposes, non-repudiation is combined with user authentication, etc. To provide the users with the previous features, a number of different methods have been developed [11,18,19]. Among them, the possibility of encoding messages within a chaotic carrier has received considerable attention in the last few years [3, 4,5,6,12,15]. In this scheme, both the transmitter and the receiver are chaotic systems. The chaotic output of the transmitter is used as a carrier in which the message is encoded (masked). The amplitude of the message is much smaller than the typical fluctuations of the chaotic carrier, so that it is very difficult to isolate the message from the chaotic carrier. Decoding is based on the fact that coupled chaotic systems are able to synchronize their output under certain

M.H. Shafazand and A M. Tjoa (Eds.): EurAsia-ICT 2002, LNCS 2510, pp. 135–144, 2002.

conditions [14]. To decode the message, the transmitted signal is coupled to the chaotic receiver, which is similar to the transmitter. The receiver synchronizes with the chaotic carrier itself, so that the message can be recovered by removing the receiver output from the transmitted signal.

This scheme has been applied to secure communications with electronic circuits [5,6] and lasers [4,12]. Unfortunately, many of these models exhibit shortcomings that dramatically restrict their application to secure communications. The main one is that, as shown in [17,22,23], messages masked by low-dimension chaotic processes, once intercepted, are sometimes readily extracted even though the channel noise is rather high [24]. This fact explains why this kind of scheme has not been extensively developed for commercial purposes yet.

Until now, the previous limitation was considered to be overcome by employing either high dimensional chaotic systems [16] or high frequency devices, such as lasers. However, some recent results have reported extraction of messages with very high dimensions and high chaoticity [25], thus limiting the applicability of these systems. On the contrary, very high frequency systems are still seen as the optimal candidates for chaotic cryptography.

In this context, the present paper introduces a new scheme based on semiconductor lasers for cryptographic communications. The proposal consists of an optical fiber communication network in which the transmitter and the receiver are both (identical) semiconductor lasers subjected to phase-conjugate feedback. The laser parameters are carefully chosen in such a way that they exhibit a chaotic behavior, which is used to mask the message. It should be noticed that the exact knowledge of the parameters of the system in the receiver is necessary to recover the information. Thus, the laser parameters serve as the *encryption key*. As described in Section 3 the receiver synchronizes with the chaotic component of the transmitter system, recovering in this way the original message. This new system provides all the features listed above, it is neither hard to make nor very expensive, it is portable, it supports much more information than any other communication system and finally it is impossible to duplicate with current technology. Each of these statements will be reasoned as the paper progresses.

The structure of the paper is as follows: Section 2 describes all the steps needed for the implementation of the cryptographic communication system presented in this paper. Then, Section 3 illustrates the chaotic masking scheme for cryptographic communications. Finally, the main conclusions and some further remarks close the paper.

2 The Based-on-Lasers Communication System

In many communication systems, it is very important to establish the clear and unquestionnable identity of the communicating parties. The process of establishing and verifying the identity of a party is usually referred to as *authentication*. On the other hand, it is important to prevent the sender from later denying that he/she sent a message. This is called *non-repudiation*. Due to some advantages (ease of deployment and lower costs) these features are often checked by software. However, the highest level of reliability for (among others) authentication and non-repudiation involves *hardware* that must be associated with authorized users and that is not easy to duplicate.

On the other hand, most of the current development of digital communications is motivated by laser technology, in which optical fibers are used for data transmission [1]. By means of the optical fiber our computers can effectively support real-time video, multimedia, Internet, etc. and enormous amounts of information can be easily transmitted. For a based-on-lasers cryptographic communication system to be developed we need to walk through four main steps that will be carefully analyzed in the following paragraphs:

1. Choice of the laser and its components.
2. Determination of the laser parameters.
3. Choice of an accesible parameter for chaoticity.
4. Synchronization of the chaotic transmitter and receiver systems.

2.1 Choice of the Laser and Its Components

In this section we discuss the choice of the kind of laser and its components. This process is somewhat similar to choosing the hardware components of a computer. Here, our decision is determined by the requirements of communication and security. The first requirement implies the use of *semiconductor lasers* (or *diode lasers*). The reason is that in this kind of laser the medium for light amplification is given by semiconductor materials that emit light with wavelengths specially attractive for optical communication purposes. In fact, the role of the diode lasers in telecommunications is analogous to the role of transistors in electronics. Due to these reasons, semiconductor lasers have played a key role in the replacement of the "old" copper-wire based networks with optical fiber networks. In addition, diode lasers are relatively easy to make and are very reliable and efficient, exhibiting a very small size: the typical length of a semiconductor laser is about 0.25 millimeters. All these facts explain why these lasers are so popular as light sources in optical communication networks.

Of course, the diode lasers also have their disadvantages: they exhibit a very complex dynamics when they are subjected to an external signal, for example, light obtained by reflection from an external mirror. Such a reflection is usually referred to as *optical feedback* [21] and it has been intensively studied since the late 70s from both the experimental and the theoretical points of view [8,10, 13]. Depending on the type of mirror (conventional or phase-conjugate), we talk about conventional optical feedback (COF) or phase-conjugate feedback (PCF). PCF mirrors exhibit interesting advantages with respect to conventional mirrors. The most important one is related to its distorsion-undoing property. Any phase distortion between source and phase-conjugator is fixed on the way back by the fact that light beams retrace exactly their path to their respective sources. This property implies that PCF mirrors are a cheaper alternative to the conventional ones, which must be controlled previously.

There is, lastly, an additional major advantage of PCF lasers: the perfect synchronization (zero error) is only possible for diode lasers with a phase-conjugate mirror. Motivated by these attractive advantages, we will focus on semiconductor lasers subjected to PCF.

2.2 Determination of the Laser Parameters

In this step, the laser parameters for the transmitter and the receiver are determined. To this end, we need to introduce the rate equations of the laser. They describe the behavior of the electric field $E(t)$ and the inversion $N(t)$ (the total number of electron-hole pairs in the cavity) in a single-mode laser and are usually based on the Lang-Kobayashi equations [9]. These rate equations are:

$$\frac{dE(t)}{dt} = \frac{1}{2}(1 - i\,\alpha)\left[G(t) - \frac{1}{\tau_n}\right] E(t) + \kappa\, E^*(t-\tau) + \sqrt{2\,\beta\, N(t)}\,\xi(t) \tag{1}$$

and

$$\frac{dN(t)}{dt} = \frac{I}{q} - \frac{N(t)}{\tau_e} - G(t)\,|E(t)|^2 \tag{2}$$

where

$$G(t) = \frac{g(N(t) - N_0)}{(1 + s\,|E(t)|^2)}. \tag{3}$$

The terms of the right-hand side of (1) can be interpreted as follows: the first term describes the behavior of the electric field for the *solitary* laser (the semiconductor laser isolated from the outside world), the term $\kappa\, E^*(t-\tau)$ is associated with the optical feedback (the superscript * of the electric field stands for the complex conjugate) which is accounted for two parameters: the delay time τ, and the feedback-rate κ. Finally, the last term $\sqrt{2\,\beta\, N(t)}\,\xi(t)$ is included here to give an account of the spontaneous emission noise, which naturally arises in real applications. Such a noise is modeled by a complex Gaussian white noise term $\xi(t)$ of zero mean, $<\xi(t)>= 0$, and correlation $<\xi_i(t)\xi_j(t')>= \delta_{ij}(t-t')$. The parameter β stands for the spontaneous emission rate.

All the remaining parameters, their symbols and the values of the model used in this paper are listed in Table 1. We remark that it is not possible to create, *a priori*, lasers with the same parameters as those listed in Table 1. What is really done in practical settings is to construct a new laser, then to determine its parameters (which are only approximately chosen by the manufacturer) and finally to analyze its dynamical behavior when varying an accesible parameter (for example, the distance between the laser and the mirror, L_{ext}). Other laser and mirror parameters cannot be chosen but only determined after construction. Especially interesting is the case of the mirror, because it is extracted from the central part of a "wafer" whose dynamical behavior changes as a function of the points of the piece. Thus, usually only a few (typically two or three) small mirrors with the same parameters can be obtained from the same piece. These mirrors can be used for the transmitter and the receiver lasers, so the assumption that both systems are identical is feasible in practice. After finishing this process, no other mirror with the same parameters can be made anymore. This is a very important question, because very strong authentication requires hardware

Table 1. Parameter values of the semiconductor laser with phase-conjugate feedback

Parameter	Symbol	Value
Linewidth enhancement factor	α	5
Photon lifetime	τ_n	2 ns.
Feedback coefficient	κ	0.0238 ps^{-1}.
External cavity round-trip time	τ	200 ps.
External cavity length	L_{ext}	1 cm.
Spontaneous emission rate	β	1.5 10^{-9} ps^{-1}.
Bias current	I	44 mA.
Electron charge	q	1.602 10^{-19} C.
Carrier lifetime	τ_e	2 ns.
Gain parameter	g	1.510^{-8}
Nonlinear gain parameter	s	5 10^{-7}
Transparency carrier number	N_0	1.5 10^{8}

components that can be in the possesion of only a few people at a time. This situation differs radically from electronic circuits, which can be easily duplicated with current engineering.

2.3 Choice of Some Accessible Parameter for Chaoticity

For quantifying the laser dynamics the use of bifurcation diagrams is well established [13]. It is a powerful tool, since the diagram shows at a glance for which strengths of feedback the laser operates stably, periodically or chaotically. For diode lasers with PCF, the bifurcation diagrams are obtained by generating a time series for each feedback level and noting the carrier number N when the laser power crossed the solitary-laser value N_{sol} vs. $\kappa\tau$, this last term meaning the feedback level. For the choice of the laser parameters given in Table 1 and $\tau = 200$ ps., we vary κ such that $\kappa\tau \in [1, 5]$ (i.e., $\kappa \in [0.005, 0.025]$). The corresponding bifurcation diagram is displayed in Fig. 1. The figure has been split up into two parts: on the left, the diagram for the whole interval $\kappa \in [0.005, 0.025]$ and, on the right, a zoom of the chaotic region $\kappa \in [0.0235, 0.025]$.

As the reader may appreciate from Fig. 1(left) the bifurcation diagram for diode lasers with PCF is very complicated. Roughly speaking, a single crossing in the bifurcation diagram implies a periodic output whereas multiple crossings indicate period doublings, quasiperiodicity or chaos. In Fig. 1(left) the laser exhibits a steady state (corresponding to the blank regions without any dots) on an interval for κ starting at the initial value $\kappa = 0.005$, which becomes unstable at a critical value of κ, and the laser becomes periodic. After a period-doubling region, quasiperiodicity (for example, for $\kappa = 0.0187$) and chaotic behaviors appear. However, chaos is interrupted by regions of nonchaotic output although, in general, this output does not become completely stable. Figure 1(right) shows a zoom of the chaotic region corresponding to $\kappa \in [0.0235, 0.025]$.

For the synchronization process to be possible we need a chaotic behavior for the laser. From Fig. 1, we have considered a feedback coefficient $\kappa = 0.0238$, for

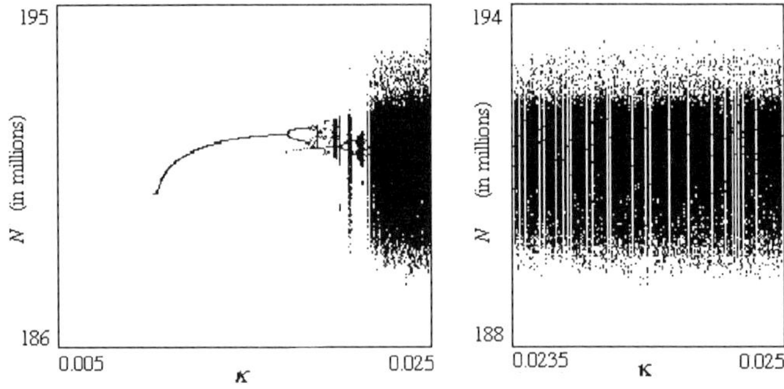

Fig. 1. Bifurcation diagram for a semiconductor laser subjected to PCF: (left) diagram for $\kappa \in [0.005, 0.025]$; (right) a zoom of the chaotic region $\kappa \in [0.0325, 0.025]$

which the laser behaves chaotically. To check this, in Fig. 2 we have represented both the trajectory of the laser intensity (left) and its Fourier spectrum (right) for this parameter value.

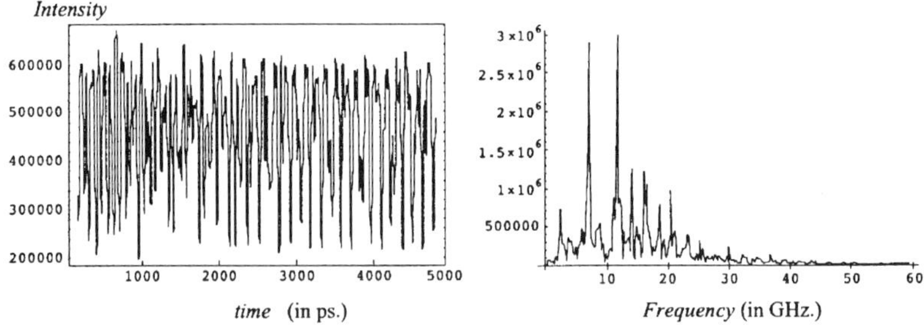

Fig. 2. Chaotic behavior of the diode laser with PCF for $\kappa = 0.0238$: (left) Intensity of the chaotic orbit during 5 ns.; (right) Fourier spectrum

2.4 Synchronization of the Chaotic Transmitter and Receiver Systems

In this section we explore the synchronization of two semiconductor lasers subjected to PCF, that is, described by:

$$\frac{dE_{t,r}(t)}{dt} = \frac{1}{2}(1 - i\alpha)\left(G_{t,r}(t) - \frac{1}{\tau_{t,r}}\right) E_{t,r} + \kappa E^*_{t,r}(t-\tau) + K_r E_t \tag{4}$$

$$\frac{dN_{t,r}(t)}{dt} = \frac{I}{q} - \frac{1}{\tau_n} N_{t,r}(t) - G_{t,r}(t)|E_{t,r}(t)|^2 \tag{5}$$

where:

$$G_{t,r}(t) = \frac{g(N_{t,r}(t) - N_0)}{(1 + s|E_{t,r}|^2)} \tag{6}$$

and the subscripts t and r are used to indicate respectively the transmitter and the receiver. On examining these equations, the receiver is seen to be similar to the transmitter except for the fact that it incorporates a new term $K_r E_t$, meaning that we inject a small amount (given by the coupling parameterK_r) of the transmitter signal E_t to the receiver. Thus, by varying this parameter K_r we modify the amount of the injected signal.

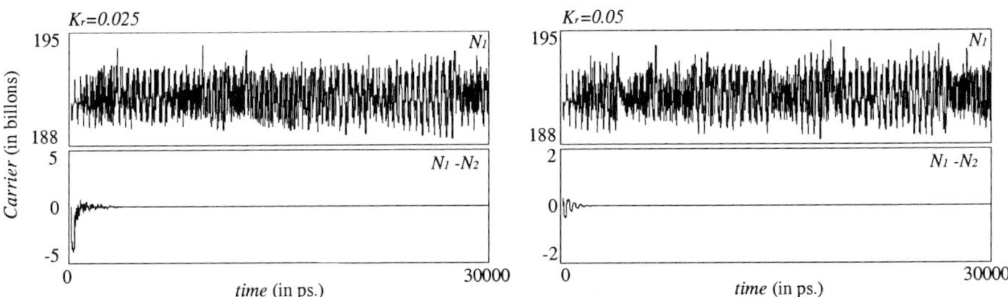

Fig. 3. Syncronization of chaotic lasers for different values of K_r: (left) 0.025; (right) 0.05. Each figure represents the transmitter (above) and the difference between the transmitter and the receiver (below)

Figure 3 shows the synchronization for two different values of K_r: 0.025 and 0.05. The figure represents the number of carriers of the transmitter laser N_1 (above) and the difference between the transmitter and the receiver $N_1 - N_2$ (below). Of course, the difference $N_1 - N_2$ should vanish for a perfect synchronization. This occurs for both values of K_r, meaning that synchronization is robust to small perturbations of this parameter. From these figures, it becomes clear that this kind of laser allows there to be a very accurate synchronization with a very brief transient, two remarkable features for communications.

3 Applications to Cryptographic Communication

Most of the interest in chaotic synchronization has been motivated by its potential application to cryptographic communication [4,5,6,12]. Here, the information message is "coded" by means of a chaotic signal and then transmitted. The receiver contains a copy of the chaotic system that generates the cipher key and produces the decoding signal by synchronizing with it. This process can be accomplished through the so-called *chaotic masking* scheme (see Figure 4). In this scheme the optical carrier modulated by a small signal (the message E_M) is added to the transmitter laser output E_D and transmitted to a response system, which is a replica of the laser that generates the chaotic signal (the *drive*). Decoding is performed by the chaotic receiver, providing that it synchronizes only on the chaotic component of the received signal, which is therefore available on the receiver laser output. A substraction of the synchronized signal $E_R \approx E_D$ from the transmitted signal $E_T = E_D + E_M$ results in the recovery of the message.

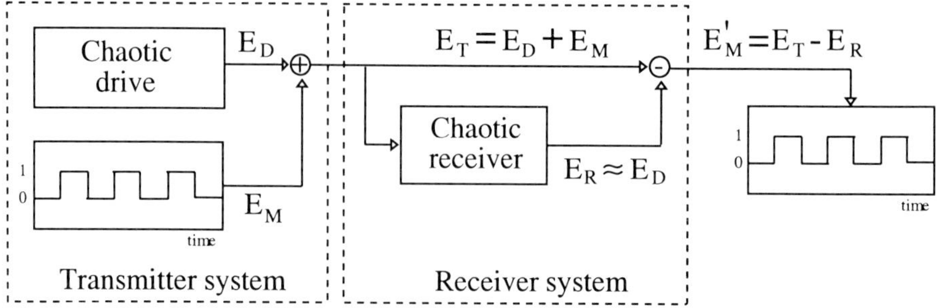

Fig. 4. Scheme of secure communication based on chaotic masking

We have applied this scheme to digital communications, where both the transmitter and the receiver are semiconductor lasers with PCF. Figure 5 summarizes the different steps in this process: the message M consists of a digital signal generated by a table of bits 1 and -1 as shown in Fig. 5(a). The message is encoded by a small variable attenuation of the chaotic ouput (for example, the electric field) of a laser subjected to PCF (transmitter) by: $|E'_t(t)| = |E_t(t)|(1 + w\, M(t)|,\ w \in \mathbb{R}$. Of course, the amplitude of the signal to be transmitted $M(t)$ must be small in comparison to the chaotic signal. Otherwise, information could be extracted directly from the waveform. The transmitted signal $||E'||$ (Fig. 5(b)) is chaotic and virtually indistinguishable from that of the chaotic drive (for instance, they both have a similar power spectrum). Therefore, no information can be extracted from the encrypted signal. Then, it is coupled to another chaotic laser (receiver) that is similar to the transmitter. The receiver synchronizes with the chaotic carrier itself $|E_r(t)| \approx |E'_t(t)|$. The message can be recovered from the transmitted signal as $D(t) = \sqrt{\dfrac{|E'_t(t)|^2}{|E_r(t)|^2}} - 1$ (Fig. 5(c)) and then filtered (Fig. 5(d)). A simple visual comparison of the original and the filtered messages (Fig. 5(e)) shows the excellent performance of the method.

4 Conclusions and Further Remarks

In this paper a new hardware implementation for cryptographic communications has been described. The proposed method substantially improves many other based-on-hardware methods for tasks such as access control, integrity, authentication and non-repudiation. In addition, the system is digital, which makes possible perfect matching of the receiver to the transmitter (at the coding level).

In addition, chaotic masking has been shown to be more secure than masking with noise, since in the later case data bits can be extracted by using a correlator. This is not possible for chaotic lasers since the correlation time of the bits and of the chaos are almost identical. Finally, fiber dispersion and nonlinearities can deteriorate the transmitted signal and make decoding impossible [2]. Although not included here because of space limitation, we have checked that, under the appropriate circumstances (such as a fiber length not longer than 50 km. and a loss coefficient $\nu = 0.2$ dB/km., among others), this does not occur here.

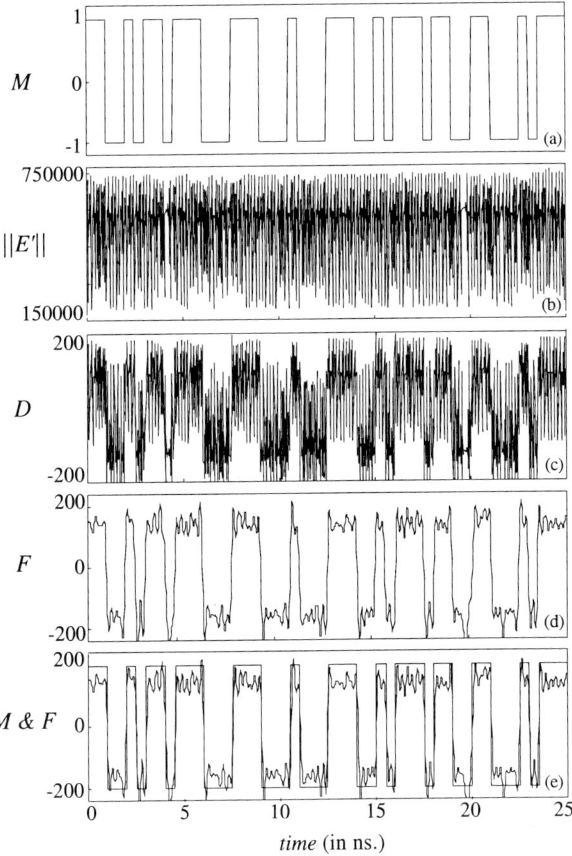

Fig. 5. Chaotic masking scheme: (a) a digital message; (b) transmitted signal encoded; (c) recovered signal; (d) recovered signal filtered; (e) original and filtered messages

References

1. G.P. Agrawal and N.K. Dutta: Long Wavelength Semiconductor Lasers, 2nd. ed., Van Nostrand Reinhold, New York, 1993
2. G.P. Agrawal: Nonlinear Fiber Optics, Academic Press, San Diego, 1989

3. P. Celka: Chaotic synchronization and modulation of nonlinear time-delayed feedback optical systems. IEEE Trans. Circuits Syst. **42** (1995) 455–463
4. P. Colet and R. Roy: Digital communications with synchronized chaotic lasers. Optics Lett. **19** (1994) 1056-2058
5. K.M. Cuomo and A.V Oppenheim: Circuit implementation of synchronized chaos with applications to communications. Phys. Rev. Lett. **71** (1993) 65-68
6. K.M. Cuomo, A.V Oppenheim and S.H. Strogatz: Synchronization of Lorenz-based chaotic circuits with applications to communications. IEEE Trans. Circuits Syst., **40** (1993) 626–633
7. D. Davies and W. Price: Security for Computer Networks, John Wiley & Sons, New York, 1989
8. G.R. Gray, D. Huang and G.P. Agrawal: Chaotic dynamics of semiconductor lasers with phase-conjugate feedback. Phys. Rev. A **49** (1994) 2096–2105
9. R. Lang and K. Kobayashi: External optical feedback effects on semiconductor injection laser properties. IEEE J. Quantum Electron., **16** (1980) 347–355
10. D. Lenstra: Theory of a diode laser with phase-conjugate feedback. Opt. Lett., **17** (1992) 1590–1592
11. A.J. Menezes, P.C. van Oorschot and S.A. Vanstone: Handbook of Applied Cryptography, CRC Press, Boca Raton, Florida, 1996
12. C.R. Mirasso, P. Colet and P. García-Fernández: Synchronization of chaotic semiconductor lasers: application to encoded communications. IEEE Phot. Tech. Lett., **8**(2) (1996) 299–301
13. J. Mork, B. Tromborg and J. Mark: Chaos in semiconductor lasers with optical feedback: theory and experiment. IEEE J. Quantum Electron., **28** (1992) 93–108
14. L.M. Pecora and T.L. Carroll: Synchronization on chaotic systems. Phys. Rev. Lett. **64** (1990) 821–824
15. L.M. Pecora and T.L. Carroll: Driving systems with chaotic signals. Phys. Rev. A, **44** (1991) 2374–2383
16. J.H. Peng, E.J. Ding, M. Ding and W. Yang: Synchronizing hyperchaos with a scalar transmitted signal. Phys. Rev. Lett., **76** (1996) 904–907
17. G. Pérez and H.A. Cerdeira: Extracting messages masked by chaos. Phys. Rev. Lett., **74** (1995) 1970-1973
18. M.Y. Rhee: Cryptography and Secure Data Communications, McGraw-Hill, Boston, MA, 1994
19. B. Schneier: Applied Cryptography, Second Edition, John Wiley & Sons, New York, 1996
20. J. Seberry and J. Pieprzyk: Cryptography: An Introduction to Computer Security, Prentice-Hall, Englewoods Cliffs, N.J., 1989
21. G.H.M. van Tarwijk and D. Lenstra: Semiconductor lasers with optical injection and feedback. Quantum Semiclass. Opt., **7** (1995) 87–143
22. T. Yang: Recovery of digital signals from chaotic switching. Int. Journal of Circuit Theory and Appl., **23** (1995) 611–615
23. T. Yang, L.B. Yang and C.M. Yang: Application of neural networks to unmasking chaotic secure communication. Physica D, **124** (1998) 248–257
24. C.S. Zhou and T.L. Chen: Extracting information masked by chaos and contamined with noise: some considerations on the security of communication approaches using chaos. Phys. Lett. A, **234** (1997) 429–435
25. C.S. Zhou and C.H. Lai: Extracting messages masked by chaotic signals of time-delay systems. Phys. Rev. E, **60** (1999) 320–323

Parallel Algorithm and Architecture for Public-Key Cryptosystem

Hyun-Sung Kim[1] and Kee-Young Yoo[2]

[1] Kyungil University, Computer Engineering,
712-701, Kyungsansi, Kyungpook Province, Korea
kim@kiu.ac.kr

[2] Kyungpook National University, Computer Engineering,
702-701 Daegu, Korea
yook@knu.ac.kr

Abstract. This paper proposes a new parallel algorithm and architecture for two modular multiplications over GF(2^m). The algorithm uses the property of irreducible all one polynomial as a modulus and computes two modular multiplications in parallel. The architecture is based on cellular automata and has smaller area and time complexity than previous architectures. Since the proposed architecture has regularity, modularity and concurrency, it is suitable for VLSI implementation. The proposed architecture can be used as a basic architecture for the public-key cryptosystems.

1 Introduction

Finite field $GF(2^m)$ arithmetic is fundamental to the implementation of a number of modern cryptographic systems and schemes of certain cryptographic systems[1][2]. Most arithmetic operations, such as exponentiation, inversion, and division operations, can be carried out using just a modular multiplier or using modular multiplier and squarer. Therefore, to reduce the complexity of these arithmetic architectures, an efficient architecture for multiplication over $GF(2^m)$ is necessary.

In 1984, Yeh, *et al.* [7] developed a parallel systolic architecture for performing the operation $AB+C$ in a general GF(2^m). A semi-systolic array architecture in [8] was proposed using the standard basis, whereas architecture to compute multiplication and inversion were represented using the normal basis [9]. A systolic power-sum circuit was presented in [10], and many bit-parallel systolic multipliers have been proposed. However, these multipliers still have some shortages for cryptography applications due to their system complexity. For the better complexity, Itoh and Tsujii [11] designed two low-complexity multipliers for the class of GF(2^m), based on an irreducible all one polynomial (AOP) of degree m and irreducible equally spaced polynomial of degree m. Later, Fenn, *et al.* in [12] and Kim in [13] developed linear feedback shift register (LFSR) based multipliers with low complexity of hardware architecture using the property of AOP.

Cellular automata (CA), first introduced by John Von Neumann in the 1950s, have been accepted as a good computational model for the simulation of complex physical

M.H. Shafazand and A M. Tjoa (Eds.): EurAsia-ICT 2002, LNCS 2510, pp. 145–153, 2002

systems [3]. Wolfram defined the basic classification of cellular automata using their qualitative and functional behavior [4], whereas Pries, *et al.* focused on the group properties induced by the patterns of behavior [5]. Choudhury proposed an LSB-first multiplier using CA with low latency [14]. However, all such previously designed systems still have certain shortcomings.

Accordingly, the purpose of this paper is to propose a parallel algorithm and architecture for two modular multiplications over GF(2^m). The algorithm uses the property of irreducible AOP as a modulus and computes two modular multiplications in parallel. There exists a common part to compute two multiplications at the same time. The algorithm computes the common parts only once, thereby reducing the required operations compared to related algorithms. The proposed architecture is based on cellular automata and has smaller area and time complexity than previous architectures. The proposed architectures can be used as a kernel circuit for exponentiation, inversion, and division architectures. These architectures are very important part for the public key cryptosystems. If we use the proposed architectures to implement cryptosystem, we can get a great cryptosystem with low hardware complexity. It is easy to implement VLSI hardware and use in IC cards as the proposed structures have a particularly simple architecture.

2 Background

This section provides a brief description of the cellular automata and finite fields. These properties will be used to derive a new algorithm and architecture.

2.1 Cellular Automata

Cellular automata are finite state machines, defined as uniform arrays of simple cells in *n*-dimensional space. They can be characterized by looking at four properties: the cellular geometry, neighborhood specification, number of states per cell, and algorithm used for computing the successor state. Cells are restricted to local neighborhood interaction and have no global communication. A cell uses an algorithm, called its computation rule, to compute its successor state based on the information received from its nearest neighbors. An example is shown below for 2-state 3-neighborhood 1-dimensional CA [3-5].

Neighborhood state	:	111	110	101	100	011	010	001	000	
State coefficient	:	2^7	2^6	2^5	2^4	2^3	2^2	2^1	2^0	
Next state	:	0	1	0	1	1	0	1	0	(rule 90)
Next state	:	1	1	1	1	0	0	0	0	(rule 240)

In the above example, the top row gives all eight possible states of the 3-neighboring cells at time *t*. The second row is the state coefficient, while the third and last rows give the corresponding states of the *i*th cell at time *t*+1 for two illustrative CA rules. If the next state function of a cell is expressed in the form of a truth table, then the decimal equivalent of the out column in the truth table is conventionally called the rule number for the cell. The next state of the CA is determined by the current state and the rules that govern its behavior. Let the previous, current, and next state be s_{i-1}, s_i, and s_{i+1}. The two rules 90 and 240 result in the following:

$$\text{rule 90 } : s_i^+ = s_{i-1} \oplus s_{i+1}, \qquad \text{rule 240 : } s_i^+ = s_{i-1}$$

where s_i^+ denotes the next state for cell s_i, and '$\oplus$' denotes an XOR operation.

In addition, CA is composed of Linear CA, Non-Linear CA, and Additive CA according to the operation rules applied between cells. Linear CA are restricted to linear rules of operation, which means that the next state of the machine can only be computed from the previous state using a linear operation, i.e., an XOR operation. Non-Linear CA is composed of XOR operation plus other operations, while additive CA is composed of only XOR and/or XNOR operations. CA is also classified as Uniform or Hybrid based on the rules applied. Uniform CA only has one rule, whereas Hybrid CA has two or more rules. Finally, CA can be referred to as named 1-dimensional, 2-dimensional, and 3-dimensional based on the structure of the array.

There are various possible boundary conditions, for example, a Null-Boundary CA (NBCA), where the extreme cells are connected to the ground level, a Periodic-Boundary CA (PBCA), where extreme cells are adjacent, etc.

The present state of a CA having m cells can be shown in terms of an m vector v=($v_0\ v_1\ v_2 \ldots v_{m-1}$), where v_i is the value of cell i, and v_i is an element of GF(2). The next state of a linear CA can be determined by multiplying the characteristic matrix with the vector in the present state, where the characteristic matrix denoted by T show all rules of the CA. If v^t is a column vector representing the state of the automata at the t-th instant of time, then the next state of the linear CA is given by $v^{t+1}=T\times v^t$.

$$T = \begin{bmatrix} 0 & 1 & 0 & 0 \\ 1 & 0 & 0 & 0 \\ 0 & 1 & 0 & 1 \\ 0 & 0 & 1 & 0 \end{bmatrix}$$

Fig. 1. Characteristic matrix for NBCA with rules <90, 240, 90, 240>.

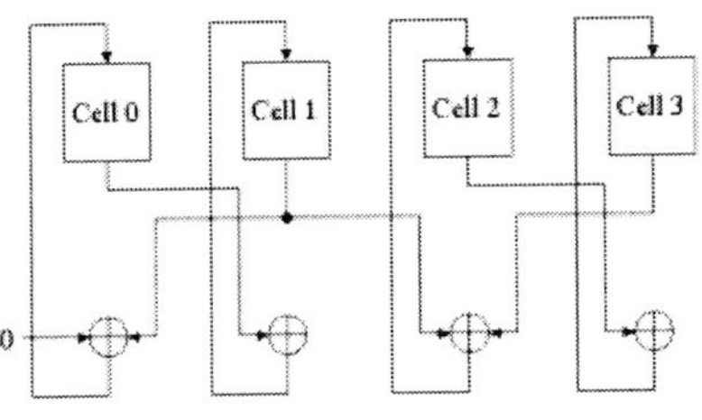

Fig. 2. One-dimensional 4-cell NBCA using rules <90, 240, 90, 240>.

Fig. 2 shows a 4-cell 2-state 3-neighborhood 1-dimensional CA architecture using rules <90,240,90,240>.

2.2 Finite Fields

A finite field $GF(2^m)$ contains 2^m elements that are generated by an irreducible polynomial of degree m over $GF(2)$. A polynomial $f(x)$ of degree m is said to be irreducible if the smallest positive integer n for which $f(x)$ divides $x^n + 1$ is $n = 2^m-1$ [6]. It has been shown that an all one polynomial (AOP) is irreducible if and only if m+1 is a prime and 2 is a generator of the field $GF(m+1)$ [11]. The values of m for which an AOP of degree m is irreducible are 2, 4, 10, 12, 18, 28, 36, 52, 58, 60, 66, 82, and 100 for $m \leq 100$. Let $f(x) = x^m + x^{m-1} + \ldots + x + 1$ be an irreducible AOP over $GF(2)$ and α be the root of $f(x)$ such that $f(\alpha) = \alpha^m + \alpha^{m-1} + \ldots + \alpha + 1 = 0$. Then we have

$$\alpha^m = \alpha^{m-1} + \alpha^{m-2} + \ldots + \alpha + 1, \ \alpha^{m+1} = 1$$

The reduction is often performed using the polynomial $\alpha^{m+1} + 1$. This property of irreducible polynomial is very adaptable for PBCA architecture.

Then any field element $a \in GF(2^m)$ can be represented by a standard basis such as $a = a_{m-1}\alpha^{m-1} + a_{m-2}\alpha^{m-2} + \ldots + a_1\alpha^1 + a_0$, where $a_i \in GF(2)$ for $0 \leq i \leq m$-1. $\{ 1, \alpha, \alpha^2, \alpha^3, \ldots, \alpha^{m-1} \}$ is the standard basis of $GF(2^m)$. If it is assumed that $\{ 1, \alpha, \alpha^2, \alpha^3, \ldots, \alpha^m \}$ is an extended standard basis, the field element A can also be represented as

$$A = A_m\alpha^m + A_{m-1}\alpha^{m-1} + A_{m-2}\alpha^{m-2} + \ldots + A_0 \tag{1}$$

where $A_m = 0$ and $A_i \in GF(2)$ for $0 \leq i \leq m$. Here, $a = A \pmod{f(x)}$, where $f(x)$ is an AOP of degree m, then the coefficients of a are given by $a_i = A_i + A_m \pmod 2$, $0 \leq i \leq m$-1.

Let C and M be elements of $GF(2^m)$, the exponentiation of M is then defined as

$$C = M^E, \ 0 \leq E \leq n \tag{2}$$

For a special case, $M = \alpha$, the exponent E, which is an integer can be expressed by $E = e_{m-1}2^{m-1} + e_{m-2}2^{m-2} + \ldots + e_1 2^1 + e_0$. The exponent also can be represented with vector representation $[\, e_{m-1}\, e_{m-2} \cdots e_1\, e_0 \,]$. A popular algorithm for computing exponentiation is the binary method proposed by Knuth [15]. The exponentiation of M can be expressed as

$$M^E = M^{e_0}(M^{2^1})^{e_1}(M^{2^2})^{e_2}\ldots(M^{2^{m-1}})^{e_{m-1}} \tag{3}$$

Based on equation 3, an algorithm for computing exponentiations is presented as following algorithm.

[Algorithm 1] Knuth's Binary Method

Input : $M, E, f(x)$

Output : $C = M^E \bmod f(x)$

```
T=M
if (e0==1) C=T else C=α^0
for i=1 to m-1
        T=TT mod f(x)
        if (ei==1) C=CT mod f(x)
```

The algorithm shows that the exponentiation can be performed with squaring in line 4 and multiplication in line 5. Modular squaring can be considered as a special case of modular multiplication, which has the same input for operand and operator. Next section presents a new parallel algorithm and their architecture for two modular multiplications with significantly low complexity of operations.

3 Parallel Algorithm

This section presents two new parallel algorithms over $GF(2^m)$. First, we propose a new modular multiplication algorithm from the ordinary modular multiplication. Then, devise a new parallel algorithm for two modular multiplications. The algorithm uses the property of irreducible AOP as a modulus. Therefore, all operations are performed over extended standard basis. There exists a common part to compute two multiplications at the same time. The algorithm computes the common parts only once, thereby reducing the required operations significantly.

3.1 Modular Multiplication Algorithm

Let A and B be the elements over $GF(2^m)$ and $f(x)$ be the modulus, which hsa the property of an irreducible AOP. Then each element over extended standard basis is expressed as follows:

$$A = A_m\alpha^m + A_{m-1}\alpha^{m-1} + A_{m-2}\alpha^{m-2} + \ldots + A_0$$
$$B = B_m\alpha^m + B_{m-1}\alpha^{m-1} + B_{m-2}\alpha^{m-2} + \ldots + B_0$$
$$f(x) = \alpha^{m+1}+1$$

The following shows the comparison between ordinary and proposed algorithms.

[Algorithm 2] Ordinary Multiplication

Input : $A, B, f(x)$
Output : $P = AB \bmod f(x)$
Initial value : $P^{(0)} = (0, 0, \ldots, 0)$
$A^{(0)} = (a_{m-1}, a_{m-2}, \ldots, a_0)$
$A^{(i)}_{-1} = 0, 0 \le i \le m$

1: for $i = 1$ to m do
2: for $j = 1$ to m do
3: $P^{(i)}_{m-j} = P^{(i-1)}_{m-j} + b_{i-1}A^{(i-1)}_{m-j}$
4: $A^{(i)}_{m-j} = A^{(i-1)}_{m-j-1} + f_{m-j}A^{(i-1)}_{m-1}$

[Algorithm 3] Proposed Multiplication

Input : A, B
Output : $P = AB \bmod \alpha^{m+1}+1$
Initial value : $P^{(0)} = (0, 0, \ldots, 0)$
$A^{(0)} = (0, a_{m-1}, a_{m-2}, \ldots, a_0)$

1: for $i = 1$ to $m+1$ do
2: for $j = 1$ to $m+1$ do
3: $P^{(i)}_{m+1-j} = P^{(i-1)}_{m+1-j} + b_{i-1}A^{(i-1)}_{m+1-j}$
4: $A^{(i)}$ = Circular_Right_Shift($A^{(i-1)}$)

Modular multiplication is divided into two operations in line 3 and 4. These operations can be processed in parallel since there is no dependency between operations. Proposed algorithm reduces the complexity of the second operation very efficiently. In line 4, the operation of $A\alpha \bmod f(x)$ is carried out by circular shifting 1-bit to the right of A.

3.2 Algorithm for Two Modular Multiplications

Algorithm 1 shows that the exponentiation can be performed with squaring and multiplication operations. Modular squaring can be considered as a special case of modular multiplication, which has the same input for operand and operator. The computation steps for the modular multiplication $P = AB \bmod f(x)$ and square $S = AA \bmod f(x)$ are follows as:

$$\begin{aligned} P &= AB \bmod f(x) \\ &= b_0A+b_1[A\alpha \bmod f(x)]+\ldots+b_m[A\alpha^m \bmod f(x)] \end{aligned} \tag{4}$$

$$\begin{aligned} S &= AA \bmod f(x) \\ &= a_0A+a_1[A\alpha \bmod f(x)]+\ldots+a_m[A\alpha^m \bmod f(x)] \end{aligned} \tag{5}$$

In the above two equations 4 and 5, [ ] is common to both. Thus, it is much more efficient to compute the common parts only once when two multiplications are carried out simultaneously. The proposed algorithm computing two modular multiplications, denoted by TMM, can be derived from the above equations 4 and 5:

[Algorithm 4] Proposed Algorithm for Two Modular Multiplications

Input : A, B

Output : $P = AB \bmod \alpha^{m+1}+1$, $S = AA \bmod \alpha^{m+1}+1$

Initial value : $P^{(0)} = (0, 0, \ldots, 0)$, $S^{(0)} = (0, 0, \ldots, 0)$, $A^{(0)} = (0, a_{m-1}, a_{m-2}, \ldots, a_0)$, $C = (0, a_{m-1}, a_{m-2}, \ldots, a_0)$

1 : for $i = 1$ to $m+1$ do

2 : for $j = 1$ to $m+1$ do

3 : $P^{(i)}{}_{m+1-j} = P^{(i-1)}{}_{m+1-j} + b_{i-1}A^{(i-1)}{}_{m+1-j}$

4 : $S^{(i)}{}_{m+1-j} = S^{(i-1)}{}_{m+1-j} + c_{i-1}A^{(i-1)}{}_{m+1-j}$

5 : $A^{(i)} = \text{Circular\_Right\_Shift}(A^{(i-1)})$

Two modular multiplications are divided into three operations in line 3, 4, and 5. These operations can be processed in parallel since there is no dependency between operations. In line 5, the common operation, $A\alpha \bmod f(x)$, between two modular multiplications is carried out just once thereby it reduces the complexity of operations significantly.

4 Cellular Automata Architecture

This section presents a new CA architecture based on TMM over $GF(2^m)$. First, we design a new CA multiplier from the proposed modular multiplication. Then, design a new CA architecture for two modular multiplications based on proposed CA multiplier.

4.1 CA Multiplier

The following is the basic operation for performing modular multiplication in Alg. 3 :

Operation 3-1: Multiply b_i to $(m+1)$-tuple of A and add it to $(m+1)$-tuple of P as follows: $(p_0, p_1, \ldots, p_{m-1}, p_m) <= (p_0, p_1, \ldots, p_{m-1}, p_m) + b_i(a_0, a_1, \ldots, a_{m-1}, a_m)$

Operation 3-2: $(m+1)$-tuple of A is circular shifting 1-bit to the right as follows: $(a_m, a_0, \ldots, a_{m-2}, a_{m-1}) <= (a_0, a_1, \ldots, a_{m-1}, a_m)$

In order to perform operation 3-2, the 1-dimensional PBCA having m+1 cells is used which is the upper part named with cell # in Fig. 3. A is input into m+1 cells of CA and CA have a characteristic matrix with all rules 240 for the operation 3-2. Fig. 3 shows the proposed modular multiplier using PBCA with rules 240. It is possible to perform multiplication in m+1 clock cycles over $GF(2^m)$.

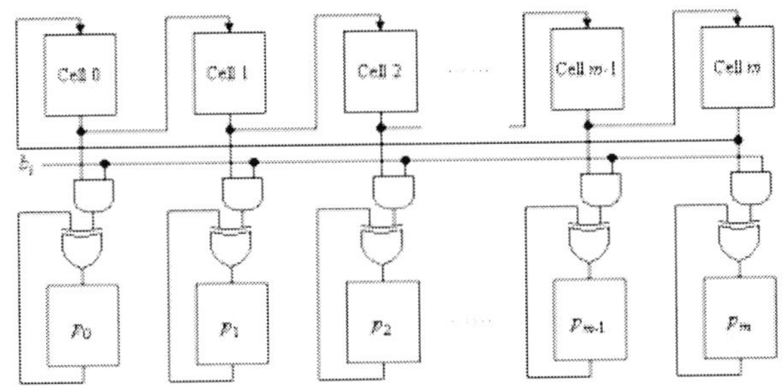

Fig. 3. PBCA Modular Multiplier.

4.2 CA Architecture for Two Multiplications

To implement algorithm for TMM, we derive a new PBCA from the Alg. 4. The following is the basic operation for performing TMM in Alg. 4 :

Operation 4-1: Multiply b_i to $(m+1)$-tuple of A and add it to $(m+1)$-tuple of P as follows: $(p_0, p_1, \ldots, p_{m-1}, p_m) <= (p_0, p_1, \ldots, p_{m-1}, p_m) + b_i(a_0, a_1, \ldots, a_{m-1}, a_m)$

Operation 4-2: Multiply a_i to $(m+1)$-tuple of A and add it to $(m+1)$-tuple of S as follows: $(s_0, s_1, \ldots, s_{m-1}, s_m) <= (s_0, s_1, \ldots, s_{m-1}, s_m) + a_i(a_0, a_1, \ldots, a_{m-1}, a_m)$

Operation 4-3: $(m+1)$-tuple of A is circular shifting 1-bit to the right as follows: $(a_m, a_0, \ldots, a_{m-2}, a_{m-1}) <= (a_0, a_1, \ldots, a_{m-1}, a_m)$

Operations 4-1 and 4-3 are the same with operations in modular multiplier. Therefore, we can implement a new CA architecture using the same PBCA and characteristic matrix with the proposed multiplier. For the operation 4-2, the same module with operation 3-2 can be added. Fig. 4 shows the proposed CA architecture for two modular multiplications. It is possible to perform two multiplications in m+1 clock cycles over $GF(2^m)$. The proposed CA architecture for TMM computes operation 4-1, 4-2, and 4-3 in parallel.

Table. 1 shows a comparison between modular multipliers. For the comparison, it is assumed that n AND and n XOR represent n number of 2-input AND gate and XOR gate, respectively and REG represents 1-bit register. There is no CA based architecture for TMM. For the forth architecture in Table 1, it is assumed the case of that using two CA architectures in [14]. Comparison shows that the proposed architectures hybrid the advantages from previous architectures. It reduces hardware and time complexity, significantly, compared to the previous architectures.

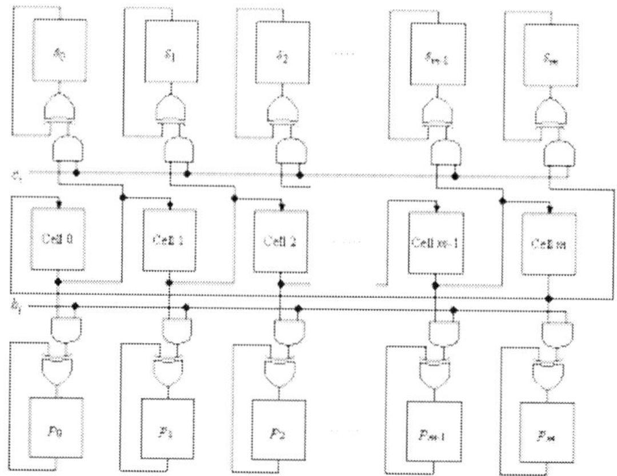

Fig. 4. CA architecture for TMM.

Table 1. Comparisons for arithmetic architectures.

Item / Circuit	Function	Number of cells	Latency	Hardware Complexity	Critical Path
Chaudhury in [14]	*0AB+C*	m	m	$2m$ AND $2m$ XOR $4m$ REG	AND+XOR
Fenn in [12]	*1AB*	m+1	$2m$+1	(m+1) AND m XOR 2(m+1) REG	AND +XOR($\log_2 m$)
Fig. 4	*2AB+C*	m+1	m+1	(m+1) AND (m+1) XOR 3(m+1) REG	AND+XOR
CA based on Choudhury in [14]	*AB*+C and A^2+C	m	m	$4m$ AND $4m$ XOR $5m$ REG	AND+XOR
Kim in [13]	*AB* and A^2	m+1	$2m$+1	(m+1) AND m XOR 3(m+1) REG	AND +XOR($\log_2 m$)
Fig. 5	*AB*+C *3*and A^2+*C*	m+1	m+1	$2m$ AND $2m$ XOR 5(m+1) REG	AND+XOR

5 Conclusion

This paper proposed a new parallel algorithm and architecture for two modular multiplications over GF(2^m). The algorithm uses the property of irreducible AOP as a modulus and computes two modular multiplications in parallel. There exists a common part to compute two multiplications at the same time. The algorithm computes the common parts only once, thereby reducing the required operations compared to related algorithms. The proposed architecture is based on cellular automata and had smaller area and time complexity than previous architectures. The proposed architectures can be used as a kernel circuit for exponentiation, inversion, and division architectures. Since it includes regularity, modularity, and concurrency, it is suitable for VLSI implementation.

Acknowledgement. This work was supported by the research fund of Kyungil University.

References

1. I. S. Reed and T. K. Truong, "The use of finite fields to compute convolutions," *IEEE Trans. on Information Theory*, IT-21, pp. 208–13, Mar. 1975.
2. B. Schneier, *Applied Cryptography – second edition*, John Wiley&Sons, Inc. 1996.
3. V. Neumann, *The theory of self-reproducing automata*, Univ. of Illinois Press, Urbana &London, 1966.
4. S. Wolfram, "Statistical mechanics of cellular automata," *Rev. of Modern Physics*, vol. 55, pp. 601–644, 1983.
5. W. Pries, A. Thanailakis, and H. C. Card, "Group properties of cellular automata and VLSI applications," *IEEE Trans. on Computers*, C-35, vol. 12, pp. 1013–1024, Dec. 1986.
6. E. R. Berlekamp, *Algebraic Coding Theory*, New York: McGraw-Hill, 1986.
7. C. S. Yeh. S. Reed, and T. K. Truong, "Systolic multipliers for finite fields $GF(2^m)$," *IEEE Trans. on Computers*, vol. C-33, pp.357–360, Apr. 1984.
8. S. K. Jain and L. Song, "Efficient Semisystolic Architectures for finite field Arithmetic," *IEEE Trans. on VLSI Systems*, vol. 6, no. 1, pp. 101–113, Mar. 1998.
9. J. L. Massey and J. K. Omura, *Computational method and apparatus for finite field arithmetic*, U. S. Patent application, submitted 1981.
10. S. W. Wei, "A systolic power-sum circuit for $GF(2^m)$," *IEEE Trans. on Computers*, vol. 43, pp. 226–229, Feb. 1994.
11. T. Itoh and S. Tsujii, "Structure of parallel multipliers for a class of finite fields $GF(2^m)$," *Info. Comp.*, vol. 83, pp. 21–40, 1989.
12. S. T. J. Fenn, M. G. Parker, M. Benaissa, and D. Taylor, "Bit-serial Multiplication in $GF(2^m)$ using irreducible all one polynomials," *IEE. Proc. Comput. Digit. Tech.*, vol. 144, no. 6, Nov. 1997.
13. H. S. Kim, *Bit-Serial AOP Arithmetic Architecture for Modular Exponentiation*, Ph. D. Thesis, Kyungpook National Univ., 2002.
14. P. Pal. Choudhury and R. Barua, "Cellular Automata Based VLSI Architecture for Computing Multiplication and Inverses in $GF(2^m)$," *IEEE 7th International Conference on VLSI Design*, Jan. 1994.
15. D. E. Knuth, *The Art of Computer Programming. Volume 2: Seminumerical Algorithms*, Addison-Wesley, Reading, Massachusetts, 2nd edition, 1998.

Specification and Verification of Security Policies in Firewalls

Rasool Jalili[1] and Mohsen Rezvani[1]

Department of Computer Engineering
Sharif University of Technology
Tehran, Iran
{jalili, mrezvani}@ce.sharif.edu

Abstract. Rules are used as a way of managing and configuring firewalls to fulfill security requirements in most cases. Managers have to specify their organizational security policies using low level and order-dependent rules. Furthermore, dependency of firewalls to the network topology, frequent changes in network topology(specially in dynamic networks), and lack of a method for analysis and verification of specified security policy may reduce to inconsistencies and security holes. Existence of a higher level environment for security policy specification can rectify part of the problems. In this paper we present a language for high level and formal specification of security policy in firewalls. Using the language, a security manager can configure its firewall based on his required security policy independent of the network topology. The language is used as a framework for analysis and verification of security policies. We designed and implemented a tool based on theorem proving for detecting inconsistencies, coverage, as well as applying a query on the specified policy. Results of analysis can be used to detect security vulnerabilities.

Keywords: Firewall, Formal Specification, Security Policy

1 Introduction

Although using firewalls is a useful method to satisfy security policy of an organization, precise specification of security requirements is vital for accuracy of firewall operations. This may occur due to the notations and language which is used to define the requirements; which is currently based on an ordered set of rules. Experiments show that complexity of working with rules, due to their low level of abstraction, results in miss-configuration of firewalls and potentially security holes for the organization.

This paper presents a security model with an inherent compatibility with overall design of most firewalls. Applying the security model does not force any change in the firewall design. A language is designed to express security requirements. Due to common applications of deontic logic[10] in security models, it

M.H. Shafazand and A M. Tjoa (Eds.): EurAsia-ICT 2002, LNCS 2510, pp. 154–163, 2002.

is used as a basis for syntax and semantics of the language. Security policy is specified using deontic formulas.

On the other hands, analysis and verification is one of the key management challenges to specify policy. We present an analysis tool based on theorem proving. The tool can be used to prove the consistency and coverage properties as well as applying queries on security policy. A theorem is generated for each property and is proved using our proof assistant.

Remainder of the paper is organized as follows. In section 2, basic concepts and related works in security policy specification are reviewed. In section 3, our security policy specification language is presented. Section 4 describes our tool for formal and automatic analysis of the specified policy. Finally conclusion and future works are presented.

2 Motivation and Related Work

Considering security issues is critical in planning an Internet connectivity. A usual method for protecting internal network of an organization is using a firewall[14]. A firewall provides the means for implementing and enforcing network access policies. In the other words, a firewall provides access control to users and services. The access control is specified as conditions named security policy.

Typical components of a firewall include:

- Stateful Packet Filter, which inspects packets based on their belonging to successfully initiated connections. Based on connections information, security policy is specified as a sequence of rules. Each rule may allow, deny, or redirect packets belonging to a connection matching the rule.
- Network Address Translation(NAT): Using NAT, an organization can hide its internal addresses and network topology, as well as overcome its limitation on IP addresses. This is specified in security policy using a set of specific rules (NAT rules).
- Application Level Gateways, which inspect connections based on their particular protocols.Rules are usually used to specify this aspect of security policy.
- Content Filter, which inspects transferring data regardless of their conveying protocol. Categorizing all possible contents is a way of expressing this aspect of specifying security.
- Log Registerer: Regardless of the action applies on a connection, logging can be used to monitor the previous activities of the organization and tune the firewall if needed. Logging can be specified as additional attribute in each filtering policy.

While firewalls are accepted as a tool for applying security policy in computer networks, they suffer from management and security problems[14]. We will focus on a different category of problems when firewalls are deployed in an organization.

Rule based nature of firewalls is an important problem which may produce inconsistency in applying a security policy[2]. A list of typical problems are:

- Adding an inappropriate rule, can influence all the rule base.
- Respecting the sequential application of the rule base, adding, removing or even interchanging of a rule can lead to inconsistency in security policy.
- Existence of different rule bases for several aspects of security policy in firewalls, makes maintaining all of rule bases consistent very complicated.
- Security policy is defined based on and dependent to the network topology.

2.1 Formal Methods

Rectifying the problem mentioned above, raises the need for an accurate method for specifying security policy. Formal methods are considered as a good alternative in this regard. Formal methods provide a mathematical framework for specifying, verifying, and developing systems. A formal approach provides the following components[1]:

- a model language to describe the system,
- a specification language to describe the correctness requirements, and
- the analysis technique to verify that the system meets its specification.

The model specifies the possible behavior of the system, and the specification describes the desired behavior of the system. Theorem proving and model checking are known methods for formal verification of computer systems[1, 5].

Advantages of using formal methods in specifying security policy are[10]:

- Accuracy and unambiguity of security policy definition.
- Possibility of verifying security policy properties such as consistency.
- Provision of verification on applying the security policy.

2.2 Related Work

Existing security policy specification methods can be broadly categorized into four classes:

- Temporal logic based specification: In this approach[11], security policy is specified formally based on the information flow. A security policy consists of a set of information classes and constraints on flow of information. The constraints are specified by a specific type of logic called branching time temporal logic[1, 8].
- Modal logic based specification: In this approach[4], a security policy is specified as an specific case of a regulation. The system to be regulated consists of agents which can execute actions on some objects. Each role is associated with a set of norms(permissions, obligations, and prohibitions). An agent can play one or more roles. In this approach, regulation is specified using a logic based on SDL(Standard Deontic Logic).
- LaSCO based specification: LaSCO(the Language for Security Constraints on Objects)[7] is a language for specifying policy as a directed graph. Semantics of the language represented by a first ordered logic.

- Applicable methods: An approach introduced in Bell-Labs[3, 9], specifies security policy in stateful packet filtering independent of network topology. The specification is based on the concept of role. Once the security policy is expressed, the packet filter rule base is automatically derived from the specification. The model is appropriate for packet filter firewalls, and does not represent any method for verification of security policy. Another approach for security policy specification is Guttman work on filtering postures[6]. In the method, a Lisp-like language is represented to specify policies, services, and network. Separation of the security policy from the network topology is not provided in this method.

3 Security Policy Specification Language

Main requirements in selecting a method for security policy specification in firewalls are:

- Separating the security policy from the network topology.
- A high level language for defining security policy considering all functionalities of all firewalls.
- Automatic generation of firewall rule bases using the specified security policy.
- A method for verification of security policy.

Based on the above requirements, we propose a method for security policy specification in firewalls. A language based on the deontic logic is the core of our method. The language supports separation of the security policy from the network topology as well as automatic generation of an existing firewall rules.

3.1 Syntax of the Language

The syntax of our proposed security policy specification language covers two parts:

- Security policy specification
- Network topology specification

At first, security policy is specified regardless of network topology and then network topology is specified. A security policy in firewalls defined as a hexad (R, Op, S, T, P, A) such that:

- R is a finite set of roles.
- Op is a set of relations on roles. For example, inheritance is a relation. If role x inherits from role y, then all hosts playing role x will do play role y.
- S is a finite set of services in network.
- T is a finite set of time periods.

- P is a finite set of primitive propositions. The propositions specify connections. On the other hand, each proposition introduces a set of connections between two roles. For example, a proposition can be $x \overset{st}{\rightarrow} y$; where x and y are roles, s is a service, and t is a time period. This proposition represents all connections from role x to role y with service s at time period t. Operator $\rightarrow$ specifies direction of connections. Other operators for this concept are $\leftarrow$ and $\leftrightarrow$.
- A is a sequence of security propositions; each proposition is a deontic logic formula. The propositions are main part of security policy specifications.

Masquerading is supported by a specific service, we call it *Masq*. A primitive proposition can be $x \overset{Masq}{\rightarrow} y$ which presents translation of role x addresses into the address of role y.

In order to support security policy in application level gateways, three more concepts are introduced. The language syntax is extended accordingly.

- Content security, which includes URL filtering, Content Filtering, and command filtering.
- Authentication.
- Logging.

Content security extends the scope of data inspection to higher levels of a service protocol. This is provided through an extra field *Resources*in the syntax. A resource is composed of three components:

- A set of commands.
- A set of URL categories.
- A set of content categories.

To satisfy the requirements of authentication and logging in application level gateways, another field *Actions* is added to the syntax. This field can be assigned to each security proposition and has the following components.

- Authentication type: This component represents the type of authentication and may have one of the values *UserAuth*, *SessionAuth*, *ClientAuth*, and *NoAuth*.
- User: This component specifies a user name or a group name whose connection is inspected.
- Log type: This component specifies type of logging. and may have one of the values *Full*, *Summary*, *WithContent*, and *NoLog*.

Network topology is specified as a set of tuples $< r, I >$ where:

- r is an element from R; defined in the security policy specification.
- I is a range of IP addresses. Those hosts with IP addresses in I, play role r.

3.2 Semantics of the Language

To simplify the language semantics, we assume the functionality of the firewall is restricted to a stateful packet filter, saving the state of all active connections. Such a firewall can simply be modelled as an state transition machine. Variables corresponding to the states are vectors saving information of active connections. Each connection can be represented by a state transition machine[13].

An active connection C can be modelled as a triple $(\Sigma_C, \tau_C, I_{0C})$, where:

- Σ_C is a finite set of states for a connection.
- τ_C is a transition relation. $\Sigma_C \to \Sigma_C$.
- I_{0C} is an initial state of a connection.

A firewall can be modelled as a composition of active connections. Accordingly, the firewall F is a state transition machine $(\Sigma_F, \tau_F, I_{0F})$ such that:

$F = \| \; C_i \qquad (1 \leq i \leq n)$
$\Sigma_F = \prod \Sigma_{C_i} \qquad (1 \leq i \leq n)$
$\tau_F = \bigcup \tau_{C_i} \qquad (1 \leq i \leq n)$
$I_{0F} = \bigcup I_{0C} \qquad (1 \leq i \leq n)$

As security policy in firewalls is specified at the lowest level based on connections, semantics of the language can use only the first level of the state transition machine. To represent the semantics, first we define the matching function ρ, followed by defining how to apply security policy via a single state. Finally applying security policy by a state transition machine is defined.

Assume that the set of truth-values is $\mathbf{B} = \{Accept, Reject, NoMatch\}$. Function π maps a primitive proposition to the set of its matching connections. The matching function ρ returns result of matching a connection with a security proposition. Matching a connection c with security proposition ϕ denoted as $\rho(c, \phi, \pi)$ and is defined as:

- if $\phi = \mathbf{P}X$ and $c \in \pi(X)$, then the result is *Accept*, otherwise is $NoMatch$.
- if $\phi = \mathbf{F}X$ and $c \in \pi(X)$, then the result is $Reject$, otherwise is $NoMatch$.

A state s of the state transition machine $(\Sigma_F, \tau_F, I_{0F})$ satisfies security policy SP, if and only if for each connection c in s, there exists a security proposition ϕ in SP such that $\rho(c, \phi, \pi) = Accept$, and there not exist any security proposition ϕ in SP such that $\rho(c, \phi, \pi) = Reject$. A state transition machine $(\Sigma_F, \tau_F, I_{0F})$ satisfies SP, if and only if all states in the machine satisfy the policy.

The semantics described so far can be extended to cover all features provided by the language including application gateways, time, and NAT. We used the semantics to show that specifications accepted by the language are applicable and consistent.

4 Analysis and Verification of Security Policy

Analyzing and verification of security policy in firewalls includes:

- proving consistency in security policy,

- proving coverage among security propositions, and
- applying a query on the policy.

Based on the syntax of our language, security propositions are translated into propositional logic formulas. Then a theorem is defined for each property in the formulas. Each theorem is also a propositional logic formula. For proving each property in policy, we prove validity of the appropriate theorem using a proof assistant.

4.1 Translation of Security Propositions

Consider the complication of automatic proof in deontic logic formulas [5], we translate security propositions into propositional logic formulas. The result of this translation is used for generating theorems.

Considering the security policy for a stateful packet filtering firewall, the key components in security propositions are:

- the role
- the service
- the time period
- the connection direction

The idea is representing numbers as a bit sequence. For example, a port number in *TCP* protocol is a number between 0 and 65535; just a sequence of bits. A number of boolean variables and expressions are introduced to represent the information in security propositions. Each field is assigned to a number between 0 and $n-1$. The number can be represented in $m = \log_2 n$ bits, and so we introduce m variables $v_0, \cdots, v_{m-1}$ to encode the field.

According to the syntax of the proposed language, There are two roles in the security proposition. Each role is assigned to a set of IP addresses. Each IP address includes 32 bits. We introduce 32 variables of the form $t_0 \cdots t_{31}$ for the first role (remind the first role) and 32 variables of the form $s_0 \cdots s_{31}$ for the second role (remind it). A range of addresses can be translated using the disjunction operator.

In many cases, the address range is defined using masks. The following algorithm is used for translating the range address $s[0 \cdots 31]$ with mask $m[0 \cdots 31]$.

```
1. Translate s into s[0]...s[31]
2. Translate m into m[0]...m[31]
3. result=TRUE
4. for i=0 to 31 do
     if m[i]=1 then result=result & s[i]
     if m[i]=0 and s[i]=1 then result=FALSE ; return
5. end of for
```

The service component contains the protocol type and the port number. Protocol type can be either *TCP*, *UDP*, *ICMP*, or *IGMP*. Two variables named

l_0 and l_1 are introduced for translating the protocol type. Port numbers can be specified using 16 boolean variables of the form $p_0 \cdots p_{15}$.

The time period is translated as disjunction of its valid periods. Each valid period can contain ranges of clock, day of week, day of month, month, and year. These parts are conjuncted for translating a valid period.

The direction component is translated into a single bit presented by the boolean variable c. The component has three values and is translated as follows:

- For value $\rightarrow$, this is translated to c.
- For value $\leftarrow$, this is translated to $\neg c$.
- For value $\leftrightarrow$, this is translated to $TRUE$.

Using the described method, the security proposition can be presented by a boolean expression. The expression is defined by conjunction of its components. General form in translation of a security proposition is:

$$P \; : \; C \Rightarrow A$$

where C is the condition part and A is the action part of a translated security proposition. The condition part is defined as conjuncted boolean expressions of translating roles, service, time period, and direction of the security proposition.

In the simplified case of a stateful packet filtering firewall, a bit named g in introduced for specifying the action part. For a security proposition in the form $\mathbf{P}X$, action part will be considered as g and for a security proposition in the form $\mathbf{F}X$, the action part will be considered as $\neg g$.

The following sub-section describes using of this method for defining the theorem and verifying properties in security policy.

4.2 Consistency in Security Policy

In an overall view, a security policy is consistent if and only if there exist no state in which the policy leads to a contradiction. Contradiction is occurred when a security proposition satisfies permission of a connection and another proposition satisfies forbidden of the same connection.

The trivial case of security policy inconsistency is when there are two security propositions with the same condition part and different action type.

Definition 1. *Let SP be a security policy; a sequence of security propositions. The security policy SP is consistent if and only if the following formula is satisfied:*

$$\forall P_1, P_2 \in SP \bullet \neg(A_1 \Leftrightarrow A_2) \Rightarrow \neg C_1 \vee \neg C_2$$

In order to prove consistency between each pair of security propositions P_1 and P_2 in security policy, exploiting their translation form, the expression $\neg(A_1 \Leftrightarrow A_2) \Rightarrow \neg C_1 \vee \neg C_2$ is generated. Then the expression is given to a proof assistant for validity. If the expression is valid then the pair of propositions have no contradiction and so are consistent. We say that a security policy is consistent if for each pair of its propositions, there is no between them.

4.3 Coverage among Security Propositions

Coverage between two propositions indicates redundancy in security policy. If the action parts of two security propositions are alike and all correct states in one of the propositions is satisfied in the other one, then the first proposition is covered by the second one.

Definition 2. *The proposition P_1 is covered by the proposition P_2 if and only if the following formula is satisfied.*

$$(A_1 \Leftrightarrow A_2) \wedge (C_1 \Rightarrow C_2)$$

The general case of coverage occurs when a proposition is covered by more than one propositions.

Definition 3. *The proposition P_n is covered by propositions $P_1 \cdots P_k$ if and only if the following formula is satisfied.*

$$(A_n \Leftrightarrow A_1) \wedge \cdots \wedge (A_n \Leftrightarrow A_k) \wedge (C_n \Rightarrow (C_1 \vee \cdots \vee C_k))$$

4.4 Querying on the Security Policy

The aim of querying from the security policy is to provide a tool for administrator to easily manifest his firewall policies. The tool interacts with the user through very high level queries. This aspect is used in [9] for analyzing a packet filter policy. A query specifies a set of connections.

It is usual that some connections specified by a query is rejected by policy and can't be established. Therefore, the answer of a query is a refined list of new queries. The semantics of the answer are that for each response query, the corresponding source can establish a connection with the service to the destination.

According to the syntax of queries in the form of a set of connections, we can translate queries into propositional logic formulas. Accordingly, the methods described in section 4.1 are used. Assume the propositional formula *query* is resulted from translating a query. We define that a query is matched with the security proposition $P : C \Rightarrow A$, if and only if the formula $query \Rightarrow C$ is valid. The result of a matched query is the action specified by A. A query is matched with a security policy if and only if it is matched with a security proposition in the policy.

In summary, we present how to generate theorem for verifying consistency, covering, and applying the query on policy. We translate security propositions to propositional logic formulas. Thus each theorem will be a propositional logic formula. We design a proof assistant for proving these theorems. The proof assistant can prove validity and satisfiability of a propositional logic formula.

A successful idea for proving propositional formulas that come from semantics of the logic, is that of binary decision diagrams, or BDDs. We might say that they are a recent invention, as the originator of BDDs as we know them today was Randall E. Bryant in 1986[5]. In our proof assistant, we use the BDD idea that is described in [12] by detailed.

5 Conclusion and Future Works

A language for high level and formal specification of security policy is presented in this paper. Once the security policy and network topology are described, it is given to the language compiler and in addition to syntax and semantics checking, rule bases of an existing firewall is automatically generated. Formalism of the language provides a framework for analyzing and verification of the security policy. A tool is presented for verifying consistency and covering in the security policy. The framework is used for applying a query on the policy. Verification of safety problem in security policy of firewalls is under investigation.

References

1. Alur R. and Henzinger T.A. *Computer-Aided Verification: An Introduction to Model Building and Model Checking for Concurrent Systems.* Draft, 1999.
2. Anderson J.P., Brand S., Gong L., and Haigh T. Firewall: An expert roundtable. *IEEE Software*, 14(5):60–66, September/October 1997.
3. Bartal Y., Mayer A., Nissim K., and Wool A. Firmato: A novel firewall management toolkit. In *Proceedings of the IEEE Symposium on Security and Privacy*, pages 17–31, May 1999.
4. Cholvy L. and Cuppens F. Analyzing consistency of security policies. In *Proceedings of the IEEE Symposium on Security and Privacy*, pages 103–112, May 1997.
5. Goubault-Larrecq J. and Mackie I. *Proof Theory and Automated Deduction*, volume 6 of *Applied Logic Series.* Kluwer Academic Publishers, Dordrecht, 1997.
6. Guttman J.D. Filtering postures: Local enforcement for global policies. In *Proceedings of the IEEE Symposium on Security and Privacy*, pages 120–129, Los Alamitos, 1997.
7. Hoagland J. *Specifying and Implementing Security Policy using LaSCO, the Language for Security Constraints on Objects.* PhD thesis, The University of California Davis, Department of Computer Science, March 2000.
8. Manna Z. and Pnueli A. *Temporal Verification of Reactive Systems: Safety.* Springer-Verlang, New York, 1995.
9. Mayer A., Wool A., and Ziskin E. Fang: A firewall analysis engine. In *Proceedings of the IEEE Symposium on Security and Privacy*, pages 177–190, May 2000.
10. Ortalo R. Using deontic logic for security policy specification. Tech. Report 96380, LAAS-CNRS Toulouse, France, October 1996.
11. Peri R.V. *Specification and Verification of Security Policies.* PhD thesis, The University of Virginia, School of Engineering and Applied Science, January 1996.
12. Rezvani M. *High Level Security Policy Specification in Firewalls.* Master's thesis, Dept. of Computer Engineering, Sharif University of Technology, September 2001, In Persian.
13. Stevens W.R. *TCP/IP Illustrated, Volume 1: The Protocols.* Addison-Wesley Professional Computing Series. Addison-Wesley, Reading, MA, USA, 1994.
14. Wack J.P. and Carnahan L.J. Keeping your site comfortably secure: An introduction to Internet firewalls. NIST special publication Computer security 800-10, U.S. Dept. of Commerce, Technology Administration, National Institute of Standards and Technology, U.S. G.P.O., Gaithersburg, MD, USA, 1994.

Image Segmentation Based on Shape Space Modeling

Daehee Kim and Yo-Sung Ho

Kwangju Institute of Science and Technology
1 Oryong-dong Puk-gu, Kwangju, 500-712, Korea
{kimdh, hoyo}@kjist.ac.kr

Abstract. In this paper, we propose a new image segmentation method based on the active contour. If we define a shape space as a set of all possible variations from the initial curve and we assume that the shape space is linear, it can be decomposed into the column space and the left null space of the shape matrix. In the proposed method, the shape space vector in the column space describes changes from the initial curve to the imaginary feature curve, and a dynamic graph search algorithm describes the detailed shape of the object in the left null space. Since we employ the shape matrix and the SUSAN operator to outline object boundaries, the proposed algorithm can ignore unwanted feature points generated by low-level image processing operations and is therefore applicable to images of the complex background. We can also compensate for limitations of the shape matrix with the dynamic graph search algorithm.

1 Introduction

The MPEG-4 visual coding standard [1] enables content-based functionalities by introducing the concept of the video object plane (VOP). A VOP is defined as a coding unit in the MPEG-4 natural visual coding, determined by the shape of the predefined video object, at a certain frame over the entire sequence. However, since it is not easy to define a mathematical model or a similarity measure for extracting video objects adequately, automatic segmentation algorithms cannot provide satisfactory segmentation results over various image sequences. If the user can define VOPs in the first frame in a user-assisted manner, we may obtain better segmentation results in the subsequent picture frames. Therefore, the user-assisted segmentation approach is more practical in generating VOPs of moving objects. In this paper, we propose a new image segmentation method based on an active contour algorithm to define objects in a given image.

In the active contour or snake algorithm, we try to find an energy-minimizing curve from the initial curve indicated by the user. Performance of the active contour algorithm depends mainly on the definition of the energy function and the shape of the active contour is controlled by internal, external and constraint forces. While the internal force is the smoothness constraint on the curve, the external force guides the active contour towards image features. The constraint force allows interactivity in manipulating the active contour. The energy functional $E^*(s)$ is represented as a parametric curve $\mathbf{r}(s)=(x(s), y(s))$, where s is the parameter for the given interval [2].

A functional of the active contour can be defined by

M.H. Shafazand and A M. Tjoa (Eds.): EurAsia-ICT 2002, LNCS 2510, pp. 164–171, 2002.

$$E^*(s) = \int_0^1 E_{snake}(\mathbf{r}(s))ds \tag{1}$$
$$= \int_0^1 \left[E_{int}(\mathbf{r}(s)) + E_{ext}(\mathbf{r}(s)) + E_{con}(\mathbf{r}(s))\right]ds$$

where E_{int}, E_{ext} and E_{con} represent the internal energy of the contour, the external image force and the constraint force, respectively. The final location of the active contour corresponds to the local minimum of the energy functional.

However, the minimization operation of $E^*(s)$ by variational calculus may cause a problem of numerical instability [3]. Since dynamic programming techniques [3] can provide more stable results, most active contour algorithms adopt dynamic programming. Unfortunately, dynamic programming techniques are extremely slow because of computational complexity of $O(nm^3)$, where n is the number of points to be processed on the active contour and m is the number of points in the search window. Greedy algorithms [4] are sensitive to the distance between adjacent points on the initial contour. Those conventional algorithms are designed for objects in simple and homogeneous backgrounds.

In this paper, we define a shape space as a set of all possible variations from the initial curve. This shape space is divided into two subspaces by the shape matrix, which is used to represent variations of the initial curve with a few parameters. Therefore, the shape space consists of a subspace defined by the shape matrix and the other space cannot be represented by a few parameters. In this paper, we first describe changes from the initial contour with a few parameters and employ a graph search algorithm that compensates for limitations of the shape matrix.

2 Shape Space Model

2.1 B-Spline Curve

The parametric curve $\mathbf{r}(s)=(x(s), y(s))$ is a particular function of the parameter s. A B-spline function $x(s)$ can be constructed as a weighted sum of N_B basis functions $B_n(s)$, where $n=0,\ldots,N_B$-1. If we set the order $d=3$, the curve has a continuous gradient. Therefore, the constructed spline function satisfies the internal energy requirement in its nature.

The spline function can be represented by

$$x(s) = \mathbf{B}(s)^T \mathbf{Q}_x \tag{2}$$

$$\mathbf{B}(s) = [B_0(s), B_1(s), \ldots, B_{N_B-1}(s)]^T \; \mathbf{Q}_x = [x_0, x_1, \ldots, x_{N_B-1}]^T$$

where x_n is the weight for a basis function $B_n(s)$. Therefore, the parametric curve $\mathbf{r}(s)$ is also represented in the matrix form.

$$\mathbf{r}(s) = \mathbf{U}(s)\mathbf{Q} \tag{3}$$

where

$$\mathbf{U}(s) = \mathbf{I}_2 \otimes \mathbf{B}(s)^T = \begin{pmatrix} \mathbf{B}(s)^T & 0 \\ 0 & \mathbf{B}(s)^T \end{pmatrix}, \; \mathbf{Q} = \begin{pmatrix} \mathbf{Q}_x \\ \mathbf{Q}_y \end{pmatrix} \tag{4}$$

In Eq. (4), **I** is a 2 × 2 identity matrix and the operation of ⊗ is the Kronecker product.

2.2 Shape Matrix

A shape space is defined by a set of all possible variations from the initial curve. If we assume that the universal shape space is a linear space, it can be composed of the column space $\mathcal{R}(\mathbf{W})$ and the left null space $\mathcal{N}(\mathbf{W}^T)$ of the shape matrix which will be defined in this section. The left null space is orthogonal to the column space [6].

The column space $\mathcal{R}(\mathbf{W})$ is constructed from a set of vectors of dimension N_X. It is desirable to restrict the displacement of control points to a lower dimensional shape space if it preserves the frame of the shape. An unconstrained control vector **Q** may lead to unstable active contours [7]. A change of the curve in $\mathcal{R}(\mathbf{W})$ is a linear mapping of the shape space vector **X** into a control vector **Q**

$$\mathbf{Q} - \mathbf{Q}_0 = \mathbf{W}\mathbf{X} \tag{5}$$

where **W** is the shape matrix of size $N_Q \times N_X$, and $\mathbf{Q}_0$ is a control vector of the initial curve. The shape space vector **X** describes the change of the initial curve.

In this paper, we describe contour changes by the 6-parameter affine model (N_X= 6). This class can be represented by the initial curve $\mathbf{Q}_0$ and the shape matrix [7]:

$$\mathbf{W} = \begin{pmatrix} \mathbf{1} & \mathbf{0} & \mathbf{Q}_{x0} & \mathbf{0} & \mathbf{0} & \mathbf{Q}_{y0} \\ \mathbf{0} & \mathbf{1} & \mathbf{0} & \mathbf{Q}_{y0} & \mathbf{Q}_{x0} & \mathbf{0} \end{pmatrix} \tag{6}$$

Each column of **W** forms a basis vector of the column space $\mathcal{R}(\mathbf{W})$ of the shape matrix **W**, but it is not necessary to be orthogonal to the other vectors.

2.3 Projection onto Column Space of Shape Matrix

In order to find the boundaries of video objects, we need to define a distortion measure that can be expressed by the curve norm of shape differences between the estimated curve $\mathbf{r}(s)$ and the desired curve $\mathbf{r}_d(s)$.

$$\left\|\mathbf{r}(f(s)) - \mathbf{r}_d(s)\right\|^2 = \frac{1}{L}\int_0^L \left|\mathbf{r}(f(s)) - \mathbf{r}_d(s)\right|^2 ds \tag{7}$$

where L is the length of the interval of s and $f(s)$ is the adjustment function which is necessary to match the corresponding points of the two curves. If two curves are very similar to each other, $\|\mathbf{r}(f(s)) - \mathbf{r}_d(s)\|$ can be replaced by $[\mathbf{r}(s) - \mathbf{r}_d(s)]\cdot\mathbf{n}(s)$ [7]. If the integral of Eq. (7) is approximated by a summation and the normal vector $\mathbf{n}(s)$ is used to measure the shape difference, our minimization criterion becomes

$$\| \mathbf{r} - \mathbf{r}_d \|^2 \approx \frac{1}{N} \sum_{i=1}^{N} \left[\left(\mathbf{r}_d(s_i) - \mathbf{r}(s_i) \right) \cdot \mathbf{n}(s_i) \right]^2 \tag{8}$$

where N is the number of regularly-spaced points in the interval of s. Eq. (8) can be rewritten in terms of the initial estimated curve $\mathbf{r}_0$ and the shape space vector $\mathbf{X}$.

$$\| \mathbf{r} - \mathbf{r}_d \|^2 \approx \frac{1}{N} \sum_{i=1}^{N} [(\mathbf{r}_d(s_i) - \mathbf{r}_0(s_i)) \cdot \mathbf{n}(s_i) - \mathbf{n}^T \mathbf{U}(s_i) \mathbf{W}(\mathbf{X} - \mathbf{X}_0)]^2 \tag{9}$$

In order to minimize $\|\mathbf{r}(s) - \mathbf{r}_d(s)\|^2$ in Eq. (9) and to find the estimated shape space vector $\mathbf{X}^*$, we employ the least square approach in this paper. The minimal $\mathbf{X}^*$ can be estimated by setting $\partial\|\mathbf{r}\text{-}\mathbf{r}_d\|^2/\partial\mathbf{X} = 0$, which leads to

$$\mathbf{X}^* = \left(\sum_{i=1}^{N} \rho_i \mathbf{W}^T \mathbf{U}^T \mathbf{n} \mathbf{n}^T \mathbf{U} \mathbf{W} \right)^{-1} \left(\sum_{j=1}^{N} \rho_i \mathbf{W} \mathbf{U}^T \mathbf{n} (\mathbf{r}_d - \mathbf{r}_0)^T \mathbf{n} \right) \tag{10}$$

where ρ_i is a weighting factor instead of $1/N$ to compensate for non-uniform sampling. Once we find $\mathbf{X}^*$, we can obtain the optimal control vector $\mathbf{Q}$ by Eq. (5). For more accurate results, we can repeat the same procedure several times by setting the previously fitted curve $\mathbf{r}(s)$ as the initial curve $\mathbf{r}_0(s)$.

Mathematically, $\mathbf{X}^*$, which describes the change from the initial curve, is the projection onto $\mathcal{R}(\mathbf{W})$. The column space of $\mathbf{W}$ is too restrictive for an active contour that tries to model arbitrary shape deformations. In order to address this problem, we consider the left null space $\mathcal{N}(\mathbf{W}^T)$ in Section 3.

2. 4 Estimate of Desired Curve

Since the aim of the active contour algorithm is to fit the initial curve to image features, $\mathbf{r}_d(s)$ of Eq. (7) should be estimated from the given image. In this paper, we employ the smallest univalue segment assimilating nucleus (SUSAN) operator [8] to extract image features. Although the SUSAN edge detector requires lower computational complexity than morphological gradient tools or the Canny edge detector, it works well for images of complex background.

The SUSAN edge detector consists of three steps. In the first step, we place the nucleus of a circular mask around a pixel. In the second step, we calculate the number of pixels that have similar brightness to the nucleus within the circular mask. This number is defined as the area of the univalue segment assimilating nucleus (USAN). In the third step, we subtract the size of USAN from the geometric threshold to produce an image of edge strength.

After the SUSAN edge detection, we utilize the edge image to extract the imaginary feature curve $\mathbf{r}_d(s)$. Since we try to fit the initial curve to image features by the active contour algorithm, we need to find feature points in the search region which includes the neighboring area of the initial curve. Features should be detected by scanning along each of the sampled normal lines to the initial curve, because of the approximation in Eq. (8). Therefore, if normal lines are constructed at points along $\mathbf{r}(s)$, we obtain a sequence of $\mathbf{r}_d(s_i)$ along the imaginary feature curve $\mathbf{r}_d(s)$.

3 Left Null Space of Shape Matrix

In Section 2, we have projected the change of the initial contour onto $\mathcal{R}(\mathbf{W})$. Here, we can ignore effects of unwanted feature points generated by low-level image processing operators and describe the overall change of the contour. However, it is not easy to describe detailed changes of the contour because the column space of the shape matrix consists of only six basis vectors.

In this section, we incorporate the change of the contour in $\mathcal{N}(\mathbf{W}^T)$ to compensate for the fitted contour and obtain the final object boundaries. However, we do not know the actual dimension of the shape space because we cannot describe the object in a finite-dimensional space due to subjectivity of the object definition. Thus, we do not employ general linear algebra approaches to solve problems of the left null space. We consider that feature points on the fitted contour obtained in Section 2 are represented well in $\mathcal{R}(\mathbf{W})$ and other feature points on the fitted contour are distorted by the projection operation onto $\mathcal{R}(\mathbf{W})$ due to the absence of the information about $\mathcal{N}(\mathbf{W}^T)$. Here, we define a seed point sequence as a sequence of points that are represented well in $\mathcal{R}(\mathbf{W})$ without the information about $\mathcal{N}(\mathbf{W}^T)$. In order to estimate the final object boundaries, we modify the distorted points of the fitted contour using the seed point sequence.

Boundary definition can be formulated as a graph search problem where the key idea is to form an image by a weighted graph where pixels represent nodes with weighted directed edges connecting each pixel and its eight adjacent neighboring pixels [9]. We set the first point and the second point of the seed point sequence to be the start point and the end point, respectively. Once we find the optimal path, we replace the start point with the end point and set the third point to a new end point. This procedure is performed repeatedly until we arrive the last point of the seed point sequence. The optimal path is defined as the minimum cumulative cost path from the start point to the goal point.

Since the minimum cost path should correspond to boundaries of the object, pixels of strong edge features should have low local costs and vice versa. In order to reflect various edge features to local costs, we define the local cost in terms of the static and dynamic costs and the total cost as the weighted sum of the static and dynamic costs.

3.1 Static Cost

First, we define the static local cost function as the weighted sum of cost functions derived from Laplacian zero-crossings and gradient magnitudes. We use multiple kernel widths because multiple kernels provide good performance psychologically [10]. Each kernel is generated from the 2-D Gaussian distribution with a different standard deviation. We define cost functions so that strong feature points produce low costs and vice versa.

3.2 Dynamic Cost

Some parts of the object boundary may have weak gradient magnitudes relative to nearby strong gradient edges. If the nearby strong edge has a relatively lower cost, the optimal path moves to the strong edge rather than the desired edge. Thus, the gradient magnitude cost function should be modified dynamically to resolve this problem.

We can define the dynamic cost function from the segmentation results along the path from the previous start point to the previous end point of the seed point sequence. In order to adapt gradual changes in edge characteristics, we can update the dynamic cost function as the interval of seed points is changed. The dynamic cost function can be obtained from the histogram of costs in the previous optimal path.

4 Experimental Results

In order to evaluate performance of the proposed algorithm, we run computer simulations on MPEG-4 test images of the CIF format. Figure 1 demonstrates the effects of the dynamic cost functions.

(a) Static Cost

(b) Dynamic Cost

Fig. 1. Effect of Dynamic Costs

As we mentioned earlier, when a section of the desired object boundary has weak gradient magnitudes relative to strong gradient magnitudes in the neighborhood of the object boundary, the section of the optimal path is set on the strong magnitudes, as shown in Figure 1(a). The desired boundary is the mother's cheek. However, since its part is so close to her high contrast lip, the optimal path is set on her lip. In order to avoid this problem, we consider the dynamic cost defined in Section 3.2. Figure 1(b) demonstrates how a segment of the optimal path latches onto the edge that is similar to the previous segment. In Figure 1(b), the white line of her shoulder is obtained only by static costs, but the white line of her cheek is obtained by dynamic costs derived from her shoulder line.

Figure 2 shows simulation results of the proposed algorithm applied to AKIYO image. Figure 2(a) shows the initial contour provided by a user-pointing device. Image features obtained by the SUSAN operator are displayed in Figure 2(b). The search region displayed in Figure 2(c) is formed by sweeping normal vectors along the initial curve. In Figure 2(c), points on white lines are candidates of $\mathbf{r}_d(s_i)$ corresponding to $\mathbf{r}(s_i)$. Figure 2(d) is the fitted curve by the repeated least square method. Figure 2(e) is the final segmentation result that is compensated with the graph search algorithm. In Figure 2(d), the fitted curve is misaligned with AKIYO's ears and parts of her shoulder connected to her arm due to the smoothness property of the 6-parameter approximation and the B-spline curve. The final curve in Figure 2(e) tracks the object boundary in more detail.

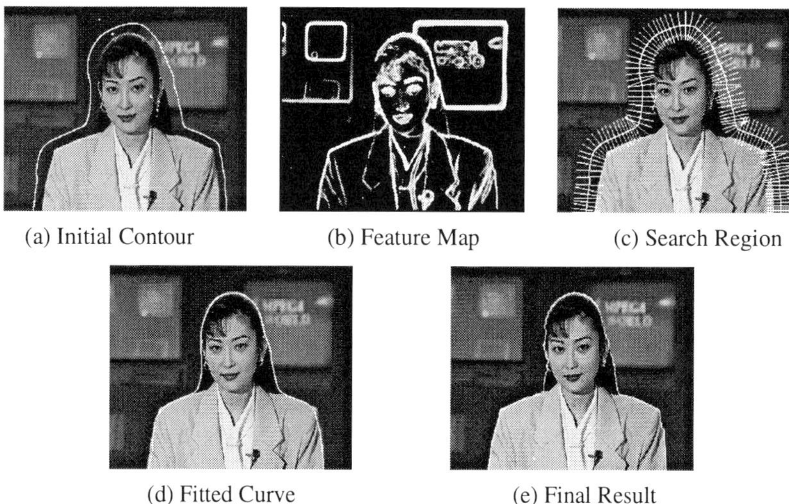

(a) Initial Contour (b) Feature Map (c) Search Region

(d) Fitted Curve (e) Final Result

Fig. 2. Results for AKIYO Image

Figure 3 shows segmentation results for MOTHER AND DAUGHTER image that has more complex shapes than AKIYO image. Figure 3(e) demonstrates a more accurate shape of the object boundary than Figure 3(d), particularly in the area of the mother and daughter's hair. In general, since conventional active contour algorithms are mainly designed for images with homogeneous backgrounds, they may not work well for objects having complex backgrounds.

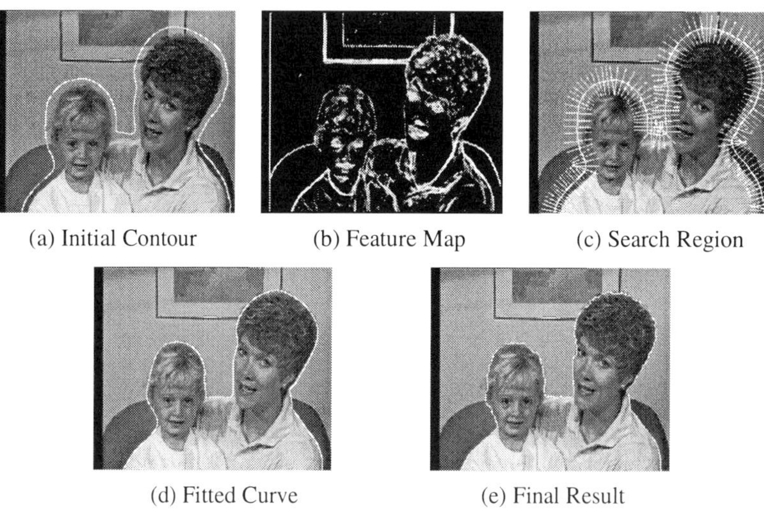

(a) Initial Contour (b) Feature Map (c) Search Region

(d) Fitted Curve (e) Final Result

Fig. 3. Results for MOTHER AND DAUGHTER Image

5 Conclusions

In this paper, we have proposed a new user-assisted active contour algorithm to extract objects from a given image. Since we employ the shape matrix and the SUSAN operator to outline object boundaries, the proposed algorithm can ignore some outliers or unwanted feature points generated by low-level image processing operations and is applicable to images with complex backgrounds. These outliers make it difficult for conventional active contour algorithms to find object boundaries in non-homogeneous backgrounds. We also use a dynamic graph search algorithm to overcome limitations of the 6-parameter affine model by describing changes of the curve in the left null space of the shape matrix. The dynamic graph search algorithm can describe the detailed shape of the object, which cannot be represented by the shape matrix.

Acknowledgement.
This work was supported in part by the Korea Science and Engineering Foundation (KOSEF) through the Ultra-Fast Fiber-Optic Networks (UFON) Research Center at Kwangju Institute of Science and Technology (K-JIST), and in part by the Ministry of Education (MOE) through the Brain Korea 21 (BK21) project.

References

1. ISO/IEC FDIS 14496-2: Information Technology - Generic Coding of Audio-Visual Objects, Part 2: Visual. ISO/IEC JTC1/SC29/WG11 (1998)
2. Kass, M., Witkin, A., Terzopoulos, D.: Snakes: Active Contour models. First International Conference on Computer Vision (1987) 259-269
3. Amimi, A., Weymouth, T., Jain, R.C.: Using Dynamic Programming for Solving Variational Problems in Vision. IEEE. Trans. Patt. Anal. Mach. Intel., Vol. 12, No. 9 (1990) 855-867
4. Williams, D. J., Shah, M.: A Fast Algorithm for Active Contours and Curvature Estimation. CVGIP:Image Understanding, Vol. 55, No. 1 (1992) 14-26
5. Foley, J. D., Dam, A., Feiner, S. K., Hughes, J. F., Phillips, R. L.: Introduction to Computer Graphics. Addison-Wesley, New York (1995)
6. Strang, G.: Linear Algebra and Its Applications. 3rd edn. Harcourt Brace Jovanovich (1988)
7. Blake, A., Isard, M.: Active Contours. Springer-Verlag, Berlin Heidelberg New York (1998)
8. Smith, S.M, Brady, J.M.: SUSAN – A New Approach to Low Level Image Processing. Int. Journal of Computer Vision, 23(1) (1997) 45-78
9. Mortensen, E.N., Barrett, W.A.: Interactive Segmentation with Intelligent Scissors. Graphical Models and Image Processing (1998) 349-384
10. Marr, D., Hildreth, E,: Theory of Edge Detection. Proc. R. Soc. Lond. B 270 (1980) 187-217
11. Kim, D., Ho, Y.S.: A User-Assisted Segmentation Algorithm Using B-Spline Curves. SPIE Visual Communications and Image Processing (2001) 734-744

HERMES: File System Support for Multimedia Streaming in Information Home Appliance*

Youjip Won[1], Jinyoun Park[1], and Sangback Ma[2]

[1]Div. of Electrical and Computer Engineering, Hanyang University, Korea
{yjwon|jypark}@ece.hanyang.ac.kr
[2]Department of Computer Science, Hanyang University, Korea
sangback@cse.hanyang.ac.kr

Abstract. The HERMES file system is state-of-the-art file system designed to handle multimedia streaming workload in consumer electronics platform. The design objective of HERMES is to minimize the *delay* and the *delay variance* of I/O request in the sequential workload. File organization, meta data structure, unit of storage, or etc. are elaborately tailored to achieve this objective. Further, HERMES provides a number of API's specific to mpeg-4 file format. It can greatly facilitate the development of multimedia applications. For the seamless integration with the existing application, HERMES file system is developed under *virtual file system(VFS)* layer. Prototype of HERMES file system is implemented on Linux operating system. Our benchmark test shows that HERMES file system exhibits suprior performance than EXT2 file system.

Keywords: Multimedia, Streaming, File System, UFS, Scheduling

1 Introduction

1.1 Motivation

Information Appliance for digital video can be thought as a lightweight computer system designed to store incoming high quality digital video stream at the local hard disk and/or to play the recorded video clips at user's convenience. Unlike the general-purpose computer, which has abundant computing resources and storage capacity, this type of consumer electronics has stringent resource constraints due to its restriction on power consumption, pricing, acoustic, reliability, etc. The disk drive in this device is not an exception. It is not feasible to use high performance disk in this type of device. The realtime playback and the retrieval of multimedia data puts intense bandwidth demand on the storage device. It is mandatory that the underlying file system is elaborately designed to fully utilize the physical performance of the disk by exploiting the characteristics of multimedia workload. Unfortunately, the fundamental design philosophy of the

* This work was supported by Korea Research Foundation Grant(KRF-2000-003-E00322.

M.H. Shafazand and A M. Tjoa (Eds.): EurAsia-ICT 2002, LNCS 2510, pp. 172–179, 2002.

most commodity file systems, e.g. Unix File System, NTFS, EXT2, FAT32, or etc. is ill-suited for meeting the real-time performance requirement of the audio and video data retrieval. One of the reasons is that navigating through multi-level tree structured file entails non-trivial amount of the disk head movement overhead in visiting internal nodes of the tree.

Efficiency of the underlying file system plays a critical role in providing the streaming service in cost effective manner. To effectively exploit the physical bandwidth of the disk, it is important that the file system layout, meta data structure, file organization, file placement, etc. are elaborately tailored so that disk fragmentation is avoided and the time to locate the data blocks is minimized. Special care needs to be taken in designing the file system to meet the requirement of the underlying workload. There is not much debate that the Unix file system is a landmark achievement in modern file system design. However, regarding the multimedia data retrieval, there are two major issues in Unix file system which requires further elaboration: (i) file structure and (ii) data abstraction. Unix file system abstracts the file as a sequence of byte stream. Mpeg-4 compressed file consists of a set of atoms and each atom may contain video data, audio data, text data, file meta data, or etc. Since HERMES has well defined target workload, it is possible to provide more specific sets of API's for the target application. We provide a set of file system level API's which can extract or record mpeg-4 specific information in HERMES In this work, we present the novel file system which effectively address the above mentioned issues: (i) file system layout, (ii) file organization, and (iii) data abstraction.

1.2 Related Works

Since legacy SCAN, FIFO, and their bifurcations do not provide bandwidth guarantee, it is not possible to provide continuous flow of data blocks from the disk to the end system. A number of works address these issues and propose the disk scheduling algorithms for multimedia data retrieval[1, 3, 7, 2, 5].

There are a number of prototype file systems which are designed to handle multimedia data[8, 12]. MMFS[10] improves interactive playback performance by supporting intelligent pre-fetching, state-based caching, prioritized real-time disk scheduling, and synchronized multi-stream retrieval. Minorca Multimedia file system[11] proposed (i) a new disk layout and data allocation techniques called MOSA which offers a high degree of contiguous allocation for large continuous media files and allows the coexistence of small, non-CM files, and (ii) a new read ahead method to optimize the input of the I/O request queue. These techniques aim at increasing disk access locality and at reducing disk seek overhead. Presto File System[4] introduces the idea of storing the data based on the semantic unit. The unit of placement is extent which consists of fixed number of semantic units. The size of file is limited to one extent. SMART file system[6] maintains a file as a linked list of extents and thus improves the file size limitation in Presto[6]. Symphony[9] also allows each video file to be accessed either as a sequence of bytes or as a sequence of frames. To support two different abstractions in accessing the file, they use two level index structure: index for frame which maps the frame index to byte offset and index for byte which maps the byte offsets to disk block addresses. In Minorca file system and Symphony file system, file is organized using index block and has tree like structure. Particularly, Minorca file system clusters the index block and data block together, which may look like B tree.

2 Synopsis: Unix File System

In Unix file system, the management information is kept strictly apart from the data and is collected in a separate structure for each file. This structure is called "i-node" and stores the metadata information of each file as well as data block references for actual data location. Data references consist of twelve direct references, one indirect reference, two-step indirect reference and three-step indirect reference. Given that multimedia file can easily go beyond a tens of mega bytes, data block retrieval in multimedia streaming operation entails the retrieval of the intermediate pointer blocks as well. While tree structure based file organization gives greater flexibility in handling wide variety of file sizes, retrieval of pointer blocks gives substantial overhead in the streaming operation.

Since disk needs to access the i-node block and possibly a number of indirect blocks to access the data block, non trivial amount of additional overhead occurs. Even though the i-node and pointer blocks are in the buffer cache, memory access time can consume significant fraction of CPU cycle. It is very unlikely that pointer blocks and data blocks are stored consecutively, especially, when a number of files co-exist in the file system. Disk head needs to travel across the platter to retrieve i-nodes and pointer blocks.

Most file systems of modern Unix family operating systems, e.g. Linux, Solaris, NetBSD, etc. adopt the mechanism to place the data blocks consecutively or as closely as possible. EXT2 file system, which is the most widely used file system in Linux operating system, uses the concept of block group. Using the concept of block group, file system can cluster the data blocks for a file within relatively closer cylindrical position. Unfortunately, block group based placement policy still splits the file into different block groups when the size of files exceeds certain limit and may suffer from significant overhead in disk seek. EXT2 file system consists of multiple block groups. Each group contains the copy of file system superblock(for file system consistency's sake), group descriptor, block bitmap, i-node bitmap, i-node table, and finally data blocks, with the respective order.

3 HERMES: Multimedia File System

UFS (Unix File System) manages both directory file and multimedia file equally. It is possible that multimedia files and the directory files are placed in the disk in interleaved fashion. This can cause substantial overhead in sequential scanning operation on multimedia file. To resolve this issue, HERMES file system maintains the directory block and data block seperately. Fig. 1 illustrates the file system layout of HERMES file system. Its partitions consist of super block, extent bitmap, i-node bitmap, i-node tables, directory extents and multimedia extents.

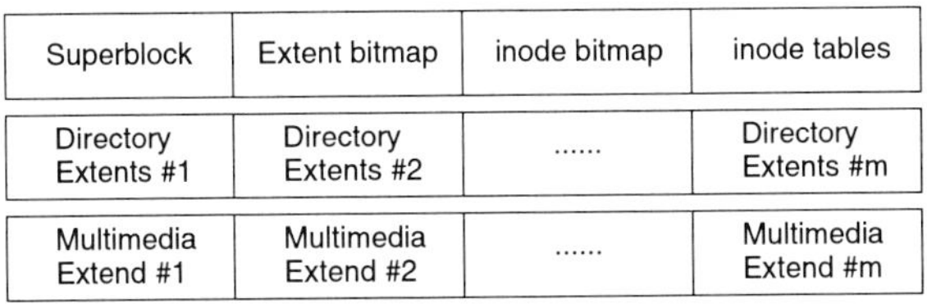

Superblock	Extent bitmap	inode bitmap	inode tables
Directory Extents #1	Directory Extents #2	......	Directory Extents #m
Multimedia Extend #1	Multimedia Extend #2	......	Multimedia Extend #m

Fig. 1. Layouts of HERMES file system

Superblock is located at the first block of the file system partition and stores general information of the file system. It contains the information about the number of extent, the number of multimedia extent, the number of free extent, size of extent, the number of i-node, creation time, and etc. Extent is the smallest allocation unit and consists of consecutive data blocks. Extent size is determined when formatting file system and cannot be altered unless the file system is reformatted. The directory entry is stored in directory extent and the multimedia file is stored in multimedia extent. By separating directory entry from multimedia data region, the HERMES file system can reduce the disk seek time and also store and retrieve multimedia data very efficiently. Extent bitmap is used to denote whether the respective extent is in use or not. Free extent is allocated using first fit algorithm. The i-node table consists of a predefined number of i-nodes. All i-nodes have the same size, 128 bytes. Each i-node maintains the file meta-data information: owner ID, group ID, file mode and references of allocated extent, etc. With this design approach, we like to achieve the followings: (i) efficient handling of multimedia playback workload, (ii) minimizing I/O latency, and (iii) supporting wide range of file size.

Fig. 2 illustrates the i-node structure of HERMES file system. We improve the i-node design proposed in [11]. The block reference pointer in UFS i-node points to single block, which can be either data block or pointer block. UFS adopts skewed tree-like file structure to cover large size file. In HERMES, we like to avoid using multi-level indirect reference in locating the data block. In HERMES, each block reference pointer can point to cluster of data blocks. Block reference pointer is augmented with `i_count` which denotes the number of consecutive data blocks as in Fig. 2.

4 Operating System Support for Streaming Operation

4.1 VFS and HERMES

HERMES file system is implemented under VFS layer so that the existing application can use HERMES file system partition without any modification. Fig. 3

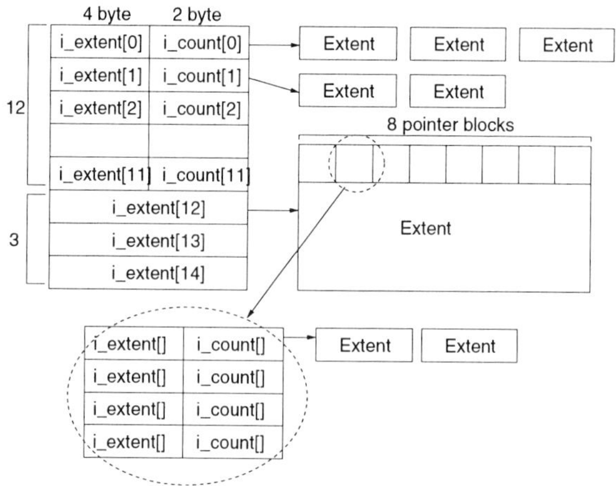

Fig. 2. i-node structure of HERMES file system

illusrates the relationship between operation system kernel, VFS and HERMES. As can be seen in Fig. 3, user can manipulate the files in HERMES in vari-

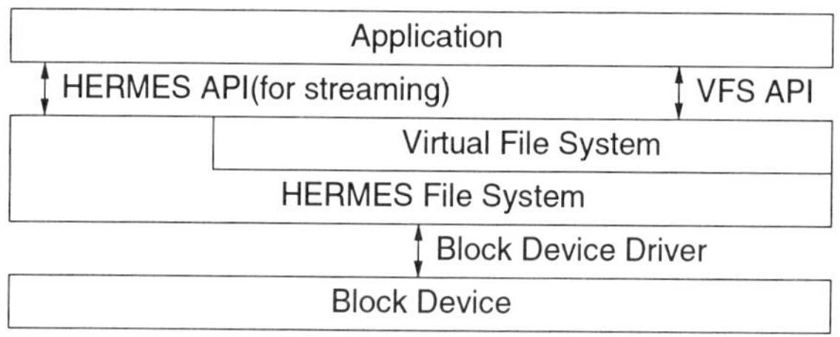

Fig. 3. Structure of HERMES file system

ous ways. It can use the existing API's provided by VFS layer. Also, HERMES provides a number of API's which is dedicated to handle multimedia specific information. The application can directly access these API's to manipulate the multimedia file.

4.2 Interface for Multimedia Streaming

HERMES defines a set of application programming interfaces to handle multimedia streaming workload. These interfaces facilitate the handling of multimedia data in more efficient manner. It achieves the following objective: (i) organizing the file as a collection of LDU(logical data unit), (ii) supporting QoS related data structure and (iii) management of streaming operation specific meta-data.

HERMES defines `mminfo` structure to convey the meta information related to streaming operation. The information includes the direction(forward or backward) or speed(x1, x2,...) of playback. The information of `mminfo` is used for scheduling the data retrieval operation in HERMES file system. The `mp4track` structure stores the information about the multimedia data track. The information shows track ID, creation time, modification time, the duration and time scale. The HERMES file system offers a set of kernel level API's. Table ?? illustrates the API's. When opening a file, `mp4open` enables the application to specify the information related to mutlimedia playback, e.g. QoS, playback rate, playback speed, etc. This information is carried in `mminfo` structure. `mp4GetMovieIOD` retrieves *IOD(Initial Object Descriptor)* in MPEG-4 file and store this information into `iniitalOD`. The size of IOD is stored into `pIoDLenght`. `int mp4GetTrackCount` retrieves the number of tracks in MPEG-4 file and stores into nCount. `mp4GetMovieTrack` retrieves information of the specified track in MPEG-4 file. `mp4read` reads the samples in a track. The sample can be video frames or the sequence of audio samples. The sample data and its size are read into `buffer` data structure. `mp4write` stores the multimedia samples in a track.

5 Performance Experiment

In this section we examine the performance behavior of the proposed file system. The HERMES file system is implemented on the Linux operating system. Performance of the file system is measured via experiments with a streaming workload. The experiment is performed on dual Pentium III (Coppermine) 746MHz processor with 256 Kbyte cache. The system has the 4 Ultra-Wide SCSI hard disks, each with 9.1 Gbytes disk space. The disk model is IBM Ultra star 36LP. A simulation program is written to sequentially scan the entire file. We vary the number of concurrent streams in the experiments.

We compare the I/O latency between HERMES and EXT2 file system. Since the streaming workload usually scans the file sequentially, it exhibits higher degree of spatial locality. Fifty MPEG-4 files with 50Mbyte each are created in both EXT2 file system partition and HERMES file system partition. In EXT2 file system, 570Mbytes file consist of 12 direct reference and 143 single indirect pointer blocks. We measure the the I/O latency between HERMES file system and EXT2 file system varying the number of concurrent streams. Fig. 4 illustrates the result of experiment. As is shown, I/O latency in the HERMES file system is approximately 70% of the latency in EXT2 file system. This is because as the number of concurrent sessions increases, the overhead of reading the indirect block constitutes more dominant fraction of the elapsed I/O time in the Linux file system. On the contrary, HERMES file system minimizes the disk seek overhead by prohibiting the usage of multi-level indirect reference.

In EXT2 file system, complex i-node structure along with multi-level data block organization and block group oriented placement strategy can make the latency of data block vary widely. In contrast, HERMES file system has relatively flat structure in organizing the data blocks. Thus, I/O latency remains relatively uniform. We measure the variance of I/O lantecy under different number of

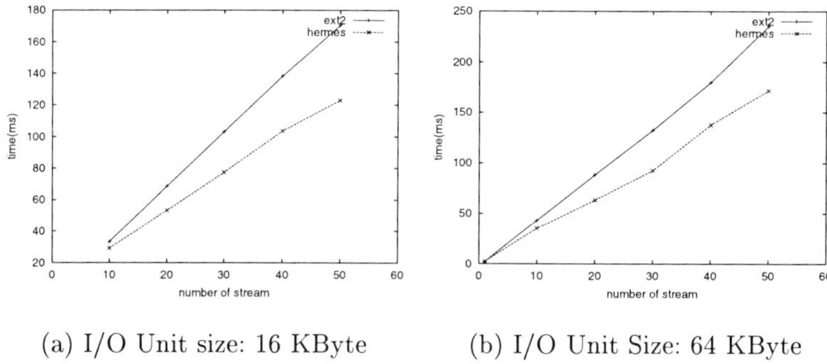

(a) I/O Unit size: 16 KByte (b) I/O Unit Size: 64 KByte

Fig. 4. Scalability Test: I/O latency of the streaming operation

concurrent streaming session. Fig. 5 illustrates the result of experiment. We can observe that the variance of HERMES is much smaller than EXT2 file system.

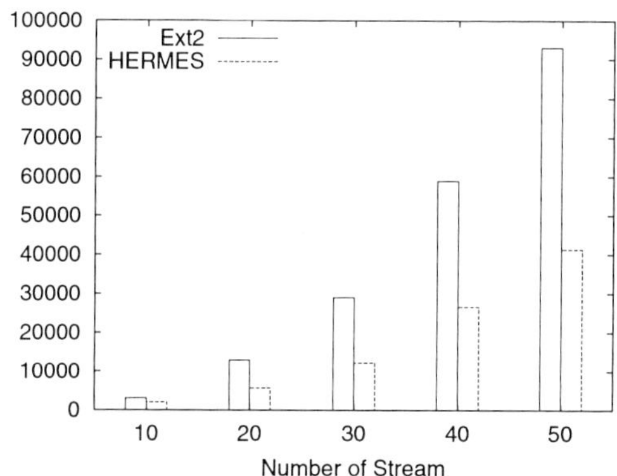

Fig. 5. Variance of I/O latency, 64 KByte I/O

6 Conclusion

In this work, we focus our effort in devising efficient file system for streaming operation and analyze its performance behavior under streaming workload. There are two design objectives in our file system. First, it should be able to support sequential access efficiently. To achieve this objective, we avoid using multi-level tree like structure. Single file is organized as a collection of data unit groups. Data unit group contains a fixed number of data units. Instead of using small size data unit(block), each data unit group consists of a collection of semantic data units. Second, the file organization is developed to handle relatively large files(tens of Mbyte), which are commonly found in multimedia applications. Our file system is implemented on the Linux platform. We examine the performance of the given file system under streaming workload and compares it with the performance of the EXT2 file system. There are a number of distinctive features which deserves attention: HERMES file system has more predictable behavior and thus is more robust against jitter. It services the I/O request with more evenly distributed latency compared to legacy UFS. We found that HERMES file system is more scalable compared to legacy UFS. The performance gap between HERMES and EXT2 becomes more dominant as there are more number of concurrent. The result of performance experiments indicates that the HERMES file system prototype successfully meet the file system constraints for multimedia streaming application.

References

1. Mon-Song Chen, Dilip D. Kandlur, and Philip S. Yu. Optimization of the grouped sweeping scheduling(gss) with heterogeneous multimedia streams. In *ACM Multimedia '93*, pages 235 – 242, 1993.
2. D. Gemmell, H. Vin, D. Kandlur, P. Rangan, and L. Rowe. Multimedia Storage Servers: A Tutorial. *COMPUTER*, 28(5):40–49, May 1995.
3. D.R. Kenchammana-Hosekote and J. Srivastava. Scheduling Continuous Media on a Video-On-Demand Server. In *Proc. of International Conference on Multi-media Computing and Systems*, Boston, MA, May 1994. IEEE.
4. Wonjun Lee, Difu Su, Duminda Wijesekera, Jaideep Srivastava, Deepak Kenchammana-Hosekote, and Mark Foresti. Experimental evaluation of pfs continuous media file system. In *Proceedings of CIKM*, pages 246–253, Las Vegas, Nevada, USA, 1997.
5. B Ozden, A. Biliris, R. Rastogi, and Avi Silberschatz. A Low-Cost Storage Server for Movie on Demand Databases. In *Proc. of VLDB '94*, 1994.
6. Jinyoun Park, Youjip Won, and Jaideep Srivastava. Smart: Yet another file system for multimedia streaming. In *Proceedings of International Conference on Distributed Multimedia Systems*, Taipei, Taiwan, Sep. 2001.
7. P. Rangan, H. Vin, and S. Ramanathan. Designing an on-demand multimedia service. *IEEE Communication Magazine*, 30(7):56–65, July 1992.
8. R.L.Haskin. Tiger shark-a scalable file system for multimedia. *IBM Journal of Research and Development*, 42:185–197, 1998.
9. Prashant J. Shenoy, Pawan Goyal, Sriram S. Rao, and Harrick M. Vin. Symphony: An integrated multimedia file system. In *Proceedings of SPIE/ACM Conference on Multimedia Computing and Networking(MMCN'98)*, pages 124–138, San Jose, CA, USA, Jan 1998.
10. T. Chiueh T.H. Niranjan and G. A. Schloss. Implemenation and evaluation of a multimedia file system. In *Proceedings of International Conference On Multimedia Computing and Systems*, 1997.
11. C. Wang, V. Goebel, and T. Plagemann. Techniques to increase disk access locality in the minorca multimedia file system. In *Proceedings of the* 7th *ACM Multimedia*, 1999.
12. R.P.Fitzgerald W.J.Bolosky and J.R.Douceur. Distributed schedule management in the tiger video fileserver. *ACM SIGOPS Operating Systems Review*, 31, 1997.

Motion Vector Recovery for Error Concealment Based on Macroblock Distortion Modeling

Jae-Won Suh and Yo-Sung Ho

Kwangju Institute of Science and Technology,
1 Oryong-Dong Puk-Gu, Kwangju, 500-712, Korea
{won, hoyo}@kjist.ac.kr
http://vclab.kjist.ac.kr/index.html

Abstract. If channel errors are introduced during the video transmission, we can apply error concealment techniques to reduce the transmission error effects by exploiting spatial and temporal redundancies of the video signal. Recently, motion vector recovery and motion compensation with the estimated motion vector has been one of the critical issues for the temporal-domain error concealment. In this paper, we prove that it is reasonable to use the estimated motion vector for concealing the lost macroblock by providing new macroblock distortion models. Based on the proposed distortion models, we develop several motion vector recovery algorithms.

1 Introduction

Fast growth of digital transmission services has generated a great deal of interest in digital transmission of video signals over a bandlimited channel. The bandwidth limitation necessitates the efficient coding schemes to compress an enormous amount of digitized video data, such as H.261, H.263, MPEG-1, and MPEG-2. Applications of digital broadcasting require about 50:1 or greater compression ratios. The MPEG-2 video coding standard [1] successfully satisfies the requirement of compression ratio using a hybrid algorithm of motion compensation (MC) and discrete cosine transform (DCT).

Due to the complex coding structure of the MPEG-2 video compression algorithm, compressed bitstreams are very sensitive to channel disturbances. Even one bit error can degrade not only the current frame but also succeeding frames. Therefore, we need error resilient coding techniques both to protect the transmission data and to reduce the transmission error effects. If we consider one layer bitstream without the network information, there are mainly two different approaches. First, we can apply channel coding techniques at the transmission level, such as forward error correction (FEC) codes. However, these approaches increase transmission bit rates because of the parity bits for error detection and correction. Therefore, if we consider the limited channel bandwidth, FEC may not always be the best solution.

As an alternative approach, we can apply error concealment techniques to hide the lost macroblock (MB) data. Error concealment techniques can be classified into two main groups by exploited redundancies in the video sequence:

M.H. Shafazand and A M. Tjoa (Eds.): EurAsia-ICT 2002, LNCS 2510, pp. 180–187, 2002.

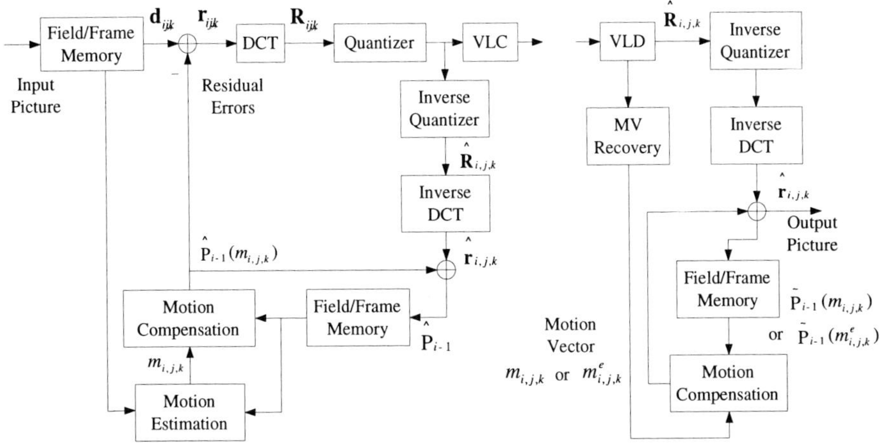

Fig. 1. Modified MPEG-2 Video Codec

spatial-domain error concealment techniques and temporal-domain error concealment techniques. Spatial-domain error concealment techniques interpolate the lost MB data using adjacent luminance and color values of the corrupted MB [2][3]. Temporal-domain error concealment techniques estimate lost motion vector (MV) of the corrupted MB and compensate it by the estimated MV under the assumption that MVs of spatially adjacent MBs are highly correlated [4]-[7]. Although we cannot recover the lost data perfectly, we can get similar MB data without increasing transmission bit rate.

In this paper, we propose new MB distortion models for the predictive frame (P-frame) to effectively conceal the corrupted MB data and increase the performance of error concealment. Based on the proposed MB distortion models, we prove that MC with the estimated MV is reasonable for temporal-domain error concealment techniques. There is a similar analysis for overall distortion of the decoded frame for H.263 [8]. In order to evaluate the proposed MV recovery algorithms, we compare and analyze simulation results with the previous works.

2 MPEG-2 Video Codec

Fig. 1 shows a block diagram of the MPEG-2 video codec. For intra frame (I-frame) coding, the first frame of group of pictures (GOP) is partitioned into nonoverlapping blocks of 8×8 pixels. After two-dimensional DCT is applied to each block independently, DCT coefficients are quantized and encoded using variable length coding (VLC). In the succeeding frames, temporal redundancies are removed by motion estimation (ME) and MC. The residual errors are the difference between the MB to be encoded of the current frame and the best matching block in the reference frame. The residual errors and MV information are encoded and transmitted. The decoding procedure is just opposite of the encoding procedure. In our modified MPEG-2 video decoder, if we receive a corrupted packet, we estimate MV of the lost MB in the MV recovery block and compensate the lost MB using the estimated MV.

The next problem is how to detect the transmission error because error concealment is largely dependent on the capability of error detection. We solve the error detection problem by checking the MB address (MBA). MBA is the unique address of the absolute position of the MB. Whenever we decode correctly received packets, we store the last decoded MBA. If we cannot decode the received packet data and resynchronization is established in the next slice header, we assume that (stored MBA + 1) is the beginning position of the erroneous MBs.

3 MB Distortion Models for P-frame

The MPEG-2 video codec is shown in Fig. 1, where $\{i, j, k\}$ indicates the k-th MB in the j-th slice of the i-th frame and bold character means MB data. $\mathbf{r}_{i,j,k}$ denotes residual errors, and $\mathbf{R}_{i,j,k}$ is DCT transformed data of the residual errors. $\hat{\mathbf{R}}_{i,j,k}$ means quantized data of $\mathbf{R}_{i,j,k}$ and $m_{i,j,k}$ represents a MV. $\hat{\mathbf{P}}_{i-1}(m_{i,j,k})$ and $\tilde{\mathbf{P}}_{i-1}$ indicate reference frames that are used for ME and MC at the encoder and the decoder, respectively, where i is the coding order. They should be differently notified because some parts of the reference frame at the decoder might be concealed.

First of all, we define the slice error probability by Eq. (1). Let $S_{i,j}$ be the event that the j-th slice of the i-th frame is corrupted.

$$\begin{aligned} Pr(S_{i,j}) &= \sum_{n=1}^{N_{i,j}} {}_nC_k P_b^n (1-P_b)^{N_{i,j}-n} \\ i &= 1, 2, \cdots, N, \quad k \leq n, \quad and \quad j = 1, 2, \cdots, J \end{aligned} \tag{1}$$

where P_b is bit error probability, $N_{i,j}$ is the number of bits of the j-th slice of the i-th frame, J is the number of the slices in the frame, and N is the number of frames.

The distortion $\mathbf{e}^I_{i,j,k}$ for the intra MB of the P-frame can be represented by

$$\mathbf{e}^I_{i,j,k} = E\big[(\mathbf{d}_{i,j,k} - \hat{\mathbf{d}}_{i,j,k})^2\big](1 - Pr(S_{i,j})) + E\big[(\mathbf{d}_{i,j,k} - \tilde{\mathbf{d}}_{i,j,k})^2\big] Pr(S_{i,j}). \tag{2}$$

In Eq. (2), if we receive the corrupted MB data, concealed data $\tilde{\mathbf{d}}_{i,j,k}$ replaces quantized MB data $\hat{\mathbf{d}}_{i,j,k}$ by error concealment operation. The first term represents the quantization error and the second term means the concealment error. If we assume that the quantization error and the concealment noise are uncorrelated, the second term in Eq. (2) can be decomposed into

$$E\big[(\mathbf{d}_{i,j,k} - \tilde{\mathbf{d}}_{i,j,k})^2\big] = E\big[(\mathbf{d}_{i,j,k} - \hat{\mathbf{d}}_{i,j,k})^2\big] + E\big[(\hat{\mathbf{d}}_{i,j,k} - \tilde{\mathbf{d}}_{i,j,k})^2\big]. \tag{3}$$

By Eq. (3), we can rewrite Eq. (2) as

$$\mathbf{e}^I_{i,j,k} = E\big[(\mathbf{d}_{i,j,k} - \hat{\mathbf{d}}_{i,j,k})^2\big] + E\big[(\hat{\mathbf{d}}_{i,j,k} - \tilde{\mathbf{d}}_{i,j,k})^2\big] Pr(S_{i,j}). \tag{4}$$

$\tilde{\mathbf{d}}_{i,j,k}$ can be obtained by $\tilde{\mathbf{P}}_{i-1}(m^e_{i,j,k})$, that is, we can employ ME of the lost MB and MC algorithms to conceal the the lost MB.

The distortion $\mathbf{e}^P_{i,j,k}$ for the inter MB of the P-frame can be expressed as

$$\begin{aligned}\mathbf{e}^P_{i,j,k} &= E\big[\{(\mathbf{r}_{i,j,k} + \hat{\mathbf{P}}_{i-1}(m_{i,j,k})) - \tilde{\mathbf{P}}_{i-1}(m^e_{i,j,k})\}^2\big]Pr(S_{i,j}) \\ &+ E\big[\{(\mathbf{r}_{i,j,k} + \hat{\mathbf{P}}_{i-1}(m_{i,j,k})) - (\hat{\mathbf{r}}_{i,j,k} + \tilde{\mathbf{P}}_{i-1}(m_{i,j,k}))\}^2\big](1 - Pr(S_{i,j})).\end{aligned} \tag{5}$$

If we receive corrupted MB data, we estimate the MV of the lost MB and compensate with the estimated MV to conceal the lost MB, as in the first term of Eq. (5). In order to separate meaningful terms, if we assume that the quantization noise of the residual errors and the concealment noise are uncorrelated, the first term of Eq. (5) can be rearranged as

$$\big(E\big[(\mathbf{r}_{i,j,k} - \hat{\mathbf{r}}_{i,j,k})^2\big] + E\big[\{\hat{\mathbf{r}}_{i,j,k} + \hat{\mathbf{P}}_{i-1}(m_{i,j,k}) - \tilde{\mathbf{P}}_{i-1}(m^e_{i,j,k})\}^2\big]\big)Pr(S_{i,j}). \tag{6}$$

If we use another assumption that the quantization noise of the residual errors and the mismatch between reference frames are uncorrelated, the second term of Eq. (5) can be represented as

$$\big(E\big[(\mathbf{r}_{i,j,k} - \hat{\mathbf{r}}_{i,j,k})^2\big] + E\big[\{\hat{\mathbf{P}}_{i-1}(m_{i,j,k}) - \tilde{\mathbf{P}}_{i-1}(m_{i,j,k})\}^2\big]\big)(1 - Pr(S_{i,j})). \tag{7}$$

By Eq. (6) and Eq. (7), we can rewrite Eq. (5) as

$$\begin{aligned}\mathbf{e}^P_{i,j,k} &= E\big[(\mathbf{r}_{i,j,k} - \hat{\mathbf{r}}_{i,j,k})^2\big] \\ &+ E\big[\{\hat{\mathbf{P}}_{i-1}(m_{i,j,k}) - \tilde{\mathbf{P}}_{i-1}(m_{i,j,k})\}^2\big](1 - Pr(S_{i,j})) \\ &+ E\big[\{\hat{\mathbf{r}}_{i,j,k} + \hat{\mathbf{P}}_{i-1}(m_{i,j,k}) - \tilde{\mathbf{P}}_{i-1}(m^e_{i,j,k})\}^2\big]Pr(S_{i,j}).\end{aligned} \tag{8}$$

In Eq. (8), we cannot treat the first term to improve the performance of error concealment since it is the quantization noise of the residual errors. If there is no error in the reference frame, we can remove the second term. From Eq. (4) and Eq. (8), it seems reasonable to conclude that we estimate the lost MV from $\tilde{\mathbf{P}}_{i-1}$ and compensate for the lost MB using the estimated MV $m^e_{i,j,k}$ for error concealment.

4 Error Concealment Techniques

In order to conceal the lost MB, we use upper and lower MB data of the lost MB. A simple estimate value for the lost MV is zero with an assumption that no motion has occurred between the previous reference frame and the current frame. Use of the zero MV produces a reasonably good approximation in the static scene. However, we cannot expect good result in the dynamic scene. Therefore, we should consider motion compensated error concealment algorithms.

In order to estimate MV, we can exploit neighboring MVs of the lost MB. MV of the lost MB can be obtained by taking the average value of MVs of the vertically adjacent MBs (AVG) [4]. In this scheme, if vertically neighboring

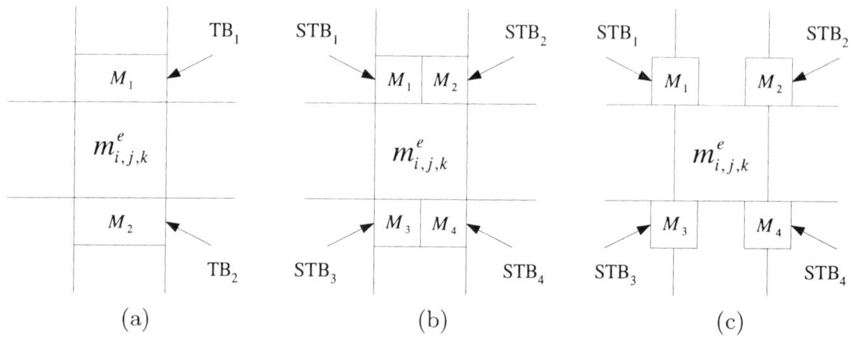

Fig. 2. Modified Average Algorithm

MBs have MVs, we can obtain reasonably good reconstruction quality for the lost MB. However, if only one or none of the vertical neighbors has a valid MV, quality of the reconstructed image is not satisfactory. For this reason, we devise the following modified average (MAVG) schemes for recovering the lost MV [6].

As shown in Fig. 2(a), if vertically adjacent MBs are coded by intra mode, we define 16 × 8 target blocks (TBs), TB_1 and TB_2, above and below the lost MB. The MV of each TB is computed by the block matching algorithm at the decoder. We substitute the lost MV by the average of the MVs of the two TBs. In order to obtain more accurate MV for the lost MB, we separate each 16 × 8 TB into two 8 × 8 small target blocks (STBs), as shown in Fig. 2(b). We can take average of the MVs of the four STBs for the lost MV. Although these two methods produce good performance, they entail a considerable amount of processing complexity at the decoder. In order to reduce the processing time, we define alternative STBs, as shown in Fig. 2(c). Because estimated MVs of STB_2 and STB_4, M_2 and M_4, can be used as M_1 and M_3 in the next MV recovery process, we can reduce the computational complexity by half.

Other MV recovery algorithms use boundary pixels of the lost MB to estimate the lost MV. Within the given search range (SR), the boundary matching algorithm (BMA) [5] calculates the squared sum of differences (SSD) between outer one pixel boundary line of the above, below, and left sides of the lost MB in the current frame and inner one pixel boundary line of the target block in the previous reference frame. BMA replaces the lost MB with the target block data that has the smallest total SSD. The decoder motion vector estimation algorithm (DMVE) [7] is very similar to BMA. Different thing is that DMVE uses outer several pixel boundary lines (two to eight) of the lost MB in the current frame and the previous reference frame to calculate the SSD. However, these algorithms, BAM and DMVE, have significant limitation: left outer boundary pixels of the lost MB are not available when successive MBs are lost. Nevertheless, if left outer boundary pixels are used in computing the SSD, it means that the MV estimation process includes error concealment mismatch from the second MB of the corrupted MBs. In order to resolve this problem, we propose an extension matching algorithm (EMA).

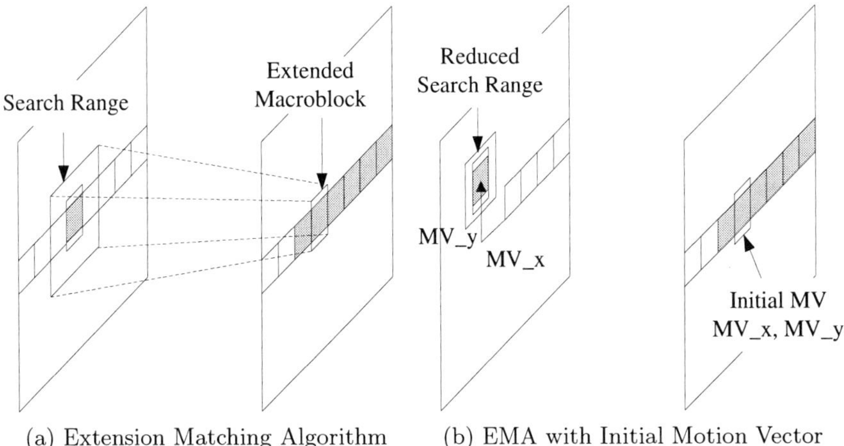

(a) Extension Matching Algorithm (b) EMA with Initial Motion Vector

Fig. 3. EMA with Initial Motion Vector

As shown in Fig. 3(a), we form an extended MB by including pixels above and below the lost MB. The MV is then searched to minimize the SSD between extended pixels in the current and the previous frame. In this algorithm, the extended width (EW) and the SR should be selected appropriately. Since the best matching block should be searched for within the given SR, this technique may entail a considerable amount of processing complexity at the decoder. In order to reduce the computational complexity of EMA, we modify the algorithm with an initial estimate of the MV, as shown in Fig. 3(b). We first set the initial MV for the lost MB by AVG. The initial MV establishes a starting point of the SR and enables to reduce the SR. If none of the vertical neighbors has a valid MV, we use normal SR. As a result, we can effectively reduce the processing time and get good performances.

5 Simulation Results

In order to evaluate the performance of the error concealment algorithms, we perform computer simulations on various test video sequences: FOOTBALL, BICYCLE, and BALLET. They have the 4:2:0 format of 720 × 480 pixels. These are encoded using the pattern of IBBPBBPBBPBB at a frame rate of 30 frames/sec. We consider TS packet transmission system for considering noisy channel. We assume that the first P-frame has lost one TS packet.

We compare the performance of conventional algorithms including AVG [4], BMA [5], and DMVE [7] and proposed algorithms, MAVG [6], EMA, and IEMA. In order to estimate MV of the lost MB, BMA, DMVE, and EMA use [-25, 24] SR. While BMA takes one pixel boundary line, DMVE, EMA, and IEMA exploit variable pixel boundary lines from one to eight. From the simulation, we observe that DMVE produces best results when it takes two pixel boundary lines. EMA and IEMA generate the best performance if they use one pixel boundary line. As

Table 1. Concealed PSNR for Various MV Recovery Algorithms

	FOOTBALL	BICYCLE	BALLET
Original	32.59	26.57	29.12
AVG	30.62	25.15	28.41
BMA	31.25	23.37	28.53
DMVE	31.23	23.30	28.52
MAVG	31.50	24.16	28.76
EMA	31.31	23.76	28.70
IEMA	32.21	25.42	29.01

described earlier, IEMA has [-5, 4] reduced SR and [-25, 24] normal SR. MAVG has alternative STB structure to reduce computation time and uses [25, 24] SR for ME of the STB.

Table 1 summarizes the peak-signal-to-noise ratio (PSNR) for the test sequences. BMA, DMVE, and the proposed EMA produce very similar PSNR results. If objects have dynamic motion, such as FOOTBALL and BALLET, BMA, DMVE, and EMA generate good PSNR results. However, if objects have static motion, such as BICYCLE, AVG and MAVG generate good performance. Because the proposed IEMA takes the advantages of each algorithm, it provides the best PSNR results. The subjective quality test of reconstructed frames is very important for the performance comparison of error concealment. As shown in Fig. 4, IEMA shows a good subjective video quality.

6 Conclusion

In this paper, we have shown that motion vector estimation and motion compensation with the estimated motion vector is reasonable for concealing the lost macroblock in temporal-domain error concealment based on proposed MB distortion models. We have reviewed merits and demerits of conventional motion vector recovery algorithms and proposed motion vector recovery algorithms to improve the performance of error concealment. Experimental results of PSNR and subjective quality lead us to the conclusion that the proposed extension matching algorithm with an initial motion vector is one of the best solution for error concealment.

Acknowledgement.
This work was supported in part by the Korea Science and Engineering Foundation (KOSEF) through the Ultra-Fast Fiber-Optic Networks (UFON) Research Center at Kwangju Institute of Science and Technology (K-JIST), and in part by the Ministry of Education (MOE) through the Brain Korea 21 (BK21) project.

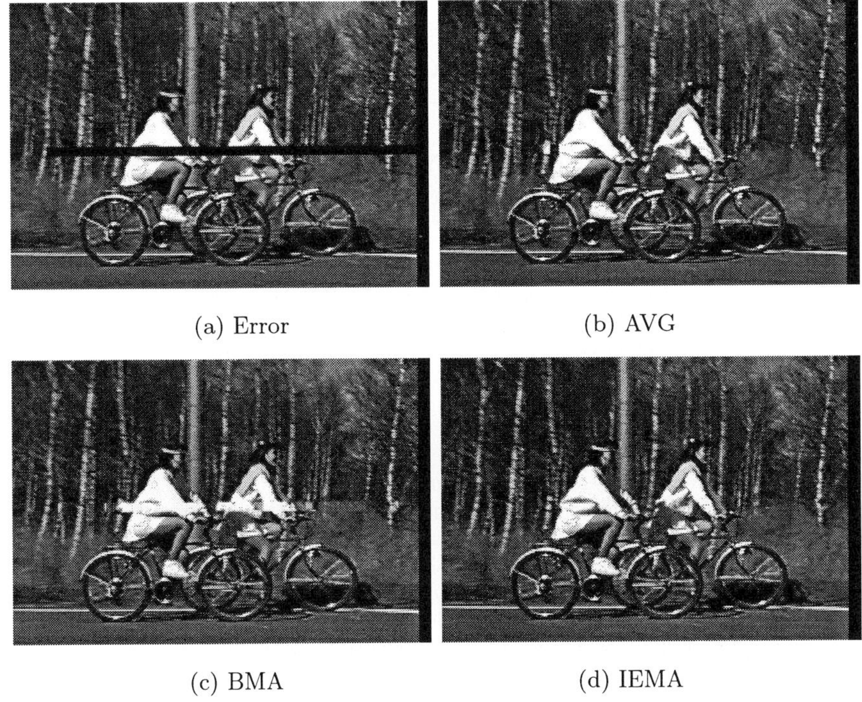

(a) Error (b) AVG

(c) BMA (d) IEMA

Fig. 4. Concealed Frames of BICYCLE

References

1. ISO/IEC IS 13818-2 (MPEG-2 Video): Information Technology - Generic Coding of Moving Pictures and Associated Audio Information, April 1996.
2. Aign, S. and Fazel, K.: Temporal and Spatial Error Concealment Techniques for Hierarchical MPEG-2 Video Codec, IEEE International Conference on Communication, Vol. 3, June (1995) 1778-1783.
3. Suh, J.W. and Ho, Y.S.: Error Concealment Based on Directional Interpolation, IEEE Transactions on Consumer Electronics, Vol. 43, No. 3, Aug. (1997) 295-302.
4. Sun, H., Challapali, K., and Zdepski, J.: Error Concealment in Digital Simulcast AD-HDTV Decoder, IEEE Transactions on Consumer Electronics, Vol. 38, No. 3, Aug. (1992) 108-116.
5. Lam, W.M., Reilbman, A.R., and Liu, B.: Recovery of lost or erroneously received motion vectors, Proc. ICASSP, April (1993) V417-V420.
6. Suh, J.W. and Ho, Y.S.: Recovery of Motion Vectors for Error Concealment, Proc. IEEE Region 10 TENCON, Sep. (1999) 750-753.
7. Zhang, J., Arnold, J.F., and Frater, M.R.: A Cell-Loss Concealment Technique for MPEG-2 Coded Video, IEEE Transactions on Circuits and Systems for Video Technology, Vol. 10, No. 4, June (2000) 659-665.
8. Zhang, R., Regunathan, S.L., and Rose, K.: Video Coding with Optimal Inter/Intra-mode switching for packet losss resilience, IEEE Journal on Selected Areas in Communications, Vol. 18, No. 6, June (2000) 952-956.

A Memory Copy Reduction Scheme for Networked Multimedia Service in Linux Kernel

JeongWon Kim[1], YoungUhg Lho[2], YoungJu Kim[3], KwangBaek Kim[3], and SeungWon Lee[4]

{[1]Dept. of Computer & Information Engineering, [2]Dept. of Computer Education, [3]Dept. of Computer Engineering} Silla Univ., Korea
{jwkim, yulho, yjkim, gbkim}@silla.ac.kr
[4]Dept. of Computer Science, Pusan National Univ., Korea
bluecity@melon.cs.pusan.ac.kr

Abstract. While Multimedia streams need an efficient support of kernel, the current buffer cache mechanism of Linux kernel originally based on the Unix operating system is designed apt for small files, which are aperiodically requested and time-uncritical. But, in case of continuous media, the overhead of CPU occurs for large copying memory from kernel address space to user address space. This overhead both degrades system throughputs and cannot guarantee QOS. In this paper, we've designed and implemented two memory copy reduction schemes in Linux kernel, direct I/O and one-copy. The direct I/O path skips the buffer cache layer of Linux kernel and directly copies the disk blocks to the user buffer. And, the one-copy provides fast disk-to-network data path without copying to user address space. These enhancements should increase the throughputs of VOD server. The experimental results demonstrate throughput improvements and show considerable reduction of CPU overhead.

1 Introduction

Multimedia services such as VOD (Video on Demand) will be indispensable components of the ongoing multimedia information super-highway. The multimedia data transferred by these services requires QoS (Quality of Service): large storage space, high network capacity and bounded delay. For example, the MPEG-II stream needs 2GB storage and 4~60Mbps bandwidth [1]. A VOD server should guarantee this QoS. So, it needs an efficient support of OS kernel for efficient networked multimedia service. There are many efforts to solve these problems: placement and caching strategies for multimedia files, real-time transmission and deadline guarantees of time-critical data, etc [2,3,4]. Of these uncountable researches, our work focuses on the fast data delivery from the disks to the network interface.

The Linux kernel is convenient for application-driven kernel development because it is free-ware, open, and stable. But, in case of multimedia file service, the kernel must copy from data on disks to kernel address space, and again to user address space over one hour. In this point, we must consider the CPU overhead for data copy. The copy overhead from disk to kernel buffer cache system may be reduced by the DMA

M.H. Shafazand and A M. Tjoa (Eds.): EurAsia-ICT 2002, LNCS 2510, pp. 188–195, 2002.

(Direct Memory Access) support, but the copy from kernel address space to user address space causes the heavy works of CPU. The CPU overhead for small and non-periodic data is trivial. But, The overheads in case of continuous media must be significant issue for QoS guarantee.

This paper describes two kernel-supported schemes. The first path is the direct I/O that lets the disk block driver to directly copy physical disk blocks on disks to user buffer space without using the kernel buffer cache system. The second is the one-copy path that provides fast data path between file system and network protocol stack. This one-copy does not need the user buffer and passes the kernel buffer pointer to the network protocol stack avoiding data copies. We have designed and implemented these schemes in Linux 2.2.12 kernel [15]. This paper has augmented our works and performed lots of extended experiments.

The rest of this paper is organized as follows: Section 2 describes in detail the related works. Section 3 describes the direct I/O mechanism. In Section 4, we present the one-copy scheme. Section 5 shows the development environments and performance evaluation results. Finally, we present the conclusions.

2 Related Works

- Zero-copy [5]: Milind designed and implemented the zero-copy to decrease the data copy overheads in 4.4BSD Unix. This method uses a new buffer system (mmbuf), which provides zero-copy path between file system and network protocol stack. The kernel file system maintains a cluster of kernel buffer blocks for each multimedia stream and passes the blocks to the protocol stack avoiding data copies.
- IRIX direct I/O [6]: the XFS of SGI IRIX provides direct I/O path between file system and user address space. The direct I/O locks the user buffer pages in memory during the I/O transfer, since the disk driver will be asked to transfer directly to or from those pages. And, the buffer cache module allows XFS to bind a buffer object to a range of user memory, and does I/O on the buffer object in the usual way.
- Raw I/O [7]: A Unix-style OS supports the raw I/O data path that provides I/O for device driver files But, this raw I/O is applied to only device files, not general files. The current Linux kernel provides this raw I/O, but uses the buffer cache module. Stephen Tweedy [8] have developed the original raw I/O patches, which will be included in Linux 2.4.x kernel.

3 Direct I/O

The direct I/O path that skips the buffer cache module and directly copies disk blocks to user buffer can decrease copy overhead of CPU and provide fast response time. Also, It must be effective in case of continuous retrieval of multimedia data. The direct I/O in IRIX proves these points [9, 10]. In generic buffer cache I/O path, two copies occur on the buffer cache I/O path. If it is added to a copy to network protocol buffer, total number of copy is three. But, the direct I/O path needs only one copy that disk driver copies physical disk blocks to user buffer. So, the copy overhead can considerably be decreased.

The core problem to provide the direct I/O path in Linux kernel for i386 architecture is to translate the user allocated buffer address (virtual address) into the corresponding physical address. We note that the block device driver must transmit physical disk blocks to real address on main memory. The address delivered to the device driver must be not virtual address but physical address. A kernel function for direct I/O path should provide this virtual-to-physical address translation functionality.

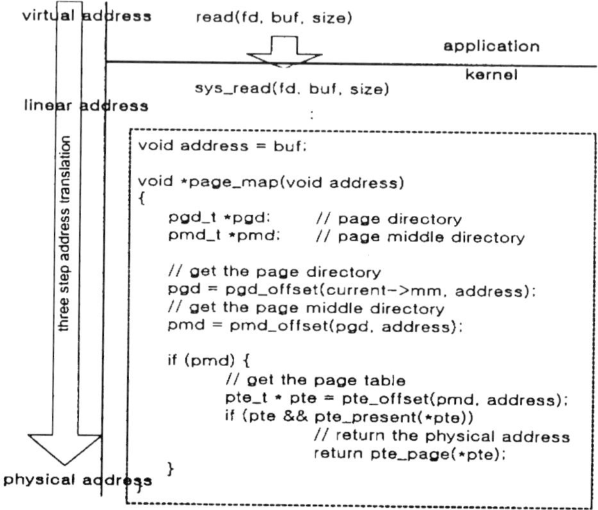

Fig. 1. Address translation: The X86 processor only supports a two-level conversion of the linear address. This X86 processor defines the size of the page middle directory as one and interprets the entry in the page directory as a page middle directory [13].

3.1 Direct I/O Implementation

The address translation scheme of i386 CPU is segmentation with pagination [11]. There is detail explanation for from virtual to physical address translation [15]. The direct I/O is implemented on the Linux 2.2.12. It refers the raw I/O patches developed by Stephen Tweedy [8]. We provide a direct I/O path for normal file using a *direct_rw()* function. In this subsection, we present the scenario and the implementation of the direct I/O path.

3.1.1 Scenario of Direct I/O

Our direct I/O doesn't need a new system call. Application programs only specify the flag O_DIRECT on a file open call. For example, when opening a new file, let's assume O_DIRECT flag is inserted.

```
Open (filename, O_RDONLY | O_DIRECT);
```

Then, a file system identifies that this open file is not normal buffer cache I/O path but direct I/O path. This information is recorded in the open file descriptor tables of the calling process. The *sys_read()* interface determines each path. For the direct I/O path, it calls the *direct_rw()* function developed by this work. If not, it calls the *generic_file_read()* function. So, we don't need a new file pointer, inode information and system call interface.

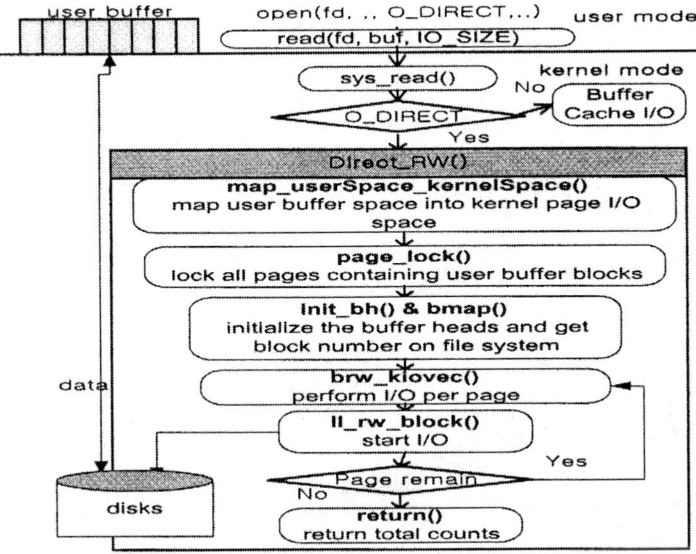

Fig. 2. Flow of direct_rw():step1-mapping user buffer space into kernel page, step 2-locking all pages, step 3-initializing buffer heads and block mapping, step 4-issuing the page I/O, step 5-performing the block device driver I/O, step 6-returing total counts

3.1.2 Direct_rw()

The read() and write() procedures for the direct I/O are almost same except for the R/W flag to the device driver. For our goals, we describes only the read () operation flow in this work. The *direct_rw()* branched in the *sys_read()* is composed of six steps. Figure 2 illustrates this flow.

4 One-Copy

We've found that the direct I/O has the following characteristics: 1) all I/Os should be 512-byte aligned both in memory and on disk. It's because the disk drivers transfer physical blocks a sector unit. 2) The generic buffer cache module does prefetching through readahead, but the direct I/O cannot do this because it cannot allocate buffer memory for the prefetching. 3) The direct I/O path is effective on a large block size of file system rather than a small block size of file system such as 4KB of Linux ext2 in i386 architecture.

One-copy means that only one copy occurs from disk to kernel address space while a VOD server transmits the physical disk blocks to the network interface. In the zero-copy scheme, an independent buffer cache system to support multimedia stream is insulated in the BSD kernel. The core concept is to combine memory inode and socket for a multimedia stream. Figure 3 is the block diagram of the one-copy scheme. The one-copy maintains the buffer blocks of the generic buffer cache module as it is, and the modified memory inode uses these buffer blocks for disk read and the modified socket structure also uses these for transmission. Therefore, the same buffer block is pointed by both memory inode and socket structure. The memory inode and the socket are one-to-one mapped.

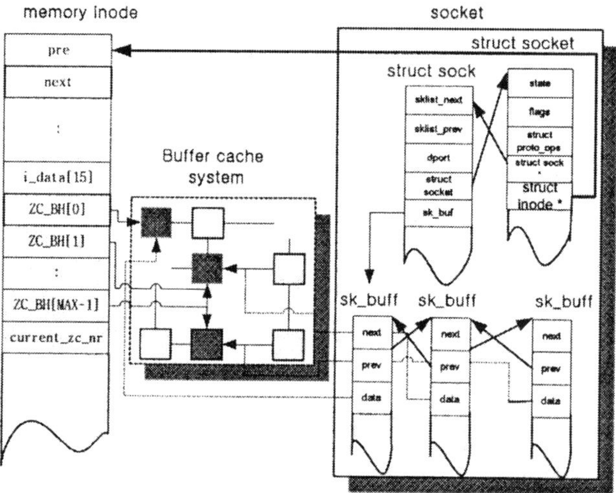

Fig. 3. Diagram of one-copy

4.1 One-Copy Implementation

Main changes to kernel are described in this subsection.

• memory inode

Two variables on the memory inode are inserted. In Figure 3, the pointer array *ZC_BH []* points to the just used buffer blocks for the previous read and the *current_zc_nr* records the number of the buffer blocks. The request blocks per a period in a VOD server can determine the maximum number of the *current_zc_nr.*

• socket structure

The *struct inode pointer* field of the new socket structure points to the associated memory inode. An application program passes the file descriptor to the kernel interface using the *fcntl()* interface. Then, the socket handling routine associates the socket with the inode pointed by the file descriptor of the calling process.

• generic_readpage()

The *generic_readpage()* function is a common interface to the read operations of all local file systems under the VFS(virtual file system). This function copies the buffer block data to the user buffer using the in-kernel function *copy_to_user()*. In our one-copy, this function just returns the number of retrieved data bytes.

- socket buffer pointer

The socket protocol stack checks whether the socket is a one-copy path or not. If the socket is a normal path, the protocol routine copies the data of user buffer space to the socket buffer named *sk_buff*. But, If a one-copy, the protocol function just points to the buffer blocks maintained by the associated inode.

5 Experiment Results

This section describes the implementation environment and experiment results to evaluate the performance gains of the direct I/O and the one-copy. There is detail experiment environment in previous work [15].

The experiments are performed on both a single media file and a VOD testbed. The ext2 file system created in this work used a 4KB-block size and a fragment size of 1KB. Total 30 video files with a size of 300 MB are stored in this file system. Each file is striped across the three disks concatenated by the software raidtools. For the experiment to a single file, we measured the overheads of CPU, response times to read () operation, and throughputs to a read()/send() combination. The normal operation and no readahead operation are inserted in each experiment to compare the proposed schemes with the other cases. For measuring the performance benefits of our solutions on a MOD server, we constructed a virtual MOD server program. This program retrieves the disk files through all I/O paths: direct I/O, one-copy, normal, and no readahead path. Before our experiments, we generated a log file that has the arrival time of a job thread and its video file name. The interarrival time between job threads is calculated with mean number of requests per total experiment time and follows a Poisson distribution. The file retrieved by a job thread is selected by a Zipf distribution (• = 0.271) [14]. Each job thread has a 1.5 Mbps of data rates and records the elapsed time per each single operation in a log file. We analyzed all log files created by all job threads.

The purpose of experiment in Figure 4 is to demonstrate that the data paths proposed in this paper show lower overhead of CPU than the normal I/O path. We performed large continuous reads for a single file on the Linux ext2 file system. This experiment was repeated 100 times and the CPU loads were logged each time. Figure 4 illustrates the results of average CPU loads vs. file size. The normal path shows a worst performance due to the CPU works for copying from kernel memory to user buffer. And, the direct path shows slightly higher CPU loads than the one-copy. The overheads for the address translation and the mapping user address space into kernel buffer structure may bring about this gap.

We also measured read throughputs of the direct and the one-copy path over the normal and the no readahead path. Figure 5 illustrates the results of total read time vs. file size in all paths. We repeated this experiment 100 times and calculated the average read time for all paths. We observed that the no readahead path also showed the worst throughput. This is because the no readahead path must issue disk accesses every read operation. The normal path is 9~10 % faster than the direct path, but 5 ~ 10 % slower than the one-copy path. We note that the direct I/O also cannot support the readahead and shows lower throughput than the normal path. The one-copy path shows the best throughput of all paths because this does not copy disk blocks to user buffer space.

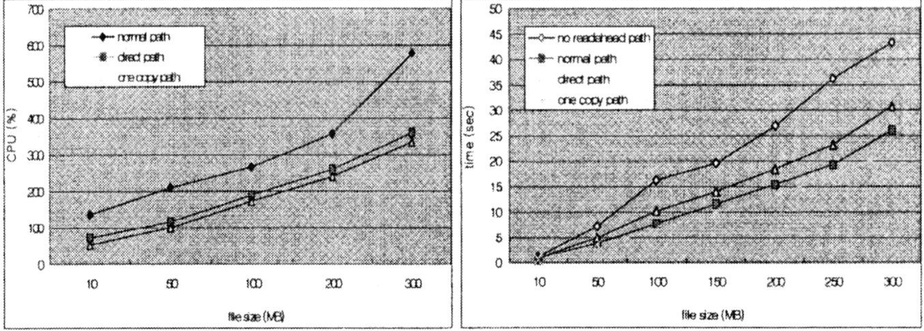

Fig. 4. CPU loads

Fig. 5. Total read time

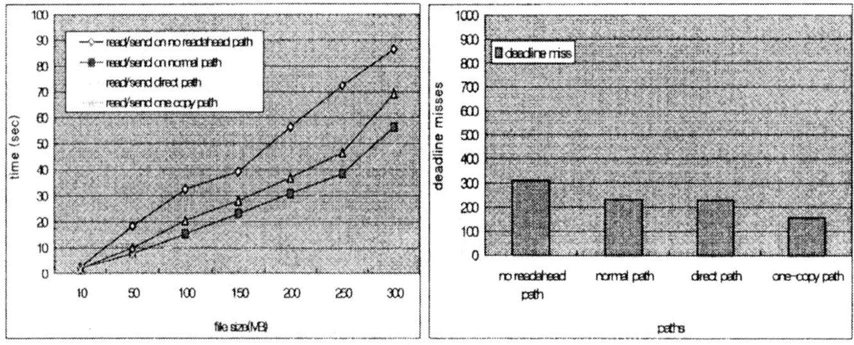

Fig. 6. Read / send throughputs

Fig. 7. Deadline misses

The read/send throughputs about all paths also were measured in same environment with Figure 5. Figure 6 shows the results. We observed consistently lower completion times for read/send operation on the one-copy path than the others and improvements varied from 3% to 12 %.

Finally, the performances about all paths on a MOD server were measured in Figure 7. This experiment was conducted under the following conditions: each request arrives with a Poisson distribution (100 users per an hour), has 1.5Mbps data rates, and retrieves a video file under a Zipf distribution (the skew for 20 videos is 0.271). One round on our MOD server is assumed three second and the deadline misses are recorded each round. We observed that the one-copy path also has the least misses but the direct and the normal path has almost same misses. This can be analyzed that the normal path under a heavy load such as MOD server cannot frequently call the readahead routine as same as a normal loads.

6 Concluding Remarks

This paper presents two schemes (direct I/O, one-copy) that can decrease the memory copy overheads of CPU for multimedia data and increase the throughputs of a VOD server. We've observed the direct I/O could decrease CPU overhead for memory copy over a single read operation, but cannot utilize caching benefits. The one-copy over continuous read for single file has shown better performance. The page size and the block size on the current Linux kernel for i386 machine are too small to upgrade the performance of the direct I/O. And, the one-copy that provides fast I/O path from the disks to the network shows better performance over the other paths because it needs not only one copy while delivering data from disks to network but also can use the readahead functionality. For future research, we need to perform our schemes on 64bit machine.

References

1. Prabhat K. and Leigh, Kiran Thakrar.: Multimedia Systems Design, pp.112, Prentice Hall PTR, 1996
2. Yuewei Wang, David H.C. Du.: Weighted striping in multimedia servers, Proc.of IEEE on multimedia computing and systems. pp 102–109, June, 1997
3. Yuewei Wang, Johathan C.L. Liu, David H.C. Du and Jenwei: An efficient video file allocation schemes for video on demand services, ACM multimedia systems journal, vol.5, no.5, 1997
4. Renu Tewari, Daniel M. Dias, Ajit Mukherjee, Harrick M. Vin: High availability in clustered multimedia servers, Proc. of the USENIX annual technical conference, Jan, 1996
5. M.M.Buddihikot, X,J.Chen, D.Wu, and G.M.Parulkar.: Enhancements to 4.4 BSD Unix for efficient networked multimedia in project MARS, IEEE ICMCS, pp.326–337, 1998
6. http://techpubs.sgi.com/library/manpages/open.html
7. Maurice J. Bach: The design of the Unix operation system, Englewood Cliffs, NJ 07632: Prentice-Hall, Inc., 1986
8. Stephen Tweedie, PATCH: Raw device I/O for 2.1.131, http://www.linuxhq.com/lnxlists/linux-kernel/lk_9812-02/msg00686.html, December 1998
9. Steen R. Siltis, Thomas M.Ruwart, Matthew T.O'Keefe.: The global file system, Proc.of the fifth NASA goodard space center conference on mass storage systems and technologies, sept 17–19, 1996
10. Jim Mostek, William Earl, and Dan Koren.: Porting the SGI XFS File System to Linux, white paper, http://oss.sgi.com/projects/xfs/, 1999
11. Silberschatz, Galvin, Operating systems concepts, fifth edition, pp. 304, 1998
12. M. Beck, H. Bohme, M. Dziadzka, U. Kunitz, R. Magnus, D. Vervorner.: Linux Kernel Internals, pp.148–151, Addison Wesley, 1998.
13. Remy Card, Eric Dumans, and Frank Mevel.: The Linux kernel book, Wiley, pp 286~298, 1999
14. D.E. Knuth.: The art of computer programming, vol.3: Sorting and searching. Addison-Wesley, 1973
15. J.W Kim, S.W Lee, K.D Chung.: Implementation of Zero-copy with caching for Efficient Networked Multimedia Service in Linux Kernel, ICME2001

Hidden Markov Model and Neural Network Hybrid

Dongsuk Yook

Speech Information Processing Laboratory
Department of Computer Science and Engineering, Korea University
Sungbookgoo Anamdong 5-1, Seoul, Korea 136-701
http://voice.korea.ac.kr
yook@voice.korea.ac.kr

Abstract. When there is a mismatch between training and testing environments, statistical pattern classification methods may suffer from severe degradation in their performance because the parameters in the classifiers do not represent the testing data well. The mismatch is typically due to the interference or noises from operating environments. In this paper, a neural network based transformation approach is studied to handle the distribution mismatches between training and testing data. The probability density functions of the statistical classifiers are used as the objective function of the neural network. The neural network maximizes the likelihood of the data from a testing environment, and allows global optimization of the network when used with the statistical pattern classifiers. The proposed approach is applied to the area of automatic speech recognition to recognize noisy distant-talking speech and it reduces the error rate by 52.9%.

1 Introduction

When a statistical pattern classifier is used in an environment that is different from the one where the training data is collected, the performance of the classifier may be degraded because the data distribution that the classifier has seen is different from the actual testing data distribution. This is a typical robustness issue in generic statistical pattern recognition problems. In this paper, we consider a popular statistical pattern recognition method; hidden Markov models (HMMs) which show high performance when applied to automatic speech recognition. A typical distribution mismatch in speech recognition is the case that a speech recognizer is trained on clean speech and there are convolutional and additive noises in a testing environment. In this paper, a neural network is applied to the parameter adaptation of HMM-based recognizers. The conditional probability of adaptation data for a given recognizer is used as an objective function of the neural network [9][12][13]. Since the likelihood of the data is maximized by the neural network, it is called *maximum likelihood neural network* (MLNN). The advantages of neural network based transformation methods are as follows. It avoids retraining of a classifier, which is an expensive task in terms of training data collection and computation time. By using a neural network, the adaptation process requires a small amount of adaptation data. Without any knowledge about the operation environments, it automatically learns the mapping function between training and testing environments from examples. For speech recognition case, it can handle ambient noise, reverberation, channel mismatches, and their combinations [8].

M.H. Shafazand and A M. Tjoa (Eds.): EurAsia-ICT 2002, LNCS 2510, pp. 196–203, 2002.

Neural networks have been used in conjunction with speech recognizers in various ways for robust speech recognition. Tamura and Waibel [7] are one of the first to use neural networks to reduce noise in noisy speech signals. Bengio *et al.* [1], Biem *et al.* [2], and Rahim *et al.* [5] used neural networks as front-end for HMM-based speech recognizers for feature transformation. Previously, the MLNN is used for mean vector transformation [9][12], and variance transformation [13]. This paper reviews the theory of MLNN, and its application to robust speech recognition. The experimental results are analyzed in terms of distribution transformation and speech recognition accuracy. In Section 2, the motivation behind the MLNN is described. In Section 3 and 4, it is applied to the transformation of mean vectors and covariance matrices to best match the testing environments. The experimental results are analyzed in Section 5.

2 Maximum Likelihood Neural Networks

The error criterion of neural networks is traditionally represented by a mean squared error (MSE) [6]. For example, the feature transformation neural network [10], which transforms distorted feature vectors to the corresponding target vectors, can be trained to minimize the mean squared error E,

$$E = \frac{1}{2}\sum_{i}(\overline{x}_i - x_i)^2 \quad , \tag{1}$$

which is the summation of the squared difference between the i-th dimension target value, $\overline{x}_i$ (e.g., clean speech), and the corresponding network output, x_i (e.g., approximated clean speech). On the other hand, continuous speech recognition is accomplished by finding the word sequence, $\overline{U}$, which gives the highest probability for a given feature vector sequence X;

$$\overline{U} = \arg\max_{U \in U^*} P(U|X) \tag{2}$$

$$= \arg\max_{U \in U^*} \frac{P(X|U)P(U)}{P(X)} \tag{3}$$

$$= \arg\max_{U \in U^*} P(X|U)P(U) \quad , \tag{4}$$

where U^* is a set of all possible word sequences, $P(X|U)$ is the acoustic score of the vector sequence X for a given word sequence U, and $P(U)$ is the score for the word sequence, which is usually computed independently using a stochastic language model [3]. The likelihood $P(X|U)$ is the only term in equation (4), which can be affected by a feature or model transformation. The acoustic score of the word sequence is the product of each word's acoustic score. In continuous speech recognition, the likelihood of the best word sequence is usually approximated by the likelihood of the best state sequence using the *Viterbi* algorithm [11];

$$P(X|U) = \sum_{S \in S^*} P(X, S|U) \tag{5}$$

$$\approx \max_{S \in S^*} P(X, S|U) \tag{6}$$

$$= \max_{S \in S^*} P(X|S,U)P(S|U) \tag{7}$$

$$= \max_{S \in S^*} \prod_{s \in S} P(x|s)P(s|r) \quad , \tag{8}$$

where S^* is a set of all possible state sequences for the given utterance U, and $P(X,S|U)$ is the acoustic score of X coming from a state sequence S for the given utterance U. $P(s|r)$ is the state transition probability from a state r to a state s, where the state r is the predecessor of the state s in the state sequence S. Assuming that the state transition probabilities are not changing, the feature or model transformation affects only $P(x|s)$, which is the likelihood of an observation vector $x \in X$ given the state $s \in S$. $P(x|s)$ is usually represented using a mixture of Gaussian distributions with diagonal covariances;

$$P(x|s) = \sum_p g_{s,p} \, e^{-\frac{1}{2}\sum_i \frac{(x_i - m_{s,p,i})^2}{v_{s,p,i}}} \quad , \tag{9}$$

$$g_{s,p} = c_{s,p} \frac{1}{\sqrt{(2\pi)^n \prod_i v_{s,p,i}}} \quad , \tag{10}$$

where $c_{s,p}$ is the weight of the p-th distribution, and $m_{s,p,i}$ and $v_{s,p,i}$ are the i-th dimension mean and variance of the p-th distribution in the state s, respectively. Minimizing the mean squared error in equation (1) does not necessarily maximizes the likelihood in equation (8).

One way to increase the likelihood in equation (8) is to maximize equation (9) using a neural network that transforms the parameters of the distributions to match the test data. The conditional probability, equation (9), can be used as an objective function of the neural network, and the weight update rule can be derived as in the standard *error back-propagation* algorithm [6];

$$w \leftarrow w + \Delta w \quad , \tag{11}$$

$$\Delta w = \eta \frac{\partial P(X|U)}{\partial w} \quad , \tag{12}$$

$$\frac{\partial P(X|U)}{\partial w} \propto \sum_{s \in \overline{S}} \frac{\partial \ln P(x|s)}{\partial w} \quad , \tag{13}$$

where w is a weight of the network, and $\overline{S}$ is the best state sequence of the vector sequence X for the given word sequence U. The neural network is trained to increase the likelihood in equation (8). The HMM-based speech recognizers that use mixture Gaussian distributions have three types of parameters; state transition probabilities, mean vectors, and covariance matrices. In the following sections, the MLNN is applied to mean and variance transformations.

3 Mean Transformation MLNNs

In the mean transformation, the observation x_i is fixed, and the mean $m_{s,p,i}$ becomes a variable (i.e., input and output of the neural network). For an output layer's weight update rule, the logarithm of equation (9) is differentiated with respect to the output neuron's weight $w_{i,j}$;

$$\frac{\partial \ln P(x|s)}{\partial w_{i,j}} = \sum_p \frac{\partial \ln P(x|s)}{\partial m_{s,p,i}} \frac{\partial m_{s,p,i}}{\partial \sigma_{p,i}} \frac{\partial \sigma_{p,i}}{\partial w_{i,j}} \quad , \tag{14}$$

where $\sigma_{p,i}$ is an input to the sigmoid function of the i-th neuron for the p-th probability density function (PDF) in a state. Note that σ is dependent on p because all mean vectors in a state are provided to the network sequentially, producing a different σ value for each PDF. The weight update rule for a hidden layer can be derived by first expressing the likelihood $P(x|s)$ in terms of hidden layer's weight, $w_{j,k}$, then differentiating it with respect to the weight;

$$\frac{\partial \ln P(x|s)}{\partial w_{j,k}} = \sum_p \frac{\partial \ln P(x|s)}{\partial h_{p,j}} \frac{\partial h_{p,j}}{\partial \sigma_{p,j}} \frac{\partial \sigma_{p,j}}{\partial w_{j,k}} \tag{15}$$

$$= \sum_p \sum_i \left[\frac{\partial \ln P(x|s)}{\partial m_{s,p,i}} \frac{\partial m_{s,p,i}}{\partial \sigma_{p,i}} \frac{\partial \sigma_{p,i}}{\partial h_{p,j}} \right] \times \frac{\partial h_{p,j}}{\partial \sigma_{p,j}} \frac{\partial \sigma_{p,j}}{\partial w_{j,k}} \quad , \tag{16}$$

where $h_{p,j}$ is the output of j-th hidden node for the p-th PDF. Note that $h_{p,j}$ is a function of hidden layer's weights. Equation (14) and (16) are plugged into equation (13) to complete the weight update rule for the mean transformation MLNN.

For a diagonal covariance matrix case, equations (14) and (16) can rewritten as follows;

$$\frac{\partial \ln P(x|s)}{\partial w_{i,j}} = \sum_p \frac{P(x|s,p)}{P(x|s)} \frac{m_{s,p,i} - x_i}{v_{s,p,i}} m_{s,p,i}(1 - m_{s,p,i}) h_{p,j} \quad , \tag{17}$$

$$\frac{\partial \ln P(x|s)}{\partial w_{j,k}} = \sum_p \sum_i \left[\frac{P(x|s,p)}{P(x|s)} \frac{m_{s,p,i} - x_i}{v_{s,p,i}} m_{s,p,i}(1 - m_{s,p,i}) w_{i,j} \right] \times h_{p,j}(1 - h_{p,j}) \dot{m}_{s,p,k} \quad , \tag{18}$$

where $\dot{m}_{s,p,k}$ is the input to the 1-hidden-layer network. The amount of the weight change is proportional to how important a PDF is, i.e., $P(x|s,p)/P(x|s)$, at current iteration, and how far the current mean is from the observation, i.e., $(m_{s,p,i} - x_i)/v_{s,p,i}$, in *Mahalanobis* distance space. It can be considered that there is a separate network for each PDF, and their weights are shared among them. This is mathematically equivalent to feeding the mean vector of each PDF sequentially and update the weights according to the accumulated weight change in a batch mode.

4 Variance Transformation MLNNs

The conditional probability, $P(x|s)$, can be applied to a variance transformation as well. Following the same procedure as in the mean transformation case, the weight update rule can be derived for the output layer's weights as follows;

$$\frac{\partial \ln P(x|s)}{\partial w_{i,j}} = \sum_p \frac{\partial \ln P(x|s)}{\partial v_{s,p,i}} \frac{\partial v_{s,p,i}}{\partial \sigma_{p,i}} \frac{\partial \sigma_{p,i}}{\partial w_{i,j}} \quad , \tag{19}$$

and for the hidden layer's weights;

$$\frac{\partial \ln P(x|s)}{\partial w_{j,k}} = \sum_p \frac{\partial \ln P(x|s)}{\partial h_{p,j}} \frac{\partial h_{p,j}}{\partial \sigma_{p,j}} \frac{\partial \sigma_{p,j}}{\partial w_{j,k}} \tag{20}$$

$$= \sum_p \sum_i \left[\frac{\partial \ln P(x|s)}{\partial v_{s,p,i}} \frac{\partial v_{s,p,i}}{\partial \sigma_{p,i}} \frac{\partial \sigma_{p,i}}{\partial h_{p,j}} \right] \times \frac{\partial h_{p,j}}{\partial \sigma_{p,j}} \frac{\partial \sigma_{p,j}}{\partial w_{j,k}} \quad . \tag{21}$$

Note that the variable is now $v_{s,p,i}$. Equation (19) and (21) are substituted into equation (13) for the weight update rule of the variance transformation MLNN.

For a diagonal covariance matrix case, equations (19) and (21) become as follows;

$$\frac{\partial \ln P(x|s)}{\partial w_{i,j}} = \sum_p \frac{P(x|s,p)}{P(x|s)} \frac{((x_i - m_{s,p,i})^2 - v_{s,p,i})}{v_{s,p,i}^2} \times v_{s,p,i}(1 - v_{s,p,i}) h_{p,j} \quad , \tag{22}$$

$$\frac{\partial \ln P(x|s)}{\partial w_{j,k}} = \sum_p \sum_i \left[\frac{P(x|s,p)}{P(x|s)} \frac{((x_i - m_{s,p,i})^2 - v_{s,p,i})}{v_{s,p,i}^2} v_{s,p,i}(1 - v_{s,p,i}) w_{i,j} \right] \times h_{p,j}(1 - h_{p,j}) \dot{v}_{s,p,k} \quad , \tag{23}$$

where $\dot{v}_{s,p,k}$ is the input to the network. Even though one neural network can be used to transform both means and variances, separate neural network is used for each transformation in this paper. The variance transformation is done after mean vectors are transformed.

5 Experimental Results

The baseline speech recognizer is trained using 3,979 clean speech utterances (3.8 hours) from the Resource Management (RM) speech corpus [4]. The recognizer has 8,808 Gaussian PDFs and 2,253 triphone models. The proposed method has been evaluated on a noisy distant-talking version of the RM database [8]. The SNR of the noisy speech data is about 20dB, and the distance between a microphone and a speaker is about 5.4~5.8 meters. The baseline clean speech recognizer is adapted to the noisy environment by the MLNNs using 100 utterances (4.9 minutes) of noisy speech.

Figure 1 shows the trajectories of *mel-frequency cepstral coefficients* (MFCCs) of an example utterance. The solid line is for the first dimension cepstral coefficients of noisy speech. The example utterance is aligned against the clean speech recognizer using the Viterbi algorithm and the corresponding mean vector sequence is plotted as the dotted

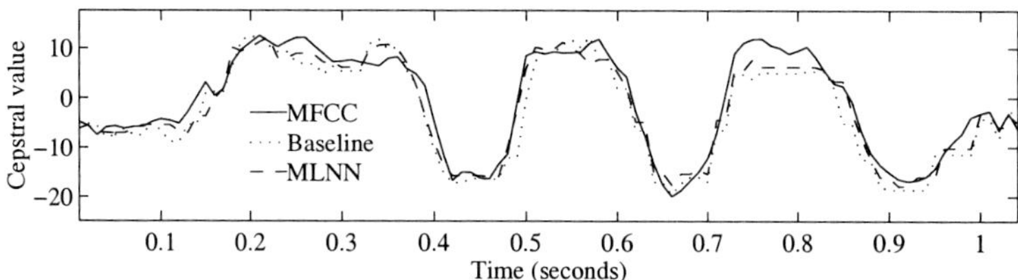

Fig. 1. The trajectories of cepstral coefficients and aligned mean vectors of an example utterance "*Find sensors*".

line in the figure. The dashed line is for the aligned mean vectors of the transformed recognizer by the MLNN. It can be seen from the figure that the transformed mean vectors (dashed line) are closer to the testing speech data (solid line) than the baseline system's mean vectors (dotted line) are.

Figure 2 shows the empirical and transformed distributions of a sound "*g*". The distributions of the middle state of a triphone model of which base phone is "*g*" are plotted in the figure. Only the first dimension of MFCCs is shown in the figure. Figure 2 (a) is a clean speech distribution. Figure 2 (b) is the result of mean transformation, and Figure 2 (c) is the result of mean and variance transformation. Figure 2 (d) is a noisy distant-talking speech distribution of the same sound. It can be observed in these figures that the original clean speech distribution, Figure 2 (a), is successfully converted to Figure 2 (c) which looks more similar to the noisy distant-talking speech distribution, Figure 2 (d), than the original distribution does.

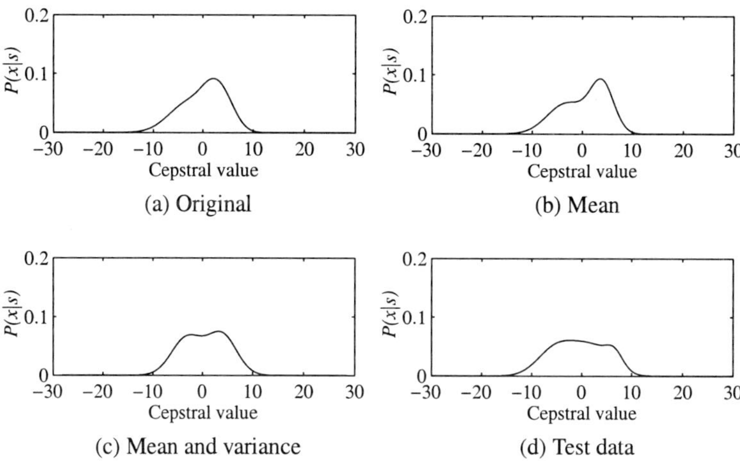

Fig. 2. The empirical and the transformed distribution of clean and noisy speech sounds "*g*".

Table 1 compares the word recognition accuracies of the MLNNs. Except in the retraining case, the clean speech recognizer or its transpormations are used for the evaluation. The original clean speech recognizer tested on the clean speech data shows

Condition	Accuracy (%)
Clean (matched)	94.5
Retraining (matched)	79.0
Distorted (mismatched)	64.3
MLNN_M	73.0
$\text{MLNN}_{M\&V}$	78.7
$\text{MLNN}_{M\&V+US}$	83.2

Table 1. Word recognition accuracies (%); "Clean" is for clean speech training and testing, "Retraining" is for noisy speech training and testing, and "Distorted" is for clean speech training and noisy speech testing. The rest are using clean speech recognizer after transformation by MLNNs; "MLNN_M" is for mean transformation, "$\text{MLNN}_{M\&V}$" is for mean and variance transformation, and "$\text{MLNN}_{M\&V+US}$ is for unsupervised speaker adaptation on top of the $\text{MLNN}_{M\&V}$.

94.5% word recognition accuracy. When the same system is used for noisy speech, the performance drops to 64.3% even after some robust feature processing techniques [10]. The mean transformation MLNN improves the word recognition accuracy to 73.0%, and the mean and variance transformation MLNN to 78.7%. This result is quite competitive with a retrained recognizer (79.0%). However, note that the retrained recognizer requires a lot more data than the MLNNs do. Using the hypotheses of 78.7% correct, one mean transformation MLNN is built for each speaker in unsupervised way. The unsupervised speaker adaptation improves the performance (83.2%) beyond the retrained recognizer.

6 Conclusions

For a typical statistical pattern recognition framework such as HMMs, the conditional probability density functions can be used as the new objective function of the neural network. The resulting MLNN can transform the distributions in the HMMs and produce parameters that are consistent with testing environments. The mean and variance transformation MLNNs learn the distortion function that degrades clean speech to noisy speech. The training can be done in unsupervised way by making use of recognized output of unknown speech. The proposed technique has been applied to large vocabulary continuous speech recognition under adverse acoustical environments, which involves background noise, reverberation, and difference in microphones. The model transformation MLNN experiment done in this research uses only one network to transform all mean vectors (or covariance matrices) of a recognizer. This can be modified to be state-dependent, where each state has its own MLNN to transform its parameters. The states can be grouped using tree structure so that those states that do not have enough training data can share an MLNN. The MLNN is a good candidate for discriminative training, because alternative hypotheses or confusing pairs can be provided from a speech recognizer. In this case, *mutual information* or *maximum a posteriori probability* criteria may be considered for the objective functions. Current unsupervised adaptation methods use the output of a mismatched recognizer. When the adaptation algorithms run iteratively, the recognizer tends to make same mistakes repeatedly. More intelligent techniques to exploit the multiple hypotheses from possibly multiple recognizers need to be investigated.

References

1. Y. Bengio, R. DeMori, G. Flammia, and R. Kompe. Global optimization of a neural network-hidden Markov model hybrid. *IEEE Transactions on Neural Networks*, 3(2):252–259, March 1992.
2. A. Biem and S. Katagiri. Feature extraction based on minimum classification error/generalized probabilistic descent method. *IEEE International Conference on Acoustics, Speech, and Signal Processing*, 2:275–278, April 1993.
3. S. Katz. Estimation of probabilities from sparse data for the language model component of a speech recognizer. *IEEE Transactions on Acoustics, Speech, and Signal Processing*, ASSP-35(3):400–401, March 1987.
4. P. Price, W. Fisher, J. Bernstein, and D. Pallett. The DARPA 1000-word resource management database for continuous speech recognition. *IEEE International Conference on Acoustics, Speech, and Signal Processing*, 1:651–654, April 1988.
5. M. Rahim and C. Lee. Simultaneous ANN feature and HMM recognizer design using string-based minimum classification error (MCE) training. *International Conference on Spoken Language Processing*, 3:1824–1827, October 1996.
6. D. Rumelhart, G. Hinton, and R. Williams. Learning internal representations by error propagation. In J. McClelland D. Rumelhart, editor, *Parallel Distributed Processing: Exploration in the Micro-Structure of Cognition*, volume 1, pages 318–362. MIT Press, 1986.
7. S. Tamura and A. Waibel. Noise reduction using connectionist models. *IEEE International Conference on Acoustics, Speech, and Signal Processing*, 1:553–556, April 1988.
8. D. Yuk. *Robust Speech Recognition Using Neural Networks and Hidden Markov Models.* PhD thesis, Rutgers University, 1999.
9. D. Yuk, C. Che, and J. Flanagan. Robust speech recognition using maximum likelihood neural networks and continuous density hidden Markov models. *IEEE Workshop on Automatic Speech Recognition and Understanding*, pages 474–481, December 1997.
10. D. Yuk, C. Che, L. Jin, and Q. Lin. Environment-independent continuous speech recognition using neural networks and hidden Markov models. *IEEE International Conference on Acoustics, Speech, and Signal Processing*, 6:3358–3361, May 1996.
11. D. Yuk, C. Che, P. Raghavan, S. Chennoukh, and J. Flanagan. N-best breadth search for large vocabulary continuous speech recognition using a long span language model. *136th meeting of Acoustical Society of America*, page 1819, October 1998.
12. D. Yuk and J. Flanagan. Telephone speech recognition using neural networks and hidden Markov models. *IEEE International Conference on Acoustics, Speech, and Signal Processing*, 1:157–160, March 1999.
13. D. Yuk, J. Flanagan, M. Krishnamoorthy, and K. Dayanidhi. Adaptation to environment and speaker using maximum likelihood neural networks. *Eurospeech*, pages 2531–2534, September 1999.

Neural Network Based Algorithms for IP Lookup and Packet Classification

Mehran Mahramian[1,2], Nasser Yazdani[2], Karim Faez[1], and Hassan Taheri[1]

[1]Amirkabir University of Technology, Electrical Engineering Department, 424 Haafez Ave, Tehran 15914, Iran
m_mahramian@isc.iranet.net, {htaheri, kfaez}@aku.ac.ir
[2]Tehran University, Engineering Department, North Kargar Ave, Tehran, Iran
nasyaz@sofe.ece.ut.ac.ir

Abstract. IP routers need lookup tables to forward packets. They also classify packets to determine which flow they belong to and to decide what quality of service they should receive. Increasing rate of communication links is in contrast with practical processing power of routers and switches. We propose a few neural network algorithms to solve the IP lookup problem. Some of these algorithms, gives promising results, however, they have problems in training time. Parallel processing of neural networks provide a huge processing power to do IP lookup. The algorithm can be implemented in hardware on a single chip. Our method can perform an IP lookup in 4.5 nanoseconds, which implies supporting 60 Gbps link rate. Pipelining and parallel processing can be used to increase the link rate up to 400 Gbps and decrease the learning time.

1 Introduction

IP lookup and packet classification are the bottlenecks of the high speed networks. The communication links have increased their baud rates up to terra bit per second, but the processing speed has not been improved at the same rate. However, increasing the communication line speed reduces the processing and search times since the packets must be switched at the wire speed. This implies the forwarding engines need a more efficient IP lookup technique [24].

Routers perform a forwarding decision on an incoming packet to determine the packet's next-hop. This is achieved by examining destination address of the packet, finding the longest matching prefix in the forwarding lookup table and then forwarding packets to the corresponding next hop. Increasing packet arrival rates at higher speed links, the complexity of the lookup mechanism and the size of forwarding tables have made IP lookups a bottleneck [10].

The objective of packet classification is to identify the highest priority rule that applies to an incoming packet. Most of the present solutions of packet classification make a transformation from packet classification domain to IP lookup domain ([2], [3], [4], [8], [9], [12], [13] and [16]). Thus, an extendable fast algorithm for IP lookup can solve the packet classification problem.

M.H. Shafazand and A M. Tjoa (Eds.): EurAsia-ICT 2002, LNCS 2510, pp. 204–211, 2002.

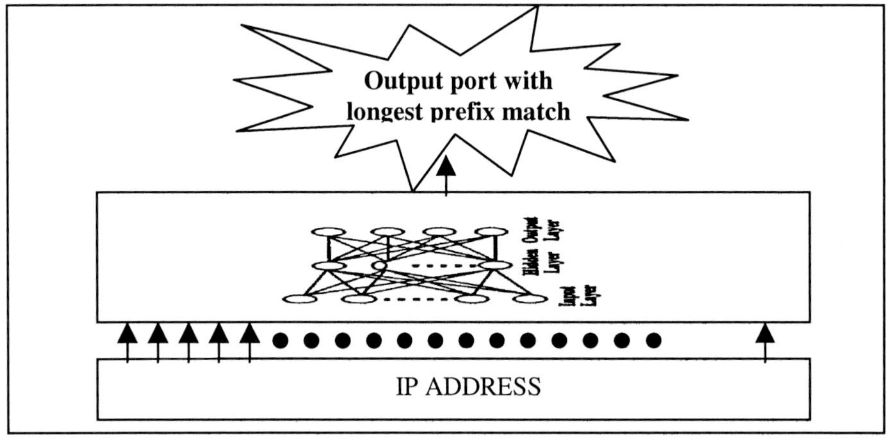

Fig. 1. The block diagram of a neural network IP lookup System.

Any lookup or classification algorithm must satisfy some requirements [10]. First, it must be fast in search to support high speed communication links. Implementation comes next. It also should have low update and initialization time.

Neural network algorithms are generally flexible and fast. They can be implemented using parallel processors. They require distributed memory architectures, which reduce the access time to the memory since they can be accessed simultaneously. By increasing the number of hidden layer neurons, neural networks can learn large routing tables [18]. Training time is the main issue in neural networks. Usually, it takes a long time for a neural network to learn all input patterns. The update rate of neural networks is another problem [7].

The rest of the paper is organized as follows. Section 2 describes IP lookup problem. Section 3 continues with back propagation networks and simulation results. We applied the architecture to the different types of back propagation networks in this section. Implementation issues including architecture of the VLSI circuit and neurons are investigated in section 4. Section 5 concludes the paper and contains the future work.

2 Background

The proposed approaches for the IP lookup problem are usually classified in four categories [22]. First category includes methods which usually modify exact matching schemes and apply them to the prefix matching problem ([24], [6], [20], [17], [11] and [5]).

Second category contains hardware based solutions. They usually use CAM (Content Addressable Memory). In CAM, the content of each memory location is compared in parallel to the input key and data in the matching location is put on the output. This is relavtive fast, but an expensive solution [19, 21].

Third category contains protocol-based solutions like MPLS and TAG switching. These methods translate the prefix matching problem into the exact matching. In these methods, we need to modify the packet headers. Furthermore, not all of the existing networks support these protocols and we will face difficulties in applying those to all points of the global network. Besides, still, we need IP lookup at the ingress or entry points to the network. Finally, the last category of solutions is caching. This solution is also expensive like CAM and cannot be used for the large IP lookup Table.

In the neural network algorithm, we train the neural network and apply all of the prefixes to the network. After the learning phase, each IP address enters the network and the next hop address appears at the output of the network. Figure 1 shows the block diagram of the system. This is a simplified view of the network. IP packets enter from input ports of the switch and the corresponding destination address, which specifies the output port that the packet should be forwarded to, enters the neural network. There is a module for training and a memory for managing updates.

Here, we try back propagation algorithm, which is a supervised neural network, i.e. each new IP address in training is trained to the neural network as well as its next hop. Simulation results show that back propagation algorithm works well, however, it is slow in initialization and updating (i.e. the insertion and deletion of IP prefixes). We tune the algorithm to make it suitable for IP lookup at very high speeds. Links running at 60 Gbps rates need the router to process 200 million packets per second (assuming minimum-sized 40 bytes TCP/IP packets) [10]. Our algorithm can accept a new packet every 4.5ns. So it can forward 200 million packets in a second.

3 IP Lookup and Back Propagation

Back propagation (BP) is simply a gradient descent method to minimize the total squared error of the output computed by the net. The BP neural networks used in this study are three layered feed forward. The length of the IP address, 32 bits, dictates the size of the input layer (X). The number of available next hops represented in the IP switch, each output unit representing one output port, determines the output layer size (Y). The hidden unit size in the first step is ten (Z). We use the Hecht-Nielsen theorem [7] for an optimized and reliable number of hidden neurons. We chose 2 * 32 + 1 = 65 for the number of hidden neurons. Bipolar input and output and bipolar sigmoid activation function are used. If input IP address is C4(H), the input of BP network will be +1+1-1-1-1+1-1-1 which is applied directly to X layer of the BP network. Nguyen-Widrow initialization [7] is used for initialization of the first and second layer weights. Biases in both layers are assumed. The operations of the present neural IP lookup system consist of two stages: the learning stage, during which the system learns IP prefixes; and the retrieving stage, during which the system retrieves IP addresses after being presented.

At the learning stage, IP prefixes are presented to the input channels of the system. For example, let the IP prefix length be 2 (n = 2) bits and number of units in hidden layer be 2 (p = 2) and number of output units be 2 (m = 2). Weight matrices are V and W. V is a (n+1)*p matrix. It connects input layer to the hidden layer. W is a (p+1)*m matrix which connects each unit of hidden layer to output layer. If IP address = [+1 – 1] appears at the input of the neural network and V and W having the following initialized values:

$$V = \begin{bmatrix} 1.2 & -.3 \\ 1.9 & .6 \\ .03 & -1.7 \end{bmatrix}, W = \begin{bmatrix} -.2 & .85 \\ -1.5 & -.9 \\ 2.1 & -1.4 \end{bmatrix}$$

The input of the hidden layer will be a Z matrix calculated by:

$$Z_i = [1 \ \ IP] * V \tag{1}$$

Outputs of the hidden units are determined by a built-in activation function f(.) like a bipolar sigmoid function:

$$Z_o = f\,(Z_i) \tag{2}$$

O_i vector represents inputs of output layer

$$O_i = [1 \ \ Z_o] * \ W \tag{3}$$

Another activation function brings the output vector O. If the input packet should be forwarded to the first output, then we expect a [+1 –1] for the output vector. During learning phase, weights in V and W are changed, so we can have the desired output vector for each input IP address. In retrieving phase, the incoming packets' IP addresses are applied to the neural network and the network decides the next hop using the multiplication, summation and activation functions (Equations 1 to 3).

Figure 2 shows the block diagram of application network. We proposed different schemes for structure and configuration of back propagation neural network. Table 1 shows simulation results. Configuration number 1 is a back propagation neural network in which the inputs and outputs are binary. All IP addresses matching a prefix are used as inputs to the neural network for training. For example, if we have an IP prefix, 192.168.18.*, all of IP addresses from 192.168.18.0 to 192.168.18.255 are inputs to the network. Configuration number 2 is similar to the first one except that the IP prefixes are sorted in the IP lookup table so that the longest prefix comes at the bottom of the table. The error is decreased in this case as shown in Table 1 and Figure (3.a). Mean square error in the same configuration, however, in a random order for IP prefixes, is shown in figure (3.b). In the next configuration, bipolar inputs and outputs are used and we place *s by 0s in prefixes. The training time is decreased in this configuration but the network routes 255 packets to a wrong next hop. There were 2304 packets, so we have more than 10 percent errors and this is not acceptable. In configuration 4, bipolar inputs and outputs are used and all IP addresses matching a prefix are used as inputs to the network during the training phase (Like config. number 1). This configuration has the best results i.e. only 6 errors in 2304 packets (Figure 3.c and 3.d). This implies less than 0.2% error in routing. Of course, errors in routing are not acceptable. Fortunately, with optimizations in the algorithm, we can improve the error rate.

The configuration number 4 is selected for the final circuit, so it has the minimum error. The IP lookups are updated almost every 30 seconds. For compensation of high training time, we duplicated the architecture. One of the neural networks is used for

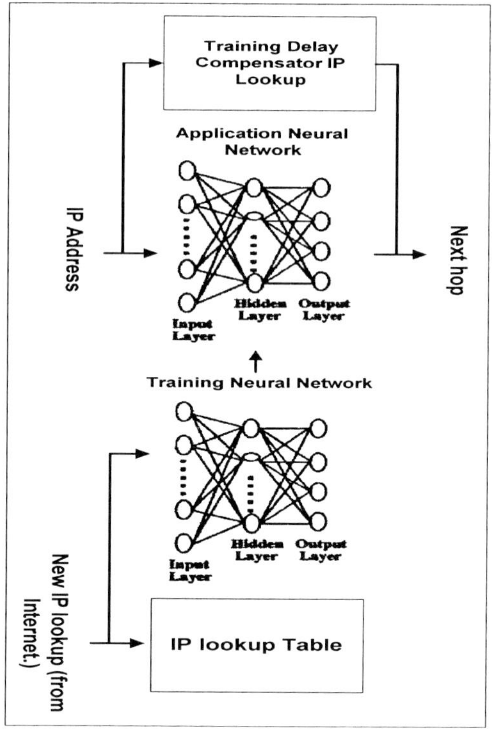

Fig. 2. Block Diagram of fast IP switch.

Table 1. Simulation results for back propagation network in different configurations (number of IP addresses in each configuration is 2304).

cfg.	N	p	m	No. of Iterations	Mean Squared Error.	Critical IP Routing Error.	IP routing Errors.
1	32	10	8	70	1625.4	*	*
2	32	10	8	70	450.2130	*	*
3	32	10	8	50	1658.2	255	255
4	32	10	8	100	562.4514	6	6

the application and another for training (Figure 2). During the training, new IP addresses may enter the IP lookup, but the application neural network is not trained for these IP addresses. A small traditional IP lookup is used in parallel with the application neural network. During update and training, a traditional IP lookup scheme routes the IP addresses that match the new IP prefixes. Application neural network routes these IP addresses too, but the output of traditional IP lookup has the highest priority and can override the result of the application network. After training, the traditional IP lookup will be disabled.

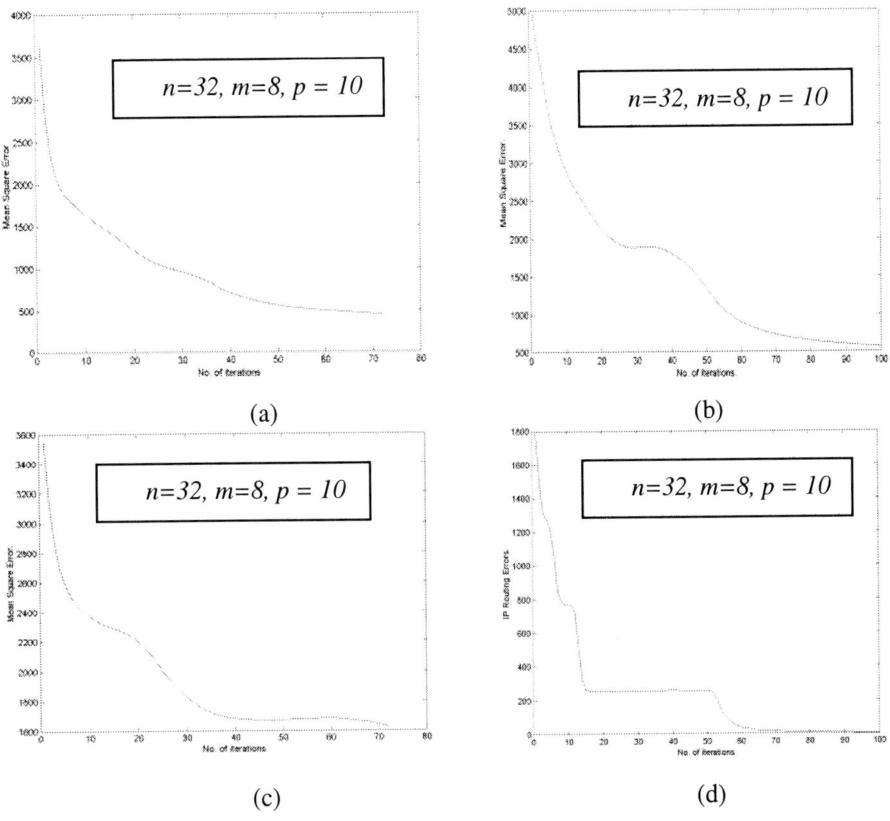

Fig. 3. Mean Square Error during training phase versus number of iterations. (a) IP prefixes were sorted before training. (b) Order of inputs is randomized (c) Configuration number 4 (d) Number of IP routing Errors during training versus number of iterations. (Configuration number 4)

4 Implementation Issues

The most important part of a neuron is the multiplier since the time needed for a multiplication is the bottleneck [15]. The transfer function can be implemented with a small lookup table or another neural network. We use the bipolar sigmoid function.

A fast multiplier used in our architecture can be found in [23]. The multiplier has a delay AND gate delay + log n * ADDER delay. Using this multiplier in a pipelined architecture, a multiplication of two numbers can be performed in only one clock cycle. The delay of a pipelined adder is only one clock cycle which can be assumed two times of an AND gate delay since a full adder can be realized by a two-level circuit [14]. Then, the delay of the multiplier is AND delay * (1 + 2* log n) where n is the number of bits used for the weights. Assuming n is 16 and delay for a fast AND gate is 0.5 ns, we will have 4.5ns delay for a multiplier. The pipelined architecture of back propagation neural network enables the network to have an output in each clock cycle. Therefore, our neural network can forward each IP packet in one clock cycle. With a 200 Mega Hertz clock, the circuit can forward 200 million packets in each second. Assuming average of 256 bytes for IP packet length, our algorithm can support a 400 Gbps link rate. Using high technologies can increase the clock rate so advancing the rate of forwarded packets in one second. For example, assuming a one Giga Hertz clock, the architecture can support a 2000 Gbps link rate.

The logic area for each multiplier is n * (n-bit AND) + (n-1) * (2n-bit ADD). Nine NAND gates are considered for each full adder, 13n gates are needed for each n bit adder. Thus the area of the silicon chip will be n^2 + (n-1) * (2n * 13) = 27 * n^2 - 26 * n. Assuming n to be 16, 6500 NAND gates is needed for each neuron. In the proposed neural network, the number of inputs is 32 and the number of outputs is 8. Choosing 10 as the number of hidden neurons, we will need 80 multipliers for the second layer, which adds up to 16 * 16 multipliers totally.

For the first layer, 1 * 16 bit multipliers are needed because inputs are 1 or –1. These multipliers can be ignored comparing with the second layer multipliers. Therefore we need 80 * 6500 = 520000 NAND gates for implementation of the neural network. Assuming 3 transistors for each NAND gate, the network will contain 1.5 million-transistor, which can be implemented on a single chip.

5 Conclusion and Future Work

Four back propagation neural networks are proposed for the IP lookup problem. Our results show we can forward one packet in each clock cycle. We also proposed architecture for the on chip implementation of our algorithm. According to our primary calculation, the method can be implemented in a chip with roughly 1.5 million-transistors. Regarding the average packet size of 256 bytes, our method can easily handle a 400 Gbps link rate.

References

1. Baraldi, A., Parmiggiani, F.: Neural Network for Unsupervised Categorization of Multivalued Input Patterns: an Application to Satellite Image Clustering, IEEE Transactions on Geoscience and Remote Sensing, Vol. 33 (1995) 305–316
2. Borg, N., Svanberg, E., Schelen, O.: Efficient Multi-Field Packet Classification for QoS Purposes, Seventh International Workshop on Quality of Service (1999) 109–118
3. Chaskar, H.M., Dimitriou, E.; Ravikanth, R.:Service Guarantees in the Internet: Differentiated Services Approach, Eighth International Workshop on Quality of Service (2000) 176–178
4. Decasper, D., Dittia, Z.; Parulkar, G.; Plattner, B.: Router Plugins: A Software Architecture for Next-Generation Routers, IEEE/ACM Transactions on Networking (2000) 2–15
5. Degermark. M., Brodlink. A., Carlsson. S., Pink. S.: Small Forwarding Tables for Fast Routing Lookups, Proceedings of SIGCOMM (1997) 3–14
6. Doeringer. W., Karjoth. G., Nassehi. M.: Routing on Longest-Matching Prefixes, IEEE/ACM Trans. Networking, vol. 4, no. 1 (1996) 86–97

7. Fausett, L.: Fundamentals of Neural Networks. Architectures, Algorithms and Applications, Prentice Hall International, Inc. (1994)
8. Fulu, L., Seddigh, N., Nandy, B., Matute, D.: An Empirical Study of Today's Internet Traffic for Differentiated Services IP QoS, Proceedings of Fifth IEEE Symposium on Computers and Communications (2000) 207–213
9. Feldman, A., Muthukrishnan, S.: Tradeoffs for Packet Classification, Proceedings of Nineteenth Annual Joint Conference of the IEEE Computer and Communications Societies, Vol. 3 (2000) 1193–1202
10. Gupta, P.: Algorithms for Routing Lookups and Packet Classification, Degree of Doctor of Philosophy Thesis Submitted to the Department of Computer Science of Stanford University (2001)
11. Gupta. P., Lin. S., McKeown,. N.: Routing Lookups in Hardware at Memory Access Speeds, Proceedings of Seventeenth Annual Joint Conference of the IEEE Computer and Communications Societies, vol.3 (1998) 1240–1247
12. Gupta P., McKcown, N.: Packet Classification on Multiple Fields, Proceedings of SIGCOMM (1999) 147–160
13. Hari, A., Suri, S., Parulkar, G.: Detecting and Resolving Packet Filter Conflicts, Proceedings of Nineteenth Annual Joint Conference of the IEEE Computer and Communications Societies, Vol. 3 (2000) 1203–1212
14. Hill. F. J., Peterson. J. R.: Introduction to Switching Theory and Logical Design, John Wiley & Sons, New York, 3rd edn. (1981)
15. Jabri, M. A., Coggins, R. J., Flower. B. G.: Adaptive Analog VLSI Neural Systems, Chapman & Hall (1996)
16. Jun, X., Singhal, M., Degroat, J.: A Novel Cache Architecture to Support Layer-Four Packet Classification at Memory Access Speeds, Proceedings of Nineteenth Annual Joint Conference of the IEEE Computer and Communications Societies, Vol. 3 (2000) 1445–1454
17. Lampson, B., Srinivasan, V., Varghese, G.: IP Lookup Using Multiway and Multicolumn Search, IEEE/ACM Transactions on Networking, Vol. 7 (1999) 324–334
18. Lawrence, S., Lee Giles, S., Ah Chung, T.: What Size Neural Network Gives Optimal Generalization? Convergence Properties of Back propagation, Technical Report, Institute for Advanced Computer Studies University of Maryland (1996)
19. McAuley, A. J., Francis, P.: Fast Routing Table Using CAMs, Proceedings of Twelfth Annual Joint Conference of the IEEE Computer and Communications Societies. Networking: Foundation for the Future, vol.3 (1993) 1382–1391
20. Nilsson, S.: IP-Address Lookup Using LC-Tries, IEEE Journal on Selected Areas in Communications, Vol. 17 (1999) 1083–1092
21. Pi-Chung, W., Chia-Tai Chan, Yaw-Chung, C.: A Fast IP Routing Lookup Scheme, IEEE International Conference on Communications, Vol. 2 (2000) 1140-1144
22. Waldvogel, M., Varghese. G., Turner. J., Plattner. B.: Scalable High Speed IP Routing Lookups, In Computer Communication Review, Vol. 27, no. 4 (1997)
23. Web site of Henryk Niewodniczanski Institute : http://chall.ifj.edu.pl/~szczygie/slides
24. Yazdani, N., Min, P.S.: Fast and Scalable Schemes for the IP Address Lookup Problem, Proceedings of the IEEE Conference on High Performance Switching and Routing (2000) 83–92

Non-linear Prediction of Speech Signal Using Artificial Neural Nets

K. Ashouri, M. Amini, and M.H. Savoji

Electrical and Computer Engineering Faculty
Shahid Beheshti University
Evin Square, Tehran 1983963113, Iran.

Abstract. Speech technology is one of the key technical issues involved in Information Technology as it constitutes an important aspect of Human Computer Interaction. Prediction of speech signal has applications in speech technology, especially in coding. Conventionally, linear prediction is used. However, non-linear phenomena exist in speech production and, considering this non-linearity should lead to lower signal dynamics during coding with a consequent reduction in bit-rate and the needed bandwidth. The non-linear prediction of speech segments, as long as a whole vowel, using neural nets is studied in this paper. It is shown that non-linear speech prediction does not lead to an appreciable further reduction in the residual signal in this case.

1 Introduction

The prediction of speech has applications in speech technology i.e. speech recognition, synthesis and coding. Linear prediction is used conventionally to reduce the redundancy of speech signal and decrease the coding bit-rate. The reduction in bit-rate is achieved by coding the residual signal or what remains from the speech once its predictable part has been removed. However, it is known that radiation effects from the lips and turbulences of the air flow from the lungs cause non-linear phenomena in speech production [1]. Therefore, considering the non-linearity in speech prediction is believed to result in lower dynamics of the residual signal to be coded. The non-linear prediction of speech can be achieved using Artificial Neural Nets. The studies carried out about a decade ago [2], [3] in this regard showed that 2 to 3dB reduction in the coder gain factor was possible when short frames of speech were analyzed. This has resulted, since, in different coding schemes using non-linear prediction. Most recent ones include [4] to name but one. Nevertheless, there are applications where segments as long as whole vowels are used e.g. speech synthesis using waveform concatenation. The non-linear prediction of long speech segments is studied in the work reported here.

M.H. Shafazand and A M. Tjoa (Eds.): EurAsia-ICT 2002, LNCS 2510, pp. 212–219, 2002.

2 Artificial Neural Nets and Non-linear Prediction

Neural nets have been used extensively in non-linear problems [5] for the mere fact that they can automatically generate an optimum solution if it exists and their high speed of execution [6].

Hundreds of structures have been proposed for neural nets [7], [8] that can be divided into two main groups: Feed Forward Neural Nets (FFNN) and Recurrent Neural Nets (RNN). In FFNNs, the mapping between the input and output remains unchanged once the training is completed and the network is stationary. On the other hand, RNNs are complex non-linear dynamic nets with cyclic connections.

In most problems it is convenient to use multi-layer FFNNs (MLFN), due to their simplicity, when supervised learning is possible. In our study, MLFNs both with and without cyclic connections have been used. This choice was motivated by the fact that MLFNs can be used for non-linear Auto Regression (AR) modeling whilst the same structure with cyclic connections can be viewed as non-linear Auto Regression Moving Average (ARMA) modeling [9].

3 Non-linear Prediction of the Speech Residual Signal

It is usually suggested that all linear trends of the input be removed before it is applied to a neural net. This can be done explicitly by calculating the linear prediction coefficients and the excitation signal and using this signal as the input of an MLFN or, implicitly by loading the calculated values as the fixed weights of the direct connections between the first layer neurons and the single neuron of the last layer of the network, as shown in figure 1, where the speech waveform is input to the structure arranged as Time Delay MLFN (TD-MLFN). The delay line is used here to convert the temporal pattern to a vector of values i.e. the input spatial pattern of the neural net.

Alternatively, instead of loading the pre-calculated linear prediction coefficients as weights of the connections shown in the figure as dashed lines, the net can be let to calculate them in combination with all other weights. This is linear and non-linear prediction combined. When the network is to carry both linear and non-linear prediction, the neuron in the last layer must have a linear characteristic instead of the usual Sigmoid or Tansigmoid function. When the prediction of the input signal is sought, the output of the TD-MLFN is $z(t)=x(n+1)$ where the network carries the prediction of the input one sample interval ahead.

4 The Back Propagation Training Algorithm and Its Generalization

The well known BP algorithm used for training of FFNNs can be generalized for training of MLFNs that include connections between a layer and any layer above it. This is shown as an example in [7]. The generalization of the BP algorithm to recurrent networks is given in [10]. This generalization as outlined in [7] has been used in this work. Other recurrent algorithms exist and can be found in [9].

The BP algorithm is the basis of training algorithm used in MATLAB NN Toolbox whose routines have been extensively used in our work.

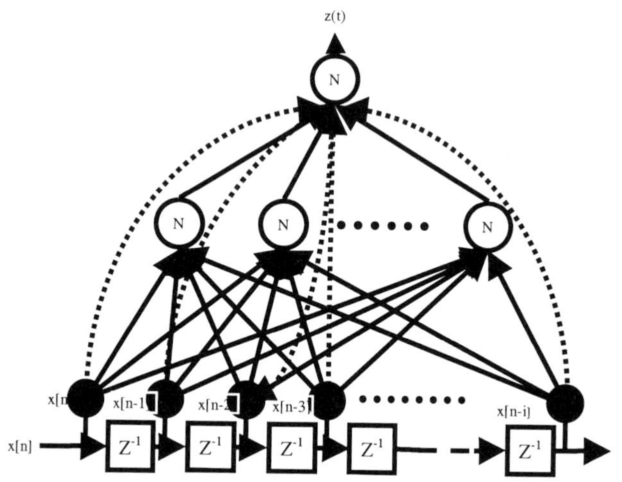

Fig. 1. TD-MLFN with direct connections between the first and last layer

We developed, however, our own package of routines in Visual C++ using the BP algorithm and its generalizations mainly, to overcome the limitations of recurrent routines in MATLAB that are rather approximations.

4.1 The Selection of the Best Training Algorithm

Many different variants of the BP algorithm exist for the training of simple MLFNs. Choosing the best algorithm is difficult and depends on many factors. MATLAB suggests the Levenberg-Marquardt (LM) algorithm as the best for networks with less than 100 eights [11]. This algorithm has been used here in batch training mode when using MATLAB Toolbox. Batch Gradient Descent with Momentum was the method used for training in our developed package. In all cases the mean square error has been used as the optimization criterion.

4.2 Initialization

When using MATLAB, the Ngugen-Widrow algorithm [11] has been used for initialization. This algorithm is known to initialize the weights and biases so that fewer neurons are left inactive in the network and all neurons participate efficiently, increasing the speed of processing. Simple random initialization has been used in our package for simplicity.

4.3 Overtraining and the Ways to Avoid It

One of the problems encountered during training is over-training or over-fitting. In this condition, the error in the training set is very low but when new data are employed the output error increases. In this situation the network is not generalized for new inputs. One of the methods used to avoid this problem is early stopping.

The Early Stopping of the Training Phase. In this method the available data is divided in three sets: Training, validation and test. The training set is used for calculating weights and gradients during training. The validation set is also used during training and the error is calculated for this set although it is not used for network calculation. In normal conditions, continuing the training will reduce both the training set and the validation set errors but, when over-fitting is occurred the error for the validation set starts to grow. That is when training should be stopped. Naturally, the test set is only to assess the performance of the network once it is trained. Early stopping has been used in this work.

5 Post-regression Analysis of Results

The performance of a trained network can be assessed considering the errors corresponding to the above three sets. However, a regression analysis can be applied between the actual and the desired outputs. A linear regression is sought and the analysis is carried out calculating three parameters; namely m,b,r. The parameters m and b are respectively the slope and the distance from the zero origin in the best linear regression if there is any. If the network performs ideally m=1 and b=0. The parameter r is the correlation coefficient between the desired and actual output of the net. If r=1 then there is a complete correlation between the two.

6 The Data-Base

The waveforms of Farsi (Persian) phrases and words uttered by two male speakers were recorded at 11 and 22 KHz sampling frequencies and digitized with 8 and 16 bits. Then words were segmented into syllables to be saved in separate files as items of our data-base. The phonetic description of the files' contents and other characteristics such as the speaker code and the code of microphone used were attached to each file. A search engine permits to extract all files with a specific phonetic content and other needed characteristics such as the sampling frequency or bit representation for different experiments.

7 The Experimental Results

The results obtained using TD-MLFNs are first reported. It is the network used unless otherwise specified.

Since the initialization of network parameters was random it was necessary to repeat each experiment many times to ensure that local optimizations were avoided and a correct solution was achieved. When assessing the performance of the net on different inputs, such as in generalization, each experiment was repeated ten times and the best result saved for later comparison with similar experiments.

7.1 Determining the Network Structure and Dimension

One of the important issues in neural computing is the selection of appropriate network structure and dimension. This problem has been dealt with in the literature [6]. Usually, as few as possible neurons are used with only one hidden layer.

In some neural structures like TD-MLFNs the correlation between input samples and other information such as the sampling frequency can be used for determining the minimum number of neurons needed in the first layer called also the sensors layer.

As for the speech signal sampled at almost twice the maximum frequency (i.e. 8 to 10 KHz), it is well known that considering neighboring samples more than 10 samples distant does not produce further reduction in the linear prediction error suggesting that the correlation drops to a negligible value after 9 to 10 samples [12]. Therefore, from the point of view of the net's dimension, the 9-3-1 structure seemed appropriate to start with.

The Optimum Structure. For input speech sampled at 11 KHz increasing the number of neurons in the hidden layer from 3 to 4 resulted in reducing the output error in most cases. But, for speech sampled at 22 KHz this reduction was insignificant. Increasing the number of neurons in the input layer from 9 to 15 reduced the output error for 11 KHz speech but again, this reduction was much less in the case of speech sampled at 22 KHz. It was concluded that the 9-3-1 structure was appropriate especially for speech inputs sampled at 22 KHz sampling frequency.

As for the neuron type, specified by the neuron's excitation function, changing it from tansigmoid (tansig) to pure linear (purelin) for the output neuron was without noticeable effect. Therefore, it was assumed that the network for combined linear-non-linear prediction, with linear function for the output neuron, was structurally no different from the non-linear prediction network making the comparison easier.

7.2 The Effect of Bit Resolution

The results obtained on 22 KHz speech files at 8 and 16 bit resolutions showed that the training errors, for an equal number of training epochs, were very close and the network performed almost equally for either resolution. This showed the relative immunity of the network to quantization noise; a well known property.

7.3 The Effect of Sampling Frequency

Different network structures and inputs with both resolutions were used, trained for the same number of epochs, with inputs sampled at 11 and 22 KHz. It was observed that a lower output error could be achieved for inputs sampled at 22 KHz as compared with those sampled at 11 KHz. This was interpreted as neural nets being capable of using more information presented in a wider input bandwidth.

7.4 Comparison of Linear and Non-linear Prediction of Speech

Despite it was hoped that non-linear prediction being more general should result in lower output error, the error using a neural net and the linear prediction error calculated in a conventional manner were almost equal in all studied cases which were all long segments, mostly as long as a whole vowel in a syllable. This suggested that non-linear prediction did not lead to an appreciable reduction in prediction error in this case. This observation can be seen in the figure 2.

The Non-linear Prediction of the Speech Excitation Signal. When the linear prediction error signal called also the excitation signal was used as input, the network was trained in the first few epochs showing that no further training was possible. In all cases the output error was only slightly smaller than the excitation input where the peaks were attenuated to some degree. This observation was confirmed by the post regression analysis as shown in figure 3.

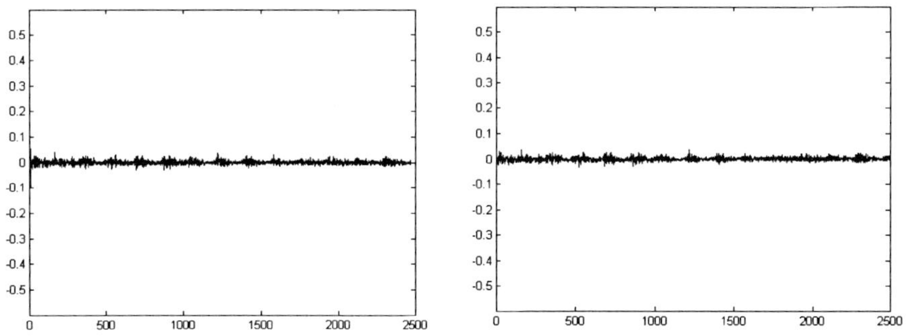

Fig. 2. Prediction error signals; linear (left) and non-linear (right)

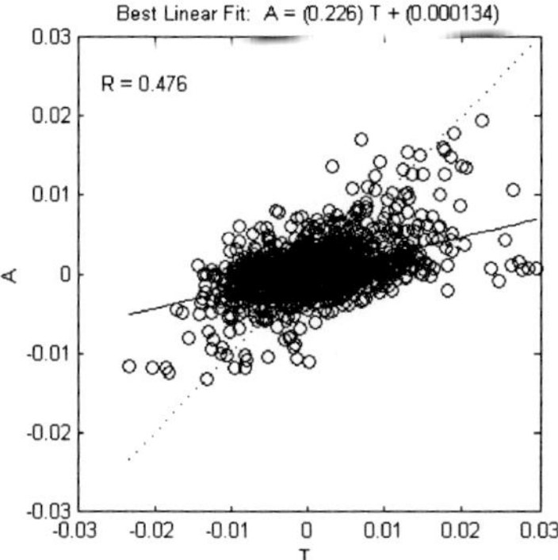

Fig. 3. Post-regression analysis of non-linear prediction of speech excitation signal

7.5 The Generalization of the Net

The Generalization for the Same Vowel. The generalization was carried out first by training a net, for a given vowel, on the first of a sequence of files with the same content and using the trained net as the starting point for continuing the training with other files in the sequence. The procedure was repeated several times by changing the order of appearance of files in the sequence. The generalization depended on the set output error. For a moderate level of error the generalization was good even for a mixture of files from different speakers, resolutions and sampling frequencies. But, for low output errors the network did not generalize for 11 KHz speech inputs. Excluding these files permitted good generalization on 22 KHz speech files.

This observation was confirmed when using the early stopping of training as a means of controlling the generalization process. It can then be concluded that the generalization for the same vowel is quite good for inputs with 22 KHz sampling frequency. The bit resolution did not have much effect on this generalization.

The Generalization for Two or More Vowels. The training procedure was the same as above and two sets of experiments were conducted: One using a sequence of different inputs and using a trained net as the starting point for the next file and second using the early stopping for controlling the generalization. The result was almost the same. The network did not generalize well when 11 KHz files were included; but generalization was good with 22 KHz speech files.

7.6 Results with Recurrent Networks

The experiments reported above were conducted with TD-MLFN in MATLAB. Nevertheless, they were confirmed using our own developed package. As for recurrent networks, the problem with MATLAB is that all recurrent connections include one sample delay and there is no obvious way avoiding these delays if a true instantaneous interaction is required. In our developed algorithm loops with and without delays were envisaged. Results using MATLAB recurrent nets were first confirmed employing our algorithm and then extended to true recurrent nets.

There was no difference in the above results when simple TD-MLFNs were replaced with recurrent substitutes where hidden layer neurons were connected two by two together. However, it is important to note that recurrent nets were trained, for the same output error goal, in less number of epochs; but because of the nature of the algorithm, the training took much longer.

8 Conclusion

The ensemble of experiments conducted in this work shows that when long segments of speech are considered the non-linearity of speech production does not warrant non-linear prediction for further reduction of the error signal and the reduction achieved with short frames are not observed here. The only advantage of non-linear prediction of speech using neural nets is perhaps its generalization power which can be achieved only if the input bandwidth is not limited in the sampling process.

References

1. H.M. Teager: Some Observations on Oral Air Flow Vocalization, IEEE Trans. ASSP, Vol. 28 (5), PP 599–601
2. N. Tishby: A dynamical systems approach to speech processing, Proc. ICASSP, 1990, PP365–368
3. B. Townshend: Non-linear prediction of speech, Proc. ICASSP, 1991, PP 425–428
4. G. DAlessandro etal.: A new sub-band non-linear prediction coding algorithm for narrowband speech signal – The NADPCMB-MLT coding scheme, Proc. ICASSP, 2002, (NEURAL-L03, paper 2066)
5. A.S. Weigend: Time Series Analysis and Prediction, www.cs.colorado.edu/~andreas /home.html
6. T. Masters: Signal and Image Processing with Neural Networks, John Wiley & Sons (1994)
7. N.K. Bose & P. Liang: Neural Network Fundamentals with Graphs, Algorithms and Applications, Mc Graw Hill (1996)
8. Limin Fu: Neural Network in Computer Intelligence, Mc Graw Hill (1994)
9. D.P. Mandic & J.A. Chambers: Recurrent Neural Networks for Prediction: Learning Algorithms, Architectures and Stability, John Wiley & Sons (2001)
10. F.J. Pineda, Generalization of Back Propagation to Recurrent Neural Networks, Physics Review Letters, 59, PP 2229–2232
11. MATLAB NN Toolbox User's Guide; The MATH WORKS INC, www.mathworks.com
12. S. Haykin & S. Kesler, Prediction Error Filtering and Maximum Entropy Spectral Estimation, in Non-linear Methods of Spectral Analysis, Springer-Verlag (1983)

Web Document Access Control Using Two-Layered Storage Structures with RBAC Server

Won Bo Shim[1] and Seog Park[2]

[1] Dept. Computer Science, Sogang University,
121-742, Seoul, Korea
cool96@chch.ac.kr

[2] Dept. Computer Science, Sogang University,
121-742, Seoul, Korea
spark@dblab.sogang.ac.kr

Abstract. Role-based Access Control (RBAC) appears to be the most appropriate technique for access control to minimize the errors likely to occur in managing users and network resources. It can also reduce management costs. In this paper, we show a method for implementing access control for Web documents without modification of the Web server or Web browser, unlike other methods. The access control of Web documents in existing Web servers is based on directories and files, and depends on Access Control Lists defined in the configuration files of the Web servers. This method cannot realize access control according to the user access permission, based on the Web document content. We also propose a Public Layer and a Protected Layer for more secure Web document storage. Finally, we achieve a fine-grained Web document access control method according to the access permissions granted to the user's role in each Web server in environments of multiple Web servers.

1 Introduction

Unlike other data structures, documents on the Web are characterized by the absence of schema, information connections between different documents, and hypertext navigation. [1]

The conventional access control methods for Web documents include the following problems. Access is limited at the directory, or file levels, using access control lists (ACLs). Thus, there is no method for access control to a part of a file, or according to the content of the file. If a business program or content is distributed over several Web servers, authentication problems can occur when accessing each Web server, because authentication must be obtained from every Web server. There is also no consideration of different permissions at each server for a single user.

M.H. Shafazand and A M. Tjoa (Eds.): EurAsia-ICT 2002, LNCS 2510, pp. 220–227, 2002.

A method that makes it possible for general users to manage their Web documents in their own domain in accordance with the management policy of the current organization cannot be provided and Web servers or Web browsers should be modified to implement the proposed method. Thus, a method that can solve those problems and complement them should be proposed.

Research on access controls for the Web can be divided into two areas: methods using the role-based access control (RBAC) concept, which has recently been highlighted as an access control model, and methods not using the concept.

The capability-based authorization model proposed by J. Kahan, the Phoenix system proposed by M.G. Lavenant and J.A. Kruper, and the DCE proposed by S. Lewontin are examples of access control methods not using the RBAC concept on the Web. [5][7][8]

The Intranet RBAC (IRBAC) proposed as an RBAC model to apply to Intranet environments by Zahir Tari at Royal Melbourne Institute of Technology, the NIST RBAC/Web implemented by John Barkley at NIST, and the Secure Cookie Method and the Smart Certificate proposed by Joon S. Park at George Mason University are access controls for the Web that use the RBAC concept. [13][3][12] [4]

Securing Web documents becomes very difficult due to their characteristics, and the existing models cannot solve the problem. Therefore, a new security method is required.

The rest of this paper is organized as follows. We firstly discuss access control using embedded Role-based Access Control (RBAC) units. Our Web document access scheme for multiple Web servers and the complete system configuration to realize this scheme is described. In Section 3, we show the prototypical implementation of our access control model in an environment of multiple Web servers. Finally, we offer our conclusions in Section 4.

2 Access Control Using Embedded RBAC Units

2.1 Embedded RBAC Unit to Access Control

Web document access control using ACLs in a Web server cannot overcome the limitations of being directory or file based. Its management is also cumbersome, as ACLs should be modified to change the access control of a Web document when its location in the system is changed. In this section, we will introduce a method that can identify the user role information at each Web document when the user accesses the document and accomplish access control of Web documents according to the permission granted to each role. Fig. 1 shows the method.

In Fig. 1, the authentication process checks whether the user has a proper session key when accessing a Web document through a Web browser. If the user does not provide an adequate session key, the user authentication process is performed by referring to information in the RBAC server after obtaining the ID and password from the user.

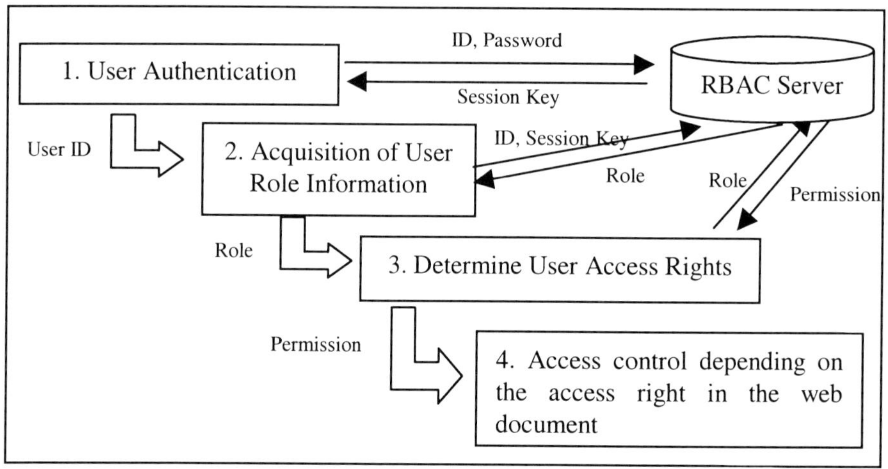

Fig. 1. Embedded RBAC Unit to Access Control

If the authentication is successful, the Web server passes the session key in cookie format to the client's Web browser. The issued session key, which includes user ID information, is valid until the Web browser is terminated. The session key disappears from the client side and cannot be used when the Web browser is finished. This is to prevent any security risk from leaving the user information on the hard disk in the client side. The user can access other Web documents with the session key until the Web browser is terminated. Until that time, no further authentication process is required. The session key can also be applied when accessing Web documents in other Web servers within the same domain. In other words, using the session key, documents in other Web servers in the same domain can be accessed with no additional redundant authentication process.

2.2 Web Document Storage Structure

Generally, Web documents open to the public are located in specific open directories such as 'htdocs'. To solve the security problem, we divided the storage locations of Web documents into two hierarchies: one is the public layer and the other is the protected layer.

The public layer is the directory location where Web documents are open to the public, whereas the protected layer is a safe directory that cannot be accessed by general users. By using the two storage hierarchies, it is possible to prevent general users from accessing confidential or classified Web documents. While the Web documents in the public layer can be freely accessed on the Web, those in the protected layer can only be accessed through the Web documents in the public layer. Through such two-layered storage structures, Web documents can be safely protected in the Web server.

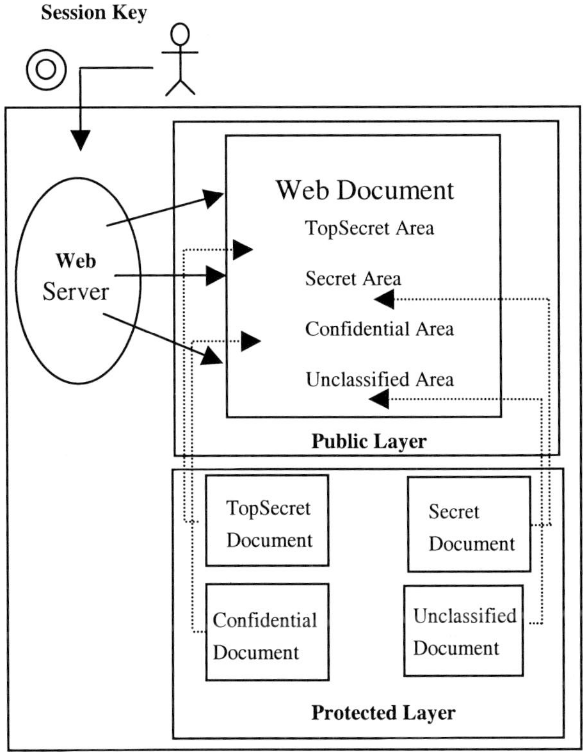

Fig. 2. Tow-Layered Storage Structures of Web Documents

2.3 Web Document View Control According to User Access Permission

We can make access control possible on a part or a line basis of a Web document, by controlling the access to the Web documents in the protected layer according to the user's access permission inside Web documents in the public layer, as shown in Fig. 2. Assume that a user with a Director's role has access permission to see all document contents including Top Secret level and that a user with an Engineer's role has access permission to see the document contents below Confidential level. Document security can then be more powerful by differentiating the document contents that can be viewed by a user having Director's role and a user having Engineer's role.

We used PHP scripts to control access to Web documents in the public layer, and used MySQL for the RBAC database. For example, when a user requests a Web document in site A, the Web document including its PHP script checks the validity of the user. PHP scripts in Web documents can check if the accessing user has a valid authenticated cookie. If the user has no valid cookie, the script passes control to the login process page. The following steps of the access control process show how we can access a Web document in the protected layer from a Web document in the public layer.

Step 1. $needlogin = true;
Step 2. obtain user id cookie and session id cookie;
Step 3. if user id and session id are correct then
    $needlogin = false;
Step 4. if $needlogin is true, move to the login process
    else move to the next step;
Step 5. select a work;
Step 6. obtain information of the roles related to the selected work;
Step 7. perform permission check for the roles;
Step 8. load files from the protected layer according to the level permission;
Step 9. the Web document shows its contents according to the user's role permission.

2.4 Access to Multiple Web Servers in the Same Domain

We will now closely examine Web document access control when there are multiple Web servers in a domain. For example, when a user accesses a document on Site A, the Web document checks whether he or she has a valid session key through the authentication process. If the user has an appropriate session key, the Web document being accessed will be shown on the Web browser. Otherwise, the user will be automatically connected to the authentication process.

To be authenticated for the Web document, the user should supply ID and password in this process. If the authentication is successful, the user will be given the session key required for subsequent connections in the form of an authentication cookie from Site A.

Generally, a cookie is stored in the hard disk and reused for subsequent connections. However, this can cause a security problem, so we chose to save the cookie not in the hard disk but in main memory. The cookie is therefore only effective while the Web browser is opened. The situation is the same even if the user visits Site B first.

As shown in Fig. 3, given a session key from any site, the user can provide authentication using the cookie stored in main memory when connecting other servers in the same domain.

The user can therefore access Web documents without additional authentication processes, and will be under the access control in accordance with his or her role in each Web server in the domain.

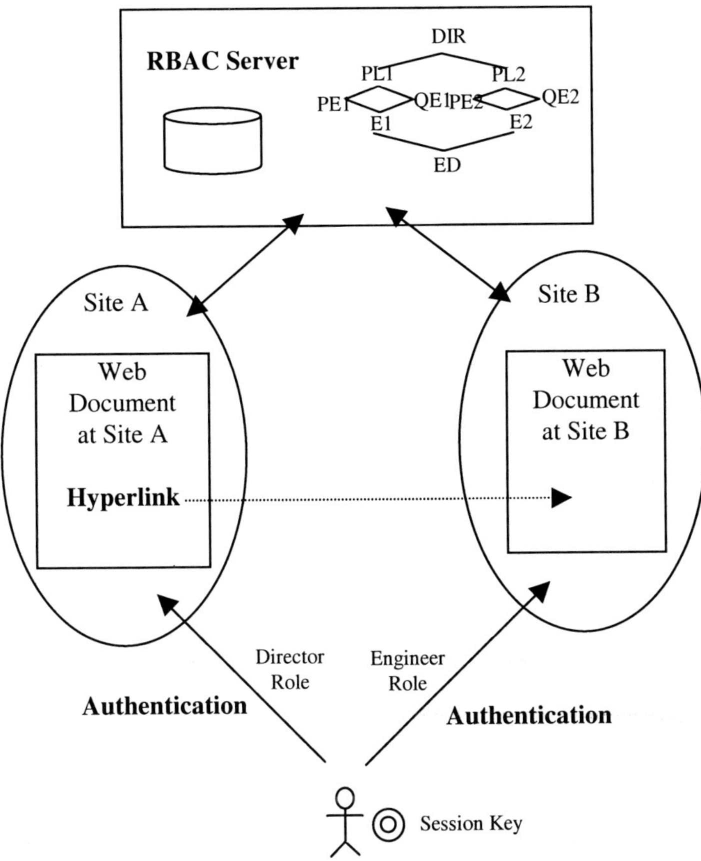

Fig. 3. Access to Multiple Web Servers in the Same Domain

3 Implementing Multiple Web Server Access Control in the Same Domain

Our method's approach to realizing access control to help the user complete a task by allowing access to multiple Web servers within the domain will be described in the following. As shown in Fig. 4, a user accessing a Web document in a 'multi' server is asked to provide certification, and upon the successful completion of the certification process, the list of projects assigned to the user is shown. Then the user selects an item in the work list. Suppose that the user selects a work, Project 1. As soon as the project is selected, its sub work list will be displayed on the screen. Because the items in the list are mapped to the roles in each server, the user will have the necessary access rights of the roles for each server needed to complete the task. The role server will deliver the information about the corresponding user's role and access control in cookie format. In Fig. 4, the user has the highest access rights by taking a Director role within the multi server, so documents including the top secret one are shown to the user through the accessed document. The user can access the other server, 'multint', (multint.chch.ac.kr), via a hyperlink within the domain after successfully logging onto the initial 'multi' Web server (multi.chch.ac.kr). The 'multint' server checks whether the user has already been verified, based on the cookie, and because the user has already received a valid cookie when accessing the initial 'multi' server, the documents in 'multint' server are made available without another verification process. In this situation, the role server provides the appropriate access rights for the user, because the user's role varies in the 'multint' server and the user's access rights vary correspondingly.

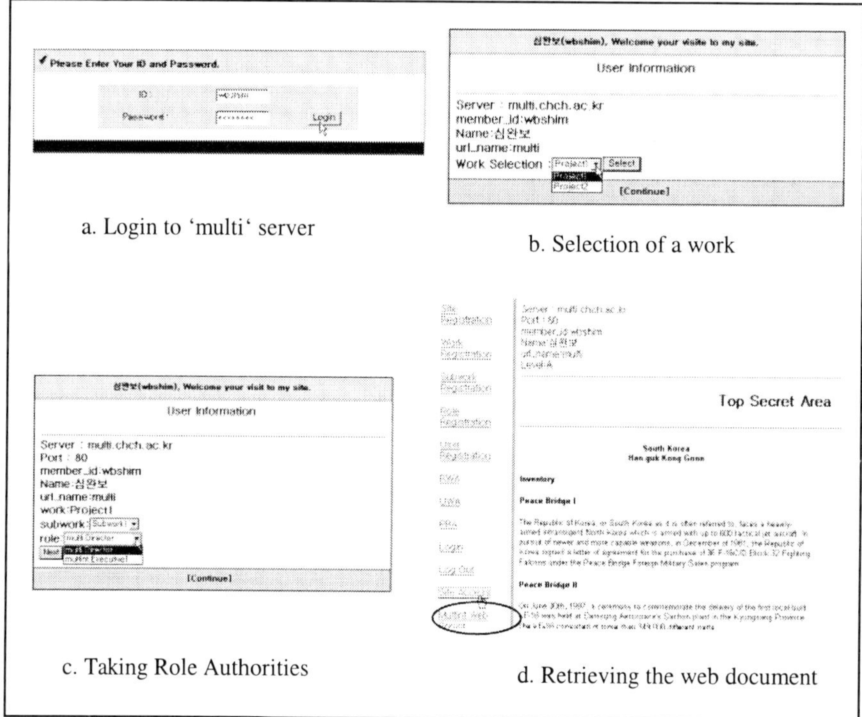

a. Login to 'multi' server

b. Selection of a work

c. Taking Role Authorities

d. Retrieving the web document

Fig. 4. Process of taking the role authority for the user

4 Conclusions

It is quite natural for the numbers of Web servers and users to increase in the distributed Web environment. Access to the resources provided by the Web servers must be controlled appropriately. However, access control for Web documents in existing Web servers is based on directories and files, depending on ACLs defined in the configuration file of the Web servers. This method cannot realize access control according to user access permissions and Web document content. In this paper, we described a method that identifies the role information of the user to enable access to Web documents through the role-based access control technique. The method can allow independent access control for Web documents in accordance with the permissions granted to each role in each Web server. If this method is employed, general users can independently manage their Web documents in their own domains in accordance with the management policy of the current organization. We also proposed public layers and protected layers for more secure Web document storage. Finally, we achieved a fine-grained Web document access control method according to the access permissions granted to the user's role in each Web server in an environment of multiple Web servers.

Acknowledgements. This work was supported by grant No.2000-1-303-001-3 from Basic Research Program of the Korea Science and Engineering Foundation.

References

1. James B.D. Joshi, Walid G. Aref, Arif Ghafoor, and Eugene H. Spafford, "Security Models for Web-based Applications", Communications of the ACM, 2001
2. James B.D. Joshi, Walid G. Aref, Arif Ghafoor, and Eugene H. Spafford, "Digital Government Security Infrastructure Design Challenges", IEEE Computer, 2001
3. J. F. Barkely, A. V. Cincotta, D. F. Ferraiolo, S. Gavrilla and D. R. Kuhn, "Role Based Access Control For the World Wide Web", 20th NCSC, 1997.
4. Joon S. Park and Ravi Sandhu, "RBAC on the web by smart certificates",Proc. of 4th ACM Workshop on Role-Based Access Control, 1999.
5. Kahan, J., "A capability-based authorization model for the World Wide Web", Proc. of the 3rd WWW Conference, 1995
6. Larry S. Bartz, "hyperDRIVE: Leveraging LDAP to implement rbac on the web", 2nd ACM Workshop on Role-Based Access Control., 1997.
7. Lavenant, M.G. and Kruper, J.A., "The Phoenix Project: distributed hypermedia authoring", Proc. of the 1st WWW Conference, 1994
8. Lewontin, S., "The DCE Web Toolkit: enhancing WWW protocols with lower-layer service", Proc. of the 3rd WWW Conference, 1995
9. Ravi Sandhu and Joon S. Park, "Secure Cookies on the Web", IEEE Internet Computing, July-August, 2000.
10. Ravi Sandhu, David Ferraiolo and Richard Kuhn, "The NIST Model for Role-Based Access Control: Toward A Unified Standard", Proc. of the Fifth ACM Wrokshop on Role-Based Access Control, 2000.
11. Ravi Sandhu, E. Coyne, H. Feinstein, and C. Younman, "Role-Based Access Control Models", IEEE Computer Magazine Vol. 29, 1996.
12. Ravi Sandhu and Joon S. Park, "Secure Cookies on the Web", IEEE Internet Computing, July-August, 2000.
13. Zahir Tari and Shun-Wu Chan, "A role-based access control for intranet security", IEEE Internet Computing, September/October 1997.

Development of UML Descriptions with USE

Martin Gogolla and Mark Richters

University of Bremen, Computer Science Department, Germany

Abstract. The Object Constraint Language OCL is part of the Unified Modeling Language UML. Within software engineering, UML is regarded today as an important step towards development of high-quality object-oriented systems. OCL allows to describe system structure by invariants and system behavior by pre- and postconditions. This paper explains the functionality of the UML Specification Environment USE which allows to validate and verify UML and OCL descriptions. The paper also uses a new approach to handle UML statecharts by OCL pre- and postconditions.

1 Introduction

The Unified Modeling Language UML [OMG01] is regarded today as an important standard for the development of software systems. Many commercial tools for UML are available. The Object Constraint Language OCL [OMG01,WK98] is part of standard UML, but up to now OCL is not supported by most commercial tools. OCL is a specification language supporting and enriching the UML with textual details which cannot be expressed in diagramatic form. OCL allows to precisely describe system structure by invariants and system behavior by pre- and postconditions.

Tool support for OCL is beginning to develop. Among the first available tools was our system USE (UML Specification Environment). USE is based on conceptional work on the formal semantics of the OCL and the needed relevant UML features [RG01] and work on the metamodel of OCL [RG99]. The main task of USE is to validate and verify specifications developed within UML and OCL. By validation we mean that the developer can animate the specification by providing a number of test cases and check whether the USE responses meet the intuition. Especially for specifications involving formal aspects we regard such validation systems as extremely helpful because they give feedback to developers in early development stages. By verification we mean, that the test cases are formally checked with respect to invariants and pre- and postconditions. Other tools for OCL include the commercial tool from Boldsoft, an OCL compiler [HDF00], and the KeY tool [ABB+00].

USE is now available as version 2.1 and has achieved more and more functionality since it was first introduced. USE is the only OCL tool allowing interactive monitoring of OCL invariants and pre- and postconditions. USE has been successfully applied in various larger case studies(e.g. [ÁES01,GB01]) and in teaching UML. It was employed in submissions for the upcoming new version

M.H. Shafazand and A M. Tjoa (Eds.): EurAsia-ICT 2002, LNCS 2510, pp. 228–238, 2002.

of UML and has been used for large specifications like the UML metamodel with about 100 classes and hundreds of OCL expressions. USE offers the following functionalities:

1. syntax check for and browsing through textual UML and OCL descriptions,
2. generation of system states through object, attribute and link manipulation,
3. representation of system states as object diagrams,
4. monitoring model inherent and explicit invariants in class diagrams,
5. operation execution and monitoring of pre- and postconditions,
6. representation of operation call sequences as sequence diagrams, and
7. querying system states with OCL expressions.

For realizing these functionalities, the USE system covers central UML language features. The input to USE are UML class diagrams in textual form, OCL invariants, OCL operation descriptions through pre- and postconditions, and UML statecharts which are encoded as OCL pre- and postconditions. USE produces as output UML object and sequence diagrams as well as reports on the validity of invariants and pre- and postconditions in table and text form.

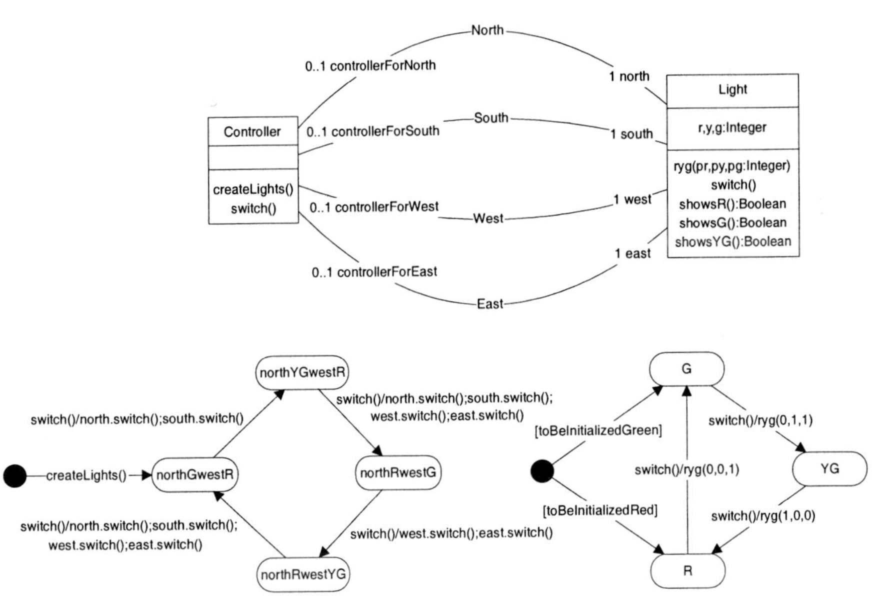

Fig. 1. Class Diagram and Statecharts for Crossing System

The central idea of the USE tool is to check for software quality criteria like correct functionality of UML descriptions already on the design level in an implementation-independent manner. This approach takes advantage of descriptive design level specifications by expressing properties shorter and in a more abstract way. Such properties are given by invariants and pre- and postconditions, and these are checked by the USE system against the test scenarios, i.e.

object diagrams and operation calls given by sequence diagrams, which the user provides. These abstract design level tests are expected to be also used later in the implementation phase.

The rest of this paper is structured as follows. In Sect. 2 we present a small case study which will be used throughout the paper. A small case study was choosen in order to be self-contained, but also to demonstrate the principal idea of specification development with USE. These ideas work for larger descriptions as well. In that case study we also employ a new approach how to translate UML statecharts into OCL pre- and postconditions. The central Sect. 3 explains the functionalities offered by USE through various screenshots. The paper ends with concluding remarks in Sect. 4.

2 Running Example: The Crossing System

Our running example describes a simple traffic control application where a controller is supposed to supervise four traffic lights on a street crossing. Figure 1 shows a UML class diagram with the classes `Controller` and `Light` and two UML statechart diagrams, the left one for class `Controller` and the right one for class `Light`.

Fig. 2. USE Command Windows

A `Controller` object can be linked to the four `Light` objects which are then accessible with the role names `north`, `south`, `west`, and `east`. The controller possesses an operation `createLights` for creating and initializing the lights and an operation `switch` for switching the crossing system. The controller is responsible for transmitting a `switch` operation to the respective lights. A `Light` object possesses three attributes `r`, `y`, and `g` for the red, yellow, and green bulb, respectively, an internal operation `ryg` for manipulating the three attributes and also an operation `switch` for switching. Furthermore, for convenience, it has three Boolean-valued operations for checking the displayed signal.

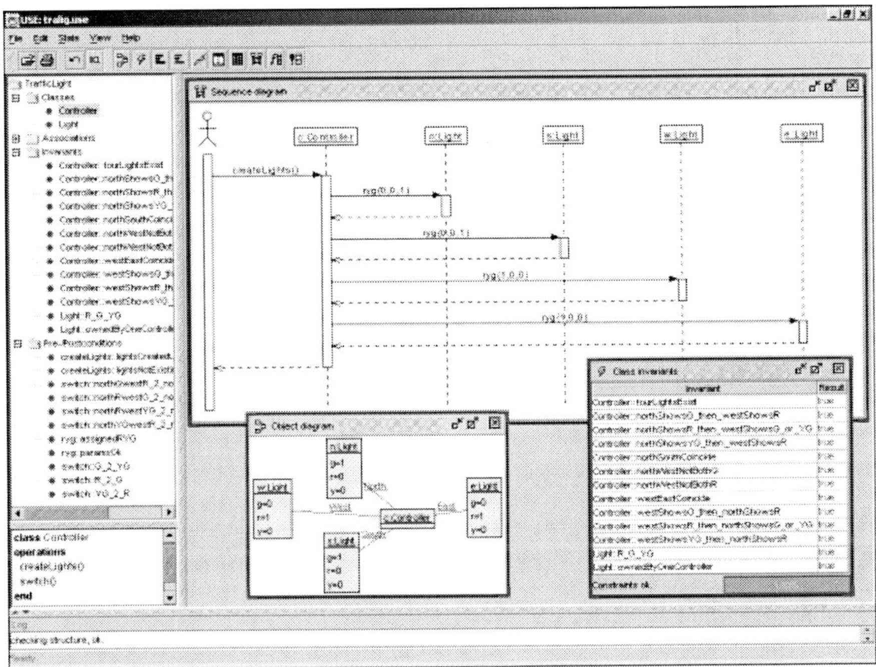

Fig. 3. USE Project and Snapshot Windows-Initialization

The `Light` statechart shows that we describe Italian traffic lights with three states `G`, `YG`, and `R`. In state `G` only the green bulb is on, in state `YG` only the yellow and green bulb, and in state `R` only the red bulb. The shows predicates characterize these states. The `switch` operation in class `Light` switches between these states in a cyclic way. The `Controller` statechart has four states which are named according to the signal on the `north` and `west` light. We require that the `north` and `south` lights coincide as well as the `west` and `east` lights. When receiving a `switch` operation, the task of the controller is to transmit the `switch` to the proper `Light` objects. Certain properties have to be satisfied. For example, for safety reasons, the `north` and the `west` light are not allowed to display only green at the same time, and they are not allowed to show only red at the same time. The purpose of USE is to bring life into this description.

3 Use of USE

We now explain how a UML description like the one in Fig. 1 can be analyzed with the USE tool. We do so by showing typical screenshots taken from sessions working with USE. But before going into details we want to give an overview on the content of the displayed screenshots in Figs. 2, 3 and 4.

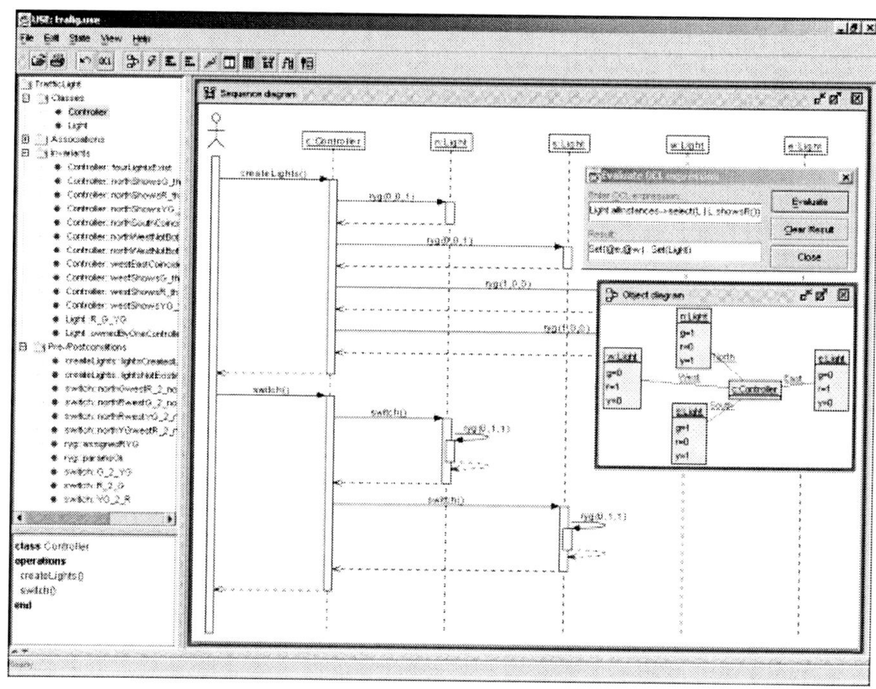

Fig. 4. USE Project and Snapshot Windows-Switching

The starting point for working with USE is to load a USE specification like the one given in [GR02]. After successful syntactical analysis, the specification is available in a project browser window as shown in the left of Fig. 3. By selecting an element of the specification, like for example the class `Controller`, this element is shown in detail in a window below the project browser window. In Fig. 3, the details of class `Controller` are pictured.

In Fig. 2 we see commands which are given to the USE system and the responses of the USE system. Roughly speaking, the commands in the left part of Fig. 2 generate the situation in Fig. 3, and the commands in the right part of Fig. 2 subsequently generate the situation in Fig. 4. Such commands construct system states which can be inspected through various windows as shown in Figs. 3 and 4. On the bottom of Fig. 3, a log window gives information on the status of the model inherent constraints, namely whether the specified multiplicities are currently valid or not. To the right of the project browser window, we see three further windows: A sequence diagram window, an object diagram window, and a class invariant window. The sequence diagram window shows which operations have been applied to which object and displays the operation order. The object diagram window shows the current system state, i.e. the objects currently existing and their attributes as well as their links. The class invariant window shows the value of the specified invariants, i.e. it reports on whether the respective invariant is false or true.

Figure 4 shows in the sequence diagram the development after some more commands have been entered. The object diagram in Fig. 4 shows the modified system state. Figure 4 also explains through the OCL evaluation window shown in the upper right part how system states can be queried with OCL expressions. Let us now go through the USE functionalities as mentioned in the introduction and discuss how they are realized in USE.

3.1 Syntax Check

UML class diagrams, invariants, pre- and postconditions as well as UML statecharts are given to USE in textual form. [GR02] shows the complete specification for the crossing system.

Class and association descriptions are written down straightforward. Classes possess attributes and operations. Query operations, like for example `showsR`, can be characterized by OCL expressions. Associations (as well as aggregations and compositions) are described with multiplicities and role names.

As usual in OCL, an invariant possesses a class as its context and a name. The crossing system has two invariants for class `Light` and eleven for class `Controller`. For example, the `Light` invariant `R_G_YG` requires that either only red is displayed or only green or only yellow and green. The unwanted case that all three shows predicates become true (this would be allowed by the `xor`) is excluded due to the non-overlapping definition of the shows predicates. Considering an example for a `Controller` invariant, the invariant `northWestNotBothG` disallows system states where both the `north` and `west` lights show green. We do not discuss all invariants in detail but emphasize that a well chosen name for invariants is useful because these names are employed in the invariant window to be discussed further down.

As usual in OCL, operations can be described by pre- and postconditions. The `Controller` operation `createLights` is specified with pre- and postconditions. The precondition requires that the current controller is not linked to any `Light` object. The postcondition ensures that four new `Light` objects are created and linked to the controller and that the `Light` objects are correctly intitialized. The `Light` operation `switch` is characterized by three postconditions which encode the `Light` statechart. The idea is that each transition in the statechart becomes a postcondition where the pre-state of the transition is described by means of the OCL operator `@pre`. The operator `@pre` allows in postconditions access to the values of attributes at the precondition time. Thus for example, the postcondition `R_2_G` is true provided the following holds: If the operation `switch` was called in the state where only red is displayed, then this call of `switch` results in the state where only green is displayed. Therefore the three postconditions for the `Light` operation `switch` realize the `Light` statechart.[1] Analogously to the `Light` statechart, the four postconditions for the `Controller` operation `switch` implement the `Controller` statechart. We refrain from explaining the details of the remaining pre- and postconditions.

3.2 System State Generation

System states can be manipulated in USE by entering commands in the command window as shown in Fig. 2. Alternatively such manipulations can also be achieved interactively by using the graphical user interface, but for documentation purposes we here use the command window. The basic manipulation commands are the commands `!create` and `!destroy` for object creation and deletion, the command `!set` for attribute modification, and the commands `!insert` and `!delete` for link modification. Commands can also be read from a command file. In Fig. 2, the two command files `createLights.cmd` and `ryg.cmd` are employed which can be considered as the USE implementation of the respective operation. The content of those command files is as follows.

```
createLights.cmd:
!create n:Light
!create s:Light
!create w:Light
!create e:Light
!insert (self,n) into North
!insert (self,s) into South
!insert (self,w) into West
!insert (self,e) into East
!openter c.north ryg(0,0,1)
read ryg.cmd
!opexit
!openter c.south ryg(0,0,1)
read ryg.cmd
!opexit
!openter c.west ryg(1,0,0)
read ryg.cmd
!opexit
!openter c.east ryg(1,0,0)
read ryg.cmd
!opexit
```

```
ryg.cmd:
!set self.r:=pr
!set self.y:=py
!set self.g:=pg
```

3.3 Object Diagrams

The system state achieved after execution of the commands in the left part of Fig. 2 is displayed by the object diagram in Fig. 3. The four `Light` objects with their attribute values, the one `Controller` object, and their links are shown.

3.4 Invariants

Invariants in USE are divided into model inherent and explicit ones. The model inherent invariants are given through the UML class diagram with the multiplicities specified therein. The log window in the bottom of Fig. 3 reports that the

[1] Due to the current semantics of pre- and postconditions for operations in UML, which requires that all preconditions have to be true at invocation time, one cannot describe the `Light` statechart more intuitively with three pairs where each pair consists of a precondition and a postcondition.

multiplicity constraints are satisfied by the current system state. The evaluation of the explicit invariants is displayed by the class invariant window. There, all invariant names together with their corresponding classes are pictured. For each invariant the result of the evaluation is given. We see that in the current system state all invariants are valid, and therefore the achieved system state has all desirable properties expressed as invariants. In case of an invariant evaluating to false, double-clicking the invariant would open an evaluation browser window which allows to analyse which parts of the invariant and which objects in the current system state contribute to the undesired result. This will be detailed in Sect. 3.8. As mentioned above, an expressive name for an invariant is important at this point in order to understand the result displayed.

3.5 Pre- and Postconditions

As exemplified in the left part of Fig. 2, operations in USE can be invoked by the command `!openter` and closed by the command `!opexit`. Upon operation invocation the precondition is checked, and after exiting the operation the postcondition is evaluated. For example, after submitting the command line `!openter c createLights()` which starts the respective operation on the `Controller` object `c`, the USE system responds with the message that the precondition of the operation evaluates to true. Thereafter, the implementation of that operation is read from a command file and executed step by step. This implementation also invokes other operations, for example through the call `!openter c.north ryg(0,0,1)`. The invocation `!openter c createLights()` is closed by the last `!opexit` command in the left part of Fig. 2, and the USE system reports at this point that the postcondition `lightsCreatedLinkedInitialized` is satisfied. This means that the chosen implementation of `createLights` meets the desired and specified behavior.

In the right part of Fig. 2 we show how the first call of the `switch` operation on the `Controller` object is handled. After invoking that operation and the induced invocations on the respective `Light` objects connected to the `Controller` object, the last three `!opexit` commands close those invocations, and USE responses that all postconditions of all operations are valid. This means that the implemented behavior through the entered commands is correct with respect to the specification, i.e. the operation calls and their implementation meet the behavior described in the statecharts which were encoded in the pre- and postconditions of the operation `switch`.

3.6 Sequence Diagrams

Sequence diagrams are employed in USE in order to capture complex operation calls. The sequence diagram in Fig. 3 shows that a user object symbolized by the strawman icon invokes the operation `createLights` on the `Controller` object and that the `Controller` object reacts by submitting respective initialization calls to the four `Light` objects. The sequence diagram in Fig. 4 points out that the first `switch` directed to the `Controller` object is transmitted only to the north and south light and that a switch operation on Light objects induces a call of the operation ryg modifying the bulb attributes. The evaluation of the invariants in this stage would result in the same table as before, i.e. all invariants are valid. Together with the positive results of evaluating the pre- and postconditions presented in Fig. 2, the sequence diagram in Fig. 4 captures the proof that the implemented behavior is correct with respect to specified invariants and the specified pre- and postconditions.

3.7 Querying

Figure 4 also shows in the upper right part how system states can be queried through OCL expressions. The OCL expression evaluator allows to compute the result of complex OCL expressions in the current system state.

3.8 Further Windows

Figure 5 shows further windows supporting the functionalities for operation invocation and invariant monitoring. The situation shown is an intermediate step reached during execution of the commands given right part of Fig. 2 before the first !opexit is issued. The sequence diagram in Fig. 5 displays this fact because the dashed return arrows present in Fig. 4 do not yet appear here.

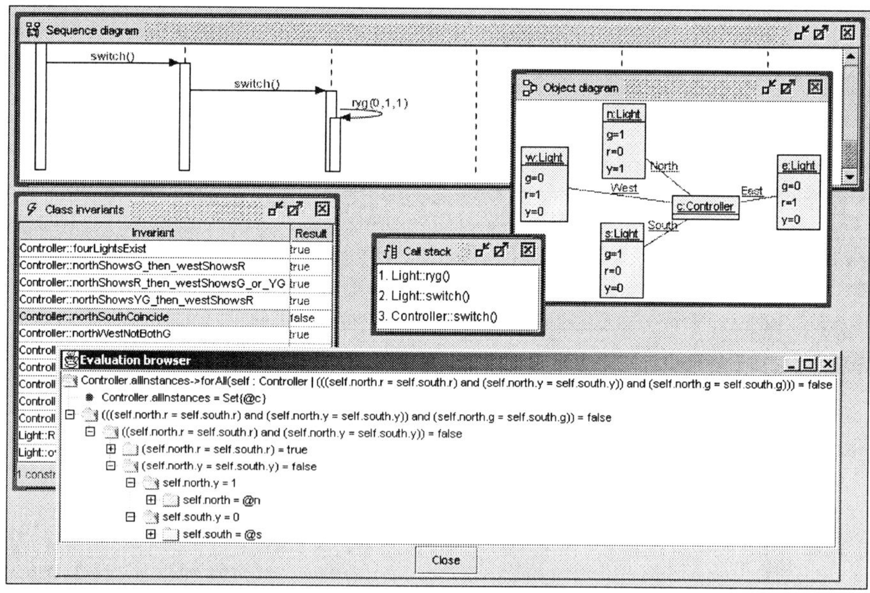

Fig. 5. USE Call Stack and Evaluation Browser Windows

The fact, that three operation calls are not yet closed, is reflected in the call stack window in the middle of Fig. 5 which exactly shows these three open operation calls. At this point, the class invariant window in the left part also shows that the invariant `northSouthCoincide` evaluates to false. Double-clicking that invariant in the class invariant window opens the evaluation browser window shown in the bottom of Fig. 5. That window traces which parts of the invariant and which part of the current system state contribute to the invariant evaluating to false. That window reveals that the `Light` objects `n` and `s` are responsible for the fact that the invariant's subexpression `north.y=south.y` is false and with this the complete invariant is false. This non-coincidence of the `Light` attribute y on objects `n` and `s` is also visually represented by the object diagram.

4 Conclusion

We have described the functionality of the UML Specification Environment USE. USE allows to validate formal specifications by providing test cases and checking whether the system's responses meet the intuition. USE also allows to verify formal specifications because it checks whether the operation implementation, namely the chosen operation commands, meet the operation specification. Thus USE covers validation and verification aspects. Future work will consider the further development of USE. The results of evaluating pre- and postconditions in the command window could be represented in the GUI. Invoking and finishing operations could also be supported by interactively. In order to increase the computational power of operations one could consider programming language-like features, i.e. conditionals and loops, for the command interpretation.

References

[ABB+00] W. Ahrendt, T. Baar, B. Beckert, M. Giese, E. Habermalz, R. Hähnle, W. Menzel, and P. H. Schmitt. The KeY approach: Integrating object oriented design and formal verification. In M. Ojeda-Aciego, I.P. de Guzmán, G. Brewka, and L. M. Pereira, editors, *Proc. 8th Europ. Workshop Logics in AI (JELIA)*, LNCS 1919, pages 21–36. Springer, 2000.

[ÁES01] J. Álvarez, A. Evans, and P. Sammut. Mapping between levels in the metamodel architecture. In M. Gogolla and C. Kobryn, editors, *Proc. 4th Int. Conf. UML (UML'2001)*, pages 34–46. Springer, LNCS 2185, 2001.

[GB01] G. Georg and J. Bieman. Using Alloy and UML/OCL to Specify Run-Time Configuration Management: A Case Study. In A. Evans, R. France, A. Moreira, and B. Rumpe, editors, *Proc. UML'2001 Workshop on Rigorous Development*, pages 69–70. LNI, German Informatics Society, 2001.

[GR02] M. Gogolla and M. Richters. USE Specification Text for the Traffic Light Case Study. `ftp://ftp.informatik.uni-bremen.de/local/db/papers/trali.use`, 2002.

[HDF00] H. Hussmann, B. Demuth, and F. Finger. Modular architecture for a toolset supporting OCL. In A. Evans, S. Kent, and B. Selic, editors, *UML 2000*, LNCS 1939, pages 278–293. Springer, 2000.

[OMG01] OMG, editor. *OMG Unified Modeling Language Specification, Version 1.4*. OMG, September 2001. `www.omg.org`.

[RG99] M. Richters and M. Gogolla. A Metamodel for OCL. In R. France and B. Rumpe, editors, *Proc. 2nd Int. Conf. UML (UML'99)*, pages 156–171. Springer, LNCS 1723, 1999.

[RG01] M. Richters and M. Gogolla. OCL – Syntax, Semantics and Tools. In T. Clark and J. Warmer, editors, *Advances in Object Modelling with the OCL*, pages 43–69. Springer, Berlin, LNCS 2263, 2001.

[WK98] J. Warmer and A. Kleppe. *The Object Constraint Language: Precise Modeling with UML*. Addison-Wesley, 1998.

FPGA Implementation of Digital Chaotic Cryptography

Dewi Utami, Hadi Suwastio, and Bambang Sumadjudin

Electrical Engineering Departement – STT Telkom
Tel. 62.22.7565933, Fax. 62.22.7565933
dwu@sttelkom.ac.id

Abstract. In this paper, we present the digital chaotic cryptography implementation on FPGA. The system realization uses AR filter and modulo function as a non - linear component. The hardware implementation of FPGA-based which consumed 288 CLBs has been successfully developed for 24 bits fixed point system using 4^{-th} order filter. The maximum clock frequency used in the experiment is 6.747 MHz.

Keywords. Chaos, digital filter overflow, cryptography, fixed point, FPGA

1 Digital Chaotic Cryptography

In this research, we used digital filter overflow to generate a chaotic signal. The filter has modulo function as a non linear component [4, 5, 7, 9]. Furthermore, we can find the system equation and its structure as:

$$y_i = \mathrm{mod}\left(z_i + \sum_{i=1}^{n} c_i y_{i-1} \right) \qquad y_i \in [-1,1) \tag{1}$$

$$\tilde{z}_i = \mathrm{mod}\left(\tilde{y}_i - \sum_{i=1}^{n} c_i \tilde{y}_{i-1} \right) \qquad \tilde{z}_i \in [-1,1) \tag{2}$$

where, the modulo function is:

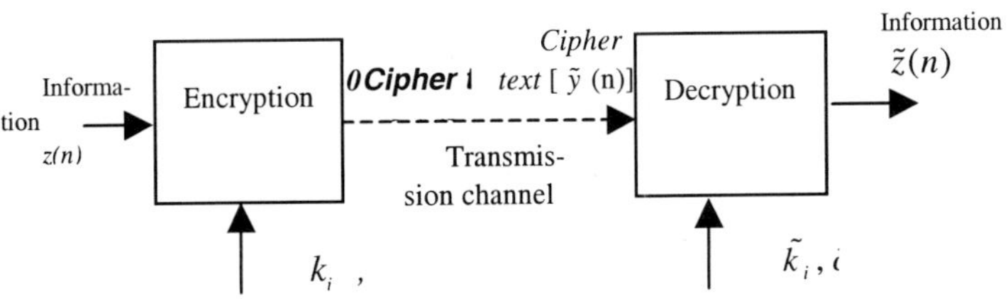

Fig. 1. Block diagram of chaotic cryptography system

M.H. Shafazand and A M. Tjoa (Eds.): EurAsia-ICT 2002, LNCS 2510, pp. 239–247, 2002.

$$f(y) = \mathrm{mod}(y)$$
$$f(y) = y - 2\left\lfloor \frac{y+1}{2} \right\rfloor \tag{3}$$

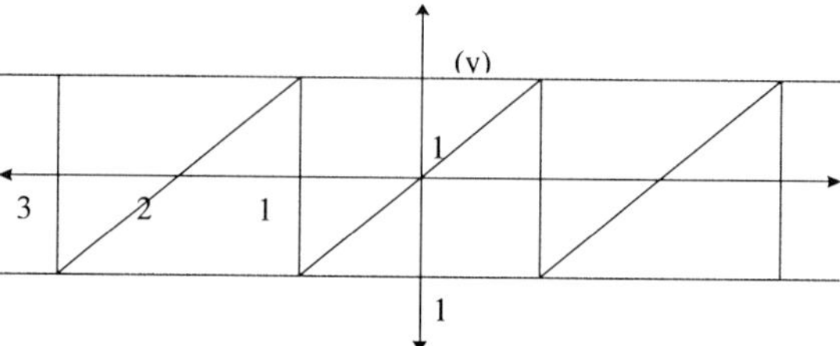

Fig. 2. Modulo function. [4]

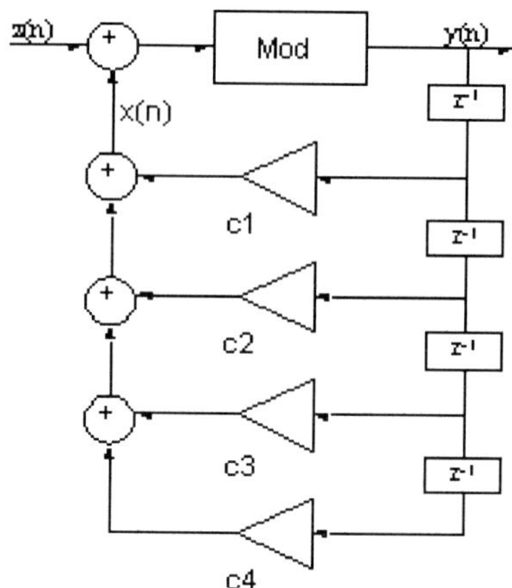

Fig. 3. Encryption system structure

The modulo function can be categorized as a piecewise linear map, that only has one gradient for x=[-1,1]. Thus the PMF (Probability Mass Function) generated by modulo function will be uniform. Then, the function (PMF) can be defined as [5]:

$$f_y(y) = \begin{cases} 2^{-1} & y \in I \\ 0 & y \notin I \end{cases} \tag{4}$$

for I = [-1,1).

Next, let we prove that the plain text can be retrieved back by decryption formula:

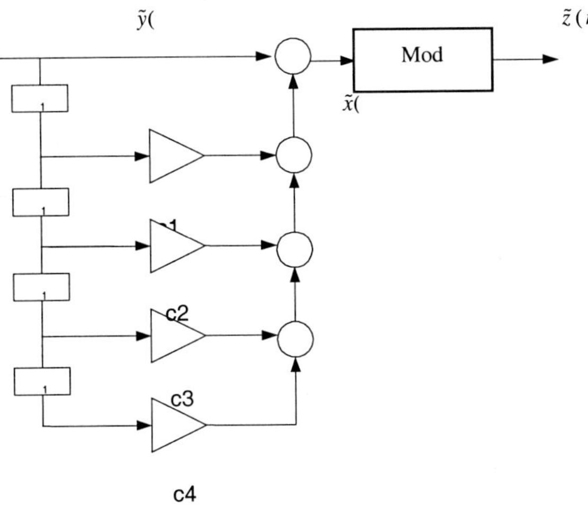

Fig. 4. Decryption system structure

$$\begin{aligned}\tilde{z} &= \text{mod}[\tilde{y} - \tilde{x}] \\ &= \text{mod}[\text{mod}[\tilde{z} + \tilde{x}] - \tilde{x}] \\ &= \text{mod}\left[\tilde{z} + \tilde{x} - 2\left\lfloor \frac{\tilde{z} + \tilde{x} + 1}{2} \right\rfloor - \tilde{x}\right] \\ &= \text{mod}\left[\tilde{z} - 2\left\lfloor \frac{\tilde{z} + \tilde{x} + 1}{2} \right\rfloor\right] \\ &= \tilde{z} \qquad \text{untuk nilai } \tilde{z} = [-1,1)\end{aligned} \tag{5}$$

The problem arises that is why does this formula always equal to z? Analytically, it is caused by $2\left\lfloor \frac{\tilde{z} + \tilde{x} + 1}{2} \right\rfloor$ that is always even integer that has a zero value for modulo function. If the condition of no error quantization was fulfilled, then we will get $\tilde{z} = z$. To get a chaotic behavior, we select the filter coefficients, which are greater than one. This selection will lead us to get unstable region of filter operation. Consequently, if we choose the coefficients that are smaller or equal to one, the initial condition dependence of system will be convergent (equation (6)). It is the opposite of chaotic signal generation theorem.

$$\left| f^{[n]}(z+\delta) - f^{[n]}(z) \right| \approx \delta\text{-}2.\text{floor}\left(\frac{z+\delta+1}{2}\right) - 2.\text{floor}\left(\frac{1+z}{2}\right) \tag{6}$$

2 Quantization Size on Fixed Point Realisation

In our paper, we defined a quantization size as a number of bits that are allocated in arithmetic processing and input-output data storage. We used 24 bits quantization size and realized in the fixed -point system. The system is designed to use integer values of filter coefficient, and the initial condition as an integer multiplying of smallest step from the level of quantization. This condition is designed to get cipher text that does not increase the bit rate, or in other words – we maintained that both of the plain text and cipher text have the same number (lengths) of bits.
To prove the above hypothesis, the following proof can be described:
Let we start from equation (1), assumed:

- $\{y_n, z_n\} \in [-1,1)$, which was represented in L signed bit
- $c_i \in$ Integer, represented in T signed bit.

So, the smallest step of $\{y_n, z_n\}$ is $2^{-(L-1)} = q$, and the smallest step from c_i is 1.
From the assumptions, we can determine the following equations:

$$\left.\begin{aligned} y_i &= sq \\ z_i &= wq \end{aligned}\right\} \text{ where s and w are an integer number with max. value is } 2^{L-1}$$

$$\begin{aligned} y_i &= \operatorname{mod}\left(z_i + \sum c_i y_{i-1}\right) \\ &= \operatorname{mod}\left(wq + \sum c_i sq\right) \\ &= \operatorname{mod}\left(\left(w + \sum c_i s\right) q\right) \end{aligned} \tag{7}$$

and, the results are :

1. if ($w + \sum c_i s$) $\leq 2^{(L-1)}$, then yi=($w + \sum c_i s$)q. (8)
2. if ($w + \sum c_i s$) $> 2^{(L-1)}$, then y_i can be broken down as an adder of an integer number (=d) with a fraction number (=pq).

So:

y_i=mod(d+pq) $\Rightarrow$ p is an integer, which maximum value of $2^{(L-1)}$.
y_i=mod(mod(d)+pq), and the final result are:

- ❑ if d=even number, then y_i=mod(pq)=pq. (9)
- ❑ if d=odd number, then :
  y_i=mod (-1+pq) =mod((-$2^{(L-1)}$+p)q). (10)

From equation 9 and 10, if there were no error quantization, then all of the results can be represented exactly as L bit.
In this research, the Keys are defined by a combinational of coefficient filters (C) and initial conditions (K). Thus, for mth order system, it will give $C^m K^m$ combination of keys. So, if we use 4th order system, which has 5 bits for coefficient filter, and 24 bits for initial condition, we can get 8,3x10^{34} combinations of Key! Moreover, if we as-

sume that the maximum number of user is a half of the number of Keys combinations [6], the system can be employed for $4{,}15\times10^{17}$ user!

3 FPGA Implementation

The hardware realization based on FPGA (Field Programmable Gate Array). The FPGA used is from Xilinx production – IC XC4010XL and 40XS development board. This FPGA has 10000 gates or equal to 400 CLBs (Configurable Logic Blocks).

From the synthesis result, we know that we have consumed 288 CLBs to realize the system. It also used 63 I/O pads. The schematic diagram of the integrated circuit is shown in figure 5.

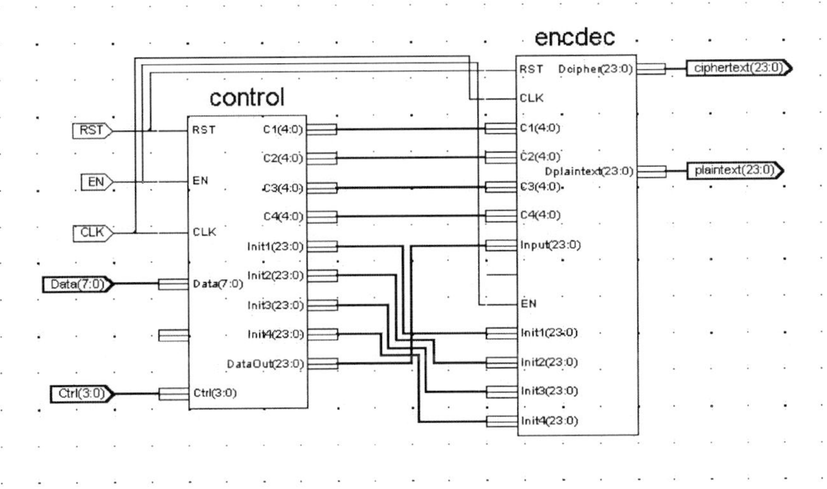

Fig. 5. Implementation schematic of cryptosystem

4 Experiment Result

In the first step of the research, we predict that all of the coefficient filter combination will give us an unstable operation of filter. If the bit number of coefficient is 5 (signed bit), the integer values of coefficient be $[-16 \le c_i \le 15]$. Now, we plot the maximum pole for every combination of coefficient IIR filter (figure 4). In the figure we can see that all of the poles created are on the outer of unit cycle, so it gives us a proof that the entire filters are unstable.

In addition, figure 4 shows the error quantization of cipher text. From this figure we notice that an error has been memorized up to m+1 sequence samples. So, the number of error will be m+2 samples, where m is order of filter.

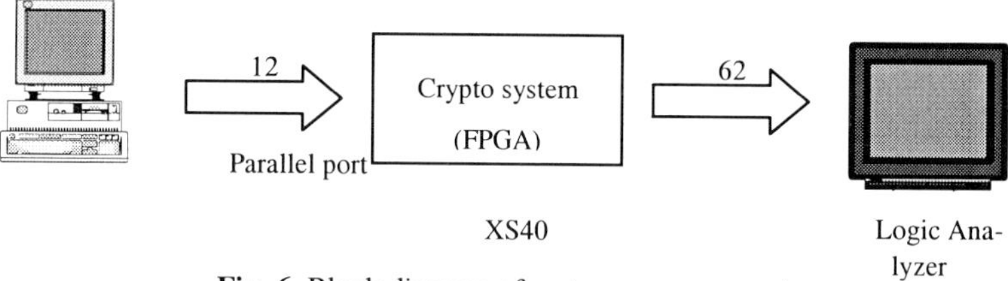

Fig. 6. Block diagram of system measurement

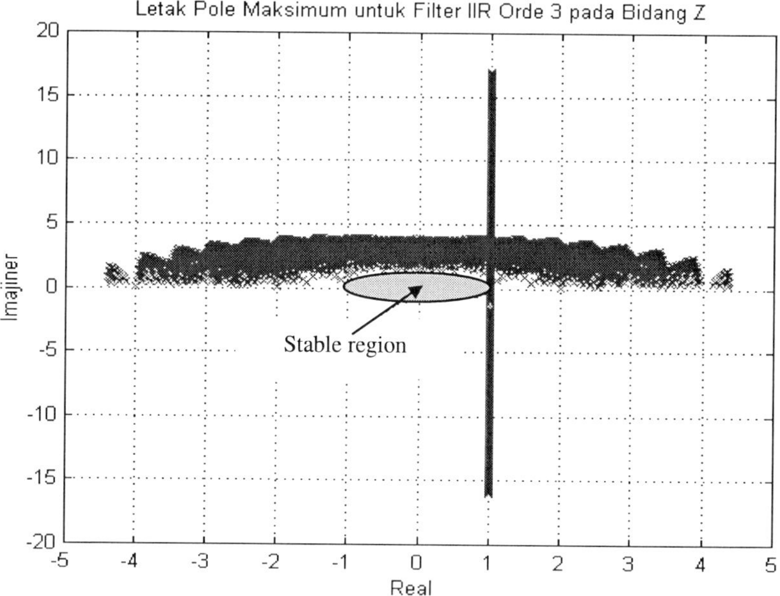

Fig. 7. Maximum pole location of integer values of IIR filter coefficient ($-16 \leq c_i \leq 15$).

5 Statistical Analysis

There are some statistical properties that must be shown in this paper, to prove some advantages of cipher text result. So it will lead us to some conclusions of chaotic cryptosystem performance.

5.1 Probability Mass Function

The uniform of PMF is the most important advantage of chaotic cryptosystem. With this property the chaotic cryptosystem cannot be cracked by statistical analysis, because all of plain text will be successfully mapped to uniform PMF cipher text. Let we describe with an empiric result of experiment that is shown on figure 9 and 10. The form of plain text causes the difference of those figures. Figure 9 is formed by a uniform plain text, but in figure 10 it is generated from a random plain text.

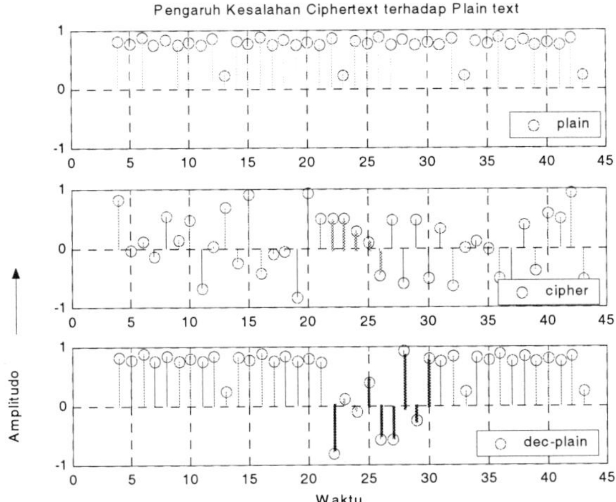

Fig. 8. The error quantization of cipher text, it will make m+2 errors of plain text.

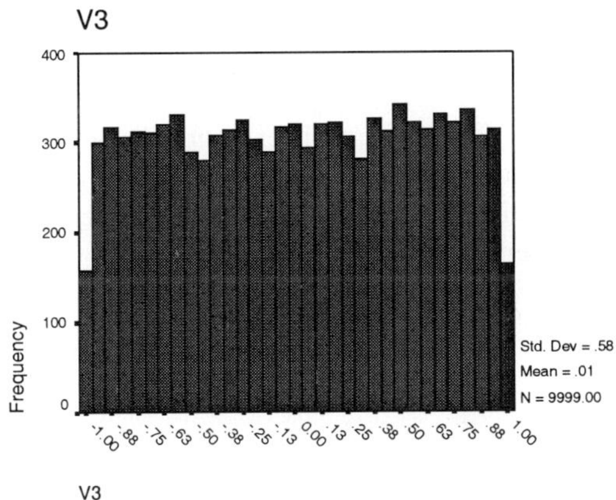

Fig. 9. The histogram of cipher text probability for a uniform plain text

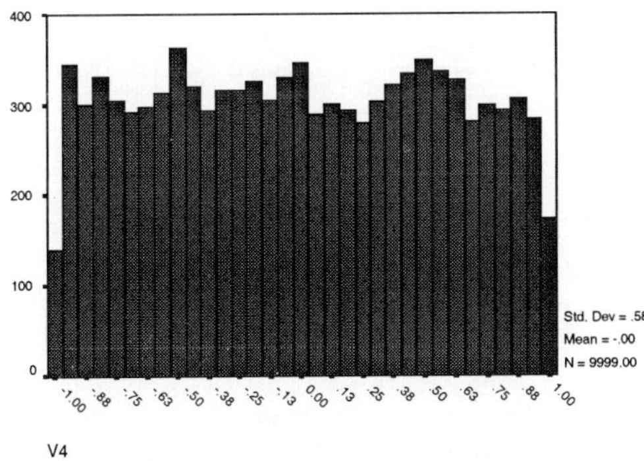

Fig. 10. The histogram of cipher text probability for a random plain text

5.2 Autocorrelation

The test on the auto-correlation is aimed to examine how far the correlation between the-nth sampling signals and the previous and next signal being generated. By applying the normalized auto-correlation to the generated cipher text signal will result in Acf graphic as shown in this picture below, The results show the values of correlation between the nth signal with the surrounding signal.

If we notice the Acf graphic of cipher text signal, the normalized autocorrelation can be written as a following equation:

$$\text{Rxx}(\tau) = \begin{Bmatrix} 1 & \text{untuk } \tau = 0 \\ \Delta & \text{untuk } \tau \neq 0 \end{Bmatrix} \tag{11}$$

where, Δ is a small value that is dependent on coefficient filter used. If the location of poles is far away from the unit cycle, Δ is smaller and closed to zero. But in the other side, the graphics plot is not moving when the input is changing (see: figure 11).

5.3 Power Spectral Density

The power spectral density (PSD) of cipher text signal is supposed to be a constantly continue wideband and is not depending upon their input. In the figure 12, the simulation result of PSD has relatively constant values with the average of 0 to 2 dB, that is closed to white noise property.

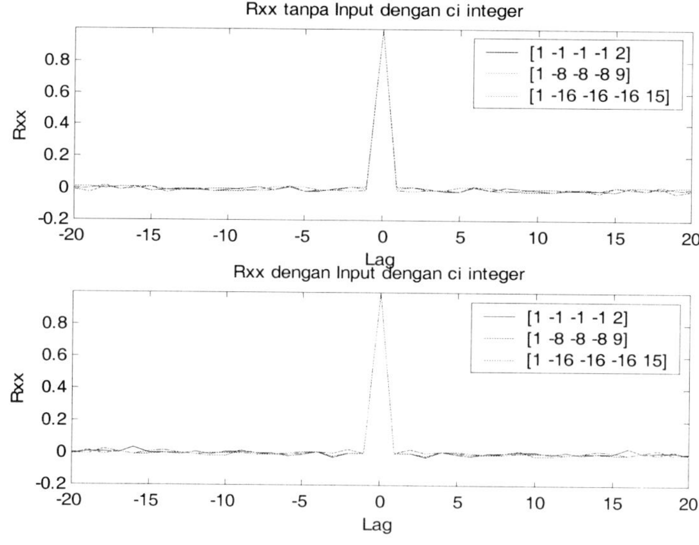

Fig. 11. Act of cipher text for two different inputs.

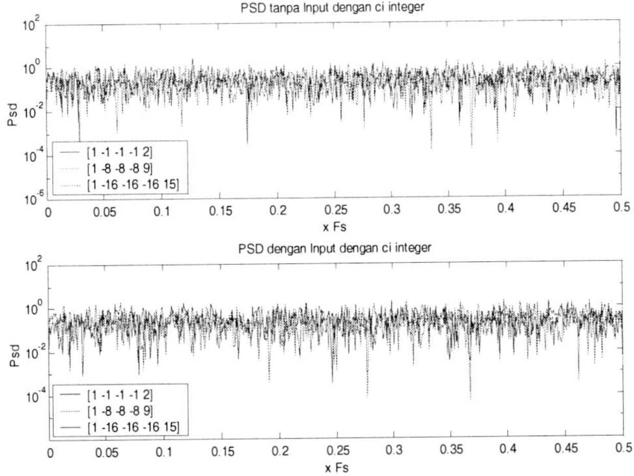

Fig. 12. The PSD graphics of cipher text

6 Conclusions

In this research we have successfully realized crypto-graphics system with chaotic method on FPGA. The system implementation used fixed point with integer values of filter coefficients. It makes the cryptography need no additional of bit rate for their cipher text. One important point of this experiment results show that the statistical properties of cipher text is independent to input signal. Due to this result, the system could not be cracked by the statistical analysis.

References

1. Bianco, Mark E.,Reed, Dana A., *Encryption System Based on Chaos Theory,* Patent Number: US5048086, September 1991
2. Conover W.J, *Practical Non Parametric Statistic*, John Wiley & Son, 1980.
3. Denny Gulick, *Encounters with Chaos*, Mac Graw Hill, 1992.
4. Kristina Kelber, Wolfgang Schwarz, *Digital Realization of Discrete Time Chaos Generators,* IEEE Transactions on Circuits and Systems, 1998.
5. Kristina Kelber, *N-dimensional Uniform Probability Distribution in Nonlinear Auto-Regressive Filter Structures*, IEEE Transactions on Circuits and Systems, September 2000.
6. Man Young Rhee, *Cryptography and Secure Communications*, Mc Graw Hill, 1994.
7. Marco Gotz, *Non Linier Digital Waveform Coding of Chaotic Signals*, Technische Universität Dresden, 1998.
8. *Runs Test for Detecting Non-Randomness*, http://www.itl.nist.gov/div898/handbook/eda/section3/eda35d.htm, 2001.
9. Wolfgang, Kutzer, *Chaotic Signal Generated by Digital Filter Overflow*, Technische Universität Dresden, 1998.

Stereo for Recovering Sharp Object Boundaries

Jeonghee Jeon[1], Choongwon Kim[2] and Yo-Sung Ho[1]

[1] Kwangju Institute of Science and Technology (K-JIST),
1 Oryong-dong Puk-gu, Kwangju, 500-712, KOREA
{jhjeon, hoyo}@kjist.ac.kr
[2] Chosun University,
375 Seosuk-dong Dong-gu, Kwangju, 501-758, KOREA
{cwkim}@chosun.ac.kr

Abstract. In this paper, we propose a stereo matching algorithm using multiple windows to recover sharp object boundaries. In order to achieve this goal, we employ commonly used techniques, such as left-right consistency, uniqueness constraint, and expansible multiple windows. By adding a new idea of multiple windows over these techniques, we can minimize the boundary overreach, which is usually caused by the unsuitable window size or shape. We demonstrate that the algorithm using a new multiple windows can display the dense and sharp disparity map and the windows adaptively used according to features of boundary edges. A problem generated by the disposition order of multiple windows is also presented.

1 Introduction

Stereo matching based on correlation or sum of squared differences (SSD) is a basic technique to obtain a dense map from images [1] [3] [4] [5] [7] [8] [10] [11]. Although this technique yields dense depth maps, it fails within occluded areas and/or poorly textured regions. Any window-based stereo matching in these regions has problems of low reliability. In general, the fattening or thinning operation along the object boundary has a problem that a window can contain both foreground and background surfaces with different disparities [2]. Kanade and Okutomi addressed the problem of choosing the right support region for high reliability with adaptive window [3]. At each point, a rectangular window is grown to an optimal size based on an estimate of disparity uncertainty in the current window. A greedy algorithm is used to select the best of the four possible directions to grow the window at each step. They also presented a multiple-baseline stereo to determine a single point in a region with repetitive patterns [4]. Fuseiello, et al. proposed a method for choosing the right support region by the multiple window approach [5]. For each pixel, they perform the correlation operation with nine different windows, and obtain the disparity from the window of the smallest SSD value. The basic idea of this scheme is that a window yielding a smaller SSD is more likely to cover a constant depth region.

In this paper, we propose a stereo matching algorithm using multiple windows to recover sharp object boundaries. Relevant techniques and a new algorithm are de-

M.H. Shafazand and A M. Tjoa (Eds.): EurAsia-ICT 2002, LNCS 2510, pp. 248–255, 2002.

scribed in Section 2. In Section 3, we are shown experimental results using synthetic and real stereo pairs, and we summarize the proposed algorithm in Section 4.

2 The Algorithm

2.1 New Multiple Windows

In recent years, stereo techniques using multiple windows have been proposed to recover precise object boundaries and display disparity maps efficiently [5] [8]. However, they do not use windows of diagonal edges that are often encountered in the image. They use only symmetric windows. The stereo algorithm using multiple windows can compute more accurate similarity than one using a single window. However, it is difficult to calculate similarity exactly in the boundaries of diagonal edges because the shapes of the windows are suitable for detecting horizontal and vertical edges. We introduce new windows of diagonal shapes, as shown in Fig. 1.

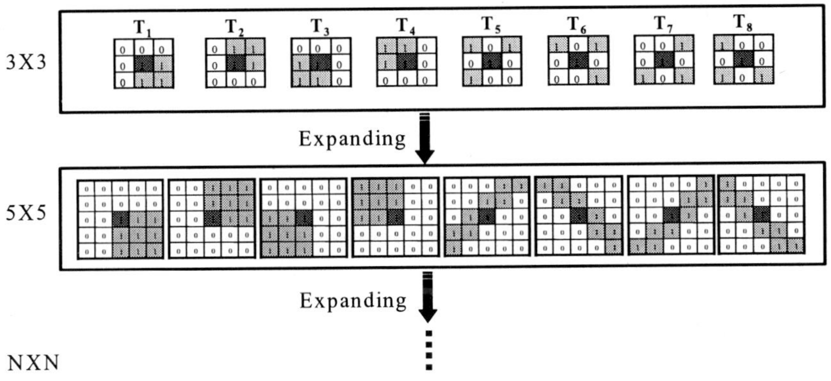

Fig. 1. Multiple windows

In Fig. 1, we show four windows T_1~ T_4 to detect horizontal, vertical, and corner edges that are borrowed from the literature [5] [8]; however, windows T_5~ T_8 for diagonal edges are newly introduced. All windows shown in Fig. 1 can be regularly expanded for stereo matching. We notice that the multiple windows have the same number of gray cells if their sizes are equal. All gray cells are considered to calculate similarity of each window and the darker gray cells are the pixels that we want to find in other images. In this paper, we assume that all stereo pairs are fully rectified by camera parameters [14].

2.2 Similarity Measure

In the area-based stereo algorithm, the similarity measure of each pixel usually uses the well-known SSD.

$$S_T(x, y, d) = \sum_{i, j \in w} [I_L(x + i, y + j) - I_R(x + i + d, y + j)]^2 \tag{1}$$

where I_L, I_R, d, w, and T are left, right images, disparity, gray cells within window, and one of multiple windows, respectively. The best match for each point in the image using SSD can be found by comparing the square window centered at this point against the window of equal size centered at points that lie on the corresponding scanline in the other image. The SSD across the window is used as the similarity measure. The location that minimizes this measure is selected as the best match, and the disparity is stored. From the similarity of each window, we determine the disparity from the window generating the smallest SSD.

$$S(x, y, d) = \arg \min_{T=1}^{8} S_T(x, y, d) \tag{2}$$

We can determine the disparity by Eq. (1) and Eq. (2); however, there are a few problems, as shown in Fig. 2(b).

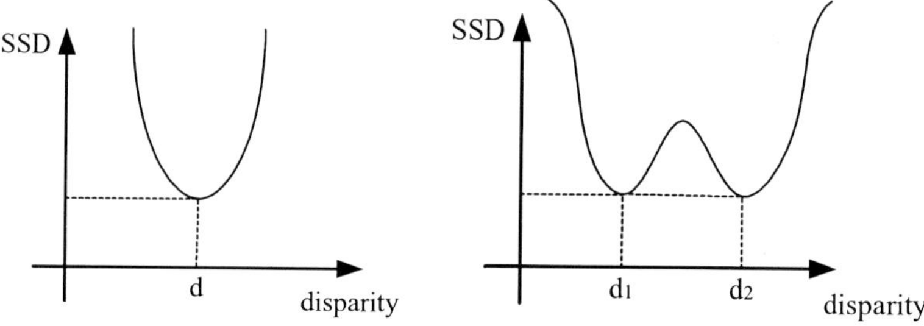

(a) In the case of using single window

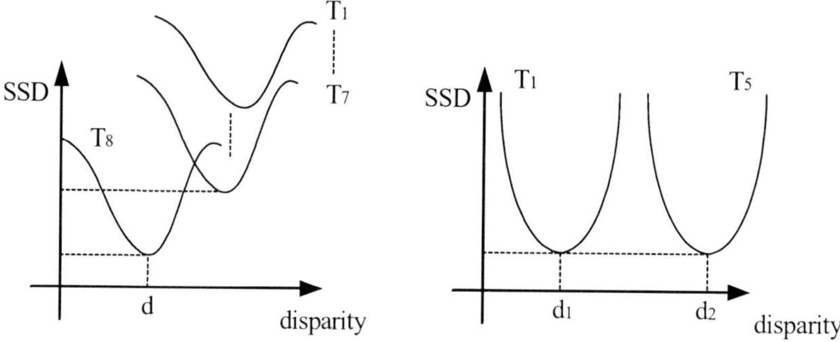

(b) In the case of using multiple windows

Fig. 2. Problems in determining a disparity

In the left graph of Fig. 2(a), the disparity can be determined uniquely at d. Marr and Poggio described uniqueness constraint to minimize false matches [9]. The constraint is that a given pixel or feature from one image can match no more than one pixel or feature from the other image. However, if the number of pixels having the same minimum value is two or more, as shown the right graph of Fig. 2(a), we cannot determine disparity exactly. It was mainly happened in image regions of regular patterns or uniform intensity values. Okutomi and Kanade presented a multiple-baseline stereo to select a single point in the region [4]. We have described a method to estimate single disparity at a region of uniform intensity by extending the window size [10]. In this paper, our method simply expands the window size to four directions to refer more pixels if there are multiple local minima within the search range. With the operation, the bulk of multiple local minima are disappeared as the SSSD function of multiple-baseline stereo [4].

In case of multiple windows, a decision of disparity is faced with a new problem. In the left graph of Fig. 2(b), we know that the disparity can be easily estimated as d if the minimum of SSD in each window is different. However, as shown in right graph of Fig 2(b), we cannot decide the disparity of window with the smallest SSD accurately. The disparities are usually determined by priority of the windows. We call the problem disposition order. As shown in right graph of Fig. 2(b), disposition order of the windows can be an important factor if the SSD of each window is same but disparities are different each other. Disposition order is priority of windows to estimate a disparity from the SSDs. If priority of T_1 is higher than T_5, disparity is estimated as d_1. However, if priority of T_1 is lower, disparity is determined as d_2. For the reasons, any algorithm using multiple windows cannot avoid the problem, which is originated by disposition order of multiple windows. Thus, it should be carefully determined to estimate right disparity. We have selected the disposition order as Fig. 1. This selection is made from the assumption that object's boundaries in the scene of real world have mostly vertical and horizontal lines.

In order to detect occlusion, we use left-right consistency. The principle is that a valid match point should be equally matched in left-right and right-left direction [7]. Each point on one image can match at most one point on the other image, and the matched points have the same disparity in both directions, respectively. Therefore, we can easily predict an occluded pixel or region by checking consistency.

2.3 Proposed Algorithm

We described relevant techniques and characteristics of the multiple windows. In order to estimate single disparity, we firstly use a smaller window, which can be expanded to four directions according to a situation of uniqueness constraint. The windows can be also expanded according to condition of left-right consistency that a valid disparity should be equally existed in both directions of left-right and right-left direction. Disposition order of multiple windows is considered.

All points of image are basically examined whether uniqueness constraint or left-right consistency are satisfied or not. If even one of two conditions is not satisfied, the size of window is enlarged to find a unique match point by referencing more pixels. Through stereo matching by using uniqueness constraint, left-right consistency, and expanding multiple windows, we know that the algorithm has a possibility to minimize boundary overreach. The proposed stereo matching procedure to minimize boundary overreach is as following.

Step 1: Set all the points to FALSE.
Step 2: If a point has FALSE, start to stereo matching using the Eq. (1) and Eq. (2).
Step 3: Check uniqueness constraint and left-right consistency.
Step 4: If two conditions are simultaneously satisfied, save the disparity.
Step 5: Mark the point as TRUE.
Step 6: Process Step 2 ~ Step 5 until the last point of image with FALSE.
Step 7: Expand window size for the point with FALSE and iterate Step 2 ~ Step 6.
Step 8: Display disparity map or windows map, that indicates which window is used for searching disparity from the proposed eight windows.

3 Experimental Results

We described stereo matching algorithm using multiple windows. The central problems for recovering precise object boundaries are to select appropriate window size and shape as described in Section 1 and Section 2. These problems are important issues in the field of stereo matching irrespective of using single or multiple windows. The issues are found in many literatures [3] [6] [7] [8] [11]. We do not test the problems originated by window size or shape of square or rectangular because the problems have excellently described in many papers [3] [6] [11].

In this section, we perform simulation on various kinds of gray-level image to evaluate performance of the proposed algorithm. The images used are classified into synthetic and real image as shown in Fig. 3(a): synthetic images are Random dot, Corridor, and Microsoft and real image is Tsukuba downloaded from web site [15][16]. The Random dot stereo pairs have features that square object of foreground and background are shifted to right direction with six and two, respectively. The Corridor stereo is clean synthetic image without noise and has various objects including such as straight lines, curves, and circle. The stereo pairs of Microsoft are image that object is slanted, and the Tsukuba stereo pairs have many features such as narrow objects, regions of uniform intensity, and so on.

The algorithm is firstly evaluated by disparity map and the influence of disposition order is secondly assessed by disparity map and window maps, that represents which window is selected out of the proposed eight windows. Fig. 3 presents various disparity maps by using our algorithm and Symmetric Multi-Window (SMW) stereo algorithm, respectively. The SMW algorithm by Fusiello, et. al uses nine windows [5]. The SMW algorithm is an adaptive, multiple windows scheme using left-right consistency to compute disparity and its associated uncertainty. From Fig. 3, it is found that the proposed algorithm produces better results in the aspect of sharp boundaries but the SMW algorithm gives smoother disparity map. It is noticed that disparity maps by our algorithm presents gray-level to only the points with "TRUE" and black to other points with "FALSE".

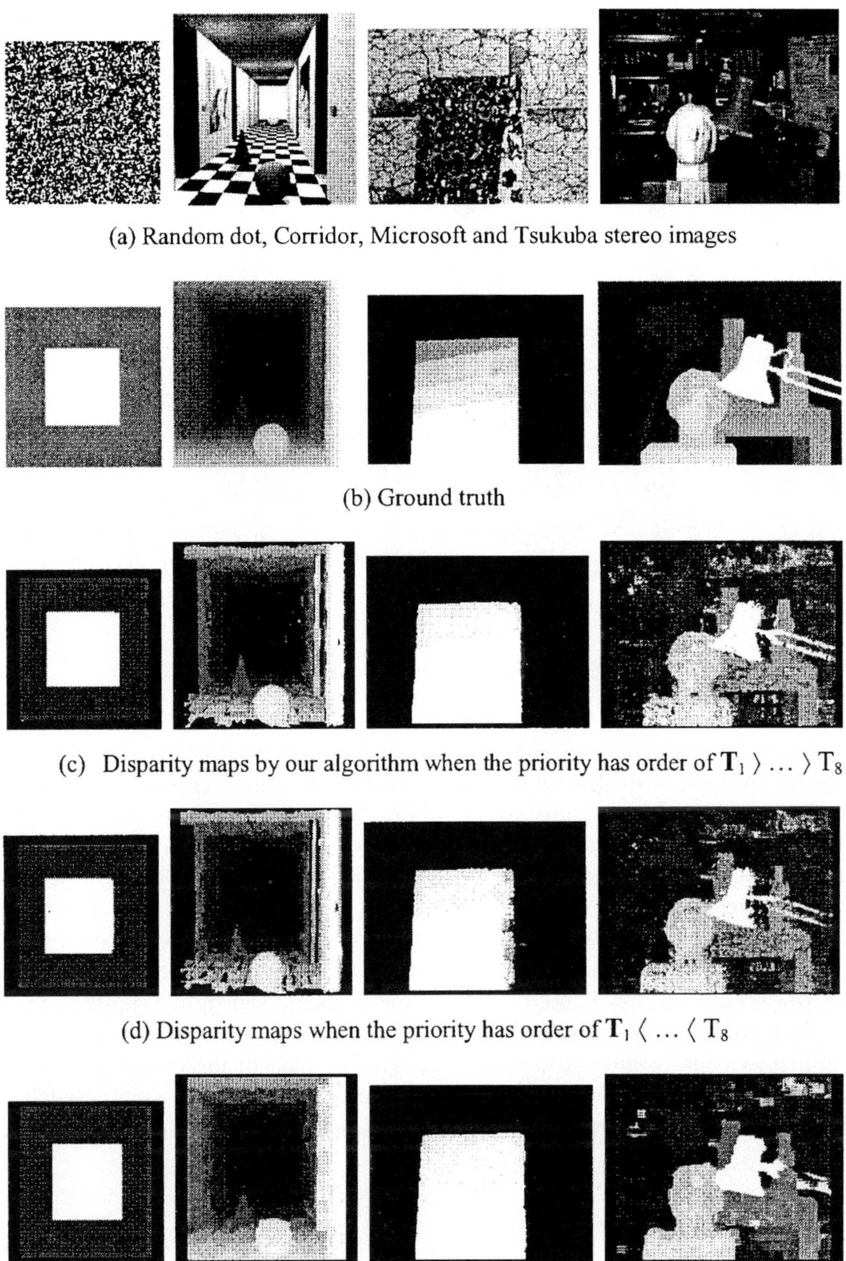

(a) Random dot, Corridor, Microsoft and Tsukuba stereo images

(b) Ground truth

(c) Disparity maps by our algorithm when the priority has order of $\mathbf{T}_1 \rangle \ldots \rangle \mathrm{T}_8$

(d) Disparity maps when the priority has order of $\mathbf{T}_1 \langle \ldots \langle \mathrm{T}_8$

(e) Disparity maps by SMW

Fig. 3. Stereo image, ground truth, and disparity maps

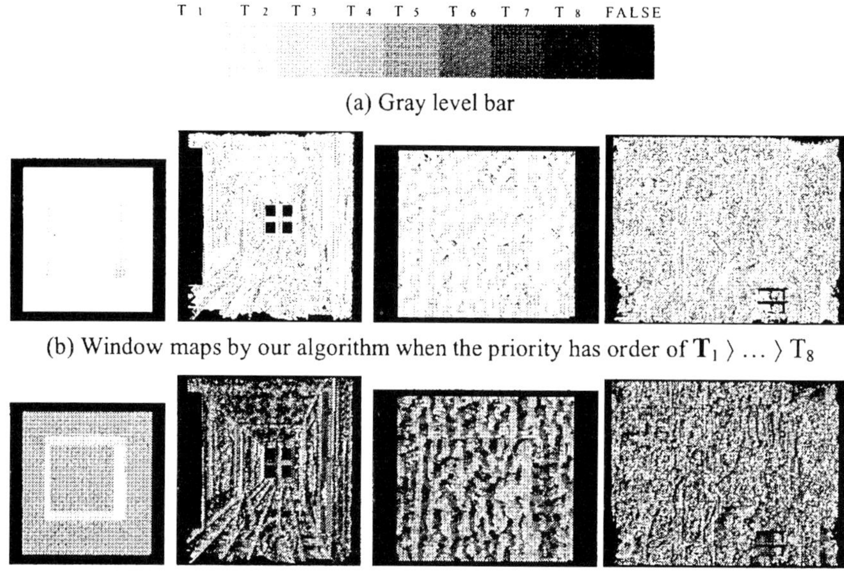

(a) Gray level bar

(b) Window maps by our algorithm when the priority has order of $\mathbf{T}_1 \rangle \ldots \rangle T_8$

(c) Window maps when the priority has order of $\mathbf{T}_1 \langle \ldots \langle T_8$

Fig. 4. Gray level bar and window maps

From the disparity maps of Fig. 3, we can find that our algorithm minimizes the boundary overreach. In case of object's boundary of Fig. 3(c) and (e), our results show steeper boundaries than those of SMW. However, performance of two algorithms cannot be simply compared because the window size of SMW is not known. We can show an influence of disposition order using window maps that indicate which window is used among the multiple windows. A careful comparison between Fig. 3(c) and Fig. 3(d) shows that disposition order has a great influence on the disparity map and out assumption about disposition order is valid. In Fig. 4(a), gray level bar represents which each window is used. The window maps are shown in Fig. 4(b) and Fig. 4(c). From the window maps, we know that the proposed windows are correctly used according to edge features. Especially, the window map of the Corridor containing objects of diagonal edges demonstrates the proposed algorithm uses appropriately the windows of T_5~T_8.

4 Conclusions

In this paper, we have proposed stereo matching algorithm using multiple windows to recover sharp object's boundary. The algorithm assumes that qualities of depth are affected by an appropriate window size, shape, and disposition order of multiple windows. We have made new multiple windows of horizontal, vertical, and diagonal edges. The windows are to be expanded according to uniqueness constraint and left-right consistency. We carefully determined disposition order of the windows because of an effect on estimation of disparity. The experimental results have shown that our algorithm presents not only dense disparity maps but also sharp at the boundaries of object. We also demonstrated that the proper window is used depending on the properties of boundaries like straight line, curve, circle and diagonal edge. These windows can be used in image retrieval, segmentation, motion estimation, image processing, etc.

Acknowledgement.
This work was supported in part by the Korea Science and Engineering Foundation (KOSEF) through the Ultra-Fast Fiber-Optics Networks (UFON) Research Center at Kwangju Institute of Science and Technology (K-JIST), and in part by the Ministry of Education (MOE) through the Brain Korea 21 (BK21) project.

References

1. Dhond, U. R., Aggarwal, J.K.: Structure from Stereo-A Review, IEEE Transactions on System, Man, and Cybernetics, Vol. 19, No. 6, Nov./Dec. (1989) 1489-1510.
2. Otha, Y., Tamura, H.: Mixed Reality, Springer-Verlag (1999).
3. Kanade, T., Okutomi, M.: A Stereo Matching Algorithm with An Adaptive Window: Theory and Experiment, IEEE Transaction on Pattern Analysis and Machine Intelligence, Vol. 16, No. 9, Sept. (1994) 920-932
4. Okutomi, M., Kanade, T.: A Multiple-Baseline Stereo, IEEE Transaction on Pattern Analysis and Machine Intelligence, Vol. 15, No. 4, (1993) 353-363.
5. Fusiello, A., Roberto, V., Trucco, E.: Efficient Stereo with Multiple Windowing, Proceeding of CVPR, (1997) 858-863
6. Scharstein, R., Szeliski, R.: Stereo Matching with Nonlinear Diffusion, International Journal of Computer Vision, Vol. 28, No. 2, July (1998) 155-174.
7. Fua, P.: A Parallel Stereo Algorithm that Produces Dense Depth Maps and Preserves Image Features, Machine Vision and Application, Vol. 6, No.1, (1993) 35-49.
8. Okutomi, M., Katayama, Y., Oka, S.: A Simple Algorithm to Recover Precise Object Boundaries and Smooth Surface, IEEE Computer Society Conference on Computer Vision and Pattern Recognition (CVPR'2001), Vol. 2, (2001) 138-144.
9. Marr, D., Poggio, T.: Cooperative Computation of Stereo Disparity, Science, Vol. 194, Oct. (1976) 283-287.
10. Jeon, J., Kim, K., Kim, C., Ho, Y. S.: A Robust Stereo-Matching Algorithm Using Multiple-Baseline Cameras, IEEE Pacific Rim Conference on Communications, Computers and Signal Processing (PACRIM 2001), Vol. I, Aug. (2001) 263-266.
11. Kang, S.B., Szeliski, R., Chai, J,: Handling Occlusions in Dense Multi-View Stereo, Technical Report MSR-TR-2001-80, Microsoft Research, Sept. (2001).
12. Gonzalez, G. C., Woods, R. E.: Digital Image Processing, Addison Wesley (1992).
13. Frei, W., Chen, C. C.: Fast Boundary Detection: A Generalization and a New Algorithm, IEEE Transaction on Computers, Vol. C-26, No. 10, Oct. (1977) 988-998.
14. Tsai, R.Y.: A versatile Camera Calibration Technique for High-Accuracy 3D Machine Vision Metrology Using Off-the-Self TV Cameras and Lenses, IEEE Journal of Robotics and Automation, Vol. RA-3, No. 4, Aug. (1987) 323-344.
15. http://www.middleburry.edu/stereo.
16. http://www-dbv.cs.uni-bone.de/stereo_data.

Priority Vantage Points Structures for Similarity Queries in Metric Spaces

Cengiz Celik

Department of Computer Science, University of Maryland at College Park
cengiz@cs.umd.edu

Abstract. Similarity search structures for metric spaces have different performance characteristics depending on the properties of the data, construction cost, and space consumption. Nonetheless, recent experiments seem to favor vantage points-based methods such as LAESA, Spaghettis, and FQA if they are allowed to use enough pivots. By using more pre-processing time, these methods can produce superior query performance in terms of distance computations. Unfortunately this also causes them to use more space and CPU time than other structures. In this paper we explore ways to organize the basic structure according to distance relations between database objects, pivots and query objects. We introduce the priority vantage points method, which reduces the CPU overhead without adding extra space requirements. The Kvp structure is also introduced as an improvement, which stores less distance values than other vantage points algorithms. Kvp needs one sequential scan over the index data, making it very suitable to be stored on disk. We show that Kvp is superior to the other methods given same amount of storage.

1 Introduction

The simplicity of distance-based similarity search structures makes them suitable for a wide variety of application areas involving complex objects, such as audio, image, video, text files, fingerprints, protein sequences [E. Chávez et. al. 2001a]. In many cases, it is the most natural manner to approach a problem. For example the similarity between two strings may easily be determined by the edit distance. The alternative approach of embedding the objects in a vector space is not straightforward.

A *metric space* is defined to be a set of objects X together with a distance function d on pairs of objects that satisfies the triangle inequality.

Two important queries used in similarity matching are *range queries* and *k-nearest neighbor queries*. A *range query*, given a query object q and a real r, returns all the objects that are within distance r of q. A *k-nearest neighbor query* returns the k closest objects to a given query object. In this paper we focus on range queries, which can be used to process k-nearest neighbor queries as well.

Metric-based methods have the simplicity of not needing to make any assumptions about the internal structures of the objects. Vector-based methods can handle the dimensions of the vector one at a time. Despite this handicap, metric based methods have been reported to outperform R* trees at high dimensions [P. Ciaccia et. al. 1997]. The metric framework can also use domain specific features when possible,

M.H. Shafazand and A M. Tjoa (Eds.): EurAsia-ICT 2002, LNCS 2510, pp. 256–263, 2002.

such as terminating the distance function computation if it exceeds a given limit, as done by R. F. Sproull [1991].

2 Proposed Methods

There has been considerable work on similarity searching in metric spaces from different disciplines, sometimes unaware of each other. A very inclusive survey can be found at [E. Chávez et. al. 2001a].

2.1 Tree Based Methods

J. Uhlmann [1991] defined the *gh-tree*, short for generalized hyperplane tree where the idea is to pick two points from the current subset, and recursively dividing the rest into two depending on which representative point they are closer to.

The GNAT tree, presented by S. Brin [1995] can be categorized as a generalization of the gh-tree, where there are more than two representatives.

One possible piece of information to keep in the node along with the representative points is the radius of the associated region. This method was used in the *M-tree* [P. Ciaccia et. al. 1997]. Another way would be to include the distances between the representatives as well. An even more precise way is used in the GNAT tree. Every representative keeps its shortest and longest distances to points in every other subset.

GNAT tree in the best case has been reported to make more than a factor of 6 less distance computations than vp-tree spending about a factor of 14 more distance computations for construction. It was reported to be beaten at some cases to vp-tree. The original study also showed that GNAT was outperformed by a variant of vantage points structure although no data was given about its parameters. Recently experiments in [E. Chávez et. al. 2001a, 2001b] show that GNAT is consistently inferior to variants of vantage points structures.

M-tree is designed to be a dynamic structure, emphasis being paid on the ability to keep its querying effectiveness despite data manipulations and to optimize IO performance. Keeping only radius of the representative objects allows it to easily reorganize disk blocks.

Slim-tree [C. Traina et. al. 2000] is shown to have improvements over M-tree with different insertion and splitting algorithms.

Vp-tree [1991] uses a *vantage point* to partition objects into k based on their distances. A nice feature of this is that it is possible to divide the space into many divisions by one single distance computation. However, as the dimensionality grows, the objects tend to cluster around a distance value [K. Beyer et. al. 1999]. Thus the distance to the vp point loses its discriminating power among the objects.

Mvp-tree [T. Bozkaya and M. Ozsoyoglu 1997] uses two vantage points per node. After partitioning the points with one primary vantage point, the partitions are further divided by using a second vantage point. The value of this partitioning approach is, instead of dividing the space into very thin shells, it strives to produce more tightly clustered subsets.

Mvp-tree stores distances to two vantage points at the leaf nodes, making it a hybrid of vantage points structures. It is reported to make up to 80% less distance calculations compared to vp-tree. Experiments in [E. Chávez et. al. 2001a, 2001b] show that pure form of Mvp-tree is consistently beaten by vantage points variants in terms of distance computations.

2.2 Vantage Points Methods

In vantage points methods, the pivots govern over the whole set of objects instead of having local scope. A subset of the objects of size k is selected as vantage points. The distances between the pivots and the rest of the objects are computed at initialization time and stored in the database. At query time these pre-computed distances are used to eliminate candidates. Consider a range query for object q with radius r, the distance of q to the vantage point being d. Triangle inequality implies that the only points that can qualify are the ones that have a distance between $d - r$ and $d + r$ to the vantage point. This elimination step brings in some extra processing compared to local methods where determination of partitions at a node is done once for all the objects covered by the node. Finally, a pass through all objects not eliminated by use of pivots is performed.

A powerful aspect of these methods is that it is possible to use as many pivots as desired at the cost of construction time, storage space and extra CPU cycles. This yields progressively better query performance in terms of distance computations. Vantage points structure can perform up to a factor of 335 less distance computations than vp-tree by spending just 2.4 times more at construction time. This is a huge improvement and is not reported by any other structure.

To our best of knowledge, the first vantage points structure that appeared in literature was LAESA [L. Mico et. al. 1996], as a special case of AESA [E. Vidal 1986]. There have been some improvements over basic LAESA algorithm, such as keeping distances to vantage points sorted and doing binary searches to identify which objects can be eliminated from search [S. Nene and S. Nayar 1997]. Spaghettis [E. Chávez et. al. 1999] is designed to lower the CPU cost as well, by using some extra pointers.

FQA [E. Chávez et. al. 2001b] sorts the points according to their distances to the first vantage point, then on the second, and so on. During a query execution, the algorithm can do binary searches within each distance range. It does not require any extra storage. However it does not work well if too many bits are used for distance values, since that would render the structure to be sorted only by the first pivot. Using fewer bits weakens the effectiveness of pivots. Their experiments show that at 20 dimensions FQA takes only 37.6 % less time than naive approach.

3 Prioritized Vantage Points

There are $k.n$ pieces of information for pivot-based algorithms, where k is the number of pivots and n is the database size. We show ways to prioritize or ignore some of this data to improve space or CPU consumption at the cost of relatively few more distance computations.

3.1 Prioritizing Vantage Points

We would expect a pivot's effectiveness to depend on its distance to the query point. Figure 1 shows the number of objects eliminated versus the pivot's distance to query point for a radius of 0.6 in 20 dimensions. We observe that the pivot is more effective as it is close or far from the query object.

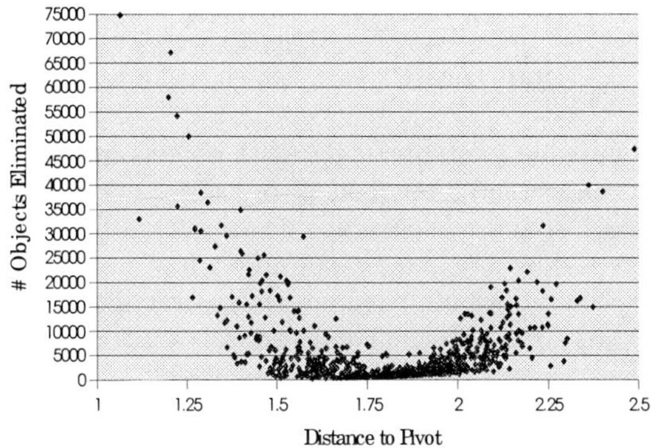

Fig. 1. Distribution of Pivot Cuts by Distance to Pivot

Figure 2 illustrates the intuition behind this observation. Here we have range queries of radius 1 on a database of English words. As the query object is closer to the bulk of the population, more database objects can hide behind it.

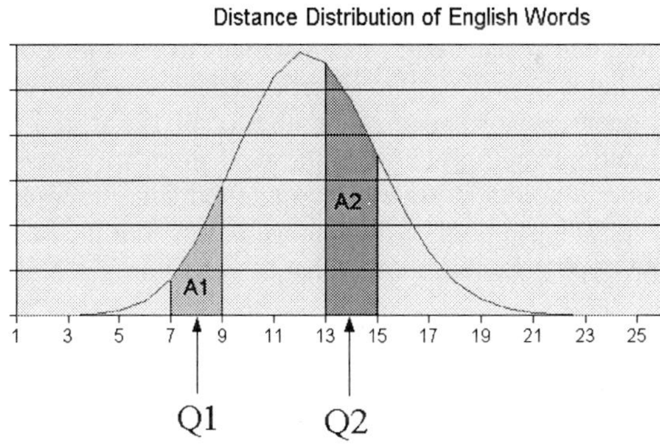

Fig. 2. Query objects Q1 and Q2 have distances of 8 and 14 to the pivot. The areas represent the set of objects that cannot be eliminated by this pivot.

Based on this observation, we can alter the pivot-based search scheme by first computing the distance of query point to all the vantage points, which is eventually done anyway. Based on these distances we can order the processing of pivots.

Figure 3 compares the reduction in search space. Here, the y-axis show the number of objects not eliminated from the candidate list. As more pivots are processed, this number drops.

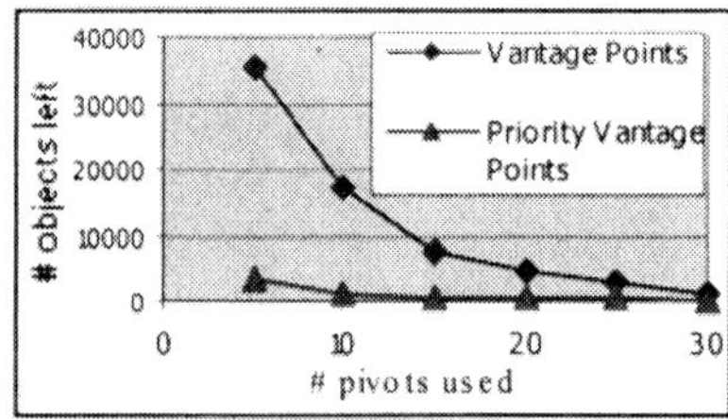

Fig. 3. Priority Vantage Points Uses Pivots More Efficiently

We see that it is possible to eliminate database objects much quicker with priority vantage points. We can use large numbers of pivots for our structure, but only use a portion of them in the search time depending on the query object. This would cut back the CPU processing considerably at the expense of a few more distance computations.

Like FQA, this scheme does not need any extra structures other than *k.n* distance values. Unlike FQA, its CPU efficiency does not rely on coarsening the distance values, it is possible to use as much precision as possible.

3.2 Kvp Structure

It was shown that pivots work better if they are close to the query point. Similarly, we can expect that a close pivot is more effective for a database object. This observation leads to *Kvp* structure, where only *k* most promising pivots are stored per object.

There are two ways this can be implemented. One way could be the usual layout where every pivot keeps an array of distances to objects. The objects can be sorted to do binary search for range queries. Another way is to have an unordered collection of object entries, where each object entry has distances to its selected pivots. The benefit of this structure is that it is very easy to insert or delete objects from the database.

Figure 4 show query performance by number of pivots stored for radius 0.4 in 20 dimensions. The random method chooses the next pivot to be used randomly, simulating a classic vantage points structure. Kvp method first processes close and distant vantage points. Kvp 50 has a pool of 50 vantage points that it prioritizes. As the number of pivots in the pool is increased, the chances of finding a better suited pivot also get better.

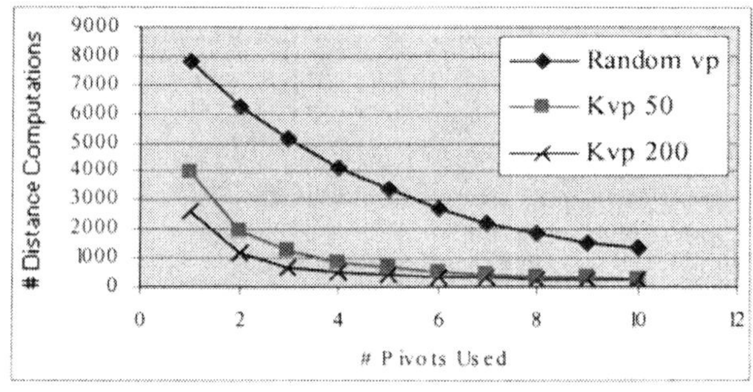

Fig. 4. Query Performance of Kvp Structure

Access patterns of pivot-based structures are targeted toward minimizing CPU processing; they are not suitable to be directly mapped into disk. For example, doing binary searches on secondary storage is expensive as it involves seek operations.

Kvp structure is very suitable to be stored on disk. It only requires a sequential scan of distance values. It does not need to keep any global data.

Kvp is the only structure to store less distance values than usual to provide similar pruning ability. Another way of storing less information is discretization of the distance values, so that fewer bits are used for the distance values.

Experiments were performed where the number of bits used per distance value is varied and the number of pivots per object was adjusted accordingly to keep the total size below a threshold. In the following figures, a label <*r, b, Vps*> stands for classical vantage points algorithm using *b* bits per database object for a query radius of *r*. <*r, b, Kvp k*> stands for Kvp with *k* pivots.

We can see from figure 5 that using more bits is much more effective than using more pivots for values below 7 bits. After that, we observe that using more bits are almost as important as using more pivots.

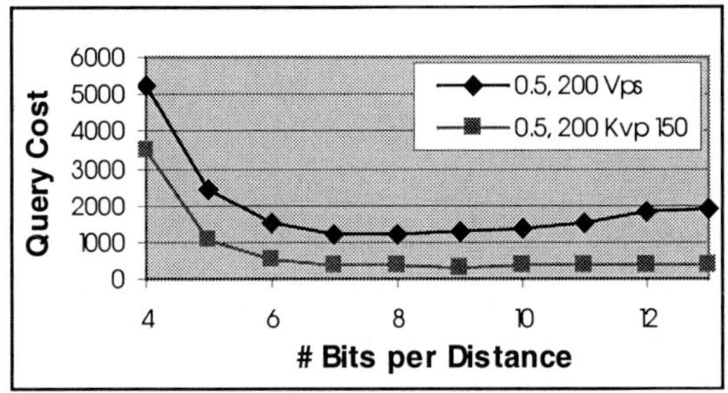

Fig. 5. Query performance observed when the total memory usage is kept at a constant and the number of bits per distance value and the number of pivots stored is varied.

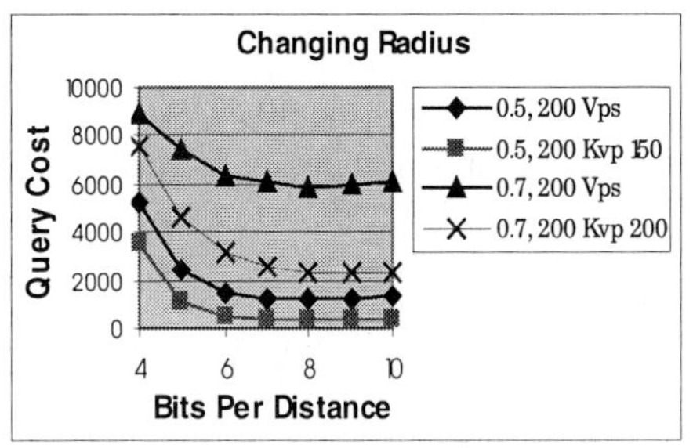

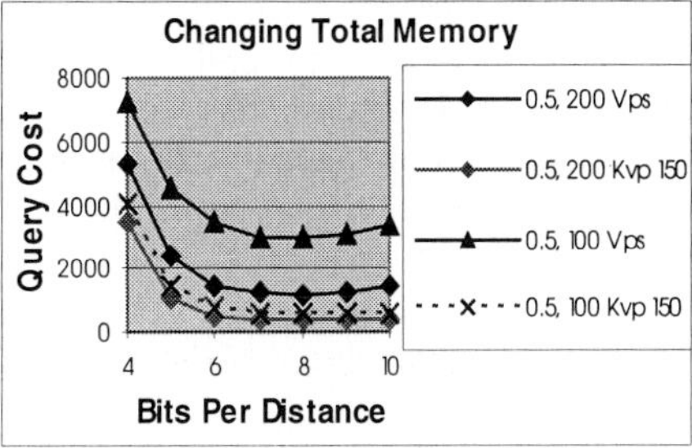

Fig. 6. Affects of changing search radius and amount of total memory on performance

Figure 6 compares two methods under varying radius and memory parameters, where we observe that Kvp is consistently better than Vps. The following table summarizes the speedups Kvp provide over Vps in terms of distance computations:

Table 1. A sample of improvements Kvp provides over Vps

Radius	# Bits	Cost Ratio Vps/Kvp
0.7	200	2.56
0.5	100	5.22
0.5	200	3.53
0.5	400	1.73
0.4	200	1.84

In figure 7 we compare vp-tree to Kvp limited to using same space and less construction cost. We assume an array implementation of vp-tree that uses one distance value per two objects.

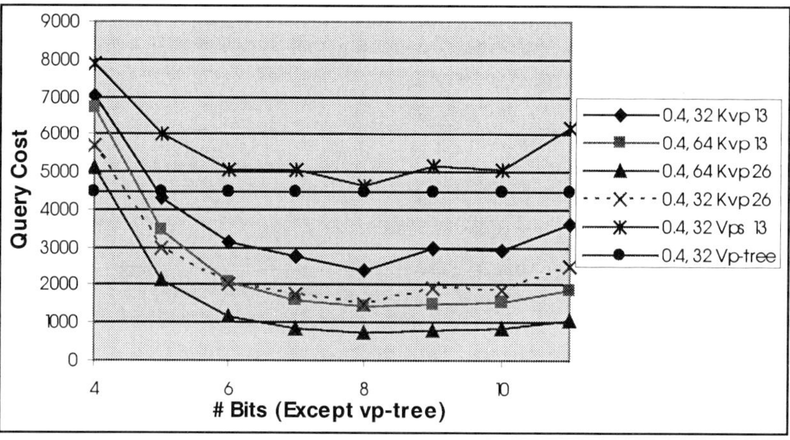

Fig. 7. Comparison of Kvp to Vp-tree

We observe that optimal Vps performs almost identically to vp-tree even though it has to work under pretty limited conditions, imposed by vp-tree structure rather than application requirements. Kvp on the other hand manages to perform 46% less distance computations. The affects of using more memory and more pivots is also briefly studied in the graph to emphasize the power of Kvp to improve its query cost.

4 Conclusions

Experiments show that Kvp offers the ability to spend less CPU cycles and less storage while putting a negligible extra distance computation overhead. It gives applications the ability to tune their resources according to the characteristics of the domain. It is also very easy to manipulate dynamically.

For applications where index can be stored in memory, Kvp offers an important advantage in requiring much less space. Therefore it enables the structure to use more vantage points than the classic method. Kvp is suited to behave well on secondary storage since it does not require seek operations.

Acknowledgement. I am very grateful for the guidance and help I received from Prof. David Mount while working on this project.

References

C. Traina Jr., Agma J. M. Traina, B. Seeger, C. Faloutsos. Slim-Trees: High Performance Metric Trees Minimizing Overlap Between Nodes. EDBT: 51–65. **2000.**

E. Chávez, J. L. Marroquín, R. A. Baeza-Yates. Spaghettis: An Array Based Algorithm for Similarity Queries in Metric Spaces. SPIRE/CRIWG : 38–46. **1999**.

E. Chávez, G. Navarro, R. Baeza-Yates, J. L. Marroquín. Proximity Searching in Metric Spaces. ACM Computing Surveys 33(3):273–321. **2001a**.

E. Chávez, J. L. Marroquín, G. Navarro. Fixed Queries Array: A Fast and Economical Data Structure for Proximity Searching. Multimedia Tools and Applications 14(2): 113–135. **2001b.**

E. Vidal. An Algorithm for Finding Nearest Neighbors in (approximately) Constant Average Time. Pattern Recognition Letters 4:145. **1986**.

J. Uhlmann. Satisfying General Proximity/Similarity Queries with Metric Trees. Information Processing Letters 40:175–179. **1991.**

L. Mico, J. Oncina, E. Vidal. A New Version of the Nearest-neighbor Approximating and Eliminating Search (AESA) with Linear Preprocessing Time and Memory Requirements. Pattern Recognition Letters 17:731–739. **1996.**

P. Ciaccia, M. Patella, and P. Zezula, M-tree: An Efficient Access Method for Similarity Search in Metric Spaces. Proceedings of the 23rd VLDB. 1997.

R.F. Sproull. Refinements to Nearest-Neighbor Searching in k-Dimensional Trees. Algorithmica 6:579–589. **1991.**

S. Brin. Near Neighbor Search in Large Metric Spaces. Proc. VLDB 574–584. **1995**.

S. Nene, S. Nayar. A Simple Algorithm for Nearest Neighbor Search in High Dimensions. IEEE Transactions on Pattern Analysis and Machine Intelligence 19(9):989–1003. **1997.**

T. Bozkaya, M. Ozsoyoglu. Distance-based Indexing for High-dimensional Metric Spaces. SIGMOD 357–368. **1997.**

W. Burkhard, R.Keller. Some Approaches to Best-Match File Searching. CACM 16(4): 230–236. **1973.**

A High Performance Image Coding Using Uniform Morphological Sampling, Residues Classifying, and Vector Quantization

Saeid Saryazdi[1] and Mostafa Jafari[2]

[1] Ass. Prof., Department of Electrical Engineering, Shahid Bahonar University of Kerman, Kerman, Iran
saryazdi@mail.uk.ac.ir
[2] Student, Department of Electrical Engineering, Shahid Bahonar University of Kerman, Kerman, Iran

Abstract. In this paper a new image coding scheme based on the uniform morphological sampling is presented. In the proposed algorithm, the image is sub-sampled uniformly using a sampling grid of squares of size 4 in Heijmans method. The sampling process is equivalent to decomposing the image into 4×4 blocks and each block is represented by its minimum intensity (sample value). The residual blocks are then classified into uniform and non-uniform blocks according to a discrete gradient. The uniform blocks are represented by their mean value. Each non-uniform block is represented by its minimum value and a block (vector) chosen among a predetermined codebook blocks (vectors). The uniform and non-uniform blocks are coded by a different number of bits. Also, a hierarchical version is proposed which provides a higher compression ratio for an approximately equivalent visual quality. Several experiments are made, and compression ratios of 22.64 to 25.19 for a good visual quality of reconstructed images are obtained.

1 Introduction

Digital image coding algorithms attempt to represent images as compactly as possible. This is essential for applications such as transmission or storage of the images. In the framework of image coding techniques, there is an increasing interest in the second- generation image coding techniques [1]. Such a technique eliminates the less important information in the image according to human visual system, and therefore, provides a high compression ratio. In the second-generation schemes, the images are converted into a set of symbols according to some subjective image model. So, the selection and the extraction of visual features in the image play an important role in any second generation coding scheme.

Recently, there has been a great interest in morphological methods in coding applications [2–8]. Mathematical morphology was originally proposed for binary images by Matheron and Serra, then it was generalized to gray scale cases by Sternberg [9]. A morphological operator modifies the geometrical features of the image using a set called "Structuring Element". The simplest morphological

M.H. Shafazand and A M. Tjoa (Eds.): EurAsia-ICT 2002, LNCS 2510, pp. 264–271, 2002.

operations are Erosion, Dilation, Opening and Closing. Erosion and Dilation operations shrink and expand the image respectively, however, Opening and Closing are non-linear smoothing filters. These are defined by:

$$
\begin{aligned}
D(F,G) &= \max_{\underline{Y}} \left[F(\underline{X}-\underline{Y}) + G(\underline{Y})\right] \quad \text{(Dilation)} \\
E(F,G) &= \min_{\underline{Y}} \left[F(\underline{X}-\underline{Y}) - G(-\underline{Y})\right] \quad \text{(Erosion)} \\
F \circ G &= D(E(F,G),G) \quad \text{(Opening)} \\
F \bullet G &= E(D(F,G),G) \quad \text{(Closing)}
\end{aligned}
\tag{1}
$$

where D and E stand for Dilation and Erosion. F and G represent image and structuring element. The vectors $\underline{X}$ and $\underline{Y}$ are the coordinates of pixels in F and G.

Mathematical morphology provides a well known method to capture geometric information and as a result the applications of morphological method in image representation become popular. Well known examples of these image representations are morphological sampling based pyramid [8], morphological skeleton [3,4], loss less morphological sampling [7], and optimal non-uniform morphological sampling [2]. Regular morphological sampling was introduced by Haralick [10]. Later, Heijmans and Toet [11] described a new regular sampling strategy based on mathematical morphology. The sampled image obtained by their approach is a restriction of image erosion by a structuring element "K", called the sampling element. It is assumed that K covers the concrete space (Z^2 in our case), i.e.:

$$S \oplus K = Z^2, \tag{2}$$

where $S \subset Z^2$ is the sampling grid, and, $\oplus$ represents the morphological dilation. Moreover, the reconstructed image (function) is the dilation of sampled image by K. The reconstructed image $\hat{F}$, obtained by this method is limited by the following upper and lower bounds:

$$F \ominus K \le \hat{F} \le F \circ K < F, \tag{3}$$

where F, $\ominus$ and $\circ$ represent original image, morphological erosion and morphological opening respectively. The quality of reconstructed image obtained by the regular morphological sampling scheme is not suitable for coding purpose due to the blocking effect.

2 Proposed Algorithm

The proposed method contains two steps. First, the image is uniformly sampled using Heijmans scheme with a square shape of 4×4 binary structuring element, K. The sampling grid, "S", is considered to be a grid of the squares of size 4, i.e. 4×Z. So the sampling process is equivalent to decomposing the image into 4×4 blocks and to represent each block by its minimum intensity (sample value). The reconstructed image will be obtained by replacing the pixel values by their minimum in all blocks.

Therefore, the image, $I = \{I_{i,j}\}$, is divided into non-overlapping blocks of size 4×4 of the form :

$$b_{i,j} = [I_{m,n} : 4i \le n \le 4i+3, 4j \le m \le 4j+3]. \tag{4}$$

Each block is designated by its minimal intensity, $m_{i,j}$, and its residual block $r_{i,j}$:

$$m_{i,j} = \min_{m,n}(b_{i,j}(m,n)) \tag{5}$$

$$r_{i,j}(k,l) = b_{i,j}(k,l) - m_{i,j}, \quad (k,l) \in \text{supp}(b_{i,j}) \tag{6}$$

Note that according to Eq. (**3**), the residual $r_{i,j}$ is always non-negative.

In the second step, the residual blocks are classified into uniform and non-uniform blocks according to a discrete gradient magnitude:

$$|\nabla b| = \sqrt{\Delta x^2 + \Delta y^2}, \tag{7}$$

where Δx and Δy represent respectively the mean difference between the right hand half-block and the left hand half-block and mean difference between the top half-block and the bottom half-block.

If the gradient magnitude of a block is smaller than a threshold value $|\nabla I_{min}|$, then the block $b_{i,j}$ is a uniform block, otherwise it is considered as a non-uniform block. The threshold value can be chosen based on Weber law [12]. The residues of non-uniform blocks are then coded by a vector quantization scheme [13,14].

Vector quantization coding schemes require the construction of a codebook containing a set of image blocks (vectors) deemed in some manner to be typical of images occurring in the coding application. The codebook is computed by finding a number of vectors that best represent the blocks. In the proposed scheme, the residues of non-uniform blocks are classified by a L.B.G algorithm [15] into N classes, each class is represented by a block (code vector). Clearly, the number of vectors of the codebook greatly affects the visual quality of the coded blocks. Note that the residual blocks (vectors) have a smaller length and therefore are more closely distributed in the space than the primary blocks used in an ordinary V.Q. coding scheme. So, to achieve a similar visual quality of the reconstructed blocks, the proposed algorithm needs a smaller number of code vectors.

In this new scheme, the uniform blocks and the non-uniform blocks are coded by a different number of bits:

– If $b_{i,j}$ is a uniform block, its mean intensity, $\overline{m}_{i,j}$ is quantized and transmitted through the communication channel using $n_u = n_{mu} + 1 = 7$ bits and the additional bit (flag bit) indicates the type of block.

– For a non-uniform block, its minimal intensity $m_{i,j}$ is quantized using $n_{me} = 4$ bits ($n_{me} \le n_{mu}$), and the index of the selected code vector is coded by $n_s = \log_2(N)$ bits. So, to transmit a non-uniform block, $n_e = n_{me} + n_s + 1$ bits are required, whereas, to code a uniform block $n_u = n_{mu} + 1 = 7$ bits are needed ($n_u \le n_e$). The obtained compression ratio, CR, depends on the original image. For images coded with 8 bits (intensity between 0 and 255), CR is limited to:

$$\frac{16\times 8}{n_e} \le CR \le \frac{16\times 8}{n_u} \tag{8}$$

The decoding process is simple: If the currently block $b_{i,j}$ is based on the type of the block uniform (according to its flag bit value), its reconstructed block will be:

$$\hat{b}_{i,j}(m,n) = \overline{m}_{i,j}, \quad m,n = 0,1,2,3. \tag{9}$$

Whereas for a non-uniform block the reconstructed block will be:

$$\hat{b}_{i,j}(m,n) = m_{i,j} + V_{i,j}. \tag{10}$$

3 Hierarchical Structure

To increase the compression ratios in the previous section, the block size must be increased. However, this will degrade the visual quality of the decoded image. The hierarchical structure presented in this section can increase the compression ratio without degrading image quality.

The hierarchical structure consists of merging the suitable uniform blocks, obtained by the proposed algorithm, into uniform blocks of size 8×8. If four blocks of size 4×4:

$$b_{i,j}, b_{i+1,j}, b_{i,j+1}, b_{i+1,j+1}, \quad \text{for } i,j = 0,2,...N-2 \tag{11}$$

are uniform, and the difference between the block value (mean intensity) of each of them is smaller than a threshold T, here in T has been chosen equal to 8. These 4 blocks will be merged into a uniform block of size 8×8. In addition, a flag bit is used to indicate that the 8×8 block $b_{i,j}$ must be considered as a uniform block of size 8×8 or as a four blocks of size 4×4.

In the hierarchical version, the uniform and the non-uniform blocks are coded by different number of bits. If $b_{i,j}$ is an 8×8 uniform block, its average sample values,

$$(m_{i,j} + m_{i+1,j} + m_{i,j+1} + m_{i+1,j+1})/4 \tag{12}$$

is coded using $n_{u8} = 6$ bits, otherwise, first it is divided into four 4×4 sub-blocks,

$$b^k_{i,j}, \ k = 1,...,4, \tag{13}$$

each sub-block is then coded as follows. If $b^k{}_{i,j}$ is a 4×4 uniform block, its mean value $\overline{m}_{i,j}$ is quantized and transmitted through the communication channel using $n_u = n_{mu} + 1 = 7$ bits, where the additional bit (flag bit) indicates the block type. For a non-uniform block, its minimum value $m_{i,j}$ is quantized using $n_{me} = 4$ bits $(n_{me} < n_{mu})$, and, the index of the corresponding code vector is coded by $n_s = \log_2(N)$ bits. Thus, to transmit a non-uniform block, $n_e = n_{me} + n_s + 1$ bits are required, whereas, to code a uniform block $n_u = n_{mu} + 1 = 7$ bits are needed $(n_u \le n_e)$. Thus the overall bit rate, for coding an image of size M×M by the hierarchical version is:

$$\left((M^2/64) + N_{u8} \times n_{u8} + N_{u4} \times n_{u4} + N_{e4} \times n_{e4} \right) \Big/ M^2 \quad \text{bits/pixel} \tag{14}$$

where N_{ui} and n_{ui} are the number of uniform blocks of size $i \times i$ and their relative number of allocated bits, respectively. N_{e4} and n_{e4} represent the number of non-uniform blocks (edge blocks) and their relative number of allocated bits, respectively.

4 Experimental Results

The new algorithm is examined for two images, Fruits and Kid. Each image has 512 × 512 pixels and 256 gray levels, and the proposed algorithm and hierarchical version are applied to these images. The gradient magnitude threshold is set as $|\nabla I_{min}| = 13$, and, the number of the code vectors is considered to be 8, therefore $n_s = \log_2(N) = 3$. Note that both $|\nabla I_{min}|$ and N play an important role in compression ratio and visual quality.

Table 1. Coding results obtained by the proposed algorithms

		Fruits	Kid
Proposed Algorithm	Number of uniform blocks	12737	11655
	Number of non-uniform blocks	3647	4729
	Bit/pixel	0.45	0.456
	CR	17.72	17.56
	PSNR	32.06	27.35
Hierarchical version	Number of 8×8 uniform blocks	1632	1403
	Number of 4×4 uniform blocks	6209	6043
	Number of non-uniform blocks	3647	4729
	Bit/pixel	0.317	0.35
	CR	25.19	22.64
	PSNR	31.9	27.30

The coded images using the new algorithm and hierarchical version are illustrated in Fig. 1. Also, the quantitative results are summarized in Table 1. The results show that the hierarchical scheme merged 51% and 48% of the uniform blocks, respectively

Fig. 1. Coding of 512×512 test images "Fruit" and "Kid": (a,b) original images; (c,d) images obtained by proposed algorithm; (e,f) coded images obtained by the hierarchical structure.

in Fruits and Kid. The proposed algorithm provides a compression ratio of 17.72 for Fruits and 17.56 for Kid, and, the hierarchical scheme achieves a compression ratio of 25.19 for Fruits and 22.64 for Kid, which corresponds to an increment of 44% and 38% respectively, compared to the proposed algorithm.

Obviously, the two proposed methods provide a high compression ratio with a good visual quality.

5 Conclusions

In this paper, we have proposed a new algorithm for image coding based on the morphological sampling. In the proposed algorithm, the image to be coded is sub-sampled uniformly using Heijmans method by a sampling grid of the squares of size 4. Then the residual blocks are classified in to uniform and non-uniform blocks according to a discrete gradient. The uniform blocks are represented by their mean value. Each non-uniform block is represented by its minimum value and a block (vector) chosen among a predetermined codebook blocks (vectors). In addition, a hierarchical version is proposed which provides a higher compression ratio for an approximately equivalent visual quality. These techniques have been experimented on several images and the results yield compression ratios 17.56 and 17.52 for ordinary and 22.64 and 25.19 for hierarchical version.

References

1. Kunt M., Ikonomopoulus A., Kocher M.: Second-Generation Image-Coding Techniques. Proceedings of IEEE, Vol. 73, No. 4 (1985) 549–574
2. Saryazdi S., Haese-Coat V., Ronsin J.: Image Represntation by a New Optimal Non-Uniform Morphological Sampling. Patten Recognition, Vol. 33, No. 6 (2000) 961–977
3. Maragos P., Shafer R.: Morphological Skeleton Representation and Coding of Binary Images. IEEE Trans. on ASSP, Vol. 34 (1986) 1228–1244
4. Maragos P.: Pattern Spectrum and Multi-scale Shape Representation. IEEE Trans. on Pattern Analysis and Machine Intelligence, Vol. 11. , No.7 (1989) 701–716
5. Salambier P.: Morphological Multi-scale Segmentation for Image Coding. Signal Processing, Vol. 38 (1994) 359–386
6. Salambier P., Torres L., Meyer F., Gu C.: Region-Based Video Coding Using Mathematical Morphology. Proceeding of IEEE, Vol. 83, No. 6 (1995) 843–857
7. Wang D., Labit C.: Segmented Images Compression Based on a Lossless Morphological Sampling Scheme. Proceeding of ICIP95 (1995).
8. Kong X., Goustias J.: A Study of Pyramidal Techniques for Image Representation and Compression. Journal of Visual Communications and Image Representation, Vol. 5, No. 2 (1994) 190–203
9. Sternberg S. R.: Grayscale Morphology. Computer Vision, Graphics, and Image Processing Vol. 35 (1986) 333–355
10. Haralick R. M., Zhuang X., Lin C., Lee J.: The Digital Morphological Sampling Theorem. IEEE Trans. on ASSP, Vol. 37, No. 12 (1989) 2067–2090

11. Heijmans H., Toet A.: Morphological Sampling. CVGIP Image Understanding, Vol. 54, No. 3 (1991) 384–400
12. Chen D., Bovik A. C.: Visual Pattern Image Coding. IEEE Trans. on Communications, Vol. 38, No. 12 (1990) 2137–2145
13. Nasrabadi N. M., King R. A.: Image Coding Using Vector quantization: A Review. IEEE Trans. on Communications, Vol. 36, (1988) 957–951
14. Netravali A. N., Haskell B. G.: Digital Picture Representation and Compression. Plenum Press, New York (1988).
15. Linde Y., Buzo A., Gray R. M.: An Algorithm for Vector Quantizer Design. IEEE Trans. on Communications, Vol. 28, No. 1 (1980) 84–95

A Genetic Algorithm for Steiner Tree Optimization with Multiple Constraints Using Prüfer Number

A.T. Haghighat[1], K. Faez[2], M. Dehghan[3], A. Mowlaei[2], and Y. Ghahremani[2]

[1] Atomic Energy Organization of Iran (AEOI), Tehran, Iran.
[2] Dept. of Electrical Engineering, Amirkabir University of Technology, Tehran 15914, Iran.
kfaez@aut.ac.ir
[3] Iran Telecommunication research Center (ITRC), Tehran, Iran.
dehghan@itrc.ac.ir

Abstract. The bandwidth-delay-constrained least-cost multicast routing is a challenging problem in high-speed multimedia networks. Computing such a constrained Steiner tree is an NP-complete problem. In this paper, we propose a novel QoS-based multicast routing algorithm based on the genetic algorithms (GA). In the proposed method, Prüfer number is used for genotype representation. Some novel heuristic algorithms are also proposed for mutation, crossover, and creation of random individuals. We evaluate the performance and efficiency of the proposed GA-based algorithm in comparison with other existing heuristic and GA-based algorithms by the result of simulation. This proposed algorithm has overcome all of the previous algorithms in the literatures.

1 Introduction

In the past, most of the applications were unicast in nature and none of them had any QoS requirements. However, with emerging distributed real-time multimedia applications such as video conferencing, distance learning, and video on demand, the situation is completely different now. These applications will involve multiple users, with their own different QoS requirements in terms of throughput, reliability, and bounds on end-to-end delay, jitter, and packet loss ratio. Accordingly, a key issue in the design of broadband architectures is how to efficiently manage the resources in order to meet the QoS requirements of each connection. The establishment of efficient QoS routing schemes is, undoubtedly, one of the major building blocks in such architectures. Supporting point to multi-point connections for multimedia applications requires the development of efficient multicast routing algorithms. Multicast employs a tree structure of the network to efficiently deliver the same data stream to a group of receivers. In multicast routing, one or more constraints must be applied to the entire tree. Several well-known multicast routing problems have been studied in the literatures. The Steiner tree problem [1] tries to find the least-cost tree, the tree covering a group of destinations with the minimum total cost over all the links. It is also called the least-cost multicast routing problem, belonging to the class of tree-optimization problems.

M.H. Shafazand and A M. Tjoa (Eds.): EurAsia-ICT 2002, LNCS 2510, pp. 272–280, 2002.

Finding either a Steiner tree or a constrained Steiner tree is NP (Non-deterministic Polynomial)-complete [2]. In this paper, we consider a bandwidth-delay-constrained least-cost multicast routing. For the purpose of clarity, in this paper we assume an environment where a source node is presented with a request to establish a new least-cost tree with two constrained: bandwidth constraint in all the links of the tree and end-to-end delay constraint from the source node to each of the destinations. In other words, we consider the source routing strategy, in which each node maintains the complete global state of the network, including the network topology and state information of each link. Most of the proposed algorithms for Steiner tree (without constraint) are heuristic. Some of the well-known Steiner tree heuristics are the RS heuristic [8], the TM heuristic [9], and the KMB heuristic [7]. Several algorithms based on neural networks [10] and genetic algorithms (GA) [13-17] have been also proposed for solving this problem.

Recently, a lot of delay-constrained least-cost multicast routing heuristics such as the KPP heuristic [4], the BSMA heuristic [3] and so on ([5], [6], and [11]) have been proposed. However, the simulation results given by Salama et al. [12] have shown that most of the heuristic algorithms either work too slowly or cannot compute delay-constrained multicast tree with least cost. The best deterministic delay constraint low-cost (near optimal) algorithm is BSMA ([12], [18], [22]). Note that the above algorithms have designed specifically for real-time applications with only one QoS constraint without mentioning how to extend these algorithms to real-time applications with two or more QoS constraints.

In this paper, we propose a novel QoS-based multicast routing algorithm based on genetic algorithms (GA). In the proposed method, the Prüfer number is used for genotype representation. Some novel heuristic algorithms are also proposed for mutation, crossover, and creation of random individuals. We evaluate the performance and efficiency of the proposed GA-based algorithm in comparison with other existing heuristic and GA-based algorithms by the result of simulation. This proposed algorithm has overcome all of the previous algorithms in the literatures.

2 Problem Description and Formulation

A network is modeled as a directed, connected graph $G = (V, E)$, where V is a finite set of *vertices* (network nodes) and E is the set of *edges* (network links) representing connection of these vertices. Let $n = |V|$ be the number of network nodes and $l = |E|$ be the number of network links. The link $e = (u, v)$ from node $u \in V$ to node $v \in V$ implies the existence of a link $e' = (v, u)$ from node v to node u. Three non-negative real value functions are associated with each link e $(e \in E)$: cost $C(e): E \rightarrow R^+$, delay $D(e): E \rightarrow R^+$, and available bandwidth $B(e): E \rightarrow R^+$. The link cost function, $C(e)$, may be either monetary cost or any measure of the resource utilization, which must be optimized. The link delay, $D(e)$, is considered to be the sum of switching, queuing, transmission , and propagation delays. The link bandwidth, $B(e)$, is the residual bandwidth of the physical or logical link. The link delay and bandwidth functions, $D(e)$ and

B(e), define the criteria that must be constrained (bounded). Because of the asymmetric nature of the communication networks, it is often the case that $C(e) \neq C(e')$, $D(e) \neq D(e')$, and $B(e) \neq B(e')$.

A multicast tree $T(s, M)$ is a sub-graph of G spanning the source node $s \in V$ and the set of destination nodes $M \subseteq V - \{s\}$. Let $m = |M|$ be the number of multicast destination nodes. We refer to M as the *destination group* and $\{s\} \cup M$ the *multicast group*. In addition, $T(s, M)$ may contain relay nodes (Steiner nodes), that is, the nodes in the multicast tree but not in the multicast group. Let $P_T(s, d)$ be a unique path in the tree T from the source node s to a destination node $d \in M$.

The total cost of the tree $T(s, M)$ is defined as the sum of the cost of all links in that tree and can be given by

$$C(T(s,M)) = \sum_{e \in T(s,M)} C(e)$$

The total delay of the path $P_T(s, d)$ is simply the sum of the delay of all links along $P_T(s, d)$:

$$D(P_T(s,d)) = \sum_{e \in P_T(s,d)} D(e)$$

The bottleneck bandwidth of the path $P_T(s, d)$ is defined as the minimum available residual bandwidth at any link along the path:

$$B(P_T(s,d)) = \min\{B(e), e \in P_T(s,d)\}$$

Let Δ_d be the delay constraint and B_d the bandwidth constraint of the destination node d. The bandwidth-delay-constrained least-cost multicast problem is defined as minimization of $C(T(s, M))$ subject to

$$\begin{cases} D(P_T(s,d)) \leq \Delta_d, \forall d \in M \\ B(P_T(s,d)) \geq B_d, \forall d \in M \end{cases}$$

3 The Proposed GA-Based Algorithms

In general, a genetic algorithm has five basic components as follows: 1) An encoding method, that is a genetic representation (genotype) of solutions to the program. 2) A way to create an initial population of individuals (chromosomes). 3) An evaluation function, rating solutions in terms of their fitness and a selection mechanism. 4) The genetic operators (crossover and mutation) that alter the genetic composition of offspring during reproduction. 5) Values for the parameters of genetic algorithm. A general structure of the genetic algorithms is as follows:

3.1 Genotype: Modified Prüfer Numbers

The Prüfer number has been used in the GA-based algorithm proposed by Gen et al. [24] to represent a spanning tree. According to the above definition, a spanning tree T has n nodes, $n \geq 3$, and its Prüfer number, $P(T)$, is an n-2 digit number.

Encoding of the Steiner tree by the Prüfer number is more difficult than encoding of the spanning tree. Special difficulty arises because:

- The Steiner trees contain a variable number of nodes in the range from m+1 to n, and their associated Prüfer numbers include between m-1 and n-2 digits.
- In the spanning case, the set of eligible nodes for consideration in decoding algorithm is the set of all nodes that are not appeared in the Prüfer number. In the Steiner case, this rule is not applicable.

We adopt the encoding/decoding algorithms of the Prüfer numbers to be suitable for the Steiner tree problems. Let i be the lowest numbered leaf (node of degree 1) in T and j be the predecessor of i. The Prüfer number is built up by appending j to the right of $P(T)$ and removing i and the edge (i, j) from T. Thus i is no longer considered at all and if i was the only successor of j, then j has become a leaf. This process is repeated, until only two nodes remain in T to be considered. Thus, $P(T)$ is built and read from left to right. Let P be the set of nodes that are part of the Prüfer number, $P(T)$. In our modified Prüfer number decoding algorithm, we consider that the set of eligible nodes, R, be all nodes in the multicast group, $\{s\}\cup M$, that are not member of P, i.e., $R=(\{s\}\cup M)\cap P'$.

The Prüfer encoding establishes a one-to-one correspondence (non-redundancy property) between k-node trees and the set of all string of k-2 digits. This means that we can use only (k-2)-digit permutation (short encoding property) to uniquely represent a tree where each digit is an integer between 1 to k inclusive. The transformation back and forth between edges and Prüfer numbers can be carried out in $O(n \log n)$ with the aid of a heap.

3.2 Pre-processing Phase

Before starting the genetic algorithm, we can remove all the links, which their bandwidth are less than the minimum of all required thresholds (Min $\{B_d \mid \forall d \in M\}$). If in the refined graph, the source node and all the destination nodes are not in a connected sub-graph, this topology does not meet the bandwidth constraint. In this case, the source should negotiate with the related application to relax the bandwidth bound. On the other hand, if the source node and all the destination nodes are in a connected sub-graph, we will use this sub-graph as the network topology in our GA-based algorithms.

3.3 Initial Population

The creation of the initial population in this study is based on the randomized depth-first search algorithm [20],[23]. We proposed a modified randomized depth-first search algorithm (*Random individual creation algorithm*) for this purpose. In this algorithm, a linked list is constructed from the source node s to one of the destination nodes. Then, the algorithm continues from one of the unvisited destinations and at each node the next unvisited node is randomly selected until one of the nodes in the previous sub-tree (the tree that is constructed in the previous step) is visited. The algorithm terminates when all destination nodes have been mounted to the tree.

3.4 Fitness Function

The fitness function in our study is an improved version of the scheme proposed in [23]. We define the fitness function for each individual, the tree $T(s, M)$, using the penalty technique, as follows:

$$F(T(s,M)) = \frac{\alpha}{\sum_{e \in T(s,M)} C(e)} \prod_{d \in M} \phi(D(P(s,d)) - \Delta_d) \prod_{d \in M} \phi(B(P(s,d)) - B_d)$$

$$\phi(z) = \begin{cases} 1 & z \le 0 \\ \gamma & z > 0 \end{cases}$$

where α is a positive real coefficient, $\phi(z)$ is the penalty function and γ is the degree of penalty (γ is considered equal to 0.5 in our study).

3.5 Selection

The selection process used here is based on spinning the roulette wheel *pop-size* times, and each time a single chromosome is selected as a new offspring. The probability P_i that a parent T_i is selected is given by:

$$p_i = \frac{F(T_i)}{\sum_{j=1}^{pop-size} F(T_j)}$$

Where $F(T_i)$ is the fitness of the T_i individual.

3.6 Crossover

Several crossover operators are described in the literatures [13–23] for Steiner tree and constrained Steiner tree problems. Some of them have used the traditional well-known crossover operators, such as the following schemes:

- One point crossover operator [18]
- One point crossover operator, with a fixed probability P_c(≈0.6-0.9) [17]
- Two point crossover operator [22]
- One point crossover operator plus "and" and "or" logic operations with a fixed probability P_c [19]

Unfortunately, according to the genotype representation in these papers, the above crossover operators are not suitable for recombination of two individuals (the crossover operation mostly leads to illegal individuals). However, Ravikumar et al. [20] have proposed a new interesting approach for crossover of Steiner trees and Wang et al. [23] have used the same scheme with some modifications. In this scheme, two multicast trees, $T_F(s, M)$ and $T_M(s, M)$, are selected as parents and the crossover operation produces an offspring $T_O(s, M)$ by identifying the links that are common to both parents. The operator selects the same links of two parents for quicker convergence of the genetic algorithm. However, these common links may be in some separate sub-trees, and some edges may have to be added in order to transform them into a multicast tree.

We propose two novel crossover schemes for recombination of two individuals, which represent Steiner trees:

Crossover I: Let $\{P_F(s, d_1), P_F(s, d_2), \ldots, P_F(s, d_m)\}$ be the set of paths from the source node s to all destination nodes in T_F and $\{P_M(s, d_1), P_M(s, d_2), \ldots, P_M(s, d_m)\}$ be the same set in T_M. Since, we have found these paths for all individuals in the current population for calculating the fitness function of them, the proposed algorithm will not be complex. We define a fitness function for the path $P(s, d_i)$ based on the total cost, the total delay and the minimum bandwidth of the path using the penalty technique, as follows:

$$F(P(s,d_i)) = \frac{\alpha}{\sum_{e \in P(s,d_i)} C(e)} \phi(D(P(s,d_i)) - \Delta_{d_i}) \phi(B(P(s,d_i)) - B_{d_i})$$

$$\phi(z) = \begin{cases} 1 & z \le 0 \\ \gamma & z > 0 \end{cases}$$

where α is a positive real coefficient, $\phi(z)$ is the penalty function and γ is the degree of penalty (γ is considered equal to 0.5 in our study). According to the crossover probability of P_c, two multicast trees $T_F(s, M)$ and $T_M(s, M)$ are selected as parents and the crossover operation produce an offspring $T_O(s, M)$. Each individual may be recombined with its right individual and its left individual through the crossover operator. For each destination node d_i, we compute the fitness of $P_M(s, d_i)$ and $P_F(s, d_i)$ and select the better path. Finally, we compose all selected paths and construct a new Steiner tree.

Crossover II: In this scheme, we first use a simple one-point crossover operator, with a fixed probability P_c. The constructed offspring do not necessarily represent Steiner trees. Then, the effective and fast *check and recovery* algorithm proposed in [21] is used to connect the separate sub-trees in the offspring and also connecting the absent nodes of multicast group to the final tree.

3.7 Mutation

Ravikumar et al. [20] have proposed a new scheme for mutation of Steiner trees and Wang et al. [23] have used the improved version of it in their study. In this scheme [23], according to the mutation probability P_m, the mutation procedure randomly selects a subset of nodes and breaks the multicast tree into some separate sub-trees by removing all the links that are incident to the selected nodes. Then, it re-connects those separate sub-trees into a new multicast tree by randomly selecting the least-delay or the least-cost paths between them.

However, the result of this complex heuristic algorithm is not necessarily a multicast tree including the source node and all destination nodes. In this paper, we propose two following algorithms for mutation operator:

Mutation I: First, we propose an improved version of the scheme presented in [23]. The mutation procedure randomly selects a subset of nodes and breaks the multicast tree into some separate sub-trees by removing all the links that are incident to the selected nodes. Then, the effective and fast *check and recovery* algorithm proposed in [21] is used to connect the separate sub-trees and also connecting the absent nodes of multicast group to the final tree.

Mutation II: According to the mutation probability P_m, the mutation procedure randomly selects an infeasible chromosome from one of the following class (If the first class is empty, a chromosome is selected from the second class and so on)

Class 1: The chromosomes, which do not satisfy the delay and the bandwidth constraints.

Class 2: The chromosomes, which do not satisfy the delay constraint.

Class 3: The chromosomes, which do not satisfy the bandwidth constraint.

If all chromosomes in the current population satisfy both of the QoS constraints, we exit from the mutation procedure. We select only the paths that satisfy both of the QoS constraints in the selected chromosome. We reconnect these selected paths by our proposed algorithm of *crossover I*. Finally, the disconnected destination nodes will be mounted to the sub-tree by our proposed algorithm of *random individual creation*.

4 Experimental Results

We have used the simulation experiments to compare the performance of the proposed GA-based algorithms with the heuristic BSMA heuristic algorithm and some existing GA-based algorithms. All simulation experiments are run on a Pentium III 800, 256 MB RAM. The experiments are run repeatedly until confidence interval of less than 5%, using 95% confidence level, are achieved for the simulation results. A random graph generator based on the Salama [12] graph generator is used. The average degree of each node in the random generated graphs is 4. The multicast group is randomly selected in the graph. The size of multicast group is considered 5%, 15%, and 25% of the number of network nodes. We have tuned the proposed GA-based algorithms and the following parameter settings are achieved: population size *pop-size* = 20, crossover probability $P_c = 0.4$ for *crossover I*, crossover probability $P_c = 0.4$ for *crossover II*, mutation probability $P_m = 0.01$ for *mutation I*, and mutation probability $P_m = 0.01$ for *mutation II*. The experiments mainly test the convergence ability, the convergence speed, and the tree cost of the achieved solutions.

Figure 1 shows the percentage tree cost of BSMA [3], Sun GA-based algorithm [18], and Wang GA-based heuristic algorithm [23] in comparison with our proposed GA-based heuristic algorithm for different network sizes. This Figure shows that our proposed GA-based heuristic algorithm can result in a smaller average tree cost than the mentioned existing algorithms.

Figure 2 shows a typical example of the execution time of our proposed GA-based heuristic algorithm in comparison with the mentioned existing algorithms. This Figure shows that our proposed GA-based heuristic algorithm can result in a smaller execution time than the mentioned existing algorithms.

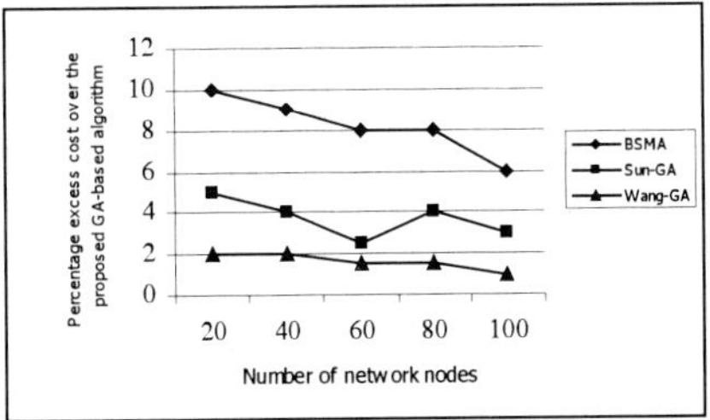

Fig. 1. Percentage excess cost over the proposed GA-base algorithm versus number of network node (Multicast group size is 30% of the number of network nodes).

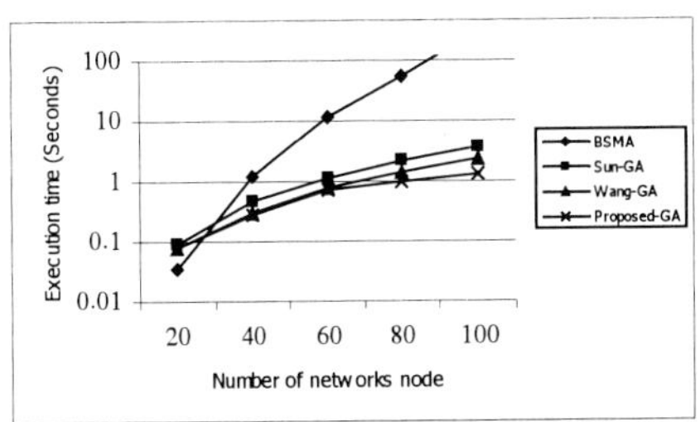

Fig. 2. Execution time of the proposed algorithm in comparison with other existing algorithm.

5 Conclusions

In this study, we have proposed a GA-based heuristic algorithm to solve the bandwidth-delay-constrained least-cost multicast routing problem which is known to be NP-complete. We have proposed Prüfer number for representation of the Steiner trees. In our study, the following new algorithms have been proposed to increase the performance of the genetic algorithm:

- An algorithm for creation of a random individual: *random individual creation*
- Two heuristic algorithms for mutation operator: *mutation I, II*
- Two heuristic algorithms for crossover operator: *crossover I, II*

We have used the penalizing strategy in the proposed fitness function to deal with the infeasible chromosomes and also the repairing strategy in the *mutation I* and *crossover II* algorithms to deal with the illegal chromosomes. On the other hand, we have proposed the *avoidance* strategy to avoid of creating illegal chromosomes in the *crossover I, mutation II*, and *random individual creation* algorithms. The simulation results have shown that the proposed GA-based algorithm has overcome all of the previous algorithms in the literatures.

In this study, we have focused on the source routing and the future work should focus on mechanisms to apply the proposed algorithms to the hierarchical routing.

References

1. S. L. Hakimi: Steiner problem in graphs and its implications, Networks, Vol. 1, (1971) 113–133.
2. R. Karp: Reducibility among combinatorial problems, in: R. E. Miller, J. W. Thatcher, *Complexity of computer computations*, Plenum Press, New York, (1972) 85–103.
3. M. Parsa, Q. Zhu, J.J. Garcia-Luna-Aceves: An iterative algorithm for delay-constrained minimum-cost multicasting, IEEE/ACM Transactions on Networking, Vol. 6, No. 4, (1998) 461–474.
4. V.P. Kompella, J.C. Pasquale, G.C. Polyzos: Multicast routing for multimedia communication, IEEE/ACM Transactions on Networking, Vol. 1, No. 3, (1993) 286–292.

5. R. Widyono: The design and evaluation of routing algorithms for real-time channels, Technical Reports TR-94-024, Tenet Group, Dept. of EECS, University of California at Berkeley, (1994).
6. A. G. Waters: A new heuristic for ATM multicast routing, 2nd IFIP Workshop on Performance Modeling and Evaluation of ATM networks, (1994).
7. L. Kou, G. Markowsky L. Berman: A fast algorithm for steiner trees, Acta Informatica, Vol. 15, (1981) 141–145.
8. V. Rayward-smith: The computation of nearly minimal steiner trees in graphs, International Journal of Mathematical Education in Science and Technology, Vol. 14, No. 1, (1983) 15–23.
9. H. Takahashi, A. Matsuyama: An approximate solution for the Steiner problem in graphs, Mathematica Japonica, Vol. 22, No. 6, (1980) 573–577.
10. E. Gelenbe, A. Ghanwani, V. Srinivasan: Improved neural heuristics for multicast routing, IEEE Journal of selected Area in Communication, Vol. 15, No. 2, (1997) 147–155.
11. Q. Sun, H. Langendörfer: An efficient delay-constrained multicast routing algorithm, Journal of High-Speed Networks, Vol. 7, No. 1, (1998) 43–55.
12. H.F. Salama, D.S. Reeves, Y. Viniotis: Evaluation of multicast routing algorithms for real-time communication on high-speed networks, IEEE Journal on Selected Areas in Communications, Vol. 15, No. 3, (1997) 332–345.
13. J. Hesser, R. Männer, O. Stucky: Optimization of Steiner trees using genetic algorithms, Proceedings of the Third International Conference on Genetic Algorithms, San Mateo, CA, (1989) 231–236.
14. B.A. Julstrom: A genetic algorithm for the rectilinear Steiner problem, Proceedings of the 5th International Conference on Genetic Algorithms, (1993) 474–480.
15. A. Kapsalis, V.J. Rayward-Smith, G.D. Smith: Solving the graphical Steiner tree problem using genetic algorithms, Journal of the Operational Research Society, Vol. 44, No. 4, (1993) 397–406.
16. H. Esbensen: Computing near-optimal solutions to the Steiner problem in a graph using a genetic algorithm, Networks, Vol. 26, (1995) 173–185.
17. Y. Leung, G. Li, Z.B. Xu: A genetic algorithm for the multiple destination routing problems, IEEE Transactions on Evolutionary Computation, Vol. 2, No. 4, (1998) 150–161.
18. Q. Sun: A genetic algorithm for delay-constrained minimum-cost multicasting, Technical Report, IBR, TU Braunschweig, Butenweg, 74/75, 38106, Braunschweig, Germany, (1999).
19. F. Xiang, L. Junzhou, W. Jieyi, G. Guanqun: QoS routing based on genetic algorithm, Computer Communications, Vol. 22, (1999) 1394–1399.
20. C.P. Ravikumar, R. Bajpai: Source-based delay-bounded multicasting in multimedia networks, Computer Communications, Vol. 21, (1998) 126–132.
21. Q. Zhang, Y.W. Lenug: An orthogonal genetic algorithm for multimedia multicast routing, IEEE Transactions on Evolutionary Computation, Vol. 3, No. 1, (1999) 53–62.
22. J. J. Wu, R. H. Hwang, H. I. Lu: Multicast routing with multiple QoS constraints in ATM networks, Information Sciences, Vol. 124, (2000) 29–57.
23. Z. Wang, B. Shi, E. Zhao: Bandwidth-delay-constrainted least-cost multicast routing based on heuristic genetic algorithm, Computer Communications, Vol. 24, (2001) 685–692.
24. G. Zhou, M. Gen: An effective genetic algorithm approach to the quadratic minimum spanning tree problem, Computers and operations research, Vol. 25, No. 3, (1998) 229–247.

A New Technique for Participation of Non-CORBA Independent Persistent Objects in OTS Transactions

Mohsen Sharifi[1], S.F. Noorani[2], and F. Orooji[2]

[1]Computer Engineering Department, Iran University of Science and Technology, Tehran, Iran
mshar@iust.ac.ir
[2]Computer Engineering Department, Shahid Beheshti University, Tehran, Iran
{s-noorani, f-orooji}@ce.sbu.ac.ir

Abstract. The OMG's *Object Transaction Service (OTS)* is an important CORBA Service that provides transaction processing facilities on top of *Object Request Broker (ORB)*. This paper proposes a technique that enables a persistent object, called *Independent Persistent Object (IPO)* to participate in OTS and keep its data consistency by itself. The IPO objects are independent objects that are not under the control of any resource manager. An IPO object can participate in OTS through interfaces that are provided by the OTS. Since the proposed technique is independent of CORBA Persistent Object Service, it can be readily used in any CORBA ORB that has CORBA OTS.

A prototype implementation of the OTS with the proposed technique on a CORBA ORB validates the feasibility of the proposed technique.

Keywords. Distributed Transaction, Object Transaction Service (OTS), CORBA, Recoverable Object, Shadow Object, Persistent Object.

1 Introduction

Heterogeneity in computing platforms and environments has caused so much complications that motivated the definition of CORBA, by the *Object Management Group (OMG)*. CORBA has set the standard for interoperability between objects in cross language, cross operating system, cross platform and cross network communication [1].

Given the requirements of reliable distributed applications for distributed transaction support [2], OMG has specified *Object Transaction Service (OTS)* [3]. OTS consists of interfaces for object involvement in transactions and their interactions with Transaction Manager. So OTS can be considered to act in two layers: *Transaction Management* and *Transactional Object Interface*. The Transaction Management provides services and management functions for transaction demarcation, transactional resource management and transaction context propagation. This paper is only concerned with the transactional object interface layer.

On the other hand, there are persistent objects that are not supported by resource manager, for example the objects that contain shared data in distributed systems [4]. Consistent updates and references to persistent objects can be attained by the

M.H. Shafazand and A M. Tjoa (Eds.): EurAsia-ICT 2002, LNCS 2510, pp. 281–288, 2002.

introduction of a kind of Transaction Manager mechanism. Each object can use the interfaces that the Transaction Manager has provided for participating in transactions while being confident about its data consistency at the end of transactions. Such objects are called Independent Persistent Object (IPO) in this paper. The IPO objects participate in transactions and OTS ensures their data consistency.

This paper's proposition for participating IPO objects in OTS uses the *shadowing* mechanism. Shadowing is a state-based technique [2]. With respect to object behavior, state-based approaches are more suitable than operation-based approaches [2] for persistent objects. The real persistent object is in stable storage and intermediate transaction updates are done on the shadow of the object. After transaction commitment, the shadow is copied to the real persistent object. The proposed technique's features can be summarized as follows:

- It introduces a mechanism for participating IPO objects in OTS.
- It leaves OTS specification intact. Therefore, OTS treats these objects quite similarly to other transactional objects.
- The transaction manager does not need to use additional functions for involving IPO objects.
- The completion protocol applies similarly to both the shadow of IPO objects and other transactional objects.
- The technique is independent of other CORBA services such as *Persistent Object Service* [5]. This property relieves us from weaknesses in persistent object service implementation, and the complexity of its integration with other key services [6].

The rest of paper is organized as follows: a brief overview of the CORBA OTS is presented in section 2. Section 3 provides details of the proposed technique. Section 4 describes the prototype implementation of the technique. Section 5 provides a comparative overview, and section 6 concludes the paper.

2 CORBA Object Transaction Service

OTS provides operations to control the scope and duration of a transaction. It allows multiple objects potentially residing at different resource managers to participate in a global atomic transaction and allows objects to associate their internal state changes with the transaction. OTS coordinates the completion of a distributed transaction by implementing presumed abort two phase commit (2PC) protocol across heterogeneous, autonomous, and distributed objects-based systems [3]. The components of OTS-based applications are:

- Transactional Object: A *transactional object* either includes persistent data or has references to persistent data, which can be manipulated by the object's methods. This object should be called in transaction boundaries. Therefore, the *Transactional interface* has no methods and can be defined as follows:

```
Interface Transactional {
        };
```

- Transaction Context: The *transaction context* includes a transaction identifier and a reference to the transaction coordinator and is shared by all participating objects.

- Recoverable Object: A transactional object capable of being committed or rolled back by a transaction is called a *recoverable object*. It does so by registering an object called a *Resource* with the Transaction Service. The Resource interface is defined as follows:
```
Interface Resource {
        Vote prepare ()
        Void rollback ()
        Void commit ()
        Void commit_one_phase ()
        Void forget ()
        };
```
- Transactional Client: A *transactional client* initiates transaction by obtaining a transaction context from OTS.
- Transactional Server: A *transactional server* is a collection of one or more objects whose behavior is affected by transactions, but have no recoverable states or resources of their own.
- Recoverable Server: A *recoverable server* is a collection of one or more recoverable objects whose data (or state) is affected by committing or rolling back a transaction.

In the proposed technique, an independent persistent object uses the OTS interface and can become a transactional object or recoverable object that participates in OTS.

3 Proposed Technique

In the proposed technique, the shadow of an IPO object participates in OTS instead of the object itself. The proposed scheme is shown with dotted outlined boxes and arrows in Fig. 1; they show the real contribution of this paper.

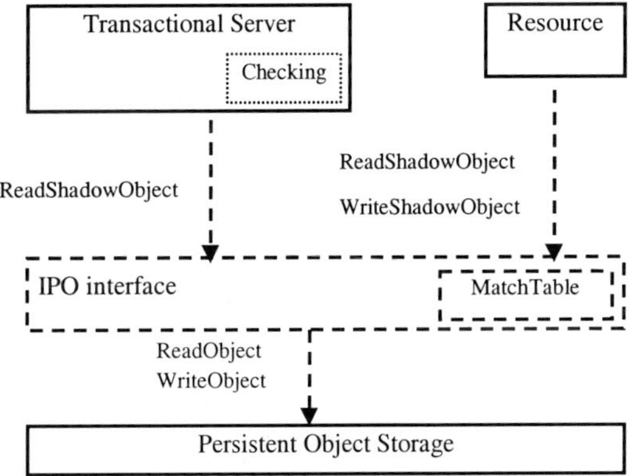

Fig. 1. The architecture of an IPO object participating in OTS.

In this architecture, the *Checking* component function, is checking object type and content of the transaction context. The *IPO Interface* layer stands between Transactional Server and Resource on the one hand and the *Persistent Object Storage* on the other hand. This interface manages the persistency of both the IPO object and its shadow by its own *MatchTable*. The Transactional Server and Resource objects interact with *IPO Interface* using two methods: *ReadShadowObject* and *WriteShadowObject*. The IPO Interface reads and writes an IPO object or IPO shadow object by *ReadObject* and *WriteObject* calls from the *Persistent Object Storage*. The implementation of these methods depends on the *Persistent Object Storage* mechanism.

The proposed scheme is applied both during object creation, and Prepare and Commit phase of 2PC, in the recoverable (independent) persistent objects. These operations and stages are described in sections 3.1, 3.2 and 3.3.

3.1 Creating IPO Objects

When a transactional client begins a transaction, it gets transaction context from transaction service. Afterwards it can ask about object reference from transactional server. After checking transactional context and object type (*Checking* component in figure 2), the transactional server may create an object or an IPO shadow object, and return its reference to the client. If the object is recoverable, it must first be registered with the coordinator of transaction. The transactional server behavior for creation and registration of objects and IPO shadow objects is briefed in Table 1.

Table 1. Creating an object in server side.

(IPO=Independent Persistent Object, Other=Non-Independent Persistent Object)

Requested Object Type		Content of Client Transaction Context	
		Null	**Not Null**
Non-Transactional	**Other**	1. Create object 2. Return object reference to client	1. Create object 2. Return object reference to client
	IPO	1. Create object 2. Return object reference to client	1. Create object 2. Return object reference to client
Transactional	**Other**	1. Don't create object	1. Create object 2. Return object reference to client
	IPO	1. Don't create shadow of object	1. Create shadow of object 2. Return shadow reference to client
Recoverable	**Other**	1. Don't create object	1. Create object 2. Register object in Coordinator 3. Return object reference to client
	IPO	1. Don't create shadow of object	1. Create shadow of object 2. Register shadow in Coordinator 3. Return shadow reference to client

The scenario for the creation of IPO shadow objects by the transactional server is shown in Fig. 2, using Unified Modeling Language (UML) sequence diagram notation.

3.2 Receiving Prepare Message

Resource objects implement recoverable objects, and their involvement in transaction completion. To do so, they must follow the two-phase commit protocol initiated by

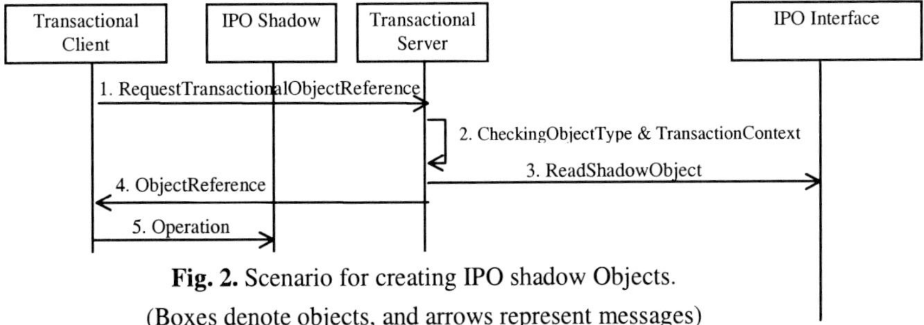

Fig. 2. Scenario for creating IPO shadow Objects. (Boxes denote objects, and arrows represent messages)

their coordinator and maintain certain elements of their state in stable storage. The responsibilities of a Resource object with regard to a particular transaction depends on how it will vote [3]. In our scheme, if a Resource object replies *VoteCommit*, in addition to functions presented in [3], it must record the IPO shadow object. IPO shadow objects can be recorded in persistent object storage in the same way as persistent objects. Fig. 3. shows the sequence diagram for receiving prepare message.

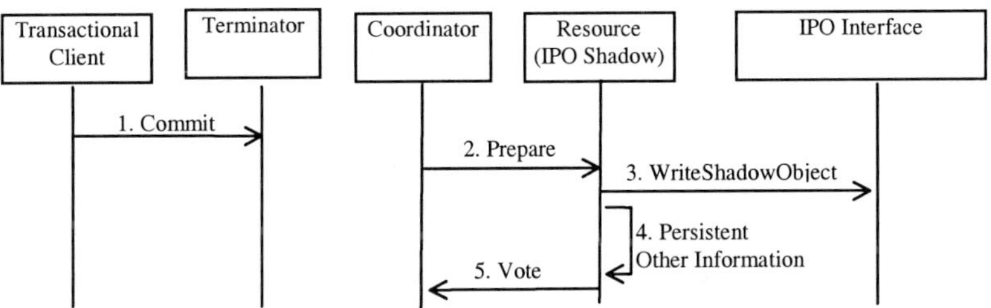

Fig. 3. Scenario for putting shadow in persistent storage.

When a Resource object calls *WriteShadowObject*, the IPO interface imports ShadowId and its related to ObjectId in MatchTable. The entries of MatchTable are used to find main persistent object identifier (ObjectId) in Commit phase.

3.3 Receiving Commit Message

After receiving all resources' vote, the coordinator decides whether to commit or rollback the transaction, and sends the outcome of its decision to resources.

After receiving *Commit* message, an IPO object should copy IPO shadow object on the main IPO object in stable storage. The IPO shadow object related, ObjectId is obtained from MatchTable. If it receives *Rollback* message, it does not need to do anything on IPO object. It should remove the IPO shadow object that has been recorded in Prepare phase (Sect. 3.2). The sequence diagram for receiving commit message is shown in Fig. 4. If a failure occurs, the IPO object remains consistent, because the intermediate changes have been made only to the IPO shadow object.

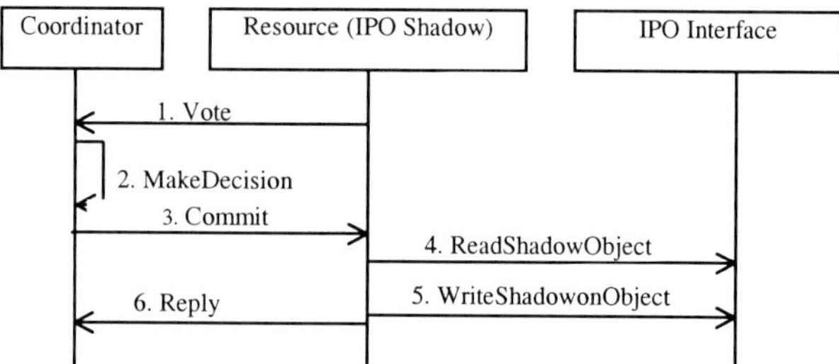

Fig. 4. Scenario for copying shadow on IPO object.

4 Prototype Description

To show the feasibility of our scheme, a prototype OTS supporting the IPO participation in transaction is implemented on Visibroker [7]. An example Bank application transacts with this prototype. Each Bank Server has an AccountManager. The AccountManager's responsibility is to create Account objects. After connection to the Bank Server, the transactional client requests the creation of an Account object by the AccountManager. The Account object is a persistent and recoverable object.

The IDL code of Account, AccountManager and some transaction service elements are defined as follows:

```
module Bank {
   interface Account :: Resource {
      float Balance ();
      Void WithDraw (in float amount);
      Void Deposit (in float amount);
   };
   interface AccountManager {
      Account Open (in string coordinatorreference, in
string name);
      Void EndWork (in string name);
   };
};
module TransactionManager {
   interface Coordinator{
      ...
      Void RegisterResource (in Bank:: Account account);
      ...
   };
};
```

A transaction can involve multiple objects performing multiple requests. In this example, the client calls the WithDraw method from account-A and the Deposit method from account-B (account-A and account-B are two instances from Account object). In our scenario, a client first begins a transaction by issuing a request to transaction service. Then the client issues a request to AccountManager for creating account-A. Account object is a persistent (IPO object) and recoverable object.

Therefore the AccountManager creates a shadow from Account, and registers it with the coordinator, and then returns its reference to the client.

Account-B follows a similar scenario. The client calls the WithDraw method from account-A and the Deposit method from account-B. Finally, the client issues commit request to transaction service. At this time, the coordinator initiates and leads 2PC.

Fig. 5 shows a representation of the above scenario. The dark outlined box named "client", and the dotted outlined boxes named "account-A" and "account-B" are run by three separate threads. The dotted outlined boxes denote IPO shadow objects.

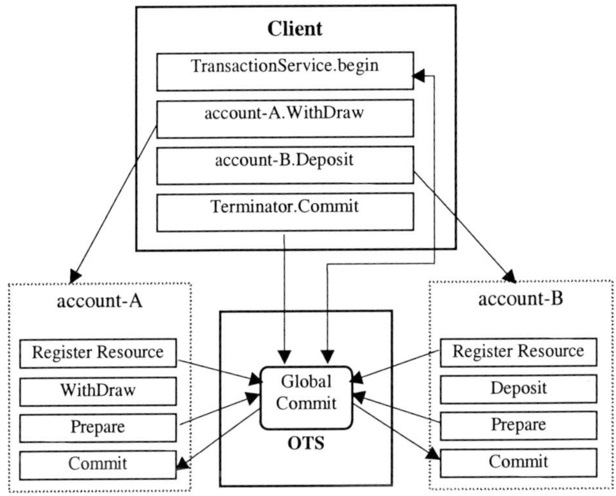

Fig. 5. Prototype architecture.

5 Comparison with Other OTS-Based Implementations

For the sake of comparison with our scheme, three important OTS implementations, and one Java Transaction Service (JTS) [12] implementation, and their interactions with persistent data, are noted here. The OTS implementations are *OrbixOTS* [8], *TPBroker* [9], and *Integrated Transaction Service* (ITS)[10]. The JTS implementation is *Total-e* Transactions from *Bluestone* [11].

In all above implementations the data maintenance container is *database*, and transaction service should interact with *database* to ensure consistency of data. But IPO objects in our scheme contain data within themselves. These objects can participate in transactions in order to ensure their consistency. There is no need for DMBS or XA-aware DBMS support.

Some of the advantages of our proposed scheme are as follows:

- Persistent objects need not to pay for the XA overhead.
- The transaction manager does not need to know about Redo-Undo mechanism.
- It extricates from Log overhead. For example in Redo-Undo method, for each data with n bytes size, at least 3n bytes must be stored in Log (for data itself, its after image and before image).

- The recovery process before recording transaction result is quite simple and does not entail any work, and after recording transaction result, only involves copying shadow object to real persistent object in stable storage.
- As is shown in Fig. 1, the *IPO Interface* layer is an interface. So, the proposed scheme can deploy any object persistence mechanism in *Persistent Object Storage* layer. For example it can use Total-e object state storage mechanism.

6 Conclusion

A new approach was proposed for independent persistent objects (IPO) participating in OTS. IPO objects participate in OTS using *Transactional* and *Resource* interfaces that ensure consistency of data objects at the end of transaction. This approach guarantees object consistency, since all updates within a transaction are performed on the shadows of objects. So even after transaction failure, the real IPO objects remain intact and are correct. It is only after transaction commitment that updates are stored on the persistent storage.

The presented technique allows the shadows of IPO objects to participate in transactions. Behaviors of IPO shadow objects are the same as IPO objects; that is if an IPO object is transactional or recoverable, its related IPO shadow object is transactional or recoverable too.

The technique is independent of other CORBA services such as Persistent State Service, and has no impact on existing systems. Therefore it does not deviate from CORBA standard. It has no additional overheads, and it does not require any changes to recovery protocols and other Transaction Services. It is also suitable for any persistent object that has no Resource Manager or OTS-aware (XA interface) Resource Manager; the persistent object needs only to participate in OTS to ensure its consistency, without paying for additional overheads.

References

1. OMG, The Common Object Request Broker: Architecture and Specification. (1998)
2. Gray, J., Reuter, A.: Transaction Processing. Morgan Kaufmann (1993)
3. Object Management Group, CORBA Services: Common Object Services Specification, Transaction Service Specification. Rev. 1.2 (2001)
4. Hirotsu, T., Tokoro, M.: Object-Oriented Transaction Support for Distributed Persistent Object. IEEE (1992)
5. Object Management Group, CORBA Services: Common Object Services Specification. Chapter 5, Persistent Object Service Specification (1997)
6. Ram, P., et als.: Object Transaction Service: Experiences and Open Issues. IEEE (1999)
7. Inprise Corporation, VisiBroker 4.1, Programmer's and Administrator's Guide, (1999)
8. IONA Technologies PLC, Orbix2000 Administrator's Guide (2001)
9. HITACHI, TPBroker, Object-Oriented Transaction Processing for Mission-Critical Applications, White Paper (1998)
10. Inprise Corporation, VisiBroker Integrated Transaction Service, Data Access Guide (1999)
11. Bluestone Inc., HP Bluestone Total-e-Transactions 2.1, Programmer's Guide, (2001)
12. Sun Microsystems Inc., Java Transaction Specification. Version 1.0 (1999)

Compositional Modelling of Workflow Processes

Khodakaram Salimifard[1,2] and Mike B. Wright[2]

[1] Department of Industrial Management, Persian Gulf University, Bushehr 75168, Iran
k.salimifard@lancaster.ac.uk

[2] Department of Management Science, Lancaster University, Lancaster LA1 4YX, UK
m.wright@lancaster.ac.uk

Abstract. Modelling workflow processes includes transformation of the process logic into a formal representation. Petri net-based modelling is popular in academia, but it suffers from the modelling difficulties and complicated structure which make the technique less practical. In this paper we introduce a component-based approach for modelling of workflow processes using Coloured Petri nets. The proposed method introduces workflow components that are the transformation of execution constructs into Petri net models. Components are process-independent models that can be used to design workflow processes. This approach accelerates the modelling process and model validation. In addition, it increases the modelling power of Coloured Petri nets for workflow applications.

1 Introduction

Business processes and workflows are process models. They are partial order of tasks or activities. The term workflow modelling implies the requirement to represent the processes through which work is accomplished. This requirement becomes a vital issue in designing and re-designing a workflow system. There are differences between workflow modelling and other types of modelling in manufacturing and computer science. Rather than solely focusing on data and functions within the system, workflow modelling also focuses on co-ordination and interacting behaviour among participating agents. On one hand, a workflow model defines what has to be done by representing the functional perspective [7]. On the other hand, the model frames the behaviour and the organisational perspectives [4] of the system.

Petri nets [9] have been widely used in modelling business processes [11]. However, the practical application of Petr nets is limited. One of the restrictions of the existing approaches is that either resources are not considered or, if considered, resources are inseparable-parts of the model. In the first case, the model designer assumes the infinite availability of the required resources which is not true. In the second case, the model gets very complicated and any change, even a small one, in either the process or the resources requires lots of corrections on the whole model, which makes the model inflexible.

M.H. Shafazand and A M. Tjoa (Eds.): EurAsia-ICT 2002, LNCS 2510, pp. 289–296, 2002.

Many of the existing approaches to workflow modelling are based on using Petri nets [2, 13, 14]. The standardisation of model elements has captured less attention, although the compositionality of Petri nest has been discussed in [16]. Recently, van der Aalst et. al. [17] investigated requirements for modelling workflow. The requirements have been classified into modelling patterns which represent comprehensive functionality necessary for a workflow software.

In this paper, we use a different methodology based on the concept of components with clearly defined interfaces. A component is a special Petri net model with only one input place and only one output place. Moreover, a component is somehow open and it can be modified. The paper is mainly focused on the compositional approach for modelling workflow processes. The rest of the paper is organised as follows. Section two contains a short reference on Coloured Petri nets that we use in this paper. The third section covers a very brief introduction to modelling requirements for workflow processes. The proposed methodology is discussed in the fourth section. The model validation is discussed in section five. The paper is finally concluded in section six.

2 Coloured Petri Nets

Coloured Petri nets (CPN) [5] is a class of high level Petri nets. In addition to the basic elements of the classical Petri nets, CPN extends the formalism by allowing a token to be of a specific *colour* and distinguishable.

A CPN model composed of the seven elements. *Colour sets* are analogous to types in programming languages determining characteristics and attributes of entities in the model. *Places* are containers for tokens of certain colour set. *Tokens* are tuples holding data values which are contained in places and consistent with the place colour. *Transitions* are active elements of the model. They change the set of tokens in neighboring places. A transition may be augmented with a code to control the modification and creation of tokens in output places of the transition. *Arcs* are arrows joining places and transitions. *Inscriptions* are expression containing constants, variables and functions associated with transitions and arcs. *Declaration* is a set of constants, variables, and functions defining colour sets and values in the initialisation of the model. A function, so called *marking*, assigns to each place the set of tokens of the same colour set currently held in the place. The set of all place markings at a given time describes the global system status. More details on CPN is out of the scope of this paper. A good introduction to CPN is covered in [5] and links to a wide range of applications and examples is provided at the CPN web site [1].

3 Process Modelling Requirements

Large and complex workflow models are expensive to build and to validate. To facilitate modelling complex workflows and to hide unnecessary details, the concepts of activity refinement and compositional modelling are required. Task refinement allows a layered modelling. Each layer can be refined into a lower layer representing a detailed structure. In the same vein, a lower layer can be composed into an upper layer avoiding unnecessary details.

The behaviour perspective [7] of a workflow is represented by control flows [13, 3]. A *control flow* specifies the execution order [4] and dependencies between workflows. It is modelled using control flow constructs [13, 3]. Typical control flow constructs [3] are *And-split*, *And-join*, *OR-split*, *OR-join*, and *repetition*. AND-constructs are used to model parallel execution and synchronisation of workflows. Alternative routing and decisions are modelled using OR- constructs. Repetitive execution of a part of a workflow is modelled using a repetition construct. These control flow constructs are very common in a workflow model, though they are not sufficient to specify all possible control flows in a complex workflow model [4]. In order to capture the complex functionality of workflow process models, more control flow constructs so-called *execution instructions* is proposed in [8]. However, not all required control flow constructs can be anticipated prior to realisation.

The proposed constructs [17, 3, 8] can be used as a basis for standardisation of workflow models. In order to have a standard definition, formal specification of the semantics of control flow constructs is required. In this paper, we define the semantic of control flow constructs using Coloured Petri Nets (CPNs) [5]. Each control flow construct is modelled as a CPN model, namely a *workflow component*. A control flow construct may compose of other constructs. In the same vein, a component may contain other components. This feature supports both top-down and bottom-up modelling. It also extends the applicability of the proposed methodology for modelling convoluted workflows.

4 The Workflow Components

A *workflow model* illustrates all possible execution flows of a particular workflow process. The model also represents required workflow activities and the logical order of their execution. As discussed in [4], a workflow model is constituted of either elementary or composite workflow activities. An *elementary* activity is a basic task or activity for a workflow participant. A *composite* activity consists of other activities. In contrast to other approaches, where a workflow activity is represented as a single transition [16, 13], we model an activity at the lowest level using two transitions linked by an internal place.

4.1 The Atomic Activity Component

An atomic activity cannot be decomposed. It has the smallest granularity. In fact, it represents a step in the processing of a workflow, where organisational resources are actually used to process the corresponding business operation. We propose a generic model for an atomic activity, namely *Atomic Activity Component* (AAC). It is illustrated in Fig. 1. An AAC is a special component. In addition to input and output places, i.e. `cmpIn` and `cmpOt`, it has a reference to organisational resources via two places `toOrg` and `frOrg`. The component also contains two transitions. `actSt` and `actEnd` represent the beginning and the end of the execution, respectively.

4.2 Components for Non-parallel Workflows

Two workflows are in non-parallel relationship if they cannot be performed simultaneously. A *sequence* is a series of workflows performed one after the other [8]. There may be a situation in which more than one workflow is enabled but only one of them can be performed. The is analogous to *exclusive OR* in programming languages. A workflow may consist of repetitive execution of one or more workflows. Depending on the execution policy, instances of the repeated workflow can be executed either sequentially, concurrently, or in any arbitrary order.

The sequential repetition of a workflow can be modelled using the proposed *Sequential Loop Component* (SeqLop). The generic model of the component is shown in Fig. 1. The number of repetitive execution, i.e. `n`, is given as the initial marking of `cmpIst`. The circuit containing `M2`, `LopSt`, `iNo`, and `LopEn` is used to model the loop. Each occurrence of `LopSt` consumes a `bp` token from `M1` and creates a new one in `M3`. Transition `Wf1` is the abstract representation of the workflow which represents the workflow to be executed repeatedly. `cmpEnd` occurs if $m = n$.

In a workflow, it is also possible to have a situation where two or more workflows are executed in any arbitrary order, but not in parallel. In a *recruitment* workflow, for example, both workflows *interview_the_applicant* and *medical_examination* should be executed. Obviously, these two workflows cannot be executed in parallel and their execution order is not important. We propose the *Arbitrary Order Component* (ArbOrd) as depicted in Fig. 1. An ArbOrd has a very complex behaviour. Occurrence of `cmpSt` creates `n` `bp` tokens in `cmpF`. Due to the fact that transitions share an `e` token from `cmpMD`, either of `T11` or `T21` occurs. Consequently, only transition `T11` (or `T21`) occurs. The initial marking of `cmpIst` implies the number of activities, say 2, can be executed in any arbitrary order. The workflow designer can replicate the sequential part of the component, shown by dashed-dotted lines, `n` times.

4.3 Modelling the Milestone

A milestone is a situation where the processing of a workflow is restricted to another workflow. In a *travel-booking* workflow, for example, as long as the invoice has not been printed, flight and hotel can be booked. It is an issue where a so-called *deadline* [4] or *milestone* [17] construct is required. The proposed *Milestone Workflow Component* (MilStn) is shown in Fig. 1.

Basically, *MilStn* represents a point where a workflow is restricted by a restricting workflow. The *restricting* part is shown by doted line. The part shown by dotted-dashed line is the restricted activity. The *restricted* activity, e.g. *flight_booking*, can be repeated as long as the restricting activity, e.g. *invoice_printing*, is not started. The restricting workflow is represented by `Wf1`

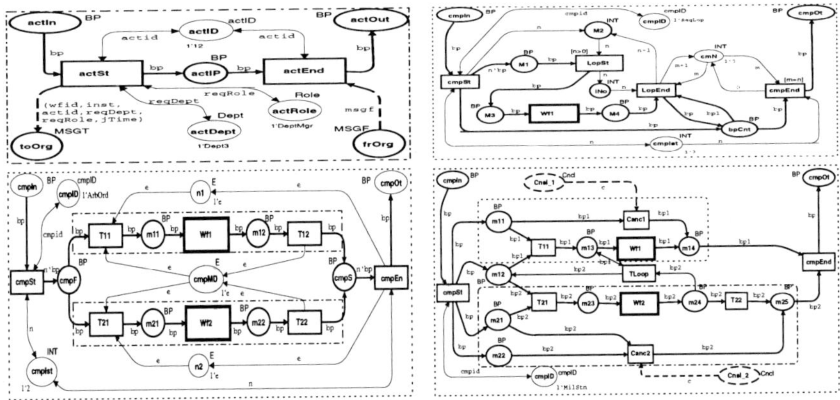

Fig. 1. Component models for parallel workflows: *AAC* (top-left), *SeqLop* (top-right), *ArbOrd* (bottom-left), and *MilStn* (bottom-right).

whilst `Wf2` models the restricted workflow. Both restricting and restricted workflow can be cancelled. Transitions `Canc1` and `Canc2` model the cancellation of restricting and restricted workflow, respectively. The restricted workflow can be repeatedly executed if the restricting workflow has not been executed.

4.4 Components for Parallel Execution of Workflows

Parallel execution is a very common building block in workflow models. It represent a situation in which more than one workflow may or must be concurrently performed, provided that they do not share the same set of resources. Depending on the starting and ending conditions, different type of parallelism can be distinguished.

Parallel Workflows with the Same Start Time. Workflows `WA1` and `WA2` must be started at exactly the same time but they may end at different points in time. This situation is a class of parallel execution in which synchronisation is not needed. The proposed component, i.e. *ParStr*, is depicted in Fig. 2. Upon occurrence of `cmpSt` both `Wf1` and `Wf2` are enabled and occur concurrently. The expression of incoming arc between Join and `cmpEn`, i.e. `bp`, conveys only one `bp` token. Therefore, in order to `cmpEn` being enabled, only a `bp` token in `Join` is required.

The Trigger and Overlapped Workflow. This execution construct represents a situation in which `Wf1` is started and triggers `Wf2` to be started. In fact, `Wf1` triggers a sub-workflow `Wf2` and waits for it to be finished. Fig. 2 shows the proposed component, i.e. *ParDur*, for modelling such a situation.

The Trigger and Overlapping Workflow. This construct represents a situation in which the start and the end of a workflow conditioned to another workflow. That is, Wf1 is started and triggers Wf2 being started. Moreover, Wf2 must be finished after Wf1 is finished. In fact, Wf2 is a pure sub-workflow for Wf1. Fig. 2 shows the proposed modelling component *ParOvr*.

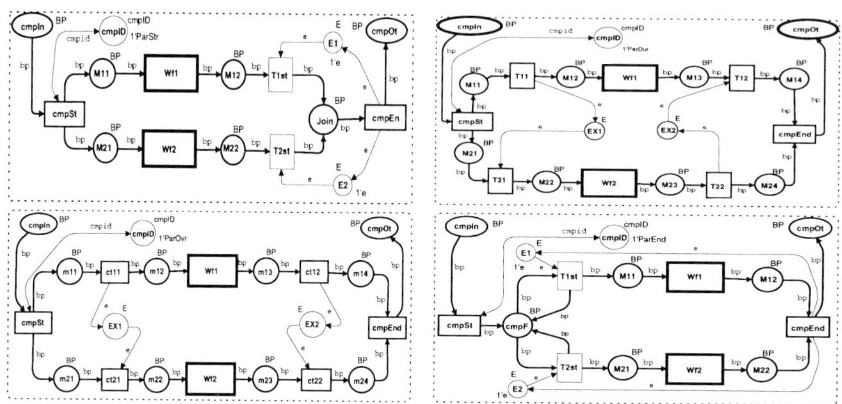

Fig. 2. Component models for parallel workflows: *ParStr* (top-left), *ParDur* (top-right), *ParOvr* (bottom-left), and *ParEnd* (bottom-right).

Parallel Workflows with Synchronisation. This is a situation where two workflows must end at the same time, while they may be started at any arbitrary time. That is, the upper level workflow cannot proceed until both workflows have finished. A CPN-based model for this execution construct so called *ParEnd*, is depicted in Fig. 2. Upon receiving a bp token in its input place, ParEnd allows two transition T1st and T2st to be enabled. The initial marking of shared input place E1 is set to $1'e$. It prevents T1st and T2st be occurred concurrently. Therefore, either of Wf1 or Wf2 occurs. CmpEnd has two input places M12 and M22. It exemplifies a synchronisation which wait until one bp token is available in both input places. Occurrence of cmpEnd consumes two bp tokens and creates a bp token in the output place of the component allowing the upper level workflow to precede.

5 Model Validation

Deriving a model is a process that involves a great deal of communication and interactions with people who are the experts of the real system. The developed model must be validated in order to justify its accuracy and consistency. The model validation considers the correct behaviour of the model based on the expected behaviour in the real situation. It is a process by which the modeller assess whether a developed model posses almost the same input-output relations as the real system does have [10].

In the context of Petri net based modelling, *qualitative analysis* considers the correct function of a Petri net model. It can be used to check the structural correctness of the model. Several structural properties of workflow nets have been discussed in [12, 13, 15]. We constrain our proposed model to adopt the modelling framework suggested in [13], therefore the proposed validation approach can be applied. In our approach, each component has only one input place and only one output place. One of the properties of such an approach is that, for any case, the model terminates with only one token in the output place, eventually. All the other places are empty.

In the proposed approach, model validation is performed using the notion of well-formed model. A component is *well-formed*, if its corresponding cyclic Petri net model is live, bounded, and also its initial marking is the home state. In order to validate a component model, it is required to prove its boundedness, liveness, and home properties.

It is customary to investigate a CPN model by means of interactive simulation option. This, however, cannot be used to prove the correctness of a model. it must be complement with the construction of state space. A state space is a directed graph. Each reachable marking is presented as a node. An arc joining two nodes represents the binding element which occurs in the source node and ends to destination node. State space is usually referred to as *occurrence graph* (OG) [6]. The OG tool is fully integrated with Design/CPN. The OG tool generates a so-called *state space report* (SSR) which provides essential information about the behaviour of the CPN model. Using the OG tool embedded in the Design/CPN, boundedness, liveness, and home state properties of component models could be decided.

6 Summary

In this paper, we presented a compositional approach for workflow process modelling. Our approach differs from the existing modelling methodologies and it is based on CPN. An important merit of this approach is that the soundness property of the resulting model is guaranteed.

The compositional characteristic of the proposed approach helps the modelling process in different aspects. Firstly, modeller can start the design of the workflow process using the most abstract workflow component. It then can be extended to a very complex workflow process model by refining each transition `Wf` in the upper component. Secondly, the proposed atomic activity component separates the workflow process model from the organisational resources. The modeler can design, validate, and analyse the workflow process independent of resources. At the final step, a sound workflow process model will be linked to the organisational resources via atomic activity components.

In this paper we mainly focused on the designing a workflow process regardless of the performance of the model. To finalise a workflow process model, the performance of the model have to be studied. The performance of a workflow model depends on the WfMS and the management of resources. Therefore, it is required to integrate a workflow process model into organisational model, in order to study the performance of the workflow process.

References

1. CPN Web Site, University of Aarhus, Denmark. *http://www.daimi.au.dk/CPnet*, 2002.
2. C.A. Ellis and G.J. Nutt. *Modelling and Enactment of Workflow Systems, Application and Theory of Petri Nets*, volume 691 of *LNCS*, pages 1–16. Springer, 1993.
3. D. Hollingsworth. Workflow management coalition: The workflow reference model. Technical Report TC00-1003, Issue 1.1, Workflow Management Coalition, November 1994.
4. S. Jablonski and C. Bussler. *Workflow Management: Modeling, Concepts, Architecture and Implementation.* International Thomson Computer Press, 1996.
5. K. Jensen. *Coloured Petri Nets, Basic Concepts, Analysis Methods and Practical Use*, volume 1. Springer, 1997.
6. K. Jensen, S. Christensen, and L.M. Kristensen. *Design/CPN Occurrence Graph Manual, Version 3.0.* Computing Science Department, University of Aarhus, Denmark, 1996.
7. T.M. Koulopoulos. *The Workflow Imperative: Building Real World Business Solution.* Nostrand Reinhold, New York, USA, 1995.
8. K. Meyer-Wegener and M. Bohm. Conceptual workflow schemas. In *Proceedings of the* 14th *IFCIS International Conference on Cooperative Information Systems*, pages 234–242. IEEE Press, 1999.
9. C.A. Petri. Kommunikation mit automaten. In *Technical Report RADC-TR-65-377*, volume 1 of *Suppl. 1, English translation.* Air Force Base, New York, Griffiss, 1966. Bonn: Institut für Instrumentelle Mathematik, Schriften des IIM Nr. 2.
10. M. Pidd. *Tools for Thinking: Modelling in Management Science.* John Wiley & Sons, 1996.
11. K. Salimifard and M. Wright. Petri net-based modelling of workflow systems: An overview. *Eurpean Journal of Operational Research*, 134(3):664–676, 2001.
12. W.M.P. van der Aalst. *Verification of Workflow Nets*, volume 1248 of *LNCS*, pages 407–426. Springer, Berlin, 1997.
13. W.M.P. van der Aalst. The application of Petri nets to workflow management. *The Journal of Circuits, Systems and Computers*, 8(1):21–66, 1998.
14. W.M.P. van der Aalst and K. van Hee. *Workflow Management: Models, Methods and Systems.* MIT Press, 2002.
15. W.M.P. van der Aalst, Hees M.W. Verbeek, and D. Hauschildt. A Petri net-based tool to analyze workflows, September 1997.
16. M. Voorhoeve. *Compositional Modelling and Verification of Workflow Processes*, pages 184–200. Business Process Management: Models, Techniques and Emperical Studies. Springer, 2000.
17. Workflow Patterns, Minicom, Australia. *http://tmitwww.tm.tue.nl/research/patterns/*, 2002.

EDMIS: Metadata Interchange System for OLAP

In-Gi Lee, Minsoo Lee, and Hwan-Seung Yong

Dept. of Computer Science and Engineering
Ewha Womans University
11-1 Daehyun-Dong, Seodaemoon-Ku
Seoul, 120-750 Korea
lig@ktf.com, {mlee, hsyong}@ewha.ac.kr

Abstract. The management of enterprise-wide information requires that metadata should be shared and globally accessible by all the heterogeneous products found in today's information technology environment. OLAP metadata includes various information about objects and the relationships among them in an OLAP system. OLAP tools are provided by different vendors and run on different hardware and software platforms. But there still is no standard for a logical OLAP model. To use OLAP tools efficiently, users need to be able to move metadata between tools or between tools and a repository. In this paper, we have designed a metadata interchange model that can be shared among OLAP systems and have implemented a prototype called EDMIS as the OLAP metadata interchange system. The OLAP metadata interchange model that we propose in this paper is designed using XML DTDs. This model lists typical mismatches among the data models of commercial OLAP tools and proposes methods to overcome these differences. The OLAP metadata interchange system enables users to browse and search OLAP metadata on the Web. In addition, if a user wants to use the metadata in another OLAP system, the implemented system can automatically create new cubes using metadata in the repository. In order to validate the usefulness of the proposed system, we have used OLAP products such as MS SQL Server Analysis Services, Pilot DSS and Oracle Express.

1 Introduction

As companies are trying to find a way to more easily administrate their databases and enable a rapid decision making process, the concept of data warehousing which enables end users to access data that is distributed within the company is rapidly gaining popularity. The core part of data warehousing is OLAP (On-Line Analytical Processing) technology, which allows the end user to directly access and analyze the multi-dimensional information in a user-friendly manner and make decisions based on the analysis. This technology, in addition to data mining, is one of the most popular uses of the data warehouse [1].

Despite the hype in data warehousing and OLAP, users are still having problems in obtaining consistent and well-organized information regarding OLAP. Currently a large number of OLAP products are available on the market, but in reality there still does not exist a common standard for a logical multi-dimensional model. The absence

M.H. Shafazand and A M. Tjoa (Eds.): EurAsia-ICT 2002, LNCS 2510, pp. 297–305, 2002.

of a standard among the OLAP products has made it very difficult to exchange data and interface with the products. Even though a user has knowledge of a specific OLAP product, it is very difficult to communicate with other users if they have used different products. To solve this problem, standards efforts are being carried out at the API level. As an example, MD-API of the OLAP Council and OLE DB for OLAP API from Microsoft have been proposed [2]. In addition to this, many vendors are working on OLAP standards but it is very difficult due to the differences in the various OLAP products that have been developed.

This paper explains a common metadata model that can be used among different OLAP products, and shows the design of an XML-based OLAP metadata interchange model based on the metadata model. The proposed OLAP metadata interchange model is an integration of the solutions obtained from the experience of extracting metadata from several OLAP products and creating new cubes. We focus on OLAP metadata required for cube creation and management. We have implemented an OLAP metadata interchange system called EDMIS (Ewha Data warehouse Metadata Interchange System). The OLAP metadata interchange system is used for exchanging cube data that includes dimension, hierarchy and variables etc. It extracts metadata from OLAP servers, stores it, and if needed will automatically create cubes in other OLAP products. This system allows users to interface with different products without going through the complex cube creation process for each product, and enable users to select the OLAP product that can satisfy their purpose and required query performance. It is also developed as a Web-based system to allow users to conveniently manage the metadata through the Web.

The organization of the paper is as follows. Section 2 discusses several standardization efforts. In section 3, an OLAP metadata model is explained and the design of the OLAP metadata interchange model is shown. Section 4 explains the implementation of the OLAP metadata interchange system and gives an example on how the metadata from the OLAP products are extracted and can be used to recreate cubes in other products. Section 5 concludes the paper with a summary of the results and future work.

2 Related Research

It has been very difficult to exchange data among OLAP products due to the absence of a commonly agreed standard for a logical multi-dimensional model. Currently about 40 OLAP products are using different OLAP metadata, vendor specific technology, and specialized user interfaces. Therefore users who have only limited experience with a single OLAP product obtain biased knowledge about OLAP technology. In order to solve this problem, standardization efforts at the API level have been recently carried out. MD-API of the OLAP Council and Microsoft OLE DB FOR OLAP API are such efforts.

MD-API is an object-oriented database-independent interface to a multi-dimensional data source. This API enables clients to select a multi-dimensional data schema, connect to it, and query the metadata. A set of objects is used to represent a query rather than using a text based language to form a query. Therefore, there is no specific textual language used by the API. All query results are returned as a cube object, and applications can obtain cell data from the cube. OLE DB for OLAP API was designed by extending OLE DB so that users can access multi-dimensional data sources. The API is a collection of COM objects and interfaces to enable efficient communication between producers and consumers of multi-dimensional data. OLAP products can easily communicate independently of the multi-dimensional data environment via the API. A multi-dimensional query language called MDX (Multidimensional Expressions) is also provided to connect the OLAP clients and servers. This query language is similar to SQL and has a large variety of functionalities.

Although the above two standard APIs have been proposed, MD-API is only implemented by a small number of vendors such as Gentia and is currently not being supported by a large number of OLAP product vendors, which is resulting in poor influence on the community. On the other hand, OLE DB for OLAP API is receiving support from almost all of the OLAP vendors, but does not define a detail OLAP model and cannot be considered as a real standard.

We propose an XML-based approach for modeling the OLAP metadata. Other approaches that have recently gained interest regarding modeling of metadata include ontology-based approaches such as DAML, OIL, and topic maps [3]. The use of ontologies provides a very powerful way to describe objects and their relationships to other objects.

3 Design of the Metadata Interchange System

Research on metadata standards includes two areas. One is the metadata model which is used to apply the metadata on the repositories. The other is the metadata exchange model which is used to exchange metadata among different tools.

A metadata model defines the structures and semantics of the metadata in order for vendors to share the metadata in a common way. Figure 1 shows an example of how the metadata can be shared and used by various users and applications. There are several international standards for metadata. The most interesting ones are the Open Information Model (OIM) proposed by the MetaData Coalition (MDC) and the Common Warehouse Metamodel (CWM) proposed by the Object Management Group (OMG). Recently, as the MDC has become part of OMG, these two efforts are now converging.

The implemented OLAP metadata exchange system is based on an extensible OLAP metadata exchange model that can be used as a standard among different OLAP products that have different hardware and platform requirements.

3.1 OLAP Metadata Model

One of the current standards for the metadata models is the Open Information Model proposed by MDC. OIM is based on an object model that can store specific types of information and is also flexible enough to support new types of information. It is an open standard that can be extended to satisfy the requirements of specific

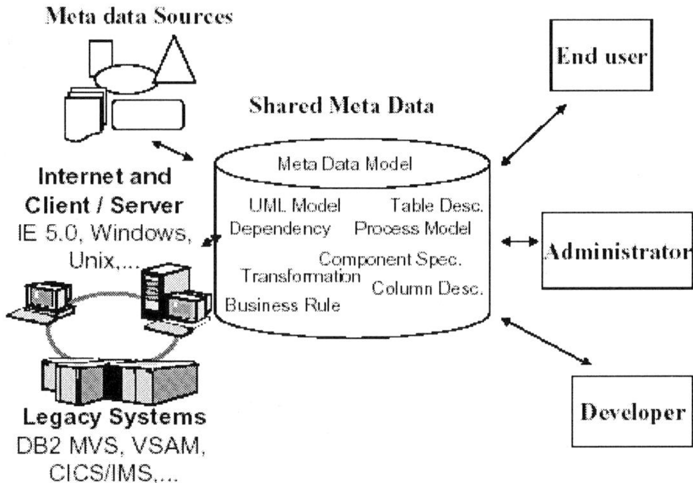

Fig. 1. A shared metadata model

users or vendors. OIM includes five areas: Analysis and Design, Component Description and Specification, Database and Data Warehousing, Business Engineering, Knowledge Management. We discuss the OLAP schema in the Data Warehousing area in the following.

The OLAP metadata model provides a common place to store the multi-dimensional schema information of the various OLAP products. The model is composed of cubes, dimensions, hierarchies, levels, dimension items, variables, and attributes as basic elements. The store is an abstract class that generalizes the different multi-dimensional storage objects. A store can represent a cube, virtual cube, physical cube, partition, and aggregation. The cube is a basic element of multi-dimensional analysis and is generally composed of a fact table and more than one dimension tables. Fact tables contain the measure variables that are related to a combination of the items in the dimension tables. The dimension tables define the item values for a dimension. As an example, there could be the three dimensions such as store, product, and time that each contain specific detail information regarding all of the stores for a company, the products that are sold for the company, and the time information of the fiscal years. The fact table will then record the transactions when a product is sold from a store at a specific time. The fact table would record the number of products sold and the margin from the transaction. Dimension hierarchies are associated with levels and mappings. Dimension levels are used to roll-up or breakdown on details. As an example, a geography dimension can have states, regions, and countries as levels. Mappings are used to indicate the use of a dimension hierarchy by a store.

The MD-API uses its own OLAP data model and has been unsuccessful in receiving support from the vendors. The OLE DB FOR OLAP API follows a standard metadata model. Because the different OLAP products use different OLAP metadata models, it is very important to define a standard OLAP metadata model. This paper proposes a flexible OLAP metadata interchange model that enables exchanging of metadata among vendors and also could be extended as a metadata model.

3.2 An XML-Based OLAP Metadata Interchange Model

The OLAP metadata exchange model is described in detail in terms of the DTDs of the XML in the following subsection. Afterwards, the application method of this model on several commercial products is investigated.

3.2.1 Model Design

The OLAP metadata interchange model is defined based on the metadata model explained in the previous section. There are several reasons for using XML in the interchange model. XML is vendor independent, programming language independent, and its structure can be defined via DTDs and XSDs. The OLAP metadata interchange model was designed by observing and factoring out the common characteristics of several products and enabling the model to handle the differences among those products. The model includes common metadata that is shared by several products (i.e., the least common denominator) and special metadata that is dependent on specific products. The special metadata can be stored in the metadata store and can be used when necessary. As an example, Attributes are used to show the query results or used as characteristics to select the set of members. However, not all OLAP products can support this Attribute. Even though some products support it, each product supports it in its own way. In this case, all of the supported Attributes were organized together to prevent the loss of data. The differences are supported by approaches discussed in the next section on how the model is applied to each product. The OLAP metadata interchange model can include and interchange the characteristics of several products in this manner.

The interchange model is composed of the basic elements such as DATABASE, CUBE, DIMENSION, MEASURE, HIERARCHY, LEVEL, MEMBER, and ATTRIBUTE.

The DATABASE is the highest level structure which is composed of PROPERTY, DATASOURCE, and CUBE. The CUBE is composed of PROPERTY, MEASURE and DIMENSION, and an optional element called STORE has the information about the storage structure. The PROPERTY element contained in each element is very similar but is organized in different ways according to the circumstances. For the CUBE, the PROPERTY includes the cube type which shows if it is a virtual cube or not, the description of the cube, and other optional items.

The DTD of the MEASURE shown in Figure 2 includes the values and the types of the real data used for analysis. The data type is usually a numeric type and has attributes such as numeric_precision, numeric_units, and numeric_scale. The properties of MEASURE include the measure_aggregator which is the basic operation used for aggregation, the caption, and unique_name. Each FIELD of MEASURE includes the dimensions that are used for the analysis with consideration of the series multi-cube approach.

Figure 3 shows that a DIMENSION is composed of HIERARCHY, DATATYPE, and ATTRIBUTE elements. Although there usually is just a single hierarchy in a dimension, some products support several hierarchies. Many products do not support DATATYPE. Some products support ATTRIBUTE at the dimension while others support it at the level or member. Looking at PROPERTY, the cube name is a required element. If the metadata is extracted from a product supporting multi-cubes

```
<!ELEMENT MEASURE(PROPERTY,FIELD+,DATATYPE)*>
<!ATTLIST MEASURE
        measure_name        CDATA #REQUIRED
        created_on          CDATA #IMPLIED
        last_schema_update CDATA #IMPLIED
        schema_updated_by   CDATA #IMPLIED
>
<!ELEMENT PROPERTY(database_name+, cube_name, measure_unique_name?,
measure_caption?, measure_guid?, measure_desc*, measure_aggregator )>
```

Fig. 2. MEASURE element of the OLAP metadata interchange model

```
<!ELEMENT DIMENSION(PROPERTY, HIERARCHY*,DATATYPE?,ATTRIBUTE*)>
<!ATTLIST DIMENSION
        dimension_name      CDATA #REQUIRED
        created_on          CDATA #IMPLIED
        last_schema_update CDATA #IMPLIED
        schema_updated_by   CDATA #IMPLIED
>
<!ELEMENT PROPERTY(database_name+, cube_name*,
                                                 dimension_unique_name?,
 dimension_desc*, dimension_ordinal, dimension_type,
                                              dimension_cardinality,
default_hierarchy, is_virtual, is_drillthrough_enabled?,
dimension_unique_settings?, is_SQL_enabled? )>
```

Fig. 3. DIMENSION element of the OLAP metadata interchange model

instead of hyper-cubes, there can be multiple cube names. Dimension_ordinal is a number that represents the order among the dimension sets that form the cube. Dimension_cardinality is the number of members in the dimension. Dimension_type indicates that the dimension is either a time dimension, quantitative dimension, regular dimension, or other dimension.

Figure 4 shows that the HIERARCHY is composed of LEVEL, ATTRIBUTE, DATATYPE elements and has a similar structure as the dimension.

The XML DTDs shown here only represent the basic structure of the OLAP metadata interchange model. The next subsection explains how to actually apply this model to several products.

3.2.2 Model Application Methodologies

The OLAP metadata interchange system was designed to handle the special characteristics of the products by extracting the metadata from each of the products. The model needs to be applied in different ways based on how the product organizes its metadata. These methods are incorporated as a module of the OLAP metadata interchange system. When the OLAP metadata interchange system creates a cube for specific products, transformations are necessary to map it to the model of the product.

```
<!ELEMENT HIERARCHY(PROPERTY, LEVEL*, DATATYPE?, ATTRIBUTE*)>
   <!ATTLIST HIERARCHY
          hierarchy_name       CDATA #REQUIRED
          created_on           CDATA #IMPLIED
          last_schema_update  CDATA #IMPLIED
          schema_updated_by   CDATA #IMPLIED
>
```

Fig. 4. HIERARCHY element of the OLAP metadata interchange model

The following explains the transformations that are required to overcome the differences in the products shown in Table 1. Cubes, Attributes, and Dimensions are mapped accordingly.

Table 1. Analysis of OLAP product models

Component / Products	**Storage**	**Cube**	**Attribute**	**Dimension**
Cognos Powerplay	MOLAP	Hypercube	Not Supported	Multi level (Special category)
Hyperion Essbase	MOLAP	Hypercube	Dimension attribute	Single Hierarchy Multi level
Informix Metacube	ROLAP	Block Multicube	Supported	Single Hierarchy Multi level
MS OLAP Services	HOLAP	Block Multicube	Level attribute	Single Hierarchy Multi level
Oracle Express	MOLAP (HOLAP)	Series Multicube	Not Supported	Multi Hierarchy Single level

4 System Implementation

An OLAP metadata interchange system was developed to exchange cube data among different OLAP products. The extracted metadata from the OLAP products is transformed according to the OLAP metadata interchange model and stored. Users can search and edit the metadata in a Web environment. Additionally, users can automatically create cubes in the desired OLAP products without needing to go through the complex process relevant to each product. The user interface is a Web-based interface so that users can easily interchange data among the OLAP products anytime anywhere.

The implementation environment is as follows. The OLAP metadata interchange system uses Object Design eXcelon as the XML data server running on Windows 2000. Excelon can efficiently store and manage the dynamic structure of the XML data. OLE DB FOR OLAP API was used for the metadata extraction, and C++ was used as the programming language. The MS Internet Information Server was used as

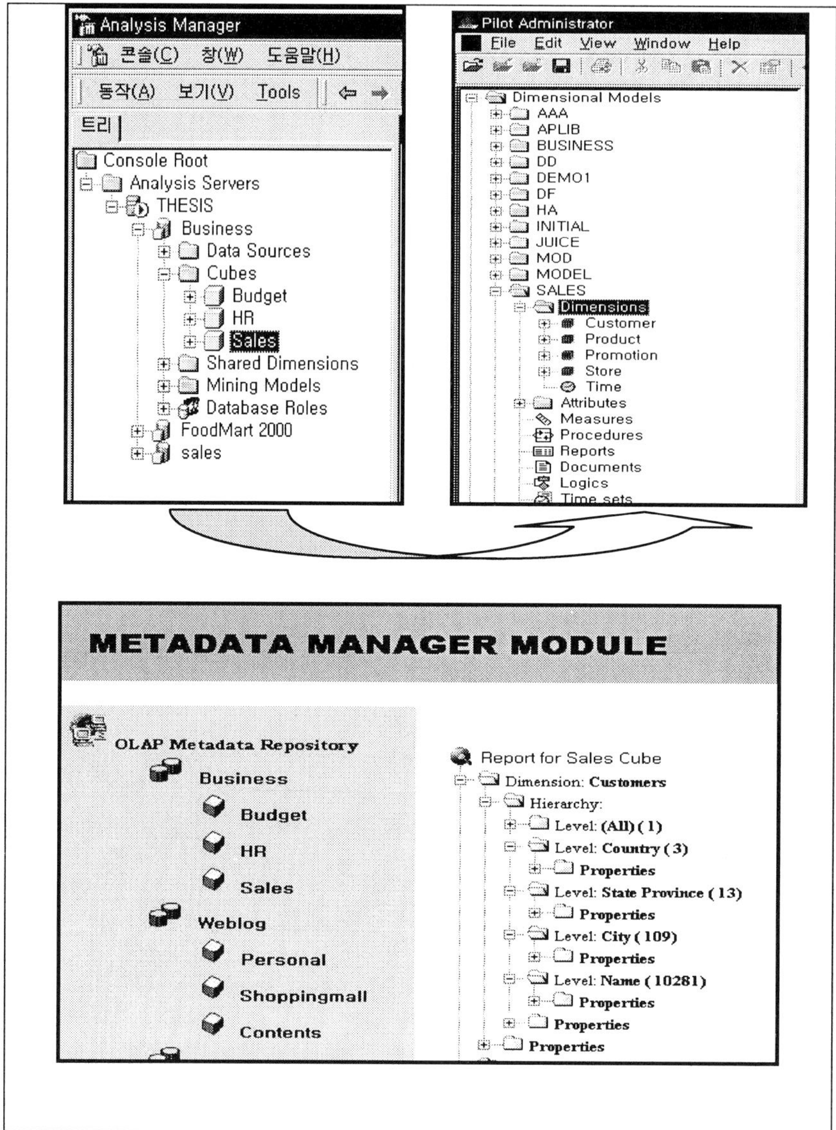

Fig. 5. Cube creation in Pilot

the Web server and MS Internet Explorer was used as the client. ASP, XQL, and XSL was used for searching and editing the metadata in a Web environment.

The OLAP metadata interchange system is composed of four modules: the metadata extraction module, the metadata transformation and storage module, the metadata manager module, and the cube creation module. The metadata extraction module will extract the metadata that is desired by the user from various OLAP products. The metadata transformation and storage module applies the OLAP metadata interchange model to the extracted metadata information and transforms it into XML and stores it into the eXcelon Server. The metadata manager module enables users to search and edit the metadata in the eXcelon Server through a Web-based interface. The cube creation module enables users to automatically create cubes based on the metadata in a specific OLAP product.

The system has been used with Microsoft OLAP Services, Pilot's Decision Support Suite, and Oracle's Express for evaluation. Figure 5 shows the metadata being exported from OLAP Services and the cube being created in Pilot. The extension of the system to support other products is easy and straightforward.

5 Conclusion

On-Line Analytical Processing(OLAP) systems is one of the key technologies for data warehouses. However, the approximately 40 OLAP products were developed with different OLAP metadata, making it difficult for OLAP products to exchange data or interface with each other. This paper proposes an OLAP metadata interchange model along with an implemented OLAP metadata interchange system.

The following points differentiate the proposed OLAP metadata interchange system with other research results. First, by analyzing many of the existing OLAP products, several problems that occur during the standardization process in the OLAP area were identified and incorporated into the model. Second, the design of the OLAP metadata interchange model can incorporate the models of the currently used products and supports the metadata model transformation process. Third, the metadata interchange model uses XML, making it easier to transmit structured documents over the Web. Fourth, users need not learn about the complex cube creation process of each product. With just a single cube, the user can experience the different analysis environments provided by several different products.

Currently, the most debated area of OLAP technology is the standardization effort. OLAP products will be able to talk to each other in the near future by extending the results of this paper to design a common query language as well as common metadata and interfaces.

References

[1] E. Thomsen, OLAP Solution, John Wiley &Sons, 1997

[2] Cho Jae Hee, Park Sung Jin, OLAP Technology, Sigma Consulting Group, 1999

[3] DAML.org, DARPA Agent Markup Language (DAML), http://www.daml.org/

[4] "Putting Metadata to Work in the Warehouse", http://www.cai.com/products/platinum/wp/wp_meta.htm

[5] Pilot Software, *White Paper*, "An Introduction to OLAP: Multidimensional Terminology and Technology", http://www.pilotsw.com/olap/olap.htm

[6] Oracle Express Server, http://www.oracle.com/ip/analyze/warehouse/servers/index.html

[7] Microsoft, OLE DB for OLAP 2.0 Beta Specification, http://www.microsoft.com/data/oledb/olap/spec/

[8] OLAP Council, MDAPI specification version 2.0,http://www.olapcouncil.org/

An Efficient Method for Controlling Access in Object-Oriented Databases

Woochun Jun[1] and Le Gruenwald[2]

[1] Dept. of Computer Education, Seoul National University of Education, Seoul, Korea
wocjun@ns.seoul-e.ac.kr

[2] School of Computer Science, University of Oklahoma, Norman, OK 73069, USA
ggruenwald@ou.edu

Abstract. In this paper, we present a locking-based concurrency control scheme for object-oriented databases (OODBs). The proposed scheme deals with class hierarchies in OODBs. Our scheme is based on implicit locking but designed to require less locking overhead than implicit locking for all types of accesses to OODBs. Our scheme makes use of intelligent method to reduce locking overhead. Especially, our scheme utilizes only structural information of OODBs so that extra information to reduce locking overhead is not necessary. We also prove theoretically that our scheme performs better than implicit locking.

1 Introduction

OODBs have been used for many advanced applications such as multimedia software engineering and XML databases due to their advanced modeling power [9, 14]. The three most fundamental aspects of the OODBs are abstract datatype, inheritance, and object identity [9,10]. These aspects are summarized as follows. In OODBs, the sets of objects that share a common definition are implemented through classes. Also, a class defines and implements all the methods that capture the behavior of its instances. Classes implement abstract datatypes that hide the implementation of the user-defined operations (methods) associated with the datatype. That is, the implementation is totally hidden within the class. Using inheritance, a new class can be developed by reusing all the definitions of an existing class and extending it by adding new definitions. In OODBs, each object has an identity that is independent of name, state and address of the object. With object identity, objects can contain or reference other objects.

In OODBs, there are two types of inheritance: single inheritance and multiple inheritance. In single inheritance, a class can inherit the class definition from one superclass. On the other hand, a class can inherit the class definition from more than one superclass in multiple inheritance. With multiple inheritance, existing classes are combined to produce a class that uses multiple superclasses in a variety of definitions. Fig. 1 (a) and (b) show examples of single inheritance and multiple inheritance, respectively.

M.H. Shafazand and A M. Tjoa (Eds.): EurAsia-ICT 2002, LNCS 2510, pp. 306–313, 2002.

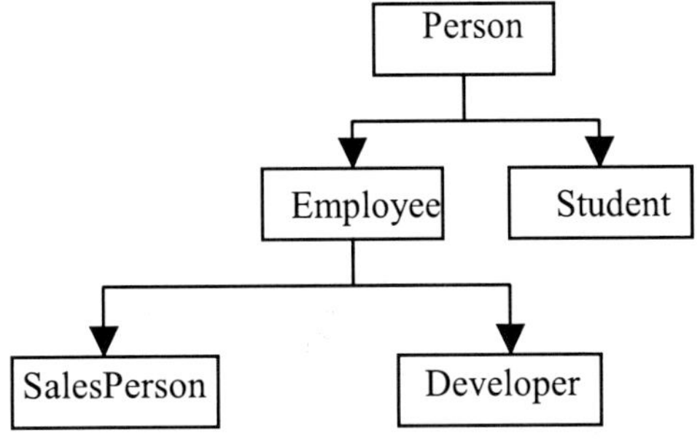

(a). Single inheritance

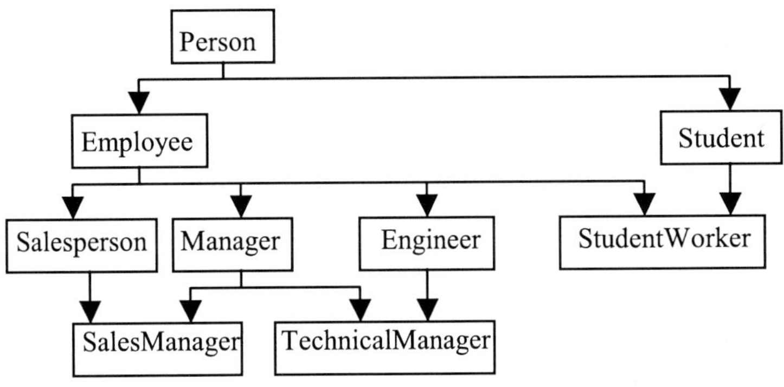

(b) Multiple inheritance

Fig. 1. Example of inheritance [9]

In general, there are two types of accesses to an object: instance access (instance read and instance write) and class definition access (class definition read and class definition write) [2]. Especially, there are two types of accesses on a class hierarchy: class definition write and instance access to all or some instances of a given class and its subclasses (also called IACH, meaning Instance Access to Class Hierarchy) [5,6,7]. A query is an example of IACH where a query is defined as instance reads to a given class and its subclasses [5]. Due to inheritance hierarchy, while a class and its instances are being accessed, the definitions of the class' superclasses should not be modified. Also, due to the is-a relationship between classes, the search space for a query against a class, says C, may include the instances of all classes on the class hierarchy rooted at C as well as all instances of C. For convenience, we call MCA (Multiple Class Access) for class definition writes and IACHs and SCA (Single Class Access) for other accesses such as class definition read and instance access to a single class.

In a database management system, transactions usually run concurrently. In order to ensure database and transaction consistency, database systems require a serializable order of executions among transactions [1]. This means that the results of the transactions are the same as if the transactions were executed in some serial order. Concurrency control schemes are used to ensure serializability of transactions. A concurrency control scheme allows multi-access to a database but incurs an overhead whenever it is invoked. This overhead may affect the performance of OODBs where many transactions are long-lived. Thus, reducing this overhead is critical to improve the overall performance. For OODBs, locking-based concurrency control schemes have been used for controlling current accesses [5,6,7,12,13].

In this paper, an efficient concurrency control scheme for OODBs will be proposed. This paper is organized as follows. In Section 2, the related work is discussed. In Section 3, a new scheme is proposed. The scheme incurs less overhead than implicit locking does. In Section 4, the correctness of the proposed scheme is proved. Finally, conclusions and future work are described.

2 Related Works

In the literature, there are two major locking-based approaches dealing with inheritance: explicit locking [2,13] and implicit locking [5,6,7,8,12]. In explicit locking, for an MCA access on a class, C, a lock is required not only on the class C, but also on each subclass of C in the class hierarchy. On the other hand, for an SCA access, a lock is required for only the class to be accessed (called target class). Thus, for an MCA access, transactions accessing a class near the leaf in a class hierarchy will require fewer locks than transactions accessing a class near the root in the class hierarchy. Also, another benefit is that it can treat single inheritance and multiple inheritance in the same way.

On the other hand, implicit locking is based on intention locks [10,11]. The purpose of an intention lock on a class indicates that some lock is already set on a subclass of the class. Thus, when a transaction needs to set a lock on a class, say C, the transaction also requires intention locks on a path from C to its root as well as on C. In implicit locking, when an MCA operation is accessed on a class, C, locks are not required for every subclass of the class C. It is sufficient to set a lock only on the class C (in single inheritance) or locks on C and its subclasses that have more than one superclass (in multiple inheritance) [5,6,7]. Thus, for an MCA access, it incurs less locking overhead than explicit locking. But, implicit locking requires more locking overhead when a target class is near the leaf in a class hierarchy due to intention lock overhead.

For example, consider the following simple class hierarchy as in Fig. 2. Note that the lock modes are based on those proposed in Orion, an OODB system [5,10]. In order to update the class definition in class, say H, the explicit locking scheme works as in Fig. 2.a. On the other hand, for the implicit locking, intention locks IWs corresponding to W locks are required for each superclass on the path from H to the root A. Fig. 2.b shows locks required by implicit locking

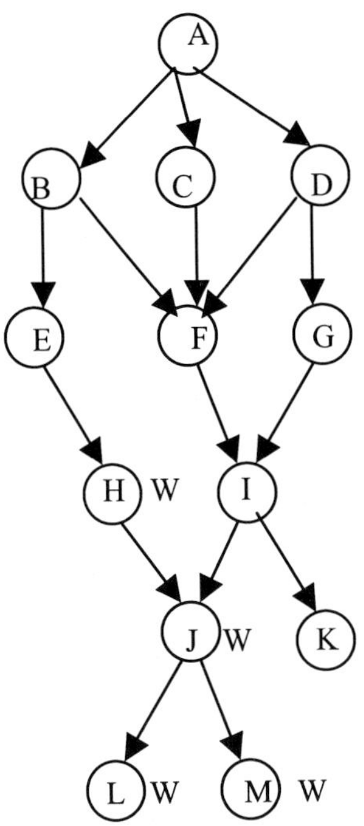

Fig. 2.a Locks by the explicit locking

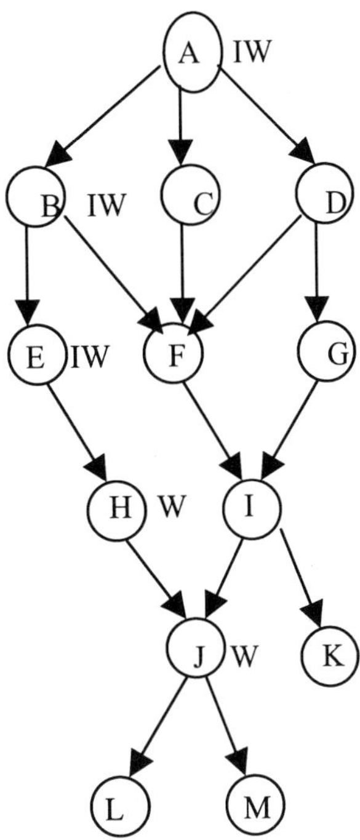

Fig. 2.b Locks by the implicit locking

3 The Proposed Scheme

As we discussed in Section 1, it is very important to reduce concurrency control overhead for performance improvement. In this work, some intelligent information is used to reduce locking overhead. When a transaction requests a lock on some object, the transaction needs to check possible conflict with locks by other transactions. The depth first search is widely used to check conflicts in search space [3]. The basic idea in this work is equivalent to reducing number of visits in finite search space, when the depth first search is applied, as follows.

Our idea is that, in implicit locking, some redundant locks can be removed without affecting the correctness of the scheme. Assume that a class C is accessed. For implicit locking, an intention lock is set on every superclass of C. On the other hand, the proposed scheme does not have to set an intention lock on every superclass of C. That is, the proposed scheme sets intention locks only on 1) the root, 2) from the root, each superclass of C up to either the first superclass that has more than one immediate subclass or superclass or the immediate superclass of the target class (if there is no such first class), and 3) the superclasses of C that have more than one immediate subclass or superclass.

Based on the above idea, the proposed scheme is summarized as follows. Assume that a lock is requested on class C. Also, for simplicity, we assume that the strict two-phase locking is adopted [1,11].

Step 1) Locking on the root class

1.a) For the root class of C, check conflicts and set an intention lock.

1.b) From root, for each superclass of C up to the first class that has more than one immediate subclass or superclass, check conflicts and set an intention lock. If there is no such first class, check conflicts and set an intention lock for each superclass of C.

Step 2) Locking on superclasses

For each superclass, which has more than one immediate subclass or superclass of C, check conflicts and set an intention lock.

Step 3) Locking on the target class

3.a) For SCA access: Check conflicts and set a lock on only the target class C.

3.b) For MCA access: Check conflicts and set locks for all subclasses of the target class, which have more than one superclass

Consider the following class hierarchy in Fig. 2. Assume that a class definition needs to be changed in class J. Assuming that a path A-> B->E->H->J is chosen for intention locks, the implicit locking adopted in Orion [5,10] needs to get locks as in Fig. 3.a. On the other hand, locks are required as in Fig. 3.b if the proposed scheme is applied. In the proposed scheme, classes E and H need not be locked.

For the same access, assume that a path A->D->G->I->J is chosen for intention locks. The implicit locking approach requires locks as in Fig. 4.a while the proposed scheme requires locks as in Fig. 4.b. In our scheme, class G needs not be locked.

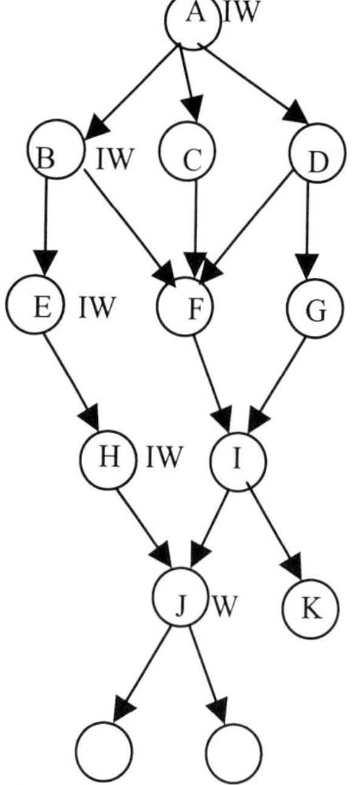

Fig. 3.a Locks by the implicit locking

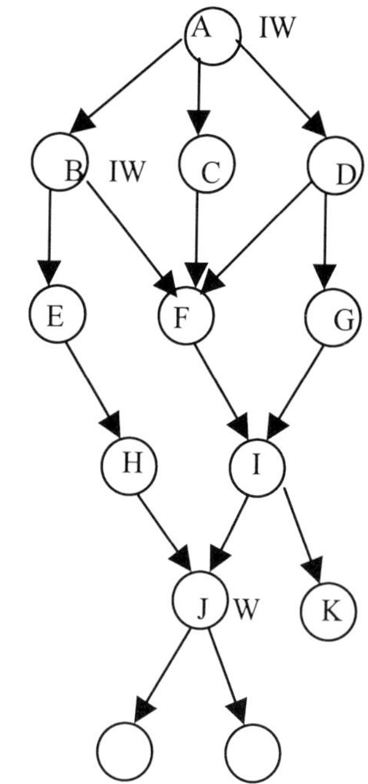

Fig. 3.b Locks by the proposed scheme

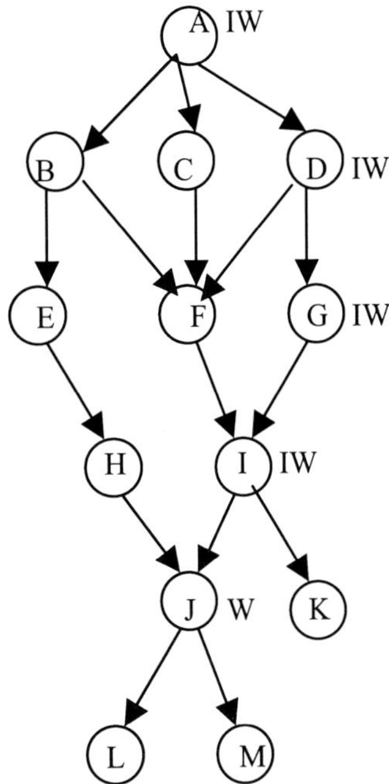

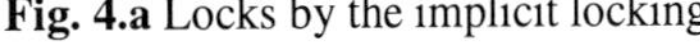
Fig. 4.a Locks by the implicit locking

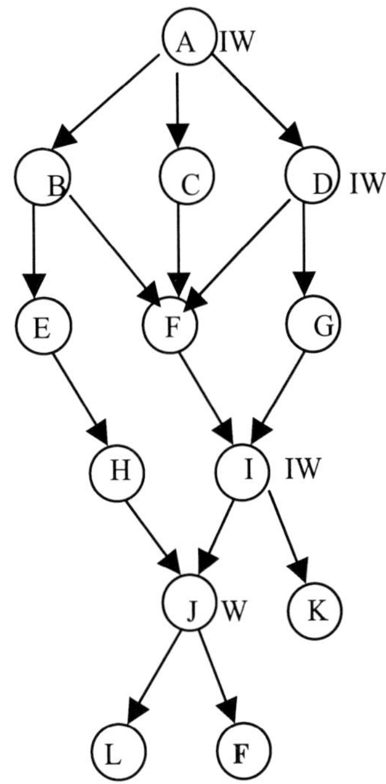

Fig. 4.b Locks by the proposed sheme

4 The Proof of Correctness of the Proposed Scheme

In this section, we prove that the proposed scheme is correct, that is, it satisfies serializability. Specifically, we prove that, for any requesters, any conflicts with a lock holder is always detected. With this proof, since the proposed scheme is based on two-phase locking, it is guaranteed that the proposed scheme satisfies serializability [4,11].

Claim) For any lock requester, possible conflict with a lock holder is always detected.

We prove that there exists at least one common class on which both the lock requester and the lock holder set a lock.

Case I) The lock requester and the lock holder have the same target class.

In this case, regardless of the intention locks set by both requester and holder, the conflict can be detected on the target class.

Case II) The lock requester and the lock holder have the different target class.

Case II.1) Single Inheritance

Assume that the target classes of the holder and the requester are C_H and C_R, respectively. In single inheritance, the conflict can be detected on the nearest common superclass (to be locked) from both C_H and C_R [8].

Case II.2) Multiple inheritance

Without loss of generality, assume that a class C_H has subclasses with more than one superclass and is locked in MCA mode by a lock holder. Then, there are two cases depending on the lock mode set by the requester.

Case II.2.a) A lock requester needs an SCA access.

If C_R is a superclass of C_H, there is no conflict. Also, there is neither superclass nor subclass relationship between C_H and C_R, there is no conflict. Assume that C_R is a subclass of C_H. In this case, if C_H is on the same path on which the requester sets intention locks, the conflict is detected on C_H. If not, the requester must get a lock on one of the subclasses of C_H, say, C_K, which has more than one superclass. Otherwise, C_R would not be a subclass of C_H. Thus, the conflict will be detected on C_K.

Case II.2.b) A lock requester needs an MCA access.

If C_R is neither a superclass nor a subclass of C_H, possible conflicts are detected on the one of subclasses of C_H and C_R that has more one superclass. Assume that C_R is a superclass of C_H. Then, if C_R is on the same path with C_H, the conflict is detected on C_H. If not, C_H must get a lock on one of the subclasses of C_R, say C_L, which has more than one superclass. Otherwise, C_R would not be a superclass of C_H. Thus, the conflict is detected on C_L. On the other hand, assume that C_R is a subclass of C_H. If C_H is on the same path on which the requester sets intention locks, the conflict will be detected on C_H. If not, C_R must get a lock on one of the subclasses of C_H, say C_M, which has more than one superclass. Otherwise, C_R would not be a subclass of C_H. That is, the conflict is detected on C_M.

From case I and II, we can conclude that our scheme is correct. This means that our scheme has less locking overhead than the implicit locking scheme does.

5 Conclusions and Further Work

In this paper, we present a locking-based concurrency control scheme for OODBs. The proposed scheme is based on the implicit locking scheme and is developed to reduce locking overhead for all applications. We adopt intelligent method to reduce search space for conflict checking. Finally, we prove theoretically that the proposed technique has less locking overhead than the implicit locking scheme does.

We are developing a locking-based concurrency control scheme dealing with composite object hierarchy, which is a major aspect in OODBs. The composite object locking scheme will be combined with the proposed scheme in this paper. There have been some implicit locking schemes for controlling composite object hierarchies [5,10]. We also plan to compare the proposed work with the existing implicit locking schemes by means of simulation.

Since most OODBs have long duration transactions, aborting or delaying a transaction due to conflicts may waste lots of system resources and valuable time.

Thus, it is also a good research topic to develop a locking-based scheme that deals with such long duration transactions.

References

1. Bernstein, P., Hadzilacos, V. and Goodman, N.: Concurrency Control and Recoveryin Database Systems, Addison-Wesley (1987)
2. Cart, M. and Ferrie, J.: Integrating Concurrency Control into an Object-Oriented Database System, 2nd Int. Conf. on Extending Data Base Technology, Venice, Italy, Mar. (1990) 363 - 377
3. Charmiak, E. and McDermott, D.: Introduction to Artificial Intelligence, Addison-Wesley, 1987
4. Eswaran, K., Gray, J., Lorie, R. and Traiger, I.: The notion of consistency and predicate locks in a database system, Communication of ACM, Vol. 19, No. 11, Nov. (1976), 624 - 633
5. Garza. J, and Kim, W.: Transaction Management in an Object-Oriented Database System, ACM SIGMOD Int. Conf. on Management of Data, Chicago, Illinois, Jun. (1988) 37 - 45
6. Jun, W. and Gruenwald, L.: An Effective Class Hierarchy Concurrency Control Technique in Object-Oriented Database Systems, Journal of Information And Software Technology, Vol. 40. No. 1, Apr. (1998) 45-53
7. Jun, W. and Gruenwald, L.: An Optimal Locking Scheme in Object-Oriented Database Systems, First International Conference on Web-age Information Management (Lecture Notes in Computer Science 1846) Shanghai, China, Jun. (2000) 95-105
8. Jun, W. and Kim, K.: A Revised Implicit Locking Scheme in Object Oriented Database Systems, The proceedings of 8th International Conference on High Performance Computing and Network Europe 2000, Amsterdam, The Netherlands, May (2000), 618 – 622.
9. Khoshafian, S. et al.: Jasmine Object Database System: Multimedia Applications for the Web, Morgan Kaufmann Press (1999)
10. Kim, W.: Introduction to Object-oriented Databases, MIT press (1990)
11. Korth, H and Silberschartz, A.: Database System Concept, 2nd edition, McGraw Hill (1991)
12. Lee, L. and Liou, R.: A Multi-Granularity Locking Model for Concurrency Control in Object-Oriented Database Systems, IEEE Trans. on Knowledge and Data Engineering, Vol. 8, No. 1, Feb. (1996) 144 - 156
13. Malta, C. and Martinez, J.: Controlling Concurrent Accesses in an Object-Oriented Environment, 2nd Int. Symp. on Database Systems for Advanced Applications, Tokyo, Japan, Apr. (1992) 192 - 200
14. Maresca, P., et al.: Transformation Dataflow in Multimedia Software Engineering Using TAO_XML: A Component-Based Approach, 2nd International Workshop in Multimedia Databases and Image Communication (LNCS 2184), Amalfi, Italy, Sep. (2001) 77-89.

Extracting Information from XML Documents by Reverse Generating a DTD*

Jong-Seok Jung, Dong-Ik Oh, Yong-Hae Kong, and Jong-Keun Ahn

Division of Information Technology Engineering, SoonChunHyang University
Shinchangmyun, Asan, Korea
{jungjs, dohdoh, yhkong, jkahn}@sch.ac.kr

Abstract. Information contained in XML documents cannot properly be interpreted without an appropriate DTD. However, XML documents collected from the web may not always be accompanied by the corresponding DTD, so that extracting information from such sources may not be easy. In this study, we reverse construct a DTD from DTD-unknown XML sources, and use it to extract information from XML inputs. The DTD construction module developed is designed to scan input XML files in 1-path, where most other implementations use 2-path approach. Developed modules provide clean Java programming interfaces as well, so that it can be integrated with other web applications seamlessly.

1 Introduction

The DTD constructor developed in this study is a part of Soonchunhyang E-COmmerce System (SECOS) [1]. SECOS is a component-based web information provision system, whose purpose is to provide a model for developing other web-oriented information management systems. Figure 1 depicts the SECOS architecture.

There are 3 component divisions in SECOS. Among them, Gathering Division is in charge of collecting information from remote sites. There are three gatherers in the division. They are Affiliated, Regular, and Meta Gatherer. Through the gatherers, SECOS performs Internet information collection activities. With Affiliated Gatherer, a SECOS site exchanges information with other SECOS sites. With Regular Gatherer, SECOS collects information from other web sites through DTD-provided XML documents. Meta Gatherer is the third gatherer SECOS provides. Its purpose is to collect information from unstructured web sources, such as HTML files. Considering the fact that majority of data on the web are still in unstructured format, gathering information from such sources is necessary for SECOS to be a genuine general-purpose web information provision system. The main focus of this paper is on the development of Meta Gatherer, especially on the part of gathering information from DTD-unknown XML document sources.

---

* This works is supported in part by the Ministry of Information & Communication of Korea

M.H. Shafazand and A M. Tjoa (Eds.): EurAsia-ICT 2002, LNCS 2510, pp. 314–321, 2002.

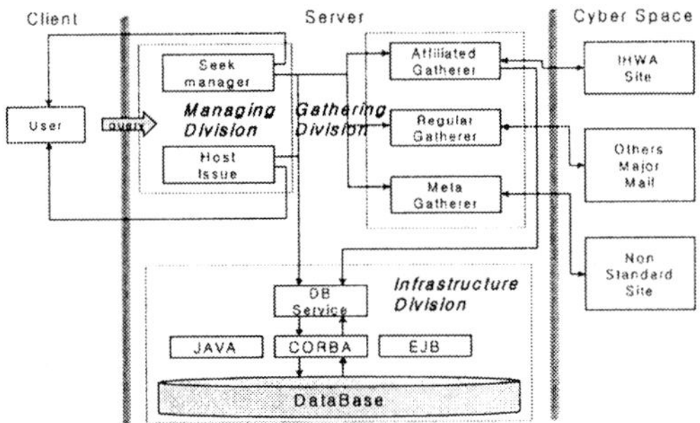

Fig. 1. Architecture of SECOS

Researches closely related to our work (namely, extracting data from DTD-unknown XML sources) can be found in recent literatures [2,3]. However, direct application of these modules to SECOS was not possible, for they either do not provide programming language level interfaces or do not cover complete DTD grammars. There was another reason for building our own DTD constructor. Conventional implementations try to generate a DTD from XML inputs by expressing XML data with trees and by incrementally placing them into a merged tree. Then a DTD that validates all XML inputs is driven from the merged tree. By generating a tree for each XML input and by later merging them to produce a common tree, they need at least 2 scans through the input. However, our module needs only one path through the input, by placing parsed information directly into the merged tree. Therefore, our implementation is more efficient than the conventional ones.

The remainder of the paper is organized as follows. In Section 2, we explain how we express XML structure using the n-ary tree. Section 3 gives an actual example and explains the algorithm for constructing the tree. We also explain how to extract a DTD out of it. Section 4 discusses how the developed DTD constructing modules can be integrated with other parts of the gathering system. It then concludes.

2 Structuring XML Information

In order to produce a DTD, which validates multiple XML inputs, we need to represent the data in trees, because XML data is inherently nest-structured. Such structures can be well expressed with n-ary trees, and examining data structure from the tree is convenient. In this section we present the data structures and algorithm to express XML information using the n-ary tree.

2.1 Data Structures

A tree consists of nodes and links. Basically, a node represents an element in the XML input, and the link represents the nested structure of the XML element. Each node may express DTD relationships among elements such as Mandatory, Optional, OR, and Repetition. There are several attributes associated with a node. Table 1 summarizes the attributes a node can have.

Table 1. Node attributes and their meaning

Attribute Name	Meaning
ele_name	Name of the XML element the node represents
isTemp	Indicates the node is a temporary one
isMandatory	Tells if the node represents a mandatory or an optional XML element
isRepeat	The node represents repetition of XML elements
isOR	Indicates that the node is used for expressing OR relationship
isChecked	Checks if the node is visited at least once during the processing
isNewCreated	Tells if the node is created while processing the current definition

In order to check repeated patterns of the XML inputs, we need to express the relationship among XML elements. To do this we need additional pointers similar to the ones used for state transition diagrams. First, there is the *refNodeId* pointer, which is used to avoid defining redundant nodes. Second, the *chain* pointer is used to indicate repeated XML patterns. Namely, we connect repeated XML patterns using the pointer and later extract the pattern by detecting a cycle. Third, there is a pointer called *link*, which is used to point to the opposite direction of the *chain* pointer. *chain* is used to detect a cycle, but *link* is used to extract names of elements processed so far.

Along with a node prepared for an XML element, we need to keep track of where in the tree the node should be placed or processed. For the purpose, we have three global pointers. *startPointer* points to the first node in the *chain* cycle. *currentPath* points to the parent node, under which the element being investigated is to be placed. *insertPointer* points to a node under *currentPath,* at which a search of an element is initiated.

In the tree, we use temporary nodes to express Repetition, Optional, and OR relationship of elements. For the reason, we may have multiple entries of the same element, and each one of them can have different descendent nodes. However, we would not want to define identical nodes multiple times. Therefore, we maintain a single copy for each node in a table (tree node table) and use references to it through the *refNodeID* pointer. Insertion, modification, and search to the tree are performed on this table.

With previously mentioned node definition and the tree structure, we can express each DTD operators as in Table 2.

2.2 Tree Construction Algorithm

The tree construction module converts nested XML data into a tree. There are 3 main steps involved.

Step1, Initialization. The module creates a node called "root", which is the ancestor of all other tree nodes. *currentPath*, which indicates the search point of tree and *insertPointer* which points to the node where an insertion is to be made, are initialized to point to the "root" node.

Table 2. Tree expression of DTD operators

Symbol	Purpose	Example	Tree Expression
|	Selection	<!ELEMENT a (b | c)>	a; OR, OR; b, c — *isTemp=true* *isOR=true*
+	Repetiton (at least 1)	<!ELEMENT a (b, c)+>	a; REP; b, c — *isTemp=true* *isRepeat=true* *isMandatory=true*
*	Repetition	<!ELEMENT a (b, c)*>	a; REP; b, c — *isTemp=true* *isRepeat=true* *isMandatory=false*
?	Optional	<!ELEMENT a (b, c)?>	a; OPT; b, c — *isTemp=true* *isRepeat=false* *isMandatory=false*
,	Ordering	<!ELEMENT a (b, c)>	a; b, c — *isTemp=false* *isMandatory=true*

Step2, Tree Insertion and Modification. Now, we extract elements from XML documents one by one. For each element extracted, we check to see if the element definition is already in the tree. Depending on the search result, a new node may be placed in the tree, or one or more existing nodes may be modified. Table 3 and 4 summarizes what actions will be taken under what contexts.

Step3, Simplification of tree. Tree generated in Step 2 may be complicated. In this step, simplification of the tree is performed to enhance readability of the tree.

The example XML input in Figure 2 will produce a tree on the left-hand side through Step 2. In this tree, two subtrees representing OR relationship have common definitions. However, they exist in different subtrees. In Step 3, such multiply defined definitions of OR relations are combined. Through the process, the tree can be simplified to the right-hand side tree of Figure 2.

```
<?xml version="1.0"?>
<test>
   <a>....</a>
   <b>....</b>
   <c>....</c>
</test>
<test>
   <b>....</b>
   <c>....</c>
</test>
```

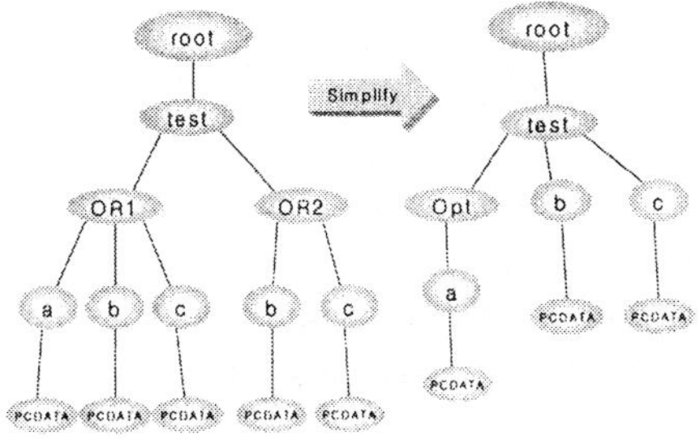

Fig. 2. An example of tree simplification

Table 3. Tree construction algorithm

Existence of children	*isNewCreated field of the currentPath node*	*Is currentPath node a temporary node?*	*Type of temporary node*	*Comparison to the insert-Pointer node*	*Is the startPoin-ternewly set?*	*Comparison of insertPointer node with the next sibling*	*Comparison of insertPointer with the previous sibling node*	*Act ion*
Terminal	True							①
	False							②
Non-terminal	True	Temporary	OR					③
			Optional					③
			Repetition		Newly set	Identical		④
						Different		⑤
					Initial value	Does not exist	Node exists	⑥
							Does not exist	⑦
		Not temporary		Same	Newly set			⑤
					Initial value			⑧
				Different	Newly set	Identical		④
						Different		⑤
					Initial value	Does not exist	Node exists	⑥
							Does not exist	⑦
	False	Temporary	OR					⑨
			Optional					⑨
			Repetition		Newly set	Identical		④
						Different		⑤
					Initial value	Identical		⑩
						Different		⑪
						Does not exist	Node exists	⑥
							Does not exist	⑦
		Not temporary		Same	Newly set			⑤
					Initial value			⑧
				Different	Newly set	Identical		④
						Different		⑤
					Initial value	Identical		⑩
						Different	Node exists	⑥
							Does not exist	⑪
						Does not exist	Node exists	⑥
							Does not exist	⑦

3 An Example of Tree Construction and DTD Extraction

In this section, we explain the tree construction algorithm in more detail using the XML document given in Table 5.

3.1 An Example of Tree Construction

In Step1, we create the "root" node and initialize *currentPath* and *insertPointer* to point to this node. When the "books" element is extracted, we search the tree for the element name. In this case, *currentPath* is pointing at "root", and the "root" node is a terminal with its *isNewCreated* attribute set to *true*. Therefore, according to the classification given in Table 3, we take action ① of Table 4. This leads us to create a new node called "books." We then insert it into the tree under the "root" node. Since a new element definition is created and inserted, *currentPath* and *insertPointer* are set to point to the inserted "books" node. Next, the "book" and "title" elements are extracted and inserted into the tree one at a time. For PCDATA of the "title" element, the same thing happens. When the EndTag of the "title" element is encountered, *currenPath* and *insertPointer* are set to point to "book" and "title" respectively.

Table 4. Tree construction algorithm (actions taken by the contexts in Table 3)

Action	Explanation
①	Place a new node under *currentPath*
②	This can happen only for the node representing PCDATA
③	Same as the non-temporary node
④	Connect the *chain* pointer of the *insertPointer* node and the *link* pointer of the next sibling
⑤	Place the node pointed by the *link* as a child of the *currentPath* node. If the *currentPath* node is a temporary node, place it as a child of the parent of the *currentPath* node
⑥	Set the value of *startPointer* using the *insertPointer* value. Connect the *chain* pointer of the *insertPointer* node with the *link* pointer of the searched node
⑦	Create a temporary node with *isMandatory* set to *false*. Place it as a child of searched node
⑧	Set the *isRepeat* attribute of the *insertPointer* node to *true*
⑨	Same as the case for the non-temporary node
⑩	Set the *isChecked* attribute of the next sibling node to *true*
⑪	Create two temporary nodes with *isOR* attribute set to *true*. Separate already processed child nodes of searched node using temporary nodes

Table 5. An example XML file

```
<?xml version="1.0" encoding="EUC-KR"?>
<books> <book>
  <title>Harry Potter and the Sorcerer's Stone</title>
  <author>J. K. Rowling </author>
 </book>
 <book>
  <title> Hamlet </title>
  <author>
   <first-name> William </first-name>
   <last-name> Shaksphere </last-name>
  </author>
 </book> </books>
```

Then comes the "author" element. According to the classification in Table 3, it is the case for non-terminal, *isNewCreated* field of the *currentPath* node is *true,* it is not a temporary node, name of the node is different from the element name, and do not have the next sibling node. Therefore action ⑤ of the Table 4 will be taken. In other words, the *isNewCreated* attribute of the "book" node - the parent of the extracted element - is *true*, the *insertPointer* is pointing at the "title" node, and there is no preceding sibling for the node pointed by the *insertPointer*. Therefore, we know that the "title" and "author" are siblings under the same parent. So, we create a new node for "author" and insert it under the "book" node, next to "title." When a new node is inserted, *currentPath* and *insertPointer* are set to point to the newly inserted node. When the EndTags of "author" and "book" are encountered, *currentPath* and *insertPointer* are set to point to "books" and "book" respectively. Figure 3 summarizes the tree construction process.

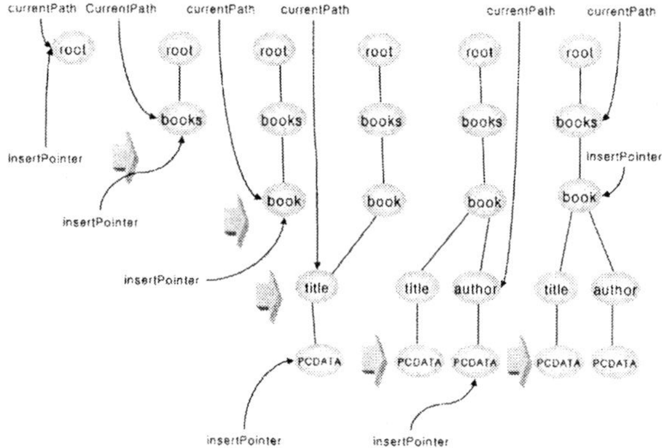

Fig. 3. Tree construction process and tree table entries for the first definition of Table 5

Next, the second definition of "book" element starts. At this time, action ⑧ of Table 4 is selected, and the *isRepeat* attribute of the "book" node is set to *true*. Namely, we indicate that the "book" definition is repeated. When the "title" element is processed, we search the tree and find that the node exists already under the node pointed by *currentPath*. Therefore, we simply adjust *currentPath* and *insertPointer* to point to the "title" node. When the EndTag of the "title" element is met, we move *insertPointer* to point to the "book" node. Next, "author" element will be extracted, but this node exists in the tree already. So, we adjust *insertPointer* to point to the "author" node. Now, the "first-name" element of the XML document needs to be processed, and it will be inserted as a child of the "author" node. In this case the *isNewCreated* attribute of the "author" node is *false*, therefore, action ⑪ of Table 4 will be taken. A new temporary node now needs to be created and inserted. When the EndTag of "first-name" is extracted, *insertPointer* points to the newly created temporary node, and *currentPath* stays to point to the "author" node. For the next "last-name" element, a new node is created and inserted under the temporary node, as a next sibling of the "first-name" node. Figure 4 summarizes the construction process.

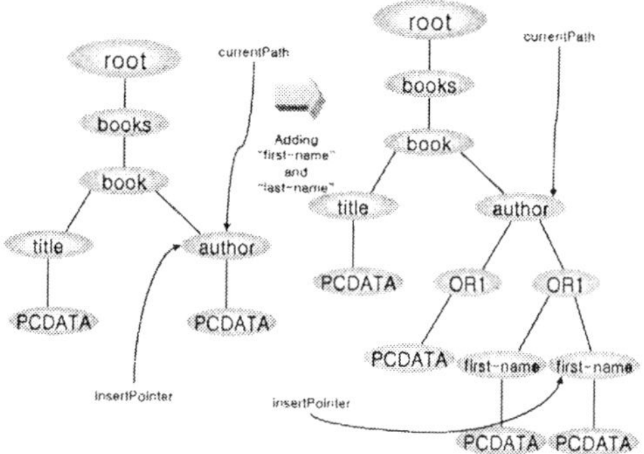

Fig. 4. Tree construction process and tree table entries for the second definition of Table 5

Table 6. DTD constructed from the tree in Figure 4

```
<!ELEMENT root books>
<!ELEMENT books (book)+>
<!ELEMENT book (title, author)>
<!ELEMENT title #PCDATA>
<!ELEMENT author ((#PCDATA) | (first-name, last-name))>
<!ELEMENT first-name #PCDATA>
<!ELEMENT last-name #PCDATA>
```

3.2 DTD Extraction

Once the tree is constructed reflecting all XML definitions, we can easily walk through the tree and construct a DTD reflecting DTD relationships. Temporary nodes used for expressing OR, Optional, and Repetition are not directly translated to a DTD definition. For them, subtree definitions are used to construct definitions instead. The DTD produced from the tree in Figure 4 is given in Table 6.

4 Conclusion

For SECOS to be able to acquire wide variety of information from the Internet, the role of Meta Gatherer is important. In this study we designed and implemented a DTD extraction module of the gatherer.

In order to automate the process of storing collected XML data into DB, we identify differences between the constructed DTD and the system DTD, which represent the underlying master DB. "XML2XML Mapper" developed by another group of SECOS development team does the operation [4]. It converts source XML files into target XML file suitable for the system DB.

For seamless operation of the system, we used several commercial tools as well. We used Oracle xmlparserv2.0 XMLTokenizer to extract tokens from XML documents and Oracle XDK utilities for storing data into DB.

References

1. Oh, D., Jung, J.: Effective Web-Based Information Gathering Services of IHWA. Proceedings of ICEIC'2000 International Conference, Shenyang, China (2000) 202-205
2. Garofalakis, M., Gionis, A., Rastogi, R., Seshadri, S., Shim, K.: XTRACT-A System for Extracting Document Type Descriptors from XML Documents. Bell Labs Tech. Memorandum (1999)
3. Moh, C.-H., Lim, E.-P., Ng, W.-K.: Re-engineering Structures from Web Documents. Proceedings of the 5th ACM International Conference on Digital Libraries (DL2000), San Antonio, Texas, USA (2000)
4. Ha, S.: The Effective Exploitation of Heterogeneous Product Information for E-Commerce. Submitted for Publication (2002)

Mapping XML-Schema to Relational Schema*

Sun Hongwei, Zhang Shusheng, Zhou Jingtao, and Wang Jing

National Specialty Laboratory of CAD/CAM, Northwestern Polytechnical University, Xi'an, China, 710072
shw2000cn@yahoo.com.cn

Abstract. XML is fast emerging as the dominant standard for representing data in Internet. One efficient path to store it is transforming XML data into relational database. Exiting XML-to-RDB algorithms focus only on the structure and ignore semantic constrains, in addition, their inputting is not XML-Schema but DTD. In this paper, we present an algorithm for mapping XML-Schema to relational schema. Our main ideas are as follows: 1) On the basis of regular tree grammar, propose a concise and precise formalization representing method named FD-XML for XML-Schema; 2) Extend the traditional ER model to Extended ER model (EER); 3) Map FD-XML to EER and then EER to relational schema. With the above procedures, the mapping algorithm comes into being, where both correct data structure and integrated data constrains are translated.

1 Introduction

XML is becoming the standard data format in Internet. One potential way to manage XML data is to reuse the effective and mature relational database techniques. Several XML-to-RDB algorithms have been proposed for this end (see [1][2][3]). But their input is DTD rather than XML-Schema, which is much more complex and powerful than DTD and recommended by W3C to replace DTD as a standard for XML Schema languages. For the much differences between them (see [4]), the algorithms mapping DTD to relational schema can't be used in mapping XML-Schema to relational schema. Therefore, there is an imperious need to study a new algorithm whose mapping input is not DTD but XML-Schema. In addition, the existing DTD-to-relational mapping algorithms mostly focus on the data structure and ignore the data constraints that contain abundant semantic information.

Aiming at the above matters, in this algorithm, we firstly propose a formal representation method named FD-XML for XML-Schema based on the regular tree grammar. FD-XML can integrally represent the information of XML-Schema. At the same time, we put forward the extended ER model, EER. FD-XML is then converted to EER model whose diagram represents the data structure and accessories containing

* We differentiate two terms, XML Schema(s) and XML-Schema; the former is a general term for a schema for XML, while the latter refers to one of the XML Schemas proposed by W3C.

M.H. Shafazand and A M. Tjoa (Eds.): EurAsia-ICT 2002, LNCS 2510, pp. 322–329, 2002.

the integrated data constraints. Besides, we adopt the equivalence transformation method of graph theory in simplifying the EER diagram by reducing entities. At last, we convert the EER model to the general relational schema.

The remainder of this paper is organized as follows. In Section 2, we introduce EER model. From section 3 to 6, the mapping algorithm is introduced step by step. Finally some test results, concluding remarks and future works are given in section 7.

2 Extend E-R Models to EER

E-R is a powerful and widely used approach to describe the real world, and it has become the most commonly used expression tool for concept model ever since. But there exist many problems in directly mapping XML-Schema to E-R because of the many differences between them. So we extend E-R model in three aspects (see figure 1). Firstly, because there doesn't exit clear parent-child relationship, which is the main relationship in XML-Schema. We use the arrowhead starting from the parent element and ending at the sub-element to express the parent and subelements in the parent-child relationship; (n, m) is then used to represent the occurrence time in XML-Schema, where n means the minimum occurrence time and m means the maximal. At last, as E-R model can't express data semantic constraints perfectly while XML-Schema holds a powerful data constraints representing mechanism, which is to be represented in FD-XML by some sets, accessories are given to EER in order to preserve the integrality of data constraints.

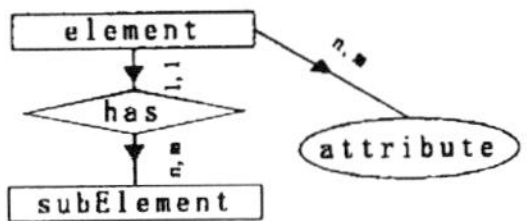

Fig. 1. Basic structure of EER diagram. The diagram is composed of elements, their relationships and attributes along with their occurrence time. Where the parent-child relationship is expressed by "has" in diamond box. In addition, EER accessories are not represented in this figure.

3 Formal Descriptions for XML-Schema: FD-XML

In this section, as a mechanism for describing permissible XML instance document, we borrow the definitions of regular tree languages and tree automata in [6], but unlike the definitions in it, we allow trees with "infinite arity"; that is, allow a node to have any number of sub-nodes, and allow the right-hand side of a production rule to have a regular expression over non-terminals. Now a RTG based formalization expression for XML-Schema, FD-XML can be given.

Definition 1. (FD-XML) An *FD* is denoted by an 8-tuple *FD*=(*N*, *T*, *S*, *E*, *A*, *C*, *U*, *K*), where,

1) N is a set of non-terminal symbols, where $N \subseteq \check{N}$;
2) T is a set of terminal symbols, where $T \subseteq \check{T}$;
3) S is a set of start symbols, where $S \subseteq \check{N}$;
4) E is a set of element production rules of the form "$X \longrightarrow a\ RE$", where $X \in N$, $a \in T$, and RE is the content model of this production rule; it is: $RE ::= \varepsilon \mid n \mid \tau \mid (RE) \mid RE + RE \mid RE,RE \mid RE? \mid RE^* \mid RE+$, where $n \in \check{N}$, $\tau \in \check{A}$
5) A is a set of attribute production rules of the form "$X \longrightarrow a\ RE$", where $X \in N$, $a \in T$, and $RE ::= \varepsilon \mid a \mid (RE) \mid RE,RE$, where a is an attribute;
6) C is a set of terminal symbol data types and their constraints, $\forall\ c_i \in C$, c_i is an 18-tuple, C_i (B_i, *Li, Lmaxi, Lmini, Pi, Ei, WSi, MAXIi, MINIi, MAXEi, MINEi, TDi, FDi, Fixi, Defi, Opti,* Pro_i*, Reqi*), where B_i denotes the basic user-defined data type of C_i, meanwhile, L_i for the *length* constraint, L_{maxi} for the *maximal length* constraint, L_{mini} for the *minimum length* constraint, P_i for the *character pattern*, E_i for the *enumerate set*, WS_i for the *whiteSpace* constraint, MAX_{Ii} for the *maximal value* including itself, MIN_{Ii} for the *minimum value* including itself, MAX_{Ei} for the *maximal value* not including itself, MIN_{Ei} for the *minimum value* not including itself, TD_i for the *totalDigits* attribute, FD_i for the *fractionDigits* attribute, Fix_i for the *Fixed* attribute, Def_i for the *Default* attribute, Opt_i for the *Optional* attribute, Pro_i for the *prohibited* attribute and Req_i for the *Required* attribute;
7) U is a set of primary key and unique constraint, $\forall\ pk_i \in U$, pk_i is a 3-tuple, pk_i (K_{mi}, *XPsi, XPfi*), where K_{mi} denotes the name of pk_i, X_{Psi} denotes the *selector* domain expressed by Xpath of pk_i, and X_{Pfi} for the field domain expressed by Xpath of pk_i.
8) K is a set of the foreign key, $\forall\ fk_i \in K$, fk_i is an 4-tuple fk_i (K_{fi}, *Kmi, XPsi, XPfi*), where K_{fi} denotes name of fk_i, K_{mi} denotes the referenced primary key in U, and X_{Psi} denotes the *selector* domain expressed by Xparth of fk_i, and X_{Pfi} denotes the field domain expressed by Xpath of fk_i.

```
<xsd:schema xmlns:xsd="http://www.w3.org/2001/XMLSchema">
<xsd:element name="purchaseOrder" />
<xsd:complexType >
<xsd:sequence>
<xsd:element name="USAdress" type="string" fixed="L.A." minOccurs="0" />
<xsd:element name="Items"  type="ItemsType"/>
</xsd:sequence>
<xsd:attribute name="orderDate" type="xsd:date" use="required" fixed="US"/>
</xsd:complexType>
</xsd:element>
<xsd:complexType name="ItemsType">
<xsd:sequence>
<xsd:element name="item" minOccurs="0" maxOccurs="unbounded">
    <xsd:complexType>
<xsd:sequence>
<xsd:element name="productName" type="xsd:string" fixed="paper"/>
<xsd:element name="quantity" default="1000">
<xsd:simpleType>
<xsd:restriction base="xsd:Integer">
<xsd:minInclusive value="1000"/>
<xsd:maxInclusive value="9999"/>
<xsd:pattern value="\d{3}[0]"/>
</xsd:restriction>
```

```
    </xsd:simpleType>
    </xsd:element>
    </xsd:sequence>
    </xsd:complexType>
    </xsd:element>
    </xsd:sequence>
    </xsd:complexType>
</xsd:schema>
```

Fig. 2. An example of XML-Schema. This is a simplified version of the example in [5]

The XML-Schema in Figure 2 is formalized as FD_I=(*N, T, S, E, A, C, U, K*), where:

N={*purchaseOrder, USAddress, Items, item, productName, quantity*}

T={<*purchaseOrder*>, <*Items*>, <*orderDate*>, <*USAddress*>, <*item*>,<*productName*>, <*quantity*>}

S={*purchaseOrder*}

E={*purchaseOrder* → <*purchaseOrder*>(*USAddress*?, *Items*),*USAddress* → <*USAddress*>(ε), *Items* → <*Items*>(*item**),*item* → <*item*> (*productName,quantity*), *productName* → <*productName*>(ε),*quantity* → <*quantity*>(ε)}

A={*purchaseOrder* → <*purchaseOrder*>(@<*orderDate*>),*USAddress* → <*USAddress*>(ε),*Items* → <*Items*>(ε),*item* → <*item*>(ε), pro*ductName* → <*productName*>(ε),*quantity* → <*quantity*>(ε) }

C={<*USAddress*> (*string*, ε , ε , ε , ε , ε , ε , ε , ε , ε , ε , ε , ε , "L.A."ε,ε,ε,ε),<*orderDate*>(*date*, ε , ε , ε , ε , ε , ε , ε , ε , ε , ε , ε , ε , ε , "20020501", ε , ε , ε), <*productName*> (*string*, ε , ε , ε , ε , ε , ε , ε , ε , ε , ε , ε , ε , "paper", ε , ε , ε , ε),<quantity> (*Integer*, ε , ε , ε , "\d{3}[0]", ε , ε , "9999", "1000", ε , ε , ε , ε , ε , "1000", ε , ε , ε) }

U ={Φ}

K={Φ}

Obviously, from the example above, FD-XML can wholly describe both the data structure and the data constraints. In *FD* =(*N, T, S, E, A, C, U, K*), tuples of *N, T, S, E, A* represent the XML-Schema structure information and tuples of *C, U, K* represent the information of data type and data constraints.

4 From FD-XML to EER

In this section we convert FD-XML to EER. The detailed procedure is as follows:

1) Represent every element in the tuple *N* of FD-XML as an entity, which is denoted by rectangle box in EER model.
2) Build up the relationships between entities according to the tuple *E* of FD-XML, hereinto the relationships are denoted by diamond boxes, the parent and subordi-

nate elements in parent-child relationship are distinguished by arrowhead's starting and ending point.

3) Build up the entities' attributes and confirm their occurrence times according to the tuple *A* of FD-XML.

4) Keep the tuples *C*, *U* and *K* of FD-XML as the accessories of EER.

After the above four steps, now we can get the resultant EER diagram of the illustrated FD_1 (see figure 3). EER diagram and it's accessories constitute the integrated EER model, where EER diagram is mainly produced by *N, T, S, E, A* and represents the data structure; the accessories include the three set of *C, U, K* and represent the data constraints.

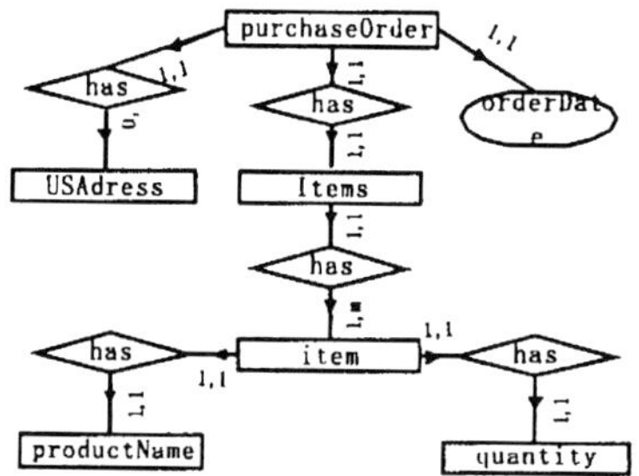

Fig. 3. EER diagram of the illustrated FD_1. The original five elements in *N* are converted to the entities in rectangle box, the parent-child relationships is build up with the "has" symbols along with the arrowhead according to *E* and the entities' attributes are built up according to *A*

5 Simplify EER

The direct mapping from EER to relational schema is unscientific, for it will result in many piecemeal and exiguous relations possessing few or even no attributes. To avoid it, we propose a simplification method for EER diagram. With this simplification method, we can simplify EER to fit for the mapping and keep semantic equivalence at the same time. The simplifying means is mainly in two aspects:

1) An entity is converted to its parent entity's attribute if it satisfies the following conditions: a) the entity has a unique parent entity; b) the entity possesses no sub-entity; c) the parent-child entities meet the qualification of Element (1,1) $\longrightarrow$ sub-Element (1,1) or the qualification of Element (1,1) $\longrightarrow$ sub-Element (0,1). After conversion, the former gets the attribute occurrence time of (1, 1), the latter gets (0, 1).

2) The sub-entity is removed from its parent entity that satisfies the following conditions: a) the entity possesses no attribute; b) the entity has no or just one parent entity; c) the entity possesses only one sub-entity and meet the qualification that Element (1,1) $\longrightarrow$ sub-Element (1,1) or the qualification Element (1,1) $\longrightarrow$ sub-Element (0,1).

After the above two procedures, the EER diagram illustrated in figure 3 can be simplified to the EER diagram in figure 4.

Besides, the accessories of EER are simplified. According to the primary key and unique constraint set *U* in EER accessories, the XPath expression is parsed to found the entity's primary key, which is marked as *primary key* $\xrightarrow{pkey}$ *entity*; Similarly, According to the foreign key constraint set *K* in EER accessories, the XPath expression is parsed to found the entity's foreign key, which is marked as *referencing entity, foreign key* $\xrightarrow{fkey}$ *referenced entity, primary key.*

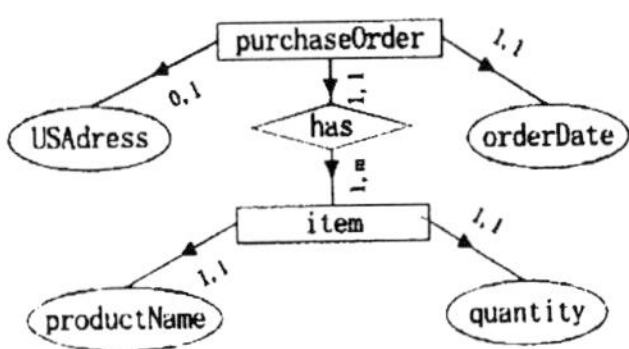

Fig. 4. Simplified EER diagram of Fig.3. The Entities *USAdress, productName* and *quantity* in Fig.3 are converted to purchaseorder's attributes in step 1; the entity *Items* in Fig.3 is removed in step 2

6 From EER to General Relational Schema

The pending problems we have to face when mapping EER model to relational schema are: 1) how to convert the entities as well as their relationships to relations; 2) how to confirm the attributes and keys of the entities; 3) how to create the data constraints from the EER accessories.

EER diagram can fully represent the original data structure of XML-Schema with concision and precision, and the accessories amply denote the apparent data constraints in XML-Schema, hence the two parts of EER are mapped respectively during the whole mapping procedure. That is, convert the EER diagram to the data structure of relational schema; then convert the constraint information of the accessories to the corresponding data constraints of relational schema, where the primary key set *U* in EER accessories is converted to entity integrality constraint, the foreign key set *K* to the reference integrality constraint and the data type along with its constraints to the user-defined integrality constraint.

One thing to be explained is that some entities possess keys in the EER accessory *U*, while some not, which fall short of the entity integrality constraint, so we must supplement the integrality constraint for them. Besides, for the latent reference integrality between parent-child entities, there are only apparent constraints without concealing references in the EER accessory *K*, so we have to dig the reference constraints to represent the constraints between entities integrally. At last, the data structure and constraints set by XML-Schema are single instance document satisfying, and if we merge two or more instance documents together, the resultant document won't always satisfy the constraints. For instance, an element possesses a "be unique" constraint, and so the data item of this element is nonrecurring in a single instance document, but there may exit same data items in two or more documents. These documents merged, the element's "be unique" constraint is surely broken.

Accounting for the above troubles, the method to convert EER model to relational schema is:

1) Map an entity to a relation, and the entity attributes are converted to the relation's attributes.
2) Add an identification code of *long* type for every entity relation to ensure the entity's integrality. *Identification code name::= entity name + ID*, and this attribute is merged into *U* as the primary key of the entity.
3) Map a relationship between entities in EER to a relation, and convert all kinds of entity keys as well as attributes of the EER relationship to the attributes of the relation. The relationship may appear in three cases: a) 1:1, both the keys of the parent entity and the sub-entity can serve as the candidate key for the relation, and we select the key of the parent entity; b) 1: n, we set the key of the n entity as the relation key; c) n: m, we set the combination of all the entity keys as the relation key.
   From the above three steps we get the following three relations:
   (***PurchaseOrderID***, *USAdress*, *orderDate*), (***ItemID***, *productName*, *quantity*) and (***ItemID***, *purchaseOrderID*).
4) Merge the relations possessing same primary key. Then the above relations remain two: (***purchaseOrderID***, *USAdress*, *orderDate*), (***ItemID***, *productName*, *quantity*, *purchaseOrderID*), mark it as *RM*.
5) According to the primary key constraint set *U* in EER, establish the primary key constraint and add primary key of every entity to *U*, then we can get *U* {(*purchaseOrderID* $\xrightarrow{pkey}$ *purchaseOrder*), (*ItemID* $\xrightarrow{pkey}$ *item*)}.
6) According to the foreign key constraint set *K* in EER, establish the foreign key constraint. For the relations meeting some given requests, we can dig the foreign keys and establish the reference integrality constraint. The given requests are: a) A key A_1 is added for relation *A* to ensure the entity integrality; b) Another relation *B* takes the A_1 as an attribute. So we can establish the reference constraint from the A_1 in relation *B* to the A_1 in relation *A*, and mark it as (B, $A1 \xrightarrow{fkey} A$, $A1$), then add it to *K*, and we get *K* {(*item*, *purchaseOrderID* $\xrightarrow{fkey}$ *purchaseOrder*, *purchaseOrderID*)}.
7) Establish all the attribute data types as well as their constraints according to the set *C* in EER, and present the data type of *ID* for the relations key automatically generated in step 2), then merge them into *C*.
   Thus, we get the relational schema as follows, which is made up of *RM*, *U*, *K*, and *C*.

RM=(***purchaseOrderID***, *USAdress*, *orderDate*), (***ItemID***, *productName*, *quantity*, *purchaseOrderID*)

U= (*purchaseOrderID* $\xrightarrow{pkey}$ *purchaseOrder*), (*ItemID* $\xrightarrow{pkey}$ *item*)}

K= {(*item*, *purchaseOrderID* $\xrightarrow{fkey}$ *purchaseOrder*, *purchaseOrderID*)}

C={<*purchaseOrderID*>(*long*, ε , ε , ε , ε , ε , ε , ε , ε , ε , ε , ε , ε , " ε ", ε , ε ,
ε , ε),<*itemID*>(*long*, ε , ε , ε , ε , ε , ε , ε , ε , ε , ε , ε , ε , " ε ", ε , ε , ε ,
ε),<*USAddress*> (*string*, ε , ε , ε , ε , ε , ε , ε , ε , ε , ε , ε , ε , "L.A.", ε , ε , ε ,
ε),<*orderDate*>(*date*, ε , ε , ε , ε , ε , ε , ε , ε , ε , ε , ε , ε , ε , "20020501", ε , ε ,
ε), <*productName*> (*string*, ε , ε , ε , ε , ε , ε , ε , ε , ε , ε , ε , ε , "paper", ε , ε ,
ε , ε),<*quantity*> (*integer*, ε , ε , ε , "\d{3}[0]", ε , ε , "9999", "1000", ε , ε , ε ,
ε , ε , "1000", ε , ε , ε) }.

7 Test Result and Conclusion

Based on section 3 to 6, the mapping algorithm from XML-Schema to general relational schema has been achieved and the experimental results indicate that our algorithm works well. Still, In our future work, we will improve FD-XML to make it fit for any other XML Schema such as DTD, RELAX etc, and then get a universal XML-to-relational mapping algorithm.

References

1. Florescu, D., Kossman, D.: Storing and Querying XML Data Using a RDBMS. IEEE Data Engineering Bulletin, Vol. 22. No. 3 (1999)27–4
2. Shanmugasundaram, J., Gang, H, et al.: Relatilnal Databases for Querying XML Documents: Limitations and Opportunities.VLDB'99, Proceedings of 25th International Conference on Very Large Data Bases, Edinburgh, Scotland (1999) 302–04
3. Lee, D. W., Chu, W. W.: CPI: Constraints-preserving Inlining Algorithm for Mapping XML DTD to Relational Schema. Data Knowledge Engineering (2001) 3–25
4. Lee, D. W., Chu, and W. W.: Comparative Analysis of Six XML Schema Languages.ACM SIGMOD Record (2000) 76–87
5. Fallside, D. C.: XML Schema Part 0: Primer. http://www.w3.org/TR/xmlschema-0 (2001.5)
6. H. Comon, M. Dauchet, R. Gilleron, F. Jacquemard, D. Lugiez, S. Tison, and M. Tommasi. "Tree Automata Techniques and Applications", 1997.

Flexible Modification of Relational Schema by X2RMap in Storing XML into Relations

Jaehoon Kim and Seog Park

Department of Computer Science, Sogang University
1-1 Shinsu-Dong Mapo-Gu Seoul Korea 121-742
{chris3, spark}@dblab.sogang.ac.kr

Abstract. Among many attempts to effectively store and query massive quantity of XML data, one method using RDBMS is drawing attention. It is to create appropriate relational schema by using DTD and then divide and store the XML document according to the schema. XML query for data stored as above can be implemented by rewriting it into SQL. Among this series of processes, in fact, intermediate mapping concept becomes necessary due to the difference between DTD and relational schema. It can help a user to compose a XML query transparently from a virtual physical structure, a relational schema. So, though the relational schema is properly modified for query performance, it doesn't have an effect on XML query. In this paper, we introduce some effective methods for flexible modification of relational schema using mapping structure, X2RMap.

1 Introduction

In addition to standardization, XML can be applied to a variety of areas connected to web and especially in such areas as e-Business, bioinformatics, medicine and multimedia, XML applications with a great deal of data have been made. Creating this mass data requires efficiently devised storage mechanism and a query process. So, efforts to store and query vast XML data have been made from diverse aspects until now. It can be broadly classified into designs of definitely new XML DBMS such as LORE of Stanford Univ., ObjectDesign's eXcelon and Software AG's Tamino, and utilization of existing OODBMS or RDBMS. Our interest lies in processing XML data efficiently by the capacity of mass data processing of existing RDBMS and utilizing many tools supporting it, and the researches concerned with it already has been examined in STORED, Shanmugasundaram et al.[1, 2], XPERANTO and SilkRoute and so on.

However, such methodologies are focused on designing an efficient relational schema for storing the given XML data. Contrary to this, our research is intended to observe a method for achieving the improvement by analyzing the XML query statistics and properly modifying the relational schema even if it has been defined. Designing a new XML DBMS middleware wrapping RDBMS without rebuilding it is just similar to producing a RDBMS application but gives us very interesting thought. The thought is that from the viewpoint of XML DBMS RDBMS is considered as a kind of a physical layer and then a relational schema can be freely modified, whenever necessary, in order to enhance a query performance. It is because when in fact

M.H. Shafazand and A M. Tjoa (Eds.): EurAsia-ICT 2002, LNCS 2510, pp. 330–338, 2002

XML data is divided and stored in relational schema, XML schema and relational schema become decoupled through mapping information between them. That is to say, relational schema can be freely modified since this mapping information is maintained and XML query can be rewritten into appropriate SQL statement by using such mapping.

This paper is organized as follows. Section 2 addresses comparison of our research with other similar researches. Section 3 introduces some methods about the flexible modification of relation schema and section 4 introduces X2RMap. Some performance analysis through tests of the methods is carried out in section 5 and section 6 describes our conclusions.

2 Related Works

Direction of our basic research is toward division of XML data and storing it by using RDBMS and this research has been performed in [1, 2, 3, 4, 5]. In fact, these related researches became basis for embodiment of our system. However, what we suggest in this paper is that relational schema and DTD can be decoup-led through mapping concept and relational schema can be freely modified for a performance enhancement.

STORED is perhaps the first trial of modification of mapping XML data into RDBMS from the viewpoint of storage cost and query cost. It classifies the schemaless XML data as high cost structure and the low one, using the apriory algorithm. And the modification by the mapping concept is that the high cost structure is stored into RDBMS and the low one into semi-structured data object repository for utilizing the high performance of RDBMS. Therefore, it is different from our approach in which the relational schema is modified by analyzing the query statistics on the XML data stored in RDBMS in which only DTD exists.

Phil Bohannon et al.[5] attempts to freely modify the relational schema when XML data with a schema is stored into RDBMS. But it has several differences from our strategy of schema modification. And although it utilizes the greedy method for system to search automatically the optimal relational schema during certain period, about this problem we are placed in a position that we can make full use of monitoring and tuning tools which support RDBMS. Another difference is that our system saves the modified schema information in mapping structure, X2RMap and efficiently performs the query rewriting into SQL using the information of X2RMap.

In commercial RDBMS, there exist mapping concepts, e.g. IBM DB2's DAD, MS SQL Server's SQL extension. However, this mapping structure is mere mapping of relational schema and then it is difficult to support the flexible modification of relational schema for a performance enhancement.

3 Flexible Modification of Relational Schema

In here, first we introduce the example that helps to illustrate some methods of modifying a relational schema and an initial mapping between XML and relations.

An initial mapping schema: Given the DTD such as (a) in Fig. 1, (d) is to be obtained by expanding the DTD graph of (b). And we generate the relational schema like (c) by *, +, | symbols from (d). We name (d) as X2RMap and it is implemented

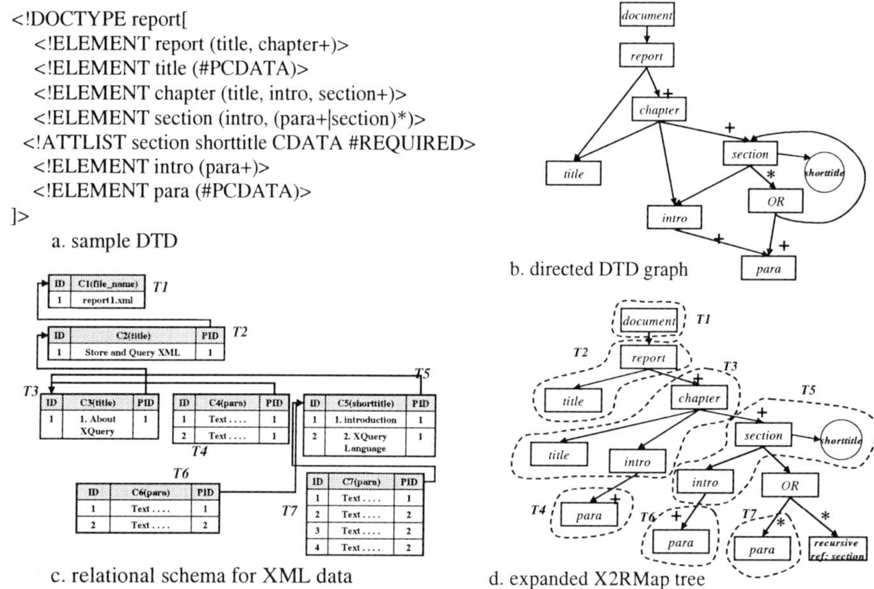

Fig. 1. An initial mapping between XML and relations

by tree data structure. It saves the basic mapping information between tables(columns) and each node of the tree.

In fact, this mapping is similar to the method suggested in [1]. About XML read query, our basic rewrite method models [2]'s *sorted outer union* method and about update query, it references [4]. The reason why we do not consider the other mapping methodology like *edge table* of [3] is that we intend to store large scale of XML data with the schema and in that case many self-joins of the edge table are inefficient.

3.1 Table Merge about Parts with the Same Subtree Structure on DTD

We introduce the following as first modification method. We expanded parts with more than two fan-ins on DTD graph of Fig. 1 and mapped them to separate tables although they have the same subtree structure. However, dividing these parts into separate tables and merging them into a single table can be appropriately adjusted for a rapid query response time. For example, PARA nodes can be mapped to one table unlike (c) of Fig. 1, and in case following query is given,

FOR $S IN //para WHERE contains($S, "XQuery") RETURN $S

processing the query only in the merged table has an advantage. If a proper index is set additionally to PARA column in the condition clause, query processing can be carried out more quickly. The merged table should additionally have a parent table code column with respect to each parent table. For PARA node, T4, T6 and T7 can be merged into new table T8(<u>ID</u>, *PCODE*, C8, PID). We will be able to process T3 ⋈ T8 or T5 ⋈ T8 join by the parent table code, PCODE. PCODE is generated automatically by system and X2RMap tree manages its information.

3.2 Maintaining the Extent of Recursion

Our system manages recursion as followings. If a document in which SECTION is recurred three times like '//chapter/section/section/section' is inserted, X2RMap tree is modified as in Fig. 2. That is to say, whenever a document with recursion is inserted, X2RMap tree maintains the extent of recursion until now. Of course, the modification of mapping information requires a dynamic creation and deletion of tables only in the first case. We may be able to manage recursion structure in each table as in Fig. 2 or merge nodes with the same structure into a single table.

Structuring the degree of recursion into X2RMap tree supports much efficient query processing. Let us consider the following query.

FOR $S IN document("report1.xml")/report/chapter/section/section
WHERE contains($S/@shorttile/text(), "XQuery") RETURN $S

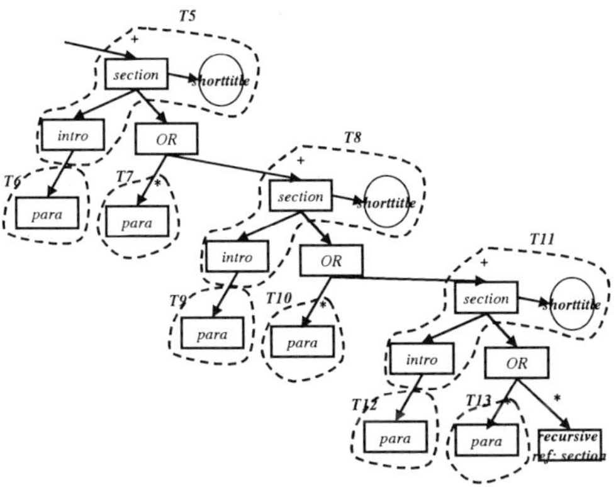

Fig. 2. Maintaining the extent of recursion by X2RMap

If recursion is not structured, we cannot but use the method such as Naïve or Semi-Naïve[6] which repeats join until the point of fixpoint since we are not sure of the extent that SECTION repeats. It means that *outer union method*[2] cannot be used for rewriting XQuery into single SELECT query and it makes overhead on the process of query. However, if the recursion is structured, it can be used. Moreover, to evaluate the path such as 'section/section', non-structured case needs extra join, but once structured the rewritten query may be a SQL statement which directly refer to T8 table using the information of X2RMap.

The point to ponder in this recursion processing is that the consistency of X2RMap tree shared by clients should be controlled and though not likely, recursion with extremely high depth may occur.

3.3 Using Replication

It is known that performance enhancement can be accomplished by appropriate replication of data in classical RDB design even though some anomalies arise. In this subsection, several replication methods will be presented.

3.3.1 ID Column Replication. If XPath, 'document("report1.xml")/report/chapter/section/intro/para' is given in Fig.1, T1 ⋈ T2 ⋈ T3 ⋈ T5 ⋈ T6 should be carried out in order to evaluate the path. But if ID of T1 or T2 is replicated in T6 table, the operations will be reduced to T1 ⋈ T6 or T1 ⋈ T2 ⋈ T6. This replication approach can be applied in two forms, IN-PLACE and SEPARATE. IN-PLACE is that ID column of an ancestor table is replicated in a descendent table and SEPARATE is that ID columns to be replicated are maintained in a separate table. The reason why replication is maintained in SEPARATE is related with update cost and storage cost.

3.3.2 SUBTREE Fragment Replication. As for a following XQuery query,

FOR $S IN document("report1.xml")/report/chapter/section
WHERE contains($S/@shorttitle/text(), "XQuery") RETURN $S

Once SUBTREE fragment with its root at SECTION node was composed beforehand and then replicated, join and outer union operations and re-composition for XML results might be relieved. This XML SUBTREE fragment can be stored in a CLOB column of the table to which the subtree root node belongs as in Fig.3.

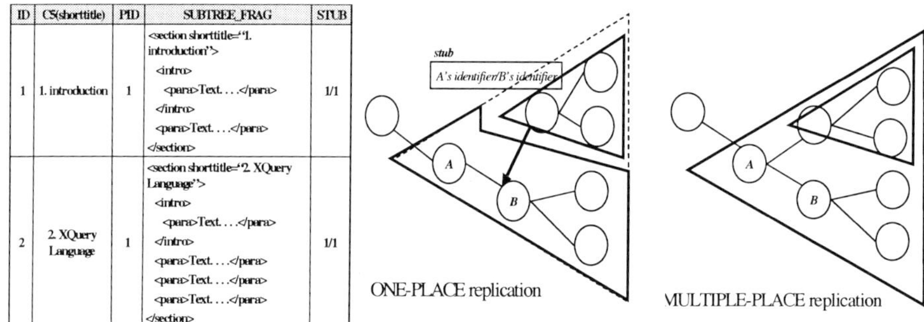

ID	C5(shorttitle)	PID	SUBTREE_FRAG	STUB
1	1. introduction	1	<section shorttitle='1. introduction'> <intro> <para>Text. . . .</para> </intro> <para>Text. . . .</para> </section>	1/1
2	2. XQuery Language	1	<section shorttitle='2. XQuery Language'> <intro> <para>Text. . . .</para> </intro> <para>Text. . . .</para> <para>Text. . . .</para> <para>Text. . . .</para> </section>	1/1

Fig. 3. SUBTREE fragment replication

The replication method of SUBTREE fragment may be classified into ONE-PLACE replication in which a fragment is once replicated and MULTIPLE-PLACE replication in which it is replicated in several places. That is, ONE-PLACE in Fig. 3 shows that an upper fragment is divided when a fragment replication is carried out for a lower subtree in order that the replication may be carried out once. Also, it is also shown that in case of MULTIPLE-PLACE replication, a fragment with respect to the subtree is newly replicated regardless of existing fragments.

In ONE-PLACE replication, all fragments of the SUBTREE that were divided should be recombined when results of the query are returned. As a divided lower fragment has stub information on an upper fragment, this connection is made in query rewrite.

In MULTIPLE-PLACE replication, a replicated fragment of a relevant node has only to be returned. Therefore, a query processing in MULTIPLE-PLACE replication is faster than that in ONE-PLACE replication but MULTIPLE-PLACE replication has such shortcomings as large quantities of replication and update maintenance.

4 X2RMap

In the previous section, we have examined several modification of relational schema by mapping concept. In fact, all information related with such methods are kept in X2RMap tree and so query rewrite and optimization are processed by means of these information. These information of X2RMap are saved in a XML file as like Fig.4 (it shows the case of Fig.1.).

```
<?xml version="1.0"?>
<!DOCTYPE X2RMAP SYSTEM "x2rmap.dtd">
<X2RMAP>
 <!-- table_count, column_count... save the number of table, column... created until now. -->
 <DATABASE_INFO  table_count="11"  column_count="11"  replication_count="6"  pcode_count="5"
stub_count="0"><TABLES>
<!-- In here, the dependency information among tables are saved through parent. Also the dependency by ID
replication is saved through REP_TO and REP_FROM. Rep_type represents IN-PLACE or SEPARATE. Col-
umn_name represents the replication column. Block_num is used for query cost evaluation. -->
   <TABLE name="T1" block_num="1">
    <REP_TO rep_type="IN" table_name="T6" column_name="R1"/>
         :         :        :        :
   <TABLE name="T6" block_num="1" parent="T5">
    <REP_FROM rep_type="IN" table_name="T1" column_name="ID"/></TABLE>
   <TABLE name="T7" block_num="1" parent="T5">
    <REP_FROM rep_type="IN" table_name="T1" column_name="ID"/></TABLE>
  </TABLES></DATABASE_INFO>

 <NODE name="document" cardinal="null" type="E">  <!-- From here, X2Map tree is represented. Cardinal
means '+', '*'... and type means Element, Attribute... -->
  <INFORMATIONS>
   <MAP table_name="T1" column_name="C1"/>
   <FRAGMENT_REP>
    <MULTIPLE column_name="R2"/>  <!-- This means that SUBTREE fragment replication by unit of a
document is done. R2 is a clob column for saving the fragment. -->
   </FRAGMENT_REP>
  </INFORMATIONS>
    <CHILDREN>
     <NODE name="report" cardinal="1" type="E">
         :         :        :        :
             <NODE name="intro" cardinal="+" type="E">
              <INFORMATIONS>
               <MAP table_name="T5" column_name="null"/>
               <FRAGMENT_REP><RNODE path="document"/></FRAGMENT_REP>
              </INFORMATIONS>
              <CHILDREN>
               <NODE name="para" cardinal="+" type="E">
                <INFORMATIONS>
<!-- Table merge about the same sub-structure --><MAP table_name="T11" column_name="C6" pcode="1"/>
     | :         :        :        :
     |       <NODE name="OR" cardinal="1" type="S">
     | :         :        :        :
     |             <NODE name="para" cardinal="*" type="E">
     |              <INFORMATIONS>
     L-----►         <MAP table_name="T11" column_name="C9" pcode="5"/>
  :         :        :        :
```

Fig. 4. The XML file saving X2RMap information

5 Some Experiments

In this section, we will show that flexible modification of relational schema is very useful through some experiments. Our test was embodied on windows 2000 server with 1 GB memory and dual 866 MHz Pentium III CPU, and oracle 8i as a commercial RDBMS was adopted and all codes were written in java.

5.1 Experiment 1

Advantages mentioned in section 3.1 are experimented in here. A simple relational schema of Fig. 5 in which some XML data are divided and stored is taken into account for this experiment. Here, T2 and T3 have the same structure and are mapped to separate tables. And CT1 is a merge case.

T2: T12 T13
T3: T14 T15
CT1: CT2 CT3

We have implemented following XQuery query under these two cases.
Case 1: XQuery extracting $field_1$s of T2, T12, T13, T3, T14 and T15 on condition of $field_2$s of T2 and T3 (that is, a multi-path query as like 'report/chapter//para')
Case 2: XQuery extracting $field_1$s of T2, T12 and T13 on condition of a $field_2$ of T1 (that is, a single-path query as like 'report/chapter/intro/para')

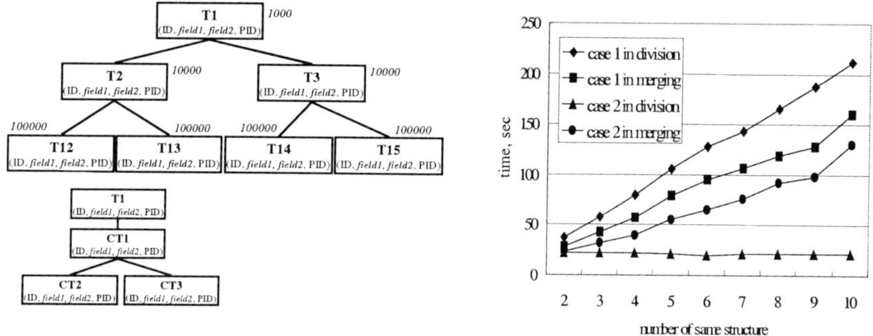

Fig. 5. Table merge about parts with the same structure

Here, for the purpose of comparison under the best situation with fast retrieval, a non-clustered index was set in $field_2$ of each table whereas a clustered index for PCODE in merged case. In the test model, numbers of tuples of T1, T2 and T3 are 1000, 10000 and 100000 respectively and accordingly average number of child table's tuples per parent table's one tuple is 10. We performed experiments while the number of cases with above structure was varied from 2 to 10. The result graph shows that table merge is advantageous for a multi-path query while division is more advantageous than table merge for a single-path query. This difference results from the fact that not only join costs of two case but also read costs of each index are different.

5.2 Experiment 2

In order to analyze the mixed query cost of read and update queries regarding ID replication, we make 10 client threads run on our XML DBMS server and each thread is made to select random queries among prepared sets of read queries and update queries and request them in the server. Selection of queries in a client is implemented by means of a probability, P_u. Here, P_u = the number of update queries of a client /the number of total queries of a client. We have set the number of total queries of a client to 10. Each client thread returns the execution times measured for read queries and update queries, and then average execution times of read queries and update queries of one client thread, C_{read} and C_{update}, are calculated. Accordingly, the average query cost of a client, C_{total}, is as follows; $C_{total} = C_{read} * (1 - P_u) + C_{update} * P_u$.

* Average number of child table's tuples per parent table's one tuple(average fan-out of one node in XML document) : 3

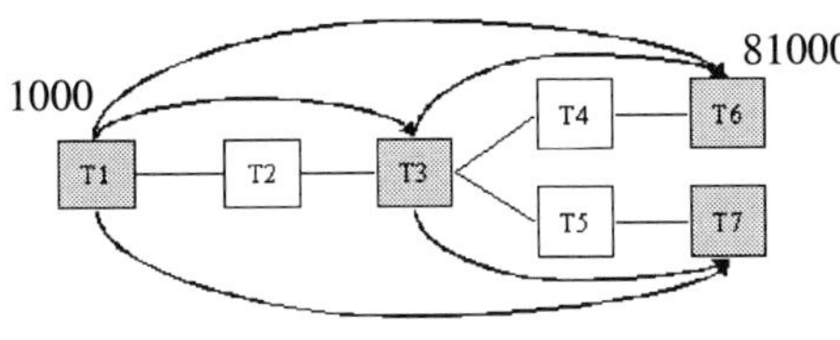

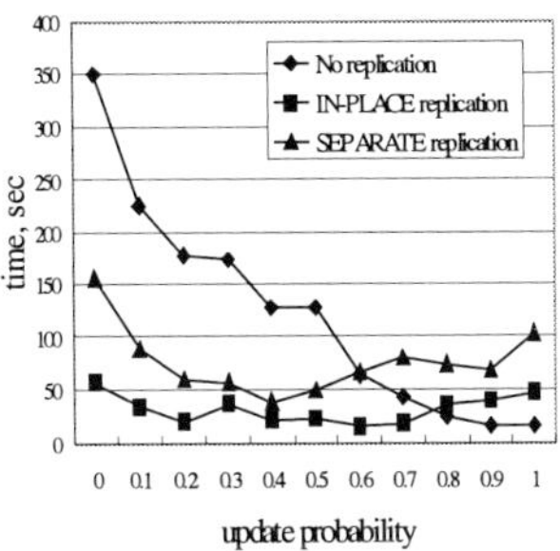

Fig. 6. ID column replication

Let's consider a test model of Fig. 6. In case of IN-PLACE T1(ID) is replicated into tables, T3, T6 and T7 and T3(ID) is replicated into tables, T6 and T7 while regarding SEPARATE, T1(ID), T3(ID) and T6(ID) are all replicated into one separate table and T1(ID), T3(ID), T7(ID) are all replicated into another separate table. As for each of these cases, some XQuery read queries are processed about T6 and T7 on certain field of table T1(f_r = the number of a table's tuples to which a read query is applied/total number of the table's tuples = 5%). And random update queries are processed with the update probability for T1, T3, T6 and T7(f_u = 5%). Graph of Fig. 6 shows that using properly replication may lower the query cost as much as three times. However, if update query becomes frequent, query cost should become higher in replications. IN-PLACE should much better performance than SEPARATE. It is related with the features of storing XML into RDBMS. But we will skip the reason about this, because this paper is intended about the usefulness of mapping concept.

6 Conclusions and Future Works

When XML data is divided and stored in RDBMS, designing a mapping structure between DTD and relational schema, as examined in this research, provides several advantages. That is, modifying relational schema without affecting XQuery query can enhance query performance. Maybe, advantages resulting from the mapping concept would be more diverse. For example, we have skipped the basic vertical/horizontal division of a table. But it can perform XML query rewrite that takes advantage of parallel query processing provided by commercial RDBMS. Although we have not presented the systematic algorithm of query rewrite which uses the mapping information, we actually have embodied a query processor which can operate for several simple XQuery syntax using X2RMap and utilized it through the test. In fact, it was not difficult one. But if the XQuery query is complex, it was not an easy problem to actually convert the query into SQL statement appropriate to it[1]. Therefore, our future work is to make more systematic the query rewrite using mapping information of X2RMap.

Acknowledgements. This work was supported by grant No.R01-2000-000-00272-0 from Basic Research Program of the Korea Science and Engineering Foundation.

References

1. J. Shanmugasundaram, K. Tufte, G. He, C.Zhang, D. De-Witt, and J. Naughton. Relational databases for querying XML documents: limitations and opportunities. In *Proceedings of VLDB*, pages 302-314, Edinburgh, UK, September 1999.
2. J. Shanmugasundaram, E. Shekita, R. Barr, M. Carey, B. Lindsay, H. Pirahesh, and B. Reinwald. Efficiently publishing relational data as XML documents. In *Proceedings of VLDB*, pages 65-76, Cairo, Egipt, September 2000.
3. D. Florescu and D. Kossmann. Storing and Querying XML data using an RDBMS. *IEEE Data Engineering Bulletin*, 22(3), 1999.
4. I. Tatarinov, Z. G. Ives, A. Y. Halevy, D. S. Weld. Updating XML. In *Proceedings of the ACM SIGMOD International Conference on Management of Data,* Pages 413–424
5. P. Bohannon, J Freire, P. Roy, and J. Simeon. From XML Schema to Relations: A Cost-Based Approach to XML Storage. *Proceedings of International Conference on DATA ENGINEERING*, Pages 64-75, San Jose, California, February-March 2002
6. Francois Bancilhon and Raghu Ramakrishnan. An Amateur's Introduction to Recursive Query Processing Strategies. In *Proceedings of the ACM SIGMOD International Conference on Management of Data*, Washington. D.C., May 1986

B2B Integration – Aligning ebXML and Ontology Approaches

Birgit Hofreiter and Christian Huemer

Institute for Computer Science and Business Informatics
University of Vienna, Liebiggasse 4, 1010 Vienna, Austria
{birgit.hofreiter,christian.huemer}@univie.ac.at

Abstract. In B2B e-commerce, XML provides means to exchange data between applications. It does not guarantee interoperability. On the syntactic level, this requires an agreement on an e-business vocabulary. Even more important, on the semantic level, business partners must share a common view unambiguously constraining the generic document types. In this paper, we present a framework that brings together work in the area of ontologies and work in the area of XML-based data interchange, namely ebXML. The framework uses an ontology based on ebXML corecomponents expressed in RDF to allow for bridging between different e-business vocabularies. Since a bridging mechanism is required, but not specified within ebXML, our approach complements ebXML. The integration of the ontology-based approach into ebXML is realized in four major steps. In this paper we exactly identify the requirements and the architecture of each step. This provides exact guidelines for future research towards implementing these steps.

1 Introduction

XML is said to overcome the most significant obstacles of traditional electronic data interchange (EDI) standards. After the first hype, people realized that XML provides means to exchange data between applications, but does not guarantee interoperability. XML only provides a syntax that could be used for data transfer in B2B, which is only one level of interoperability that must be met in B2B. Fig. 1 gives an overview of all B2B levels that must be agreed upon, or that a mapping between different protocols must be realized for. On the lowest level, interoperability on the level of the transport protocol (e.g. HTTP, SMTP, X.400) must be reached. Using a message-oriented middleware approach in a B2B setting requires reliable messaging and additional messaging envelope mechanisms, e.g. SOAP. It should be noted that even if both business partners use SOAP, interoperability is not guaranteed, since they might use incompatible SOAP variants, e.g. ebXML SOAP vs. BizTalk SOAP.

The third level has to ensure interoperability on the syntax used to encode business documents, like XML or UN/EDIFACT. In this paper, we only consider XML-based middleware and ignore all other syntaxes as well as the document transport on the lower levels. XML provides syntax, not semantics, since tags have no predefined meaning [2]. The meaning of XML languages is defined by the document designer. This resulted in a proliferation of XML-based e-business vocabularies within the first few years of XML in existence [12]. Although we expect vocabularies disappearing

M.H. Shafazand and A M. Tjoa (Eds.): EurAsia-ICT 2002, LNCS 2510, pp. 339–349, 2002.

and merging, a certain number of well known "standard" vocabularies will co-exist. This means that on the fourth level business partners must either agree on a certain e-business vocabulary, or a mapping between their preferred vocabularies must be realized. We expect that companies will prefer a single interface that automatically maps to the different e-business vocabularies over implementing an interface to the in-house information system for each e-business vocabulary. Therefore, this paper emphasizes interoperability between different e-business vocabularies.

However, document types of e-business vocabularies are much too ambiguous including a lot of optionality and covering much more semantics than an involved application is able to process. In other words, a valid XML business document does not guarantee that the business partner is able to process the document. It has to follow the shared view of the business partners on the business content. This shared view on a document's semantics is an agreement to be met on the fifth level of interoperability. Implementing and maintaining these agreements - called message implementation guidlines (MIGs) in EDI - makes EDI expensive [10].

Level 6	Business Process Semantics	order management
Level 5	Document Semantics	purchase order
Level 4	E-Business Vocabulary (incl. generic document types)	xCBL order
Level 3	Transfersyntax	XML
Level 2	Messaging Envelope	SOAP
Level 1	Transport Protocoll	HTTP

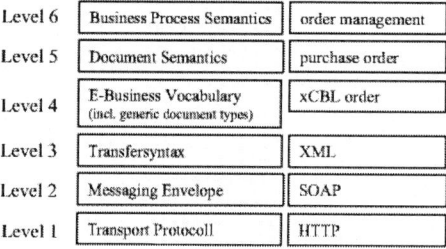

Fig. 1. Levels of Interoperability in B2B

Instead of standardizing voluminous and vague business document types, ebXML is based on unambiguous business collaborations that represent an agreement on the sixth level of interoperability. For this purpose, a clear choreography of business activities including an unambiguous and context-specific definition for the business content exchanged in each activity is defined. In other words, this paper concentrates on a framework for context-specific views into the core components-based ontology for each single business activity. Furthermore, the framework aims to automatically transform a corresponding view into various e-business vocabularies, provided an existing core component binding for the respective vocabulary.

The remainder of the paper is structured as follows: In Section 2 we give a brief introduction into the concepts we have adopted from related work, namely e-business vocabularies, ontologies, Open-edi, UMM, and ebXML. The main contribution of our work is elaborated in Section 3, where we present the necessary steps to integrage the basic ideas on ontologies into the ebXML framework. These steps are the following: definition of a document ontology, languange binding for e-business vocabularies, definition of contex-specific views into documents to support a specific business activity, and representing these views in different e-business vocabularies. Each of these steps is presented in its own subsection. The paper concludes with a short summary.

2 Related Work

The work presented in this paper does not by itself create any new B2B technology, rather does it interlink and coordinate already existing technologies to ensure B2B interoperability. This section introduces the key concepts the paper refers to.

E-business vocabularies: XML became the preferred way to exchange business data over the Internet. A lot of organizations developed their own vocabulary. Lacking guiding standards for interoperability, those solutions use different data structures and tagging to encode the same business concept. Although some of the vocabularies have become the first choice within a vertical, there are still some competing efforts. Popular e-business vocabularies include languages of market place providers, like Commerce One's xCBL and Ariba's cXML, the Open Application Group's OAGI to interconnect ERP systems, and domain-specific solutions like RosettaNet in the IT sector. Getting them all to interoperate is still a challenge for the B2B community. An overview of E-business vocabularies is provided in [13].

Ontologies: An ontology is defined by Gruber as a "formal, explicit specification of a shared conceptualization" [7]. According to this definition, even a DTD or an XML schema for a business document type can be regarded as a very primitive ontology.

However, both DTDs and XML schemas are basically just a set of terms and do not define relationships between different terms (cf. [6]). More advanced applications require more expressive ontology languages, like RDF/RDF schema [3], DAML+OIL [4], SHOE [8], or OML [15]. A lot of ontology approaches are directed towards the semantic web [1]. It is the goal to define rules and meanings of web data that precisely enough that machines can correctly interpret them. Similar to the problem of web data is that of e-business vocabularies. For interoperability of different vocabularies a shared set of terms and their interrelationships with a common understanding is needed. An approach to develop an ontology for business documents based on reverse engineering existing e-business vocabularies is described in [14].

Open-edi: The idea of separating the business semantics and its representation in a certain e-business vocabulary was alreay a key concept of the Open-edi initiative started in 1988. Open-edi distinguishes between a business operational view (BOV) and a functional service view (FSV). The BOV is defined as ´a perspective of business transactions limited to those aspects regarding the making of business decisions and commitments among organizations, which are needed for the description of a business transaction', while the FSV focuses on implementation-specific technological aspects of Open-edi. The Open-edi reference model [11], which became ISO standard 14662, guides B2B standard works to ensure the coherence and integration of related standardized modeling and descriptive techniques, services, service interfaces, and protocols.

UMM: UN/CEFACT's Modelling Methodology (UMM) is a modeling technique to describe the BOV aspects of Open-edi. The UMM meta model describes the business semantics that allows trading partners to capture the details for a specific business scenario using a consistent modeling methodology that utilizes UML [17]. A business process describes in detail how trading partners take on shared roles, relationships and responsibilities to facilitate interaction with each other. An interaction between roles follows a choreographed set of business transactions, whereby each transaction is expressed as an exchange of electronic business documents. Business partners will be able to communicate with each other if they support the same unambiguously defined choreography of transactions using unambiguously defined business document types.

ebXML: In order to provide an FSV layer that takes full advantage of the Open-edi concept and UMM, UN/CEFACT joined with OASIS in the ebXML initiative [5]. ebXML offers a modular suite of specifications. These specifications provide a standard method to exchange business messages, conduct trading relationships, communicate data in common terms, and define and register business processes [9]. In the context of this paper the ebXML specifications for business processes and core components are of particular relevance. The ebXML business process specification schema (BPSS) adopts a subset of UMM needed to configure ebXML-compliant software. An ebXML-compliant software will then be able to control a business process from the corresponding business partners view by monitoring state changes resulting from document exchanges. In ebXML, a document type does not correspond to the union set of all possibly required data structures needed for anyone's version of a given transaction type. A document is defined by an unambiguous data structure exactly meeting the business requirements to reach the business goals of a single activity in a business process. However, ebXML does not use its own e-business vocabulary to describe business documents. Instead, ebXML document types are assembled from so-called core components which are syntax-neutral descriptions of semantically meaningful business concepts. Currently, ebXML does not specify any methododology to represent syntax-neutral core components in targeted e-business vocabularies. Thus, the framework presented in this paper will complement the ebXML approach.

3 ebXML Core Component-Based Ontology Framework

In this section we present our framework to extend ebXML by ontology concepts. In order to develop a document ontology, two main approaches are introduced by Ontoprise's Semantic B2B Broker [16]: a *top-down* and a *bottom-up* approach. In a top-down approach business experts will first define a document ontology that describes their shared understanding of a business document type. The conceptual model of this document ontology builds the foundation to develop a new e-business vocabulary (represented as DTD or XML schema). Vice versa, a bottom-up approach takes DTDs or XML schemas from existing e-business vocabularies to analyze their semantic content. The result is an harmonized ontology of all considered e-business vocabularies have to be harmonized in order to define a unified document ontology.

The ebXML initiative is currently the strongest supported initiative by industry with respect to development of vocabulary-independent components, so-called core components. It is our goal to take advantage of a future pool of core components. Hence, our ontology layer is not defined by reverse engineering of existing e-business vocabularies like in the bottom-up approach. Instead, our ontology layer is based on ebXML core components. However, we do not use a pure top-down approach, because we will not develop a new e-business vocabulary. We have to mediate the ebXML-based ontology layer with existing e-business vocabularies. Since we are coming from top as well as from bottom, we call our approach "meet in the middle".

The presented framework is based on 4 major steps depicted in Fig. 2. The building of a document ontology starting from ebXML core components constituting the first one. The second step covers the definition of language bindings for various e-business vocabularies. The third step requires the definition of a view into the ontology that exactly meets the requirements of the document exchange supporting an ebXML business activity. Finally, the fourth step, which can be done automatically, takes on the language binding and the view specification and derives an implementation guideline in a certain e-business vocabulary. Each of these steps is introduced in the following subsections.

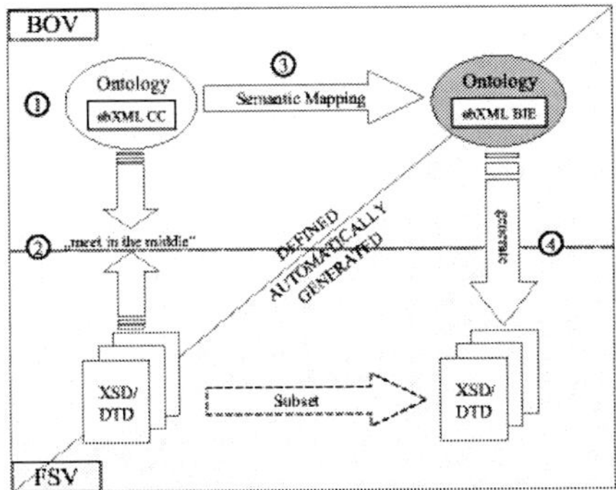

Fig. 2. ebXML Core Component-based Ontology Framework

3.1 Definition of a Core Components-Based Document Ontology

In the first step we develop an ontology that follows the latest draft of the ebXML core components specification [18]. For this purpose we have developed an RDF schema (RDFS) [3] for the core components meta model that is depicted as a graph in Fig. 3. Note that all boxes with solid lines are of the RDFS type Class and those with broken lines are of type Property.

A *core component* is defined as a semantic building block that is used as a basis to construct all electronic business messages. There exist 3 different types of core components. A *basic core component* represents a singular business concept with a unique business-semantic definition. Each basic core component is of a certain core component type. A *core component type* (e.g. amount type) consists of a content component that carries the actual content (e.g. 12) plus one or more supplementary components giving an essential extra definition to the content component (e.g. Euros). Note, that content and supplementary components are nothing else than core components. Core component types do not have business meaning. An *aggregate core component* is a bag of core components that convey a distinct business meaning.

In the RDF Schema in Fig. 3 the three different types *AggregateCoreComponent*, *Basic Core Component*, and *CoreComponentType* are represented as subclasses of the class *CoreComponent*. The property elementType is used to assign a core component type to a basic core component. The composition of a core component type is defined by the properties *contentComponent* and *supplementaryComponent*, referencing basic core components. The property coreComponentChild is used to reference the components within an aggregate.

Each ebXML core component contains the following dictionary information: A dictionary entry name is the unique official name of the core component. It corresponds to the RDFS property *label*. The definition of the unique semantic business meaning of the core component is given in the RDFS property *comment*. The property *remark* is used to further clarify the definition, to provide examples and/ or to reference a recognized standard. If there exist further synonym terms under which the core component is commonly known and used in the business, the property *businessTerm* is used to define them. We assign the properties *objectClass*, *representationTerm*, and *propertyTerm* to core components as defined in the ebXML specification.

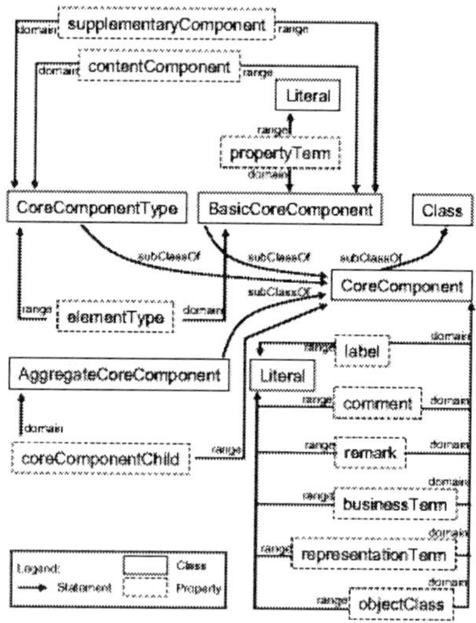

Fig. 3. RDFS meta model for CC

The RDF library of core components must be populated with all core components currently being developed by UN/CEFACT. Each core component will be expressed as a RDF model that follows the RDF (meta) schema of the core components' meta model. Fig. 4 demonstrates the population of the RDF library by the means of the aggregate core component *PostalAddress.Details*. Due to space limitations, the resource representing this aggregate core component is marked *#000005* in Fig. 4. This represents only the fragment identifier of a URI uniquely identifying core components. The unique identifications (UID) assigned to core components by ebXML are used as fragment identifier. The label of the resource is equal to the data dictionary entry name *PostalAddress.Details*. The comment states the definition of the PostalAddress.Details. The business terms *Address* and *Location* which are commonly used to refer to PostalAddress are assigned to the aggregate core component. Each of the components aggregated within PostalAddress.Details is assigned as coreComponentChild. Owing to space limitations we have detailed only the basic core component *#00027 Street.Name*. In addition to label, comment, remark and business term (*Road*), the basic core component gets assigned an object class

(*Street*), property term (*Name*), and a presentation term (*Name*). More important is the fact that a basic core component is of exactly one core component type. *Street.Name* is of type *#000090 Text.Type*. This core component type includes the content component Text.Content (which includes at one instance the name of the street) and the supplementary component Language.Code.

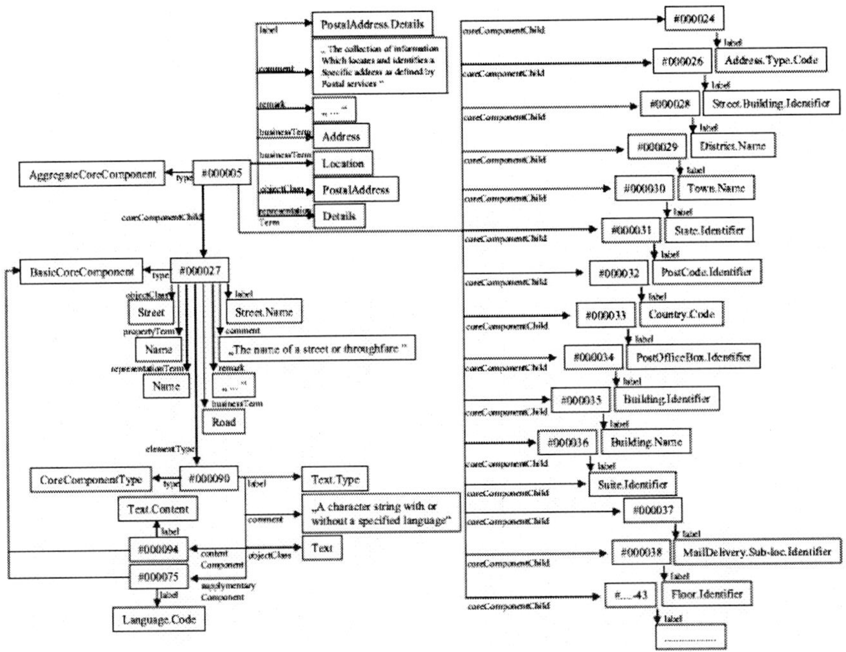

Fig. 4. RDF-Model for the Aggregate Core Component "Postal Address Details"

So far we have been considering the meta model of core components to capture the content of an ebXML core component library by the means of RDFS. A document ontology will use identified core components and their interrelationships as basic semantic building blocks for document types. The document ontology is also expressed in RDFS. Consequently, document instances are valid RDF models of the document's RDFS. Fig. 5 depicts an example of a document instance that is a valid fragment of a postal address representing the German (ISO Code 936: *DE*) name of the street *Liebiggasse*. For a better understanding, we have marked the anonymous resources in the grey boxes with a meaningful name. Each of these resources is an instance of a core component. The corresponding core component is referenced as the resource's type. This allows to identify the semantic context of the resource. Consequently, the semantic relationship between core component instances is comprehended even using an anonymous referencing mechanism via the property "references" between all instances of any types of core components.

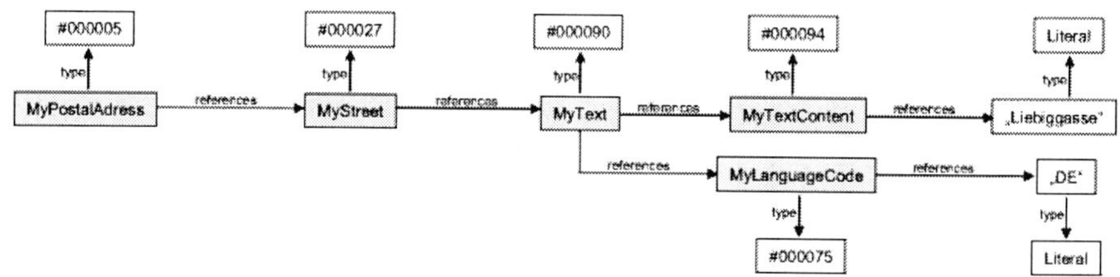

Fig. 5. RDF Instantiation using Aggregate Core Component "Postal Address Details"

3.2 Language Binding for E-business Vocabularies

Having defined a document ontology, the next step is to provide language bindings between the document ontology and the corresponding document type of an e-business vocabulary. Since our "meet-in-the-middle" approach and a bottom-up approach only differ in the way the ontology is built, the problem of defining a language binding remains the same. Thus, our framework considers the language binding defined in the bottom-up approach by Omelayenko and Fensel [14].

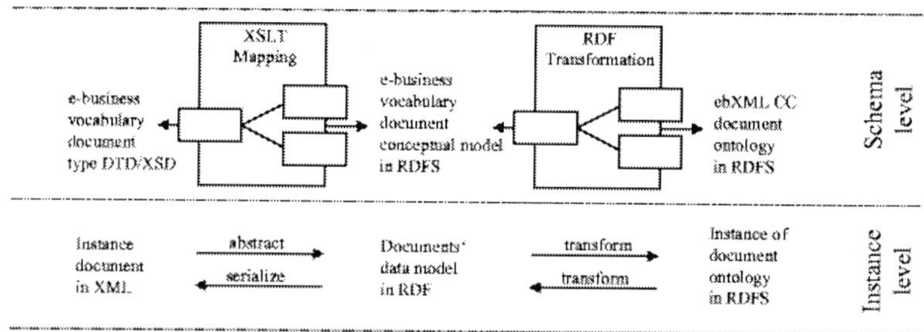

Fig. 6. Language Binding between Ontology and e-business vocabulary

The basic concept of this language binding is depicted in [6] It is based on the definition of a common conceptual model for document types of different e-business vocabularies. The conceptual models are described in RDFS. Thus, a mapping between the vocabulary's DTD or XML schema on the one side and the conceptual data model on the other side is required. This mapping on the schema level is defined by the means of XSLT. On the instance level, an incoming document is abstracted from its XML serialization and translated into its RDF data model. Vice versa, in order to create an outgoing document, the RDF data model of the target vocabulary is serialzed according to the target XML format.

Furthermore, the conceptual data model of an e-business vocabulary's document type (expressed in RDFS) must be mapped to the document ontologies data model of the same document type (also expressed in RDFS). This means that the equivalence mapping between the terminologies requires a transformation from one RDFS model to the other. The mapping must be described by the means of a RDFS mapping language.The mappings must be automatically translated into an RDF transformation language. This transformation language is applied to translate the conceptual RDF model of an incoming document into an instance of the document ontology and, vice versa, to translate an instance of the document ontology into the conceptual RDF model of an outgoing document. It follows that all mappings between different vocabularies will also be managed via mapping each standard to the core components-based document ontology. Unfortunately, there neither does exist a standard for the RDFS mapping language nor for the RDF transformation language. The specification of these languages to support the overall framework is an essential future work item.

3.3 Context-Specific View Definitions

If business partners were able to process all the semantics that are usually included in a business document type of a standard vocabulary, step 2 would already be the final step. Consider the fact that e.g., flattening an xCBL purchase order will result in about 16,000 data element types covering most probably a similar number of semantic concepts. Then it becomes evident that a company usually supports only a subset of the concepts or in other words a specific view of the general standard document.

Hence, exchanging valid XML documents is not enough to ensure interoperability, because these documents might carry data representing business concepts not supported by the business partner. It is an indispensable requirement that business partners share a common view on the semantic concepts included in a document type to collaborate in a B2B transaction. In the context of ontologies, a view setting approach as defined by Ontoprise's Semantic B2B Broker is required [16]. The view setting is used if different users have to be provided with different views on the same information source (= document type). In this case, the ontology of the information source may be semanically restricted to the conceptual view shared by all business partners.

According to UMM, the information requirements of each business activity resulting in a document exchange are analyzed to support the overall business process. The information requirements result in an unambiguous set of concepts that must be supported in the exchanged document type. Unambiguous means that there is no open space for partner-specific views agreed on at run time. The semantic concepts to be shared by all business partners supporting a business process are fixed at design time.

ebXML uses UMM artifacts to define a corresponding conceptual data model based on core components. When a core component is used in the context of a business activity, its refinement becomes a so-called *business information entity*. A business information entity is a piece of data or a group thereof with a unique business semantic definition in a given context. Accordingly, a language to define refinements on a semantic level is necessary. For this purpose, ebXML specifies a constraint language. The scope of the constraint language is to refine an assembly as appropriate. As depicted in Fig. 7, the document ontology representing the general document structure based on core components (expressed in RDFS) must be refined into an activity-specific document structure based on business information entities (also expressed in RDFS). Thus, a language allowing for a refinement of an RDFS must be devloped. This language should follow the semantics captured by ebXML's constraint language.

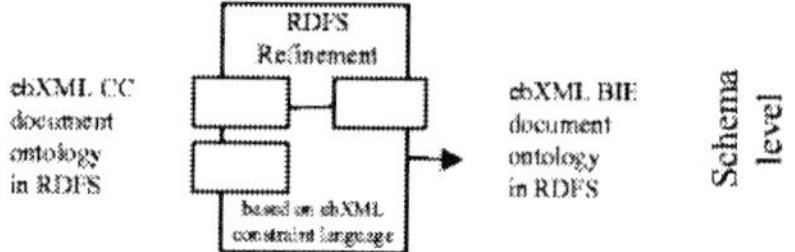

Fig. 7. Document ontology refinement

3.4 Representing Views in E-business Vocabularies

The last step of the framework covers the definition of unambiguous document types in the XML syntax of e-business vocabularies. The document ontology based on business information entities to support a specific activity is more restrictive than the general document ontology. Consequently, the DTD or XML schema for the e-business vocabulary supporting a well-defined activity must be more restrictive than the general DTD or XML schema of the corresponding document type. Taking the example of the previous subsection, a certain step in a well-defined business activity will use only a well-defined subset of the 16,000 data elements in a general xCBL purchase order.

This time the mapping from the document ontology into the e-business vocabulary's DTD or XML schema should be done automatically. This will be enabled by the information about the mapping of the gerneral document type (step 2) as well as the information about the constraint language (step 3). We expect two major types of refinements in regard to their consequences on the automatic mapping. Firstly, some semantic concepts are not used at all. In this case applying the same RDF transformation language as in step 2 will result in the desired output. This is due to the fact that missing input will simply be ignored. Secondly, a component is used differently in different situations according to more complex constraints. This can be expressed neither in the document's DTD nor XML schema. Thus, it does not have any consequences on the output of an appropriate DTD or XML schema. However, these more complex constraints should be expressed as an declarative (XML-based) language which accompanies the document type [10]. In order to check the validity of a document for a given business activity, not only conformance to the document type, but also conformance to the rules of an instance of the declarative constraint language is required.

4 Summary

In this paper we presented an ebXML core component-based ontology framework to be used in B2B e-commerce. It is based on the idea of Open-edi to separate a business operational view represented by the document ontologies and a functional service view represented by the XML schemas or DTDs of business vocabularies. It specifyies four major steps to describe semantically equivalent document types in different e-business vocabularies for the same activity in a well defined business process.

Each of these steps is described in this paper on a conceptual level and directs our future research towards the implementations of these steps. In the first step the semantics of ebXML core components are defined in a RDFS-based ontology. The second step defines a mapping between the ontology and RDF representations of e-business vocabularies as well as another one between the latter RDF representations and the DTDs or XML schemas of the e-business vocabularies. Beyond traditional ontology approaches, we take on the ebXML idea to further restrict a general document ontology to the specific needs of a certain business activity. This refinement will lead to a certain view of the general document ontology and must be specified by means of a constraint language. Steps 1 to 3 have to be done manually, whereas the last step will be derived automatically. Using the information of the previous steps a more restrictive and appropriated XML schema or DTD for a specific activity will be created. The XSDs and DTDs are a subset of their corresponding general ones. Further restictions expressed in a declarative language will accompany the document types. This framework focuses both, a mapping between different e-business vocabularies and at the same time guaranteeing their semantical equivalence in support of a specific activity.

References

1. Berners-Lee, T., Hendler, J., Lassila, O.: The semantic Web. Scientific American (2001)
2. Bosak, J.: Media-Independent Publishing: Four Myths about XML. IEEE Computer, Vol. 31, No. 10 (1998)
3. Brickley, D., Guha, R.: RDF Vocabulary Description Language 1.0: RDF Schema, W3C Working Draft. (2002), http://www.w3.org/TR/2002/WD-rdf-schema-20020430
4. DARPA: The DARPA Agent Markup Language Homepage. http://www.daml.org
5. ebXML: Homepage of the ebXML Initiative. http://www.ebXML.org
6. Erdmann, M., Studer, R.: Ontologies as Conceptual Models for XML Documents. 12th Workshop on Knowledge Acquisition, Modeling and Management, Canada, (1999)
7. Gruber, T.R.: A Translation Approach to Portable Ontology Specifications. Knowledge Acquisition Vol. 6, No. 2 (1993)
8. Heflin, J., Hendler J.: A Portrait of the Semantic Web in Action. IEEE Intelligent Systems, Vol. 16, No. 2 (2001)
9. Hofreiter, B., Huemer, C., Klas, W.: ebXML: Status, Research Issues and Obstacles, Proc. of 12th Int. Workshop on Research Issues on Data Engineering (RIDE02), San Jose (2002)
10. Huemer, C.: <<DIR>>-XML^2 -Unambiguous Access to XML-based Business Documents in B2B E-Commerce. Proc. of 3rd ACM Conference on Electronic Commerce, Tampa (2001)
11. ISO: Open-edi Reference Model. ISO/IEC JTC 1/SC30 ISO Standard 14662 (1995)
12. Kotok, A.: Extensible and More - A Survey of XML Business Data Exchange Vocabularies. O'Reilly xml.com, http://www.xml.com/pub/2000/02/23/ebiz/index.html
13. Li, H.: XML and Industrial Standards for Electronic Commerce. Knowledge and Information Systems, Vol. 2, No. 4 (2000)
14. Omelayenko, B., Fensel, D.: Scalable Document Integration For B2B Electronic Commerce. submitted to Electronic Commerce Research Journal
15. Ontology Consortium: Ontology Markup Language Project. http://www.ontologos.org/OML/OML.html
16. Ontoprise, B^3 Semantic B2B Broker, http://www.ontoprise.de, 2000
16. UN/CEFACT TMWG, UN/CEFACT Modelling Methodology, http://www.gefeg.com/tmwg
18. UN/CEFACT ebTWG, Core Component Technical Specification, Version 1.8, (2002) http://www.ebtwg.org/projects/documentation/core/CoreComponentsTS1.80.pdf

An Agent Based Service Discovery Architecture for Mobile Environments

Zhou Wang[1] and Jochen Seitz[2]

[1]Institute of Telematics, University of Karlsruhe
D-76128 Karlsruhe, Germany
Zhou.Wang@tm.uka.de
[2] Institute of Communications Technology and Measurement
Ilmenau Technical University, D-98693 Ilmenau, Germany
Jochen.Seitz@tu-ilmenau.de

Abstract. The rapid expansion of networked services and wide deployment of easily accessible wireless Internet connection pave the way to access information and services using portable devices in mobile environments. However, the inherent resource shortage of mobile devices and heterogeneous environments prevent mobile users to utilize services freely and effectively. This paper presents a mobile agent based service discovery architecture CHAPLET which provides a general-purpose, integrated and efficient way to locate services in nomadic environments. The proposed architecture supports uniformed query representation, asynchronous service lookup, flexible service selection, and transparent interoperation with different heterogonous discovery protocols, while taking the constraints of wireless communication and portable devices into consideration.

1 Introduction

The advances of network technology and decreasing cost of computer hardware promote the wide deployment of computational resources within the network. Internet is evolving from a global communication backbone to an open information infrastructure providing various types of resources and services, from simple network enabled printers to more sophisticated application-level services, e.g. multimedia services, and online financial services. Recently, the trends in network-enabled information appliances boost the growth of services available on the network.

In order to use various services on the network, the first necessary step is to find the exact address of services. As the number and diversities of services on the network increase greatly, certain service discovery mechanisms become an important foundation for users to gain access to desired services. Service discovery is especially important as computers become more portable. On the one hand, with the wide deployment of wireless or mobile communication network, e.g. wireless LAN, GSM, or UMTS in future, connection to Internet is no more an arduous task. Mobile users desire to access networked information and services anytime anywhere. On the other hand, mobile users may change their geographic locations over time. Consequently, the set of services available on the network would vary drastically with location and time. They need additional support to discover what services are available, as they come into an unknown environment.

M.H. Shafazand and A M. Tjoa (Eds.): EurAsia-ICT 2002, LNCS 2510, pp. 350–357, 2002.

In this paper we proposed a mobile agent based service discovery architecture CHAPLET which provides a general-purpose, integrated, and efficient way to discover services in nomadic environments. The CHAPLET architecture enables mobile users to transparently interact with heterogeneous services discovery mechanisms, while taking care that they use a uniformed means to represent queries. Moreover, mechanisms for asynchronous service lookup, flexible service selection, and security control are integrated in the CHAPLET architecture.

This paper is organized as follows. In section 2 we reviewed existing service discovery approaches and their suitability for nomadic users. Then we presented the mobile agent based service discovery architecture CHAPLET, followed by the description of system design and implementation. Finally we summarize the paper and give an outlook to future work.

2 Analysis of Existing Service Discovery Mechanisms

Until now, a number of service discovery solutions are developed. These solutions range from hardware-based technologies such as Bluetooth SDP, to single protocols, e.g. SLP and SDS, to frameworks such as UPnP, Jini and Salutation. Due to the space limitation we could not go to insight of these approaches. Instead, we just identify some technical highlights of several typical solutions, to illustrate that each approach has both strength and weakness.

SSDP. The UPnP Simple Service Discovery Protocol (SSDP) [6] is one of typical protocols employing announcement/listen model to discover services without any directory server. The primary advantage of SSDP is that it needs "zero" or little configuration and administration. The weak side of SSDP includes lack of a query facility that can search for services by attributes.

SLP. The Service Location Protocol (SLP) [7] from IETF is an example of central directory based solution. Service registration and lookup are performed through UDP-based unicast communication between UAs/SAs and DAs. The service description and search are based on name-value paired strings. Compared to the SSDP, the SLP has better scalability, which can operate in networks ranging from a single LAN to enterprise networks.

Jini. Sun's Jini provides a simple infrastructure for creating spontaneous interaction between devices and services [15]. Jini resembles SLP both in principle and in architecture, but it is tightly bound to the Java environment. The main strength of Jini is its ability to move code across the network. In Jini, services provide a proxy (service interface) during registration which will be moved to the client side as the service is invoked, thus it eliminates the necessity of pre-installing drivers on the client.

SDS. The Service Discovery Service (SDS) developed in Berkeley [5, 10] is a typical example of distributed directory based architecture. The significant feature of SDS is the hierarchical structure with lossy aggregation to achieve better scalability and reachability. Besides, the security mechanisms in SDS are also meaningful. However, the SDS is more favorable for applying in stationary network environments, since it requires additional overheads to maintain the hierarchical structure and to propagate index updates.

Other service discovery solutions include Salutation[14], Bluetooth SDP[3], HAVi[9], INS[1],and VIA[4]. Despite that most of them provide similar functionality, namely automatically discovering services based on service characteristics, they have different features and are not compatible with each other. The reasons for this undesirable diversity include the built-in characteristics of the application environment, inherent technical challenges and motivations to share market. By the same reasons, these diverse approaches will continue to exist for a long time. However, the incompatibility of the approaches prevents mobile users to discover and utilize networked services freely and conveniently. Mobile users may have to simultaneously use multiple service discovery protocols with considerable configuration overhead. For example, they might have the SLP protocol for daily use, but have to install the UPnP protocol which is commonly used when they have a meeting, and other protocols which are possibly needed during the business trip, since they do not know what protocols may be applied in an unknown environment. Moreover, these approaches have different ways to describe service characteristics, e.g. SSDP and SDS use XML, and SLP uses string encoded name-value pairs. Users are forced to learn different description means, in order to interact with different approaches. To know so many technical details is burdensome, if not impossible, for end users who usually have no special technical knowledge.

Apart from the heterogeneous environments, most of the existing approaches rarely take the issues of thin client and poor wireless link into consideration. For example, synchronous operation is one of the intrinsic natures of most existing service discovery approaches. This property is either revealed explicitly by protocol specifications, such as SLP, or implied by the defined APIs, e.g. Jini's RMI-based APIs, and Salutation's RPC-based APIs. Although synchronous operation simplifies protocol and application design, it is fastidious for mobile environments. The unexpected but frequent disconnections and possible long delay of wireless link greatly influence the usefulness and efficiency of synchronous calls.

These obstacles that mobiles users are facing motivate us to develop the CHAPLET system, which will be described in next sections.

3 Service Discovery with Agents – The CHAPLET Architecture

In recent years, mobile agent is treated as an innovative concept for creating distributed systems. Plenty of applications such as distributed information retrieval and E-commerce benefit from using mobile agents [11]. In [16,17] we introduced a mobile agent based service model to access resources and services in nomadic environments. From our point of view, mobile agents technology is not only feasible to surmount the above difficulties, but also useful to enhance search functionality and flexibility.

3.1 Design Goals

As design goal, we aim at providing a general-purpose integrated platform to enable mobile users easily and efficiently to discover services in nomadic environments.

Asynchronous Search. Compared to synchronous mode, asynchronous operation seems more attractive for mobile users. Clients neither await results, nor continuously keep the active connection, after they sent requests. The results will be sent back to clients when they connect into the network later. Such loose coupling relaxes the communication restrains, and thus improves efficiency and flexibility.

Uniformed Query Representation. Mobile users desire to have a uniformed and simple means to describe search requirements, without to be involved in different description methods supplied by underlying protocols. Moreover, they expect a more expressive and flexible query mechanism, in order to support inexact or dynamic search queries to a certain extent, such as search for a color printer if available, otherwise a black-and-white printer; or find a printer with the shortest print queue.

Interoperability. The incompatibility of different architectures is one of the biggest obstacles for effective service discovery. Although some solutions have been proposed to bridge different mechanisms, they are limited to pair-wise bridges, such as Jini to SLP [2,8], Salutation to Bluetooth SDP [13]. To the best of our knowledge, universal, multi-way interoperability solutions are not available yet. Rather than those individually developed, isolated extensions, we in this work aim to provide a common integrated platform to enable interoperability between different protocols.

Generality. The CHAPLET system is oriented to develop a general-purpose service discovery architecture. Hence, we neither impose any assumptions on service types nor limit clients to search for only certain types of services.

3.2 Overview of the CHAPLET System

Before we go insight into the CHAPLET architecture, some terms are defined as follows:

Service Domain (SD): a set of service offers that are governed by an instance of a particular service discovery protocol. A service domain is associated with one, and only one protocol. For example, in Fig. 1, the UPnP SSDP protocol is applied in the service domain A, while service domain B is based on the SLP protocol.

Autonomous Domain (AD): a collection of service domains that are under the authority of a particular administration. The concept of autonomous domain comes from the real life, where services and service domains are offered by some kind of organizations. The administrator wants to impose certain control policies on services that pertain to service domains within the autonomous domain.

The conceptual architecture for discovering services in CHAPLET is given in Fig. 1.

3.3 The Service Discovery Procedure

The mobile host connects into the wired network via a particular access point. The selection of an appropriate access point and hand-offs between different access points are handled in the data link layer and should not be considered in the CHAPLET.

Once a mobile user requests services in wired networks, he specifies query requirements using XML-based query schemes, and sends them together with agent chaperon into the wired network (❶ in Fig. 1.) The chaperon is prepared by the client on portable device, and sent into the wired network. The task of chaperon is to act on behalf of the mobile client to retrieve services which satisfy the needs of the client.

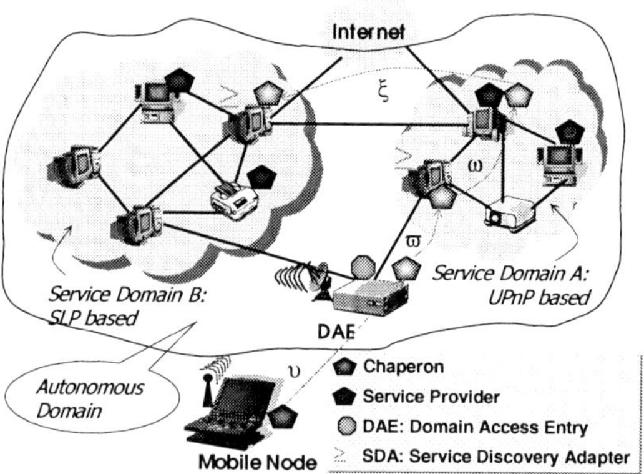

Fig. 1. Service Discovery Procedure in CHAPLET

The chaperon firstly locates the node where the agent Domain Access Entry (DAE) is running. The agent DAE, similar to access point in the data link layer, is the rendezvous point for mobile clients to interact with the autonomous domain. As a part of infrastructure services, the DAE usually resides statically on nodes in the autonomous domain. The DAE is adopted for two purposes: admission control and configuration management. The agent DAE authenticates the client firstly, and sends to the chaperon a list of service domains which the client is allowed to access if the authentication is successful. Along with the access list, an access ticket is returned, which is used to show that the chaperon is already certificated by the DAE. The access ticket has a validity period. Before it expires, the chaperon should ask the DAE to renew it.

Having the configuration of currently available service domains in mind, the chaperon designs an itinerary for traveling through them, and then starts to search for services. For each service domain, there is at least one node which is furnished with Service Discovery Adapters (SDAs). The SDA is a bridge between the chaperon and the service domain it is visiting.

The chaperon begins with the first service domain listed in its itinerary. It migrates to one of nodes equipped with an SDA in that service domain (❷ in Fig.1.). The chaperon passes the search predicates to the SDA. Upon receiving the query, the SDA transforms the XML query scheme into the form supported by the local service domain. Then the SDA calls APIs, which are specialized for the local service domain, to search for services that match the predicates. After the chaperon gets search results from the SDA, it might require the SDA to retrieve service descriptions of particular services. The reason is that some discovery protocols, e.g. UPnP, support only very limited search ability. The chaperon has to match the search predications with the given service description by itself, in order to compensate the insufficiency of these protocols. Moreover, if the chaperon thinks that the service description is inadequate to make a decision, it migrates to the service node that is of interest (❸ in Fig.1.), and negotiates there with the service provider about further service details. In this way, the chaperon is able to get to know more about services, especially some performance parameters or attributes that are changed rapidly.

After the chaperon completes the search process in a service domain, it migrates into the next one specified in the itinerary if it wants to continue (❹ in Fig.1.). After the chaperon has visited all service domains, it collates the retrieved services and sends the most appropriate services to the client on the mobile node.

So far, the service discovery procedure is accompanied through cooperation of three fundamental components: the chaperon is the representative of mobile client, the DAE works as an rendezvous point in the autonomous domain, and SDAs execute the service discovery task in particular service domains.

4 System Design and Implementation

4.1 Query Representation

In CHAPLET, we rely on XML to describe client queries. The flexible hierarchical structures and semantic-rich content of its self-describing syntax makes XML the good choice for describing and disseminating queries.

For each service type there is a predefined query schema based on W3C's XML Schema Definition. The query schema defines a set of rules to dictate which service attributes can be applied in the query. In Fig. 2.a, we illustrate a part of schema definition for printer service type.

Using schema-aware XML input tools, clients can easily specify the service-specific queries. Before queries are submitted to the chaperon, they are validated to check the conformance to the particular query schema. In Fig. 2.b, a simple example is shown. This example indicates such a query that the user wants to search for a printing service that accepts PS- or PDF-format document, and uses laser printer as output device. These two conditions are regarded as mandatory by setting the tag attribute "Priority" to "0". In contrast, other service characteristics such as "Color", "Location" and "Price" are specified as optional. The given priorities are used to select the most appropriate service.

```
<?xml version="1.0" encoding="utf-8"?>
<xsd:schema targetNamespace="http://tm.uka.de/chaplet/xml/Printer">
<xsd:element name="Query" type="prn:PrinterQueryType"/>
<xsd:complexType name="PrinterQueryType">
 <xsd:all>
   <xsd:element name="ServiceType" type="xsd:string" fixed="Printer"/>
   <xsd:element name="PrinterType" type="prn:PrnList"  minOccurs="0"/>
   <xsd:element name="DocumentType" type="prn:DocList"  minOccurs="0"/>
   <xsd:element name="Color" type="lib:TypeBoolean" minOccurs="0"/>
   <xsd:element name="Price" type="lib:IntRange" minOccurs="0"/>
 </xsd:all>
</xsd:complexType>
</xsd:schema>
```

a) The Query Schema Definition of Printer Service

```
<?xml version="1.0" encoding="UTF-8"?>
<Query xmlns=http://tm.uka.de/chaplet/xml/Printer
     xsi:schemaLocation="http://tm.uka.de/chaplet/xml/Printer prn.xsd">
  <ServiceType>Printer</ServiceType>
  <PrinterType Priority="0">Laser</PrinterType>
  <DocumentType Priority="0">PDF PS</DocumentType>
  <Color Priority="10">true</Color>
  <Price Priority="5">
     <lib:Minimum>1</lib:Minimum>
```

```
    <lib:Maximum>10</lib:Maximum>
  </Price>
</Query>
```

b) A Query Example of Printer Service

Fig. 2. Examples of Query Representation in CHAPLET

4.2 System Architecture

As seen in Fig. 3., the CHAPLET architecture needs an agent platform. Rather than develop an own agent system, we prefer to let CHAPLET run on existing agent platforms. Now the CHAPLET architecture builds on IBM Aglet System.

The CHAPLET architecture consists of the runtime system, system agents and user agents. The runtime system provides fundamental common functions for system agents and user agents. For example, the communication manager deals with issues related to possible unpredictable disconnections of wireless link. The event manager is a general asynchronous event publish/subscribe mechanism serving as a basic communication means between agents and the CHAPLET runtime system. Besides user authentication, the security manager applies encryption and digital signatures to protect the data carried by chaperon from modification by malicious hosts.

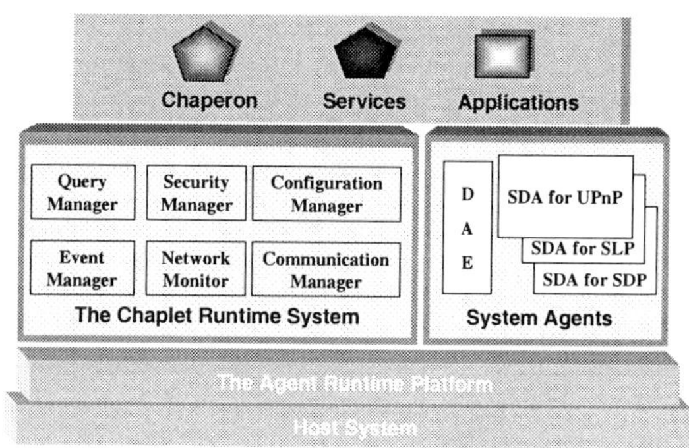

Fig. 3. The CHAPLET System Architecture

The system agents include Domain Access Entry(DAE) agents and Service Discovery Adapter(SDA) agents. The SDA translates the uniformed query language into the specialized representation of particular service discovery protocol, and then performs search process according to specification of the protocol. Hence, the SDA is protocol dependent.

5 Conclusions

In this paper we presented a mobile agent based service discovery architecture CHAPLET to facilitate mobile users easily, efficiently and effectively to locate the required services in networks. We believe that agent technology provides a feasible approach for accessing services in nomadic environments. The CHAPLET architecture contains a number of distinguished features for nomadic service discovery: it enables asynchronous service lookup to alleviate the influence of narrow vulnerable wireless link to a great extent; it provides a common middleware platform which makes clients use an uniformed means to transparently interact with heterogeneous service discovery protocols. Additionally, security control mechanisms to prevent misuse and overloading are integrated in CHAPLET.

In future, we will develop further SDAs for other popular protocols, and extend the CHAPLET architecture to locate services across autonomous domains. Besides, the mechanisms supporting mobility specific services such as location based services and personalized services will be integrated in CHAPLET.

References

[1] W. Adjie-Winoto, E. Schwartz, H. Balakrishnan, and J. Lilley. *The design and implementation of an intentional naming system.* Proceedings of the 17th ACM Symposium on Operating Systems Principles (SOSP '99), December 1999.

[2] M. Bathelt and J. Nickles, *Das Plug and Play der Automatisierung*, Elektronik, Vol. 6, March 1999, pp 54–62.

[3] Bluetooth SIG, *Specification of the Bluetooth System – Core*, Part E, February 2001.

[4] P. Castro, B. Greenstein, R. Muntz, P. Kermani, C. Bisdikian and M. Papadopouli. *Locating Application Data Across Service Discovery Domain*, 7th Annual Int. Conf. Mobile Computing and Networking, Rome, Italy, July 2001.

[5] S. Czerwinski, B. Zhao, T. Hodes, A. Joseph, R. H. Katz, *An Architecture for a Secure Service Discovery Service*, Proceedings of The Fifth ACM/IEEE International Conference on Mobile Computing, Seattle, WA, August 1999, pp. 24–35

[6] Y. Goland, T. Cai, P. Leach, et al. *Simple Service Discovery Protocol*, IETF Draft, draft-cai-ssdp-v1-03.txt, October 1999.

[7] E. Guttman, C. Perkins, J. Veizades, and M. Day. *Service Locaiton Protocol, Version 2, IETF, RFC 2608*, http://www.ietf.org/rfc/rfc2608.txt, June 1999.

[8] E. Guttman and J. Kempf, *Automatic Discovery of Thin Servers: SLP, Jini and the SLP-Jini Bridge*, Proceedings of 25th Annual Conference of IEEE Industrial Electronics Society (IECON'99), Piscataway, USA, 1999.

[9] HAVi Consortium, HAVi Specification V1.0, www.havi.org, 2000.

[10] T. D. Hodes, S. E. Czerwinski, B. Y. Zhao, A. D. Joseph, R. H. Katz, *An Architecture for Secure Wide-Area Service Discovery*, ACM Wireless Networks Journal, March-May 2002, Vol. 8, Nr 2-3, pp 213–230.

[11] D.B. Lange, M. Oshima, "Seven Good Reasons for Mobile Agents", *Communication of the ACM 42(3)*, March 1999, pp. 88–89.

[12] Microsoft. *Universal Plug and Play Device Architecture, version 1.0*, June 2000.

[13] B. Miller and R. Pascoe, *Mapping Salutation Architecture APIs to Bluetooth Service Discovery Layer*, http://www.bluetooth.com, July 1999.

[14] The Salutation Consortium. *Salutation Architecture Specification, Version 2.1*, 1999.

[15] Sun. *Jini Technology Core Platform Specification.* Sun Microsystems, Inc, October 2000.

[16] Z. Wang and J. Seitz, *An Agent-based Distributed Service Model for Nomadic Users*, The Eighth International Conference on Parallel and Distributed Systems, June 2001, Korea.

[17] Z. Wang and J. Seitz, *Accessing Distributed Services in Nomadic Environments*, The Thirteenth IASTED International Conference on Parallel and Distributed Computing and Systems, August 2001, Anaheim, USA.

Location Management Using Multicasting HLR in Mobile Networks

Dong Chun Lee

Dept. of Computer Science Howon Univ., Korea
ldch@sunny.howon.ac.kr

Abstract. We propose mobile strategy called Multicasting Home Location Register (MHLR) scheme. When a call is established, MHLR records the caller's VLR ID according to the callee. Periodically, MHLR ranks the VLRs and determine which VLRs frequently make calls to the callee. During a location registration process, MHLR sends the terminal's location information to the determined VLRs and it can eliminate HLR queries from the call tracking. The proposed method distributes messages to VLRs and effectively reduces mobility management cost. The performance analysis proves that the proposed strategy is superior to the Interim Standard 41 (IS-41) scheme.

1 Introduction

A major problem in the mobile networks(MN) is how to locate mobile callees. This is named the location management issues. The standard commonly used in North America is the IS-41 scheme, and in Europe the GSM [2]. Those standards use the two-tier database system of home location register (HLR) and visitor location register (VLR). IS-41 and GSM have a structural drawback: as the number of users increase, HLR becomes the bottleneck problem.

Many papers in the literature have demonstrated that the IS-41 scheme does not perform well. This is mainly because whenever a mobile callee moves, the VLR of a Registration Area (RA) which detected the arrival of the callee always reports to the HLR about the callee's new location. For convenience in presentation, in the rest of the paper we will simply regard VLR as a general term which represents the local hardware/software system managing mobile clients information within a RA. While a call is placed, the callee is also located by going to the HLR's database to find the callee's new location. As the HLR could be far away from a VLR, communication to the HLR is costly. Several other location management strategies have been proposed to improve the performance of the IS-41 scheme.

Among them, the Forwarding Pointer (FP) strategy and the Local Anchoring (LA) method are introduced in [1][5]. Under these schemes, signaling traffic due to location registration is reduced by elimination the need to report location changes to the HLR. Location update and paging subject to delay constraints is considered in [8]. When an incoming call arrives, the residing area of the terminal is partitioned into a number of sub-areas, and then these sub-areas are

M.H. Shafazand and A M. Tjoa (Eds.): EurAsia-ICT 2002, LNCS 2510, pp. 358–365, 2002.

polled sequentially. With increasing the delay time needed to connect a call, the cost of location update is reduced. A queuing model of three-level hierarchical database system is illustrated in [10]. These schemes can reduce both signaling traffics due to location registration and call tracking using the properties of local mobility. As for the LA and FP schemes, they are cost effective in reducing the HLR access traffic. However, there is a trade-off between the registration cost and call tracking cost.
In this paper, we propose mobility strategy where it is multicasting home location register (MHLR) which exploits receiver side call locality.

2 MHLR Scheme

The location information of terminal is multicasting from a HLR to some VLR's chosen for the location management. When a call is established, MHLR records the caller's VLR ID according to the callee. Periodically, MHLR ranks the VLRs and determine which VLRs frequently make calls to the callee. During a location registration process, MHLR sends the terminal's location information to the determined VLRs and it can eliminate HLR queries from the call tracking. The proposed method distributes messages to VLRs and effectively reduces mobility management cost. Fig. 1 and Fig. 2 show the major steps of the mobility management scheme.

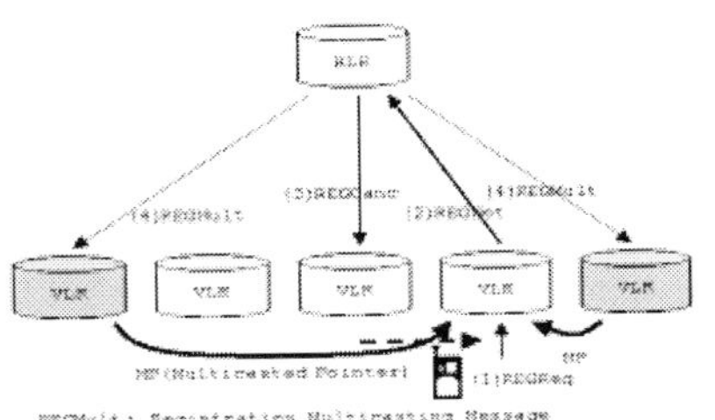

Fig. 1. Location registration (When previous VLR is not multicasting, and m=2)

Fig. 2. Call tracking (When multicasting information of the RT exists in query result

Here **m** represents the number of VLR to multicast the location information of terminal. We outline the major steps of the MHLR scheme location registration as follows:

1. The registration request(REGREQ) message from a terminal is transferred to the VLR that messages a new registration area(RA).
2. VLR transfers the registration notification(REGNOT) message to HLR and HLR performs the query by number of the received terminal(RT).
3. If previous VLR of terminal is not multicasting,
   The registration cancellation(REGCANC) message is transferred to previous VLR of terminal.

The registration multicasting message (REGMULT) is transferred to the multicast VLRs.
4. Else previous VLR of terminal is multicasting,
The REGMULT message is transferred to the multicast VLRs.

And the call tracking is outline as follows:

1. The sending terminal requests a call to the VLR. The VLR performs the query by number of the RT.
2. If a RT exists in query result, TLDN is assigned to the RT via MSC.
3. Else if the multicasting information of the RT exists in query result,
The routing request(ROUTREQ) message is transferred to the VLR of the RT. The VLR of the RT assigns temporary local directory(TLDN) via MSC, and transfers to the VLR of sending terminal.
4. Else if the multicasting information of the RT doesn't exists in query result,
The VLR of sending terminal transfers LOCREQ message to HLR. The HLR find the VLR of receiving terminal by query and transfer ROUTREQ message. The VLR of receiving terminal assigns TLDN via MSC and transfers it to HLR. The HLR transfers TLDN to the VLR of sending terminal.

In the MHLR scheme, it is explained how to determine the VLR to receive multicasting information **m**. The conceptual structure of CL field is depicted in Fig.3 (a).

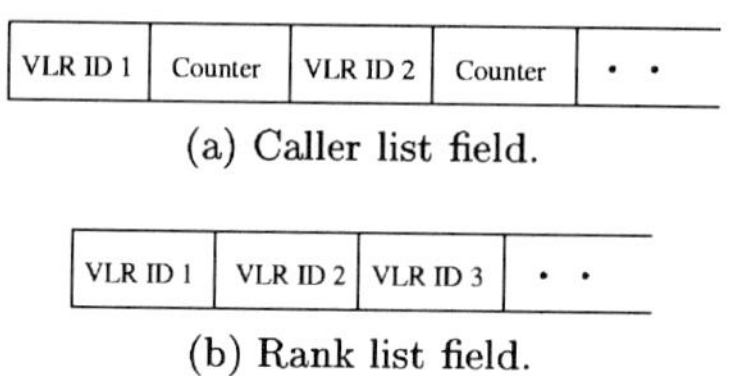

Fig. 3. Conceptual structure of the CL field and RL field.

The CL field consists of the pairs of the ID of frequently calling VLR and number of calls from it. Every established call adds a new pair or increases the number of existing calls. The CL fields are distributed over the HLR and multicast VLR recording the established calls, and merged into HLR periodically to construct the RL field. In Fig.3 (b), each element of RL field is an ID of VLR and the ID of VLR calling more frequently course before. The RL field exists only in the user profile of HLR while the CL field is dispersed over VLR and HLR. The reconstruction of RL field is done during an idle time of the network in the period of a week, and it is a little overhead to write on the CL field, which is done only for the established calls. In addition, since the fields have small sizes, they little change the size of database.

3 Analytic Models

To evaluate this end-to-end delay, we treat the database system of MN as Jackson's network. The service time of each database operation is assumed to be a major delay, and we do not consider a link cost [7][9][10]. We assume that there are n VLRs and one HLR in the system. The HLR is assumed to have an infinite buffer and single exponential server with the average service time $\frac{1}{\mu_h}$. Likewise, the VLR is assumed to have an infinite buffer and single exponential server with the average service time $\frac{1}{\mu_v}$. We assume that within a RA, the location registration occurs in a Poisson fashion with rate λ_u and the call origination occurs in a Poisson fashion with rate λ_c. With these assumptions, the MIS using MHLR scheme as a mobility management method becomes Jackson's network modeling as shown in Fig.4.
λ_{lr} and λ_{tt} represent the average arrival rate of REGCANC message and the average arrival rate of ROUTREQ message, respectively. λ_h represents the average arrival rate of messages to the HLR from other VLRs, and by the Burke's theorem it is the same as the average departure rate of messages from the HLR. P_{vo} is the probability the departure message from the VLR leaves the system. P_{vh} is the probability the departure message from the VR enters the HLR. P_{hv} is the probability the departure message from the HLR enters one of n VLRs. From the definition of λ_{lr} and λ_{tt}, we know that these messages get out of the system after going through the VLR. λ_u enters the VLR in the form of REGREQ message, and after receiving services from the VLR it is delivered to the HLR in the form of REGNOT message. After receiving services from the VLR, $\frac{1}{n}\lambda_c$get out of the system because the probability the callee is in the same RA with the caller is $\frac{1}{n}$. So, we have

$$P_{vo}=\frac{\frac{1}{n}\lambda_c+\lambda_{lr}+\lambda_{tt}}{\lambda_{lr}+\lambda_{tt}+\lambda_c+\lambda_u} \tag{1}$$

$$P_{vh}=\frac{\frac{n-1}{n}\lambda_c+\lambda_u}{\lambda_{lr}+\lambda_{tt}+\lambda_c+\lambda_u} \tag{2}$$

$$P_{hv}=\frac{\lambda_h}{n}. \tag{3}$$

By the property of Jackson's network, we have

$$\lambda_h=n\times(\tfrac{n-1}{n}\lambda_c+\lambda_u). \tag{4}$$

From $\frac{\lambda_h}{n}=\lambda_{lr}+\lambda_{tt}$, we have

$$\lambda_{lr}=\lambda_u \tag{5}$$

$$\lambda_{tt}=\frac{n-1}{n}\lambda_c. \tag{6}$$

Though Eqs. (5) and (6) could be directly inferred from the definitions of λ_c and λ_u, we lead these equations through Eqs. (1)-(4) to help understand how λ_c and λ_u are delivered to HLR and VLRs and how many messages arrive at these DBs on average using the property of Jackson's network and theorem 1. We will repeat these steps in the following Jackson's network modeling the MHLR scheme.

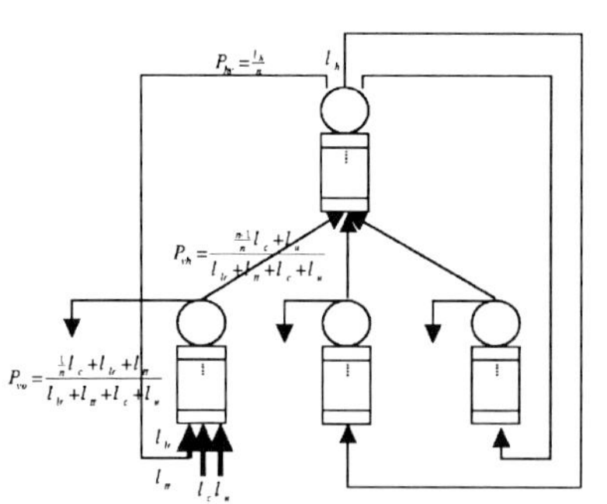

Fig. 4. Jackson's network modeling of the IS-41

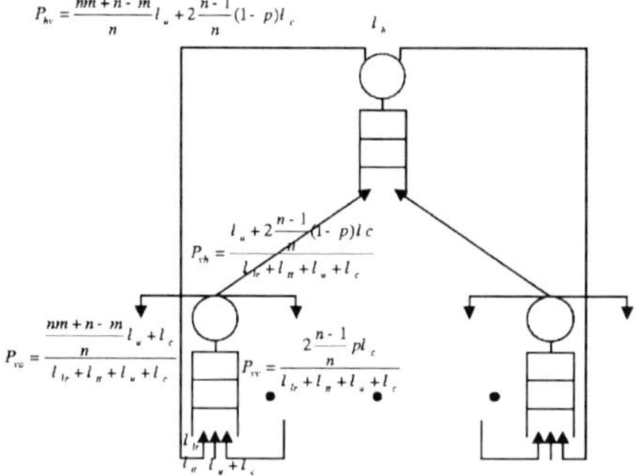

Fig. 5. Jackson's network modeling of the MHR scheme.

Now, let W_v and W_h represent the average system time (queue plus service) in the VLR and the average system time in the HLR, respectively. By the Little's Rule, W_v and W_h becomes

$$W_v = \frac{1}{\mu_v - (\frac{2n-1}{n}\lambda_c + 2\lambda_u)} \tag{7}$$

$$W_h = \frac{1}{\mu_h - n(\frac{n-1}{n}\lambda_c + \lambda_u)}. \tag{8}$$

Likewise, the MN using the MHLR scheme become Jackson's network as shown in Fig.5. The focus of message flow aims at the quantity. The wide difference between Fig.4 and Fig. 5 is that the number of message out coming HLR. When call managed in HLR, passing one more device dealing with multicasting can solve this problem. Another difference is what communicates call between VLR passing HLR. It is because of obtaining location information of the receiving terminal by multicasting information. In the Jackson's network modeling the MHLR scheme, we need to calculate the probability that the departure message from the VLR is delivered to another VLR. We denote this probability by P_{vv}. Considering the definition of λ_{lr},λ_{tt},λ_c,λ_u as follows:

$$P_{vo} = \frac{\frac{nm+n-m}{n}\lambda_u + \lambda_c}{\lambda_u + \lambda_{lr} + \lambda_c + \lambda_{tt}}. \tag{9}$$

$$P_{vh} = \frac{\lambda_u + 2\frac{n-1}{n}(1-p)\lambda_c}{\lambda_u + \lambda_{lr} + \lambda_c + \lambda_{tt}}. \tag{10}$$

$$P_{vv} = \frac{2\frac{n-1}{n}p\lambda_c}{\lambda_u + \lambda_{lr} + \lambda_c + \lambda_{tt}}. \tag{11}$$

$$P_{hv} = \frac{nm+n-m}{n}\lambda_u + 2\frac{n-1}{n}(1-p)\lambda_c. \tag{12}$$

Here P represents probability that multicasting message search to VLR of receiving terminal. By the definition of λ_{tt} and λ_{lr}, we obtain

$$\lambda_{tt} = 2\frac{n-1}{n}(1-p)\lambda_c, \tag{13}$$

$$\lambda_{lr}=\frac{nm+n-m}{n}\lambda_u, \tag{14}$$

$$\lambda_h=n(2\frac{n-1}{n}(1-p)\lambda_c+\frac{nm+n-m}{n}\lambda_u). \tag{15}$$

Let W_v' and W_h' denote the average system time in the VLR and in the HLR, respectively, From the Little's Rule, W_v' and W_h' are:

$$W_v' = \frac{1}{\mu_v-(2\frac{n-1}{n}(1-p)\lambda_c+\frac{nm+n-m}{n}\lambda_u+\lambda_c+\lambda_u)}. \tag{16}$$

$$W_h'=\frac{1}{\mu_v-n(2\frac{n-1}{n}(1-p)\lambda_c+\frac{nm+n-m}{n}\lambda_u)}. \tag{17}$$

Table 1 shows the average number of arrival messages to the VLR, and the average system time in the VLR when the MHLR scheme are used as compared to those of IS-41. Table 2 shows the average number of arrival messages to the HLR, and the average system time in the HLR when the MHLR scheme are used as compared to those of IS-41 scheme.

Table 1. VLR comparisons between the MHLR and IS-41scheme

	Average number of arrival messages	Average system time in VLR
IS-41	$\frac{2n-1}{n}\lambda_c+2\lambda_u$	$\frac{1}{\mu_v-(\frac{2n-1}{n}\lambda_c+2\lambda_u)}$
MHLR	$(2\frac{n-1}{n}(1-p)\lambda_c + \frac{nm+n-m}{n}\lambda_u+\lambda_c+\lambda_u)$	$\frac{1}{\mu_v-(2\frac{n-1}{n}(1-p)\lambda_c+\frac{nm+n-m}{n}\lambda_u+\lambda_c+\lambda_u)}$

Table 2. HLR comparisons between the MHLR and IS-41scheme

	Average number of arrival messages	Average system time in VLR
IS-41	$n(\frac{n-1}{n}\lambda_c+\lambda_u)$	$\frac{1}{\mu_h-n(\frac{n-1}{n}\lambda_c+\lambda_u)}$
MHLR	$n(2\frac{n-1}{n}(1-p)\lambda_c + \frac{nm+n-m}{n}\lambda_u)$	$\frac{1}{\mu_h-n(2\frac{n-1}{n}(1-p)\lambda_c+\frac{nm+n-m}{n}\lambda_u)}$

As shown in table 2 and table 3, the MHLR scheme distributes messages from the HLR to the VLRs, and it also reduces the average system time in the HLR with the small increase of the average system time in the VLRs. We define the costs for mobility management as follows.

Location registration rate $\times$ location registration cost + call tracking rate $\times$ call tracking cost.

Based on the delay times of the HLR and the VLR as shown in Table 2 and Table 3, we can calculate the mobility management costs for IS-41 scheme and the MHLR scheme as follows.

- Mobility management cost when using the IS-41scheme:
W_{IS-41M}(Mobility Management cost)$=\frac{\lambda_u}{\lambda_c+\lambda_u} \times W_{IS-41L} + \frac{\lambda_c}{\lambda_c+\lambda_u} \times W_{IS-41C}$.

- Mobility management cost when using the MHLR scheme:
W_{MHLRM}(Mobility Management cost)$=\frac{\lambda_u}{\lambda_c+\lambda_u} \times W_{MHLRL} + \frac{\lambda_c}{\lambda_c+\lambda_u} \times W_{MHLRC}$.

4 Numerical Results

To get numerical results, we use the same value of system parameters as those in [3], n= 128, for example. From these parameters, the average occurrence rate of location registration in an RA, λ_u, is calculated as $\lambda_u = (390 \times 30.3 \times 5.6)/3600 \times\pi = 5.85/s$. And the average call origination rate in an RA, λ_c, is calculated as $\lambda_c = (1.4 \times 57.4 \times 390)/3600 = 8.70/s$. We assume that the average service rates of HLR and VLR are $\mu_h = 2000/s$, $\mu_v = 1000/s$.
Fig.6 shows message ratio in mobility management cost. Y-coordinate presents the ratio dividing the result value of IS-41scheme into the result value of the MHLR scheme. If the ratio is 1, performance of IS-41scheme is equal to the MHLR scheme. If it has greater than 1, the performance of MHLR scheme is superior to the IS-41scheme. If it has less than 1, the performance of IS-41scheme is superior to the MHLR scheme. And also, the example of the graph can show multicasting factor, **m**. It is determined VLRs which can be multicast the location information of terminal.
If probability **p** is 0.5, and **m** is 3, the value is equal to 0.87 in short. It is fact that the MHLR scheme is 1.15 times as many messages as IS-41scheme. For the ratio of delay time in VLR, the MHLR scheme has a fewer values than IS-41scheme. In previous case that the probability p is 0.5 and **m** is 3, the value is 0.99. If probability p is 0, the value is over 56 times. If **m** is 3 and probability p is 0.5, the value is over 28 times. We can know that the MHLR scheme has prominent performance than IS-41scheme.

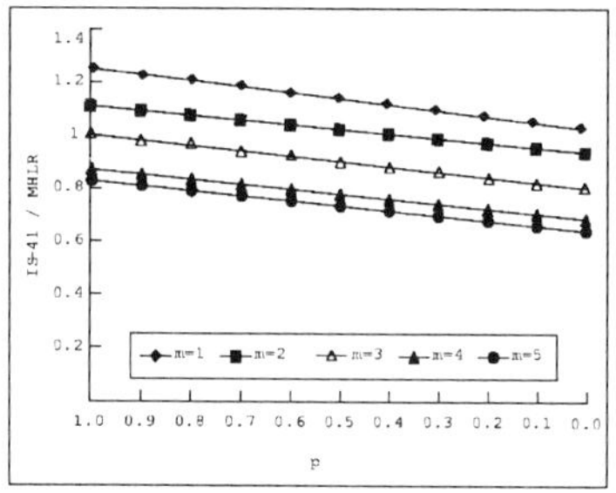

Fig. 6. Mobility management cost in the MHLR scheme

5 Conclusions

We propose new mobility management strategy to reduce the location traffic of HLR because the most important problem of IS-41 and GSM used in standard of the mobile management at present is to solve the location traffic problem of HLR. We also estimate overall mobility management cost from the MN models based on the Jackson's network model. In numerical results, in the MHLR scheme, the number of call tracking message and the delay time of HLR is decreased a little as compared to that of the IS-41scheme. Especially, the management delay time of HLR has more performance result than the IS-41scheme. However, the delay time of VLR is increased a little as compared to that of the IS-41 scheme. The proposed schemes can be expected to have the prominent performance advancing for location management in the MN.

Acknowledgments. This work is supported by Howon University, 2002.

References

1. F. Akyildiz, J, McNair, J, Ho, H. Uzunalioglu, W. Wang, "Mobility Management in Current and Future Communication Networks", IEEE Network, Jul./Aug., 1998, pp. 39–49
2. EIA/TIA IS-41.3, "Cellular Radio Telecommunications Intersystem Operations", Technical Report (Revision B), July 1991.
3. R. Jain, Y. B. Lin, C. N. Lo. and S. Mohan, "A caching Strategy to Reduce Network Impacts of PCS", IEEE Jour. on Selected Areas in Comm., Vol. 12, No. 8, 1994, pp. 1434–1445
4. C. -L. I, G. P. Plooini and R. D. Gitlin, "PCS Mobility Management using the Reverse Virtual All Setup Algorithm", IEEE/ACM Trans. Vehicle. Tech, 1994, pp. 1006–1010
5. T. Russel, "Signaling System ♯7", McGraw-Hill, 1995
6. J. S. M. Ho and I. F. Akyildiz, "Local Anchor Scheme for Reducing Location Tracking Costs in CNs", Proceeding of ACM MOBICOM'95, Nov. 1995, pp. 181–194
7. J. Z. Wang, "A Fully Distributed Location Registration Strategy for Universal Personal Communication Systems", IEEE Personal Comm., First Quarter 1994, pp. 42–50
8. X. Qui and V. O. K. Li, "Performance Analysis of PCS Mobility Management Database System", Proceeding of USC/IEEE Int'l Conf. Com. Com., and Networks, Sept. 1995
9. C. Eynard, M. Lenti, A. LOmbardo, O. Marengo, S. Palazzo, "Performance of Data Querying Operations in Universal Mobile Telecommunication System (UMTS)", Proceeding of IEEE INFOCOM '95, Apr. 1995, pp. 473–480
10. Y. Bing Lin and S. Y. Hwang, "Comparing the PCS Location Tracking Strategies", IEEE Trans. Vehicle. Tech., Vol. 45, No. 1, Feb. 1996

Packet Error Probability of Multi-carrier CDMA System in Fast/Slow Correlated Fading Plus Interference Channel

Jae-Sung Roh[1], Chang-Heon Oh[2], Heau-Jo Kang[3], and Sung-Joon Cho[4]

[1] Dept. of Information & Communication Eng., SEOIL College, Seoul, KOREA
jsroh@seoil.ac.kr

[2] School of Information Technology, Korea Univ. of Technology and Education, Chungnam, KOREA
choh@kut.ac.kr

[3] Dept. of Electrical and Electronic Eng., Dongshin Univ., Chonnam, KOREA
hjkang@dongshinu.ac.kr

[4] School of Electronics, Telecommunication and Computer Eng., Hankuk Aviation Univ. Kyonggi-do, KOREA
sjcho@mail.hangkong.ac.kr

Abstract. The probability of packet error for Multi-Carrier CDMA system with maximum ratio combining diversity is evaluated and compared in the slow and fast correlated Nakagami fading plus multiple access interference channel. From the results, the probability of packet error in fast fading is higher than that in slow fading, and the difference of packet error probability between the fast fading and slow fading diminishes with Nakagami fading parameter approaching infinity. When the error correction coding technique is used, it is observed that coding technique for Multi-Carrier CDMA system is more effective in the fast fading case than that in the slow fading case.

1 Introduction

Many urban and vehicular communication systems are subject to fading and co-channel interference caused by multipath propagation due to reflections, refractions and scattering by buildings and other large structures. Thus, the received signal is a sum of different signals that arrive via different propagation paths [1].

Several statistical models have been used in the literature to describe the fading envelope of the received signal [2]. A more versatile statistical model is Nakagami's m-distribution [3],[4], which can model a variety of fading environments including those modeled by the Rayleigh and one-sided Gaussian distributions. Also, the log-normal and Rician distributions may be closely approximated by the Nakagami distribution in some ranges of mean signal values. The Nakagami distribution is more flexible and more accurately fit experimental data for many physical propagation channels than the log-normal and Rician distributions. For this reason, there is contin-

M.H. Shafazand and A M. Tjoa (Eds.): EurAsia-ICT 2002, LNCS 2510, pp. 366–376, 2002.

ued interest in modeling a variety of propagation channels with the Nakagami distribution.

Diversity reception technique is used extensively in radio channels to reduce the effect of fading on system performance, including both fixed terminals and mobile communication systems [5]. For example, diversity in land mobile radio can be used also in mobile terminals, on cars, and even in hand held portable radios. In order to obtain the diversity gain, there must be a sufficient degree of statistical independence in the fading of the received signal in each of the diversity branches. The assumption of statistical independence between the diversity channels is valid only if they are sufficiently separated. However, there are other cases of practical interest where the assumption of statistical independence is not valid. The effect of correlated fading on the performance of a diversity reception system has received a lot of attention in the literature [4],[6].

2 Analysis Model of Multi-carrier CDMA System

2.1 Multi-carrier CDMA System Model

This section is concerned with the calculation of the error probability of the Multi-Carrier CDMA system in a fading channel that is modeled by a discrete set of Nakagami faded paths. The block diagram of Multi-Carrier CDMA transmitter of the k-th user is shown in figure 1, where $a^{(k)}$ and $C_i^{(k)}$ denote the information symbol and the i-th spreading code of length M_C of the k-th user, respectively. By using M_C orthogonal codes, the maximum number of users is equal to M_C.

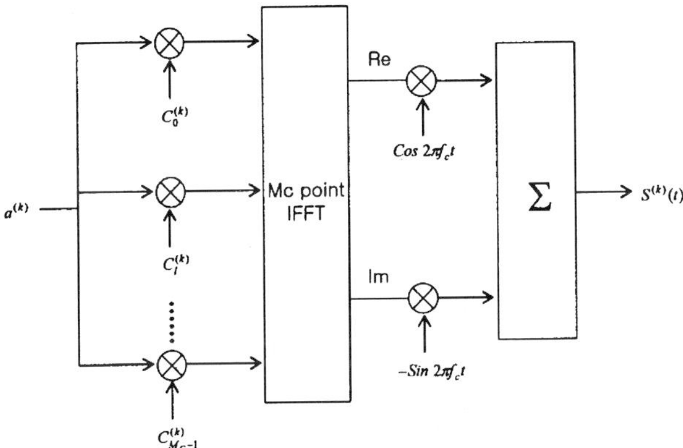

Fig. 1. Multi-Carrier CDMA transmitter for the k-th user in reverse link.

The transmitted signal of the k-th user is in general shaped in the time domain by a window function $w(t)$ to minimize excessive out-of-band emissions. The quadrature sinusoidal carriers are then modulated by the real and imaginary part of the baseband signals generated from IFFT respectively, while the former component at the j-th symbol instant can be described as

$$s_{I,j}^{(k)}(t) = \sqrt{\frac{2P}{M_C}} a_j^{(k)} \sum_{m=1}^{M_c} c_m^{(k)} \cos\left(\frac{2\pi mt}{T_s}\right) w(t - jT_s) \tag{1}$$

where $\{a_j^{(k)}\}$ is the bi-phase information sequence and $\{c_m^{(k)}\}$ is the spreading sequence over the alphabet {+1, -1}. It is assumed that the spreading gain M_C is the large and equals the number of carriers present. Also, the fundamental carrier frequency f_0, equals the symbol rate $1/T_S$ and the transmitted signal power P of every user is assumed to be same. Then, the j-th transmitted symbol $s_j^{(k)}$ of the k-th user with modulator phase shift θ_k is

$$s_j^{(k)}(t) = s_{I,j}^{(k)}(t)\cos(2\pi f_c t + \theta_k) - s_{Q,j}^{(k)}(t)\sin(2\pi f_c t + \theta_k) \tag{2}$$

where f_c to $f_c + M_C / T_S$ is the desired spectrum in used and $s_{Q,j}^{(k)}(t)$ equals (1) with $\cos(2\pi mt / T_s)$ replaced by $\sin(2\pi mt / T_s)$. For the sake of clarity, we let $u_k(t) = \sum_{m=1}^{M_C} c_m^{(k)} \cos(2\pi mt / T_S)$, $v_k(t) = \sum_{n=1}^{M_C} c_n^{(k)} \sin(2\pi nt / T_S)$ in the analysis of the following sections.

For simplicity, we assume $w(t)$ in Eq. (1) equals $p_{T_S}(t)$. Where $p_{T_S}(t)$ is the pulse waveform of each symbol, given as

$$p_{T_S}(t) = \begin{cases} 1, & 0 \le t < T_S \\ 0, & otherwise \end{cases} \tag{3}$$

where T_S is a symbol period.

For the quadrature modulator and the receiver, the transmitted data sequence, $\{a_j^{(1)}\}$ is coherently detected and multipath fading is not presented. To simplify our argument, neither the guard interval nor the influence of inter-symbol interference (ISI) of each user will be taken into account.

With a perfect power control, the j-th transmitted symbol at each receiver is given by

$$r_j(t) = \sum_{k=1}^{U} \left\{ s_{I,j}^{(k)}(t - \tau_k)\cos(2\pi f_c t + \phi_k) - s_{Q,j}^{(k)}(t - \tau_k)\sin(2\pi f_c t + \phi_k) \right\} + N(t) \tag{4}$$

where $N(t)$ is white Gaussian noise and $\phi_k = \theta_k - 2\pi f_c \tau_k$. The delay and modified phase offset $\{a_j^{(k)}\}$ of every transmitted signal are assumed to be uniformly distributed on the interval $[0, T_S]$ and $[0, 2\pi]$ respectively. It is further assumed that the parameters τ_1 and ϕ_1 equal zero.

Multi-Carrier CDMA signal consists of sum of M_C waveform. As M_C increases, its waveform will have a Gaussian distribution. That is, for asynchronous Multi-Carrier CDMA system, Gaussian approximation is then used to estimate the probability of error in the Multi-Carrier CDMA system.

In the analysis, random signature sequences are assumed in operation, and for asynchronous Multi-Carrier CDMA system, interference from other users can be well approximated as a Gaussian noise with zero mean and the average value of the k-th multiple access user's variance in the Multi-Carrier CDMA system, $E\left[Var\left\{I^{(1,k)}\right\}\right]$ is as follow [7].

$$E\left[Var\left\{I^{(1,k)}\right\}\right] = \frac{PT_S^2}{4\pi^2 M_C}\left\{\left[\sum_{n=1}^{M_C}\sum_{i=1, i\neq n}^{M_C}\frac{3n^2+i^2}{(n^2-i^2)^2}\right]+\frac{2\pi^2}{3}\right\} \tag{5}$$

By assuming the interference among all other users are independent, the total interference power would simply be the summation of every $Var\left\{I^{(1,k)}\right\}$. The equivalent signal-to-noise and interference power ratio γ_{eq} at the receiver is then approximated as follow.

$$\gamma_{eq} = \frac{M_C PT_S^2/2}{\frac{(U-1)PT_S^2}{4\pi^2 M_C}\left\{\left[\sum_{n=1}^{M_C}\sum_{i=1, i\neq n}^{M_C}\frac{3n^2+i^2}{(n^2-i^2)^2}\right]+\frac{2\pi^2}{3}\right\}+\frac{M_C N_0 T_S}{4}} \tag{6}$$

where E_b is transmitted signal energy in one symbol period, U is the number of multiple access users, and N_0 is the one-side noise spectral density.

2.2 Nakagami Fading Channel Model

The Multi-Carrier CDMA signals of k-th user are transmitted to channel, and then they are distorted by the Nakagami fading. A Nakagami characterizes channels with different fading depths through a parameter (m) called amount of fading. The received signal envelope, R is a random variable with a Nakagami probability density function (pdf) [2] i.e.

$$f_\Omega(R) = \frac{2m^m R^{2m-1}}{G(m)\Omega^m}\exp\left(-\frac{mR^2}{\Omega}\right) \tag{7}$$

where $G(\cdot)$ is the Gamma function, $\Omega/2=\overline{R^2}/2$ is the mean power of the faded signal, and m is Nakagami fading parameter ($m=\Omega^2/\left(\overline{R^2-\Omega}\right)\geq 1/2$).

In a Nakagami fading channel, the magnitude of the received signal is characterized by the Nakagami distribution. However, it is more convenient to use the square of the received signal envelope which is proportional to the signal power ($\gamma=R^2/2N_0$). The desired signal-to-noise power ratio γ is then gamma distributed with probability density given by [3]

$$f_{\Gamma}(\gamma)=\left(\frac{m}{\Gamma}\right)^m\frac{\gamma^{m-1}}{G(m)}\exp\left(-\frac{m\gamma}{\Gamma}\right),\qquad \gamma>0 \tag{8}$$

where $m\geq 1/2$ and Γ is the average signal-to-noise power ratio. The constant m is called the Nakagami fading parameter. m=1 and m=∞ correspond to the Rayleigh fading and the nonfading case, respectively.

2.3 Diversity Reception of Correlated Fading Signal

Next, we consider the reception of Multi-Carrier CDMA signal in correlated fading channel. For convenience we assume that the correlation between the fading signals is known and completely characterized by a single parameter ρ. Experimental data on correlation of the fading signals are usually given in terms of envelope correlation because of ease of measurement. Given the envelope correlation, however, the power correlation can be found, and vice versa. In fact, for many practical purposes, both correlations can be taken to be approximately equal [4].

In this section, we consider the constant correlation model which characterize the correlation between the fading signals in each of the diversity branches is presented as $\rho_{ij}=\rho$, where $i,j=1,2,\cdots,M_B$ and $0<\rho<1$. Such a constant correlation model may approximate closely placed diversity antennas. For the above correlation model, Aalo [4] has obtained the probability density function for γ from its characteristic function. It is given by

$$f_{C\Gamma}(\gamma)=\frac{\left(\frac{\gamma m}{\Gamma}\right)^{M_B m-1}\exp\left(\frac{-\gamma m}{\Gamma(1-\rho)}\right)}{(\Gamma/m)(1-\rho)^{m(M_B-1)}(1-\rho+M_B\rho)^m G(M_B m)} \cdot {}_1F_1\left(m,M_B m;\frac{M_B m\rho\gamma}{\Gamma(1-\rho)(1-\rho+M_B\rho)}\right) \tag{9}$$

where ${}_1F_1(\cdot,\cdot;\cdot)$ is the confluent hypergeometric function. For ρ=0, Eq. (9) reduces to the probability density of the sum of M_B independent random variables, each of which has the density given by Eq. (8).

3 Performance Evaluation of Multi-carrier CDMA System

3.1 Probability of Packet Error for Multi-carrier CDMA System

In the analysis of data transmission with steady-signal reception, the probability of bit error plays a fundamental role in describing the performance of a digital communication system. Digital systems employing error detection or error correction coding are generally based on the transmission of blocks of L_P sequential bits, where each block of L_P bits may be a complete message or a submessage element, such as a character.

The performance of Multi-Carrier CDMA system will then depend upon the probability of occurrence of various numbers of errors in a block of L_P bits. For Multi-Carrier CDMA modulation and steady-signal reception in white Gaussian noise, the errors in a block will be binomially distributed, i.e., the probability of exactly t errors in a block of L_P bits is given by

$$P_t = \binom{L_P}{t} p^t (1-p)^{L_p - t} \tag{10}$$

where channel error rate, $p = P_e(\gamma)$ for a given signal to noise power ratio γ in AWGN and fading channel.

For the modulation schemes to which the binomial distribution applies, the probability of more than M errors within a block of L_P bits, which we define as the probability of packet error is

$$P(M, L_P) = \sum_{t=M+1}^{L_P} P_t \tag{11}$$

In a communication system that transmits data in blocks of L_P bits, the probability $P(M, L_P)$ of more than M bit errors in a block is an important quantity. If an error-correction code capable of correcting up to M errors in each block of L_P bits is employed, system performance is governed by the probability of more than M errors in a block. If a simple automatic repeat request (ARQ) scheme is used, the throughput can be determined from $P(0, L_P)$. On the other hand, if the use of forward error correction is to be investigated, then the knowledge of $P(M, L_P)$ is required. For AWGN channel in which the bit errors are independently and identically distributed, $P_g(M, L_P)$ can be readily calculated from the bit error probability $P_{eg}(\gamma)$, namely,

$$P_g(M, L_P) = \sum_{t=M+1}^{L_P} \binom{L_P}{t} P^t{}_{eg}(\gamma)(1 - P_{eg}(\gamma))^{L_P - t} \tag{12}$$

However, for a fading channel as encountered in VHF or UHF mobile radio applications, no simple relationship exists between $P_g(M, L_P)$ and $P_{eg}(\gamma)$. One commonly used analytic model for the signal fluctuations assumes that they can be closely approximated by a Rayleigh distribution [8]-[11]. In the past, a number of authors [8],[9] have derived expression for $P_g(M,L_P)$ in the assumption that the fading is so slow that the received signal strength can be assumed constant over the duration of a block of L_P bits. We will refer to such a channel as a very slow fading channel. Unfortunately, these expressions have proved awkward to evaluate and ad hoc approximations for $P_g(0,L_P)$ have been suggested.

It is well known [7] that the standard Gaussian approximation formula of uncoded Multi-Carrier CDMA BPSK system is therefore given by

$$P_{eg}(\gamma) \approx Q(\sqrt{\gamma_{eq}}) = 0.5 \cdot erfc(\sqrt{0.5\gamma_{eq}}) \tag{13}$$

where γ_{eq} is the equivalent signal to noise plus interference power ratio at the base station receiver and $Q(x)$ is the complementary error function. If we have a very slow fading channel in which the received average SNR varies according to a probability density function (pdf) $f_{CT}(\gamma)$, then the resulting probability of packet error in fading channel $P_f(M,L_P)$ can be written as

$$\begin{aligned} P_f(M,N) &= < P_f(M,N;\gamma) > \\ &= \int_0^{\infty} P_g(M,N) \cdot f_{c\Gamma}(\gamma) d\gamma \end{aligned} \tag{14}$$

where is a pdf of correlated Nakagami fading channel.

In the case of fast fading where the signal strength varies continuously such that two adjacent bits in a block are faded independently, we have

$$P_f(M,L_P) = \sum_{m=M+1}^{L_P} \binom{L_P}{t} P^t{}_{ef}(\gamma)(1-P_{ef}(\gamma))^{L_P-t} \tag{15}$$

where $P_{ef}(\gamma)$ is the bit error probability under correlated Nakagami fading given by

$$P_{ef}(\gamma) = \int P_{eg}(\gamma) \cdot f_{C\Gamma}(\gamma) d\gamma \tag{16}$$

In the absence of error-correction coding, the probability of packet error under fast fading channel is

$$\begin{aligned} P_f(0,L_P) &= \sum_{t=1}^{L_P} \binom{L_P}{t} P^t{}_{ef}(\gamma)(1-P_{ef}(\gamma))^{L_P-t} \\ &= 1-(1-P_{ef}(\gamma))^{L_P} \end{aligned} \tag{17}$$

4 Numerical Results and Discussion

In this paper, the probability of packet error for Multi-Carrier CDMA system are evaluated as a function of the Nakagami fading parameter (m), the number of Multi-Carrier (M_C), the number of multiple access users (U), correlation parameter (ρ), error correcting capability (M), the length of packet (L_P), and the number of branch (M_B). For the correlation parameter ρ =0, the probability density reduced the sum of M_B independent random variables, each of which has the uncorrelated probability density. In this section, to compare the performance of Multi-Carrier CDMA system, we select the error correcting capability at the same bits, and same correlation parameter.

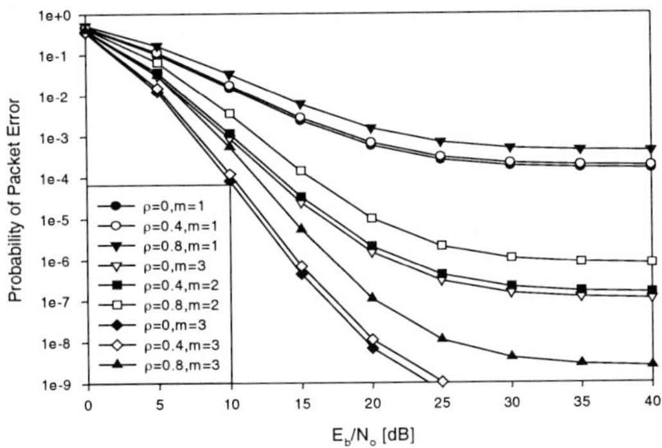

(a) U=5, M_B =2, M_C =127, L_P =100, M=0, ρ =0.4, slow fading

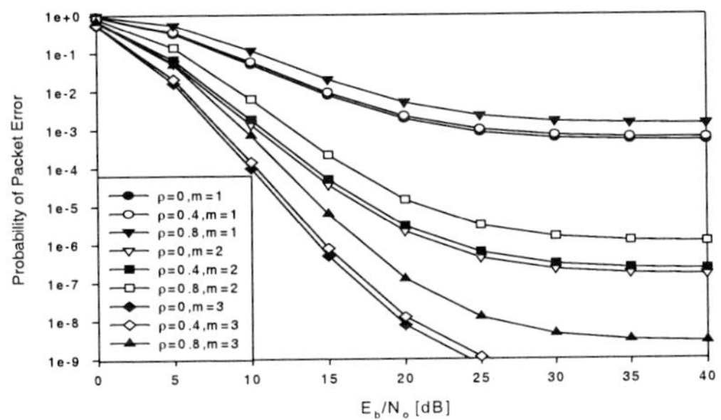

(b) U=5, M_B =2, M_C =127, L_P =100, M=0, ρ =0.4, fast fading

Fig. 2. Average y of pa cprobabilitket error for various E_b / N_0

Figure 2(a) and figure 2(b) show the effect of slow and fast fading according to the Nakagami fading parameter (m), the correlation parameter (ρ), and average E_b / N_0. Clearly, as the Nakagami fading parameter (m) increase, the packet error performance is getting better and the curves approach that corresponding to the case of non fading channel. Also, as the correlation parameter (ρ) decrease, the packet error performance is getting better and the curves approach that corresponding to the uncorrelated case. From the curves, it is noting that the performance of $\rho < 0.4$ can result in similar packet error performance of $\rho = 0$ in slow and fast fading channel.

Figure 3 shows the probability of packet error for various block length (N), fading parameters (m) and fading rate (slow/fast) in Nakagami fading channel. From the Fig. 3, it is observed that for a given value of m, the probability of packet error in fast fading is always higher than that in slow fading, especially when m is small (large amount of fading). The difference of packet error probability between the fast fading and slow fading diminishes as m approaches to infinity.

Figure 4(a) and figure 4(b) show the effects of error correction coding in slow and fast fading conditions respectively. Comparing the figure 4(a) and the figure 4(b), it is observed that error correction coding technique is more effective in the fast fading case than that in the slow fading case.

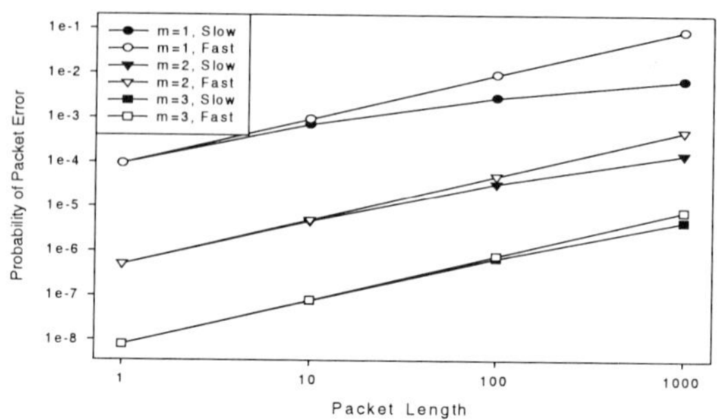

Fig. 3. Average probability of packet error for various block length (E_b / N_0=15dB, U=5, M=0, M_C=127, ρ=0.4)

5 Conclusion

In this paper, we have considered the probability of packet error for the Multi-Carrier CDMA system in correlated Nakagami fading channel that includes the Rayleigh fading channel. As a technique for the performance improvement, MRC diversity and block coding techniques have been used, and the performance of Multi-Carrier CDMA system has been compared and analyzed in slow and fast correlated Nakagami fading plus multiple access interference environments.

From the results, the probability of packet error in fast fading is always higher than that in slow fading, and the difference of packet error probability between the fast fading and slow fading diminishes as m approaches to infinity. When the error correction coding technique is used, it is observed that error correction coding technique is more effective in the fast fading case than in the slow fading case.

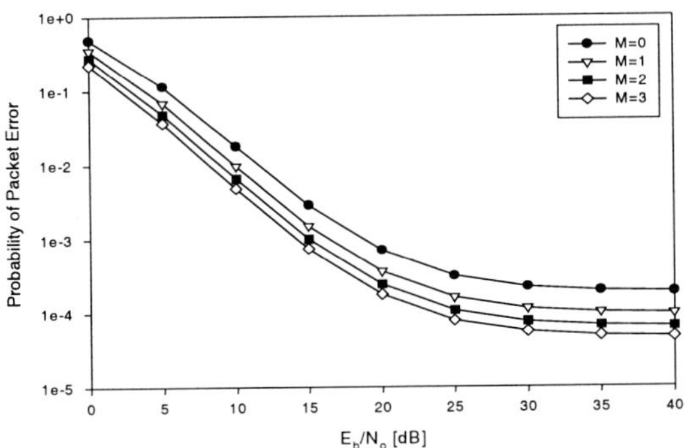

(a) $U=5,\ M_B=2,\ M_C=127,\ L_P=100,\ \rho=0.4$, slow fading

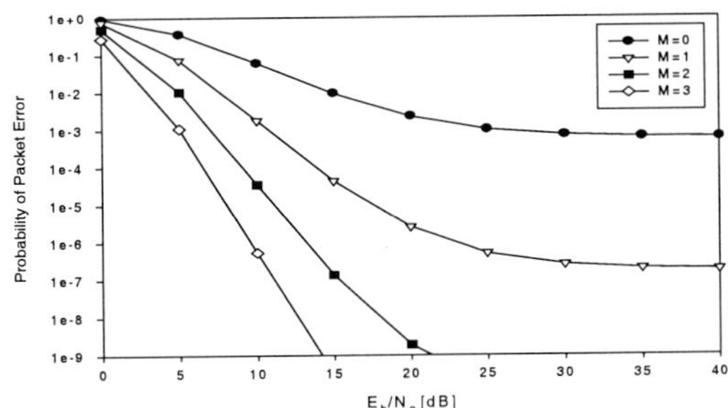

(b) $U=5,\ M_B=2,\ M_C=127,\ L_P=100,\ \rho=0.4$, fast fading

Fig. 4. Average probability of packet error for various error correcting capability

Acknowledgements. This work was supported by the Korea Science & Engineering Foundation (KOSEF) and the Kyonggi Province through the Internet Information Retrieval Research Center (IRC) of Hankuk Aviation University.

References

1. J. G. Proakis, *Digital Communications*: New York, McGraw-Hill, 2001.
2. T. Yamazaki and H. Yashima, "A modeling for multipath fading channel using a Gilbert model in coded multicarrier transmission," *Proc. IEEE Region 10 International Conference on Electrical and Electronic Technology*, pp. 565–570, 2001.
3. Y. D. Yao, T. Le-Ngoc, and U. H. Sheik, "Block error probabilities in a Nakagami fading-channel," *IEEE Trans., on Commun.*, vol. COM-29, pp, 130–133, 1993.

4. V. A. Aalo, "Performance of maximal-ratio diversity systems in a correlated Nakagami-fading environment," *IEEE Trans., on Commun*, vol. COM-43, pp, 2360-2369, Aug. 1995.
5. V. Aalo, G. Efthymoglou, and H. Helmken, "Path diversity performance of a DS-CDMA based land-mobile satellite system in a shadowed Rician-fading channel," *Wireless Communication-96*, pp. 133–140, 1996.
6. A. S. Akki, "Negatively correlated diversity branches using various modulation methods and maximal combining," *IEEE Trans., on Commun*, vol. COM-33, pp, 1323–1326, Dec. 1985.
7. T. F Ho, "Performance evaluation for multi-carrier CDMA system," *Proc. IEEE VTC'96*, pp. 1101–1105, 1996.
8. R. E. Eaves and A. H. Levesque, "Probability of block error for very slow Rayleigh fading in Gaussian noise," *IEEE Trans., on Commun,* vol. COM-25, pp. 368–373, Mar. 1977.
9. C. E. Sundberg, "Block error probability for noncoherent FSK with diversity for very slow Rayleigh fading in Gaussian noise," *IEEE Trans., on Commun*, vol. COM-29, pp. 57–60, Jan. 1981.
10. L. H. C. Lee, *Error-Control Block Codes for Communications Engineers*, Artech House, 2000.
11. F. Adachi and K. Ohno, "Block error probability for noncoherent FSK with diversity reception in mobile radio," *Electron. Lett.*, vol. 24, pp. 1523–1525, Nov. 1988.

A Distributed Low-Cost Dynamic Multicast Routing Algorithm with Delay Constraints*

Min-Woo Shin[1], Nak-Keun Joo[1], and Hyeong-Seok Lim[2]

[1] Department of Computer Science, Dongshin University,
252 Daeho-dong, Naju, Chonnam 520-714, Korea
{mwshin,nkjoo}@blue.dongshinu.ac.kr

[2] Department of Computer Science, Chonnam National University,
300 Youngbong-dong, Buk-gu, Kwangju 500-757, Korea
hslim@chonnam.chonnam.ac.kr

Abstract. Many real-time multimedia applications such as distance education, video conferencing have stringent end-to-end delay requirement and consume large amounts of network resources. In order to support these applications efficiently, multicast routing algorithms are needed for computing the least cost multicast trees that satisfy a given end-to-end delay constraint. However, finding such a tree is known to be computationally expensive. Therefore, we propose a heuristic distributed multicast routing algorithm that satisfies a given end-to-end delay constraint and minimizes the average resulting tree cost. Also, simulation results show that the proposed algorithm has better average cost performance than the other existing algorithms.

1 Introduction

Recently, the appearance of many new real-time network applications, such as distance education and video conferencing, has great influences on the network routing algorithms. These new applications usually have stringent end-to-end delay requirement and consume a lot of network resources. In order to support these new applications efficiently, multicast routing algorithms computing least cost multicast trees that satisfy a given end-to-end delay constraint are desirable. The traditional multicast routing algorithm can be thought of as the minimum Steiner tree problem and classified into two categories by means of route computation method. The first category is the centralized multicast routing algorithms. In centralized one, also called source routing, complete network topology information must be available at any node running the centralized algorithm. So, the computation is easy and fast. However, the overhead to maintain the whole network status in the route computing node can be very large. The other category is the distributed multicast routing algorithms. In distributed algorithms, each network node participates in the route computation. The route is generated by

* This work was supported by Korea Research Foundation Grant.(KRF-2000-041-E00299)

M.H. Shafazand and A M. Tjoa (Eds.): EurAsia-ICT 2002, LNCS 2510, pp. 377–386, 2002.

exchanging messages between nodes that have only partial information of the network status. The distributed one is slow and complex, but it is not needed to maintain the whole network status in each node. Also, there are two major classes of routing methods to which routing algorithms can be applied: dynamic and static. Dynamic multicast routing algorithms permit sources and receivers to join and leave a multicast session and the corresponding multicast trees at any moment. In static multicast routing algorithms, however, the multicast group is fixed, and paths from the sources to all receivers are computed at the same time, when initiating the multicast session. Although a number of delay-constrained multicast routing algorithms[8,15,14] have been proposed in the past few years, there exist few delay-constrained multicast routing algorithms that explicitly support dynamic multicast groups. Some existing multicast routing protocols used in the Internet(MBone) can very well support dynamic multicast groups. However, the underlying multicast routing algorithms[4,12,5,6,1,2] have been designed only for best-effort delivery so it can not support dynamic change of a multicast group and guarantee the QoS of new distributed real-time applications. In addition, most existing dynamic multicast routing algorithms consider only one link metric (link cost or link delay). For the first time, WAVE[10] considers more than one link metric, but it does not consider explicitly a given delay constraint. Recently, Quan Sun proposed a distributed delay-constrained dynamic multicast routing algorithm(DCDMR)[12] which is based on a distributed delay-constrained unicast routing algorithm(denoted by DCR)[11]. But DCDMR has the following drawbacks: (1) packet duplication can occur in some cases; (2) resulting tree cost is dependent on DCR entirely.

In this paper, we propose a distributed delay-constrained dynamic multicast routing algorithm which removes the drawbacks of DCDMR and perturbs the existing multicast tree as little as possible when group membership changes. The proposed algorithm can be easily incorporated into the distance vector based routing protocol. The benefits of this algorithm are that a given end-to-end delay constraint is satisfied and efficient network resource utilization can be achieved.

In the next section, our network model and problem definition are presented. In section 3, we give the previous works based on our paper. In section 4 and 5, we describe the proposed algorithm and the simulation results. Section 6 concludes the paper.

2 Network Model and Problem Definition

The network is modeled as a connected, directed graph $G = (V, E)$, where V is a set of nodes and E is a set of links. Two non-negative real value functions are associated with each link $e(e \in E)$: delay $D(e)$ and cost $C(e)$. The link delay $D(e)$ is the delay a data packet experiences on the corresponding link. The link cost $C(e)$ is a measure of the utilization of the corresponding link's resources. Links are asymmetrical, that is usually $C(e) \neq C(e')$ and $D(e) \neq D(e')$. Given a source node $s(s \in V)$, a set of destination nodes $S(S \subseteq V - s)$, a tree $T(T \subseteq G)$ rooted at s and spanning all the nodes in S (with all leaf nodes $\subseteq S$) is called a *multicast tree*. The cost of tree T is defined as follows:

$$C_T = \sum\nolimits_{e \subseteq T} C(e)$$

The nodes in S are open called multicast group members, and $|S|$ is the multicast group size. Let $P_T(s, v)$ denote the unique path from s to $v \subseteq S$ on the multicast tree T, if the delays of all source-destination paths on the tree are kept within a given bound Δ, namely:

$$\sum\nolimits_{e \subseteq P_T(s,v)} D(e) < \Delta, \quad \forall v \in S$$

then the tree T is called a *delay constrained multicast tree* for the given delay constraint Δ. The problem of finding a least cost delay constrained multicast tree is NP-complete [1]. This paper focuses on the Delay Constrained Dynamic Multicast routing problem (DCDM). This problem can be formally defined as follows: For a directed network $G = (V, E)$ with non-negative link costs C and link delay D(note that C and D may change with time), a delay tolerance Δ,the initial delay constrained multicast tree T_0 with the source node $s(s \in V)$, and a series of requests $\{r(t_1), r(t_2), \cdots, r(t_i), \cdots\}$ where $r(t_i)$ represents a request of adding a new group member or deleting an existing group member at time $t_i(t_1 \leq, t_2 \leq, t_3 \leq, \cdots)$, find a series of delay constrained multicast trees $\{T_1, T_2, \ldots, T_i, \cdots\}$ such that the members of T_i are those of T_0 modified by the requests $r(t_1), r(t_2), \cdots, r(t_i)$ and the cost of T_i is minimum among all possible choices for tree T_i.

A simple solution for DCDM is to reconstruct a new delay constrained multicast tree using a static delay constrained multicast heuristic like [15] for each request. Rebuilding a new tree in this way, however, may cause large disruptions for the current multicast sessions. This is especially unacceptable for real-time multimedia applications. Thus, there are needed efficient dynamic delay constrained multicast routing algorithms that disrupted as little as possible for the current multicast sessions. In the next section, we discuss DCDMR which is the motivation for our paper and one of the dynamic delay-constrained multicast routing algorithms.

3 Related Work

To do the route computation correctly, DCDMR's each node must have the following information: the delays of all outgoing links, a cost vector, and a delay vector. The delay vector at node $v_i \in V$ consists of $|V| - 1$ entries, one entry for each other network node. The entry for node $v_j \in V(v_j \neq v_i)$ contains the following items:

- The destination node identity : v_j
- The end-to-end delay of the least delay path $P_{ld}(v_i, v_j)$ from v_i to v_j : $D(P_{ld}(v_i, v_j))$
- The cost of the least delay path $P_{ld}(v_i, v_j)$ from v_i to v_j : $C(P_{ld}(v_i, v_j))$
- The next hop node on the least delay path $P_{ld}(v_i, v_j)$ from v_i to v_j: $id(P_{ld}(v_i, v_j))$

Similarly, the cost vector information is the same as the delay vector information, if P_{ld}(least delay path) is replaced with P_{lc}(least cost path). For the sake of convenience, we let $D(P_T(s,v))$ denote the delay of the unique path from s to v_i on the existing multicast tree T where s is the source node of T and v_i is any other node on T. In DCDMR, when a new member v_n joins an existing tree, v_n contacts node v_p on the existing tree with a request message(Req) along the shortest delay path where v_p can be any node of T.

Step1. Set $v_i \leftarrow v_p$
Step2. If $v_i = s$ and $D(P_{ld}(s,v_n)) \geq \Delta$, there exits no delay-constrained path from s to node v_n and DCDMR terminates; otherwise go to next step.
Step3. If $D(P_T(s,v_i)) + D(P_{ld}(v_i,v_n)) < \Delta$, node v_i computes a path from itself to the node v_n satisfying the delay constraint of $\Delta - D(P_T(s,v_i))$ using **DCR**; otherwise go to next step.
Step4. v_i sends the received Req to v_i's parent node v_q on the existing tree. Set $v_i \leftarrow v_q$ and go to Step 2.

As shown in the above algorithm in step 3, DCDMR constructs a path between nodes by using a distributed delay-constrained unicast routing algorithm, DCR. In the following, we give a short description of DCR.

DCR Algorithm

For simplicity, we give the notation for the description of DCR as follows.

- s : source node
- d : destination node
- $active\_node$: the node which received a path construction message
- $delay\_so\_far$: the delay of already constructed path from s to $active\_node$

To select the next node for the path construction, $active\_node$ checks

$$delay\_so\_far + D(P_{lc}(active\_node, d)) < \Delta \quad (1)$$

If (1) is satisfied, $id(P_{lc}(active\_node, d))$ is selected as the next node, otherwise $id(P_{ld}(active\_node, d))$ is selected. Then, $active\_node$ constructs a path construction message which including $(d, \Delta, delay\_so\_far)$ and sends it to the selected node. If the destination node d receives the path construction message, it means that the delay-constrained path has been successfully constructed.

The DCDMR algorithm is very simple and efficient, but the multicast tree cost is dependent on DCR wholly and packet duplication can occur in some cases. Fig. 1 shows an example of the packet duplication when a new member d joins an existing tree.

As shown in Fig. 1, the node c has the shortest delay path from node d. However, $D(P_T(s,c)) + D(P_{ld}(c,d)) < \Delta$ is not satisfied. So, c just sends the received Req message to c's parent node, b. Since the same situation happens in the nodes b and a, Req message is merely transmitted to the node s. In node s, $D(P_{ld}(s,d)) < \Delta$ is satisfied, so the node s computes a path from itself to the

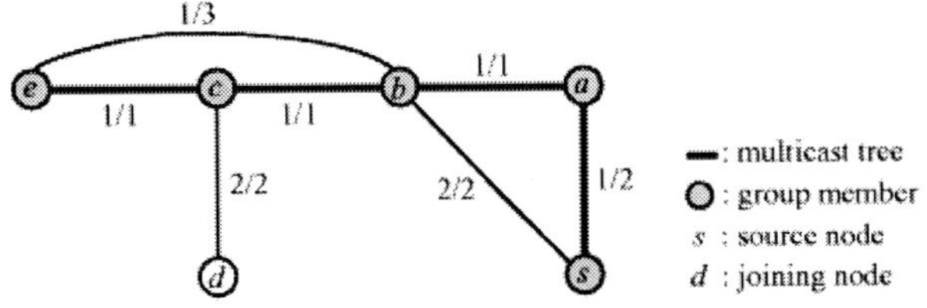

Fig. 1. An example of the packet duplication with a delay bound 5(link labels are cost/delay)

node d using DCR. As a result, the new path $s \to b \to c \to d$ is constructed and packet duplication occurs in the node b. Therefore, we present the method that can solve the packet duplication of DCDMR and propose the efficient unicast routing algorithm to improve the average cost of the multicast tree.

4 The Proposed Algorithm

DDMR(Delay Constrained Dynamic Multicast Routing algorithm) uses the same routing informations, namely, delay vectors and coat vectors which already mentioned in chapter 3. These vectors are similar to the distance vectors of some existing routing protocols[7]. So, the same procedures which are used for maintaining the distance vectors can be used for maintaining the cost vectors and delay vectors. We will not discuss these procedures, because our focus is on a routing algorithm that uses the cost vectors and delay vectors as input information. In the following, we assume that the contents of these vectors do not change during the execution of the routing algorithm.

DDMR is also based on a distributed delay-constrained unicast routing algorithm as DCDMR. Therefore, we first describe the unicast routing algorithm, DDUR(Distributed Delay constrained Unicast Routing algorithm) used in DDMR, and then consider the DDMR algorithm.

4.1 DDUR Algorithm

In Quan Sun's DCR algorithm, any *active_node* at the end of the partially constructed path can choose to add one of only two alternative outgoing links. One link is on the least cost path from *active_node* to the destination, while the other link is on the least delay path from *active_node* to the destination. This limitation reduces the amount of computation required at any node, but it increases the cost of the constructed path. Therefore, in order to get the low cost of constructed path, DDUR expands the selection criteria while deciding what outgoing links are to be added. Because of this expansion, DDUR's computation time at any node increases slightly, but the total cost of the constructed path is very satisfactory. DDUR constructs the path one node at a time, from *active_node* to the destination node d. In the following, we give a short description how to select the next *active_node* in DDUR algorithm.

- If $D(P_{lc}(active\_node, d)) \leq \Delta$, then the next *active_node* becomes the next hop node on the least cost path towards destination node, namely $id(P_{lc}(active\_node, d))$ as DCR algorithm.
- If $D(P_{lc}(active\_node, d)) > \Delta$, then the *active_node* sends Query messages to its neighbor nodes $v_i, (i = 1, \cdots, k)$. The Query message contains the destination node d and $delay\_so\_far + D(active\_node, v_i)$. For convenience, we denote $delay\_so\_far + D(activr\_node, v_i)$ as $D(temp)$. After receiving Query message, each v_i checks if

$$D(temp) + D(P_{ld}(v_i, d) > \Delta \tag{2}$$

If (2) is satisfied, there exists no delay-constrained path from v_i to d and v_i sends a "No path found message" back to the *active_node*. Otherwise, it checks if

$$D(temp) + D(P_{lc}(v_i, d) \leq \Delta \tag{3}$$

If (3) is satisfied, v_i sends a Response message back to the *active_node* with $(C(P_{lc}(v_i, d)), v_i)$. Otherwise, it sends a Response message back to the *active_node* with $(C(P_{ld}(v_i, d)), v_i)$. After receiving a set of Response messages from all v_i, the current *active_node* checks the cost value, namely, $(C(active\_node, v_i) +$ received cost value from $v_i)$, and then selects the node v_i with the least cost value. If we let this selected node be v_j,then the node v_j becomes the next *active_node*.

After deciding the new *active_node*, the previous *active_node* constructs a path construction message and sends it to the new *active_node*. Then the new *active_node* executes the same procedure until destination node d receives a path construction message. If $D(P_{lc}(active\_node, d)) \leq \Delta$, then DDUR constructs the same path as that of DCR, otherwise, our simulation results shown that the average path cost of DDUR is up to better 40% than DCR. Fig. 2 shows the percentage excess cost over DDUR when DDUR and DCR find a different path. DCR requires $O(|V|)$ messages to construct a path in the worst case, however DDUR requires $O(|E|)$ messages due to handling Query messages, where $|E|$ is the number of network links.

4.2 DDMR Algorithm

If DDMR has no packet duplication, it is similar to DCDMR except using DDUR instead of DCR. However, if packet duplication occurs, DDMR needs to modify the algorithm for reconstructing the existing multicast tree after removing packet duplication. In such case, each node on the existing multicast tree has to maintain added information about the list of destination nodes. This information can be added easily modifying the DDUR algorithm, whenever any node received a path construction message, it appends the destination node included in the path construction message to the list of destination nodes. In the following, we give a short description about the procedure for solving the packet duplication.

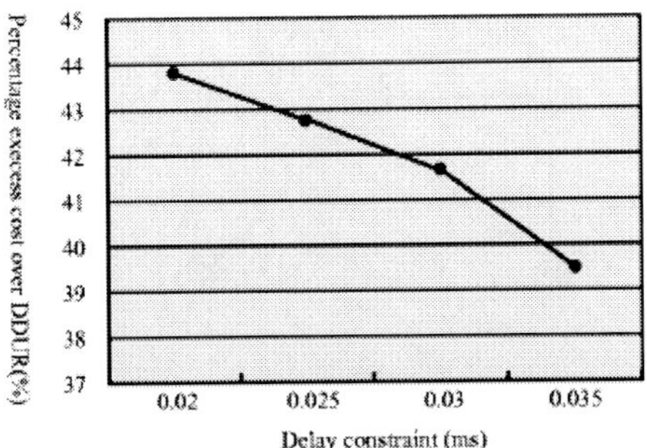

Fig. 2. Percentage excess cost over DDUR when the number of nodes equal to 50

- v_i, v_n : the same meaning as in DCDMR
- v_d : the node where packet duplication is found during the execution of the DDUR
- $D_{DDUR}(v_i, v_d)$: delay between from v_i to v_d when applied to DDUR
- *slack* : $D(P_T(s, v_d)) - (D(P_T(s, v_i)) + D_{DDUR}(v_i, v_d))$

F1) calculate the *slack* at v_d
F2) modify $D(P_T(s, v_d))$ to $(D(P_T(s, v_d)) - slack)$
F3) disconnect the link existing in the previous tree between v_d and its parent node
F4) set v_d to *active_node* and execute DDUR for each node in the list of destination nodes maintained by v_d

When we apply the above procedure to Fig. 1, packet duplication is found in the node b. For solving this duplication, first calculate the *slack* value(=1), and then disconnect the link between the node b and the node a after modifying the value of $D(P_T(s, b))$ 3 to 2. Node e and d are destination nodes maintained in the node b, so set b to *active_node* and execute the DDUR algorithm for each destination node. As a result, multicast tree is reconstructed as in Fig. 3. Because the *delay_so_far* at node b saved as *slack* value, the new cheapest path can be selected from node b to e Therefore, the total multicast tree cost is reduced.

5 Simulation

We used simulation experiments to evaluate the cost performance of DDMR. The multicast routing simulation environment described in [9] was modified to evaluate and compare the performances of DDMR, DCDMR and BSMA [15] which has the best cost performance among all the proposed delay-constrained static multicast heuristics. In the simulation, full duplex ATM networks with homogeneous link capacities of 155 Mbps were used. To create networks, we used a random graph generator based on Waxman's generator[13]. The generator first

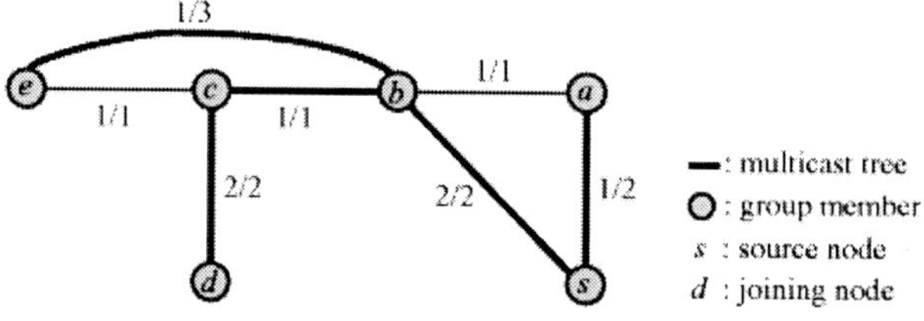

Fig. 3. Multicast tree after removing the packet duplication

creates a list of nodes(at random locations) and then creates links between these nodes. The probability of a link to exist between two nodes u and v is given by

$$P(u,v) = \beta exp \frac{-l(u,v)}{L\alpha}$$

where $l(u,v)$ is the distance between u and v, and L is the maximum distance between any two nodes, α and β are parameters between $(0,1]$. A large value of α increases the ratio of the number of long links to short links, and a large β increases the average node degree of the network. The simulated networks were always connected and had an average node degree of 5 through adjusting the parameters. Each node in the created network represented a non-blocking ATM switch. Each link was assumed to have a small output buffer. The propagation speed through the links was taken to be two thirds the speed of light. The propagation delay was dominant under these conditions, and queuing component was neglected when calculating the link delay $D(e)$. In the simulation, the link cost $C(e)$ was set to be the total currently reserved bandwidth on the link, which is a suitable measure of the utilization of both the link's bandwidth and its buffer space.

For each run of the experiment we generated a random network. Random background traffic for each link of the network was also generated. The equivalent bandwidth of each link's background traffic was a random variable uniformly distributed between 1 Mbps and 125 Mbps. We then generated an initial multicast tree T_0, which consists of only a randomly generated source node s, and a random series of Req requests $\{r_1, r_2, \cdots, r_m\}$ (note that $r_i \neq r_j \;\; if \; i \neq j \; for \; i,j = 1,2,\cdots,m$). The algorithm DDMR and DCDMR were applied to generate a new delay-constrained multicast tree T_1 for the request r_1 joining T_0, T_2 for the request r_2 joining $T_1, \cdots, T_m$ for the request r_m joining T_{m-1}. The BSMA algorithm was used directly to generate the tree T_m for the source node s and all the m members to be added without considering the request sequence of the nodes to be added. The experiment was run repeatedly until a confidence interval of less than 5% using 95% confidence level was achieved for the cost of the tree T_m. Fig. 4 shows the percentage excess cost over BSMA versus number of network nodes for the number of Req requests of 15 and delay bound of $35ms$. Fig. 5 shows the percentage excess cost over BSMA versus number of Req requests for the number of network nodes 60 and a delay bound of 35ms. From theses figures, we see that DDMR has a cost performance always within 20% worse than BSMA, although the number of *Req* requests and the number of nodes may increase. However, DCDMR is up to 40% worse than BSMA.

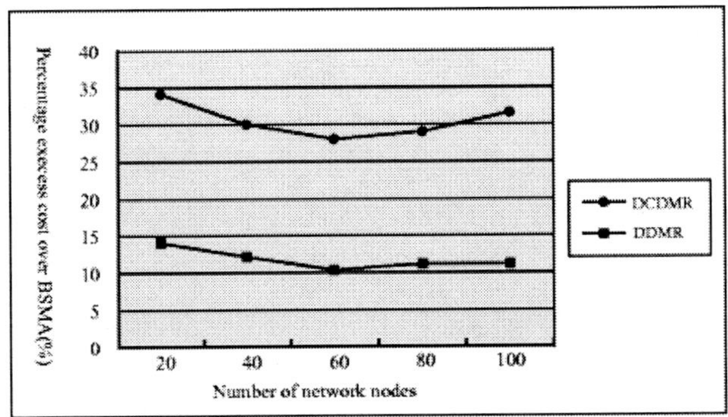

Fig. 4. Percentage excess cost over BSMA versus number of Req = 15

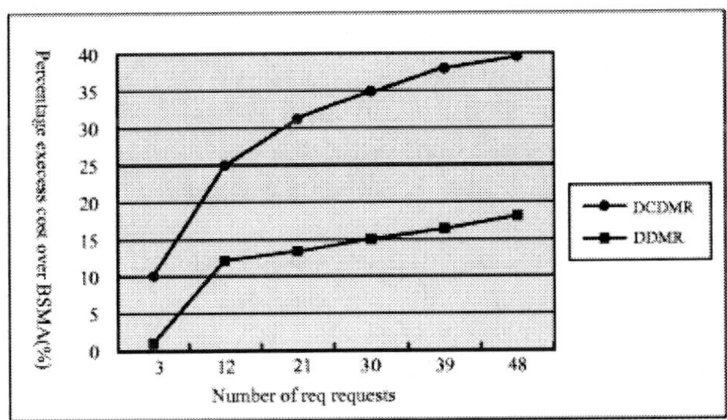

Fig. 5. Percentage excess cost over BSMA versus number of Req, for number of network nodes = 60

6 Concluding Remarks

In this paper, we proposed a distributed delay-constrained multicast routing algorithm (DDMR) supporting dynamic multicast groups. DDMR perturbs the existing multicast tree as little as possible when group membership changes. DDMR scales well because the source node may not be involved in the route computation or needs only small computation, also has very good cost performance. Simulation results show that the average tree cost of DDMR is always within 20% worse than BSMA and at least 20% better than DCDMR.

There are a few areas for further research : (a) apply the characteristics of DDUR or DDMR to broadcast routing algorithm with delay constraints, (b) consider the mobile or Ad hoc network environment based on the proposed algorithm.

References

1. Ballardie, A., Tsuchiya, P. and Crowcroft, J.: Core Based Trees(CBT) - Scalable Multicast Routing, Internet Draft, Apr., 1993
2. Ballardie, A., Francis, P. and Crowcroft, J.: Core Bases Trees(CBT) - An Architecture for Scalable Inter-Domain Multicast Routing, ACM SIGCOMM'93, Sep., 1993
3. Biersack, E. and Nonnenmacher J.: WAVE: A New Multicast Routing Algorithm for Static and Dynamic Multicast Groups, in proceedings of 5th Workshop on Network and Operating System Support for Digital Audio and Video, pp. 228–239, 1995
4. Deering, S.: Multicast routing in Datagram Internetworks and Extended LANs, ACM Transactions on Computer Systems, 8(2), pp. 85–110, May, 1990
5. Deering, S. : Host Extensions for IP Multicasting, RFC1112, Aug., 1989
6. Estrin, D. Rekhter, Y. and Hotz, S. : A unified Approach to Inter-Domain Routing, RFC1322, May. 1992
7. Hedrick, C.: Routing Information Protocol, Internet RFC 1058, June, 1988
8. Kompella, V. P., et al.: Multicast Routing for Multimedia Communication, IEEE/ACM Transactions on Networking, 1(3), pp.286-292, 1993
9. Salama, H.F., et al.: Evaluation of Multicast Routing Algorithms for Real-time Communication on High-Speed Networks, in High Performance Networking VI, IFIP 6th International Conference on High Performance Networking, pp. 27–42, 1995
10. Sun, Q. and Langendorfer, H.: A Distributed Delay-Constrained Dynamic Multicast Routing Algorithm, in European Workshop on Interactive Distributed Multimedia Systems and Telecommunication Services (IDMS'97), pp. 97–106, 1997
11. Sun, Q. and Langendorfer, H.: A New distributed Routing Algorithm for Supporting Delay-Sensitive Applications, International Report, Institute of Operating Systems and Computer Networks, TU Braunschweig, Germany, March 1997
12. Waitzman, D., Partridge, C. and Deering, S.: Distance Vector Multicast Routing Protocol, RFC1075, Nov., 1988
13. Waxman, B.M.: Routing of Multipoint Connections, IEEE Journal on Selected Areas in Communications, 6(9), pp. 1617–1622, 1988
14. Wi, S. and Choi, Y.: A Delay Constrained Distributed Multicast Routing Algorithm, 12th International Conference on Computer Communication, ICCC'95, pp. 833–838, 1995
15. Zhu, Q., et al.: A Source-Based Algorithm for Near-Optimum Delay Constrained Multicasting, IEEE INFOCOM'95, pp. 377–385, 1995

A New Bandwidth Reduction Method for Distributed Rendering Systems*

Won-Jong Lee, Hyung-Rae Kim, Woo-Chan Park, Jung-Woo Kim, Tack-Don Han, and Sung-Bong Yang

Media System Laboratory, Department of Computer Science, Yonsei University, Seoul 120-749 Korea,
{airtight, kimhr, chan, mosh, hantack}@kurene.yonsei.ac.kr
yang@mythos.yonsei.ac.kr

Abstract. Scalable displays generate large and high resolution images and provide an immersive environment. Recently, scalable displays are built on the networked clusters of PCs, each of which has a fast graphics accelerator, memory, CPU, and storage. However, the distributed rendering on clusters is a network bound work because of limited network bandwidth. In this paper, we present a new algorithm for reducing the network bandwidth and implement it with a conventional distributed rendering system. This paper describes the algorithm called geometry tracking that avoids the redundant geometry transmission by indexing geometry data. The experimental results show that our algorithm reduces the network bandwidth up to 42%.

1 Introduction

As the images become increasingly complex, the size of the model data in 3D graphics grows explosively. A variety of parallel processing techniques have been researched on designing a graphic accelerator to generate high quality images at real time frame rates. Recently scalable displays are highlighted as an important parallel rendering research area. Scalable displays are the graphics hardware/software systems that can generate high resolution(several million pixels) images on the multiple displays.

There are two approaches to build parallel rendering systems for scalable displays. One method is to employ a parallel machine with extremely high-end graphics capabilities. PowerWall [1] and InfinityWall [2] are typical parallel systems. This approach is, however, limited to the number of graphics accelerators that can fit in one computer and is quite expensive. The other method is to utilize the networked clusters of PCs each of which is equipped with a fast graphics accelerator, memory, CPU, and storage.

As compared to high-end parallel computers, there are several advantages of the networked clusters. Since the processors of PCs communicate only by a network protocol, they may be added and removed from the system easily. Because

* This work was supported by the NRL-Fund from the Ministry of Science & Technology of Korea.

M.H. Shafazand and A M. Tjoa (Eds.): EurAsia-ICT 2002, LNCS 2510, pp. 387–394, 2002.

each PC has its own CPU, memory, AGP bus driving a single graphics accelerator, the aggregated hardware computing, storage, and bandwidth capacity of a PC clusters can grow linearly.

However, a cluster system does not have fast access to the shared virtual memory space and the latencies and bandwidths of inter-processor communication in the system are significantly inferior. Thus the main challenge is to develop efficient parallel rendering algorithms that scale well within the processing, storage, communication characteristics of a PC cluster[3]. AT&T InfoLab [4], Princeton, and Stanford[5] are exploring a variety of parallel rendering techniques for clusters. Stanford developed a software remote rendering system called WireGL[5]. Because it is implemented as a driver that stands in for the system's OpenGL driver, an application can render without modification.

In an immediate-mode graphics API like OpenGL, geometry data are specified by individual function calls. In scenes with significant geometric complexity, an application can perform several millions of such function calls per frame and is limited by the available network bandwidth. Since the network traffic overhead overwhelms the computation overhead on the distributed renderers, it should be a network-bound work rather than a computation-bound work. Thus the network bandwidth reduction scheme is essential.

In this paper, a new method to reduce the network bandwidth for distributed rendering systems is proposed. The patterns of geometry data that dominate network transmission are analyzed. As a result, we have found significant redundancy in the geometry data. The proposed algorithm called *geometry tracking* avoids the retransmission of redundant geometry data by indexing geometry data in a network packet. It is implemented by modifying the WireGL's source code. The experimental results with *SPECViewperf*[6] and *Quake III* are given. They show that our algorithm reduce the network bandwidth up to 42%.

The rest of the paper is organized as follows. Section 2 reviews the WireGL's data transmission scheme and the geometry patterns of the benchmarks are analyzed for measuring its redundant occurrence rate. Section 3 explains the proposed geometry tracking algorithm. Section 4 shows the experimental results. Section 5 concludes the paper.

2 Geometric Redundancy

This section describes the WireGL's data transmission scheme for distributed rendering and analyzes the redundant geometry occurrence rate.

2.1 Data Distribution Scheme in WireGL

WireGL is implemented as a graphics driver that intercepts the application's calls to graphics hardware. WireGL then distributes the application's opcodes and the data to the multiple rendering servers that are responsible for their own tiled displays. Each buffer is placed on the client and distributed servers. The opcodes and the data are stored in each buffer separately. When the geometry buffer is

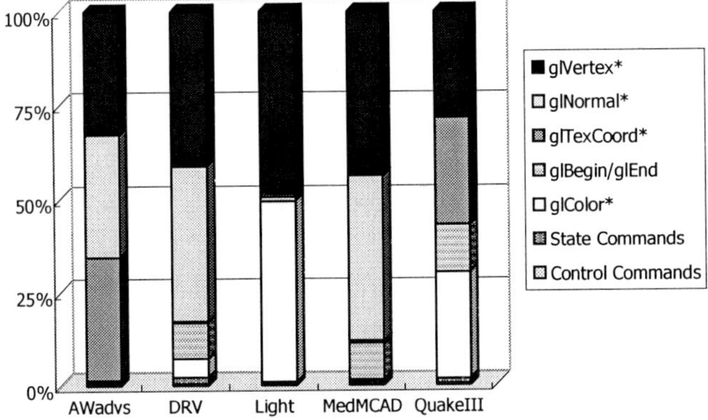

Fig. 1. Comparison of the generated data sizes by the geometry commands and other commands in the OpenGL applications renderings

full or the OpenGL state is changed, the geometry buffer is flushed. At this time the codes and the data in the geometry buffer are packed as a packet which is then transmitted to its corresponding server. That is, the network transmission is performed per packet base, where the size of the packet is the same as that of the geometry buffer.

2.2 Redundancy of the Geometry Data

In order to investigate a method to reduce the network bandwidth, we profile the patterns of the geometry data in all transmitted packets. OpenGL performance benchmarks, SPECViewperf, and the Quake III game are used for the experiment. In the case of the SPECViewperf, the several hundreds of frames of each model are rendered while the demo is rendered in the case of Quake III. Fig. 1 gives the profiling results of the transmitted data to the servers. It shows that the data generated by the geometry commands(`glVertex*`, `glNormal*`, `glTexCoord*`) are occupied from 45% to 95% and the size of the data is much larger than that of the data generated by the other commands.

Table 1 provides the average occurring rate of the redundant geometry commands in each packet, whenever the client's geometry buffer is flushed. It shows

Table 1. The average redundant geometry occurrence rate of OpenGL commands in the WireGL network packets

Benchmarks	Number of frames	Number of all vertices	Number of redundant vertices	Rate(%)
AWadvs-04	600	44,013,480	26,000,664	59.07
DRV-07	370	87,499,104	46,790,684	53.48
Light-04	100	62,312,104	38,305,636	61.47
MedMCAD-01	723	1,155,849	403,092	34.87
Quake III	1,346	13,126,642	6,027,431	45.91

that the average redundant occurrence rates of the geometry data in WireGL packets ranges between 35% and 61%. Thus if these redundant data are not retransmitted and the previous transmitted data are reused, the overall network bandwidth will be reduced considerably.

3 Geometry Tracking

In this section, the network bandwidth reduction scheme called geometry tracking algorithm is proposed. Its implementation by modifying WireGL is then described.

3.1 Geometry Tracking Algorithm

The proposed geometry tracking algorithm performs the indexing and the tracking for all geometry data in a single packet. Then, if the same arguments values are occurred repeatedly, the geometry data of the vertex need not be retransmitted. Only its index value is transmitted.

To transmit the index value for the redundant geometry commands and to detect redundant geometry commands in the servers, the following three kinds of new commands are defined. `glIndex_Vertex*`, `glIndex_Normal*`, and `glIndex_TexCoord*`. These index commands are used only within the client-servers network and are hidden to applications. The fig. 2, 3 show the geometry tracking algorithm for the client and the server.

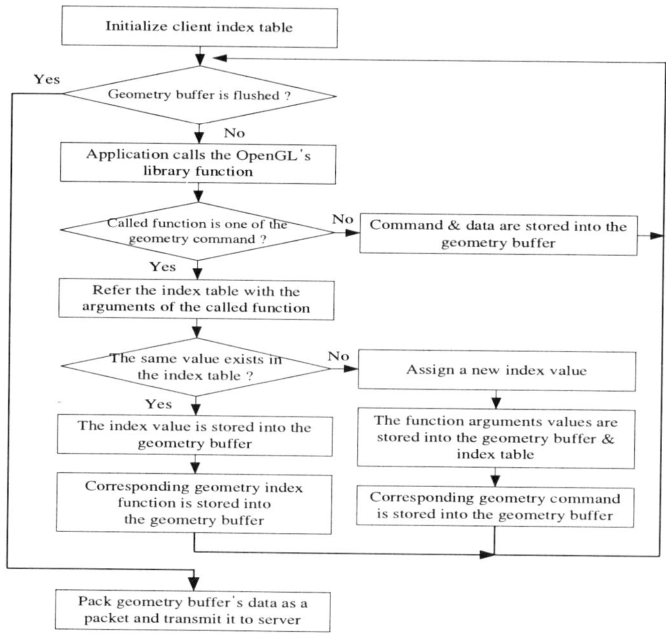

Fig. 2. The flow of geometry tracking algorithm for client behalf

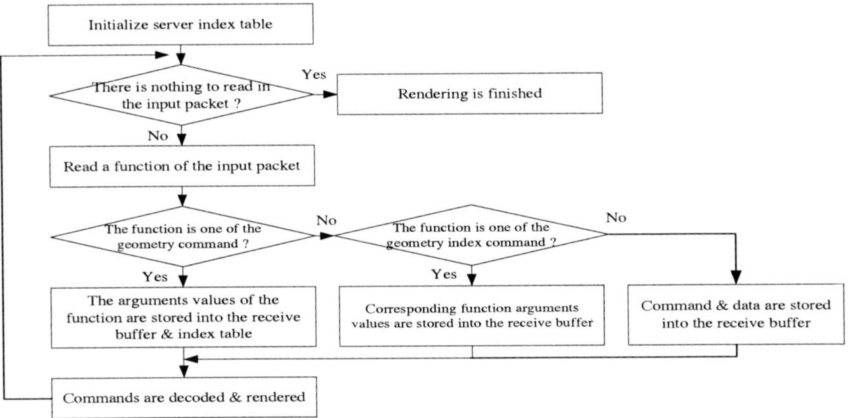

Fig. 3. The flow of geometry tracking algorithm for server behalf

3.2 Implementation on WireGL

On the client's behalf, when the application calls the OpenGL library function, WireGL intercepts application's call to the graphics hardware. If the called function is one of the geometry commands(`glVertex*`, `glNormal*`, `glTexCoord*`), the index table is referred with the corresponding function arguments values. If these values already exist in the index table, it means that the same geometry command has been called previously. Thus the corresponding index value is stored into the geometry buffer instead of the arguments values of the called function. The corresponding index command is then encoded and stored. If the arguments values of the called function do not exist in the index table, it means that the same geometry command is not called yet. The arguments values of the called function are stored into the geometry buffer directly and into the index table with the newly assigned index value. This procedure is iterated until the geometry buffer is flushed. Then the geometry data and the index values are transmitted to the distributed servers.

On each server's behalf, the opcodes and the data of the transmitted packet are read in order. If the corresponding function of the opcode is one of the geometry commands, which means that it is not a redundant opcode, thus it is stored into the server's index table and into the receive buffer directly. If the corresponding function of the opcode is one of the index geometry commands(`glIndex_*`), the server's index table is referred with the corresponding index value. Then the arguments values for this function are read from the index table and are stored into the receive buffer. Also the original opcode is restored and is stored into the receive buffer. Finally the opcodes of the receive buffer are decoded and rendered on the corresponding server.

Fig. 4 shows a snapshot after the execution of the 25th line of the application source code. The geometry buffer is flushed either when the buffer is full or when the state is changed. On the client's behalf, since the state of the color was changed in the 15th line as compared with the 2nd line, the geometry buffer was

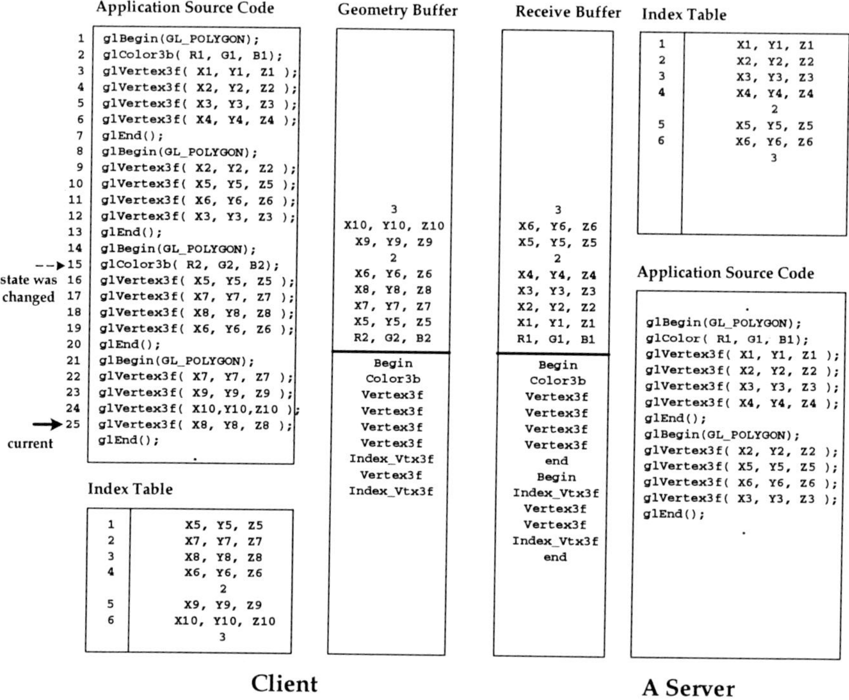

Fig. 4. A snapshot of the proposed geometry tracking scheme

flushed. Then the generated and accumulated data by application source codes between the 1st line and the 15th line were transmitted. On the server's behalf, the geometry data in the receive buffer were read in order and the index table was constructed. The application source codes was then generated by decoding with the opcodes in the receive buffer and the data of the index table.

After the packet was transmitted, the client's index table was reinitialized. In the figure, the opcodes and the data between the 16th line and the 19th line were stored into the index table and the geometry buffer in order. Because the command `glVertex3f` in the 22nd line has the same vertex data as that in the 17th line, the index value of the data in the 17th line, which is 2, was stored into the geometry buffer for the vertex data in the 22nd line. In the same manner, the index value of the vertex data in the 18th line, which is 3, was stored into the geometry buffer for the vertex data in the 25th line.

4 Experimental Results

We built an 8-node cluster that consists of 8 rendering servers as well as a client machine. Each rendering server contains a Pentium IV 1.6GHz processor and an NVIDIA GeForce3 Ti 200 graphic accelerator. The client machine contains a dual AMD Athlon XP 1800+ 1.5GHz and an NVIDIA GeForce3 Ti 200 graphic accelerator. The cluster is connected with a high speed network. Each

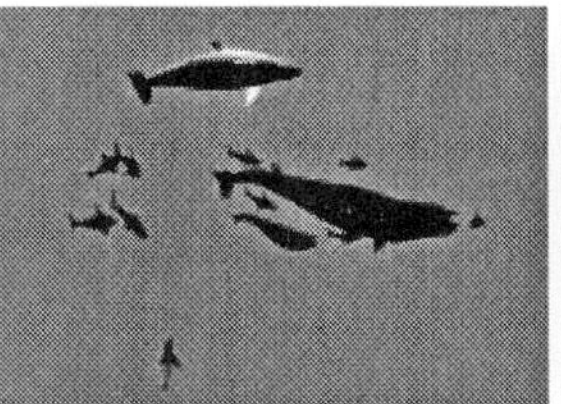
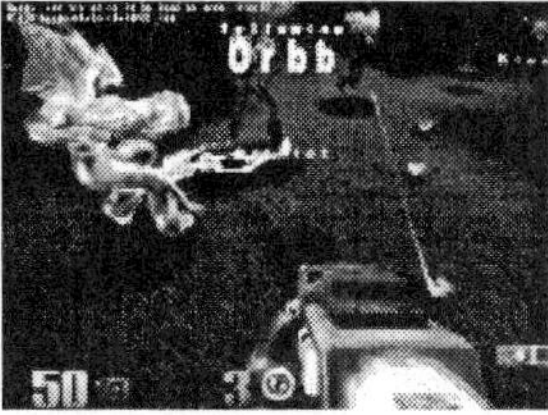

Fig. 5. Three benchmarks to be experimented. DRV-07, Atlantis, Quake III

rendering server outputs a 1024x768 resolution video signal. We experimented on this cluster with the implemented software. Fig. 5 shows the benchmarks to be tested our system for the experiment.

1. *DRV-07* is a product of the SPECViewperf. It contains 367178 vertices in 42821 primitives and its size is greater than 50 megabytes. It is used as a benchmark because of its high scene complexity compared with other SPECViewperf products.
2. *Atlantis* is one of the OpenGL demo programs. It simulates a pool of swimming sharks, whales, and a dolphin. The body poses for each object are computed in real time. Because it is rendered infinitely, it is limited by 500 frames.
3. *Quake III* is one of the typical current video games and is frequently used as a benchmark. One of its demos that contain 280 frames is tested.

The number of vertices in a packet can be increased since the redundant vertex can be stored with the small size into the geometry buffer by the geometry tracking method. In Table, the geometry tracking method and WireGL are compared in terms of the number of overall transmitted packets. They are tested the single rendering server and client. As shown in Table 2, it can be reduced up to 45%.

Each application is tested on four different server configurations: 1x1, 1x2, 2x2, and 2x4 until each rendering is finished. Fig. 6 shows that the total sizes of the transmitted data as the number of servers increases. In order to compare with WireGL's data sent to the servers, the relative transmission rate which is the ratio of the data sent to the servers to WireGL's data sent to a 1x1 configuration is measured. According to Fig. 6, the traffic reduction rate in DRV-07 ranges between 11% and 27%, the rate in Atlantis ranges between 34% and 42%, and the rate in Quake III ranges between 8% and 12%. Since Quake III has heavy texture images that should be sent to each sever, the reduction rate is not better

Table 2. Comparison of the number of overall transmitted packets

	DRV-07	Atlantis	Quake III
WireGL	774,806	2,723	10,228
Geometry tracking	494,572	1,489	9,096
Reduction rate (%)	36.17	45.32	11.08

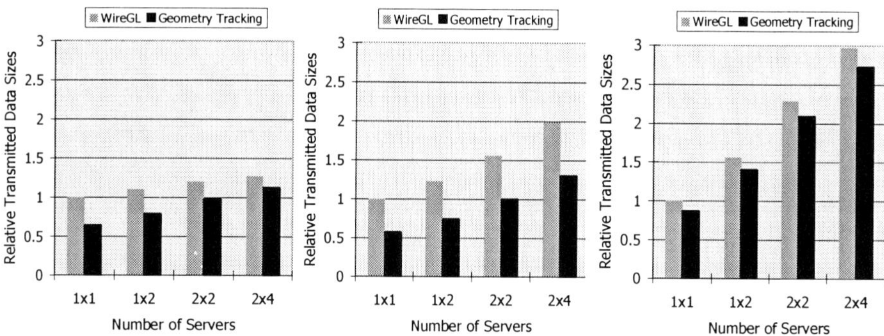

Fig. 6. Relative transmitted data sizes with increasing number of servers. DRV-07, Atlantis, Quake III

than those of other benchmarks. Consequently geometry tracking can reduce the network bandwidth up to 42%.

5 Conclusion

This paper has introduced a new algorithm called geometry tracking that can reduce the network bandwidth by indexing the geometry data. The geometry tracking algorithm can reduce the network bandwidth up to 42%. However, the texture-intensive application like a Quake III cannot reduce the rate as much as other applications can because of its heavy texture traffic.

The geometry tracking algorithm will be extended to reduce the texture traffic and will be used for other applications that have more redundancy such as volume rendering. The algorithm is expected to improve the overall rendering performance for volume rendering systems.

References

1. University of Minesota PowerWall, http://www.lcse.umn.edu/research/powerwall/powerwall.html
2. Marek, C., Dave, P., Daniel, S., Tom, D., Gregory, L.D., Maxine, D.B.: The ImmersaDesk and InfinityWall Projection-based Virtual Reality Displays, ACM SIGGRAPH Computer Graphics, Vol. 31. ACM Press, New York (1997) 46–49
3. Dirk, B., Bengt-Olaf, S., Claudio, S.: Rendering and Visualization in Parallel Environments, Proceedings of the SIGGRAPH 2000 Course Notes (2000)
4. Bin, W., Claudio, S., Eleftherios K., Shankar K., Stephen N.: Visualization Research with Large Displays, IEEE Computer Graphics and Applications, Vol. 20, No. 4, July/Aug. (2000) 50–54
5. Greg, H., Matthew, E., Ian, B., Gordan, S., Matthew, E., Pat, H.: WireGL: A Scalable Graphics System for Clusters, Proceedings of the 2001 Conference on Computer Graphics (2001) 129–140
6. SPECViewperf, http://www.spec.org/gpc/opc.static/opcview.htm

Neural Networks Based Mesh Generation Method in 2-D

Çınar Ahmet and Arslan Ahmet

Fırat Universitesi, Muhendislik Fakultesi,
Bilgisayar Muhendisligi, 23119, Elazig, Turkiye
acinar@firat.edu.tr

Abstract. This paper describes a novel method for mesh generation in 2D by means of feed forward single layer neural networks. Original values of an initially given boundary are represented by finite values instead whole of points. For this aim, b-spline control points are made up to boundary curve. The obtained control points are used as inputs of the single layer feed forward neural network and points which belongs to closed area are obtained from output of neural networks. Obtained points are meshed by proposed method. As application, some mesh samples are given.

1 Introduction

Mesh is especially used in the CFD (computational fluid dynamics) for analyzing a specific region by dividing it into sub regions and preparing data with respect to realizing method(1,2). Mesh is obtained by dividing surface into triangles or rectangles and volume into tedrahedrals or hexahedrals(4,5). The goal of meshing process is to prepare appropriate data for analyzing method. Before solving a numerical systems, a mesh which is suitable to structure of system should be prepared. Depending on physical phenomena, the middle point distribution of generated mesh can vary. Numerical mesh generation method satisfies flexible distribution along area to be solved(3,7). This process is realized by using structured or unstructured meshes. It is known that, unstructured mesh algorithms requires large memory and long time solving time requirements and it is difficult to prepare data for solving (2).

In this paper, a new unstructured mesh generation algorithm basing on neural networks is introduced. The algorithm is summarized as follows:

Step 1. To make up control points to initially given boundary curve. Thus boundary curve is represented with finite control points instead of whole points. If boundary curve is given as a mathematical function, then process on step 1 does not need. But, generally boundary curves are represented as a free form.

Step 2. To apply to inputs of single layer feed forward neural networks of control points which computes at the step 1 and obtaining inner points from out of neural networks, all of which will be meshed. At this stage, whether or not inner points belong to closed area needs to determine.

Step 3. Meshing of points that are calculated in the step 2 by means of developing a new method.

The reason of using neural network is that, complex inner points calculations do not need. Because, outputs of neural networks are directly inside points.

M.H. Shafazand and A M. Tjoa (Eds.): EurAsia-ICT 2002, LNCS 2510, pp. 395–401, 2002.

The organization of paper is as follows: Section 2 is to make up b-spline control points to boundary of the initially given closed region. In Section 3, the proposed method is described and model of neural networks is given for evaluating internal points and some 2D generated meshes are given at finally conclusion is offered.

2 Discretization of Boundary Curves

This operation is originally given in Wang (1995) (8).
For $n+1$ control points; $\{p_0,p_1,\ldots,p_n\}$, ^{ith} b-spline function is as follows (6).

$$B(t) = \sum_{i=0}^{n} p_i N_{i,k}(t) \qquad 1 \le k \le n \tag{1}$$

where, $N_{i,k}(t)$ is called as blending function and k is the degree of generated curve.

$$N_{i,k}(t) = \left(\frac{t - t_i}{t_{i+k-1} - t_i}\right) N_{i,k-1}(t) - \left(\frac{t_{i+k} - t}{t_{i+k} - t_{i+1}}\right) . (N_{i+1,k-1}(t)) \tag{2}$$

$$N_{i,1}(t) = \begin{cases} 1 & \text{if } t_i \le t \le t_{i+1} \\ 0 & \text{Others} \end{cases}$$

T *is* knot vector, and its number is $n+k+2$ and its maximum value is $n\text{-}k+1$ for open uniform b-spline curve

$$T = \{t_0, t_1, t_2, \ldots\}, \tag{3}$$

The reason of selecting b-spline curve is that, it has continuity by certain degree at the all of points

Let $f(x)$ be a function to represent given boundary curve.

$$f(x) = \sum_{i=0}^{n} p_i N_{i,k}(x), \quad 1 \le k \le n \tag{4}$$

Let $d(x_j)$ be control points of the given boundary curve. Then error between original point and its control point is as follows:

$$e(x_j) = d(x_j) - f(x_j) = d(x_j) - \sum_{i=0}^{n} p_i N_{i,k}(x_j) \tag{5}$$

J is mean squared error .

$$J = \sum_{j=0}^{n} e^2(x_j) = \sum_{j=0}^{m} \left[d(x_j) - \sum_{i=0}^{n} p_i N_{i,k}(x_j) \right]^2 \tag{6}$$

Partial derivative of J with respect to p_i should be zero to minimize J.

$$\frac{\partial J}{\partial p_i} = 2\sum_{j=0}^{m} \left[d(x_j) - \sum_{i=0}^{n} p_i N_{i,k}(x_j) \right] \left[-N_{i,k}(x_j) \right] = 0$$

$$\Rightarrow \sum_{j=0}^{m} d(x_j) N_{i,k}(x_j) = \sum_{j=0}^{m} \sum_{l=0}^{n} p_l N_{l,k}(x_j) N_{i,k}(x_j) \tag{7}$$

By letting

$$Q_i = \sum_{j=0}^{m} d(x_j) N_{i,k}(x_j), \tag{8}$$

$$N_{il} = \sum_{j=0}^{m} N_{i,k}(x_j), N_{l,k}(x_j), \tag{9}$$

Lastly, given boundaries are represented as b-spline control points (8).

$$f(x) = \sum_{i=0}^{n} p_i N_{i,k}(x) \tag{10}$$

f(x) is original values and p_i is the computed control points.

3 The Use of Neural Networks for Mesh Generation

This section describes how to compute internal points requested for meshing of closed region by means of feed forward neural networks. P_i is the set of points, all of which represent border with less points. Inputs of neural networks are points *pi.* Point that will obtain from output of neural networks will be evaluated for inner region and if the points belong to inner area, then points are belong to set which will be meshed.

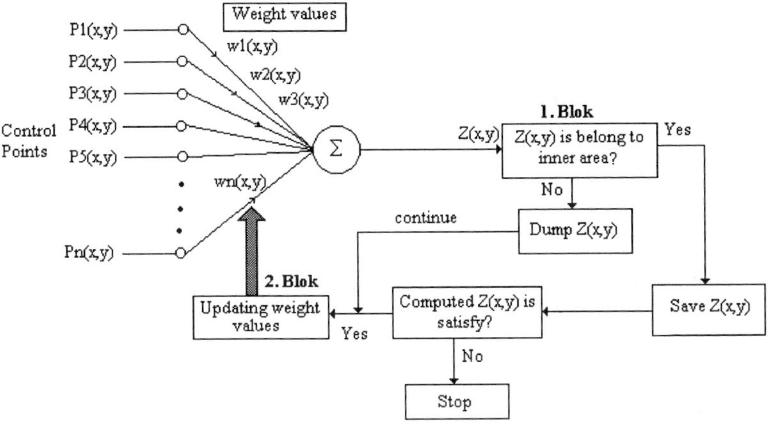

Fig. 1. Block diagram of proposed method.

Neural network model used in this study is a single layer and single-output neural network. Different points are obtained by adjusting values of neural networks. Inner points are acquired depending on whether or not points are belonging to inner area.

Figure 2 shows a sample closed area. The control points for this are *(2.0,2.0),(4.0,2.0),(5.0,4.0)* and *(3.0,4.0* object with respect to error *J=0.001*. Because shape is geometric, knowing corners of shape is satisfied to define it.

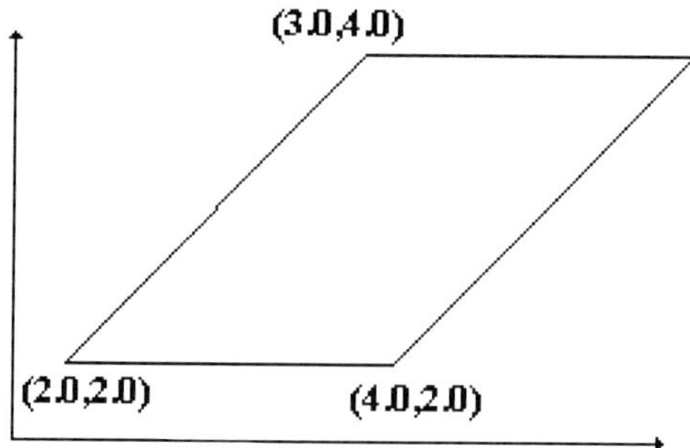

Fig. 2. A sample closed area.

These control points are inputs of neural network. Following table shows some values obtained for different weights.

Table 1. Weight values for a sample closed area and some values, which is into closed area.

Values	X=3.23	Y=3.40	X=3.87	Y=3.74	X=3.68	Y=3.69	X=3.82	Y=3.87
Values of Weights	0.0540	0.2507	0.2171	0.4095	0.3657	0.0292	0.0631	0.3249
	0.0592	0.0237	0.1320	0.0448	0.1056	0.0792	0.1620	0.5100
	0.1242	0.0339	0.2997	0.1185	0.0416	0.1556	0.1911	0.0016
	0.1415	0.6521	0.2060	0.0859	0.2884	0.4963	0.2963	0.0677
	0.1360	0.0531	0.0256	0.1396	0.0803	0.1102	0.0650	0.1458
	0.1849	0.1407	0.0591	0.0027	0.1549	0.0636	0.0861	0.0841
	0.2448	0.0998	0.3026	0.2344	0.3168	0.0430	0.0450	0.2034
	0.0185	0.0728	0.0730	0.1952	0.0062	0.2308	0.3035	0.0980

The following method is applied to find whether or not the computed values are belong to inner area. The problem of whether or not given a point is belong to closed area is commonly knowing problem in the computer graphics(6). In this study, vector product is applied to this problem. Let *z(x,y)* point be the obtained value by vector product. Figure 3 shows this process. Obtained value is belong to inner area when direction of vector *v1,Z(x,y)* is down of plane or paper, otherwise is not.

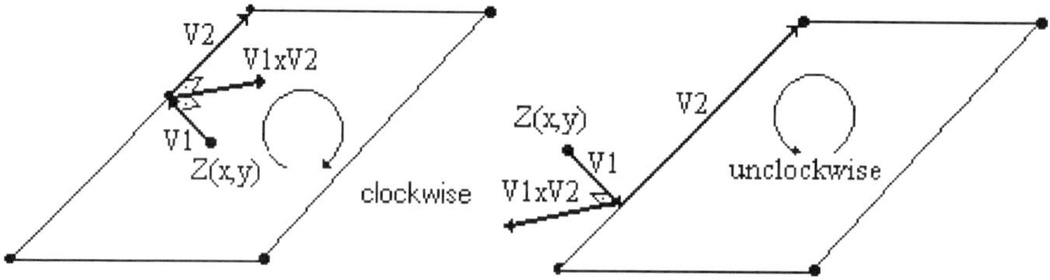

Fig. 3. Determination of whether computed *Z(x,y)* belongs to closed area or not. a) Point *Z(x,y)* belongs to closed area because *V1xV2* is the direction of down paper. b) Point *Z(x,y)*is the excluded point.

After making this computation, point in the inner area is produced by number of points. Control points computed for different weight values are given in table 1.

In this phase, scattered data is meshed by developed method. For this process, following method is applied; Let n be number of points, which will determine closed area. P_i is the point ; where $i=0,1,...,n-1$. To define border of shape is satisfy these points. Let m be number of points computed inner area and points t_i are inner points; where $i=0,1,2,...,m-1$. According to given method, figure 4 shows step-by-step meshing process of shape in the figure 2.

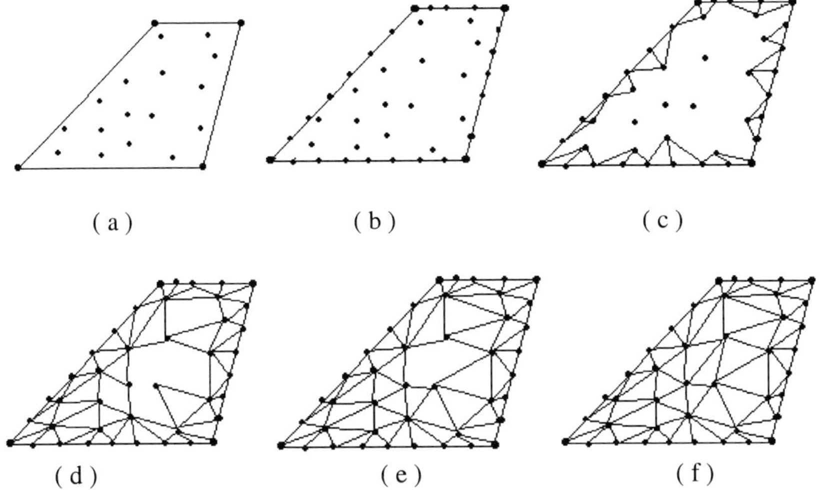

Fig. 4. Step-by-step meshing process.

Developing method is summarized as follows via figure 4;

a) Control points on the original shape and points, which will be meshed. Internal points and corner points are meshed.
b) Control points don't satisfy to be meshed fully. So, control points is divided according to shortest distance among each for other of internal points.
c) Meshing process is done by beginning from boundary. To determine near points to boundary, the nearest two points on the boundary are calculated, and then distance between these two points and the computed value by product with *1.5* this distance is accepted the inner value near to boundary. Afterwards, on the direction of unclockwise, three points are connected to generate the mesh . In this case, it must be noticed intersection situations.
d) Process in the step (c) is repeated for whole of inner points, which are near to boundary curve. In this stage, boundary of inner area is accepted new boundary.
e) After then, new boundary is taken. Step c and step d are repeated for other inner points.
f) Inside checking process is done so that unmeshed points are not allowed.

Figure 5 shows an other object meshed by means of developed method.

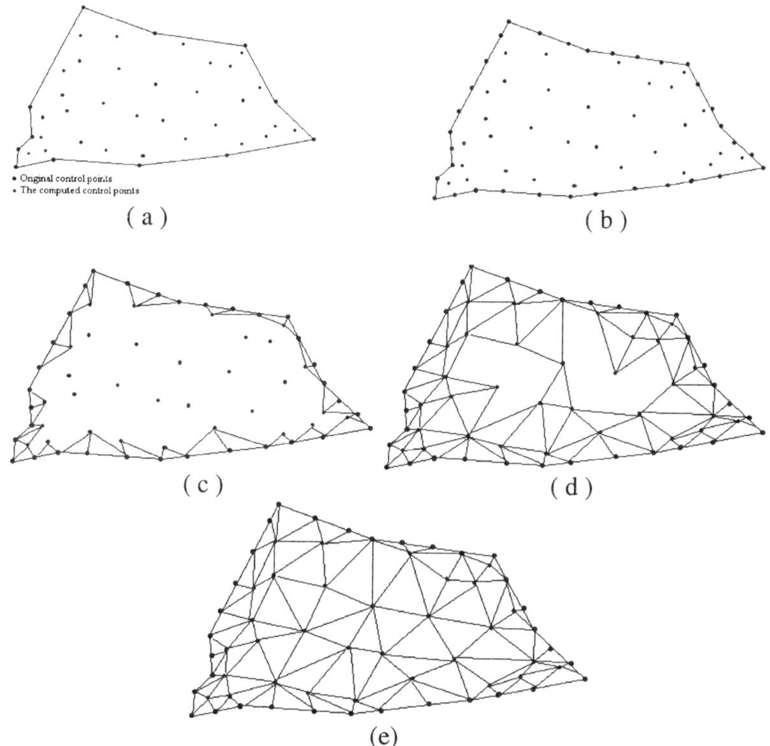

Fig. 5. Step by step meshing process of a sample object.

4 Conclusion

In this paper, preparing of data needed to mesh in 2D and their meshing are done by neural networks. The developed method differs from other methods because it doesn't need coordinate transform, and so more mathematical calculations don't requested. By using neural networks, computation time is decreased. Method perfectly works provided that, weight values are suitable. If weight values is not appropriately selected, then computed values can be outside of the shape. As future work, this anxiety (trouble) will be got ride of. For this process, a new technique based upon fuzzy logic will be used and thus problem will be directly solved.

References

1. Steven. Owen, Meshing Software Survey, URL: http:/www.andrew.cmu.edu/user/sowen /softsurv.html (1998)
2. David L. Marcum, "Unstructured Grid Generation using Iterative Point Insertion and Local Reconnection" AIAA Journal, (1995), 33 (9) 1619-1625

3. R. Lohner, "Progress in Grid Generation via the Advancing Front Technique" Engineering with Computers, (1996), 12, 186-210
4. Scott A. Mitchell, "High Fidelity Interval assignment "Proceedings 6th International Meshing Roundtable, (1997), 33-44
5. Matthew L. Staten, Steve Owen, "BMSWEEP: Localing Interior Nodes During Sweeping" Proceedings 7th International Meshing Roundtable, (1998), 7-18
6. D.F. Rogers, J.A. Adams",Mathematical Elements for computer Graphics, 2nd Edition NewYork, McGrawHill,1989
7. Robert Schneiders, "A Grid-Based algorithm for the Generation of Hexahedral Element Meshes " Engineering with Computers, (1996), 12, 168-177
8. Wang Chi-Hsu,Wang Wei-Yen, Lee Tsu-Tian, Tseng Pao-Shun, Fuzzy B-Spline Membership Function (BMF) And Its Application in Fuzzy-Neural Control, IEEE Transactions On System Man And Cybernetics, (1995), Vol. 25, No.5 May, pp: 841-851.

Image Denoising Using Hidden Markov Models

Leila Ghabeli and Hamidreza Amindavar

Amirkabir University of Technology
Department of Electrical Engineering
Hafez Avenue , 15914
Tehran, Iran
Tel: (98)(21) 64543362
Fax: (98)(21) 6406469
{g7823257, hamidami}@cic.aut.ac.ir

Abstract. In this paper a new method is proposed for image denoising based on HMM (hidden Markov modeling). In this manner the noisy imageis modeled as a hidden Markov process, we use the local statistics of the images in defining the HMM parameters, and the image restoration is achieved by computing the most likely state at each pixel. Among the features of the proposed method is the adaptive window size for different regions of the image (smooth and non-smooth regions). The adaptive window size allows us to obtain a better estimate of the local variance of the noise, therefore, the additive noise is removed more in the smooth regions while the edges are preserved in nonsmooth ones. Another feature of this method has to do with the proportionality of the execution time and the noise power; the less the noise power, the faster is the execution of the proposed algorithm. The performance of this algorithm is evaluated through subjective and objective criteria and it is shown that the restored images by HMM have higher contrast and clearness which is attributed to nearly optimal usage of the statistical properties of the image by HMM.

1 Introduction

Traditionally, there are three basic approaches to image denoising, spatial filtering methods, frequency domain filtering, and entropy methods. In special smoothing methods such as mean and the median filtering [1,2], the noise is removed without any attempt to explicity identify it. On the other hand, in frequency smoothing methods [3,4] the removal of the noise is achieved by designing a frequency domain filter and adapting a cut-off frequency when the noise components are decorrelated from the useful signal in the frequency domain. These methods are time consuming and de-pend on the cut-off frequency and the filter function behavior. Furthermore, they may produce artificial frequencies in the processed image. In the entropy based methods [5], a noise-filtering treatment can be obtained by decreasing the entropy of the local contrast in a given neighborhood and they may be useful in removing the impulse noise.

In this paper we present a new method for image denoising based on HMM. The HMM approach [6] is a well-known and widely used in characterizing the spectral

M.H. Shafazand and A M. Tjoa (Eds.): EurAsia-ICT 2002, LNCS 2510, pp. 402–409, 2002.

properties of the frames of a pattern. HMM is a useful tool for image denoising because it can draw the maximum amount of information about the nonstationary statistics of the image, hence, we have a better restored image.

In the proposed method the noisy image is modeled as a hidden Markov process, we use the local statistics of the images in defining the HMM parameters and we assume the local stationary and gaussian distribution in a neighborhood of each pixel, To estimate the local variance of the original image from the noisy image the size of the processing window is adaptively increased in smooth regions (that their variances are small). By this technique we treat with two kinds of regions (smooth and nonsmooth regions) differently and so the higher performance are acheived. finally the image res-toration is achieved by computing the most likely state at each pixel. We also present an adaptive algorithm to reduce the required time at low noise power.

To measure the performance of the proposed algorithm, the root mean square (RMS), and logarithmic mean square error (LOGMSE) are utilized. Because of the nonlinearity response of the human visual system to the light intensity the LOGMSE is a better criterion than RMS.

The restored images by the HMM method have higher clearness, contrast and lower RMS and LOGMSE in comparison with the other methods such as wiener, mean, median ,.... This is due to the nearly optimal usage of statistical properties in hidden Markov modeling of noisy images.

The paper is organized as follows: In section II, we provide the hidden Markov modeling of the noisy image and denoising using HMM method are descirbed. Results of simulations are in section III. In section VI some concluding remarks are provided.

2 Hidden Markov Modeling of Images

A noisy image model is formulated by integrating the image and the noise models into the hidden and observation layers of a hidden Markov model. To model the original image s(x,y) as a Markov process, The No. of graylevels that is used to show the pixel's brightness, is assumed to be the states of the model e.g. for images with 8 bit/pixel, we have 256 states $\mathbf{q} = (0, 1, \ldots, 255)$ and we assume that this Markov proess is hidden and is indirectly observed by the noisy image o(x,y).

$$o(x, y) = s(x, y) + n(x, y) \qquad n(x, y) \propto N(0, \sigma^2), \tag{1}$$

where σ^2 is the noise variance. By this definition the symbol probabilities are computed as follows:

$$\begin{aligned} b_i[o(x, y)] &= f[o(x, y) \mid s(x, y) = q_i] \\ &= \frac{1}{\sqrt{2\pi\sigma^2}} \times \exp\left(- \frac{(o(x, y) - q_i)^2}{2\sigma^2} \right), \end{aligned} \tag{2}$$

and also we need to determine the state transition probability matrix, the elements of this matrix, $a_{ij}(k)$, expresses the transition probability from ith state to jth state at kth pixel.

$$a_{ij}(k) = pr(s_k = q_j \mid s_{k-1} = q_i) , \tag{3}$$

where s_{k-1}, s_k denote the graylevels of adjacent pixels at (k-1), (k) th pixel respectively. In order to avoid the blurring effect in the restored image we neglect the correlation coefficient of the adjacent pixels, (because we have to estimate this value by using the small number of pixels in a processing window so we cannot acquire an accurate value for it). By this assumption we have

$$a_{ij}(k) = pr(s_k = q_i) = a_j(k) , \tag{4}$$

Most of the estimators assume local stationary statistics for the images and use the gaussian function to model the probability density function of the pixels in a neighborhood

$$a_j(k) = \frac{1}{\sqrt{2\pi\sigma_s^2}} \exp\left(- \frac{\left(q_j - \mu_{s_k}\right)^2}{2\sigma_{s_k}^2} \right), \tag{5}$$

where μ_s and σ_s^2 are the local mean and variance of the pixels in the processing window centered by the pixel $'s'$ and they are computed from the noisy image. By the assumption that the signal and noise are uncorrelated these values are computed via the following equations

$$\tilde{\mu}_{s(x,y)} = \tilde{\mu}_{o(x,y)} = \frac{1}{N_B^2} \sum_{m=-b}^{b} \sum_{n=-b}^{b} o(m,n) , \tag{6}$$

$$\begin{aligned} \hat{\sigma}_{s(x,y)}^2 &= \hat{\sigma}_{o(x,y)}^2 - \hat{\sigma}_n^2 \\ &= \frac{1}{N_B^2} \sum_{m=-b}^{b} \sum_{n=-b}^{b} \left(o(m,n) - \hat{\mu}_{o(i,j)}\right)^2 - \sigma_n^2 , \end{aligned} \tag{7}$$

N_B is the size of window and should be an odd number (N_B=2b+1).

From equation 7, we see that it is possible to have negative values for variance σ_s^2 in the regions with small value σ_o^2 but larg σ_n^2 (smooth regions), In these regions the size of window will be increased until the positive value is reached for the variance σ_s^2 in Eq. 7. Consequently, in the smooth regions (with small σ_o^2), μ_o is computed in larger window and for more number of pixels, so the effect of the noise is more reduced in these regions, and in nonsmooth regions (with large σ_o^2), the mean μ_o is computed in small window and so the effect of the edges is preserved.

2.1 Image Restoration Using HMM Method

Finally, the restored image is acquired by estimating the most likely state from each noisy pixel as follows

$$\delta_k(j) = a_j(k) \times b_j(o_k), \tag{8}$$

$$j_w = \max_{1 \le j \le N} [\delta_k(j)] , \tag{9}$$

$$\hat{s}_k = q_{j_w} , \tag{10}$$

where $b_j(o_k)$ denotes the probability that the kth noisy pixel o_k is equal to the jth state q_j and it is computed via Eq. 2. $a_j(k)$ denotes the probability that the true value of the kth pixel s_k is equal to q_j and is computed by the aid of its neighborhood pixel values according to Eq. 10. Hence, we obtain the estimated value for the kth pixel s_k is the state q_j that has the most $[a_j(k) \times b_j(o_k)]$.

All the graylevels in the restored image are integer and are in the range [0,255], and there are some errors in restored image (perfect restoration is impossible). These errors are also integer and in the range [0 255]. Also they don't extend to all pixels so they are very similar to the impulsive noise. In order to reduce the effect of these errors, we use CWM which is capable of reducing the impulsive noise. To remove the impulsive noise the CWM filter is prefered to the other methods like mean and wiener filter because it yeilds minimum smoothing to the image and it mainly influences the noisy pixels and has the least effect on the noise free pixels.

Next, we present an algorithm to reduce the computational load by refering to the normal standard table.

2.2 Reduced Computational Algorithm

We present here a reduced computational algorithm where the required time changes is proportional to the noise power. In this method by using the standard normal curve table we determine the following probabilities for the magnitude of the noise.

$$\Pr(|n| < 3\sigma_n) = \%98 \ , \Pr(|n| < 2\sigma_n) = \%94 \ , \ \Pr(|n| < \sigma_n) = \%62,$$

where σ_n denotes the noise standard deviation and |n| is its absolute value of the noise, consequently, we use these probabilities in the proposed algorithm, and then a_{ij} is computed only for the states are located in a finite duration e.g.

$$o_k - 2\sigma_n < q_j < o_k + 2\sigma_n \qquad \text{for} \qquad \Pr(|n| < 2\sigma_n) = \%94,$$

for example if σ_n =25, a_{ij} is computed nearly for 100 states (instead of 256 states). By using this technique the required time to execute the algorithm is decreased especially for small noise power. Through extensive simulations we determined the proper value for maximum noise value is $3\sigma_n$. This is appropriate in terms of execution time and the RMS value. In Fig. 1, the graghs of RMS and the required time versus k ($\times\,\sigma_n$) for cameraman image at different values of k, (0<k <5) are shown.

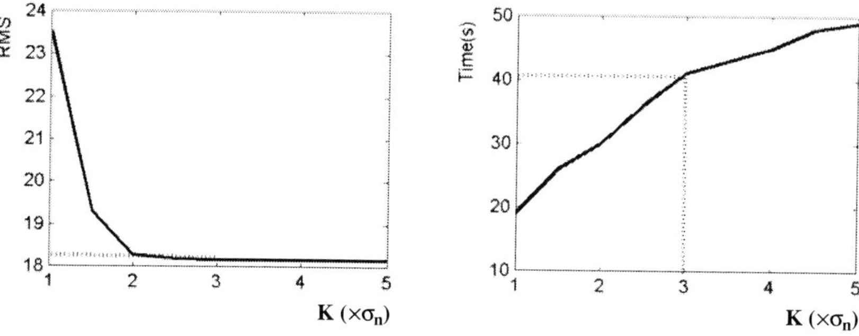

Fig. 1. Plot of RMS and time required vs. K($\times\sigma_n$) for cameraman image at SNR = 10 dB.

3 Simulation

We examined the performance of the proposed HMM to denoise images in cascade with CWM for three different images 1) camareman with high details 2) woman with high frequency 3) lake with smooth regions, at different SNR. The same scenarios were also tested for the well know wiener [3], and mean filter, these two methods have proven themselves effective in removing the white Gaussian noise.

The results are shown in figures 2–4. The restored images by this algorithm have higher contrast , clearness and lower RMS and LOGMSE value in comparison to the other methods.

Fig. 3 shows the results for the lake image at SNR = 0 dB, from this figure we observe the ability of HMM to reduce the noise effect in the smooth regions.

Fig. 4 shows the results for the woman image at SNR=20dB, among these four methods only HMM reduced the value of RMS in comparison with the noisy image.

The woman image is a high frequency image that treats like the noise and most of the methods for image denoising tend frequency image, thus it behaves similar to the additive noise. Majoriety of the image denoising methods tend to smooth out the noise by averaging, this results in the elimination of the high frequency components of the images. But in HMM, due to the optimal usage of the local statistics and adaptive selection of the processing window size the high and low frequencies of the images are preserved more in comparison with the other existing methods.

To quantitavely measure the performance of the proposed method, two criteria, RMS and LOGMSE were used. LOGMSE is defined as

$$LOGMSE = \frac{\sum_{j=1}^{N}\sum_{i=1}^{N}\{\log_{10}(1+s(i,j)) - \log_{10}(1+\hat{s}(i,j))\}^2}{\sum_{j=1}^{N}\sum_{i=1}^{N}\{\log_{10}(1+s(i,j))\}^2}, \tag{11}$$

(a) (b) (c)

(d) (e) (f)

Fig. 2. (a) Original image, Cameraman 256×256, (b) Noisy image at SNR=10dB, RMS=42.56, (c) 3×3 mean filter, RMS=25.62, (d) 3×3 wiener filter, RMS=19.62, (e) HMM method MS=18.15, (f) HMM & 3×3 CWM, RMS= 15.77

(a) (b) (c)

(d) (e) (f)

Fig. 3. (a) Original image, Lake 256×256, (b) Noisy image at SNR=0 dB, RMS = 141.52, (c) 5×5 mean filter, RMS=53.23, (d) 5×5 wiener filter, RMS=42.2, (e) HMM method RMS=44.64, (f) HMM & 3×3 CWM, RMS= 32.48

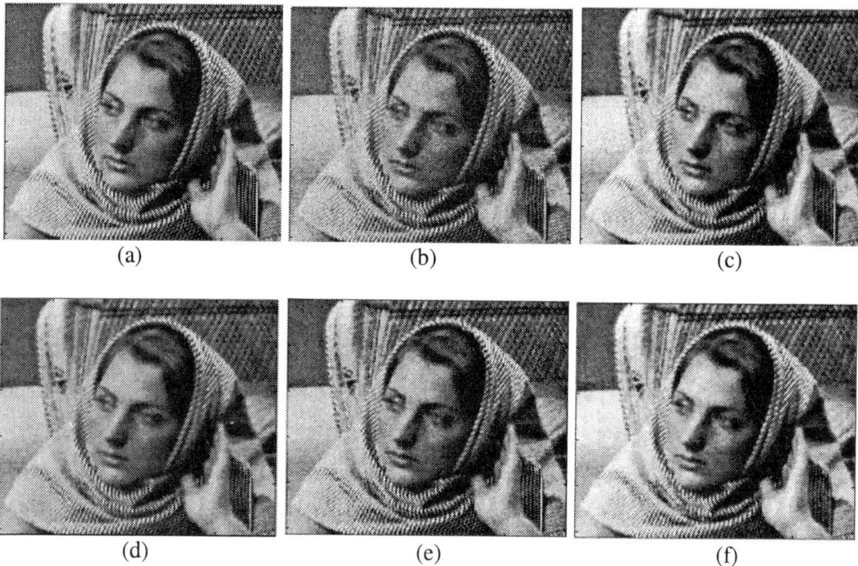

Fig. 4. (a) Original image, Woman 256×256, (b) Noisy image at SNR = 20 dB, RMS = 12.89, (c) 5×5 mean filter, RMS=21.7, (d) 5×5 wiener filter, RMS=14.7, (e) HMM method RMS=10.81, (f) HMM & 3×3 CWM, RMS= 14.73.

where s(m,n), $\hat{s}(m,n)$ denote the original image and the shifted filtered version of the noisy image o(m,n) respectively.

The plot of RMS and LOGMSE vs. SNR are shown in figure 5. From this we observe the higher performance of (HMM&CWM) method in comparison with the other methods. But because of the smoothing property of CWM filter, for low SNR ratio, HMM yield better results than (HMM &CWM).

As it is mentioned before the LOGMSE criterion is closer to the subjective criteria, and from fig. 5 it is seen that the high quality of the restored images by the HMM in comparison with the other methods is evaluated more correctly based on this subjective criterion.

The important property of the images restored by HMM method is that the pixel's graylevels of the restorted images are in the actual range [0,255], this results in the higher contrast restored images in comparison with the other methods.

4 Conclusions

In this paper, we presented HMM for image denoising. The proposed algorithm extracted the maximum amount of information about the local statistics of the original image along with the noise statistics. We used this information for the image restoration. Among features of the proposed algorithm is the adaptive window size for different regions of the image, i.e., smooth and nonsmooth regions.

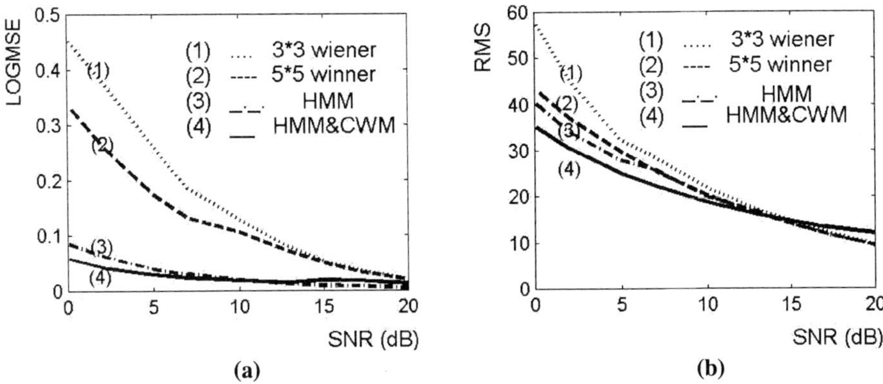

Fig. 5. (a) Plot of LOGMSE vs. SNR, (b) Plot of RMS vs. SNR, for four kinds of methods (1) 3×3 wiener (2) 5×5 wiener (3) HMM (4) HMM & CWM.

The proposed algorithm, based on the adaptive window size feature, preserves edges in nonsmooth regions while it removes noise substantially in the smooth regions. We obtained an even better performance by cascading the proposed method with CWM filter, this can compensate the errors in the initial assumptions of the algorithm.

A future work is to test the ability of other probabilities to model the local stationary statistics of the images and also test the ability of this method in removing the signal dependent noise.

References

[1] A. Rosenfeld and A. Kak, Digital Picture Processing. *New York: Academic*, 1982.

[2] P. K. Sinha and Q. H. Hong, "An improved median filter," *IEEE. Trans. Med.Image.*, vol. 9, pp. 345–346, Sept. 1990.

[3] E. L. Hall, Computer Image Processing and Recognition. *New York:Academic,* 1979.

[4] A. K. Jain, "Fundamentals of digital image processing", *Prentice-Hall,* 1989.

[5] A. Beghdadi and A. Khellaf, "A noise filtering method using the local information measure ", *IEEE Trans. On image image processing,* Vol. 6, No. 6, June 1997.

[6] L. Rabiner, B. H. Juang, "Fundamentals of speech recognition," Prentice-Hall, Englewood Cliff, NJ, 1993.

[7] S. J. Ko and Y. H. Lee, "Center Weighted Median filters and their application to image enhancement", *IEEE Trans. Circuits Syst.,* Vol.38, 984–993, 1991.

Anaphoric Definitions in Description Logic*

Maarten Marx[1] and Mehdi Dastani[2]

[1] LIT, ILLC, Universiteit van Amsterdam, The Netherlands
marx@science.uva.nl, http://www.science.uva.nl/ marx
[2] CS, Universiteit Utrecht, The Netherlands
mehdi@cs.uu.nl, http://www.cs.uu.nl/ mehdi/

Abstract. This paper investigates the possibility of adding machinery to description logic which allows one to define self-referential concepts. An example of such a concept is a narcissist, someone who loves himself. With domains in which the natural ontology is a graph instead of a tree, this extra expressive power is often desired (e.g., when writing an ontology about web pages or molecular structures). Our results show that one has to be very careful with such additions. We add self-reference to $\mathcal{ALC}$ with inverse. Then we obtain all well known difficulties of having individual concepts or nominals together with inverse relations and even worse, checking for concept consistency becomes undecidable. Most of this expressive power seems not to be needed and we can identify a useful fragment whose complexity does not exceed that of $\mathcal{ALC}$.

1 Introduction and Motivation

We investigate adding a form of self-reference to description logic. This form is inspired by the downarrow operator from hybrid logic which names the "here and now" [3]. We can call a definition anaphoric if the definition contains a pronoun which refers to the thing or object which is being defined. For example,

narcissist: *someone who loves* oneself;
stepmother: *a female who is married to a person who has a child which is not* hers.

An example from the web could be a "solipsistic page" —a page which only links to pages which link back to *it*.

Using the notions of bisimulations for description logics developed by Kurtonina and de Rijke [7] one can simply show that these concepts are not definable in $\mathcal{ALC}$ and other extensions. This is due to the fact that these concepts exploit the graph like structure of the underlying domain, while $\mathcal{ALC}$ concepts can only capture (part of) the tree like aspects of it. This part of the design makes it robustly decidable [9].

There are some domains in which the graph like nature of the relations is important and the definition of concepts makes use of it. The web is a good

* Research supported by NWO grant 612.000.106. A preliminary version appeared in the proceedings of the 2002 workshop on description logic.

M.H. Shafazand and A M. Tjoa (Eds.): EurAsia-ICT 2002, LNCS 2510, pp. 410–417, 2002.

example. One has to be careful in designing languages which may speak about the graph like nature. Once grids can be defined, undecidability is very close.

Instead of adding variables as in hybrid logic, we add here the personal pronouns **I** and **me** to description logic with the following intended meaning:

> If C is a (complex) concept and a an element of the domain, then a belongs to $\mathbf{I}.C$ if a belongs to C under the assumption that all occurrences of **me** in C denote the individual concept $\{a\}$.

Note that **me** can be seen as a kind of dynamic version of the one-of operator. With **I** and **me** we can define the earlier mentioned concepts:

narcissist	$\mathbf{I}.\exists$ `loves` **me**
stepmother	`female` $\sqcap\, \mathbf{I}.\exists$ `married-to` $\exists$ `has-child` $\neg\exists$ `has-child`$^{-1}$ **me**
solipsistic web page	$\mathbf{I}.\forall$ `has-link` $\exists$ `has-link` **me**.

The definition of stepmother can graphically be represented as below. Here the node labeled `female` is the stepmother.

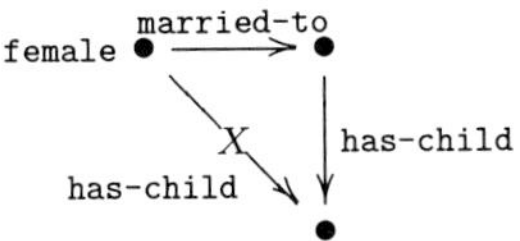

We can also define concepts which intrinsically need three variables, like sibling. (This definition goes back to Schröder.)

sibling: $\mathbf{I}.\exists$ `has-child`$^{-1}$ (`female` $\sqcap$
$\exists$ `has-child` ($\neg$ **me** $\sqcap$
$\exists$ `has-child`$^{-1}$ ($\neg$ `female` $\sqcap$
$\exists$ `has-child` **me**))).

In the picture below, the (different) nodes a and b are both siblings.

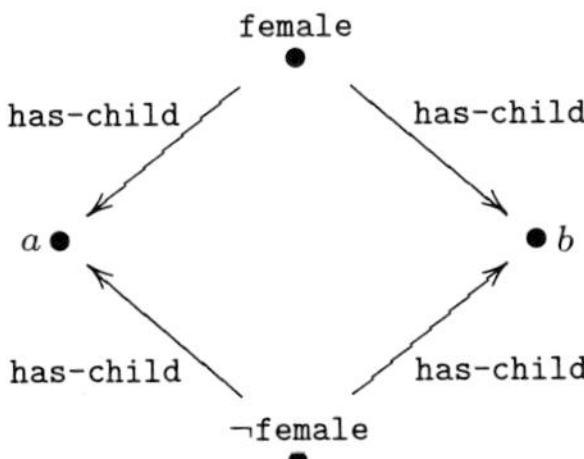

From DL and hybrid logic it is known that the combination of nominals or individual concepts together with inverse roles makes reasoning much harder (e.g., checking concept consistency for empty T-Box in $\mathcal{ALCI}$ with one nominal is EXPTIME hard). With self-reference things are much worse. Consider a T-Box $\{\top \sqsubseteq (1)\}$, with (1) the concept:

$$\begin{array}{lr} & \exists\,\texttt{has-link}\,\top\,\sqcap \\ (1) & \forall\,\texttt{has-link}\,\neg\texttt{startpage}\,\sqcap \\ & \mathbf{I}.\forall\,\texttt{has-link}\,\forall\,\texttt{has-link}^{-1}\,\mathbf{me}. \end{array}$$

The concept `startpage` can only be non empty in a model making this T-Box true if the model is infinite. This is because $\mathbf{I}.\forall\,\texttt{has-link}\,\forall\,\texttt{has-link}^{-1}\,\mathbf{me}$ expresses that *has-link* is an injective relation.

Things get even worse with "spy pages": a web page that has a direct link to all pages which can be reached by following a path of links from itself. They are defined by

$$(2) \qquad \mathbf{I}.\forall\,\texttt{has-link}\,\forall\,\texttt{has-link}\,\exists\,\texttt{has-link}^{-1}\,\mathbf{me}.$$

Now consider the concept (3), in which R is a new relation.

$$\begin{array}{lr} & \mathbf{I}.\forall\,\texttt{R}\,\forall\,\texttt{has-link}\,\exists\,\texttt{R}^{-1}\,\mathbf{me}\,\sqcap \\ (3) & \exists\,\texttt{R}\,\texttt{startpage}\,\sqcap \\ & \forall\,\texttt{R}\,(1). \end{array}$$

The previous result immediately implies that the concept defined by (3) is only non empty on infinite models.

In the remainder of the paper we show undecidability of concept consistency with empty T-Box for $\mathcal{ALC}$ with **I** and **me** added. For this result only one relation is needed and no converse. This contradicts the decidability result from Theorem 7.10 in [2]. The mistake in that proof lies in the given reduction to a problem without occurrences of **I**. The concept (1) above showed that this cannot be done in a finite way.

On the positive side, we also define a decidable existential version of the language. Here we tame the power of the **I**–**me** construction by disallowing universal quantifiers in the scope of **I** (as occurring in (1) and (2)).

Before we start let's make things precise. If $\mathcal{X}$ denotes some description language, let $\mathcal{X}\mathsf{self}$ be the language with the added clauses:

- **me** is a concept;
- if C is a concept, then $\mathbf{I}.C$ is a concept.

Here C is called the *scope* of **I**. The concept **me** occurs *free* in C if it is not in the scope of some **I**. We only consider concepts in which every occurrence of **me** occurs in the scope of some **I**. Now we can make the meaning of the new concepts precise.

> a belongs to $\mathbf{I}.C$ if a belongs to C under the assumption that all free occurrences of **me** in C denote the individual concept $\{a\}$.

In the sequel we use $C \to D$ as an abbreviation of $\neg C \sqcup D$.

2 The Existential Fragment

In this section we tame the power of self reference by restricting the type of concepts which can occur in the scope of **I**. Decidability and a matching complexity bound are obtained by a translation into the guarded fragment of first order logic with three variables [1,6].

Note that the spy-page (2) is equivalent to the first order formula $\forall y(Rxy \rightarrow \forall z(Ryz \rightarrow Rxz))$, when R is interpreted as *has-link*. This formula uses three variables in a nonreducible way and is not equivalent to a (loosely) guarded formula.

The description logic we consider is $\mathcal{ALCI}$, $\mathcal{ALC}$ with inverse roles. Every $\mathcal{ALCI}$ concept is equivalent to one in negation normal form, that is, constructed by $\sqcap, \sqcup, \exists\,\mathtt{R}, \forall\,\mathtt{R}, \exists\,\mathtt{R}^{-1}$ and $\forall\,\mathtt{R}^{-1}$ from atomic concepts and their negations. We define the *existential* $\mathcal{ALCI}\mathsf{self}$ concepts as those constructed by $\sqcap, \sqcup, \exists\,\mathtt{R}, \exists\,\mathtt{R}^{-1}$ and **I** from atomic concepts, **me**, their negations and $\neg\exists\,\mathtt{R}\,\mathbf{me}$ and $\neg\exists\,\mathtt{R}^{-1}\,\mathbf{me}$. Thus universal quantification is not allowed, except in the form of $\neg\exists\,\mathtt{R}\,\mathbf{me}$ in which form it is just an atomic statement.

The set of $\mathcal{ALCI}\mathsf{self}^{\exists}$ concepts is the smallest set such that every atomic concept name including **me** and their negations are concepts, and if C and D are concepts, then $C \sqcap D, C \sqcup D, \exists\,\mathtt{R}\,C, \exists\,\mathtt{R}^{-1}\,C, \forall\,\mathtt{R}\,C, \forall\,\mathtt{R}^{-1}\,C$ are also concepts. Moreover, if C is an existential $\mathcal{ALCI}\mathsf{self}$ concept, then $\mathbf{I}.C$ is also $\mathcal{ALCI}\mathsf{self}^{\exists}$ concept. An $\mathcal{ALCI}\mathsf{self}^{\exists}$ T–Box consists of a set of GCI's of the form $C \sqsubseteq D$, for C, D $\mathcal{ALCI}\mathsf{self}^{\exists}$ concepts with the requirement that **I** does not occur in C. Note that $\mathcal{ALCI}\mathsf{self}^{\exists}$ contains $\mathcal{ALCI}$ and that narcissist, stepmother and sibling can still be defined. Also note that even if $\mathcal{ALCI}\mathsf{self}^{\exists}$ is not closed under negation, the subsumption problem can in specific cases still be reduced to the satisfiability problem. In particular, if $\neg D$ is equivalent to an $\mathcal{ALCI}\mathsf{self}^{\exists}$ concept, then $\Sigma \models C \sqsubseteq D$ reduces to the satisfiability problem covered by the following theorem.

Theorem 1 *Let Σ be an $\mathcal{ALCI}\mathsf{self}^{\exists}$ T–Box and C an $\mathcal{ALCI}\mathsf{self}^{\exists}$ concept. The problem of checking concept consistency ($\Sigma \not\models C \doteq \bot$) is decidable in* EXPTIME.

Proof. We translate the problem using the standard translation to the universal guarded fragment[1] with three variables. In this fragment only universal formulas need to be guarded. Grädel [6] showed that the satisfiability problem for this fragment is complete for exponential time. The $\mathcal{ALCI}$ concepts are translated as usual (e.g., as in Table 2 of Borgida [5]). The new clauses are

$$\mathcal{T}^x(\mathbf{I}.C) := \exists w(x = w \wedge \mathcal{T}^x(C)) \quad \mathcal{T}^y(\mathbf{I}.C) := \exists w(y = w \wedge \mathcal{T}^y(C))$$
$$\mathcal{T}^x(\mathbf{me}) := x = w \qquad\qquad \mathcal{T}^y(\mathbf{me}) := y = w.$$

For example, $\mathcal{T}^x(\mathbf{I}.\exists\,\mathtt{R}\,\exists\,\mathtt{R}\,\mathbf{me})$ is $\exists w(w = x \wedge \exists y(Rxy \wedge \exists x(Ryx \wedge x = w)))$ which is equivalent to $\forall y(Rxy \rightarrow Ryx)$. The restriction on the scope of the **I** ensures that there are no non guarded universal quantifiers in the translation.

[1] Formulas in this fragment are constructed from atoms and their negations by conjunction, disjunction, unrestricted existential quantification and guarded universal quantification [8].

Because of the restriction on the form of the CGI's in the T–box, we may assume that they all have the form $\top \sqsubseteq D$, for D an $\mathcal{ALCI}\mathsf{self}^{\exists}$ concept. Then $\Sigma \not\models C \doteq \bot$ iff the universally guarded sentence $\forall x(x = x \rightarrow \bigwedge\{\mathcal{T}^x(D) \mid \top \sqsubseteq D \in \Sigma\}) \wedge \exists x \mathcal{T}^x(C)$ is satisfiable.

3 Undecidability

Theorem 2 *Let C be an $\mathcal{ALC}\mathsf{self}$ concept containing just one relation symbol R. The problem of checking concept consistency with empty T-Box for such C is undecidable.*

Undecidability is shown by encoding the $\mathbb{N} \times \mathbb{N}$ tiling problem (cf. [4]). The main point in such a proof is to show that two commuting functions *up* and *right* can be defined. Let Σ be a T-Box consisting of the following CGI's:

$$(4) \qquad \begin{array}{l} \top \sqsubseteq \exists\, \mathtt{up}\, \top \sqcap \exists\, \mathtt{right}\,, \\ \top \sqsubseteq \mathbf{I}.\forall\, \mathtt{up}^{-1} \forall\, \mathtt{up}\, \mathbf{me} \sqcap \mathbf{I}\forall\, \mathtt{right}^{-1} \forall\, \mathtt{right}\, \mathbf{me}, \\ \top \sqsubseteq \mathbf{I}.\forall\, \mathtt{up}^{-1} \forall\, \mathtt{right}^{-1} \forall\, \mathtt{up} \forall\, \mathtt{right}\, \mathbf{me}. \end{array}$$

If we apply the standard translation from the previous section to (4) (and simplify formulas) we obtain a theory which says that for all x,

$$\begin{array}{l} \exists y R_{\mathsf{up}} xy \wedge \exists y R_{\mathsf{right}} xy, \\ \forall yz(R_{\mathsf{up}} xy \wedge R_{\mathsf{up}} xz \rightarrow y = z) \wedge \forall yz(R_{\mathsf{right}} xy \wedge R_{\mathsf{right}} xz \rightarrow y = z), \\ \forall yz(R_{\mathsf{up}} yx \wedge R_{\mathsf{right}} zy \rightarrow \forall w(R_{\mathsf{up}} zw \rightarrow R_{\mathsf{right}} wx)). \end{array}$$

Thus $I \models \Sigma$ if and only if *I(up)* and *I(right)* are commuting total functions. Having this part it is standard to code up the tiling problem.

Now we turn to the proof of the theorem. Because we do not want to use inverse relations and only one relation symbol, we need some additional coding. In Figure 1 we present how we would model the grid. In this model there is only one relation R which is symmetric. The atomic concepts are $\{0, 1, 2\}$ and $\{u, r\}$. The nodes in the picture are given by the labels u and r and by the labels $i_{(k,m)}$. These last nodes correspond to positions in the grid. For $i \in \{0, 1, 2\}$, define $s(i) = i + 1 \bmod 3$ and $p(i) = i + 2 \bmod 3$. The idea is to model $\exists\, \mathsf{up}\, C$ by (for $i \in \{0, 1, 2\}$)

$$i \rightarrow \exists\, \mathtt{R}\, (u \sqcap \exists\, \mathtt{R}\, (s(i) \sqcap C)),$$

and $\exists\, \mathtt{right}\, C$ by $i \rightarrow \exists\, \mathtt{R}\, (r \sqcap \exists\, \mathtt{R}\, (p(i) \sqcap C))$. The corresponding relation up then is

$$\{\langle x, y\rangle \mid \exists z(x\mathrm{R}z\mathrm{R}y \wedge z \in I(u) \wedge \exists i(x \in I(i) \wedge y \in I(s(i))))\}.$$

We now present a number of concepts (5)–(9) which force an unraveled model to have the model from Figure 1 as a substructure in the case that each element

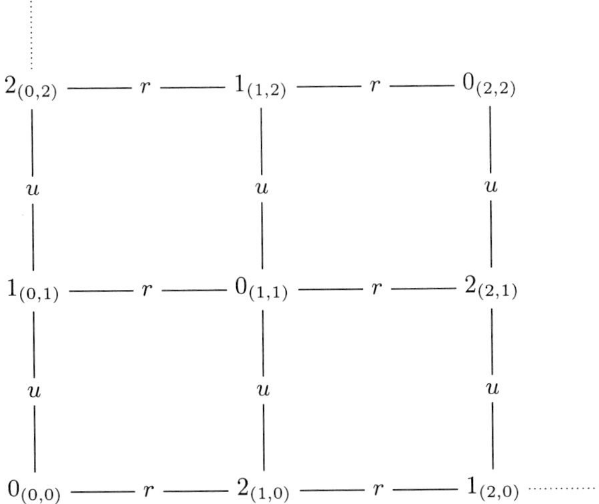

Fig. 1. Our standard model of the grid.

in the domain belongs to these concepts. The first concept (5) expresses that the relation R is symmetric.

$$\mathbf{I}.\forall\,\mathrm{R}\,\exists\,\mathrm{R}\;\mathbf{me}. \tag{5}$$

For $i \in \{0,1,2\}$, (6) expresses that every point has an *up* and a *right* successor.

$$i \to \exists\,\mathrm{R}\,(u \sqcap \exists\,\mathrm{R}\,s(i)) \sqcap \exists\,\mathrm{R}\,(r \sqcap \exists\,\mathrm{R}\,p(i)). \tag{6}$$

The next two concepts express that these successors are unique and that the relations `up` and `right` are irreflexive. For $i \in \{0,1,2\}$,

$$s(i) \to \mathbf{I}.\forall\,\mathrm{R}\,(u \to \forall\,\mathrm{R}\,(i \to (\neg\,\mathbf{me} \sqcap \forall\,\mathrm{R}\,(u \to \forall\,\mathrm{R}\,(s(i) \to \mathbf{me}))))) \tag{7}$$

$$p(i) \to \mathbf{I}.\forall\,\mathrm{R}\,(r \to \forall\,\mathrm{R}\,(i \to (\neg\,\mathbf{me} \sqcap \forall\,\mathrm{R}\,(r \to \forall\,\mathrm{R}\,(p(i) \to \mathbf{me}))))). \tag{8}$$

The last concept expresses the confluence property depicted in Figure 2 (assuming R is a symmetric relation). For $i \in \{0,1,2\}$,

$$\begin{aligned} s(i) \to \mathbf{I}.\forall\,\mathrm{R}\,(u \to \forall\,\mathrm{R}\,(i \to \forall\,\mathrm{R}\,(r \to \forall\,\mathrm{R}\,(p(i) \to \\ \exists\,\mathrm{R}\,(u \sqcap \exists\,\mathrm{R}\,(i \sqcap \exists\,\mathrm{R}\,(r \sqcap \exists\,\mathrm{R}\;\mathbf{me}))))))). \end{aligned} \tag{9}$$

Concepts (5)–(9) take care of the structural side of the encoding. Now let $T = \{T_1, \ldots, T_k\}$ be a set of tile types. For each type T_i, there is a corresponding atomic concept t_i. The next concept expresses that at every grid position exactly one tile concept holds. For $i \in \{0,1,2\}$,

$$i \to \bigsqcup_{1 \le n \le k} t_k \sqcap \bigwedge_{1 \le n \ne m \le k} \neg(t_n \sqcap t_m). \tag{10}$$

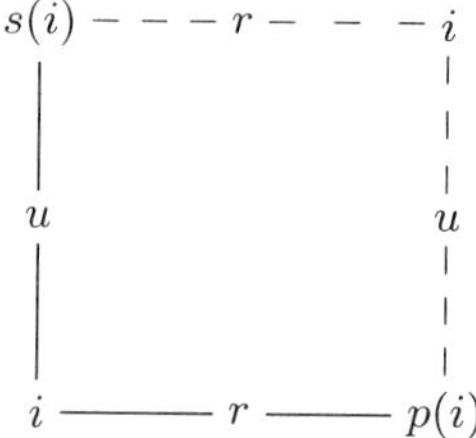

Fig. 2. (9) expresses a confluence property.

The last two concepts express that the colors of the tiles match. For $i \in \{0,1,2\}$ and $1 \leq n \leq k$,

$$(11)\quad i \sqcap t_n \rightarrow \forall\, \mathtt{R}\,(u \rightarrow \forall\, \mathtt{R}\,(s(i) \rightarrow \bigsqcup\{t_m \mid top(T_n) = bottom(T_m)\}))$$

$$(12)\quad i \sqcap t_n \rightarrow \forall\, \mathtt{R}\,(r \rightarrow \forall\, \mathtt{R}\,(p(i) \rightarrow \bigsqcup\{t_m \mid right(T_n) = left(T_m)\})).$$

Let ϕ_T be the conjunction of all concepts (5)–(12). It is straightforward to show that T tiles the grid if and only if the T–box $\{\top \sqsubseteq \phi_T\}$ is satisfiable. So we showed that checking T-Box consistency is undecidable. To obtain the result in the theorem we use the spy point technique from [3]. Consider the following concept

$$(13)\quad \begin{array}{l} \mathbf{I}.\forall\, \mathtt{R}\, \forall\, \mathtt{R}\, \exists\, \mathtt{R}\ \mathbf{me} \quad \sqcap \\ \forall\, \mathtt{R}\, \forall\, \mathtt{R}\, \mathbf{I}.\forall\, \mathtt{R}\, \exists\, \mathtt{R}\ \mathbf{me}. \end{array}$$

Applying the standard translation, the meaning becomes clearer:

$$\forall yz(Rwy \wedge Ryz \rightarrow Rzw) \wedge \forall xyz(Rwy \wedge Ryz \wedge Rzx \rightarrow Rxz).$$

Together the conjuncts imply that R is transitive from w: $\forall yz(Rwy \wedge Ryz \rightarrow Rwz)$. Thus we can use (13) to force that every element in the model belongs to a concept.

Putting everything together we obtain that T tiles the grid if and only if the concept (13) $\sqcap\ \exists\, \mathtt{R}\, 0\ \sqcap\ \forall\, \mathtt{R}\, \phi_T$ is non empty. This finishes the proof of Theorem 2.

4 Conclusion

We showed that adding a simple form of self reference to $\mathcal{ALC}$ makes it very, and indeed, too expressive: the language becomes undecidable. By restricting the concepts to which self reference can be applied we "tamed" the expressive power and obtained decidability. What is left is more or less the possibility of expressing that certain loops exist; still a useful extension of $\mathcal{ALC}$. I conjecture that existing tableau based procedures for $\mathcal{ALC}$ can be adapted to include this limited form of self reference. I even believe that the problem of checking for concept consistency with empty T-Box can still be done in PSPACE.

References

1. H. Andréka, J. van Benthem, and I. Németi. Modal languages and bounded fragments of predicate logic. *J. of Philosophical Logic*, 27(3):217–274, 1998.
2. C. Areces. *Logic Engineering.* PhD thesis, Institute for Logic, Language and Computation, University of Amsterdam, 2000.
3. C. Areces, P. Blackburn, and M. Marx. Hybrid logics: Characterization, interpolation and complexity. *J. of Symbolic Logic*, 66(3):977–1010, 2001.
4. E. Börger, E. Grädel, and Y. Gurevich. *The Classical Decision Problem.* Springer Verlag, 1997.
5. A. Borgida. On the relative expressiveness of description logics and predicate logics. *Artificial Intelligence*, 82:353–367, 1996.
6. E. Grädel. On the restraining power of guards. *Journal of Symbolic Logic*, 64(4):1719–1742, 1999.
7. N. Kurtonina and M. de Rijke. Classifying description logics. In M.-C. Rousset et al, editor, *Proc. International Workshop on Description Logics (DL'97)*, pages 49–53, LRI, CNRS, Gif sur Yvette, 1997.
8. M. Marx. Tolerance logic. *Journal of Logic, Language and Information*, 10:353–373, 2001.
9. M. Vardi. Why is modal logic so robustly decidable? In *DIMACS Series in Discrete Mathematics and Theoretical Computer Science 31*, pages 149–184. American Math. Society, 1997.

Storage and Querying of High Dimensional Sparsely Populated Data in Compressed Representation

Abu Sayed M. Latiful Hoque

Department of Computer and Information Sciences,
University of Strathclyde, 26, Richmond St, Glasgow, G1 1XH, UK
Latiful.Hoque@cis.strath.ac.uk

Abstract. Storage and querying of high dimensional sparsely populated data creates new challenge to conventional horizontal model. It requires supporting large number of columns and frequently altering of database schema. The sparsity of data degrades performance in both time and space. A 3-ary vertical representation [5] can be used. But the cardinality of the vertical table grows exponentially when the density of the non-null values increases. It is also difficult to support multiple data types using a single vertical table. In this paper, we have presented a compressed 1-ary vertical representation where schema evolution is easy and size grows linearly with non-null density. Queries can be processed on compressed form of data without decompression. Decompression is done only when the result is necessary. We have considered three alternative representations: 3-ary uncompressed vertical, 1-ary compressed bit-array and 1-ary compressed offset. Experimental results show the superiority of 1-ary offset representation in both space and time.

1 Introduction

Storage and updating of new generation e-commerce data require frequent alteration of database schema. In addition, tables data is sparsely populated because of the same structured nature of such data. Conventional horizontal representation needs large number of columns. Consider the marketplace of electronics industry in Agrawal et. al. [5]. It consolidates information about 1000 manufacturers and distributors. The current catalog contains nearly 2 million parts classified into 2000 categories. There are more than 5000 part attributes across various categories. New suppliers are expected to join the marketplace every week. They bring with them new parts, causing new attributes to be added to the current catalog. In a relational database system, data objects are conventionally stored using a horizontal scheme. A data object is represented as a row of a table. There are as many columns in the table as the number of attributes of the objects. In trying to store information such as that described above in one table, we run into problems caused by the large number of columns. Current database systems typically limit the number of columns in a table. This limit

M.H. Shafazand and A M. Tjoa (Eds.): EurAsia-ICT 2002, LNCS 2510, pp. 418–425, 2002.

is 1012 in DB2 and Oracle [5]. Where a DBMS allows the desired numbers of columns, we would have nulls in most of the fields. In addition to creating storage overhead, nulls increase the size of the index. Further problems can be caused by frequent alteration to table to accommodate new features. Schema evolution is expensive in current database systems. Queries on tables with high degree incur a large performance penalties if the data records are very wide but only a few columns are used in the query. In this paper, we have presented a number of compressed vertical representations for storage and querying sparsely populated data. Schema evolution is easy in this representation. Related work has been given in section 2. A number of storage options have been described in section 3. The organization of the dictionaries is given in section 4. Query translation and processing on compressed data are given in section 5. The result, discussion and conclusion are presented in section 6 and 7.

2 Related Work

Many compression schemes have been proposed to improve performance in storage or querying database systems. A lossy reduction scheme for very high dimensional data has been presented in [3]. Their reduction scheme is based on a statistical model and queries are performed on the model rather than the original data set. McGregor et. al. [1] have described a compressed column oriented organization for relational database systems. The method is suitable for high dimensional data and schema evolution is easy. There are however architectural limitations in the support of such sparsely populated data. A 2-ary storage model has been presented in [2]. The number of table is equal to the number of attributes. Having a large number of tables instead of one makes the system harder to manage. A 3-ary representation of data has been given in [5]. This representation offers an interesting design point between the conventional horizontal n-ary representation and the 2-ary binary representation. This representation loses data typing since all values are stored as VARCHARs in the val field. The growth of the number of 3-ary tuple is $O(mn)$, where m and n are the two dimensions of the horizontal table. We have presented an 1-ary compressed vertical representation where sparsely populated data can be stored in compact form.

3 Storage Options

In conventional implementations, relations are stored with the tuples implemented as records directly mapping the input form of the information. Though other arrangements are possible this is the simplest and we shall adopt it for ease of explanation. It is the technique used in IBM's DB2. A fixed maximum size of storage is allocated to represent each record field. Thus the database system must allocate sufficient storage space to allow for the largest storage representation required by any tuple in the relation resulting in a considerable waste of storage for all but the extreme tuple field values. We have considered both conventional horizontal and a 3-ary vertical structure given in [5] in uncompressed form.

In compressed representation, a number of storage options have been considered. The basis of the compression architecture has been described in [1]. Figure 1 shows a conventional horizontal table and corresponding 3-ary vertical, 2-ary vertical and 1-ary vertical tables. The schema of the 3-ary vertical table is {Oid, Key (attribute name), Value} where, Value is the attribute value in the horizontal table. For each data value in the horizontal table, a tuple in the 3-ary vertical table is included. In 2-ary representation, the number of tables is equal to the number of columns. In 1-ary representation, each column in the horizontal table is an 1-ary object.

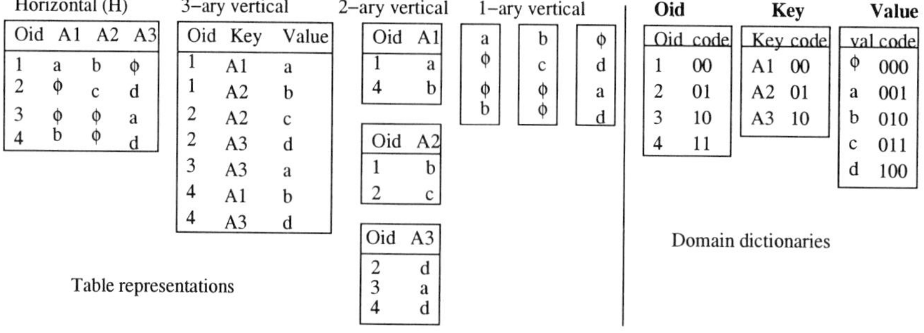

Fig. 1. Table representations and dictionaries

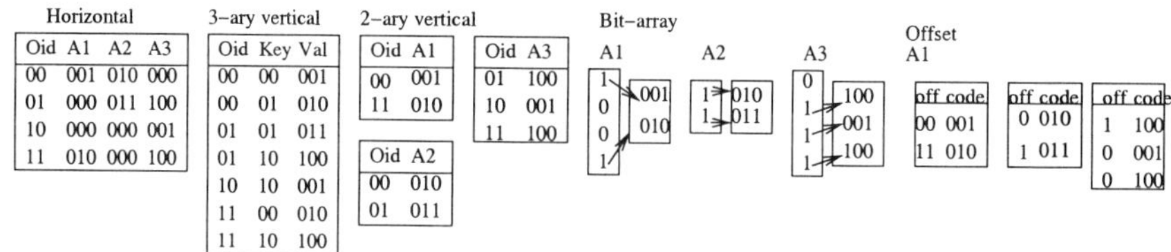

Fig. 2. Compressed representations

Using the basic compression approach [1], three dictionaries have been created for three domains e.g., Oid, Key and value (fig. 1). Fig. 2 shows the compressed representations. We have designed two alternative structures for 1-ary vertical representation. One is using bit array index and the other is using an offset method. The size of compressed tables for different options are: horizontal, 44 bits; 3-ary vertical, 49 bits; 2-ary vertical, 35 bits; 1-ary bit array index, 31 bits and 1-ary offset, 30 bits. Let n be the cardinality of the horizontal table and m be the number of columns and c the size of the code for values. The compressed size of the horizontal table is $(m-1)nc + nc_1$ where, c_1 is the code for *Oid*. The compressed size of the vertical table is $mn\rho(c_1 + c_2 + c)$ where c_2 is the code for key. The size of the bit-array index is mn bits and the total size of the compressed file is $mn + mn\rho c$. The non-null density is ρ and the density is uniformly distributed. The average distance between non-null values is d. The number of additional bits for offset is $mn\rho\log_2(d+1)$ bits, and $0 \leq d \leq (n-1)$. For uniform distribution, $d = (1-\rho)/\rho$, where $1/n \leq \rho \leq 1$. The size of the compressed file using offset is $mn\rho\log_2(1/\rho) + mn\rho c$. When $\rho = 1$, the size of the offset method approaches to the size of the compressed horizontal table. When $\rho\log_2(1/\rho) < 1$, the size of the compressed table in offset method is smaller than the bit-array index method.

4 Dictionary Organization

The values are stored in a dictionary and integer tokens are used to represent these values. A minimum number of bits is used to represent the tokens. There are different options for the dictionary: a single dictionary for the scheme; three dictionaries, one for objects, one for key (attributes) and one for values; more than three, one for objects, one for key (attributes) and one for each data type. We can also use a single dictionary for the scheme or a dictionary per domain. A dictionary per domain will reduce the size of the code. Using the compression approach in [1], three dictionaries have been created for the three domains; object identifier (Oid), Key (Attributes) and the values (fig. 1). The dictionary (fig. 3) has three important characteristics. It maps the attribute values to their encoded representation during the compression operation: $encode(lexeme) \rightarrow token$. It performs the reverse mapping from codes to literal values when parts of the relation are decompressed: $decode(token) \rightarrow lexeme$. The mapping is cyclic such that $lexeme = decode(encode(lexeme))$ and also $token = encode(decode(token))$. The operations that can be performed on a dictionary are findAString(), getAString() and insertAString(). In the former we can search a given string in the dictionary. If the string is already in the dictionary it outputs a token. If the string is not in the dictionary, it is inserted in the dictionary and a corresponding token is the output. In the later operation (getAString) a token is input to the dictionary, it outputs the corresponding string. For a given string, it always gives the same token. If a token is given to the dictionary, it always gives the same string. For insertAString, if a string is not present in the dictionary, it is inserted at the end of the string heap.

5 Processing Queries on Compressed Data

The fact that in a domain a specific token always refers to the same field value, enables many operations to be carried out by processing only the table component of the data, ignoring the dictionaries until string values are essential (e.g.

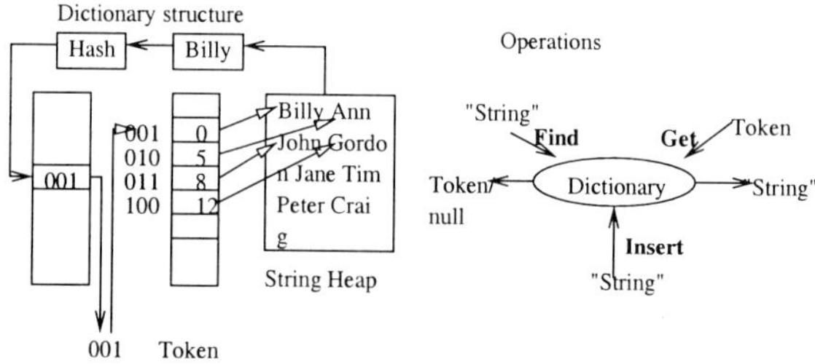

Fig. 3. The structure of the dictionary and the operations

for output). Queries are carried out directly on the compressed data, without requiring any decompression during the processing. The query is translated to compressed form and then processed directly against the compressed form of the data. This is inherently more efficient than the conventional alternative of processing an uncompressed query against uncompressed data since it requres manipulation of fewer bits. The amount of data that has to be moved into the CPU and compared for each selection operation is significantly reduced because of the compression, and because only a subset of the relation's columns are accessed. This results in increased efficiency of buffering of data both in RAM and in CPU caches. The final answer has to be converted from the internal identifier form to a normal uncompressed representation. However the computational cost of this decompression occurs only for the tuples that are returned in the result, normally a small fraction of those processed. Queries can be translated by the use of views. In uncompressed form of data, a horizontal view H is defined for a 3-ary vertical table V using an extended DDL: CREATE VIEW H ON VERTICAL TABLE V USING COLUMNS (A_1, A_2, A_n)

Where A_1, A_2, A_n are the attribute names. The same DDL is applicable to compressed representation as follows:

CREATE VIEW H_{ct} ON VERTICAL TABLE V_{ct} USING COLUMNS CODE $(C_1, C_2, ...Cn)$

Where $C_1, C_2, ...C_n$ are the corresponding codes for $A_1, A_2, ...A_n$. Now the user is free to run queries on the horizontal view. Each time, user input a query, it is translated into compressed form using the dictionary and run on H_{ct}. An example is given in the next section.

6 Results and Discussion

All experiments were run on an 800 MHz AMD DuronTM processor machine with 256 MB of physical memory. The operating system was Microsoft Widows 2000 Professional. We have implemented the model using Sun JDK version 1.3. We have used Java object serialization to provide a persistent storage of data objects. To study the performance characteristics of compressed representation, we used synthetic data. We used a dictionary of more than 15 thousands key words of table of contents database of Electronic Text Center, University of Virginia Library. We split them randomly over 1000 different categories. We have considered 20000 data objects and according to non-null density, an average of 10 to 300 key words have been used to describe each data object and place them into different categories. We included in this study, the space occupied by horizontal relational storage and different vertical options in uncompressed and compressed form, the time of projection of k columns where $k = 20, 40, ...100$ and the time of selection operation for different options. We performed all the operations on data in compressed representation. The results include the time for decompression. Fig. 4a shows a comparison of the size of the table in uncompressed and a compressed form. The size of the table in the horizontal relational storage representation is 1440 MB. The non-null density ranges from 1% to 30%. Using 1-ary offset, we have achieved a compression factor up to 710 compared to horizontal uncompressed form and a factor of 20 compared to 3-ary uncompressed vertical form. In both the cases, the non-null density is 1%. The growth of 3-ary table in both uncompressed (fig. 4a) and compressed (fig. 4b) form is very high. We have implemented three compressed representations: 3-ary vertical, 1-ary bit-array and 1-ary offset. In all range of non-null density (1% to 95%), 1-ary offset option outperforms the other two (fig. 4b). In lower range of non-null density (1% - 5%), the size in 3-ary vertical representation is smaller than 1-ary bit-array option. This is because of the fixed overhead of bit array index. But at higher non-null density region, 1-ary bit-array outperforms 3-ary vertical. We have defined a horizontal view (H_{ct}) over the vertical uncompressed/compressed representations. In compressed representation, the query engine translates the user queries into compressed form to this view. The performance of projection operation for varying number of columns is shown in fig. 4c. The number of projected columns varies from 20 to 100. In the 3-ary vertical representation, the projection of k columns requires a $k - way$ join of columns. The performance of this join operation depends on the physical level clustering of data. Agrawal et. al. [5] have shown that the clustering of data according to the attribute order offers better performance over the clustering in O-id order. We have used the best performance attribute order for the implementation of the 3-ary representation. A single level dense index on the key attribute of a 3-ary table has been used. We have compared the 3-ary uncompressed representation with a 1-ary options. The overall gain in performance using compression is a factor of 20 (fig. 4c, 4d). This is because less data is transferred for the projection operation (a factor of 12-20) in compressed form and 1-ary column organization offers a more compact clustering of data.

The selection experiment were run using the query: SELECT A_{20}, A_{60} FROM H_{ct} WHERE A_{60} = "*Communication*". "*Communication*" is a key word in the column A_{60}. The integer identifier of the key word "*Communication*" is 1025. This token was searched against the tokenized data represented by a dynamic vector. The design and implementation of the vector is given in [4]. We have used a selectivity of 5%. For very low non-null density (1%), 3-ary performs better than the other two options. This is because of the fixed access overhead

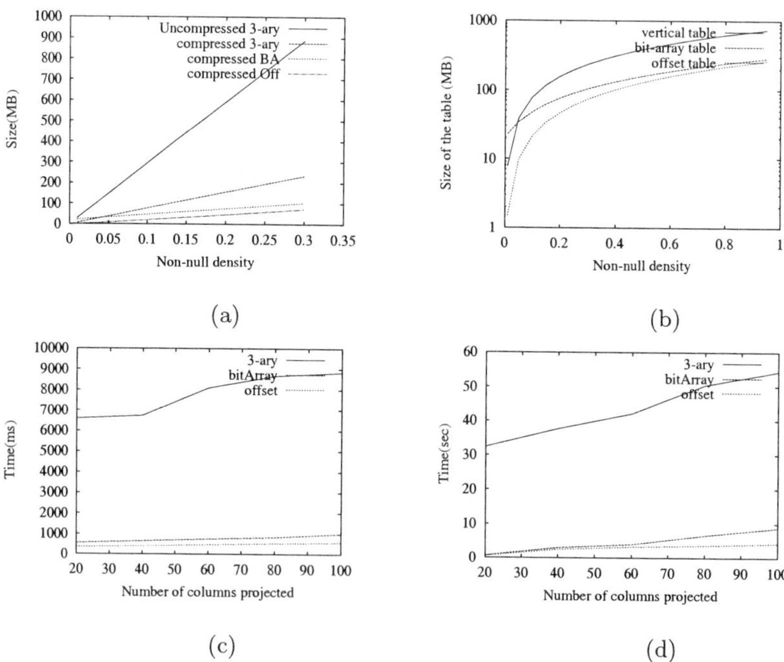

Fig. 4. (a) Uncompressed and compressed size (b) Growth of compressed table for different scheme. Projection time for varying number of columns (c) 1% non-null density (d) 5% non-null density

of the compressed vector data. When the non-null density increases, the 1-ary offset outperforms the 3-ary option even there is a burden of decompression on 1-ary offset. The decompression cost is negligible. The factor of tuple to be decompressed is of the order of 0.0005 to 0.0075. The selection performance of bit-array is less than the other two, because of the two-way access to the ith element; first to the bit array and then the vector.

Table 1. Selection performance for 5% selectivity

Option	1%	5%	10%	15%
bitArray	0.28	0.32	0.37	0.41
offset	0.12	0.12	0.13	0.14
3–ary	0.05	0.16	0.24	0.30

7 Conclusion

Storage and querying of high dimensional sparsely populated data is a challenge for conventional horizontal relational systems [5]. Emerging applications in e-commerce, digital libraries and Metadata repositories require large numbers of attributes for conventional horizontal representation. Data are also sparsely populated. Existing relational database management systems do not support such a large number of columns in a single table. In these applications, database structures are constantly evolving. Changing a schema using horizontal representation is costly. Even if the DBMS were to allow the desired number of columns, the performance of the system will be degraded because of the sparsity of data. In addition to creating storage overhead, nulls increase the size of the indexes. A 3-ary vertical format given in [5] can be used. However the cardinality of the vertical table grows in the order of $O(mn)$ when the density of the non-null values increases. It is also difficult to support multiple data types using a single vertical table. In this paper, we have presented a compressed 1-ary vertical representation where schema evolution is easy and size grows linearly with non-null density. Queries can be processed on the compressed form of data without decompression. Decompression is done only when the result is necessary. We have considered three alternative representations: 3-ary uncompressed vertical; 1-ary compressed bit-array and 1-ary compressed offset. We have achieved a compression factor of up to 20 compared to uncompressed 3-ary representation. We have shown that the effect of decompression has a very little effect while processing queries on compressed data. Finally, 1-ary offset option performs better than the other options in terms of both time and space. We have used a synthetic data set to measure the performance. We believe that the behavior of the architecture will be same if the real data sets of the same kind of the application area mentioned are used. Performing range queries on the compressed data is a difficult task. Further research is needed to develop an inverted index structure to support range queries. Using one dictionary for each data type can support multiple data types. Acknowledgements: This work was financed by Commonwealth Scholarship Commission of UK.

References

1. Wilson J. Cockshott W. P., MCGregor D. High-performance operations using a compressed architecture. *The Computer Journal*, 41.
2. G. P. Copeland and S. Khoshafian. A decomposition storage model. In *Proceedings of the 1985 ACM SIGMOD*, Austin, Texas, May 1985.
3. C. Jermaine and E. Omiecinski. Lossy reduction of very high dimensional data. In *Proceedings of the 18th International Conference on Data Engineering (ICDE (ICDE 02)*. IEEE, 2002.
4. D. R. Latiful Hoque A. S. M, McGregor. Improved compressed data representation for computational intelligence systems. In *UKCI-01*, Edinburgh, UK, September 2001.
5. Yirong Xu Rakesh Agrawal, Amit Somani. Storage and querying of e-commerce data. In *Proceedings of the 27th VLDB Conference*, Roma, Italy, 2001.

The GlobData Fault-Tolerant Replicated Distributed Object Database *

Luís Rodrigues, Hugo Miranda, Ricardo Almeida, João Martins, and PedroVicente

Universidade de Lisboa
Faculdade de Ciências
Departamento de Informática
{ler,hmiranda,ralmeida,jmartins,pedrofrv}@di.fc.ul.pt

Abstract. GLOBDATA is a project that aims to design and implement a middleware tool offering the abstraction of a global object database repository. This tool, called COPLA, supports transactional access to geographically distributed persistent objects independent of their location. Additionally, it supports replication of data according to different consistency criteria. For this purpose, COPLA implements a number of consistency protocols offering different tradeoffs between performance and fault-tolerance. This paper presents the work on strong consistency protocols for the GLOBDATA system. Two protocols are presented, that rely on the use of atomic broadcast as a building block to serialize conflicting transactions. The paper also describes the procedure to reintegrate failed nodes.

1 Introduction

GLOBDATA [1] is an European IST project started in November 2000 that aims to design and implement a middleware tool offering the abstraction of a global object database repository. The tool, called COPLA, supports transactional access to geographically distributed persistent objects independent of their location. Application programmers have an object-oriented view of the data repository and do not need to be concerned of how the objects are stored, distributed or replicated. The COPLA middleware supports the replication of data according to different consistency criteria.

This paper reports the work on strong consistency replication protocols for the GLOBDATA system that is being performed by the Distributed ALgorithms and Network Protocols (DIALNP) group at Universidade de Lisboa. Based on the previous work of [8,7], two protocols are being implemented: a voting protocol

* This work has been partially supported by the project IST-1999-20997, GLOBDATA.

[1] The GLOBDATA partners are: Instituto Tecnológico de Informática de Valencia (ITI), Spain; Faculdade de Ciências da Universidade de Lisboa (FCUL), Portugal; Universidad Pública de Navarra (UPNA), Spain; GFI Informatique (GFI), France; Investigación y Desarrollo Inforático (IDI EIKON), Spain.

M.H. Shafazand and A M. Tjoa (Eds.): EurAsia-ICT 2002, LNCS 2510, pp. 426–433, 2002.

and a non-voting protocol. The protocols are executed on top of an off-the-shelf relational database that is used to store the state of persistent objects and protocol control information. All protocols rely on the use of atomic broadcast as a building block to help serialize conflicting transactions. A specialized total order protocol is being implemented to support replication in large-scale [10]. The paper introduces the GLOBDATA architecture, and resumes the consistency protocols and the procedure for dealing with recovering nodes.

This paper is organized as follows: Section 2 describes the general COPLA architecture. Section 3 presents the consistency protocols. Section 4 describes the recovery procedure for failed nodes. Section 5 discusses related work. Section 6 concludes this paper.

2 Copla System Architecture

COPLA is a middleware tool that provides transparent access to a replicated repository of persistent objects. Replicas can be located on different nodes of a cluster, of a local area network, or spread across a wide area network spanning different geographic locations. To support a diversity of environments and workloads, COPLA provides a number of replica consistency protocols. The main components of the COPLA architecture are depicted in Figure 1. The upper layer is a "client interface" module, that provides the functionality used by the COPLA applications programmer. The programmer has an object-oriented view of the persistent and distributed data: it uses a subset of Object Query Language [2] to obtain references to distributed objects. Objects can be concurrently accessed by different clients in the context of distributed transactions.

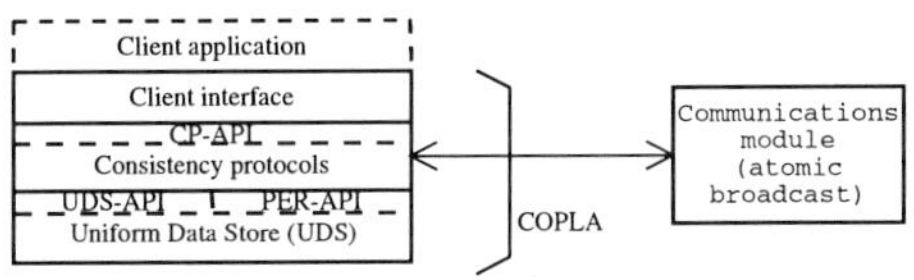

Fig. 1. COPLA architecture

For fault-tolerance, and to improve locality (i.e., transactions access locally stored objects) of read-only transactions, an object database may be replicated at different locations. Several consistency protocols are supported by COPLA; the choice of the best protocol depends on the topology of the network and of the application's workload. To maintain the user interface code independent of the actual protocol being used, all protocols adhere to the same specification (labeled CP-API). This allows COPLA to be configured according to the characteristics of the environment where it runs. The uniform data store (UDS) module (developed by the Universidad Pública de Navarra) is responsible for storing the state of the persistent objects in an off-the-shelf relational database management system

(RDBMS). To perform this task, the UDS exports an interface, the UDS-API, through which objects can be stored and retrieved. It also converts all the queries posed by the application into normalized SQL queries. Finally, the UDS is used to store in a persistent way the control information required by the consistency protocols. This control information is stored and accessed through a dedicated interface, the PER-API.

Architectural challenges. The GlobData project is characterized by a unique combination of different requirements, that make the design of the consistency protocols a challenging task. These are: *i) Large-scale.* The consistency protocols must support replication of objects in a geographically dispersed system, in which the nodes communicate through the Internet. This prevents the use of protocols that make used of specific network properties; *ii) RDBMS independence.* A variety of commercial databases should be supported as the underlying data storage technology. This prevents the use of solutions that require adaptations to the database kernel; *iii) Protocol interchangeability.* Copla must be flexible enough to adapt to changing environment conditions, like the scale of the system, availability of different communication facilities, and changes in the application's workload. Therefore it should allow the use of distinct consistency protocols, that can perform differently in several scenarios; *iv) Object-orientation.* Even if Copla maps objects into a relational model, this operation must be isolated from the consistency protocols. In this way, the consistency algorithms are not tied to any specific object representation.

3 Strong Consistency Protocols

In GlobData, the application programmer may trade fault-tolerance for performance. Therefore, a suite of protocols with different behavior in the presence of faults is being developed by different teams. Another project's partner, the ITI, is developing a suite of protocols based on the notion of object ownership [6]: Each node is the manager for the objects created in it, and is responsible for managing concurrent accesses to those objects. On the other hand, the DIALNP team at Universidade de Lisboa, is developing two protocols that enforce strong consistency even in the presence of faults. The strong consistency protocols rely extensively on the availability of an uniform atomic broadcast primitive. The protocols to be described can be further optimized, but these optimizations are not presented here due to lack of space. For more details see [9].

3.1 Interaction among Components

Both protocols cooperate with the Uniform Data Store to obtain information about which objects are read or updated by each transaction. This information, in the form of a list of unique object identifiers (OIDs), allows the protocols to have fine-grain information about which transactions conflict with each other. Since the consistency protocols only manipulate OIDs, they remain independent from the representation of objects in the database.

The COPLA transactional model. In COPLA, the execution of a transaction includes the following steps: *1)* The programmer signals the system that a transaction is about to start. *2)* The programmer makes a query to the database, using a subset of OQL. This query returns a collection of objects. *3)* The returned objects are manipulated by the programmer using the functions exported by the client interface. These functions allow the application to update the values of object's attributes, and to read new objects through object relations (object attributes that are references to other objects). *4)* Steps 2-3 are repeated until the transaction is completed. *5)* The programmer requests the system to commit the transaction.

Interaction with the consistency protocols. The common protocol interface basically exports two functions: a function that must be called by the application every time new objects are read by a transaction, and a function that must be called in order to commit the transaction.

The first function serves two main purposes: to make sure that the local copies of the objects are up-to-date (when using deferred updates, the most recent version may not be available locally); and to extract the state of the objects by calling the UDS (the access to the underlying database is not performed by the consistency protocol itself; it is a function of the UDS component).

The second function is used by the application to commit the transaction. In response to this request the consistency protocols module has to coordinate with its remote peers to serialize conflicting transactions and to decide whether it is safe to commit the transaction or if it has to be aborted due to some conflict. In order to execute this phase, the consistency protocol requests the UDS module to provide the list of all objects updated by the current transaction, and an opaque structure containing the state of the updated objects. It is the responsibility of the consistency protocol to propagate these updates to the remote nodes.

Interaction with the atomic broadcast primitive. An atomic broadcast primitive disseminates messages among a group of servers, guaranteeing atomic and ordered delivery of messages. Specifically, let m and m' be two messages sent by atomic broadcast to a group of servers g. *Atomic* delivery guarantees that if a member of g delivers m (resp. m'), then all correct members of g deliver m (resp. m'). *Ordered* delivery guarantees that if any two members of g deliver m and m', they deliver them in the same order. These two properties are used by both consistency protocols: the order property is used by the conflict resolution mechanism, and atomic delivery is used to simplify atomic commitment of transactions. COPLA uses a specialized broadcast primitive described in [10].

3.2 The Non-voting Protocol

This protocol is a modification of the one described in [8], altered to use a version scheme for concurrency control, and adapted to the COPLA transactional model. A more detailed description of this algorithm can be found in [9].

The protocol maintains, for each object, a version number. This version number is not replicated, but its kept consistent with other replicas by the protocol. When an object is created, its version number is set to zero. Each time a transaction updates an object, and that transaction commits, the object's version number is incremented by one. Version numbers are kept synchronized across replicas, since the total order ensured by atomic broadcast causes all replicas to process transactions in the same order. These version numbers are used to detect conflicting transactions. In this scheme, conflicting transactions can be defined as follows: Two transactions t and t' conflict if t' has read an object o with version v_o and when t' is about to commit, object o's version number in the local database, v'_o, is higher than v_o. That means that t' has read data that was later modified (by a transaction t that modified o and committed before t', thus increasing o's version number), and therefore t' should be aborted. Note that this definition covers only read/write conflicts, because in GLOBDATA all objects are read before they are written, therefore no write/write conflicts exist.

The general outline of the non-voting algorithm is now presented: *1)* All the transaction's operations are executed locally on the node where the transaction was initiated (this node is called the delegate node). *2)* When the application requests a commit, the set of read objects and its version numbers, and the set of written objects is sent to all nodes using the atomic broadcast primitive. *3)* When a transaction is delivered by the atomic broadcast protocol, all servers verify if the received transaction does not conflict with other local running transactions. There is no conflict if the versions of the objects read by the arriving transaction are greater or equal to the versions of those objects present in the local database. If no conflict is detected, then the transaction is committed, otherwise it is aborted. Since this procedure is deterministic and all nodes, including the delegate node, receive transactions by the same order, all nodes reach the same decision about the outcome of the transaction. The delegate node can now inform the client application about the final outcome of the transaction.

Note that the last step is executed by *all* nodes, including the one that initiated the transaction.

3.3 The Voting Protocol

This protocol is an adaptation of the protocol described in [5] adapted to the COPLA transactional model. It consists in two phases, a write set broadcast phase, and a voting (decision) phase. A more detailed description of this algorithm can be found in [9]. The general outline of the algorithm is as follows:

1) All the transaction's operations are executed locally on the delegate node, obtaining (local) read locks on read objects (note that, in order to be written, an object must be previously read). *2)* When the application requests a commit, a message $< t, \mathrm{WS}_t >$ containing the set of written objects is sent to all nodes using atomic broadcast. *3)* When the write set of a transaction t is delivered by atomic broadcast, all nodes try to obtain local write locks on all objects in the set. If there is a transaction that holds a write lock on any object of the write set of t, t is placed on hold until that write lock is relinquished. Transactions holding read locks on any object of the write set of t are aborted (sending an abort message through atomic broadcast). When the delegate node has obtained all write locks, sends a commit message to all servers, through atomic broadcast. *4)* Upon the reception of a commit message, a node applies the transaction's writes to the local database and subsequently releases all locks held on behalf of that transaction. Upon the reception of an abort message, the delegate node aborts the transaction an releases all its locks (other nodes ignore that message).

The algorithm uses the order given by atomic broadcast to serialize conflicting transactions. The final transaction order is given by the order of the $< t, \mathrm{WS}_t >$ messages. Conflict detection is done using locks.

4 Node Recovery

To support recovery of nodes, each node stores information about every committed transaction in a non-replicated persistent log (the log is maintained by the PER-API of the UDS module). Writes to the log are performed in the context of the corresponding transaction. Therefore, the UDS ensures that if the transaction commits the log update is also committed. The log consists of a table with three columns: the transaction's sequence number, the transaction's unique identifier and the set of changes performed by the transaction. Each line corresponds to a transaction log record. Since this log has finite size, two alternatives exist for recovery: one is to transfer the contents of the entire database. The other is to use the information stored in the log to replay the transactions the recovering node has missed.

Both the voting and the non-voting algorithms operate as described in the previous sections as long as the local node remains in a majority partition. If due to failures or due to a network partition a node finds itself in a minority partition it stops processing write transactions. It is possible to configure the system to let read-only transactions to read (possibly stale) data in a minority partition. The total number of servers in the system and their locations is a configuration parameter that is provided by the system manager. As such, establishing whether or not a given group of nodes is a minority (if there are n servers, a minority group is one with $n/2$ members or less) is trivial.

To support the recovery procedure, the consistency protocol maintains three complementary *views* of the server membership: *i)* The *up-to-date process group view* $P(t)$ contains all nodes that are capable of processing transactions at time t. *ii)* The *recovery group view* $R(t)$ contains all nodes that are connected with the nodes on $P(t)$ at time t. It therefore contains the process group view and all the nodes that are updating their state but noy yet capable of processing transactions. *iii)* The *static group view* S contains all the nodes in the system and is specified in the initial node configuration. This view is static and does not change with time. A majority (or minority) partition is always defined compairing the up-to-date process group view $P(t)$ with the static group view S, *i.e,* we consider that $P(t)$ has become a minority group if $\#P(t) \leq \frac{\#S}{2}$ Note that, by definition, $P(t) \subseteq R(t)$ at any given time t. In the normal case, $R(t) = P(t) = S$, i.e., all nodes are up and can process transactions.

The recovery method outlined requires that an unique sequence number is assigned to every transaction processed in the system. This sequence number can easily be derived from total order.

When a node recovers, that node must synchronize its database with the nodes that belong to $P(t)$. This is done in the following manner: *1)* The recovering member r joins the group of replicas, and is placed in $R(t)$. *2)* While recovery is taking place, processes that belong to the up-to-date process group view may continue to accept and process new transactions. Therefore, the recovering node must immediately start collecting all transactions it receives. Since it does not have its state up-to-date, it does not process these updates: they are kept in a pending state. The unique identifier (not the sequence number) of the first transaction added to the pending state is stored in a variable $pending_r$. *3)* A node $p \in P(t)$ is selected to be the *peer* node for the recovering node. The peer node will be responsible for helping the recovering node to synchronize its database. *4)* The recovering node r sends to p the sequence number of the last transaction that it processed successfully ($last_r$) and the identifier of the first transaction added to pending ($pending_r$). The transactions in this interval are all the transactions the recovering node has missed. *5)* The peer node p then checks if all the transactions missed by the recovering node are registered in p's log. If yes, the outcome of these transactions are transfered to the recovering node. If not, this means that the recovering node has been crashed or disconnected for too much time. In this case, a copy of the complete database is initiated. *6)* After receiving all missed transactions, the recovering node processes the transactions that have been stored in a pending state. At this point, r joins $P(t)$ and starts processing transactions normally.

5 Related Work

In the database literature, one can find different alternatives to enforce replica consistency, like for example using a voting scheme [3]. These and other related techniques suffer from scalability problems, which are identified in [4].

An alternative approach followed in COPLA consists in using an active replication scheme based on the use of efficient atomic multicast primitives. Systems such as [8,7], use the message order provided by atomic broadcast to aid in the serialization of conflicting transactions. An example of such a system is the Dragon project [11,5]. However, unlike our approach, the Dragon protocols are implemented at the database-kernel level, and cannot be used with off-the-shelf database systems.

Concurrently with our work, the CNDS group has developed a system [1] similar to GLOBDATA. Their approach also separates the consistency protocol from the database module. However, unlike our protocols, their system does not provide fine grain conflict detection. In COPLA, because a node knows the read and write set of the transactions that is executing, it can: a) Detect if a given query (read-only transaction) can be applied immediately (i.e., if it does not conflict with pending update transactions); b) abort transactions earlier, saving an expensive atomic broadcast message.

6 Conclusion

This paper presented the strong consistency protocols supported by the COPLA middleware, a tool that provides transactional access to persistent transparently replicated objects. These protocols are based on the use of atomic broadcast primitives to serialize conflicting transactions and to enforce the consistency in the transaction commit phase. The paper also introduced the procedure for reintegrating failed nodes. Currently we are completing the implementation of both consistency protocols and the atomic broadcast protocol. We then plan to evaluate the impact of different loads on both algorithms under varying conditions, as well as the performance of the atomic broadcast primitive in the those varying conditions.

References

1. Y. Amir, C. Danilov, M. Miskin-Amir, J. Stanton, and C. Tutu. Practical wide area database replication. Technical Report CNDS-2002-1, Center for Networking and Distributed Systems, February 2002.
2. R. Cattell. *The Object Data Standard: ODMG3.0.* Morgan Kauffmann, 2000.
3. D. Gifford. Weighted voting for replicated data. In *Proc. of the 7th ACM Symposium on Operating System Principles*, pages 150–162, USA, December 1979.
4. J. Gray, P. Helland, P. O'Neal, and D. Shasha. The dangers of replication and a solution. In *Proc. of the 1996 ACM SIGMOD International Conference on Management of Data*, pages 173–182, Montreal, Quebec, Canada, June 1996.
5. B. Kemme and G. Alonso. A suite of database replication protocols based on group communication primitives. In *Proc. of the 18th International Conference on Distributed Computing Systems (ICDCS)*, The Netherlands, May 1998.
6. F. D. Muñoz, L. Irún, P. Galdámez, J. M. Bernabéu, J. Bataller, and M. C. Bañuls. Globdata: Consistency protocols for replicated databases. In *Proc. of the IEEE YUFORIC'2001*, pages 97–104, Spain, November 2001. ISBN 84-9705-097-5.
7. M. Patiño Martínez, R. Jiménez-Peris, B. Kemme, and G. Alonso. Scalable replication in database clusters. In *Proc. of the 14th International Symposium on Distributed Computing (DISC)*, Toledo, Spain, October 2000.
8. F. Pedone, R. Guerraoui, and A. Schiper. Exploiting atomic broadcast in replicated databases. In *Proc. of EuroPar (EuroPar'98)*, Southampton, UK, September 1998.
9. L. Rodrigues, H. Miranda, R. Almeida, J. Martins, and P. Vicente. Strong replication in the Globdata middleware. In *Proc. of the Workshop on Dependable Middleware-Based Systems 2002 (part of DSN 2002)*, USA, June 2002.
10. P. Vicente and L. Rodrigues. An indulgent uniform total order algorithm with optimistic delivery. Technical report, Departamento de Informática, Faculdade de Ciências, Universidade de Lisboa, 2002.
11. M. Wiesmann, F. Pedone, A. Schiper, B. Kemme, and G. Alonso. Database replication techniques: a three parameter classification. In *Proc. of the 19th IEEE Symposium on Reliable Distributed Systems (SRDS2000)*, Germany, October 2000.

A Levelized Schema Extraction for XML Document Using User-Defined Graphs

Sungrim Kim[1] and Yong-ik Yoon[2]

[1] Computer Major, Division of Computer & Information Science
Dongduk Women's University
23-1 Wolgok-dong, Sungbuk-ku, Seoul, Korea
srkim@dongduk.ac.kr

[2] Department of Computer Science,
Sookmyung Women's University,
53-12 Chungpa-dong 2-ga, Yongsan-gu, Seoul, Korea
yiyoon@sookmyung.ac.kr

Abstract. EXtended Markup Language (XML), which are becoming new standard for representing and exchanging data in the Internet, don't have defined schema. It is not proper to directly apply XML documents to the existing SQL or OQL. Research on how to extract schema for XML documents and query language is going on actively. For users' query, the query results could be too many or too few. It is important to give the users good size results. We suggest the way to extract schema using some graphs according to the frequency of element occurrence in XML documents. The extracted schema can be reduced or extended to correspond to the users' query more flexibly. We test our proposed method and the results show that our method is a desirable extracting schema for XML documents.

1 Introduction

EXtended Markup Language(XML) is becoming a new standard for representing and exchanging data in the Internet [1,15]. XML data is a data that is tagged by XML elements. Such XML tags describe data themselves and can show documents in various forms.

XML documents are composed of groups of tag elements that show data structure. Even though it doesn't have schema that the other existing database have, it can be said that each document has a structure (DTD : Document Type Definition). The structure of XML data model is different from existing database. It is not proper to apply the existing SQL or OQL. As a result, research on how to extract schema for XML documents and query language is going on actively [5,7,9].

We suggest the way to extract levelized schemas based on elements information of XML documents using some graphs and to extract levelized schemas according to some thresholds. This method of extracting schema reduces or extends the scope of query by applying it to levelized schema when the query results are too many or too few. Then it can meet the requests of users efficiently.

M.H. Shafazand and A M. Tjoa (Eds.): EurAsia-ICT 2002, LNCS 2510, pp. 434–441, 2002.

The structure of this paper is as follows. Chapter 2 deals with related research on schema extraction. Chapter 3 deals with theories that support this paper. Chapter 4 shows step of extraction and examples. Chapter 5 explains our experiment environment. Chapter 6 concludes the paper.

2 Related Work

First, the way to find occurrence frequency pattern has been studied in the area of transaction database time series database and the other database area. Among many methods, the way to find maximum pattern has been suggested by establishing occurrence frequency pattern tree (*FP-tree*) [3].

FP-tree has information on occurrence frequency pattern. It improved the efficiency of finding frequency pattern by forming conditional FP-tree using the information.

It has the merit of extracting variety of schema based on the number of user-defined occurrence frequency. But it can extract special pattern of schema rather than the entire schema and it repeats schema extraction phases every time for the user-defined value for the special pattern and the number of occurrence frequency.

Second, there is a way to extract common schema with maximum *tree expression* according to occurrence frequency of tree expression [10, 11].

At tree expression *te*, schema extraction is done according to the following. The MINISUP of *te* is the number of documents that has weaker *te* than document *d*. If the MINISUP is greater than the MINISUP that a user defines, it can be said that *te* is frequent. If *te* is high and it is more frequent than the other *te*, the *te* can be said that it has the maximum frequency.

It has the merit of executing similar queries efficiently by making *te* for frequently occurring similar queries and extracting schema based on it. But it has a weak point that it is hard to find schema for the entire document.

Last, *Lore* is database management system for XML developed by Stanford University [13,14]. As XML documents don't have schema that was pre-defined, it is difficult to create meaningful queries if there aren't tags and attribute pattern. Query engine should understand the database structure to perform the queries efficiently. For this, Lore provides *DataGuide*.

DataGuide shows clear and dynamic arranged structure for XML database and performs the role of database schema and DTD. The user can understand the entire structure of database through DataGuide and create useful queries.

DataGuide can understand the entire schema for all documents. But as data in all documents are expressed, the created schema could be maximized and the searching area in XML documents could be widened.

3 Preliminaries

3.1 Edge-Labeled Graph

XML documents can be expressed with edge-labeled graph like semi-structured data [2,14]. Elements are expressed as nodes in edge-labeled graph. Each object has object identifier (*oid*) like &01. Edge-labeled graph defines two objects-simple object and complex object. Edge-labeled graph has edge between objects. Each edge has label with element name and has direction to express sub element.

3.2 Graphs for Schema Extraction

- Definition 1: *Data Graph*

We define edge labeled directed graph that describes all data in XML documents as *Data Graph*. Edge is made from root node to lower nodes and the label of edge becomes the name of element. Each node has *oid* (object identifier) and the node has the form of simple object or complex object.

- Definition 2: *Schema Graph*

The graph that is created to express the all paths only once using the depth first search in Data Graph for XML documents is called *Schema Graph*. Like the DataGuide[2] of Lore system[6], all the label paths are concise and all data in XML documents should be expressed(accuracy) and the structure of each node should be known (convenience).

3.3 Schema Extraction Using Bitmap Indexing

Basically, bitmap indexing uses 0 or 1 to express whether the attribute in tuple has special value or not. The merit of bitmap indexing is that it can increase the performance speed with hardware operation of bitwise-AND, OR, NOT calculation. Because of this merit, it is widely used in decision support system and dataware housing [4,8].

This paper creates levelized Schema Graph with bitmap indexing of label path at XML documents defined as follows [12].

- Definition 3: *Label Path*

The path from one node to a certain lower node in Data Graph or Schema Graph is defined as *Label Path*. Edge is existent between nodes in graph. The label with element name is existent. The middle node that appears in the path from root node and leaf node is expressed as . (dot).

3.4 Graph and Operation for Schema Extraction

- Definition 4 : *Graph Projection*

Graph Projection is the process of bitwise-AND calculation for element of schema that is extracted according to limit value and elements of each XML document. The result of Graph Projection may be XML document to which the extracted schema could be applied.

- Definition 5 : *Query Graph*

Query Graph describes elements that appear at SELECT clause and WHERE clause of user's query. Query Graph can extract the most adequate schema for query performance among many schemata created.

4 Schema Extraction Model

4.1 Structure of Schema Extraction Model

Figure 4.1 shows structure of proposed schema extraction model. It receives XML document and performs parsing operation. Then, Graph Generator creates Data Graph and Schema Graph based on analyzed element information. Then, it calculates the frequency of label path using Schema Generator and Schema-XML Graph Projector. It extracts label path that has greater frequency than some limited value. Then, it creates levelized schemas. Query Graph Generator and Schema-Query Graph Projector process users' queries using the extracted schema. It can define document area that includes all elements in schema when extracting schema.

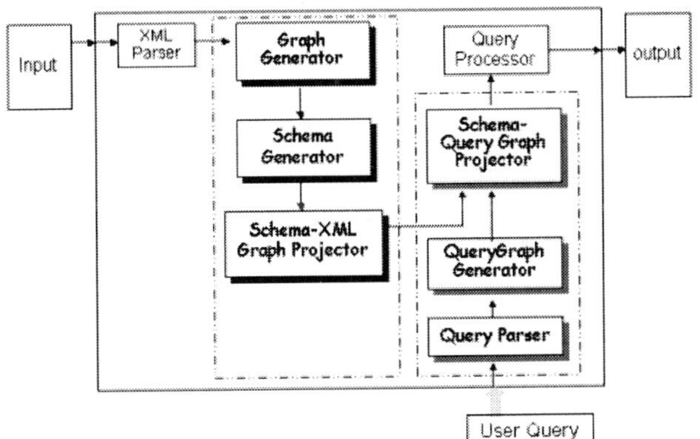

Fig. 4.1. Schema Extraction Model

4.2 Example of DTD and XML Documents

A XML document is supposed as shown in Table 4.1 with the information on movies in IMDB Top 250 films from http://www.imdb.com/.

4.3 Data Graph and Schema Graph

If the XML document in Table 4.1 is expressed in *Data Graph*, it is shown as in Figure 4.2. As defined in 3.2, the *Data Graph* expresses all elements in XML

Table 4.1. XML document

```
<movie>
  <title>Usual Suspects, The </title><year>1995</year>
  <director><firstname>Bryan</firstname><lastname>Singer</lastname></director>
  <writer>Christopher McQuarrie </writer> <genre>Crime</genre> <genre>Thriller</genre>
  <cast><name>Stephen Baldwin</name><role>Michael McManus </role>
   <spouse><name>Kennya Deodato</name>
   <occupation>(1990 - present)</occupation></spouse></cast>
  <cast><name>Benicio Del Toro</name><role>Fred Fenster </role>
     <awards><award>Independent Spirit Award</award>
     <category>Best Supporting Male</category></awards></cast>
 <language>English</language><country>USA</country><country>UK</country><color>Color</color>
    <keywords>drug-dealer</keywords><keywords>twist-in-the-end</keywords>
  </movie>
```

document. If the XML document in Table 4.1 is expressed in *Schema Graph*, it is shown as in Figure 4.3. As defined in 3.2, *Schema Graph* expresses the all label paths only once.

4.4 Label Path

We found label paths from root node to a leaf node in Schema Graph using depth first search. The result is Table 4.2. Using these label paths, we can extract levelized schema and proper schema to user's query.

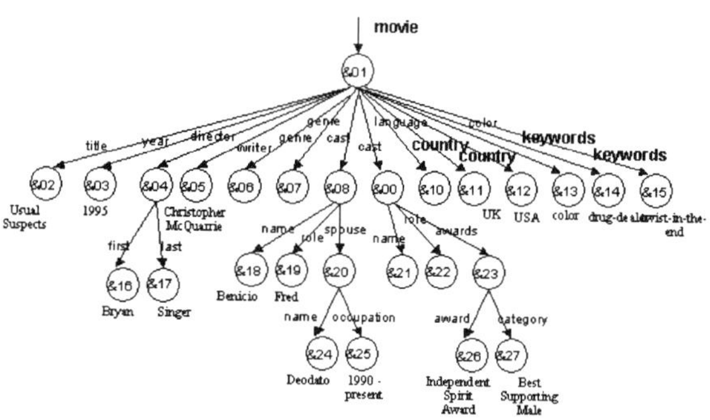

Fig. 4.2. Data Graph

4.5 Schema Extraction Using the Frequency of Label Path

It can describe label path in Table 4.2 and label path in XML document with bitmap indexing. We use bitmap indexing method and writes 1 or 0 according to existing of elements of each XNL document.

It can get the frequency of label path. Then it gives limit value for total creation frequency. It writes 1 if a label path is greater than the limit value. It writes 0 if a label path is smaller than the limit value. It can create schema by extracting only label path that is described with 1. Table 4.3 shows schema that is composed of only label paths that have creation frequency that is greater than 3.

Table 4.2. Label Paths

Label Path	Label Path
p1	movie.title
p2	movie.year
p3	movie.director.first
p4	movie.director.last
p5	movie.director
…	…
p17	movie.country
p18	movie.color
p19	movie.keywords

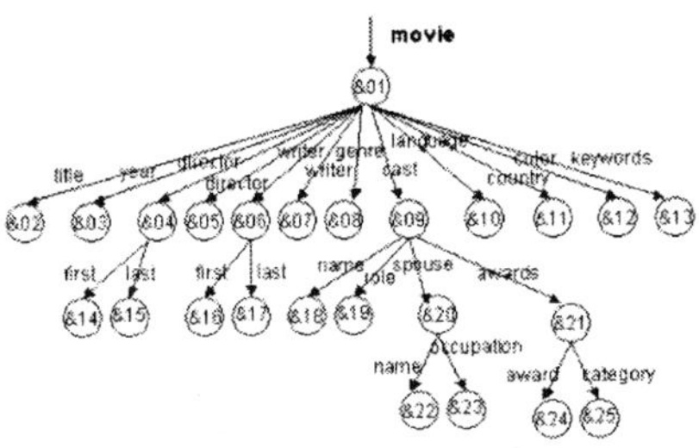

Fig. 4.3. Schema Graph

If the creation frequency of label path is small, it creates schema that has more label paths than the greater creation frequency.

4.6 Graph Projection

It performs bitwise-AND operation for label path of schema that is extracted according to limit value and label path of each XML document. If all results are 1, it can be XML document to which extracted schema is applied. Graph Projection in this paper is bitwise-AND calculation for label path of extracted levelized schema and label path of XML document.

Table 4.4 shows the result of Graph Projection for XML documents that could apply to schema that is extracted when frequency is greater than 3. We can know that XML document including all label paths of schema, of which the frequency is more than 3, is *d4*. But the result of graph projection calculation for schema, of which frequency is more than 2, is zero. It shows that there is no XML document that includes all elements of schema that is extracted when frequency is more than 2. We can know that the scope of XML document which is apply to extracted schema according to frequency can be acquired using label path to reduce the searching area before processing query.

Table 4.3. Schema Frequency of Label Path ≥ 3

	p1	p2	P3	P4	p5	p6	P7	p8	p9	…	p17	p18	p19
d1	1	1	1	1	0	0	0	1	1	…	1	1	1
d2	1	1	0	0	1	0	0	1	1	…	0	0	1
d3	1	1	0	0	1	1	1	0	1	…	1	0	0
d4	1	1	1	1	1	1	1	1	1	…	1	1	0
Frequency ≥ 3													
	1	1	0	0	1	0	0	1	1	…	1	0	0

Table 4.4. Graph Projection: Frequency ≥ 3

	p1	p2	P5	p8	p9	p10	p11	p16	P17
d1	1	1	0	1	1	1	1	1	1
d2	1	1	1	1	1	1	1	0	0
d3	1	1	1	0	1	1	1	1	1
d4	1	1	1	1	1	1	1	1	1
Bitwise-AND									
	1	1	1	1	1	1	1	1	1
d4	1	1	1	1	1	1	1	1	1

5 Experiments

The environment for testing schema extraction method is as follows. The OS is Windows 2000 and the database is Oracle9i. It uses JDK 1.3.1 and JSP. It also used Oracle JDBC thin driver to be compatible to Oracle. The web server is IIS5.0. It uses resin 2.0.1 as JSP engine. It uses Oracle Parser (version 2.0.1.0) as XML parser.

It created and tested XML documents based on text data except multimedia data among 100 movies from IMDB Top250 films of http://www.imdb.com.

6 Conclusion

This paper suggests a way to extract schema according to creation frequency by expressing elements for XML documents, which is becoming the standard of exchanging data on the Internet. It uses data, which is related to movies, to extract schema in many steps. It also acquires the scope of applicable XML documents according to extracts schema and finds that the searching area could be reduced before processing query.

To extract schema from XML document, it suggests many steps of schema according to frequency of label path appeared at each XML document by acquiring creation frequency of data graph, schema graph and label path based on input XML documents. It creates query graph according to users query and apply it to many schemata according to its frequency to find a way to extract the most adequate schema for user's query. It performed a test and analyzed the result of the method this paper suggests.

We can understand that it can expand or reduce the searching area before querying using the schema extracted according to creation frequency of label path and information of users query graph.

It is required to perform a further research for extracting schema from XML documents that include not only text data but also multimedia data.

References

[1] Jon Bosak, "XML, Java, and the Future of the Web", http://webreview.com/wr /pub/97/12/19/xml/index.html

[2] Roy Goldman, Jennifer Widom, "DataGuides: Enabling Query Formulation and Optimization in Semistructured Databases", *In Proceedings of VLDB*, 1997

[3] Jiawei Han, Jian Pei, Yiwen Yin, "Mining Frequent Patterns without Candidate Generation", *Proceedings of the 2000 ACM SIGMOD on Management of data*, 2000, pp. 1–12

[4] Theodore Johnson, "Performance Measurements of Compressed Bitmap Indices", *VLDB 1999*, pp. 278–289

[5] Alon Levy, "More on Data Management for XML", University of Washington, May 9th, 1999. http://www.cs.washington.edu/homes/alon/ widom-response.html

[6] J. McHugh, S. Abiteboul, R. Goldman, D. Quass, J. Widom, "Lore : A Database Management System for Semistructured Data", *SIGMOD Record*, 26(3), pp.54–66, September 1997

[7] Jayavel Shanmugasundaran, Kristin Tufte, Gang He, Chun Zhang, David DeWit, Jeffrey Naughton, "Relational Databases for Querying XML Documents: Limitations and Opportunities", *Proceedings of the 25th VLDB Conference 1999*

[8] M.C. Wu, A.P. Buchmann, "Encoded Bitmap Indexing for Data Warehouses", *Proc. ICDE '98*, 220–230

[9] Jennifer Widom, "Data Management for XML",Working Document, initial draft appeared April 1999, Also *IEEE Data Engineering Bulletin*, Special Issue on XML, 22(3):44–52, September 1999

[10] Ke Wang, Huiqing Liu, "Schema Discovery from Semistructured Data", *International Conference on Knowledge Discovery and Data Mining*, August 1997, pp. 271–274

[11] Ke Wang, Huiqing Liu, "Discovering Typical Structures of Documents : A Road Map Approach", *The ACM SIGR conference on Research and Development in Information Retrieval*, August 1998, pp. 146–154

[12] J. Yoon, S. Kim, "Schema Extraction for Multimedia XML Document Retrieval", in *Journal of Applied Systems Studies, Cambridge International Science Publishing*, Cambridge, UK, 2001

[13] Roy Goldman, Jason McHugh, Jennifer Widom, "Lore: A Database Management System for XML", Dr. Dobb's Journal, 25(4):76–80. April 2000, http://www.ddj.com /articles/2000/0004/0004i/0004i.htm ?topic=xml

[14] Serge Abiteboul, Peter Buneman, Dan Suciu, "*Data on the Web : From Relations to Semistructured Data and XML*", Morgan Kaufmann, 2000

[15] Tim Bray, Jean Paoli , C. M. Sperberg-McQueen , "Extensible Markup Language (XML)1.0", http://www.w3.org/TR/REC-xml#dt-xml-doc, 1998

Extracting, Interconnecting, and Accessing Heterogeneous Data Sources: An XML Query Based Approach

Gilles Nachouki and Mohamed Quafafou

IRIN, Université de Nantes – Faculté des Sciences et des Techniques,
2, rue de la Houssinière, BP 92208 44322 Nantes cedex 03, France
{nachouki,quafafou}@irin.univ-nantes.fr

Abstract. This paper describes the design of a system that facilitates ***Extracting, Interconnecting and Accessing Heterogeneous D****ata* ***S****ources*. Data sources can be static or active: static data sources include structured or semistructured data like databases, XML and HTML documents; active data sources include services which can be any executable programs such as C, C++ programs or Java classes. Users interact with the system using XQuery language. We have extended the XQuery language in order to call services. In this paper services resolve conflicts between static data sources (naming conflicts, data representation conflicts, data scaling conflicts…) in order to obtain an integrated view of data.

1 Introduction

The problem of interconnecting and accessing heterogeneous data sources is an old problem known under the name of Interoperability. As organisations evolve over time interoperability is still an important field of current research in both academic and industry. In the past interoperability was limited to structured data sources like databases. Interoperability is the ability of two or more systems (or components) to exchange information and to use the information that has been exchanged [6]. Interoperability can be classified into two approaches [11]: tightly coupled approach (or static approach) and loosely coupled approach (or dynamic approach). In the static approach, schematic and semantic heterogeneity are resolved when a new component database is incorporated in the system. The first stage of this approach uses a Wrapper that translates the database schema from its local data model into a Canonical Data Model [11]. The second stage uses a Mediator that combines multiple component schemes into an integrated schema, in which inconsistencies are resolved and duplicates are removed. In some systems, all component schema are integrated into a single global schema, while in others, multiple integrated schema exist for end-users. In the dynamic approach, the heterogeneity is resolved by the end-user at query time. In this approach the end-users are not provided with a predefined view. Instead, they are given direct access to the component schemes at query time by means of a multidatabase query language or a graphical user interface. This approach uses a metadata resource dictionary that contains descriptions of the component schemes and includes

M.H. Shafazand and A M. Tjoa (Eds.): EurAsia-ICT 2002, LNCS 2510, pp. 442–449, 2002.

conversion information. With the advent of the World Wide Web (WWW) data management has branched out from traditional framework to deal with the variety of information available on the WWW. Section 2 summarises related work in the domain of interoperability between heterogeneous data sources. The others sections describe our contribution in this domain.

2 Related Work

In the literature many systems are developed in order to connect structured and semi-structured data sources; among them we cite TSIMMIS [4], Information Mainfold [7], MOMIX [1], AGORA [8], [9], LIXTO [2] and XML-Media [5] systems:

In *TSIMMIS* Wrappers use an object model called OEM (Object Exchange Model). Mediators use a Mediator Specification Language (MSL), based on rules and functions for translating objects: the tail of a rule specifies patterns found in the sources, while the head describes patterns of the top-level objects of integrated views. Tsimmis system is query-centric: it chooses a set of queries, and for each such query the system provides a procedure to answer the query using the available sources. *INFORMATION MANIFOLD* is a system for integrating web-based information. The end-user specifies his queries against a single global view, called the World View, which is a collection of virtual relations with class hierarchies that describe the contents of the information sources. For each data source the integrator specifies source descriptions for this data source, consisting of content records and capability records. The content record of a given source identifies the world-view attributes, which can be found in that source. A capability record describes which attributes can be used as binding and selection criteria. User queries are formulated in terms of the world-view relations by means of conjunctive queries. In *MOMIX* Wrappers use a common object-oriented data model (ODL) in order to describe source schema for integration purpose. A mediator provides to users a unique view of data sources. In Momis a common thesaurus is constructed by analysing ODL descriptions of the sources. In particular, a set of terminological relationships is stored in the thesaurus, by interacting with the user and by using description logic to express the logical links existing between sources. The knowledge in the Thesaurus is then exploited for the identification of semantically related information in ODL descriptions of different sources and for their integration at the global level. *AGORA* unlike the systems described above starts from an integrated view of data sources expressed with XML. Users queries are posed using XML query language (QUILT language [3]), while all data flows inside the query processor consist of relational tuples. An extension of Agora for integrating heterogeneous data sources (including XML and relational) is proposed in [9]. In this system, relational and XML data sources are defined as view over the global XML schema, by means of an intermediate virtual, generic, relational schema modelling the generic structure of an XML document. *LIXTO* is a visual Web system for Information extraction. Lixto generates wrappers that translate pieces of HTML to XML. A Lixto wrapper is created interactively, using the Lixto browser, by creating patterns in a hierarchical order. A filter represented by Elog rule defines each pattern. An extracted instance from the HTML page must satisfy all conditions defined in a filter. The output of the extractor is translated into an XML document. This approach generates infinite numbers of wrappers in order to extract information from the web. In *XML-Media* wrapper generator uses rule-based scripting to produce XML data from various source formats. A mediator permit to integrate relational and XML sources using a DBMS extender supporting XML on the top of relational DBMS.

3 Contributions

Our work follows a loosely coupled approach where the data sources are under the responsibility of the users. Our approach is based on the XQuery language in order to provide interoperability between data sources. This approach uses metadata that contains descriptions of data sources (expressed with DTD formalism). The choice of this approach is that data sources on the web are constantly being added, modified or suppressed and the generation of an integrated view of data sources is still a difficult task:
– We have developed wrappers that extract DTDs from data sources and mediators which offer to users views of data sources expressed with DTDs. DTDs describe static and/or active data sources: static data sources include structured or semistructured data like relational databases, XML and HTML documents; active data sources are services, which can be any executable program over an operating system such as C, C++ programs or Java classes;

– We have developed a new approach [10] based on XQuery language in order to extract information from the web. Our approach shows the web as a large database containing XML documents. Users extract information from one or several documents using XQuery language;

– We have extended the XQuery language in order to call services. In this paper we consider services that resolve conflicts between data sources such as naming conflicts, data representation conflicts and data scaling conflicts. At this stage conflicts between data sources are supposed known. Users use services in their queries in order to obtain an integrated view of data.

Section 4 describes the architecture of our system that permits to extract, interconnect and access data sources. Our approach is based on XQuery language, Corba and Web technologies (XML framework). Section 5 provides experimental results.

4 System Architecture

Figure 1 illustrates the different levels composing the system: Interfaces, Mediators and T-Wrappers levels:

Interfaces: User interface offers to users a list of DTD of data sources. The system provides three distinct interfaces: the first interface permits to familiar users with XQuery language to access data sources by the formulation of queries. The generator interface permits users non-familiar with XQuery language to generate queries starting from their selected attributes. The Administrator interface consists to add (or suppress) a data source to (or from) the list of existing data sources and to create and run CORBA object servers (like DSS and DSR objects servers described below). CORBA provides location transparency that allows clients to access objects using their objects identifiers independent of their location and communication protocols between the client and the object.

T-Wrappers: A T-Wrapper is a wrapper for a data source of type T. For static data sources we have developed two wrappers denoted SQL-Wrapper and XML-Wrapper that correspond to the most important sources of information existing over the web (e.g. SQL databases, XML and HTML). Wrappers extract automatically DTD from data sources and register it beside the mediator, then execute queries on data sources and return results to the mediator.

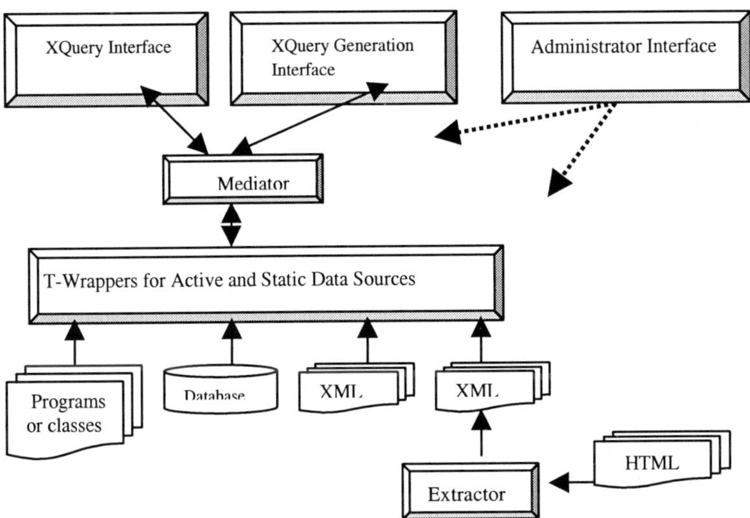

Fig. 1. The components of the system

To extract data from the web our approach is based on XQuery language. This approach follows two steps: in input we consider the selected Html web pages and a DTD that describes the structure of the extracted data (optional). The first step of this approach consists to convert each selected Html page to an XHTML [15] document. The second step consists to specify useful data to extract from the resulting XHTML document using XQuery query. The results of execution of these queries over the XHTML documents are integrated into one XML document. More detail about this approach is given in [10].

For active data sources we have developed two distinct wrappers denoted P-Wrapper and J-Wrapper. P-Wrapper permits to run any executable programs (e.g. C, C++,...)

and J-Wrapper permits to run Java classes. The design of T-Wrappers is composed of two components: Data Sources Server and T-Engine:

– Data Source Server (DSS): A DSS is designated as responsible of a data source (static or active). It represents the interface of wrappers of any type. There's no way to reach data sources directly in this model, all the communications toward data sources have to go through the DSS. The DSS hides the implementation differences between the different data source types and provides a common prototyping interface for data retrieval.

– T-Engine: T-Engine is an engine dedicated to a type T of data source (static or active). It is composed of all the pieces necessarily to extract DTDs, retrieve data or run services.

Mediator: This component offers to users views of data sources expressed with their DTDs. A mediator receives queries from users interfaces, processes queries and returns results to users as XML documents. The mediator has the knowledge necessary to forward a specific query to the concerned wrappers. Three components compose a mediator: Data Source Repository, Query Server and Cache:

– Data Source Repository (DSR): the goal of a DSR is to collect information about the DSSs (references, addresses, names, descriptions and most importantly the DTD of the associate data source). All the DSSs have to register themselves at start-up at the central DSR, providing all the needed pieces of information about the represented Data Source. A Repository therefore has the knowledge of all the main details about the connected data sources, and has also the possibility to share this information with the clients centrally.

– Query Server (QS): A Query Server receives complex queries from the client side, and processing them. Under processes, we mean analysis them first, decomposes into sub-queries, which refer only to one data source, executes those sub-queries, and after collecting the results, joins them together based on the previous analyses and returns the result to the client. Processing of a query is improved with the use of a **cache**, which contains data selected from data sources. This mechanism avoids selecting same data from data sources when they are used frequently by queries. The QS communicates with the DSR in order to obtain more information about data sources and with T-Wrappers in order to execute specific sub-queries.

5 Experimental Results

We consider five static data sources called Restaurant, Cinema, Underground, "Personne" and People. Restaurant is an Html page web selected by Google engine searcher using the key word "Restaurants Nantes" (URL is given in Appendix), Underground, Personne and People are three XML documents and Cinema is a Mysql database. We consider two services (Java classes) called "ToEuro" and "Concat". The first service converts dollar currency to Euro and the second service concatenates strings into one. We propose two distinct applications:

a) The first application uses the data sources Restaurant, Cinema and Underground. We select at the Nantes city the streets that contain a tube, a restaurant and a cinema. The XQuery query and the result is given in figure2:

```
<Query>FOR $var1 in document("Restaurant_Html")/html/body/level/.../level
FOR $var2 in document("Underground_Xml")/Underground/Line/Station
where contains($var1,$var2/Street/text())
RETURN
FOR $var3 in  document("Cinema_DataBase")/cinema/cinema
where $var3/street/text()=$var2/Street/text()
RETURN
<Result> <Restaurant>$var1</Restaurant>, <Underground>$var2/Station_name/text()</Underground>,
<Cinema>$var3/name/text()</Cinema></Result> </Query>
```

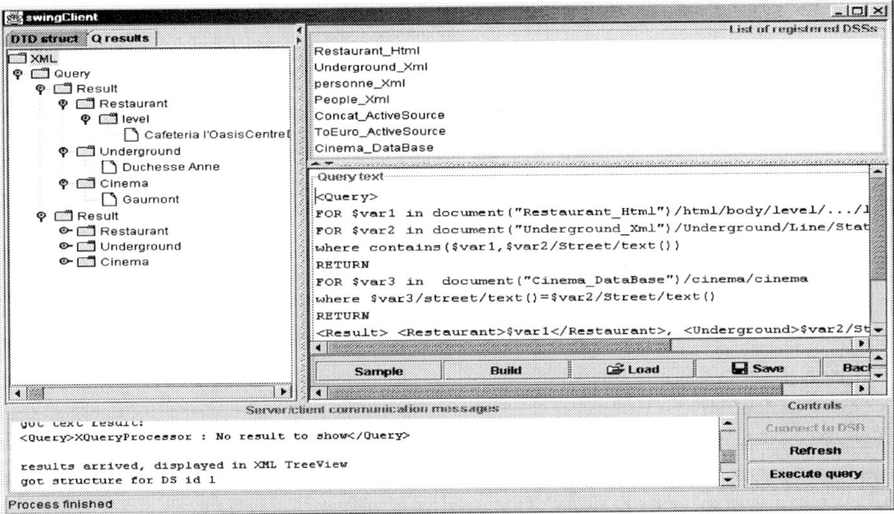

Fig. 2. Querying heterogeneous data sources

b) The second application considers the two active data sources ToEuro and Concat and the two static data sources "Personne" and People:
People and "Personne" describe the same concept but some conflicts exist between them like conflicts of names (i.e. salaire in "Personne" and salary in People), structures (i.e. "adresse" in "Personne" and address in People) and scale (i.e. "salaire" is expressed in Euro and salary is expressed in dollar). Conflicts of names are resolved with the choice of the name of tags in the same language (french or english). The structure of the element "adresse" in "Personne" is expressed in a string format using Concat service. The salary of a person is converted to Euro using ToEuro service. The following query selects data, from the data sources People and Personne. The result of this query is shown in figure 3 inside the user interface.

```
<Query>FOR $a IN document("personne_Xml")/personne/unepers
RETURN
<From_Personne> <numero>$a/ID/text()</numero>, <nom> $a/NOM/text() </nom>,
<adresse>exec("java","Concat",$a/RUE/text(),$a/CP/text(),$a/VILLE/text(),"c:/AIMS/Sources")
```

```
</adresse>,
<salaire>$a/SALAIRE/text()</salaire>, <age>$a/AGE/text()</age>
</From_Personne>,
For $b IN document("People_Xml")/people/onepeople
RETURN
<From_People><numero>$b/NUM/text()</numero>, <nom>$b/NAME/text()</nom>,
<adresse>$b/ADDRESS/text()</adresse>,
<salaire>exec("java","ToEuro",$b/SALARY/text(),"c:/AIMS/Sources")</salaire>,
<age>$b/OLD/text()</age>
</From_People></Query>
```

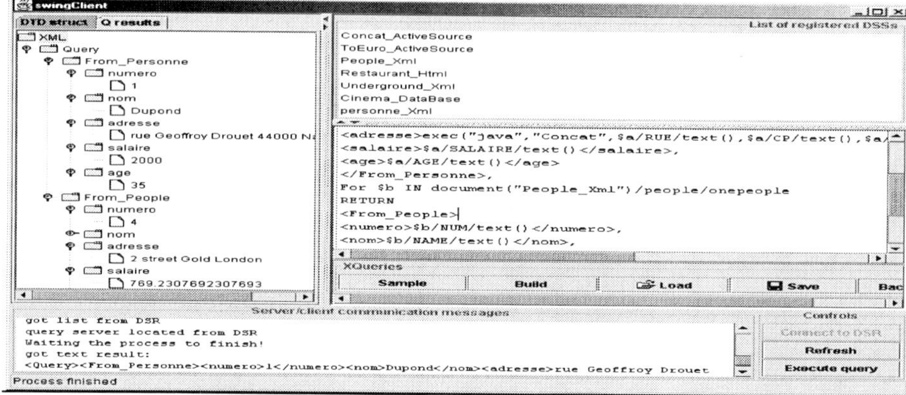

Fig. 3. Integrated view of two conflicts data sources

6 Conclusion

Our approach is based on XQuery language in order to extract, interconnect and access heterogeneous data sources. This approach follows a loosely coupled approach where the data sources are of the responsibility of users. We implement XQuery language and we have extended this language in order to call external services. Services permit to resolve conflicts existing between heterogeneous data sources. In the future, we plan to study semantics interoperability between data sources.

References

1. Bergamaschi, S., Beneventano, D. “Integration of Information from Multiple Sources Textual Data”, Springer-Verlag (editor M. Klusch), March 1999, ISBN 3-540-65112-8
2. Baumgartner, R., Flesca, S., Gottlob, G.: Visual Web Information Extraction with Lixto.Proceedings of the 27th VLDB Conference, Roma, Italy, 2001
3. Chamberlin, D., Robie, J., Florescu, D. “QUILT: An XML Query Language for Heterogeneous Data Sources”, in Proceedings of WebDB 2000 Conference in Lecture Notes in Computer Sciences, Springer-Verlag 2000.
4. Garcia-Molina, H.., Hammer, J., Ireland, H., Papakonstantinou, Y., Ullman, J., Widom, J. “Integrating and Accessing Heterogeneous Information Sources in TSIMMIS”, in Proceeding of the AAAI Symposium on Information Gathering, California, Mars 1995

5. Gardarin, G., Sha, F., Ngoc, T. D.: XML-Based Components for Federated Multiple Heterogeneous Data Sources. Lecture Notes in Computer Science, vol 1728. Springer-Verlag, Paris – France (1999)
6. IEEE Standard Computer Directory: A compilation of IEEE standard Computer Glossaires. New York, NY:1990.
7. Kirk, T., Levy, A., Sagiv, Y., Srivastava, D. "The Information Manifold" In working Notes of the AAAI Spring Symposium: Information Gathering from Heterogeneous Distributed Environments, 1995
8. Manolescu, I., Florescu, D., Kossmann, D., Xhumari, F., Olteanu, D. "Agora : Living with XML and Relational", in Proceedings of the 26th VLDB Conference, Cairo, Egypt, 2000.
9. Manolescu, I., Florescu, D., Kossmann, D., "Answering XML Queries over Heterogeneous Data Sources", in Proceedings of the 27th VLDB Conference, 2001.
10. Nachouki, G., Quafafou, M.: Extracting semistructured data from the web: An XQuery Based approach". In Shafazand, H., Tjoa A., Traunmuller, R., Wagner, R. (eds.): Workshop on Web Content Mapping. Austrian Computer Society, Eurasia-ICT 2002, Iran
11. Sheth, A., Larson, J., "Federated Databases Systems for Managing Disributed Heterogeneous and Autonomous Databases", ACM Computing Survey, Vol.19, September 1990.

Appendix : Data Sources Description

Restaurant: http://www.bestofnantes.com/files/hotel_resto/resultat.asp?fichier=resto&s=-2&type=

Personne.xml:

<personne><unepers><ID>1</ID><NOM>Dupond</NOM><ADRESSE>Nantes</ADRESSE>
<RUE>rue Geoffroy Drouet</RUE><CP>44000</CP><VILLE>Nantes</VILLE><AGE>35</AGE>
<SALAIRE>2000</SALAIRE></unepers>…</personne>

People.xml:

<people><onepeople><NUM>4</NUM><NAME>Bouleau</NAME>
<ADDRESS>2 street Gold London</ADDRESS><SALARY>5000</SALARY><AGE>35</AGE>
</onepeople>...</people>

Underground.xml:

<Underground> <Line><Line_number>1</Line_number><Line_name>ligne bleue</Line_name>
<Station><Station_name>Duchesse Anne</Station_name><Street>route de Vannes</Stree
<Town>Nantes</Town></Station>…</Underground>

Cinema:

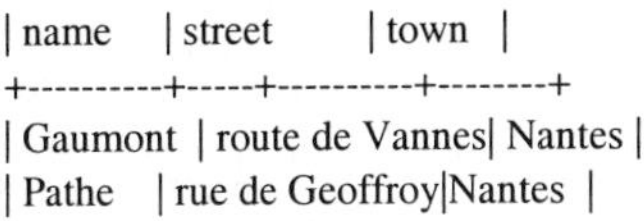

name	street	town
Gaumont	route de Vannes	Nantes
Pathe	rue de Geoffroy	Nantes

Call Admission Control in Cellular Mobile Networks: A Learning Automata Approach

Hamid Beigy and Mohammadreza Meybodi

Soft Computing Laboratory
Computer Engineering Department
Amirkabir University of Technology
Tehran, Iran
{beigy, meybodi}@ce.aku.ac.ir

Abstract. Dropping probability of handoff calls and blocking probability of new calls are two important QoS measures for cellular networks. Call admission policies, such as fractional guard channel and uniform fractional guard channel policies are used to maintain the pre-specified level of QoS. In this paper, we propose a learning automata based call admission policy in which a learning automaton is used to accept/reject new calls. This call admission policy can be considered as adaptive uniform fractional guard channel policy. In order to study the performance of the proposed call admission policy, the computer simulations are conducted. The simulation results show that for some range of input traffics, the performance of the proposed approach is close to the performance of the uniform fractional guard channel policy. The proposed policy is fully adaptive and doesn't require any information about the input traffics.

1 Introduction

Introduction of micro cellular networks leads to efficient use of channels but increases the expected rate of handovers per call. As a consequence, some network performance parameters such as *blocking probability of new calls* (B_n) and *dropping probability of handoff calls* (B_h) are affected. In order to maintain B_h and B_n at a reasonable level, *call admission control (CAC) algorithms* are used, which play a very important role in the cellular networks because directly control B_n and B_h. Since B_h is more important than B_n, CAC policies give the higher priority to handoff calls. This priority is implemented through allocation of more resources (channels) to handoff calls. A general CAC policy, called *fractional guard channel policy* (FG), accepts new calls with a probability that depends on the current channel occupancy and accepts the handoff calls as long as channels are available [1]. Suppose that the given cell has C full duplex channels. The FG policy uses a vector $\Pi = \{\pi_0, \ldots, \pi_{C-1}\}$ to accept the new calls, where $0 \leq \pi_i \leq 1$ and $0 \leq i < C$. The FG policy accepts new calls with probability of π_k when k $(0 \leq k < C)$ channels are busy. Unfortunately, there is no algorithm to find the optimal vector Π^*. A restricted version of FG is called *guard channel policy* (GC) [2]. The GC policy reserves a subset of channels,

M.H. Shafazand and A M. Tjoa (Eds.): EurAsia-ICT 2002, LNCS 2510, pp. 450–457, 2002.

called *guard channels*, for handoff calls (say $C - T$ channels). Whenever the channel occupancy exceeds the certain threshold T, the GC policy rejects new calls until the channel occupancy goes below T. The GC policy accepts handoff calls as long as channels are available. It has been shown that there is an optimal threshold T^* in which B_n is minimized subject to the hard constraint on B_h [3]. An algorithm for finding such optimal threshold is given in [3]. In order to have more control on B_h and B_n, *limited fractional guard channel policy* (LFG) is introduced [1]. The LFG can be obtained from FG policy by setting $\pi_k = 1$, $0 \leq k < T$, $\pi_T = \pi$, and $\pi_k = 0$, $T < k < C$. It has been shown that there are an optimal threshold T^* and an optimal value of π^* in which B_n is minimized subject to the hard constraint on B_h [1]. The algorithm for finding such optimal parameters is given in [1]. In [4], a restricted version of FG policy, called *uniform fractional guard channel policy* (UFG), is introduced. The UFG policy accepts new calls with probability of π independent of channel occupancy. The UFG can be obtained from FG by setting $\pi_k = \pi$, $0 \leq k < C$. In order to find the optimal value of parameter π, in [4] a binary search algorithm is given.

In this paper a reinforcement learning based algorithm is given to find the optimal value of π for the UFG policy. In context of CAC, the use of reinforcement learning techniques can lead to good solution in reasonable time. Instead of relying on a known teacher, the system is designed to learn an optimal assignment policy by directly interacting with the environment. Learning automaton (LA) is a reinforcement learning technique and has been used successfully in many applications such as telephone and data network routing [5,6], solving NP-Complete problems [7,8,9] and capacity assignment [10], to mention a few. In this paper, we propose a LA based CAC algorithm for cellular networks. In this algorithm, LA is used to determine acceptance/rejection probability (π) of new calls.

The rest of this paper is organized as follows: Section 2 presents performance parameters of UFG policy and gives an algorithm to find the optimal value of parameter of UFG policy. The LA is given in section 3 and the proposed CAC policy is given in section 4. The computer simulations is given in section 5 and section 6 concludes the paper.

2 The Uniform Fractional Guard Channel Policy

In UFG policy, handoff calls are accepted as long as channels are available and new calls are accepted with probability π, which is independent of channel occupancy. The description of UFG policy is given algorithmically in figure 1. In the next subsections, we first study the blocking performance of UFG and then give a binary search algorithm to find the optimal value of parameter π.

2.1 The Blocking Performance of UFG

In what follows, we study the blocking performance of UFG policy. The blocking performance of UFG policy is computed based on the following assumptions.

```
if (HANDOFF CALL) then
    if c(t) < C ) then
        accept call
    else
        reject call
    end if
end if
if (NEW CALL) then
    if (c(t) < C and rand (0,1) < π) then
        accept call
    else
        reject call
    end if
end if
```

Fig. 1. Unform fractional guard channel policy

1. The arrival process of new and handoff calls is poisson process with rate λ_n and λ_h, respectively.
2. The holding time for both types of calls are exponentially distributed with mean μ^{-1}.
3. The time interval between two calls from a mobile host is much greater than the mean call holding time.
4. Only mobile to fixed calls are considered.
5. The network is homogenous.

The above first three assumptions have been found to be reasonable as long as the number of mobile hosts in a cell is much greater than the number of channels allocated to that cell. The fourth assumption makes our analysis easier and the fifth one lets us to examine the performance of a single network cell in isolation.

Suppose that the given cell has a limited number of full duplex channels, C, in its channel pool. We define the state of a particular cell at time t to be the number of busy channels in that cell and is represented by $c(t)$. The $\{c(t)|t \geq 0\}$ is a continuous-time Markov chain (birth-death process) with states $0, 1, \ldots, C$. The state transition rate diagram of a cell with C full duplex channels and UFG call admission policy is shown in figure 2.

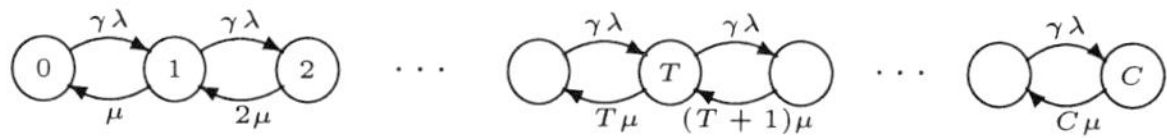

Fig. 2. Markov chain model of cell

At the state $0 \leq n < C$, new calls are accepted with probability $0 \leq \pi \leq 1$ and handoff calls are accepted with probability 1. Both types of calls are blocked in state C. Thus, the state dependent arrival rate in the birth-death process is equal to $[a + (1 - a)\pi]\lambda$, where $a = \frac{\lambda_h}{\lambda}$ is the handoff traffic in Erlangs and $\lambda = \lambda_h + \lambda_n$ is total traffic seen by the cell. Because of the structure of the Markov chain, we can easily write down the steady-state balance equations. Define the steady state probability $P_n = \lim_{t\to\infty} \text{Prob}[c(t) = n]$ for state $n = 0, 1, \ldots, C$. Then, the following expression can be derived for P_n $(n = 0, 1, \ldots, C)$.

$$P_n = \frac{(\rho\gamma)^n}{n!} P_0 \tag{1}$$

where $\gamma = [a + (1 - a)\pi]$ and $\rho = \lambda/\mu$ is the total traffic in Erlangs as seen by a cell and P_0 is the probability that all channels are free. The value of P_0 is calculated by the following expression.

$$P_0 = \left[\sum_{n=0}^{C} \frac{(\rho\gamma)^n}{n!}\right]^{-1} \tag{2}$$

Thus, the dropping probability of handoff calls, $B_h(C, \pi)$, id given by

$$B_h(C, \pi) = \frac{(\rho\gamma)^C}{C!} P_0. \tag{3}$$

Similarly, the blocking probability of new calls, $B_n(C, \pi)$ is equal to

$$\begin{aligned} B_n(C, \pi) &= \sum_{n=0}^{C-1} (1 - \pi) P_n + P_C \\ &= 1 - \pi \left[1 - B_h(C, \pi)\right] \end{aligned} \tag{4}$$

2.2 Finding of Optimal Parameter of UFG Policy

The objective of our call admission control policy is to find a π^* that minimizes $B_n(C, \pi)$ with the constraint that $B_h(C, \pi)$ must be at most p_h. The value of p_h specifies the QoS of the network. It is too complex to obtain an exact solution for this problem. Hence, a search algorithm is given to determine the optimal value of π for given traffic and constraint p_h. This algorithm is given in figure 3.

Since the proposed algorithm needs all parameters of input traffic, which are unknown and possibly time varying, we introduce a LA based algorithm for determination of parameter π.

3 Learning Automata

The automata approach to learning involves the determination of an optimal action from a set of allowable actions. An automaton can be regarded as an abstract object that has finite number of actions. It selects an action from its finite set of actions. This action is applied to a random environment. The random environment evaluates the applied action and gives a grade to the selected

```
set upper ← 1; lower ← 0
set k ← 0
if (B_h(C, 1) ≤ p_h) then
    return 1
end if
while (k < 20 and (upper - lower) < 0.0001) do
    set π ← (upper + lower) / 2
    if B_h(C, π) > p_h ) then
        set upper ← π
    else
        set lower ← π
    end if
    set k ← k + 1
end while
return π
```

Fig. 3. Algorithm for determination of π^*

action of automata. The response from environment (i.e. grade of action) is used by automata to select its next action. By continuing this process, the automaton learns to select an action with best grade. The learning algorithm used by automata to determine the selection of next action from the response of environment. An automaton acting in an unknown random environment and improves its performance in some specified manner, is referred to as *learning automaton.* LA can be classified into two main families: *fixed structure learning automata* and *variable structure learning automata* [11]. Variable structure learning automata is represented by triple $< \beta, \alpha, T >$, where β is a set of inputs actions, α is a set of actions, and T is learning algorithm. The learning algorithm is a recurrence relation and is used to modify the state probability vector. It is evident that the crucial factor affecting the performance of the variable structure learning automata, is learning algorithm for updating the action probabilities. Various learning algorithms have been reported in the literature. Let α_i be the action chosen at time k as a sample realization from probability distribution $p(k)$. In linear reward-ϵpenalty algorithm ($L_{R-\epsilon P}$) scheme the recurrence equation for updating p is defined as

$$p_j(k+1) = \begin{cases} p_j(k) + a \times [1 - p_j(k)] \text{ if } & i = j \\ & \\ p_j(k) - a \times p_j(k) \quad \text{ if } & i \neq j \end{cases} \quad \text{if} \quad \beta(k) = 0 \tag{5}$$

$$p_j(k+1) = \begin{cases} p_j(k) \times (1 - b) \quad \text{ if } & i = j \\ & \\ \frac{b}{r-1} + p_j(k)(1 - b) \text{ if } & i \neq j \end{cases} \quad \text{if} \quad \beta(k) = 1 \tag{6}$$

The parameters $0 < a < 1$ and $0 < b \ll a$ represent *step lengths* and r is the number of actions for LA. The a and b determine the amount of increase and decreases of the action probabilities, respectively. If the a equals to b the recurrence equations (5) and (6) is called *linear reward penalty*(L_{R-P}) algorithm.

4 LA Based CAC Strategy

In this section, we introduce a LA based call admission algorithm (figure 4). This algorithm is used to determine admission probability π when parameters a and ρ (or equivalently λ_h, λ_n and μ) are unknown or probably time varying. The proposed CAC strategy adjusts parameter π as long as network operates. This algorithm gives a higher probability to handoff calls by allowing handoff calls to be accepted with higher probability than new calls. This algorithm can be described as follows: The proposed algorithm uses one reward-penalty type LA with two actions in each cell. The action set of this automaton corresponds to {ACCEPT,REJECT}. The automaton associated to each cell determines the probability of acceptance of new calls (π). Since initially the values of a and ρ are unknown, the probability of selecting these actions are set to 0.5. When a handoff call arrives, it is accepted as long as there is a free channel. If there is no free channel, the handoff call is blocked. When a new call arrives to a particular cell, LA associated to that cell chooses one of its actions. Let π be the probability of selecting the action ACCEPT. Thus, the LA accepts new calls with probability π as long as there is a free channel and rejects them with probability $1 - \pi$. If action ACCEPT is selected by automaton and the cell has a free channel, then action ACCEPT is rewarded. If there is no free channel to be allocated to the arrived new call, the call is blocked and action ACCEPT is penalized. When the automaton selects action REJECT, it estimates the dropping probability of handoff calls ($\hat{B}_h$). If the current estimate of dropping probability of handoff calls is less than the given threshold p_h and there is a free channel, then the new call is accepted and action REJECT is penalized. This rule causes that the channels used more efficiently. In other case, the new call is rejected and action REJECT is rewarded.

5 Simulation Results

In this section, we compare performance of the uniform fractional guard channel and the LA based call admission policies. The results of simulations are summarized in table 1. The simulation is based on the single cell of homogenous cellular network system. In such network, each cell has 8 full duplex channels ($C = 8$). In the simulations, new call arrival rate is fixed to 30 calls per minute ($\lambda_n = 30$), channel holding time is set to 6 seconds ($\mu^{-1} = 6$), and the handoff call traffic is varied between 2 calls per minute to 20 calls per minute. The results listed in table 1 are obtained by averaging 10 runs from $2,000,000$ seconds simulation of each algorithm. The objective is to minimize the blocking probability of new calls subject to the constraint that the dropping probability of handoff calls is less than 0.01. The optimal parameter of uniform fractional guard channel policy is obtained by algorithm 3.

By carefully inspecting the table 1, it is evident that for some range of input traffics, the performance of the proposed policy is close to the performance of the uniform fractional guard channel policy.

```
if (NEW CALL) then
    if (LA.action () = ACCEPT) then
        if (c(t) < C )then
            accept call
            reward action ACCEPT
        else
            reject call
            penalize action ACCEPT
        end if
    else //LA selects action REJECT
        if (c(t) < C and B̂_h < p_h ) then
            accept call
            penalize action REJECT
        else
            reject call
            reward action REJECT
        end if
end if
```

Fig. 4. LA based call admission control algorithm

Table 1. Minimize B_n such that $B_h \leq 0.01$

Case	λ_h	Uniform Fractional Guard Channel Policy		LA Based Uniform Fractional Guard Channel Policy	
		B_n	B_h	B_n	B_h
1	2	0.023935	0.024675	0.054314	0.023388
2	4	0.089897	0.023639	0.088707	0.025217
3	6	0.15725	0.022202	0.147586	0.024847
4	8	0.223872	0.020367	0.193917	0.024625
5	10	0.289849	0.019248	0.193917	0.024625
6	12	0.356866	0.017607	0.249867	0.024294
7	14	0.424072	0.016390	0.289657	0.025141
8	16	0.489967	0.015076	0.391456	0.024460
9	18	0.557026	0.013939	0.444263	0.023861
10	20	0.623746	0.013318	0.488290	0.024800

6 Conclusions

In this paper, we introduced a LA based call admission control policy for cellular networks. In the proposed policy, a LA is used to accept/reject new calls. This call admission policy can be considered as adaptive uniform fractional guard channel policy. In order to study the performance of the proposed call admission policy, the computer simulations are conducted. The simulation results show that for some range of input traffics, the performance of the proposed approach is close to the performance of the uniform fractional guard channel policy. Since LA are adaptive and don't need any information about its environment, the proposed policy is fully adaptive and doesn't require any information about the input traffics.

References

1. R. Ramjee, D. Towsley, and R. Nagarajan, "On Optimal Call Admission Control in Cellular Networks," *Wireless Networks*, vol. 3, pp. 29–41, 1997.
2. D. Hong and S. Rappaport, "Traffic Modelling and Performance Analysis for Cellular Mobile Radio Telephone Systems with Priotrized and Nonpriotorized Handoffs Procedure," *IEEE Transactions on Vehicular Technology*, vol. 35, pp. 77–92, Aug. 1986.
3. G. Haring, R. Marie, R. Puigjaner, and K. Trivedi, "Loss Formulas and Their Application to Optimization for Cellular Networks," *IEEE Transactions on Vehicular Technology*, vol. 50, pp. 664–673, May 2001.
4. H. Beigy and M. R. Meybodi, "Uniform Fractional Guard Channel," in *Proceedings of Sixth World Multiconference on Systemmics, Cybernetics and Informatics, Orlando, USA*, July 2002.
5. P. R. Srikantakumar and K. S. Narendra, "A Learning Model for Routing in Telephone Networks," *SIAM Journal of Control and Optimization*, vol. 20, pp. 34–57, Jan. 1982.
6. O. V. Nedzelnitsky and K. S. Narendra, "Nonstationary Models of Learning Automata Routing in Data Communication Networks," *IEEE Transactions on Systems, Man, and Cybernetics*, vol. SMC-17, pp. 1004–1015, Nov. 1987.
7. B. J. Oommen and E. V. de St. Croix, "Graph Partitioning Using Learning Automata," *IEEE Transactions on Commputers*, vol. 45, pp. 195–208, Feb. 1996.
8. H. Beigy and M. R. Meybodi, "An Algorithm Based on Learning Automata for Determining the Minimum Number of Hidden Units in Three Layers Neural Networks," *Amirkabir Journal of Science and Technology*, vol. 12, no. 46, pp. 111–136, 2001.
9. M. R. Meybodi and H. Beigy, "Neural Network Engineering Using Learning Automata: Determining of Desired Size of Three Layer Feedforward Neural Networks," *Journal of Faculty of Engineering*, vol. 34, pp. 1–26, May 2001.
10. B. J. Oommen and T. D. Roberts, "Continuous Learning Automata Solutions to the Capacity Assignment Problem," *IEEE Transactions on Commputers*, vol. 49, pp. 608–620, June 2000.
11. K. S. Narendra and K. S. Thathachar, *Learning Automata: An Introduction.* New York: Printice-Hall, 1989.

An Adaptive Flow Control Scheme for Improving TCP Performance in Wireless Internet

Seung-Joon Seok, Sung-Min Hong, and Chul-Hee Kang

Department of Electronics Engineering, Korea University,
1, 5-ga, Anam-dong, Sungbuk-gu, 136-701, Seoul, Korea
{ssj, mickey, chkang}@widecomm.korea.ac.kr
http://widecomm.korea.ac.kr/~ssj

Abstract. The end-to-end TCP (Transmission Control Protocol) performance is one of important issues in wireless Internet services. In this paper, a TCP aware link layer protocol, called *Adaptive-TCP(A-TCP),* is proposed in order to improve the performance. The key idea of the protocol is that an A-TCP agent, which is located in each base station, makes a mobile host look as if it has a wired link with the base station. In this paper, this concept is referred to a *virtual host model.* In order to implement it, the A-TCP agent performs three functions; Local Retransmission, Freezing Sender and A-TCP Flow Control. In this proposal, the A-TCP Flow Control is original and is also the principal factor for improving the performance in a wireless Internet. In the A-TCP Flow Control, the A-TCP agent marks the window field of acknowledgement segment with a retransmission buffer size. Therefore, the TCP congestion controls, which happen in a TCP sender, are not caused by wireless link overflow. Performance evaluations were conducted by computer simulations. The results of the evaluation show that the A-TCP can provide near optimal performance in wireless bottleneck conditions. These also suggest that the A-TCP improves performance by at least 20% compared to other TCP approaches.

1 Introduction

In the Internet, a TCP is an important protocol that guarantees reliable transmission between end-users at the transport level. Since mobile hosts (MHs) will expect the same services that are offered to fixed hosts (FHs), it is necessary to implement a TCP for the mobile domain. If a legacy TCP designed for a wire-based carriage is used in wireless networks without modification, however, a serious drop in its performance will occur. This is because wireless environments have several characteristics that are quite different from those of wired network. Unfortunately, legacy TCP designers have not taken these characteristics into consideration. The principal characteristics of wireless communications are a high bit error rate (BER) within the wireless channel, frequent disconnections of the wireless channel, and a narrow and variable bandwidth of the wireless channel. These are result from the unreliability of wireless channels and the mobility of MH. Current TCP protocols assume that packet loss is a result of

M.H. Shafazand and A M. Tjoa (Eds.): EurAsia-ICT 2002, LNCS 2510, pp. 458–465, 2002.

network congestion, whereas in reality, the packet is corrupted in a wireless channel. This was uncovered from discussions on TCP modification so as to make it suitable for a wireless Internet.

Much effort has been made to improve the TCP performance for mobile Internet services. The solutions are classified into three broad categories; a link-layer solution, a split connection, and an extension of the TCP itself [4]. These methods for improving end-to-end TCP performance mostly focus on the packet losses caused by the BER and the channel disconnections, because the TCP sender assumes that all losses are caused by the congestions within network congestion. In addition, a wireless link generally becomes a bottleneck portion in an end-to-end TCP connection because of its narrow bandwidth, as compared to wired links. Thus, a TCP sender's congestion controls are apt to be caused by wireless link congestions, even though the wireless channels are reliable via other solutions. So far, this point has not been considered in efforts to improve TCP throughput. This paper, however, considers the narrow bandwidth characteristic of the wireless channels, as well as their high BER and frequent disconnections, in attempting to improve TCP performance and the utilization of wireless channels.

This paper proposes a conceptual model of a wireless Internet service, called a *virtual host*. As a method of improving end-to-end TCP throughput, this model supports not only a high BER and frequent channel disconnections, just as previous study does, but also the narrow and variable bandwidth of wireless channels. In addition, the Adaptive TCP (A-TCP) is proposed for the implementation of the virtual host concept. The A-TCP performs three functions: local retransmission, sender freezing, and A-TCP flow control, all of which are performed by an *A-TCP agent* at a base station. The first and second functions diminish the impact of the BER and long-term channel disconnection, respectively, and are based on previous solutions such as snoop-TCP, M-TCP and TCP-SMART [7]. The third function is a principal mechanism of our solution to improve wireless channel utilization. A-TCP flow control involves the application of the current TCP's flow control mechanism to a wireless Internet environment.

2 Description of Virtual Host Model

This section describes a new communication model to improve the performance of TCP connections with wireless links. For convenience, this model is referred to as a virtual host. The operation of the virtual host is for a wireless system to emulate a host that is connected through a wired link, although the host really has a wireless channel. The wireless system is composed of base stations, MHs and wireless links. To completely hide the mobile environment from the FH, several problems that occur in a solely wireless network should be considered. These are packet corruption within a wireless channel, wireless channel disconnection, and wireless link congestion, which had already been mentioned. Strictly speaking, an instance of the virtual host represents the part of an end-to-end TCP connection between a base station and an MH.

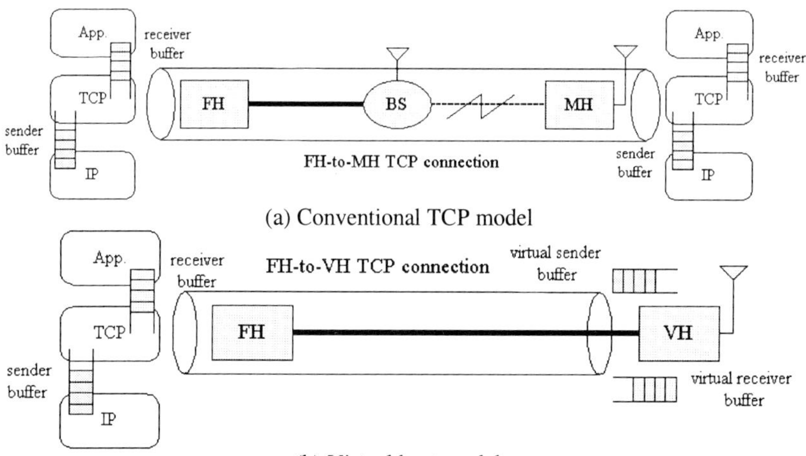

(a) Conventional TCP model

(b) Virtual host model

Fig. 1. The comparison of TCP and virtual host models

The virtual host masks the part of the TCP connection that is affected by the wireless channel from the FH's TCP sender, and so the FH's TCP come to think to connect with another FH's TCP. We thus define a virtual host as a conceptual host that supports only a TCP connection and diminishes the effect of wireless characteristics. The virtual host has two virtual buffers that are used as the receiver and the sender buffer of the normal TCP, respectively. The virtual buffer size implies the virtual host's capacity to receive and send segments. In a normal TCP, a TCP receiver informs a TCP sender of its buffer capacity or the number of packets that the receiver buffer can accept anew, using the window field of ACK segment. This is a TCP flow control behavior. Therefore, the virtual host also performs a modified flow control. The virtual host also uses the window field to inform a TCP sender of the capacity of the virtual receiver buffer. It is thus important to know how the window field is marked. On the other hand, the virtual host does not send a new packet whenever the virtual sender buffer is full. This behavior is the same as that of a normal TCP sender. The virtual host also uses the window field to inform a real TCP sender of the capacity of the virtual sender buffer.

Figure 1 illustrates the virtual host model at both the receiver and the sender sides, as compared to a normal TCP case. In the case of data transmission from an FH to an MH, data segments are buffered in the virtual receiver buffer, until the segments are confirmed to have reached the mobile host. The packets received by the normal TCP receiver are buffered at the receiver buffer, until an application reads the packets. In the case of data transmission from an MH, however, the virtual sender buffer may act as a normal sender buffer in which unacknowledged segments are buffered. These receiver and sender virtual buffers of the virtual host are embodied as two retransmission buffers that are located in the base station. One is used for connection from the MH and the other is used for connection to the MH. As the BER and the disconnection rate of the wireless channel increase, the virtual buffer occupancy should also increase. The occupancy can thus be assumed as representing the current channel status.

3 Adaptive-TCP Behaviors

To implement the virtual host model, we propose the use of a TCP-aware link layer protocol called Adaptive TCP (A-TCP). An A-TCP agent module, which exists in every base station, performs the A-TCP for all TCP connections that pass through the base station. The A-TCP of a virtual host object performs the three basic functions of local retransmission, sender freezing and A-TCP flow control in order to hide the wireless environment from the FH and emulate the FH's TCP. The local retransmission diminishes the effect of a high BER and temporal channel disconnection. Sender freezing also solves the problem of long-term channel disconnection. In addition, A-TCP flow control improves the utilization of a wireless channel with a variable and narrow bandwidth. Since the third function is a principal and original A-TCP mechanism for improving end-to-end TCP performance, this paper focuses on the A-TCP flow control mechanism. Details of these three functions are as follows.

1) *Local Retransmission*: The A-TCP locally retransmits lost packets at the wireless link level. In this, the A-TCP agent has a timer as a retransmission timer of the TCP sender. If the timer is expired for a particular segment, the A-TCP immediately retransmits the segment. Alternatively, if the A-TCP agent receives several duplicate acknowledgements, the agent filters the duplicate acknowledgements and locally retransmits the lost packet. This behavior is similar to a snoop TCP[4].
2) *Sender Freezing:* If a wireless channel disconnection condition continues for a relative long time, an A-TCP agent sends TCP sender a duplicate acknowledgement for the last acknowledged packet, which has a zero window size, to the TCP sender. The zero window caused makes the TCP sender to remain in the persist mode and start off with the previous window size, after the wireless channel is reconnected. This behavior is referred to as sender freezing and is based on previous reports [1][5][7]. For this behavior, the A-TCP has the second timer to examine the channel disconnection time. The timer starts whenever the local retransmission timer expires and this disconnection timer does not active. If no acknowledgement is received from the mobile host, until the disconnection timer has expired, the A-TCP agent recognizes that the wireless channel has been disconnected for a lengthy period of time and a freezing of the sender is needed.
3) *A-TCP Flow Control:* The A-TCP performs a flow control for improving wireless channel utilization. The basic concept of an A-TCP Flow Control is that an A-TCP agent informs a TCP sender of its wireless channel capacity using the window field of ACK segment and, then, TCP sender does not increase its sending rate over the capacity. The capacity means the number of packets that the A-TCP agent can guarantee from loss. This is the similar concept of the normal TCP flow control. In the following section, full details of the A-TCP Flow Control process are provided.

A-TCP flow control is based on two fundamental facts. The first is that the occupancy of retransmission buffer can precisely represent the current channel status. The other is that to prevent its overflow, the receiver can limit the sending rate using legacy TCP flow control. If a TCP sender sends packets merely according to the A-TCP's retransmission buffer size, the packet can be virtually guaranteed from loss caused by wireless link's overflow. In the A-TCP, the A-TCP agent has a buffer that is used like a receiver buffer. If a packet reaches the A-TCP agent, the packet is cached at a retransmission buffer for local retransmission. When an ACK of the packet is received from the MH, the cached packet is removed and the window field of the ACK is remarked with the minimum between the current receiver window value and the local retransmission buffer size of the corresponding virtual host object. Note that the buffer size is the total buffer size, not the residual size of the buffer, unlike the receiver buffer. Thus, the sender window size is not larger than the retransmission buffer size and the receiver window size. As a result, virtual host object overflow can never happen. Figure 2 illustrates the A-TCP flow control operation.

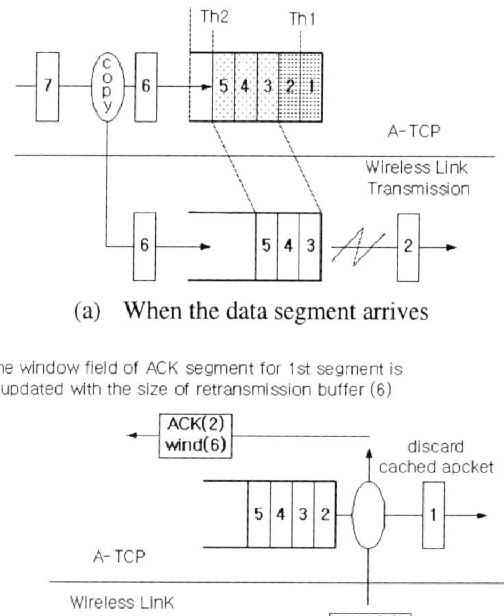

Fig. 2. The A-TCP flow control mechanism performed by an A-TCP agent at the base station

The MH's TCP receiver informs the TCP sender of the receiver buffer's remaining capacity, using the ACK segment's receiver window through the A-TCP agent. If the virtual host object's local retransmission buffer is smaller than the receiver buffer, the A-TCP agent transmits to the sender the retransmission buffer size, instead of the receiver buffer size. In addition, the sender window must be the smallest among the congestion window, the receiver window and the sender buffer. This is because the

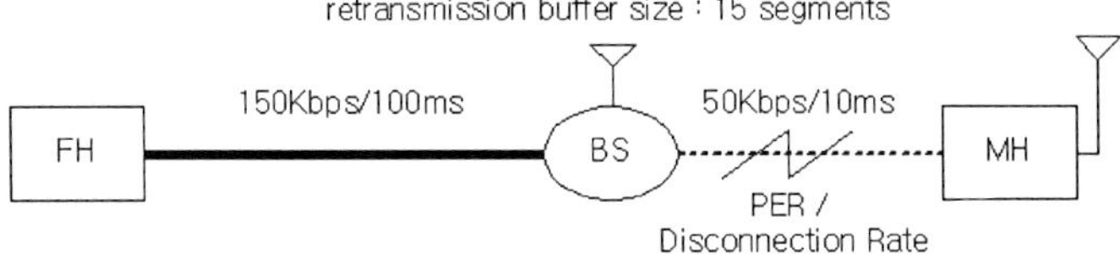

Fig. 3. Simple analysis / simulation model

sender should be able to control the network congestion, sent packets should be cached at the sender for retransmission, and the receiver buffer and the local retransmission buffer should not overflow.

4 Performance Study of A-TCP Protocol

In this subsection, we evaluate the performance of our proposal through simulations. The network model used in these simulations had a simple topology (Figure 3) in which an FH was connected to a base station through a wired link and the base station was connected to an MH through a wireless link that was experiencing channel disconnections and error conditions, and had a third part of the wired link's bandwidth. These simulations focused on data transmission from the FH to the MH. This simple simulation was implemented using C++. In this model, the BER within a segment was expressed as the Packet Error Rate (PER) of the segment, and channel disconnection was modeled using the ON-OFF model. The MSS of the TCP connection was 1000

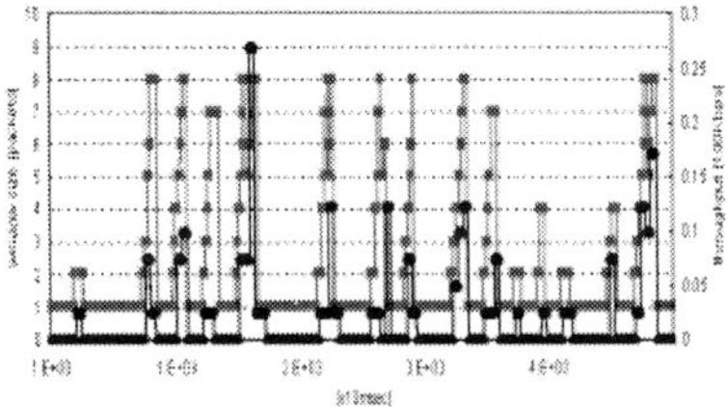

(a) Normal Reno-TCP congestion control

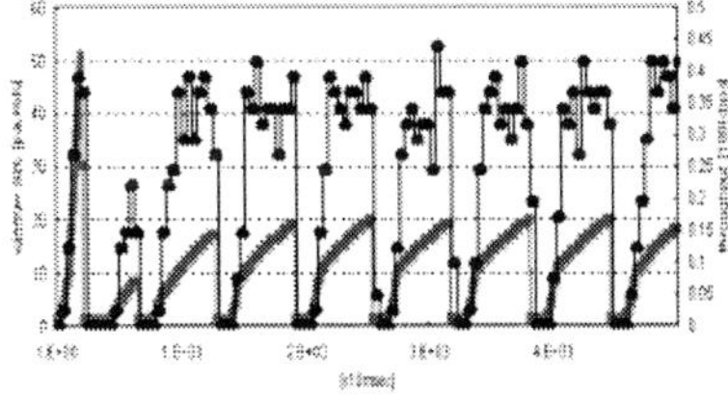
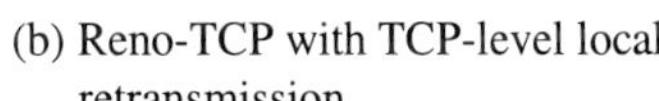

(b) Reno-TCP with TCP-level local retransmission

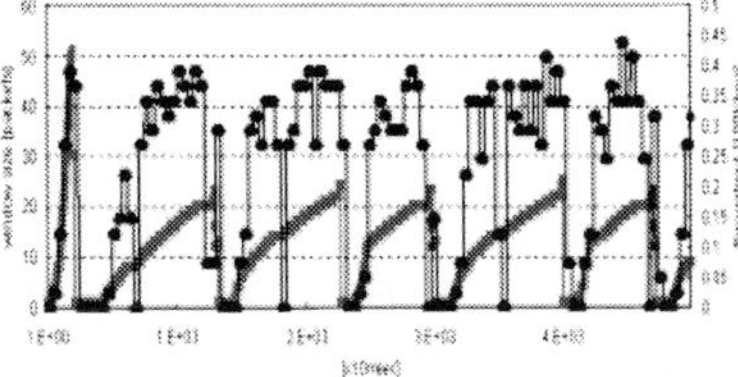

(c) Reno-TCP with local retransmission and freezing sender

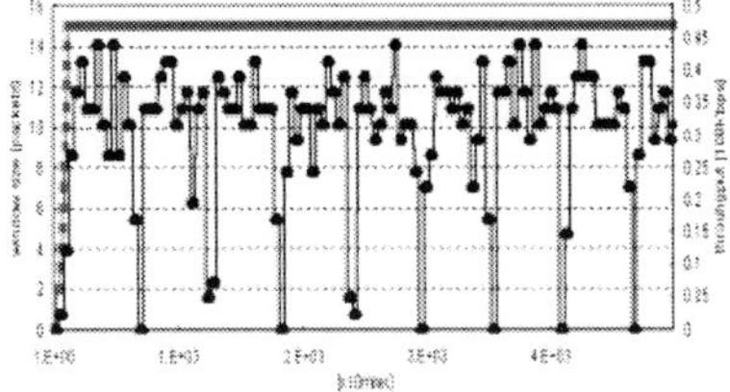

(d) A-TCP: Reno-TCP with local retransmission, freezing sender and A-TCP flow control

Fig. 4. Simulation results of several approaches in throughput and window size.

bits. The MH was also assumed to have large buffer size and enough processing capacity. This subsection considered four TCP strategies for wireless environments. The first strategy adopted a conventional TCP Reno simply to a wireless network. The second strategy performed local retransmission of packets lost in a wireless link. The third strategy adopted both local retransmission and sender freezing. The fourth strategy adopted A-TCP flow control in addition to the third strategy. For the implementation of the second and third cases, we simplified the procedures of the snoop protocol and the M-TCP. The fourth case represents the A-TCP.

Figure 4 illustrates the sender window size and the throughput of previous strategies for a 30% PER and a long channel disconnection for every 5 seconds. In these simulations, the local retransmission buffer size was set at 15 packets, with each packet size made up of 1000 bits and wireless link's and wired link's bandwidths are 150 Kbps and 50 Kbps, respectively. The average throughputs of four strategies were 1.8, 20.3, 21.6 and 29.8 Kbps, respectively. If the optimal throughput of the wireless link is considered to be approximately 30 Kbps, the A-TCP throughput is remarkable.

Next, Figure 5 shows the comparison of four strategies in different PER conditions. Other simulation conditions of this experiment are the same as are considered in the previous experiment. In this experiment result, fourth strategy using A-TCP always produces a near optimal solution regardless of the PER. Since both second and third strategies perform the local retransmission, the performance curves have a similar wireless throughput degradation pattern. It can be also deduced from the similarity of the second and third results that a possible reason for the degradation in performance is the frequent congestion controls caused by the narrow bandwidth. The first strategy using the legacy TCP-Reno experienced dramatic performance degradation as the PER increases. The effect of the long term-channel disconnection was examined next. Figure 6 shows the throughputs of four strategies in different disconnection rate. The result of this experiment is similar to that of the second experiment. The A-TCP can also provide near optimal throughput regard of the disconnection rate. The sender

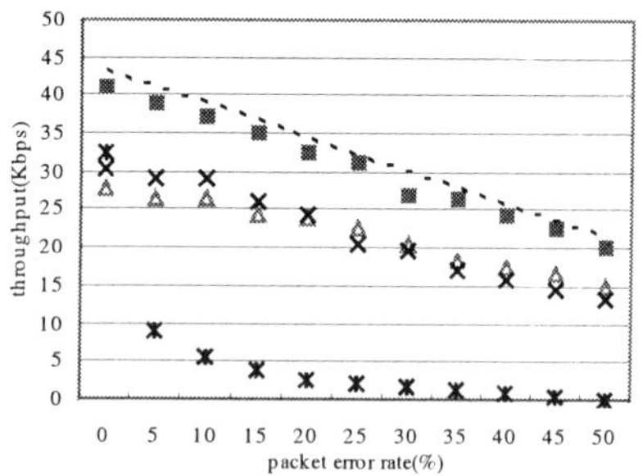

Fig. 5. Throughput curves for several packet error

-- : wireless channel throughput

•: A-TCP

Δ : Snoop+freezing TCP

X : Snoop TCP

* : Normal Reno TCP

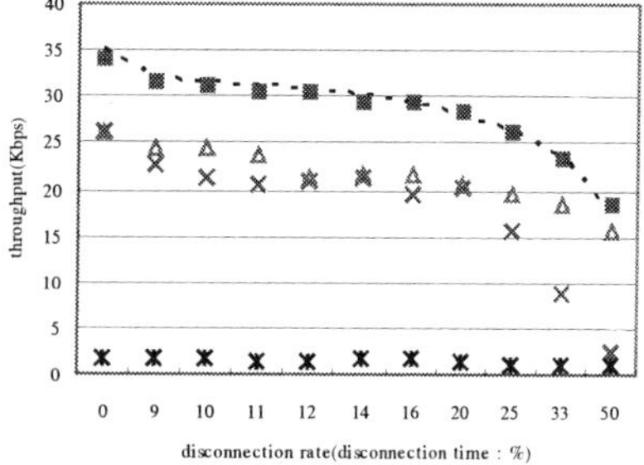

Fig. 6. Throughput curves for several packet error rates

freezing method is unnecessary when there is a low disconnection rate, but is quite effective when there is an exceedingly high disconnection rate. The local retransmission and sender freezing are core principles that snoop-TCP and M-TCP, respectively, improve the TCP performance. Previous experiment results thus demonstrate that A-TCP improves the wireless Internet more than snoop and M-TCP.

5 Conclusion

In this paper, a virtual host concept and a protocol, called an Adaptive-TCP, were proposed to improve the TCP performance in wireless Internet environments. This paper focused first on increasing the wireless link utilization through A-TCP flow control. Furthermore, the A-TCP is based on local retransmission and freezing sender for other wireless link characteristics of high BER and long time channel disconnection. The performance was evaluated through computer simulations. The simulation results show that the A-TCP protocol can provide a near optimal throughput under variable wireless conditions and A-TCP flow control is the key reason for this.

There are two reasons of packet loss in wireless environment. These are wireless channel corruption and mobility of mobile host. This paper focused on the former and. Thus the next issue is to describe the handoff procedure and protocol carefully and to evaluate these more precisely. Also another issue is to implement and experiment the virtual host concept in real experimental network based on Linux operating system.

References

1. Moronski, J.: Freeze-TCP: A true end-to-end TCP enhancement mechanism for mobile environments. In Proc. IEEE Infocom'2000(2000) pp 1537–1545
2. Perkins, C.: IP Mobility Support. IETF TFC 2002(1996)
3. Balakrishnan, H., Oadmanabhan, V.N., Seshan, S., Kata, R. H.: A Comparison of Mechanisms for Improving TCP Performance over Wireless Links. In Proc. ACM SIGCOM'96(1996) pp 256–269
4. Balakrishnan, H., Seshan, S., Kata, R. H.: Improving Reliable Transport and Hand-off Performance in Cellular Wireless Networks. ACM Wireless Networks, Vol. 1(4) (1995)
5. Brown, K., Singh, S.: M-TCP: TCP for Mobile Cellular Networks. ACM Computer Communication Review, Vol. 27(5) (1997)
6. Bakre, A., Badrinath, B. R.: I-TCP: Indirect TCP for Mobile Hosts. In Proc. 15th International Cof. On Distributed Computing Systems (1995) pp 136–143
7. Elaroud, M., Ramanathan, P.: TCP-SMART: A Technique for Improving TCP Performance in a Spotty Wide Band Environment. In Proc. ICC'2000 (2000) pp 1783–1787

An Adaptive TCP Protocol for Lossy Mobile Environment

Choong Seon Hong[1], YingXia Niu[1], and Jae-Jo Lee[2]

[1] School of Electronics and Information, Kyung Hee University
Korea 449-701
{cshong, niuyx}@khu.ac.kr
[2] Korea Electrotechnology Research Institute
Korea 437-808
jjlee@keri.re.kr

Abstract. TCP has been designed and tuned as a reliable transfer protocol for wired links. However, it incurs end-to-end performance degradation in wireless environments where packet loss is very high. TCP HACK (Header Checksum Option) is a novel mechanism proposed to improve original TCP in lossy links. It presents an extension to TCP that enables TCP to distinguish packet corruption from congestion in lossy environments. TCP HACK performs well when the sender receives the special ACKs correctly, but if many ACKs are also lost, the efficient of TCP HACK will not be prominent. In this paper we present an extension to TCP HACK, which can perform well even if the ACKs are severely corrupted. We use OPNET to simulate our proposal. The results have shown that our proposal performs substantially better than TCP HACK when corruptions occur on both data transmission path and acknowledgement path.

1 Introduction

Recent years, supporting Internet service over wireless network is a hot issue that has attracted many researchers to develop enhancements. Many applications are built on top of TCP, and will continue to be in the foreseeable future. So the performance of TCP in wireless environments has received much attention in recent years. The transmission control protocol (TCP) has been designed, improved and tuned to work efficiently on wired networks where the packet loss is very small. Whenever a packet is lost, it is reasonable to assume that congestion has occurred on the connection path. Hence, TCP triggers congestion recovery algorithms when packet loss is detected. These algorithms work reasonably well as the assumption on packet losses remains valid in most situations. However, in the wireless Internet environment, the bit error rate is much higher. As a result, the assumption that packet loss is (mainly) due to congestion is no longer valid. And the original TCP cannot work well in a heterogeneous network with both wired and wireless links.

Many protocols have been proposed to improve the performance of TCP over wireless links, but most of them need to use an intermediary node (such as base station) to modify TCP. So they cannot maintain an end-to-end TCP. TCP HACK (Header Checksum Option) [12] is a true end-to-end protocol proposed to improve the performance of TCP over lossy links. TCP HACK can work well when there are

M.H. Shafazand and A M. Tjoa (Eds.): EurAsia-ICT 2002, LNCS 2510, pp. 466–478, 2002.

bursty errors on the data transmission path but no corruption on the acknowledgement path.

In this paper, we propose an extended TCP HACK that can work well even when there are much corruptions on the acknowledgement path.

The rest of this paper is organized as follows: Section 2 gives the problems of original TCP over lossy links; In section 3, we introduce the existing proposed solutions, indicating their strengths and weaknesses; our extended TCP HACK is proposed in section 4; simulation comparisons and drawbacks of our protocol are also presented. Concluding remarks are given in section 5.

2 Problems of TCP over Lossy Links

The original TCP detects missing ACKs via three duplicate acknowledgements or time-outs. When high bit error rates occur, even if a single packet loss will be considered that congestions have happened, it just could not distinguish between the congestion and packet loss. Then the TCP sender will trigger slow start mechanism: it drops the congestion window down to 1, and first grows it by a factor of 2 each time an ACK is received, until it reaches half of the threshold of the congestion window. Fig.1-1 shows the process of slow-start mechanism after a packet loss is detected, Fig.1-2 shows if we can distinguish the packet loss from congestion and immediately retransmit the lost packet, the slow-start will be avoided. The comparison between these two figures indicates that the wrong assumption drastically decreases the performance of TCP in the cases where bit error rates are high.

Fig. 1.1 Slow-start algorithm is triggered after packet loss in detected

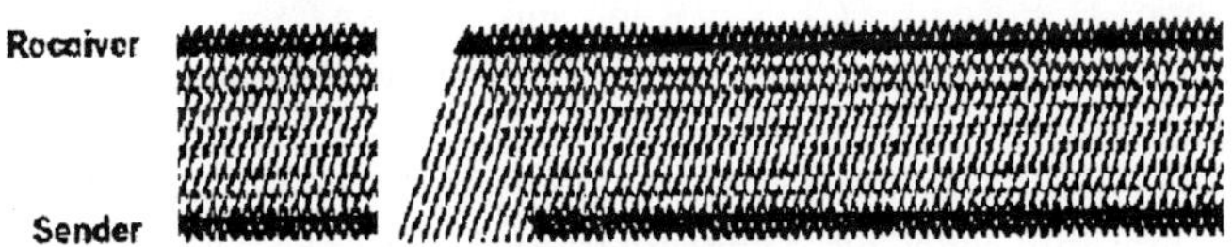

Fig. 1.2 Avoiding to trigger slow-start when detected packet loss

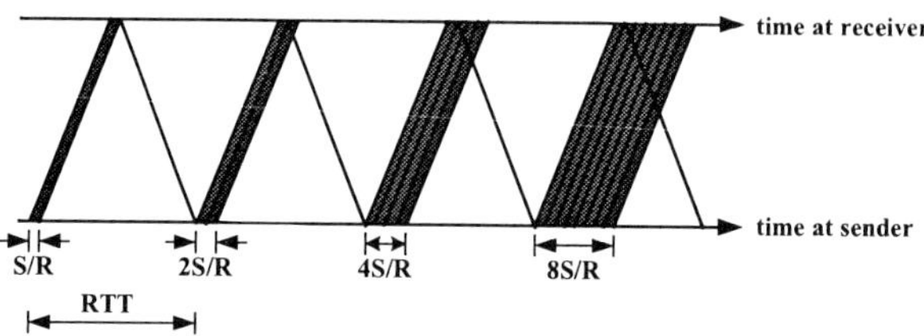

Fig. 2. TCP timing during slow-start

Fig.2. shows the TCP timing during slow start. S means the MSS (maximum size segment) is S bits; RTT is the round-trip time (including the transmission time of the packet) and R denotes the transmission rate of the link from the sender to the receiver. So the idle time (approximate) of the sender in slow start stage is given by:

Idle time=P[RTT]-(2^P-1)S/R

In which P=min {Q, K-1}, K is the number of windows that cover the object and Q is the number of times the sender would stall. For details, please read [15].

If the slow-start scheme is not triggered, then the number of extra packets that can be transferred is approximate given by [10]:

Extra Segments=W^2/8+WlgW-5W/4+1;

W is the unACKed packets can be sent in a congestion window.

It should be noted that the upper two expressions are under the assumptions that no corruptions in the traffic and ignoring all protocol header overheads.

3 Strengths and Drawbacks of Existing Solutions

Now the existing TCP implementation normally conforms to one of the four protocols: TCP Tahoe; TCP Reno; TCP NewReno and TCP SACK.

TCP Tahoe [3] is the original protocol that has Slow-Start, Congestion Avoidance, and Fast Retransmit algorithms.

TCP Reno [3] introduces a fast recovery algorithm to TCP Tahoe. After the fast retransmission algorithm sends what appears to be the missing segment, the fast recovery algorithm sets the congestion window to a half of its current window, and invokes congestion avoidance from a halved congestion window, not like TCP Tahoe that sets the congestion window to the smallest value. The reason that doesn't trigger the slow-start algorithm is because the receipt of the duplicate ACKs not only indicates that a segment has been lost, but also that the receiver can still receive the segments.

TCP NewReno [20] improves Reno with a Partial Acknowledgment algorithm. When TCP enters the Fast Recovery, it records the highest sequence number. If a new acknowledgment arrives during the Fast Recovery but does not cover the highest sequence number, TCP evaluates it as a Partial Acknowledgment and anticipates that more packets are lost.

TCP Tahoe, Reno, and NewReno will experience poor performance when multiple packets are lost from one window of data. With the limited information, A TCP sender can only learn about a single lost packet per round trip time.

TCP SACK [11], a Selective Acknowledgment mechanism, combined with a selective repeat retransmission policy, can help to overcome this limitation. The receiving TCP sends back SACK packets to the sender informing the sender of data that has been received. The sender can then retransmit only the missing data segments. But TCP SACK is still constrained by the tiny congestion window.

Recently, many protocols have been proposed to alleviate the poor end-to-end TCP performance in the heterogeneous network environments. These mechanisms can be mainly divided into two classes: one class needs to use Performance Enhancing Proxies (PEPs) [18]; the other class is end-to-end mechanisms that do not require TCP-level awareness by intermediate nodes. The first class includes Indirect TCP (I-TCP), Snoop TCP, New Snoop TCP, Mobile TCP (M-TCP), and Explicit bad-state notification (EBSN) etc; the second class includes TCP HACK etc.

In the following, we will describe the mechanisms of these proposals and see the strengths and drawbacks of each solution.

I-TCP [5] suggests that any interaction from a mobile host (MH) to a host on the fixed network (FH) should be split into two separate interactions - one between the MH and the Base Station (BS) over the wireless medium and the other between the BS and the FH over the fixed network. The data sent to MH is received, buffered and ACKed by BS. It is then the responsibility of BS to deliver the data to MH. The BS communicates with the MH on a separate connection using a variation of TCP that is tuned for wireless links and is also aware of mobility. By using I-TCP it can achieve better throughputs than standard TCP. But it does not preserve end-to-end semantics of TCP because the BS partitions the TCP connection.

The Snoop protocol [6] introduces a module, called snoop agent, at the base station. The agent monitors every packet that passes through the TCP connection in both directions and maintains a cache of TCP packets sent across the link that has not yet been acknowledged by the receiver. The snoop agent retransmits the lost packet if it has cached and suppresses the duplicate acknowledgments (ACKs). The Snoop module performs extremely well in high BER environments and maintains end-to-end TCP semantics. But it needs the intermediary (the Base Station) to do TCP modifications.

The New Snoop protocol [7] was proposed to overcome the shortcomings of Snoop protocol. It uses a two-layer hierarchical cache scheme. The main idea is to cache the unacknowledged packets at both Mobile Switch Center (MSC) and Base Station (BS), thus forming a two-layer cache hierarchy. If a packet is lost due to transmission errors in wireless link, the BS takes the responsibility to recover the loss. If the loss/interruption is due to a handoff, the MSC performs the necessary recovery. With this proposed hierarchical cache architecture, New Snoop protocol can effectively handle the packet losses caused by both handoffs and link impairments. But both Snoop protocol and New Snoop protocol need the intermediary (Such as a Base Station) to do TCP modifications and New Snoop even needs the MSC's participation.

The M-TCP [8] has the same goals as I-TCP and snoop TCP: to prevent the sender window from shrinking if bit errors or disconnection but not congestion cause current problems. M-TCP wants to improve overall throughput, to lower the delay, to maintain end-to-end semantics of TCP, and to provide a more efficient handover. It splits the TCP connection into two parts as I-TCP does. An unmodified TCP is used on the standard FH-BS connection, while an optimized TCP is used on the BS-MH connection. The BS monitors all packets sent to the MH and ACKs returned from the MH. And it retains the last ACK. If the BS does not receive an ACK for some time, it assumes that the MH is disconnected. It then shut down the TCP sender's window by sending the last ACK with a window set to zero. Thus, the TCP sender will go into persist mode. The M-TCP approach does not perform caching/retransmission of data via the BS. If a packet is lost on the wireless link, it has to be retransmitted by the original sender. This maintains the TCP end-to-end semantics. But it still requires a substantial base station involvement.

An explicit bad-state notification (EBSN) scheme [9] does not split the connection in two connections, it uses two types of acknowledgments: one is a partial acknowledgment informing the sender that the packet had been received by the base station and the other is a complete acknowledgment which has the same semantics as the normal TCP acknowledgment, i.e. the receiver (MH) received the packet. So it can distinguish the losses on the wired portion from the losses on the wireless link. Now the base station is responsible for retransmissions on the wireless link, while it delays timeout at the sender by sending a partial acknowledgement. This idea is that these explicit notifications prevent the sender from dropping congestion window. It also requires an intermediate node to modify TCP.

TCP HACK (Header Checksum Option) [12] is a solution based on the premise that when packet corruption occurs, it is more likely that the packet corruption occurs in the data and not the header portion of the packet. This is because the data portion of a packet is usually much larger than the header portion for many applications over typical MTUs. It introduced two TCP options: the first option is for data packets and contains the 1's-complement 16-bit checksum of the TCP header (and pseudo-IP header) while the second is for ACKs and contains the sequence number of the TCP segment that was corrupted. These "special" ACKs do not indicate congestion in the network. Hence, the TCP sender does not halve its congestion window if it receives multiple "special" ACKs with the same value in the ACK field. With this scheme, TCP is able to recover these uncorrupted headers and thus determine that packet corruption and not congestion has taken place in the network. TCP HACK performs substantially better than both TCP SACK and NewReno in cases where burst corruptions are frequent.

4 Our Proposed Solution: Extended TCP HACK

4.1 Requirements for Enhancing the Performance of TCP

Our goal in developing a new TCP protocol is to provide a general solution to the problem of improving TCP's efficiency for lossy links. Specifically, we want to design a protocol that has the following characteristics:

- It should preserve end-to-end TCP semantics. I-TCP and EBSN are not end-to-end-semantics. So, if a sender receives an acknowledgement, it assumes that the receiver got the packet. Receiving an acknowledgement now only means (for the mobile host and a correspondent host) that the foreign agent received the packet. The correspondent node does not know anything about the partitioning. Thus a crashing access node may also crash applications running on the correspondent node assuming reliable end-to-end delivery.
- It should not require the intermediate node to do TCP modifications. I-TCP, Snoop, New Snoop, M-TCP and EBSN all need the intermediate node to do TCP modifications. So the intermediary will become the bottleneck and add the third point of failure besides the endpoints themselves.
- It can handle encrypted traffic. As network security is taken more and more seriously, encryption is likely to be adopted very widely. Finally, all efforts for snooping and buffering data in the intermediate nodes may be useless if certain encryption schemes are applied end-to-end between the correspondent host and mobile host. Using IP encapsulation security payload the TCP protocol header will be encrypted, so that the intermediate nodes may not even know that the traffic being carried in the payload is TCP. Furthermore, retransmitting data from the foreign agent may not work any longer because many security schemes prevent replay attacks and retransmitting data from the foreign agent may be misinterpreted as replay. Encrypting end-to-end is the way many applications go. Therefore, it is not clear how these schemes (I-TCP, Snoop, New Snoop, M-TCP and EBSN) could be used in the future.
- It doesn't need a symmetric routing. The protocols that need to modify TCP in an intermediate node usually require that traffic to and from the end mobile host is routed through the same intermediate node. But in some networks, data and ACKs can take different paths, so these schemes based on intermediary involvement cannot be accomplished and may result in non-optimal routing.
- It can handle high BER. I-TCP, M-TCP, Snoop, New Snoop, and TCP HACK all can handle high BER.

From the analysis above we can get a conclusion that the existing mechanisms that need the PEPs have many drawbacks. The adoption of these protocols in the future should be considered. So we pay attention to the TCP HACK protocol. We think it is a novel mechanism to improve the TCP over lossy links; we do some modification to TCP HACK to make it more efficient. The modified TCP HACK is called Extended TCP HACK.

4.2 Our Proposal: Extended TCP HACK

In TCP HACK, when the receiver receives a corrupted packet whose header is not corrupted, it recovers the packet sequence number from the header and sends a special ACK packet including that sequence. If the return path is lossless, the TCP sender can get the information in time. But since the return path carrying ACKs and special ACKs is not lossless, the special ACK conveying the information that the packet was corrupted might be lost, now what the sender could do is waiting for the timeout.

The idea of our extended TCP HACK comes from the structure of TCP SACK. In SACK, the SACK option is defined to include more than one SACK block in a single packet. The redundant blocks in the SACK option packet increase the robustness of SACK delivery in the presence of lost ACKs. So we try to send more than one sequence number in one special ACK packet. We add a buffer in the TCP receiver (here we call it s_buffer). Then we save all the recovered sequence numbers into the s_buffer. These sequence numbers have been recovered from those packets whose data are corrupted but the sequence numbers in headers can be recovered (these packets have been transmitted in the same window). Then in the HACK special ACK option, we acknowledge all these sequence numbers saved in the s_buffer. So if the last special ACK is lost, the sender can get the sequence number from the next special ACK and retransmit the corrupted packet. It should be mentioned that sequence numbers in the s_buffer would be cleared if the corresponding retransmitted packets have been received correctly or the timers expire.

In TCP HACK, it introduced two options: one is Header Checksum option; the other is the Header Checksum ACK option. In our proposal, we don't make any change in the first option (see Fig.3.), but the second option has been extended (see Fig.4.). The packet length of the special ACK is variable and the sequence number in the option can be more than one.

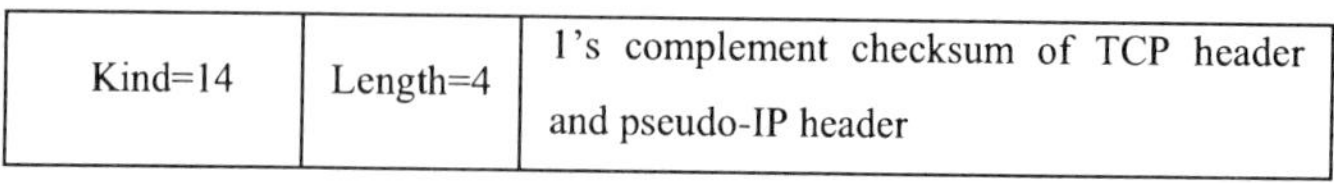

Fig. 3. TCP Header Checksum option

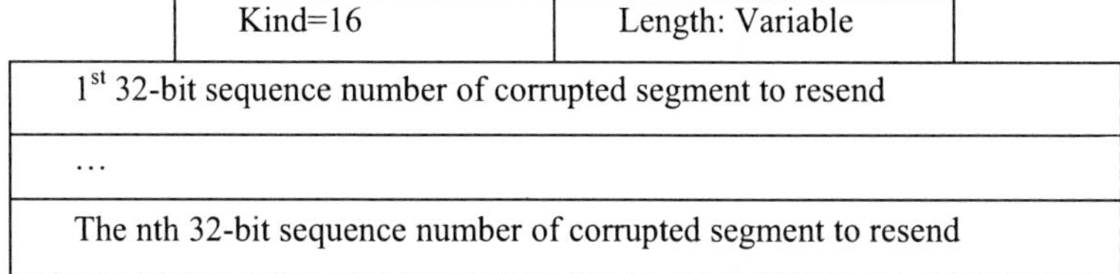

Fig. 4. Extended TCP header Checksum ACK option

The nth 32-bit sequence number means that now there are n sequence numbers in the s_buffer. When the receiver recovers first sequence number, it writes it into the s_buffer and sends a special ACK contains the 1st sequence number; when the receiver recovers the second sequence number, it writes it into the s_buffer and sends another special ACK contains the 1st and the 2nd sequence numbers, and so on.

Like the TCP HACK, with our proposed extended TCP HACK, we should also do modifications to the TCP sender, receiver and the ACK processing algorithms, but there are some differences from the TCP HACK. In the following, we will explain in details.

4.2.1 Modification to TCP Sender

When a segment is sent, the TCP sender first checks if there is header checksum option enabled. If there is not, it will continue as normal; otherwise, it calculates the header checksum of that segment and places it into the header checksum option and then continues as normal TCP. This modification is same with TCP HACK.

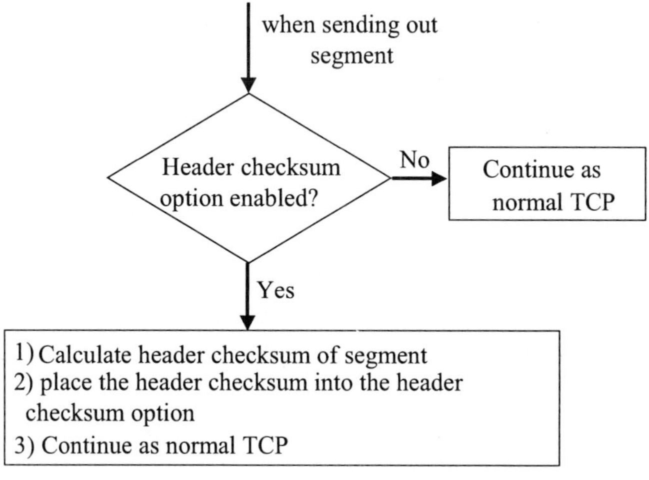

Fig. 5. Modification to TCP sender when sending a packet

4.2.2 Modification to TCP Receiver

When the TCP receiver receives a packet, first it checks the integrity of the packet using the standard checksum, if there is no error, then it works as normal TCP. If the packet is corrupted, then it will use the header checksum to check if the header portion has been corrupted. If the header portion is also corrupted, then the packet will be discarded, otherwise, the modified extended TCP does following:

1) Recover the sequence number form the packet header;
2) Save the sequence number into the s_buffer
3) Send a "special" ACK (option 16) to the sender. Its option contains all the sequence numbers that have been saved in the s_buffer. So the TCP sender can distinguish that this special ACK was generated because of packet corruption. And with these recovered sequence numbers, the TCP sender can selectively retransmit these corrupted packets.

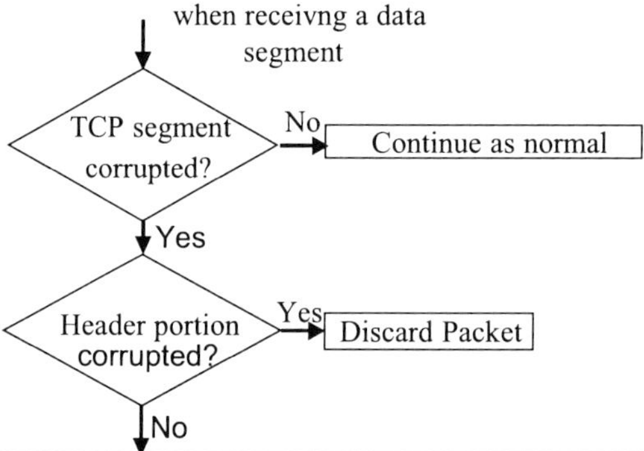

Fig. 6. Modification to TCP receiver when receiving a packet

4.2.3 Modification to the ACK processing

When the TCP sender receives an ACK, first it checks if there is a HACK option (option 15) or an extended HACK option (option 16). If the received ACK is not a special ACK, the TCP sender will perform as normal TCP; otherwise, if it's a special ACK with option 15, TCP sender will work as TCP HACK, if the option kind is 16, the TCP sender does as follows:

1) Extract all the sequence numbers from the option of the special ACK
2) Retransmit Selectively these packets,
3) Discard the special ACK without further processing

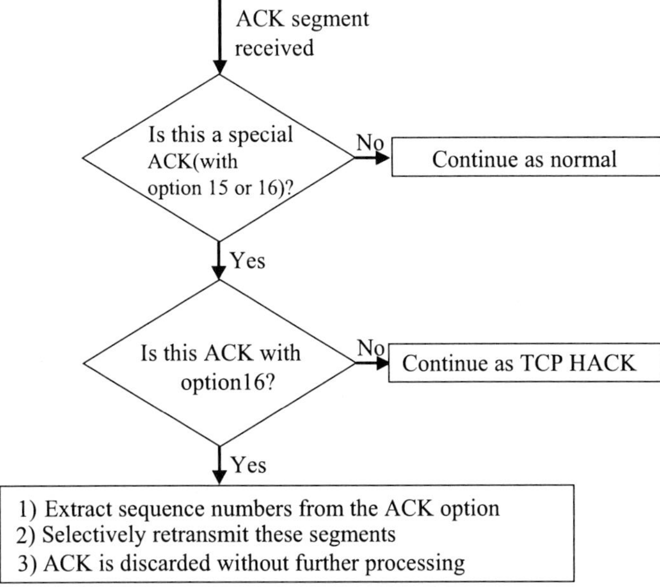

Fig. 7. Modification to the ACK processing

It should be mentioned that our extended TCP HACK could work together with normal TCP and TCP HACK. We implement the TCP HACK and extended TCP HACK with the original TCP. The user can choose to use any of them according to the transmission conditions.

4.3 Drawbacks of Our Proposal

There is one problem should be considered by using our proposed extended TCP HACK: the software overload. Using extended TCP HACK, more complex software will be on the sender and receiver sides, but while memory sizes and CPU performance permanently increase, the bandwidth of the air interface remains almost the same. Therefore the higher complexity is no real disadvantage any longer as it was in the early days of TCP.

4.4 Performance Evaluation

We carried our experiments by using OPNET modeler 8.0. The simulation model is shown as Fig.8. We ran our experiments by sending bulk data from the server to the client. The connect link between the server and client is a wired link but has been configured by different error bit rates to simulate the lossy wireless link. The link rate is set to 10M to simulate a wireless LAN and in order to avoid the congestion. The s_buffer is set to infinite not to be a limiting factor.

2%~15% packet loss was considered. The burst packets length is configured to 3 packets.

We also disabled the link layer CRC as TCP HACK did. So the corrupted TCP packet can arrive to the TCP stack.

We run experiments in two situations, one is corruption only on one direction and the other is corruption on both directions.

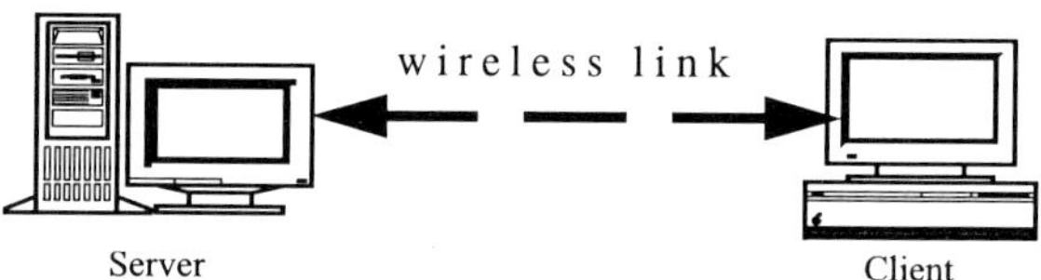

Fig. 8. Simulation model

4.4.1 Corruptions on One Direction

First, we run experiments in the condition that errors are only on the forward path; packets on the reverse path (the ACK packets) are not corrupted. Fig.9. shows the throughput for various packet loss rates. The result shows that TCP HACK and our extended TCP HACK all work better than the original TCP. When the packet loss rate is very high (more than 10%), the original TCP almost cannot receive correct packets, but using TCP HACK or extended TCP HACK, the receiver can still receive the packets

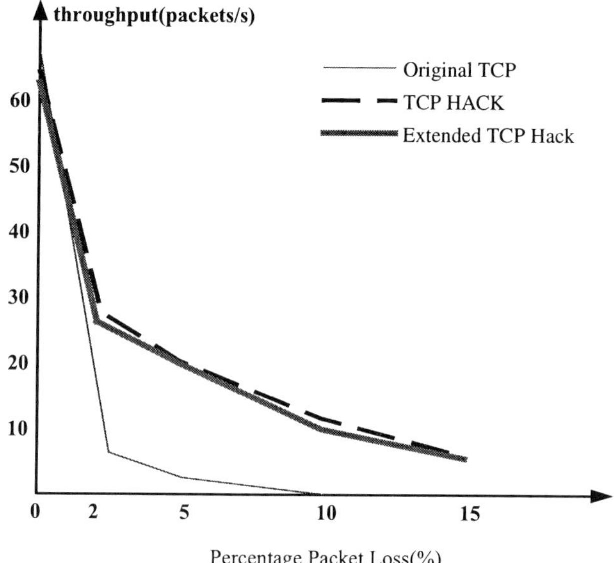

Fig. 9. Throughput for various packet loss rates when corruption on one direction

4.4.2 Corruptions on Both Directions

Next, we run experiments in the condition where corruptions on both directions (forward and reverse path). The acknowledgement packets have the same error rate as the data packets. From Fig.10 we can see that our extended TCP HACK performs much better in this condition. Because when there are many corruptions on the ACK path, the special ACKs in TCP HACK can't arrive at the sender, so the TCP HACK sender cannot get the loss information and cannot retransmit the corrupted packets.

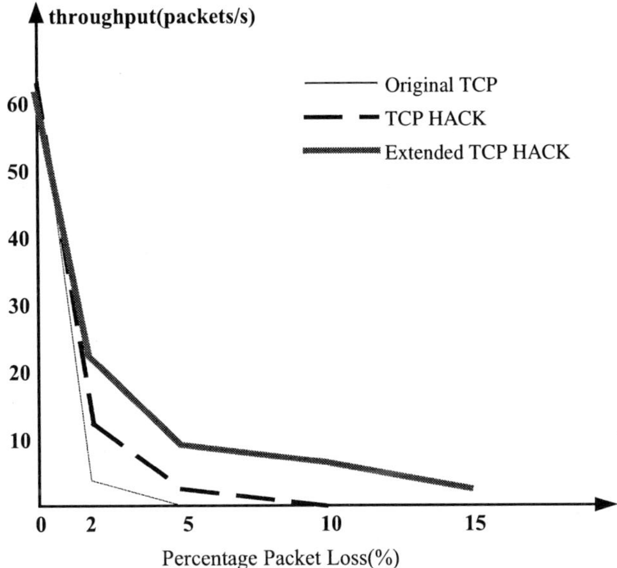

Fig. 10. Throughput for various packet loss rates when corruption on both directions

5 Conclusions and Future Works

In this paper, we analyzed the problems of original TCP over lossy links. Then we summarized the existing protocols, indicating their strengths and weaknesses. TCP HACK is one of these protocols, which has been proposed to improve original TCP over lossy links. We think that TCP HACK can be adopted in the future. But it has some drawbacks. So we proposed an extension to TCP HACK. Our proposal can enhance the TCP HACK in the situation where not only the data on the forward data are corrupted much, but also the ACKs on the inverse path susceptible to packet corruption.

Simulations have been done to test our extended TCP HACK protocol by using OPNET modeler 8.0. The results proved that our proposed protocol performs much better than original TCP in the cases that corruptions are big. In addition, the extended TCP HACK can improve the TCP HACK in the condition where packet loss not only on the forward path but also on the inverse path.

Works are being done to test the effectiveness of our proposal in situations where congestion, corruptions occur at the same time.

References

[1] Behronz A.Forouzan, "TCP/IP protocol suite" international editions 2000,271-311

[2] Jochen Schiller, "Mobile Communications" Pearson Education Limited 2000, 290-307

[3] M.Allman,V.Paxson,and W.Stevens, "TCP congestion control", IETF RFC 2581,1999

[4] Jiangping Pan,Jon W.Mark and Xuemin Shen, "TCP performance and its improvement over wireless links" GlobeCom2000

[5] Ajay Bakre and B.R. Badrinath, "I-TCP: Indirect TCP for mobile hosts", Tech. Rep., Rutgers University, May 1995.

[6] H.Balakrishnan, S.Seshan, Eamir, and R.H. Katz, "Improving TCP/IP performance over Wireless Networks", In Proc.1st ACM Conf. on Mobile computing and Networking, November 1995.

[7] Jian-Hao Hu,Kwan L.Yeung,Wiew Chee Kheong and Gang Feng, "Hierarchical Cache Design for Enhancing TCP over Heterogeneous Networks with Wired and Wireless Links" GlobeCom2000

[8] K.brown and S.Singh, "M-TCP: TCP for Mobile Cellular Networks", ACM computer Communications Review (CCR), vol.27, no.5, 1997

[9] N.Vaidya, Overview of work in mobile-computing (transparencies).

[10] Tom Goff, James Moronski, D.S.Phatak, "Freeze-TCP: a true end-to-end TCP enhancement mechanism for mobile environment" InfoCom2000

[11] M.Mathis, J.Mahdavi, S.Floyd, A.Romanow, "TCP selective acknowledgment options" IETF RFC 2018, 1996

[12] R.K.Balan, B.P.Lee,K.R.R.Kumar, L.Jacob,W.K.G.Seah, A.L.Ananda, " TCP HACK:TCP header checksum option to improve performance over lossy links", InfoCom2001

[13] Hari Balakrishnan, Venkata N. Padmanabhan, Srinvasan Seshan, Mark Stemm, Elan Amir, Randy H.Katz, "TCP Improvements for Heterogeneous Networks: The Daedalus Approach".

[14] Ramon Caceres and Liviu Iftode,"Improving the performance of reliable transport protocols in mobile computing environments" IEEE JSAC Special Issue on Mobile Computing Network,1994.

[15] James F. Kurose and Keith W. Ross, "Computer Networking" 2001 by Addison Wesley Longman, Inc. 167-260.

[16] Modeler & Radio powered by OPNET Simulation Technology. SIMUS Technologies. Inc.

[17] IT Decision Guru powered by OPNET Simulation Technology. SIMUS Technologies.Inc.

[18] IETF PILC WG homepage, http://www.ietf.org/html.charters/plic-charter.html

[19] J. Border , M. Kojo , J. Griner , G. Montenegro , Z. Shelby , "Performance Enhancing Proxies Intended to Mitigate Link-Related". IETF RFC 3135. June 2001.

Design and Implementation of Application-Level Multicasting Services over ATM Networks*

Sung-Yong Park[1], Jihoon Yang[1], and Yoonhee Kim[2]

[1] Department of Computer Science and Engineering
Sogang University, Seoul Korea
{parksy,jhyang}@ccs.sogang.ac.kr,
[2] Department of Computer Science
Sookmyung Women's University, Seoul Korea
yulan@cs.sookmyung.ac.kr

Abstract. The ACS (Adaptive Communication System) is a multi-threaded message-passing system that provides application programmers with multithreading and flexible communication services. This paper outlines the general software architecture of ACS and describes how the ACS architecture is applied to implement its flexible application-level group communication services. We provide the performance results of ACS multicasting services and compare them with those of p4, PVM, and MPI.

1 Introduction

The Adaptive Communication System (ACS) [1] is a multithreaded message-passing system that provides users with multithreading and dynamic communication services (e.g., point-to-point and group communication services). The ACS capitalizes on thread-based programming model to overlap computation and communication, and develop a dynamic message-passing environment with separate data and control paths. This leads to a flexible and scalable message-passing environment that can support multiple communication algorithms (e.g., error control, flow control, multicasting algorithms) and interfaces at runtime. This paper primarily focuses on the *flexible*, *scalable*, and *application-level* group communication services provided by ACS.

The group communication services provided by current message-passing systems have several drawbacks. First of all, some message-passing systems (e.g., PVM) implement group communication operations (e.g., collective communication) by repeatedly calling send routines for portability, which is computationally expensive and not scalable for groups with large members. Although the tree-based multicasting operations can be implemented at the source level, this process is cumbersome and prone to errors. Second, their communication primitives are static, which means that they are not able to adapt to rapidly changing

* The publication of this paper was supported in part by Institute for Applied Science and Technology of Sogang University

M.H. Shafazand and A M. Tjoa (Eds.): EurAsia-ICT 2002, LNCS 2510, pp. 479–486, 2002.

network dynamics. The ability to adapt to varying network conditions is one of the important features that need to be supported in the communication systems, especially for wide-area computing. Third, in traditional message-passing systems, the transfer of control and data are usually tightly coupled. For a large number of small groups, this results in generating a large amount of control traffic associated with group operations and potentially decreases the performance of applications. There have been several distributed computing software systems specially designed to support group communication services such as Horus [5], Totem [6] and Transis [7]. However, most of them are designed to support special functionalities (e.g., fault tolerance, message ordering, virtual synchrony, group partition) rather than to provide high throughput or scalable group communication services. They fail to address the issues mentioned above.

The group communication services in ACS are based on the dynamic grouping. Each ACS process can dynamically create, join or leave a group during the lifetime of the process. The multicasting operation in ACS is implemented by using a spanning tree (e.g., binary tree) and this is more efficient than repetitive techniques for large group size. The multicasting tree is virtually created at the application level upon unicast connections using the application specific performance metric (e.g., topology, latency, or bandwidth). The ACS architecture which separates the data and control transfer allows the multicasting operations to be implemented efficiently by utilizing the control connections when transferring status information (e.g., membership change, acknowledgment to maintain reliability etc.). This separation optimizes the data path and thus improves the performance of ACS applications.

The rest of the paper is organized as follows. We begin by providing an overview of the ACS architecture in Section 2. Section 3 presents an implementation approach to the ACS group communication services. Section 4 compares the multicasting performance of ACS with those of other message-passing systems such as p4 [2], PVM [3], and MPI [4]. Section 5 contains the summary and conclusion.

2 Overview of ACS Architecture

ACS is a multithreaded message-passing system that provides multithreading (e.g., thread synchronization, thread management) and communication services (e.g., point-to-point communication, group communication) for High Performance Distributed Computing (HPDC) applications with different Quality of Service (QoS) requirements (See Figure 1). ACS uses multiple *Compute_Threads* to implement the computations of HPDC applications. These threads use the ACS primitives to communicate and synchronize with other *Compute_Threads*. This allows ACS to provide efficient support for fine-grained applications, and to reduce the propagation delay impact on HPDC applications especially in Wide Area Network (WAN)-based distributed computing by overlapping computation and communication.

ACS decouples the control and data paths by creating different threads for both control and data functions. Moreover, the control and data information from the two paths are transmitted on separate connections. The *control threads* implement important control functions such as connection management, flow control, error control, and configuration management in an independent manner. The *data transfer threads* are spawned based on a per-connection basis to perform only the data transfers associated with a specific connection. The separation of control and data functions eliminates the process of demultiplexing control and data packets within a single connection, and allows the concurrent processing of control and data functions.

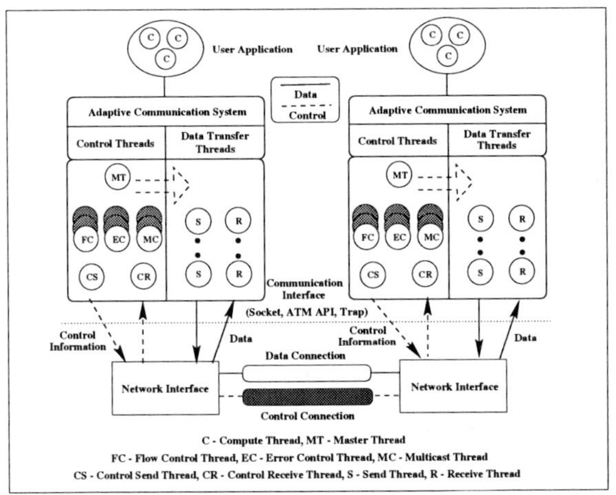

Fig. 1. ACS General Architecture

In ACS, multiple flow control (e.g., window-based, credit-based, or rate-based), error control (e.g., go-back N or selective repeat), and multicasting algorithms (e.g., repetitive send/receive or a multicast spanning tree) are provided as *control threads* and programmers activate the appropriate thread when establishing a connection to meet the QoS requirements of a given connection. This allows the HPDC programmers to select for a given HPDC application the appropriate flow control, error control, and multicasting algorithms per-connection basis at runtime.

ACS is designed to support these classes of applications by offering three application communication interfaces: 1) Socket Communication Interface (SCI); 2) ATM Communication Interface (ACI); 3) High Performance Interface (HPI). The SCI is used mainly for providing high portability over a network of computers (e.g., workstations, PCs, parallel computers). The ACI is the application communication interface that allows programmers to access the inherent features of ATM network. The HPI is built to achieve high-throughput and low-latency inter-process communications.

3 ACS Group Communication Services

ACS group communication services support *dynamic* groups as shown in Figure 2. At program startup, a default ACS group, called *ACS_GRP*, is created and each ACS process in the *hostfile* joins this group automatically. The first process specified in the *hostfile* becomes a *Master Group Server* (MGS). Each process that creates a new group with a unique name becomes a *Local Group Server* (LGS) of that group by default. The MGS represents the whole LGSs and coordinates the group communication operations between these servers. The LGS is responsible for multicasting operations within the local group and maintains the membership information of the local group only. A *Global Multicasting Tree* (GMT) is built to connect all the LGSs rooted at the MGS. All the group members within the same group are connected by a *Local Multicasting Tree* (LMT) rooted at the LGS of that group.

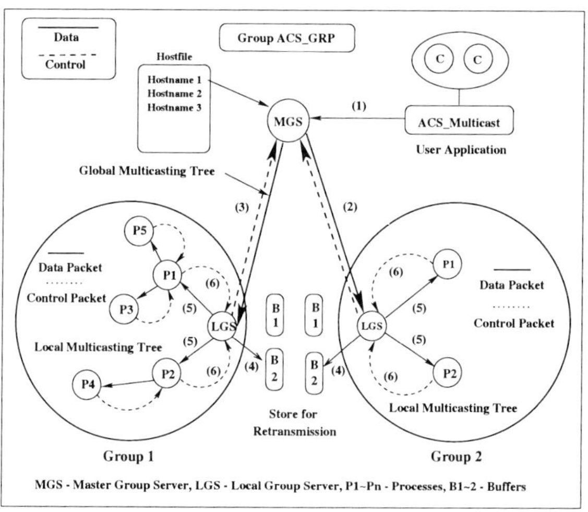

Fig. 2. Multicasting in the ACS Environment

The MGS and LGSs periodically exchange the status information of each group. By having a distributed group server in each group and making it to manage the local group only, the status traffic can be minimized and the information is managed with reliability. Each process that joins a particular group is identified by the *rank* within that group. Since ACS group communication primitives allow the overlapping of different groups, all ACS processes can join multiple groups at the same time.

When a group is created or destroyed, the MGS updates the GMT based on the application specific performance metric (e.g., topology, latency or bandwidth) and broadcasts the information to all the LGSs over the control connections. Each LGS updates the group information it is maintaining such as routing information (e.g., left child, right child) and group server information (e.g., identifier of each group server) after receiving the information from the MGS. Each LGS in turn broadcasts the group server information to all the members of its group over the control connections. On the other hand, when a group member joins or leaves a particular group, the LGS of that group also updates the LMT and broadcasts the information to all the members of the same group only. Consequently, the information of the MGS and LGSs are visible to all the members within the global group *ACS_GRP* and the information of the group members are known only to the members within the same group. Therefore, the GMT and LMT are built before the multicasting operation is invoked. This reduces the setup time to perform multicasting operations and thus improves the performance. The ACS architecture that separates the control and data path allows us to implement this scheme efficiently by using the control path to transfer the membership changes without interfering with the data traffic.

3.1 Multicasting Protocol

ACS group communication primitives support *open* groups by providing three classes of multicasting operations: (1) *Global Broadcast*; (2) *Local Broadcast*; and (3) *Global Multicast.*

The *Global Broadcast* is used to transmit messages to the entire groups defined in the global group *ACS_GRP*. The *Local Broadcast* is used to transmit messages to the all members within the same group. The *Global Multicast* is used to transmit messages to the part of the entire groups. For all three operations, the destination end-point is not the members but the group server. The configuration information of the destination group (e.g., How many members are in the group, topology of the group) need not be visible to the multicasting process and the destination group server takes care of broadcasting the message to the its members. Keeping the state information for each member of all different groups is not efficient and generates a lot of broadcasting traffic whenever the membership status of a process is changed. In the *dynamic* group where a lot of membership changes are expected, the performance of the applications can be improved by reducing the traffic associated with the transfer of the status messages.

Since the three classes of multicasting operations are implemented using similar schemes, we will provide the algorithm for the *Global Broadcasting* only. The multicasting algorithm for the *Global Broadcasting* consists of six steps as shown in Figure 2 :

1. When the *Compute_Thread* of a process invokes the *ACS_mcast()* primitive, the *Multicast_Thread* of that process activates the corresponding *Send_Thread* to transmit an actual message to the MGS.
2. The MGS transmits the received message to the other LGSs using its GMT.
3. If the *ACS_mcast()* is invoked with reliable mode, each LGS that received the message sends an acknowledgment back to the MGS. The acknowledgment is merged along the GMT and the MGS should guarantee that all the LGSs receive the message.

4. A LGS maintains two buffers. The first buffer is used to assemble the messages, which are then transferred to the second buffer. The second buffer is used to retain the messages that have not been correctly received by group members.
5. Each LGS locally multicasts the message to its group members using its LMT.
6. If the *ACS_mcast()* is invoked with reliable mode, each member that received the message sends an acknowledgment back to the LGS. Again, the acknowledgment is merged along the LMT and the LGS should guarantee that all the members receive the message. If there is any group member which has not received a message within the timeout period, the LGS of the group retransmits the message again. This reduces the retransmission traffic from the source process.

4 Benchmarking Results

In this section, we analyze and compare the performance of ACS with those of other message-passing systems such as p4, PVM, and MPI in two levels: primitive performance level and application performance level using Back-Propagation Neural Network (BPNN) learning algorithm.

4.1 Primitive Performance

Figure 3 shows the performance of broadcasting primitives (e.g., *ACS_mcast()*, *pvm_mcast()*, *p4_broadcast()*, and *MPI_Bcast()*) of four message-passing systems over an ATM network for message sizes from 1 byte to 32 Kbytes. The group size varies from two to ten.

As we can see from Figure 3, ACS primitive (*ACS_mcast()*) shows the best performance for various message sizes and group sizes. Furthermore, *ACS_mcast()* primitive shows almost similar performance for large group sizes (over six members) as we increase the message size (over 4 Kbytes). In the *ACS_mcast()* primitive where most of the information for performing group communications (e.g., setup binary tree, setup routing information) is set up in advance by using the separate connections, the start-up time for the broadcasting operations is very small. Also, the tree-based broadcasting scheme improves the performance as the group size gets bigger.

The performance of PVM primitive (*pvm_mcast()*) is not so good for small message sizes but as the message size and group size increase, it shows better performance. In the *pvm_mcast()* where the broadcasting operation is implemented by repeatedly invoking a send primitive, the performance is expected to increase linearly as we increase the group size. Moreover, *pvm_mcast()* constructs a multicasting group internally for every invocation of the primitive, which results in the high start-up time when transmitting small messages as shown in Figure 3 (message size 1 byte).

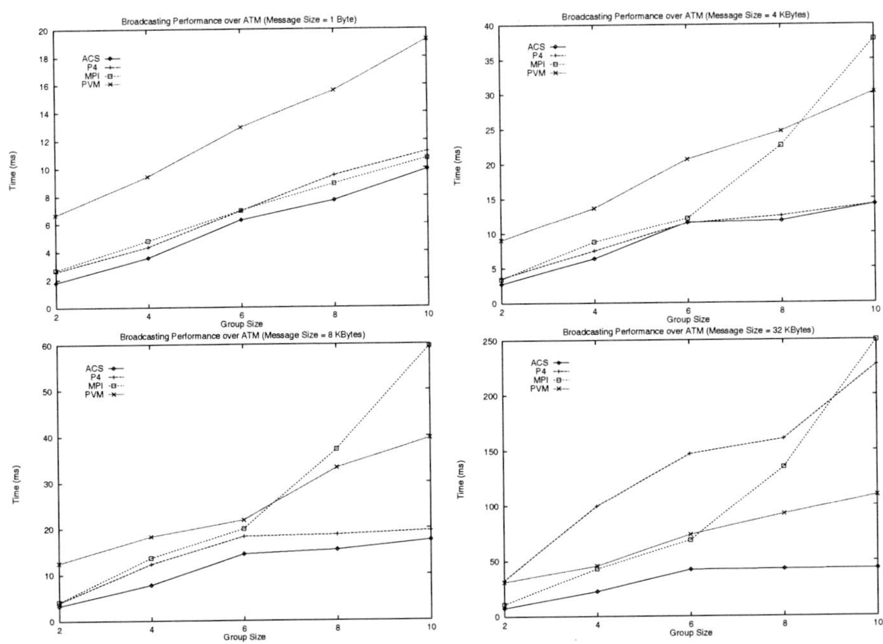

Fig. 3. Comparison of Broadcasting Performance over ATM

The p4 primitive (*p4_broadcast()*) and MPI primitive (*MPI_Bcast()*) shows comparable performance to ACS for relatively small message sizes and small group sizes but it is getting worse drastically when it is running for large message sizes and large group sizes.

4.2 BPNN Learning Algorithm

Training BPNN for character recognition is one of the problems in the Artificial Intelligence (AI) area which require highly intensive computation. We used master/slave programming model to parallelize this application. This application intensively uses the broadcasting primitives when distributing the weight vectors to all the *slave* processes. The BPNN used in this experiment has 100 input nodes, 630 hidden nodes, and 4 output nodes to train 16 input vectors which represent the hexadecimal digits from 0x01 to 0x0F.

Figure 4 shows the performance comparison of each message-passing system running over four homogeneous workstations (e.g., four SUN-4 workstations running SunOS 5.5 or four IBM/RS6000 workstations running AIX 4.1) and eight heterogeneous workstations (e.g., four SUN-4 workstations and four IBM/RS6000 workstations) interconnected by an ATM network.

As we can see from Figure 4, the ACS implementation outperforms other implementations regardless of the platform used. In the BPNN application where large messages are broadcasted repeatedly, the performance improvement is noticeable and it is widening as we increase the group size. We believe that most

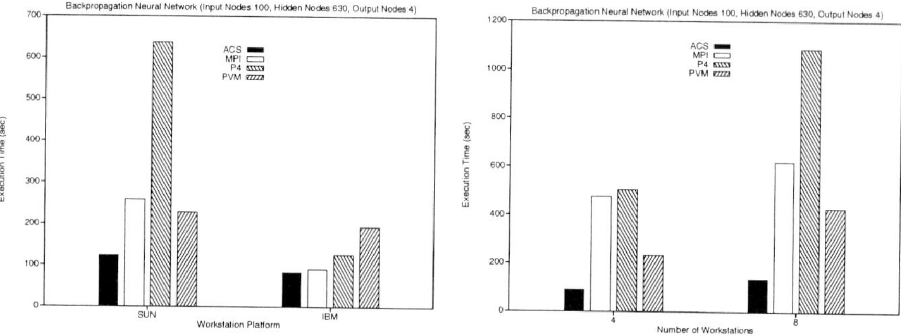

Fig. 4. Comparison of Application Performance

of the improvements of ACS are due to the overlapping of communication and computation and the tree-based broadcasting primitive.

5 Conclusion

In this paper, we have outlined the software architecture of a multithreaded message-passing system, ACS, and presented how ACS architecture can be applied to provide flexible and application-level group communication services. We have evaluated the performance of ACS group communication services and showed that ACS outperforms other message-passing systems. It is clear that the ACS novel architecture, which separates the data and control transfer and tree-based multicasting scheme played an important role in improving the performance of the communication primitives and the ACS applications.

References

1. S. Park and S. Hariri, "ACS: An Adaptive Communication System for Heterogeneous Wide-Area ATM Clusters", *Cluster Computing Journal*, pp. 229–246, 1999.
2. R. Butler and E. Lusk, "Monitors, message, and clusters: The p4 parallel programming system", *Parallel Computing*, Vol. 20, pp. 547–564, April 1994.
3. V. S. Sunderam, "PVM: A Framework for Parallel Distributed Computing", *Concurrency: Practice and Experience*, Vol. 2, No. 4, pp. 315–340, December 1990.
4. MPI Forum, "MPI: A Message Passing Interface", *Proc. of Supercomputing '93*, pp. 878–883, November 1993.
5. R. Renesse, T. Hickey, and K. Birman, "Design and performance of Horus: A lightweight group communications system", Technical Report TR94-1442, Cornell University, 1994.
6. L. E. Moser, P. M. Melliar-Smith, D. A. Agarwal, R. K. Budhia and C. A. Lingley-Papadopoulos, "Totem: A Fault-Tolerant Multicast Group Communication System", *Communications of the ACM*, Vol. 39, No. 4, pp. 54-63, 1996.
7. D. Dolev and D. Malki, "The Transis Approach to High Availability Cluster Communication", *Communications of the ACM*, Vol. 39, No. 4, pp. 64–70, 1996.

Bon: The Persian Stemmer

Masoud Tashakori [1], Mohammadreza Meybodi [2], and Farhad Oroumchian[3]

[1] Computer Engineering Department, Amirkabir University of Technology, Tehran, Iran
Tashakori@noarvar.com

[2] Computer Engineering Department, Amirkabir University of Technology, Tehran, Iran
Meybodi@ce.aku.ac.ir

[3] Electrical & Computer Engineering Department, Tehran University, Tehran, Iran
Foroumchian@acm.org

Abstract. Stemmers are softwares that find syntactic roots of the words. They play an important role in natural language processing and other fields such as information retrieval (IR). In IR using stemmed words instead of the original words, could increase as much as 15 percent to the overall performance. In this paper, we report on the development of a Persian stemmer (Bon). Bon is tested on a collection of Persian texts in the domain of computer science. In our experiments, the recall has been improved by 40 percent.

1 Introduction

Morphological analysis is part of natural language processing and linguistic. Suffix stripping and stemming is part of computational morphological analysis. Stemming is a widely used method of word standardization designed to allow the matching of morphologically related terms. If, for example, a searcher enters the term *stemming* as part of a query, it is likely that he or she will also be interested in such variants as *stemmed* and *stem*. Stemmers are softwares that extract stems of word automatically [3].

In the field of information retrieval many experiments have been conducted to determine the value of stemming in retrieval process. There is a verity of methods to build a stemmer. The widely used Porter's algorithm [1] is a rule-based system, which iteratively removes suffixes. Porter algorithm does not guarantee correct form of the words to be produced after stemming. However, his algorithm is consistent and it is shown that it increases recall by up to 15%. Karaaj et al [2] have demonstrated that Porter's algorithm compresses the index vocabulary by about 43% on English text.

David Hull [4], Harmann [5] and others almost all agree that in information retrieval (IR), stemmers play an important role. In IR using stemmed words instead of the original words, could reduce the size of vocabulary. Since a single stem typically corresponds to several full terms, by storing stems instead of terms, compression factors of around 40-50 percent can be achieved. Thus in this paper we report on the development of a Persian stemmer which is called Bon. In next section we will look

M.H. Shafazand and A M. Tjoa (Eds.): EurAsia-ICT 2002, LNCS 2510, pp. 487–494, 2002.

at some of the properties of Persian words. Section 3 presents the Bon algorithm and section 4 describes the experiment. The last section is the conclusion.

2 Persian Words

Persian is an Indo-European language. As so in this language also new words can be constructed by adding prefixes and suffixes to base forms of words. Bon is an affix removal stemmer. Affix removal algorithms remove suffixes and/or prefixes from terms leaving a normalized form of the word. These algorithms sometimes also transform the resultant word into the real linguistic stem. A simple example of an affix removal stemmer is one that removes the plurals from terms [3].

Persian verbs have inflectional property, because they include person, number, and tense. For example, the verb "مي روم" (*mi-ra-vam*) which means, "I am going", consists of three parts: "مي + رو + م" (*mi* + *ro* + *m*) that is "I + go + ing" all in one word. Moreover, the infinitive verbs in Persian can be simple, or compound, or phrasal. There is at least one space between components of a compound or phrasal infinitive. For example, the verbs "قسم خوردن"(*ghasam xordan)* that means " to oath", and " از دست دادن" *(az dast dädan)* that means "to lose" have two and three components each in that order. In order to stem these verbs all their components should be located and evaluated as one word. This problem is considered in the Bon's stemming algorithm.

In Persian plural nouns are made by adding "ان"*(än)* or "ها"*(hä)* to the end of nouns. But if any noun ends in a "ه"*(eh)*; then before adding "ان"*(än)*, "ه"*(eh)* transforms into "گ"*(ge)* as depicted in the figure 1(a). There exceptions also, such as nouns that end in with "ان"*(än)* but are not plural e.g. "قهرمان"*(ghahramän)*. Plural form of some nouns are made by adding Arabic plural signs like "ون"*(un)*, "ين"*(in)*, and "ات"(at) as shown in the figure 1(b). But if a noun ends in a "ا"*(ä)*,"و"*(u)*, "ه"*(eh)*, or "ي"*(y)*, instead of adding "ات"*(ät)*, "جات"*(jät)* is added, as shown in the figure 1(c). Moreover some nouns that are adopted from Arabic language have irregular plural forms ("Mokassar") as shown in the figure 1(d) [6].

In Persian a pronoun can be attached to the end of a noun. But if the noun ends in an "ا"*(ä)*, or "و"*(u)*; then a "ي"*(y)* is inserted before attaching the pronoun. Examples of this case are the word ("پا"(*pä*) meaning: foot) → ("پايم"(*päyam*) meaning: my foot), or the word (" چاقو"(*chäghu*) meaning: knife) → ("چاقويش"(*chäghuyaš*) meaning: his knife). Also if a singular pronoun is added to the end of a noun and the noun ends in a "ه"*(eh)*; then an "ا"*(a)* is added to the end of the noun before the pronoun. Example of this case is ("خانه"*(xäneh)* meaning: house) → (" خانه‌ام"*(xäneham)* meaning: my house).

Bon utilizes a dictionary of infinitives and present tense of infinitives for exceptional cases. For stemming words that are adopted from Arabic language, either a rule based or table look up approach can be used. Bon uses the first method.

a	خواننده ← خوانندگان (*xänandeh → xänandegän*)		
b	روحاني ← روحانيون *ruhäny → ruhänyun*	مؤمن ← مؤمنين *momen → momenin*	تدارك ← تداركات *tadärok → tadärokä*
c	دوا ← دواجات (*davä → daväjat*)	شيريني←شيرينيجات (*širini → širinijat*)	شور ← شورجات (*šur → šurjat*)
d	كتاب ← كتب (*ketäb → kotob*)		

Fig. 1. Different Cases of Plurals in Farsi (Persian)

3 Bon Algorithm

Bon uses an iterative longest-match stemming algorithm. An iterative longest match stemmer removes the longest possible string of characters from a word according to a set of rules. This process is repeated until no more characters can be removed. After all characters have been removed, the resulting stem may not be correct. For example the word "خانگي"*(xänegy)* that means, "home-made", may be reduced to the stem "خانگ"*(xäneg)* which is incorrect form of the real stem "خانه"*(xäneh)* "house". There are two techniques to handle this: re-coding or partial matching [3].

Re-coding is a context sensitive transformation of the form AxC -> AyC where A and C specify the context of the transformation, x is the input string, and y is the transformed string. In partial matching, only the n initial characters of stems are used in comparing them. Using this approach, one might say that two stems are equivalent if they agree in all but their last character [3]. Bon benefits from re-coding technique.

Bon has four major components:

1. Stemming rules that are extracted from Persian word construction rules.
2. A dictionary of Persian infinitives.
3. A dictionary of "Mokassar" words and their singular form
4. A dictionary of Persian roots.

Because of difficulties in building the last dictionary, an experimental version with almost 7000 words was built from the collection. In order to construct this dictionary, words were extracted from 450 abstracts in our collection. Then by running Bon, words that were derivative of any other word in the dictionary were eliminated. Moreover we gradually added or eliminated many words to/from roots dictionary.

In writing Persian, some of the letters of the words are attached together and some are separated. For example in word "خوردن"*(xordan)* that means eating, the first two letters are attached together but the last three letters are separated. A word boundary detection program finds the boundary of the words by looking simultaneously at each letter and its followings letters.

As depicted in the figures 2 and 3 the Bon algorithm has two major procedure: *Stem()* and *AffixRemove()*. The *Stem()* function takes a word and returns the stem of the word. This function first, checks whether the word could be a verb or not. If the word is a verb then *Stem()*will return the infinitive of the verb. But if the word is not a verb, *Stem()* will search the word in Mokassar dictionary. If the word is found in the dictionary, its corresponding singular form is returned as the stem of the word. Now, if the word is not a verb or a Mokassar plural noun, then the word is turned over to the *AffixRemove()* function.

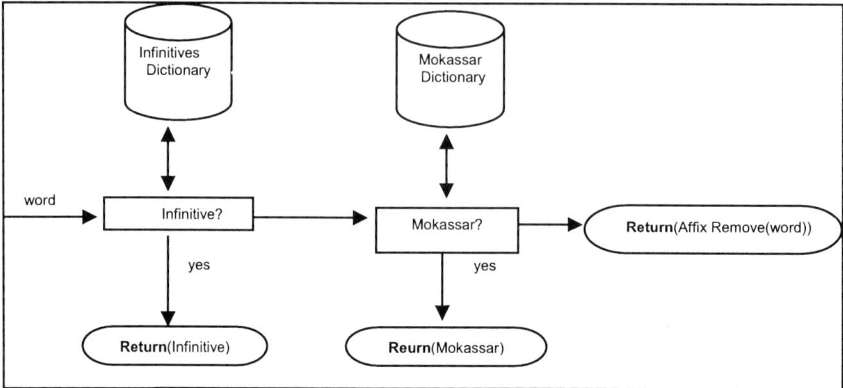

Fig. 2. Bon Algorithm: Stem() procedure

The *AffixRemove()* function starts with removing first (longest) possible affix from the word. If a possible affix could not be find, *AffixRemove()* will return the input word. But in the case of removing a possible affix, the stripped word is searched in the Persian roots dictionary. If the stripped word exists in this dictionary, *AffixRemove()* will return the stripped word. Otherwise it examines the possibility that the stripped word is a verb or Mokassar plural noun. Finally if this possibility fails, the word is restored and the *AffixRemove() function* will try to remove another affix until no more affixes could be found.

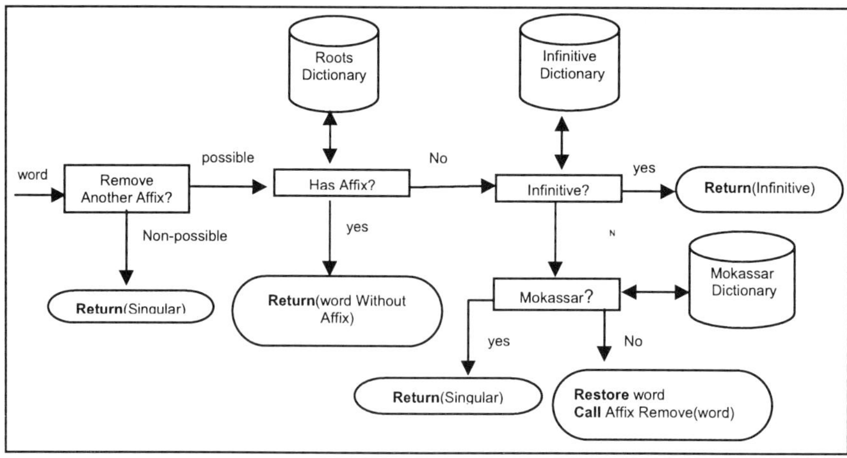

Fig. 3. Bon Algorithm: Affix remove() procedure

The reason for using Persian roots dictionary is that in some cases although the word contains the sign of an affix but it is actually a root word and should not be stemmed. For example the word "بیرون" (*birun*) ends with an "ون" (*un*) which is a sign of plural. But this word is a root word and not a plural. By having the root forms in the Persian roots dictionary, mistakes such as this would be prevented.

Next section describes experiments of running Bon on a Persian text collection.

4 Experiments

There are many ways to evaluate the performance of a stemmer. Paice has used two parameters in evaluating different stemming algorithms in English [7]. The Paice's method is independent of the context where the stemmer is going to be used. David Hull uses the context of IR to evaluate different stemming algorithms [4]. Kraaij et al have a good review of the evaluation methods of stemmers [2]. For this experiment, we looked at the improvements that can be achieved by using stems instead of the original words in a Persian information retrieval system. This is a popular method of evaluation used by Porter, David Hull and others. The retrieval effectiveness can be measured by recall and precision [8].

We have put together a corpus of 450 abstracts of Persian texts in the domain of computer science, which is called PCA (Persian Computer Abstracts). Experiments were performed on this collection using 32 queries.

Persian word boundary detection was applied first and the Persian words have detected. Then a stop list of 150 high frequency Persian words was used to filter out words with very low discrimination value. Table 1 lists our Persian stop list and their meanings.

For this experiment a Boolean retrieval system was implemented. Then computer students made up 32 queries, and then went through the collection and determined which documents are relevant to their query. To evaluate the Bon's performance, first the queries were presented to the system and the output were taken and recall and precision of each query was calculated based on the students relevance judgments. Next, Bon was used to present the students with all the terms that shared the same stem as the query words. Students could then expand their query by these newly introduced words. Then the expanded queries were run through the system again and the precision and recall of their output was calculated again. The precision and recall values were averaged over all the queries. Table 2 shows the difference in precision and recall with and without query expansion by using Bon stemming algorithm.

As depicted in table 2, query expansion by using Bon can increase the recall by 40 percent. There is also a slight drop in precision but this drop is more than offset by the increase in recall. To illustrate how a Persian stemmer such as Bon can help in searching, consider the following example:

Table 1. Typical stop list for a Persian IR System

آن(that)	آنان(they)	آنجا(there)	آنچه(whatever)
آنکه(which)	آنگاه(then)	آنها(those)	از(from)
است(is)	اگر(if)	اگرچه(although)	الان(now)
اما(but)	انجام(do)	او(he,she)	اي*
ايشان(they)	اين(this)	اينجا(here)	اينست(it's)
اينگونه(so)	اينها(these)	با(with)	باشد(be)
بايد(must,should)	بدون(without)	بر(upon,over)	براي(for)
بنابر**	بنابراين(therefore)	به(at)	بيشتر(more)
بين(between)	پس(then)	تا(until)	تايي(many)
تو(you)	توسط(via)	چرا(why)	چگونگي(manner)
چگونه(how)	چنان(that)	چنانچه(if)	چند(a few)
چندان(so much)	چندگانه(multiple)	چندين(many)	چنين(this)
چه(what)	چون(since)	چيز(thing)	چيزي(anything)
چيست(what is)	حتي(even)	خواهد(will)	خواهيم(will)
خود(itself)	خودت(yourself)	خودتان(yourselves)	خودش(himself,herself)
خودشان(themselves)	خودم(myself)	خودمان(ourselves)	خويش(self)
داده(given)	داراي(having)	دارد(has)	داشته(having)
در(in)	درباره(about)	دو(both)	ديگر(other)
ديگران(others)	ديگري(another)	را***	روي(on)
زيرا(because)	سپس(next)	شامل(including)	شايد(may)
شد(became)	شده(becoming)	شما(you)	شود(be)
صورت****	فقط(only)	كدام(which)	كرد(do)
كردن(do)	كه(that)	گرچه(though)	گرفته(taken)
لكن(however)	ليكن(however)	ما(we)	مابين(between)
مانند(as)	مختلف(several)	من(I)	مورد(case)
ميباشد(is)	ميتوان(can)	ميتواند(could)	ميدهد(give)
ميشود(becomes)	ميشوند(become)	ميكند(do)	ميكنند(do)
نظر(seem)	نميتوان(cannot)	نوع(kind)	نياز(need)
نيز(again)	نيست(isn't)	نيستند(aren't)	ها*****
هاي*****	هايي*****	هر(each)	هرچه(whatever)
هريك(every)	هست(is)	هستند(are)	هستيم(are)
هم(too)	همان(same)	همانطور(same)	همانند(like)
همچنين(also)	همچون(like)	همديگر(each other)	همه(all)
همواره(ever)	هميشگي(usual)	هميشه(always)	همين(this)
هنوز(still,yet)	هيچ(never)	هيچكدام(none)	هيچگونه(any)
و(and)	وجود(being)	وگرنه(otherwise)	ولي(but)
وي(he,she)	يا(or)	يك(a,an)	يكديگر(each other)
يكي(one)			

* This word is part of Persian present perfect verbs for the second person.

** part of "بنابر اين" (therefore)

*** particle as a sign of the definitive direct object

**** part of "بدين صورت" (so)

***** particles as plural sign

Table 2. Comparison of retrieval effectiveness with and without using Bon stemmer

	Recall	Precision
Without stemming	0.3595258	0.8974702
Using Bon stemmer	0.5421372	0.8397220

Example1. Suppose a searcher has requested the following query:

Persian Query: شبکه **یا** اینترنت**)** **و** امنیت

(*šabake* ***yä*** *internet)* ***va*** *amnyat)*

Meaning in English: (Network ***OR*** Internet) ***AND*** security

If the IR system doesn't have stemming, the following text, for example, wouldn't be retrieved. However this text is relevant to the query.

يكي از شايعترين پروتكلهاي توزيع كليد در حال حاضر پروتكلهاي KeyptoKnight است كه در اوايل دهه جاري ميلادي ارائه گرديده است. در اين مقاله پس از معرفي خانواده پروتكلهاي مذكور به ارزيابي آن اقدام ميشود. هدف از اين ارزيابي پاسخ مشروحي به اين سؤال است كه آيا KeyptoKnight به اهداف اعلام شده طراحان خود دست يافته است. در همين راستا يك مقايسه تحليلي بين اين پروتكل با پروتكلهاي مطرح و همسنگ آن انجام ميگيرد. بالاخره ضمن معرفي يك رخنه در اين خانواده نشان داده ميشود كه اهداف طراحان در تلاش براي دستيابي به حداكثر سازگاري با انواع توپولوژيهاي ايجاد ***شبكه*** به ايجاد خلل در جنبه‌هاي ***امنيتي*** پروتكل منجر شده است.

But as it is highlighted in the text the word "امنيتي" (*amnyati*) which is from the same stem as the word "امنيت"(*amnyat*) is present in the text. Therefore using a stemmer such as Bon can increase the recall by retrieving documents that could have been missed otherwise.

Also the Bon stemming algorithm is fast. The speed of stemming measured for all the words in the collection and averaged. Stemming of a word on a Pentium III 550 machine with 64Mbytes memory takes only 0.03 seconds on average.

5 Conclusions

This paper describes a Persian stemmer called Bon. Persian stemming rules have many exceptions that are considered and handled in designing this stemmer. Experiments revealed that using stems as index terms gives better retrieval results than using full words. In the experiment reported here, Bon have increased recall by 40 percent.

A by-product of our experiment is building a small Persian corpus (PCA). It seems appropriate to use this collection in future experiments on Persian retrieval models.

References

1. Porter M. F., "An Algorithm for Suffix Stripping", Program, vol. 14, no. 3, pp. 130-137, 1980.
2. Kraaij W., Pohlmann R., “Evaluation of A Dutch Stemming Algorithm”, The New Review of Document and Text Management, vol. 1, pp. 25-43, 1995.
3. Frakes W. B., Stemming Algorithms, available at: http:// matrix.nbu.bg/books/books/book5 /chap08.htm
4. Hull David A., “Stemming Algorithms: A Case Study for Detailed Evaluation”, Journal of The American Society For Information Science, vol. 47, no. 10, pp. 70-84, 1996.
5. Harman D.,”How effective is suffixing?”, Journal of the American Society for Information Science, vol. 42, no. 1, pp. 7-15, 1991.
6. Anvari H., and Ahmadi Geavi H. Persian Grammer, Fatemi, Tehran, 1995.
7. Paice C. D., “An Evaluation Method For Stemming Algorithms”, Proceedings of ACM-SIGIR94, pp. 42-50, 1994.
8. Salton G. and Mc Gill M. J., Introduction to Modern Information Retrieval, Mc Graw Hill, New York, 1983.

Current and Future Features of Digital Journals

Harald Krottmaier

Institute for Information Processing and Computer Supported new Media (IICM)
Inffeldgasse 16c, A-8010 Graz, Austria
hkrott@iicm.edu

Abstract. Features in currently available systems are often restricted to passive 'consumption' of articles stored in the corresponding digital journal. This paper classifies features into two categories (overall system structure, and content based features) and gives an outlook of planned implementations in the Journal of Universal Computer Science (J.UCS). It also shows that using the powerful Hyperwave Information Server (HIS) makes it quite easy to *implement* features and make knowledge management features (such as 'find an expert on a topic') available to users of a digital journal.

1 Introduction and Overview

First generation publishing systems were very static and inflexible. Simple pages encoded in some digital format were prepared by some editorial team, stored on a server system, and then transfered to the user on request. Systems were based on ordinary web-servers without any interactive features but 'get document XXX' were very often used as base system. With ongoing development of technologies user supporting tools like 'search through the content stored on a server' etc. were implemented. This paper is about features available in digital journals. The following systems were explored in detail:

ACM-DL: Digital library of the ACM (Association for Computing Machinery) [ACM Digital Library, 2002].
LINK: Information service published by Springer [LINK, 2002].
Xplore: Service published by IEEE (Institute of Electrical and Electronics Engineer) [IEEE-xplore, 2002].
ScienceDirect (SD): published by Elsevier [Science Direct, 2002]
JoDI: The Journal of Digital Information maintained by the IAM Research Group, University of Southampton [JODI, 2002].
JUCS: The Journal of Universal Computer Science. A publication of the Know-Center in cooperation with Springer Co.Pub., JOANNEUM RESEARCH and the IICM, Graz University of Technology [J.UCS, 2002].

To simplify the following discussion, we use the bold printed keys to refer to the systems, i.e. we write *Xplore* when we talk about the digital library system of the IEEE. *ACM-DL*, *LINK*, *Xplore* and *ScienceDirect* are serving

M.H. Shafazand and A M. Tjoa (Eds.): EurAsia-ICT 2002, LNCS 2510, pp. 495–502, 2002.

several journals, conference proceedings and books. *JoDI* and *JUCS* are serving *one* journal each. The selected libraries are very diverse in terms of amount of data provided to the user, offered features, licensing etc. Nevertheless, the selected systems are all electronic publishing systems trying to support the user in information and knowledge gathering, therefore the demands on the features are very similar.

In the following sections we take a critical look at selected qualities of systems, categorized this two categories:

Overall System Structure and Features: like exploring the content stored in the systems via searching and browsing as well as refining articles once found in the system.

Content Related Features: like document formats and interactive and active features using the content or parts of the content.

At the end of each subsection we take a look at the differences between the selected features and features available in *JUCS*. We also present some ideas of future improvements related to the corresponding topic.

2 Overall System Structure and Features

Organization and structure of *information entities* are a main issue in information systems. The digital world consists of servers, index- and content pages. Disadvantages of traditional libraries (like limited space in shelves, borrowed books are not available to other users etc.) are easily 'repaired' in the digital equivalent. But there are problems to solve: we all know the *lost in hyperspace syndrome* (e.g. [Theng and Thimbleby, 1998]) and try to avoid it in the design of an information system's user interface.

The content of almost every investigated digital libraries is available not only in electronic form but also in printed form. Articles also appear on paper, whereby the electronic edition appears usually earlier (e.g. *LINK*s 'Online First'-service). Since every page includes a page number, an additional navigation feature is added to the content.

In a printed environment space is always an issue and articles must not exceed some predefined number of pages. Electronic editions are easier to handle in this respect: They offer (nearly) unlimited storage space, therefore there is (usually) no limit to the size of an information entity. It is easy for an electronic system to add additional material to an entity, e.g. information about the author (link to the homepage of the author, curriculum vitae, email address etc.), additional software, tutorials etc. While all these features are 'nice to have', they also have massive drawbacks: just consider archiving the material. Problems related to the electronic format of the content and reliability of storage mediums arise.

Let us take a look at navigation-features of a digital publishing system. Finding the right content starting from the entry page of the system may be accomplished in two ways: by browsing (see section 2.1) or by searching for a title, author's name etc. (see section 2.2). If the right article is explored, functions like

'Remember this article' (see section 2.3) and 'Find related articles to the current one' are very useful (see section 3.2).

2.1 Browsing

Articles are organized in issues of a journal, issues are collected in volumes, and volumes are again collected into some category. Since entities of publishing systems are accessible via different index pages, all thinkable kinds of sorting (by author, title, category,...) are possible.

In *ACM-DL* it is possible to sort the content by the type of publication (journals, magazines, proceedings etc.), after following e.g. the *journal*-link, a list of all journals published in the *ACM-DL* is presented to the user. All other investigated digital libraries offer similar browsing mechanisms to the user. *LINK* and *Xplore* offer listings by title of the journal, *ScienceDirect* also generates listings clustered by the subject of the journal (e.g. Computer Science, Chemistry etc.).

Since *JoDI* and *JUCS* are serving one journal, there is no need for an index at that level. *ScienceDirect* is the only examined system where several publishers offer content. The volume- and issue-listing differ in the level of detail displayed to the user.

Very often articles are also categorized by some classification system (e.g. the ACM computing classification system, ACM-CCS, [ACM, 1998] for computer science related material). This classification of articles makes it possible to browse through categories of similar documents (see also section 3.2). Thereby it is easier to gain an overview of available documents about a specific topic. Nevertheless, if the amount of articles published in one category increases, a 'flat' listing is inappropriate. Additional features like filtering of selections and searching through collections are required. All of the investigated server systems have the browsing feature implemented by generating HTML documents.

In *JUCS* articles are categorized via the ACM-CCS and may be linked to several categories. At the moment all categories are equally important so there is no *rating* of a category. We think of an additional classification of these categories like 'most related category' and 'related category'. In the future much more sophisticated (graphically visualized) and easier to use graphical browsing techniques will be implemented too because the number of articles published in *JUCS* increases.

2.2 Searching

An expert uses a digital library much more different than an ordinary user. It is very likely that an expert prefers the search-feature for exploring the library over browsing. Searching is ideal if someone knows exactly what to look for. Publishing systems usually offer search features for bibliographic data and/or for fulltext data. More sophisticated systems also allow scoping of the search query. Please note that we have explored text based systems so fulltext search in those systems is also an issue.

Bibliographic data (also know as 'metadata', like author, title of the item, publishing date etc.) is typically stored in some database or in a separate file. There are many different formats for storing bibliographic data. Dublin Core (DC, [Dublin Core Metadata Initiative, 2001]) is one of the well-known standards for metadata-entries. DC classifies metadata (called 'Core Elements') into three categories: content, intellectual properties, and the instantiation of the document . It is possible to qualify elements. To give an example: the 'Date'-element (sub-element of the instantiation core-element) may be qualified by *created*, *issued*, *modified* etc. The concept of elements and qualifiers makes DC very powerful.

Aside from the search in metadata, fulltext search is available in most systems. Several different search engines are used and the input forms presented to the user are very different. Depending on the search engine implementation, operations like 'these words should appear in the same sentence' (e.g. implemented in *JUCS*) are possible. Some operations also use thesaurus search, i.e. if someone searches for the term 'magnetic tapes' also other 'computer storage devices' may be found.

ACM-DL, *ScienceDirect* and *JUCS* offer a combined search form for metadata- and fulltext search, whereas *LINK* offers different forms for each kind of search. At the time of writing *Xplore* did *not* support fulltext search – but this feature will be included in the future.

If some entries are found, they are presented to the user. The listing of matching articles may be sorted by users preferences like date of publication, author, relevance etc. In some evaluated systems it is possible to store search queries and work with the results of those queries.

In *JUCS* we are going to improve the search-interface on the one hand and on the other hand we are going to implement different interfaces for experts and new users. Some special queries (like all publications from a specific research institute) are under consideration but therefore a few more metadata must be added to the objects.

2.3 Refinding and Organizing Content

Once an article is explored, it is very likely that a user wants to read it again for reference purpose without storing the article on the local computer system. Stability related to storage location (e.g. via Uniform Resource Name, URN, [Moats, 1997]) and content of the information entity is therefore an issue. Some journals (like *ACM-DL* and *LINK*) provide Digital Object Identifiers (DOI, [DOI, 2002]) which are a kind of URN.

In an electronic system electronic bookmarks are the right solution for most of the users to the problem of refinding articles ([Krottmaier, 2001]). They can be created either on the client-, the server side, or on some intermediate or proxy system. The most important feature of server side bookmarking is the possibility to share bookmarks with other colleagues. Therefore one user may collect and share a selection of articles about a specific topic. Other users may use these articles via the given bookmarks and rate or comment on the articles via the bookmarks. A proper user- and group management on the server side is a prerequisite for bookmarks on the server side.

Beside static bookmarks, dynamic bookmarks should be implemented, i.e. bookmark listings are generated on the fly by performing some database search query. Therefore the list is always up to date. Users may be informed by some alert email mechanism if new articles arrive in the listing. So called 'active' bookmark listings may be arbitrary scheduled by the user.

At the time of writing *ACM-DL* was the only examined system which supports server side bookmarking (called *binders*). Binders are private by default but it it possible to share the bookmarks with other users or groups of users. It is also possible to create *active binders*, which performs some *saved search query*. Notifications via email, if the listing changes, are optional.

LINK and *JoDI* do not offer any structuring feature for users. Users of *ScienceDirect* have the option to collect journals in some *Personal Journal* region on the server. Unfortunately, it is not possible to share this collection with other users.

In *JUCS* a personal restructuring feature is currently under development. It is planned that each registered user will have the possibility to create arbitrary collections of articles, reuse existing collections (like articles structured by category), or store some active *search query*. These collections may be shared with other registered users. To provide flexible sort orders of the articles (like sorting by date, title, author of the article etc.) surrogates for the original articles will be created in the database. These surrogates will store context specific attributes of the original article. With surrogate objects it will also be possible to represent articles not stored in the database and well defined parts of articles.

3 Content Related Features

The features described in the previous section are system wide features supporting the user by searching and exploring the *structure* of a publishing system. In this section we explore features which are available as soon as the right *content* is displayed on the screen.

3.1 Format of an Article

The content of an article is very often available in some 'printer friendly' format like the Portable Document Format (PDF, [Adobe, 1993]) or PostScript (PS). Although PDF is a very popular document format and viewers are available for different operating systems, there are enormous problems with PDF:

Interacting with PDF documents requires additional software:
Interaction with a PDF document is limited to the functionality implemented in the major viewer: Adobe Acrobat Reader. Taking a closer look at the implemented features of that software uncovers the big problems when trying to work *actively* with PDF: It is not possible to change any content of

the PDF document. There are some navigation and zooming features available (like goto next/previous/first/last page and zoom-in/out etc.), as well as searching, but no *interaction* with the document is possible. It is not possible in Acrobat Reader (up to version 5) to add bookmarks or annotate/highlight parts of a document. Other implementations of viewers (e.g. Xpdf, [Noonburg, 2002]) are also very limited in functionality. Buying the full version of Adobes PDF-aware software where it is possible to interact with the document as described above is not acceptable for most of the users because of the price. Nevertheless, PDF was not designed to be heavily modified after creation.

PDF documents are not suitable for reading on screen: A study by Nielsen (see [Nielsen, 2001]) showed that PDF is not the preferred document format for reading content online.

> "Forcing users to browse PDF files makes usability approximately 300% worse compared to HTML pages. Only use PDF for documents that users are likely to print. [...] PDF is great for distributing documents that need to be printed. But that is all it's good for. No matter how tempting it might be, you should never use PDF for content that you expect users to read online."

Unfortunately, many digital journals provide the content of documents exclusively in PDF. This forces the majority of users to print out the articles and read these articles offline. When reading the article offline it is not possible to use *interactive features* of the digital journal *while* exploring the content.

These two problems are solvable: In the first place if someone does not have access to full PDF-aware software (i.e. to Adobe Acrobat) and wants to interact with the document, the document *itself* must provide additional hyperlinks to *interactive web-applications.*

Existing PDF documents may be modified on the server side by some PDF manipulation software. This application may add hyperlinks to interactive web applications. Unfortunately, no information about the structure of a PDF document (like sections, headings etc.) is available. Granularity of the added hyperlinks is therefore limited to pages. To give an example: The PDF manipulation software may take a single PDF document requested by the user, add some header with a special link to each page on the fly and deliver then the modified PDF document to the user. If the user wants to annotate the paper, it is possible to add annotations to a specific page or the whole document. Bookmarks can also be evaluated to enhance the granularity of annotations. Please note that this feature of adding interactivity to *existing* PDF documents stored in *JUCS* is currently under development and therefore there are no usability studies and user acceptance tests available at the time of writing.

If content is available in XML (Extensible Markup Language) and the PDF document is generated on the fly, the document structure may be extracted out of the XML document.

Secondly it is possible to provide content in different document formats (e.g. in *JUCS* content is currently provided in HTML, PostScript and PDF). The user may select the appropriate document format depending on the used device. If the content is stored in some generic format (like XML) transformation to arbitrary document formats is possible using the Extensible Stylesheet Language (XSL).

3.2 Features Using Content

The content of an article is obviously used in fulltext search. Most of the available digital journals index the content into a fulltext index regardless of the document format. The usage of a thesaurus and additional information of the words (e.g. position in the document) allows high sophisticated search queries like 'search for words within a sentence or paragraph' (see section 2.2). Finding similar articles is a feature often available in KM systems. In *JUCS* and *ACM-DL* this feature is available and may be applied to articles stored on the respective system.

Disassembling the content into the different parts of an article (like abstract, sections, conclusion, bibliography etc.) is a necessary task to make additional features work. Using the abstract (or all abstracts of published articles) makes it possible to support the user with an overview of the available material. The user may then decide which article to read and which article not to read before downloading.

Sections in the document may be interlinked automatically (intra-document links) and it is very common (e.g. in *ACM-DL* and *LINK*) to link even bibliography entries to the electronic versions of the cited material (extra-document links). Section headings may be used to give an additional article navigation frame to the user. Single words may be interlinked to a dictionary or thesaurus to support the user with additional information. Especially in systems supporting teaching it is very common to enable links to dictionaries.

In digital libraries it is also possible to put additional material of the content on the publishing server, including multimedia attachments, demonstration packages, glossaries, programs, additional references etc. Although this is very interesting for online-publications, it is not possible to archive some of these additional contents on paper.

An active document concept is introduced in [Heinrich and Maurer, 2000]. With this concept it is possible to ask questions at any time and any position in the document. The system will then either search for an appropriate answer in the system or will inform the author of the document.

In *JUCS* we are going to work much more with parts of documents, especially with the reference section. We want to extract information out of this very special section to support the reader with real important information like 'Which other resources should be read to understand this article?' and 'Who are the *real* experts on this topic?'.

4 Conclusion and Future Work

Features of digital publishing systems may be separated in overall features and content related features. This paper has described some currently available features of different systems. It has been shown that (inter-)active features already available for most knowledge management systems are not widely available in digital journals.

Future work is addressed to bring some of those KM features into the journal environment of *JUCS*. Currently restructuring of content objects is available for internal use. The next step is to restructure parts of a document at least for HTML formated content and make this feature available to registered users of *JUCS*. An interactive typed-annotation feature is currently available for HTML documents only. We are going to add this feature also to PDF formated documents.

References

[ACM, 1998] ACM (1998). ACM Computing Classification System.

[ACM Digital Library, 2002] ACM Digital Library (2002). http://www.acm.org/dl .

[Adobe, 1993] Adobe (1993). *Portable Document Format Reference Manual.* Addison-Wesley Longman, Inc.

[DOI, 2002] DOI (2002). Digital Object Identifier. http://www.doi.org.

[Dublin Core Metadata Initiative, 2001] Dublin Core Metadata Initiative (2001). Dublin Core Metadata Initiative. http://www.dublincore.org .

[Heinrich and Maurer, 2000] Heinrich, E. and Maurer, H. (2000). Active Documents: Concept, Implementation and Applications. *Journal of Universal Computer Science*, 6(12):1197–1202. http://www.jucs.org/jucs_6_12/active_documents_concept_implementation.

[IEEE-xplore, 2002] IEEE-xplore (2002). Digita Library of the IEEE. http://ieeexplore.ieee.org .

[JODI, 2002] JODI (2002). Journal of Digital Information. http://jodi.ecs.soton.ac.uk .

[J.UCS, 2002] J.UCS (2002). Journal of Universal Computer Science. http://www.jucs.org .

[Krottmaier, 2001] Krottmaier, H. (2001). Improving the usability of a digital library. In Hübler, A., Linde, P., and Smith, J. W., editors, *Electronic Publishing*, pages 178–182, Canterbury, Kent, United Kingdom.

[LINK, 2002] LINK (2002). Link Service. http://link.springer.de .

[Moats, 1997] Moats, R. (1997). *URN Syntax.* RFC 2141.

[Nielsen, 2001] Nielsen, J. (2001). Avoid PDF for On-Screen Reading. http://www.useit.com/alertbox/20010610.html .

[Noonburg, 2002] Noonburg, D. (2002). Xpdf. http://www.foolabs.com/xpdf/ .

[Science Direct, 2002] Science Direct (2002). Science Direct. http://www.sciencedirect.com .

[Theng and Thimbleby, 1998] Theng, Y. and Thimbleby, H. (1998). Addressing Design and Usability Issues in Hypertext and on the World Wide Web by Re-Examining the "Lost in Hyperspace" Problem. *Journal of Universal Computer Science*, 4(11):839–855. http://www.jucs.org/jucs_4_11/addressing_design_and_usability.

Solving Language Problems in a Multilingual Digital Library Federation*

Nieves R. Brisaboa, José R. Paramá, Miguel R. Penabad, Ángeles S. Places, and Francisco J. Rodríguez

Departamento de Computación. Universidade da Coruña
{brisaboa,parama,penabad}@udc.es,
{asplaces,franjrm}@mail2.udc.es

Abstract. This work presents an architecture to federate pre-existing Digital Libraries with documents written in different languages. A user will be able to ask queries to all the federated Digital Libraries using a unique and friendly user interface that will be generated in the language she chooses among the available ones. The query will be executed over all the relevant databases in the system no matter which language their documents are written in. The query will be automatically translated if it is necessary. This architecture is based on ontologies, which are used not only to represent the global schema but also to guide the execution of software modules in the system; it also includes three dictionaries to solve the inter-language barriers.

Keywords: Digital Library Federation, Cross-language search, Federated Search, Ontologies.

1 Introduction

Within the European Union, there are more than 50 autochthonous languages in use, even when only 11 are official languages [12]. Europe is a large jigsaw of languages and cultures with a cultural richness that is necessary to preserve. Nowadays the Web is becoming the medium for the preservation and dissemination of any cultural manifestation. It is clear that only languages with presence in the Web will have the opportunity to survive. All around Europe, there are efforts to create document databases, supported by the European Union. These databases are real Digital Libraries with documents and literature in different languages, but when those languages have few speakers, the isolated efforts are not sufficient, because visitors of such Web sites are scarce and its maintenance is expensive.

Federating Digital Libraries into a cross-language federation system will increase the number of visitors of all the Web sites in the federation. Such a system will take advantage of the fact that some languages are similar enough to be understood by speakers of other languages (who are able to read but not to write a query directly). On the other hand, the federation will facilitate international researchers to find documents from other societies, even if they do not understand the language in which

* This work was partially granted by CICYT (TEL99-0335-C04-02).

M.H. Shafazand and A M. Tjoa (Eds.): EurAsia-ICT 2002, LNCS 2510, pp. 503–510, 2002.

the documents are written. Those two facts made the Cross-language Text Retrieval (TR) systems gain more attention in the last years [11][16].

The Federation of Digital Libraries also helps to solve a problem, not specifically related with Multilingual Digital Libraries, but with the use of several different Digital Libraries in general. A user interested in a specific subject must know the URL where each relevant database is placed, as well as adapt the same query to the characteristics of each Digital Library user interface and write it using its own query language. This drawback makes obvious the need to work with a system that integrates a number of document databases (that agree to be asked in a federated way) under a unique user interface through a unique URL.

At the Database Lab of the University of A Coruña, we are working on the design and implementation of such a system for a Multilingual Digital Library Federation. In this paper, we present the architecture of this system. In our proposal, we assume that the Digital Libraries are pre-existing and heterogeneous in the language of their documents and type of corpus, as well as in their data model and technology. When the user starts the connection to our system, she will choose the language (among the available ones) for the user interface and then she will write the query in the chosen language over a friendly interface in Bounded Natural Language [6][21]. The query will be redirected to the appropriate Digital Libraries in the federation. The retrieved data and documents will be presented to the user in an interface in the chosen language. Documents will be kept in their original language, because our system does not translate literary works.

The rest of the paper is distributed as follows. Section 2 is an overview of the system architecture. Section 3 details the role of ontologies in the architecture as well as the dictionaries. Section 4 makes a description of the query process, including a brief overview of BNL technique. The last section contains our conclusions and directions for future work.

2 System Architecture

Several architectures have been proposed to build systems that work with heterogeneous and geographically dispersed databases [1][2][9][18][23][24]. The relevance of the Database Federation [22] subject has been increasing at the same time as the number of isolated data sources available in the Web grows. Nowadays, this is an important and current investigation area [16].

The proposed architecture is composed of four isolated layers and we define three exchange languages to communicate them. This architecture is shown in Fig. 1. To understand the system it is necessary to consider the ontological architecture described in the next section. The architecture layers and software modules are:

1. **Layer 1. User Interface:** It must allow expressing queries about both structured data and document contents of any database integrated in the system. In addition, it must be extremely intuitive and easy to use. The technique we use to facilitate the user to express restrictions about concepts/attributes is the *Bounded Natural Language (BNL)* technique [6][21].
2. **Layer 2. Overlapped Integrator System:** It includes the General Ontology as well as all the dictionaries. Its task is to follow the General Ontology to create the Query User Interface (User Interface Generator). This module presents the

concepts and the attributes to the user in order to allow her to write a query in an easy way and in her own language. After the user writes her query by expressing restrictions over the concepts, this layer decides which Databases the query must be redirected to (Query Analyzer), using the information associated to concepts and attributes in the General Ontology.

3. **Layer 3. Query Systems:** Each Digital Library in the federation has its own Query System. Its task is to query (Query generator) its associated database. This module translates the query in FQL (from layer 2) into the language of the associated database, using the information associated to each concept in its Specific Ontology. The retrieved document *Ids* are stored (Answer Store) so the user can navigate through them (Answer Manager) and through those retrieved from other Digital Libraries by the same query. Although all the Query Systems perform similar tasks, they need to be adapted to the specific associated database.
4. **Layer 4. Document Databases:** They are pre-existing and independent of the system. That is, managing the databases is not a task of our system. Therefore, if a database has TR capabilities [3][4][5][17][20], any needed pre-process has to be already performed and those TR techniques have to be already implemented.

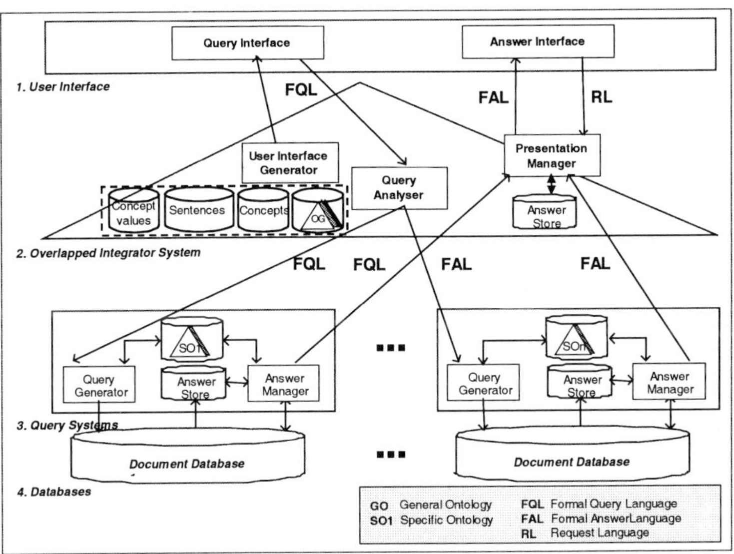

Fig. 1. System Architecture.

The communication between layers is made by means of three exchange languages defined by us in XML [25]: **FQL:** *Formal Query Language;* **RL:** *Request Language*; and **FAL:** *Formal Answer Language.*

3 Ontologies and Dictionaries

Although there are a number of definitions of ontology, most authors agree that an ontology is a specification of a conceptualisation [10][13][14][15], that is, a set of concepts and the relationships among them, that describe a domain of interest.

In our system, ontologies are conceptual models with an abstraction level higher than the schemas of any federated database. Ontologies give a homogeneous description to different schemas of databases integrated in the system, so offering a way to make the necessary schema conciliation. All concepts, as well as their attributes, in the ontologies are common notions in the domains of the integrated databases. Each concept has associated attributes, that is, properties that define a concept. We arrange ontologies in tree shapes where the nodes are concepts. Among them, we define two kinds of relationships: (1) Generalization / Specialization relationships ("is a" relationship): for instance, a *Work* can be a *Book*, a *Journal*, etc. (2) Description relationships ("has" relationship): for instance, a *Work* has an *Edition* and an *Author*.

As shown in Fig. 1, two ontological levels are considered: (1) an upper level: where a unique General Ontology is placed; (2) a lower level: where there is one Specific Ontology associated to each federated database. The General Ontology is an abstraction of the schemas of all member databases and it integrates all the concepts in the Specific Ontologies. Figs. 2 and 3 show the General Ontology for a federated system that integrates Digital Libraries with documents of journals, thesis, books and compilation books. One Specific Ontology is defined for each document database. A Specific Ontology describes the concepts of its associated document database. Therefore, it has a subset of the concepts in General Ontology. Fig. 4 shows the Specific Ontology of a database of thesis.

In both General and Specific Ontologies we distinguish two parts: (1) The *Common Part* includes general concepts that any user, even if she is not an expert in any corpus domain, can perfectly understand. The Common Part is the left subtree in any ontology in our system; (2) The *Expert Part*: For each corpus, there is a set of concepts that perfectly describes such a domain. This set of concepts that we arrange in a subtree shape, forms a "Corpus" Expert Set. The name of the corpus is the concept in the root of the subtree. Probably only experts in the domain can understand concepts in the corresponding "Corpus" Expert Set. The Expert Part of the General Ontology has a "Corpus" Expert Set for each corpus in the federation. The Specific Ontologies have only one "Corpus" Expert Set each. In Fig. 2, the Common Part of the General Ontology is shown. The root of the tree, "Work", is specialized in four "Corpus" Specific Sets. Fig. 3 shows the Journal Expert Set.

The attributes of the concepts in both General and Specific Ontologies have associated useful information. However, this information is different in the General and Specific Ontologies:

- **General Ontology:** The attributes in the General Ontology have associated information useful to generate the user interface and to redirect the query from the user to the appropriate databases. The information is: (1) The *Ids of the sentence skeletons,* needed to allow the user to express restrictions over an attribute of a concept. (See Section 5); (2) The *list of databases* where the attribute is relevant.
- **Specific Ontology:** The attributes in Specific Ontologies contain the expression to access the corresponding data in the associated database. This expression depends on the database DBMS. For example, in a relational database, an attribute can have the relation and attribute names where it is really stored. However, this expression can be more complex, like a SQL sub-query if the underlying database is relational. These statements are used to build the query to retrieve data and documents from the underlying database. On the other hand, if the database has

some kind of TR capabilities, the information associated to an attribute as *"content"* or *"subject"* will be the directions to call the TR algorithm implemented in the database.

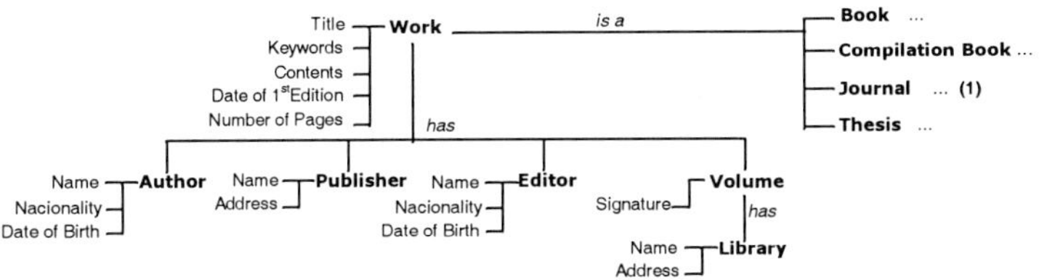

Fig. 2. General Ontology. Common Part.

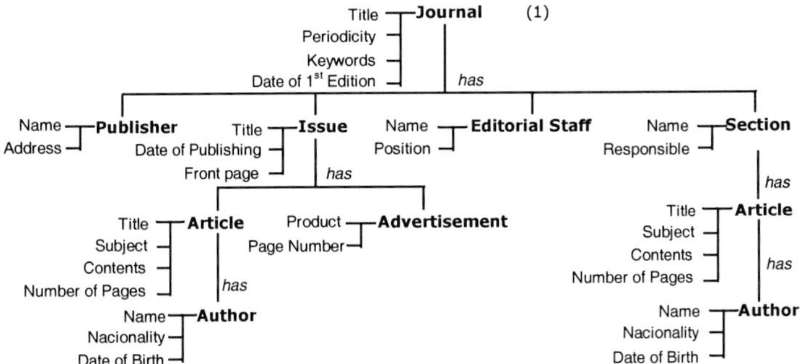

Fig. 3. General Ontology: *Journal Expert Set.*

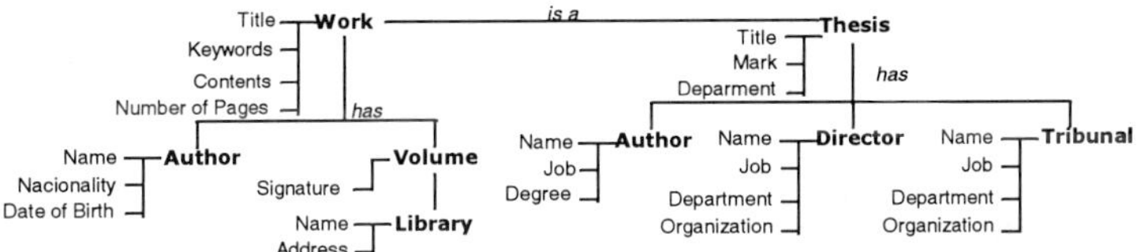

Fig. 4. Specific Ontology for a Thesis Database

Ontologies are defined using XML [8][25]. Among the possibilities to represent the ontologies [18] we have chosen XML because it is a universal language allowing us to define formats for data and metadata exchange that are human readable and computer parseable. Fig. 5 shows the two DTD's for the two ontologies.

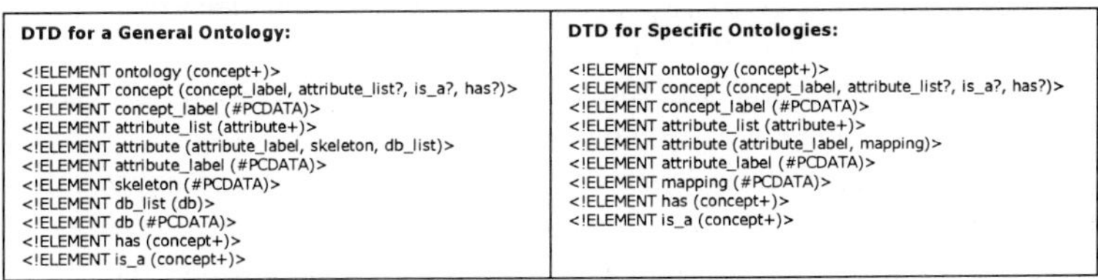

DTD for a General Ontology:

```
<!ELEMENT ontology (concept+)>
<!ELEMENT concept (concept_label, attribute_list?, is_a?, has?)>
<!ELEMENT concept_label (#PCDATA)>
<!ELEMENT attribute_list (attribute+)>
<!ELEMENT attribute (attribute_label, skeleton, db_list)>
<!ELEMENT attribute_label (#PCDATA)>
<!ELEMENT skeleton (#PCDATA)>
<!ELEMENT db_list (db)>
<!ELEMENT db (#PCDATA)>
<!ELEMENT has (concept+)>
<!ELEMENT is_a (concept+)>
```

DTD for Specific Ontologies:

```
<!ELEMENT ontology (concept+)>
<!ELEMENT concept (concept_label, attribute_list?, is_a?, has?)>
<!ELEMENT concept_label (#PCDATA)>
<!ELEMENT attribute_list (attribute+)>
<!ELEMENT attribute (attribute_label, mapping)>
<!ELEMENT attribute_label (#PCDATA)>
<!ELEMENT mapping (#PCDATA)>
<!ELEMENT has (concept+)>
<!ELEMENT is_a (concept+)>
```

Fig. 5. DTD for General and Specific Ontologies

Given that we are dealing with multilingual digital libraries, and that users must be able to choose the language of the user interface, we need some dictionaries to translate the appropriate terms. These dictionaries, located in Layer 2, are the following: (1) *Concept Values*, for the concept in the ontology and its values, if they are a finite set (see Fig. 6); (2) *Skeleton Sentences*, for the skeletons of the sentences in Bounded Natural Language; and (3) *Interface Text*, for texts or directions in the user interface.

English		Spanish		Galician		...
Genre	Male	**Sexo**	Hombre	**Sexo**	Home	...
	Female		Mujer		Muller	...
Type	Novel	**Tipo**	Novela	**Tipo**	Novela	...
	Tale		Cuento		Conto	...
	Journal		Revista		Revista	
Date of Birth ...		**Fecha de Nacimiento ...**		**Data de Nacemento ...**		...

Fig. 6. Dictionary of concepts and values into the General Ontology

4 Description of the Query Process

The technique we use to build the Query User Interface is the BNL technique [6, 7]. It lays on showing the user a set of sentences in natural language with "gaps". The user can choose the sentences he wants and fill the gaps. It is necessary to remark that these sentences will be written in the selected language. Finally, the set of selected and completed sentences expresses the query the user wants to ask.

For each attribute the user wants to restrict, the system instantiates its associated sentence skeleton, using the appropriate language, and shows the BNL sentence.

For example, the sentence skeleton in Fig. 7 could be associated to the attribute *Name* of concept *Author* in the General Ontology. If a user wants to add a restriction about this attribute, the system will take the sentence skeleton in Fig. 7(a) following the Sentence Skeleton Id of the attribute *Name* and will present it to the user after replacing the tag <ATTRIBUTE> with the correct translation for *Name*, again using the needed dictionary. So the sentence in BNL the user would see is shown in Fig. 7(b).

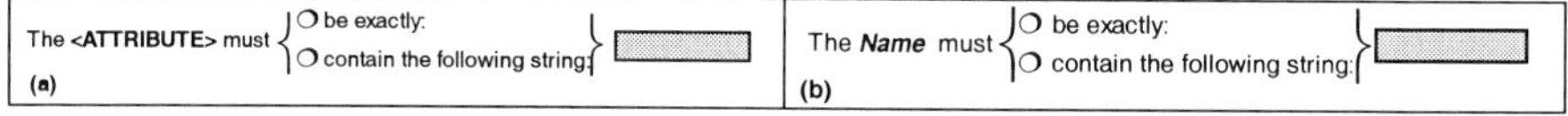

Fig. 7. Sentence skeleton for *Short String* data type and an example

The possible restrictions could be restrictions about structured data, like in the previous BNL sentence, or restrictions on non-structured data. With the BNL technique, we allow the user to express a set of words that describes the text that documents must include to be retrieved.

We already said that each attribute in the General Ontology has associated a sentence skeleton, but other two sentence skeletons are also necessary to allow the user the navigation through the General Ontology so that she can express her whole query. Both of them are used with the concepts in the General Ontology. We call these sentences: (1) *Disjunction Sentence:* It is used with concepts having a *"is a"* relationship. Filling this sentence the user will say if she is interested in the general concept or in one of its specializations; and (2) *Choice Sentence:* It is used to ask the user which attributes and concepts describing a concept he is interested in.

The User Interface Generator uses an algorithm that we call User Interface Algorithm (UIA). The first input parameter for the UIA is the concept at the root of the General Ontology. The algorithm first checks if the current concept has a *"is a"* relationship. In this case, the algorithm instantiates the *disjunction* sentence skeleton with the current concept and its specializations, takes the option the user chooses and recursively calls the UIA algorithm using as input parameter the concept chosen by the user. After checks the *"is a"* relationship, the algorithm instances the *choice* sentence skeleton for that concept, if it has a *"has"* relationship. With this sentence the user can choose as many options as she wants and two lists of concepts are built. The *AttributeList* contains the chosen attributes of concept under process. The *ConceptList*, contains concepts, related to the current concept through a *"has"* relationship, that have been chosen by the user. For example, if the current concept was *Article* (in Fig. 3), the instantiated *choice* sentence would be:

"Regarding to Article, I´m interested in expressing restrictions about:
❒ Title ❒ Subject ❒ Contents ❒ Number of Pages ❒ Author."

Assuming the user marks *Subject, Number of Pages* and *Author* then *AttributeList* will contain *Subject* and *Number of Pages* and *ConceptList* will contain *Author*. Notice that, in this step, the user is saying that the *Author*, the *Number of Pages* and the *Subject* of the *Article* are the three features she wants to restrict to describe the *Articles* she is looking for.

These two lists are used to follow the process. The algorithm will use concepts and attributes in both lists to continue its execution. For each attribute in the *AttributeList* the associated sentence skeleton is instantiated and presented to the user. Her restrictions are translated into a *Formal Query Language* (FQL) statement that will be part of the whole FQL document that will represent the query. For each concept in the *ConceptList,* the algorithm recursively calls itself using such a concept as input parameter.

Notice that all the sentences and the names of the concepts and attributes are presented in the appropriate language using the dictionaries shown in Section 3 to perform the correct translation.

5 Conclusions and Future Work

Federating and integrating different data sources into a cross-language broker system is an important and current investigation area in general, and for document databases in particular. The use of ontologies to federate databases gives an interesting and useful way to integrate different database schemas. Moreover, ontologies provide logical and physical independence to the system. Using the dictionaries allows us to present the interface in the desired language. The *User Interface* of our system is based on BNL technique, allowing the user to build queries in an intuitive manner. Using the dictionaries allows us to present the interface in the desired language.

Nowadays, we are refining the *User Interface* by including cognitive metaphors. That is, images that represent a piece of real world. So, the *User Interface* becomes more intuitive because the user will know how he has to query the system just by looking at the Interface. We are working in the implementation of a prototype, applying the ideas exposed in this paper. Initially, this system integrates three document databases that store two different corpora of historic documents in Galician and Spanish languages.

References

1. Abelló, A.; Oliva, M.; Rodríguez, E.; Saltor. F. The BLOOM Model Revisited: An Evolution Proposal. Proc. ECOOP Workshops & posters, ECOOP'99, Lisbon, June 1999.
2. Arens, Y., Hsu, C., Knoblock, C. A. Query processing in the SIMS Information Mediator. Advanced Planning Technology, Austin Tate (Ed.), AAAI Press pp. 61-69, CA, 1996.
3. Baeza-Yates, R.; Navarro, G. Integrating contents and structure in text retrieval. ACM SIGMOD Record, 25(1):67–79, Marzo 1996.
4. Baeza-Yates, R.; Navarro, G.; Vegas, J.; Fuente, P. A model and a visual query language for structured text. En Berthier Ribeiro-Neto (Eds.) Proc. of the 5th Symposium on String Processing and Information Retrieval, pp. 7–13, Santa Cruz, Bolivia, 1998. IEEE CS Press.
5. Baeza-Yates, R.; Ribeiro-Neto, B. Modern Information Retrieval, Addison-Wesley, 1999.
6. Brisaboa, N. R., Penabad, M. R., Places, A. S., Rodríguez, F. J.: A documental Database Query Language. 8th SPIRE'01. IEEE Computer Society. Chile (2001), pp. 242–245.
7. Brisaboa, N. R., Penabad, M. R., Places, A.S., Rodríguez, F. J.: Tools for the design of user friendly Web applications. LNCS 2115 (EC-WEB'2001), Munich (2001), pp. 29-38.
8. Brisaboa, N.R., Penabad, M. R., Places, A.S., Rodríguez, F. J.: Using ontologies for federation of Web accesibles databases. Procs. of SEKE'2001, Argentina (2001), pp. 87-94.
9. Kirk,T., Levy, A. Y., Sagiv, Y., Srivastava, D. The Information Manifold. In Proc. of the AAAI Spring Symposium on Information Gathering in Distributed Heterogeneous Environments.
10. Chandrasekaran, B.; Josephson, R. What are ontologies, and why do we need them? In IEEE Inteligent systems, 1999.
11. Cross- Language Evaluation Forum. http://www.iei.pi.cnr.it/DELOS/CLEF/
12. Euromosaic: The production and reproduction of the minority language groups in the European Union, ISBN 92-827-5512-6. Luxembourg (1996).
13. Gruber, T. Toward Principles for the Design of Ontologies Used for Knowledge Sharing. IJHCS, 43 (5/6): 907–928. 1994.
14. Gruber, T. http://www-ksl.stanford.edu/kst/what-is-an-ontology.html
15. Guarino, N. (ed.), Formal Ontology in Information Systems. Proceedings of FOIS'98. Amsterdam, IOS Press, pp. 3-15, Trento, Italy, 6–8 June 1998.
16. Hasselbring, W.; van den Heuvel, W.-J.; Houben, G.J.; Kutsche, R.-D.; Rieger, B.; Roantree, M.; Subieta, K. Research and Practice in Federated Information Systems. Report of the EFIS'2000. ACM SIGMOD RECORD Web Edition. Vol. 29, Núm. 4. (2000).
17. Excalibur.Informix Corporation. http://www.informix.com/
18. Mena, E., Illarramendi, A., Kashyap, V., Sheth, A. OBSERVER: An Approach for Query Processing in Global Information Systems based on Interoperation across Pre-existing Ontologies. Published in the journal Distributed And Parallel Databases (DAPD). 1998.
19. Miller, G. WordNet: A lexical database for English. ACM, 38(11), 1995.
20. Context.Oracle Corporation. http://www.oracle.com/
21. Penabad, M.; Durán, M.J.; Lalín, C.; López, J.R.; Paramá, J; Places, A.S.; Brisaboa, N.R.; Using Bounded Natural Language to Query Databases on the Web. Procs of ISAS'99. Florida, 1999.
22. Sheth, A. P., Larson, J. A. Federated databases for managing distributed, heterogeneous, and autonomous databases, Computing Surveys 22:3 (1990), pp. 183–236.
23. Subrahmanian, V.S., Adali, S., Brink, A., Emery, R.,. Lu, J., Rajput, A., Rogers, T., Ross, R., Ward. C. HERMES: A heterogeneous reasoning and mediator system. Technical report, University of Maryland, 1995.
24. Chawathe, S., Garcia-Molina, H., Hammer, J., Ireland, K., Papakonstantinou, Y., Ullman, J., Widom, J. The TSIMMIS project: Integration of heterogenous information sources. 16th Meeting of the Information Processing Society of Japan pp. 7–18, Tokyo, Japan, 1994.
25. World Wide Web Consortium. Standard XML http://www.w3.org/XML

Performing IP Lookup on Very High Line Speed[1]

Nasser Yazdani and Nazila Salimi

Tehran University and Amir Kabir Univ. of Technology
nasyaz@sofe.ece.ut.ac.ir

Abstract. IP lookup is still considered a hard and challenging problem in routers. In high speeds, around 100Gbps, and with current growth rate of lookup tables, it sounds IP lookup can be a bottleneck. Complexity of the problem stems from the fact that routers must find the longest matching prefix with a packet destination address in the lookup table in order to forward the packet. Basically, this process is slow. We are currently developing a hardware-based scheme, which can perform IP lookup in a time proportional to access time to the external memory. The method implements DMP-tree, Dynamic M-way Prefix tree, which is a superset of B-tree and initially devised for prefix matching. Implemented in a FPGA, the scheme can forward around 100 million packets per second and regarding the average packet size can support IP lookup for over 100Gbps line speed. Our technique scales well to the next generation IP addressing, IPv6.

1 Introduction

Different factors have made routing IP packets on Internet very challenging. Classless addressing requires finding the longest prefix matching a packet destination address in the lookup table in order to forward the packet. The data rates of links have increased rapidly to hundreds of gigabits per second. Clearly, the packet forwarding system must work at the wire speed. Otherwise, it will be a bottleneck. In practice, assuming the minimum IP packet length of 64 bytes implies the forwarding mechanism must determine the packet output port in 200ns for OC-48, 2.5Gbps, and in 50ns for OC-192. Meanwhile, any solution to this problem must scale well for the feature technology since the communication lines speed doubles approximately in every four months. With the current trend in technology, solutions based purely on ASIC technology cannot keep pace with increase in lines speed. In this circumstance, handling lookup in software is completely out of consideration.

According to [1], the size of lookup table grows exponentially. A lookup engine should be able to support approximately 400K-500K prefixes in coming years. On the other hand, increasing line speeds shortens the time of processing and forwarding packets. Therefore, it seems the IP lookup problem is a serious bottleneck so that the advances in the semiconductor technology and processing capability alone will not be able to solve it.

---

[1] This work has been partially supported by Iran Telecommunication Research Center (ITRC) and Control and Intelligent Processing Center of Excellency at University of Tehran.

M.H. Shafazand and A M. Tjoa (Eds.): EurAsia-ICT 2002, LNCS 2510, pp. 511–518, 2002.

Many schemes have been proposed recently for the IP lookup problem [1],[4],[5],[9],[10]. Unfortunately, these methods while working for small lookup tables are usually not scale well to the large number of prefixes, address lengths and higher speeds. In this paper, we report the design specification of a hardware-based scheme, that implements the Dynamic M-way Prefix Tree [16]. DMP-Tree is a superset of B-tree for matching string that may be prefixes of each other. According to [16], the maximum height of DMP-Tree is $\log_M N$, the same as B-tree for the large branching factors in the internal nodes (≥ 6) and large number of prefixes (≥ 500). This implies that the scheme scale well for large data set of prefixes. Implemented in hardware, the scheme can perform one lookup in each clock cycle and assuming current FPGAs, it can route more than 100 million packets per second with over 500K prefixes.

The rest of the paper is organized as follows. Section 2 is related work. Section 3 briefly introduces DMP-Tree. In section 4, current stage implementation is discussed. Section 5 discusses extension of the implementation to IP version 6. Finally, we conclude in section 6.

2 Related Work

The basic scheme for IP lookup, which inspired proposing some new approaches, is binary trie [1]. Binary trie is simple to implement, but its worst case search time is O(32) for IPv4. Patricia trie modifies binary trie by compressing the paths and eliminating unnecessary nodes [2]. Patricia trie has been implemented in the BSD kernel [3]. Another optimization to binary is level compression [4]. V.Srinivasan and G.Varghese in [5] apply a dynamic programming technique to compress the hight of the trie and optimize the memory usage. The scheme proposed in [6] is a specific case of [5] by locally optimizing memory usage in each step. DP-Trie, Dynamic Prefix Tries, [8], tries to compact the Patricia trie by keeping prefixes themselves and the index of bit position differing in the subtries of each node. The method proposed in [9] exploits the same idea, applying binary tree search, and extends it to a multiway tree by treating each prefix as a range. Authors in [17] propose a scheme which starts with binary tree and is extended to a M_way tree structure.

An alternative approach for IP routing table search is hashing [10]. Unfortunately, hashing based methods do not take advantage of the hierarchical address structure and does not well suited for Internet applications. Furthermore, [11] reports poor cache hit ratios for backbone routers.

Some hardware-based solutions have been proposed in literature. Content-addressable memories (CAMs) are used to implement best matching prefix. A scheme in [12] uses separate CAM for each possible prefix length. This requires 32 CAMs for IPv4 and 128 for IPv6, which makes the scheme expensive. It is possible to use CAMs that allow "don't care" bits in CAM entries to be masked out [13]. Such design requires a single CAM, however; the size of CAMs is still a problem for using them in backbone routers. Another method proposed in [14] needs a simple logic and each packet is routed by few memory accesses. However, it does not scale to IPv6 with 128bit addresses. Caching is another hardware-based scheme, which has not worked well in the past in the backbone routers. Lulea algorithm, proposed by Degemark et al [7], is motivated by the objective of minimizing the storage requirements of their data structure, so that it can fit in the L1 cache of a conventional general purpose processor. Finally, a good survey about IP lookup problem can be found in [18].

3 DMP-Tree

DMP-tree, which stands for Dynamic M-way Prefix tree, has been proposed for matching strings of different lengths where strings can be prefixes of each others [16]. DMP-tree utilizes the following definition to compare and sort string of different lengths assuming the characters constructing data set are ordered.

Definition 1. Assume two strings $A = a_1a_2...a_n$ and $B=b_1b_2...b_m$ and a special character $\perp$ belonging to the character set, then,

1. if n=m, two strings have the same length, the values of A and B are compared.
2. if $n \neq m$ (assume n<m), two substrings $a_1a_2...a_n$ and $b_1b_2...b_n$ are compared. The substring with bigger (smaller) value is considered bigger (smaller) if they are not identical. Otherwise, they are identical, b_{n+1} character is checked. $B \leq A$ if b_{n+1} is equal or before $\perp$ in the ordering of characters, and $B > A$ otherwise.

As an example, in the binary alphabet, {0,1}, assuming $\perp$ is 0, clearly, 1101 is greater than 1011 and smaller than 11101, and 1011 is greater than 101101. Two strings A and B are considered *disjoint* if each is not a prefix of the other.

Definition 2. String S is called an enclosure if there exists at least one data string A such that S is a prefix of A. A is called an enclosed data element. For example, 1011 is an enclosure of 1011010.

DMP-Tree is a generalization of B tree and differentiated from B-tree by the property of that no data element can be in a higher level than its enclosure in the index tree. This property enforces some restriction on node splitting and makes B-tree a special case of DMP-tree. When internal branching is big and the data set are large and random, which is true for IP prefixes in Internet [9], DMP-Tree approaches B-tree in terms of search and memory utilization. For an example of DMP-Tree readers can refer to [16].

4 Implementation

The most important factors in designing any lookup scheme are high speed and low storage requirements. To obtain the first goal, the search scheme is implemented as pipeline. By pipelining, we can have lookup in a time proportional to a (random-access) memory time in the worse case. The second goal is obtained by storing prefixes with minimum extra storage which is nearly 32N bits where N is the number of prefixes.

4.1 Overall Design

Figure 1 illustrates architecture of the system. It uses internal and external memory modules to store prefixes. Internal nodes are stored on on-chip memory and leaves in an external memory (RAM). Search and update modules do the search and updating, respectively. Insertion and delete modules are sub-modules of the update module.

This architecture considers having only one pipe for search. Then, the search time corresponds to the height of the tree. Considering 6 levels in the index tree, finding the next hop of a packet will take 6 memory accesses. Of course, this is slow. To expedite the search, we have considered 1K cash in the left bottom, Port2Addr, module. This module acts like a CAM, Content Addressable Memory, taking the destination addresses as keys and putting out the corresponding output port. Cache works in parallel with the rest of logic in order to not slow down the search. Considering even 50 percent hit ratio in the cache, the average lookup time is reduced to 3 memory accesses. Certainly, this is a good improvement. In the following, we propose a scheme to reduce the lookup time to one memory access by copying the search pipe and performing everything in parallel. In this case, we do not need the cache.

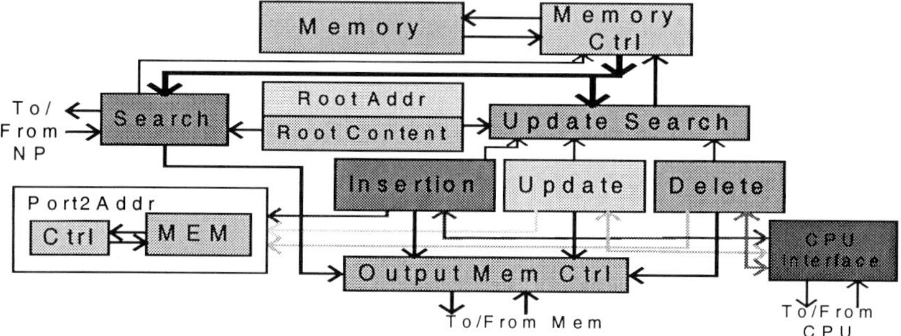

Fig. 1. A general view of the DMPTchip chip design.

4.2 Node Structures

To determine the best number for branching factor, we did some simulation. The results are shown in table 1. Intuitively, the bigger the branching factor the smaller tree height and consequently, the faster the search process. Another important parameter is the memory requirement to build the tree. Since we want to support more than 100K prefixes, then, we have to use off chip memory. According to table 1, the memory bandwidth and number of pins increase with increasing the branching factor. Based on table 1, it seems a branching factor of 9 should be optimum. We choose branching factor of 8 to save a little memory and make branching factor a power of 2.

All internal nodes are located in the on chip memory and leaves in the external memory. This is due to the fact that in large branching factors the majority of prefixes are kept in leaves. For example, more than 70 percent of nodes are located in leaves with 8 branching factor. Internal nodes are more frequently accessed than leaves, then. We also need more memory bandwidth for internal nodes since they need to keep pointer to the lower level subtrees. Then, the natural choice is to put them inside the chip. Obviously, to have a better performance, we need to read a node at once.

Prefixes are stored as 33-bit. For prefix matching, not exact matching, we need lengths of prefixes. We can keep lengths with just one extra bit. We put a 1 in the end of each prefix and make the rest bits zero. As an example, 10010110110* with length 11 is stored as 100101101101000..0.

Table 1. Simulation Result

#of Prefixes	Required Memory	Branching Factor	Memory Pins	Mem. Size (on chip) mm2	Max heights
64K	5.5Mbits	11	655	82	5
64K	5.4Mbits	9	527	81	5
100K	8.5Mbits	11	655	125	5
100K	8.3Mbits	9	527	122	6
100K	9Mbits	6	335	135	8

Addr1 []	Prefix1 [33]	Next-Hop [6]	Addr2 []	Prefix2 [33]	Next-Hop [6]	...	Addr []

Prefix1 [33]	Next-Hop [6]	Prefix2 [33]	Next-Hop [6]	Prefix3 [33]	...	Prefix [33]	Next-Hop [6]

Fig. 2. a. An internal node, b. A leaf node of DMP-tree

We use different number of bits for addressing the next level in each level. For example, in level 2, with branching factor of 8, we need 3 bits. We also need 6 bits for next hop assuming switch in the backbone has at most 64 ports. Adding up to these 33 bits for each prefix, we will have 297 bit for the root. Assuming 128K prefixes and branching factor of 8, we will have 128K/8 = 16K nodes in the tree. However, around 35% of nodes are empty [16]. This implies that we need around 5k extra nodes. Putting internal levels inside the chip and leaves, level 6, in the external memory, total memory required for the internal nodes is 1609728bits. We have chosen device XC2V4000 Virtex-II of Xilinx family, which has 2,160Kbits on-chip RAM and 912 pins which obviously meets our requirements.

4.3 Search Module

The block diagram of each stage of the pipeline is shown in figure 3. There are three main modules: Length Extraction, Find-Longest-Match and Find-Next-Level-Address. Function of each block is briefly explained in the following.

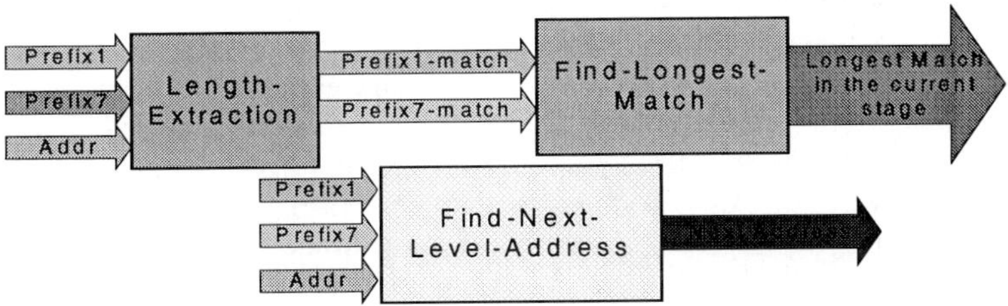

Fig. 3. Block diagram of each stage of the pipeline

Length-Extraction: This module extracts the lengths and finds prefix matching the destination address. Extracting lengths is implemented as a multiplexer with case statement (32 case statements) and can be done in one clock cycle. To find the matching prefix some XNOR gates are used. The output of this module is a series of bits for each prefix (See figure 4).

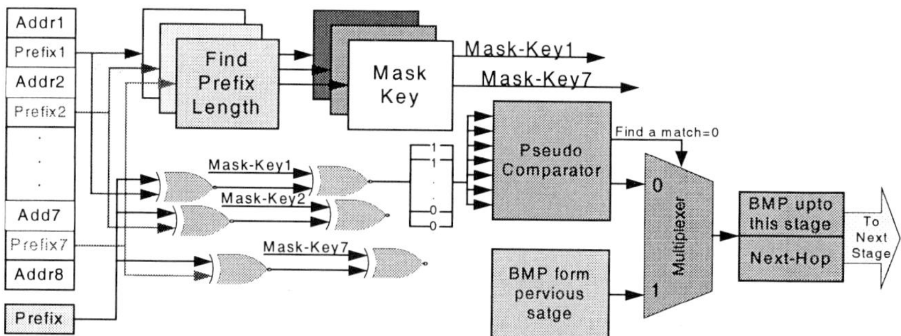

Fig. 4. The detail of the each stage of the pipeline

Find-Longest-Match: uses the result of the pervious block to find the longest matching prefix. It also takes the lengths of prefixes as inputs.
Find-Next-Level-Address: Finds the address of the next level of the tree.

Figure 4 and 5 show these blocks in details. Based on the lengths of prefixes, some masks are constructed. For instance, if "101000110100000…0" is a prefix (the formal form of prefix 101000110 with length 9), its corresponding mask key will be "11111111000000…0". These masks are helpful in checking if a prefix is matching a destination address. To find the longest match, XNOR gates are used. At the output of XNOR gates, there are eight strings of 1s and 0s. If a match is found, the one found in the pervious stage is substituted with this one. Few registers store the longest matching prefix from the previous stage in each stage of the pipeline.

Figure 5 shows the Find-Next-Level-Address module. This module compares the packet destination address with prefixes in the current node. According to the result of comparison the address of next level stored in the current node is selected.

Blocks in figures 4 and 5 are repeated 6 times in order to fully pipeline the whole search process and perform a lookup in each clock cycle.

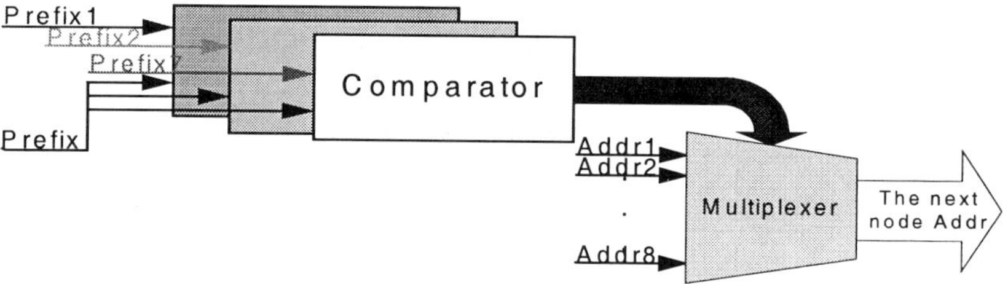

Fig. 5. Find the Address of the Next Node in the next level

4.4 Update Module

Due to the specific way of building tree and definition of enclosure, [17], updating is not as easy as the search procedure. Currently, we consider different options to implement the update module.

1. Copy whole memory space and rewrite tree in the second memory while doing search on the first one. This has few drawbacks. First, doubling on chip memory is expensive and we may not be able to implement the scheme in VirtexII without considerable modification. Second, rewriting the whole tree takes long time, whereas during this time we have to use pervious tables. This may lead to many wrong routing, which does not sound good.
2. Find better way to build the tree. This is the option we are deeply looking on it. We have found an easy way to build the tree in hardware. However, the method has long update time which we try hard to fix it.
3. Use a special core CPU to build and maintain tree. We try to modify some existing core to accommodate it for this application.

Ultimately, we can implement the existing algorithms. We consider options 2 and 3 more seriously.

5 IP Version 6

To the best of our knowledge, DMP-Tree is the only lookup scheme which scales well to the next generation of IP addressing, IPv6. The same data structure and hardware/software implementation can be applied to the new addresses and the complexity of the system grows less than linear. According to our calculations, the amount of the memory required to the last two levels of the tree is two large to accommodate inside the chip. Therefore, we have to put them off chip. In this case, the speed of lookup will be decreased by factor of two compared to IPv4. Since the number of bits is so many in each node and this requires a large memory bandwidth or pins, we have to put each node in two memory locations. In other words, to access to any node of the tree in the external memory, we need two memory access times. Considering the memory accessed time at rate 100MHz, and two levels in the external memory, we can do a lookup in four access times to the external memory. According to the amount of memory, speed and pins (about 1000pins for communicating with external environment and memory) DMP-Tree can be easily implemented on device XC2V6000 of VirtexII form Xilinx.

6 Conclusion

The IP lookup problem is still considered a bottleneck in very high speed links. We are currently implementing DMP-tree on FPGA to perform lookup for lines over 100Gpbs rate. The implementation uses pipelining to achieve a lookup in a time proportional to a memory access. It exploits a combination of on-chip and off-chip memory to build the index tree. In this way, it sharply reduces the memory bandwidth requirements for the tree. All internal nodes of the tree are stored on on-chip memory and leaves in the external memory. This well takes advantage from the fact that in the leaves we do not need to keep any address to the next level subtree. Consequently, the size of each node and required bandwidth are reduced. To support the next generation of IP, IPv6, we only need to change the data prefix length from 32 to 128 in our implementation. Furthermore, since the memory requirement grows linearly with the address length, we have to move the last two levels of the tree to the external memory. Our implementation can perform a lookup for IPv6 in a time proportional to four memory access time.

References

[1] S. Nil Nilsson and G. Karlsson, "Implementing a Dynamic Compressed Tree.", Proceedings of WAE'98, Saarbrücken, Germany, Aug. 1998.
[2] G. H. Gonnet and R.A. Baeza-Yates, "Handbook of Algorithms and Data Structures", Addison Wesley, 2th Edition, 1991.
[3] Sklower K., "A Tree-Based Routing Table for Berkeley Unix", Proceeding of the Winter Usenix Conference, 1991.
[4] S. Nilsson and G. Karlsson. "IP-address lookup using LC-tries" IEEE Journal of Selected Areas in Communications, vol. 17, no. 6, pages 1083–92, June 1999.
[5] V. Srinivasan and George Varghese, "Fast Address Lookups using Controlled Prefix", Proceedings of ACM Sigmetrics, Sep. 1998.
[6] S. Nilsson and G. Karlsson, "Fast Address Lookup for Internet Routers", Proceeding of IEEE Broadband Communication 98, Apr. 1998.
[7] Mikael Degermark, Andrej Brondnik, Svante Carlsson and Stephan Pink, "Small Forwarding Tables for Fast Routing Lookups", Proceeding of SIGDOMM 1997.
[8] W. Doeringer, G. Karjoth and M. Nassehi, "Routing on Longest-Matching Prefixes", IEEE/ACM Trans. Networking, vol. 4, no. 1, pp. 86–97, Feb. 1996.
[9] B. Lampson, V. Srinivasan and G. Varhese, "IP Lookups Using Multiway and Multicolumn Search", 1998.
[10] Torrent Networking Technologies Corporation, "High-Speed Routing Table Search Algorithms", Atechnical paper, http:// www.torrentnet.com.
[11] P.Newman, G, Minshall and L. Huston, "IP Switching and Gigabit Routers", IEEE Communications Magazine, Jan. 1997.
[12] A.J. McAuley, P. Tsuchya and D. Wilson "Fast multilevel hierarchical routing table using content-addressable memory," U.S. Patet serial number 034444, 1995.
[13] A.J. McAuley and P. Francis. "Fast routing table lookup using CAMs," *Proceedings of IEEE Infocom*, vol. 3, pages 1382–91, April 1993.
[14] P. Gupta, S. Lin and N. McKeown. "Routing lookups in hardware at memory access speeds," *Proceedings of IEEE Infocom*, vol. 3, pages 1240–7, April 1998.
[15] P. Gupta "Algorithms for Routing lookups and packet classification", Stanford University, Dec. 2000.
[16] Nasser Yazdani, "DMP-Tree: A Dynamic M_way Prefix Tree Data structure for Strings Matching," submitted to Iranian Journal of Science and Technology, July 2002.
[17] Nasser Yazdani and Paul Min, "A Fast and Scalable schemes for the IP address Lookup Problem", Proceeding of IEEE Conference on High Performance Switching and Routing, Heidelberg Germany, June 2000.
[18] Pankaj Gupta, "Algorithms for Routing Lookups and Packet Classification", A PhD thesis, Stanford Univ., Dec. 2000.

A Study of Marking Aggregated TCP and UDP Flows Using Generalized Marking Scheme

Seung-Joon Seok, Sung-Min Hong, and Chul-Hee Kang

Department of Electronics Engineering, Korea University,
1, 5-ga, Anam-dong, Sungbuk-gu, 136-701, Seoul, Korea
{ssj, mickey, chkang}@widecomm.korea.ac.kr
http://widecomm.korea.ac.kr/~ssj

Abstract. End-to-end Assured Service may be provided by the combination of assured service models over multiple domains through service level agreements (SLA) between two neighbor domains. Many studies revealed that the current assured service model does not meet the target rate of large-profile TCP flows in the presence of numerous small-profile TCP flows and that it does not equably distribute the profile rate of an SLA with multiple flows that are included in the SLA. We proposed a marking rule, called G-Marking (Generalized Marking) in order to diminish these problems simultaneously. The G-Marking consists of an Ingress marking scheme (I-Marker) and an Egress marking scheme (E-Marker) that are performed in ingress and egress routers respectively. In addition, there is another unfairness among UDP flows and TCP flows. In Internet, USP flows and TCP flows are mixed but TCP flows are seriously affected by USP flows. Thus, the G-Marking should effectively support this Internet traffic distribution and so this point is focused in this paper. Two experiments with the ns-2 simulator were performed: with E-Marker and I-Marker respectively. Simulation results show that compared to other schemes, G-Marking can more greatly correct the current problems.

1 Introduction

Assured Service is an example of end-to-end service that can be built from the proposed differentiated services enhancements to IP using an AF PHB (Assured Forwarding PHB), as first defined in [1]. With Assured Service, packets are forwarded with high probability on the condition that the traffic from the customer site does not exceed the subscribed rate. It is therefore necessary that a user or user group of an assured service establishes a contract with the provider, which defines a profile for the service expected to be obtained; SLA (Service Level Agreement). This contract can be made only for a TCP (Transmission Control Protocol) flow or a UDP (User Datagram Protocol) flow, for an aggregation of TCP flows, or for a combination of multi-protocol flows and includes a Traffic Profile element indicating reservation rate.

Previous work reports that there are two types of unfairness in Assured Service. Several studies [2][3][4] have found that TCP flows with large profile rate and/or long round trip time (RTT) barely meet their target rate in the presence of numerous flows

M.H. Shafazand and A M. Tjoa (Eds.): EurAsia-ICT 2002, LNCS 2510, pp. 519–526, 2002.

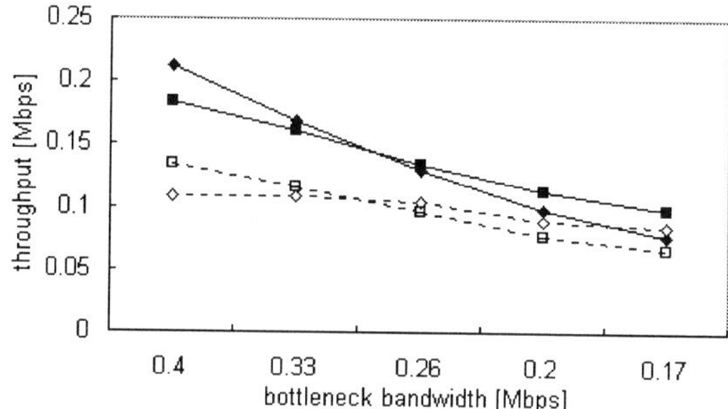

Fig. 1. Effectiveness for non-responsive flows.
* : TCP (s1) + UDP (s2)
□ : TCP (s1) + TCP (s2)
dotted line : average throughput of flows from s2
solied line : average throughput of flows from s1

with small profile rate when these two flows are competed in a common bottleneck link. The reason of this problem is that flows with large profile rate to have longer time to reach to its target rate after a packet loss or not to reach to. For convenience, this problem is herein referred to "*inter-SLA unfairness*". The unfairness is defined as excess bandwidth of a bottleneck link not being distributed equably among flows that go through the link. A SLA may cover a set of flows. In this case, however, there exists unfair sharing reserved bandwidth or profile rate within aggregated flows [5][6][7]. This unfairness between the aggregated flows is a serious problem in Assured Service model. The unfairness can be caused by differences in RTTs, in link capacities, or in congestion levels experienced by flows within the network, though the total throughput of the aggregation still reaches the target rate. This type of unfairness is herein referred to as "*intra-SLA unfairness*".

When responsive flows with TCP and non-responsive flows with UDP competes the bottleneck link bandwidth each other, responsive flows are seriously affected by the non-responsive flows. The non-responsive flow's sender does not reduce the sending rate in response to the network congestion. As a result, the non-responsive flow worsens the congestion state of the bottleneck queue. Thus, the OUT packet drop probability of the other responsive (i.e., TCP) flows will increase, and consequently, the throughput of the TCP flows will decrease. The impact of the non-responsive flow has been demonstrated by other studies. Figure 1 shows the experimental result demonstrating this impact. This simulation considered two flow sources of S1 and S2. S1 is a long-lived FTP source with 0.16 Mbps profile rate and S2 becomes a CBR (Constant Bit Rate) source using UDP protocol or a long-lived FTP sources with 0.1 Mbps profile rates. The CBR sources have 0.11Mbps generating rate and 0.1 Mbps profile rate. Figure 1 shows that an unfairness problem happens between TCP flows and UDP flows and hardly happen between two TCP flows with similar reservation rates (0.16Mbps and 0.1Mbps). In this result, TCP performance is reduced as congestion grows worse while the throughput of UDP flows does not seriously affected by link congestion like TCP flows. Thus, this unfairness becomes a serious problem in the case when network is over-subscribed. This denotes new inter-SLA unfairness between two SLAs with UDP flow and TCP flow respectively. If a SLA include some TCP flows and some UDP flows and if UDP flows are not congested and UDP sources generate much traffic, the UDP flows is allocated much higher *IN* rate by edge marker than TCP flows. Therefore, the intra-SLA unfairness problem more severe for the mixture of UDP and TCP flows. If UDP flows are, however, congested or UDP sources generate few traffic, this unfairness are dominated by TCP flows.

In this paper, we first describe a marking architecture, called G-Marking [8], to simultaneously correct the intra-SLA unfairness and inter-SLA unfairness. G-Marking strategy is based structurally on the REDP [9]. G-Marking also has two operation phases: demotion and promotion. The common idea behind G-Marking is that the marking results of previous edge routers are used as information representing the state of flow and arriving packets are remarked according to this information. Next, G-Marking strategy is evaluated through several simulations. In particular, the robustness of G-Marking scheme against the non-responsive flow's effect is tested. Finally this paper is concluded.

2 Generalized Marking Scheme for Assured Service

G-Marking strategy [8] consists of an “Ingress marking scheme” (I-Marker) and an “Egress marking scheme” (E-maker) that are located in ingress routers and egress routers respectively. The upstream E-Marker and the downstream I-Marker are associated with each other to mark packets effectively according to the SLA between the two domains. The role of the upstream E-Marker is to distribute a reservation rate fairly among flows that are included in the SLA to reduce intra-SLA unfairness, while the downstream I-Marker guarantees the target rate of the SLA to reduce inter-SLA unfairness.

E-Marker should determine the drop precedence (colors) of packets that arrived from several different ingress routers within its domain so as to distribute traffic profile rate fairly to the flows aggregated in the SLA. However, it is difficult because all flows experience different path conditions such as different congestion level and round trip delay, and in particular because the flows have different source reservation rates. In these conditions, equal throughput or the strictly equal sharing profile rate, that are pursued by other studies, does not represent the fairness among aggregated flows. This paper considers, thus, “fairness” for aggregated flows as *how equably all flows have the excess throughput*. E-Marker operates in color-aware mode and has two operation phases: promotion phase and demotion phase. The promotion phase may occur when aggregated *IN* profile rates is less than the traffic profile rate of the SLA with downstream domain. Otherwise, the E-Marker may be in the demotion phase. In promotion phase some of *OUT* (*RED or YELLOW*) packets are promoted to the *IN* (*GREEN*) packets, and in demotion phase some of *IN* packets are remarked as *OUT* (*RED*). The E-Marker's basic methodology is increasing *IN* marking rate in reverse-proportion to the *OUT* profile marking rate of each aggregated flow in the promotion phase, but decreasing *IN* marking rates in proportion to the *OUT* profile marking rate of each aggregated flow in the demotion phase. Therefore, this E-Marker mechanism prevents TCP flows from UDP flows because UDP traffic is generally CBR traffic. E-Marker is implemented using a token bucket and several leaky buckets. The token bucket is configured according to the traffic profile rate of the SLA and enforces *IN* marking rate of the aggregated flows to the profile rate. The presence of the token causes the promotion behavior of E-Marker. Conversely, the absence of the token represents the demotion phase. The leaky buckets are used to provide fair marking among the flows, and thus a leaky bucket is set up corresponding to each flow. E-Marker generates a trace packet for each *OUT* (*YELLOW or RED*) packet and inputs it into corresponding leaky bucket. Therefore, the occupancy of a leaky bucket can represent the state of the related flow. All leaky buckets have same output rates and bucket sizes. The leaky bucket rate is "*average out rate*" (aggregated out packet rate / the number of aggregated flows) and the leaky bucket size is as large as the burst packets absorbed.

I-Marker aims to guarantee the target rate of SLA. As mentioned in section 1, the throughput of flow with large reservation rate hardly achieve its target rate because the small-profile flows greedy the excess bandwidth of bottleneck link. I-Marker modifies the marking result of upstream E-Marker using bgTCM (bandwidth guarantee three color marking) scheme in order to guarantee the flows' target rates. bgTCM is a three-color marking scheme. bgTCM scheme marks arrival packets as *GREEN* by the profile rate, *YELLOW* between profile rate and 4/3 profile rate or *RED* over 4/3 profile rate which is called as extended profile rate. In addition, I-Marker has the added characteristic of avoiding non-responsive flows impact on the responsive flows. I-Marker first identified the flow type of arriving packets and does not apply the three-color marking rule to non-responsive flows, thus reducing the impact of non-responsive flows. However, how to detect the non-responsive flows is a current issue. I-Marker is implemented using two token buckets. The first, called the profile bucket, is set up with the traffic profile of the SLA. The other, called the bg-bucket, is set up with extended profile rate. The extended profile rate has 4/3 profile rate and 4/3 profile bucket size.

3 Performance Study

In this section, we analyze the effectiveness of G-Marking scheme using ns-2 simulator in several network environments. In particular, we focus on some cases when non-responsive flows coexist with responsive flows. Because G-Marking is composed of the I-Marker and the E-Marker, each playing a different role in the marking scheme, we qualify the effectiveness of these two components through two experiments to other marking schemes: Proportional marking and Per Flow marking. In Proportional marking, *IN* marking rate of a flow is allocated in proportion to its sending rate. In Per Flow marking, the profile rate is distributed to all flows equally.

3.1 Test of Intra-SLA Unfairness

E-Marker is designed to correct the intra-SLA fairness. We first test the robustness of the E-Marker against the non-responsive effect. Figure 2 depicts the experimental

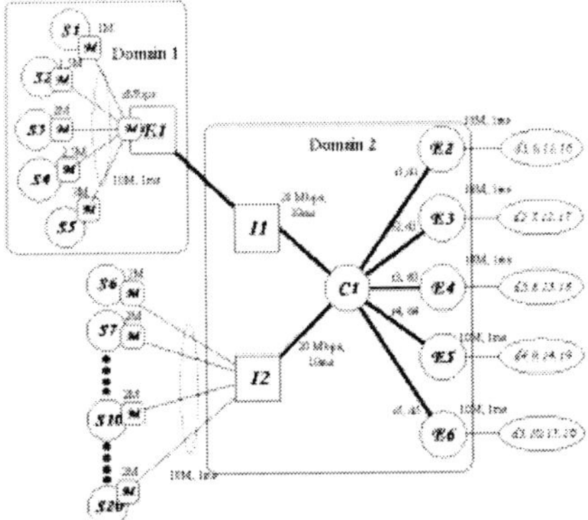

Fig. 2. Experimental topology used to study the E-Marker

topology used to test the performance of the E-Marker. In this topology, all sources are assumed to be long-lived FTP sources using TCP Reno protocol. Flows originated from sources s1, s2, s3, s4 and s5 pass through domain1 and domain2 sequentially. Thus, these flows should be remarked by the SLA between domain1 and domain2 at the E1. The other flows are used for background traffic. All sources have a source-integrated marker assigned its profile rate. Every router uses a RIO as multi-level dropping rule. The RIO parameter is [10/50/0.5, 60/80/0.02]. Flow1, 2, 3, 4 and 5 are routed through different routes in domain2 with different congestion levels and different RTT delays. Source s3 is CBR (Constant Bit Rate) source using UDP protocol and the other sources are long-lived FTP sources using TCP protocol. The CBR source has a 3.5Mbps generating rate and a 2Mbps source profile rate. To limit the investigation to E-Marker in this experiment, the egress node of domain1 conducts three aggregated marking schemes and the ingress nodes of domain2 do not remark arrived packets. We use a fairness index (equation 1) in order to compare simulation results among these aggregated marking schemes.

$$\frac{Max(Th./\,Src.profile)}{Min(Th./\,Src.profile)} \tag{1}$$

where *Th.* and *Src.profile* denote a flow's throughput and source profile rate respectively. This index represents how fairly the excess bandwidth is distributed. The lesser the index value, thus, the greater the fairness among aggregated flows.

Two simulations were conducted: a promotion case simulation with 15Mbps SLA profile rate and a demotion case simulation with 7.5Mbps SLA profile rate. Simulation results are shown in Figure 3 in which legends denote the combinations of bottleneck link bandwidths – protocols and short lines are source marking rate. From these results, the throughput of the CBR source is almost always at its generating rate of 3.5Mbps, because bottleneck link load of 85% and 73% is not severally congested. To evaluate the effect of the non-responsive source, we calculated the Max/Min value for TCP flows in the three marking schemes: Per Flow marking, Proportional marking and E-Marker. In the promotion simulation, the Max/Min values of the Proportional marking, E-Marker and Per Flow marking are 6.31, 3.52 and 4.48 respectively. The E-Marker has the minimum value among three schemes. It is natural that the Per Flow marking does not affected by the non-responsive source. In demotion simulation, the

Max/Min values are 9.89, 3.90 and 4.71. In this demotion case, the E-Marker provides the flow5 with about 1.7Mbps but the Proportional marking provides with about 1Mbps, in spite of 3Mbps source profile rate. In result, as usual, the E-Marker is effective in the case where non-responsive flow is included in the aggregated flows.

3.2 Test of Inter-SLA Unfairness

In this experiment, we evaluate the effectiveness of I-Marker for the inter-SLA unfairness. Figure 4 depicts the experimental topology used for this experiment. In this topology, there are 10 aggregations, each aggregation composed of four flows. All

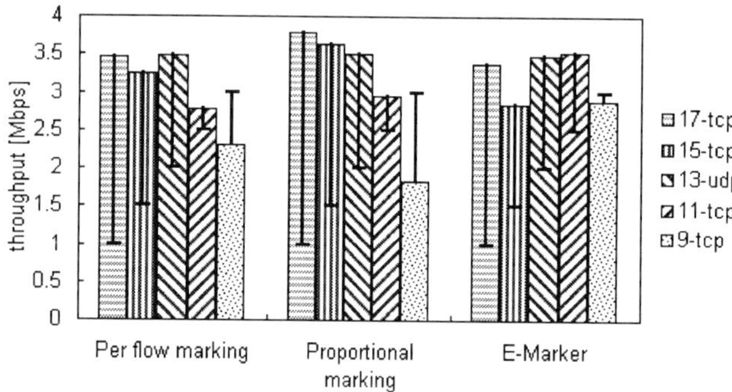

(a) Promotion Phase : 15Mbps SLA Profile Rate

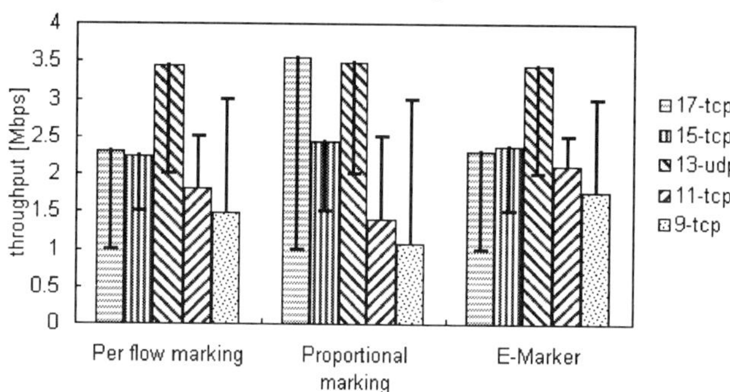

(b) Demotion Phase : 7.5Mbps SLA Profile Rate

Fig. 3. Impact of non-responsive flow3

aggregations, from a1 to a10, have its profile rate of 0.4Mbps, 0.8Mbos, 1.2Mbps, 1.6Mbps, 2Mbps, 2.8Mbps, 4Mbps, 6Mbps, 8Mbps and 12Mbps. Aggregations a1, a3, a5, a7, and a9 consist of four CBR UDP flows and Aggregations a2, a4, a6, a8, and a10 consist of four long lived TCP flows. All sources have a source integrated marker assigned a profile rate of aggregate profile rate / 4. I-Marker classifies flows into non-responsive flows and responsive flows and does not apply the bgTCM to the non-responsive flows in order to reduce the impact of the non-responsive flows. In stead, sr2CM(single rate two-color marking) is applied to the flows. Here, to limit investigation to I-Marker, the ingress nodes of domain conduct the only I-Marker scheme and all flows are assumed to arrive at the ingress node from different domains. In this topology, thus, I-Markers remark arriving packets from different upstream domains. In this experiment, the case using I-Marker is compared with using no ingress marking.

Figure 5 shows another experimental result for I-Marker. As the bottleneck link load is increased in normal case, responsive flows with high profile rate are seriously affected but non-responsive flows get throughput over its reservation rate. However, if I-Marker is used, the gradients of throughput lines of non-responsive flows are steeper than the Source-only marking and those of responsive flows are gentler than Source-only marking. From this experiment, it is observed that I-Marker can help to improve the impact of non-responsive flows.

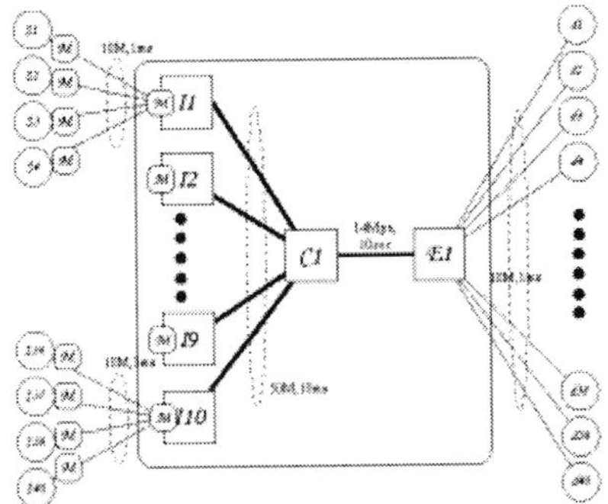

Fig. 4. Experimental topology used to study the I-Marker

4 Conclusion

In this paper, we first described an inter-domain marking scheme, named G-Marking, for the assured service. The G-Marking strategy consists of Ingress marking scheme (I-Marker) and Egress marking scheme (E-Marker) that are carried out in the ingress routers and egress routers respectively. These two markers have different roles for the SLA at the inter-domain interface. E-Marker is responsible for distributing the marking rate fairly to the aggregated flows to improve the intra-SLA unfairness. The role of I-Marker is to guarantee the target rate of the SLA under competing with other flows for the excess bandwidth of the bottleneck link. I-Marker carries out bgTCM and is implemented with two token buckets. In particular, this paper focused on the effectiveness of G-Marking on the unfairness between UDP flows and TCP flows caused by different behaviors for the packet losses each other. We have simulated the E-Marker and I-Marker schemes using the ns-2 simulator to test this effectiveness. Results were compared with other marking schemes. From the simulation results, we have been convinced that E-Marker scheme and I-Marker scheme can alleviate intra-

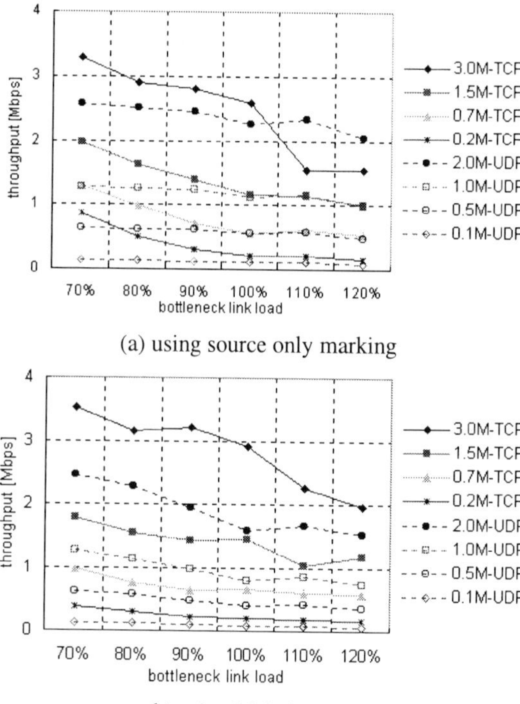

(a) using source only marking

(b) using I-Marker

Fig. 5. Effectiveness on non-responsive flows

SLA unfairness and inter-SLA unfairness respectively although UDP and TCP flows are mixed within a SLA or included in different SLAs.

References

1. Heinanen, J., Baker, F., Weiss, W., Wroclawski, J.: Assured Forwarding PHB Group. RFC2597, IETF(1999)
2. Yeom, I., Reddy, A. L. N.: Realizing Throughput Guarantees in a Differentiated Service Network. In Proc. ICMCS'99 (1999) pp 372–376
3. Rezende, J. F. : Assured Service Evaluation. In Proc. GLOBECOM'99(1999) pp 100–104
4. Ibanez, J., Nichols, K.: Preliminary Simulation Evaluation of an Assured Service. Internet Draft, IETF(1998)
5. Yeom, I., Reddy, A. L. N.: Impact of marking strategy on aggregated flows in a differentiated services network. In Proc. IWQoS' 1999(1999)
6. Kim, H.: A Fair Marker. Internet Draft, IETF(1999)
7. Andrikopoulos, L., Pavlou, G.: A fair traffic conditioner for the assured service in a differentiated service Internet. In Proc. IEEE ICC 2000(2000)
8. Seok, S. J., Lee, S. H.,, Kang, C. H.: G-Marking: Improved Marking Method for Multiple-Domain Differentiated Services. In Proc. INC'2002(2002) pp 271–278
9. Wang, F., Mohapatra, P., Mukherjee, S.: A Random Early Demotion and Promotion Marker for Assured Service. IEEE Journal of Selective Area in Communications, Vol. 18(12) (2000) pp 2640–2650

Towards the Global Information Society: The Enactment of a Regulatory Framework as a Factor of Transparency and Social Cohesion

Panagiotes S. Anastasiades

University of Crete, Department of Education,
GR-74100 Rethymno Crete, Greece
panas@ucy.ac.cy

Abstract. The new emerging Information Society will bring about significant changes in the way in which we perceive and experience both our working and social environment. Frameworks, everyday functions and practices, which, for decades, constituted the obvious reality around which the whole of our life revolved, are starting to look effete and unable to respond to the new electronic reality. The challenges of the new era, must become an object of attention for all citizens who will be called upon to enjoy its advantages, but also to confront the consequences which it will bring about to their life and the life of their children. The construction of an integral statutory framework, which will prepare the access to the new electronic reality for all citizens, under as less as possible disturbances, comprises the highest priority of the current era, since it is a factor of political clarity and cohesion of our society.

1 Introduction

Undoubtedly, we are in a transitional era, where the end of the 20th century is identified with the end of the industrial society as it has been formed during the passage of time [1].

At the beginning of the third millennium, the explosion of new technologies is leading to a totally different, and let me say, exciting era, which will be marked by the revolution of digital information and the numerous technological and communicational applications which will accompany it [2].

The new technological applications and the rapidity in which they are being introduced in our social and professional lives, tend to sweep along with them structures and functions, which currently form an inseparable part of the social and economic body of countries [3,4].

A characteristic feature of our era is the effort of the co-ordinated government's legislative bodies [5] to follow breathless the rapid changes which are taking place on the whole of our professional and social environment.

The continuous introduction of new applications and services within the bounds of the new digital era does not merely affect the existing balances, but also generates a totally different operational model in the frame of which the existing status quo is

M.H. Shafazand and A M. Tjoa (Eds.): EurAsia-ICT 2002, LNCS 2510, pp. 527–535, 2002.

disdained with an unbelievable speed, seeming most of the times unable to act according to the new conditions which are being created [6].
In the area of consumption and market, where the introduction and rapid expansion of e-commerce undoubtedly creates a totally new framework, the issues of trade competition, consumer protection, procedure of selling goods and services, structure and even operation of marketing businesses, are placed in a totally different framework from that which we have known until now.
In the area of employment, the introduction of Telework [7] creates totally different facts for both the businesses and employees. The concepts of workplace, working hours, leave, working rights, insurance, acquire a totally different substance, which needs to be imprinted in a new statutory framework [8].
The rapid expansion of Distance Education changes the map in the area of education [9], since structures, functions, and practices, which had been established for a large period of time, are suddenly seen as old-fashioned and unable to lift on their shoulders the heavy weight of the new era [10].
Serious issues, which developing countries had in the recent past, confronted efficiently, are now coming to the fore with more intensity [11].
Protection of our privacy is threatened from the possible uncontrolled use of technologies [12]. Man seems to be unable to harness the new digital typhoon, which is approaching, since the means he holds in his hands are probably not enough [13].
The spread of illegal and damaging content in all types of networks child - molesting, electronic fraud, and other relevant criminal activities, modernising their way of functioning, have been transferred from the streets and cities to the electronic highways of information, rendering their action as uncontrollable.
Telecommunications [14], networks [15], satellites [16], digital applications [17], do not constitute merely the puzzle of a new fictitious reality, which although visible it is far away from us. The Information Society is at the eve of our generation, it will constitute part of our reality, existence, and daily routine [18].
Therefore, it is noticed that the upcoming changes do not concern a selected and specialised number of people who elaborate on them [19], nor will their consequences be restricted presumably to some particular sectors. The upcoming changes concern all of us and they will break through the whole of social, working, and human activities and functions [20].

2 The Formation of a Unified Statutory Framework

In the course of human evolution, the statutory framework's degree of adaptation to the new facts of each era, gave the comparative advantage to both the organised government and its members, to exploit, on the one hand, with the most efficient way all the new opportunities which were created, and to absorb, on the other hand, with the most possible minimum cost the shocks from the acute changes and realignments [21].
In an opposite case, either functional and legislative idles are created having as a result the manifestation of uncontrolled situations, or the disdained legal regime and its structures act as a brake to technological and social progress [22].

Such an environment is in favour of powerful economic centres, since in the name of unimpeded evolution they can move freely within the limits of law creating the conditions and rules which will be later on legislated by the institutionally consolidated legislative bodies. A recent example is the establishment of "free television" in Greece [21].

The preparation of the new digital reality cannot be left to the implicit powers of the market and its laws [8], nor can it become an object of worm-shafted discussions and analyses between exclusive teams and clubs of speculation [21].

The need to elevate a unified regulatory and legal environment, with the enactment of measures and rules which are obliged to be in harmony with the modern social and economic needs, comprises the most urgent priority in the European Union [23,24,25,26,27,28].

The new regulatory framework is obliged to obey to five main priorities.

a. Regulations and equivalent rules ought to respect, more than anything else, the human being and its existence. Protection of privacy and respect to human rights are the highest priorities .

It is now well known that any type of personal information can be found under the possession of ordinary or legal faces who have the capability of breaking into security codes of various files with the aim of using private and confidential information for their own interest. Therefore, the issue of electronic and legal protection [29] is the most significant issue, which needs to be dealt with immediately. The huge interest, which the private sector has shown for solving this issue, has its explanation to the fact that, apart from the problems that it might bring to the private life of citizens, lack of electronic protection can also function as a suspending factor in vital sectors of technological applications, such as the electronic market.

b. In the frame of a modern and effective regulatory framework, assuring citizens' free access to public interest networks as well as the unimpeded use of the new services by all those interested comprise of the second main priority. The establishment of a universal service [30] serves the big necessity of social cohesion, which is, in our era of digital information, one of the most important issues breaking in with a most dangerous manner.

c. The creation of an open and transparent action framework [31], in which all those interested will have the capability to offer their services, is believed to be an essential element for the rapid and effective development of the peak technological applications. In the frame of a competitive environment, all the contributors ought to be properly equipped with the necessary tools, knowing with clarity the frame in which they have the potential to act.

d. The telecommunications sector [32] has already become the apple of discord, since it stands as the necessary means for the effusion of the marketing product of information. Telecommunications must be competitive without, however, losing their character of common benefit which governs them.

The competitive environment may be necessary for the release of private capitals and the encouragement of investment efforts, but it is also likely to create side-effects for which confrontation cannot rely on the laws of the market [33].

Removal of political obstacles which during periods of time were imposed and served the necessities of an era with different characteristics and priorities, is now

acknowledged as a necessary and desired progress. All facts are showing that the practice of adapted monopoly invoices is been abandoned and harmonised in the frames of appointing proper regulatory frameworks which will have as an aim the release of telecommunication infrastructures [34].
However, in order to realise something like this, we need to trace down and establish, on a European level, the minimum required preconditions for the security of a balanced service and information infrastructure development.

e. The issue of mass media ownership presents a unique interest, since centralisation of the media's power to isolated individuals or groups of self-interest hides many dangers.Moreover, this issue will play a determinant role in the free distribution of programmes and information, a fact which will contribute to the multilingual and multicultural character of vision and sound programmes, irrespective of the transmission method [35].
The issues which need to be confronted in the vision and sound sector are numerous and significant.The granting of licences, interconnection of networks, management of common finite resources, such as distribution of radio frequencies, and enumeration of subscribers, are at the stage of settlement [36].
It would be a huge mistake to confront all the above significant open issues fragmentarily. On the contrary, they must be approached through collective elaboration on a European level, with the enactment of a unified regulatory framework which will have universal validity, ensure the feeling of security to the average citizen, scientist, intellectual, artist, building all the necessary mechanisms of protection.
Simultaneously, it must be open to the challenges of our days, flexible in its structure, and ready to embody anything new while isolating it from the undesired side-effects which may accompany it.
The legal framework must function as a factor of enactment, certifying the above mentioned priorities, which are necessary to be enriched by practical, clear and understandable for the average citizen rules, for solving any conflicts and applying immediate remedial actions, in cases of abusing the actually unlimited possibilities which the new technology has to offer us.

3 The Need to Construct a Modern Regulatory Framework

It has been mentioned many times that the Information Society lies in the distant future, but it is starting slowly-slowly to become a part of the modern man's life [37]. However, many of the applications, which promise the change in our life, are still not visible to the average citizen.
It is true that one of the obstacles which are encountered in the effort to expand the use of new communication applications, is the high telecommunication cost which arises for those interested from the phone bills, international or long-distance.
The monopoly conditions, which prevail in many parts of the world, do not allow any adjustment of bills to lower levels.

Reduction of bills is an urgent priority which must be accompanied by the justified distribution between the exploitation bodies of duties having to do with rendering of common utility services [38].

Competitiveness on its own cannot cope with the cost demanded for the introduction of new applications, if simultaneously the extend of demand to consume the offered services is not sufficient.

Particular attention is given to the release of the telecommunications sector [39] since, in such a case, it is estimated that with the infiltration of private investments, all the necessary construction works will be accomplished much more rapidly, while on the other hand, in the frame of an intense competitiveness which will be developed, the much desired reduction of bills will be achieved.

The existence of low bills will attract more users having as a result the increase in demand for the specific services. Therefore, what is needed is the creation of a critical mass to which adjustment of bills contributes to, without however being enough.

Abolition of all monopolies, in theory and practice, does not concern only the public sector. It also concerns private initiative. In the mass media and vision and sound sectors, the creation of monopoly situations, which create, different kinds of distractions and dependencies have been noticed.

The above fact in combination with the conflicts between the producers of information and the producers of goods and services, who most of the times are the same individuals, create uncontrolled situations of adulteration in the parliamentary system itself.

Similar practices may be tolerable in the common sense of the "free market", but, of course, they are not consistent with declarations on clear rules of competitiveness for the interest of consumers.

From the above mentioned it is understandable that co-operation between the involved bodies is judged as imperative:

- In the private sector, encouragement of co-operation between the competitors will contribute to the creation of the needed size and emphasis will be given to specific sectors of the market. A nice example is the mobile phone services sector.
- In the public sector, public administrations must form a common framework of requirements and specifications, in relation to the extensive promotion and best use of existing and future networks and services.
- The need to ensure a balanced participation in the new satellite systems of personal communication marks the importance of effective access to the global market for the new worldly mobile technology.

The formation of an open and competitive environment is a necessary precondition for the speeding-up of procedures, which will contribute to the development of the Information Society. In a healthy competitive environment the public and private sector must operate under clear rules, distribution of roles and duties, open field of action, and enterprising initiatives.

This procedure can be left to the unrestrained powers of the market, since the one and only criterion, which satisfies them - profit - does not constitute the best guarantee for achieving the creation of an open, clear, and competitive environment.

Therefore, the creation of a regulatory framework is believed to be necessary, since it will constitute the embankment to the uncontrolled, anarchic, and beyond measure pursuits of those bodies who will attempt to become wealthy, ignoring any social and human cost.

In no case, however, should this mean the building of an inflexible, inefficient, and eventually hostile regulatory framework, which will hinder the citizens' access to the new era, placing continuous obstacles to technological evolution itself.

With the term continuous obstacles what is meant is bureaucracy, public corruption, ineffectiveness and lack of organisation in many public services, and not a real interest for issues having to do with human existence and dignity.

Essentially, an inflexible and hostile regulatory framework will be by-passed by reality itself and its needs, having as a result the creation of uncontrolled and dangerous situations.

Therefore, whoever desires to express in practice the logic of the market and social automatism, has to choose between two pathways: either to leave the market at the mercy of its own powers, or to build hostile bureaucratic regulatory frameworks. The result will be the same. The difference lies in the fact that in the second case, "people's justification" coexists.

The creation of a regulatory framework, which will prepare the grounds for the transition to the Information Society, is imperative. On a European level [40], a great effort has already begun with the aim of laying down measures and rules which will, on the one hand, facilitate the development of new applications, and, on the other hand, act protectively in relation to the citizens' rights, and in the frame of a free and competitive environment.

Every country-member is trying to lay down a framework of measures and rules towards that direction.

It is necessary to organise informative campaigns having as receivers the public administration, the mass media, educational institutions, and particularly the new generation with the aim of making more sensitive wider population groups in relation to the benefits but also the dangers, which the advent of the Information Society will bring about. A modern, functional, and under the above-mentioned preconditions regulatory framework, will be in the position to maximise the benefits and minimise the costs from the upcoming change.

The unified regulatory framework constitutes the smooth underbelly of the Information Society. Its significance is enormous and from the content and mainly the political will of National and European authorities for the strict observance of conditions and rules which will be enacted, the smooth course of our transition to the new electronic era of networks and progressive technological and communicational applications will be judged.

4 Conclusions

The new emerging society probably will not cause human and material catastrophes, but definitely the transition from the given model of social organization which we are experiencing today to the new technologically digital environment, will cause huge and significant turbulence in the social body of countries. The hard core of "revolution" is located in the redevelopment of the meaning and significance of the term "information".

Rapid development of new technologies has liberated information from geographical, time, and quantity limitations giving it a highly added value. Digitalisation of information, has allowed information to flow with incredible speed through electronic venues (networks) and towards every direction.

The huge overturning of our era is recorded exactly at this point. If, until today man had to open passages in order to reach to the source of marketing and economic information, in the frame of the new digital environment, search and regaining is possible through networks without the need of human locomotion. Now information travels to man, just as long as he is connected to the net and equipped with the appropriate tools to trace it down. Therefore, man's capability to connect the net comprises the necessary and apt condition for access to the new era. Gratification of the above condition demands the existence of appropriate telecommunication infrastructures, necessary knowledge for their exploitation and of course a reasonable purchasing cost of the necessary equipment and of the use of nets. It is characteristic that even if only one of the above preconditions is not in effect, then the much desired access to the new era will be particularly difficult.

References

1. Anastasiades, S.P.: In the Information Age. A.A.Livanis editions, Athens Greece (2000)
2. Dutton, William (ed) Information and Communication Technologies, Oxford University Press, 1996
3. Pestel, R; Johnston, P. (2000) "Technology driving change: scenarios for Europe in a global information society", Foresight, Volume 2 No. 2.
4. Murahashi, K. Duties of persons who live in information-oriented society Journal of Information Processing and Management, 2001, vol. 44, no. 8, pp. 581-582
5. Marsden, C. T. Cyberlaw and International Political Economy: Towards Regulation of the Global Information Society Detroit College of Law At Michigan State University Law Review, 2001, no. 2, pp. 355–422
6. Dertouzos M, Gates B. What Will Be: How the New World of Information Will Change Our Lives. Harper San Francisc, 1997
7. Bertin I., Denbigh A.,The Teleworking Handbook. New Ways to Work in the Information Society, TCA, Warwickshire, 1996.
8. Anastasiades P.: «A Unified Regulatory Framework On A European Information Society: Suggested Building Levels». Proceedings of the International Federation for Information Processing – IFIP/SEC 2002 Int.Conference: Security In the Information Society, May 6-8, 2002. Cairo, Egypt.
9. Holmberg, B. (1986). Growth and Structure of Distance Education, Croom Helm, London
10. P.S. Anastasiades and S.Retalis. «The Educational Process in the Emerging Information Society: Conditions for the Reversal of the Linear Model of Education and the Development of an Open Type Hybrid Learning Environment». Republished paper to Computers in the Social Studies Journal, Volume10 Number1, January 2002. ISSN 1090-8595
11. Thomas Ruddy. (202) "The World Summit on the Information Society", Solothurn Univ. Switzerland.

12. Branscomb, A.W. Who Owns Information? From Privacy to Public Access. New York: Basic Books, 1994
13. Anastasiades P.: «Netizen: The concept of isolation and loneliness in the emerging Internet society». Proceedings of the Computing and Philosophy @ Oregon State University Conference, January 24, 25, 26, 2002. Oregon, USA
14. Dutton, William (ed) *Information and Communication Technologies*, Oxford University Press, 1996
15. Tapscott Don.(1999). Growing Up Digital: The Rise of the Net Generation. McGraw-Hill Professional Publishing
16. Bradley, S.P., Hausman, J.A. & Nolan, R.L. (1993). "Global Competition and Technology," in *Globalization, Technology and Competition: The Fusion of Computers and Telecommunication in the 1990s*, Harvard Business School Press, Boston, MA
17. Blais Pamela (1996). How the Information Revolution Is Shaping Our Communities. Planning Commissioners Journal, 14(16).
18. Bangemann,M. The Europe and the global information society Recommendations to the European Union, 1994.
19. Guidi, L., Towards an Information Society for All. Managing Information, 2001, no. MAY, pp. 46–51
20. Allen, J. C (2001) Ductael, Webster, and Herrmann: The Information Society in Europe: Work and Life in the Age of Globalization. Rural Sociology, vol. 66, no. 4, pp. 636–637
21. Anastasiadis, P. The European Information Society – The Need to Establish an Integrated Regulatory Framework, Nea Synora, A.A. Livani, Athens, 1999.
22. James Slevin (2000) The {Internet} and society. Blackwell, Malden, MA, USA
23. Official Journal- European Communities Information and Notices C, 2001. 2001/C 311/04 Opinion of the Economic and Social Committee on the `Communication from the Commission to the Council, the European Parliament, the Economic and Social Committee and the Committee of the Regions – Creating a Safer Information Society by Improving the Security of Information Infrastructures and Combating Computer-related Crime' , vol. 44, no. 311, pp. 12–18
24. Official Journal- European Communities Information and Notices C, 2001. A5-0043/2001 European Parliament legislative resolution on the Council common position for adopting a European Parliament and Council directive on the harmonisation of certain aspects of copyright and related rights in the Information Society (9512/1/2000 – C5-0520/2000 – 1997/0359(COD)), vol. 44, no. 276, pp. 121–123
25. Official Journal – European Communities Information and Notices C, 2001. Copyright in the Information Society ***II (vote), vol. 44, no. 276, pp. 51
26. Official Journal – European Communities Information and Notices C, 2001. 2001/C 207/03 Statistics relating to technical regulations notified in 2000 under the Directive 98/34/EC procedure - Information supplied by the Commission in accordance with Article 11 of Directive 98/34/EC of the European Parliament and of the Council of 22 June 1998 laying down a procedure for the provision of information in the field of technical standards and regulations and regulations applying to information society services, vol. 44, no. 207, pp. 5–7
27. Official Journal- European Communities Legislation l, 2001, Directive 2001/29/EC of the European Parliament and of the Council of 22 May 2001 on the harmonisation of certain aspects of copyright and related rights in the information society, vol. 44, no. 167, pp. 10–19
28. Official Journal- European Communities Information and Notices C. 2001. A5-0106/2000 European Parliament legislative resolution on the Council common position for adopting a European Parliament and Council directive on certain legal aspects of Information Society services, in particular electronic commerce, in the Internal Market (14263/1/1999 – C5-0099/2000 - 1998/0325(COD)) vol. 44, no. 41, pp. 38

29. European commission,' status report on European union telecommunication policy, dg xiii, Brussels 26-1-1998 (pp 47).
30. European commission,' status report on European union telecommunication policy, dg xiii, Brussels 26-1-1998 (pp 15–17)
31. Communication to the European Parliament of the European Union and the economic and social committee, ' Regulatory transparency in the internal market for information society services', 24-7-1996,(pp 9–15).
32. European commission,' status report on European union telecommunication policy, dg xiii, Brussels 26-1-1998 (pp 7)
33. European commission, growth, competitiveness, employment the challenges and ways forward into the 21st century white paper, com (93) 700 final.
34. European commission, dg xiiia/1' 1998 interconnection tariffs in member states', Brussels 11-3-1998.
35. http://europa.eu.int/pol/av/en/info.htm, the audiovisual policy.
36. European commission,' status report on European union telecommunication policy, dg xiii, Brussels 26-1-1998 (pp 42–46)
37. European Commission, Towards the information society Communication from the Commission to the Council, the European Parliament, the Economic and Social Committee and the Committee of the Regions on a Methodology for the implementation of information society applications Proposal for a European Parliament and Council Decision on a series of guidelines for Trans- European telecommunications networks (presented by the Commission), COM (95) 224 final.
38. European comission,' status report on European union telecommunication policy, dg xiii, Brussels 26-1-1998 (pp 13–15).
39. European Commission, Green Paper on the convergence of the telecommunications, media and information technology sectors, and the implications for Regulation Towards an information society approach (presented by the Commission), COM (97) 623 final.
40. European Commission, Green Paper on the convergence of the telecommunications, media and information technology sectors, and the implications for Regulation Towards an information society approach (presented by the Commission), COM (97) 623 final.

On the Application of the Semantic Web Concepts to Adaptive E-learning

Juan M. Santos, Luis Anido, Martín Llamas, and Judith S. Rodríguez

Departamento de Enxeñería Telemática, Universidade de Vigo
E.T.S.E. Telecomunicacións, Vigo, Spain
{jsgago, lanido, martin, jestevez}@det.uvigo.es

Abstract. Adaptive learning means offering customized educational environments for each individual. For this to be possible in virtual environments, like the Internet, actual standardized data models for e-learning must be enriched with explicit "machine-processable" semantic information and appropriate inference rules. This would allow Learning Management Systems (LMS) to automatically process and integrate these models meaningfully in order to offer alternative behaviours depending on each learner and his particular needs. This paper analyses the impact of the upcoming Semantic Web, an extension of the current Web in which information is given an explicit meaning, on the consecution of real high-quality adaptive e-learning environments.

1 Introduction

In the last years more and more institutions have adopted the Information and Communication technologies, particularly the Internet, for delivering education and training. During these few years, Internet-based e-learning systems have progressively evolved from the very basic repositories of simple learning documents accessible via Web to the advanced learning environments that include sophisticated tools like pedagogical simulators or complex communication facilities, enabling different educational approaches and methods, like "learning by doing" [1] or "collaborative learning". Adaptive learning seems to be the next step in this evolving process.

Future e-learning systems should provide each apprentice with learning experiences that are unique and tailored to his needs, interests, preferences and learning style in order to maximize the effectiveness of learning. For this personalized learning to be possible a set of complex data structures are needed to express alternative behaviours for the Learning Management System (LMS) depending on each learner and his particular needs. Currently, an active learning technology standardization process is dealing with the lack of interoperability among heterogeneous systems. Nevertheless, expressing complex behaviour (e.g. data needed to offer adaptive learning) is incompatible with standards established by consensus.

The organization of the paper is as follows. Section 2 gives an overview of the adaptive learning requirements and briefly describes some systems that deal with

M.H. Shafazand and A M. Tjoa (Eds.): EurAsia-ICT 2002, LNCS 2510, pp. 536–543, 2002.

these requirements. Section 3 introduces the learning standardization process, illustrating some meaningful proposed models. Some intermediate conclusions are explained in section 4. Section 5 deals with the concepts involved in the Semantic Web. In section 6 it is showed how those concepts can be applied in Learning Managing Systems to provide better personalization of the learning experiences of students. Section 7 concludes and summarizes the paper.

2 Adaptive Learning

The personalised attention from teachers to their students is, too often an impossible to reach goal in the classic educational environments, where a teacher has to convey his/her classes to groups, habitually very populated, of students. However, in the modern Internet-based electronic learning environments that homogeneity among students is not so feasible. The access to the flexible "virtual classrooms" can be done from anywhere in any time and by heterogeneous students. Due to this heterogeneity, traditional teaching methods, which have been systematically reflected in the e-learning environments, miss their efficiency and they have difficulties to adapt to these new settings.

Thus, the future of e-learning is in the adaptive e-learning systems. These systems are characterized by offering educational experiences dynamically customized to the real and particular needs of each student at any instant and therefore maximizing the effectiveness of learning. This new generation of e-learning systems will be possible with the introduction of the inferential technologies, which will allow the development of intelligent systems that model the behaviour and profile of the student and personalize the educational session according principally to his learning style and his background knowledge in the subject of study. Other parameters can be considered too to obtain a fully comfortable environment, but they are not exclusive of the learning domain. Hence, it is desirable to get:

- Personalization in content presentation or rendering according to the personal taste of the individual.
- Adaptability in the presentation of contents according to the characteristics of the visualization device. This kind of personalization is mainly emerging because of the proliferation of handheld computers having small displays and reduced computational capacity.
- Accessibility for people with disabilities. For example, making text fonts bigger for short-sighted people.

The e-learning systems that overpower the current market don't present these possibilities of personalization or else they tend to present them in a very limited way. However, several platforms, in an experimental state, have been developed by the research community. Here, some of them are briefly described:

- IDEAL [2]: IDEAL is a prototype of an intelligent agent assisted environment for active learning. In the system, students' learning-related profiles, such as

learning styles and background knowledge, are used for selecting, organizing and presenting the learning materials to individual students and in supporting active learning. The student models are inferred from the performance data using a Bayesian belief network.
- SAC [3]: SAC is a self-paced and adaptive courseware system developed at the Hong Kong Polytechnic University. This system provides dynamic navigational guidance to students taking online courses. Specialized data models not compatible with those presented in the next section are used to perform the adaptation.
- MATS [4]: The Multi-Agent Tutoring System (MATS) is an agent based educational platform that models a "one student - many teachers" learning situation. Each MATS agent represents a tutor, capable of teaching a distinct subject. All MATS tutors are also capable of collaborating with each other for solving learning difficulties that their students may have.

3 The E-learning Standardization Process

Recently, several organizations and institutions (e.g. IMS, ADL, AICC, IEEE's LTSC, CEN/ISSS/LT) have been working towards the development of standards and recommendations aimed to solve the interoperability problems currently found in the e-learning domain. The broad adoption of these proposals by the e-learning platform manufacturers will ensure, in the near future, the development of educational resources that can be exploited by a wide variety of compliant platforms. This feature will significantly reduce the time and the costs related to the development of on-line courses. Some of the most outstanding fields of standardization are showed in the next subsections.

3.1 Educational Metadata

Metadata is one of the most prolific fields in the e-learning technology standardization process. Generally speaking, metadata can be defined as "data about data". The educational metadata provide descriptions and additional information about learning resources. This information can be used not only for characterizing the resources but also for searching, cataloguing and use improvement.

One of the main contributors in this field has been the IEEE's LTSC with the *Learning Object Metadata* (LOM) specification [5]. LOM defines a total amount of 60 elements grouped into nine meaningful categories that try to reflect all aspects that must be considered in a pedagogical environment.

3.2 Learning Resources Organization and Packaging

The need to exchange educational resources between e-learning platforms and authoring tools has caused the development of content packaging formats and procedures. In this way, the definition of a single entity (e.g. a file) that encapsulate the educational content together with its organization and its related metadata will make easier the course transfer among different systems.

The most outstanding recommendation in this field is that proposed by the ADL initiative: the *SCORM Content Aggregation Model (CAM)* [6]. CAM is an extension of the IMS *Content Packaging specification* [7] including some results from the *AICC CMI Guidelines for Interoperability* [8]. A CAM *Package* is composed of two elements: the *Manifest*, an XML file that describes the encapsulated resources and their organization, and the *Resources Collection*, i.e. the physical files that store the resources. The *Organization* element of the *Manifest* defines the static relations among the resources of the aggregation (often a whole course) in a hierarchical tree of *items*. The *Prerequisites* subelement of each *item* defines what other parts of the content aggregation must have been completed before starting the *item*. The LMS is responsible for interpreting these prerequisites and controlling the actual sequencing of the learning resources at run-time.

3.3 Student Profiles

As it happens in conventional educational environments, e-learning platforms must handle information about students. The proposals in this field define information models whose aim is to provide not only a structured way to represent this information but a standardized way to interchange it among different systems. The first recommendation made in this field is the one from the IEEE's LTSC, the *Public and Private Information for Learners* (PAPI Learner) [9]. PAPI includes a subset of useful elements organized into five categories: *Personal*, *Preference*, *Performance*, *Portfolio*, *Relations* and *Security*. The IMS consortium has made its own proposal based on the PAPI work. The IMS recommendation, *Learner Information Package* (LIP) [10], identifies eleven categories to describe student-related information: *Identification*, *Securitykey*, *Transcript*, *Goal*, *Qcl*, *Activitity*, *Interest*, *Competency*, *Relationship*, *Affiliation* and *Accessibility*.

4 Interim Conclusions

First proposals from the e-learning standardization process are currently in scene and some manufacturers have begun to adopt them. These proposals sound useful in order to solve the interoperability problem existent in the e-learning domain. However, they present some troubles for using them in the next generation of adaptive e-learning systems:

- They are not flexible enough. Particularly, the proposed educational content organizations are too rigid, being difficult the delivery of alternative contents to the student depending of his particular needs.
- They are hardly scalable. Changes on them, for example to include a new element in an educational metadata specification, may imply important modifications at software level in the LMS. It is due to the lack of explicit machine-processable semantic in the recommendations.
- Relations among the different proposals are very poor. For example, it is not easy to match the *Preferences* category from the PAPI specification to

the LOM metadata elements. It makes difficult searching the most appropriate content for a particular leaner. So, such relations should be made more explicit.

5 Semantic Web

Currently, the web is a large and decentralized source of information designed primarily for human consumption. The coming Semantic Web [11] is an extension of the current web in which information in a machine-processable form can coexist and complement the existing human-readable information, better enabling computers and people to work in cooperation.

For this emerging Web to function, facilities to put machine understandable data must be developed. Ontologies [12] figure prominently (c.f. Figure 1) as a way of representing the semantics of documents and enabling that semantics to be used by web applications and intelligent agents. An ontology defines the terms used to describe and represent an area of knowledge (like medicine, tool manufacturing, automobile repair, financial management, etc.), including computer-usable definitions of basic concepts in the domain and the relationship among them. They encode both knowledge in a domain and also knowledge that spans domains. In this way, they make that knowledge reusable.

It must be clear that ontologies declare not just hierarchical categories or taxonomies but they may also include other logical rules. Logical rules in ontologies can be used by inference engines to derive new knowledge from current data. For the sake of illustration below we include a very simple example:

A particular ontology about Natural Sciences can state:

```
CARNIVORE "is subclass of" ANIMAL; FOX "is subclass of" CARNIVORE
HERBIVORE "is subclass of" ANIMAL; RABBIT "is subclass of" HERBIVORE
```

A property for the class CARNIVORE, defined as a logical rule, can be:

```
CARNIVORE "eat" ANIMAL
```

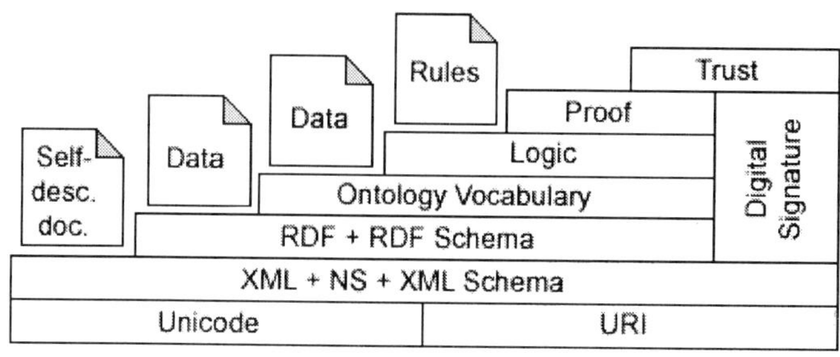

Fig. 1. The Semantic Web layers

From these declarations, an inference engine can reason that a fox eat rabbits (if no other rule contradicts this deduction).

In order for ontologies to fulfill their role in the semantic integration of the Web, there will need to be some standardization of Web ontology languages. The W3C is already moving in this direction with languages such as RDF and RDFS. However, in order to achieve the widest possible acceptability, these languages have deliberately been kept very simple and have relatively weak semantics. Much richer ontology specification languages [13] will be needed in order to exploit the full potential of the Semantic Web.

6 Semantic Web Concepts in E-learning

The LTSA [14] states that, from a conceptual viewpoint, every Learning Management System is composed by the elements showed in Figure 2. Briefly, the overall operation has the following form [14]: (1) the learning styles, strategies, methods, etc., are negotiated among the learner and other stakeholders and are communicated as learning preferences; (2) the learner is observed and evaluated in the context of multimedia interactions; (3) the evaluation produces assessments and/or learner information; (4) the learner information is stored in the learner history database; (5) the coach reviews the learner's assessment and learner information, such as preferences, past performance history, and, possibly, future learning objectives; (6) the coach searches the learning resources, via query and catalog info, for appropriate learning content; (7) the coach extracts the locators from the available catalog info and passes the locators to the delivery process, e.g., a lesson plan; and (8) the delivery process extracts the learning content from the learning resources, based on locators, and transforms the learning content to an interactive multimedia presentation to the learner.

The Coach is the process that incorporates information from several sources to search and select learning content for appropriate learning experiences. In essence, it is an inference engine, i.e. an automatic control mechanism that applies the axiomatic knowledge present in the knowledge base to determine: the adequate learning contents to be delivered to the learner.

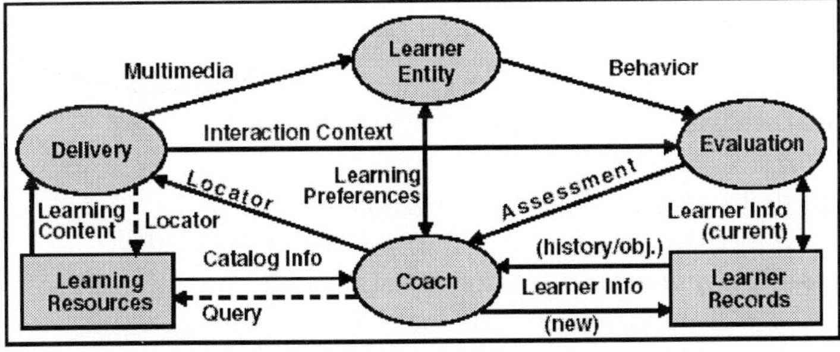

Fig. 2. The LTSA architecture components

The Coach work is usually facilitated by the definition of a specific organization for the resources each course is compliant with (c.f. Section 3.2). So, the searching of the next resource to be delivered to the learner in a concrete learning session consists just in finding the next item in hierarchical tree containing the course organization. One of properties of this *item* immediately specifies the resource to deliver. After that, the Coach only has to check that the learner fulfills the prerequisites established for that *item*. This checking is based on the data stored in the learner profile. However, as it was stated in Section 4, this approach is very restrictive because it does not allow the choice among different resources depending on the learner characteristics, knowledge and preferences.

A more open and flexible approach consists in not linking each *item* with a particular resource. On the contrary, each *item* should explicitly define the characteristics of the resource to be delivered in a machine-readable way. In this sense, inference rules constructed with the terms and vocabularies from one or more ontologies become the most useful tool. Such rules will allow the Coach to automatically carry out the search of the suitable resources from the LMS repository and then to select the one that best fits to the learner needs. This approach also permits changing already existing resources and even adding new ones without updating the course structure.

The selection process also needs explicit inference rules to meaningfully carry out the automatic matching between the learner profile data (c.f. Section 3.3) and the resource descriptions (c.f. Section 3.1). The definition of these logic rules in an ontology-based format would avoid the implementation of complex and subjective (i.e. conceived by every implementor) software algorithms in the Coach that would have to be modified on changes in the underlying data models.

Obviously, for this approach to be possible it is necessary to properly annotate the resources in a machine-readable way. The metadata schemes presented in Section 3.1 must include a vocabulary with a huge amount of terms to reduce ambiguity. Such a complex vocabulary would be expensive to construct and difficult to maintain as an only and isolated entity. A most practical approach is to define several small specific ontologies and to establish the possible relations among the identified terms on them through logic rules. Those specialized ontologies developed by different institutions and developers would be populated in one or several repositories accessible to the whole Internet Learning Community. This is precisely the basis of the emerging Semantic Web.

7 Conclusions

The notion of the Semantic Web, as thought by Tim Berners-Lee, inventor of the classic Web, is to transform the current World Wide Web so that the information can be understood and used both by computers and humans. The Semantic Web will provide an environment where software agents can perform advanced tasks on behalf of humans. This notion seems very interesting in the e-learning domain. As the Web is in essence a large and decentralized knowledge base, building that knowledge in a "machine-readable" format would permit the development of specialized agents (tutor agents) that asynchronously explore, gather, extract, combine and transform information from several sources in order to present adequate and personalized contents depending on each individual particular learning needs.

Perhaps, this will be possible in a distant future. However, at present, the concepts involved in the Semantic Web can be used to improve the current e-learning systems. In this paper some restrictions of the proposed recommendations from the learning standardization process were identified to deal with adaptive e-learning. Such restrictions would be relaxed by the introduction of distributed ontologies and appropriate inference rules allowing the meaningful integration of different models.

Acknowledgments. We want to thank "Xunta de Galicia" and "Ministerio de Ciencia y Tecnología" for their partial support to this work under grants "Arquitecturas distribuidas para Teleservicios" (PGIDT00TIC32203PR) and "CORBALearn: Interfaz de Dominio guiada por Estándares para Aprendizaje Electrónico" (TIC2001-3767).

References

1. Anido, L., Llamas, M., Fernández, M.J.: Internet-based Learning by Doing. IEEE Transactions on Education, vol. 44, n. 2 (2001).
2. Shang, Y., Hongchi, S., Su-Shing, C.: An Intelligent Distributed Environment for Active Learning. Proc. of 10th Int. WWW Conference, Hong Kong, May 1–5 (2001).
3. Chan, A., Chan, S., Cao, J.: SAC: A Self-paced and Adaptive Courseware System. Proc. of IEEE Int. Conf. on Advanced Learning Technologies, August (2001).
4. Solomos, K., Avouris, N.: Learning From Multiple Collaborative Intelligent Tutors: An Agent Based Approach. Journal of Interactive Learning Research, vol. 10, n. 3.4. (1999).
5. Hodging, W.: Draft Standard for Learning Object Metadata V. 6.4. IEEE LTSC Technical Report, March (2002).
6. Dods, P.: SCORM Content Aggregation Model. ADL Tech. Report, October (2001).
7. Anderson, T., McKell, M.: IMS Content Packaging Information Model V. 1.1.2. IMS Technical Report, August (2001).
8. Hyde, J.: AICC/CMI Guidelines for Interoperability V. 3.5. AICC Technical Report, April (2001).
9. Collet, M., Farance, F.:Draft Standard for Learning Technology - Public and Private Information (PAPI) for Learners (PAPI Learner) V. 7. IEEE's LTSC Technical Report, November (2000).
10. Smythe, C., Tansey, F., Robson, R.: IMS Learner Information Packaging Information Model Specification. IMS Technical Report, March (2001).
11. Berners-Lee, T., Hendler, J., Lassila, O.: The Semantic Web. In: Scientific American, May Issue (2001).
12. W3C Web Ontology Working Group at http://www.w3.org/2001/sw/WebOnt
13. Bechhofer, S.: Ontology Language Standardisation Efforts. OntoWeb Rep (2002).
14. Farance, F., Tonkel, J.: Learning Technology Systems Architecture (LTSA). IEEE LTSC Technical Report, November (2001).

An Integrated Programming Environment for Teaching the Object-Oriented Programming Paradigm

Stelios Xinogalos and Maya Satratzemi

Dept. of Applied Informatics, University of Macedonia
Egnatia 156, P.O. Box 1591, 54006 Thessaloniki, Greece
{stelios, maya}@uom.gr

Abstract. In this paper we propose a new integrated programming environment for teaching the object-oriented programming paradigm. The environment is based on the microworld approach to teaching programming and the programming language of Karel++. Its main features are: a series of e-lessons, a special kind of structure editor, run-time error detection, program animation and recordability of students' actions. In this paper we present the programming environment, the results and our own experiences in using the integrated programming environment to teach object-oriented programming to undergraduate students.

1 Introduction

Programming is without doubt one of the most difficult topics in Computer Science. The extended research that has been carried out over the past three decades showed that one of the most important factors that makes programming difficult to learn is the fact that *students are taught the principles of programming by the classic approach* [3], which is based on:

- A *general purpose programming language* that is too big and too idiosyncratic. The volume of the language makes it difficult for the student to understand the conceptual basis of the language together with the main principles of programming properly. As a result students fail to assimilate what they are taught. Furthermore, special attention is paid from the students to learning the syntax of the language and not to obtaining problem solving skills.
- A *professional programming environment* for the chosen programming language that does not support students neither in understanding the semantics of the control structures and the flow of control nor in the process of debugging their programs.
- A set of *problems from the area of number and symbol processing* that are far from the students' everyday experiences and are not attractive to them.

The fact that the classic approach of teaching programming to novices does not fulfil their didactic needs resulted in many efforts to develop special methodologies, languages and tools. Each one of these approaches differs from the classic approach in at least one of its three main characteristics. The main proposed approaches to teaching programming [9] are:

M.H. Shafazand and A M. Tjoa (Eds.): EurAsia-ICT 2002, LNCS 2510, pp. 544–551, 2002.

Microworlds. A representative educational tool that has been developed in the context of this approach is Karel Genie [7].
Compilers with Improved Diagnostic Capabilities. Representative educational tools of this category are THETIS [5] and CAP [8].
Iconic Programming Languages. A representative environment of this type is BACCII-BACCII++ [4].
Program Animation. A good example of a program animator is Dynalab [2].

Motivated by the research about the difficulties of novice programmers while learning to program, the small number of papers that explore the difficulties of object-oriented programming [6], and the efforts of researchers to develop educational tools that can help students, we developed an educational environment for teaching object-oriented programming. This environment extends the environment Karel++ [1] and attempts to incorporate the results of the research about the teaching of programming. The features that differentiate it from the existing environments are: a series of e-lessons that introduce the student from the beginning to the principles of object-oriented programming, a structure editor that makes the development of programs easier, and the ability to record students actions. The last feature provides invaluable assistance to the teacher, so as to detect students' conceptions about programming, and especially object-oriented programming, and the techniques they use during program development. In this paper we present the programming environment and its use with students of an Informatics department.

2 Overview of the Programming Environment

The extended research that we carried out about the various approaches to teaching programming and the educational tools that have been developed in the context of each approach [9] resulted in our decision to adopt the microworld approach, as the main approach, for our programming environment.

The programming language of our environment is based on Karel++ [1], which is closely related to C++ and Java. The metaphor used is that of a world of robots. The actor of the microworld is a one or more robot (object) that is assigned various tasks in a world that consists of: crisscrossing horizontal streets and vertical avenues forming one block intervals, wall sections between adjacent streets or avenues used to represent obstacles (hurdles, mountains, mazes etc), and beepers – small plastic cones that emit a quiet beeping noise – placed on street corners. Students write programs that instruct robots how to perform their tasks. If the available class of robots – that can move forward a block, turn in place (90° to the left), pick beepers from the current corner or put beepers from their bag to the corner – is not appropriate for a task, then students can create a new class of robots that inherit the properties and methods of the basic or a previously declared class and extend them.

The sections that follow describe the most important features of our programming environment.

E-lessons. The new integrated programming environment incorporates a series of e-lessons for supporting students in understanding the *basic principles of object-oriented programming* and the most common *control structures*. The main purpose of these lessons is to minimize the possibility of developing misconceptions and acquiring wrong or insufficient knowledge. For this purpose each lesson contains:

- \vspace*{-2mm}*theory* intended to teach a specific concept of object-oriented programming or a control structure. The most important aspects of the paradigm are disclosed before the control structures and are emphasized throughout the lessons. Students manipulate objects (robots) from the beginning and write their own classes almost immediately. Great emphasis is placed on eliminating wrong conceptions that students usually form.

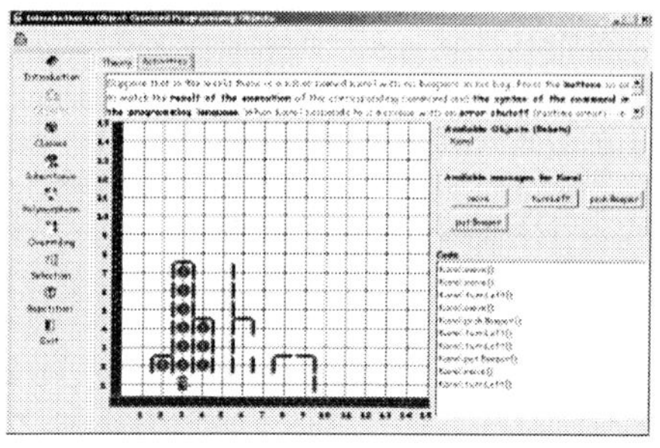

Fig. 1. E-lesson Window

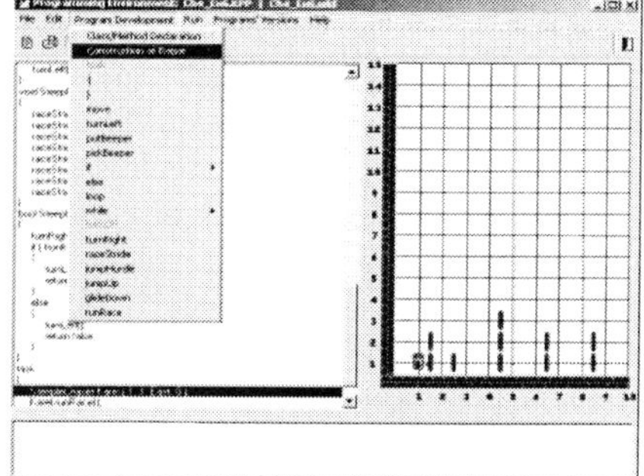

Fig. 2. Main Window and Program Animation.

- an *activity* that lies in the heart of the concept that the lesson is intended to teach. For example, the student can click on the buttons, labelled with the messages that the available object (robot Karel) understands; and watch (i) the robot execute the message, (ii) the syntax of the command in the programming language, as shown in Figure 1. We believe that these activities will help students internalise object-oriented programming concepts better and easier than if they were only presented with verbal descriptions.

A special kind of Structure Editor. A structure editor, in contrast with traditional text editors, operates directly on an abstract syntax tree while the program is being developed. That means that the structure editor provides a menu of the syntactically correct transformations for any location in a partially complete program, yielding always syntactically correct programs [7].

The editor of our programming environment is a special kind of structure editor:

- Writing a program is accomplished by: 1) choosing the appropriate action (class/method declaration, construction of object) or choosing a message to send to an object from a single *menu* as shown in Figure 2. This menu is automatically updated whenever the user deletes or edits a class/method or declares a new class/method. 2) interacting with the system through *dialog boxes*. For example, when the "Construction of robot" menu's choice is clicked the dialog box shown in Figure 3 appears, so as to specify the name and the parent class of the robot and initialise its properties.
- The *source code is automatically indented*, as shown in Figure 2, in order to avoid misleading appearances.
- Logic and some syntax errors are deliberately not prevented.

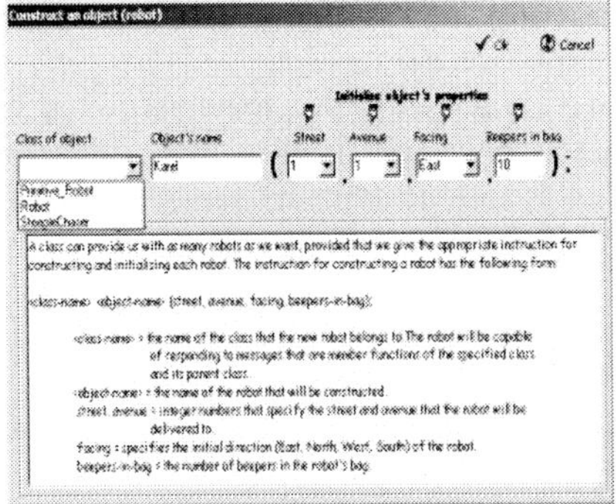

Fig. 3. "Construction of Object" dialog box.

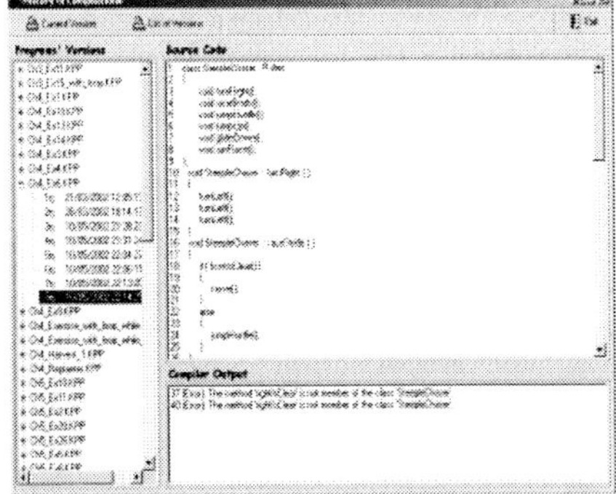

Fig. 4. History of Compilations

The design of the editor is the result of a didactic rationale that focuses on fulfilling specific didactic needs of students:

- Development of programs becomes easier and the common problem of focusing on the syntactic details, even in the case of a simple language, is eliminated. Students can *focus on acquiring problem solving skills and assimilating the principles of programming*.
- Long sessions of searching for and correcting syntax errors are eliminated.
- The programming environment can be used for teaching programming to *young students*.
- The editor's partially compiled knowledge of the program gives us the chance to provide students with valuable information, such as *run-time error detection* and *understandable error messages*.

Run-time Error Detection – Understandable Error Messages. The most difficult and frustrating phase of debugging is finding the logic errors of a program. Unfortunately commercial compilers are designed for experts rather than novices and so they do not really help students in debugging their programs [5]. Our programming environment intends to reduce this burden by detecting logic errors and reporting *understandable and highly informative error messages*:

- The *line number* reported is the actual line of the mistake.
- Messages report not only what is wrong but also *explain why it is wrong*.
- The error messages use *physical language* and not codes.

In this way the misconceptions that usually account for the errors are revealed to students. Furthermore, the run-time error detection *saves on teachers' time and effort* while studying student's errors and misconceptions because they do not have to check the programs themselves (for run-time errors), as is usually the case.

Executing Programs - Program Animation. Since a program has been successfully compiled the user has three choices of executing it:

1. *Running the program.* This choice, which is the standard way of executing programs in professional environments, presents the user with the final result of the program's execution.
2. *Tracing through the program.* The program is executed in slow motion and each command is highlighted (by changing its background colour from white to blue

and the font colour from black to white), while its execution's result on the situation of the microworld and the actor is visible on the screen.

3. *Executing the program step by step.* In contrast with the previous ways of executing a program the user takes an active role and decides when the next command will be executed. Each command is executed in the same way as in tracing. The user can terminate the execution of the program at any time.

The last two choices (2, 3) of executing a program, known as *program animation*, are of great importance to the student. Executing the program step by step in particular:

- Reveals the dynamic nature of program execution.
- Assists students in understanding the semantics of the control structures and flow of control, and supports them in finding logic errors.
- Gives instructors the chance to cover more material in less time.

When the student uses the last two choices (2, 3) of executing programs, he is also presented with explanatory messages about the semantics of the command being executed. The explanatory messages use physical language and are presented in the pane that stretches across the bottom of the window (Figure 2). This feature is known as *explanatory visualisation.*

Recordability of Students' Actions. With the term recordability we refer to the system's ability to *auto save the student's program and its errors/warnings each time the program is compiled.* The *history of compilations* is presented in a separate window with the form of a two-level tree as shown in Figure 4.

The first level presents as with the names of the programs that have been developed and the second one with the date and time of each compiled version of the selected program. When a version of a program is selected the source code and the compilation's result is presented as shown in Figure 4. Source code is prefixed with line numbers so as to make studying of versions of a program easier.

The aim of this module is to:

- assist teachers in *studying students' errors and misconceptions* when they are introduced to the object-oriented programming paradigm.
- help teachers detect students' problem-solving techniques, since each compiled version of a program represents a step in the development of the problem's solution.

3 Empirical Study

The software we developed was tested and evaluated by 20 undergraduate students from the department of Applied Informatics of the University of Macedonia in Greece. The students that participated in the empirical study had attended in the 1st semester the compulsory course "Programming", which is based on the procedural programming paradigm and uses Pascal as the programming language, and they had all failed in the exams. The students were separated into 10 groups. Six lessons were carried out, all of them at the laboratory, and the programming environment was used as the medium of instruction - occasionally the blackboard was used too. We also used Camtasia Recorder for recording (video) the actions of 2 groups. The lessons took place weekly, and their duration was 2 hours. In each one of the first 5 lessons we used the theory of the lessons that are incorporated in the programming environment in order to teach the basic concepts of each lesson. Next, students used the activities that are incorporated in the programming environment, and in some cases existing programs, so as to assimilate the concepts and become familiar with the features of the programming environment. Finally, the groups developed one or two programs without help from the teacher. The 6th lesson was devoted to the assessment of students' knowledge and the evaluation of the programming environment and the series of lessons from the students. Next, we give the subject taught in each of the five lessons.

1st Lesson – Objects-Classes.
2nd Lesson – Inheritance.
3rd Lesson – Polymorphism & Overriding.
4th Lesson – Conditional Structures.
5th Lesson – Repetitive Structures.
6th Lesson. The aim of this lesson is: (i) the *evaluation of students' knowledge*, and (ii) the *evaluation of the programming environment and the series of lessons by the students*. To be more specific, the students:

- test and debug an existing program that includes all the categories of errors.
- answer – personally – open type questions, so as to detect to what degree they understood the basic concepts of the object-oriented programming paradigm.
- fill in - personally – a questionnaire that includes open and closed type questions.

The results. For each one of the first 5 lessons:

- the teacher recorded difficulties in understanding the taught concepts, drawbacks and problems that were encountered while using the programming environment,
- the consecutive versions of the students' programs were studied,
- the recorded actions and dialogues of two groups were analysed.

The most important conclusions that were drawn from the analysis of the data are:

1st Lesson – Objects-Classes. The programming environment and the problems stimulated student's interest. Students became familiar with the programming environment very quickly, and from the study of the consecutive versions of their programs we concluded that they had no difficulties in understanding the taught concepts. The detection and correction of logic errors was immediate due to step by step execution and explanatory visualisation.

2nd Lesson – Inheritance. Students solved the proposed problems without having any particular difficulty. Only one group seems to confuse the concept of class with the concept of object. Although students assimilated the concept of inheritance without difficulty – only 15% of the programs included semantic errors – they had difficulty in developing the programs – 60% of the programs led to run-time errors.

3rd Lesson – Polymorphism & Overriding. Several students had difficulty in understanding the concepts of *polymorphism* and *overriding*. With the help of step by step execution of existing programs they all understood that when a method declared in different classes with the same name is called, then the class of the objects decides which method will execute. The thorough study of the students' programs made clear that 50% of the students assimilated the concepts of *polymorphism* and *overriding*, while 20% of them did not understand the concept of *overriding* or confuses the two concepts.

4th Lesson – Conditional Structures. The syntax errors that were detected, in the programs of nearly all the groups, were mainly related to the omission of the syntactic markers that define the scope of the *then* and *else* parts of the conditional structures. Several students seem to become tense when the programming environment reports syntax error messages, since they immediately compile the program without having made any attempt to correct it. However, the vast majority of syntax errors is corrected with the first attempt due to "good" error messages (plain and physical language is used, the actual line of the error is reported). The study of the consecutive versions of students' programs made clear that without the assistance of the programming environment – "good" error messages, step by step execution, explanatory visualisation – most of the groups would not have been able to solve the problem, or at least they would have had great difficulty.

5th Lesson – Repetitive Structures. Students had difficulty with repetitive structures. The study of the consecutive versions of programs made clear that a significant number of students finds it difficult to (i) distinguish the structures *if* and *while*, and (ii) develop programs that need both *if* and *while*.

6th Lesson. After studying the versions of the programs that were developed by the students we concluded that: (i) understanding and debugging an existing program is quite difficult and time-consuming for some students, and (ii) editing a program, with the syntax editor, is easy. The results of the test are in agreement with the results of the 5 lessons. The analysis of the questionnaires showed that students evaluated positively the programming environment and the series of lessons. There were not reported any usability problems, and generally there was not reported any kind of problem. All the students stated that step by step execution helped them. The vast majority of students mentioned that: (i) used the lessons (theory, activities) that are incorporated in the programming environment during program development, (ii) was benefited by explanatory visualisation, and (iii) the structure editor assisted them in developing their programs.

4 Conclusions

The educational programming environment that we developed was based on the results of the extended research about teaching and learning programming. Furthermore, we tried to incorporate those features that would give as the chance to record the students' difficulties when they are taught object-oriented programming, since the research that has been carried out until now has revealed important facts about teaching programming but very few about object-oriented programming. The results of the empirical study that we carried out show that the programming environment can help invaluably the teacher and the student in teaching and learning programming respectively. The most important conclusions are:

- The ability to record students' actions provides invaluable assistance to the teacher, so as to detect students' conceptions about object-oriented programming and the techniques they use during program development.
- The structure editor solves the problem of focusing on the syntactic details and gives the chance to focus on understanding the concepts and principles of object-oriented programming, structure matters and developing problem-solving skills. Furthermore, we would like to mention the ease with which all the students used the structure editor.
- The incorporation of the lessons (theory, activities) in the programming environment helped students during instruction and program development.

- Step by step execution, tracing through the program and explanatory visualisation helped students understand flow of control and the semantics of control structures, and helped them overcome their difficulties and misconceptions.

Furthermore, the pilot use of the programming environment showed that it is possible to teach object-oriented programming in a short period of time with specially designed educational environments, while this is impossible with the usual professional programming environments.

References

1. Bergin, J., Stehlik, M., Roberts, J. and Pattis, R.: Karel++ - A Gentle Introduction to the Art of Object-Oriented Programming. 2nd edn. , Wiley, New York (1997)
2. Birch, M., Boroni, C., Goosey, F., Patton, S., Poole, D., Pratt, C., Ross, R.: DYNALAB: A Dynamic Computer Science Laboratory Infrastructure Featuring Program Animation. Papers of the 26th SIGCSE technical symposium on Computer Science Education, ACM Press, Nashville, TN USA (1995) 29–33.
3. Brusilovsky, P., Calabrese, E., Hvorecky, J., Kouchnirenko, A., Miller, P.: Mini-languages: a way to learn programming principles. Journal of Education and Information Technologies, **2**, Kluwer Academic Publishers (1997) 65–83
4. Calloni, B. A., Bagert, D. J.: Iconic Programming Proves Effective for Teaching the First Year Programming Sequence. In the Proceedings of the 28th SIGCSE technical symposium on Computer Science Education, ACM Press, San Jose, CA USA (1997) 262–266
5. Freund, S. N., Roberts, E. S.: THETIS: An ANSI C programming environment designed for introductory use. In Proceedings of the 27th SIGCSE technical symposium on Computer Science Education, ACM Press, Philadelphia, PA USA (1996) 300–304
6. Holland, S., Griffiths, R., Woodman, M.: Avoiding Object Misconceptions. SIGCSE '97 CA, USA, ACM Press (1997) 131–134
7. Miller, P., Pane, J., Meter, G., Vorthmann. S.: Evolution of Novice Programming Environments: the Structure Editors of Carnegie Mellon University. Computer Science Department, Carnegie Mellon University, Pittsburgh, PA 15213–3890 (1994)
8. Schorsch, T.: CAP: An automated self-assessment tool to check Pascal programs for syntax, logic and style errors. Papers of the 26th SIGCSE technical symposium on Computer Science Education, ACM Press, Nashville, TN USA (1995) 168–172
9. Xinogalos, S., Satratzemi, M., Dagdilelis, B.: Introduction to programming: Teaching Approaches and Educational Tools. In Proceedings of the 2nd Pan Hellenic Conference with International Participation "Information and Communication Technologies in Education", Patra, (2000) 115–124 (in Greek)

The Current Legislation Covering E-learning Provisions for the Visually Impaired in the EU

Hamid Jahankhani, John A. Lynch, and Jonathan Stephenson

University of East London, East London Business School, UK,
Hamid.jahankhani@uel.ac.uk

Abstract. In the education environment many colleges, universities and training establishments are keen to exploit the e-learning environment. In the current population there are significant numbers that are visually impaired. These people tend to be excluded from the socially popular vehicle for education. This paper is to review and evaluate the e-learning provisions of educational establishment in relationship to the government guide lines and in particular look at how the university of East London tackles such an interesting and challenging opportunity.

1 Introduction

In the education environment many colleges, universities and training establishments are keen to exploit the lucrative e-learning environment. Growth in the e-learning market in Europe is expected to be around £3 billion by 2004 which represents a compound annual growth rate of 96 percent. The United Kingdom, the Netherlands and Sweden are the strongest markets in the adoption of e-learning because Internet adoption levels are already high in these countries. The number of colleges and universities offering e-learning will more than double, from 1,500 in 1999 to more than 3,300 in 2004, according to statistics and student enrolment in these courses will increase 33 percent annually during this time. With such an explosive growth it is important that the visually impaired are not neglected and will have equal opportunities to such provisions. Many blind and partially sighted people prefer particular types of information in Braille.

On 25th September 2001, European commission adopted a communication on improving the accessibility of public Web sites and their content. The aim is to make web sites more accessible to people with disabilities. Visually impaired people, 'read' web pages using software tools known as screen readers, which generate speech and/or refreshable Braille output.

2 Current Trends

Most educational organisations have to address the imperative for change. The technological conditions in which universities operate have changed in very

M.H. Shafazand and A M. Tjoa (Eds.): EurAsia-ICT 2002, LNCS 2510, pp. 552–559, 2002.

significant ways [1]. The real challenge is to discover the ways to improve learning. In order to do this we need to state what set of initial conditions might be imagined to improve learning by the use of new technologies [2]. The World Wide Web improves access to information and ideas and is becoming an important learning resource, which is potentially accessible to most people in the world. For the learner to effectively use this resource they need to become more autonomous in their learning.

The promise of an emerging trend of WBT was confirmed in Callaway article [3]. A developer of WBT should not overlook, Barron's article [4], which provides guidelines and rules for designing Web-delivered instruction. Xiaodong [5] described an innovative approach for CD-ROM training using DHTML. Mayer and Moreno [6] presents a cognitive theory of multimedia learning which draws on dual coding theory; cognitive load and constructive learning theory. Cognitive Load Theory (CLT) can provide guidelines to assist in the presentation of information in a manner that encourages learner activities that optimise intellectual performance [7].

3 Technical Innovation

Web is information medium, but too many web designers still think of it as a purely visual medium, and are unaware even that visually impaired people can access the web.

An accessible website is one that can be accessed by anyone. In this context, the essence of good design involves ensuring that a text alternative exists for every non-text element on the web page. The Internet currently connects more than 40 million workstations [8] and the user community is still expanding rapidly.

Many people with sight problems have some useful vision, and read web pages in exactly the same way as fully sighted people with their eyes. However the needs of people with poor sight vary considerably, depending on how their eye condition affects their vision.

The National Library for the Blind (Visugate) [9] in conjunction with a number of other partner organisations provides a gateway to electronic resources that address all aspects of visual impairment. Goose, S. et al, [10], researched in Vox Portal, a scalable VoxML client and a WWW Server-hosted dynamic HTML↔VoxML converter. They reported that, Interactive voice browsers offers an alternative paradigm that enables both sighted and visually impaired users to access the World Wide Web. This technology can facilitate a safe, 'hands-free' browsing environment. VoxML, is a standard mark-up language for specifying the dialogs of interactive voice response applications that feature speech synthesis and recognition technologies.

JAWS® for Windows® developed by Freedom Scientific provides speech technology that works with the Windows operating system to access popular software applications and the Internet. Connect Outloud is an easy to use tool offering Braille and speech output to the web and gives the user the ability to send and receive email. MAGic helps those with low vision view information on their computer screen with magnification up to 16 times, while hearing it through their speech synthesiser.

OPENBook, which is also developed by them, is for individuals to read, edit and manage printed media by scanning it and converting it to digital information.

There are around 23,000 people in the UK who has a severe loss of both sight and hearing. About 200,000 have less serious dual sensory loss. Some deafblind people retain enough sight to be able to use systems used by deaf people for example lip reading or British Sign Language or the Deaf Alphabet. Many visually impaired people have additional disabilities, which may affect the manner in which they communicate. Visually impaired people from ethnic minorities may face additional communication difficulties.

4 Web Accessibility, Special Educational Needs, and Disability Act 2001

The Special Educational Needs and Disability Act 2001 (SENDA) [13] received the Royal Assent in May 2001 and came into force on the 1st September of the same year. Compliance with the Act is required by September of 2002.

This Act covers a range of accessibility issues in education and is not specifically about the web, though that is included in its provisions. The SENDA is an extension of the Disability Discrimination Act of 1995 and may also be referred to as part IV of that legislation. As its title suggests the Act's target is the education sector, but similar provisions also exist under part III of the Disability Discrimination Act affecting other sectors. In some circumstances these would also apply. Web sites may also be actionable under the Human Rights Act 2002. Similar legislation is coming into, or is already in force elsewhere in Europe, in America (Section 508) and Australia. In separate contexts the laws relating to defamation and data protection may also apply to a university's use of the web. This is also an international issue [11].

The code of practice for providers of post 16 education and related services provides guidelines for the higher education on the interpretation of the law and distance learning and independent learning opportunities such as e-learning are specifically included. The law focuses on the institution's relationship with the student, but seems likely to cover prospective students too. Web sites will qualify as services and may fall into one or more categories that are covered, depending on the specific circumstances.

In practice since the duty clearly extends to the admission process and the circumstances surrounding that, any use of the web to offer information about programmes and services to prospective students is almost certainly covered too.

5 Web Content Accessibility Guidelines

The Web Content Accessibility Guidelines (WCAG) published by the Web Accessibility Initiative (WAI) which is part of the World Wide Web Consortium (W3C) and Royal National Institute for Blinds (RNIB) is a contributing member has dawn up the WAI guidelines (W3C 2001). While here we focus on the needs of visually impaired people, accessible web design is something that affects everyone, and the guidelines reflect this. The tips given here are an introduction to some of the issues that should be considered when designing for accessibility.

5.1 Images and Background, Text and Colours

The alternative text attribute (ALT text) [12] of the image tag exists to provide a textual representation of the image for people accessing the page in a non-graphic way (e.g. text only, speech or braille). The ALT text should convey what is important or relevant about the image:

If what is important is the actual content of the image, the ALT text should consist of a brief description of the image, e.g. "A photo of my house". If the image is essentially functional (e.g. a "Search Now" button) the ALT text should convey the function. In the case of the example given, the ALT text could be, simply, "Search Now" - a description of what the button looks like is unnecessary.
If the image is a bullet or a horizontal section divider, the ALT text should be a text representation of the same thing, e.g. " - " or " # " for bullets, or " - - - " for a divider. If the image is essentially "eye candy" or is used for the purpose of visual layout (e.g. "spacer" images), the ALT text should be set to read "*". This will let the user know that the image contains no information and performs no relevant function - they are not left in the position of knowing the image exists but being unsure of what it might contain or do. Longer descriptions of more complex images may be provided by using the LONGDESC attribute, or by placing a "d" next to the image and linking that to a page containing the detailed description.

Choose a background that is a single, solid colour. The choice of background and foreground colours is not as important as the contrast between the background and the text. Simple ways to check is to take a screen shot of your proposed page and use a graphic editor to convert this into a greyscale image.

Ensure that your chosen colour scheme can be over-ridden by the user's browser settings. Some people have eye conditions that mean they can read only black on white, while others can read only yellow on black - if your design is flexible, everyone will be able to read it. Don't underline large blocks of text. In addition, since underlining usually indicates hyperlink text, it can be confusing for users if it is used where no link exists. Capitalisation of whole sentences should be avoided, as many people find it difficult to read.

Use headings appropriately. Don't use headings simply to increase text size, and don't simply use bold or larger font size to simulate headings.

5.2 Frames, Links, and Navigation

Some people have difficulty navigating within frames, either because the frames are confusing or because the software they are using simply cannot read frames. When using frames, always offer meaningful NOFRAMES content for those people who cannot read framed information. The NOFRAMES section should contain meaningful content with links to the other pages in your site, so that they can be accessed without frames. Ensure that each frame has a sensible TITLE (in addition to the NAME). If possible, avoid using frames.

All links should contain enough useful information about their destination that they make sense on their own, without surrounding text or graphics. Links should not be presented directly next to each other, as some access software will interpret a group of links as being one single link. Instead, separate links with text (e.g. the bar character " | ") or a graphic. If you have used graphics of text as links, provide text based links as well to accommodate partially sighted users who are using their browser settings to increase the font size.

5.3 PDF Documents

If documents are provided in Portable Document Format (PDF), ensure that HTML or plain text versions are also available. Consider how to make any charts or graphs in the PDF file accessible. PDF documents are readable by blind people using access technology with the help of 'Access Adobe'. Access Adobe translates PDF into HTML or into a text email, making it readable by someone unable to access PDF in the usual way. If you offer PDF files on your site, ensure that you give the URL of each PDF file and that a link to http://access.adobe.com/ is available.

5.4 Shockwave, Scrolling Text, JavaScript, Plug-Ins, etc.

If you create web pages in which the content and functionality is presented in formats other than plain HTML, you may be excluding some people from your site. Not everyone has the desire or capability to download and use all scripts and plug-ins. Always provide plain HTML alternatives so that everyone can access the information and services on your site. If a Flash movie is used on your entrance page, ensure that any meaningful content is available to users who can't access Flash. Ensure that a plain text link is available to enable users to access subsequent pages of your site - embedding a "Skip intro" link in the Flash movie itself is of little use to anyone who can't access the movie!

Some browsers can't read JavaScript and may tell the visitor only that there is 'an unsupported script' on the page. Wherever JavaScript is used, ensure that the page functions correctly without the script, or that a parallel page which is JavaScript-free is available. Some access software works in conjunction with a "standard" browser but is unable to translate and present to the user everything that the browser itself is able to handle. Moving, blinking and auto-refreshing text is hard to deal with if you have poor sight - avoid all of these if you can.

5.5 Tables and User Interaction

Text based browsers generally can't display tables in the same way as graphical browsers. Usually, they will display the content cell by cell and row by row. As a result of this, it is important that you ensure that the page makes sense if the table content is presented in this "de-columnised" way. Not everyone can use a mouse. If you can't see, you can't see where the mouse cursor is positioned on the page. Don't require users to be able to click on a small or moving target in order to proceed to another page. Offer additional plain text links. Give some thought to the TAB order of the various links and form elements on the page. The default is the order in which these elements appear in the HTML code. In most cases, this is appropriate, but in some instances, it may make more sense for a different TAB order to be set up - you can do this using the TABINDEX attribute.

5.6 Testing

Validate your pages. Although writing valid HTML doesn't guarantee that your pages will be accessible, it is an important step in the process. If you use a mouse and a graphical browser, unplug the mouse and switch off graphics in your browser.

Install a text only browser (e.g. Lynx which is free and can be downloaded from the Web) and test the pages in that - do they make sense? Whenever possible, ask a range of people with various abilities and disabilities to test your pages and give you feedback. Use "Bobby" to test your pages. Bobby is a program, which can check your pages for compliance with the WAI Web Content Guidelines (it can also check for compatibility with various browsers and browser versions). Bobby is an automated program and can only test for some accessibility errors - the fact that a page passes the Bobby test does not guarantee that it is fully accessible.

6 Practical Application of Accessible Design

The current UK legislation (SENDA) does not specifically target the use of the web, but rather the provision of educational products (i.e. courses or programmes) or services that might conceivably be delivered by it. The associated Code of Practice [13] makes it clear that most, probably all, of the established and developing uses of the web in education are included. E-learning and all communication or provision of information by means of the web must therefore be made accessible. Here are some potentially complex obstacles to consider:

- Neither the legislation nor the Code of Practice (again referring to the position in the UK) actually define what an accessible web site is or give guidance on creating one. This means that an external benchmark must be found. It is generally accepted that the WAI Guidelines are the best point of reference. Given the global nature of the Internet and the world wide potential of e-learning it is perhaps advisable to consider additional resources as well. An example would be the USA's Section 508 legislation.
- There is a conflict of interest between the needs of the disabled user and the expectations of the much larger and increasingly sophisticated audience of non-disabled web users.
- Expertise in web design at the required level is very unlikely to be found amongst teaching or administrative staff, so institutions may have to review all of their established practices as part of the process of making their use of the web accessible.
- Much e-learning is delivered via bought in Virtual Learning Environments

(VLEs). These certainly have potential problems and because they are proprietary products there are technical limitations and licensing restrictions on what may be done with them.

7 The University of East London Model

The University of East London's new web site achieves a very high level of accessibility. The site's visual appeal and the impression it is intended to convey to its users, are not sacrificed to the legal requirement. Instead all criteria are met as part of a balanced relationship. Many of the new pages pass accessibility testing at AAA rating (Priorities 1, 2 and 3 of the WAI Guidelines) - and also conform to America's Section 508. In practice it relies on many interwoven details, but these essential principles strongly influenced the design process:

- Accessibility is merely an aspect of usability. Maximise usability for the benefit of the entire audience and many accessibility issues are resolved in the process.
- Embrace the design principles that are either expressed or implied in the WAI Guidelines and use them creatively. For many it will mean a complete rethink of how their sites and web pages are designed.
- Do not, when seeking to make the web accessible, lose sight of the web site's fundamental objectives, or the needs of its entire audience.

The method that underpins the UEL web site's successful combination of aesthetic quality with accessibility basically relies on the separation of design and content. Parts of this arrangement can either be ignored, or just not understood, by the user or their technology without any serious detriment to the meaning. This is how it is done:

- Use HTML 4.00/4.01 (Hyper Text Mark-up Language) or XHTML 1.0 (Extensible Hyper Text Mark-up Language) to correctly describe the content's structure. These versions of the mark-up language have features that assist accessibility.
- Then use CSS (Cascading Style-Sheets) to the maximum extent possible to display the page. This presents some technical problems because current web browsers implement CSS differently and none support them fully. It is perhaps the most challenging aspect of accessible design, but it also offers significant possibilities.
- Test the design thoroughly.

This is obviously a simplified explanation of the techniques used in the UEL web site design. Some but not all of the additional detail would be found in the WAI Guidelines or are touched on in this paper. The value of the information architecture should not be underestimated either. The result is not only accessible it is also adaptable.

8 Conclusions

The current UK legislation (SENDA) does not specifically target the use of the web, but rather the provision of educational products (i.e. courses or programmes) or services that might conceivably be delivered by it.

The University of East London's new web site achieves a very high level of accessibility, but does so within the context of a forward looking and carefully considered design. The end result is not only accessible it is also adaptable. It works at all common screen resolutions, is cross-browser compatible and degrades into a usable form in old browser versions. It can also be easily modified for different purposes. That adaptability is also an important step towards truly user centred learning and the provision of e-learning experiences that not only benefit the disabled, but also other user groups and the learning styles identified before.

References

1. Evans, T.D., Nation, D., Changing University Teaching: Reflections on Creating Educational Technologies, London (2000) Kogan Page.
2. McElhone, M.J., A methodological approach to educational innovation: a case study involving web-based learning, education on-line (2000), http://www.leeds.ac.uk/educol/documents.
3. Callaway, E., "The learning Web", PC week (1996), 13:49,53–7.
4. Barron, A., "Designing Web-based training", British Journal of Educational Technology, (1998), 29:4.
5. Xiaodong, L., Designing an interactive Web tutorial with cross-browser dynamic HTML, Library Hi Tech (2000), 18,4:369–382.
6. Mayer, R. E., Moreno R., aids to computer-based multimedia learning, Learning and Instruction, (2002), 12:107–119.
7. Kirschner, P., Cognitive load theory: implications of cognitive load theory on the design of learning, Learning and Instruction (2002), 12:1–10.
8. Zakon, R.H., Hobbes Internet timeline, (2001), http://www.personal.umd.umich.edu/~nhughes/htmldocs/timeline.html.
9. Egan, D., VisuGate: the gateway to information on visual impairment, (2001), http://www.rnib.org.uk/techshare/visugate_egan.htm.
10. Goose, S., Newman, M., Schmidt, C.,Hue, L., Enhancing web accessibility via the Vox Portal and web-hosted dynamic HTML↔VxML converter, Computer Networks (2000), 33:583-592. Motorola, VoxML: The mark-up language for voice applications, http://www.voxml.com
11. Jahankhani, H., Alexis S. A., E-Commerce Business Practices in the EU, 4th International conference on Enterprise Information Systems, ICEIS 2002, Vol. 2, 929–936, Ciudad Real, Spain.
12. Slatin, J.M., The art of ALT: toward a more accessible Web, Computers and Composition, (2001), 18:73–81.
13. Disability Discrimination Act 1995 Part 4, Code of Practice for providers of Post 16 education and related services, New duties (from September 2002) in the provision of post-16 education and related services for disabled people and students.

Monte Carlo Soft Handoff Modeling*

Alexey S. Rodionov[1] and Hyunseung Choo[2]

[1] Institute of Computational Mathematics and Mathematical Geophysics
Siberian Division of the Russian Academy of Science
Novosibirsk, RUSSIA +383-2-396211
alrod@rav.sscc.ru

[2] School of Information and Communication Engineering
Sungkyunkwan University
440-746, Suwon, KOREA +82-31-290-7145
choo@ece.skku.ac.kr

Abstract. In this paper some models for obtaining the distribution of a sojourn time in CDMA cellular systems are proposed. Knowledge on this is essential for reliable modeling of the soft handoff and for solving other related problems in the analysis of cellular systems. The proposed model is based on random walks and can be adopted to different conditions. Analytical results can be obtained that lead to a quit complicated numerical scheme so simulation models are used for Monte Carlo experiments. Main assumptions include different kinds of mobile carriers (pedestrians and transport passengers) and round shape of a cell. The scheme for simulation experiments is presented along with the discussion of simulation results.

1 Introduction

It is known that the same frequency band can be used simultaneously over neighboring cells in CDMA cellular systems. This enables so called soft handoff scheme when a new base station (BS) is assigned to a mobile while this mobile is still served by the old one and will be served by it until reaching some *outer handoff border*. The assignment process of new BS starts when a mobile reaches some *inner handoff border* or just starts a call inside the *soft handoff zone (SHZ)*. In the later case two or more stations are assigned to this mobile, but for distinctness we consider one of them as "old."

For different models of the soft handoff it is necessary to have the distribution of the sojourn time or the time period that an active (busy by call) mobile stays inside the handoff zone. Unfortunately, there is no available real data on the sojourn time, so some authors make attempts to obtain this distribution from some plausible (sometimes not very) reasoning. In last years some papers concerning this problem were published. Our attention was attracted by [1] and

* This paper was partially supported by BK21 program and grant No. 2000-2-30300-004-3 from the Basic Research Program of Korea Science and Engineering Foundation. Dr. Choo is the corresponding author.

M.H. Shafazand and A M. Tjoa (Eds.): EurAsia-ICT 2002, LNCS 2510, pp. 560–568, 2002.

we have tried to improve the model proposed by its authors. The use of round shape of a cell seems a good choice for us but we cannot agree with the model of a mobile movement that was proposed in the paper. In [2] the authors tried to simplify the problem by considering the square form of a cell instead of the round one. We think this is not a good idea as the real simplification of the problem can be made in a quite other direction. Proposed models are based on the random walks approach [3,4] and can be adopted for different conditions. Analytical results can be obtained that lead to a quit complicated numerical scheme [5] so simulation models are used for Monte Carlo experiments [6].

The rest of the paper is organized as follows. Section 2 is devoted to the main modeling assumptions, notation and basic concepts. In Section 3 we discuss about obtaining the probability of a mobile appearance on the border of the sojourn zone. Section 4 presents the description of simulation results. Section 5 is a brief conclusion.

2 Modeling Assumptions

In order to simplify the modeling of soft handoff in [1] the following basic assumptions were formulated (some notations are changed for fitting with the following text):

1. calls are being uniformly generated in a cell;
2. the base station is located in the origin of the Cartesian coordinates;
3. the aerial of BS is omni-directional;
4. only the path loss is considered,i.e. fast fading and shadowing effects are not considered;
5. a cell has round form with a radius R, the BS is located in the center;
6. SHZ is of a ring form and is surrounded by two concentric circles with radii r and R, respectively (see Fig. 1). The SHZ area is not larger than half of the cell area, i.e., $2r^2 \geq R^2$;
7. a mobile can move in the four perpendicular directions ($\rightarrow\uparrow\leftarrow\downarrow$). At the point of call generation or change of velocity the mobile chooses one over four directions with the same probability 1/4. This point is denoted as (X, Y);
8. the discrete random process $N(t)$ is the number of changes of velocity during $(0, t]$, and follows a Poisson process with parameter λ;
9. the continuous random variable V denotes the speed of a mobile which is uniformly distributed on $[5, V_{\max}]$. V_i is the constant speed of a mobile between $(i-1)$-th and i-th changes of velocity, and $\{V_i\}$ is independent and identically distributed (i.i.d.);
10. the continuous random variable T is the time between two consecutive changes of velocity or until the completion of a talk, and is exponentially distributed with parameter λ. T_i is the time between $(i-1)$-th and i-th changes of velocity, where $T_0 = 0$. $\{T_i\}$ is i.i.d.;
11. the random variables V and P are statistically independent;

12. the continuous random variable S is the distance between points of two consecutive changes of velocity or until the completion of a call from the instance of velocity change. S_i is the traveling distance between $(i-1)$-th and i-th changes of velocity and $\{S_i\}$ is i.i.d. with a following relationship: $S_i = T_i \cdot V_i$;
13. the discrete random variable M is the number of changes of velocity during a call which is geometrically distributed with parameter p;
14. the discrete random variable K is the number of changes of velocity just before leaving the call origination cell. If a mobile remains inside the cell during a call, then $K \geq M+1$;
15. the continuous random variable Γ is the total call duration. Thus $\Gamma = T_1 + T_2 + \ldots + T_M + T_{M+1}$;
16. and so on.

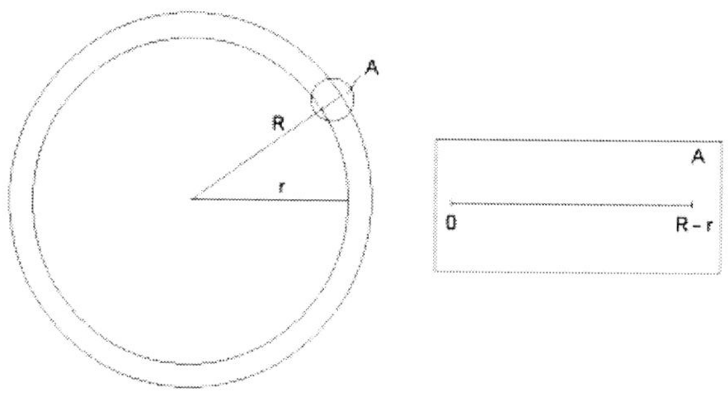

Fig. 1. Sojourn area and its one-dimensional projection

It is clear that this description of modeling is not only very complicated but also unreliable. In other words, why V_i was chosen to be uniformly distributed? Why only four directions are possible for a mobile? And what is the reason for exponential distribution between velocity changes? We can add more questions. Hence it is no wonder that these statements lead to quite complicated analytical derivations including consideration of 8 types of events between two consecutive changes of a mobile velocity.

This complexity is so obvious that the new model is proposed in [2]. In that paper the authors propose to simplify tasks by modifying assumptions 6 and 7 above: the square form is proposed for a cell shape and the SHR is now located between inner square and outer octagon, see Fig. 2. This approach simplifies analytical expressions in comparison to those from [1], however the number of possible transitions is increased up to 18. Therefore the task still is rather complex.

We think that the refusal from a round shape of a cell is a wrong decision and we are not lonely in this opinion, refer to [7] where the round shape of a

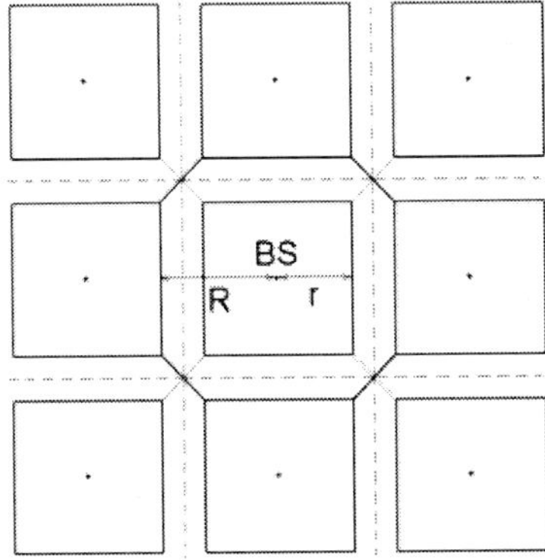

Fig. 2. Square shaped cells from [2]

cell is used successfully for the evaluation of soft handoff. In this we propose to simplify and improve reliability of the model by changing all the assumptions after 6 as follows:

- a mobile can move in *any* direction;
- observations of a mobile position are made in constant time intervals Δt, so the time for the i-th observation is $i \cdot \Delta t$;
- call durations T_i^c are i.i.d. random variables and the distribution (let us denote its cumulative function as $G(t)$) is a renewal one.

Now we present some more assumptions which require explanations on one hand and help us in the problem simplification on the other hand.

First. As *only* the time in a handoff zone is of interest, we assume that a call starts with some probability ω on the border of the zone, or uniformly at any point inside the zone. As no difference exists for sojourn between mobiles exiting from the sojourn zone into an inner zone of any neighboring cell and motions are omni-directional, we consider all mobiles incoming into the sojourn zone as penetrating through one border (say, the inner one).

Second. As inter-observation intervals are constant, we assume that a mobile's velocity is a step function unchanged on them. So the distance that a mobile moves between observations at time t_i and t_{i+1} is simply $V_i \cdot \Delta t$. For more adequacy we also assume that there is non-zero probability ν that the velocity V_i is 0 (mobile is immovable during the time step). Each time when the inter-observation interval ends the new value of velocity is being chosen based on the given probability distribution and values are i.i.d.

Third. As any movement along the circle is out of our interest from the point of sojourn time, we analyze only its projection on the radial direction. Therefore our task remains in one dimension (see Fig. 1).

3 The Probability Distribution for the Sojourn Time

In this section, we discuss about possible initial distributions needed for obtaining the distribution for the sojourn time. As usual, we assume that random walk process is renewable and thus a mobile that is detected on the border is treated as if it just starts a call.

3.1 Distribution of a Mobile Speed

Let us define the distribution of an absolute value of a velocity in the radial projection. As the most plausible variant we choose the following.

$$F_V(x) = \begin{cases} (1-\nu)F(x) & \text{if } x < 0 \\ \nu + (1-\nu)F(x) & \text{if } x \geq 0 \end{cases} . \tag{1}$$

Here $F(x)$ is a cumulative function of some distribution (it is discussed later) with non-zero expectation V and ν is the non-zero (in a common case) probability of zero speed as it was mentioned above. Negative velocity means that a mobile goes toward the inner border of a sojourn zone.

The function $F(x)$ we obtain is based on the following assumptions:

1. the mobile user can be a pedestrian or a driver (passenger) of some vehicle;
2. the mobile user can change his status during the call duration (a pedestrian takes a seat in a bus or a passenger gets off from a taxi, for example).

From this we have that

$$F(x) = \delta F_p(x) + (1-\delta)F_t(x) \tag{2}$$

where $F_p(x)$ and $F_t(x)$ are the cumulative distribution functions of velocity for cases of pedestrian and vehicle, respectively, and δ is the probability of the mobile user to be a pedestrian.

Hence, the final common expression for the cumulative distribution function of a velocity is

$$F_V(x) = \begin{cases} (1-\nu)[\delta F_p(x) + (1-\delta)F_t(x)] & \text{if } x < 0 \\ \nu + (1-\nu)[\delta F_p(x) + (1-\delta)F_t(x)] & \text{if } x \geq 4 \end{cases} . \tag{3}$$

As for the kind of $F_p(x)$ and $F_t(x)$ we choose normal (Gauss) distributions with expectations m_p and m_t, and standard deviations σ_p and σ_t, correspondingly. Definitely it is more plausible to consider the Gauss distribution for the motion in Cartesian coordinates and make transformation into polar ones, but for a narrow ring the difference is negligible.

Now the task for obtaining the distribution of the life-time of an active mobile inside the sojourn zone T_s is reduced to the task of obtaining the distribution of a minimum value from a pair $\{T_c^z, T_z\}$, where T_z is time for a mobile to be inside the sojourn zone and T_c^z is a part of call duration from the moment when a mobile enters the sojourn zone till a call end. Based on our assumption for the distribution of a call duration being renewal and from (3) we have

$$F_s(x) = P(\min\{T_v^z, T_z\} < x) = P(T_c^z < x \vee T_z < x) \\ = \begin{cases} G(x) + F_{T_z}(x) - G(x) \cdot F_{T_z}(x) & \text{if } x > 0, \\ 0 & \text{otherwise} \end{cases} . \tag{4}$$

Therefore, our task is to obtain the distribution of T_z. Analytical solution was discussed in [5,8] where the numerical scheme for obtaining $F_s(x)$ in a table form is obtained. Here we present experimentat results obtained by the Monte Carlo method.

3.2 Model Parameters

In this subsection, we employ reliable or at least reasonable initial data for simulation. First of all let us estimate parameters fof initial distributions. Average speed of a pedestrian lies obviously somewhere between 3 and 6 km, meanwhile average vehicular speed should be chosen from wider range and highly depends on the area: it may be from 20 to 40 km for city area and up to 70-80 km for a countryside. We consider the projection of a vector of a move onto a cell radius and the average absolute value of speed with which carrier goes to the border is $V_o = \frac{2}{\pi}V$, where V is linear speed of a carrier. The multiplier is simply obtained by transformation of a motion in Cartesian rectangular coordinates into motion in polar ones. So an average speed with which a pedestrian goes to a cell border is about 2-4 km in the meantime the speed for a vehicle is about 12-27 km for a city and 47-53 km for a country region. As for a reliable variance, we obtain it from a simple reason that probability of a speed being more than twice of the average must to be negligible, say not more than 1%. Thus if an average speed of a pedestrian is 3 km then the corresponding standard deviation is 1.29 km.

Next we need the probability ω that an active mobile gets into the sojourn zone through border (call does not start inside the zone but continues). For this we use the following simulation. The calls start inside the inner part of a cell (radius r), and motions are simulated according to the scheme described above and M_e – the number of mobiles that leave the inner part while calls have not yet ended – is stored. Let M be the total number of simulated calls. As a part of M in the whole number of calls that start inside the whole cell (radius R) is $\frac{r^2}{R^2}$, then ω is obtained as

$$\omega = \frac{M_e r^2}{M_e r^2 + M(R^2 - r^2)} .$$

For $R - r \ll r$ we consider the flows of mobiles through inner and outer borders to be equal and then we can simplify our simulation model and treat all the incoming calls as entering through inner border. Thus

$$\omega = \frac{2M_e r^2}{2M_e r^2 + M(R^2 - r^2)} . \tag{5}$$

Categorizing carriers into pedestrians and passengers is not a constant and greatly depends on the area: almost all are pedestrians inside a city park and almost all are passengers on the highway crossing. Hence we try $\delta = 0.0, 0.1, 0.2, \ldots, 1.0$. The intensity of a call flow is of no difference in our task. The average call duration $\bar{T}_c$ we choose is equal to 3 minutes that is rather a common choice. However we make experiments for $\bar{T}_c$ to be equal to 2 and 4 minutes as well. We need also the value of ν – the probability that a mobile is immovable. This probability can significantly vary in different areas: on a road it is obviously smaller than that in a city park or recreation zone. We have no real reliable information about its value but being guided by a common sense we simply made simulation with different $\nu \in [0, 0.3]$.

As for cell radius and width of sojourn zone we choose R to be 400,500,800,1200 and 2500 meters and r is equal to either $0.8R$ or $0.9R$.

4 Simulation Results

The simulation model is really rather simple and does not need any additional explanation. The most interesting results are presented here. Time interval Δt between changes of speed is equal to 5 sec.

Dependency of ω from a Cell Radius: We have obtained the following results for $\bar{T}_c$=3 minutes by making 100000 experiments for each variant with $r = 0.8R$ and $\nu = 0.1$: It is easy to see that value of ω decreases with the growth

Table 1. Dependency of ω from a cell radius

δ	0.2	0.5	0.8
$R = 400m$	0.684	0.677	0.652
$R = 500m$	0.682	0.673	0.644
$R = 800m$	0.673	0.660	0.618
$R = 1200m$	0.662	0.544	0.583
$R = 2500m$	0.626	0.589	0.489

of a cell radius. This is due to the fact that with the growth of a cell radius more calls have time enough for completion inside the inner part of a cell.

Distribution of a Sojourn Time: Simulation has been made for some combinations of parameters and values of ω were taken from the table presented above. As expected, the experiments show that the distribution of a sojourn time greatly depends on the zone width and average speed of a mobile carrier, the last depends on division of carriers on pedestrians and passengers. In Fig. 3 we can see this clearly (the kind of distribution is the same but the scale is quit different). In the first case the average sojourn time is about 8 seconds (maximum – 45 sec) while in second, where part of pedestrians is a majority, the average sojourn time is about 17 seconds (maximum – 135 sec). For a larger cells the dependency on

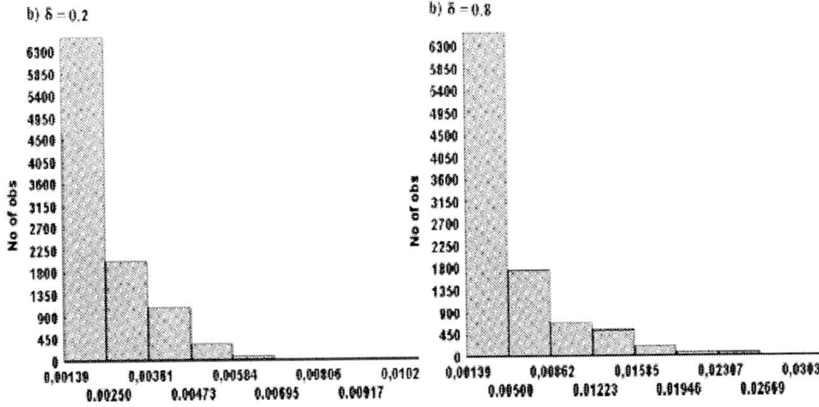

Fig. 3. Histograms of a sojourn time for $R = 400$m, $r = 320$m

average speed is even more. In the Table 2 the results of experiments with a cell outer radius equal to 2500 m and different inner radiuses are presented. The number of experiments for each radius was 100000, $\nu = 0.05$, $\delta = 0.2$. In the

Table 2. Dependency of sojourn parameters from the width of sojourn zone

r	ω	T_s	$\max\{T_s\}$
2000	0.626	0.00682	0.0278
2050	0.669	0.00628	0.0264
2100	0.712	0.00577	0.0250
2150	0.752	0.00524	0.0236
2200	0.792	0.00471	0.0225
2250	0.831	0.00413	0.0209
2300	0.868	0.00355	0.0181
2350	0.904	0.00299	0.0167
2400	0.937	0.00242	0.0139
2450	0.970	0.00176	0.0125

Fig. 4 the dependency of an average time of sojourn from the width of a zone is presented (column 3 of the Table 2). It is clear that dependency is almost linear. The same is with maximum value also.

5 Conclusions

Detailed simulation of the motions of an active mobile inside the sojourn zone based on the assumption of a round form a cell gives results that differ from results presented in [1,2]. It seems to us that our assumptions are more reliable

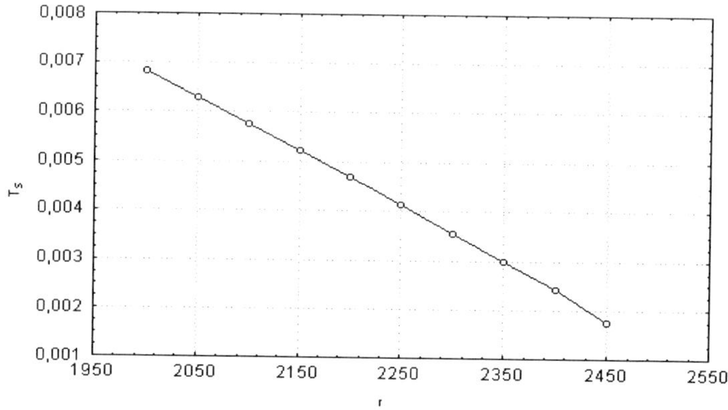

Fig. 4. Dependency of an average time of sojourn from the width of a zone

than those in these papers. The comparison of simulation results with those obtained by analytical solution for the distribution of a sojourn time in [8] shows good coincidence.

References

1. Suwon Park, Ho Shin Cho, and Dan Keun Sung, "Modeling and Analysis of CDMA Soft Handoff," *Proceedings of the VTC'96*, Atlanta, U.S.A., pp. 1525–1529, 1996.
2. Jae Kyun Kwon and Dan Keun Sung, "Soft Hadoff Modeling in CDMA Cellular Systems," *Proceedings of the Vehicular Technology Conference, IEEE 47th, Vol. 3*, pp. 1548–1551, 1997.
3. W. Feller, *An introduction to probability theory and its applications,* John Wiley&Sons, 1971.
4. F. Spizer, *Principles of random walk*, Princeton, New Jersey, 1964.
5. A.S. Rodionov, V.S. Antyufeev, H. Choo, and H.Y. Youn, "About One Problems of Soft Handoff Modeling," *Proc. of the ICS-NET'2001 Int. Workshop, Moscow*, pp. 205–209, 2001. (in Russian)
6. J.M. Hammersley and D.C. Handscomb, *Monte-Carlo Methods*, Methuen&Co, Ltd, 1964.
7. A. Nagate, M. Murata, H. Miyahara, and M. Sugano, "An Integrated Approach for Performance Modeling and Evaluation of Soft Handoff in CDMA Mobile Cellular Systems," *Proc. of the VTC'2000, IEEE-VTS 52nd, Vol. 6*, pp. 2605–2610, 2000.
8. A.S. Rodionov, V.S. Antyufeev, H. Choo, and H.Y. Youn, "Some Problems of Soft Handoff Modeling," *Proc. of the Seventh International Conference on Information Networks, Systems and Technologies, Belarus, October 2–4, 2001, Vol. 1.*, pp. 34–38, 2001.

A QoS Provision Architecture for Mobile IPv6 over MPLS Using HMAT

ZhaoWei Qu[1], Choong Seon Hong[2], Sungyoung Lee[3]

[1] School of Electronics and Information, Kyung Hee University, Korea 449-701
quzhaowei@hotmail.com
[2] School of Electronics and Information, Kyung Hee University, Korea 449-701
cshong@khu.ac.kr, sylee@oslab.khu.ac.kr

Abstract. The Resource Reservation Protocol (RSVP) provides a signaling mechanism for end-to-end QoS in Integrated Services Internet. Multi-Protocol Label Switching (MPLS) is a fast label-based switching technique that can offer new QoS capabilities for large scale IP networks. In this paper, in order to obtain more efficient use of scarce wireless bandwidth, increase data rate and reduce QoS signaling delay and data packet delay during handoff in Mobile IPv6, we propose a novel scheme that improves the efficiency using a Hierarchical Mobile Agents Tree (HMAT) based on the definition of new option called QoS Object Option (QOO) over MPLS. The HMAT can be chosen and configured in any way as the network administrator thinks appropriate. The QOO is included in the hop-by-hop extension header of certain packets that carry Binding messages. Mobile agents are required to manage QOO, resource reservation and other mobility related tasks on behalf of mobile hosts. The root mobile agent of the HMAT is also an edge Label Switch Router (LSR) in the MPLS Core Network. This paper defines the efficient signaling and control mechanisms to support QoS in Access and Core Network for Mobile IPv6.

1 Introduction

The Internet Engineering Task Force (IETF) has introduced the Mobile IPv4 [1] and Mobile IPv6 [2] to interoperate seamlessly with protocols that provide real-time services in the Internet. Multi-Protocol Label Switching (MPLS) is a fast label-based switching technology that integrates the label-swapping paradigm with network-layer routing [3]. Resource Reservation Protocol (RSVP) [4] [5] is a resource reservation setup protocol designed for a wired network. Provision of end-to-end QoS in wireless networks is more complex [6] than in wired networks because of the user mobility.

In this paper, we will propose a novel scheme that defines the signaling and control mechanisms to support QoS using the Hierarchical Mobile Agents Tree (HMAT) based on the definition of new option called QoS Object Option (QOO) [12] over MPLS to improve the efficiency during handoff in Mobile IPv6.

In section 2, we provide related works, and in section 3, we describe our scheme to

M.H. Shafazand and A M. Tjoa (Eds.): EurAsia-ICT 2002, LNCS 2510, pp. 569–579, 2002.

provide a new QoS mobility support. In section 4, we present simulation results to prove the efficiency of our scheme. Finally, we give our conclusions in section 5.

2 Related Works

Recently there have been some works [7-14] that focus on the handoff management problem. In [7], the drawback of this architecture proposed by Talukdar is that a mobile knows the addresses of all the subnets it is going to move into and which is not always possible. In [8], the proposal proposed by Mahadevan is based on the assumption that a base station knows the addresses of the base stations in all the neighboring cells. In [9], the protocol proposed by Zhang may result in triangle routing problem, and the pre-provisioned RSVP tunnels are not flexible and efficient. In [10], Chen describes another signaling protocol based on IP Multicast Tree. Figure 1 shows an example for reservations. When the handoff occurs, the Multicast Tree will be modified dynamically. After the new Multicast Tree is formed, the Predictive Reservation from Merge Point to current mobile proxy is switched into Conventional Reservation. The original Conventional Reservation from the Merge Point to the original mobile proxy is switched to Predictive Reservation, and some new Predictive Reservations along the new Multicast Tree from the source to the neighboring cells surrounding the current cell of mobile host should be set up. Then the flow of data packets can be transmitted over that new Conventional Reservation link. In this protocol, there are 8 additional messages presented to complete the functions of Multicast Tree modifying

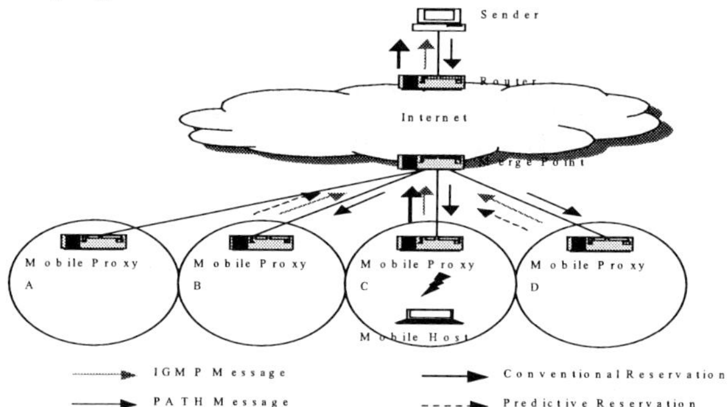

Fig. 1. RSVP Mobility Based on Multicast Tree

and RSVP setting up. Talukdar [7], Mahadevan [8] and Chen [10] have a problem that is how to predict the Mobile node's movement behavior so that pre-reservations can be done only in necessary cells to reduce the QoS signaling delay, the data packet delay. In [11], it is difficult to choose a proper router as the Nearest Common Router, and this scheme is not feasible. In [12], when the MN is receiver in Access Network, the Binding Acknowledgment has to be used so that the proposal is not efficient, and has more data packet delay. In [13], the handoff message has to be used only for Access Network, thus the flexibility is not better. In [14], the QoS provision Architecture has been considered only in the Access Network for the Mobile IPv6. In [17],

Zhong considered the supporting MPLS only in Mobile IPv4, and the QoS guarantee has not been considered. In [18], Choi does not show details of the data delivery procedure and QoS guaranteed LSP setup procedure.

3 Proposed Scheme

In this section we propose a framework using a Hierarchical Mobile Agents Tree (HMAT) based on the definition of new option called QoS Object Option (QOO) [12] to get more efficient use of scarce wireless bandwidth, minimize the QoS signaling delay, the data packet delay and losses and get higher data rate during handoff in mobile IPv6 environment.

3.1 QoS Object Option (QOO)

Table 1. Composition of a QoS Object

<table>
<tr><td></td><td>0</td><td>0</td><td>1</td><td colspan="2">Option Type 5bit</td><td>Option Data Len 8bit</td></tr>
<tr><td>Reserved</td><td colspan="5">Object Length 8bit</td><td>QoS Requirement 8bit</td></tr>
<tr><td colspan="5">Max Delay (ms) 16bit</td><td colspan="2">Delay Jitter (ms) 16bit</td></tr>
<tr><td colspan="7">Average Data Rate 32bit</td></tr>
<tr><td colspan="7">Burstiness : Token Bucket Size 32bit</td></tr>
<tr><td colspan="7">Peak Data Rate 32bit</td></tr>
<tr><td colspan="7">Minimum Policed Unit 32 bit</td></tr>
<tr><td colspan="7">Maximum Packet Size 32 bit</td></tr>
<tr><td colspan="7">Values of Packet Classification Parameters</td></tr>
</table>

This option is included in the hop-by-hop extension header of certain packets carrying Binding Update message in Mobile IPv6. The composition of a QOO [12] is shown in Table 1 by using TLV format. A QoS Object is not only an extension of RSVP QoS that can be used in Access Network, but also used in Core Network to get a better QoS support. In QOO, the QoS Requirement describes the QoS requirement of the MN's packet stream, the fields Max Delay and Delay Jitter specify the delay that packet stream can tolerate, the fields Average Data Rate, Burstiness, Peak Data Rate, Minimum Policed Unit and Maximum Packet Size describe the volume and nature of traffic that the corresponding packet stream is expected to generate, the field Packet Classification Parameters provide values for parameters in packet headers that can be used for packet classification.

3.2 Hierarchical Mobile Agents Tree (HMAT)

Our hierarchical mobile agents tree is aimed at solving the problem that the QoS signaling delay, the data packet delay will increase and the data rate will decrease and the packet losses, possible service degradation may occur due to frequent handoff. The HMAT means Hierarchical Mobile Agents Tree that contains mobile agents of several levels, and can be chosen and configured in any way as the network administrator thinks appropriate.

A mobile agent is an entity that manages QOO, resource reservations and other mobility related works. Mobile agents in a HMAT can be divided into two kinds. First type is the mobile agent in a domain and the first level of the HMAT, similar to home agents in Mobile IPv6, manages QOO for QoS support, processes the mobile related RSVP messages and maintains the mobile soft state for mobile hosts, is organized into a hierarchy to handle local movements of Mobile hosts within the domain. And the second type is the mobile agent in higher levels of the HMAT can manage QOO for QoS support, merge path message and reservation message. The first type mobile agent's function includes the second type mobile agent's function. Specially the root of the HMAT, as a second type mobile agent, is also a Label Switch Router (LSR) in the MPLS Core Network.

3. 3 Mobile IPv6 Support in MPLS Core Network

In our scheme, we integrate Mobile IPv6 with MPLS, and the QOO [12] is used to interoperate with RSVP-TE [15][16] to make the performance of QoS-sensitive applications running on the MN maintained at a desired level. We also use the HMAT in the Access Network where the Mobile Node can roam freely and the root of the HMAT is as an edge LSR in the Core MPLS network. Thus the LSP is established only between the root mobile agent of the HMAT and CN's edge LSR. And the design principle is that only active data is supposed to traverse over QoS guaranteed LSP. This would be efficient to save the bandwidth on network and to reduce end-to-end delay. Our scheme considers the smooth handoff support in Mobile IPv6 over MPLS to reduce the signaling load and the latency or interruption due to handoff. There are no additional messages on legacy MPLS signaling to setup QoS guaranteed LSP. There are no obligation of MPLS signaling on the Mobile Node because of the use of the HMAT so that MN does not need to install RSVP-TE at all, this can reduce memory cost and complexity of a MN device

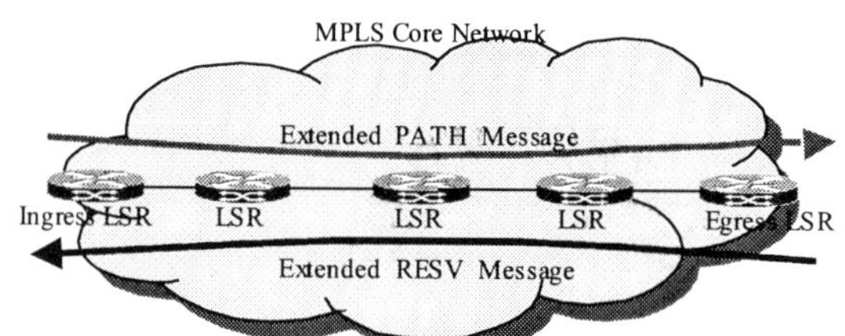

Extended PATH Message (Session, Label-Request Object, Explicit-Route Object, Session-Attribute Object, Record-Route Object)

Extended RESV Message (Session, Label Object, Record-Route Object)

Fig. 2. LSP Setup Using RSVP-TE

3. 3. 1 LSP Setup Using RSVP-TE

We use RSVP-TE to extend the RSVP, allow the establishment of the explicitly routed Label Switched Path (LSP) Tunnel which can be automatically routed away from network failures, congestion and bottlenecks, and distribute label-binding information using the RSVP-TE as a signaling protocol. We extend the RSVP to setup the LSP because of already existing RSVP implementations in our HMAT and it would provide a unified signaling system within the whole network. Figure 2 shows this procedure of LSP setup using RSVP_TE

3. 3. 2 Smooth Handoff

When handoff occurs, the MN sends the Binding Update with QOO along the HMAT to the root of the HMAT. The root of HMAT, as an edge LSR in MPLS Core Network, receives the Binding Update with QOO from the MN, examines the QOO, then sends the PATH message of RSVP-TE to the CN's edge LSR. The CN's edge LSR receives the PATH of RSVP-TE and Binding Update with QOO, sends the RESV message of RSVP-TE back to the root of the HMAT to setup the QoS guaranteed LSP, also sends the Binding Update with QOO to the CN. The CN receives the Binding Update with QOO, sends back the Binding Acknowledgement to the MN. The root of the HMAT receives the RESV of RSVP-TE, then the new QoS guaranteed LSP has been established between the root of the HMAT and the edge LSR of the CN before the MN receives the Binding Acknowledgment. Figure 3 shows this smooth handoff support in MPLS Core Network

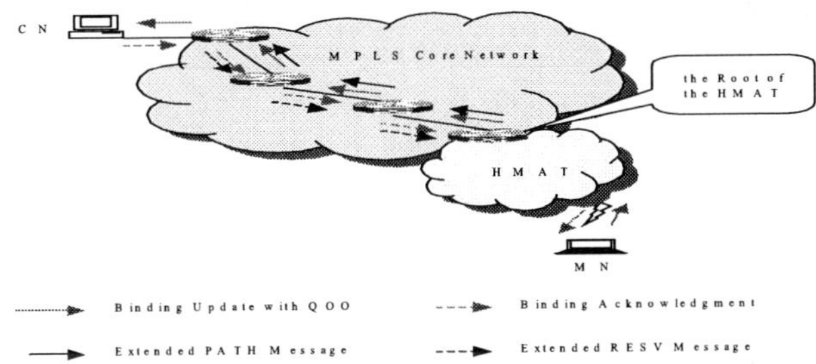

Fig. 3. Smooth Handoff Support in MPLS Core Network

3. 3. 3 Data Delivery Procedure

The packets from the CN to the MN are forwarded to the edge LSR of CN. Then the edge LSR of CN receives the packet from CN, searches the IP routing table, sends the packet to the MPLS layer, then looks up its label table to find the out label and out port, inserts the label to the packet and sends the packet along the LSP that has been setup to the root of the HMAT. The root of the HMAT receives the packet, and then looks up its label table. The out label and out port are empty, therefore the root of the

HMAT striped off the label and sends the packet to the IP layer, finally looks up its routing table and forwards the packet to the MN along the HMAT.

The data delivery procedure from the MN to the CN is the same with the data delivery procedure from the CN to the MN.

3.4 QoS Support Using HMAT

There are two scenarios in QoS support based on our HMAT scheme.

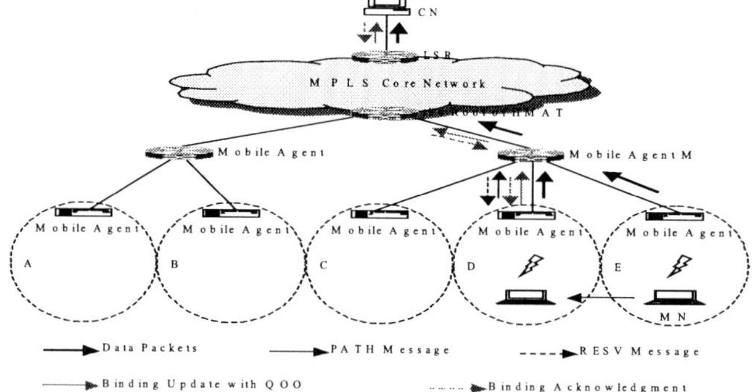

Fig. 4. MN as Sender in HMAT Model

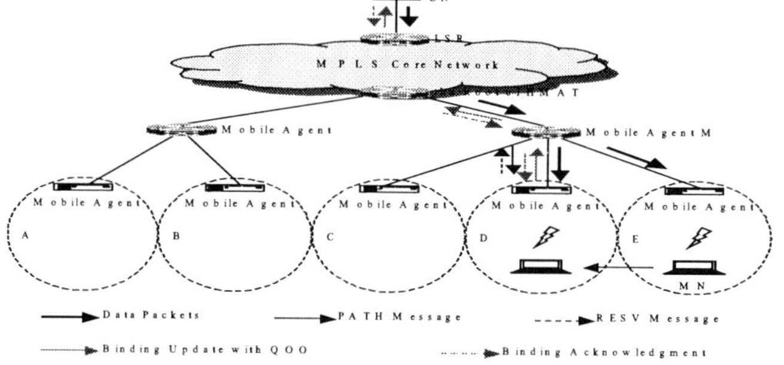

Fig. 5. MN as Receiver in HMAT Model

When the MN is sender, the CN is receiver, after a handoff, the MN sends a Binding Update with QOO to CN along HMAT, in the first level mobile agent, this agent examines QOO and immediately performs the resource reservation, sends the new PATH message to CN with the same source flow identity as the one before handoff, and also sends the Binding Update with QOO to the CN. Then the PATH message can be merged at the merging mobile agent M that has already a path state in HMAT for that flow which is created before. This will make RSVP to have a Local Repair for sender route. Therefore the mobile agent sends a RESV message associated with the flow along the new path in HMAT to the MN upstream at once, also sends the Binding Update with QOO to the CN downstream. The flow path reserved resources previously from the mobile agent to the CN can be reused. After the CN receives the Binding Update with QOO, the CN will send the Binding Acknowledgment to the MN's current location through the HMAT. Then the data packets will be sent from the MN's new location to the CN. Figure 4 shows this scenario.

When the MN is receiver and the CN is sender, the difference with previous scenario is that when the Binding Update with QOO gets to the merging mobile agent M, the M examines QOO and immediately sends the new PATH message to the MN downstream and at the same time sends the Binding Update with QOO upstream to the CN. When the MN receives the new PATH, it sends RESV to the M. And after the CN receives the Binding Update with QOO, the CN sends the Binding Acknowledgment to the MN's current location. Figure 5 shows this scenario.

We use Rational Rose 2000 to show the Sequence Diagram of the Multicast Tree scheme in Figure 6, the Sequence Diagram of our HMAT scheme in Figure 7 and Figure 8

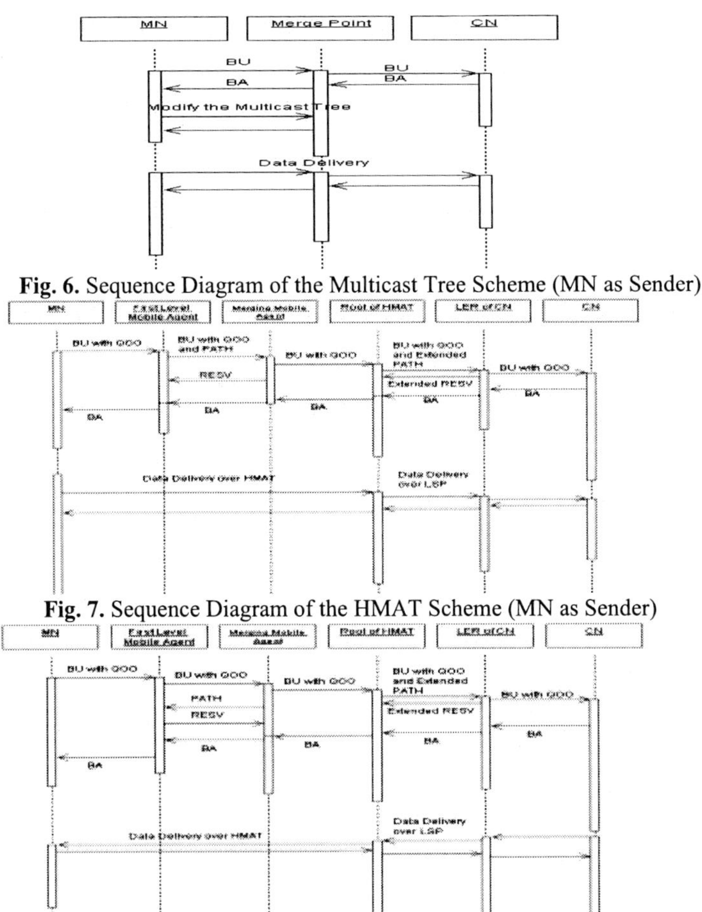

Fig. 6. Sequence Diagram of the Multicast Tree Scheme (MN as Sender)

Fig. 7. Sequence Diagram of the HMAT Scheme (MN as Sender)

Fig. 8. Sequence Diagram of the HMAT Scheme (MN as Receiver)

4 Simulation Results

We use the OPNET Modeler v8.0 to simulate our scheme and compare our scheme with RSVP Mobility Based on Multicast Tree [10]. For simplicity our simulation is based on an assumption that the capacity of the links between the mobile agents is not limited. And we only considered the unicast data flows from a single mobile sender roaming freely at a rate of 5km/hr in wireless domain to a fixed static receiver for simplicity. Figure 9 shows the network topology used for our simulation. There are two cells in this network, we set the radius of each cell 1km, and each cell has a mobile agent as the first type mobile agent in HMAT that has two levels.

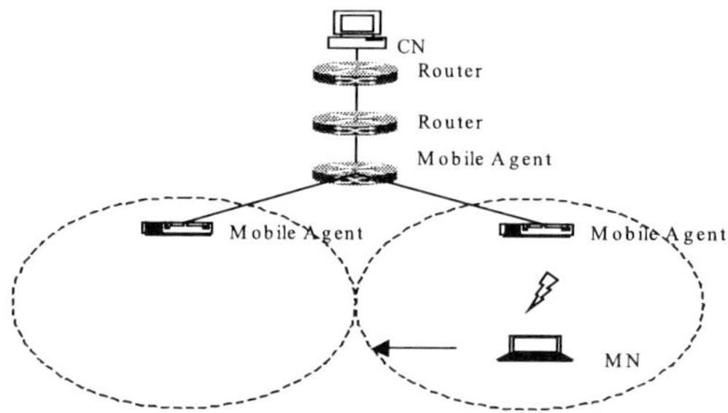

Fig. 9. Network Configuration for Simulation

The goal of our simulation work is for evaluating the QoS, such as data rate, packet loss ratio and packet delay, by using our scheme and comparing with RSVP Mobility Based on Multicast Tree [10] when the handoff occurs. We use a real time traffic source at a peak rate of 1Mbps to get the variations of data rate, packet loss ratio and packet delay received by the fixed static receiver CN due to the handoff.

Figure 10 shows the simulation results of the data rate using our HMAT scheme and Multicast Tree scheme over simulation time. In the figure, the X-axis represents the simulation time (minute) and the Y-axis represents the relative data rate (kbps). We can see that the data rate of Multicast Tree scheme is obviously decreased, especially at the moment a handoff takes place, and our HMAT scheme has a real smooth handoff. Figure 11 shows the packet loss ratio of the Multicast Tree scheme is more than that of the HMAT scheme especially when handoff takes place. Figure 12 shows the Multicast Tree scheme has more packet delay than our HMAT scheme especially when handoff takes place.

These simulation results have the following reasons. First, in the Multicast Tree scheme, when the handoff occurs, the Binding Update is sent to the CN and the MN receives the Binding Acknowledgment, then the Multicast Tree should be modified dynamically and some QoS signaling messages should be used, the new Conventional Reservation and all of the Predictive Reservations should be made again, then the data packets is be sent continually. Second, in our HMAT scheme, there are no any extra QoS setting up delay not only in the Access Network but also in the Core Network due to handoff and all the packets will be offered QoS as desired. Therefore the Multicast Tree scheme has more QoS signaling delay, packet loss ratio, data packet delay and lower data rate than our HMAT scheme whenever a handoff occurs.

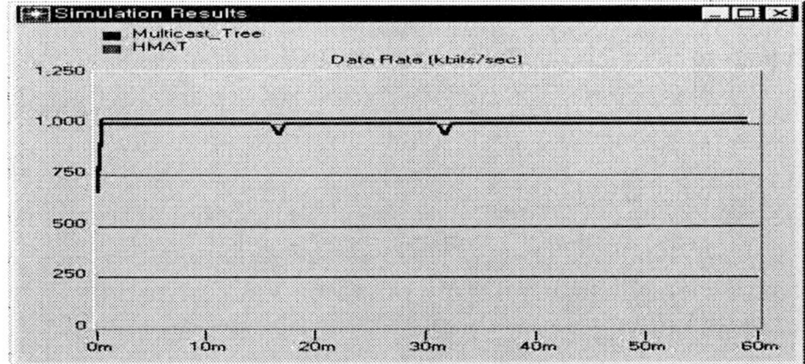

Fig. 10. Simulation Results (Data Rate)

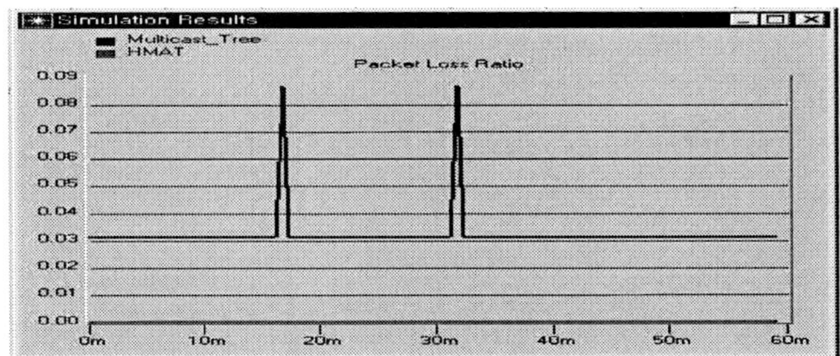

Fig. 11. Simulation Results (Packet Loss Ratio)

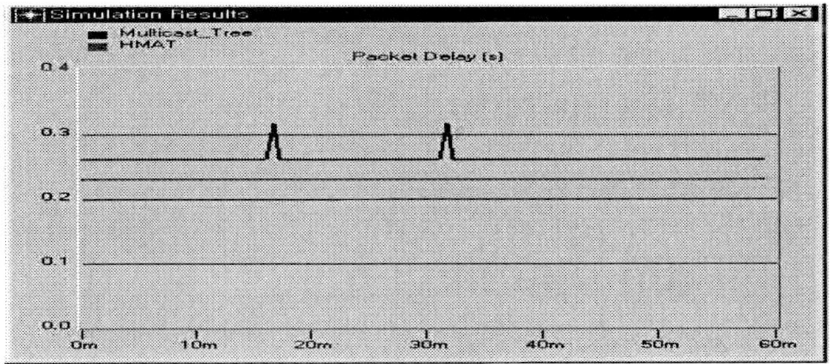

Fig. 12. Simulation Results (Packet Delay)

5 Conclusions

In this paper, a novel framework for QoS support in Mobile IPv6 using the HMAT based on QoS Object Option (QOO) [12] over MPLS has been proposed. When a handoff occurs, the RSVP can be made a Local Repair in the HMAT based on the QOO and the LSP can be setup using RSVP-TE and QOO during the time when the Binding Update with QOO and Binding Acknowledgement message are transmitted between the MN and the CN through the HMAT Access Network and the MPLS

Core Network. There are no any extra QoS setting up delay not only in the Access Network but also in the Core Network due to handoff and all the packets will be offered QoS as desired. The RSVP and RSVP-TE have been used because there is a unified signaling system applied in the whole network.

Moreover, we use OPNET Modeler to simulate our scheme and compare with RSVP Mobility Based on Multicast Tree [10]. The simulation results prove that our scheme can provide higher data rate, lower packet delay and packet loss ratio, and improve the efficiency by using HMAT based on QOO over MPLS when handoff takes place in Mobile IPv6.

Acknowledgement

This work was supported by grant (No.R02-2001-00976) from the Basic Research Program of the Korea Science & Engineering Foundation.

References

1. C.Perkins, " IP Mobility Support, " RFC 2002, October 1996.
2. D.B.Johnson and C.Perkins, " Mobility Support in IPv6, " IETF Internet-Draft, work in progress, November 2000.
3. E.Rosen, "Multiprotocol Label Switching Architecture, " RFC3031, January 2001.
4. R.Braden, L.Zhang, S.Berson, S.Herzog, and S.Jamin, " Resource ReSerVation Protocol (RSVP) Version 1 Functional Specification, " RFC 2205, September 1997.
5. J.Wrclawski, " The Use of RSVP with IETF Integrated Services, " RFC 2210, September 1997.
6. A.Terzis, J.Krawczyk, J.Wroclawski, L.Zhang, " RSVP Operation Over IP Tunnels, " RFC 2746, January 2000.
7. A.K.Talukdar, B.R.Badrinath, and A.Acharya, " MRSVP: A Reservation Protocol for an Integrated Services packet Network with Mobile Hosts, " Tech. Rep. Dcs-tr-337, Department of Computer Science, Rutgers University, U.S.A., 1997.
8. I.Mahadevan and K.M.Sivalingam, " An Experimental Architecture for providing QoS guarantees in Mobile Networks using RSVP, " IEEE PIMRC, Boston, September 1998.
9. A.Terzis and M.Srivastava and L.Zhang, " A Simple QoS Signaling Protocol for Mobile Hosts in the Integrated Services Internet, " INFOCOM 1999.
10. W.Chen and L.Huang, " RSVP Mobility Support: A Signaling Protocol for Integrated Services internet with Mobile Hosts, " INFOCOM 2000.
11. Qi.Shen and W.Seah and A.Lo, " Flow Transparent Mobility and QoS Support for IPv6-based Wireless Real-time Services, " IETF Internet-Draft, work in progress, February 2001.
12. H.Chaskar and R.Koodli, " A Framework for QoS Support in Mobile IPv6, " IETF Internet-Draft, work in progress, March 2001.
13. ZhaoWei Qu, ChoongSeon Hong, " QoS Provision Architecture for Mobile IP using RSVP, " Proceedings of KICS National Conference, Vol.23, pp.1551, Cheju, Korea, July 2001.
14. ZhaoWei Qu, ChoongSeon Hong, "A QoS Provision Architecture for Mobile IPv6 Using RSVP, " ICOIN16, January 2002.
15. D.Awduche, L.Berger, D.Gan, T.Li, V.Srinivasan, G.Swallow, "RSVP-TE: Extensions to RSVP for LSP Tunnels, " RFC3209, December 2001.

16. D.Awduche, A.Hannan, X.Xiao, " Applicability Statement for Wxtensions to RSVP for LSP-Tunnels, " RFC3210, December 2001.
17. Zhong Ren, C.K.Tham, C.C.Foo, C.C.Ko, "Supporting MPLS in Mobile IP, " ICT2000, May 2000.
18. J.K.Choi, M.H.Kim, Y.J.Lee, "Mobile IPv6 Support in MPLS Network, " IETF Internet-Draft, work in progress, December 2001.

A New Propagation Model for Cellular Mobile Radio Communications in Urban Environments Including Tree Effects

Reza Arablouei and Ayaz Ghorbani

Faculty of Electrical Engineering, Amirkabir University of Technology
P.O. Box 15875-4413, Hafez Ave., Tehran, Iran
Fax: +98 21 640 6469
rezaa219@yahoo.com

Abstract. An advanced model for radio wave propagation in vegetated urban areas is proposed. The new model is based on uniform theory of diffraction (UTD) and so appropriate for numerical calculation. We have recently proposed a new UTD-based model for multiple diffractions by buildings and in this paper we have endeavored to extend the new model to involve tree effects. We have supposed that apart from the high-rise urban core, a city's buildings are of nearly the same height and spacings and are organized by street systems into rows. In addition, trees are assumed to be situated adjacent to the buildings. The canopies of trees are modeled as discrete ensembles of leaves and branches. It is revealed that the new model including tree effects provides reasonable and convincing predictions.

1 Introduction

Radio signal characteristics force fundamental limits on the design and performance of the cellular mobile communication systems. Therefore, without considering effects of buildings and trees on radio wave propagation in vegetated urban environments, an accurate prediction for the path loss and coverage area will not be available. The more rigorous model for buildings and trees will conclude the more precise prediction.

A prior work tried by Torrico *et al.* [1] proposes a theoretical model for investigating the influence of trees on radio wave propagation. Their model is an extension of Walfisch and Bertoni's one [2] to include tree effects. In their approach, a row of buildings and the adjacent canopy of trees are represented by an absorbing screen and a partially absorbing phase screen, respectively. They have applied the physical optics and multiple Kirchhoff-Huygens integration [1].

In this study, the propagation environment is assumed as in [3], [4], which is shown in Fig. 1. Thus, the expressions derived and propounded in [3], [4] are developed in order to include effects of trees in the model. Furthermore, a heuristic UTD diffraction coefficient and higher-order diffracted fields introduced by Holm [5] are employed.

M.H. Shafazand and A M. Tjoa (Eds.): EurAsia-ICT 2002, LNCS 2510, pp. 580–587, 2002.

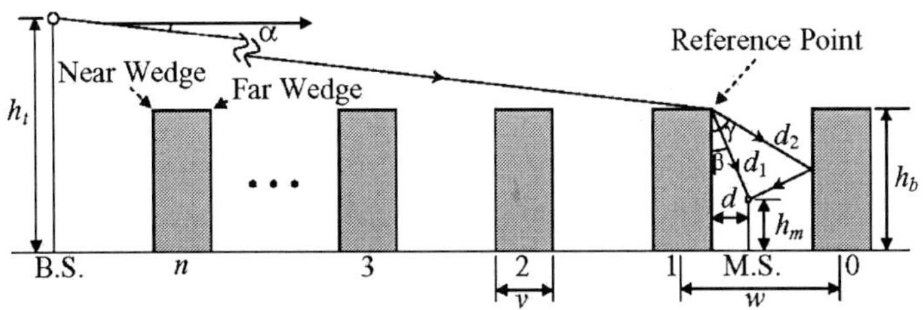

Fig. 1. Radio wave propagation in presence of buildings

Finally, the predicted results by the new model are presented and compared with some existing measurements. A good agreement between them is observable.

2 Applied Diffraction Coefficient

In the new model, we use a heuristic UTD diffraction coefficient for finitely conducting wedges, offered by Holm [5]:

$$D(\varphi,\varphi',L,n) = D^{(1)} + R_0 R_n D^{(2)} + R_0 D^{(3)} + R_n D^{(4)} \tag{1}$$

where R_0 and R_n are the reflection coefficients for the zero- and n-face, respectively (see Fig. 2(a)). The constituents $D^{(l)}$ (l=1, 2, 3, 4) are defined in [5].

At the case of double diffractions by two consecutive wedges, which grazing incidence happens at the second wedge, it is evident that first-order diffracted fields cannot precisely explain the multiple diffraction events. Hence, higher-order diffracted fields must be emphasized on account of ensuring accuracy of the prediction.

The field of the doubly diffracted ray in Fig. 2(b) can be written as

$$E_{UTD} = \frac{E_0 e^{-jks_T}}{s_T} A_s(s_1,s_2,s_3) \sum_{m=0}^{\infty} \frac{1}{m!}\left(\frac{-1}{jks_2}\right)^m \frac{\partial^m D_a}{\partial \varphi_1^m} \frac{\partial^m D_b}{\partial \varphi_2'^m} \tag{2}$$

where D_a and D_b are the diffraction coefficients relevant to the first and second wedges, respectively. The function A_s is the proper spreading factor and s_T is the total path length [5]. We rewrite (2) as

$$E_{UTD} = \frac{E_0 e^{-jks_T}}{s_T} A_s(s_1,s_2,s_3) \cdot DD(\varphi_1,\varphi_1',\varphi_2,\varphi_2',L_1,L_2,n_1,n_2,s_2) \tag{3}$$

in this manner, we define the new function DD, which facilitates the formulation. We take into account up to third-order diffracted fields (up to m=2 in (2)).

At the case of right-angled consecutive wedges, interior angles of all wedges equal to $\pi/2$, this implies that $n_1=n_2=3/2$, $\varphi_1=3\pi/2$, and $\varphi'_2=0$. Moreover, s_2 is assumed identical for all building rows. Therefore, we will suppress known variables throughout.

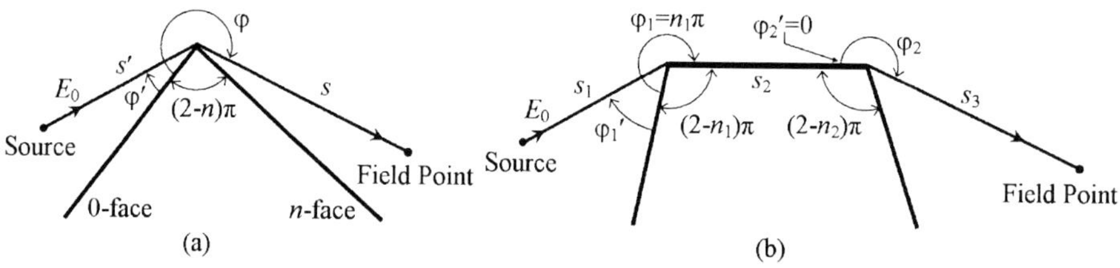

Fig. 2. a) Ray geometry for diffraction by a noncurved wedge b) Ray geometry for diffraction by two joined noncurved wedges

3 UTD-Based Enhanced Model for Multiple Diffractions by Buildings

Recently, we have proposed an enhanced model based on UTD for multiple diffractions by buildings [3], [4]. In this model, the most realistic geometry for buildings and multiple diffractions by buildings has been assumed. We use this model and try to modify it in order to include effects of trees.

3.1 Multiple Building Diffraction Loss

Path loss contribution due to multiple diffractions by buildings is defined as

$$L_{MBD} = -20log_{10}\left(\frac{E_n}{E_0}\right) \tag{4}$$

where E_0 is the transmitted field and E_n is the received electric field intensity at the reference point with n existing building rows. The reference point is defined just top of the last forwardly diffracting wall close to the mobile station as shown in Fig. 1.

Assuming n building rows and according to [3], [4], the received field normalized to the transmitted field E_0 can be written as

$$\frac{E(n)}{E_0} = e^{-jk[v+(n-1)w\cdot cos(\alpha)]}\left(D_1 + e^{-jkv[cos(\alpha)-1]}\right) + D_3 e^{-jk[w(1+(n-2)cos(\alpha))+v]} \cdot\left(D_4 + D_2 e^{-jkv[cos(\alpha)-1]}\right)\left(\frac{1-\left(D_5 e^{-jkw[1-cos(\alpha)]}\right)^{n-1}}{1-D_5 e^{-jkw[1-cos(\alpha)]}}\right) \tag{5}$$

where v is the thickness of the individual building row, w is the spacing of the building rows, and α is the incident plane wave's angle with the horizontal axis (see Fig. 1), also D_1 to D_5 are defined in [4].

3.2 Excess Path Loss

Excess path loss is achieved by (4) when substituting E_n with E_n^m, which represents the received field at the mobile station with n existing building rows.

In our scenario shown in Fig. 1, the normalized received field at the mobile station is found according to [4] as

$$\frac{E^m(n)}{E_0} = D_9 e^{-jknw} + e^{-jk[v+(n-1)w\cdot cos(\alpha)]}\left(D_7 + D_6 e^{-jkv[cos(\alpha)-1]}\right) + D_8 e^{-jk[w(1+(n-2)cos(\alpha))+v]}\left(D_4 + D_2 e^{-jkv[cos(\alpha)-1]}\right)\left(\frac{1-\left(D_5 e^{-jkw[1-cos(\alpha)]}\right)^{n-1}}{1-D_5 e^{-jkw[1-cos(\alpha)]}}\right) \quad (6)$$

where D_6 to D_9 are defined in [4].

4 Including Effects of Trees in the UTD-Based Enhanced Model

Trees are modeled as discrete random ensemble of leaves and branches all having prescribed orientation and location statistics [1]. Due to the randomness associated within the medium of discrete scatterers, the wave treat in a tree is more properly described by a modern stochastic model [6], [7]. Therefore, we are provided the basis for computing the mean field and the propagation constant in a multiple scatterer, i.e. a tree.

We avoid including all details here and are contented with using the previously derived results in our model.

If we consider a plane wave of unit amplitude and polarization $\hat{q}$ is incident in the direction $\hat{i}$, the mean field in the canopy of a tree can be written as

$$\langle E(x,z;\hat{q})\rangle = \hat{q}e^{jkx-jk_0\cos(\theta_i)z} \quad (7)$$

where

$$k = k_0\sin(\theta_i) + \frac{2\pi}{k_0\sin(\theta_i)}\sum_t \rho_t \left\langle f_{qq}^{(t)}(\hat{i},\hat{i})\right\rangle \quad (8)$$

and $\sin(\theta_i) = \hat{i}\cdot\hat{x}$. The x and z directions are revealed in Fig. 3.

Here, k is the propagation constant in the x direction of polarization $\hat{q}$ and $\left\langle f_{qq}^{(t)}(\hat{i},\hat{i})\right\rangle$ is the mean forward scattering amplitude over the scatterers orientation and the sum is over scatterer type t (leaves or branches). It can be written as

$$\left\langle f_{qq}^{(t)}(\hat{i},\hat{i})\right\rangle = \frac{1}{2\pi}\int f_{pq}^{(t)}(\hat{i},\hat{i})p(\theta)d\theta \quad (9)$$

where $p(\theta)$ is the probability density function of the inclination angle. The parameters f_{pq} and ρ_t are defined in [1].

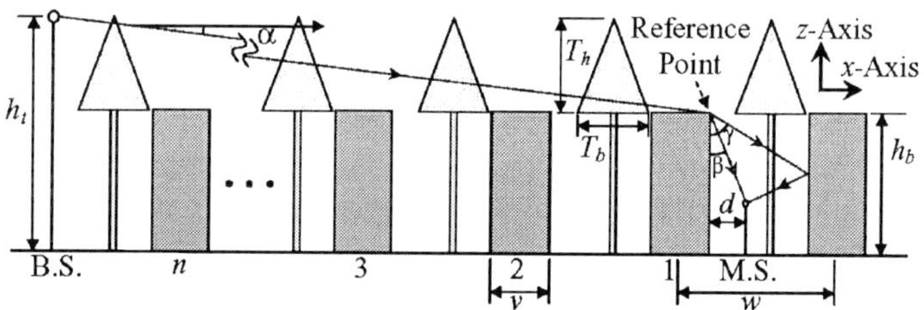

Fig. 3. Radio wave propagation in presence of buildings and trees

The approach used in [1] is based on physical optics, whilst our approach is based on UTD, so we only use the propagation constant expression (i.e. (8)), derived by Torrico *et al.* We should note that the propagation constant in a tree usually has a complex value.

In fact, in a UTD-based propagation model we deal with rays that connect the transmitter to the receiver. Thus, the influence of trees on the propagation process can be interpreted by the effects of trees on the rays. Knowing that a ray passing through a tree incurs a different propagation constant from the free space's one motivates us to modify (5) and (6) in order to take into account the effects of trees. Doing so, the length of the path that each ray passes through trees l_T is found, and then multiplying extra term $e^{-jk'_T l_T}$ modifies the received field intensity contribution transmitted via that ray, which is already known. We know $k'_T = k_T - k_0$, where k_T and k_0 are the propagation constants for tree and free space, respectively.

To be able to estimate l_T distances for all rays, we define L_L and L_R as

$$L_L = \sin(A)\cos(\alpha)(1+n_L)\left(\frac{T_b}{\sin(A+\alpha)} + w\frac{n_L}{2}\left(\frac{1}{\sin(A+\alpha)} - \frac{1}{\sin(A-\alpha)}\right)\right) \tag{10}$$

and

$$L_R = \sin(A)\cos(\alpha)(1+n_R)\left(\frac{T_b}{\sin(A+\alpha)} + \left(w\frac{n_R}{2}+v\right)\left(\frac{1}{\sin(A+\alpha)} - \frac{1}{\sin(A-\alpha)}\right)\right) \tag{11}$$

where L_L is the path length that a ray incident to the near wedge of the building row (closer to the transmitter) passes through the trees. Likewise, L_R is associated to the far wedge (farther from the transmitter). The near and the far wedges are shown in Fig. 1.

Parameters n_L and n_R are the number of trees that are passed through by the rays corresponding to L_L and L_R, respectively. Accomplishing some geometrical calculations, we can simply show

$$n_L = \left\lceil \frac{T_h / \tan(\alpha) - T_b / 2}{w} \right\rceil \tag{12}$$

$$n_R = \left\lceil \frac{T_h / \tan(\alpha) - T_b / 2 - v}{w} \right\rceil \tag{13}$$

where T_h and T_b are the height and the base of the tree canopies, respectively, as shown in Fig. 3. In addition, $\tan(A) = 2T_h / T_b$ and $\lceil \ \rceil$ means the integer part.

4.1 Multiple Building Diffraction Loss

According to the findings above, modifying (5) gives us

$$\frac{E(n)}{E_0} = e^{-jk[v+(n-1)w\cdot\cos(\alpha)]}\left(D_1 e^{-jk_T' L_L} + e^{-j(kv[\cos(\alpha)-1]+k_T' L_R)}\right)$$
$$+ D_3 e^{-j(k[w(1+(n-2)\cos(\alpha))+v]+k_T' T_b)}\left(D_4 e^{-jk_T' L_L} + D_2 e^{-j(kv[\cos(\alpha)-1]+k_T' L_R)}\right)$$
$$\cdot\left(\frac{1-\left(D_5 e^{-j(kw[1-\cos(\alpha)]+k_T' T_b)}\right)^{n-1}}{1-D_5 e^{-j(kw[1-\cos(\alpha)]+k_T' T_b)}}\right) \tag{14}$$

where, putting (14) in (4) yields us the multiple building diffraction loss in presence of trees.

4.2 Excess Path Loss

In a similar manner to the preceding subsection applied to (6), we are provided that

$$\frac{E^m(n)}{E_0} = D_9 e^{-j(knw+k_T' L_L)} + e^{-jk[v+(n-1)w\cdot cos(\alpha)]}\left(D_7 e^{-jk_T' L_L} + D_6 e^{-j(kv[\cos(\alpha)-1]+k_T' L_R)}\right)$$
$$+ D_8 e^{-j(k[w(1+(n-2)\cos(\alpha))+v]+k_T' T_b)}\left(D_4 e^{-jk_T' L_L} + D_2 e^{-j(kv[\cos(\alpha)-1]+k_T' L_R)}\right)$$
$$\cdot\left(\frac{1-\left(D_5 e^{-j(kw[1-\cos(\alpha)]+k_T' T_b)}\right)^{n-1}}{1-D_5 e^{-j(kw[1-\cos(\alpha)]+k_T' T_b)}}\right) \tag{15}$$

Then, putting (15) in (4) provides us the excess path loss in a vegetated urban area.

5 Results and Discussion

In Fig. 4, the settled multiple building diffraction loss (for a large n) is plotted versus α with and without including tree effects. The leaves of trees are assumed to have a radius a=4 cm and a thickness of t=0.4 mm, a dielectric constant of ε_r=26+j7 [1], and a density of ρ_l=250/m^3. The branches are supposed to have a radius a=1 cm and a branch length l=40 cm, a dielectric constant ε_r=20+j7, and a density ρ_b=2/m^3. The probability density for the leaves and branches in the azimuth coordinate φ is assumed to be uniformly distributed from 0° to 360°. The probability density in the θ coordinate is dependent on vegetation type. For the branches and the leaves it is considered to be uniformly distributed as $p_\theta(\theta) = 1/(\theta_2 - \theta_1)$, where for the leaves, θ_2=180° and θ_1=0°

and for the branches θ_2=60° and θ_1=0°. The tree canopy dimensions are taken as T_b=2.8 m and T_h=3 m.

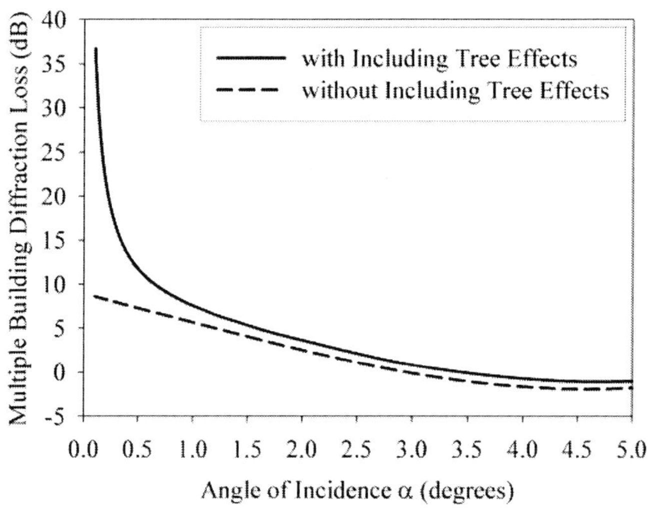

Fig. 4. The settled multiple building diffraction loss versus α predicted by the new model with and without including tree effects. Here f=920 MHz, w=40 m, v=20 m, ε_r=5.5, σ=0.023 S/m, T_b=2.8 m and T_h=3 m

From Fig. 4, it is evident that the multiple building diffraction loss in a vegetated area gives rise suddenly when α grows less than 0.5°.

In Fig. 5, we plot the excess path loss as a function of α predicted by the new model with and without including tree effects and Walfisch's physical-optics-based model [2]. Moreover, some measured data treated by Okumura [2] are comprised in Fig. 5.

Examining Fig. 5, one can conclude that the excess path loss predicted by the new model with including tree effects has a better agreement with Okumura's measurements. The agreement gets worse at α<0.3°. This can be explained in this fashion: when α decreases about 0.3° or less, slightly overestimation can take place at estimating the loss due to the rays, which pass through trees.

6 Conclusion

A UTD-based propagation model for predicting the path loss in the urban cellular mobile environments has been developed. This new model provides most rigorous expressions for path loss contributions with considering effects of trees. Utilizing this enhanced model makes available more precise prediction for the path loss in the

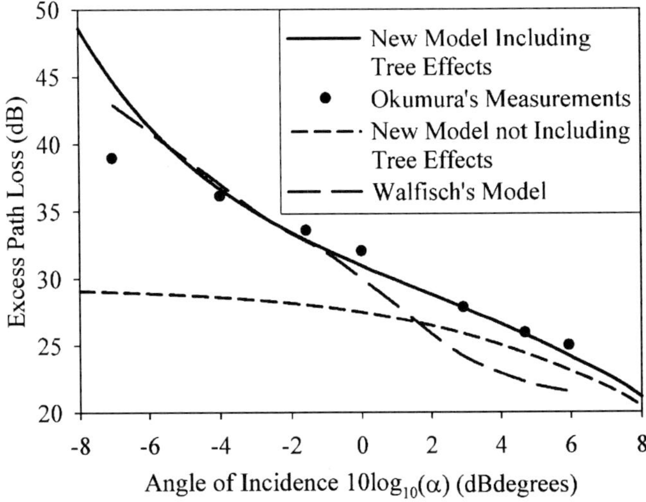

Fig. 5. The excess path loss versus α predicted by the new model with and without including tree effects and Walfisch's model [2] along with Okumura's measurements [2]. Here, f, w, v, ε_r, σ, T_b, and T_h are as in Fig. 4, also h_b=9 m, h_m=1.8 m, and d=7 m

vegetated urban mobile environments. Moreover, the new expressions are so appropriate for the numerical calculation. The new model can be used for predicting the path loss, coverage area, intercellular interference and other propagation characteristics for the cellular mobile radio communications in the vegetated residential areas.

References

1. S. A. Torrico, H. L. Bertoni, and R. H. Lang, "Modeling tree effects on path loss in a residential environment," *IEEE Trans. Antennas Propagat.*, vol. 46, pp. 872–880, June 1998.
2. J. Walfisch and H. L. Bertoni, "A theoretical model of UHF propagation in urban environments," *IEEE Trans. Antennas Propagat.*, vol. 36, pp. 1788–1796, Dec. 1988.
3. R. Arablouei, A. Ghorbani, "A New UTD-Based Model for Multiple Diffractions by Buildings," in *Proc. 3rd International Conference on Microwave and Millimeter Wave Technology ICMMT 2002*, Beijing, China.
4. R. Arablouei, A. Ghorbani, "An Improved Model Based on UTD for Multiple Diffractions by Buildings," in *Proc. International Seminar Day on Diffraction 2002*, St.Petersburg, Russia.
5. P. D. Holm, "A new heuristic UTD diffraction coefficient for nonperfectly conducting wedges", *IEEE Trans. Antennas Propagat.*, vol. 48, pp. 1211–1219, Aug. 2000.
6. M. X. Lax, "Multiple scattering of waves," *Rev. Mod. Phys.*, vol. 23, no. 4, pp. 287–310, 1951.
7. V. Twersky, "Multiple scattering of electromagnetic waves by arbitrary configurations," *J. Math. Phys.*, vol. 8, no. 3, pp. 589–610, 1967.

A Secure Mobile Agent System Applying Identity-Based Digital Signature Scheme

Seongyeol Kim[1] and Ilyong Chung[2]

[1] School of Computer and Information, Ulsan College, Ulsan, 682-090, Korea
kimsy@mail.ulsan-c.ac.kr
[2] Dept. of Computer Science, Chosun University, Kwangju, 501-759, Korea
iyc@mail.chosun.ac.kr

Abstract. Even though a mobile agent system contributes largely to mobile computing on distributed network environment, it has a number of significant security problems. In this paper, we analyze security attacks to this system presented by NIST[3]. In order to protect it from them, we suggest a security protocol for a mobile agent system by employing Identity-based key distribution and digital multi-signature scheme. To solve these problems described on NIST, securities of mobile agent and agent platform should be accomplished. Comparing with other protocols, our protocol performs both of these securities, while other protocols mentioned only one of them. Furthermore, it is designed to guarantee the liveness of agent, and to detect message modification immediately by verifying each step of agent execution.

1 Introduction

Due to the progress of distributed technology[2] languages and paradigms for drawing up network-based application have been gradually developed. Among these paradigms, the mobile agent paradigm extending the concept of Code-on-Demand model has drawn considerable attentions since it has many advantages from the aspects of system label, middleware, and user-level.

In a mobile agent system, its executable code that is not depended upon a specific system performs tasks during traversal between systems. This system has characteristics of a mobile code executed in JAVA applet, and furthermore, has characteristics of travelling lots of systems and of being mobile in its own when needed. So it is superior to previous models by providing low cost of communication, better asynchronous interaction, and improved flexibility in terms of distributed computing. Thus, many advantages are given when a mobile agent is used in information retrieval, network management, electronic commerce, and mobile computing[3].

However, since it has to execute an outside code in its own system, it poses problems, in which it can not provide system resources with blind confidence about the codes arrived from outside. Also for the mobile code exposed thoroughly to the executing environment, information exposure or information modification can occur. This mobile agent system has been faced with serious secu-

M.H. Shafazand and A M. Tjoa (Eds.): EurAsia-ICT 2002, LNCS 2510, pp. 588–596, 2002.

rity problems since the birth, and therefore many studies have been performed to solve them.

For protection of agent platforms, the Sandbox model sets the memory area where an agent can access only separated memory area in order to prevent abuse of host support against malicious agents. To protect the agent from malicious hosts is somewhat more complicated. A host manipulating agent codes and data can attack the agent, can ignore migration request received from the agent, or can destroy the agent.

Vigna et al[4] presents a structure of protecting an agent by transmitting the execution status of agent to a verifier and this verifier would audit it. However, this method would be expensive since a significant amount of network resources would be utilized and execution codes would be expanded. Bennet et al[5] also proposes a Proof Verification, but it was evaluated to be inappropriate for application. In Baek et al[6], the problems seen in [4] and [5] can be solved. An agent is protected by performing a security system using routing information on a agent and information on execution status obtained from digital signatures and audit tools. However, a drawback of this method is that modification activities done by agents can be audited only at the final stage. Since immediate detection of these activities is not performed, a system can take unnecessary overheads and lengths of signature becomes longer owing to repeated usage of signature.

A few access methods to protect a mobile agent have been suggested - the first method that does not allow a sufficient amount of time for attackers to analyze the code[7], the second that encrypta an agent where it is located on the perfect environment TPE(tamper-proof environment)[8], the third that enables the execution of the agent encrypted with the key generated by time and environment of server[9]. Although a security protocol for a server against unauthorized agents was suggested while ignoring the danger of server to an agent[10], these systems do not provide mutual authentication between an agent and a server.

In this paper, we propose a new protocol to solve the security issues for mobile agent systems by employing the digital multi-signature and Identity-based key distribution scheme[11]. In Identity(ID)-based cryptographic system, there are a public key ID and a secrete key that corresponds to this ID, where a secret key can be generated only at the key distribution center. Since an ID is used as a public key, authentication of the public key is not needed. Furthermore, the electronic signature that plays an important role in an exchange of electronic document satisfies fixing signature length, allowing verification, detection of tampering, maintaining confidentiality and commonality, and has developed into a multi-signature scheme to solve an issue of inefficient single signature scheme.

2 Design of the Security Protocol for Mobile Agent Systems

The security protocol proposed in this paper obtains secure communications by using one-time password between hosts for each section through Identity-based key distribution rather than maintaining a public key directory. It generates multi-signature on the mobile code and verifies results of the previous step when migrating to next server. Then, the executable code and resulting data can be protected and unauthorized tampering can be detected in real time. Moreover, malicious disposal of agent and unauthorized copying can be detected by monitoring the migration condition of agent at the agent management center. Fig.1 illustrates an interaction of factors required for executing this protocol and Table 1 shows the notation used in this chapter.

2.1 Registration (Agent-Server Registration and Key Distribution)

All agent platforms should be registered ahead at the AMC in order to execute services in mobile agent systems. When a request of registration from an AP is called, the AMC generates a key according to the following method, distributes it to agent platforms and registers it.

(1) Select two large prime numbers, p and q, and calculate N = p × q.
(2) Compute $\varphi(N)$ = (p-1)(q-1) and e with gcd(e, $\varphi(N)$) = 1.
(3) Compute d with ed = 1 mod $\varphi(N)$.
(4) Select a primitive root g, g $\in GF(p)$, g $\in GP(q)$.
(5) Compute the values S_i and S_{ij} $(1 \leq j \leq k)$, k_{ix} for AP_i.

$$S_i = {AP_i}^d \text{ mod N}$$
$$I_{ij} = \text{f}(\text{AP}_i, j)\ (j = 1, 2, ..., k)$$
$${I_{ij}}^{-1} = {S_{ij}}^2 \text{ mod N, } (\exists\, I_{ij},\ \ I_{ij} \in QR_N)^1$$
$$k_{ix}\ \text{, where } 1 \leq k_{ix} \leq k\ ,\ I_{ij} \in QR_N$$

(6) Register AP_i at the agent-sever DB and store p, q securely.
(7) Distribute the key using smart card containing the following factors

$$(N, e, g, S_i, f, h, S_{i1}, ..., S_{ik}, k_{i1}, ..., k_{ix}, AP_1, ..., AP_m, AMC)$$

$k_{i1}, .., k_{ix}$ is a series of j that ${I_{ij}}^{-1}$ is a quadratic residue. When the key is renewed in the future, a new key is received using this session key between AP_i and the AMC and the smart card key is updated.

2.2 CreateAgent

When a agent is created at the homeplace, the AP_H let the AMC inform this creation and conduct security management. For these purposes, the procedure is designed as follows.

(1) The migration path of agent is generated as follows.
A_route = $\text{AP}_1 \parallel \text{AP}_2 \parallel ... \parallel \text{AP}_n$

[1] quadratic residue for modulus N

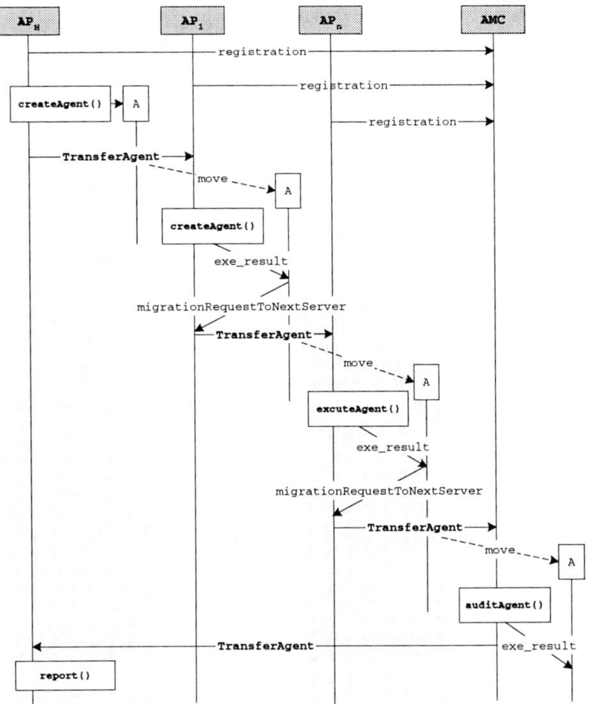

Fig. 1. Interaction of factors required for execution

(2) A sign value is setup with executable code of agent and multi-signature which are yet null state.
A_sign = H_H = f(A_code) A_code, A_sign $\in$ MA

(3) $k_{H,AMC}$ is created by generating a session key shared with AMC according to registration process.

(4) $k_{H,1}$ shared with AP_1 in which an agent is executed for the first time, is created.

(5) Agent code and path information are sent to the AMC as follows.
$AP_H \rightarrow AMC : E_{k_{H,AMC}}(MA\ ,\ t1_H)$

(6) A signature is generated as follows.

$$X_H = {R_H}^2 \bmod N$$
$$(e_{H1}\ , \ldots ,\ e_{Hk}\ ) = h\ (\text{A\_route, A\_exe\_results, } H_H\ ,\ X_H)$$
$$Y_H = R_H \prod_{e_{Hj}=1} S_{Hj} \bmod N\ , \quad j = 1, 2, \ldots, k$$
$$\text{A\_sign} = (H_H, X_H, k_H, Y_H)\ , \quad \text{A\_sign} \in MA$$

(7) AP_H transmits the data to AP_1.
$AP_H \rightarrow AP_1 : E_{k_{H,1}}\ (MA\ ,\ t1_H)$

Table 1. Notation used in this protocol

Notations	Explanation
AMC	Agent Management Center
AP_i	ID of the i th agent platform on migration path of agent
AP_H	Home Place of Agent
MA	Mobile Agent. This is a set of { A_name , A_code , A_route , A_exe_results , A_sign }
A_route	Migration Path of Agent
A_name	Name of Agent
A_exe_results	Result of executing an agent
A_code	Executable code of Agent
A_sign	multi-signature for result of executing an agent
f , h	public unilateral function
Hm	Hash values of agent code created by APm
$k_{i,j}$	Session key between AP_i and AP_j
Ek(M)	Message M is encrypted with the key k
R_i	The random number generated by AP_i
$R_{i,j}$	The random number generated by AP_i to transmit to AP_j
$t1_i$, $t2_i$	Timestamp generated by AP_i

Generation of a session key between AP_i and AP_j: The session key of AP_i and AP_j is generated by applying Diffie-Hellman key exchange method.

(1) AP_i selects a random number $R_i \in Z_N$.
(2) AP_i calculates C_i and sends it to AP_j.
$C_i = g^{R_i} \bmod N$
(3) After AP_j selects a random number R_j, calculates C_j and sends it to AP_i.
$C_j = g^{R_j} \bmod N$
(4) AP_i computes the session key shared with AP_j
$k_{i,j} = (C_j)^{R_i} \bmod N = g^{R_i \cdot R_j} \bmod N$
(5) AP_j computes the session key shared with AP_i.
$k_{i,j} = (C_i)^{R_j} \bmod N = g^{R_i \cdot R_j} \bmod N$

2.3 ExecuteAgent

When an agent migrates to AP_{i+1} , a host treats this agent as one thread.

(1) AP_i executes a mobile agent and then it makes log_i, and renews A_exe_results

$log_i = E_{k_{i,AMC}}$ ($A\_exe\_results_i$, $t2_i$)
$A\_exe\_results_i = A\_exe\_results_{i-1} \| log_i$

The results of execution should be protected from other agent platforms and finally be verified by the AMC.
(2) A signature is made as follows.

$$R_i \in Z_N$$
$$X_i = {R_i}^2\ X_{i-1} \bmod N\ ,\ (i = 1, 2, \ldots, k)$$
$$(e_{il}, \ldots, e_{ik}) = h(\text{A\_route, A\_exe\_results}_i,\ X_i,\ H_H)$$
$$Y_i = Y_{i-1} R_i \prod_{e_{ij}=1} S_{ij} \bmod N\ ,\ (j = 1, 2, \ldots, k)$$
$$\text{A\_sign} = (H_H, X_H, k_H, X_1, k_1, X_2, k_2, \ldots, X_i, k_i, Y_i)$$

2.4 TransferAgent

When an agent is transmitted at AP_H, or a migration request made at AP_i and an agent migrates to AP_{i+1}, TransferAgent procedure is accomplished between AP_i(or AP_H) and AP_{i+1}. It is designed to guarantee that no security problems have occurred in agent's performance.

(1) $AP_{i,i+1}$, the session key of AP_i and AP_{i+1}, is generated.
(2) AP_{i+1} to receive an agent generates the session key $k_{i+1,AMC}$ shared with the AMC.
(3) AP_i transmits the sign for results of execution and agent code to AP_{i+1}.
$AP_i \rightarrow AP_{i+1} : E_{k_{i,i+1}}(\text{MA,t1}_i)$
(4) AP_{i+1} can verify the signature written by previous AP as follows.

$$I_{xj} = f(AP_x, j),\ (x = 1, 2, \ldots, i\ ,\ j = 1, 2, \ldots, k)$$
$$(e_{i1}, \ldots, e_{ik}) = h(\text{A\_route, A\_exe\_results}_i,\ X_i,\ H_H)$$
$$X_i = {Y_i}^2 \prod_{x=1}^{i} \prod_{e_{ij}=1} I_{xj} \bmod N,\ (j = 1, 2, \ldots, k)$$

If X_i obtained from the fourth step of this procedure is equal to X_i in A_sign, the result would be effective.

(5) When the verification is correct, AP_{i+1} executes this agent, but if incorrect, AP_{i+1} reports immediately to the AMC.

2.5 AuditAgent

The AMC knows the migration path according to $AP_H \rightarrow AMC : E_{k_{H,AMC}}(\text{MA,t2}_H)$ performed at the procedure of CreateAgent. Moreover, each time an agent migrates, a session key is shared with AP_i , the AMC can detect it until an agent stop. When an agent normally travels the planned path and arrives at the AMC, the entire processes of signing can be verified as follows.

$$I_{ij} = f(AP_i, j)$$
$$(e_{i1}, \ldots, e_{ik}) = h(\text{A\_route, A\_exe\_results}_i,\ X_i,\ H_H)$$
$$X_i = {Y_i}^2 \prod_{x=1}^{i} \prod_{e_{ij}=1} I_{ij} \bmod N,\ ,\ (x = 1, 2, \ldots, n\ ,\ j = 1, 2, \ldots, k)$$

Table 2. Table of session key in AMC

Agent-server ID	Session key
AP_1	$g^{e \cdot R_1 \cdot R_{AMC}}$
AP_2	$g^{e \cdot R_2 \cdot R_{AMC}}$
⋮	⋮
AP_{n-1}	$g^{e \cdot R_{n-1} \cdot R_{AMC}}$
AP_n	$g^{e \cdot R_n \cdot R_{AMC}}$

If the signature is valid, the result of execution can be obtained referring to a table of session keys.

Since A_exe_results = $\log_1 \| \ldots \| \log_n$ and $\log_i = E_{k_{i,AMC}}$(A_exe_results,t1_i), the result of execution recorded by each server can be deciphered. And then this deciphered data is ciphered with $k_{H,AMC}$ and transmitted to home. Finally, the mobile agent reports to home and terminates execution.

3 Analysis of the Proposed Protocol

The proposed protocol employs Identity-based key distribution and digital multi-signature on a mobile agent system. Identity-based digital signature is based on discrete logarithm problem and it is just as safe as the Fiat-Shamir scheme. It is designed to solve security threat of the system and conventional security problems by distributing Identity-based keys, signing and ciphering information. It also accomplishes mutual authentication between mobile agent and platform, but it does not allow tampering of the agent and does not reveal results of execution. Since this structure can guarantee the liveness of agent, fault-tolerance on this system can be done. Furthermore, because the protocol includes a function signed multiplely at each agent platform when an agent passes through, authentication of agent platform, confidentiality, integrity, and protection of reputation are satisfied.

Table 3 shows analysis of our protocol compared with other schemes such as Hole method[7], Whlhelm-Stamann method[8], Riordan-Schneier method[9] and Baek method[6] in terms of security features.

4 Conclusions

In this paper, we propose the security protocol for a mobile agent system using Identity-based on key distribution and Fiat-Shamir digital signature in order to authenticate between agent and server, to protect the resulting data, and to guarantee liveness of an agent. It is able to do interim verification so that it would not have unnecessary overheads. Identity-based cryptography can be applied in key distribution and digital signature, and it has advantages of simplification of key management and of fast signature compared with

Table 3. Table of session key in AMC

schemes / matters	[7]	[8]	[9]	[6]	proposed protocol
protection method	time limited and code mixture	hardware	environmetal key	repeated simple signature	information hiding and muli-signature
cryptography method	asynchronous	asynchronous	asynchronous	asynchronous	synchronous by ID key
authentication method	unidirectional	unidirectional	unidirectional	unidirectional	bidirectional
liveness	×	×	×	partially supported	○
confidentiality	○	○	○	not described but supported	○
integrity	×	×	×	not described but supported	○
preventing repudiation	×	×	×	○	○
The result of execution is opened	oepn	open	open	open	secret

the public key mechanism. The followings are summary of characteristics of the proposed protocol. The first characteristic is a structure of simplifying key management. This not only can overcome problems related to directory management for the public key mechanism but also has an advantage to setup a new onetime session key every time. Secondly, it provides security services - authentication, confidentiality, integrity, and nonrepudiation and prevention of replay attack. In this system, one's own data is never exposed to any other APs by ciphering execution result. Each AP can confirm integrity of agent code, path information, and status information through a verification process before it executes the arrived agent. It immediately reports to the AMC in case that some problems occur. And the repudiation service can be offered by using digital signature. Thirdly, liveness of an agent can be guaranteed. The AMC receives A_route generated by each AP at every hop of agent migration and then the AMC monitors unauthorized termination done by an arbitrary AP. Fourth, the result of execution should be protected. Problems related to agent behaviors by reading the results of execution obtained from previous steps or by modifying them illegeally are solved. This security model premises that an agent will migrate along the planned path. Later this method, if the mobile platform is determined at the time of migration, should consider management of locations for remote access and recording an active list of APs.

Acknowledgement. This study was supported by research funds of Chosun University, 2002.

References

1. Dale, J. and Mamdani, E., "Open Standards for Interoperating Agent-Based Systems," In Software FOCUS, Wiley, 2001.
2. Poslad, S. and Calisti, M., "Towards Improved Trust and Security in FIPA Agent Platforms," Autonomous Agents 2000 Workshop on Deception, Fraud and Trust in Agent Societies, Spain, 2000.

3. Jansen, W. and Karygiannis, T., "Mobile Agent Security," NIST Special Publication 800-19, 1998.
4. Vigna, G., "Protecting Mobile Agents through Tracing," Mobile Object Systems ECOOP Workshop, 1997.
5. Bennet, S. Y., "A Sanctuary for Mobile Agents," DARPA Workshop on Foundations for Secure Mobile Code Workshop, pp. 26–28, 1997.
6. Baek, J. and Lee, D., "Security of Mobile Agent Using Digital Signature and Audit trail," Proc. of KISS Fall Conference, Vol. 24, No.2, KISS, 1997.
7. Hohl, F., "An approach to solve the problem of malicious hosts," Universitat Stuttart, Fakultat 3nformatik, Fakultatsbericht Nr., 1997.3.
8. Wilhelm, U. G. and Stamann, S., "Protecting the Itinerary of Mobile Agents." Proc. of the ECOOP Workshop on Distributed Object Security, pp. 135–145, INRIA, France, 1998.
9. Riordan, J. and Schneier, B., "Environmental Key Generation towards Clueless Agents," Mobile Agents and Security, pp. 15–24, Springer-Verlag, 1998.
10. Ordille, J., "When agents roam, who can you trust?," Proc. of the First Conference on Emerging Technologies and Applications in Communications, Porland, May 1996.
11. Shamir, A., "Identity-based cryptosystem and signature scheme," Advances in Cryptology, Springer-Verlag, pp. 47–57, 1985.

Transmission Time Analysis of WAP over CDMA System Using Turbo Code Scheme

Il-Young Moon[1], Jae-Sung Roh[2], and Sung-Joon Cho[1]

[1] School of Electronics, Telecommunication and Computer Eng., Hankuk Aviation Univ. Kyonggi-do, KOREA
{iymoon21, sjcho}@mail.hangkong.ac.kr

[2] Dept. of Information & Communication Eng., SEOIL College, Seoul, KOREA
jsroh@seoil.ac.kr

Abstract. In this paper, we have analyzed transmission time for wireless application protocol (WAP) over code division multiple access (CDMA) using turbo code scheme. In order for segmentation and reassembly (SAR) to improve the transfer capability, the transmission of messages have been simulated using a fragmentation that begins with the total package and incremental fragmentation for each layer using the wireless transaction protocol (WTP) to define the resultant packet size and the level of fragmentation for each proceeding layer. This turbo code scheme decreases transmission time of radio link protocol (RLP) baseband packets by sending packets. From the results, we were able to obtain packet transmission time and optimal WTP packet size for WAP over CDMA in a Rician fading channel.

1 Introduction

The WAP is an open standard for the presentation and delivery of wireless information and telephony services on mobile phones and other wireless terminals enabling information access from handheld devices requires a deep understanding of both technical and market issues that are unique to the wireless environment. Wireless devices represent the ultimate constrained computing device with limited CPU, memory, battery life and a simple user interface. Wireless networks are constrained by low bandwidth, high latency and unpredictable availability and stability.

In addition, WAP and hypertext markup language (HTML) offers an interoperable presentation platform for end-user interfaces. Applications that will be increasingly popular in the future, such as man-machine and machine-machine interfaces, will drive WAP in CDMA system. But a problem of WAP is limited transmission time by severe mobile environment. So it takes much time to transmit WAP packet and causes packet loss.

In this paper, it is analyzed that WAP packet transmission time to improve performance of WAP using SAR algorithm and turbo code scheme in CDMA system. In order for SAR to progress the transfer capability, the whole messages are fragmented

M.H. Shafazand and A M. Tjoa (Eds.): EurAsia-ICT 2002, LNCS 2510, pp. 597–606, 2002.

in WTP layer and segmented further as it passes through each layer towards the physical layer, where the actual packets are sent sequentially [1],[2].

Also, it is analyzed the transmission time of WAP packet with variable RLP layer size using SAR algorithm on the CDMA wireless channel for next generation systems.

2 WAP Architecture

2.1 WAP Model

WAP is an open specification that offers a standard method to access Internet based content and services from wireless devices such as mobile phones and personal digital assistants (PDAs).

The WAP model is very similar to the traditional desktop Internet. The mobile device has embedded browser software that connects to a WAP Gateway (software infrastructure residing in the Operator's Network that optimizes the transmission of content for the wireless network) and makes requests for information from web servers in the normal form of a URL. The content for wireless devices can be stored on any web server on the Internet. Content must be formatted suitably for the mobile phone's small screen and low bandwidth/high latency connection. Content is written in a markup language called wireless markup language (WML). WMLScript enables client side intelligence.

2.2 WAP Protocol Stack

The WAP architecture provides a scalable and extensible environment for application development for mobile communication device. With the open system interconnection (OSI) in mind, the WAP stack basically is divided into five layers. They are a wireless application environment (WAE) as transaction layer, wireless session protocol (WSP) [3] as session layer, WTP as transaction layer, wireless transport layer security (WTLS) as security layer and wireless datagram protocol (WDP) as transport layer. Figure 1 shows the WAP protocol stack and how it relates to the protocols on the Internet. Each layer of the WAP protocol stack specifies a well-defined interface, meaning that a certain layer makes lower layers invisible in figure 1. The layered architecture allows other applications and services to utilize the features provided by the WAP stack as well. This makes it possible to use the WAP stack for service and applications that currently are not specified by WAP.

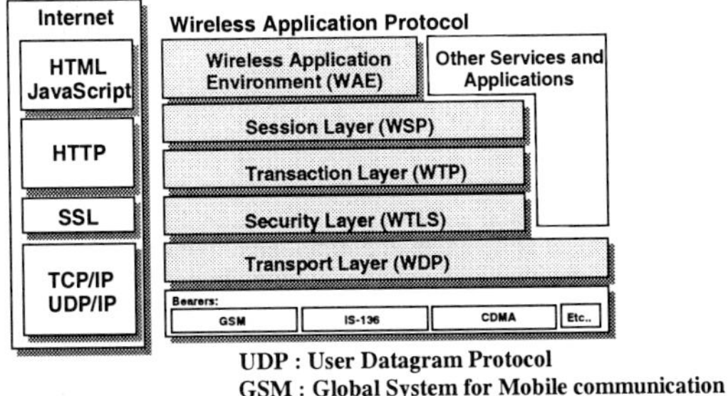

Fig. 1. WAP Protocol Architecture

3 Transmission Time for WAP over CDMA

3.1 SAR (Segmentation and Reassembly)

SAR operations are used to increase efficiency by supporting a maximum transmission unit (MTU) larger than the maximum allotment of a single packet [4]. This reduces overheads by spreading the packets used by higher layer protocols over several packets. In order for SAR to progress the transfer capability, the whole messages are fragmented in WTP layer and segmented further as it passes through each layer towards the physical layer, where the actual packets are sent sequentially.

3.2 Turbo Code Scheme

Turbo codes are obtaining a large success and their introduction in many international standards is in progress. It is known that turbo codes have exceptional performance at low/medium signal-to-noise power ratio (SNR), very close to the Shannon limit.

Due to the poor error resilience of some source-coding schemes, some applications can require very low values of bit error rate (BER). Unfortunately, turbo code performance may be significantly worse at these error rates. In fact, their free distance d_{free} (the minimum Hamming distance between different codewords) can be low, even with very large interleaver lengths. This cause their BER curves to flatten following the "error floor" imposed by d_{free}, after the "water-fall" decreases at low SNR

To analyze the code performance, simulation could be employed. For a $C(n, k)$ code, the knowledge of the free distance d_{free}, its multiplicity N_{free}, and its information bit multiplicity w_{free} (defined as the sum of the Hamming weights of the N_{free} information sequences generating the codewords with weight d_{free}), allows to determine the error floor slope. The performances of any binary code at high SNR are well approximated by the expression of the union bound, truncated to the contribution of the free distance. It is worthwhile to note that, for turbo codes, a small penalty must be also taken in account, due to the sub-optimality of the iterative decoding.

As a result, it can be written as

$$BER \cong \frac{1}{2}\frac{w_{free}}{k} erfc\left(\sqrt{d_{free}\frac{k}{n}\frac{E_b}{N_0}}\right) \quad (1)$$

In this paper, it is analyzed that rate 7/8 and 1/3 turbo code with 8-state constituent encoders to improve WAP packet transmission time [5].

3.3 Rician Fading Channel Model

To analyze WAP packet transmission time, it is simulated in Rician fading channel. Rician distribution and probability density functions can be written as

$$f(r) = \frac{r}{b_o}\exp\left[-\frac{r^2 + A^2}{2b_o}\right] I_0\left(\frac{A\,r}{b_o}\right) \quad \text{for } A \geq 0, r \geq 0 \quad (2)$$

$$K_R = \frac{A^2}{2b_o} \quad (3)$$

where, Rician factor K_R is defined as the ratio of the specular power A^2 to scattered power $2b_o$. When $K_R = 0$ the channel exhibit Rayleigh fading, when $K_R = \infty$ the channel does not exhibit any fading at all.

4 Simulation of Packet Transmission Time using Turbo Code Scheme

4.1 Simulation Model of WAP Packet

To be fragmented to WTP of WAP, the number of total message packet (K) follows by equations

$$K = \left\lceil \frac{M_{TOTAL}}{M_{SEG}} \right\rceil, \quad (4)$$

where, $\lceil x \rceil$ is integer less than $x+1$ and M_{TOTAL} is size of total message of upper layer WTP.

The total size of WTP packet, M_{WTP}, consist of the size of message to be fragment at upper layer, $M_{SEG,}$ and the size of WTP header, H_{WTP}, and can be expressed as

$$M_{WTP} = M_{SEG} + H_{WTP}, \quad (5)$$

Also, the size of last packet to be fragmented to WTP, L_{WTP} is obtained by

$$L_{WTP} = M_{TOTAL} - (K-1)M_{SEG} + H_{WTP}, \tag{6}$$

The size to be transfer from upper layer to RLP layer M_{RLP} is

$$M_{RLP} = M_{WTP} + H_{UDP} + H_{IP} + H_{PPP}, \tag{7}$$

where, H_{UDP} is size of UDP header, H_{IP} is size of IP header, H_{PPP} is size of PPP header, and a number of be fragmented in RLP layer, N, is defined as

$$N = \left\lceil \frac{M_{RLP}}{F_D} \right\rceil, \tag{8}$$

where, F_D is size of data to be transfer per one slot in RLP layer. A probability to be successfully transfer at total slot $E(F)$ can be expressed as

$$E(F) = \sum_{m=1}^{\infty} m \cdot P(F = m) = \frac{1}{p}, \tag{9}$$

where, $P(\cdot)$ is a probability of data frame to be successfully transfer and p is a probability to be successfully transfer. And, the number of average time slot to transfer WAP packet is as follows

$$E(P) = N \cdot E(F) = \frac{N}{p}, \tag{10}$$

Using the equation (9) and slot time S_{TIME}, the transmission time of WAP packet to be fragment is obtained as

$$T_{PKT}(N) = E(P) \cdot S_{TIME} = \frac{S_{TIME} N}{p} (ms) \tag{11}$$

Therefore, we are achieved the total message transmission time $T_{MSG}(ms)$, and is expressed as

$$\begin{aligned} T_{MSG} &= (K-1)T_{PKT}(q) + T_{PKT}(r) \\ &= (K-1)\frac{S_{TIME} \times q}{p} + \frac{S_{TIME} \times r}{p}, \end{aligned} \tag{12}$$

$$q = \left\lceil \frac{M_{WTP} + 36}{F_D} \right\rceil, \quad r = \left\lceil \frac{L_{WTP} + 36}{F_D} \right\rceil, \tag{13}$$

where, q and r are the number of time slot to computed by F_D.

Also, we value for the BER of the payload part in the receiver part has been calculated. In order to simulate the BER performance, an independent, static Rician fading and AWGN channel were assumed for every packet in CDMA/BPSK environment. A BER of CDMA/BPSK can be written as

$$p_e = \frac{1}{2} erfc(\sqrt{SNR}) \tag{14}$$

$$SNR = \frac{1}{\left(\frac{E_b}{N_0}\right)^{-1} + \frac{2(U-1)}{3PG}} \tag{15}$$

U : Number of multiple access user,
PG : Processing gain of system,
E_b/N_o : Signal-to-noise ratio per bit.

It is analyzed considering reverse link, cause it passes through WAP client towards the WAP proxy/server. Figure 2 depicts a protocol model of WAP packet in this paper. The simulation model consists of WAP client, wireless channel, and WAP proxy/server. To find the transmission time of WAP, we must transmit the total message by segmenting it into data packets.

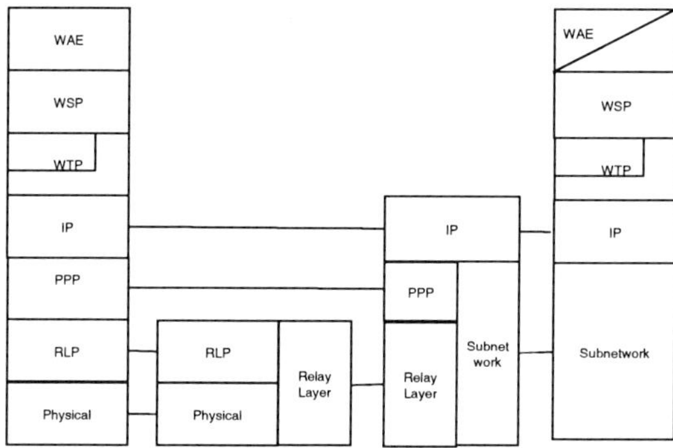

Fig. 2. A protocol model of WAP packet

4.2 Analysis of Transmission Time for WAP Packet

By changing E_b/N_o in the wireless channel, we gain transmission time of packets and analyze BER performance of the WAP over CDMA system using SAR algorithm and turbo code scheme in AWGN and Rician fading channel [6],[7]. It is considered the transmission time of WAP packet with variable RLP layer size using SAR algorithm on the CDMA wireless channel next generation systems. So, it is analyzed RLP layer size from $F_D = 27$ to $F_D = 31$. Figure 3 shows the transmission time of WAP packet with variable RLP layer size using turbo code scheme in Rician fading channel of wideband code division multiple access (WCDMA). M_{WTP} is analyzed fragmenting from 100 to 2000 bytes. In addition, in the case of CDMA and WCDMA system, S_{TIME} is set to 20ms, 10ms, respectively. For achieving transmission time of packets for WAP over CDMA system, the total message transmission time is simulated at M_{TOTAL}= 5000 byte, E_b/N_o = 2 dB and 4 dB, U = 20, PG = 64 in AWGN channel and at M_{TOTAL}= 5000 byte, E_b/N_o = 4 dB and 6 dB, U = 20, PG = 64, K_R = 10 dB in Rician fading channel.

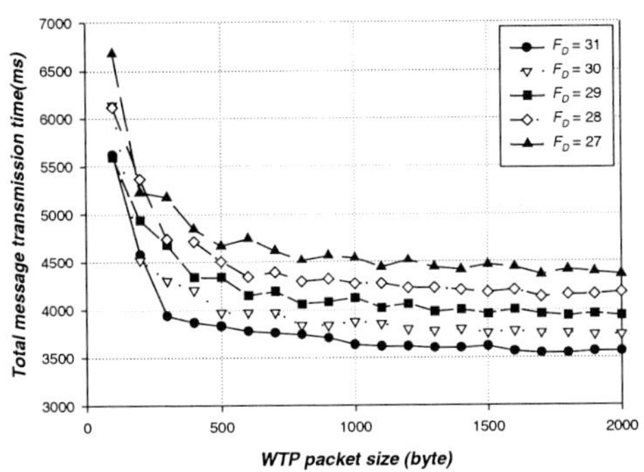

Fig. 3. Total message transmission time for WCDMA

(E_b/N_o = 4 dB, U = 20, PG = 64, K_R= 10 Rician)

In figure 4 and 5, the parameter E_b/N_o is set at 2 dB using turbo code scheme in CDMA and WCDMA AWGN channel, respectively. In figure 4, Furthermore, when the packet size increases from 100 to 2000 bytes, transmission time is less than the transmission time of a 100 bytes. Also, in the case of CDMA, transmission time is increased more about 2 times than WCDMA transmission time. In figure 6 and 7, the parameter E_b/N_o is set at 4 dB using turbo code scheme in CDMA and WCDMA in Rician fading channel, respectively. In figure 6 and 7, the result is approximately the same as figure 4 and 5, but one is AWGN channel and the other is Rician fading channel.

From these results of figure 4,5,6 and 7, we find that the WTP packet size should increase in order for the transmission time in wireless channel to decrease. Besides, when E_b/N_o in AWGN and Rician fading channel increases, transmission time is decreased. To obtain an appropriate WTP packet size, we also considered the BER in a wireless channel. In the case of optimal WTP packet size (about 500 byte) in AWGN and Rician fading channel, the WAP packet transmission time (considering trade-off between total message transmission time and WTP packet size) is about4300~7900 ms (CDMA), 2200~3900 ms (WCDMA) in AWGN and is about 4200~7600 ms (CDMA), 2100~3700 ms (WCDMA) in Rician fading channel.

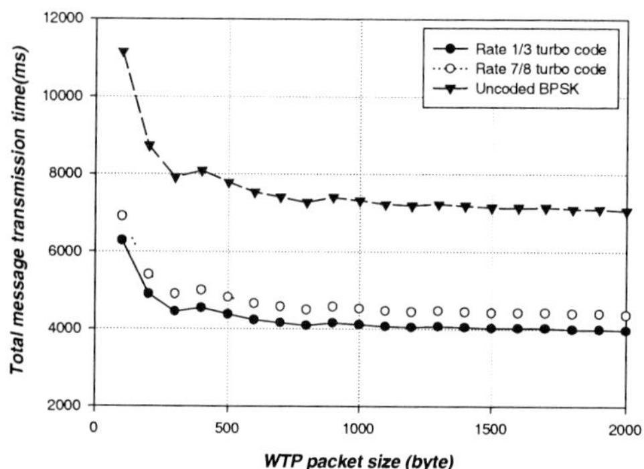

Fig. 4. Total message transmission time for CDMA ($E_b/N_o = 2$ dB, $U = 20$, $PG = 64$, AWGN)

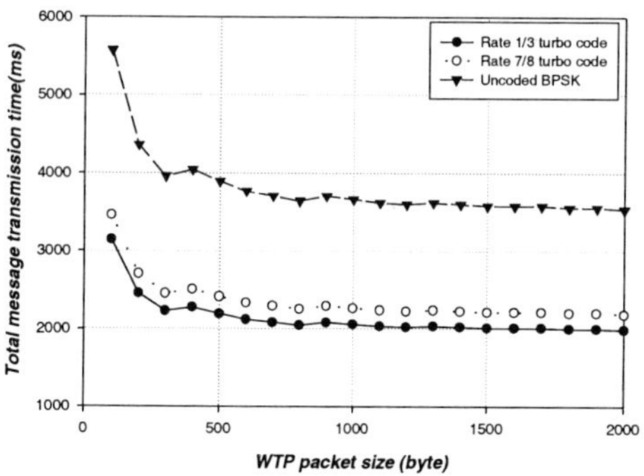

Fig. 5. Total message transmission time for WCDMA ($E_b/N_o = 2$ dB, $U = 20$, $PG = 64$, AWGN)

5 Conclusion

This paper has analyzed WAP packet transmission time using turbo code scheme. It is considered the transmission time of WAP packet with variable RLP layer size using SAR algorithm on the CDMA wireless channel. Also, we analyzed the WAP packet transmission time by changing E_b/N_o in AWGN and Rician fading channels. The combined SAR algorithm and turbo code scheme decreases WAP packet transmission time of baseband packets in wireless channel.

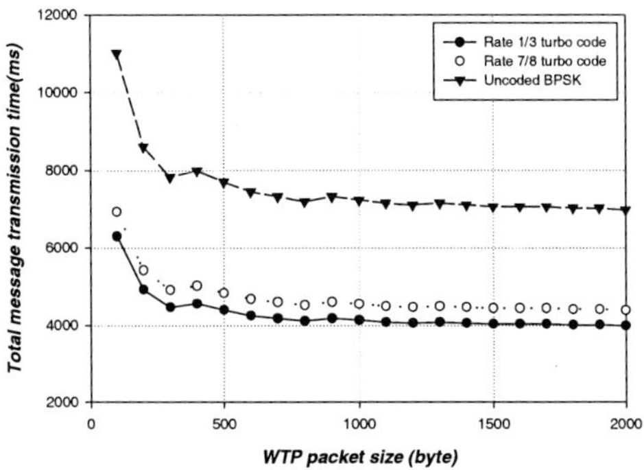

Fig. 6. Total message transmission time for CDMA (E_b/N_o = 4 dB, U = 20, PG = 64, K_R = 10 Rician)

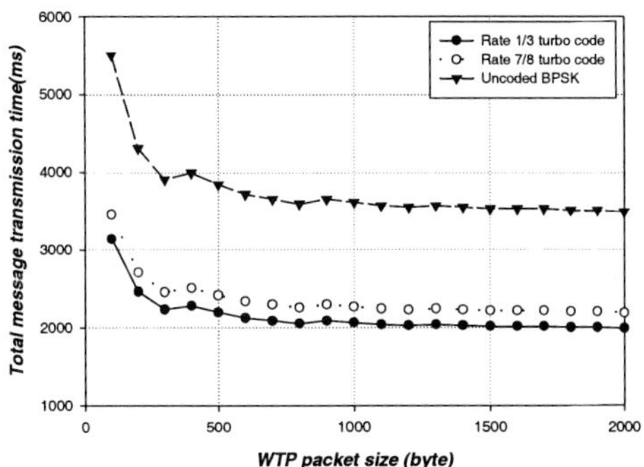

Fig. 7. Total message transmission time for WCDMA (E_b/N_o = 4 dB, U = 20, PG = 64, K_R = 10 Rician)

From the results, in CDMA channels, we can get that the transmission time in wireless channel decreases as the WTP packet size increases. As a result, based on the data collected, we can infer the correlation between packet size and the transmission time, allowing for an inference of the optimal packet size in the WTP layer. This paper will be able to give a hint on improvement of quality of service (QoS) and performance improvement of WAP over CDMA system.

Acknowledgements. This work was supported by the Korea Science & Engineering Foundation (KOSEF) and the Kyonggi Province through the Internet Information Retrieval Research Center (IRC) of Hankuk Aviation University.

References

1. WAP forum, Wireless Application Protocol: Wireless Datagram Protocol Specification, Version 14-June-2001.
2. WAP forum, Wireless Application Protocol: Wireless Transaction Protocol Specification, Version 10-July-2001.
3. WAP forum, Wireless Application Protocol: Wireless Session Protocol Specification, Approved Version 5-July-2001.
4. H. S. Park and K .W. Heo, "Performance evaluation of WAP-WTP," The Jornal of the Korean Institute of communication sciences, vol. 26, no. 1A, pp. 67–76, Jan. 2001.
5. R. Garello, F. Chiaraluce, P. Pierleoni, M. Scaloni, and S. Benedetto, "On Error Floor and Free Distance of Turbo codes," IEEE International Conference on Communications, vol.1, pp. 45–49, 2001.
6. J. S. Lee and L. E. Miller, CDMA systems engineering handbook, Artech House, 1998.
7. T. Ojanperä and R. Prasad, Wideband CDMA for Third Generation Mobile Communications, Artech House, 1998

On the Use of New Technologies in Health Care

Luis Anido[1], Fernando Aguado[2], Olga Folgueiras[3], Judith S. Rodríguez[1], Juan M. Santos[1], and Manuel Caeiro[1]

[1] Departamento de Enxeñería Telemática, Universidade de Vigo
Lagoas-Marcosende s/n, Vigo E-36200, SPAIN
{lanido, jestevez, jsgago, mcaeiro}@det.uvigo.es
[2] Departamento de Teoría do Sinal e da Comunicación
Lagoas-Marcosende s/n, Vigo E-36200, SPAIN
faguado@tsc.uvigo.es
[3] Área de Tecnología de la Información y las Comunicaciones
Lagoas-Marcosende s/n, Vigo E-36200, SPAIN
ofolgueiras@uvigo.es

Abstract. We present a Web-based management Tool for Health Care Services with appointment required. This tool lets patients request appointments with doctors and specialists via the Internet. It changes the previous situation where the only way to get an appointment was via a telephone call. Now, with the new system, it is possible to use a desktop computer or a WAP-enabled cellular phone to request appointments for health care services in addition to normal telephone booking. Moreover, the system eases daily work to doctors and administrative staff, reducing previous paper work significantly and improving the whole service. According to the users, transparency, flexibility and availability are the main advantages of the new system.

1 Introduction

Currently, Internet technologies are widely used all over the world. Almost every aspect of our daily life is affected, and usually improved, by the use of the WWW. Nevertheless, this is not the case for the Public National Health Care System in Spain. The Spanish Health Care System, with a total budget of 87.732 millions of Euros, is continually taking advantage of new technology improvements, for example, the telemedicine. This year, more than 26.940 million Euros of the Spanish national budget has been dedicated to support public hospitals and medical centers. Nevertheless, this policy, which has been maintained by different governments since 1960's, had no effect up till now in the incorporation of the WWW based services for the end user, in this case, the patient. The Spanish Health Care System WWW site [1] is composed of a set of static pages that only provide general information about the services offered, geographical location for several centers, etc.

In this paper, we present the first experience of the use of WWW technologies to offer an appointment service for the Spanish National Health Care System. Patients access through the Web to get and appointment with their doctor,

M.H. Shafazand and A M. Tjoa (Eds.): EurAsia-ICT 2002, LNCS 2510, pp. 607–614, 2002.

who can consult his workload or request a specialist doctor appointment for the patient. Both computer desktops and WAP-enabled [2] cell phones can be used. Administration duties for the whole system, which is geographically distributed, can also be carried out through a desktop computer using a Web interface. Security to protect patients' privacy is handled using SSL [3] and WTLS [4].

The rest of this paper is organized as follows: section 2 shows the previous situation in the National Health Care System in Galicia, which was exactly the same for the rest of Spain. Then, we introduce our solution using WWW/WAP technologies. Section 4 describes the whole functionality of the system presented in this paper. We end the paper with some conclusions and proposals for the future.

2 The Previous Situation in the Galician Health Care System

Apart from medical instruments, computerized systems were used in the SERGAS Health Care system to support the administrative tasks in hospitals and local health centers. Nevertheless, they usually were standalone systems with no compatibility among them.

Patients' complaints were centered on three main topics: the procedure to get an appointment with their doctor, long waiting queues for many medical services and the paperwork required to access some services. The only way to get an appointment with a doctor was to make a telephone call to an appointment center, where administrative staff gathered requests from patients. This procedure presented several drawbacks:

- During rush hours, especially early in the morning telephone lines were very busy. It was hard to get in touch with the appointment center. Several attempts were usually needed and sometimes patients gave up.
- The appointment telephone center was only available at office hours.
- Waiting queues were not clear for the public.
- The telephone was the only available medium to get an appointment.
- It was hard to get information about a previously requested appointment.

On the other hand, there were some medical services that required too much paperwork. For example, prescriptions were always handed over by the family doctor even to those patients that suffered chronic diseases and required the same prescription periodically. Appointments with a specialist doctor need the family doctor approval. Referrals to the specialist were always delivered in paper and used by the patient to get a new appointment. These processes would be eased if all these formalities could be done electronically.

3 A WWW/WAP-Based Solution

We implemented a WWW-based management tool for Health Care Services with appointment required. This tool is used by patients to request an appointment with their doctors and make later modifications if needed. It is important to stand out that the implementation that was carried out maintains the current appointment service by telephone calls integrated with the new one. In fact, there are three different possibilities to get an appointment with the doctor:

1. *A telephone call to an appointment center.* Administrative staff at this center use a Web interface to manage requests from patients.
2. *Web-based request through a desktop computer.* In this case, a Web browser is used to support the appointment request.
3. *WAP-enabled cell phone.* Using the mobile phone it is possible to request or modify an appointment with the doctor.

We believed it was important to maintain the three interfaces. The telephone call system represents the traditional system. It is essential because there is still an important percentage of the population in Spain that does not have Internet access. This is especially true for elderly people, who, by the way, are an important group of users for Health Care Services.

The Web browser interface has been actively used during the last months. As we will check in the next section, it provides a very easy-to-use tool that makes it very quick to get an appointment. Internet users are now estimated in 4 million people in Spain, that is a 10 % of the whole population. In 1997, there were only 600,000 Internet users. Numbers are increasing exponentially. With the recent adoption of a plain tariff of 16.4 Euros per month, it is expected that during the year 2001 the number of users will be increased over 25%.

Doctors are provided with both a WAP and WWW interface to access the following services: consult their assigned workload, manage patients' records and request referrals to specialists. Administrative staff is provided with a Web interface on a desktop computer to manage all data stored in a centralized database. They are responsible for tasks like registering new doctors or patients, set up every doctor's schedule and slot characteristics, move patients from their Health Care center to a new one, etc.

3.1 Advantages over the Previous System

From the patient point of view, the introduction of the new system offers:

1. *Transparency.* The appointment request is managed by the patient by himself, who is allowed to access all the information about his doctor availability. Therefore, previous confusion on the mechanism that generated so long waiting queues is now clarified. The SERGAS's willingness to disclose information has been reflected in the information offered to the end user.
2. *Efficiency.* There are no long phone waiting times at busy hours. Both the Internet and cell phone interfaces offer an always available appointment service. Typical service times range from 1 to 3 minutes depending on network delays. Previous studies showed that service time reached 15 minutes at rush hours. Moreover, appointments with a specialist are more agile since, with the new system, the referrals are directly requested by the family doctor and confirmed by the system automatically.

3. *Availability.* The new service is offered 24 hours per day, 365 days per year. There are no office hours never again.

Benefits for the SERGAS are twofold: firstly, the image offered to its users has been improved, secondly, the new system implies financial savings. Administrative staff needed to maintain the appointment service is reduced. Data is now gathered around an integrated service that eases patient movements among health centers, change of assigned doctors, management of patient records, etc.

4 Functional Description

This section includes a brief description of the functionality provided by our system. This presentation is organized according to the different roles involved: patients, doctors and administrative staff.

4.1 Patients

Patients access the system both from a Web desktop browser and from a WAP-enabled phone.

Web Interface. The Web interface is shown in Figure 1. The user interface can be displayed in Spanish or Galician. The options menu is on the left side frame and it is displayed in the form of a thermometer. As the user moves the mouse through the thermometer, the mercury inside moves with it showing which option is being selected, from top to bottom: request appointment, consult appointment, cancel appointment, change personal data, consult appointment with the specialist, help and go to the main page.

Patients are authenticated using the traditional National Health Care Service personal identification number. In order to ease this task for the user, especially for those not familiar with computers, we resembled the SERGAS Health Care Service card in the user interface, see Figure 1.

After being authenticated, the user is allowed to request a new appointment. There are two options:

1. *Request the next available.* In this case patients request the next available free slot for their doctor. This slot is reserved for this patient and, additionally, the next five days with any free slot available are returned, just in case the patient wants to select any other available slot.
2. *Request an appointment on a particular date.* It is possible to select any particular day suitable for our needs. Five days with free slots are returned: the two previous to the selected date (if possible), the selected day, and the two following dates. Then, the patient would choose a free slot from those available.

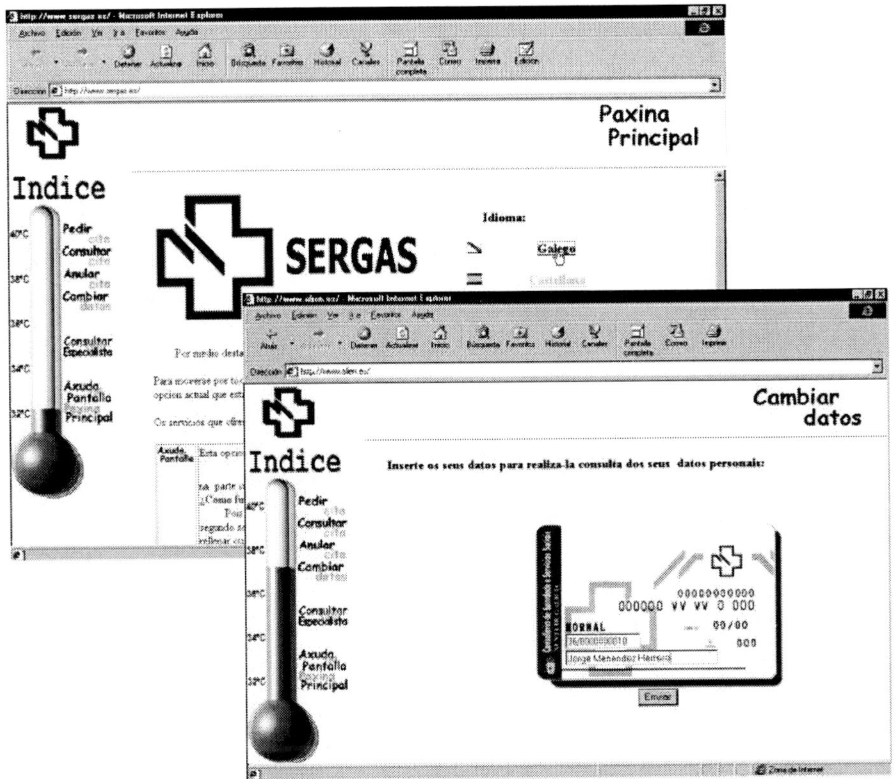

Fig. 1. WWW interface

In any case, the patient chooses the slot desired on an agenda like interface, see Figure 2. A slot is represented as a black box and is selected clicking on the box. The selection is marked with a tick on the box. A small red dot on the right bottom corner of the box indicates that this slot is not available. If the dot is green, it means the slot is free and therefore, the appointment can be made at the time situated on the left of the box. If the next-free-slot-available option is used, then this slot would be marked as the default option.

Every day in the agenda contains as many slots as assigned to this doctor. To display other slots in the same day the patient must use the buttons "+" and "-" on the bottom of the right side page. Buttons "<" and ">" on the top of that page allows the user to display other days availability. Let us recall here that any request will return five days with at least one free slot. Anyway, if the user moves to a date not yet delivered from the server, the request for the next five days would be done at that moment. The left side page of the agenda includes the following information: Patient's name, doctor's name and type of appointment (appointments could be requested just to ask for a prescription).

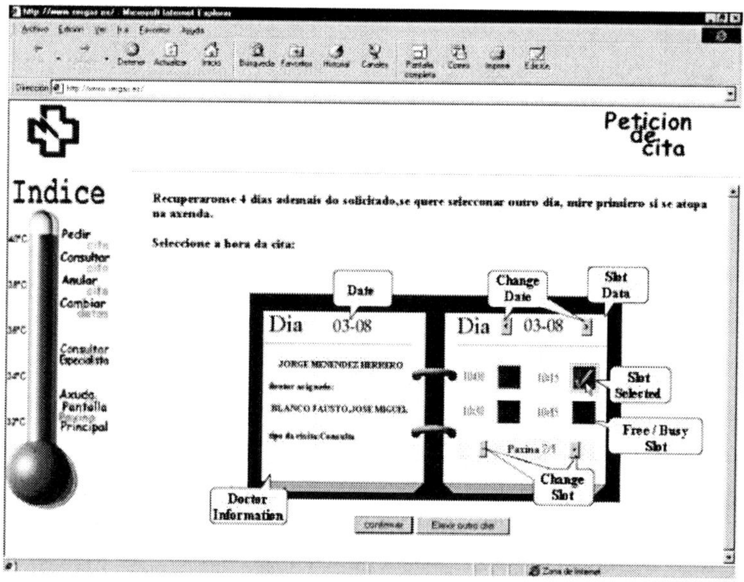

Fig. 2. Selection of slot using an agenda like interface

Other options available are: consult or cancel a previous appointment and consult an appointment with a specialist (these types of appointments are requested by the family doctor through an electronic referral to the specialist). Appointment data includes the concrete day and hour, health center or hospital address and location of room where the appointment will take place.

WAP interface. It is possible to request, consult and cancel an appointment using a WAP-enabled cell phone. In this case, the interface was redesigned taking in mind the small display of a cell phone. The number of steps required to select a free slot was reduced to three for the next available slot option: request appointment, display appointment data and confirm it. If the user selects a particular date, the required steps can be reduced to four (best case): request appointment, select hour, display appointment data and confirm it.

During the WAP interface design process we used the Nokia's WAP toolkit [5], see Nokia WAP-enabled cell phone simulator. Figure 3 shows the WAP interface on an actual Nokia 7110 WAP cell phone. Those pictures present the SERGAS logo (left picture) and main menu (right picture).

Other requirement was to optimize the bandwidth available as the Spanish mobile phone network providers offer WAP service over SMS through their GSM networks, which means that the service is not as fast as desirable. Due to this, for the next available slot option or if the requested date has any free slot, no additional data is sent to the client. Only if there is no available free slot in the date chosen by the patient, five days with free slots are sent to the cell phone.

Fig. 3. WAP interface on a Nokia's 7110.

4.2 Doctors

Doctors use the system both from a desktop computer connected to the Internet and a WAP-enabled cell phone. The main functionality offered is as follows:

- Management of patients assigned to the doctor.
- Patients' records updating. Patients' records are stored electronically by the system and can be accessed by authorized users from anywhere. A patient's record is stored as plain text to allow compatibility with databases.
- Referrals to specialists. Now, it is possible to manage appointments with specialists automatically using stored data about the specialist's schedule and availability. An electronic referral to the patient's specialist is used instead of the old referral in hard paper.
- Consult Workload assigned: who is going to be examined in each slot, number of prescriptions expected, etc. Doctors are also allowed to make some rescheduling tasks like changing the duration of slots.

4.3 Administrative Staff

Administrative staff is provided with a Web-based interface to manage all users and Health Care centers data. Among others, their responsibilities are: registering new doctors, patients and specialists, assigning family doctors to patients and specialists to family doctors, updating personal data for every user, scheduling doctor's working days, setting up local holidays for each Health Care center, etc.

In addition to these tasks, members of the administrative staff team receive and process telephone calls to the new appointment center. In this case, a new appointment can be fixed in just two steps. If the preferences of the patient cannot be satisfied, the system provides alternatives according to the user requirements. Design of the new call center took in mind the need of easing and fastening the process.

5 Conclusions and Work for the Future

The previous situation in the SERGAS Health Care System regarding the appointment telephone call centers made it essential to reduce waiting times and system availability. The tool presented in this paper brought a new appointment system to the Galician Public Health Services. While keeping the previous system, which was essential to cope with the habits of many people, especially the elderly, two new possibilities were offered: access through a Web browser and WAP-enabled cell phones.

According to the users the main advantages provided by the new appointment service were: flexibility (users do not depend upon a telephone call center), transparency (doctors schedule and available/busy slots are at anyone's disposal) and availability (the service can be used whenever needed).

Drawbacks are mainly related to the fact that some users were reluctant to change the previous system they were used to, despite of its disadvantages. This was the reason to maintain a telephone call center integrated with the new system. Also, there are not enough computers in the local health centers to allow every doctor to use the new system. The same computer must be used by several doctors making it difficult to take full advantage of it (for example, doctors have to share the same computer to access patients' records or request appointments with specialists). We hope this situation will change in a near future. In fact, the success of the new appointment service is encouraging Galician politicians to invest more on Web technologies.

Acknowledgments. We want to thank "Xunta de Galicia" and "Ministerio de Ciencia y Tecnología" for their partial support to this work under grants "Arquitecturas distribuidas para Teleservicios" (PGIDT00TIC32203PR) and "CORBALearn: Interfaz de Dominio guiada por Estándares para Aprendizaje Electrónico" (TIC2001-3767) respectively.

References

1. Spanish National Public Health Care System Web site. http://www.seg-social.es/. [Last accessed July 2002]
2. Wireless Access Protocol (WAP). Forum Web site. http://www.wapforum.com. [Last accessed July 2002]
3. Netscape's SSL page. http://home.netscape.com/security/techbriefs/ssl.html. [Last accessed July 2002]
4. WAP Forum: Wireless Transport Layer Security. WAP Forum Technical Report. Web page: http://www1.wapforum.org/tech/terms.asp?doc=WAP-261-WTLS-20010406-a.pdf. [Last accessed July 2002]
5. Nokia Mobile Internet Toolkit. http://www.nokia.com/corporate/wap/sdk.html. [Last accessed July 2002]

Hybrid Queuing Strategy to Reduce Call Blocking in Multimedia Wireless Networks

Dong Chun Lee[1], Il-Sun Hwang[2], and Robert Young Chul Kim[3]

[1] Dept. of Computer Science Howon Univ., Korea
ldch@sunny.howon.ac.kr
[2] R&D Network Manag. Supercom. Center KISTI, Korea
[3] Dept. of CIC Hoogik Univ., Korea

Abstract. In this paper, we propose a hybrid queuing strategy to reduce the blocking rate of channel allocation for multiple priority calls in Multimedia Wireless Networks (MWN). The proposed scheme is provided with an analytic model, wherein a Two- Dimension Markov Process. In numerical results, our method show correct analytic model and has better performance result than previous work in MWN.

1 Introduction

For delivering the desired levels of Quality of Service (QoS) in Wireless Networks (WN) to multiple types of mobile users, an improved channel allocation mechanism is required. It is to obtain a high-admitted traffic and to reduce blocking rate while guaranteeing the protection of calls in restricted channels [2], [8]. More recently, to reduce the blocking rate of channel allocation in WN, previous work has been proposed the schemes that make use of queuing method [5]. The schemes are classified calls in WN as voice call and data call. Intuitively, when channels are busy, waiting only the hand-off data call in a queue can reduce the blocking rate of the hand-off data call. The blocking rate of voice call can be also reduced by giving priority to voice call in a queue environment [8], [9]. However, the previous works consider only two types of traffic and have limitations for reducing the high blocking rate of multiple priority calls by allocating channels efficiently in WN. An efficient queuing scheme for multi-class calls in MWN can reduce the blocking rate of calls in restricted channels in a cell. So we have been introduced two queuing schemes that reduced the blocking rate of multiple priority calls in channel allocation in MWN [6],[7].
In order to obtain the better QoS for channel allocation, we propose an improved queuing scheme for multiple calls, which is investigated the hybrid queuing strategy in MWN.

2 Related Work

The previous scheme has been considered two types of traffic (voice calls and data packets), which is supported by a set of C channels plus a buffer of size K-C

M.H. Shafazand and A M. Tjoa (Eds.): EurAsia-ICT 2002, LNCS 2510, pp. 615–623, 2002.

[9]. Any type of arrival has access to any facility but voice call can preempt the service of data packet which return to the queue next to the last voice call arrival. Thus this scheme has a system with preemptive priority in the C channels and Head-of-the-Line (HOL) priority in the queue, where voice has a priority over data packet. A call in such a system is blocked only if there are already K calls in the system while a data packet is blocked if the system is full. Moreover, any type of traffic must leave the queue after a finite time because the vehicle has to leave the cell. This scheme is depicted in [9] and the state diagrams are given in two dimensional case. In schemes, the blocking probabilities and the mean waiting times with two types of traffic calls are given in [9].

We have been proposed two queuing schemes with n priority calls which are based on priority control methods [6],[7]. First scheme shows the queuing model to be proposed in our work using Head of Line priority control [1].

The queue is logically divided by threshold values T_i. When multiples calls arrive in a queue, each call waits in intervals of each threshold value, which high priority call prior to low priority call waits in the queue. When each call arrives in a queue of which state is over its threshold values, it is terminated by force and blocked. These calls cannot allocate channels or the portable must makes out of the coverage area. If the highest priority call λ_1 arrives in the queue which state is over the first threshold value T_1, it is terminated by force and allocated channels in a cell. If the queue size M is full, all of n+1 class calls are terminated by force irrespective of their priorities.

The second scheme also shows the basic queuing model to be proposed in our work using Partial Buffer Sharing(PBS) priority control. When multiple calls arrive in the queue, each call waits in the queue which is divided by threshold values and shares the queue partially irrespective of priority. But when each call arrives in the queue which state is over its threshold values T_i, it is terminated by force. If the lowest priority call λ_{n+1} arrives in the queue which state is over the first threshold value, it is terminated by force and blocked. The channel cannot be allocated for this call because it shares the queue by the first threshold value T_1. If multiple calls arrive in the queue which state is over the second threshold value T_2, the lowest priority call λ_{n+1} and the second lower priority call λ_n are also terminated by force. The channels cannot be allocated for those calls because they only share the queue by the first threshold value and the second threshold value. If queue size M is full, all of n+1 class calls are terminated by force irrespective of their priorities. In two schemes, analytic model are also given in [6],[7].

3 Hybrid Queuing Strategy

Fig. 1 shows a hybrid queuing model that integrate two previous methods with multiple priority calls.

In this queuing scheme, when multiple calls λ_i, $1 \leq i \leq n+1$ arrive in a queue,each call shares the queue partially without regard to priority, and waits

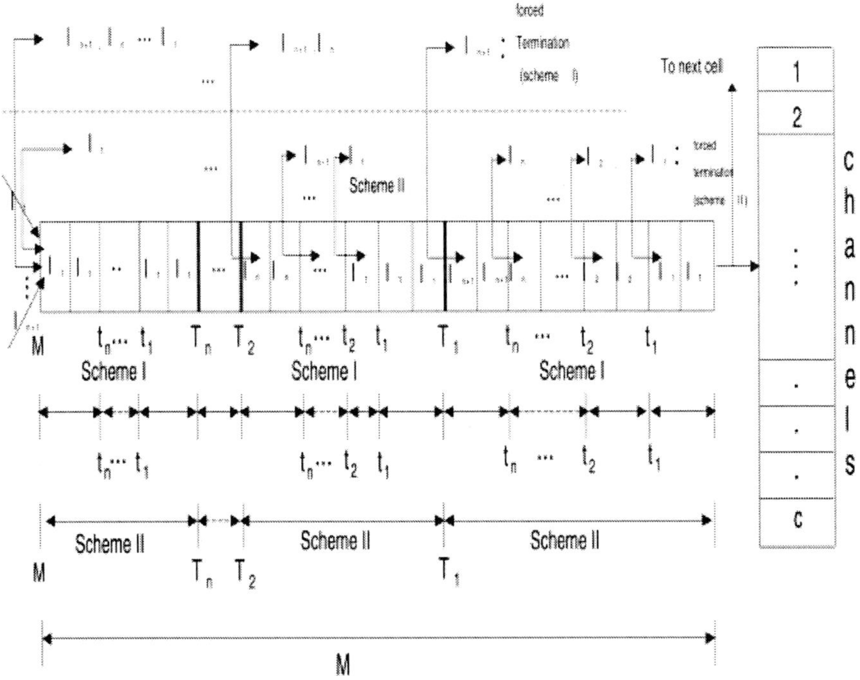

Sub-threshold value of scheme I : t_i, $1 \leq \text{i} \leq n$
Threshold value of scheme II:T_I,1≤i≤n
Threshold value of hybrid scheme:$T_i \leq t_i \leq t_2 \cdots \leq t_n \leq T_{i+1}, 1 \leq i \leq n-1$

Fig. 1. The hybrid queuing model with (n+1) priority calls in a cell.

in the queue as intervals of each threshold value T_i of scheme II. And also multiple calls in threshold value T_i wait in the queue according to sub-threshold value t_i of scheme I depending on priority i. When each call arrives in the queue of which state is over its threshold value T_i and sub-threshold value t_i of scheme I, it is terminated by force and blocked. These calls cannot allocate channels or the terminal must leave to next cell. If the second lowest priority call λ_n is over its threshold value T_n of scheme II, and If multiple calls are over its sub-threshold value tn of scheme I while the call waits in threshold value T_i of scheme II, they are terminated by force and blocked. The calls must also leave the cell because the channels cannot be allocated for them.

To analyze the proposed queuing model, we assume the followings: (a) Arrival calls are modeled as Poisson process with arrival rates λ_i of i class traffic. (b) The waiting time and the channel holding time of multiples calls have the exponential distributions with service rate μ_q and service rate μ, respectively. (c) Queue size M is finite (K-C) and FIFO discipline is served in each threshold area. (d) A cell is equipped with C permanently assigned channels. (e) The model for multiple calls is M/M/C/M/K.

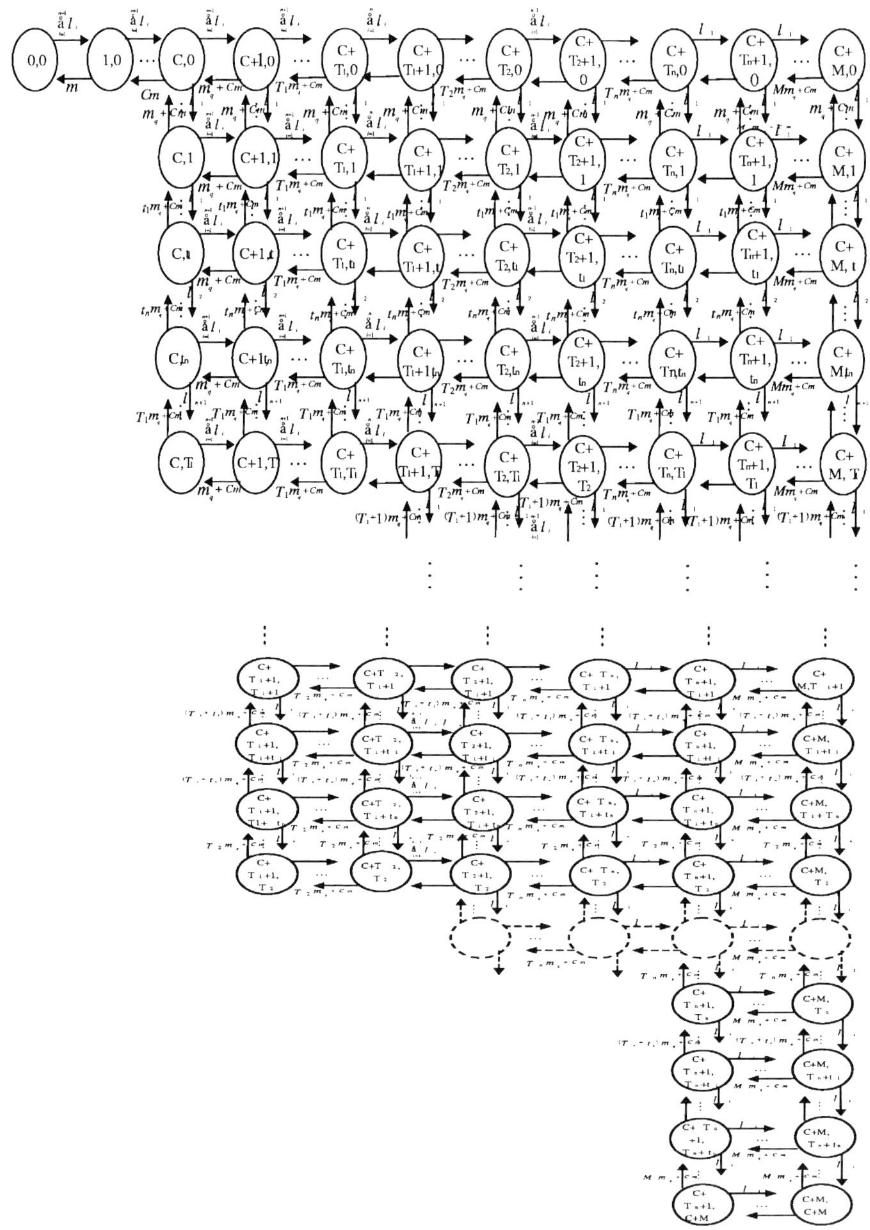

Fig. 2. The state transition diagram of hybrid queuing strategy

An analytic model is carried out by resolving a two- dimensional Markov Chain, where each state (i, j) represents the number i of multiple calls in any threshold values T_i of scheme II and the number j of multiples calls in any sub-threshold value t_i of scheme I within any threshold value T_i of scheme II. Let P_{ij} (t) = Pr [I (t) = i,J(t) = j] denotes the probability that the process is in state (i, j) at time t and let $P_{ij} =$ $_{ij}$ (t) denote the steady state probability that the process is in state (i, j). From the state transition diagram of Fig. 5, let us classify the state diagram into (n+1) parts as follows:

(part 1) $0 < i < C, \quad j = 0$
(part 2) $C < i \leq C + T_1, \quad j = 0, 1, \cdots, t_1, t_1 + 1, t_n, \cdots, T_1,$

$\vdots$

(part n+1)
$C + T_n < i \leq C + M, \quad j = 0, 1, \cdots, t_1, t_1 + 1, t_n, \cdots, T_1, T_1 + 1, \cdots,$
$T_1 + t_1, \cdots, T_{n-1}, \cdots, T_n \cdots M, t_n < j \leq T_1$

It can be observed that the state space can be partitioned into (n+1) parts of state to obtain the steady state probability P(i,j).

part 1 0<i<C, j=0

$$p(i,j) = \frac{(\sum_{i=1}^{n+1}\lambda_i)^i}{i!\mu^i}P(0,0) \tag{1}$$

part 2 C≤ i≤ C+T$_1$,

$$\mathbf{p(i,j)} = \begin{cases} \frac{(\sum_{i=1}^{n+1}\lambda_i)^c}{C!\mu^c}\frac{1}{\prod_{k=1}^{i}(K\mu_q+C\mu)}(\sum_{i=1}^{n+1}\lambda_i)^{i-C}\frac{1}{\prod_{m=1}^{j}(m\mu_q+C\mu)}(\lambda_1)^j p(0,0) \\ \quad ,0\leq j\leq t_1 \\ \frac{(\sum_{i=1}^{n+1}\lambda_i^)c}{C!\mu^c}\frac{1}{\prod_{k=1}^{i}(k\mu_q+C\mu)}(\sum_{i=1}^{n+1}\lambda_i)^{i-c}\frac{1}{\prod_{m=1}^{t_1}(m\mu_q+C\mu)}(\lambda_1)^{t_1} \\ \quad \{\frac{1}{\prod_{m=t_1+1}^{j}(m\mu_q+C\mu)}(\lambda_2)^{j-t_1}\}p(0,0) \ ,t_1\leq j\leq t_2 \\ \quad \vdots \qquad \vdots \\ \frac{(\sum_{i=1}^{n+1}\lambda_i)^c}{C!\mu^c}\frac{1}{\prod_{k=1}^{i}(K\mu_q+C\mu)}(\sum_{i=1}^{n+1}\lambda_i)^{i-c}\frac{1}{\prod_{m=1}^{t}n(m\mu_q+C\mu)}(\lambda_1)^{t_1} \\ \quad \{\frac{1}{\prod_{m=t_n+1}^{j}(m\mu_q+C\mu)}(\lambda_2)^{j-t_{n-1}}\} \\ \quad \times\{\prod_{k=2}^{n}(\lambda_k)^{t_k-t_{k-1}}\}p(0,0) \ ,t_{n-1}<j\leq t_n \\ \frac{(\sum_{i=1}^{n+1}\lambda_i)^c}{C!\mu^c}\frac{1}{\prod_{k=1}^{i}(K\mu_q+C\mu)}(\sum_{m=1}^{n+1}\lambda_i)^{i-c}\frac{1}{\prod_{m=1}^{t_{n-1}}(m\mu_q+C\mu)}(\lambda_1)^{t_1} \\ \quad \{\frac{1}{\prod_{m=t_{n-1}+1}^{j}(m\mu_q+C\mu)}(\lambda_n)^{j-t_{n-1}}\}\{\prod_{k=2}^{n-1}(\lambda_k)^{t_k-t_{k-1}}\}p(0,0) \\ t_n<j\leq T_1 \end{cases} \tag{2}$$

$\vdots$

Part n+1 $C + T_n < i \leq C + M$

$$\mathbf{p(i,j)} = \left[\begin{array}{c} \frac{(\sum^{n+1}\lambda_i)^c}{C!\mu^c} \frac{1}{\prod_{k=1}^{T_n}(K\mu_q+C_\mu)} (\sum_{i=1}^{n+1}\lambda_i)^{T_1-C}(\sum_{i=1}^{n}\lambda_i)^{T_2-T_1} \\ \cdots(\lambda_2+\lambda_1)^{T_n-T_{n-1}} \\ \times\{\frac{1}{\prod_{k=T_n+1}^{i}(m\mu_q+C\mu)} \times (\lambda_1)^{i-T_n}\}\times(\lambda_2)^{T_n-(T_{n-1}+t_n)} \\ (\lambda_3)^{T_{n-1}-(T_{n-2+t_n})}\cdots(\lambda_n)^{T_2-(T_1+t_n)}(\lambda_{n+1})^{T_1-t_n} \\ \times\{\frac{1}{\prod_{m=1}^{T_n}(m\mu_q+C\mu)}(\lambda_1)^{t_1+t_2\cdots+t_{n-1}+t_n}\}\times\{\frac{1}{\prod_{m=T_n+1}^{j}(m\mu_q+C\mu)}(\lambda_1)^{j-T_n}\} \\ \times\{\prod_{k=2}^{n}(\lambda_k)^{t_k-t_{k-1}}\} \\ \times\{\prod_{k=2}^{n-1}(\lambda_k)^{t_{k+1}-t_k}\}\cdots\times\{\prod_{k=2}^{3}(\lambda_k)^{t_{k+n-3}-t_{k+n-4}}\} \\ \{\prod_{k=2}^{2}(\lambda_k)^{t_{k+n-2}-t_{k+n-3}}\}P(0,0), \quad T_n < j \le M \end{array} \right. \tag{3}$$

Here, initial probability P (0, 0) is obtained from equations (1) using the normalization condition.

$$\textstyle\sum_{i=o}^{M}\sum_{j=0}^{M}P(i,j)=1 \tag{4}$$

From equations (1) - (4), we can obtain blocking probability ($P_b^{(i)}, 1 \le i \le n+1$) depending on priority i can be obtained by

$$P_b^{(1)} = \left(\sum_{j=0}^{t_1}P(M,j) + \sum_{j=T_1+1}^{T_1+T_2}P(M,j)\cdots + \sum_{j=T_n+1}^{M}P(M,j)\right),$$

$$P_b^{(2)} = \sum_{i=T_n+1}^{M}\left(\sum_{j=t_2+1}^{T_1}P(i,j) + \sum_{j=T_1+t_3+1}^{T_2}P(i,j)\cdots + \sum_{j=T_{n-2}+t_{n-1}+1}^{T_{n-1}}P(i,j)\right)$$

$$\vdots \qquad\qquad \vdots$$

$$P_b^{(n+1)} = \textstyle\sum_{i=T_1+1}^{M}\left(\sum_{j=T_1+1}^{M}P(i,j)\right), \tag{5}$$

where $P_b^{(1)}$ is the blocking probability with which the highest priority callλ_1 occurs when it arrives in the queue whichever state is over the first sub-threshold value t_1 within each threshold value T_1 and over the queue size M, and $P_b^{(n+1)}$ is the blocking probability with which the lowest priority call λ_{n+1} occurs when it arrives in the queue whichever state is full from n^{th} threshold value t_n within each threshold value T_i and over the first threshold value T_1.
From equations (2)–(3), the mean waiting time $W_q^{(i)}$ depending on priority i can be obtained using Little's Rule [4].

$$W_q^{(1)} = \frac{1}{\sum_{i=1}^{n+1}\lambda_i}\left[\sum_{i=0}^{M} i\left(\sum_{j=0}^{t_1}P(i,j) + \sum_{j=T_1+1}^{T_1+t_2}P(i,j)\cdots + \sum_{j=T_n+1}^{M}P(i,j)\right)\right],$$

$$W_q^{(2)} = \frac{1}{\sum_{i=1}^{n}\lambda_i}\left[\sum_{i=T_1+1}^{M} i\left(\sum_{j=T_1+1}^{t_1} P(i,j) + \sum_{j=T_1+t_2+1}^{T_1+t_3} P(i,j)\cdots + \sum_{j=T_{n-1}+t_n+1}^{T_n} P(i,j)\right)\right],$$

$$\vdots \qquad \vdots$$

$$W_q^{(n+1)} = \frac{1}{(\lambda_1)}\left[\sum_{i=0}^{T_1} i\left(\sum_{j=t_n+1}^{T_1} P(i,j)\right)\right], \tag{6}$$

where $W_q^{(1)}$ is the mean waiting time that the highest priority call λ_1 waits in the queue until the first sub-threshold value t_1 within each threshold value T_i, and W_q^{n+1} is the mean waiting time that the lowest priority call λ_{n+1} waits in the queue until n^{th} sub-threshold value t_n within each threshold value T_i.

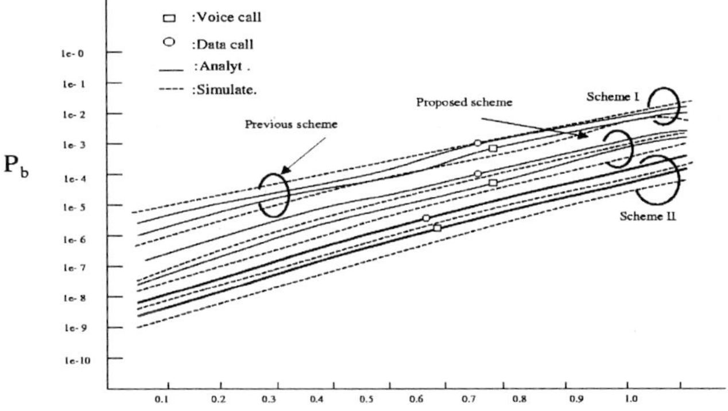

Fig. 3. Blocking probability vs. traffic intensity p.

4 Numerical Results

For the performance of the proposed hybrid queuing scheme, we verify the proposed scheme using the computer simulation and simulation model can be characterized as follows: (a) Simulation is performed by SIMSCRIPT II.5 Package using discrete time event scheduling. (b) The calls are arrived $10^7 \sim 10^8$ times in a cell. (c) Modules of simulation have PREAMBLE,MAIN, INITIAL, ARRIVAL, DEPARTURE, and STOP.SIM.

In Fig. 3, the proposed method is compared to the previous schemes with two types of traffic (voice call and data call) in aspect of the blocking probability. In numerical results, we are given for the following parameters: high priority callλ_1 is voice call, low priority callλ_1 is data call, queue size M is 40, threshold value T is 20, first sub- threshold value t_1 is 10, and number of channel per-cell C is 40.

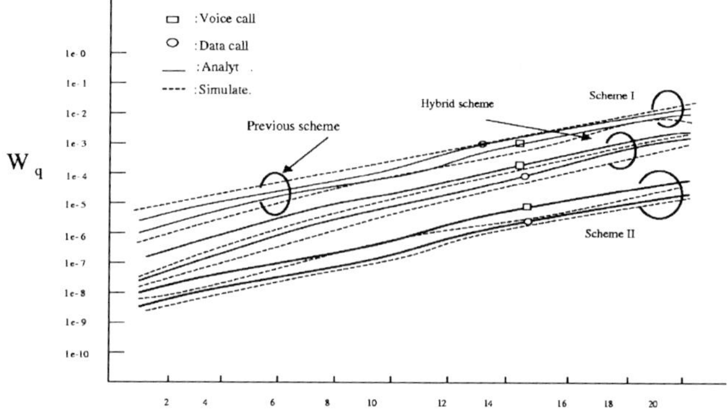

Fig. 4. Mean waiting time vs. arrival rate λ

Compared to the previous scheme, the proposed method shows the better performance result than previous scheme or our previous scheme I in terms of the blocking probability with two types of traffic because the method is based on hybrid priority control.

For the mean waiting time in Fig.4, our previous scheme I also shows the same performance result compared to previous scheme because our previous scheme I is based on HOL control traffic. However, the proposed hybrid strategy shows better performance result than the previous scheme because the scheme is based on hybrid priority control with two types of traffic.

5 Conclusions

We proposed on hybrid queuing strategy with traffic control to reduce the high blocking rate of multiple priority calls for channel allocation in MWN. In numerical results, the proposed method is better performance than our previous scheme I in terms of the blocking probability, and than our previous scheme II with respect to the mean waiting time under same priority. Proposed method must be getting better analytic model than previous schemes.

References

1. H. Cobham, "Priorities Assignment in Waiting Line Problem", Operation Research, No. 2, pp. 70–76, 1964.
2. R. Guerin, "Queuing Blocking System with Two Arrival Streams and Guard Channels", IEEE Trans.Comm., 1988.
3. H. Kroner, "Comparative Performance Study of Space Priority Mechanism for ATM Networks", Proceeding of IEEE INFOCOM'90, 1990.
4. L. Kleinrock, Queuing Systems, Vol. I, Wiley interscience, 1975.

5. Y. Jun, S. Cheng, “A Novel Priority Queue Scheme for Hand-off Procedure”, Proceeding of IEEE ICC’94
6. “Multi Service Personal Communications Services”, Journal of Com.Comm.,Vol.23, No.11, pp.1069–1083, 2000
7. D.C. Lee, S.J. PARK, and J.S. Song, “Performance Evaluation of Queuing Schemes for Multiple Priorities in Multiservice PCS”, Proceeding of IEEE GLOBECOM’98, 1998.
8. Y.B. Lin, S. Mohan, and A. Noeopel: “Queuing Priority Channel Assignment Strategies for PCS Hand-off and Initial Access”, IEEE Trans. on Vehicular Technology. Vol. 43, No. 3, PP. 704–712, 1994.
9. F.N.Pavlzdou: “Two-dimensional Traffic Models for Cellular Model System”, IEEE Trans. on Communication Vol.42, No. 2/3/4, pp. 1505–1511, 1994.

A Dynamic Backoff Scheme to Guarantee QoS over IEEE 802.11 Wireless Local Area Networks

Kil-Woong Jang, Sung-Ho Hwang, and Ki-Jun Han

Dept. of Computer Engineering, Kyungpook National Unversity, Daegu, Korea
{jangkw, sungho}@netopia.knu.ac.kr and {kjhan}@bh.knu.ac.kr

Abstract. In this paper, we propose a dynamic backoff scheme to support real-time traffic with quality of service (QoS) over IEEE 802.11 wireless local area networks. According to the type of traffic transmitted in a station, it is designed to carry out backoff procedures with the different size of contention window. Additionally, as collisions increase, we dynamically change the window size for backoff. The proposed scheme supports advanced QoS for real-time stations as well as being compliant with the 802.11 standard. We evaluated the performance of the proposed scheme using Markov model analysis. The numerical results indicate that the proposed scheme may offer better performance than the conventional 802.11 scheme in terms of saturation throughput.

1 Introduction

Currently, wireless local area networks (WLANs) are an emerging technology that provides high bandwidth and real-time multimedia applications to users in next generation networks. This technology is supported by two standards: the IEEE 802.11 standard [3,4] and High Performance Radio LAN (HIPERLAN) Type 1 [5].

The 802.11 standard provides two service types: asynchronous and delay bound. The asynchronous type of service is provided by the distributed coordination function (DCF), which implements the basic access method of the 802.11 media access control (MAC) protocol. It is also known as the carrier sense multiple access with collision avoidance (CSMA/CA) protocol. The delay bound type of service can be provided by the point coordination function (PCF), which is a centralized MAC protocol able to support collision free and access channel in a round-robin based on a polling mechanism.

In CSMA/CA, each station seeking access to the medium selects a random time slot within the contention window (CW). The station that selects the shortest random time will gain access for transmission; the others freeze their backoff times until the transmission is finished and wait for the remaining time in the following cycle. However, this contention-based MAC protocol cannot guarantee transfer delay for real-time traffic. In WLANs, all stations must compete for access to the shared medium. Therefore, the competition may cause a longer transmission delay and a lower throughput due to collisions. In this paper, we propose a new backoff scheme to support delay-bound stations with QoS by dynamically resizing the contention window.

M.H. Shafazand and A M. Tjoa (Eds.): EurAsia-ICT 2002, LNCS 2510, pp. 624–631, 2002.

2 IEEE 802.11 Backoff Scheme

The IEEE is developing an international WLAN standard identified as IEEE 802.11. The 802.11a standard describes support for a 54 Mb/s WLAN and the 802.11b standard describes support for a 11 Mb/s WLAN. Mandatory support for asynchronous data transfer is specified as well as optional support for distributed time bounded services. However, the 802.11a and 802.11b standards do not support traffic that is bounded by a specified time delay to achieve an acceptable QoS (i.e., priority). Currently, the IEEE is developing a QoS-supported WLAN standard identified as 802.11e. In the 802.11e networks, it supports an enhanced station (QSTA) that contains an 802.11e conformant MAC that supports QoS and an 802.11 conformant physical interface to the wireless medium.

In general, a QSTA may transmit a pending packet when it is operating under the DCF access method, either in the absence of a Point Coordinator or in the Contention Period of the PCF access method. In 802.11e, the priority value is provided with each packet at the medium access control service access point (MAC SAP). By default, priority 7 is the highest priority while priority 2 is the lowest priority. While priority 0 is used for best effort traffic while priority 1 (spare) is ordered between priority 3 and priority 2. The resulting default ordering is then {7, 6, 5, 4, 3, 0, 1, 2}, which matches the recommended priority mapping in IEEE Standard 802.1d [6]. A QSTA operates according to the same general rules defined for DCF by providing separate output queues. Each queue initiates a DCF state machine that contends for the wireless medium with AIFS[i] rather than DIFS. In addition, it employs a $CW_{min}[i]$ rather than a CW_min between queues within an QSTA is resolved within that station.

The backoff scheme shall be invoked whenever a QSTA desires to transfer a frame and finds the medium busy, indicated by either the physical or virtual carrier sense mechanism. The backoff scheme shall also be invoked when a transmitting station infers a failed transmission. To begin the backoff scheme, the QSTA shall set its Backoff Timer to a random backoff time using the following equation:

$$T_{backoff}[i] = Random(i) \times SlotTime \ . \tag{1}$$

where $Random(i)$ is a pseudo random integer drawn from a uniform distribution over the interval [0,$CW[i]$]. $CW[i]$ is an integer within the range of values of the MIB attributes a $CW_{min}[i]$ and a $CW_{max}[i]$.

To compute the new $CW[i]$ value, denoted $CW_{new}[i]$, from the old $CW[i]$ value, denoted $CW_{old}[i]$, in the event of a collision, a station shall choose a value of $CW_{new}[i]$ that meets the following criterion:

$$CW_{new}[i] = ((CW_{old}[i] + 1) \times PF) - 1 \ . \tag{2}$$

where the persistence factor, PF, is computed using the procedure described in [4]. The values of CW_{min}, AIFS and PF are transmitted by an enhanced access point (QAP) using the management frame with the QoS parameter set element [4].

3 Dynamic Backoff Scheme

We propose a new backoff scheme to increase saturation throughput for real-time traffic, which is called a dynamic backoff scheme (DBS), over IEEE 802.11 WLANs. Our basic idea is to allow the backoff scheme to dynamically resize the contention window under a number of stations (or amount of packet loss rate) in the QoS basic service set (QBSS).

To support the QSTA, the 802.11e standard includes management frames. Of these frames, the management frame with the QBSS load element [4] contains information on the current station population and traffic levels in the QBSS. The station count field in the frame indicates the total number of STAs and QSTAs currently associated with this QBSS. In addition, the frame loss rate field indicates the portion of transmitted packets that require retransmission or are discarded as undeliverable. A QAP knows how it is transmitted within each QBSS. Therefore, the QAP can control the size of the contention window under the certain status of transmission using the management frame with the QoS parameter set element.

```
Backoff Scheme ()
  if (IsQSTA() == True) // if a station is QSTA
     if (Rc > 1) // retransmission
        if (Rc <= Rt)
           if (Priority == Real-time Traffic)
              if (CWold < CWmax) {
                 if (FrameLossRate < Threshold)
                        Set PF > Default PF;
                 else   Set PF < Default PF;
                 CWnew = ((CWold+1)*PF)-1; }
              else CWnew = CWmax;
           else  // for a non-real-time traffic
              if (CWold < CWmax) {
                 Set PF = Default PF;
                 CWnew = ((CWold+1)*PF)-1; }
              else CWnew = CWmax;
        else Discard Packet; // if over Rt
     else CWnew = CWmin; // if first transmission in QSTA
  else CWnew = ((CWold+1)*Default PF)-1;  // if not QSTA
```

We first assume that the traffic is divided into two types: real-time (i.e., priority 7, 6 and 5) and non-real-time traffic (i.e., priority 4, 3, 0, 1 and 2). According to the type of the traffic, DBS carries out a retransmission procedure using the different contention window. In DBS, we first distinguish whether a station can support the QoS or not. If a station is a STA, the backoff procedure is conducted according to the conventional backoff procedure. The set of CW values shall be sequentially ascending integer powers of the value of default PF, minus 1, beginning with a CW_{min} value and continuing up to and including a CW_{max} value. In this paper, we assume that the value of the default PF is 2.

If a station is a QSTA, the backoff procedure carries out the DBS. In DBS, for real-time traffic, we use a boundary is called a threshold, ϕ, to distinguish the size of the contention window under two states of transmission in QBSS: the idle and busy states. If a number of stations (or the frame loss rate) is lower than ϕ, it is called an idle state. Otherwise, it is called a busy state.

In an idle state, to increase the waiting delay during the backoff procedure, a high-priority station with real-time traffic has a larger PF value than the default PF value. However, if traffic occurs heavily, the throughput will be decreased because the CW for real-time traffic has a high value. Therefore, the amount of the transmitted packet is reduced due to low throughput. Therefore, in a busy state, the value of the CW is reduced as the AP decreases PF. The following are some of the significant features of DBS:

1. An important feature of DBS is that the size of CW is dynamically changed under a state of transmission in a QBSS. In other words, when a QSTA number is increasing due to collision and transmission error, the QSTA with a real-time traffic may have a higher probability of selecting the backoff time than a non-real-time traffic due to the resizing CW.
2. The DBS is compliant with the 802.11 standard as well as supports advanced QoS for real-time stations. To carry out DBS, we use a management frame with an information element defined as the 802.11e standard.

4 Performance Evaluation

In this section, we present a Markov model to obtain saturation throughput, γ. This is the most important QoS factor in which we are interested. A Markov model for backoff time was proposed in [1,2] to analyze the performance of the 802.11 that only employs the DCF, as shown in Fig. 1. In order to accurately analyze the performance of DBS, we adopt the Markov model proposed in [1,2].

We considered a fixed number n of contending stations. Let $c(t)$ be the stochastic process representing the backoff time counter for a station at time t and $s(t)$ be the stochastic process representing the backoff stage of the station at time t. In addition, let m be the maximum value of the backoff stage [1]. We assume that each packet collides with constant and independent probability P_c. Once independence is assumed and P_c is supposed to be a constant value, the tuple $\{c(t), s(t)\}$ is a discrete-time Markov chain with transition probabilities [1]. In this Markov chain, we set $\mathrm{P}\{i_1, j_1 \mid i_0, j_0\}=P\{s(t+1)=i_1, c(t+1)=j_1 \mid s(t)=i_0, c(t)=j_0\}$. To describe the decrement of the backoff time counter, we have

$$P\{i,j \mid i,j+1\} = 1 \qquad (0 \le i \le m, \quad 0 \le j \le W_i - 2) \; . \tag{3}$$

Next, a new packet following a successful packet transmission starts with backoff stage 0. Then, the backoff is uniformly chosen in the range $(0,W_0 - 1)$. To describe this state, we have

$$P\{0,j \mid i,0\} = \frac{(1 - P_c)}{W_0} \qquad (0 \le i \le m, \quad 0 \le j \le W_0 - 1) \; . \tag{4}$$

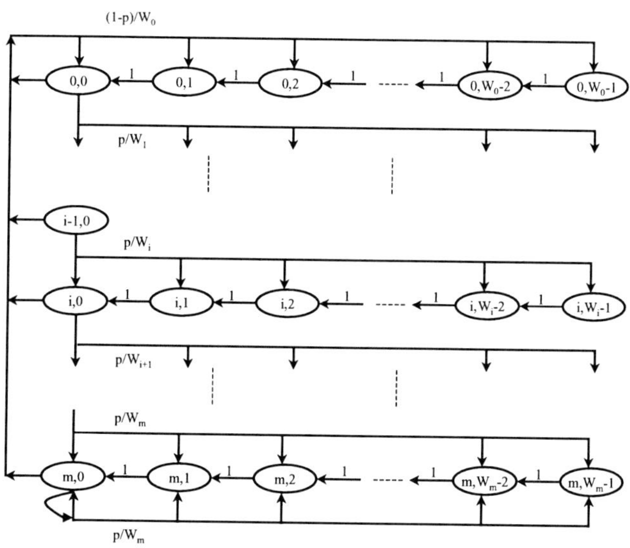

Fig. 1. Markov model for backoff time

When an unsuccessful transmission occurs at backoff stage i-1, the transition probability that the new initial backoff value is uniformly chosen in the range (0, W_i) are described as

$$P\{i,j \mid i-1,0\} = \frac{P_c}{W_0} \qquad (1 \le i \le m,\ \ 0 \le j \le W_i - 1)\ . \tag{5}$$

After the backoff stage reaches the value m, the transition probability that the size of the backoff window is uniformly chosen in the range (0,$W_m - 1$) are described as

$$P\{m,j \mid m,0\} = \frac{P_c}{W_m} \qquad (0 \le j \le W_m - 1)\ . \tag{6}$$

Here, we let $c_{i,j}$ be the stationary distribution of the Markov chain and we have

$$c_{i,j} = \lim_{t=\infty} \{s(t) = i, c(t) = j\} \qquad (1 \le i \le m,\ \ 0 \le j \le W_i - 1)\ . \tag{7}$$

Due to the Markov chain regularities, we can obtain a solution for this Markov chain. We can obtain the value of $c_{0,0}$ by imposing the normalization condition

$$1 = \sum_{i=0}^{m} \sum_{j=0}^{W_i - 1} c_{i,j} = c_{0,0} \left[\frac{W_0}{2} \left(\sum_{i=0}^{m-1} (2P_c)^i + \frac{2P_c}{1-P_c} \right) + \frac{1}{1-P_c} \right]\ . \tag{8}$$

Thus, we can obtain

$$c_{0,0} = \frac{2(1-2P_c)(1-P_c)}{(1-2P_c)(W_0+1) + P_c W_0 (1-(2P_c)^m)}\ . \tag{9}$$

Now, we can obtain ξ, which is the probability that a station transmits in a slot time. As any transmission occurs when the backoff size is equal to zero, we have

$$\xi = \sum_{i=0}^{m} c_{i,0} = \frac{c_{0,0}}{1-P_c} = \frac{2(1-2P_c)}{(1-2P_c)(W_0+1)+P_cW_0(1-(2P_c)^m)} . \tag{10}$$

The probability P_c that a transmitted packet collided in a slot time can be computed by the following equation:

$$P_c = 1-(1-\xi)^{n-1} . \tag{11}$$

where n is the number of the active stations and ξ is given in (11) as function of the unknown P_c. Numerically solving the previous equations, the values of P_c and ξ can be found. Let P_t be the probability that there is at least one transmission in a slot time and let P_s be the probability that a packet is transmitted successfully. Once ξ is known, P_t and P_s can be obtained as follows:

$$P_t = 1-(1-\xi)^n . \tag{12}$$

$$P_s = \frac{n\xi(1-\xi)^{n-1}}{P_t} = \frac{n\xi(1-\xi)^{n-1}}{1-(1-\xi)^n} . \tag{13}$$

In addition, the mean number of consecutive idle time slots between two consecutive packet transmissions, $E[\tau]$, is given by

$$E[\tau] = \frac{1}{P_t} - 1 . \tag{14}$$

We let $E[a]$ be the average amount of payload information successfully transmitted in a slot time and $E[l]$ be the length of a renewal interval. Finally, we can determine the normalized saturation throughput, γ, defined by

$$\gamma = \frac{E[a]}{E[l]} = \frac{P_sE[P]}{E[\tau]S_t+P_sT_s+(1-P_s)T_c} . \tag{15}$$

where $E[P]$ is the average packet length and S_t is the size of a slot time. Moreover, T_s is the average time that the channel is sensed busy due to a successful transmission and T_c is the average time that channel is sensed busy by the stations during a collision. We assume for simplicity that all the stations use the RTS/CTS access method for all the transmitted packets over the 802.11 networks. In such a case, collisions can occur only on RTS frames, so we have

$$T_s = RTS + 3SIFS + 4\delta + CTS + H + E[\tau] + ACK + AIFS[i] . \tag{16}$$

$$T_c = RTS + AIFS[i] + \delta . \tag{17}$$

where H is the length of packet header, which is the total of the MAC header and PHY header.

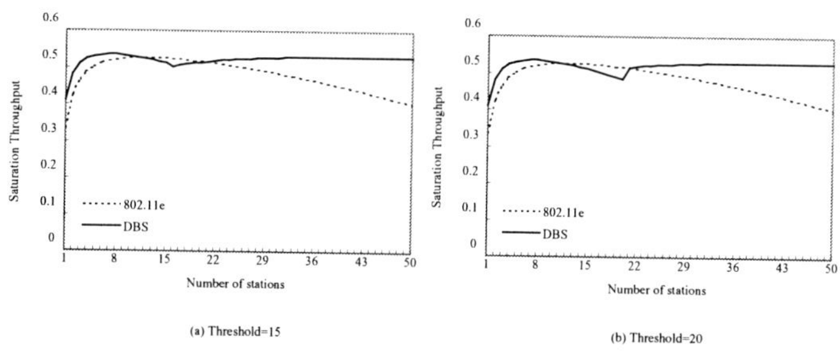

Fig. 2. Saturation throughput (case 1)

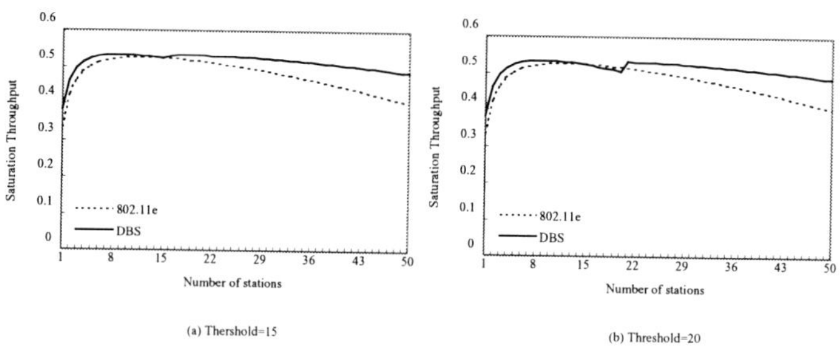

Fig. 3. Saturation throughput (case 2)

According to the type of traffic, we apply a value to the contention window size in the above equations. We define the PF value for the real-time in an idle state and a busy state, denoted by w_1 and w_2, respectively. Additionally, the non-real-time traffic in a busy state are denoted by w_3.

The performance of DBS was evaluated by analysis. The results of this analysis were obtained assuming the OFDM PHY system parameters [4]. We provide numerical results, based on analysis, to compare the performance of the 802.11e and DBS. To obtain the results, we used the Markov model. We then obtained values for the saturation throughput under various system parameters. In the analysis, we assumed that the traffic priority is uniformly distributed with a mean 1/8.

Figs. 2 and 3 show the saturation throughput theoretically achievable by the backoff scheme in both the cases of the 802.11e and DBS. From these figures, we can see that the DBS performs better than the 802.11e under most traffic loads. In Figs. 2 and 3, since we change the backoff scheme at this point, which can be called the change point, we can find steep lines in the DBS at n=15 and n=20. Around the change point, the performance of DBS is partially lower than 802.11e. Nonetheless, when the traffic load of the stations becomes heavy, the DBS performs much better than the 802.11e.

In DBS, according to the PF value during the backoff procedure, the saturation throughput can be affected by the PF. Thus, we considered two cases: case

1 (w_1=1, w_2=3, w_3=2) and case 2 (w_1=1.5, w_2=2.5, w_3=2), as shown in Figs. 2 and 3, respectively. In these figures, we can see that the results of case 1 are much better than case 2 in terms of difference between the idle and busy states. Under heavy traffic loads, case 1 offers much better performance than case 2. Conversely, around the change point, case 1 performs lower than case 2.

Finally, we can show that an adequate dynamic backoff scheme using threshold and PF values performs better than static backoff schemes. In addition, DBS performs much better than 802.11e under heavy traffic loads.

5 Conclusions

In this paper, we proposed a dynamic backoff scheme to support traffic with QoS over IEEE 802.11 WLANs. Our basic idea is to dynamically resize the contention window under a pre-defined threshold in QBSS. We developed a Markov model to analyze the DBS and the 802.11e for backoff time. We then evaluated the saturation throughput for the DBS and 802.11e. Numerical results demonstrate that the DBS performs better than the conventional 802.11e standard with DCF in WLAN.

References

1. G. Bianchi: Performance analysis of the IEEE 802.11 distributed coordination function. IEEE J. Select. Areas Commun., vol. 18, no. 3, Mar. (2000) 535–547
2. S. T. Sheu, T. F. Sheu: A bandwidth allocation/sharing/extension protocol for multimedia over IEEE 802.11 ad hoc wireless LANs. IEEE J. Select. Areas Commun., vol. 19, no. 10, Oct. (2001) 2065–2080
3. IEEE Standard for Wireless Medium Access Control and Physical Layer Specifications, Aug. (1999)
4. IEEE Standard for Wireless Medium Access Control and Physical Layer Specifications, Medium Access Control Enhancements for Quality of Service, Mar. (2001)
5. Broadband Radio Access Networks HIPERLAN Type 2, Apr. (2000)
6. IEEE Standard for Media Access Control Bridges, June (1998)
7. T. S. Ho and K. C. Chen: Performance evaluation and enhancement of the CSMA/CA MAC protocol for 802.11 wireless LAN's. Proc. IEEE PIMRC, Oct. (1996) 392–396
8. G. Bianchi, L. Fratta and M. Oliveri: Performance analysis of IEEE 802.11 CSMA/CA medium access control protocol. Proc. IEEE PIMRC, Oct. (1996) 407–411

Performance Evaluation of Serial/Parallel Block Coded CDMA System with Complex Spreading in Near/Far Multiple-Access Interference and Multi-path Nakagami Fading Channel

Jae-Sung Roh[1], Choon-Gil Kim[2], and Sung-Joon Cho[3]

[1] Dept. of Information & Communication Eng., SEOIL College, Seoul, KOREA
jsroh@seoil.ac.kr

[2] Basics Science, KAIST, Taejon, KOREA
kimcg@kaist.ac.kr

[3] School of Electronics, Telecommunication and Computer Eng., Hankuk Aviation Univ. Kyonggi-do, KOREA
sjcho@mail.hangkong.ac.kr

Abstract. The BER performance of concatenate block coded CDMA signal with coherent RAKE reception and complex spreading are considered in this paper. General multi-path intensity Nakagami fading and multiple-access near/far interference channel are assumed. A dedicated pilot channel for coherent demodulation is used for the purpose of channel estimation. Pilot channel estimation error, due to multiple-access interference and multi-path fading is studied. The analysis for system performance shows that the error of channel estimation significantly degrades BER performance and can be effectively suppressed by serial concatenate block code (SCBC) and parallel concatenate block code (PCBC) schemes. Also in this paper, an attempt for comparing the BER on different concatenate block coding schemes has been made. And a discussion on the multiple-access near/far interference is also included, which illustrates that it can be effectively limited by power control and SCBC/PCBC schemes.

1 Introduction

To provide higher data rates for end users, as well as to accommodate more users over wireless channels in the next generation communication systems, W-CDMA has become the focus of current research interests. Two of the important features of W-CDMA systems are the use of complex spreading and user-dedicated pilot channel [1]-[5]. In the W-CDMA standard of the 3GPP, a time-multiplexed pilot is used in the forward-link, whereas an in-phase/quadrature (I/Q) code-multiplexed pilot is used in the reverse-link [3]. In this paper, I/Q code-multiplexed pilot is adopted. In addition, the effects of multiple-access near/far interference and serial/parallel concatenate block code schemes have been analyzed in order to achieve reliable system performance.

M.H. Shafazand and A M. Tjoa (Eds.): EurAsia-ICT 2002, LNCS 2510, pp. 632–642, 2002.

I/Q code complex spreading can be implemented either by a complex valued sequence [4],[5], e.g., poly-phase sequence, or by two binary sequences. It is claimed that complex spreading can reduce the peak-to-average power ratio of modulated signals, thus improving RF power amplifier efficiency.

Coherent reception outperforms noncoherent by 3 dB in AWGN channel. In multi-path fading channels, a coherent RAKE receiver with maximum ratio combining (MRC) has the optimal performance. Eng and Milstein [6] presented the performance of coherent DS-CDMA in Nakagami fading channels. Coherent reception requires the knowledge of channel characteristics, which are time varying in fading environments. A conventional and effective method to accomplish this task is to use a separate pilot channel or insert pilot symbols in data symbols. Choi [7] proposed an adaptive method to estimate channel parameters by jointly utilizing pilot and data channels. However, in this paper, the channel estimation error and multiple-access near/far interference are analyzed in multi-path Nakagami fading channel. The system performance improvement due to the serial/parallel concatenate block coding schemes is also investigated.

2 System Model

2.1 Transmitter Model

In the reverse link of CDMA system, where an I/Q code-multiplexed pilot is utilized, spreading consists of two operations. The first is channelization operation, in which pilot and data symbols on I- and Q-branches are independently multiplied with an orthogonal variable spreading factor code and transformed into a number of chips. The second operation is scrambling, where the resultant signal are further multiplied by a complex valued scrambling code. This spreading scheme is called complex spreading, as illustrated in figure 1.

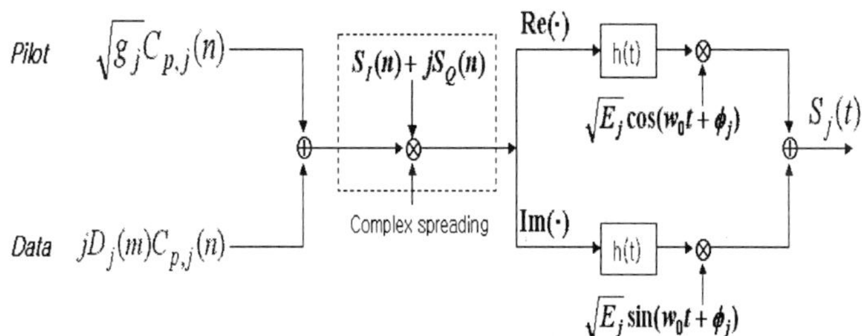

Fig. 1. Transmitter model of the *k*-user

The transmitted signal of the *k*-th user can be written as [1]

$$
\begin{aligned}
S_j(t) = \sqrt{E_{D,j}} \sum_{n=-\infty}^{\infty} \Big\{ & \\
& \cdot \left[\sqrt{g_j} C_{P,j}(n) S_I(n) - D_j(m) C_{D,j}(n) S_Q(n)\right] \\
& \cdot h(t - nT_c)\cos(\omega_o t + \phi_j) \\
& \cdot - \left[\sqrt{g_j} C_{P,j}(n) S_Q(n) - D_j(m) C_{D,j}(n) S_I(n)\right] \\
& \cdot h(t - nT_c)\sin(\omega_o t + \phi_j) \Big\}
\end{aligned}
\tag{1}
$$

where $E_{D,j}$ is chip energy of data channel, g_j is power ratio of pilot channel to the data channel, $D_j(m)$ is data symbols for the j-th user, m is integer part of n/P_N , P_N is spreading factor, ϕ_j is carrier phase, $C_{P,j}(n)$ and $C_{D,j}(n)$ are orthogonal channel codes for pilot and data symbols, respectively, $S_I(n)$ and $S_Q(n)$ are real and imaginary parts, respectively, of the cell-specific scrambling sequence, $h(t)$ is impulse response of the pulse shaping filter truncated by the length of AT_c for practical systems, where $A > 1$, and T_c is the chip interval. And, the power of the transmitted signal is expressed as

$$
P_j = (1 + g_j) E_{D,j} / T_c = E_j / T_c \tag{2}
$$

where E_j is the chip energy for both data channel and dedicate pilot channel of the j-th user.

2.2 Channel Model and Receiver Structure

The complex low-pass equivalent impulse response of multi-path fading channel can be written as

$$
h_j(t) = \sum_{l=0}^{L-1} \alpha_{j,l}(t)\delta[t - \tau_{j,l}(t)]e^{j\theta_{j,l}(t)} \tag{3}
$$

where L ($L \geq 1$) is the number of resolvable propagation paths. For the sake of simple notation, it is assumed that all users have the same number of multi-paths. $\alpha_{j,l}(t)e^{j\theta_{j,l}(t)}$ and $\tau_{j,l}(t)$ are the complex fading factor and propagation delay of the l-th path of the j-th user, respectively. Note that $\alpha_{j,l}(t)$ can be Rayleigh-, Rician- or Nakagami-distributed, depending on a specific channel model. All random variables in (3) are assumed independent for j and l. Assuming that there are J active users in the system, the received signal is given by

$$r(t) = \sum_{j=1}^{J} \sum_{l=0}^{L} \alpha_{j,l}(t) S_j[t - \tau_{j,l}(t)] e^{j\theta_{j,l}(t)} + \eta(t) \tag{4}$$

where $\eta(t)$ is the AWGN with double-side power spectrum density $\eta_0/2$.

In order to mitigate multi-path effect, a RAKE receiver with pilot symbol aided coherent demodulation and maximum ratio combining (MRC) is employed. The RAKE receiver structure is shown in figure 2, where the number of branches is less or equal to the number of resolvable paths. The received signal is multiplied by local carriers and passed through the pulse-matching filters and then sampled once per chip duration. The samples are fed to each branch of the RAKE receiver, where the pilot and data symbols are separately demodulated, and pilot symbols are used to eliminate the phase error of different path and achieve MRC. Assuming that the *i*-th path delay $\tau_{1,i}$ can be accurately estimated for the reference user (*j*=1), each path that corresponds to a RAKE branch gives an output component. The outputs of all branches are added together to form the decision statistic.

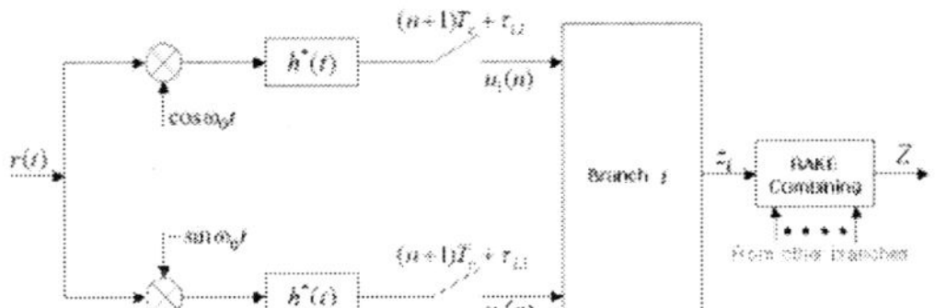

Fig. 2. RAKE receiver structure

3 Performance Evaluation

3.1 Effect of Channel Estimation

From the analysis results [1], it has been seen that in such W-CDMA system, the channel estimation using pilot symbols suffers from multiple-access interference and multi-path fading. In most publications on coherent demodulation, perfect channel estimation is assumed for the purpose of simplification. Therefore, such channel estimation error is considered in the following performance analysis. Assuming that the fading of each path is independent of each other and the output in each branch of the RAKE receiver is independent of each other, the summed output $Z(m) = \sum_{i=0}^{R_f - 1} z_i(m)$ is a Gaussian variable with mean and variance given by [1]

$$\begin{aligned} &E\left\{Z(m)\middle|_{\alpha 1,i(mP_N),\, i=0,1,\ldots,R_f-1}\right\} \\ &= E\left\{Z(m)\middle|_{\zeta}\right\} \\ &= \sum_{i=0}^{R_f-1} E\{z_i(m)\} \end{aligned} \tag{5}$$

and

$$\begin{aligned} &\operatorname{var}\left\{Z(m)\middle|_{\alpha 1,i(mP_N),\, i=0,1,\ldots,R_f-1}\right\} \\ &= \operatorname{var}\left\{Z(m)\middle|_{\zeta}\right\} \\ &= \sum_{i=0}^{R_f-1} \operatorname{var}\{z_i(m)\} \end{aligned} \tag{6}$$

where R_f is the number of RAKE finger.

Therefore, conditioned on the instantaneous multi-path fading amplitude ζ of the reference user, the equivalent signal-to-noise plus interference ratio (*SNIR*) is expressed as

$$\gamma_{eq1} = \left\{ \begin{aligned} &\frac{(1+g)}{\zeta}\left(1+\frac{1}{gN_P}\right)\left(\frac{1}{E_b/N_o}+\frac{\Delta\Omega_T J}{2P_N}\right) \\ &+\left(\frac{2M(1+g)^2}{\zeta^2 gN_P}\right)\left(\frac{1}{E_b/N_o}+\frac{\Delta\Omega_T J}{2P_N}\right)^2 \end{aligned} \right\}^{-0.5} \tag{7}$$

where $\Delta = 2/3$ for rectangular pulse shaping filter, Ω_T is channel parameter given by $\Omega_T = \sum_{l=0}^{L-1} E\{\alpha_{j,l}^2(mN)\}$, and E_b/N_o is the signal-to-noise ratio per bit.

3.2 Effect of Multiple-Access Near/Far Interference

As well known, DS/CDMA system is susceptible to near/far interference, which occurs when the base station input include one or more other CDMA signals that are stronger by adjusting the transmitted power of mobile users so that the base station gets the same power from the received signal of each transmission. In this paper, we consider the effects of multiple-access near/far interference on performance of the CDMA system with complex spreading scheme that has the channel estimation error. Modifying the previous equation (7), the equivalent *SNIR* include multiple-access near/far interference is expressed as

$$\gamma_{eq2} = \left\{ \begin{array}{l} \frac{(1+g)}{\zeta}\left(1+\frac{1}{gN_P}\right)\left(\frac{1}{E_b/N_o}+\frac{\Delta\Omega_T}{2P_N}\sum_{j=1}^{J}\frac{P_j}{P_o}\right) \\ +\left(\frac{2M(1+g)^2}{\zeta^2 gN_P}\right)\left(\frac{1}{E_b/N_o}+\frac{\Delta\Omega_T}{2P_N}\sum_{j=1}^{J}\frac{P_j}{P_o}\right)^2 \end{array} \right\}^{-0.5} \tag{8}$$

where P_o is the power controlled reference signal.

3.3 Serial/Parallel Concatenate Block Coding

In this section, we will present the high-performance coding schemes, show how to analyze their average performance, and finally give some design guidelines for concatenate codes with interleaver [8]-[10]. The main ingredients of a concatenated code with interleaver are two constituent codes (CCs) and one interleaver. They can be connected in serves, like in figure 3, or in parallel, as in figure 4. In figure 3 we show the example of a SCBC, composed by two linear cascaded CCs, the outer (N, mk) code C_o with rate $R_c^o = mk/N$ and the inner (mk, N) code C_i with rate $R_c^i = N/mk$, linked by an interleaver of size N, an integer multiple of the length of code words of the outer code, $N = mk$. The overall SCBC is then a linear (mn, mk, N) code, denoted by C_s, with rate $R_c^s = R_c^o R_c^i = k/n$.

Parallel concatenate codes are obtained as in figure 4, which refers to the case of a PCBC. Two linear block codes C_1 with parameters (n_1, k) and rate $R_c^{(1)} = k/n_1$, and C_2 with parameters (n_2, k) and rate $R_c^{(2)} = k/n_2$, the constituent codes, having in common the length k of the input information words, are linked through an interleaves of size $N = mk$. The block of m input words to the second encoder is a permuted version of the corresponding input block of the first one. The PCBC code word is formed by concatenating the two code words generated by the first and second encoder, The PCBC, that we denote as C_p, is then an $(n_1 + n_2, k, N)$ linear systematic code with rate

$$R_c^p = \frac{R_c^{(1)} R_c^{(2)}}{R_c^{(1)} + R_c^{(2)}} \tag{9}$$

We can have more than two CCs and one interleaver, but we will consider only this case for simplicity.

To obtain a union bound to the bit error probability of the SCBC of figure 3, we need to compute the following equation

$$P_b(e) \le \sum_{w=1}^{k} \frac{w}{k}\left[B_w^{C_s}(D)\right]_{D=\exp\{-R_C E_b/N_o\}} \tag{10}$$

where $B_w^{C_S}(D)$ is the conditional weight enumerating function (CWEF) of the SCBC, i.e., the weight enumerating function of the code words of the SCBC generated by information words of weight w. Assuming for simplicity that the interleaver size equals the length of the outer code words, i.e., $N=k$, the number $B_{w,d}^{C_S}$ of code words of the SCBC of weight d associated with an input word of weight w is given by

$$B_{w,d}^{C_S} = \sum_{j=0}^{N} \frac{B_{w,j}^{C_O} \times B_{j,d}^{C_i}}{\binom{N}{j}} \tag{11}$$

From equation (11), we easily derive the expressions of the conditional weight enumeration function of the SCBC

$$B_w^{C_S}(D) = \sum_{j=0}^{N} \frac{B_{w,j}^{C_O}(D) \times B_j^{C_i}(D)}{\binom{N}{j}} \tag{12}$$

From which we see that, to obtain the enumerators of the SCBC, we only need to know the two constituent codes.

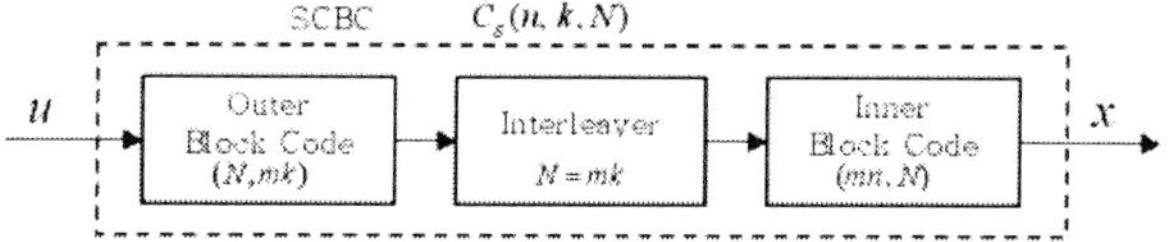

Fig. 3. SCBC encoder

In this paper, we consider the $(7,3,4)$ SCBC code obtained by concatenating the $(4,3)$ parity-check code with a $(7,4)$ Hamming code through an interleaver of size $N=4$. Using the CWEFs of the outer code and inner code for SCBC, we can be expressed the union bound to the bit error probability as follow

$$P_e \le 1.75\exp\left(-\frac{9}{7}\gamma_{eq}\right) + 1.75\exp\left(-\frac{12}{7}\gamma_{eq}\right) + \exp\left(-3\gamma_{eq}\right) \tag{13}$$

where γ_{eq} is equivalent signal-to-noise plus interference ratio.

As for the case of SCBCs, to obtain an upper bound to the bit error probability of the PCBC shown in Fig. 4, we will use Eq. (10), here rewritten with the replacement of C_S with C_P to denote that we are dealing with parallel instead of serial concatenate codes

$$P_b(e) \le \sum_{w=1}^{k} \frac{w}{k} \left[B_w^{C_P}(D)\right]_{D=\exp\{-R_C^P E_b/N_o\}} \tag{14}$$

where $B_w^{C_P}(D)$ is the CWEF of the PCBC, i.e., the weight enumerating function of the code words of the PCBC generated by information words of weight w. Also, we need to compute the CWEF $B_w^{C_P}(D)$. Using, as for SCBCs, a uniform interleaver, a given word of weight w at the input of the interleaver is mapped into all its permutations, which are then encoded by the code C_2. As a consequence, all data words of the same weight generate the same set of code words C_2, so that the CWEFs of C_1 and C_2 become independent, and can be multiplied and suitably normalized to yield the CWEF of the PCBC. In formulas, we have, for $N = k$

$$B_w^{C_P}(D) = \frac{B_w^{C_1}(D) \times B_w^{C_2}(D)}{\binom{N}{w}} \tag{15}$$

In PCBC, we consider a $(10,4,4)$ PCBC obtained through the use of a $(7,4)$ Hamming code C_1 and a $(4,3)$ code C_2 whose code words are the three parity-check bits of the $(7,4)$ Hamming code. Using the CWEFs of PCBC, we obtain the bit error probability as follow

$$\begin{aligned} P_e \le\; & 0.25\exp\left(-\frac{6}{5}\gamma_{eq}\right) + 3\exp\left(-\frac{8}{5}\gamma_{eq}\right) + 7.5\exp\left(-2\gamma_{eq}\right) \\ & + 3\exp\left(-\frac{12}{5}\gamma_{eq}\right) + 0.25\exp\left(-\frac{14}{5}\gamma_{eq}\right) + \exp\left(-4\gamma_{eq}\right) \end{aligned} \tag{16}$$

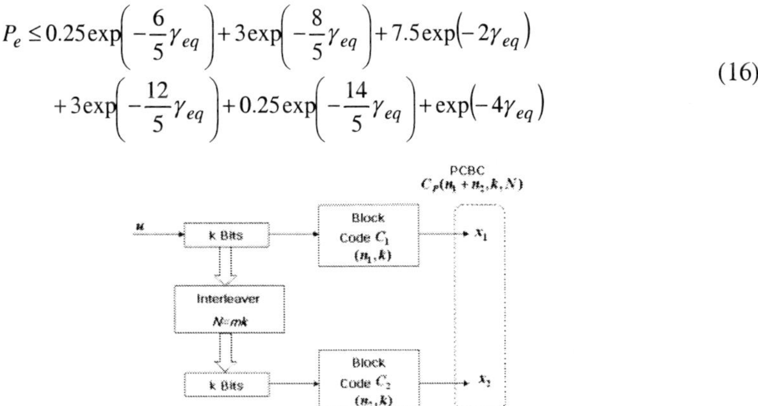

Fig. 4. PCBC encoder

3.4 Multi-path Intensity Nakagami Fading Model

The probability density function is unobtainable if the fading in each path is to be chosen arbitrarily. In order to simplify the numerical evaluation, it is assumed that each user has independent but the same fading characteristics. The amplitude fading in each path is assumed to be Nakagami distributed since the Nakagami distribution is more versatile and more adequate to describe different fading situations. For example, Nakagami fading can closely represent Rayleigh and Rician fading under appropriate parameters. The Nakagami pdf of ζ is given by

$$f(\zeta) = \left(\frac{m_T}{\Omega_T}\right)^{m_T} \frac{\zeta^{m_T - 1}}{\Gamma(m_T)} \exp\left(\frac{-m_T \zeta}{\Omega_T}\right) \tag{17}$$

where $\Omega_T = \sum_{i=0}^{R_f - 1} \Omega_o e^{-i\delta}$, $m_T = \sum_{i=0}^{R_f - 1} m_i$, and δ is the exponential decay ratio of the multi-path intensity profile (MIP).

4 Numerical Results

In this section, the effects of different parameters on the BER performance of the CDMA system with complex spreading, imperfect channel estimation by dedicated pilot symbols, multiple-access near/far interference, and serial/parallel concatenated block coding schemes are investigated by numerical calculations. The multi-path fading is assumed to be Nakagami distributed, and the exponential decay ratio of MIP model is considered. Unless noted otherwise, the number of multi-path is 8, the decay ratio of MIP is 0.2, the LPF length is 32, the power ratio of in-phase and quadrature branches is 0.3, and the spreading factor, i.e., number of chips per data symbol, is fixed to be 64. Figure 5 illustrates the BER as a function of E_b/N_o for different number of RAKE fingers, i.e., $R_f = 1,2,4,8$, and shows the coding gain of SCBC CDMA system. It can be seen that the RAKE receiver takes advantage of the multi-path and gives better performance as the number of RAKE fingers increases. For the case of $R_f = 4$, $BER \approx 10^{-3}$ of serial concatenated block coded CDMA system can be obtained when $E_b/N_o = 10$ dB. Figure 6 and figure 7 show the effects of multiple-access near/far interference according to the coding scheme. Figure 6 illustrates the BER performance of SCBC CDMA system with multiple-access near/far interference according to E_b/N_o, different number of RAKE fingers, and near/far interference model. In this figure, we select the two kinds of near/far interference model. Table 1 shows the distribution of multiple-access near/far interference according to the number of multiple-access users and interference power. In order to simplify the performance analysis, we assumed that multiple-access near/far interference model has discrete power and users distribution. According to figure 6, model 1 gives better performance than model 2. This is because model 2 has a large variation of power control error than the model 1. Figure 7 shows the effects of coding gain on PCBC CDMA system in multi-path Nakagami fading with MIP model. Comparing figure 6 and figure 7, we know that PCBC scheme gives better BER performance than SCBC scheme about 2 dB at $BER \approx 10^{-5}$.

Table 1. Distribution of near/far interference

P_j/P_o	- 3 dB	0 dB	+ 3 dB
Model 1	3 users	4 users	3 users
Model 2	4 users	2 users	4 users

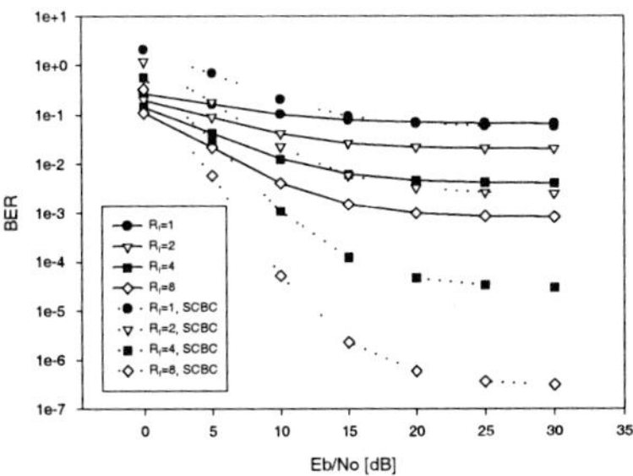

Fig. 5. BER performance of uncoded and SCBC CDMA system according to E_b/N_o and different number of RAKE fingers

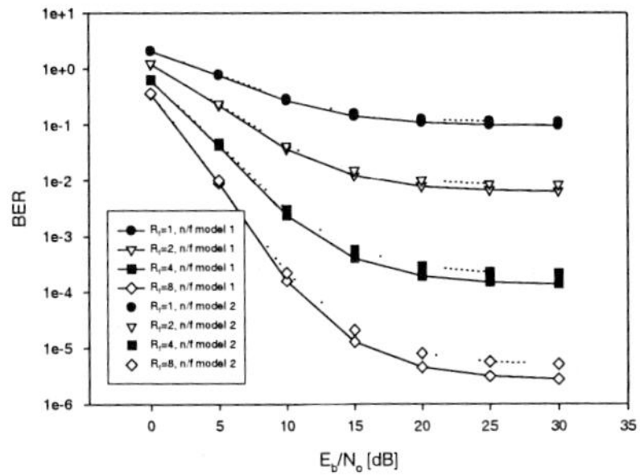

Fig. 6. BER performance of SCBC CDMA system with multiple-access near/far interference according to E_b/N_o , different number of RAKE fingers, and near/far interference model

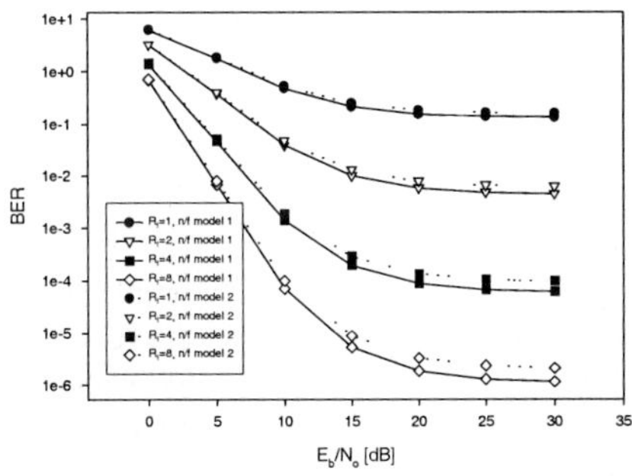

Fig. 7. BER performance of PCBC CDMA system with multiple-access near/far interference according to E_b/N_o , different number of RAKE fingers, and near/far interference model

5 Conclusion

In this paper, the performance evaluation of the CDMA system with complex spreading, imperfect channel estimation, multiple-access near/far interference, and serial/parallel concatenate block coding scheme is considered. The following conclusions have been drawn. 1) The effects of multi-path fading, imperfect channel estimation error, and multiple-access interference can be suppressed by RAKE receiver, I/Q complex spreading and serial/parallel concatenate block coding scheme 2) The system performance degrades significantly due to the multiple-access near/far interference.

However, this interference can be effectively limited by power control and serial/parallel concatenate block coding scheme.

Acknowledgements. This work was supported by the Korea Science & Engineering Foundation (KOSEF) and the Kyonggi Province through the Internet Information Retrieval Research Center (IRC) of Hankuk Aviation University.

References

1. J. Wang and J. Chen, "Performance of wideband CDMA systems with complex spreading and imperfect channel estimation," *IEEE J. Select. Areas Commun.*, vol. 19, no. 1, pp. 152–163, Jan. 2001.
2. F. Adachi, M. Sawahashi, and H. Suda, "Wideband DS-CDMA for next generation mobile communications systems," *IEEE Commun. Mag.*, vol. 36, pp. 56–69, Sept. 1998.
3. 3G TS 25.213 version 3.3.0, "Spreading and modulation (FDD)," 3GPP TSG-RAN, 2000-06.
4. L. Staphorst, M. Jamil, and L. P. Linde, "Performance evaluation of a QPSK system employing complex spreading sequences in a fading environment," in *Proc. IEEE VTS 50th Vehicular Technology Conf.*, pp. 2964–2968, Sept. 1999.
5. T. G. Macdonald and M. B. Pursley, "Complex processing in quaternary direct-sequence spread-spectrum receivers," in *Proc. IEEE Military Communications Conf.*, pp. 494–498, Oct. 1998.
6. T. Eng and L. B. Milstein, "Coherent processing in quaternary direct-sequence spread-spectrum receivers," *IEEE Trans. Commun.*, vol. 43, pp. 1134–1143, Feb./Mar./Apr. 1995.
7. J. Choi, "Multipath CDMA channel estimation by jointly utilizing pilot and traffic channels," *Proc. Inst. Elec. Eng. Commun.*, vol. 146, no. 5, pp. 312–318, Oct. 1999.
8. S. Benedetto and E. Biglieri, *Principles of Digital Transmission with Wireless Applications*, KA/PP, 1999.
9. S. Benedetto and G. Montorsi, "Serial concatenation of block and convolutional codes," Electron. Lett., 1996, 32, (10), pp. 887–888.
10. S. Benedetto and G. Montorsi, "Average performance of parallel concatenated block codes," Electron. Lett., 1995, 31, (3), pp. 156–158.

A Learning Automata Based Dynamic Guard Channel Scheme

Hamid Beigy and Mohammadreza Meybodi

Soft Computing Laboratory
Computer Engineering Department
Amirkabir University of Technology
Tehran, Iran
{beigy, meybodi}@ce.aku.ac.ir

Abstract. Dropping probability of handoff calls and blocking probability of new calls are two important QoS measures for cellular networks. Call admission policies, such as fractional guard channel and uniform fractional guard channel policies are used to maintain the pre-specified level of QoS. Since the parameters of network traffics are unknown and time varying, the optimal number of guard channels is not known and varies with time. In this paper, we introduce a new dynamic guard channel policy, which adapts the number of guard channels in a cell based on the current estimate of dropping probability of handoff calls. The proposed algorithm minimizes blocking probability of new calls subject to the constraint on the dropping probability of handoff calls. In the proposed policy, a learning automaton is used to find the optimal number of guard channels. The proposed algorithm doesn't need any a priori information about input traffic. The simulation results show that performance of this algorithm is close to the performance of guard channel policy for which we need to know all traffic parameters in advance. Two advantages of the proposed policy are that it is fully autonomous and adaptive. The first advantage implies that, the proposed policy does not require any exchange of information between the neighboring cells and hence the network overheads due to the information exchange will be zero. The second one implies that, the proposed policy does not need any priori information about input traffic and the traffic may vary.

1 Introduction

With increasing popularity of mobile computing, demand for channels is on the rise. Since number of allocated channels for this purpose is limited, the cellular and micro cellular networks are introduced, in which the service area is partitioned into regions called cells. Introduction of micro cellular networks leads to improvement of network capacity but increases the expected rate of handoff. When a mobile host moves across the cell boundary, handoff is required. If an idle channel is available in the destination cell, then the handoff call is resumed; otherwise the handoff call is dropped. Dropping probability of handoff calls (B_h) and blocking probability of new calls (B_n) are important quality of service (QoS)

M.H. Shafazand and A M. Tjoa (Eds.): EurAsia-ICT 2002, LNCS 2510, pp. 643–650, 2002.

measures of the cellular networks. Since the disconnection in the middle of a call is highly undesirable, B_h is more serious than B_n. In order to control B_n and B_h, *call admission control* (CAC) policies are introduced. The call admission policies determine whether a new call should be admitted or blocked. Both B_n and B_h are affected by call admission control policies. The simplest CAC policy is called *guard channel* policy (GC) [1]. Suppose that the given cell has C full duplex channels. The guard channel policy reserves a subset of channels, called *guard channels*, allocated to a cell for sole use of handoff calls (say $C - T$ channels). Whenever the channel occupancy exceeds the certain threshold T, the guard channel policy rejects new calls until the channel occupancy goes below T. The guard channel policy accepts handoff calls as long as channels are available. As the number of guard channels increased, B_h will be reduced while B_n will be increased [2]. It has been shown that there is an optimal threshold T^* in which B_n is minimized subject to the hard constraint on B_h [3]. Algorithms for finding the optimal number of guard channels are given in [3,4]. These algorithms assume that the input traffic is a stationary process with known parameters. The GC policy reserves an integral number of guard channels for handoff calls. In order to have more control on B_n and B_h, *limited fractional guard channel* (LFG) policy is introduced, which reserves a non-integral number of guard channels [3]. It has been shown that there is an optimal threshold T^* and an optimal value of π^* for which B_n is minimized subject to the hard constraint on B_h [3]. An algorithm for finding such optimal parameters is given in [3]. Since the input traffic is not a stationary process and its parameters are unknown a priori, the optimal number of guard channels is different for different traffic. In such cases the *dynamic guard channel* policy can be used. In dynamic guard channel policy, the number of guard channels varies during the operation of the cellular network.

Learning automaton (LA) is a reinforcement learning technique and has been used successfully in many applications such as telephone and data network routing [5,6], solving NP-Complete problems [7,8,9,10] and capacity assignment [11], to mention a few. In this paper, we propose an adaptive and autonomous call admission control algorithm, which uses LA. This algorithm uses only the current channel occupancy of the given cell and dynamically adjusts the number of guard channels. The proposed algorithm minimizes the blocking probability of new calls subject to the constraint on the dropping probability of handoff calls. Since the learning automaton starts its learning without any priori knowledge about its environment, the proposed algorithm does not need any a priori information about input traffic. One of the most important advantage of the proposed algorithm is that no status information will be exchanged between neighboring cells. The exchange of such status information increase the performance of the proposed algorithm. The simulation results show that the performance of this algorithm are near to performance of GC policy that knows all traffic parameters.

The rest of this paper is organized as follows: The section 2 presents the performance parameters of guard channel policy. The LA briefly is given in section 3. The proposed LA based dynamic guard channel policy is presented in section 4. The computer simulations is given in section 5 and section 6 concludes the paper.

2 The Blocking Performance of Guard Channel Policy

The blocking performance of guard channel policy is computed based on the following assumptions.

1. The arrival process of new and handoff calls is poisson process with rate λ_n and λ_h, respectively. Let $\lambda = \lambda_n + \lambda_h$.
2. The call holding time for both types of calls are exponentially distributed with mean μ^{-1}.
3. The time interval between two calls from a mobile host is much greater than the mean call holding time.
4. Only mobile to fixed calls are considered.
5. The network is homogenous.

The above first three assumptions have been found to be reasonable as long as the number of mobile hosts in a cell is much greater than the number of channels allocated to that cell. The fourth assumption makes our analysis easier and the fifth one lets us to examine the performance of a single network cell in isolation. Suppose that the given cell has a limited number of full duplex channels, C, in its channel pool. We define the state of a particular cell at time t to be the number of busy channels in that cell, which is represented by $c(t)$. The $\{c(t)|t \geq 0\}$ is a continuous-time Markov chain (birth-death process) with states $0, 1, \ldots, C$. The state transition rate diagram of a cell with C full duplex channels and dynamic guard channel policy is shown in figure 1.

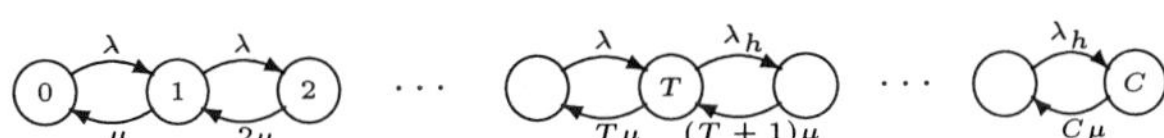

Fig. 1. Markov chain model of cell

Because of the structure of the Markov chain, we can easily write down the steady-state balance equations. Define the steady state probability $P_n = \lim_{t\to\infty} \text{Prob}[c(t) = n]$ for state $n = 0, 1, \ldots, C$. Then, the following expression can be derived for P_n $(n = 0, 1, \ldots, C)$.

$$P_n = \begin{cases} \frac{\rho^k}{k!} P_0 & k = 0, 1, \ldots, T-1 \\ \frac{\rho^k \alpha^{k-T}}{k!} P_0 & k = T, \ldots, C, \end{cases} \tag{1}$$

where $\rho = \lambda/\mu$, $\alpha = \lambda_h/\lambda$ and P_0 is the probability that all channels are free and given by the following expression.

$$P_0 = \left[\sum_{k=0}^{T-1} \frac{\rho^k}{k!} + \sum_{k=T}^{C} \frac{\rho^k \alpha^{k-T}}{k!} \right]^{-1} \tag{2}$$

Thus, dropping probability of handoff calls, $B_h(C,T)$, ie equal to $B_h(C,T) = \frac{\rho^C \alpha^{C-T}}{C!}$ and blocking probability of new calls is equal to $B_n(C,T) = \sum_{k=T}^{C} P_k$.

The objective of call admission control policies is to find a T^* that minimizes the $B_n(C,T^*)$ given the constraint $B_h(C,T^*) \leq p_h$. The value of p_h specifies by the quality of service of the network. In order to find the optimal value of T^*, in [3] a binary search and in [4] a linear search algorithms are given. These algorithms assumes that the all parameters of input traffic are known in advance.

3 Learning Automata

The automata approach to learning involves the determination of an optimal action from a set of allowable actions. An automaton can be regarded as an abstract object that has finite number of actions. It selects an action from its finite set of actions. This action is applied to a random environment. The random environment evaluates the applied action and gives a grade to the selected action of automata. The response from environment (i.e. grade of action) is used by automata to select its next action. By continuing this process, the automaton learns to select an action with best grade. The learning algorithm used by automata to determine the selection of next action from the response of environment. An automaton acting in an unknown random environment and improves its performance in some specified manner, is referred to as *learning automaton* (LA). Learning automata can be classified into two main families: *fixed structure learning automata* and *variable structure learning automata* [12].

Variable structure learning automata are represented by triple $< \beta, \alpha, T >$, where β is a set of inputs actions, α is a set of actions, and T is learning algorithm. The learning algorithm is a recurrence relation and is used to modify the state probability vector. It is evident that the crucial factor affecting the performance of the variable structure learning automata, is learning algorithm. Various learning algorithms have been reported in the literature. Let α_i be the action chosen at time k as a sample realization from probability distribution $p(k)$. In linear reward-ϵpenalty algorithm ($L_{R-\epsilon P}$) scheme the recurrence equation for updating p is defined as

$$p_j(k+1) = \begin{cases} p_j(k) + a \times [1 - p_j(k)] \text{ if } & i = j \\ & \\ p_j(k) - a \times p_j(k) \quad\quad \text{ if } & i \neq j \end{cases} \quad \text{if} \quad \beta(k) = 0 \tag{3}$$

$$p_j(k+1) = \begin{cases} p_j(k) \times (1-b) \quad\quad \text{ if } & i = j \\ & \\ \frac{b}{r-1} + p_j(k)(1-b) \text{ if } & i \neq j \end{cases} \quad \text{if} \quad \beta(k) = 1 \tag{4}$$

The parameters $0 < a < 1$ and $0 < b \ll a$ represent *step lengths* and r is the number of actions for learning automata. The a and b determine the amount of increase and decreases of the action probabilities, respectively. If the a equals to b the recurrence equations (3) and (4) is called *linear reward penalty*(L_{R-P}) algorithm.

```
if (NEW CALL) then
    set g ← LA.action ()
    if (c(t) < C − g )then
        accept call
        if (B̂_h < p_h ) then
            reward action g
        else
            penalize action g
        end if
    else
        reject call
        if (B̂_h < p_h ) then
            penalize action g
        else
            reward action g
        end if
    end if
end if
```

Fig. 2. LA based dynamic guard channel algorithm

4 LA Based Dynamic Guard Channel Policy

In this section, we introduce a new LA based algorithm (figure 2) to determine number of guard channels when the parameters λ_n, λ_h, and μ are unknown and possibly time varying. In this algorithm, LA is used to adjust number of guard channels. Assume that the cell has C full duplex channels. Let the number of guard channels at time instant t denoted by $g(t)$ is in interval $g(t) \in [g_{\min}, g_{\max}]$, where $0 \leq g_{\min} \leq g_{\max} \leq C$. In the proposed algorithm, each base station has a LA with $g_{\max} - g_{\min} + 1$ actions, where action α_i denotes that the base station must use $g(t) = g_{\min} + \alpha_i - 1$ guard channels. The proposed algorithm can be described as follows. When a handoff call arrives at the given cell and a channel is available, then the call is accepted; otherwise it is dropped. When a new call arrives at the given cell, LA associated to the cell selects one of its actions, say α_i. If the cell has at least $g_{\min} + \alpha_i - 1$ free channels, then the incoming call is accepted; otherwise it is blocked. Then the base station computes the current estimate of dropping probability of handoff calls $(\hat{B}_h)$ and then compare this quantity with the specified level of QoS (p_h). If the incoming new call is accepted and the current value of $(\hat{B}_h)$ is less than p_h then action α_i is rewarded; otherwise penalized. If the incoming new call is blocked and the current value of $(\hat{B}_h)$ is greater than p_h then the action α_i is rewarded; otherwise the action α_i is penalized. The comparison of current estimate of dropping probability of handoff calls and the specified level of QoS (p_h) is done to guarantee the specific level of QoS. The proposed algorithm requires less resources (bandwidth of the wired-line network) than other distributed call admission algorithm for which the status of all neighboring cells are needed for determination of guard channels. In other distributed call admission algorithms, status information must

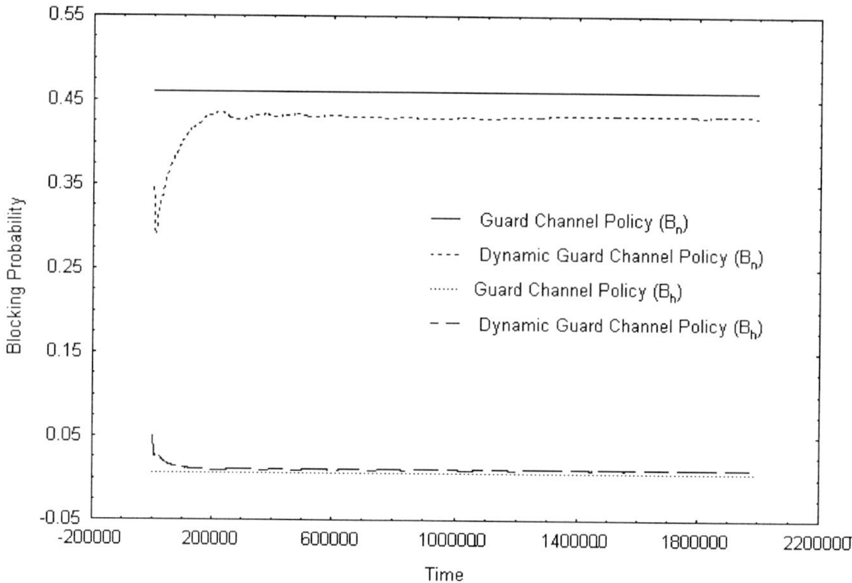

Fig. 3. The comparison of guard channel policy and dynamic guard channel policy

be exchanged between neighboring cells in the case of arrival of a call, departure of a call, and handoff of a call. However, the exchange of status information can be used to speed up the convergence of the proposed algorithm, which results an improvement of the proposed algorithm. Since the learning automata begin their learning without a priori knowledge about its environment, the proposed algorithm does not require any information about input traffic. Even though the priori information about input traffic is not needed by the algorithm, availability of such information may be used to find a better learning algorithm in order to choose a better learning algorithm for adaptation of traffic parameters. The use of a priori information in the proposed algorithm needs to be investigated. The proposed algorithm at the beginning does not perform well but as it proceeds, the performance of the algorithm approaches to its optimal performance. Initially, the proposed guard channels randomly.

5 Simulation Results

In this section, we compare performance of the guard channel [1], the limited fractional guard channel [3], and the dynamic guard channel algorithms proposed in this paper. The results of simulations are summarized in table 1. The simulation is based on the single cell of homogenous cellular network system. In such network, each cell has 8 full duplex channels ($C = 8$). In the simulations, new call arrival rate is fixed to 30 calls per minute ($\lambda_n = 30$), channel holding time is set to 6 seconds ($\mu^{-1} = 6$), and the handoff call traffic is varied between 2 calls per minute to 20 calls per minute. The results listed in table 1 are obtained by averaging 10 runs from $2,000,000$ seconds simulation of each algorithm. The objective is to minimize the blocking probability of new calls subject to the constraint that the dropping probability of handoff calls is less than 0.01. The optimal number of guard channels for guard channel policy is obtained by algorithm given in [4] and the optimal parameters of limited fractional guard channel policy is obtained by algorithm given in [3].

Table 1. The simulation results of the LA base dynamic guard channel policy

		GC		LFG		DGC	
Case	λ_h	B_n	B_h	B_n	B_h	B_n	B_h
1	2	0.063507	0.001525	0.031609	0.023283	0.053433	0.010619
2	4	0.077080	0.003538	0.051414	0.020675	0.080966	0.010039
3	6	0.091013	0.005923	0.071632	0.018707	0.125500	0.009964
4	8	0.105002	0.008380	0.092138	0.016706	0.154861	0.010031
5	10	0.120260	0.011877	0.114445	0.015572	0.207490	0.010067
6	12	0.231559	0.004309	0.147902	0.014044	0.245842	0.010017
7	14	0.255346	0.005975	0.204217	0.012675	0.290619	0.009960
8	16	0.275489	0.007999	0.250642	0.011554	0.331478	0.009983
9	18	0.296834	0.010518	0.294441	0.010877	0.377334	0.009953
10	20	0.459183	0.006081	0.384157	0.010182	0.427894	0.010005

By inspecting table 1, it is evident that the performance of dynamic guard channel policy is close to the performance of guard channel policy. One reason for the difference in performances of the guard channel policy and the proposed policy is due to the fact that transient behavior of the proposed algorithm. Since, the performance parameters (the blocking probability of new calls and the dropping probability of handoff calls) in the early stages of simulation are far from their desire value, they affect the long-time calculation of the performance parameters. However, such effect can be removed by excluding the transient behaviors of the proposed algorithm, which is shown in figure 3. Figure 3 shows the evolution of the performance parameters for the guard channel policy and the proposed dynamic guard channel policy. The traffic parameters used for figure 3 corresponds to case 10 in table 1. By carefully inspecting figure 3 and ignoring the transient behavior of the proposed algorithm, it can be concluded that the dropping probability of handoff calls approaches its prescribed value (p_h), while the blocking probability of new calls is less than the corresponding performance parameter of guard channel policy.

6 Conclusions

In this paper, a dynamic guard channel policy based on learning automata is given. The proposed algorithm adapts the number of guard channels in a cell using current estimate of dropping probability of handoff calls. This algorithm minimizes blocking probability of new calls subject to the constraint on dropping probability of handoff calls. The simulation results show that the performance of this algorithm is very close to the performance of guard channel policy that knows all traffic parameters in advance. The proposed policy has three advantages: 1) doesn't require any exchange of information between the neighboring cells leading to less network overheads. 2) doesn't need any a priori information about the input traffic. 3) the algorithms works for time varying traffics.

References

1. D. Hong and S. Rappaport, "Traffic Modelling and Performance Analysis for Cellular Mobile Radio Telephone Systems with Priotrized and Nonpriotorized Handoffs Procedure," *IEEE Transactions on Vehicular Technology*, vol. 35, pp. 77–92, Aug. 1986.
2. S. Oh and D. Tcha, "Priotrized Channel Assignment in a Cellular Radio Network," *IEEE Transactions on Communications*, vol. 40, pp. 1259–1269, July 1992.
3. R. Ramjee, D. Towsley, and R. Nagarajan, "On Optimal Call Admission Control in Cellular Networks," *Wireless Networks*, vol. 3, pp. 29–41, 1997.
4. G. Haring, R. Marie, R. Puigjaner, and K. Trivedi, "Loss Formulas and Their Application to Optimization for Cellular Networks," *IEEE Transactions on Vehicular Technology*, vol. 50, pp. 664–673, May 2001.
5. P. R. Srikantakumar and K. S. Narendra, "A Learning Model for Routing in Telephone Networks," *SIAM Journal of Control and Optimization*, vol. 20, pp. 34–57, Jan. 1982.
6. O. V. Nedzelnitsky and K. S. Narendra, "Nonstationary Models of Learning Automata Routing in Data Communication Networks," *IEEE Transactions on Systems, Man, and Cybernetics*, vol. SMC-17, pp. 1004–1015, Nov. 1987.
7. B. J. Oommen and E. V. de St. Croix, "Graph Partitioning Using Learning Automata," *IEEE Transactions on Commputers*, vol. 45, pp. 195–208, Feb. 1996.
8. H. Beigy and M. R. Meybodi, "Backpropagation Algorithm Adaptation Parameters using Learning Automata," *International Journal of Neural Systems*, vol. 11, no. 3, pp. 219–228, 2001.
9. M. R. Meybodi and H. Beigy, "New Class of Learning Automata Based Schemes for Adaptation of Backpropagation Algorithm Parameters," *International Journal of Neural Systems*, vol. 12, pp. 45–68, Feb. 2002.
10. M. R. Meybodi and H. Beigy, "A Note on Learning Automata Based Schemes for Adaptation of BP Parameters ," *Accepted for Publication in the Journal of Neuro Computing, To Appear.*
11. B. J. Oommen and T. D. Roberts, "Continuous Learning Automata Solutions to the Capacity Assignment Problem," *IEEE Transactions on Commputers*, vol. 49, pp. 608–620, June 2000.
12. K. S. Narendra and K. S. Thathachar, *Learning Automata: An Introduction.* New York: Printice-Hall, 1989.

Dynamic System Simulation on the Web

Khaled Mahbub and M.S.J. Hashmi

Faculty of Engineering and Design
Dublin City University
Dublin 9, Ireland
Khaled.mahbub2@mail.dcu.ie

Abstract. The objective of this paper is to present a Java-based simulation package (SimDynamic) for modelling, simulating and analysing dynamic systems on the Web. The package is a collection of several functional nodes and these nodes serve as the basic building blocks of a model. A model is constructed by building a directed graph with nodes and edges. A model of moderate complexity can be constructed with minimal of effort and without writing any code. Finally the package allows maximum user control of model execution and provides both graphical and numerical output.

1 Introduction

Research and the development work in the area of Web-based simulation is growing rapidly since its inception around mid 1990s. Web-based simulation is an attempt to exploit Web technology to support the future of computer simulation. Existing computer simulation support is either language-based or a library approach. In either case, they suffer from lack of portability to other environments. Also markets in educational software are small. There are thus little commercial interests in the production of simulation software for the educational market. Any simulation software, which is produced is aimed at the industrial market and is often too expensive for the educational purchaser. Moreover the demand for consulting in modelling and simulation has grown faster than the consulting companies can offer. Use of the Internet and its supporting tools such as the virtual environment and the interactive distributed simulation has the potential to overcome these factors limiting the wider use of simulation. The rapid advances in Web technology, most notably Java, enable to execute highly interactive and dynamic simulation/animation on the Web. The facilities provided by Java [1] programming language to support Web technology is the power behind Web-based simulation. Java benefits from being simple, secure, object-oriented, multi-threaded and architecture neutral. More and more researchers develop their simulation packages based on Java.

M.H. Shafazand and A M. Tjoa (Eds.): EurAsia-ICT 2002, LNCS 2510, pp. 651–658, 2002.

2 Literature Review

Fishwick [2] focuses on three important aspects of simulation that might be affected by incorporating it in the Web: a) Education and training b) Publications and c) Simulation Programmes. With respect to education and training, the Web offers storage and retrieval of supporting materials far exceeding the capacity of CD-ROMs or diskettes that may be packaged with a textbook. Regarding publications, the Web offers new and convenient mechanism for submitting, refereeing and disseminating research result. However, it is the last category in his discussion that has received the most attention in the Web-based simulation area and arguably, represents the predominant use of the term.

Simkit, created by Buss and Stork [3] a set of Java classes for creating discrete event simulation models. They use the event graph design approach to build models. SimKit permits user interaction through a detailed, model entry form. Additionally, it is combined with a java-based graphic package to allow a useful output of statistics and graphs.

Miller, Nair, Zhang and Zhao [4] present JSIM, a java-based simulation and animation environment. It demonstrates Web-based simulation through a unique combination of java applets and query driven databases. A model is constructed by building a graph with nodes and edges. For the purpose of storing and retrieving simulation models and results, JSIM incorporates database connectivity with simulation classes.

SimJava, presented by McNab and Howell [5] is a process based discrete event simulation package for building working models of complex systems, with animation facilities. It provides a set of simulation foundation classes useful for constructing discrete event models. A simulation model is a collection of entities, connected together by ports, each running in its own thread.

Veith, Kobza and Koelling [6] present Netsim, a discrete event simulation package based on the event graph approach. Netsim provides a maximum amount of user interaction with the simulation model. A programming interface provides a blank template with text fields for the various parameters of a simulation model., such as event name and state variables. A second interface allows user interaction with running the simulation model.

Cole and Tooker [7] develop Web-based physics tutorials to assist physics students. It allows students to see interesting cases of a given simulation model without requiring prior knowledge of the parameters defining theses cases or of the background programming languages involved. The tutorials use Apple's OpenDoc Frameworks to provide a basic simulation environment. The main attribute of this environment is its extensibility. One can extend the power of the environment by adding parts. This extensibility means it is functional for a broad spectrum of users.

Marr, Storey, Biles and Kleijnen [8] discuss a Web-based simulation manager program, called SimManager, which executes simulation study in a parallel-replications mode., utilising a set of slave processors (engine) available to it. Engines can belong either to an intranet, to the Internet, or to a combination thereof. An engine is simply known to SimManager by its IP address. SimManager also knows the performance characteristics of each engine and it serves as the interface between a user and the simulation system. SimManager receives request for work and assigns this workload to the available engine processors. Once all jobs are completed, an email message is sent to the user stating that the results are ready for pick up.

Research in the area of WWW-based simulation is developing rapidly as WWW programming tools develop. But most of this research is only devoted to discrete event simulation. SimDynamic is an attempt to introduce dynamic system simulation on the Web and provides a positive, unique contribution to Web based simulation. In the following sections SimDynamic has been introduced and discussed in greater detail. A comparison of SimDynamic to Simulink [9] has also been shown.

3 Overviews of SimDynamic

SimDynamic provides a simulation environment for modelling, simulating and analysing dynamic systems. It supports linear and non-linear systems, modelled in continuous time, sampled time or a hybrid of the two. SimDynamic is actually a collection of several functional nodes, where each node carries out a specific function such as integration, generation of sine wave, multiplication etc. These nodes serve as the basic building blocks of a model.

The package was designed with an aim to achieve as much modelling ease as possible and to provide maximum user flexibility in model execution. For modelling the package provides a graphical user interface (GUI) for building models as block diagram, using simple mouse operations. This interface allows users to draw all the required nodes in a simulation model window by selecting nodes from the given node set followed by drag and drop operations. All the nodes will be added to the model with their default properties/parameters. These values can be checked and redefined by double clicking the node and refilling property/parameter dialog box for the node.

Simulating a dynamic system with this package is a two-step process. First, the user has to create a model of the system to be simulated. The model graphically depicts the time-dependent mathematical relationships among the inputs, states and outputs of the system. After the model creation user has to select an integration method and then can use SimDynamic to simulate the behaviour of the system for a specified period of time. Finally the results of the simulation can be viewed as both graphical and numerical output.

4 The Design

The software consists of two packages, each package containing a set of related classes that work together to provide functionality in a key area to support a simulation process. The package hierarchy of this software is shown in fig 1.

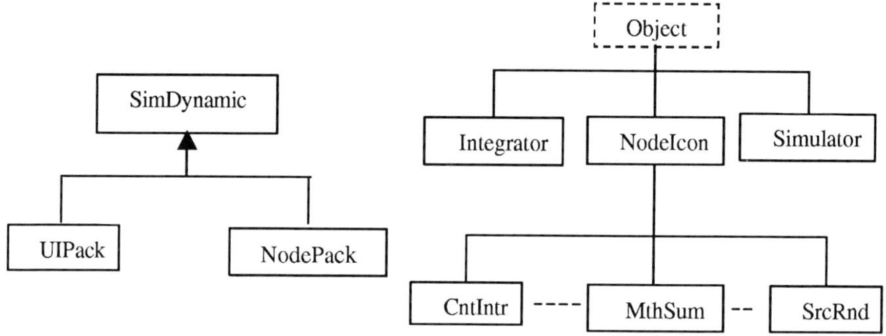

Fig. 1. Package hierarchy diagram for SimDynamic

Fig. 2. Class hierarchy diagram for the NodePack package

4.1 NodePack

NodePack is the core of the software and provides all the functional nodes and also contains the main simulation engine (the *Simulator* class). The class hierarchy of the NodePack package is shown in fig 2. In this hierarchy diagram foundation classes are represented by dashed rectangle and software classes is shown by regular rectangles.

NodeIcon. The *NodeIcon* class is defined as an abstract base class, that encapsulates the features common to the classes that appear as nodes in a model. All the functional nodes have been derived from this class.

```
public abstract class NodeIcon{
        . . . . . .
        public abstract boolean action(double simT,
              double step);
        public abstract void updateOutput(double simT,
              double step)
        . . . . .
}
```

NodeIcon class contains several abstract methods and the most important of them is the *action(double simT, double step)* method. All the functional nodes implemented this abstract method to perform its own specific function at the simulation time *simT* with a time step *step*. This method returns false in case it fails to carry out its task.

Integrator. The dynamic models are generally described by differential equations. The solutions of these equations display the operations of the system and this is the fundamental property of dynamic models. This class provides a number of numerical methods to solve Ordinary Differential Equations (ODE). Here is a brief description of each method that has been implemented

- **RK_DP:** It uses an explicit Runge-Kutta formula, the Dormand-Prince pair.
- **RK_O4:** It is based on fourth order explicit Runge-Kutta formula.
- **RK_BS:** It uses an explicit Runge-Kutta pair of Bogacki and Shmpine.
- **RK_HE:** It uses Heun's method or modified Euler formula.
- **RK_EU:** It uses simple Euler's method for integration.

Simulator. This is the main simulation engine and it controls the whole simulation process of a model. The *simulator* class makes the nodes to interact as a system and generate desired output. The way it accomplishes this, is described in section 5.

4.2 UIPack

This package contains all the necessary classes to provide the graphical user interface (GUI) of the software. The main components of this package are node set browser window and the simulation model sheet window. Node set browser will let the user to choose the required node set and then appropriate nodes for his/her model. The user will use the simulation model sheet to build his/her model. This package also contains classes to display both numerical and graphical outputs.

5 How It Works

Simulating a dynamic system is nothing but the process of computing the states and outputs of the system over a particular period time. In SimDynamic each functional node calculates its own output and state at a particular time step through the *action* method. So the only requirement to simulate a model is to co-ordinate the functional nodes to make them interact as a system, and *simulator* class does this. When user click the start from the simulation menu, the *simulator* class takes over the control and the simulation starts. It accomplishes its task in the following two phases

Initialisation Phase

Simulator performs following tasks in this phase

- It sorts the nodes of the model into the order in which the nodes need to be executed during the execution phase.

- Adjusts the sample time of each node. Because some node may have implicit sample time and some nodes may have explicit sample times specified by user.

Execution Phase

In this phase *simulator* simply calls the *action* method of each of the functional node in the order determined in the previous step. After all the ndoes have been executed it advances the simulation clock to the next time step.

6 Comparison of SimDynamic with Simulink

SimDynamic was compared to Simulink using a simple damped pendulum (fig 3). It is assumed that the mass is m kg, length of the weightless rod is L metre, and there is a frictional torque at the pivot point that is related to the angular velocity through the coefficient c N-m-s/rad.

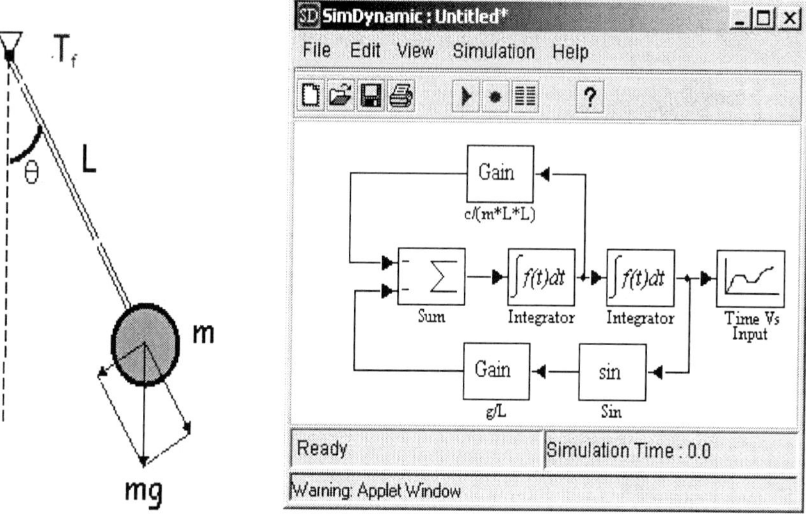

Fig. 3. A simple damped pendulum **Fig. 4.** Pendulum model developed using SimDynamic

The equation of motion of a simple damped pendulum is,

$$\frac{d^2\theta}{dt^2} = -\frac{g}{L}\sin\theta - \frac{c}{mL^2}\frac{d\theta}{dt} \tag{1}$$

From equation 1, above (fig 4) model is built with SimDynamic. Similar model is also built in Simulink [10], but it is not shown here. In this model, the right most integrator node outputs θ, on which trigonometric function node is applied to have sinθ. Gain node is used to multiply sinθ by g/L. Similarly the left integrator node outputs dθ/dt, which is multiplied by c/m*L*L (upper gain node). Outputs from these two gain nodes are fed to the sum node to complete the equation.

Simulation was carried out for 10 seconds with the following parameter values, L=0.5 m, g = 9.8 m/s^2, m = 1 kg, c = 0.1 N-m-s/rad. In both Simulink and SimDynamic we used a fixed step Runge-Kutta (Dormand-Prince pair) integrator with initial value *pi/3* and a step size 0.2. Figure 5 shows the graphical result followed by the numerical result in table 1. In figure 5, simulation time is shown along x-axis and angle (θ) is shown along y-axis.

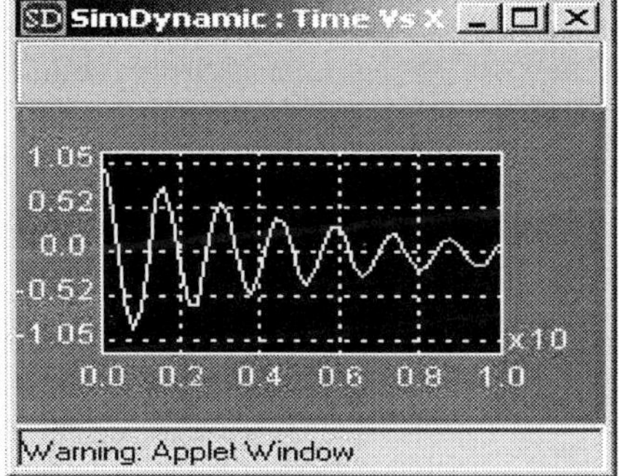

Fig. 5a. Graphical results for pendulum model from SimDynamic

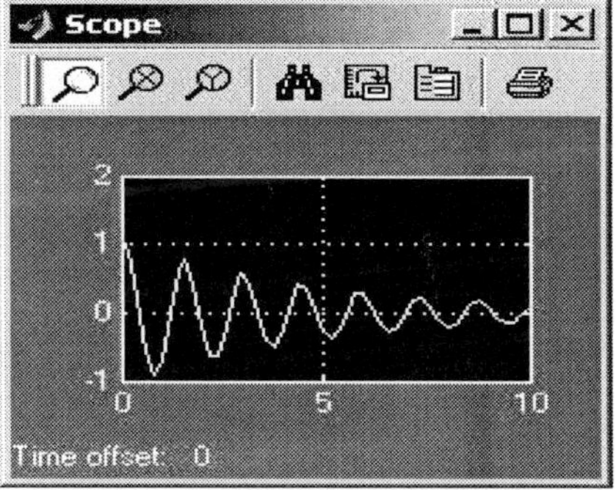

Fig. 5b. Graphical results for pendulum model from Simulink

Table 1. Numerical results for the pendulum model. (first 5 seconds)

Time	SimDyna	Simulink	Time	SimDyna	Simulink	Time	SimDyna	Simulink
0.0	1.04719	1.0472	1.8	0.43198	0.4320	3.6	-0.3983	-0.3983
0.2	0.91565	0.9157	2.0	-0.17691	-0.1769	3.8	-0.48989	-0.4899
0.4	0.23019	0.2302	2.2	-0.61732	-0.6173	4.0	-0.24031	-0.2403
0.6	-0.55600	-0.5560	2.4	-0.61236	-0.6124	4.2	0.15672	0.1567
0.8	-0.91581	-0.9158	2.6	-0.19949	-0.1995	4.4	0.41391	0.4139
1.0	-0.67531	-0.6753	2.8	0.31889	0.3189	4.6	0.36628	0.3663
1.2	-0.01451	-0.0145	3.0	0.57817	0.5782	4.8	0.06929	0.0693
1.4	0.60502	0.6050	3.2	0.42820	0.4282	5.0	-0.25325	-0.2533
1.6	0.77629	0.7763	3.4	-0.00256	-0.0026			

7 Conclusions and Future Work

SimDynamic is an attempt to introduce dynamic system simulation on the Web. It provides the basic elements necessary for building dynamic system simulation. This version of SimDynamic provides around 85 functional nodes, but it is extendable. As an ongoing project, still much work has to be done. Current implementation involves only fixed step integrators and integration of stiff problems have not been considered. Another area of focus is to provide 3D animated illustration to the output wherever applicable. Please visit the site http://student.dcu.ie/~mahbubk2/simdynamic.html for more information and examples.

References

1. Sun Microsystems, Inc. © 1995 – 2002, The Source For Java Technology. http://java.sun.com/
2. Fishwick, P.A.: Web-Based Simulation: Some Personal Observations. Winter Simulation Conference, Coronado, CA, pp 772–779, 1996
3. Buss, A.H., Stork, K.A.: Discrete Event Simulation on the World Wide Web Using Java. Winter Simulation Conference, Coronado, CA, pp 780–785, 1996
4. Miller, J.A., Nair, R.S., Zhang, Z., Zhao, H.: JSIM: A Java-Based Simulation and Animation Environment. Proceedings of the 30th Annual Simulation Symposium, Atlanta, Georgia, pp. 31–42. 1997
5. Howell, F., McNab, R.: SimJava: A Discrete Event Simulation Package for Java With Applications in Computer Systems Modelling. Proceedings of First International Conference on Web-based Modelling and Simulation, San Diego CA, Society for Computer Simulation, Jan 1998.
6. Veith, T.L., Kobza, J.E., Koelling, C.P.: NetSim: Java ™ – Based Simulation for the World Wide Web. Computer & Operation Research 26 (1999), pp 607–621
7. Cole, R., Tooker, S.: Physics To Go: Web-based Tutorials for CoLoS Physics Simulations. Proceeding of the Frontiers on Education FIE'96 26th Annual Conference.
8. Marr, C., Storey, C., Biles, W.E., Kleijnen, J.P.C.: A Java-Based Simulation Manager for Web-Based Simulation. Winter Simulation Conference, Orlando, Florida, pp 1825–1822, 2000
9. The MathWorks Inc. © 1994 – 2002, http://www.mathworks.com/products/simulink/
10. Dabney, J.B., Harman, T.L.: Mastering Simulink 2. Prentice-Hall, Englewood Cliffs, NJ, 1998.

Using Proximity Information for Load Balancing in Geographically Distributed Web Server Systems*

Dheeraj Sanghi, Pankaj Jalote, and Puneet Agarwal

Department of Computer Science and Engineering
Indian Institute of Technology Kanpur, UP, INDIA
{dheeraj,jalote}@iitk.ac.in, puneet.agarwal@alumni.cse.iitk.ac.in

Abstract. Many popular web sites get millions of hits everyday. To service a large number of requests, clusters of fully replicated web servers are used. In such a setup, the client's request has to be directed to a cluster and then to a server within the cluster in a manner that the client receives the response in minimum time. In this paper, we propose an adaptive policy of selecting the *nearest* cluster for a request. Proximity is assessed by the round trip delay between the cluster and the client. An innovative idea is to measure this delay only for those clients who are sending a large number of requests. We have implemented this scheme, and using a test-bed which simulates the world wide web environment, compared the performance of the scheme with that of some existing schemes. The results indicate that the proposed scheme performs better, both in terms of average response time, as well as throughput.

1 Introduction

Multiple clusters of servers placed in different geographical regions is a very common architecture for web server systems. Deploying such a system requires a proper strategy for selecting a cluster, and then selecting a server within the cluster for handling a client request. The strategy should result in low response time for the client. In many existing strategies, a server or cluster is selected without taking system state into account, e.g., random, round robin, etc. Some policies use weighted capacity algorithms to direct more percentage of requests to more capable servers or clusters but most policies do not take into account the proximity between clients and clusters.

When clients are geographically far apart, network conditions play an important role in the latency perceived by clients. In such a situation, we can provide better service to clients by taking proximity information into account. A few schemes that do try to use proximity information, either use only the IP address allocation table, which is not very reliable, or use probes between client and DNS/server, which has high overhead.

* This research was partially supported by a grant from Avaya Labs, NJ, USA.

M.H. Shafazand and A M. Tjoa (Eds.): EurAsia-ICT 2002, LNCS 2510, pp. 659–666, 2002.

In this paper, we propose a scheme for load balancing that employs proximity information, as well as information about server loads. But proximity is estimated only when some client has several requests for the website (e.g., proxy server of a large organization.) Proximity is estimated by round trip delay, which is measured by different clusters sending probes to the client. We have used a test-bed for evaluating the performance of this approach and comparing it with other approaches. The experimental results show that our proximity based approach performs better than other approaches.

2 Related Work

Cardelini et al [6] classify web server architectures based on the entity which distributes the incoming requests among the servers.

In *client-based approach*, the client side entity is responsible for selecting the server. The selection can be done by the client software (browser) or Java Applet run by the client [16] or client-side DNS [5] or proxy servers [4]. Evaluation of several approaches for server selection has been done by [15].

In *DNS-based approach*, server side authorized DNS maps the domain name to IP address of one of the servers. Several DNS based approaches are discussed in References [6] and [7]. An example of this type is the round robin approach (DNS-RR) [13]. The selection of the server can also be based on the server load or based on proximity with clients or a combination of both.

In *dispatcher-based approach*, DNS returns the address of a *dispatcher* that routes all the client request to other servers in the cluster. Thus it acts as a centralized scheduler at the server side. Dispatching of requests can be done using various techniques, like *Packet single-rewriting*, or *Packet double-rewriting* [2], or *Packet forwarding using MAC address* [12], or using *ONE-IP address* [9], or using *HTTP redirection.*

Many server side approaches for load balancing have been proposed. Guyton et al [11] focus on hop count and cost of collection of information for server selection. In Cisco's Distributed Director, server selection at DNS is done based on geographical proximity estimated using client IP address or hop count information obtained from routers. Given that clients are distributed geographically, static metrics like hop count for proximity information were not found good in study by [8]. Ammar et al [10, 17] propose local anycast resolver that is near a large number of clients, to which servers push their performance information.

3 Load Balancing Using Proximity Information

We assume a two level architecture. The web server system consists of multiple clusters, each cluster having a dispatcher and multiple servers, In this system, the response time depends on the load at selected server and the path characteristics between the client and the server. the DNS selects the cluster (returns IP address of its dispatcher), and the dispatcher selects the server within the cluster.

3.1 Using Proximity Metrics for Cluster Selection

We aim to use proximity information to provide a better response time. There are various metrics to measure proximity. Some of these metrics are: round-trip time, number of hops, geographical distance, bandwidth availability, etc.

For minimizing the response time, RTT is perhaps the best metric to use along with the cluster (or server) load information. However, use of RTT requires its periodic measurement. But this overhead is within limits. When used at the client side, results in [8] indicate that the overhead was less than 1% in terms of additional network traffic. Also use of three ping messages for measurement of RTT gave better results. They found that RTT has high correlation with latency perceived by the client.

For a server side approach, measuring RTT for each client can result in a huge overhead as a heavily loaded cluster may get requests from thousands of client. Arlitt et al [3] found out that 75% of total HTTP requests to a server come from 10% of networks. So if we measure the RTT for only the high load generating clients, we can minimize the overhead.

Our approach is to dynamically distribute requests based on the current system state information. All servers in a cluster report state information to the dispatcher. Dispatcher reports aggregate cluster information to the DNS. Servers also count the number of requests coming from each IP address. This information is also sent to the dispatcher, which aggregates this information and reports IP addresses of clients having very high request rate to the DNS. Using the request rate information, DNS asks a subset of clusters to measure round-trip delay with only those clients which generate the heavy load. Round-trip delay is measured by sending ping messages to clients.

While selecting the cluster, DNS ensures that clusters have enough free capacity to serve the client requests.

4 Implementation

We have implemented the scheme on PCs running Linux.

We run *Apache* web server on all the server machines. Another process collects state information like load average, CPU and memory utilization, number of active connections, etc. The system load is obtained periodically, and sent to dispatcher. Using the web server access log, the number of requests from each IP address is determined and this information is also propagated to the dispatcher.

As discussed earlier, DNS-based schemes for load-balancing require that DNS returns the IP address of server/dispatcher, based on the state information like load, and information about the client like its round-trip time from various clusters. We have extended BIND to permit this flexibility.

Dispatcher receives regular load updates from the servers. It also receives client request rates from each server. The aggregate information is sent to the DNS. Dispatcher may also receive requests from DNS to measure RTT of select clients. It can send ICMP echo request messages to those clients and report RTTs between clients and itself.

5 Experiments

We compare the performance of the proposed scheme with some existing schemes using our test-bed [1]. The results of these experiments are presented below.

5.1 Setup

For conducting our experiments, we have setup a test-bed as shown in Fig. 1. It has three clusters on different logical networks representing three different geographical regions. Each cluster has one dispatcher (or front node, represented by 'F' node) and two servers ('S' nodes).

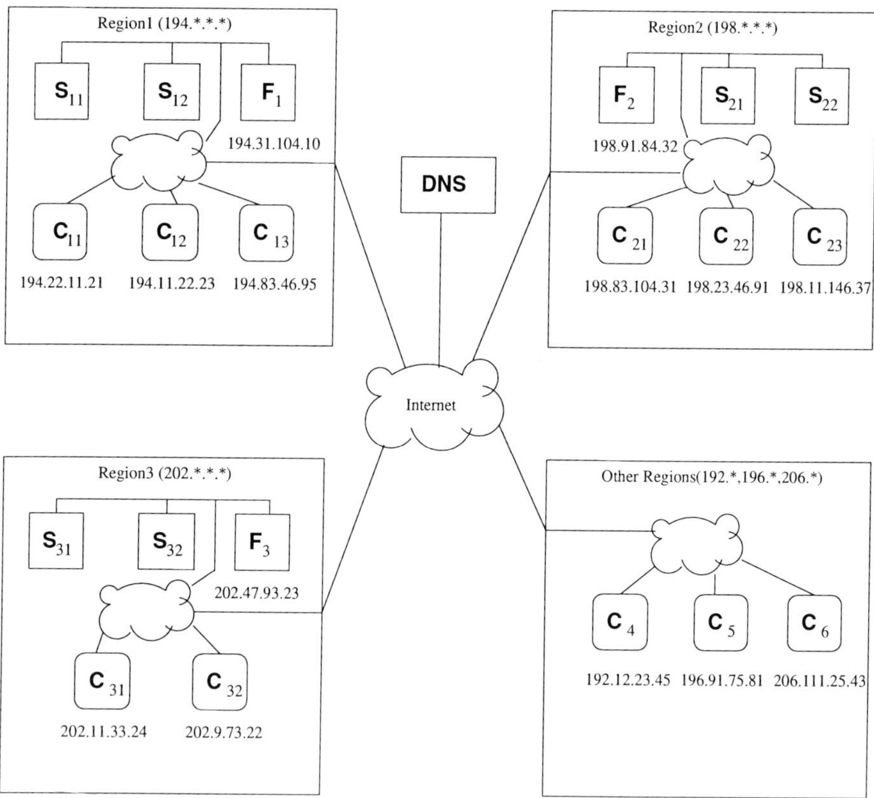

Fig. 1. Test bed used in experiments

In this setup, we have eleven clients ('C' nodes) to generate requests. Note that each client-node in the test-bed actually generates a very large number of requests in the simulation, and is equivalent to more than 10 clients in real environment. Three clients each were present in region 1 and region 2 and two clients in region 3. We also had three clients in other regions which represent mix of clients not close to any of the three clusters. A DNS was also setup to resolve IP addresses of clusters.

To model WAN effects, artificial delays and packet losses were introduced using Nistnet software [14]. We have introduced smaller delays for IP packets sent and received between clients and servers in the same region and higher delays for packets between clients and servers in different regions. We have assumed less packet loss (2%) while they travel between the same region. For packets traveling across the regions, the packet loss is assumed to be higher (5%).

To generate load and measure performance in experiments, we have used *WebStone*. This is a standard software used to benchmark web servers. Different schemes were tested with everything kept identical except policy for cluster selection at DNS.

5.2 Experimental Results

Since minimizing the response time was the primary objective of the scheme, we use average response time as the metric for performance evaluation. We have compared the proximity-based policy with three other policies for load balancing, viz., round robin (RR), random (R) and Weighted capacity (WC).

We conducted experiments to measure performance under two different conditions for RTT on links. In one condition, there was less variations in RTT (delays were mostly in the range of 20 to 45 milliseconds with a mean of 30 milliseconds for links within the same region and in the range of 200 to 500 milliseconds with a mean of 300 milliseconds). In the other condition, there were high variations in RTT (delays were mostly in the range of 10 to 70 milliseconds, with a mean of 30 milliseconds for links in the same region, and in the range of 100 to 600 milliseconds with a mean of 300 milliseconds). We have performed experiments with file sizes of 5KB and 50KB.

We varied the number of client processes from 20 to 100 in steps of 20. We also measured the request rate with the number of client processes. The request arrival rate increases linearly with the number of client processes.

Average response time of the different schemes for file sizes 5KB and 50KB are shown in Figures 2 and 3, when there were high variations in round-trip time on links. As seen in the figure, the RR, R, and WC schemes have almost similar performance, while the proximity-based scheme has a significantly lesser response time.

For 5KB files, while the average response time with other three policies was between 0.98 to 1.1 seconds, the average response time with our policy was between 0.49 to .51 seconds. For 50KB files, the average response time with other three policies was between 1.96 to 2.75 seconds, and the average response time with our policy was between 1.02 to 1.13 seconds. The average response time was reduced to half when our policy was used as compared to other policies.

The average response times of different schemes when there was low variations in round-trip time on links are shown in Figures 4 and 5. In this case also, our scheme performed better than the other schemes. For 5KB files, the average

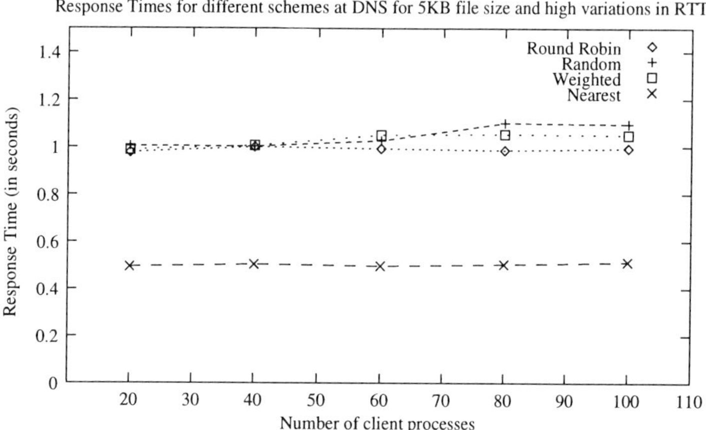

Fig. 2. Average response time with 5KB file size and high variations in RTT

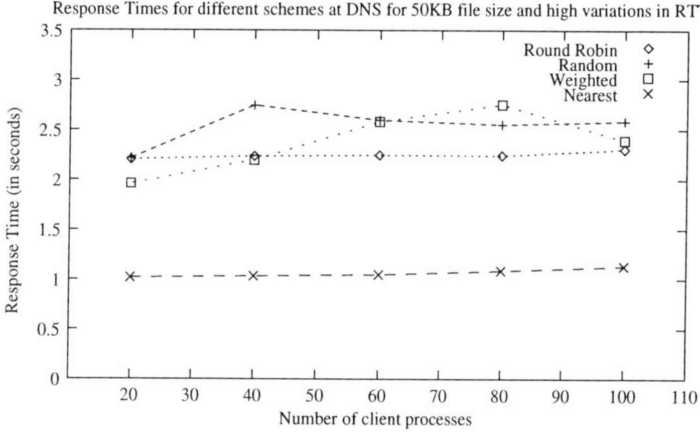

Fig. 3. Average response time with 50KB file size and high variations in RTT

response time with first three policies was between 0.86 to .99 second, the average response time with our policy was between 0.45 to .48 second. For 50KB files, the average response time with other three policies was between 1.7 to 2.27 seconds, and the average response time with our policy was between 0.8 to 0.93 seconds. Thus we see that the response time is reduced by a large factor when our policy is used and clients are distributed geographically across the globe.

We have also measured throughput (in Mbps) in all simulation experiments. We find that the throughput achieved by our policy is approximately twice in most cases.

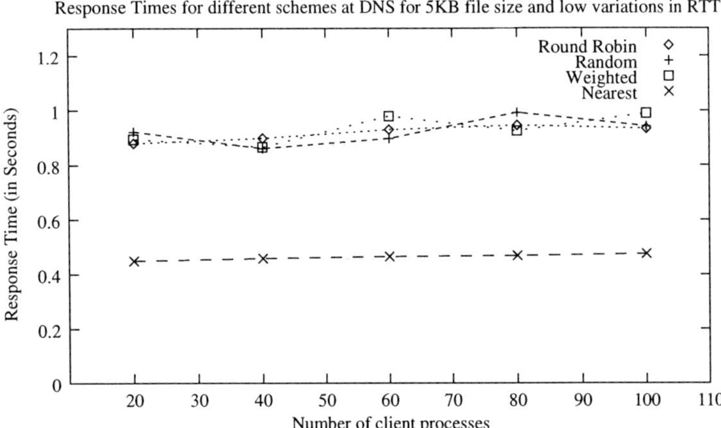

Fig. 4. Average response time with 5KB file size and low variations in RTT

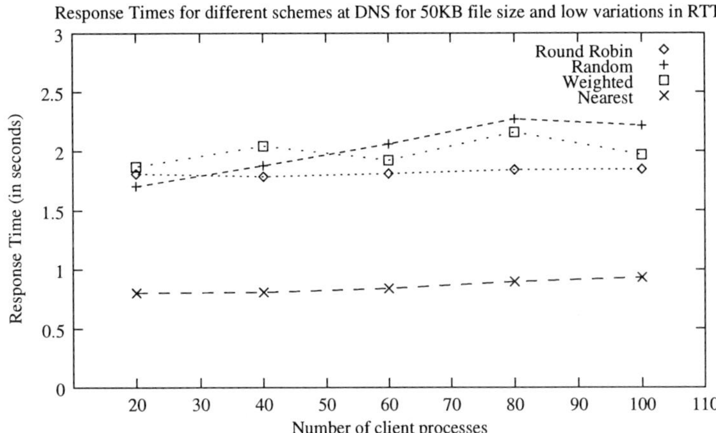

Fig. 5. Average response time with 50KB file size and low variations in RTT

6 Conclusions

We have proposed a proximity-based scheme for request distribution for a very large website. The scheme also takes server load into account. In the proposed scheme, the round trip time (RTT) for the client is measured, and the request is sent to the cluster that has the least RTT. This is done only for clients generating a lot of requests.

We have implemented the scheme on Linux machines. Using a test-bed, which simulates the delays of a real network, we compared the performance of our proposed policy with round robin, random, and weighted capacity. We found that the proposed policy gives better performance in terms of response time and throughput as compared to other policies.

References

[1] P. Agarwal. A test-bed for performance evaluation of load balancing strategies for web server systems. Master's thesis, CSE Dept., IIT Kanpur, India, May 2001.

[2] E. Anderson, D. Patterson, and E. Brewer. The Magicrouter: An application of fast packet interposing. http://www.cs.berkeley.edu/~eanders/projects/-magicrouter/osdi96-mr-submission.ps, May 1996.

[3] M. F. Arlitt and C. L. Williamson. Internet web servers: Workload characterization and performance implications. *IEEE/ACM Trans. on Networking*, 5(5):631–644, Oct. 1997.

[4] M. Baentsh, L. Baum, and G. Molter. Enhancing the web's infrastructure: From caching to replication. *IEEE Internet Computing*, 1(2):18–27, March-April 1997.

[5] M. Beck and T. Moore. The Internet-2 distributed storage infrastructure project: An architecture for Internet content channels. In *Proc. of Third Int'l WWW Caching Workshop*, Manchester, UK, June 1998.

[6] V. Cardelini, M. Colajanni, and P. S. Yu. Dynamic load balancing on web server systems. *IEEE Internet Computing*, 3(3):28–39, May-June 1999.

[7] M. Colajanni, P. S. Yu, and V. Cardelini. Dynamic load balancing on geographically distributed heterogenous web servers. In *IEEE 18th Int'l Conference on Distributed Computing Systems*, pages 295–302, May 1998.

[8] M. E. Crovella and R. L. Carter. Dynamic server selection in the internet. In *Proc. of the 3rd IEEE Workshop on the Architecture and Implementation of High Performance Communication Subsystems (HPCS '95)*, June 1995.

[9] O. P. Damani, P. Y. Chung, and C. Kintala. ONE-IP: Techniques for hosting a service on a cluster of machines. In *Proc. of 41st IEEE Computing Society Int'l Conference*, pages 85–92, Feb. 1996.

[10] Z. Fei, S. Bhattacharjee, E. W. Zegura, and M. Ammar. A novel server selection technique for improving the response time of a replicated service. In *Proc. of IEEE INFOCOMM '98 Conf.*, 1998.

[11] J. Guyton and M. Schwartz. Locating nearby copies of replicated internet servers. In *Proceedings of ACM SIGCOMM '95 Conference*, pages 288–298, Oct. 1995.

[12] G. D. H. Hunt, G. S. Goldzsmit, R. P. King, and R. Mukherjee. Network dispatcher: A connection router for scalable internet services. In *Proc. of 7th Int'l World Wide Web Conference*, Apr. 1998.

[13] T. T. Kwan, R. E. McGrath, and D. A. Reed. NCSA's world wide web server: Design and performance. *IEEE Computer*, pages 68–74, Nov. 1995.

[14] NIST. NistNet network emulator. http://snad.ncsl.nist.gov/itg/nistnet/.

[15] M. Sayal, Y. Breitbart, P. Scheuermann, and R. Vingralek. Selection algorithms for replicated web servers. In *Proc. of the Workshop on Internet Server Performance*, 1998. http://www.cs.wisc.edu/~cao/WISP98/final-versions/mehmet.ps.

[16] C. Yoshilakawa, B. Chun, P. Eastham, A. Vahdat, T. Anderson, and D. Culler. Using smart clients to build scalable services. In *Proc. of Usenix Ann. Tech. Conf.*, Anaheim, CA, Jan. 1997.

[17] E. W. Zegura, M. H. Ammar, Z. Fei, and S. Bhattacharjee. Application-layer anycasting: A server selection architecture and use in a replicated service. *IEEE/ACM Transactions on Networking*, 8(4):455–466, Aug. 2000.

Strategic Tool for Assessment of the Supply and Demand Relationship between ASPs and SMEs for Competitive Advantage

Babak Akhgar[1], Jawed Siddiqi[1], and Mehrdad Naderi[1]

School of Computing and Management Sciences,
Sheffield Hallam University, City Campus,
Sheffield, S1 1WB, United Kingdom
B.Akhgar@shu.ac.uk

Abstract. Competitive organisations must be able to manage and implement successful IS strategies. In an effort to create more effective information systems organisations are turning to Application Service Providers (ASPs) because they claim to be able to offer cheap, effective applications and services for many companies irrespective of their size. Many companies have been drawn by these claims only to be disappointed with the results. This paper aims to give clarity and structure to the process of selecting an ASP by employing an ASP Wheel Metaphor that assists in determining the necessity or otherwise of choosing an ASP. Two companies are provided as case studies for the application of this wheel to give an insight into how an organisation may decide on how to change its IS strategy.

Keywords
Application Service Provider, IS Strategy, Strategic Assessment Tool

1 Application Service Providers

More and more Organisations are beginning to realise the benefits of using an Application Service Provider (ASP). ASPs can offer organisations a number of advantages. Partly or wholly outsourcing the role of the I.T Department enables the organisation to posses a degree of freedom and flexibility. Having a large I.T Department means higher fixed costs. Information Systems (IS) evolve at a similar pace, with technology allowing more complex and thorough IS.

This evolution of systems does of course increase the expenditure of an organisation. This expenditure is not limited to equipment purchases; it includes personnel costs, research and development costs, and purchasing costs.

An ASP also can be viewed as a means of providing an organisation with a way in which to outsource IT and IS requirements. This can take various forms and an ASP can handle the whole operation or just an element of it.

M.H. Shafazand and A M. Tjoa (Eds.): EurAsia-ICT 2002, LNCS 2510, pp. 667–676, 2002.

A commonly outsourced task is technical support. ASPs also act in a more conventional way, by providing applications such as E-mail, Office Applications, and other productivity Software by connecting an organisation to a remote data centre via telephone or private network (www.aspnews.com/#1435.asp). This form of service can benefit almost any SME that uses desktop computers. Indeed many companies use ASPs for E-mail routing and provision without realising it.

Reducing the number machines reduces failures, staffing costs and replacements, allowing an organisation to allocate these resources elsewhere.

The multitude of services offered by ASPs will continue to grow. Given that more companies are moving to the ASP model it is conceivable that many more SMEs will face the decision as part of their IT and IS strategies. This paper will examine two organisations for which the ASP model appears attractive. By examining their business objectives, IT, and IS Strategies, it will be possible to analysis what type of ASP model would best suit each company and how they should select an ASP.

2 Case Study

For the purposes of this paper the companies will be referred to as Company A and Company B. Company A is a multinational charitable organisation having offices throughout the UK and each office is allowed a certain degree of autonomy. There is no centralised IT or IS strategy, each office manager makes the decisions for strategy formulation and implementation. The particular element of the organisation this paper will focus upon is a addition to the service portfolio offered by the charity that is 5 years old. It has experienced significant growth in each year of operation. However, as the service continues its growth the current specially developed software is ready to be replaced. To date, IT and IS management has been an ad-hoc affair, with little or no forward planning given to strategic choices. Currently IT is critical to the operation of the business. Should any failures occur, the consequences would be catastrophic as the service is responsible for the lives of it clients. This service employs 200 people and has a turnover of 1.5 million.

Company B is smaller in size, but has a greater turnover at 15 million, and employ 54 people. This organisation is an advertising agency operating within a major UK city. It has clients throughout the UK and places numerous press advertisements each week. The entire operational structure is based upon a singular computer program that controls every aspect of the business. The program integrates each department and logs time spent working on each job (from initial design through to payment for the finished product).

IT strategy within this company is largely governed by the industry standard applications used by most media and advertising businesses. To ensure compatibility with suppliers the predominant packages have a significant market share within this industry. Similarly the hardware used to run these packages is largely Apple Macintosh, and enjoys a similar percentage of market share. Given these factors, the company's strategy has to take account of limitations within the technology, and longer development cycle when compared with the PC.

Both companies have IT departments, however CB outsourcers its software development. CA has a national IT department to supply hardware but each service is autonomous with regard to operating procedures, software and strategy. For companies A and B, like many SMEs, financial considerations are placed highest. If the current strategies of companies A and B were mapped onto the Balanced Scorecard (Kaplan and Norton, 1996) it would be clear that the main focus is on appearance to the financial stakeholders. These stake holders differ for each company. However the precedence these stake holders take in formulation of strategy does not. For both of the companies, internal political issues often influence and shape strategy. This causes decisions to be based upon political factors rather than actual strategic thinking.

Both companies manage the Internal and Customer perspectives of Balanced Scorecard (Kaplan and Norton, 1996), in the same way. Each of them has a similar reliance upon the interaction between their internal perspective and customer perspective. These two areas are directly linked, by concentrating on productivity and quality both companies, by the nature of their industries, offer better service, and therefore raise the level of customer focus and care.

By investing in IT company B ensures that they are able, technically at least, to deliver customer satisfaction. Since advertising is based around advances within technology it is imperative that this investment continue. Possessing the latest hardware enables company B to run the latest software. It is this that provides the latest innovations within advertising, therefore enabling the company to compete. The cycle of new software releases ensures the company will be forced to purchase the latest release in order to offer the latest features to clients. This upgrade cost represents 64

Within both industries the margins for profit are being reduced. One way in which both companies can combat this is with a review of their IS strategy.

3 A Tool for Assessing the Necessity for an ASP

For both Company A and B the ASP model is not an option either had considered. This is due to a number of factors; the primary reason is the limited knowledge within the IT departments of each company. The companies are primarily concerned with day-to-day operations rather than long-term planning and strategy formulation. This lack of knowledge compounds the problems associated with the implementation of the ASP model.

The level of implementation of the ASP model depends upon the IS needs, as previously mentioned, ASPs can provide a varied range of solutions. The challenge for an organisation lays in two areas, firstly selecting what services to have an ASP provide, secondly, and most importantly which ASP to use.

When selecting ASPs, organisations often focus upon which particular ASP to use. Following their selection companies can be disappointed with their choice. While the choice of ASP is important it is vital that the company has successfully analysed its own internal structure and has a feasible IS strategy. Only by doing this can a company begin to select what services are needed and which ASP should supply them. The use of a framework would make the process of ASP selection simpler and more readily understandable. Based on our own industrial experience and abstraction of the learning outcomes from the action research conducted into the companies in the case studies noted above, we submit the model described below as a tool in the delineation of key issues in the decision making process of whether a candidate SME is firstly ready for such strategic move, if so, what are the internal factors that should be in place in readiness of such embarkation and lastly the quantification of the selection and evaluation criteria for aspirant ASPs.

The ASP wheel above requires both internal and external inputs. For correct use each segment should be considered in turn and all segments should result in satisfactory conditions. All the elements of the ASP Wheel are interdependent; the analysis of the ASP considerations will depend upon the analysis of the Internal Factors. In the following sections we will apply our proposed tool to the case studies discussed above, by first defining a base line definition followed by the application of the latter to the case studies.

Internal Factors:	ASP Considerations:
S - Strategy	D - Delivery
E - Expectations	L -Location
R - Results	S - Security
A - Action	R - Reliability

Fig. 1. ASP Wheel

3.1 Strategy (S) and Expectations (E)

The starting point in an evaluation using the ASP Wheel is the Strategy segment. This segment deals with current or proposed strategy. In the case of companies A and B it will be current strategy. This segment should be used in tandem with the Expectations segment both of these are vitally important to the outcome of the analysis. They can be seen as the foundation for the segments that follow. The user should be aware of current strategy and have in depth knowledge of its implementation. However, when considering if current strategy is meeting the criteria expected of it, analysis maybe become more difficult. More factors will need considering than the sole effect of strategy. Have any unforeseen, unavoidable circumstances taken place? Such as shift in market conditions, action by competitors (Porter, 1998), or more recently events. Following the analysis of the first two segments the user should have a clear idea of what the current strategy is and what the expectations are of that strategy. The users should be aware of where the strategy is succeeding and where the failures lie. The user should be able to identify areas of concern and critical success factors (CFS) (Johnson and Scholes 2002) needed to ensure delivery of the strategy.

Company A has problems in that current strategy was poorly implemented and market conditions have constricted the IT budget. Therefore a change in this strategy needs to be made. Company A has the goal to reduce IT costs while retaining and improving their propriety software. The company also need to improve IT support, again without increasing costs. This set of objectives and strategic shifts can be seen as the most desirable outcome. It provides an initial platform for understanding and determining what is obtainable.

Company A must achieve these goals without compromising its Critical Success Factors (CSF). The critical success factors within company A are:

- Quality of the service delivered is the most important CSF as so many other elements of the business strategy depend upon it. The high quality of service allows a higher price to be charged to the clients. To allow a high quality of service the company has to have accurate information on each client and what their needs are.
- Employees. The quality and training of the employees is paramount as this enables good customer care and a high standard of work.
- Understanding of External Political Trends. The industry is highly regulated and standards change frequently, failure to manage this change would result in loss of permits and clients.

Having identified where the company stands and where it needs to be, it is possible to determine what course of action to take and which, if any, ASP to use. Company B has become concerned that strategy implementation is costing far more than was expected and requires a method of reducing this cost. Like company A, cost reduction is a prime concern and the strategic goals reflect this. Although costs must be reduced, it is vital that the CSFs are met. Company B has to ensure that:

- It has good relationships with its clients and provides the best service it can. To facilitate this, it has to produce high quality product that is better than the competition. This is only possible through the latest IT equipment and skilled employees.
- It has to provide prices for advertising that are lower than its competitors. The management information system is crucial to ensuring this. The clients are based throughout the UK, the company charges different prices to its various clients, often for the same service. For this reason it is vital this information is kept confidential and that access to it available for use when negotiating prices.

As with company A, once this process of examination is complete the company can determine its course of action.

3.2 Results (R) and Action (A)

Once the strategy and expectations segments have been analysed it is possible examine the results and decide upon the action to be taken. If for example the first phase highlighted poor performance of the IT system, then action should be taken to improve this. The issues highlighted my not be addressed by merely moving those areas to an ASP. Performance could be linked to poor training or incorrect configuration of hardware, these problems would not be solved by using an ASP. The key questions in this phase are can our needs be met internally? Can this be done cost effectively? What problems could occur? These questions could be answered by restructuring an element of the IS strategy, altering processes, improving training, or changing hardware and software. These questions should be considered against the service an ASP can offer, considering the issues of cost, downtime, resistance to change (Paton and McCalman 2000), culture (Graves 1986), and the possible staff changes that may occur and the effects these will have. All these elements must be considered as their effects will have direct results on business strategy and culture within the organisation. For example, the move to an ASP may result in some staff changes, certain roles may become unnecessary. This is an important consideration as this may cause resentment amongst remaining staff. This action may lower overall fixed costs but would the detrimental effect to staff morale adversely affect productivity or the service offered? Given these factors, it is necessary to evaluate the consequences of a certain course of action, as the most obvious solution may not suit the business needs in the long term.

Following a decision on what action to take, objectives should be set. These objectives should have a timeframe and be realistic (Preace 1995). In the case of ASP selection, if a company requires a greater service than is actually possible then this will be unachievable and will fail. It is important for the objectives to be manageable and attainable. If the objectives are unobtainable then problems with motivation, strategy formulation, and analysis will occur (Mager 1990). By this stage the organisation should have a clear picture of what is required from an ASP. The company should be aware of what is on offer and what they need to implement, the organisation should also be clear regarding the time frames they wish to work to, and any staffing and organisational changes which are needed. Only by being fully aware of what is required can an organisation accurately asses a given ASP and evaluate if it meets the requirements. If no clear focus is present then a company cannot select the correct ASP. This may result in total failure to select an ASP or selection of an ASP that does not meet requirements and the process has to occur again at extra cost and disruption. Ensuring strategic planning is thorough and sustainable at this stage is vital as it underpins all future decisions regarding ASP selection and implementation.

Following the consideration of the first phase of the ASP Wheel, company A has decided the best course of action is to examine the feasibility of using an ASP to provide standard desktop productivity packages, to take over and development of the propriety program while providing this application from the data centre, and to provide technical support. This option is a relatively comprehensive use of an ASP and requires fundamental changes to the way the IS strategy is implemented and developed. This examination and conclusion of the first Phase of the ASP Wheel, allows the company to examine potential ASPs with more clarity. Having identified its strategic and organisational needs the company can best assess which ASP can help realise those needs. The following phases of the wheel will examine what factors should be considered and what criteria should be used when selecting and ASP.

Having evaluated the impact of change upon the business it was decided to continue with their existing systems and forgo the use of an ASP. Given the massive change required to implement an ASP, the company evaluated the impact of this change and decided the cost would be too great in the initial phase. The problems of increasing IT costs are still present; however, the company believes that by improving its existing practises and re-structuring its design department, it can manage the increase. Without the initial evaluation the company may have decided to commit to an ASP, and in this case that decision would have resulted in failure.

3.3 Delivery (D)

Following the full analysis of the internal phase of the wheel, it is possible to move on to the ASP Considerations. This phase primarily concerns the ASP and what it can offer. However, as with the first phase, each segment is dependent on the others and each segment must result in positive and obtainable outcomes if a successful choice is to be made. The user should also be aware of the links between what an ASP offers and what is needed, constantly referring to the strategic choices and analysis made during the first phase of the wheel. The first segment of the ASP phase is Deliverability. This refers to the ASP's ability to deliver the services identified within the first phase. However, this concerns a greater number of elements than whether or not the ASP can provide the service. Within this segment the user should give careful consideration to the requirements of working with an ASP. the points following should be considered:

Ability to Scale

Can the ASP cope as organisational needs change? If possible, making the correct choice of ASP the first time is preferable, changing ASPs would cause further disruption and the process of selection would require repeating. It is essential that the ASP can handle increased demand from an organisation it serves. Associated with this consideration is the size of the ASP, should the ASP take on new customers it is imperative that the service to its existing customers does not suffer. Companies should ensure that their contract with the ASP not only states what service they will provide, but also states what quality of service they will provide. It is tempting for ASPs, as with many Internet Service Providers (ISPs), to over subscribe their service thus reducing the quality of service. The culture of the ASP is also important, once an organisation agrees to work with an ASP it is vital that both companies can communicate effectively and their respective staff can work together. This is important as the ASP will be responsible for vital aspects of the business operation and management systems. Small details are important, such as a regular point of contact at the ASP, someone who understands business and the service require. Should conflict occur it would result in poor performance and hinder the implementation of strategy. An ASP has to understand its customer's needs, especially if the ASP takes on the role of development. It is also desirable that the ASP is familiar with a clients business; however this familiarity could be due to hosting competitor's applications. The ASP has a number of areas where a conflict of interest may arise.

Applications and Services Offered

At the time of selection the ASP must host all the applications an organisation requires. However, over time, an organisations requirements change. The ASP should not only offer the applications needed now, but posses the capacity to deliver applications needed in the future. The ASP should also have favourable links with its suppliers, this ensures the ASP has high product knowledge and can offer competitive prices on applications needed in the future.

3.4 Location (L)

Should the ASP be located in another city, or even another country, then certain considerations should be made. Depending upon the type of applications and services required bandwidth maybe an issue. It may not be possible to deliver the bandwidth required over normal telephone line, should broadband or other fast connection be unavailable. Support from the ASP is also location dependant, should the ASP be a great distance away then support may take time to arrive. This could be problematic should there be a serious failure. If the ASP is located abroad the economic climate of that country should be examined. How would a downturn affect the ASP and the service they provide?

3.5 Security (S)

Security is one of the key concerns regarding ASPs. They are an important supplier, trusted, in some cases with an organisations vital and confidential data. They are also responsible for providing systems that enable the organisation they supply to function. They are however external to the organisation. This raises fears about the integrity of the ASP and raises questions about how safe the data they store is. It is vital that any contract entered into with an ASP clearly states who owns the data held by the ASP (www.aspnews.com), and what the responsibilities and liabilities are regarding that data and its use. When selecting an ASP the organisation should examine closely what procedures are in place to secure data. An organisation should be certain a disgruntled employee or unauthorised person cannot access or manipulate the data. Physical security should also be examined, how secure is the site? Where is backed up data held? The security of the data between the ASP and organisation is also important. Data should be secure and encrypted so as to avoid interception. It is also important to examine whether competitors use the same ASP and what level of integrity the ASP has. For example could the ASP divulge data to a competitor or an employee accidentally leak or lose data?

3.6 Reliability (R)

This segment concerns the ability of the ASP to provide an uninterrupted service. Examination of this aspect should focus upon the quality of the hardware and the skill of the employees. The ability of the ASP to attract and retain skilled employees is also important. High staff turnover could point to ineffective recruitment, poor morale and inconsistent training; these factors would impede the ability of the ASP to provide a service. Externally the ASP is faced with events that can cause disaster. These include natural disasters, fire, crime and severe equipment failure. Any of these could cause total failure of the service the ASP provides. Ideally an ASP should be located away from any potential hazards, i.e. earthquake zones or industrial hazards. The ASP should have clear functional contingency plans in the event of disaster.

4 Summary

In an SME context the strategic feasibility of adopting an ASP as a solution to many problems, outsourcing, rising costs and expensive software was explored. A conceptual tool was introduced to analyse the supply and demand relationship between internal factors related to strategic objectives of the organisation and required ASP considerations. Application of the tool was evaluated on a case study with two different companies. It was concluded that selecting an ASP is a difficult task with profound strategic implications. Our proposed approach that begins with the supply side (i.e. an examination of the internal issues within the organisation) and then attempts to match this with the demand characterisation of potential ASP's by following the ASP Wheel Metaphor. It is claimed that following such enable an organisation to be able to foresee problems before they occur and have a clear selection criterion for an ASP. The application of the wheel (tool) clearly depends on what level of importance the ASP holds for a company. For example if a company outsource email via an ASP it will be less concerned about certain elements of the wheel than a company which is highly dependent on an ASP. However, unless the organisation implements an ASP as part of a comprehensive and clear IS strategy, organisations will continue to be disappointed with the results from the ASPs they select. The future direction of this research is to establish a roadmap that addresses not only a tool for determining the necessity or otherwise for an ASP, but also to explore the necessary stages as part of process integration across the supply chain. Early indications of this research show that elements such as the identification of business requirements, infrastructure requirements and service level agreements form pre-requisite necessary activities for effective deployment of an ASP.

References

1. Anthony, R (2000) "Cyber Supply", Electronic Business, June, http://www.findarticles.com
2. Martin, C (1998); Logistics and supply chain management: Strategies for reducing costs and improving services, Financial Time Management
3. Graves, D (1986) Corporate Culture: Diagnosis and change, , Pinter , Review of Corporate Culture
4. Johnson, G, and Scholes, K (2002); Exploring Corporate Strategy 6th Edition, 2002, Prentice Hall, Harlow, England
5. Kaplan, R and Norton, D (1996); The balance scorecard: Translating strategy into action. Harvard Business Review 74, pp: 75-85
6. Mager, R (1990) Goal Analysis, Kogan Page, A guide on System View, Vol 12 SHU re-print
7. Paton, R and McCalman, J (2000); Change Management: A guide to effective implementation, Sage Publications,
8. Porter, M(1998) On Competition, Harvard Business School Publishing, Boston, Mass, HBR Re-print
9. Preece, D(1995); Organisations and Technical Change: Strategy and Involvement, Routledge

Trust and Commitment in Dynamic Logic

Jan Broersen[1], Mehdi Dastani[2], Zhisheng Huang[1], and Leendert van der Torre[1]

[1] Faculty of Mathematics and Computer Science, Vrije Universiteit Amsterdam
{broersen,huang,torre}@cs.vu.nl
[2] Institute of Information and Computing Sciences, Utrecht University
mehdi@cs.uu.nl

Abstract. Trust and commitment have been identified as crucial concepts in electronic commerce applications. In this paper we are interested in the relation between these social concepts. We introduce a dynamic logic in which violations of stronger commitments result in a higher loss of trustworthiness than violations of weaker ones. We illustrate how the logic can be used to analyze some aspects of a well known example of trust within reason.

1 Introduction

In advanced applications of multi agent systems agents interact more frequently, deliberate more extensively, and in general act more autonomously. State of the art computer programs are capable of searching the web for the cheapest books, advising users on movies, negotiating bandwidth, participating in auctions, etc. Moreover, experiments with the contract net protocol have revealed that more flexible protocols based on levelled commitment lead to better global results, because agents can engage in several interactions simultaneously [13]. Researchers envision a continuation of this trend of increasing complexity of agent interactions and discuss washing machines negotiating the purchase of micro units of electricity with electricity companies [5]. One promising approach to build such complex agents introduces methods and concepts from the social sciences, such as organization, negotiation, commitment and trust [2,4,11].

A high level of trustworthiness is normally beneficial for the long term profits of agents, and a low level has a negative effect on them. However, whereas the short term profits are usually easy to calculate, these long term profits are much more difficult to quantify. This leads to a problem for an agent that has to balance its short term profits with its long term ones. The question is, how to balance the short term profit of violating a commitment with its cost in the long run due to the decrease in trustworthiness?

In order to formalize some of these concepts and reasoning mechanisms, we introduce a dynamic logic in which the violation of stronger commitments results in higher loss of trustworthiness than the violation of weaker ones. This logic describes an agent that proposes commitments, accepts proposals to engage in commitments, and violates commitments by performing actions other than the ones committed to.

This article is organized as follows. In Section 2 we motivate our work with an example of reasoning about trust. In Section 3 we introduce our dynamic logic. In Section 4 we show how to apply the logic to the motivating example. Finally, in Section 5 we discuss some formal properties of trust and commitment.

M.H. Shafazand and A M. Tjoa (Eds.): EurAsia-ICT 2002, LNCS 2510, pp. 677–684, 2002.

2 Trust and Commitment in Strategic Decisions

The pennies pinching example is a problem discussed in philosophy that is also relevant for advanced agent-based computer applications. It is related to trust, but it has been discussed in the context of game theory, where it is known as a non-zero sum game. Hollis [8,9] discusses the example and the related problem of backward induction as follows.

> A and B play a game where ten pennies are put on the table and each in turn takes one penny or two. If one is taken, then the turn passes. As soon as two are taken the game stops and any remaining pennies vanish. What will happen, if both players are rational? Offhand one might suppose that they emerge with five pennies each or with a six-four split – when the player with the odd-numbered turns take two at the end. But game theory seems to say not. Its apparent answer is that the opening player will take two pennies, thus killing the golden goose at the start and leaving both worse off. The immediate trouble is caused by what has become known as backward induction. The resulting pennies gained by each player are given by the bracketed numbers, with A's put first in each case. Looking ahead, B realizes that they will not reach (5,5), because A would settle for (6,4). A realizes that B would therefore settle for (4,5), which makes it rational for A to stop at (5,3). In that case, B would settle for (3,4); so A would therefore settle for (4,2), leading B to prefer (2,3); and so on. A thus takes two pennies at his first move and reason has obstructed the benefit of mankind.

Game-theory and its backward induction reasoning do not offer the intuitive solutions to the problem, because agents are assumed to be rational in the sense of economics and consequently game-theoretic solutions do not consider an implicit mutual understanding of a cooperation strategy [1]. Cooperation results in an increased personal benefit by seducing the other party in cooperation. The open question is how such 'super-rational' behavior can be explained.

Hollis considers in his book 'Trust within reason' [9] several possible explanations why an agent should take one penny instead of two. For example, taking one penny in the first move 'signals' to the other agent that the agent wants to cooperate (and it signals that the agent is not rational in the economic sense). Two concepts that play a major role in his book are trust and commitment (together with norm and obligation). One possible explanation is that taking one penny induces a commitment that the agent will take one penny again in his next move. If the other agent believes this commitment, then it has become rational for him to take one penny too. Another explanation is that taking one penny leads to a commitment of the other agent to take one penny too, maybe as a result of a social law. Moreover, other explanations are not only based on commitments, but also on the trust in the other party.

In this paper we do not want to sum up and classify all the analyses of the pennies pinching example discussed in the literature. We want to introduce a language in which some aspects of these analyses can be represented. In Section 4 we discuss these aspects as well as scenarios of pennies pinching with communication.

3 A Dynamic Logic of Trust and Commitment

Our logic formalizes a variety of examples such as the pennies pinching example as well as examples in electronic commerce. First it formalizes the discussed notions of trust and commitment. Second, it formalizes complex actions that enable the specification of protocols and communication acts. For example, the protocol of the pennies pinching game states that the only possible actions are to take one or two pennies at a time. Trust and commitments can be created by communication. Our logic therefore consists of a dynamic logic for actions and modal operators for trust and commitment.

The dynamic logic is an extension of standard propositional dynamic logic [6,7] that contains operators $\cup$ for choice, $*$ for iteration and ; for sequence. The formula $\langle\alpha_i\rangle\varphi$ expresses that agent i is able to perform action α and by doing so it possibly reaches a state where φ holds. Our extension incorporates a concurrency operator $\cap$ and an action negation operator $-$. Concurrency is needed to synchronize processes or agents, and negation is needed to formalize obligations (for details see below and [3]).

In this dynamic logic we introduce a modality for commitment $C_{i,j}(\alpha \geq \beta)$, whose intended meaning is 'agent i, towards agent j, is more committed to perform α than to perform β', and we introduce a modality $T_{i,j}(\alpha \geq \beta)$, whose intended meaning is 'agent i trusts agent j more after the performance of α than after the performance of β'.

Definition 1. *Given a set G of agent identifiers, a set $\mathcal{A}$ of action symbols (that may be indexed by individual agents or sets of agents and that may include actions for speech acts), and a set $\mathcal{P}$ of proposition symbols. The well-formed formula $\varphi, \psi, \ldots$ are defined through the following BNF with $i, j \in G$, $a \in \mathcal{A}$ and $p \in \mathcal{P}$.*

$$\begin{array}{l} \varphi, \psi, \ldots ::= p \mid \neg\varphi \mid \varphi \wedge \psi \mid \langle\alpha\rangle\varphi \mid C_{i,j}(\alpha \geq \beta) \mid T_{i,j}(\alpha \geq \beta) \\ \alpha, \beta, \ldots ::= a \mid any \mid -\alpha \mid \alpha \cup \beta \mid \alpha \cap \beta \mid \alpha;\beta \mid \alpha^* \end{array}$$

For formulas φ we use the usual abbreviations $\vee, \rightarrow, \top$ and $\bot$.

The semantics is defined using modal action structures, supplemented with orderings $\succeq^C_{i,j}$ and $\succeq^T_{i,j}$ that interpret respectively $C_{i,j}(\alpha \geq \beta)$ and $T_{i,j}(\alpha \geq \beta)$. $\succeq^C_{i,j}$ orders levels of commitment of agent i with respect to agent j, and $\succeq^T_{i,j}$ is a reflexive and transitive ordering over S that orders levels of trust of agent i in agent j.

Definition 2. *Let $\mathcal{A}$ be a set of action symbols, G a set of of agent identifiers, and $\mathcal{P}$ a set of proposition symbols. A structure is a tuple $\mathcal{S} = (S, R, \pi, \succeq^C, \succeq^T)$, where S is a nonempty set of possible states, R defines for each action $a \in \mathcal{A}$ and agent $i \in G$ an accessibility relation over S, π is a valuation function $\pi : \mathcal{P} \rightarrow 2^S$ that interprets propositions $p \in \mathcal{P}$, and $\succeq^C$ and $\succeq^T$ each return for every pair of agents $i, j \in G$ a reflexive and transitive ordering over S.*

The semantics for the comparative commitment operator $C_{i,j}(\alpha \geq \beta)$ is, that an agent is more committed to choose α than to choose β if and only if the best possible outcome can be reached by α and the worst possible outcome can be reached by β. The semantics of the operator $T_{i,j}(\alpha \geq \beta)$ has a similar definition.

Definition 3. *The meaning of well-formed formulas in a state s of a structure $\mathcal{S}$ is given by:*

$$\begin{array}{lll}
R_{\alpha\cap\beta} & = & R_\alpha \cap R_\beta \\
R_{\alpha\cup\beta} & = & R_\alpha \cup R_\beta \\
R_{-\alpha} & = & R_{any} \setminus R_\alpha \\
R_{\alpha;\beta} & = & R_\alpha \circ R_\beta = \{(s,s'') \mid (s,s') \in R_\alpha \textit{ and } (s',s'') \in R_\beta\} \\
R_{\alpha^*} & = & (R_\alpha)^* = Id \cup R_\alpha \cup R_\alpha \circ R_\alpha \cup \ldots \textit{ with } Id = \{(s,s) \mid s \in S\} \\
R_{any} & = & (R_a \cup R_b \cup R_c \ldots)^* \textit{ with } \{a,b,c,\ldots\} = \mathcal{A} \\
\mathcal{S}, s \models P & \textit{iff} & s \in \pi(P) \\
\mathcal{S}, s \models \neg\varphi & \textit{iff} & \textit{not } \mathcal{S}, s \models \varphi \\
\mathcal{S}, s \models \varphi \wedge \psi & \textit{iff} & \mathcal{S}, s \models \varphi \textit{ and } \mathcal{S}, s \models \psi \\
\mathcal{S}, s \models \langle\alpha\rangle\varphi & \textit{iff} & \exists s' \textit{ such that } (s,s') \in R_\alpha \textit{ and } \mathcal{S}, s' \models \varphi \\
\mathcal{S}, s \models C_{i,j}(\alpha \geq \beta) & \textit{iff} & \textit{for all } (s,s') \in R_{\alpha\cup\beta}, \textit{ there is} \\
 & & (1)\ \textit{a } (s,s'') \in R_\beta \textit{ such that } s' \succeq^C_{ij} s'' \\
 & & (2)\ \textit{a } (s,s''') \in R_\alpha \textit{ such that } s''' \succeq^C_{ij} s' \\
\mathcal{S}, s \models T_{i,j}(\alpha \geq \beta) & \textit{iff} & \textit{for all } (s,s') \in R_{\alpha\cup\beta}, \textit{ there is} \\
 & & (1)\ \textit{a } (s,s'') \in R_\beta \textit{ such that } s' \succeq^T_{i,j} s'' \\
 & & (2)\ \textit{a } (s,s''') \in R_\alpha \textit{ such that } s''' \succeq^T_{i,j} s'
\end{array}$$

Validity of a formula on a structure and general validity are defined as usual.

For some applications the additional constraint may be added that commitments are restricted to the agent's own actions. However, for some mechanisms such as delegation it may be useful to express that an agent is committed to the actions of other agents. For example, a boss in an organization may be committed to actions of his employees.

In this logic, several other operators are available as syntactic definitions. First we define:

$$\begin{array}{ll}
[\alpha]\varphi & =_{def} \neg\langle\alpha\rangle\neg\varphi \\
C_{i,j}(\alpha > \beta) & =_{def} C_{i,j}(\alpha \geq \beta) \wedge \neg C_{i,j}(\beta \geq \alpha) \\
C_{i,j}(\alpha = \beta) & =_{def} C_{i,j}(\alpha \geq \beta) \wedge C_{i,j}(\beta \geq \alpha)
\end{array}$$

Traditional deontic notions can be defined in terms of the operator $C_{i,j}(\alpha \geq \beta)$. This expresses that an agent that has made commitments has put himself in a normative position with obligations, permissions and prohibitions. First we define intention $I_i(\alpha \geq \beta)$ as self-commitment (as in agent oriented programming [14]). Second we define obligation $O_{i,j}(\alpha)$ as the commitment to perform α rather than its complement $-\alpha$. The strict version of the commitment operator for obligation is justified by the observation that an obligation for α cannot be complied to in any way by performing an action that possibly brings us to a state not reachable by α. Prohibition is defined in terms of the obligation operator as the obligation to do $-\alpha$. Permission is defined as the negation of prohibition.

$$I_i(\alpha \geq \beta) =_{def} C_{i,i}(\alpha \geq \beta) \qquad O_{i,j}(\alpha) =_{def} C_{i,j}(\alpha > -\alpha)$$

$$F_{i,j}(\alpha) =_{def} O_{i,j}(-\alpha) \qquad P_{i,j}(\alpha) =_{def} \neg F_{i,j}(\alpha)$$

A further discussion of these deontic notions and a comparison with alternative definitions is beyond the scope of this paper.

4 The Pennies Pinching Example in Dynamic Logic

In this section, we illustrate the dynamic logic by formalizing aspects of the pennies pinching example. In all examples we accept the following relation between trust and commitment, which denotes that violations of stronger commitments result in a higher loss of trustworthiness than violations of weaker ones.

$$C_{i,j}(\alpha > \beta) \rightarrow T_{j,i}(\alpha > \beta)$$

We first consider the example without communication. The set of agents is $G = \{1, 2\}$ and the set of atomic actions $A = \{take_i(1), take_i(2) \mid i \in G\}$, where $take_i(n)$ denotes that the agent i takes n pennies. The following formula denotes that taking one penny induces a commitment to take one penny later on.

$$[take_1(1); take_2(1)]C_{1,2}(take_1(1) > take_1(2))$$

The formula expresses that taking one penny is interpreted as a signal that the agent 1 will take one penny again on his next turn. When this formula holds, it is rational for agent 2 to take one penny.

The following formula denotes that taking one penny induces a commitment for the other agent to take one penny on the next move.

$$[take_1(1)]C_{2,1}(take_2(1) > take_2(2))$$

The formula denotes the implications of a social law, which states that you have to return favours. It is like giving a present to someone's birthday, thereby giving the person the obligation to return a present for your birthday.

More complex examples involve besides the commitment operator also the trust operator. For example, the following formula denotes that taking one penny increases the trust.

$$T_{i,j}((\alpha; take_j(1)) > \alpha).$$

The following formulas illustrate how commitment and trust may interact. The first formula expresses that each agent intends to increase the trust (=long term benefit). The second formula expresses that any commitment to itself is also a commitment to the other agent (a very strong cooperation rule).

$$T_{i,j}(\beta > \alpha) \rightarrow I_j(\beta > \alpha).$$

$$C_{j,j}(\beta > \alpha) \leftrightarrow C_{j,i}(\beta > \alpha).$$

From these two rules, together with the definitions and the general rule, we can deduce:

$$C_{i,j}(take_i(1) > take_i(2)) \leftrightarrow T_{j,i}(take_i(1) > take_i(2))$$

In this scenario, each agent is assumed to act to increase its long term benefit, i.e. act to increase the trust of other agents. Note that the commitment of i to j to take one penny increases the trust of j in i and vice versa. Therefore, each agent would not want to take two pennies since this will decrease its long term benefit.

We now consider the extension of the set of primitive actions with the communication actions or speech acts $\text{propose}_{i,j}(\alpha \text{ for } \beta)$ and $\text{accept}_{i,j}(\alpha \text{ for } \beta)$, that denote a proposal of agent i to agent j to perform α in return for β, and the acceptance of the proposal from agent j by agent i. For example, an agent may propose to the other agent to take one, if the other agent will take one afterwards too:

$$\text{propose}_{1,2}(take_1(1) \text{ for } take_2(1))$$

Moreover, the agent may propose that the other agent will take one, and that he in return will take one instead of two.

$$\text{propose}_{1,2}(take_2(1) \text{ for } take_1(1))$$

The following formula expresses that a propose followed by an accept action creates a commitment for both agents. To make the formula fit the page, we abbreviate propose by p, accept by a, and *take* by t.

$$[\text{p}_{1,2}(t_1(1) \text{ for } t_2(1)); \text{a}_{2,1}(t_1(1) \text{ for } t_2(1))](C_{1,2}(t_1(1) > t_1(2)) \wedge [t_1(1)]C_{2,1}(t_2(1) > t_2(2)))$$

Finally, properties of the protocol can be specified in the logic. Due to space limitations we are brief. The first formula says that first agent i and then agent j in turn take one or two pennies, and that no other actions are involved. The second formula further constrains the allowed actions by stipulating that the decision to take one penny each time cannot be repeated more than five times. The third formula gives the additional constraint that at any stage, after taking two pennies no more actions can be performed, which means that the game has stopped.

$$[-(((take_i(1) \cup take_i(2)); (take_j(1) \cup take_j(2)))^*)]\bot$$

$$[(take_i(1); take_j(1))^5; (take_i(1) \cup take_i(2))]\bot$$

$$[(take_i(2) \cup take_j(2)); (take_i(1) \cup take_j(1) \cup take_i(2) \cup take_j(2))]\bot$$

Other properties may require further extensions of the logic. The following formulas say that taking one penny increases the number of pennies an agent possess with one, and that taking two pennies increases the number of pennies an agent possesses with two. To formalize these formulas we either have to introduce a first order language, in which we can quantify over the variable k in the formulas, or we may read the formulas as representing the finite set of formulas that we get by instantiating k with the finite set of values relevant for the example.

$$((Possess_i = k) \rightarrow [take_i(1)](Possess_i = k + 1))$$

$$((Possess_i = k) \rightarrow [take_i(2)](Possess_i = k + 2))$$

Further possible extensions of the logic are a decision model, for example based on the fact that agents have as a goal to maximize the value of $Possess_i$.

5 Some Properties of the Operators

In this section we mention some properties of the logic to give the reader a feeling of it. A full account of the logic is beyond the scope of this paper.

To distinguish between both cases we will use the symbols $\models$ and $\not\models$. With the transitivity and reflexivity of the orderings $\succeq^C_{i,j}$ and $\succeq^T_{i,j}$ correspond the following properties:

$$\models C_{i,j}(\alpha \geq \beta) \wedge C_{i,j}(\beta \geq \gamma) \rightarrow C_{i,j}(\alpha \geq \gamma) \qquad \models C_{i,j}(\alpha \geq \alpha)$$
$$\models T_{i,j}(\alpha \geq \beta) \wedge T_{i,j}(\beta \geq \gamma) \rightarrow T_{i,j}(\alpha \geq \gamma) \qquad \models T_{i,j}(\alpha \geq \alpha)$$

We now show that we avoid the counter-intuitive properties of the normal preference logics [12,15], like disjunction expansion: if getting an apple is better than getting an orange, then getting an apple or losing million dollars is better than getting an orange. The construction with best and worst choices guarantees to the following requirements:

$$\not\models C_{i,j}((\alpha \cup \beta) \geq \gamma) \rightarrow C_{i,j}(\alpha \geq \gamma)$$
$$\not\models C_{i,j}(\alpha \geq \gamma) \rightarrow C_{i,j}((\alpha \cup \beta) \geq \gamma)$$
$$\not\models C_{i,j}((\alpha \cap \beta) \geq \gamma) \rightarrow C_{i,j}(\alpha \geq \gamma)$$
$$\not\models C_{i,j}(\alpha \geq \gamma) \rightarrow C_{i,j}((\alpha \cap \beta) \geq \gamma)$$

Within the set of actions an agent can be obliged to perform, it may distinguish certain levels concerning the relative commitment to perform actions. The following properties show how the logic behaves with respect to these situations:

$$\models O_{i,j}(\alpha \cup \beta) \wedge C_{i,j}(\alpha \geq \beta) \rightarrow O_{i,j}(\alpha) \quad \models F_{i,j}(\alpha \cup \beta) \wedge C_{i,j}(\alpha \geq \beta) \rightarrow F_{i,j}(\alpha)$$
$$\not\models O_{i,j}(\alpha \cup \beta) \wedge C_{i,j}(\alpha \geq \beta) \rightarrow O_{i,j}(\beta) \quad \not\models F_{i,j}(\alpha \cup \beta) \wedge C_{i,j}(\alpha \geq \beta) \rightarrow F_{i,j}(\beta)$$

The following properties hold for sequence of actions:

$$\models O_{i,j}(\alpha; \beta) \rightarrow \langle\alpha\rangle O_{i,j}(\beta) \quad \models F_{i,j}(\alpha; \beta) \rightarrow \langle\alpha\rangle F_{i,j}(\beta)$$

6 Concluding Remarks

An autonomous agent can decide to violate its commitments and obligations. A rational autonomous agent has to balance short term effects like paying penalties with long term effects like loosing trustworthiness and reputation. Resource bounded agents cannot quantify this balance and therefore base their decisions on qualitative decision models. This paper shows how trust and commitment can be related to each other in such qualitative models. We illustrate how the logic can formalize various aspects of the pennies pinching example. We think that reasoning about trust and commitment is highly relevant for advanced agent applications for the following reason. Agents may imagine flexible and realistic negotiation protocols which, beside buying or selling, allow reservations with (or without) a deadline in such a way that retracting a reservation commitment implies less penalty than retracting a buy or a sell commitment. In general, more flexible protocols allow agents to achieve different levels of agreements at different stages of negotiation and therefore allow agents to make what is called levelled commitments [13].

An agent's reputation is based on the degree of trust other agents have in his behavior, in particular the degree in which he fulfills his commitments. Whether agents trust other agents, and whether they use this trust in making decisions, depends on the application at hand. The extension of trustworthiness to a full agent profile is left for further research. Another interesting issue for further research is the resolution of conflicts between desires and obligations. Often an agent prefers a state which is forbidden; how to act? In our system, this becomes a trade-off between loss in utility versus loss of trustworthiness, i.e. between short term and long term effects. To resolve this kind of conflicts additional machinery has to be introduced in the logic, such as qualitative preferences between these two items, or quantitative measures. One proposal can be found in [10].

References

1. R. Auman. Rationality and bounded rationality. *Games and Economic behavior*, 21:2–14, 1986.
2. C. Basu, H. Hirsh, and W. Cohen. Recommendation as classification: Using social and content-based information in recommendation. In *Proceedings of the AAAI-98*, pages 714–720, 1998.
3. J. Broersen. Relativized action negation for dynamic logics. In *Advances in Modal Logic*, 2002.
4. A. Chavez, P. Maes, and Kasbah. An agent market-place for buying and selling goods. In *Proceedings of the PAAM'96*, pages 75–90. The Paractical Application Company Ltd, 1996.
5. M. Dastani, Z. Huang, and L. van der Torre. Dynamic desires. In S. Parsons, P. Gmytrasiewicz, and M. Wooldridge, editors, *Game Theory and Decision Theory in Agent-Based Computing*. Kluwer, to appear.
6. M.J. Fischer and R.E. Ladner. Propositional dynamic logic of regular programs. *Journal of Computer and System Sciences*, 18(2):194–211, September 1979.
7. D. Harel, D. Kozen, and J. Tiuryn. *Dynamic Logic*. MIT Press, 2000.
8. M. Hollis. Penny pinching and backward induction. *Journal of Philosophy*, 88:473–488, 1991.
9. M. Hollis. *Trust within Reason*. Cambridge University Press, 1998.
10. N.R. Jennings and J.R. Campos. Towards a social level characterisation of socially responsible agents. In *IEEE proceedings on software engineering*, pages 144:11–25, 1997.
11. C. Jonker and J. Treur. Formal analysis of models for the dynamics of trust based on experiences. In *Proceedings of MAAMAW'99. LNAI 1647*, 1999.
12. N. Rescher. The logic of preference. In *Topics in Philosophical Logic*. D. Reidel Publishing Company, Dordrecht, Holland, 1967.
13. T. Sandholm and V. Lesser. Issues in automated negotiation and electronic commerce. In *Proceedings of the ICMAS'95*, 1995.
14. Y. Shoham. Agent oriented programming. *Artificial Intelligence*, 60:51–92, 1993.
15. G.H. von Wright. *The Logic of Preference*. Edinburgh University Press, 1963.

Modelling Heterogeneity in Multi Agent Systems

Stefania Bandini, Sara Manzoni, and Carla Simone

Dipartimento di Informatica, Sistemistica e Comunicazione
Università di Milano Bicocca
Via Bicocca degli Arcimboldi, 8
20126, Milano, Italy
{bandini,manzoni,simone}@disco.unimib.it

Abstract. The aim of this paper is to consider the issue of heterogeneity in modelling complex systems through a multi agent perspective. Complex systems are often characterized by a high degree of *heterogeneity of the involved components* and by their natural aggregation in communities representing different views of the same system. These communities are autonomous and heterogeneous, in turn, since different views of a same system usually imply different types of components and relations among them. Since the interaction among the identified system components is often influenced by the *spatial relationships* among them, the model has to consider *heterogenous components situated in dedicated heterogenous spaces*, i.e. showing different topological structures. The paper discusses heterogeneity by presenting the Multilayered Multi Agent Situated System model (MMASS). The approach will be illustrated with an example in the framework of Computer Supported Cooperative Work, specifically for the design of a technology promoting awareness in cooperation[1].

1 Introduction

This paper proposes a model whose main features match the requirements identified in the different application domains we have considered in our research activities. Complex systems are often characterized by a high degree of *heterogeneity of the involved components* and by their natural aggregation in communities representing different views of the same system [1]. Moreover the interaction among the identified system components is often influenced by the *spatial relationships* among them, e.g, because it can take place through a specific medium or because the mutual distance of the interacting components changes the contents, intensity and perception of the interaction [2]. Hence, the modelling approach has to incorporate the notion of *space* whose nature, in terms of topological

[1] This work has been partially funded by the Italian Ministry of University and Research within the project 'Cofinanziamento Programmi di Ricerca di Interesse Nazionale'.

M.H. Shafazand and A M. Tjoa (Eds.): EurAsia-ICT 2002, LNCS 2510, pp. 685–692, 2002.

structure and medium, provides components with interaction capabilities based on a flexible set of communication modalities. The topological structure representing space captures the physical and/or logical relations binding the involved components and allows their interactions. This consideration, in combination with the above described heterogeneity of components and communities, implies that a modelling approach should be able to represent *heterogenous components situated in dedicated heterogenous spaces*, i.e. showing different topological structures.

In order to illustrate the MMASS model its application in the Computer Supported Cooperative Work (CSCW) [3] domain will be shown. In CSCW framework, empirical studies [4] have highlighted how people use forms of coordination based on their perception (the so called *awareness*) of the presence and distance of the other people both in the physical space and in the 'logical space' identified by the considered application. Here distance plays a fundamental role in identifying a suitable technological solution supporting this recently uncovered form of cooperation, as distance is one of the main parameters that can be used to tune and filter the amount of awareness information to be sent or perceived by an entity occupying a site of the logical space. The CSCW literature proposes models of coordination based on a space (called Spatial Models of Awareness since [5]) that focus on solutions to specific problems and give a limited formalization of the underlying model. The approach proposed here allows system designers to construct the functionalities devoted to promote awareness in cooperative applications, thus overcoming the above limits. The work presented in this paper is part of a larger project aimed to develop a language and the related infrastructure for the development of Multi Agent Systems that are characterized and influenced by their spatial position. Specifically, we will focus on primitives manipulating the space topology to extend our application framework to the domains characterized by mobility and dynamic topologies on the World Wide Web. This paper outlines the MMASS model and its application to the CSCW domain. The preliminary language specification and a discussion on the mechanisms needed to implement the model can be found respectively in [6] and [7].

2 The Multilayered Multi-agent Situated System Model

The *Multilayered Multi Agent Situated System (MMASS)* model describes Multi–Agent Systems (MAS) [8] situated in an environment whose topological structure (i.e. *space*) is a multilayered network of sites. Each layer of the MMASS space reproduces a physical or logical space in which a system of agents is situated. The spatial structure of the environment in which the system of agents is situated allows the representation of the relationships existing among agents. The heterogeneity of this relationships can be represented by the representation of the entities of the modelled domain as situated in more than one layer.

Agents situated in this environment are strongly influenced by their position, that is, the site of the space in which they are situated. In particular, the position in the environment defines a potentially complex combination of internal and external events and states that the agent has to take into account for its actions. A system of situated agents is made of reactive agents, which performs their actions as a consequence of the perception of influences coming either from other agents or from the environment. Being situated, agents are sensitive to the spatial relationships that determine their constraints, abilities and cooperation relationships. The MMASS model defines a set of influences (named *fields*) generated by agents and propagating along the edges of the multilayered structure representing agent environment and leaving on sites information about their presence. Field emission–propagation–perception models asynchronous and at–a–distance interaction among agents. Moreover, in order to model local interaction within the MAS, the MMASS approach defines *reaction*, according to which, a set of agents situated in adjacent sites synchronously influence each other and, as a consequence, change their state.

2.1 Heterogeneous Communities

A MMASS can be defined as a constellation of interacting *Multi Agent Situated Systems* (MASS) through the primitive: **Construct**($MASS_1 \ldots MASS_n$). Each MASS is defined by the triple $< Space, F, A >$ where *Space* models the structure of the environment in which a set A of agents are situated, act autonomously and interact via the propagation of a set F of fields. In order to allow MASS interaction the MMASS model introduces the notion of *interface*. The interface of a MASS specifies fields imported into and exported from each MASS. Field emission–propagation–perception is the mechanism defined for asynchronous interaction among agents situated in the same or different MASS's: an agent emits a field that propagates throughout the *Space* and can be perceived by other agents. The environment where the agents of a MASS are situated is named *Space* and it is defined as made up of a set P of sites arranged in a network. In this way the Space can be considered as a undirected graph of sites. Each *site* $p \in P$ can contain at most one agent and it is characterized by the agent situated in it ($a_p \in A$), the set of fields active in it ($F_p \subset F$) and the set of its adjacent sites ($P_p \subset P$), that is $p =< a_p, F_p, P_p >$.

2.2 Heterogeneous Interactions

The MMASS model allows the representation of two types of interactions among agents. Asynchronous interaction among agents may involve agents situated in spatially distance sites and takes place through a field emission–propagation–perception mechanism. When an agent state is such that it can be source for a field, it executes an *emit action* generating and defining parameters for a field (see section 2.3 for more details about agent actions). Field propagation occurs according to a diffusion function that specifies how its values propagate throughout the space according to its spatial structure. A field $f \in F$ is characterized by the set of values (W_f) that the field can assume during its propagation in the Space, and a *diffusion function* ($Diffusion_f : P \times W_f \times P \rightarrow (W_f)^+$)

computing the value of a field on a given site taking into account in which site and with which value it has been generated. Since the structure of a Space is generally not regular and paths of different length can connect each pair of sites, the field diffusion function returns a number of values depending on the number of paths connecting the source site with each other site. A *composition function* ($Compose_f : (W_f)^+ \rightarrow W_f$) associated to each field expresses how field values have to be combined in order to obtain, for instance, the unique value of field f at a site. Finally, a *comparison function* ($Compare_f : W_f \times W_f \rightarrow \{True, False\}$) is the function that compares field values. For instance, in order to verify whether an agent can perceive a field value, the value of a field at a site and agent sensitivity threshold are compared (see the definition of agent perception in the section 2.3). The perception function that characterizes each agent type defines the second side of an asynchronous interaction among agents: that is, the possible reception of broadcast messages conveyed through a field, if the sensitivity of the agent to the field is such that it can perceive it. This means that a field can be neglected by an agent of a given type if its value at the site where the agent is situated is less than the sensitivity threshold computed by the second component of the perception function.

Synchronous interaction is defined in terms of *reaction* among a set of agents characterized by given states and types and pair–wise situated in adjacent sites (i.e. *adjacent agents*). Synchronous interaction is a two–step process. Reaction among a set of adjacent agents takes place through the execution of a protocol introduced in order to synchronize the set of autonomous agents. When an agent wants to react with the set of its adjacent agents since their types satisfy some required condition, it starts an *agreement* process whose output is the subset of its adjacent agents that have agreed to react. An agent agreement occurs when the agent is not involved in other actions or reactions and when its state is such that this specific reaction could take place. The agreement process is followed by the synchronous reaction of the set of agents that have agreed. See Section 2.3 for more details about reaction and other actions that agents can undertake.

2.3 Heterogeneous Agents

The Space of each MASS is populated by a set of individuals called *agents*. Each agent $a \in A$ is characterized by a *state* ($s \in \Sigma_\mathcal{T}$), the *site* of the Space where the agent is situated ($p \in P$) and a *type* τ, that is $a =< s, p, \tau >$.

An agent type τ specifies the *set of states* the agent can assume ($\Sigma\mathcal{T}$), a *perception function* to express agent sensitivity to fields ($Perception_\mathcal{T} : \Sigma_\mathcal{T} \rightarrow [\mathbf{N} \times W_{f_1}] \ldots [\mathbf{N} \times W_{f_{|F|}}]$), and the *set of actions* that the agent can perform ($Actions_\mathcal{T}$). Thus, an agent type τ is defined by $< \Sigma_\mathcal{T}, Perception_\mathcal{T}, Actions_\mathcal{T} >$. $Perception_\mathcal{T}$ is a function associating to each agent state a vector of pairs in which, for the i–th pair, the first element expresses a coefficient to be applied to the field value (f_i), and the second one expresses the agent sensibility threshold to f_i in the given state (let $c^i_\mathcal{T}(s)$ and $t^i_\mathcal{T}(s)$ be their names). This means that an agent of type τ in state $s \in \Sigma_\mathcal{T}$ can perceive a field f_i only when it is verified $Compare_{f_i}(c^i_\mathcal{T}(s) \cdot w_{f_i}, t^i_\mathcal{T}(s))$, that is, when the first component of the i–th pair of the perception function ($c^i_\tau(s)$) multiplied for the received field value w_{f_i} is greater than the second component of the pair ($t^i_\tau(s)$).

The set of actions that agents of a given type can perform specifies whether and how agents of that type change their state (*trigger* action) or position (*transport* action), and how they interact with other agents. As introduced in Section 2.2 interactions can occur both synchronously and among adjacent agent (*reaction* action) and asynchronously and at–a–distance (*emit* action). In order to define agent action set ($Actions_\tau$) let us consider an agent $a =< s, p, \tau >$ that is, an agent of type τ whose current position is site p and whose current state is s. Moreover, $p =< a, F_p, P_p >$ where F_p is the set of fields active in p and P_p is the set of sites adjacent to p. To define the four actions above outlined, we will use operators of the form *action–condit–effect*. *action* is an expression of the form $f(x_1 \ldots x_n)$ where f specifies the action name and x_i are variables which can appear in *condit* and *effect* expressions. *condit* and *effect* express respectively the set of conditions that must be verified in order to let the agent execute the action, and the set of effects deriving from its execution. They are sets of atomic formula $p(a_1 \ldots a_k)$ where p is a predicate of arity k and a_i are either constants or variables. According to this syntax, the four basic agent actions are:
$action : trigger(s, f_i, s')$
$condit : state(s), perceive(f_i)$
$effect : state(s')$
where $state(s)$ and $perceive(f_i)$ are verified when the agent state is s, and $f_i \in F_p$ and $Compare_{f_i}(c^i_\tau(s) \cdot w_{f_i}, t^i_\tau(s)) = True$ (that is, the field f_i is active in p and agents of type τ in state s can perceive it). The effect of a trigger action is a change in state of the agent according to the third parameter.
$action : transport(p, f_i, q)$
$condit : position(p), empty(q), near(p, q), perceive(f_i)$
$effect : position(q), empty(p)$
where $perceive(f_i)$ has the same meaning as in $trigger$, while $position(p)$, $empty(q)$ and $near(p, q)$ are verified when the agent position is p, $q \in P_p$ and $q =< \bot, F_q, P_q >$ (q is a site adjacent to p and no agent is situated in it). The effect of the execution of a transport action is the change in position of the agent undertaking the action and, as a consequence, the change of the local space where the agent is situated.
$action : emit(s, f, p)$
$condit : state(s), position(p)$
$effect : added(f, p)$
where $state(s)$ and $position(p)$ are verified when the agent state and position are s and p. The effect of a emit action is a change at each site of the space according to $Diffusion_f$. In particular the set of fields in p where the emitting agent is situated changes to $p =< a, F_p \diamond f, P_p >$, where the operator $F \diamond f$ simply adds a field f to a field set F if the field does not already belong to the set. Otherwise it composes its already present field value(s) and the new one.
$action : reaction(s, a_{p_1}, a_{p_2}, \ldots, a_{p_n}, s')$
$condit : state(s), agreed(a_{p_1}, a_{p_2}, \ldots, a_{p_n})$
$effect : state(s')$

where $state(s)$ and $agreed(a_{p_1}, a_{p_2}, \ldots, a_{p_n})$ are verified when the agent state is s and agents situated in sites $\{p_1, p_2, \ldots, p_n\} \subset P_p$ have previously agreed to undertake a synchronous reaction. The effect of a reaction is the synchronous change in state of the involved agents; in particular, agent a changes its state s'.

3 MMASS Application

This section illustrates how the features of the MMASS model can be used to construct mechanisms supporting the promotion of awareness within and across cooperative applications [9]. In order to endow existing applications with awareness mechanisms, we designed a dedicated module, called AW–Manager, that can be associated to each application to provide human actors with the desired awareness functionality. Here heterogeneity plays a basic role, as illustrated by the situation depicted in Figure 1. In a given work setting, several actors cooperate by using (1) a combination of specialized applications. The behavior of human actors interacting with each application generates a flow of facts (2) that are passed to the pertinent instance of the AW–Manager (3). The latter elaborates them in order to produce awareness information (4) that is presented to the human actors at their user interface. Awareness information is the outcome of elaboration in each awareness space as well as of the information crossing the various spaces (5).

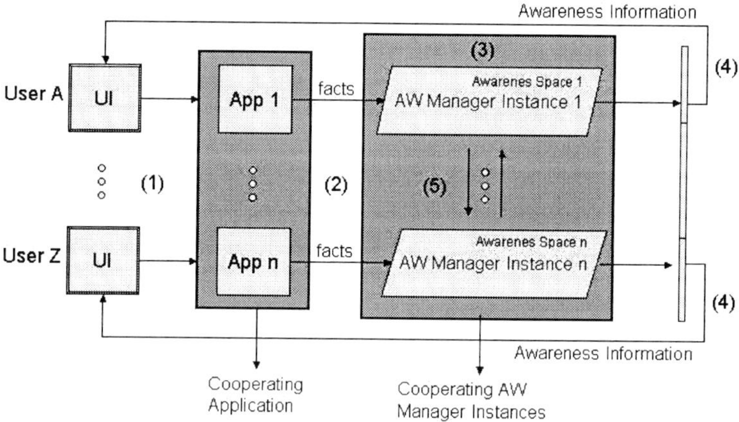

Fig. 1. A multi–layered structure of awareness spaces.

Each instance of the AW–Manager incorporates the main features of the MMASS model: a dedicated space and a set of heterogeneous agents that stand proxy for the relevant entities of the application. For example, if the application is a workflow management system typical agents are Tasks to be accomplished, the Resources they use, the Actors performing the related Activities, the Role actors play, and so on. The topology of the space (called here, awareness space) can be defined, for example, by the relations expressing causal dependencies between tasks, assignment of responsibility for tasks to roles and allocation of actors to tasks. Each agent is endowed with a behavior, expressed in terms of MMASS actions, that is peculiar of the type of entity it stand proxy for, and that specifies how the agent contributes to the promotion of awareness in cooperation with the other agents. The MMASS model incorporated in the AW–Manager specifies a MAS that is open to the influence of the environment through the incoming facts: the latter are interpreted by the model as standard imported fields the involved entity are always sensitive to. In this way, human actors behavior is immediately considered in the computation of the awareness information reaching the user interfaces.

In the MMASS model fields are characterized by a set of values and a diffusion function that take into account the requirements of the specific domain. In the case of awareness promotion, fields are pairs <*strength, content* >. The first component allows for the "modulation" of the awareness information specified by the second component. Modulation is governed by the diffusion function that specifies how the field value is modified during its propagation in the space. Hence, the same message issued by a sender agent reaches different receiving agents with values that depend on their (logical) distance from the source. If the receiving agents have associated a user interface (typically, if they stand proxy for human actors), then the representation of the awareness information takes into account both components of the field value. Specifically, the strength can determine the degree of obtrusiveness by which the message content is presented at the user interface (e.g., level of sound, dimension of graphical cues, or their combination). Going on with the above example, a typical fact is generated when an actor activates a task. In the awareness space, it can be interpreted by a *transport* action performed by the Actor agent bringing it close to the Task agent and by a *reaction* between them changing their state synchronously: the Actor becomes *busy* and the Task becomes *active*. If for some reasons, the task gets in troubles, the related fact can be interpreted in the following way: the Task agent emits a field that notifies the problem to other agents (other Tasks, Roles, Actors) with different strength. The concurrently active Tasks agents receive stronger messages than Tasks agents related to Tasks already concluded or close in the future. Each notified agent propagates awareness information to other agents, depending on its behavior, the structure of the awareness space and the state/location of the other agents leaving in it. On the other hand, each receiving agent can be differently sensitive to this field (*perception function*): future Tasks agents are likely to be more sensitive than Tasks agents corresponding to concluded tasks. More details of how the above behaviors can be formally defined in terms of MMASS primitives can be found in [2].

Field diffusion function specifies also some temporal aspects characterizing field values at each site. Typically, a field can be persistent, transient, or decrease its strength in the long run. This property affects the overall behavior of the various agents since their actions are influenced by the fields they perceive, and again influences the way in which the associated awareness information is possibly represented at the user interfaces. For example, the above field is likely to be persistent, that is, perceivable until the problem generating it is solved. Fields persistency allows agents to promote awareness among themselves in an asynchronous way. The AW–Manager interface allows the designer of the awareness functionality to specify (and modify at run time) all the component of the incorporated MMASS instantiation, and among the others fields properties. This is possible thanks to the fact that the MMASS provides primitives that the AW–Manager makes visible and accessible to the users (application designers and/or end–users). This goes far beyond the possibility provided by other approaches to awareness promotion, that limit adaptability to the tuning of predefined parameters. Another original feature of our approach is based on the multi–layered structure of MMASS and the possibility to import/export fields across spaces. In fact, cooperative applications incorporating awareness functionalities limit their action to a single target application. In our case, cooperative applications can exchange awareness information and contribute to the construction of a unified awareness functionality at the user interface. For example, the above described field can be exported to another awareness space, associated to another application supporting, e.g., a meeting (a video–conference), where the actor responsible for the workflow process is involved. In this way, the related Actor agent receives the notification and reacts according to its specified behavior.

References

1. Bar-Yam,Y.: Dynamics o Complex Systems. Addison Wesley, Reading, MA (1997)
2. Bandini, S., Manzoni, S., Simone, C.: Dealing with space in multi-agent system: a model for situated MAS. In: Proc. of the 1st International Joint Conference on Autonomous Agents and MultiAgent Systems (AAMAS 2002), Bologna, Italy. (2002)
3. Beaudouin-Lafon, M.: Computer Supported Co-operative Work. John Wiley & Sons, New York (1999)
4. Heath, C., Luff, P.A.S.: Reconsidering the virtual workplace: Flexible support for collaborative activity.In:Proc.of the 4th European Conference on CSCW. (1995) 83–99
5. Greenhalg, C., Benford, S.: Massive:a collaborative VE for tele-conferencing. ACM TOCHI (1995) 239–261
6. Bandini, S., Manzoni, S., Pavesi, G., Simone, C.: L*MASS: A language for situated multi-agent systems. In Esposito, F., ed.: AI*IA 2001: Advances in Artificial Intelligence. Volume 2175 of LNCS, Berlin, Springer (2001) 249–254
7. Bandini, S., DePaoli, F., Manzoni, S., Simone, C.: Mechanisms to support situated agent systems. Proc. of the 7th IEEE Symposium on Computers and Communications (ISCC2002), 1–4 July 2002, Taormina, Italy (2002)
8. Ferber, J.: Multi-Agents Systems. Addison-Wesley, Harlow (UK) (1999)
9. Simone, C., Bandini, S.: Integrating awareness in cooperative applications through the reaction-diffusion metaphor. CSCW, The international Journal of Collaborative Computing 11 (2002), to appear.

Pricing Agents for a Group Buying System

Yong Kyu Lee, Shin Woo Kim, Min Jung Ko, and Sung Eun Park

Dept. of Computer Engineering, Dongguk University
Pil-dong, Jung-gu, Seoul 100-715, Rep. of Korea
yklee@dgu.edu

Abstract. Internet group buying systems have been widely used recently. In those systems, because the reserve price is provided by the buyer, the success rate can be decreased if the reserve price is set too low compared with the normal price. Otherwise, an unsuitable successful bid can be made if the reserve price is set too high based on inaccurate information. Likewise, the seller's providing too high a bid price can deteriorate his/her own successful bid rate, whereas a successful bid with too low a price may make no profit in the sale. Therefore, pricing agents that recommend adequate prices based on the past buying and selling history data can be helpful. In this paper, we propose two kinds of agents. One suggests reserve prices to buyers based on the past buying history database of the system. The other recommends bid prices to a seller based on the past bidding history data of the company using the cost accounting theory. Through performance experiments, we show that the successful bid rate can increase by preventing buyers from making unreasonable reserve prices. Also, we show that, for the seller, the rate of successful bids with appropriate profits can increase. Using the pricing agents, we design and implement an XML-based group buying system. Because it is based on XML standards, it has advantages such as interoperability and extendibility compared with previous proprietary electronic commerce systems.

1 Introduction

Group buying or joint buying, a kind of reverse auction, is used to lower the unit price of an item by bulk purchasing with a group of buyers [1][4][7][8][16]. In commercial group buying systems, buyers set reserve prices when purchasing items are registered for bidding. The reserve price is hidden from bidders and tells the real amount one is willing to accept for an item [11]. The group buying can be unsuccessful if the reserve price is unreasonably low compared with the normal price. Also a successful bid with an unreasonable price can be made when a buyer carelessly provides a quite higher reserve price. These can occur when buyers do not have accurate information about normal prices or when they make mistakes during registration. Therefore, an unsuitable reserve price is one of the main reasons that purchasing becomes unsuccessful.

Similarly, if a seller provides too high a bid price compared with the normal price, it can decrease his/her own successful bid rate. Also, a successful bid with too low a

M.H. Shafazand and A M. Tjoa (Eds.): EurAsia-ICT 2002, LNCS 2510, pp. 693–700, 2002.

price may make no profit or a loss in the sale. Therefore, pricing agents that recommend adequate reserve prices for buyers and bid prices for sellers based on the past buying and selling history data can be helpful [2][10].

In electronic commerce systems, recommendation systems have been used to recommend new items to expected buyers according to their interests [9][12][13][14]. The system handles only items that are likely to be purchased by people and does not deal with pricing. Even though there can be found a program for group buying that automatically generates the estimated price with a given quantity of items, it simply multiplies the unit price by the quantity, not considering that the unit price may drop through bulk purchasing [15]. Therefore, those approaches cannot be directly applied to automatic pricing for group buying systems.

In this paper, in order to solve the problems, we present a pricing agent that automatically recommends appropriate reserve prices of purchasing items to buyers based on the past group buying history records of the system. We use the vector space model [3] in information retrieval to retrieve most similar cases to the present purchasing item from the database. By adjusting the weighted average of the reserve prices of the retrieved similar cases, an appropriate reserve price is generated for the item. Through experiments, we show that it can decrease the unsuccessful purchasing rate by preventing buyers from making unreasonably low or high reserve prices compared to the normal prices.

Also, we propose another pricing agent that automatically suggests appropriate bid prices of group buying items to a seller using the past bidding history data of his/her company and the cost accounting theory [6]. The agent calculates the unit price based on the cost equation using the company cost DB and generates a candidate for the bid price. Also, it generates another candidate using past similar cases retrieved from the bidding history DB. Adjusting the weighted average of two candidates, it finally suggests the bid price. We analyze performance through experiments and show that the rate of successful bids with appropriate profits can increase by using it.

Using the pricing agents, we implement a prototype of an XML-based group buying system. The system is based on XML standards [5] such as XML Schema, DOM, XSL, SOAP, etc. Therefore, it has advantages such as interoperability and extendibility compared with previous proprietary electronic commerce systems.

2 Reserve Pricing Agent

Figure 1 explains the procedure that the reserve price for a group buying item is generated. The history database stores the records of the past group purchases. Similar cases to the current purchasing item are retrieved from the database based on the vector space model. The reserve price is obtained by adjusting the weighted average of the retrieved reserve prices.

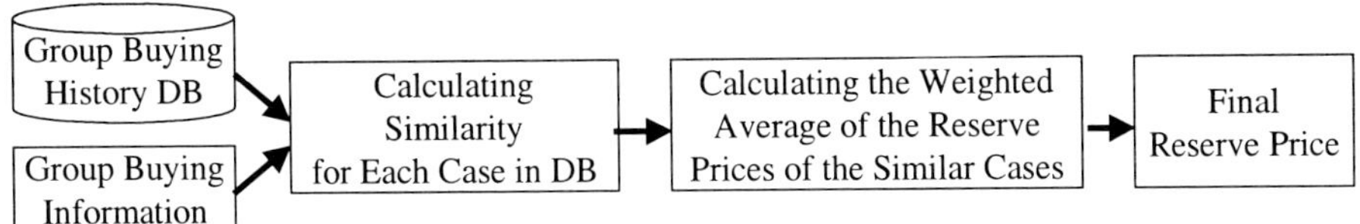

Fig. 1. Reserve Pricing Procedure

2.1 Similarity Measure

The vector space model [3] is used to retrieve ranked similar cases from the history database. The record consists of the attributes describing a case. It also contains information such as bidding date, quantity, reserve price, bid prices, winning bid price, etc. The similarity measure is as follows:

$$similarity(r_j, q_k) = \frac{\sum_{i=1}^{n}(tr_{ij} \times tq_{ik})}{\sqrt{\sum_{i=1}^{n} tr_{ij}^2 \times \sum_{i=1}^{n} tq_{ik}^2}} \quad (1)$$

where tr_{ij} = i-th term in the vector for record j, tq_{ik} = i-th term in the vector for query k, and n = number of fields in the record.

2.2 Generating the Reserve Price

From the search results, we calculate the reserve price using the following formula.

$$P_{reserve} = \left(\frac{1}{n}\sum_{i=1}^{n} w_i \cdot r_i\right) \times q \quad (2)$$

where n = number of similar cases, w_i = weight of i-th case, r_i = unit reserve price of i-th case, and q = quantity.

3 Bid Pricing Agent

Figure 2 illustrates the procedure that an appropriate bid price for a group buying item is generated. In order to retrieve similar cases from the bidding history database, we use the same similarity measure that we have used to get the reserve price. The cost equation for the item is obtained using the company cost DB. One candidate is generated from the most similar cases and the other is obtained using the cost equation. Finally, the bid price is generated by adjusting two candidates.

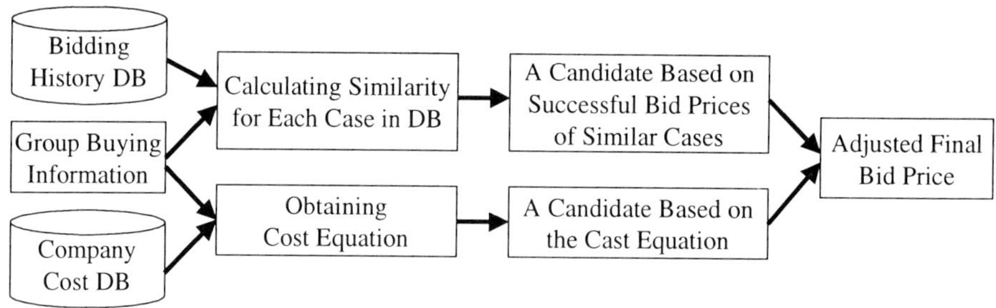

Fig. 2. Bid Pricing Procedure

3.1 Cost Equation

We use the cost equation in cost accounting [6]. From Eq. (3), the total cost according to the quantity can be obtained and it becomes a candidate for the bid price.

$$P_{\cos t-equation} = (f + v) \times q \tag{3}$$

where f = fixed unit cost, v = variable unit cost, and q = quantity.

To get the fixed unit cost f and the variable unit cost v, we have modified the high-low method [6] as shown in Eq. (4) and (5). That is, we use the average high(low) point value instead of using the single high(low) point value to loosen the possibility of bias. The cost and the quantity values are obtained from the company cost DB.

$$v = (p_{high} - p_{low}) / (q_{high} - q_{low}) \tag{4}$$

$$f = (p_{high} / q_{high}) - v = (p_{low} / q_{low}) - v \tag{5}$$

where p_{high} = average total cost of mass selling, p_{low} = average total cost of small-scale selling, q_{high} = average quantity of mass selling, and q_{low} = average quantity of small-scale selling.

3.2 Generating the Bid Price

Another candidate for the bid price is obtained by Eq. (6) using the history database.

$$P_{bid-history} = \left(\frac{1}{n} \sum_{i=1}^{n} w_i \cdot b_i \right) \times q \tag{6}$$

where n = number of similar cases, w_i = weight of i-th case, b_i = unit bid price of i-th case, and q = quantity.

The final bid price is obtained using Eq. (7).

$$P_{bid-price} = w_1 \times P_{cost-equation} + w_2 \times P_{bid-history} \qquad (7)$$

where w_1 and w_2 are weights ($w_1 + w_2 = 1$).

4 Experimental Evaluation

Performance experiments have been performed using a purchasing history database of a commercial group buying system and a bidding history database of a participating computer vendor. The purchasing history database contains 200 cases (154 successful purchases and 46 unsuccessful purchases) with quantities between 4 and 20, whereas the bidding history database has 188 cases (47 successful bids and 141 unsuccessful bids). All the computers have the same options. The company cost database contains the cost information required to get the cost equation.

For the comparison, we use the normalized absolute error as follows:

$$E = |b_i - r_i| / b_i \qquad (8)$$

where b_i = winning bid price for successful purchasing or lowest bid price for unsuccessful purchasing (or bid price for bidding), and r_i = reserve price.

The E values for the test database are shown in Figure 4 (sorted ascending by the E values of the no agent approach). The agent uses the most similar three successful purchases and weights are given at the rate of 3:2:1. It shows that the agent approach has smaller E values.

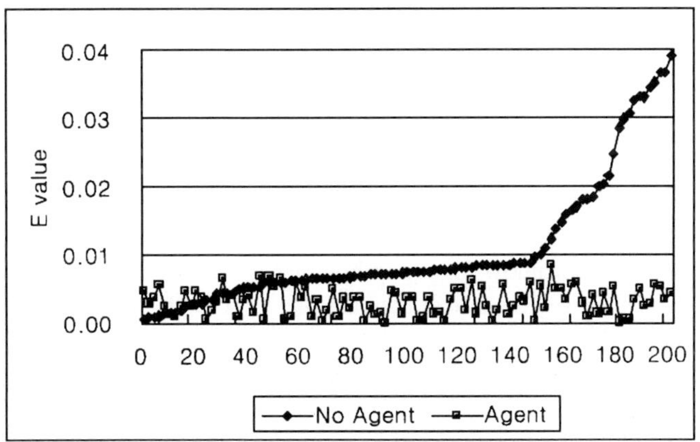

Fig. 3. Comparison of Reserve Prices

The unit prices according to the quantity are represented in Figure 4. The left graph shows the bid prices generated by the agent when $w_l = 0.5$ and the right shows those of the past bidding records. The graphs also show the average reserve prices for comparison. The bid prices of the agent converge near the reserve prices, whereas those of the past bidding records do not.

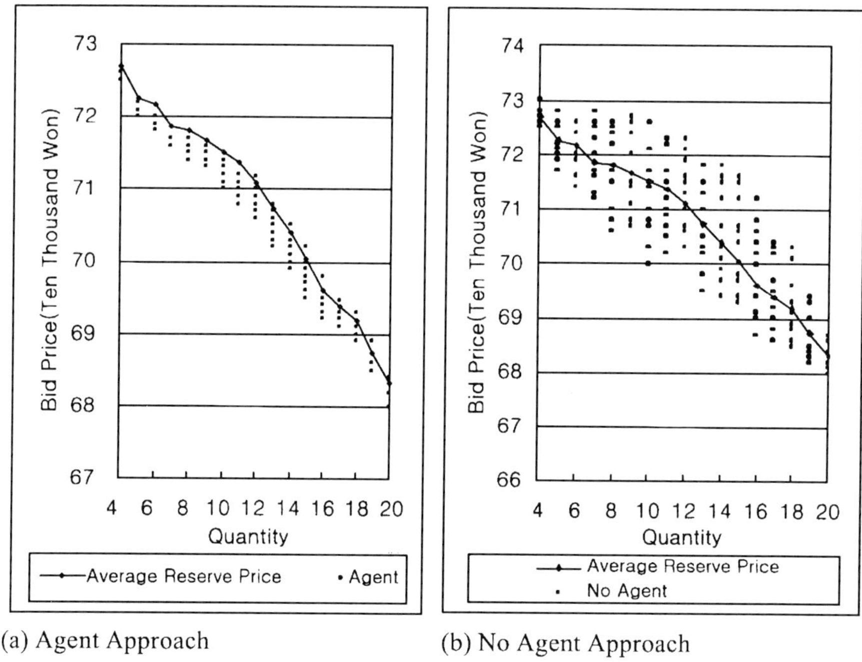

(a) Agent Approach (b) No Agent Approach

Fig. 4. Comparison of Bid Prices

5 Implementation of an XML-Based Group Buying System

A prototype group buying system has been developed on the web using JavaScript, ASP, and MS-SQL. We have used XML standards [5] including XML schema, DOM, XSL, XPath, and XQL.

Figure 5 shows the window for group buying registration. Here, sellers can give the reserve price for themselves, or can activate the reserve pricing agent to make it automatically generated.

The bidding window shows the information on the item to be purchased as illustrated in Figure 6. One can participate in the group buying by typing a bid price and pushing the 'Bidding' button. Here, the bid price can be given by the bidder or can be generated by the bid pricing agent.

Purchaser Name	Shin Woo Kim			E-MAIL	purian@dgu.edu	
Member ID	purian			Phone	011-123-4567	
Product	Computer			Quantity	10	
Bid Openning Day	Year 2002 Month 4 Day 20 Hour 12 Minite 00					
Bid Closing Day	Year 2002 Month 4 Day 22 Hour 12 Minite 00					
CPU Type	Pentium4-1.7GHz	Memory Type	256MB SDRAM	Disk Type	40GB UDMA	More Options
Recommended Reserve Price	Reserve Pricing Agent		790,000			

Registration | Reset | Cancel

Fig. 5. Group Buying Registration Window

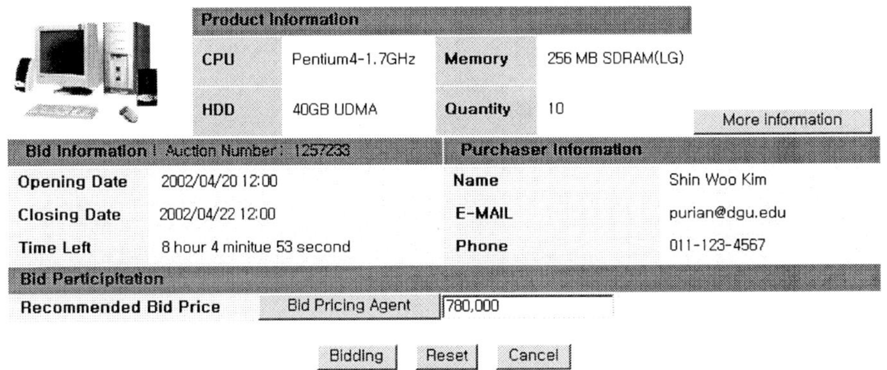

Product Information

CPU	Pentium4-1.7GHz	Memory	256 MB SDRAM(LG)
HDD	40GB UDMA	Quantity	10

More Information

Bid Information | Auction Number : 1257233		Purchaser Information	
Opening Date	2002/04/20 12:00	Name	Shin Woo Kim
Closing Date	2002/04/22 12:00	E-MAIL	purian@dgu.edu
Time Left	8 hour 4 minitue 53 second	Phone	011-123-4567

Bid Participation

Recommended Bid Price | Bid Pricing Agent | 780,000

Bidding | Reset | Cancel

Fig. 6. Bidding Window

6 Conclusions

In this paper, we have proposed a reserve pricing agent that recommends reserve prices to sellers for a group buying system based on the past purchasing records. Also, we have presented a bid pricing agent that suggests bid prices to sellers for the group buying based on the past selling records and the cost accounting theory.

According to the results of performance experiments, the successful bid rate can be increased by preventing sellers from making unreasonable reserve prices. Through experiments using 200 past records obtained from a computer group buying system, we have known that the average E value of the agent for unsuccessful cases decreases into 12.5% of that of the previous approach, and all the reserve prices generated by the agent are within the boundaries of the successful cases. It means that the unsuccessful bid rate can decrease by preventing group buying from becoming unsuccessful due to the too high a reserve price compared to the normal price. Also, it can stop a successful group buying with too low a price. Also we have shown that the bid price generated by the bid pricing agent is superior to the previous approach. The average E value of the agent for unsuccessful cases decreases into 25% of that of the previous approach. Therefore, it can prevent from abnormal bid prices using the cost accounting theory and the bidding history database.

Using the agents, we have developed a prototype group buying system. Because it is based on XML standards, it offers advantages in extendibility and interoperability compared with other proprietary group buying systems. In the future, experiments with a large database of various purchasing items are required.

References

1. Bapna R., Goes P., Gupta A.: Online Auctions: Insights and Analysis. Communications of the ACM, Vol. 44 No. 11 (2001) 42–50
2. Dasgupta, P., Das, R.: Dynamic Service Pricing for Brokers in a Multi-Agent Economy. 4th Int'l Conf. on MultiAgent Systems, Boston USA (2000) 375–376
3. Frankes, W. B., Ricardo B. Y.: Information Retrieval: Data Structures & Algorithms. Prentice Hall (1995)
4. Goes, P., et. al.: Simulating Online Yankee Auctions to Optimize Sellers Revenue. 34th Hawaii Int'l Conf. on System Sciences, Hawaii USA (2001) 2453–2462
5. Harold, E. R., Means W. S.: XML In a Nutshell. O'Reilly (2001)
6. Horngren, C. T., et. al.: Cost Accounting. Prentice Hall (2000)
7. Kauffman, R. J., Bin W.: New Buyers' Arrival under Dynamic Pricing Market Microstructure: The Case of Group-Buying Discounts on the Internet. 34th Hawaii Int'l Conf. on System Sciences, Hawaii USA (2001) 2443–2452
8. Kauffman, R. J., Riggins, F. J.: Information Systems and Economics. Communications of the ACM, Vol. 41 No. 8 (1998) 32–34
9. Kumar, R., et. al.: Recommendation Systems: A Probabilistic Analysis. 39th Int'l Conf. on IEEE Symposium on Foundations of Computer Science, Palo Alto USA (1998) 664–673
10. Maes, P. M.: Agent that Buy and Sell. Communications of the ACM, Vol. 42 No. 3 (1999) 81–91
11. McAfee, R. P., McMillan, J.: Auctions and Bidding. Journal of Economic Literature, Vol. 25 No. 2 (1987) 699–738
12. Sarwar, B., et. al.: Analysis of Recommendation Algorithms for e-Commerce. 2nd ACM Int'l Conf. on Electronic Commerce, Minneapolis USA (2000) 158–167
13. Schafer, J. B., Konstan, J., Riedl, J.: Recommender System in E-Commerce. 1st ACM Int'l Conf. on Electronic Commerce, Denver USA (1999) 158–166
14. Yukawa, T., et. al.: An Expert Recommendation System Using Concept-Based Relevance Discernment. 13th Int'l Conf. on Tools with Artificial Intelligence, Dallas USA (2001) 257–264
15. Poweruser Computer Shopping Mall, http://www.poweruser.co.kr (2002)
16. Korea Auction Site, http://www.auction.co.kr (2002)

Evolution of Cooperation in Multiagent Systems

Brian Mayoh

Department of Computer Science, University of Aarhus, Ny Munkegade, bldg. 540, 8000 Aarhus C, Denmark
brian@daimi.au.dk

Abstract. Many multiagent systems are simulations of aspects of real world societies. In any such simulation of a real world society one must either evolve or design an appropriate balance between cooperation and competition among the individual agents. The question of when a laborious design approach can be replaced by a selforganising evolutionary approach is illuminated by the investigation of a simple game theoretical model in this paper.

Many multiagent systems are simulations of aspects of real world societies. Some are simulations of natural or ecological societies, while others are simulations of economic or communal activity in human societies. One can find examples of both kinds of multiagent systems in our earlier paper (Mayoh & Junping 1999) on the requirements for an agent behaviour language. In any such simulation of a real world society one must either evolve or design an appropriate balance between cooperation and competition among the individual agents. Cooperation is widespread in nature; there are many theories as to why it is beneficial and some theories why it has evolved. For an example of evolution of cooperating robots see (Watson et al.2002). In a prizewinning paper (Jennings 2000) N. Jennings advocates designing an organisational level in multiagent systems. The question of when such a laborious design approach can be replaced by a selforganising evolutionary approach is illuminated by the investigation of a simple game theoretical model in this paper.

In any multiagent system an agent A may benefit from actions done by the other agents in the system and agent A may do actions that benefit the other agents. Often agent A's willingness to do actions for another agent depends only its previous interactions with that agent, so one can separate the multiagent cooperation policy problem to the two agent cooperation policy problem. Thus our model is the game of repeated interactions between two individuals, known as the reciprocal altruism game. Usually the payoff matrix in the repeated games does not change, but some have followed May's suggestion (May, 1987) and allowed the game players to vary their investment in the repeated games. This much widens the players' choice of strategies and in the tournaments described in (Roberts & Sherratt, 1999) the RTS, "Raise The Stakes", strategy does very well. However Killingback and Doebeli (Killingback &

M.H. Shafazand and A M. Tjoa (Eds.): EurAsia-ICT 2002, LNCS 2510, pp. 701–710, 2002.

Doebeli, 1999) claim that RTS inevitably evolves into "defection", so strategies like RTS are not the reason why reciprocal altruism has evolved. The tournaments described in this paper show that there are strategies that lead to reciprocal altruism even faster than RTS. This complements the results in (Killingback et al., 1999) that altruism arises in spatially distributed ecosystems even when individuals cannot recognise each other so repeated games are inappropriate. For an example of evolution of cooperating robots using RTS and other strategies see (Birk & Wiernik.2002).

The main innovation in this paper is that we allow the two game players to have very different cost-benefit ratios. All previous studies assume identical cost-benefit ratios, but this assumption is not realistic for multiagent systems.

1 Reciprocal Altruism Game

In the reciprocal altruism game there are R rounds between Leader and Follower. In round i each player benefits from an altruistic payment of the other player and the payoffs are

$$P_L(i) = k_L \times pay_F(i) - pay_L(i)$$

$$P_F(i) = k_F \times pay_L(i) - pay_F(i)$$

where k_F and k_L are the players' altruistic cost-benefit constants. The strategies of the players determine the amounts pay_L and pay_F; these amounts cannot be negative but they can be zero- corresponding to deception in the much studied Prisoners Dilemma game. Our game reduces to that in (Roberts & Sherratt, 1999) when the altruistic benefit is real and the same for all players: $1 < k_F = k_L$.The total payoffs for the players are:

$$Reward_L = k_L \times Cost_F - Cost_L$$

$$Reward_F = k_F \times Cost_L - Cost_F$$

where $Cost_L = \Sigma\, pay_L(i)$ and $Cost_F = \Sigma\, pay_F(i)$. Altruism is rewarding for both players if $k_F > Cost_F/Cost_L > 1/k_L$, so it will be interesting to see if altruistic strategies evolve when $k_F \times k_L > 1$ but k_F or k_L is under 1.

Example: Imagine two agents alternatively delegating tasks to one another. $Cost_L$ measures the resources used by the Leader agent in doing tasks for the Follower agent.

The reciprocal altruism game reduces to the much studied Iterative Prisoners' Dilemma when $k_F = k_L = k > 1$ and all $pay_L(i)$ and $pay_F(i)$ are either 0 or 1. The payoff to two cooperators is $R = k-1$, the payoff to two deceivers is $P = 0$, the traitor payoff is $T = k$ and the sucker payoff is $S = -1$. Clearly the crucial inequalities $T > R > P > S$ and

R >(S + T)/2 are satisfied. Our distinction between Leaders and Followers only matters if a follower strategy is allowed to use pay_L (i) in determining pay_F (i).

Estimating partner quality seems to be biologically realistic and it is incorporated in our new strategies for the reciprocal altruism game in the form:
λ = kOwn / kOpponent.

2 Strategies in the Reciprocal Altruism Game

The strategy of a game playing agent can be considered as its "personality" (Gmytrasiewicz& Lisetti.2002), so the strategy names indicate the personality and the *labels* indicate the "emotions" of the agent.The strategies we consider are

NA) Non-altruism:
pay = 0 always

GGG) Give-as-good-as-you-get:
if pay_L (1) then a else opponent's last payment

SC) Short-changer:
if pay_L (1) then 1 else opponent's last payment - 1

RTS) Raise-the stakes:
if pay_L (1) then a
else if opponent undercut then *sad:* opponent's last payment
else if opponent matched then own last payment + b
else *happy;* own last payment + 2 b

OSC) Occasional-short-changer:
chance determines either RTS or RTS -1

OC) Occasional-cheat:
chance determines either RTS or 0

AWD) Anything-will-do:
pay = a always

AON) All-or-nothing
if opponent undercut then 0 else a

RTG) Generous raise-the stakes:
if pay_L (1) then a
else if opponent undercut then *sad:* opponent's last payment
else if opponent matched then *dubious:* own last payment + b
else *happy;* opponent last payment + 2 b

OSG) Generous occasional-short-changer:
chance determines either RTG or RTG -1

OCG) Generous occasional-cheat:
chance determines either RTG or 0

RTL) Flexible Mean raise-the stakes: λ = kOwn / kOpponent
if pay_L (1) then a
else if own last payment > λ opponent last payment
then *sad:* opponent's last payment
else if own last payment = λ opponent last payment

then *dubious:* own last payment + b
else *happy;* own last payment + 2 b

OSL) Flexible raise-the stakes: λ = kOwn / kOpponent
if pay_L (1) then a
else if own last payment > λ opponent last payment
then *sad:* λ opponent's last payment
else if own last payment = λ opponent last payment
then *dubious:* own last payment + b
else *happy;* own last payment + 2 b

OCL) Flexible Generous raise-the stakes: λ = kOwn / kOpponent
if pay_L (1) then a
else if own last payment > λ opponent last payment
then *sad:* own last payment
else if own last payment = λ opponent last payment
then *dubious:* own last payment + b
else *happy;* own last payment + 2 b

The first eight of these strategies are taken from (Roberts & Sherratt, 1999). Some of them are familiar from Iterative Prisoners'Dilemma Game; NA is "always deceive", AWD is "always cooperate" and the others are like "Tit For Tat". In the robot experiment (Birk & Wiernik, 2002) we find some of our strategies but also "Justified snobbism " and several others.

In our coevolutionary implementation there was a population of 50 followers and a population of 50 leaders. In each generation every follower (leader) plays at least 5 randomly chosen leaders (followers); the average number of opponents is 10. We separated followers and leaders to see if this distinction has any influence on which strategy that evolves. The parameters of the leaders were: strategy type, a,b; the parameters of the followers were:strategy type,b and the number of iteration rounds. We fixed follower's a to 1 as we wanted to see how the number of iterations evolved.

3 Results

To run an experiment one must fix the cost-benefit ratios, k_F and k_L, choose a variety of strategies and run the simulation for a number of generations. During the simulation not only can the distribution of follower and leader strategies be tracked, but one can also plot the fitness and other parameters of individuals in various ways.
The results of our first experiments are given in table 1 and illustrated in fig.1.

The last row is not a separate run; it gives the averages over the last 10 generations so it checks that the stability of the final strategy distributions and fitnesses in rowL. The table gives no information on the evolution of strategy parameters because the only patterns observed were

– initial investment a increases but large variation between 2 and 8,
– increment b keeps wide variation between 0 and 4.

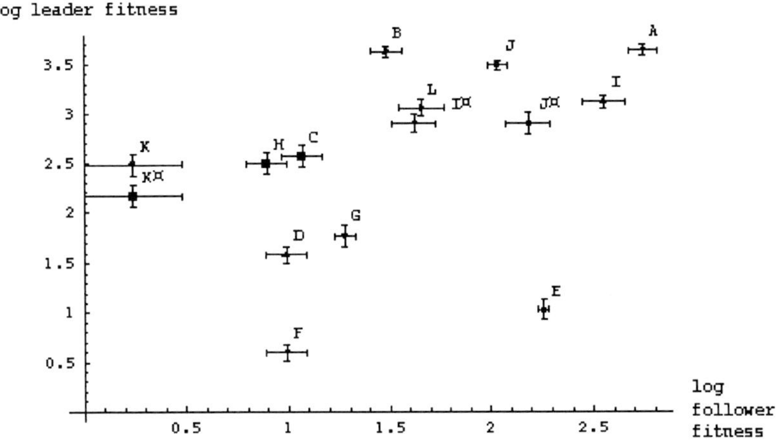

Fig. 1. Experimental Results-error bars shortened and K-fitness made positive

The last column of the table shows that the number of iterations does evolve, increasing appreciably from the initial 25. This is surprising as the number of iterations is determined by the follower, who is unlikely to benefit from more iterations when k_F is low.

It is satisfying that altruism usually evolves when it should, but it is surprising that the specially designed adjustable payment strategies RTL,OSL and OCL were never the most popular follower or leader strategy when more altruistic strategies were available. We shall see why in the next section.

In (Bazzan *et al.*, 1999) experiments with the Iterated Prisoner's Dilemma game were interpreted as showing that moral sentiments, being benevolent to build up trust, play a role in altruism. Some of our experiments can also be interpreted like this, the player who benefits most from mutual altruism tolerates shortchanging strategy SC and cheating by the other player. Mathematical investigations of a simple altruism game (Eshel *et al.* 1999) show that sometimes altruism can be an "unbeatable" strategy, just as if one was interacting with close kin.

4 Analysis of Altruism Game Experiments

The ideal result of a game model is that players' strategies evolve to evolutionary stable strategies, but this rarely happens with realistic models. Even with such a

Table 1. Experimental Results – for each experiment upper row gives follower results, lower row gives leader results

	k	RTG	OSG	OCG	RTS	OSC	OC	RTL	OSL	OCL	GGG	AWD	AON	NA	SC	fitness	rounds	
A)	1.1	26	16	6	-	-	-	-	-	-	2	0	0	0	0	557+353	41+8	
	2	23	23	4	-	-	-	-	-	-	0	0	0	0	04424+2369			
B)		1	11	21	16	-	-	-	-	-	-	2	0	0	0	0	31±21	40±9
	2	28	17	5	-	-	-	-	-	-	0	0	0	0	04318±2417			
C)		0.95	0	3	12	-	-	-	-	-	-	13	5	7	3	7	12±31	38±8
	2	20	8	12	-	-	-	-	-	-	0	0	0	0	0	386±457		
D)		0.8	0	0	2	-	-	-	-	-	-	9	12	7	4	16	10±23	38±10
	2	16	20	12	-	-	-	-	-	-	1	1	0	0	0	40±31		
E)		0.5	0	0	0	-	-	-	-	-	-	2	4	12	3	29	181±44	35±15
	2	0	6	10	-	-	-	-	-	-	11	10	7	2	4	11±34		
F)		0	0	0	1	-	-	-	-	-	-	5	9	10	1	24	10±80	30
	2	4	16	14	-	-	-	-	-	-	7	4	3	2	0	4±72		
G)		1	0	0	0	0	0	0	-	-	-	5	26	16	4	0	19±9	32±4
	2	0	1	6	24	16	3	-	-	-	0	0	0	0	0	62±109		
H)		0.95	0	1	1	2	1	7	-	-	-	10	6	9	3	10	8±37	39±8
	2	16	9	8	11	3	1	-	-	-	2	0	0	0	0	323±350		
I)	0.95	0	0	6	12	2	2	8	12	8	0	0	0	0	0	365±335	44±3	
	2	0	0	0	8	7	3	6	14	11	1	0	0	0	0	1350±783		
I?)		0.95	3	3	5	2	7	3	2	3	7	9	5	1	0	0	43±123	32±9
	2	6	8	10	4	3	2	5	6	3	3	0	0	0	0	824±650		
J)		1	0	0	3	0	14	25	0	0	0	8	0	0	0	0	109±47	47±2
	2	37	13	0	0	0	0	0	0	0	0	0	0	0	03106±1099			
J?)		1	1	2	8	3	3	4	5	4	7	5	7	1	0	0	158±193	23±9
	2	10	9	8	1	3	3	4	2	6	1	3	0	0	0	826±879		
K)		0.8	0	0	0	2	5	2	0	0	0	7	10	8	4	12	-36±125	34±10
	2	7	8	9	7	2	4	4	5	4	0	0	0	0	0	316±364		
K?)		0.8	0	1	0	1	3	2	1	0	1	6	9	9	9	8	2±87	32±9
	2	8	7	7	7	3	1	1	8	4	3	1	0	0	0	151±177		
L)		0.8	-	-	-	-	-	-	7	24	9	6	4	0	0	0	47±367	43±7
	2	-	-	-	-	-	-	8	18	16	5	2	1	0	0	1171±876		
L?)		0.8	-	-	-	-	-	-	6.1	18.6	16.5	6.2	2.4	0.2	0	0	112±361	42±7
	2	-	-	-	-	-	-	7.5	17.8	17.6	4.9	1.8	0.4	0	0	1182±824		

classical simple game as Hawk-Dove simulations show that players' strategies do not evolve to evolutionary stable strategies (Fogel *et al*, 1997). However one often has self-organising criticality (Bak, 1996) in that the player's strategies evolve to a critical level where each player has a reasonable fitness. The pattern of strategies played varies while at this critical level and more or less serious avalanches happen from time to time; during an avalanche one or more players get low fitness and change their strategies appropriately. Sometimes evolution leads to "mediocre stable states" which are adequate but far from optimal for all players. Rarely does evolution lead to periodic oscillation or random wandering of strategies. There is some biological evidence for oscillation in lizard mating strategies (Sinervo & Lively, 1997) , and some altruism studies show cycling from mild altruism to too-generous altruism to cheating to mild altruism again.

All of these patterns occur in our experiments but space limitations prevent us presenting the detailed "Theory of moves" analysis of our experiments here (see publication "The evolution of altruism" at www.evalife.dk). The evolutionary behaviour of games is complex and depends strongly on the possible strategies available to each player, but analysis and understanding is not impossible. However one should distrust

experimental results when they are not repeatable. To check for this the experimental results for 12 runs in the case $k_F= 0.8$, $k_L= 2$ and all 14 strategies are shown in table 2 and figure 2.The case $k_F= 0.8$, $k_L= 2$ was chosen because it gave the most different results in earlier runs when we had varying numbers of strategies (compare rows D,K,L in table 1or fig.1).

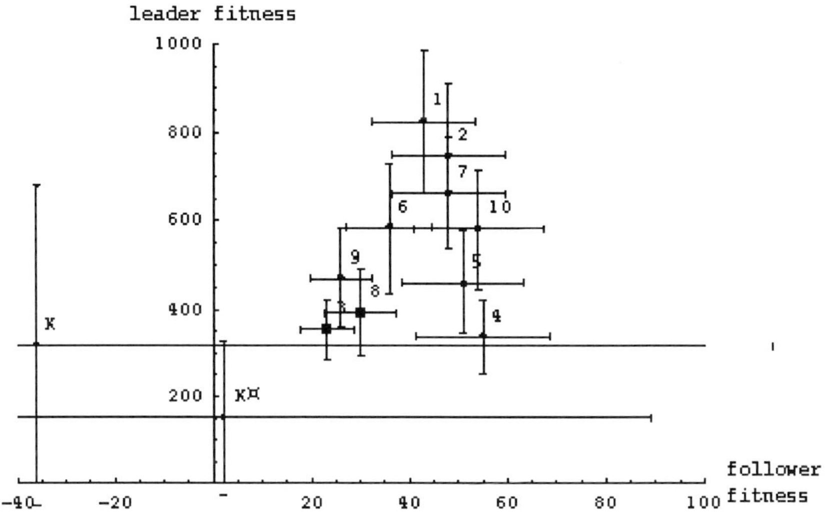

Fig. 2. Experimental Results-error bars quartered (except for K and K?)

Figure 2 suggests that our experimental fitnesses are mutually consistent and this is confirmed by the fact that each follower fitness has [-19,65] and each leader fitness has [174,328] within their fitness error bounds. Table 2 shows that leader strategy distributions are mutually consistent, follower strategy distributions are mutually consistent, but follower strategy distributions are very different from leader strategy distributions. Figure 3 shows the crosscorrelation of the followers strategies against the leaders strategies. In figure 3 there are two regions of high positive cross-correlation:

(Mediocre) top corner peak -leader strategies 1-3 (RTG, OSG, OCG)
follower strategies 10-14 (GGG,AWD, AON, NA, SC)

(Fair) middle peak – follower & leader strategies 7–9 (RTL, OSL, OCL)

Both followers and leaders get higher payoffs in the Fair region, but then leaders become greedy, the followers retaliate by not cooperating and both end in the Mediocre region. Evolution selforganises the delicate balance between the Mediocre and the Fair.

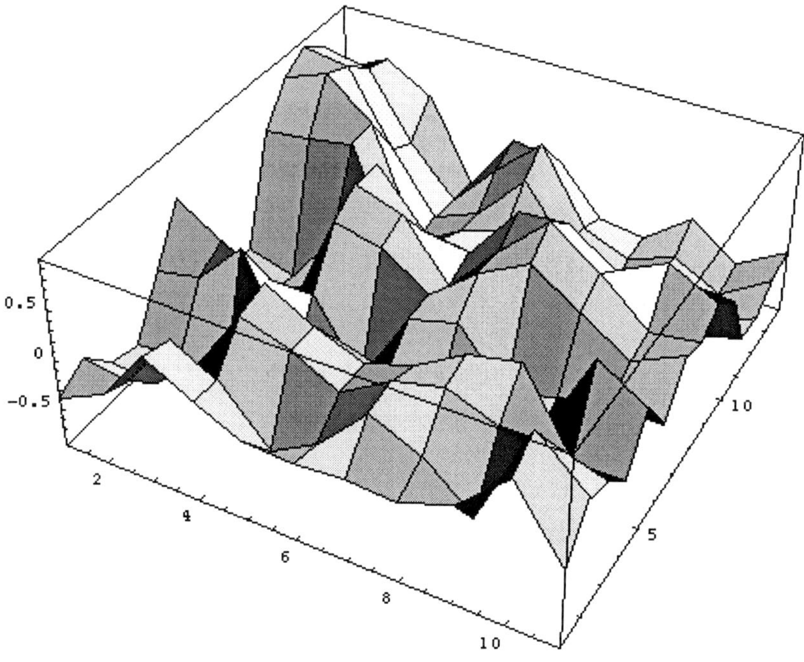

Fig. 3. Crosscorrelation (z-axis) of follower strategies (y-axis) against leader strategies (x-axis)

5 Conclusions

Spatial distribution of interacting individuals encourages the development of altruistic strategies, but it is not a necessary precondition. Our experiments show that cooperation can evolve, even when partners have very different cost-benefit ratios and must make very different investments. Our experiments give surprising insights into the advantages and disadvantages of the strategy families we tried:

Table 2. Experimental Results K10 is the same run as K9 but the results are *after 500 generations*

RTG OSG OCG RTS OSC OC RTL OSL OCL GGG AWD AON

NA SC fitness rounds

K1)		3	3	5	2	7	3	2	3	7	9	5	1	0	0	43±123	31±10
	6	8	10	4	3	2	5	6	3	3	0	0	0	0	824±650		
K2)		1	1	7	7	8	2	6	5	6	4	2	1	0	0	48±119	36±8
	5	8	6	2	5	7	5	5	2	5	0	0	0	0	747±666		
K3)		0	1	2	5	4	9	2	2	4	7	5	8	1	0	23±42	29±9
	15	12	5	3	0	2	5	4	3	1	0	0	0	0	354±273		
K4)		0	0	4	1	4	4	2	4	4	9	8	7	1	2	55±91	35±9
	14	16	11	2	5	1	4	3	4	0	0	0	0	0	337±341		
K5)		0	1	1	2	5	4	3	5	5	14	3	3	1	3	51±79	32±11
	7	7	8	1	5	1	4	7	3	3	4	0	0	0	460±511		
K6)		0	2	5	4	4	5	6	7	8	5	3	1	0	0	36 + 70	34+10
	8	5	9	4	1	2	7	6	2	2	4	0	0	0	583±684		
K7)		2	1	4	6	6	5	3	3	6	7	3	2	0	2	48±113	35±10
	4	7	11	4	3	2	5	2	9	2	1	0	0	0	661±510		
K8)		0	0	1	7	4	2	4	3	0	6	9	6	1	7	30±59	34±8
	11	15	10	1	3	1	4	2	3	0	0	0	0	0	393±419		
K9)		0	0	3	3	5	4	3	1	3	8	6	8	2	4	26±50	37±8
	13	11	6	2	4	3	5	1	3	0	0	0	0	0	470±439		
K10)		0	0	2	2	3	3	1	3	4	11	9	7	3	2	54±82	32±9
	10	10	12	1	2	0	5	4	4	2	0	0	0	0	824±650		
K)		0	0	0	2	5	2	0	0	0	7	10	8	4	12	-36±125	34±10
	7	8	9	7	2	4	4	5	4	0	0	0	0	0	316±364		
K?)		0	1	0	1	3	2	1	0	1	6	9	9	9	8	2±87	32±9
	8	7	7	7	3	1	1	8	4	3	1	0	0	0	151±177		

NA never evolved even when it was best

SC often evolved (benefits more than NA in low kF)

Leaders often prefer the greedy strategies to the fair RTL, OSL and OCL so the followers choose shortchanging strategies OC and OCG. They also suggest an explanation of why followers should be more cooperative than leaders, even when followers are not coerced by leaders.

Our game results indicate that designers of multiagent systems should not expect to find an optimal autonomous agent strategy for responding to task requests and they should consider allowing individual agents to choose their strategy on the basis of their experience with each of the other agents in the system.

References

BAK, P. (1996) How Nature works: the science of self-organised criticality . *Springer*

BAZZAN, A.L .C. *et al.* (1999) Moral sentiments in multi-agent systems, *ATAL98* Springer LNAI 1555, pp 113–131

BIRK, A. & WIERNIK, J. (2002) An N-player prisoner's dilemma in a robotic ecosystem, *Robotics and Autonomous Systems 39,* pp 223–233

ESHEL, I. et al. (1999) The emergence of kinship behaviour in structured populations of unrelated individuals". *Int.J.Game Theory* 28 , pp. 447–463

FOGEL, D.B. et al. (1997) On the instability of evolutionary stable strategies. *Biosystems* 44, pp. 135–152

GMYTRASIEWICZ, P.J. & LISETTI, C.L. (2002) Emotions and personality in agent design and modelling, *Intelligent Agents VIII*, Springer LNAI 2333 pp.21–31

JENNINGS, N.R.. (2000)On agent-based software engineering. *Artificial Intelligence.*117 pp. 277–296

KILLINGBACK, T. & DOEBELI, M. (1998) Raise the stakes' evolves into a defector. *Nature* 400 p. 518

KILLINGBACK, T. et al. (1999) Variable investment, the continuous prisoner's dilemma, and the origin of cooperation. *Proc. R. Soc. Lond. B* 266 pp. 1723–1728

MAY, R.M. (1987) More evolution of cooperation. *Nature* 327 pp. 15-17

MAYOH, B.H. & JUNPING, DU. (1999) How flexible should agent behaviour languages be?. *Proc. IAT99,*World Scientific Pub.

ROBERTS, G. & SHERRATT, T.N. (1998) Development of cooperative relationships through increasing investment. *Nature* 394 pp. 175–178

ROBERTS, G. & SHERRATT, T.N. (1999) The emergence of quantitatively responsive cooperative trade. *J. Theor. Bio.* 200 pp. 419–426

SHERRATT, T.N. & ROBERTS, G. (1998) The evolution of generosity and choosiness in cooperative exchanges. *J. Theor. Bio.* 193 pp. 167–177

SINERVO, B & LIVELY, CM (1997) The rock-paper-scissors game and the evolution of alternative male strategies. *Nature* 380, pp. 240–243

WATSON, R.A, FILICI, S.G. & POLLACK, J.B. (2002) Embodied Evolution: distributing an evolutionary algorithm in a population of robots. *Robotics and Autonomous Systems* 39 pp. 1–18

A Dynamic Window-Based Approximate Shortest Path Re-computation Method for Digital Road Map Databases in Mobile Environments[1]

Jaehun Kim and Sungwon Jung

Department of Computer Science, Sogang University
Seoul, 121-742, Korea
freeso@mclab.sogang.ac.kr, jungsung@ccs.sogang.ac.kr

Abstract. In this paper, we have studied the shortest path re-computation problem that arises in the dynamic route guidance system (DRGS) in ATIS where the cost of topological digital road map is frequently updated as traffic condition changes dynamically. Previously suggested methods are trivial in that they do not intelligently utilize the previously computed shortest path information. We have developed an efficient approximate shortest path re-computation method based on the dynamic window scheme. The proposed method re-computes an approximate shortest path very quickly by utilizing the previously computed shortest path information. We first show the theoretical analysis of our methods and then present an in-depth experimental performance analysis by implementing it on grid graphs.

1 Introduction

One of commercial applications of mobile computing is Advanced Traveler Information Systems (ATIS) in Intelligent Transport Systems (ITS). In ATIS, a primary mobile computing task is to compute the shortest path from the current location to the destination. One of the major problems in ATIS is that a significant amount of computation time is required to find the shortest path when the digital road map is large. For example, it requires about 2.4 Gbytes of storage to store a small 100 *mi* × 100 *mi* map discretized at 100 *feet* intervals [1,2]. Since ATIS are real time mobile systems, it is critical that the path be computed while satisfying a time constraint.

In this paper, we have studied the shortest path re-computation problem. This problem arises in the dynamic route guidance system (DRGS) in ATIS, where the *cost* of topological digital road map is frequently updated as traffic condition changes dynamically. As a result, the shortest path planned at the beginning of the trip may not be the path finally taken due to the unacceptable path cost. Thus, an efficient shortest path re-computation method is essential for DRGS to be successful. The two methods currently being used either re-compute the shortest path from scratch or they re-

[1] This research was sponsored in part by Institute for Applied Science and Technology of Sogang Univ.

M.H. Shafazand and A M. Tjoa (Eds.): EurAsia-ICT 2002, LNCS 2510, pp. 711–720, 2002.

compute the shortest path just between the two end points where the cost change occurs. Usually, after some changes of the cost, the new shortest path recomputed does not differ significantly from the old one. In this respect, above methods are clearly trivial in that they do not re-use the previously computed shortest path information.

We have developed an efficient approximate shortest path re-computation method based on dynamic window scheme. Our method takes advantage of the previously computed shortest path information and produces paths that are fairly close to an optimal shortest path. The basic assumption behind our approach is that for many real traffic situations, drivers may not be that interested in getting the optimally recomputed shortest path at the cost of large path re-computation time. Instead, they rather have approximate shortest paths generated rapidly.

The rest of the paper is organized as follows. Section 2 briefly discusses A^* and OTO algorithms. In section 3, we experimentally analyze the update effects for one edge on the shortest paths on two dimensional grid graphs. In section 4, we propose a dynamic window-based approximate shortest path re-computation method and give the performance analysis of our method. Finally, section 5 gives concluding remarks.

2 A* and OTO Shortest Path Algorithms

The main performance overhead of the shortest path computation comes from the size of the search space it needs to explore. A^* algorithm is a modification of Dijkstra algorithm in that it utilize the semantic information about road maps such as latitude and longitude to reduce the explored search space. It uses an estimator function *f(u,d)* to estimate the cost (i.e., Euclidean Distance) of shortest path between node *u* and *d*. The pseudo code in Fig. 1 describes the SPSP algorithm between the nodes *s* and *d*.

```
1.  For each node u ∈ V, C(u) = ∞.
2.  Let C(s) = 0, frontierSet = {s}, exlporedSet = ∅.
3.  Select a node u in frontierSet for which C(u) + f(u,d) is minimum
4.  frontierSet = frontierSet – {u}, exploredSet = exploredSet ∪ {u}
5.  If (u==d) then C(u) is the shortest path cost and stop
6.  For every edge (u,v) in E
        if C(v) > C(u) + L(u,v) then
        C(v) = C(u) + L(u,v)
        frontierSet = frontierSet ∪ {v} if v ∉ (frontierSet ∪ exploredSet)
7.  Go to step 3
```

Fig. 1. A^* algorithm

In contrast to A^* algorithm, Mohr and Pasche proposed a scheme called OTO which finds the shortest path by alternatively building two trees, one rooted at the source and the other rooted at the destination [3]. The two trees in OTO algorithm are expanded alternatively. That is, if one tree has the next shortest path than the other, the one with the next shortest path is expanded. Note that OTO uses A^* algorithm to expand each tree. For the detailed description of OTO algorithm, readers can refer to the paper in [3]. Since OTO algorithm explores the search space from the both ends, the size of its explored search space becomes smaller than that of A*.

3 Analyses of Update Effects on the Shortest Paths

In this section, we analyze the update effects of a single edge on the shortest paths. For this analysis, we use OTO algorithm and two dimensional grid graphs G(V,E,L). Two dimensional grid graphs are considered as typical examples of road maps [2,4].

3.1 Measuring the Degree of Path Similarities

After updating an edge on the shortest path, we measure the degree of path similarities between a new shortest path recomputed from scratch and the old one. We first present the following two definitions.

Definition 1. Let $P_s(s,d)$ represent the shortest path from source s to destination d. Then, $P_s(s,d)$ can be represented by the node sequence, $(z_0, z, ..., z_{m-1}, z_m)$ where $z_0 = s$ and $z_m = d$. Let $P_n(s,d)$ represent the newly recomputed shortest path from source s to destination d after the updating the cost of any edge (z_i, z_{i+1}) on $P_s(s,d)$ where $0 \le i \le m-1$. $P_n(s,d)$ can also be represented by the node sequence, $(w_0, w_1, ..., w_{n-1}, w_n)$.

Definition 2. Let $Q_0, Q_1, \ldots, Q_t$ be the distinct and non-overlapping subsequences satisfying the following condition for $P_s(s,d)$ and $P_n(s,d)$: $Q_k = (x_i, x_{i+1}, ..., x_{i+z}) = (y_j, y_{j+1}, ..., y_{j+z})$ such that $x_{i-1} \neq y_{j-1}$ and $x_{i+z+1} \neq y_{j+z+1}$ where $0 \le k \le t$, $1 \le z$, $i+z<m$ and $j+z<n$. Let $|Q_k|$ be the number of nodes in the subsequence Q_k. Then, the degree of path similarities $DPS(P_s, P_n)$ between $P_s(s,d)$ and $P_n(s,d)$ is defined as:

$$DPS(P_s, P_n) = \sum_{k=0}^{t} | Q_k | \times 100/m \ (\%).$$

Before analyzing the update effects, we first define the following three notations:

- Δ : the variations of edge costs for grid graph $G(V,E,L)$. For example, the edge costs of the grid graph in $\Delta=x\%$ are allocated randomly in the range $[100,100+x]$.
- θ : the angle between the source node s and the destination node d.
- $f(s,d)$: the Euclidean Distance from the node s to the node d.

For this analysis, we consider 6 different types of grid graphs $G(V,E,L)$ where $|V| = 50\times 50$ nodes and $|E| = 4 \times 50 \times 49$ directed edges. Their edge costs in L are generated from 7 uniform distributions corresponding Δ=20%, 40%, 80%, 160%, 240%, 320% respectively. For each grid graph, we compute the shortest paths of the five nodes pairs (s, d) with $f(s,d)=30$ having θ=0, 15, 30, 45, and 90 respectively. We then update the cost of the edge at 5%, 10%, 15%...90%, 95% of the location on each shortest path. We measure 19 $DPS(P_s, P_n)$ at 5%, 10%, 15%, ... ,90%, 95% of the position on the shortest path. Note that $DPS(P_s, P_n)$ computed at each position are then averaged over grid graphs generated with 100 seeds having the same value of Δ. Fig. 2 shows the result of our experiment.

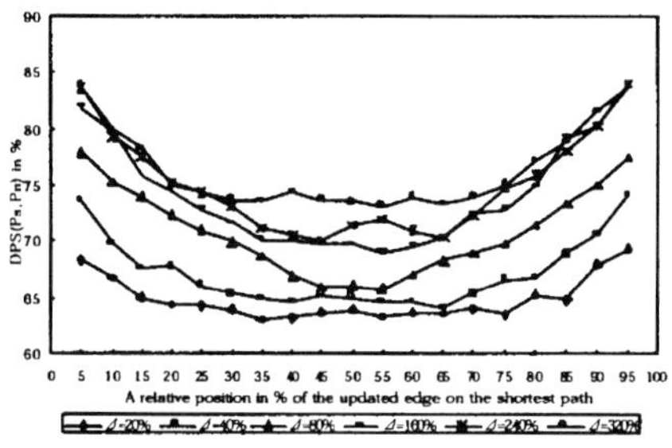

Fig. 2. *DPS*(P_s, P_n) representing Update effects

From Fig. 2, we can observe the following three interesting properties:

Property 1. *DPS*(P_s, P_n) varies from 60% to 90% depending on the value of Δ.

Property 2. As Δ is increased from 20% to 320%, *DPS*(P_s, P_n) is also increased.

Property 3. *DPS*(P_s, P_n) is the maximum at the both end nodes of the shortest path and begin to decrease to the minimum towards the middle node.

From the *property 1*, we can conclude that the most of the nodes on the previously computed shortest path are re-used in the newly recomputed shortest path after an edge cost update. We discuss *property 2* and *3* in the following section 3.2.

3.2 Correlation between the Search Spaces and *DPS*

In this section, we study how Δ and θ affect *DPS*(P_s, P_n). Assume 2 grid graphs *G*(*V*,*E*,*L*) with |*V*|=50×50 and |*E*|=4×50×49 whose *L* is from Δ=20%, 320% respectively. For each grid graph, we generated 100 grid graphs with 100 different seeds. We then compute the shortest paths for five node pairs (*s, d*) when θ =0, 15, 30, 45, 90 and *f*(*s*,*d*) is fixed to 30. Similar to the experiment given in Fig. 2, we measure the average *DPS*(P_s, P_n) over 100 grid graphs. Fig. 3 shows the results of this experiment.

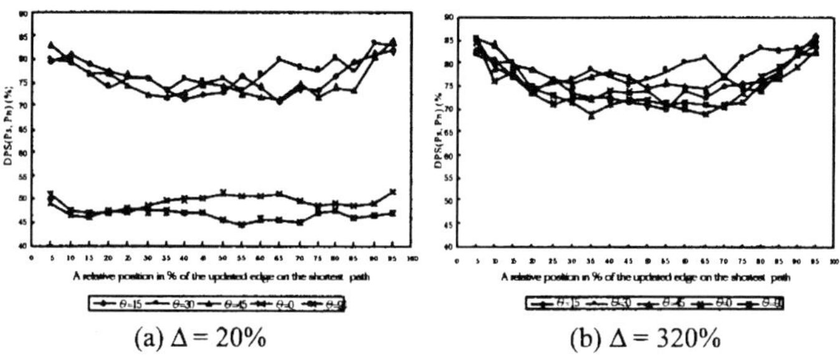

(a) Δ = 20% (b) Δ = 320%

Fig. 3. *DPS*(P_s, P_n) according to the values of Δ and θ

Fig. 3 shows the following two properties which in turn explain the *property 2*:

1. When θ=0 and 90, *DPS*(P_s, P_n) is dependent on Δ.
2. When θ=15, 30 and 45, *DPS*(P_s, P_n) is not dependent on Δ.

From the above two properties, we can conclude that the two parameters Δ and θ affect $DPS(P_s, P_n)$. To study how Δ and θ affect the size of *frontierSet*, we measure it for the shortest. Fig. 4 illustrates that Δ and θ affect the size of *frontierSet*.

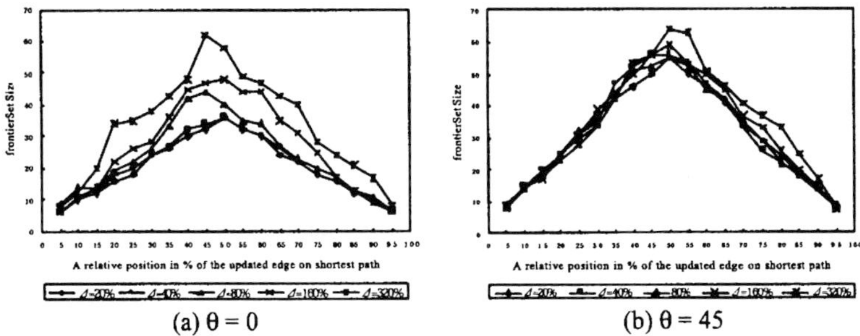

Fig. 4. The size of *frontierSet* according to the values of Δ and θ

Note that the size of *frontierSet* is the number of nodes adjacent to the explored search space. This means that the explored search space can also be described by the nodes in *frontierSet*.

Theorem 1. Δ and θ determine the size of the explored search space.
Proof. The proof is given in [5].

We now show in *Theorem 2* that the size of explored search space affects $DPS(P_s, P_n)$. By proving this, we can utilize the size of explored search space to analyze $DPS(P_s, P_n)$ instead of using Δ and θ.

Theorem 2. $DPS(P_s, P_n)$ is linearly dependent on the size of explored search space.
Proof. The proof is given in [5].

By *Theorem 2*, we can also verify the *property 3*. This verification is given in [5].

4 A Dynamic Window-Based Approximate Shortest Path Re-computation Method

4.1 Basic Idea

To illustrate our method, consider the shortest path from node *s* to *d* in Fig. 5 where the cost of edge *(p,q)* is updated.

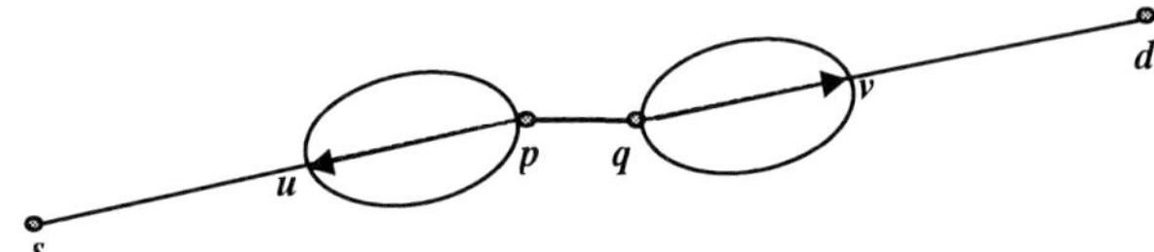

Fig. 5. Shortest Path $P_s(s,d)$ from source node s to destination node d

After the cost of edge (p,q) is updated, we know that $DPS(P_s, P_n)$ varies from 60% to 90%. This means that the cost update of edge (p,q) makes an effect on the parts from p to u and from q to v on $P_s(s,d)$. Thus the edges from s to u and from d to v will be reused in the re-computed shortest path $P_n(s,d)$. If we can correctly find the nodes u and v, we can re-compute the shortest path as follow:

1. Compute the shortest path from node u to v, $P_s(u,v)$ by using either A* or OTO.
2. $P_n(s,d)$ is the concatenation of the paths $P_s(s,u)$, $P_n(u,v)$, and $P_s(v,d)$. That is, $P_n(s,d) = P_s(s,u) \,||\, P_n(u,v) \,||\, P_s(v,d)$ where || operator represents the path concatenation.

By the above two procedure, after the cost of one edge on $P_s(s,d)$ is updated, we can compute only the shortest path from u to v (only the part affected by the update) but re-compute from scratch to yield $P_n(s,d)$.

However, it is difficult to identify the exact nodes u and v, which makes the cost of $P_s(s,d)$ often approximate to the actual shortest path cost. We proposed "a dynamic window-based approximate shortest path re-computation method" that is basically the same as the above. The term "windows" indicates the two parts of shortest path that is affected by the cost update of an edge. We use also the term "dynamic" in the sense that we dynamically determine the size of windows based on the parameters.

4.2 A Dynamic Method for Determining Window Size

The most important thing for our objective is to determine the proper window size. As indicated in Section 4.1, a window size is shown to be the size of congestion effect which is closely related to $DPS(P_s, P_n)$. That is, as $DPS(P_s, P_n)$ is increased, the congestion effect is decreased thus making window size smaller. In this respect, the proposed method determines the proper window size based on the parameters affecting $DPS(P_s, P_n)$. We know that there will be two windows created after an edge cost update like Fig. 5. Our method determines separately two windows as such:

- W_s: the window toward the source node s. For example, $W_s=[p,u]$ in Fig 5.
- W_d : the window toward the destination node d. For example, $W_d=[q,v]$ in Fig. 5.

The following *theorem 3* and *4* will be discussed before we explain our method.

Theorem 3. Given a fixed value of $f(s,d)$, the size of explored search space is the smallest when $\Delta=0\%$ and $\theta=0$(or $\theta=90$).

Proof. By *Theorem 1*, *Theorem 3* is trivially true.

Theorem 4. Given a fixed value of $f(s,d)$, $DPS(P_s, P_n)$ is 50% on the average for the smallest size of explored search space for the shortest path from node s to d.
Proof. The proof is given in [5].

Now, we illustrate our method that determines dynamically the proper window size using *Theorem 4.* Suppose that the cost of an edge (p,q) at a distance of n from node s on $P_s(s,d)$ is updated. Fig. 6 shows the smallest search space explored and the larger from s to p. Here the ellipse in dotted line (let π_s^p) represents the smallest search space, and one in solid line (let Π_s^p) represents the larger one. Besides, α specifies congestion effect toward the source node at π_s^p, and β at Π_s^p.

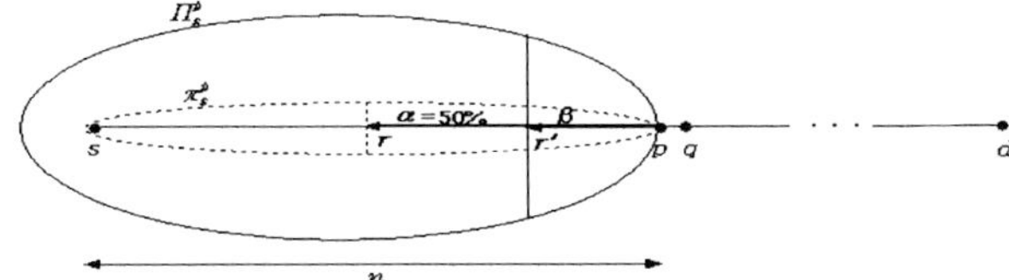

Fig. 6. Congestion effect according to the size of explored search space.

From *Theorem 4*, we know that average α=50%. Next, we determine congestion effect β. From *Theorem 2*, congestion effect is decreased as explored search space is larger, thus it is clear that β<α. We then determine how degree β is smaller than α. From *Theorem* 2, β can be get by the proportion of Π_s^p and π_s^p. That is to say, we determine β from α as (1).

$$\beta = \alpha \times \frac{\text{Size of } \pi_s^p}{\text{Size of } \Pi_s^p} \ (\%) \tag{1}$$

Formula (1) means that we get β decreasing from α as much as Π_s^p is larger than π_s^p (i.e. how much times). Because the size of explored search space is in direct ratio to *frontierSet* size, we can rewrite formula (2) from formula (1).

$$\beta = \alpha \times \frac{|\,frontierSet_p^{\pi}\,|}{|\,frontierSet_p^{\Pi}\,|} \ (\%) \tag{2}$$

where $|frontierSet_p^{\pi}|$ is the size of search space explored from s to p at π_s^p, and $|frontierSet_p^{\Pi}|$ is at Π_s^p. *frontierSet* at π_s^p is to be adjacent to each node on the shortest path. Thus formula (2) is simply expressed as follows:

$$\beta = \alpha \times \frac{n \times 2 + 2}{|\,frontierSet_p^{\Pi}\,|} \ (\%) \tag{3}$$

We know the values of α, n and $|frontierSet_p^{\Pi}|$, thus can get the value of β. Consequently, $W_s=[p,r]$.

However, we cannot figure out W_s and W_d at all the positions on the shortest path using the above method, because $|frontierSet_p^{\Pi}|$ is the search space explored from s and $|frontierSet_q^{\Pi}|$ from d. OTO algorithm finds the shortest paths by alternately exploring search space from the source and the destination. Part 1 and Part 2 represent respectively [0%, 50%] and [50%, 100%] of the shortest path. And then *frontierSet* in Part 1 is explored from node s and that in Part 2 is explored from node d. If (p,q) is located in Part 2, using formula (3) to obtain W_s is undesirable because $|frontierSet_p^{\Pi}|$ is the size of search space explored from node d instead of node s. Similarly, if the updated edge (p,q) is located in Part 1, we can not get W_d. Accordingly, we need a method to determine W_d when the updated edge (p,q) is located in Part 1 and W_s when (p,q) is located in Part 2. Fig. 7 explains this method.

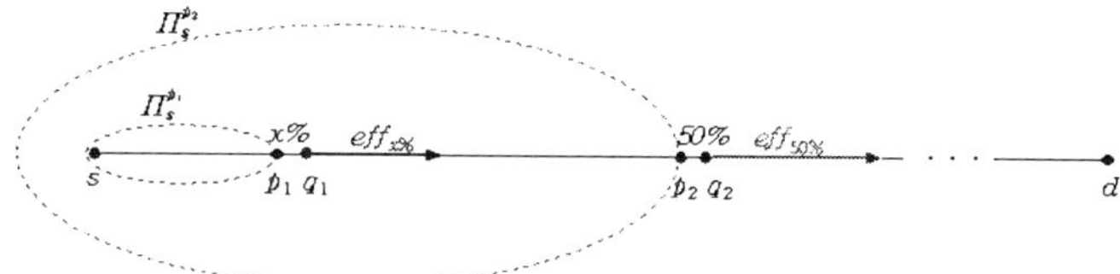

Fig. 7. Determining the congestion effect

In Fig. 7, the two updated edges, (p_1, q_1) and (p_2, q_2) are at respectively 50% and $x\%(0 \le x \le 50)$ of the location on the shortest path. $\Pi_s^{p_1}$ and $\Pi_s^{p_2}$ represent explored search space measured respectively at node p_1 and p_2. Let $eff_{50\%}$ be congestion effect of the edge (p_2, q_2) toward the destination after (p_2, q_2) is updated. Moreover, let $eff_{x\%}$ be that of the edge (p_1, q_1) toward the destination after (p_1, q_1) is updated. Then we can easily figure out $eff_{50\%}$ because $|frontierSet_{q_2}^{\Pi}|$ measured at node q_2 is explored from node d. However, we cannot obtain $eff_{x\%}$ by formula (3) because $|frontierSet_{q_1}^{\Pi}|$ measured at node q_1 is the size of search space explored from node s instead of node d. Consequently, we cannot determine W_d of the edge (p_1, q_1).

Now, we explain a method to obtain $eff_{x\%}$. According to *Theorem 2*, we know $eff_{x\%} < eff_{50\%}$ because the size of *frontierSet* from d to q_1 is larger than $|frontierSet_{q_2}^{\Pi}|$. Then we have to find how much smaller $eff_{x\%}$ is when compared to $eff_{50\%}$. We compute $eff_{x\%}$ by reducing $eff_{50\%}$ as much as $\Pi_s^{p_1}$ is smaller than $\Pi_s^{p_2}$. We give the formal definition of $eff_{50\%}$ as follows:

$$eff_{x\%} = eff_{50\%} \times \frac{|\, frontierSet^{\Pi}_{p_1} \,|}{|\, frontierSet^{\Pi}_{p_2} \,|} \tag{4}$$

The smaller $|frontierSet^{\Pi}_{q_2}|$ is over $|frontierSet^{\Pi}_{q_1}|$, the larger the size of *frontierSet* from *d* to q_1 is over $|frontierSet^{\Pi}_{q_2}|$. By taking advantage of this fact, we can establish formula (4) from *Theorem 2*. Using formula (4), we can obtain W_d when the updated edge (*p*,*q*) is located in Part 1 and W_s when (*p*,*q*) is located in Part 2.

4.3 Experimental Analysis

In this section, we experimentally analyze our proposed method. Let $P_{app}(s,d)$ be the approximate shortest path re-computed by our method. We show the efficiency of our method by comparing $P_{app}(s,d)$ with $P_n(s,d)$. The comparison is performed in the following three respects:

- node difference: we evaluate $DPS(P_{app}, P_n)$.
- cost difference: we evaluate P_{app} cost - P_n cost / P_n cost $\times 100$
- computation time difference: the comparison of the computation time of $P_{app}(s,d)$ and $P_n(s,d)$.

For this analysis, we consider a grid graph *G*(*V*,*E*,*L*) where |V|=50×50, |E|=4×50×49. Note that the edge costs in *L* is generated from Δ=80%. We generated 100 grid graphs with 100 different seeds. We then chose two node pairs: respectively, θ=0 and 45 (*f*(*s*,*d*)=30). For each grid graph, we compute 2 shortest paths and then updated the cost of the edge at 5%, 10%, 15%, … , 90%, 95% of the location on each shortest path. For the update on the cost of edge at each location, we measured node difference ($DPS(P_s, P_n)$), cost difference and computation time difference. These measured values are then averaged over 100 grid graphs. Fig. 8 and 9 show the results. They clearly indicate the efficiency of our proposed method.

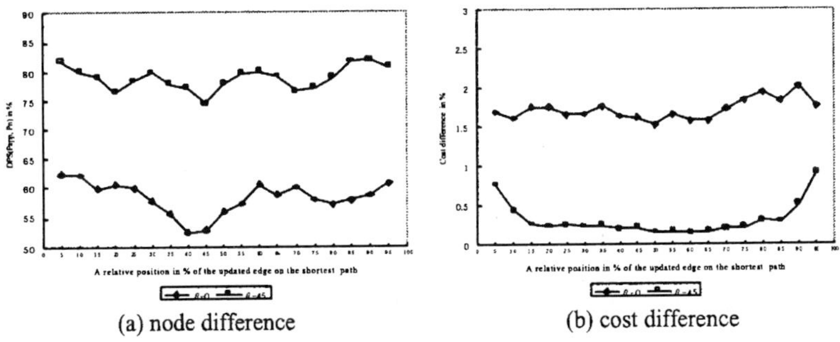

(a) node difference (b) cost difference

Fig. 8. The comparison between $P_{app}(s,d)$ and $P_n(s,d)$.

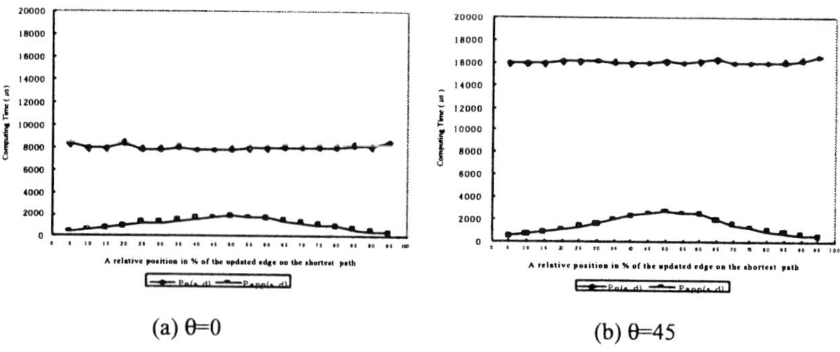

(a) θ=0 (b) θ=45

Fig. 9. The computation times for $P_{app}(s,d)$ and $P_n(s,d)$.

5 Conclusions

In this paper, we developed an efficient dynamic window-based approximate shortest path re-computation method that utilizes the previously computed shortest path information. Our method can be used to re-compute efficiently and rapidly an approximate shortest paths after the cost update of edges on the shortest path.

Above all, it is the most important to determine the proper window size in our method. In this respect, we dynamically determined the size of windows based on the analysis of update effects on the shortest paths. We also presented an in-depth experimental analysis of our method by implementing it on various grid graphs. Experimental results showed the efficiency of our proposed method in terms of the computation time as well as the cost of the approximate shortest path. We are currently investigating how to handle multiple updates on the shortest path.

References

1. R. Agrawal and H. Jagadish, "Algorithms for Searching Massive Graphs", In *IEEE Transactions on Knowledge and Data Engineering*, Vol. 6, No.2, pp. 225-238, April 1994.
2. R. Kung, E. Hanson, Y. Ioannnidis, T. Sellis, L. Shapiro, and M. Stonebraker, "Heuristic Search in Data Base System", In *Proc. 1st Int. Workshop Expert Database Systems*, pp. 96–107, Oct. 1984.
3. T. Mohr and C. Pasche, "A Parallel Shortest Path Algorithm", In *Computing*, Vol. 40, pp. 281–292, 1988.
4. S. Shekhar, A. Kohli, and M. Coyle, "Path Computation Algorithms for advanced Traveler Information System (ATIS)", In *Proc. IEEE 9th Int'l Conf. Data engineering*, pp. 31–39, 1993.
5. J. Kim, S. Jung, "A Dynamic Window-based Approximate Shortest Path Re-Computation Method for Digital Road Map Databases in Mobile Environments", Technical Report, Dept. of Computer Science, Sogang University, July 2002

Web-Based Process Control Systems: Architectural Patterns, Data Models, and Services

Mykola V. Tkachuk[1,3], Heinrich C. Mayr[1], Dmytro V. Kuklenko[2,3], and Michail D. Godlevsky[3]

[1] Institute of Business Informatics and Applications Systems, University of Klagenfurt, Universitätsstr. 65-67, A-9020 Klagenfurt, Austria
{mayr | nikolay}@ifit.uni-klu.ac.at
[2] eBusiness Institute, University of Klagenfurt
Universitätsstr. 65-67, A-9020 Klagenfurt, Austria
dmytro.kuklenko@biztec.org
[3] National Technical University "Kharkiv Politechnical Institute"
Frunze Str. 21, 61002, Kharkiv, Ukraine
god_asu@kpi.kharkov.ua

Abstract. The paper discusses results gained from a real-life project in the domain of Information Systems for Technological Process Control (TPC-IS) in the gas-transport and gas-production branches. It emphasizes the crucial requirements which have to be satisfied by TPC-IS and, therefore, investigated when analyzing and designing Web-based TPC-IS. In addition, *architectural design patterns* reflecting a multi-level distributed topology and a real-time operating mode are presented. The concept starts from an *Integrated Node Database* model, which includes process-oriented storage structures and appropriate functionality. An *Advanced Services Collection* provides the functionality needed for supporting process coordination and control. Using these pattern solutions and data modeling concepts, reusable software components may be developed, some of which already being implemented and operative in our projects.

1 Introduction

The automated control of complex technological processes in the real-time mode is still an actual and challenging issue. Strong time and cost constraints, strong requirements w.r.t. reliability and stability, as well as the need for integration with decision support systems and all other company information systems (e.g. ERP) may be seen as the main reasons for that actuality: Information systems for Technical Process Control (TPC-IS) have to support such actions as receiving the data from technical controllers and devices, performing the control commands and actions, representing the actual state of controlled technological processes, supporting the decision making process of operating staff, etc.

The main distinguishing features of TPC-IS can be considered as the following:

- distributed architecture, characterized by many and multi-dimensional relationships between several subsystems,
- continuous working cycle (so called "24x7 mode"),

M.H. Shafazand and A M. Tjoa (Eds.): EurAsia-ICT 2002, LNCS 2510, pp. 721–729, 2002.

- heterogeneous data resources,
- inclusion of legacy software applications combined with additional modules and subsystems,
- several and heterogeneous user groups and roles,
- reliability of action and user support in extreme operating cases (the analysis of failures, devices trouble-shooting, etc).

One of the main international trends in the domain of TPC-IS development is the usage of the SCADA (Supervisory Control Access and Data Acquisition) system concept [1-4] which extends the CAD/CAM functionality by the hard-ware devices level to be controlled. A typical SCADA system is a special software solution, which facilitates a lot of programmable logical controllers (PLCs). The main goal of a SCADA system is to collect the technological process data, to represent them in proper time and form, and to accomplish the necessary control actions.

According to our investigations and practical experience, the usage of Web-technologies substantially eases the problem of complex TPC-IS design because of the following reasons and considerations:

- The TCP/IP and HTTP protocols support distributed applications design and development, the latter mainly due to the stable functioning of these protocols, even in situations, where communication links or channels have low capacity and work unstable,
- Internet environment shows the best correlation with the paradigm of TPC-IS functioning in the "24x7 mode",
- Web-standards and technologies as XML, DOM, RDF, and modern browsers support integrated processing of heterogeneous data,
- Server-side and component programming technologies, like ActiveX and ASP (Microsoft Corp.), or EJB and JSP (OMG and Sun Microsystems) can be used as effective platforms for legacy software integration.

The paper is organized as follows: Section 2 outlines the main aspects of an TCP-IS which has been developed for the Kharkiv Region Gas Production Enterprise (KRGPE) in a couple of projects. In section 3 we present some architectural patterns that we found out to be effective for TCP-IS within that field. Two of them are discussed in some more detail, namely the Integrated Node Database (Section 4) and the Advanced Services Collection (Section 5). Section 6 gives some hints on TCP-IS implementation issues. The paper ends with concluding remarks and some references.

2 TPC-IS in the Kharkiv Region Gas Production Enterprise

During the last 3 years our research team has carried out a number of projects aimed at the re-engineering and the construction of TPC-IS for KRGPE. Some particular results of these projects were presented in [5-8]. Let's consider now our generalized approach to designing the architecture of that Web-based TCP-IS based on the following requirements:

- continuous and quick access to the required information related to all technological processes and technical objects in the regional system. Such access must be possible in an authorized mode from every system node, as well as from mobile communication devices,
- hot link between system users (e.g. an operator in a gas-production node must be able to obtain technical advises from an expert in the regional management center),
- support of storing, systematizing and analyzing of all data, which is necessary for day-to-day and strategic decision making.

For design and implementation we applied the fundamental principles of modern information systems development, such as OSI standard, object-oriented analysis, holistic and components structure design (based on the OMG and ODMG standards [9]), heterogeneous information resource integration. MS COM+/MS .Net and ASP were used as component technologies for Web-programming (their alternatives, such as J2EE and JSP were implemented for comparative studies), MS SQL Server 2000 was used as primary DBMS (Oracle 9i for comparative studies).

The system has distributed multilevel network architecture of functional nodes, including the Web-oriented software package (WSP), dedicated to storing, representing and analyzing technological information. We distinguish the following architectural levels (see figure 1):

- *Gas Field level*: the WSPs of the gas-fields,
- *Sectional level*: organizational union of several neighboring WSPs,
- *Regional level*: includes the WSP for top-level branch management; the main goal of this node is to support the decision making process for the whole distributed technological system control.

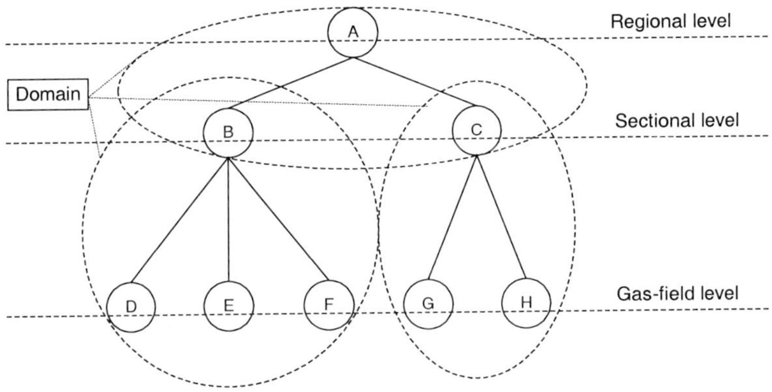

Fig. 1. Hierarchy of level nodes and domain concept

We call a set of level nodes having the same parent node a *domain*, e.g., the set of gas field level nodes belonging to a particular sectional level node form the domain of that sectional level node. For instance, in the fig. 1 the nodes of gas-field level *D,E,F* form the domain of sectional level node *B,* etc.

The system architecture described here is open – it means that it is possible to add new logical levels for both, to cooperate with national and international branch management organizations, and to include some additional intermediate nodes for technological data processing.

3 Architectural Design Patterns

Dealing with such complex TPC-IS it proves useful to elaborate *architectural patterns* which may be used for design, and for developing reusable software components. Thus, architectural patterns provide a standardized and structured way of designing and implementing the whole distributed TPC-IS. From our experience, the following architectural patterns are adequate:

1. *Integrated Node Database (INDB)*: dedicated to storing and processing all data needed for WSP's operating.
2. *Data Exchange Server (DES)*: supports the data exchange with the technical processes in the various system nodes.
3. *Data Visualization Service* (*DVS*): provides the visualization of the technical processes.
4. *Data Synchronization Service* (*DSS*): is aimed to synchronize the data in the different system nodes.
5. *User Accounts Manager* (*UAM*): provides the data access rights for the system users.
6. *Advanced Services Collection* (*ASC*): provides simulation models and appropriate software tools for decision making.

Figure 2 shows the interrelationships between the components built from these architectural patterns and their aggregation on the afore-mentioned architectural levels.

The components in figure 2 are grouped in three packages, which aggregate the functionality to the following types of system nodes:

- *Base level* node (Gas field level): its main functionality stems from the controllers of external devices (PLC-Controllers), which serve as the data sources for whole the system. Further components of this package are: DES, INDB, DVS, and DSS. This DSS acts an interface service (IServer) to the DSS of the higher level nodes.
- *Sectional level* node: includes as an additional component the UAM, which supports the definition of the appropriate user access policy.
- *Regional level* node: is the root node in the hierarchical system. Its structure is the same as the structure of sectional level's node, only extended by the ASC component.

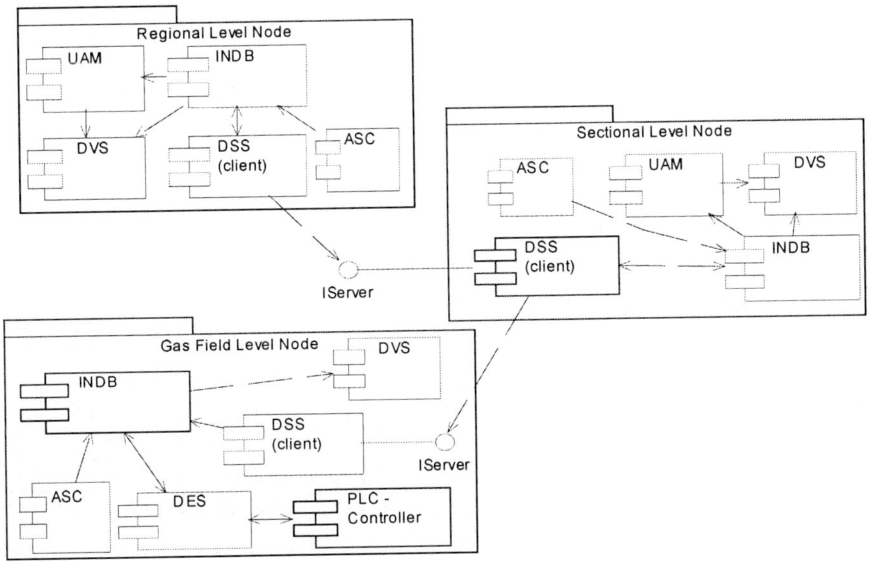

Fig. 2. TPC-IS components and their aggregation on architectural levels

4 Node Database Concept and Domain-Dependent Data Presentation

For designing the *Integrated Node Database* structure we considered the modern trends in relational and post-relational databases (e.g. [10]) with special emphasis on temporal data models (e.g. [11]). The INDB schema consists of sub-schemas for the following (sub-)databases:

- *Retrospective Database (RDB)*: stores and operates retrospective data about the technological process.
- *Technological Database (TDB)*: stores the configuration information needed by the Data Exchange Server.
- *Account Database (ADB)*: stores and manipulates information about user profiles, access rights etc.
- *Synchronization Database (SDB)*: stores the information needed by the Data Synchronization Service.

Implementations of INDB can exclude some of the sub-schemes. E.g., TDB can be omitted in nodes that do not collect its own technological data but serve as pure data bridges or storages. Other INDB might come without ADB, in simple cases UAM might be omitted at the Gas field level.

The domain concept provides the localizability of all process data. The user's access rights and privileges are managed in UAM and may be delimited by domains or single nodes.

If INDB is not only to store the necessary technological data but also to provide means for decision making, a special tool, called *Supervisory Rules Engine* (*SRE*), has to be added to INDB. The rule concept was introduced in order to extend the functionality of RDB and to support the automatic processing of technological data in non-regular situations. *Rule* R can be defined as a triple *R={C,A,P}*, where

- C is a set of conditions (logical expressions) C_i related to current values of technological parameters,
- A is a set of Actions A_j performed if the conditions evaluate to true; these actions store the data relevant to decision making,
- P is a set of rule parameters P_k which define the rule scheduling, termination, and others.

5 Advanced Services Collection

The Advanced Services Collection (ASC) component is dedicated to support decision making. Commonly, at the top level of the system decisions are made based on the strategies and interests that combined with the whole system. They may be defined by a number of factors (such as profit, social problems, ecology, safety of the system, etc.). Within a scheduled period $[1,T]$ we will present them as a trajectory $\{P_i^t, i \in I, t \subseteq [1,T]\}$ where P_i^t is a vector, the components of which define the parameters of factor *i* valid in time interval *t*.

Vector function (1) defines the degree of deviation from the given factors, where $\hat{u}_j^t$ is a specific value vector corresponding to the parameters defined by of vector u_j^t, that defines the set output nomenclature and quantity of node $j \in J$ during subperiod *t*.

$$\Phi = \{\Phi_i(\{P_i^t, \{\hat{u}_j^t\}\})\}, i \in I, j \in J, t \in [1,T] \tag{1}$$

For optimization control actions are used that are defined by the vector

$$\gamma = \{\gamma_j^t\} = \{(\bar{\gamma}_j^t, V_j^t), j \in J, t \in [1,T]\} \tag{2}$$

where *J* is a set of the coordinated nodes at the sectional level. The component $\bar{\gamma}_j^t$ is a vector defining control actions for the *j*-th node in the period *t*. The vector V_j^t defines the parameters of the recommended functioning area $D_{oj}^{dt}(V_j^t)$ of the *j*-th node in the *t*-th period in the space of its functionality *U* defining the output nomenclature and quantity. Assume that $\gamma \in \Gamma$ where Γ is a range of possible control actions changes.

This approach is discussed in more detail in [12], where the problem of coordinating technological processes at the level of a separate node is solved. Here we would like to notice, that such formal methods can be used in the process of complex distributed technological system control, so appropriate software components which realize them may be included in ASC.

6 Some Implementation Issues

By now, our research team accomplished the implementation of the key components for base level node WSP, which was successfully implemented for the Ulyanovka gas field (Kharkiv region). Fig. 3 presents a fragment of the user interface which allows the system operator of this gas-field to check all the main technological parameters, receive and process the system messages and so on. The data exchange between system components is realized with the means of HTTP by transporting XML-documents which include the parameter values. The fragment of such a XML document is shown in Fig. 4.

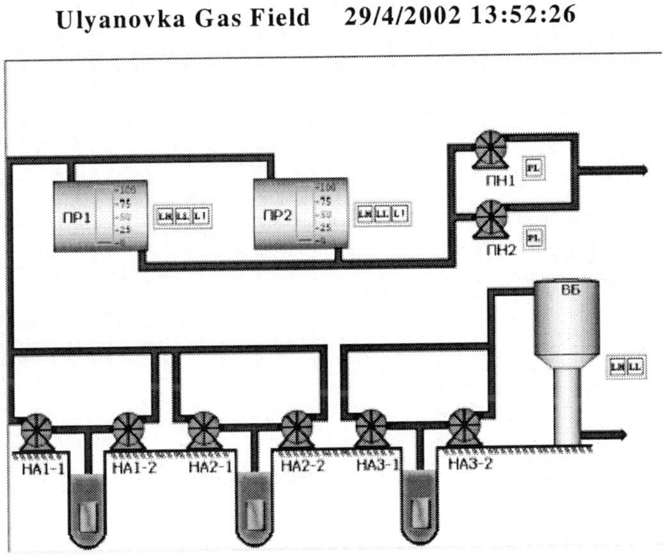

Ulyanovka Gas Field 29/4/2002 13:52:26

Fig. 3. User interface screenshot

7 Conclusions

In this paper we discussed some results we have achieved in TPC-IS development for the branch of gas-production enterprises in the Ukraine. The main features of TPC-IS

```
<?xml version="1.0" encoding="WINDOWS-1251"?>
<facility name="Ulyanovka Gas Field">
      <message id="1">
        <number>121</number>
        <title>The tap #13 is open</title>
        <type>Ordinal message</type>
      </message>
...
      <parameter id="1">
        <name>Gas pressure in separator</name>
        <value>15,2</value>
        <dimension >MPa</dimension >
      </parameter>
...   </facility>
```


Fig. 4. Fragment of a XML document containing technological parameters values

were discussed, and it was proved, that it is advantageous to realize such systems using Web technology. Architectural patterns were introduced as mean for structuring and easing the design and development of complex TPC-IS. Implementation issues for some of the described patterns solution were also presented.

Acknowledgments. We would like to express our thanks to our industrial partners in Ukraine, especially to Mr. Y.M. Chrapach ("KharkivGas-Vidobuvannya" company), Mr. S.V Ovasapov and Mr. A.I. Kuzmin (design firm «PromAvtomatika», Kharkiv, Ukraine), for their fruitful project cooperation, and to the graduate students participating in our R&D group Mr. K.Shchekotykhin and Mr. P. Pochuyev.

References

1. AdAstra Group, Homepage,.URL: www.tracemode.ru, accessed on Apr, 2002
2. Motorola MOSCAD Homepage, URL: http://www.moscad-systems.com/, accessed on Apr, 2002
3. "Attention: SCADA!", technical paper, URL: www.scada.com.ua/article2.html, accessed on Apr, 2002
4. V. Kanninen, "PC-based Automation Systems in Chemical Research", URL: http://www.hut.fi/Yksikot/Auttieto/rap2/kanninen/kanninen.htm, accessed on Apr, 2002
5. M. Tkachuk, D. Kuklenko, S. Ovasapov, K. Shchekotykhin, "Knowledge-based Maintenance Environment for Large Information Handling Systems", Lecture Notes in Informatics (LNI), GI-Edition, P-2, Bonn 2001, pp. 139–154.

6. D.V. Kuklenko, H.C. Mayr, M. V. Tkachuk, K. M. Shchekotykhin, "Web-Based Information Systems For Technological Process Control: Architectural Framework And Software Solutions", "Problems of programming", the magazine of Science Academy of the Ukraine, Kyiv, #1-2 (special bulletin), pp.317-325.
7. M. Tkachuk, R. Kaschek, S. Popov, A. Habboush, "A Gas – Compressor Station Case Study in Software Re-engineering", ReTIS-2000: 6-th International Conference on Re-Technologies for Information Systems (Preparing to E-Business), Oesterreichische Computer Gesellschaft,. Band 132, 2000, pp. 142–153
8. D.Kuklenko, M.Tkachuk,."Retrospective Database in TPC-IS: Concept and Implementation Experiense", "System Analysis, Management and Information Technologies", Kharkiv, Ukraine, #8, 2001, pp. 62-68 (in Russian).
9. W3C Homepage, URL: http://www.w3.org/, accessed on May, 2002
10. S. Purba, "High Performance Web-Databases: Design, Development, and Deployment",. CRC Press LLC, USA, 2001
11. A.Kaiser. "Time-dependent Data Modeling", Peter Lang GmbH, Europäischer Verlag der Wissenschaften, Frankfurt am Main, 2000
12. M. D. Godlevsky, Ya. N. Ghamlouche. "Principle of Construction of a MIS Subsystem of the Distributed Hierarchical Technical-Economic System Development Management", Proceedings of the International Conference on Automatic Control "Automation 2000", Lviv, 2000, pp.103–106.

A Comparison of Techniques to Estimate Response Time for Data Placement*

Shahram Ghandeharizadeh[1], Shan Gao[1], and Chris Gahagan[2]

[1] University of Southern California, LA, CA 90089
[2] BMC Software Inc., Houston, TX 77042

Abstract. Technological advances in networking, mass storage devices, processor and information technology have resulted in a variety of data services in diverse applications such as e-commerce, health-care, scientific applications, etc. While the cost of purchasing technology is becoming cheaper, the same cannot be stated about the cost of *managing* an information infrastructure. In order to reduce this cost, one needs tools that empower system administrators to explain and reason about a storage subsystem's past performance, e.g., response time. Ideally, an administrator would employ these tools to speculate on both physical organization of data and hardware changes. With a hypothetical change, one may use the previously observed response times to quantify the expected enhancements. In this study, we investigate linear regression, a M/D/1 queuing model and SEER as three alternative techniques to estimate response time. All techniques enable an administrator to speculate on changes to the placement of data and its expected impact on response time. A choice between these techniques is a tradeoff between accuracy and space/computational complexity to estimate response time. In our experimental studies, SEER provides a higher accuracy by using more storage space and computational cycles.

1 Introduction

An economical trend in the area of mass storage is cheaper, faster devices that provide higher capacities. The same trend does not apply to the cost of *managing* an information infrastructure [6,7]. Management cost includes (a) the overhead of an administrative staff to maintain systems and (b) the cost associated with down-times that render a service (data) unavailable. With the fast pace of technological changes, it is unrealistic to develop a tool that fully automates the management of a computing infrastructure. Instead, it is more appropriate to develop tools that empower administrators to reason about a system's behavior, identify limitations, speculate on solutions to address these limitations, and evaluate the impact of their proposed solutions. These tools must be extensible in order to incorporate new hardware solutions as they become available.

Response time is an important performance criterion for many applications. A slow system might appear as unavailable to the end user, contributing to the down-time costs.

* This research was supported in part by an unrestricted cash gift from BMC Software Inc. The email addresses for S. Ghandeharizadeh and S. Gao are {shahram,sgao}@cs.usc.edu. C. Gahagan's email address is chris_gahagan@bmc.com.

M.H. Shafazand and A M. Tjoa (Eds.): EurAsia-ICT 2002, LNCS 2510, pp. 730–738, 2002.

The focus of this paper is to evaluate the alternative techniques that estimate system response time. These techniques are intended for those environments where: a) a request employs only one resource, and b) the environment gathers and stores data about user requests and system behavior over a period of time.

A simple approach to computing response time is to maintain a simulator that models the critical components of an application. The system collects a trace of when a request arrives and how much resources it consumes. When the system administrator inquires about a proposed change, the system invokes the simulator with the gathered traces in order to estimate the change in response time. This approach, termed Absolute, would provide a fairly accurate estimate of the anticipated response time (assuming the simulator models system components accurately). It would require some storage space for the trace data and CPU cycles to invoke the simulator. With the current technological advances, Absolute might become feasible in the near future.

The focus of this paper is on techniques to estimate response time of a system. Even with Absolute becoming feasible, these techniques continue to be of use for several reasons. First, they will always be faster than Absolute because they have a lower complexity. Second, they provide a user with an approximate view of the anticipated response time improvements. The user may then invoke Absolute on those temporal regions that are of interest. This is similar to an art critic previewing low-resolution images of many paintings. For those paintings (regions) of interest, the critic retrieves full-resolution images (invokes Absolute).

In order to have a focused description of the alternative techniques, and without loss of generality, assume a multi-disk computer as our hardware platform, see Figure 1a. We assume the relations that constitute the database are horizontally declustered [5, 4, 1] into fragments, with one or more fragments assigned to a single disk. We consider requests at the granularity of block references that retrieve a single block from a fragment. Thus, a request references a single disk drive. One objective of this study is to observe the past response times and reason about changes that would improve these observed response times. With this objective, we break time into intervals, termed time slices. For each time slice, we store enough information to: a) reconstruct the observed response time, and b) reason about changes to the placement of fragments and how it would impact the observed response time. For example, for a given time slice, we strive to answer the following hypothetical question: "How would the observed response time change if fragments f_1 and f_2 were assigned to disk number 6?" Obviously, we want to provide an accurate response with minimal storage, bandwidth, and computational resources.

In Sections 2, 3, and 4, we describe linear regression, queuing models, and SEER as alternative techniques for our objective. In each section, we present experimental results from a 9 disk configuration that services requests based on a trace of requests gathered from an Oracle database management system. The pattern of request arrival is shown in Figure 1b. In this figure, the x-axis consists of time slices where each time slice is six minutes long. Clearly, this arrival pattern does **not** correspond to a Poisson distribution. In the reported experiments, the requested block size is fixed, resulting in a 6 millisecond service time. With these experiments, linear regression stores the following for each time slice: a) a pair of values, i.e., (load, average response time),

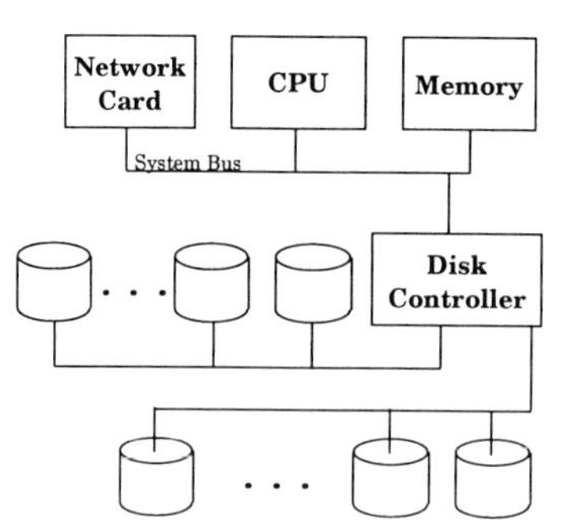

1a. A multi-disk hardware platform

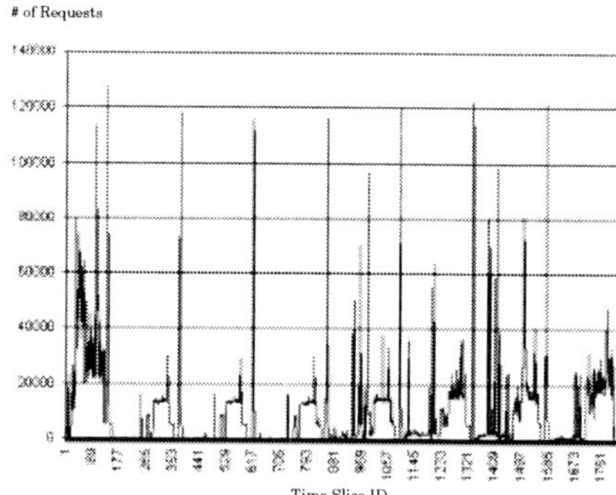

1b. Number of requests as a function of time

Fig. 1. Target hardware and a trace driven load.

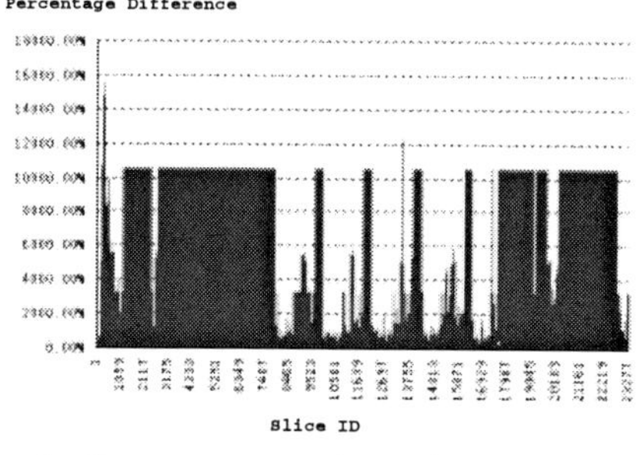

2a. Percentage error for each time slice

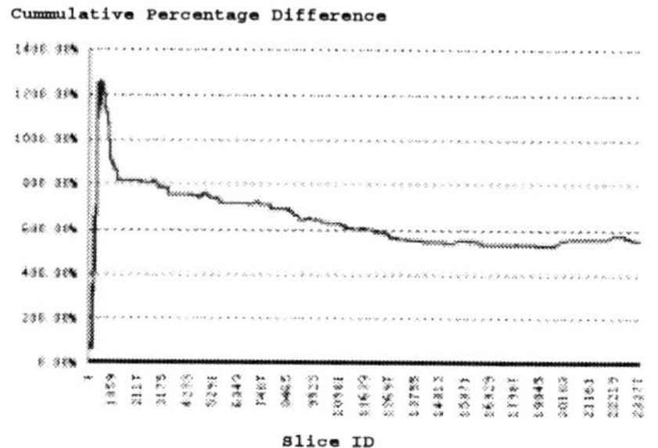

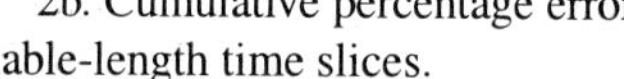

2b. Cumulative percentage error

Fig. 2. Linear regression with variable-length time slices.

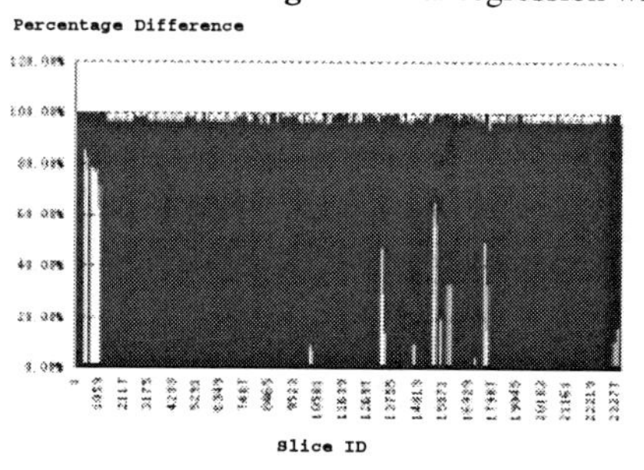

3a. Percentage error for each time slice

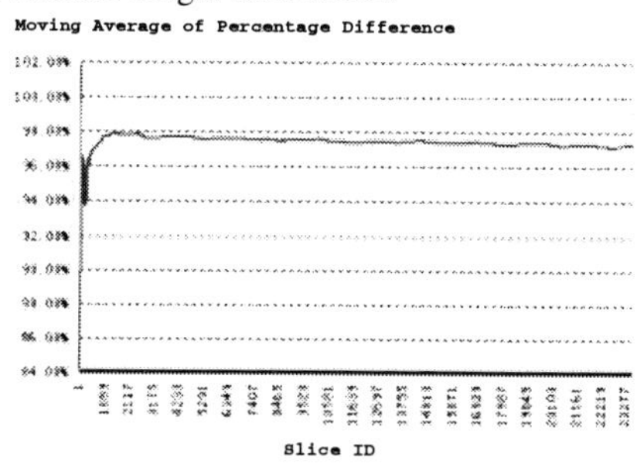

3b. Cumulative percentage error

Fig. 3. M/D/1 with fix-length time slices.

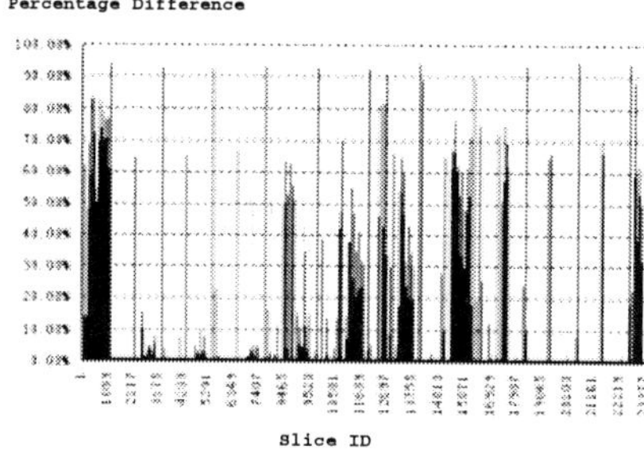

4a. Percentage error for each time slice

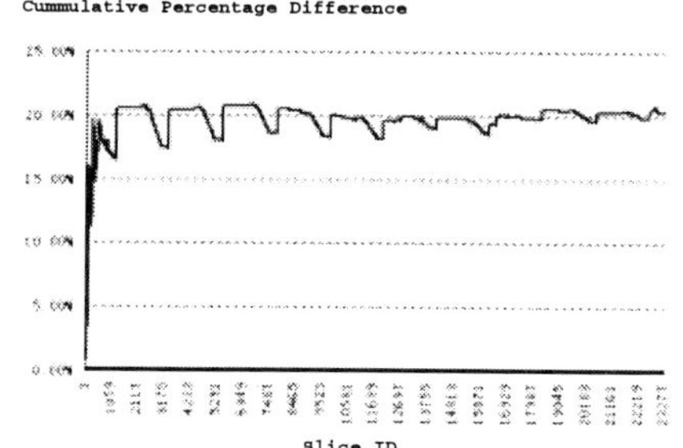

4b. Cumulative percentage error

Fig. 4. SEER with variable-length time slices.

and b) a vector of N values pertaining to the load imposed by each fragment during a time slice. It computes an equation that can estimate response time for a) each time slice, and b) a hypothetical data placement. The queuing models store a similar data set with one difference. Instead of a pair of values, it only stores system utilization per time slice. SEER is a new technique that stores significantly more data, an $N \times N$ matrix. Its computational complexity is also higher. However, it provides a higher degree of accuracy. The current technological trends justify the use of SEER. We conclude with brief remarks in Section 5.

2 Linear Regression

A simple statistical approach to estimate response time is linear regression. With this approach, we record the response time and system load for each time slice. This produces a scatter plot of response time as a function of system load. Thus, there might be several observed response times for each load. The load can be quantified in different ways. We quantify it as the number of bytes retrieved during a time slice.

Linear regression inputs the points on the scatter plot to compute a function for response time:

$$RT = a + (b \times load) \tag{1}$$

where a is an offset relative to the response time, resembling the minimum response time, and b is the slope of the line specified by the linear function. Each observed response time has a certain distance from the line specified by equation 1. This might be a positive or negative value. It is positive when it exceeds that of the point predicted by Equation 1. It is zero if it matches the point predicted by Equation 1. Otherwise, it is a negative value. The following equations for a and b ensure that the sum of these positive and negative values is zero:

$$b = \frac{N \times \sum(RT \times load) - (\sum load \times \sum RT)}{N \sum load^2 - (\sum load)^2} \tag{2}$$

$$a = \frac{\sum RT - b \times \sum load}{N} \tag{3}$$

N denotes the total number of observed response times.

In order to reason about the placement of data, the system can store the number of bytes retrieved from a fragment per time slice. This is the load imposed by that fragment on a given disk during a time slice. With a hypothetical assignment of fragments to a disk, we sum the total load attributed by these fragments and use Equation 1 to estimate an average system response time.

We first quantified the accuracy of linear regression with fixed-length time slices. We used a simulator to measure both the observed response time and system load for each time slice. We stored this information in a database management system. Next, we computed the b and a values using Equations 2 and 3. Using these, we invoked a program that employs linear regression to estimate the response time for each time slice (with the observed load). We observed a high percentage error between the observed and estimated values for each time slice. A factor contributing to this error is the fixed-length time slices. This is because requests may arrive in one time slice (say S_1) and finish during the subsequent time slice (say S_2). With our methodology, the load and response time of these requests are attributed to S_2. We adjusted our methodology by using variable-length time slices. With this methodology, a time slice does not end until the system is idle, i.e., there are no active requests in the system. Thus, both the load and observed response times are attributed to a single time slice. Figure 2 shows the percentage error with this methodology. In these figures, the first 9 ticks on the x-axis correspond to the percentage error for each of the 9 disks for the first time slice of Figure 1b. Hence, there are 9 times as many ticks on the x-axis of these figures when compared with Figure 1b. Figure 2.b show the cumulative average percentage difference as a function of time. This value is computed as a function of the x-axis by computing the average for the first x values. For example, with x-value equal to 200, the reported y value is the percentage error observed for the first 200 observed x-values. In essence, the error in one time slice is carried over to the computation of the percentage error for the next time slice. In these figures, the minimum time slice is six minutes long. These results demonstrate that the accuracy of linear regression is improved using variable length time slices. In these experiments, the maximum time slice is 995848 milliseconds (16.59 minutes) long.

The obtained results demonstrate that our use of linear regression is inaccurate for estimating response time. In particular, our measure of load fails to capture the bursty nature of request arrivals, see Figure 1b. To illustrate, assume that the service time of each request is 1 time unit and consider the following two scenarios. In the first, twelve requests arrive one time unit apart during one time slice. In the second, the same twelve requests arrive at the beginning of a time slice. Both time slices report the same observed load. However, while the average response time in the first case is 1 time unit, the average response time in the second scenario is 6.5 time units. The accuracy of linear regression might be improved if response time is described in terms of both the imposed load and burstiness of requests.

3 Queuing Models

There is an existing body of work from queuing theory [3] and operation research to estimate response time. M/G/N is Kendall's notation for representing a queuing model. The first part represents the input process to the model, the second is the service distribution, and the third is the number of servers. For example, a M/M/1 model has a Poisson arrival rate (M is an abbreviation for Markovian), exponential service time, and consists of one server. A M/D/1 is similar except that its service time is deterministic. The Pollaczek-Khinchin mean formula states that the mean response time of a one server system that employs a first-come-first-serve (FCFS) policy is:

$$(1 + \frac{1 + C_v^2}{2} \times \frac{\rho}{1 - \rho}) \times E[S]$$

E[S] is the expected service time, ρ is the utilization of the server, and C_v^2 is the squared coefficient of variation on service time. C_v^2 is defined as, $C_v^2 = \frac{V[S]}{E[S]^2}$, where V[S] is the variance of the service time.

With a M/D/1 queuing model, the service time is constant. Hence, V[S] equals zero, and C_v^2 in turn is zero. For this model, the average response time is:

$$(1 + \frac{1}{2} \times \frac{\rho}{1 - \rho}) \times E[S]$$

With a M/M/1 queue, the service time is exponentially distributed. Hence, the variance in service time, V[S], equals the square of the expected service time, E[S]^2. This means C_v^2 equals 1. Thus, the estimated response time is: $\frac{E[S]}{1 - \rho}$. These models of response time assume a Poisson arrival rate. A departure from this assumption changes average wait time and queue lengths.

For each time slice, the system may store the utilization of the server, the service time of the disk and its variance in order to estimate the response time. For example, if the utilization of the disk drive is 40% in a time slice and its service time is deterministic and fixed at 6 milliseconds then the expected response time is 8 milliseconds. Moreover, it may store a vector that maintains the utilization attributed to each fragment per time slice. Based on this, it can reason about placement of fragments and compute an answer to the hypothetical question posed in Section 1.

Figure 3 shows the accuracy of Equation 3 to estimate response time. Similar to the discussion of Section 2, we ran our simulator with a fixed service time of 6 milliseconds and 6 minute (fix-length) time slices, and stored the observed response time and utilization. Utilization is the percentage of time the system is busy during a time slice. If T is the duration of a time slice and the system is busy for B time units then utilization is $\frac{B}{T}$. Next, we used Equation 3 to estimate response time for each time slice (using the observed utilization) and compare it with the observed response time. Figure 3.a shows the percentage error for each time slice. Figure 3.b shows the cumulative percentage error.

One factor contributed to the observed inaccuracies is the arrival pattern of requests with the trace; it does not correspond to a Poisson distribution. Another contributing factor is the fix-length time slices. Similar to the discussion of Section 2, we configured the system to use variable length time slices with a minimum length of 6 minutes. In our experiments, the variable length time slice had no impact on the observed percentage error (and is eliminated from presentation due to lack of space). Generally speaking, one must use M/M/1 model when the service time is not deterministic. In this case, if the distribution of observed service time is not exponential then one might observe a higher percentage of error.

When compared with linear regression, the queuing model appears more accurate because its x-axis has a much smaller scale. One reason for this is that equation 3 always under-estimates the average response time. Thus, the percentage error is always below 100%. Of course, the parameters of linear regression, a and b, could also be adjusted to always underestimate response time (by using the minimum observed response time for each load).

4 SEER

One approach to estimate response time with 100% accuracy is to store the average wait time and service time per time slice. (We assume variable-length time slices where a time slice has a minimum duration and ends only when the system is idle.) However, this cannot estimate the expected response time when the placement of fragments across the disks is modified.

In this section, we describe SEER, a new approach to estimate the average response time. Our experimental results demonstrate that this technique is more accurate than the other two alternatives. This new approach stores the following per time slice: a) the average service time for requests that reference a fragment, S(f_i), b) the total number of requests referenced during that time slice, NumReq, and c) an SEER matrix that is used to estimate the wait-time between the requests that reference fragments. SEER is also a measure of burstiness for request arrivals. It is a $N \times N$ matrix where N is the number of fragments. Assuming that the fragments in the system are uniquely numberd as f_1 to f_N, the value $f_{i,j}$ in the matrix denotes SEER(f_i, f_j). These values are initialized to zero at the beginning of a time slice. When a request that references fragment f_j arrives, it has a service time, S_{f_j}, and an arrival time, T_{arr,f_j}. The sum of these two values is the departure time of f_j, $T_{depart,f_j} = T_{arr,f_j} + S_{f_j}$. The same holds true for each request referencing fragment f_i. (Note that we do not consider the wait time in queues.) For two requests that reference f_i and f_j, if $T_{arr,f_i} \leq T_{depart,f_j} \leq T_{depart,f_i}$ then SEER(f_i, f_j) is a positive value and equal to $T_{depart,f_j} - T_{arr,f_i}$, see Figure 5. Otherwise, it is zero. This value is added to SEER(f_i, f_j) in the matrix. Note that SEER(f_i,f_j) is defined for all fragments, i.e., there is a value for two fragments that reside on different disks. This information is useful because it enables the system to hypothesize about scenarios when these two fragments are assigned to the same disk.

The computation of the SEER matrix is as follows. For each fragment, the system maintains an arrival time and departure time. Upon the arrival of a request referencing fragment f_i, the system checks to see if the arrival time of this request is greater than the current departure time for f_i. If this is the case then the departure time of f_i is set to the departure time of f_i. Otherwise, the system (a) computes the SEER between fragment f_i and itself[3], and (b) the departure time of f_i is incremented with the service time of the new request. Next, it computes the temporal overlap between this fragment and every other fragment with a departure time greater than the arrival time of this request. The obtained value is added to SEER(f_i, f_j).

The value of SEER defines an optimistic wait time between requests referencing two different fragments. For a time slice, if one wants to speculate on the average response time with K fragments, numbered f_1 to f_K, assigned to the same disk drive then we add the observed average service time for the requests referencing these K fragments, $S_{avg}(K)$, to the wait time expected between these K fragments, $W_{avg}(K)$. These quantities are defined as follows: $S_{avg}(K) = \sum_{i=1}^{K} S_{avg}(f_i)$, Where $S_{avg}(f_i)$ is the average service time observed for requests that reference fragment f_i. The wait time between these K fragments is defined as:

$$W_{avg}(K) = \frac{\sum_{i=1}^{K-1} \sum_{j=i+1}^{K} SEER(f_i, f_j) + \sum_{i=1}^{K} SEER(f_i, f_i) + \sum_{i=2}^{K} \sum_{j=1}^{i-1} SEER(f_i, f_j)}{NumReq}$$

[3] This is the self temporal overlap. See the next paragraph for further discussion.

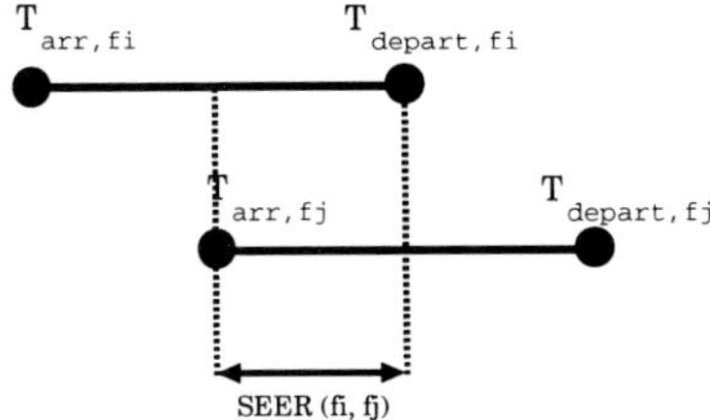

Fig. 5. SEER between two requests referencing fragments f_i and f_j.

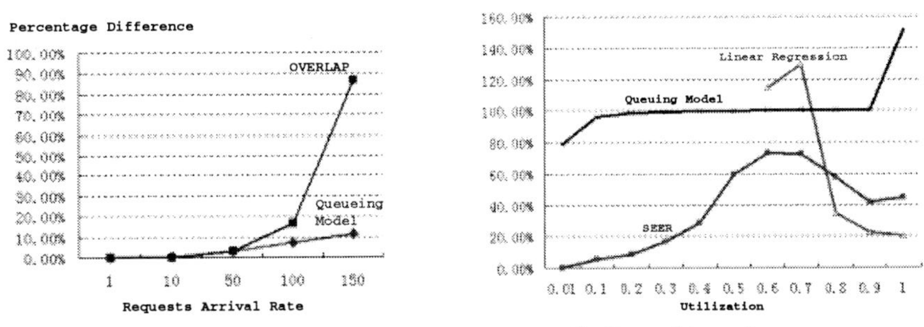

6a. Poisson arrival rate, 1 server 6b. Trace driven, 9 servers

Fig. 6. Comparison of M/D/1 and SEER

The self overlap, SEER(f_i, f_i), defines how long the requests that reference the same fragment overlap with one another in the system.

Figure 4 shows the percentage difference between the predicted and observed response times for variable length time slices[4]. The obtained results are a significant improvement when compared with linear regression and M/D/1 queuing models. Similar to the M/D/1 queuing model, SEER under-estimates the average response time of a request. This is because it does not consider the wait-time in queues to service requests. To illustrate assume that three requests (r_a, r_b and r_c) reference different fragments (f_a, f_b and f_c) assigned to a single disk. Assume a) the service time of each request is 1 time unit, 2) r_a and r_b arrive at time T_1, and 3) r_c arrvies at time $T_1 + 1$. Because of queueing delays r_c might overlap either r_a or r_b. However, our current model ignores such delays. Hence, it sets SEER(f_b,f_c) to zero because it employs only service time to estimate how long these two requests are in the system. This explains why the percentage error is higher when the system load is higher, compare Figures 4a and 1b.

5 Discussions and Future Research Directions

This paper details three alternative techniques to estimate response time and reason about the placement of data across multiple disks. We employed a trace driven eval-

[4] With fix-length time slices, the x-axis of Figure 4.a would have a scale that ranges in value from 0 to 1600%. We eliminated these results because the previous sections motivated the importance of using variable length time slices.

uation study to compare these techniques with one another. SEER is more accurate when compared with linear regression and M/D/1 queuing model because it maintains state information pertaining to burstiness of requests. Note that SEER outperforms the queuing model because pattern of request arrivals is not Poisson. With an environment consisting of a single server and a Poisson arrival rate, the queuing model and overlap provide the same accuracy with low system loads (utilization less than 20%), see Figure 4a. With high system utilization (high arrival rates), the M/D/1 queuing model is significantly more accurate than SEER.

We also analyzed the percentage of error for different system utilizations using the trace driven study, see Figure 4b. In this figure, the y-axis is the percentage error for each observed system utilization. With a low system utilization, the percentage error is so high for linear regression that it eclipses the difference between the queuing models and SEER. Hence, we show linear regression starting at 60% utilization. In our experiments, linear regression predicts response time with a high accuracy when the system load is high (more than 80%). SEER predicts response time with a high accuracy with a low system load because it ignores queuing delays. While its percentage error increases with a higher system load, it is more accurate than queuing model.

An immediate future research direction of this effort is to analyze extensions to SEER to further increase its accuracy. In addition, this study assumed a deterministic response time for all block requests, i.e., 6 milliseconds. While this might be reasonable for database applications because they reference the storage subsystem with fixed block sizes, it does not hold true for all data intensive applications, e.g., continuous media servers [2]. We intend to investigate extensions of this study in support of variable length blocks. A long term objective is to develop a re-configurable simulator that collects traces and reconfigures itself to dynamically model its target environment. This simulator should be extensible to include new solutions. It should empower a system administrator to reason about the system's past behavior and estimate enhancements with proposed hardware changes, e.g., purchase of newer disk models, replacement disks, etc.

References

1. S. Ghandeharizadeh and D. DeWitt. A Multiuser Performance Aanalysis of Alternative Declustering Strategies. In *Proceedings of the 6th IEEE Data Engineering Conference*, 1990.
2. S. Ghandeharizadeh and R. Muntz. Design and Implementation of Scalable Continuous Media Servers. *Parallel Computing*, pages 91–122, 1998.
3. L. Kleinrock. *Queuing Systems*. Wiley Interscience, 1975.
4. M. Livny, S. Khoshafian, and H. Boral. Multi-disk management algorithms. In *Proceedings of the 1987 ACM SIGMETRICS Conference on Measurement and Modeling of Computer Systems*, pages 69–77, 1987.
5. D. Ries and R. Epstein. Evaluation of Distribution Criteria for Distributed Database Systems. Technical Report UCB/ERL Technical Report M78/22, UC Berkeley, May 1978.
6. A. Veitch, E. Riedel, S. Towers, and J. Wilkes. Towards global storage management and data placement. Technical Report HPL-SSP-2001-1, Hewlett Packard Laboratories, March 2001.
7. Gerhard Weikum. The Web in 2010: Challenges and Opportunities for Database Research. In *Informatics*, pages 1–23, 2001.

Using a Real-Time Web-Based Pattern Recognition System to Search for Component Patterns Database

Sung-Jung Hsiao[1], Kuo-Chin Fan[1], Wen-Tsai Sung[2], and Shih-Ching Ou[2]

[1] Artificial Intelligence and Pattern Recognition Lab. Department of Computer Science & Information Engineering, National Central University, Taiwan
song1208@ms5.hinet.net

[2] Automation & CAD Lab., Department of Electrical Engineering, National Central University, Taiwan
sung.wentsai@msa.hinet.net

Abstract. Faraway engineers are able to sketch direct the shape of engineering components by the browser, and the recognition system will proceed with search for the component database of company by the Internet. In this paper, component patterns are stored in the database system. Component patterns with the approach of database system will be able to improve the capacity of recognition system effectively. In our approach, the recognition system adopts distributed compute, and it will raise the recognition rate of system. The system uses a recurrent neural network (RNN) with associative memory to perform the action of training and recognition. The final phase joins the technology of database match in process of the recognition except distributed compute, and it will solve the problem of spurious state. In this paper, our system will be carried out in the Yang-Fen Automation Electrical Engineering Company. The plan of experiment has gone through four months, and their engineers are also used to take advantage of the way of Web-Based pattern recognition.

Keywords. Web-Based, Pattern Recognition, engineering components, component database, RNN

1 Introduction

Several recognized procedures, with limited capacity, are considered. Such technologies could be partially improved but they have not yet yielded an optimal solution to the restricted capacity when many data are involved, Kak, Duin, and Kraaijveld [2][3][4].Associative memory is critical in a neural network used as an approach for pattern recognition. Many studies of pattern recognition have focused on the structure of associative memory, TAN and Perus [5][6]. The recurrent neural network (RNN) possesses the function of non-linear associative memory. The RNN is very effectively used in pattern recognition, Brouwer [7][8].

In many sample patterns, our paper proposes using the shape of engineering component and circuit sign to take as a recognized sample pattern. The user is able to input

M.H. Shafazand and A M. Tjoa (Eds.): EurAsia-ICT 2002, LNCS 2510, pp. 739–746, 2002.

the pattern which will be searched in the handwritten region of client-end. The system begins to perform the recognition task after we click the searching button. In the recognized process for training phase, the system uses the method of distributed compute to improve the capacity of stored pattern. On the other hand, in the retrieval stage, the Web-Based system will use the technology of database contrast to easy improve the problem that the RNN produces spurious states. A simulation experiment is also discussed to clarify and corroborate the above Web-based PR technology.

2 System Calculation and Phase Analysis

In the classical approach, Kamp and Hasler [9], an RNN is a discrete-time discretely valued dynamic system which, at any given time, t, is characterized by a binary state vector

$$x(t) = \left[x_1(t),\ldots,\ x_i(t),\ldots x_n(t)\right] \in \{1,-1\}^n \tag{1}$$

The behavior of the system is given by a dynamic equation of the type,

$$x_i(t+1) = \operatorname{sgn}\left[\sum_{j=1}^{n} W_{ij} X_j(t) - \theta_i\right] \qquad i = 1,2,\ldots\ldots.n. \tag{2}$$

A point, x, is fixed for any pattern prototype vectors, $\xi^1, \xi^2, \ldots. \xi^p$,Gimenez [10]

$$\xi^u = \left[x_1(t),\ldots., x_{i(t)},\ldots x_n(t)\right] \in \{1,-1\}^n \tag{3}$$

In our approach, x (t) is a record in the pattern database. $x_1(t)$or $x_i(t)$ is a field of any data record. Furthermore, bipolar data are between 1 and -1. A "1 "represents a black point in the pattern, and a "-1" represents a white blank in the pattern.

Initially, in the training stage, the records of a pattern database are cut; a distributional computation is employed to determine the W and θ values of every segment, and afterwards W and θ of every segment are again used to Eq.(2) and, thus, determine the most similar pattern records of retrieval in every segment.

2.1 Storage Phase

According to the outer product rule of storage, Hebb's postulate concerning learning the synaptic weight from neuron i to neuron j is generalized as,

$$W_{ji} = \frac{1}{p}\sum_{\mu=1}^{N} \xi_{\mu \cdot j} \times \ \xi_{\mu \cdot i} \tag{4}$$

1/p is taken as the constant of proportionality to simplify the mathematical description of information retrieval. Simon [12], Notably, the learning rule in Eq. (4) is a "one shot" computation.

In the normal operation of the Hopfield network, the following is set.

$$W_{ii}=0 \text{ for all } i\ ,\ i=1,\ldots,p \tag{5}$$

$W_{ii}=0$, prevents positive feedback, Hopfield [13].

Let W denote the P by P synaptic weight matrix of the network, with Wji as its jith element. Equations (4) and (5) can then be combined into a single equation written in matrix form:

$$W = \frac{1}{P}\sum_{\mu}^{N} \xi_{\mu} \xi_{\mu}^{T} - \frac{N}{P} I \tag{6}$$

I is the P x P identity matrix, and W is a symmetric matrix of which the diagonal line is zero in all places.

$$W = \begin{bmatrix} W_{11} & \cdots & W_{1p} \\ \vdots & \ddots & \vdots \\ W_{p1} & \cdots & W_{pp} \end{bmatrix} \tag{7}$$

$$= \begin{bmatrix} 0 & \cdots & \cdots & W_{1p} \\ \vdots & 0 & & \vdots \\ \vdots & & \ddots & \vdots \\ W_{p1} & \cdots & \cdots & 0 \end{bmatrix} \tag{8}$$

The threshold of the jth neuron has two modes:

$$\theta_j = 0, j = 1,..., p$$

or (9)

$$\theta_j = \sum_{i=1}^{P} W_{ij}, i = 1,..., p \tag{10}$$

The threshold of Eq.(10) can increase the memory capacity of the network , Mueller Reinhardt, and Strickland [14].

2.2 Retrieval Phase

If a recognizing pattern vector X is input, then the initial output value is X (0). Every neuron follow-up output is computed by Eq.(11)

$$\begin{aligned} X_j(n+1) &= \operatorname{sgn}(\sum_{i=1}^{P} W_{ji} X_j(n) - \theta_j) \\ &= \operatorname{sgn}(u_j(n) - \theta_j) \\ &= \begin{cases} 1 & \text{if} \quad u_j(n) > \theta_j \\ X_j(n) & \text{if} \quad u_j(n) = \theta_j \\ -1 & \text{if} \quad u_j(n) < \theta_j \end{cases} \end{aligned} \tag{11}$$

In Eq. (11), n is the number of iterations. Importantly, the discrete Hopfield network used an asynchronization method to alter the output of each neuron, and the complete process of associative memory employed Eq. (12) to describe the chain-state relationship:

$$X(0) \rightarrow X(1) \rightarrow X(2) \rightarrow \ldots \rightarrow X(k) \rightarrow X(k+1) \rightarrow \ldots \tag{12}$$

Although an asynchronization method is used here to change the output of the network, X converges on the stable state, sometimes also on the incorrect recall, Simon [18]. The X state of final convergence is therefore used to match the original pattern database. Computing each Hamming distance determines the minimal value of the dH , Jinwen [19]. With n pattern records, the Hamming distance is computed by,

$$dH = \sum_{i=1}^{P} \left| X_i - \xi_i^u \right| \quad , \quad u = 1,2,...n \tag{13}$$

And the minimal value is,

$$dHmin = \min\{\sum_{i=1}^{p}|X_i - \xi_i^1|, \sum_{i=1}^{p}|X_i - \xi_i^2|, \cdots\cdots, \sum_{i=1}^{p}|X_i - \xi_i^n|\} \tag{14}$$

If the convergent result of the $\underline{X}$ equals a vector of the sample pattern, ξ^u, then $dHmin = 0$. If the convergent result of the $\underline{X}$ does not equal a vector of the sample pattern, ξ^u, then $dHmin > 0$. In such a case, $\underline{X}$ is similar to the sample pattern, ξ^u.

3 Storage Capacity Analysis and Improvement

As an important model of associative memory, the Hopfield network has been comprehensively researched and applied to pattern recognition using the sum-of-outer products, Hopfield [11]. Further research has addressed asymmetric or generalized Hopfield model with other learning algorithms, since the memory capacity of the Hopfield network using the sum-of-outer products scheme, is very low, McEliece, Abbott,and Venkatesh [21][22][23].

The capacity of a Hopfield RNN is the number C of stable states it has. Obviously, C depends on the weight matrix, which is taken to be symmetric with zeros on the diagonal. McEliece et a1. [21] showed that,

$$P/[(4)\ln(P)] < C < P/[(2)\ln(P)] \tag{21}$$

For example, for 100 neurons, C satisfies 5<C<10, where C is the number of data records in the stable state.

The memory capacity of a discrete Hopfield network has an upper limit. If the number of neurons is P, Eq. (21) yields,

$$M_{max} = \frac{P}{2\ln(P)}, M_{max} \text{, which is the maximum memory capacity.} \tag{22}$$

D.J.Amit[24] stated that the number, P, of neurons is 99% correct in the retrieval phase ,and the number of the stored data records is limited by the following formula.

$$M \leq \frac{P}{4\ln(P)}, \quad \text{M: memory capacity} \tag{23}$$

Notably, M in the Eq.. (22) and Eq.(23) becomes the basis of the divided segment in the pattern database: M is the number of distributed computation .

4 Implementing the Web-Based Pattern Recognition System

In Figure 1, the right of the web page is the client-end and the left of the web page is the server-end. If the user inputs the pattern at the client-end, the correct recognized result will be presented at server-end even if in the case that the source pattern of client-end be interfered with some noise. The new recognition system has already overcome many problems which previously existed.

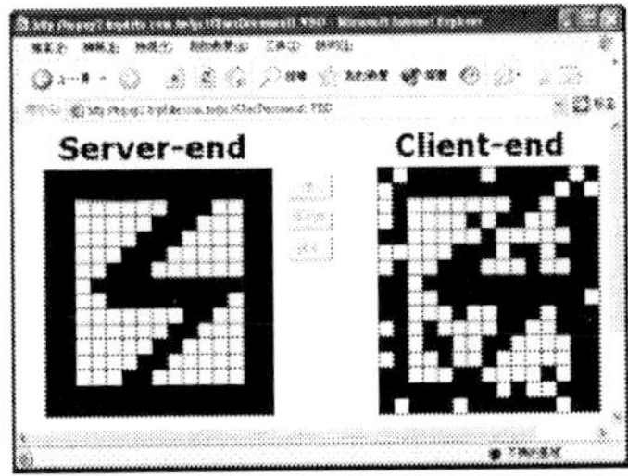

Fig. 1. Patterns are inputted at the client-end, and they are displayed at the server-end

Next, for a case of spurious states, the noisy pattern is input at the client-end, and the partial pattern of recall is recognized at the server-end. The pattern is not correctly recalled because such a sample pattern was not input, as shown in Figure 2.

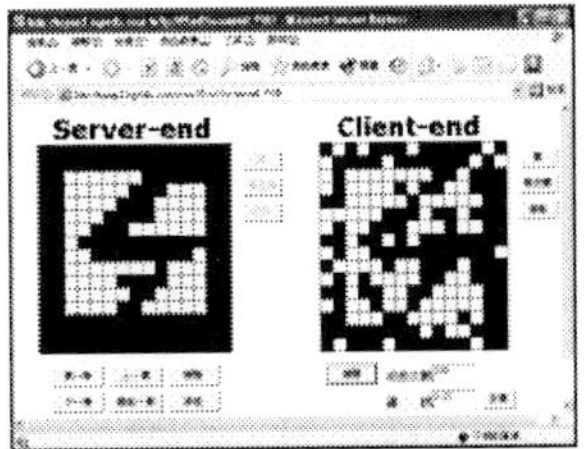

Fig. 2. Partially incorrect recollections

The system increased the matching of the pattern database. Accordingly, the partial incorrectness in Figure 2 is not again arisen. After the recognized result matched the pattern database, it converges to an accurate pattern, as in Figure 3.

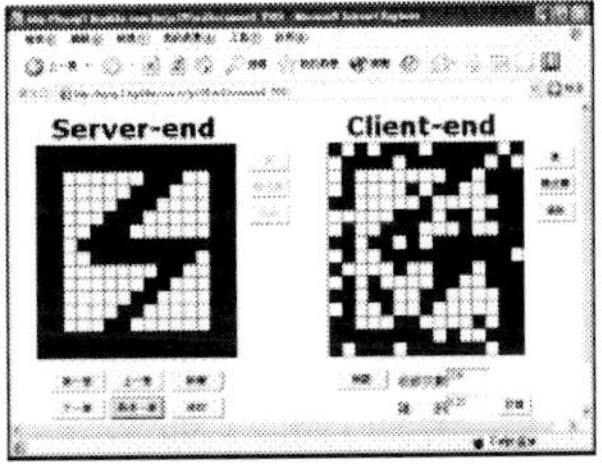

Fig. 3. Accurate condition for convergence after the recognized result matched the pattern database.

5 Cooperative Example

According to our Web-Based recognition system, our laboratory and Yang-Fen Automation Electrical Engineering Company had performed the plan of technologic cooperation during the time of January; 2002.Yang-Fen Automation Electrical Engineering Company operates in the installation of the power distribution equipment of the factory. Formerly these engineers query the head office about the stock amount of these components by telephone, and the pronunciations through telephone make mistakes of inquiry easily. Afterward Yang-Fen Company uses the method of web page to do these tasks of query by Internet. Unfortunately, there are some engineers forget these name of engineering component usually, and it will bring a persecution about the query of stock amount.

Next, Yang-Fen Company and we cooperate to do the experiment which uses Web-Based and real-time way to search for these patterns of component database by Internet, and the search is a recognized task namely recognized search. First, we build the pattern of the shape of each component in the server of component database. The pattern of each component uses their shape to become the pattern of component database. These component patterns are shown in the following figures from the Web page of Yang-Fen Company, such as figure 4.

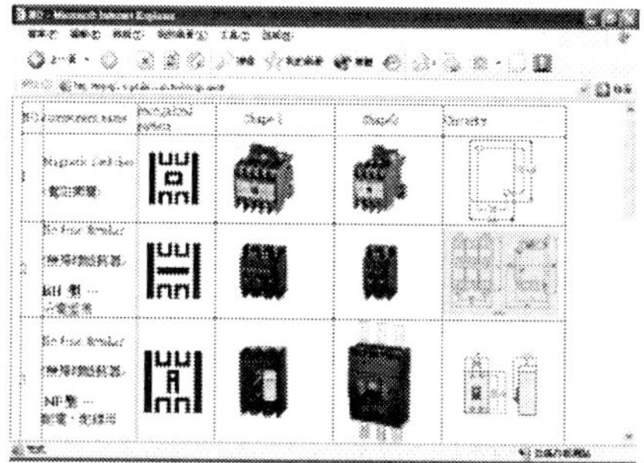

Fig. 4. Component patterns are list in the Yang-Fen's Web page.

According to the recognized statistics of Yang-Fen Company from January to April in the 2002, their engineers weren't used to be familiar with the operation of Web-Based recognition system in January; therefore, the recognition rate was low, and these conditions were improved until February. In the cooperative process, we modify the database of original component patterns frequently. We didn't let these component patterns too alike, and the recognition rate of system will be raised.

Next, we list these data that each engineer login the recognition system for the number of times of success and failure from January to April.

6 Conclusions and Future Work

A pattern database overcome many defects in recognition technology. This work provides three new solutions.

1. Using the pattern database to establish the learning pattern, and solves the problem of the capacity of RNN.

2. Adopting the matching technology of a pattern database to determine the most similar patterns reduces the prevalence of the spurious states of RNN; relatively and raise the recognition rate of a neural network.

The program presented here is built in the Web-server environment. The performance of the program is without delay because the system is real-time in learning and recognition. This method is new, and can ensure the completeness and security of these patterned data

Table 1. Recognized statistics for cooperative example

Month	Total recognition times	Correct recognition times	Incorrect recognition times	Recognition ratio
1	232	163	69	70.26%
2	256	228	28	89.06%
3	247	230	17	93.12%
4	269	262	7	97.40%

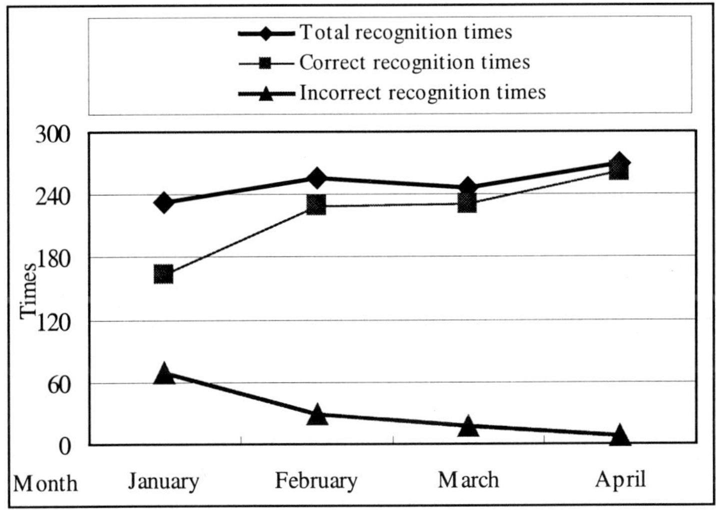

Fig. 5. The cooperative plan is analyzed by their histories.

References

1. Singh, S., "A Long Memory Pattern Modeling and Recognition System for Financial Forecasting," Pattern Analysis and Applications, vo1.2, no.3, (1999) 264–273.
2. S. Kak, "Better Web Searches and Prediction with Instantaneously Trained Neural Networks" IEEE Intelligent Systems, vo1.14, no.6, (1999) 78–81.

3. R.P.W.Duin, "Superlearning and neural network magic," Pattern Recognition Letters, vol.15, 1994, pp.215–217.
4. M.A.Kraaijveld and R.P.W.Duin, "The effective capacity of multilayer feedforward network classifiers,"Proc.12th Int'l Conf. on Pattern Recognition.(ICPR 94), Israel, vol.B,(1994). 99–103.
5. Z.TAN and M.K.ALI, "Pattern recognition with stochastic resonance in a generic neural network," International Journal of Modern Physics C, vo1.11, no.8, (2000)1585-1593.
6. M.Perus, "Neural networks as a basis for quantum associative networks," Neural Network World, vol.10, no.6, (2000) 1001–1013.
7. Brouwer, R.K., "An Integer Recurrent Artificial Neural Network for classifying Feature Vectors," International Journal of Pattern Recognition and Artificial Intelligence, vol.14, no. 3, (2000) 339–335.
8. Brouwer, R.K., "A Fuzzy Recurrent Artificial Neural Network for Pattern classification, "International Journal of Uncertainty, Fuzziness and Knowledge-Based Systems, vo1.8, no.5, (2000) 525–538.
9. Kamp, Y. and Hasler, M., Recursive Neural Networks for Associative Memory,: Wiley-Interscience Series in Systems and Optimization, England , (1990) 10–34.
10. V.Gimenez, L.Aslanyan, J.Catellanos, and V.Ryazanov. "Distribution Functions as Attractor for Recurrent Neural Networks," Pattern Recognition and Image Analysis. vol. 11, no. 3, (2001) 492–497.
11. J.J. Hopfield, "Neural networks and physical systems with emergent collective computational abilities," Proc. the National Academy of sciences, USA,vol. 79, (1982) 2554–2558..
12. Simon Haykin, Neural networks a comprehensive foundation, 2nd, Macmillan College Publishing Company, Inc., New York (1999).
13. J.J.Hopfield and D.W. Tank, "Computing with neural circuits: a model," Science, vol.233, (1986) 625–633.
14. B.Mueller, J.Reinhardt, and M. T. Strickland, Neural Networks, Springer-Verlag, Berlin Heidelberg (1995).
15. Zurada, J.M., Artificial Neural Systems, West Publishing, St. Paul, UN. (1992).
16. Lippmann, R.P., "An Introduction to Computing with Neural Nets," IEEE ASSP Mag., (1987)4-22, also reprinted in neural networks: Theoretical Foundations and Analysis, edited by C. Lau, IEEE Press, New York, (1992) 5–23.
17. W. A. Little and G.. L. Shaw, "Analytical study of the memory storage capacity of a neural network, "Mathematical Biosciences, vo1.39,no.1, (1978) 281–290.
18. Simon Haykin, Neural networks a comprehensive foundation, Macmillan College Publishing Company, Inc., New York(1994).
19. Jinwen.Ma, "A Neural Network Approach to real-time pattern recognition," International Journal of Pattern Recognition and Artificial Intelligence, vol.15, no. 6, 2001, pp. 934–947.
20. Ma, J.W., "The stability of the generalized Hopfield networks in randomly asynchronous mode," Neural Networks, vol.10, no.6, (1997) 1109–1116.
21. R.E.McEliece, E.C. Posner, E.R.Rodernich and S.S. VenKatesh, "The capacity of the Hopfield associative memory, "IEEE Trans. Inform. In.IT, vol.33, no.2, (1987) 461–483.
22. L.F.Abbott and T.B.Kepler, "Optimal learning in neural network memories, "J.Phys. A:Math. General, vol.22, (1989) 711–717.
23. S.S. Venkatesh and D. Pitts, "Linear and logarithmic capacities in associative memory, "IEEE Trans. Inform. Th. IT, vol.35, (1989) 558–568.
24. D.J. Amit, Modeling Brain Function: The World of Attractor Neural Networks, Cambridge University Press, Net York (1989).

An Adaptive Call Admission Control to Support Flow Handover in Wireless Ad Hoc Networks

Joo-Hwan Seo and Ki-Jun Han

Department of Computer Engineering, Kyungpook National University, Daegu,Korea
jhseo@netopia.knu.ac.kr and kjhan@bh.knu.ac.kr

Abstract. In this paper, we proposes a Call Admission Control(CAC) scheme to support the burst handover calls occurring due to the movement of a mobile base station(MBS) and mobile nodes in a wireless ad hoc network. The proposed scheme is a dynamic guard channel scheme that adapts the number of guard channels in each MBS coverage area according to the current estimate of the potential handover call rate derived from the number of ongoing calls within the coverage of an MBS before it initiates a flow handover. The parameters used include the new call blocking probability, handover call blocking probability, and resource utilization when an MBS initiates a flow handover in a wireless ad hoc network. Simulation studies are performed to compare the proposed scheme with other CAC schemes.

1 Introduction

A wireless ad hoc network differs from a conventional wireless network in that it does not have a fixed infrastructure or wired base station. The recently proposed BAHAMA is a wireless ad hoc ATM network consisting of Portable Base Station and mobile nodes. However, in this network, although the mobile nodes have mobility, the Portable Base Stations do not[1]. In this paper, the proposed wireless ad hoc network consists of two types of network element: Base Stations with mobility, MBSs, and mobile nodes. Each MBS includes several mobile nodes and combines a switching functionality with wireless access interfaces, plus it communicates with other mobile nodes or MBSs via a wireless link.The geographic area within which mobile nodes can communicate with a particular MBS is referred to as the MBS coverage area. In a wireless ad hoc network that does not have a backbone network or fixed base station, developping a CAC scheme to support flow handover initiated by MBS is a challenging problem.

Recently, various approaches to support mobility in a wireless ad hoc network have been studied. However, none of these approaches are directly applicable to a wireless ad hoc network with group mobility as they do not consider the flow handover caused by the situation where the MBS itself moves from one coverage area to another while calls are in progress through the MBS. An MBS that can initiate a flow handover is identified as an MBS_{fh}.Since a wireless ad hoc network does not have a backbone network and its MBSs can move randomly, a new MBS can not be provided with information on the potential handover call

M.H. Shafazand and A M. Tjoa (Eds.): EurAsia-ICT 2002, LNCS 2510, pp. 747–754, 2002.

rate through a fixed base station or backbone network. In addtional, because a handover resulting from the movement of an MBS generates burst traffic, conventional handover and CAC policies developed for wireless networks are not feasible in a wireless ad hoc network with group mobility.

Accordingly the present paper proposes an adaptive CAC scheme to solve this problem in a wireless ad hoc network. The proposed scheme adjusts the number of guard channels based on the current wireless ad hoc network traffic conditions. As such, the new scheme is a dynamic guard channel scheme which adapts the number of guard channels in each MBS coverage area according to the current estimate of the handover call rate derived from the number of ongoing calls within an MBS_{fh} coverage area. Although estimating the instantaneous burst handover call rate due to the movement of an MBS in a wireless ad hoc network is a difficult proposition, the problem is considerably eased by using information from ongoing calls within an MBS_{fh} coverage area.

2 Flow Handover

In the proposed wireless ad hoc network, the neighboring coverage areas of an MBS overlap with each other to ensure the continuity of communications when the mobile nodes and MBS move from one coverage area to another. A set of channel is allocated to each MBS. In the proposed wireless ad hoc network, when an MBS with several mobile nodes moves from its current coverage area to another coverage area, a flow handover occurs. As such, a flow handover is initiated by an MBS while calls are in progress through that MBS. These ongoing calls are identified as burst handover calls. It is assumed that the flow handover is equal to the handover of the MBS. To support a burst handover call resulting from an MBS_{fh}, new and handover calls then compete for the connenction resources at the new MBS, MBS_n. Handover calls require a higher congestion performance relative to new calls because the forced termination of a burst handover call due to a blocking flow handover call is generally more objectionable than blocking a new call from a mobile node of the new MBS[3,4,5]. In this paper, we assume that a MBS use hint message that include the information of the number of ongoing calls within a MBS coverage area and the mobility information about a MBS or mobile nodes. Also, we assume that a MBS broadcast hint message to neighboring MBS by two hops for predicting the number of necessary guard channel when a MBS initiate flow handover in future. When a mobile node wants to communicate with another mobile node and MBS; it must first obtain a channel from one of the MBSs that hears it. When a MBS moves from one coverage area to another, while several calls are in progress, the MBS requires the MBS_n to flow handover. Using the handover procedure, the MBS requires that the MBS_n that it moves into will allocate it a channel so many as the total number burst handover call. If no channel is available at the MBS_n, the handover call is blocked. If the number of channels is less than that of burst handover call, the available channels are allocated to the burst handover call with the highest priorities and the others are blocked. The mobile nodes release the channel when the mobile node completes the call or the MBS moves to another coverage area before the call is completed.

3 The Proposed Adaptive CAC Scheme

3.1 Model for the Proposed CAC Scheme

In this paper, we consider channel capacity and congestion control discipline based on the blocked-calls-dropped discipline for our adaptive dynamic guard channel scheme. Thus, call admission control is based on the current channel usage and the respectively assigned capacity limit of each call class[3,4,5,6,7].

The proposed adaptive dynamic guard scheme adapts the number of guard channels in each MBS coverage area according to the current estimate of the instantaneous burst handover call rate derived from information of ongoing calls within the MBS_{fh} coverage area. Here, the number of guard channels will be dynamically determined so as to keep the handover call blocking probability close to the handover blocking threshold. The guard channels depend on traffic parameters, network traffic condition changes, and the number of burst handover calls. In a wireless ad hoc network, the proposed scheme can be implemented in a MBS_n. A MBS_n that is formed by a set of C_t channels can be treated as an M/M/c/c priority queuing system in which a handover call has priority over a new call. C_t is the total number of available channels at a MBS_n. In addition to the call admission and congestion control functions as required by the fixed scheme, the proposed scheme also requires the instantaneous estimation about a burst handover calls occurring in future and capacity limits adaptation functions.

In Fig.1, we show a state-transition-rate diagram of this scheme. $C_g(t)$ is the number of guard channels for handover call out of C_t at time t.

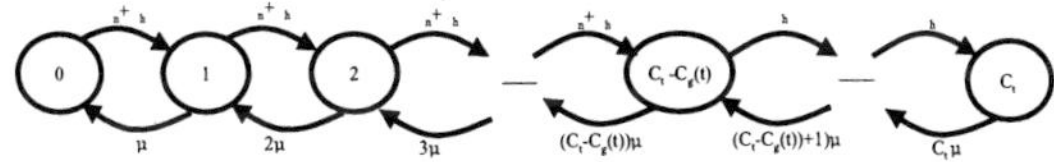

Fig. 1. The state-transition-rate diagram for the proposed scheme

Essentially, the system at a single cell can be modeled by using a Markov process whose state space describes the number of used channels. Assuming $C_g(t)$, guard channels out of C_t, are made accessible exclusively to handover calls, the state probabilities that channels i, i = 1, 2, ... , C_t, are occupied are given as follows:

$$P_j(t) = \frac{\lambda_n(t) + \lambda_{bh}(t)^j}{j!\mu} * P_0(t) \quad if \;\; j \leq C_t - C_g(t) \tag{1}$$

$$P_j(t) = \frac{\lambda_h(t) + \lambda_{bh}(t)^{C_t - C_g(t)}(\lambda_{bh}(t))^{j-(C_t - C_g(t))}}{j!\mu} * P_0(t)$$

$$if \;\; j \leq C_t - C_g(t) \qquad (2)$$

Here, $\lambda_{bh}(t)$ is identified as $\lambda_h(t) * N_H(t)$. We assume that the new and handover call arrival process at a MBS_N can be modeled by the Poisson Process with time-varying rates of $\lambda_n(t)$ and $\lambda_h(t)$, respectively. $N_h(t)$ is the number of the potential burst handover calls within a MBS_{fh} coverage area and can be modeled uniform process with time-varying rates. Channel holding time of two type calls is assumed to follow an exponential distribution with mean $1/\mu$. For a given C_t, $C_g(t)$ is employed as a control parameter which can be set within the range $0 \leq C_g(t) \leq C_{g-max}$. Here, the number of guard channels is determined by instantaneously estimating the burst handover call before a MBS initiates handover at time t. $P_0(t)$ is the probability that all channels are unoccupied at time t. $P_0(t)$ can be determined from the normalization condition by $\sum_{j=0}^{C_t} P_j(t) = 1$ at time t.

The congestion-related performance parameters at time t are the new call blocking probability $B_N(t)$ and the handover call blocking probability $B_H(t)$. According to the CAC policy which is defined in the proposed scheme at time t, a burst handover call in a MBS_n is blocked only when all channels out of C_t are occupied. Similarly, new calls are blocked when $C_g(t)$ are free at most channels. The handover and new call blocking probability is described as follows:

$$B_H(t) = P_0(t)\frac{(\lambda_n(t) + \lambda_b h(t))^{C_t - C_g(t)} \; (\lambda_{bh}(t))^{C_g(t)}}{C_t!\mu_t^C} \qquad (3)$$

$$B_N(t) = P_0(t)[(\frac{\lambda_n(t) + \lambda_{bh}(t)}{\mu})(\frac{\lambda_{bh}(t)^{C_t - C_g(t)}}{j!\mu^{h-(C_t - C_g(t))}})] \qquad (4)$$

3.2 The Proposed Adaptive CAC Scheme

Since randomly roaming of MBS_{fh} will occur flow handover and the burst handover call of mobile node, a wireless ad hoc network traffic conditions may change. Therefore, a CAC policy for handover will adjusts the number of guard channels based on the current network conditions. Because a MBS and mobile node roams randomly, the concept of a dynamic guard channel used in a wireless network can not be directly applied to reserve guard channel for a burst handover call occurring whenever a MBS initiates flow handover. This is because a backbone network and fixed base station cannot provide information of a handover call and mobility about a mobile node and MBS in a wireless ad hoc network. In this paper, the proposed scheme is based on the per-call equivalent bandwidth and deterministic channel assignment in a wireless ad hoc network. To extend the dynamic guard channel scheme to a wireless ad hoc network, instantaneous estimation of the burst handover call rate within MBS_{fh} coverage area is required to adapt the channel access priority to changing handover call arrival rates. $N_{cn}(t)$ is the total number of the burst handover call at a time t.

C_t is the total number of available channels at a MBS. $C_{na}(t)$ is the number of busy channels and $C_a(t)$ is the number of idle channels at time t. R_n is the number of rejected new calls and R_{cn} is the number of rejected handover calls. This part of the proposed CAC scheme is summarized and illustrated as follows:

```
 If handover call requests
     If Cna(t) < Ct and Ncn(t) < Ci(t),then
     allocate Ncn channels and Cna(t) = Cna(t) + Ncn(t)
     Else if Cna(t) < Ct and Ncn(t) > Ca(t),then
     allocate min(Ncn, Ca(t)) and Rcn = Ncn(t)-Ca(t)
     Else reject Ncn(t) and Rcn = Ncn(t)
If new call requests
     If Cna(t) < Cg(t),then
        allocate idle channel and Cna(t) = Cna(t) + 1
     Else reject new call and Rn  = Rn + 1
If a new and handover call is completed, then Cna(t)=Cna(t)-1
```

We note that the adaptation of the number of guard channels take place if a MBS initiate flow handover in wireless ad hoc network. This is too late in a wireless ad hoc network. In this paper, we assume that a MBS broadcast hint message to neighboring MBS for predicting the number of necessary guard channels in a MBS_n. The MBS_n will predict the potential handover call by monitoring the ongoing call state information within a MBS coverage area. The MBS_n will monitor the handover blocking probability and adapt the number of necessary guard channels to satisfy a given handover blocking threshold before a MBS initiates handover. In the proposed scheme, the handover blocking probability is the primary measure used to adjust the number of guard channels and it is necessary to keep it below the threshold when a MBS initiates handover in future. After obtaining the burst handover calls at an estimation interval, the number of guard channels $C_g(t)$ that is required to satisfy a given handover threshold for B_H can be determined via (3). An adaptation call control policy can be formulated under the given handover blocking threshold. Let N_{g-max} be the greatest value that can be assumed to satisfy $B_H(\lambda_n(t), \lambda_h(t), N_h(t), \mu, C_g(t)) \leq T_H$ according to (3) and N_{g-min} is the smallest value that can be assumed to satisfy $B_H(\lambda_n(t), \lambda_h(t), N_h(t), \mu, C_g(t)) = \alpha_{down} * T_H$ according to (3). Before MBS initiates handover, each MBS may estimate the expected handover call blocking probability, $B_E(t)$, according to (5).

$$B_E(t) = \frac{the\ number\ of\ blocked\ handoff\ calls(t)}{the\ number\ of\ burst\ handoff\ calls(t)} \tag{5}$$

The proposed scheme for determining adaptively the number of guard channels $C_g(t)$ according to (6) and (7) by using $B_E(t)$ is described as follows:

$$If\ B_E(t) > \alpha_{up} * T_H,\ then\ C_g(t) = min(N_{g-min}, C_{g-max}); \tag{6}$$

$$If\ B_E(t)\ <\ \alpha_{down} * T_H,\ then\ C_g(t)\ =\ N_{g-min}; \qquad (7)$$

4 Simulation Results

In our simulation study, a MBS with a total of $C_t = 50$ channels and an adapted number of guard channels, $C_g(t)$, is set within the range $0 \leq C_g(t) \leq C_{g-max}$ according to (6) and (7) at estimation time interval t. The threshold for handover call blocking probability, T_H, is 0.2. In the simulation, we assumed that the new call and handover call arrival time is modeled by the Poisson Process with mean new call arrival rate λ_n / $\lambda_h = 5/1$ and $1/\mu$= 180 seconds. The total simulation period is chosen to be 10 hours. The conventional channel allocation schemes are no-priority scheme, static and dynamic guard channel scheme. In figures 2, 3, and 4, we compare the proposed scheme with the conventional channel allocation schemes. Our simulations are run for call arrival rates from 5 calls/minute to 35 calls/minute. The new call blocking rate and handover blocking rate are both estimated according to the total simulation period. Handover blocking rate denotes the blocking rate of the burst handover call occurring whenever a MBS initiates handover.

Fig. 2 and Fig. 3 compares the proposed scheme with the conventional channel allocation schemes in terms of blocking probability for varying the call arrival rate. Fig. 2 shows the blocking probability of handover calls against the new call arrival rate. In Fig. 2, we note that the handover blocking probability of the proposed scheme is lower than that of the conventional channel allocation schemes. This is because the proposed scheme can be used to automatically determine the optimal number of guard channels to be reserved in the coverage area of new MBS according to (6). Here, No-Priority is the no-priority channel scheme and S-G is the static guard channel scheme. D-G is the dynamic guard channel scheme and P-D-G is the proposed scheme.

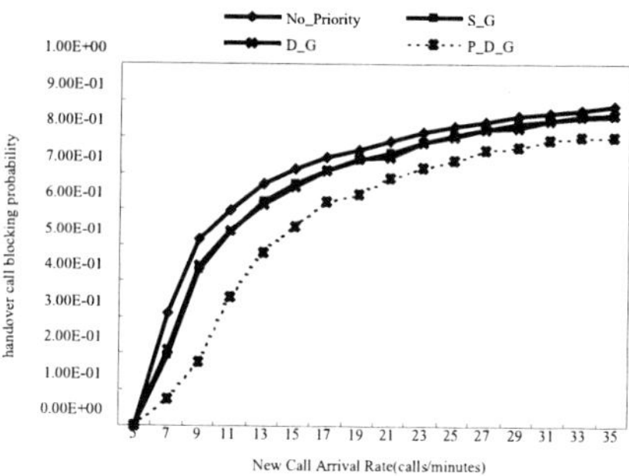

Fig. 2. Handover call blocking verse New call arrival rate

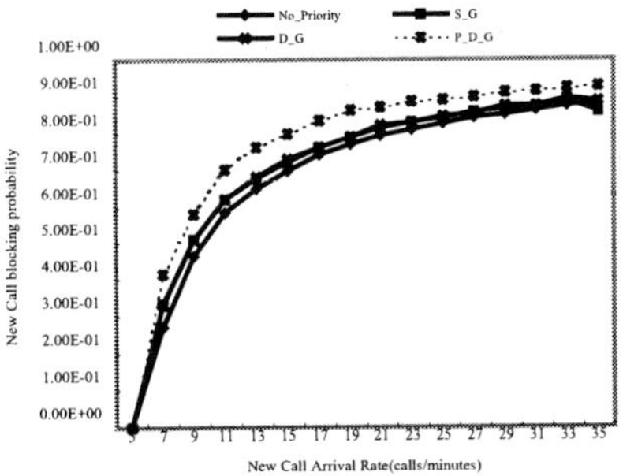

Fig. 3. New call blocking verse New call arrival rate

In Fig. 3, we note that the new blocking probability of the proposed scheme is higher than that of the conventional channel allocation schemes. In Fig. 4, the resource utilization of the proposed scheme is lower than that of No-priority scheme and is higher than that of S-G and D-G schemes when new call arrival rate is low. But the resource utilization is similar to the conventional channel allocation schemes when the new call arrival rate is high.

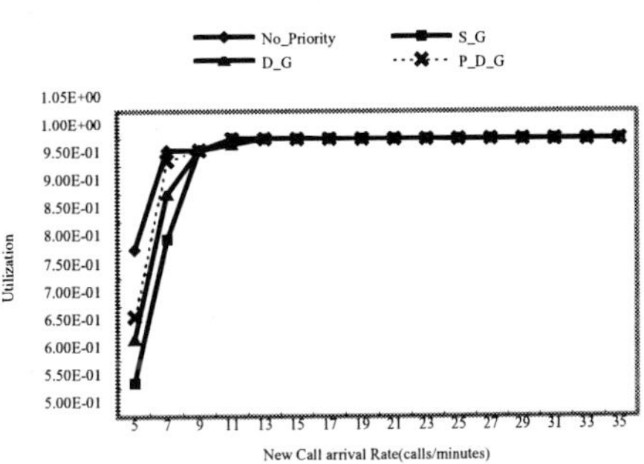

Fig. 4. Utilization

5 Conclusions

Since a wireless ad hoc network does not have a backbone network and the MBS moves randomly, CAC for supporting the burst handover call in a wireless ad hoc network is a complex problem. In this paper, we proposed an adaptive CAC scheme to solve this problem. The proposed scheme estimates the burst handover call occurring in future by monitoring the state information of ongoing calls within a neighboring MBS. And this scheme determine the number of guard channels to satisfy the handover blocking threshold. In a wireless ad hoc network, the proposed scheme can search automatically the optional number of guard channels to be reserved at a MBS under changing traffic conditions such as the burst handover call arrival rate. We note that the proposed scheme uses the information of the number of ongoing calls within MBS_{fh} coverage before a MBS_{fh} initiates handover. By using this information, it is shown through simulations that the proposed scheme provides a lower handover call blocking rate than does the conventional channel allocation schemes. Our proposed scheme is applicable to channel allocation for supporting mobility of MBSs and is extended to support group mobility in a wireless ad hoc network.

References

1. Veeraraghavan M; Karol M.J; Eng K.Y, "Mobility and Connection management in a Wireless ATM LAN," IEEE Journal on Selected Areas in Communications, Vol. 15 Issue:1., Page(s): 50–68 , Jan. 1997.
2. William Su, Gerla M., "Ipv6 flow handover in ad hoc wireless networks using mobility prediction, " Global Telecommunications Conference,1999. GLOBECOM '99.Vol: 1a, Page(s): 271–275.
3. Toh, C.-K, Shih C.-K, Vassiliou V., Delway M., "Emerging and Future Research Directions for Mobile Networks," Wireless Communications and Networking Conference, vol.1, WCNC. 1999 IEEE, Page(s): 363–367.
4. Yi Zhang and Derong Liu, "An Adaptive Scheme for Call Admission Control in Wireless Networks," Global Telecommunications Conference, 2001. GLOBECOM '01. IEEE, Vol. 6, 2000, Page(s): 3628–3632.
5. O. T. W. Yu and V. C. M. Leung, "Adaptive Resource Allocation for Prioritized Call Admission over ATM-based wireless PCN," IEEE Journal on Selected Areas in Communications, vol. 15, Page(s). 1208–1225, Sept. 1997.
6. M. H. Chiu and M. A. Bassiouni, "Predictive Schemes for Handover Prioritization in Cellular Networks Based on Mobile Positioning," IEEE Journal on Selected Areas in Communications, vol. 18, pp. 510–522, Mar. 2000.
7. S. Boumerdassi and A. L Beylot, "Adaptive Channel Allocation for Wireless PCN," Mobile nwtwork and Applications, vol. 4, pp. 11–116, 1999.
8. Grace, T.C. Tozer, and A.G. Burr, "Reducing call dropping in distributed dynamic channel assignment schemes by incorporating power control in wireless ad hoc networks," IEEE Journal of Selected Areas Communications, vol.18, pp. 2417–2428, Nov. 2000.

Design of Optimal LA in Personal Communication Services Network Using Simulated Annealing Technique

Madhubanti Maitra[1], Ranjan Kumar Pradhan[2], Debasish Saha[3],

Amitava Mukherjee[4]

[1,2] Department of Electrical Engineering, Jadavpur University, Calcutta 700 032, India.
m_madhubanti@hotmail.com
[3]MIS and Computer Science, Indian Institute of Management Calcutta, Joka, Calcutta 700 104, India.
ds@iimcal.ac.in
[4]PricewaterhouseCoopers Software (P) Ltd, Sector 5, Salt Lake, Calcutta 700 091,India.
Amitava.mukherjee@in.pwcglobal.com

Abstract. In this work we have addressed the issue of inherent trade-off between location update cost and paging cost for proper location area (LA) planning and have attempted to find out an optimal LA size such that the total cost comprising of location update and paging can be minimized. We have formulated a constrained cost optimization problem under the known patterns of call arrival and terminal mobility. The present work is divided into two parts. First, we have tried to validate the model itself; next we have proposed an algorithm based on simulated annealing technique as a solution methodology for the hard constrained optimization problem.

1 Introduction:

In a personal communication service network, because of uncertainty of location of a mobile terminal (MT), location update and paging process, both consume sufficient amount of radio resource. It is apparent that both these processes are coupled in a sense that there is an inherent trade-off between these two cost components (location update cost and paging cost), and these two together determine the total network cost. Whenever a terminal is within one LA it is instantly traceable because the location information is stored in the Visitor Location Register Databases (VLRDB) dedicated for the location area (LA) and Home Location Register Database (HLRDB. If the size of one LA is large enough, there is a high possibility that an MT with moderate velocity (for example a pedestrian or a person using his cell within one building) will roam inside the LA and as a result they will hardly cross the LA boundary. However if a mobile terminal crosses the boundary of LA where it was residing previously, and comes under a separate mobile switching center (MSC), that demands for a database update (new VLR-HLR-old VLR). So it is obvious that if the LA size is large, frequency of location update and as a result, associated cost will reduce drastically. However, the paging cost directly depends upon the number of cells to be polled at a time in one polling cycle and also upon the number of polling cycles required for paging. If we assume that in one single polling cycle, the terminal has to be paged, the paging cost depends only upon the size of LA. Hence it signifies that the size of the LA affects the signaling load generated due to paging and location update. As a

M.H. Shafazand and A M. Tjoa (Eds.): EurAsia-ICT 2002, LNCS 2510, pp. 755–765, 2002.

designer's point of view it is required to find out optimum size of an LA so that the total cost for location update and paging could be minimized.

Various location update schemes are proposed in the literature. Some of these are selective LA update[1], Profile based [2], Distance based [3], Timer based [4], Lezi-Update[5] and Movement based [6]. In this present work we have adopted Movement based location update scheme, where a mobile terminal executes discrete movements and location update is only performed when the number of movements reaches a predefined threshold value d. Instead of predefining the threshold value, in our work, we have aimed to find out optimum threshold value, which will determine the number of cells to be crossed by a mobile terminal before a location update is performed. That will give us the size of an LA if we assume a particular cell configuration.

Like wise several algorithms are proposed in the literature for performing effective paging. In our work we have considered blanket paging, where all the cells within an LA are polled simultaneously. Hence from the above discussion it is obvious that the paging cost will also depend upon the value of d and if the value of d is high, cost for paging will increase as more number of cells are to be polled to search an user and vice-versa.

Partha et al attacked the problem of designing optimum LA where they optimized paging and location update cost. To do so, they used the fluid flow model to capture the macroscopic mobility pattern. In our work we have attempted to solve the same problem of finding optimum LA considering microscopic mobility model [7], [8]. To depict the mobility pattern of the mobile terminal, we have used per-user basis movement based 2-D random walk model developed by Akyildiz et al [6]. In our work we borrowed the basic analytical framework proposed by Akyildiz et al and modified it to accommodate a moderate population size of mobile users within an LA. Moreover, we have assumed that a terminal moves in a given radial direction. Next, we have attempted to formulate a constrained cost optimization problem to find out the optimum value of movement threshold. Present work is divided into two parts. First, we have tried to validate the model itself; next we have formulated a constrained cost optimization problem to find out an optimum size of an LA. After formulating the problem we have found that the nature of the constrained optimization problem is NP-complete. Several methods have been suggested in the literature for solving this kind of non-linear optimization problems. For our problem, we tried to use some of the traditional methods but lastly decided for a heuristic solution due to the non-differentiable nature of the problem. Finally, we have presented an algorithm based on simulated annealing technique as a solution methodology [8]. Moreover, we have also developed another heuristic solution methodology based on Real-Coded Genetic Algorithm [11]. In this work, we have provided a few sample comparisons of the results that we obtain using the two heuristics. Detailed comparisons and analysis of the performance of the heuristics is not within the present scope of work.

The rest of the paper is organized as follows. In section 2 we describe the system as well as our model. In section 3 constrained cost optimization problem for LA planning is presented. In section 4 we discuss the simulated annealing technique in general and then we propose an algorithm based on simulated annealing technique for solving the formulated problem. In section 5, some representative results are presented. Section 6 concludes the present work.

2 System and Model Description

We assume that the PCS coverage area comprises of equal-sized hexagonal shaped cells. An MT resides in each cell it enters for a generally distributed time interval and then moves to any of the neighboring cells. Let us assume that the probability distribution of cell residence time of a mobile user is $P_m(t)$. When a mobile user leaves a cell, there is an equal probability that any one of the immediate neighboring cells is selected as the destination.

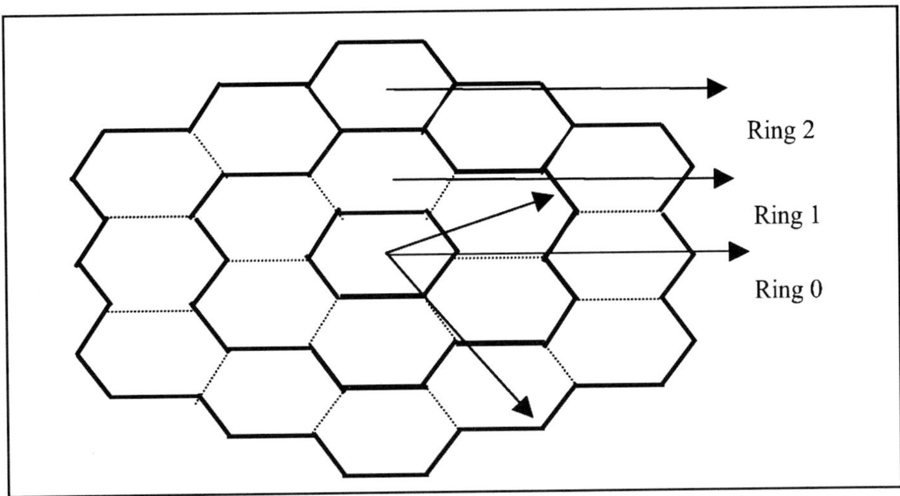

Fig. 1. Hexagonal Cell Configuration

We assume center cell is the cell where the last location registration occurred and a location update is performed by a mobile user, when the number of cell boundary crossing in a particular radial direction since the last location update registration equals a threshold value 'd'. As per our formulation, if a mobile terminal makes say d movements in one particular radial direction, then the LA will be defined as the area within (d-1) rings from the center cell (Fig. 1). If the LA consists of D rings then,

$$D=(d-1) \tag{1}$$

The number of cells within the LA L_D, will be:

$$N(D)= 3*(D+1)*D+1 \tag{2}$$

The perimeter L(D) of the LA L_D, can be calculated as

$$L(D) = (12D+6)*R \tag{3}$$

Where R denotes the cell radius and the area of each cell is $(6.\sqrt{3}/4)*R^2$ And

$$d = 2.\sqrt{3}.R \tag{4}$$

The area S(D) of the LA L_D, is

$$S(D) = [3D(D+1)+1]\,(2.6)\,R \tag{5}$$

We have assumed the incoming call arrivals to each mobile user follow a Poisson process. The paging area is the area within (d-1) rings from the center cell. For our optimization purpose we have assumed paging delay to be one. Under this broadcast paging scheme, the cost for the paging will be maximum. This assumption allows us to consider the worst-case possibility for minimizing total cost.

Let α (k) be the probability that there are k boundary crossings performed by a mobile terminal between two successive call arrivals. If the probability density function of cell residence time t_m has the Laplace-Stieltjes transform fm*(s) and mean 1/λm, the call arrival to each mobile terminal follows Poisson process with rate λc. Based on these parameters the expression of α (k) will be as follows [6]. Here, θ is the call-to mobility ratio (CMR) [10],

$$\alpha(k) = \begin{cases} 1-1/\theta\,[1-fm^*(\lambda c)], & k=0; \\ 1/\theta[1-fm^*(\lambda c)]^2[fm(\lambda c)]^{K-1} & k>0; \end{cases} \tag{6}$$

Considering that cell residence time t_m follows Gamma distribution, we get

$$fm^*(s) = (\lambda_m.^*\gamma/s+\lambda_m)^{\gamma}, \quad \text{where, } \gamma = 1/(v^*.\lambda_m^2) \tag{7}$$

Let β (k, K) denote the probability that the MT is k rings away from the center cell, given K number of cell boundary crossings are performed. To depict the mobility pattern of a mobile terminal, we have considered 2-D Random Walk model. Let us assume that P_K denotes the K X K state transition matrix, where an element $p_{i,j,k}$ in P_K gives us the probability that a mobile terminal moves from one ith ring cell to one jth ring cell in single step. Then the probability β (k, K) comes out to be [6]

$$\beta(k, K) = \begin{cases} P_K & \text{single step state transition matrix} \\ P_K * P_K^{(n-1)} & \text{n step state transition matrix} \end{cases} \tag{8}$$

Using equations 6,7,8, we express location update cost *Cu* as a function of d and the expression is as follows (detail derivation is given in [12]):

$$Cu(d) = \frac{U * \frac{Y[1-fm(\lambda c)]}{\theta} * ([fm(\lambda c)]^{d-1})}{(1-[fm(\lambda c)]^{d})^{2}} \tag{9}$$

Similarly, we derive paging cost *Cv* as a function of d and it is as follows:

$$Cv(d) = V^* \sum_{k=0}^{d-1} \rho_k N(k) \Phi_k \tag{10}$$

Where Y is the number of mobile terminal attached to the network within the LA and U and V are respective cost coefficients for performing a location update and paging and Φ_k is the probability of finding the mobile terminal within the LA L_k, the density of the mobile terminal in that area is denoted by ρ_k.
So, the total cost is

$$Ct(d) = Cu(d) + Cv(d) \tag{11}$$

3 Problem Formulation

The optimization problem can be stated mathematically as

$$\textit{Minimize (P)}: \quad Ct(d) = Cu(d) + Cv(d) \tag{12}$$

Subject to :

$$0<p<1 \tag{13}$$

$$q_k.\ S(k).\ \rho_k \geq N_k .p \tag{14}$$

$$R<R_{max} \tag{15}$$

In constraint equation (13), p denotes the penetration factor. Constraint equation (14) gives us the total number of mobile terminals in a service area where ρ_k is the density of the mobile terminals in a LA L_k, and there are q_k switches. This number must be greater than or equal to the total number of attached mobile users. In this equation N_k denotes population size within the kth LA. The maximum radius of a cell should be less than maximum allowable radius so that the constraint on system power budget is not violated (equation 15). We have attempted to solve this non-linear constrained optimization problem with help of Simulated Annealing method.

4 Simulated Annealing

There are two possible approaches to solve combinatorial optimization problems: one is via exact optimization algorithms and other is through approximation algorithms. The traditional methods give global optimum solution in a possibly prohibitive amount of time and present day evolutionary computing methods yield an approximate solution in an acceptable amount of computation time. Thus, though the quality of solution is traded off against computation time one can always reach to a near-optimal solution within bounded computation time. Such a powerful general-purpose algorithm was introduced by Kirkpatrck [9] and is commonly known as simulated annealing (SA) due to the analogy with the annealing process of solids.
The design of the algorithm based on simulated annealing consists of four important elements.
(1). A set of allowed system configurations. (Configuration space).
(2). A cost function.
(3). A set of feasible moves. (Generation or Perturbation mechanism).
(4). A cooling schedule.
At each temperature feasibility test is done to monitor any violation of equality or inequality constraints. If these constraints are satisfied, then SA proceeds to step 3. Otherwise, the move is discarded and the configuration before this move is used for next step.

4.1 Configuration Space

A configuration space is a set of allowed system configurations and over this range of feasible configurations an optimal configuration is to be searched for. So, we define, the configuration space as the set of feasible movements made by a mobile terminal after which a location update will occur.

Let us define the configuration space as

$\Omega := \{d1, d2, d3, \ldots\ldots\ldots dn\}$ $\quad\quad \forall\, di \subset \Omega$

A new configuration space could be generated by any of the following two types of moves, which would change the LA topology.

1. Add/Subtract move: Add or subtract a realistic step size to or from the movement threshold value chosen randomly from the set Ω, with the help of a random number generator. The value of the realistic step size is arbitrarily determined in each step using a random number generator.
2. Multiplicative move: add or subtract a realistic step size which is positive integer multiple of a movement threshold that is chosen arbitrarily from the set Ω, by using a random number generator. The multiplication factor is also determined randomly with the help of a random number generator.

4.2 Cooling Schedule

Next, we develop the cooling schedule, which consists of two important components: setting the initial temperature, and developing a scheme for lowering the temperature. For this problem, we have started from the high initial temperature (900° C), which we have set such that the ratio between accepted moves and total moves is 55%, and then decreased it gradually according to the formula:

$$T_{k+1} = T_k \cdot \alpha(T_k)\,. \quad \textbf{(16)}$$

The cooling factor, α, has got the direct bearing with the quality of solutions and the run time involved. In our problem we have considered a range for α(.) within the interval [0.85, 1].
In addition to the above four elements the following two elements are also important, which are
Acceptance Criteria:
Let us assume that c and ĉ be the value of cost function before and after one particular move. We find dc= ĉ-c, if the differential cost decreases i.e. dc<0, the move is accepted and new configuration is retained. When dc>0, (i.e. the move is uphill) acceptance is treated probabilistically. We first calculate Boltzman factor, which is exp *(-dc/T)*, then take a random number r from a uniform distribution in the interval [0,1]. If $r \leq exp(dc/T)$, the new configuration is retained else the move is discarded. In this context we define acceptance ratio, which actually determines the number of moves at each temperature, the ratio being defined as $n_k = c^k n_0$, where n_0 is the number of moves at initial temperature, n_k is the number of moves at k-th stage and c=1. The temperature at k-th stage T_k is lowered when the acceptance ratio reaches a n_k generated in each iteration randomly.
Stop Criterion:
The algorithm stops when no significant improvement in the cost function has been found for a number of consecutive iterations.

4.3 Simulated Annealing Algorithm Developed For Constrained Optimization Problem (*P*)

Proposed simulated annealing based solution algorithm for our problem is described as follows:

Step1: Input system and network data:
Input the system and network data, the control parameters such as the initial temperature T, an initial number of moves, and random number.

Step 2: Obtain a feasible configuration by checking the constraints:
- *i)* *Select an initial configuration from the configuration space by a random number generator.*
- *ii)* *Check the feasibility for the selected configuration. If any constraint is violated go to i).*
- iii) *Calculate the total cost (Ctold).*

Step 3: Cooling schedule:
- *i)* *Apply the cooling schedule discussed in sec. 4.2*
- *ii)* *At each temperature T_k perform nv moves. For nv =1,2,3...n_k do steps 4 and 5. Else go to step 6.*

Step 4: Generate new feasible configuration.
- *i)* *Generate a new configuration using perturbation mechanism discussed in sec 4.1*
- *ii)* *Check constraints. If any constraint is violated go to step i).*
- *iii)* *Calculate the total cost (Ctnew).*
- *iv)* *Calculate the differential cost ΔC (=Ctnew-Ctold).*

Step 5: *Update system configuration:*
i) *Accept the move and retain the new configuration if $\Delta C < 0$*
ii) *Accept the move and retain the new configuration if $\Delta C \geq 0$ and if $r \leq \exp(-\Delta C/T)$; Else discard the move.*
Step 6: *Check the stop criterion:*
i) *If the stop criterion is not satisfied go to step 3.*
Step 7: ***Obtain the optimal solution.***

5 Performance Analysis

In this section we have varied some system parameters to observe their reflections on the model behavior, without considering any constraints.

5.1 Model Validation

To validate the analytical model, we have run our SA heuristic to solve the unconstrained problem and varied the value of movement threshold 'd'. We observe corresponding changes on location update cost (Cu), paging cost (Cv), and on the total cost (Ct), [fig.(2),fig.(3) and fig.(4)]. We take three different values of CMR to demonstrate the effect of changing mobility and call arrival patterns. For a very small value of 'd', the location update cost is very high as after these d number of boundary crossings there will be one location update and lesser number of d implies that the frequency of location update is higher and vice versa. This fact can be corroborated from fig. 2. From the fig.2, it has also been seen that location update cost varies with CMR values. Low CMR means the probability of boundary crossing is high by a mobile terminal between two successive calls, which results in higher location update cost. For higher CMR values the situation is just the reverse.

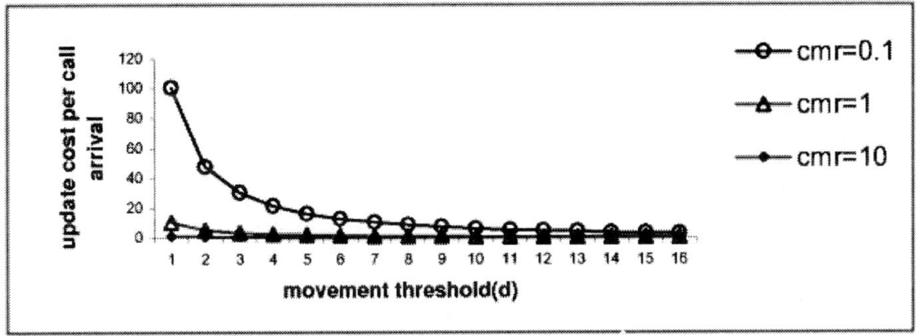

Fig. 2. Change of update cost for different CMR

Paging cost also varies as 'd' changes. If the value of 'd' is small, number of cells within the LA will be small. It has been observed in fig.3, that, for small values of 'd', paging cost is minimal. As 'd' increases ,naturally, the paging cost increases.

The paging cost also varies with CMR. If the CMR is high, call arrival rate is high and that will raise the paging cost significantly.

The total cost Ct varies as d changes and the variation is convex in nature (fig.4). For certain range of values of 'd' the total cost attains global minimum. From figure 4, we find that for the unconstrained problem for three different values of CMR, the optimum value of d lies in the interval [2,4] when we have run SA algorithm. However, to compare the quality of the solution, we have provided one plot (CMR=0.1) that we obtained after running GA based heuristic. In this case the optimum value of d comes out slightly less and lies within the interval [1.75,3].

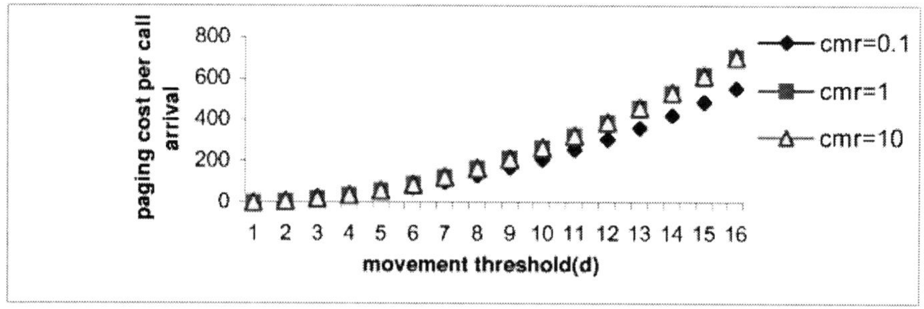

Fig. 3. Change of paging cost with movement threshold

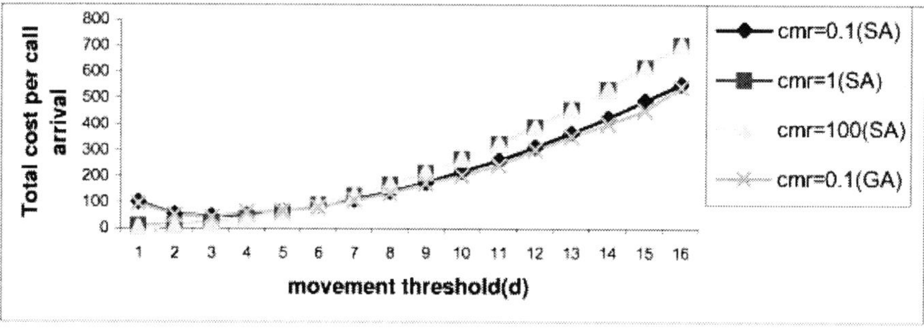

Fig. 4. Change of total cost with movement threshold

5.2 Results on Optimization of (*P*)

In this section we present the results obtained after solving the constrained optimization problem (*P*). Figure 5 shows the variation of perimeter of LA with population size. If the population size is increased, optimum value of d increases and hence the size of the LA is increased. It is obvious that, if, larger number of mobile terminals, serviced by a single MSC, are to be accommodated in a single LA, number of mobile terminals residing closer to the boundary of a location area will be high. Due to this, possibility of number of boundary crossings by mobile terminals also increases. To minimize the total cost, the movement threshold value d should also increase so that after each boundary crossing one location update does not take place. Similar result is also obtained using GA. However result shows that GA results in slightly smaller location area.

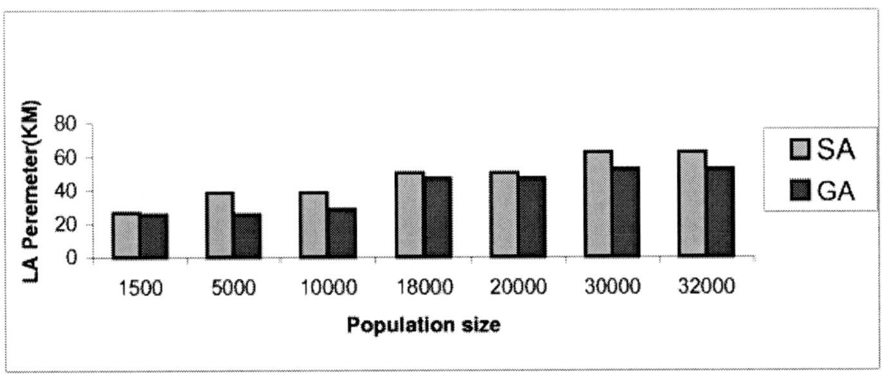

Fig. 5. Comparative Analysis of Variation of perimeter of LA with N(Q=3, p=1.0, ρ=50 km^{-2}, λc =1, R=1)

Figure 6 shows the nature of variation of the normalized total cost with number of cells per LA. Total cost is normalized with respect to average total cost value. It is observed from the figure that paging cost increases with number of cells per LA, where as location update cost gradually decreases. We observe from figure that the *normalized total cost is minimum for (217) number of cells.*

The variation of optimum cost per call arrival with CMR is given in figure 7. It is apparent that for low CMR, call arrival rate is low and the mobility rate of the mobile terminal is high. In this situation the possibility of boundary crossings by a mobile terminal increases. This would increase the location update cost significantly whereas paging cost will be minimal. This is reflected in figure 7, where the optimum cost per call arrival decreases with increase in CMR.

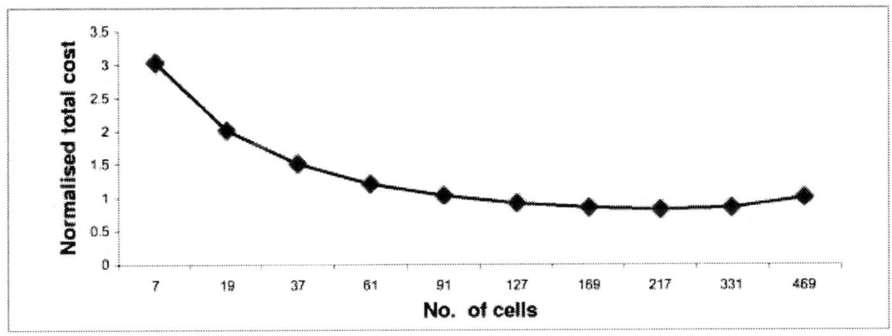

Fig. 6. Variation of normalized total cost with no of cells per LA(R=1 km, λc =0.1, CMR=0.1)

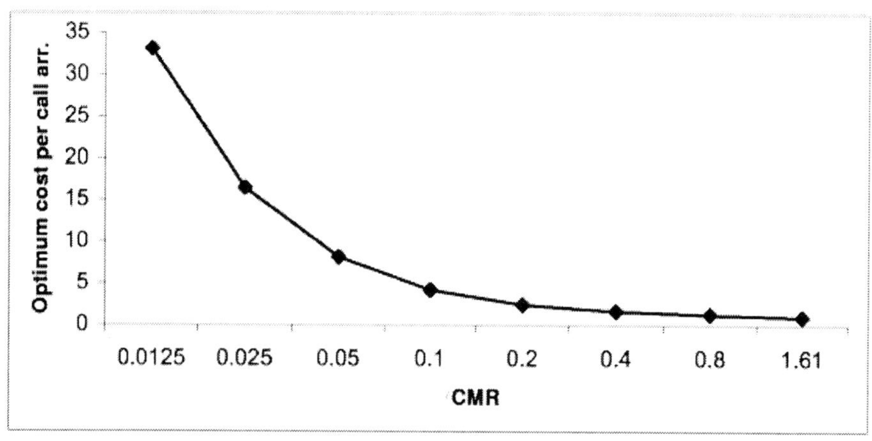

Fig. 7. Variation of optimum cost per call arrival with CMR (R=1Km,ρ=50km^{-2}.)

Fig 8 and 9 show that even as the network size grows or population size increases, SA based algorithm proposed by us tends to find a near-optimal solution within tolerable run time.

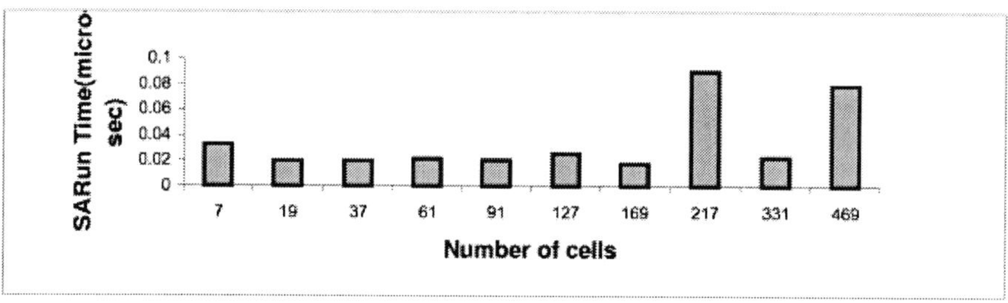

Fig. 8. Convergence time required for SA based heuristic with number of cells

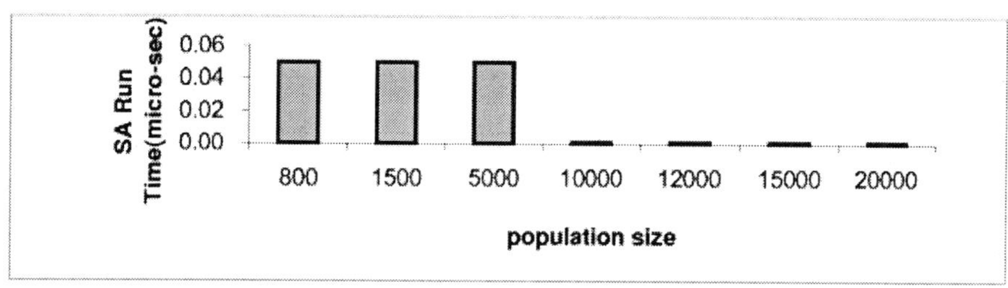

Fig. 9. Convergence time required for SA based heuristic Vs. Population size

6 Conclusion

In the present work, we have attempted to find out optimum LA size in a PCSN environment so as to minimize the signaling load generated in the system due to location update and paging. We have formulated a constrained cost optimization problem and proposed a heuristic solution based on simulated annealing technique. Promising results have been obtained (some of which are presented in the work) which indicates that evolutionary computing methods offer themselves as strong candidate for solving hard constrained combinatorial optimization problems. However in this work we considered simple blanket paging scheme. Our future work will be devoted towards developing some intelligent paging strategy so that the signaling load generated for paging can further be minimized.

References

1 Sen, S. K. Bhattacharya, A., and Das, S. K.:"A Selective Location Update Strategy for PCS users", ACM/Baltzer J. Wireless Networks, vol. 5, no. 5, pp 313-326, Sept.99.

2 Pollini, G.P., I, C. L.: "A Profile –Based Location Strategy and Its Performance", IEEE JSAC, Vol. 15, no 8. pp. 1415-1424, Oct. 97.

3 Ho, J., Akyildiz, I. F.: " Mobile User Location Update and Paging under Delay Constraints", ACM/Baltzer J. Wireless Networks, vol. 1, no. 4, pp. 413-425, Dec. 95.

4 Liang, B., Haas, Z.: " Predictive Distance-based Mobility Management for PCS Networks", Proc. IEEE INFOCOM'99, NY, March 99.

5 Bhattacharya, A., Das, S. K. :" LeZi-Update: An Information-Theroretic Approach to Track Mobile Users in PCS networks", Proc. of ACM/IEEE MobiCom '99, Settle, WA, pp. 1-12, Aug. 99.

6 Akyildiz, I. F., Ho, J. S. M., Lin, Yi-Bing.: "Movement-Based Location Update and Selective Pageng for PCS Networks", IEEE/ACM Transactions on Networking, Vol_4., No. 4, Aug,96.

7 Saha,D., Mukherjee, A., Bhattacharya, P. S.:" Optimization of Location Area Size in a Hierarchical PCSN", Conference on Distributed Processing and Networking, 97, pp 21-26, IIT, Kharagpur, India.

8 Bhattacharya, P.S., Saha, D., Mukherjee, A.." Determination of Optimum Size of a Location Area", COMM_SPHERE 99, France.

9 van Laarhoven, P. J. M., Aarts, E. H. L.:" Simulated Annealing Theory and Applications", Kluer Academic Publishers, London /Boston 87.

10 Maitra, M., Saha, D., Mukherjee, A.:"Cost Minimization for a Personal Communication Services Network Using Random-Walk Mobility Management Model", International Conference on Systematic, Cybernetics and Informatics (SCI-2000), Orlando, Florida.

11 Goldberg, D. E.:" Genetic Algorithms, in Search, Optimization & Machine Learning", PHI, Fourth Indian Reprint, 2001.

12 R. K. Pradhan, "Some Aspects of Design of Location Area in Personal Communication Services Network" , Post Graduate Dissertation, 2002

Secure Bluetooth Piconet Using Non-anonymous Group Key

Dae-Hee Seo, Im-Yeong Lee, Dong-ik Oh, and Doo-soon Park

Division of Information Technology Engineering, SoonChunHyang University, #588, Eupnae-ri, Shinchang-myun, Asan-si, Choongchungnam-Do, 336-745, KOREA
{patima, imylee, dohdoh, parkds}@sch.ac.kr
http://sec-cse.sch.ac.kr

Abstract. In accordance with the changes in the wireless communication environment, there has been a great need to satisfy the demand for diverse modes of information exchange. However, Bluetooth has weaknesses in its security features when its in -house security services are applied to networks. The purpose of this study is to propose a secure Bluetooth scenario with an upgraded security feature. This paper first reviews the general characteristics and security features of Bluetooth together with an analysis of its weaknesses, and presents the formation and maintenance process of Bluetooth piconet with is created by using ECDSA and group key in the ACL (Asynchronous Connection-less Link) connection through a secure Bluetooth inquiry process.

1 Introduction

Bluetooth is a wireless communication protocol that sends information to a device that is fixed or mobile, and it is expanded into Scatternet by creating piconet through one major master, which controls two or more devices that share the same channel.

However Bluetooth has had many security problems and weak points, e.g. If it were applied to a real network, users wouldn't enjoy secure communication because of an invasion of privacy. The most serious of these is the problem of an inquiry. To upgrade Bluetooth, a better process of the inquiry is necessary. Piconet should be studied as a secure formation as well as to maintain Bluetooth. The second section deals with the formation of a security system furnished in Bluetooth standard program version 1.1. The third section includes a new proposal that is analyzed in the fourth section. The final section expresses concluding remarks. [1]

[1] This work is supported in part by the Ministry of Information & Communication of Korea ("Support Project of University Information Technology Research Center" supervised by KIPA)

M.H. Shafazand and A M. Tjoa (Eds.): EurAsia-ICT 2002, LNCS 2510, pp. 766–777, 2002.

2 Standard Method of Bluetooth

2.1 The Analysis of Weakness of Bluetooth

Bluetooth has two major weak points. One happens in the technical specifications. The other happens when actual Bluetooth is applied.

A. Actual Bluetooth Is Applied

Bugging communication between ciphered devices or attacking PIN

Communication through Bluetooth requires that users should have a cipher that they have agreed on. While devices are sharing a cipher or used together, there might be a chance of bugging the shared information. For instantly, if cryptography hasn't been done successfully by Bluetooth and application program, Man-in-the-middle-attack will be possible. The first protocol is important because it shares a lot of important information between Bluetooth devices. The second protocol will work if the first one cannot find suitable information. The second protocol creates and repeats the first process with a different link key. Even though the Pin consisted of 8-128bit, there is a weak point, i.e., it cannot treat the value zero. An attacker can calculate a link key with a device that can create random numbers. It means that transmitting and receiving devices can disguise an attacker wiretapping for information. In case where the attacker knows the unit key of a device, it seems that the attacker is enabled to be camouflaged.

B. Analysis of the Weaknesses of Bluetooth Technical Specifications

If a Piconet or a Scatternet is formed, there might be security loopholes for eavesdropping, camouflaged attacking, as well as attacking PIN off-line, which will be attacks on Bluetooth.

Wiretaps and camouflaged attacks

There is a method of Bluetooth for creating a protocol. This method calculates random numbers for a link key and PIN addresses for Bluetooth devices. If PIN is not valid or is transmitted to other units, an attacker will recognize it promptly. To protect this kind of weakness, users should make the PIN longer than now.

3 The New Proposed Scheme

This paper proposes that a mobile device should have WPKI certificate such is renewed every 25 hours. The device offers a securer inquiry in the application class than in the Bluetooth link class, because the mobile device creates a secure piconet and a process of mutual authentication.

3.1 System Parameters

Each system factor of an entity is followed below.

* : H - A Central server, I - A Personal computer, M - Bluetooth master, S - Blue tooth slave

p_*, q_* : Prime number ($p, q \geq 512bit$)

x_*, y_* : Based on RSA algorithm * of private and public key

a_*, c_* : Based on ECC algorithm * of public and private key {($a \geq 128bit$), ($a = cP$, $P \in E(Z_P)$)}

g, η : System coefficient that center server exhibits

h_* : *'s secure hash value

H : Secure hash functions

r_* : *'s Pseudo random number

$PINLg_*$: PIN code lengths of *

$PINcode_*$: PIN code of *

T_* : Time stamp of *

EC : Elliptic Curves Cryptography

ER : RSA algorithm

E : Block Cipher algorithm

G : Base Point

M : Message

Ψ_*, Ξ_* : ESIGN Signature of *

$V_{*\_@_\#}$: Encryption value that is Encrypted by # and is transmitted from * to @

R_*, S_* : ECDSA signature of *

$ID_{\inf o_*}$: Information of *

Z : System coefficient that Bluetooth Master exhibits to piconet group circle ($Z = g^{c_m} \bmod \eta$, $[g \in GF(a_M)]$)

i : Number of key that Bluetooth Master is allocated to piconet group circle ($i \in (1, \ldots, a_M - 1)$)

BD_ADDR_* : 48bit Bluetooth address of *

S_j : Bluetooth piconet group key signature value

$(\xi)_*$: *'s connection value that create

$S_{\inf o}$: Bluetooth Slave information that require connection to Bluetooth piconet

$S_{\inf o_{res}}$: Bluetooth Slave information that require connection to Bluetooth piconet

R_{B_S}, S_{B_S} : Bluetooth slave signature value that require connection to Bluetooth piconet

(p_{Mi}, n_{Mi}) : Free pair that creates in Bluetooth Master for group key creation

ζ : Bluetooth slave number that form Bluetooth piconet

3.2 Secure Protocol of a Bluetooth Piconet by Using a Open Group Key

In order to form a secure Bluetooth piconet, the following processes should be finished.

- This preliminary stage 1 : The creation of session keys
- This preliminary stage 2: The creation of signature for a central server and a per sonal computer
- Mutual authentication between central server and personal computer
- Communication between Bluetooth master and slave
- Mutual authentication between Bluetooth master and slave
- Exchange of signature value and setting up the group key
- Closing of event
- Piconet protocols

Preliminary stage1 : Session key setup stage

This preliminary stage 1- A creation of a session key is not only able to create a session key, but also it sets up

- A central sever: After creating a random number, it sends the number to a personal computer

$$K = g^{r_H} \bmod \eta$$

- A personal computer: After creating a random number, it sends the number to a central server.

$$Y = g^{r_I} \bmod \eta$$

- A central sever calculates (see below) and sends it to personal computers

$$X = Y^{r_H} \bmod \eta$$

- A personal computer calculates the following (see below) and verifies if the answer it true or not. The computer creates a session key after that.

$$r_I^{-1} = y$$

$$K' = X^{y} \bmod \eta$$

Preliminary stage 2: Creation of a signature for a central server and a personal computer

At the preliminary stage 2 – The signature of sign is created for a central server

- A central server creates an ESIGN signature.

$$\Psi_H \geq (H_M(m) - (r_H)^K \bmod \eta)/(p_H * q_H)$$

$$\Xi_H = r_H + ((\Psi_M / K * (r_H)^{K-1})^{-1} * \bmod p_H) * (p_H * q_H)$$

- A personal computer creates an ESIGN signature.

$$\Psi_I \geq (H_I(m) - (r_I)^K \bmod \eta)/(p_I * q_I)$$

$$\Xi_I = r_I + ((\Psi_I / K * (r_I)^{K-1})^{-1} * \bmod p_I) * (p_I * q_I)$$

– To use a signature in the next stage, a receiver calculates $(\Xi_*)^K \bmod \eta$. If the answer is $H_*(M) \leq \Xi_*{}^K \bmod \eta$ and $H_*(m) \leq \Xi_* \leq H_*(m) + 2^{(2/3\log\eta)}$, the signature will be valid.

Stage 1: The mutual authentication between a central server and an personal computer.

Stage 1 aim for mutual authentication with a session key, shared between a central server and a personal computer.

Step ①: A central server creates a secure hash value, encrypts it by using a session key K and finally *sends* $v_{H\_I_K}$ to a personal computer.

- $(\xi_1)_H = (\Xi_H \| \Psi_H \| T_H)$
- $Z_H = ((\xi_1)_H)^{r_H} \bmod \eta$
- $V_{H\_I_K} = E_K(Z_H \| r_H)$

Step ②: A personal computer creates a secure hash value, encrypts it by using a session key K, and sends $v_{I\_H_K}$ to a center server.

- $(\xi_1)_I = (\Xi_I \| \Psi_I \| T_I)$
- $Z_I = ((\xi_1)_I)^{r_I} \bmod \eta$
- $V_{I\_H_K} = E_K(Z_I \| r_I)$

Through this process, the mutual authentication is carried out between a central server and a personal computer, which is the initial process for Bluetooth communication.

<In case of requiring a link key between a central server and a Bluetooth master.>

Step ③: Bluetooth master creates a secure hash value, encrypts it by using a public key of a central server and sends $(V_{M\_Hy_H} \| h_M)$ to a central server.

- $(\xi_1)_M = (BD\_ADDR_M \| r_M \| T_M)$
- $h_M = H((\xi_1)_M)$

$$V_{M\_Hy_H} = ER_{y_H}((\xi_1)_M)$$

Step ④: A central server verifies the BD_ADDR_M of a Bluetooth master, calculates $(V_{H\_Mr_M} \| V_{H\_M_K} \| h_H)$ and sends it to the Bluetooth master.

- $(\xi_2)_H = (BD\_ADDR_H \| T_H)$
- $h_H = H((\xi_2)_H)$
- $V_{H\_M_K} = E_K(g^{r_H r_M} \bmod \eta \| \Xi_H \| \Psi_H)$
- $V_{H\_Mr_M} = E_{r_M}((\xi_2)_H)$

The Bluetooth verifies the BD_ADDR_H of the Bluetooth master from the transmitted hash value and saves $V_{H\_M_K}$ in the temporary buffer. In order to certify a Bluetooth slave certified by a personal computer, saved $V_{H\_M_K}$ must be created through the mutual authentication of a Bluetooth master and a Bluetooth slave.

Step ⑤: A message about the consummation of an event should be sent to a central server.

<In case of requiring a code of PIN between a central server and a Bluetooth master>

Step ③': A Bluetooth master creates a secure hash value, encrypts it with a public key of a central server and sends $(V_{M\_Hy_H} \| h_M)$ to the central server.

- $(\xi_2)_M = (BD\_ADDR_M \| r_M \| T_M)$
- $h_M = H((\xi_2)_M)$
- $V_{M\_Hy_H} = ER_{y_H}((\xi_2)_M)$

Step ④': A central server saves the value transmitted form a Bluetooth master and verifies the BD_ADDR_M of a Bluetooth master, calculates $(V_{H\_M_K} \| V_{H\_Mr_M} \| h_H)$ and sends it to the Bluetooth master.

- $(\xi_3)_H = (BD\_ADDR_H \| PINLg_H \| PINcode_H \| T_H)$
- $h_H = H((\xi_3)_H)$
- $V_{H\_M_K} = E_K(g^{r_H r_M} \bmod \eta \| \Xi_H \| \Psi_H)$
- $V_{H\_Mr_M} = E_{r_M}((\xi_3)_H)$

The Bluetooth master verifies $BD\_ADDR_H, PINLg_H, PINcode_H$ and saves $V_{H\_M_K}$ in the temporary buffer. In order to certify a Bluetooth slave certified by a personal computer, saved $V_{H\_M_K}$ must be created through the mutual authentication of a Bluetooth master and a Bluetooth slave.

Step ⑤: A message about the consummation of an event should be sent to a central server.

<In case that a user wants to PIN into a device>

Step ⑥': A Bluetooth slave creates a secure hash value, encrypts it with a public key for a central server and sends $(V_{S\_Iy_I} \| h_S)$ to the personal computer.

- $(\xi_1)_S = (BD\_ADDR_S \| r_S \| T_S)$
- $h_S = H((\xi_1)_S)$
- $V_{S\_Iy_I} = ER_{y_I}((\xi_1)_S)$

Step ⑦': A personal computer verifies the BD_ADDR_S of a Bluetooth slave then it saves, calculates $(V_{I\_S_K} \| V_{I\_Sr_S} \| h_I)$ and sends it to the Bluetooth slave

$$\bullet\ (\xi_2)_I = (BD\_ADDR_I \| PINLg_I \| PINcode_I \| T_I)$$

$$\bullet\ h_I = H((\xi_2)_I)$$

$$\bullet\ V_{I\_S_K} = E_K(g^{r_I r_S} \bmod \eta \| \Xi_I \| \Psi_I)$$

$$\bullet\ V_{I\_Sr_s} = E_{r_S}((\xi_2)_I)$$

A Bluetooth slave verifies the $BD\_ADDR_I, PINLg_I, PINcode_I$ of a personal computer with the transmitted value and saves it in the temporary buffer.

Step ⑧': A message about the consummation of an event should be sent to a personal computer.

Stage 2 : A communicative stage between a Bluetooth master and the slave

This stage is aimed at a secure inquiry between a Bluetooth master and the slave. It creates a session key, encrypts $V_{H\_M_K}$, $V_{I\_S_K}$ that are shared by a central server and a personal computer and sends the results somewhere.

Step ⑥: A Bluetooth master calculates a secure hash value and $(V_{M\_S_\rho} \| \rho)$ and sends it to a Bluetooth slave.

$$\bullet\ (\xi_3)_M = (V_{H\_M_K} \| r_M \| T_M)$$

$$\bullet\ h_M = H((\xi_3)_M)$$

$$\bullet\ \rho = g^{r_M} \bmod \eta$$

$$\bullet\ V_{M\_S_\rho} = E_\rho((\xi_3)_M \| h_M)$$

Step ⑦: A Bluetooth slave saves the transmitted $(V_{M\_S_\rho} \| \rho)$, calculates $(V_{S\_M_\tau} \| \tau)$ and sends it to a Bluetooth master.

$$\bullet\ (\xi_2)_S = (V_{I\_S_K} \| r_S \| T_S)$$

$$\bullet\ h_S = H((\xi_2)_S)$$

$$\bullet\ \tau = g^{r_S} \bmod \eta$$

$$\bullet\ V_{S\_M_\tau} = E_\tau((\xi_2)_S \| h_S)$$

Step ⑧: A Bluetooth master saves the transmitted $(V_{S\_M_\tau} \| \tau)$ temporary, calculates the following (see below) and sends $(\Lambda \| T_M)$ to a Bluetooth slave.

$$\bullet\ \Lambda = \tau^{r_M} \bmod \eta$$

The Bluetooth slave will verify that the following $\rho' = \rho$ is right if all above expressions are correct.

$$\bullet\ r_S^{-1} = \omega$$

$$\bullet\ \rho' = \Lambda^{\omega} \bmod \eta$$

Then the master and the slave will codify temporarily saved value, verify r_M , r_S and saves $V_{H\_M_K}$, $V_{S\_I_K}$.

Stage 3 : The mutual authentication between a Bluetooth master and a slave

This stage is aimed at the mutual authentication between a central server and a personal computer for the secure inquiry with a protocol. A master and a slave create new session key.

Step ⑨: A Bluetooth master creates a secure hash value, encrypts with a public key of the central server and sends $V_{M\_Hy_H}$ to a central server.

$$\bullet\ (\xi_4)_M = (V_{H\_M_K} \| T_M \| ID_M)$$

$$\bullet\ h_M = H((\xi_4)_M)$$

$$\bullet\ V_{M\_Hy_H} = ER_{y_H}((\xi_4)_M \| h_M)$$

Step ⑩: A Bluetooth slave creates a secure hash value, encrypts with a public key of a personal computer and sends $V_{S\_Iy_I}$ to a personal computer.

$$\bullet\ (\xi_3)_S = (V_{S\_I_K} \| T_S \| ID_S)$$

$$\bullet\ h_S = H((\xi_3)_S)$$

$$\bullet\ V_{S\_Iy_I} = ER_{y_S}((\xi_3)_S \| h_S)$$

Step ⑪: When a central server creates an incipient session key K with the value transmitted from a Bluetooth master, it calculates the following by using both the used random number r_H , r_M and the thing created in the Bluetooth master. After finishing the above process, the computer encrypts a secure hash value into a public key and sends $(V_{H\_Iy_I} \| h_H)$ to a personal computer.

$$\bullet\ (\xi_4)_H = (V_{H\_M_K} \| r_H \| r_M \| T_H)$$

$$\bullet\ h_H = H((\xi_4)_H)$$

$$\bullet\ V_{H\_Iy_I} = ER_{y_I}((\xi_4)_H \| h_H)$$

Step ⑫: When a personal computer creates an incipient session key K with the value transmitted from a Bluetooth, it calculates the following by using both the used random number r_I , r_S and the thing created in the Bluetooth slave. After finishing the above process, the computer encrypts a secure hash value into a public key and sends $(V_{I\_Hy_H} \| h_I)$ to a central server.

$$\bullet\ (\xi_3)_I = (V_{I\_S_K} \| r_I \| r_S \| T_S)$$

$$\bullet\ h_I = H((\xi_3)_I)$$

$$\bullet\ V_{I\_Hy_H} = ER_{y_H}((\xi_3)_I \parallel h_I)$$

Step ⑬: A central server calculates the following and sends $v_{H\_Mr_M}$ to a Bluetooth master.

$$\bullet\ (\xi_5)_H = (K \parallel T_H)$$

$$\bullet\ h_H = H((\xi_5)_H)$$

$$\bullet\ V_{H\_Mr_M} = E_{r_M}((\xi_5)_H \parallel h_H)$$

Step ⑬: A personal computer calculates the following and sends $v_{I\_Sr_S}$ to a Bluetooth slave.

$$\bullet\ (\xi_4)_I = (K \parallel T_I)$$

$$\bullet\ h_H = H((\xi_5)_H)$$

$$\bullet\ V_{H\_Mr_M} = E_{r_M}((\xi_5)_H \parallel h_H)$$

Step ⑭: After Bluetooth master and a slave decrypt $V_{H\_M_K}$ and $V_{I\_S_K}$, they check a central server and a personal computer, and send a message of consummation after the verification of signature. After the full process, a Bluetooth master and a slave calculate $K \oplus \rho = \chi$ and start the encrypts communication with a session key.

Stage 4 : The exchange of signature values and the creation of a group key.

Stage 4 is an incipient step for setting up exchange a signature and forming a group key in order to create a Bluetooth Piconet.

Step ⑮: After a Piconet master calculates the following (see below); It creates a pair of optional keys (p_{Mi}, n_{Mi}) and the ECDSA signature of R_M, S_M and calculates a secure hash value. Then, a Blutooth master encrypts $V_{M_A\_S_\chi}$ with a session key χ and sends it to a Bluetooth slave.

$$\bullet\ (\xi_1)_{M_A} = ((p_{Mi}, n_{Mi}) \parallel R_M \parallel S_M)$$

$$\bullet\ h_{M_A} = H((\xi_1)_{M_A})$$

$$\bullet\ V_{M_A\_S_\chi} = E_\chi((\xi_1)_{M_A} \parallel h_{M_A})$$

Step ⑯: After a Piconet slave verifies the integrity and confidentiality with the transmitted $V_{M_A\_S_\chi}$, it chooses a pair of optional keys and carries out the process of signature ECDSA, encrypts $V_{S\_M_A\chi}$ and sends it to the Bluetooth master.

$$\bullet\ (\xi_4)_S = (S_j \parallel R_S \parallel S_S \parallel T_S)$$

$$\bullet\ h_S = H((\xi_4)_S)$$

$$\bullet\ V_{S\_M_{A\chi}} = E_\chi((\xi_4)_S \parallel h_S)$$

– The ECDSA signature will be used to certificate a Bluetooth slave in a Bluetooth piconet later.

Step 5 : The closing step of an event

This stage is the step that an event is closed after exchanging a group key and a signature.

Step ⑰: A master verifies the integrity and confidentiality of the value transmitted from a slave with a session key χ after verifying $V_{S\_M_A\chi}$, h_S. It sends a closing message to a Bluetooth slave.

Step ⑱: The slave sends an event-closing message to a master.

3.3 Piconet Protocol

There have been two kinds of connective methods considered as the connection of a slave since the piconet that was needed a group key was formed. The Request for the connection of the slave that is included in a piconet.

A. Request for Connection from a Slave Within the Piconet (When There Is No Authentication between Masters)

The next is about the protocol that is needed in the case of the following. On condition that there are two more piconets, a mobile slave requests that the piconet not authenticated should be connected

Step (a): A slave of Master$_B$ makes a request for connection by sending its own in formation $S_{\inf o}$ to master$_A$.

Step (b): The Master$_A$ calculates $V_{M_A\_M_B\chi}$, the following and sends it to the mas ter$_B$.

- $(\xi_2)_{M_A} = (ID_{M_A} \| R_A \| S_A \| T_{M_A} \| S_{\inf o} \| ID_{M_A})$
- $h_{M_A} = H((\xi_2)_{M_A})$
- $V_{M_A\_M_B\chi} = E_\chi((\xi_2)_{M_A} \| h_{M_A})$

Step (c)': The Master$_B$ saves the ECDSA signature of the master$_A$, verifies ID_{M_A} calculates a secure hash value, and encrypts it with a public key of the master$_B$ then sends $V_{M_B\_M_A a_{M_A}}$ to the master$_A$.

- $(\xi)_{M_B} = (R_B \| S_B \| R_{BS} \| S_{BS} \| T_{M_B} \| S_{\inf o_{res}})$
- $h_{M_B} = H((\xi)_{M_B})$
- $V_{M_B\_M_A a_{M_A}} = EC_{a_{M_A}}((\xi)_{M_B} \| h_{M_B})$

Step (c): The slave requested for being connected calculates $V_{S\_M_A a_{M_A}}$ and sends it to the master$_A$.

$$\bullet\ (\xi_5)_S = (R_S \parallel S_S \parallel T_S)$$

$$\bullet\ h_S = H((\xi_5)_S)$$

$$\bullet\ V_{S\_M_A a_{M_A}} = EC_{a_{M_A}}((\xi_5)_S \parallel h_S)$$

Step (d): After the master$_A$ verifies the integrity and confidentiality of the value transmitted from a slave and the master$_B$, it checks the signature transmitted from the slave and the signature of the slave transmitted from the master$_B$. Finally it authenticates the slave requested for being connected

B. The Request for the Connection of the Slave Included in a Piconet (After It Is Authenticated by Masters)

If masters finish authenticating, step 4 and 5 will be started with the slave requested for being connected.

4 Discussion and Security Analysis

- The management of a key : The secure Maintenance of PIN by using session distribution and Hughes Keys on the negotiatory table for the security of application.
- Authentication : An offer to certificate a user by using a users signature based on PIN, Personal Information Number
- An area attack : A secure control device through the communication authenticated between slaves by a pertinent piconet. The piconet uses a group key.
- Authentication attack : The Protection against randomized attack in order to keep a piconet slave secure on the basis of a session key and a group
- A PIN attack: Conquering security against a PIN attack. To achieve this, a class of application should use all kinds of ciphers like a hash function, a public key and the upgrade of cryptography against Man-in-the-middle-attack.
- Confidentiality: The assurance of confidentiality by using ciphers like a session key and a public key
- Integrity: The assurance of integrity by using a secure hash value and a time stamp
- Users authentication : Because of forming the network standing on the basis of users PINs, a device as well as a user can be certificated.

5 Conclusion

Since communicative technology is being developed rapidly, the demand on personal news-communication is also increasing day by day. As a result, many studies about close range wireless communication have been progressing upon the demand from users. It seems that Bluetooth is one of the studies. Now the technology is receiving a lot of attention from all over the world.

However it has some weak points, especially in the field security. Even though many companies keep studying it, it still cannot be sufficient. That is the reason why this paper proposes a method of forming and maintaining a secure and reliable Piconet by upgrading the poor field security. It is based on Bluetooth's standard program version 1.1, which resolves the problems anticipated in its actual application in the home network.

The next proposed method will cope with the titanic volume of communication in the future and it is believed that more reliable Bluetooth will be working for AD-hoc network and IMT-2000 successfully.

References

[1]. http://www.bluetooth.com (Bluetooth White Paper)

[2]. http://www.bluetooth.org/specifications.htm (Bluetooth v1.1 Core Specifications)

[3]. http://www.niksula.cs.hut.fi/~jiitv/bluesec.html (Juha T. Bluetooth security)

[4]. http://www.cs.hut.fi/Opinnot/Tik-86.174/sectopics.html (Ullgren T. Security in Bluetooth: Key management in Bluetooth)

[5]. http://www.bell-labs.com/usermarkusj/bt.html (Jakobsson M., Wetzel S. Security Weakness in Bluetooth: RSA 2001)

[6] E. Hughes, "An Encrypted Key Transmission Protocol," presented at the rump session of CRYPTO '94, Aug 1994.

[7]. http://www.intel.com/mobile/technology/wireless.htm

Differentiated Bandwidth Allocation and Power Saving for Wireless Personal Area Networks

Tae-Jin Lee[1] and Yongsuk Kim[2]

[1] School of Information and Communication Engineering, Sungkyunkwan University, Suwon, KOREA

[2] i-Networking Lab, Samsung Advanced Institute of Technology, Suwon, KOREA

tjlee@ece.skku.ac.kr, yongsuk@samsung.com

Abstract. We consider differentiated bandwidth allocation for short-range wireless personal communication systems: Bluetooth. Since bandwidth requirements may vary among applications/services, and/or it may change over time, it is important to decide how to allocate limited resources to various service classes to meet their service requirements. We propose a simple and efficient bandwidth allocation mechanism which meets bandwidth requirements of various service types while saving power consumption by a power saving mode, *i.e.*, sniff mode. We compare our proposed mechanism with a conventional (weighted) round-robin polling scheme and show that it achieves significant improvement of throughput, delay, and power consumption.

1 Introduction

As the demand of seamless communications is increasingly growing, wireless personal area networks (PAN) are expected to provide such connection between numerous personal devices to other long range communication networks. Bluetooth is considered as one of the promising technologies for such wireless PAN systems and is expected to provide ubiquitous wireless services with inexpensive cost in personal areas. Numerous applications/services using Bluetooth systems are envisioned, *e.g.*, cable replacement, data synchronization, and networking between devices [8]. In order to support such services, it is required to efficiently use limited bandwidth and power of Bluetooth systems. Currently, Bluetooth can support maximum data rate of 723 kbps which needs to be shared in a piconet comprising of up to seven active Bluetooth-enabled devices [9]. Recently, bandwidth allocation and polling schemes for Bluetooth have been researched [5], [3], [9], [6], [10], in which fair scheduling mechanisms under the normal connection mode are studied.

We notice that bandwidth requirements may vary among applications/services, or it may change over time. In this sense, it is important to decide how to allocate limited resources to various service classes to meet those requirements. In the prior research on piconet scheduling [5], [3], [6], [10], polling algorithms based on queue status of master-slave pairs are mainly considered, in which bandwidth allocation is conducted "implicitly" based on the queue status. In order to truly differentiate various service classes, it needs to "explicitly" allocate bandwidth differently according to service requirements. We propose an explicit and adaptive bandwidth allocation mechanism to provide such differentiated and adaptively varying resource allocation.

M.H. Shafazand and A M. Tjoa (Eds.): EurAsia-ICT 2002, LNCS 2510, pp. 778–785, 2002.

In addition to the efficient bandwidth allocation, power consumption needs to be limited to the minimum to meet the small battery requirements of mobile devices in wireless PAN. To tackle this limitation, Bluetooth Specification [7] defines three types of power saving modes, *i.e.*, hold, sniff and park mode. A hold mode can be used to interrupt the normal data exchange temporarily, which causes the waste of bandwidth during hold interval. A park mode requires control of many parameters and it is not efficient in terms of bandwidth utilization since bandwidth may be wasted during park interval, although it can control many devices (up to 255 devices) with low bandwidth requirements. A sniff mode has only two control parameters and it can achieve efficient bandwidth utilization with less power consumption. In [11], policies under the park mode has been presented. The sniff mode has been considered in [4] to reduce the power consumption and packet delays.

In this paper, we propose an efficient bandwidth allocation mechanism using a sniff mode. Our mechanism jointly considers power saving and differentiated bandwidth allocation. Thus it can meet bandwidth requirements of various service types while saving power consumption compared with a normal connection mode. In addition, our proposed mechanism is simple to leverage the requirement on less complex control algorithm for wireless personal area network devices. In our proposed mechanism, each connection in a network is assigned a priority determined from the service characteristics/requirement. The parameters of the sniff mode are then decided differently based on its priority. By appropriately selecting the parameters according to the priority and traffic load, one can allocate bandwidth with differentiation and improve the performance of priority services. We have shown that our mechanism can provide differentiated bandwidth allocation according to service types, and efficient resource utilization in terms of bandwidth and power consumption.

This paper is organized as follows. In §2, we briefly describe features of the Bluetooth system. We then propose our bandwidth allocation mechanism in §3. Simulation results of the proposed mechanism are followed in §4. In §5, conclusion is presented.

2 Bluetooth

Bluetooth is a low power and low cost wireless communication system covering a small personal area within 10m. In a piconet, two or more units sharing the same channel, slave devices with ACL (Asynchronous Connection-Less) connections are allowed to transmit data packets right after the TX slot of a master device, only when they are explicitly "polled" by the master. Thus ACL connections provide best-effort service. The maximum effective data rate one can achieve from an ACL connection between a single master and slave is 723 kbps. The number of slots for a packet can occupy varies among one, three, and five according to the amount of data to be transmitted. For SCO (Synchronous Connection Oriented) connections, periodic slots are strictly reserved for SCO packets between a master and slaves resulting in a guaranteed rate of 64 kbps. Since Bluetooth is expected to be employed in small mobile devices like PDAs and mobile phones, battery power consumption needs to be minimized. In this sense, power saving modes, *i.e.*, hold, sniff and park mode, are defined in the specification.

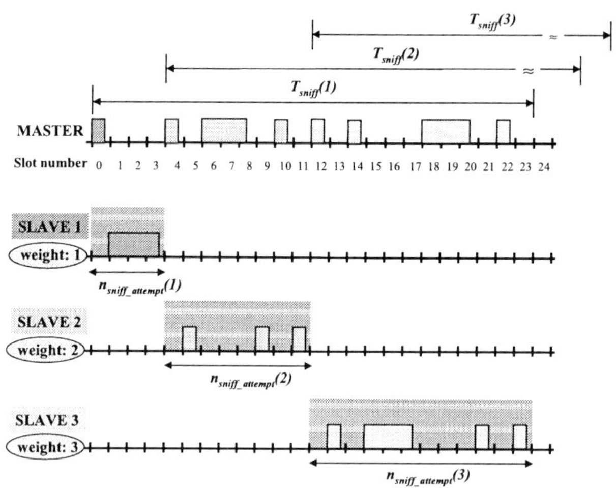

Fig. 1. Proposed Weighted Sniff (WS) mechanism based on service priorities.

3 Efficient Bandwidth Allocation

If a slave device operates on a regular connection mode, it always has to listen to packet headers to check whether they are indeed destined to the slave. This causes unnecessary power consumption, which can be saved by employing a sniff mode. Operating in a sniff mode, a slave only needs to listen to packets during its own *sniff attempt interval.* Otherwise it can sleep and does not have to listen to any packet headers, which saves RX power.

Suppose there are N_s slave devices in a piconet, two or more units sharing the same channel. Each slave is assigned a priority $w(i), i = 1, \cdots, N_s$ according to the type of service requirement. We propose a method to decide the priority based on the polling interval, which is reserved for QoS (Quality of Service) option in the Bluetooth Specification [7].

Let $p_{int}(i)$ denote the polling interval of slave i, *i.e.,* the time interval during which a master should poll the slave i at least once. We note that the polling interval re¡ ects the priority to the slave, *i.e.,* the priority is inversely proportional to the polling interval. Thus, we use the polling interval $p_{int}(i)$ to decide the priority $w(i)$ of slave i:

$$w(i) = \left\lceil \frac{1}{\frac{p_{int}(i)}{\max_j p_{int}(j)}} \right\rceil = \left\lceil \frac{\max_j p_{int}(j)}{p_{int}(i)} \right\rceil, \quad i = 1, \cdots N_s, \tag{1}$$

where $\lceil x \rceil$ denotes the closest integer which is greater than or equal to x. One can show that the priority $w(i)$ is a non-negative integer, so $w(i) \in W \subset \mathbb{N}$, where W is a small subset of non-negative integers. We note that the higher the priority, the larger the weight $w(i)$ is.

In order to use a sniff mode, two parameters, sniff interval T_{sniff} and sniff attempt interval $n_{sniff\_attempt}$, have to be decided prior to the sniff mode. For a sniff attempt interval $n_{sniff\_attempt}(i)$ of slave i, u_{min} times of its priority is given to each slave, which will provide the sufficient number of slots to each slave for packet TX and RX:

$$n_{sniff\_attempt}(i) = u_{min}w(i) \quad \text{(slots)}, \quad i = 1, \cdots N_s, \tag{2}$$

where u_{min} is the minimum number of slots to be assigned to all the slaves. By this way, more slots are allocated to a slave with higher priority. A sniff interval parameter $T_{sniff}(i)$ is determined as sum of sniff attempt intervals of the slaves:

$$T_{sniff}(i) = \sum_{j=1}^{N_s} n_{sniff\_attempt}(j) \quad \text{(slots)}, \quad i = 1, \cdots N_s. \tag{3}$$

We note that $T_{sniff}(i) = T_{sniff}(j)$ for all i, j in the same piconet to synchronize the sniff attempt intervals (see Fig. 1). Note also that the sniff interval $T_{sniff}(i)$ must be smaller than the polling interval $p_{int}(i)$ to satisfy the requested and negotiated QoS. That is

$$T_{sniff}(i) \leq \min_j \ p_{int}(j), \quad i = 1, \cdots, N_s. \tag{4}$$

The parameters can be negotiated between a master and slaves, and adaptively changed according to the varying bandwidth requirements. Hence our proposed mechanism WS (Weighted Sniff) is given by the equations (1) - (4), which is summarized in the following theorem.

Theorem 1. *The sniff attempt intervals $n_{sniff\_attempt}(i)$ and sniff intervals $T_{sniff}(i)$, $i \in N_s$ can be decided to satisfy (1) - (4).*

Proof: Given polling intervals, one can find sniff parameters according to (1) - (3). If the negotiated polling intervals between a master and slaves meet (4), the polling intervals for the sniff parameters can be used. Now suppose that this is not the case, *i.e.*, $T_{sniff}(i) > \min_j \ p_{int}(j)$. Then we let

$$T_{sniff}(i) = \min_j \ p_{int}(j) + \Delta, \quad \Delta > 0. \tag{5}$$

Suppose we can negotiate (increase) the polling intervals without much sacrificing the QoS as required initially. Increasing the polling intervals by $\alpha > 0$, one can show the followings:

$$\frac{\max_j p_{int}(j)}{p_{int}(l)} - \frac{\max_j p_{int}^{\alpha}(j)}{p_{int}^{\alpha}(l)} = \frac{\max_j p_{int}(j)}{p_{int}(l)} - \frac{\max_j (p_{int}(j) + \alpha)}{(p_{int}(l) + \alpha)}$$
$$= \frac{\alpha(\max_j p_{int}(j) - p_{int}(l))}{p_{int}(l)(p_{int}(l) + \alpha)} = \gamma(l) > 0, \quad \forall l \in N_s, \tag{6}$$

where $p^{\alpha}_{int}(l)$ is an adjusted polling interval. After the adjustment of polling intervals, the sniff interval becomes

$$T^{\alpha}_{sniff}(i) = \sum_{l=1}^{N_s} u_{min} \left\lceil \frac{\max_j p^{\alpha}_{int}(j)}{p^{\alpha}_{int}(l)} \right\rceil = \sum_{l=1}^{N_s} u_{min} \left\lceil \frac{\max_j (p_{int}(j) + \alpha)}{(p_{int}(l) + \alpha)} \right\rceil$$
$$= \sum_{l=1}^{N_s} u_{min} \left\lceil \frac{\max_j p_{int}(j)}{p_{int}(l)} - \gamma(l) \right\rceil = T_{sniff}(i) - \beta, \quad \beta > 0, \qquad (7)$$

by (6). Then, from (5) and (7), one can find α such that

$$T^{\alpha}_{sniff}(i) = T_{sniff}(i) - \beta = \min_j \; p_{int}(j) + \Delta - \beta$$
$$\leq \min_j \; (p_{int}(j) + \alpha) \leq \min_j \; p^{\alpha}_{int}(j).$$

Thus, α can be chosen to satisfy (4). □

We call our proposed mechanism WS in the sequel. For example, consider a piconet of three active slaves (*i.e.*, $N_s = 3$). We assume the negotiated polling intervals between the master and the slave1, 2 and 3 are 40, 30 and 20 slots, respectively. Then the master determines the weight $w(i)$ and sniff attempt interval $n_{sniff\_attempt}(i)$ of slave i, $i = 1, 2, 3$, from (1) and (2). That is, $w(i)$ = 2, 2 and 1 slot(s), and $n_{sniff\_attempt}(i)$ = 8, 8, and 4 slots, for i = 1, 2 and 3, respectively, assuming $u_{min} = 4$. And the maximum sniff interval T_{sniff} becomes 20 slots (10.25 msec).

The implementation of our proposed mechanism requires LMP message exchange between a master and slaves. The proposed mechanism can operate properly without any modification to the current Specification of the Bluetooth system [7].

3.1 Power Consumption

Let's examine the power consumption of our mechanism and other schemes. Let p_{TX} be the power consumption to transmit a single slot packet, p_{RX} be the power consumption to receive a single slot packet, p_{RX_H} be the power consumption to check the header of a single slot packet, and p_{SL} be the power consumption during a sleep state. Assuming TX/RX of single slot packets (DM1 or DH1), we compute the power consumption of slave i as follows. In order to normalize power consumption, we compute power consumption per unit bit.

First, suppose we use a normal connection mode with weighted round-robin bandwidth allocation, *i.e.*, a master and a slave i communicate during $n_{sniff\_attempt}(i)$, which depends on the priority/weight given to the slave, under a normal connection mode. Then the power consumption during T_{sniff} will be

$$pw_n(i) = (p_{TX} + p_{RX})(n_{sniff\_attempt}(i)/2) + \sum_{j=1, j\neq i}^{N_s} (p_{SL} + p_{RX\_H})(n_{sniff\_attempt}(j)/2).$$

The normalized power consumption per unit bit will be

$$p_n(i) = pw_n(i)/(n_b n_{sniff\_attempt}(i)/2). \qquad (8)$$

Note that we use the intervals $n_{sniff\_attempt}(i)$ and $T_{sniff}(i)$ to compare the power consumption of weighted round-robin scheme with that of our proposed scheme during the same interval.

In order to see the impact of our proposed scheme on power consumption, we compute power consumption $p_s(i)$ as follows.

$$pw_s(i) = (p_{TX} + p_{RX})(n_{sniff\_attempt}(i)/2) + \sum_{j=1, j \neq i}^{N_s} p_{SL} n_{sniff\_attempt}(j).$$

Note that $p_{RX\_H} = 0$ due to a sniff mode. The normalized power consumption per unit bit is

$$p_s(i) = pw_s(i)/(n_b n_{sniff\_attempt}(i)/2). \quad (9)$$

The power consumption during T_{sniff} can be computed similarly for a normal connection mode with a conventional round-robin scheme. Since TX consumes more power than RX [1] (*e.g.*, WaveLAN cards as in [6]), we assume $p_{TX} = 2, p_{RX} = 1, p_{RX_H} = 1/6$ and $p_{SL} = 0$, and $N_s = 4$ with $u_{min} = 2$, $w(1) = 1, w(2) = 2, w(3) = 3$, and $w(4) = 4$. Let's denote a normal connection mode with round-robin as RR, and a normal connection mode with weighted round-robin as WRR. Then total power saving of our scheme (WS) compared with that of WRR is 18.9 %. When the number of slaves is seven, we can achieve 34.2 % and 25 % of power saving over WRR and RR, respectively.

4 Simulation Results

In order to explore the actual performance (throughput and delay) of WS algorithm, we have conducted simulations. A piconet consisting of a master and three slaves is considered. We build a simulation environment, where system behaviors (*e.g.*, realistic traffic, bit errors, ARQ, error correction, etc.) are modeled to evaluate actual performance. We have modeled the Bluetooth system using BONeS Designer [2] and simulated the scheme with a connection mode and round-robin policy (RR), and our proposed scheme (WS). We consider a piconet of a master and three slaves, in which slave3 requires premium service since the traffic load on 3 is higher than the other slaves (slave1 and 2). The simulation time for each scenario was 1200 seconds (20 minutes). DM packets are used assuming that data packets are vulnerable to bit errors due to error-prone wireless channel characteristics. To ensure reliable data transmission, FEC and ARQ [7] are employed in the simulation. The BER (Bit Error Rate) of wireless channel is assumed to be 0.001.

As traffic sources, two-state MMPP (Markov-Modulated Poisson Process) is used which is shown to be well matched with multimedia traffic [12]. If we assume symmetric TX/RX, the maximum rate for a single slave using DM packets would achieve is 286.7 kbps. So we let average service rate of each connection $\bar{\mu} = 286.7$ kbps. We used the parameters summarized in Table 1 to explore the impact of scheduling mechanisms on different traffic load conditions. The symmetric traffic is applied to the master-to-slave and slave-to-master traffic sources. The sniff parameters assigned to the slave 1 and 2 is $n_{sniff\_attempt}(i)$ = 4 for $i = 1, 2$. In order to provide sufficient bandwidth required for different traffic load conditions of slave3, we use $n_{sniff\_attempt}(3)$ = 8, 12, 16, 20, and 24 as traffic load increases from 0.2 to 0.6. We have simulated a round-robin scheduling mechanism as a comparison with our proposed scheduling mechanism. In

Table 1. Parameters used in the simulation.

Parameter	Slave1,2	Slave3
Load (ρ)	0.1	0.2, 0.3, 0.4, 0.5, 0.6
Rate	28.7	57.4, 86.1, 114.8, 143.5, 172.2 kbps
$n_{sniff\_attempt}(i)$	4	8, 12, 16, 20, 24 slots
$T_{sniff}(i)$	16 ~ 32	16 ~ 32 slots

order to provide sufficient bandwidth required for different traffic load conditions of slave3, the maximum possible number of slots allocated for TX or RX of a packet was set to 5 (*i.e.*, DM1 ~ DM5).

We have collected average throughput and delay of packets received in each slave for each of mechanisms. The simulation results are shown in Fig. 2 and Fig. 3. In the RR scheme, average delay of slave1 gradually increases from 0.5 to 0.8 sec as the traffic load on slave3 increases. The delay of slave3 under RR abruptly increases as the traffic load on slave3 exceeds 0.4, *i.e.*, more than 10 sec, which might be intolerable to the service on slave3. On the other hand, the slave3 experiences significantly low delays under our WS mechanism, *e.g.*, 86 msec at load 0.2 (54.4 kbps) and 15 msec at load 0.6 (172.2 kbps). The slave1 under WS shows slight degradation of delay performance, *i.e.*, less than a few seconds under the traffic loads from 0.2 to 0.6.

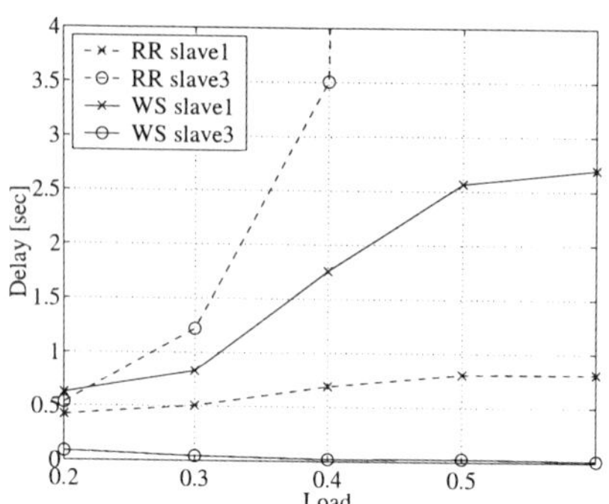

Fig. 2. Average delay of slave1 and slave3 as traffic load on slave3 increases.

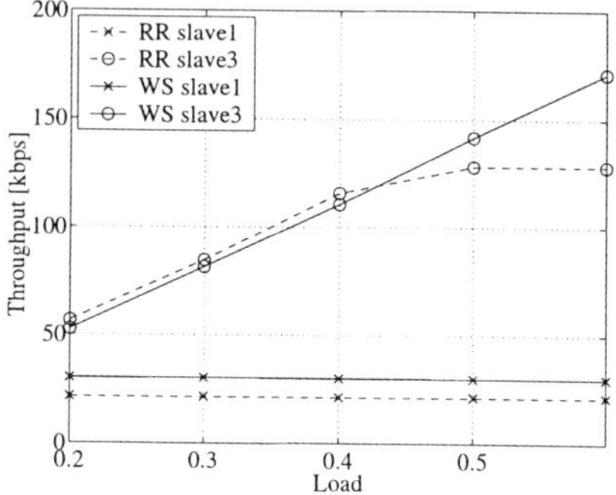

Fig. 3. Average throughput of slave1 and slave3 as traffic load on slave3 increases.

We then investigate the throughput performance under RR and WS policies. Slave1 achieves 21 kbps under RR while it shows 30 kbps under WS since WS tends to poll slave1 for longer duration than that of RR resulting in slightly more bandwidth allocation to slave1. As for slave3, RR mechanism is unable to allocate sufficient bandwidth required for the traffic source, thus the allocation of bandwidth saturates at around 128 kbps as shown in Fig. 3. However, in our WS scheme, allocated bandwidth increases by expanding the sniff attempt duration as the traffic load on slave3 increases, so that the premium service on slave3 does not have to suffer from the lack of bandwidth.

5 Conclusion

We have proposed an efficient and adaptive bandwidth allocation mechanism using a sniff mode, *i.e.,* a weighted sniff (WS) mechanism. In our mechanism, each connection in a network has a priority assigned to it and determined from the service characteristics/requirement. The sniff attempt duration is then decided based on the priority. By appropriately selecting the sniff attempt duration according to the priority and traffic load, we have shown that average delay and throughput performance can be significantly improved. Our policy can also save power consumption, which can not be achieved by a conventional connection mode using a round-robin polling mechanism. Therefore, our policy can provide differentiated bandwidth allocation while efficiently utilizing the limited resources of Bluetooth, *i.e.,* power and bandwidth.

References

1. L. Bonomi and L. Donatiello. A distributed contention control mechanism for power saving in random-access ad-hoc wireless local area networks. In *Proc. of 6th IEEE MoMuC'99*, pages 114–123, Nov. 1999.
2. Cadence. *BONeS Designer ver. 4.0.* 1999.
3. A. Capone, M. Gerla, and R. Kapoor. Efficient polling schemes for Bluetooth picocells. In *Proc. of IEEE ICC 2001*, pages 1990–1994, 2001.
4. I. Chakraborty, A. Kashyap, A. Kumar, A. Rastogi, H. Saran, and R. Shorey. Policies for increasing throughput and decreasing power consumption in Bluetooth MAC. In *Proc. of IEEE ICPWC*, pages 90–94, 2000.
5. A. Das, A. Ghose, A. Razdan, H. Saran, and R. Shorey. Enhancing performance of asynchronous data traffic over the Bluetooth wireless ad-hoc network. In *Proc. of IEEE INFOCOM*, pages 591–600, 2001.
6. S. Garg, M. Kalia, and R. Shorey. MAC scheduling policies for power optimization in Bluetooth: A master driven TDD wireless system. In *Proc. of IEEE VTC*, pages 196–200, 2000.
7. Bluetooth Special Interest Group. *Core, Specification of the Bluetooth System ver. 1.1.* Nov. 2000.
8. Bluetooth Special Interest Group. *Profiles, Specification of the Bluetooth System ver. 1.1.* Nov. 2000.
9. J. C. Haartsen and S. Mattisson. Bluetooth - a new low-power radio interface providing short-range connectivity. *Proceedings of the IEEE*, 88(10):1651–1661, Oct. 2000.
10. M. Kalia, D. Banjal, and R. Shorey. Data scheduling and SAR for Bluetooth MAC. In *Proc. of IEEE VTC*, pages 716–720, 2000.
11. M. Kalia, S. Garg, and R. Shorey. Efficient policies for increasing capacity in Bluetooth: An indoor pico-cellular wireless system. In *Proc. of IEEE VTC*, pages 907–911, 2000.
12. T.-J. Lee, K. Jang, H. Kang, and J. Park. Model and performance evaluation of a piconet for point-to-multipoint communications in Bluetooth. In *Proc. of IEEE VTC 2001*, 2001.

Combining Extreme Programming with ISO 9000*

Jerzy R. Nawrocki, Michał Jasiński, Bartosz Walter, and Adam Wojciechowski

Poznan University of Technology, ul. Piotrowo 3A, 60-965 Poznan, Poland
{Jerzy.Nawrocki, Michal.Jasinski, Bartosz.Walter, Adam.Wojciechowski}
@cs.put.poznan.pl
http://www.cs.put.poznan.pl

Abstract. The main drivers of the growing ICT market are software products. European Information Technology Observatory estimates, that in year 2002 the total value of ICT software products in Western Europe will be more than 70 billions Euro. Unfortunately very few people are satisfied with quality of the software products and processes. Software Process Improvement tools, like CMM and ISO 9000 were to cure this situation, but some people complain that they are too bureaucratic and inflexible. As a result new, so-called *agile*, methodologies appeared. One of them is Extreme Programming (XP) - a lightweight, change-oriented and customer-oriented approach to software development. Although XP proposes many interesting practices, it has some limitations. Moreover, it is not clear how to introduce XP to an organization certified to ISO 9001:2000. The aim of the paper is to present a modified version of XP that would be acceptable from the point of view of ISO 9000.

1 Introduction

European Information Technology Observatory (EITO) predicts that the total value of Western Europe ICT market will reach 678 billion Euro in 2002 [7] and 11% of it (more than 70 billions Euro) will be spent on software products. However, the market is getting more and more demanding. To be successful, software companies have to attract customers in various ways. One of possible steps is certification to ISO 9001:2000.

ISO 9000 is a series of international standards concerning establishment and maintenance of a quality management system. They are general-purpose standards comprising vocabulary [3] requirements [4] and recommendations for improvement [5]. The only standard an organization can be certified to is ISO 9001. They can be used in a private factory as well as in a government institution, in a big shipyard and in a small software company. ISO 9000 originated in UK and it is getting more and more popular. In 1997 there have been approximately 102 000 registrations worldwide and three years later the number was over 250 000 [15].

* This work has been financially supported by the State Committee for Scientific Research as a research grant 4 T11F 001 23 (years 2002-2005)

M.H. Shafazand and A M. Tjoa (Eds.): EurAsia-ICT 2002, LNCS 2510, pp. 786–794, 2002.

Opinions on ISO 9000 vary significantly. For instance, the Director General of the British Standard Institute claimed that ISO 9000 *"will save your money"*, *"it will ensure satisfied customers"*, and *"it will reduce waste and time-consuming reworking of designs and procedures"* [15]. At the opposite pole is the opinion of John Seddon. According to him *"by being labelled a quality standard, ISO 9000 has only succeeded in steering quality into troubled waters. Far from being a first step to quality it has been a step in the wrong direction. The hope is that it hasn't conditioned management to lose interest in the subject"* [15]. That opinion is somehow confirmed by the fact, that almost 10% of Australian companies have decided to discontinue registration to ISO 9001 [15]. What is wrong with ISO 9000? The general impression is that ISO 9000 standard requires too much documentation and it is too bureaucratic. Moreover, some people consider ISO 9000 too general and different types of software-oriented maturity models have been proposed [16,14,6]. Nevertheless, some software companies decided to go through the ISO 9001 certification process. The danger is that certification to ISO 9000 will result in a well-documented but still inefficient system. The initial enthusiasm of workers for the software process improvement will soon be dissipated if they find out that the ISO 9001 is just a marketing subterfuge, and the company has no intention to introduce a real process improvement.

On the other hand, a few years ago so-called *lightweight* (or *agile*) software development methodologies have appeared. The most popular is Extreme Programming (*XP* for short). XP emphasizes the importance of on-site customer, oral communication, product quality, short feedback from customer and end-users, simplicity, minimal documentation and avoidance of overtime. In the context of Information and Communication Technology it is important that XP is a change-oriented methodology and it tries to deliver maximum functionality at a minimum cost in a short time. A typical reaction of a programmer to XP is: *"At last a methodology for people, not people for methodology"*. Our idea is to use this enthusiasm as a starting point for real software process improvement and to combine it with the requirements of ISO 9000. Unfortunately, that merge is not straightforward. What we propose is a modified methodology, based on XP practices and conformant to ISO 9001:2000.

In the next section we will present most important features of Extreme Programming. Then, in section 3, we will discuss the Quality Management System of an ISO-9000 software company. We will show how to put together a general Quality Management System proposed by ISO 9000 and a software-oriented knowledge base which contains documents, artifacts, and data specific to a software organization. Our focus will be on two parts of ISO 9001:2000: product realization, and monitoring and measurement. Product realization, which is based on XP practices and conformant to ISO 9001:2000, will be described in section 4. To gain flexibility, we propose a four-level improvement schema resembling the Capability Maturity Model [14]. In section 5 we will describe measurements which are necessary from the point of view of ISO 9000 and useful in the context of XP practices. In the last section our early experience concerning the proposed approach will be presented.

2 XP Overview

Extreme Programming [1,2,8] represents a new wave in software development known as the *approach*. Tom de Marco, the father of structural analysis, calls XP the most important movement in software engineering (see the foreword to [2]). The strong points of XP in the ICT context are as follows:

- *Risk minimization.* ICT is developing very fast. To catch up with current developments it is necessary to make investments in new technologies and try new tools out. On the other hand, new tools and technologies are immature and one cannot depend on them. The best approach is to make some (preferably small) investment now and after some time invest more or give up, depending on the developments (it is like buying an option on the stock exchange). XP is based on incremental software development and its suites the strategy very well.
- *Customer orientation.* In XP all the business decisions are made by the customer and he has the full control over the development process.
- *Lack of excessive paperwork.* In XP programmers concentrate on programming, not on writing documentation. The only artifacts they have to produce are test cases and code.
- *Quality assurance through intensive testing.* In XP programmers first create test cases then they write code. Automated tests and integration are performed several times a day and they drive the development process.
- *Lack of overtime.* Short releases and increments allow to gain experience very fast. This makes planning easier and more dependable. As a result programmer do not have to (always) work overtime.

XP has also weak points. The most important are problems with software maintenance. Since the only artifacts are test cases and code, after some time it can be very difficult to maintain the software. It would be also the problem from the ISO 9000 point of view. In the remaining part of the paper we propose how to solve that problem.

3 Software Development in an ISO 9000 Company

ISO 9001:2000 standard defines requirements for a process-oriented Quality Management System (QMS for short). This means that desired results are achieved more efficiently when the related resources and activities, together with encompassing customer needs and satisfaction, are managed as a process. QMS is specified in a Quality Manual document featuring a three-tier structure, which consists of Quality Processes (including Quality Policies), Quality Procedures and Work Instructions. This structure is presented in Fig. 1.

The problem is that Work Instructions are sometimes too bureaucratic. A good example of that approach is Tricker's book on ISO 9000 [18]. According to it, a Work Instruction takes about 16 pages. Half of them contains purely administrative data (document data sheet, distribution list, amendments, list of

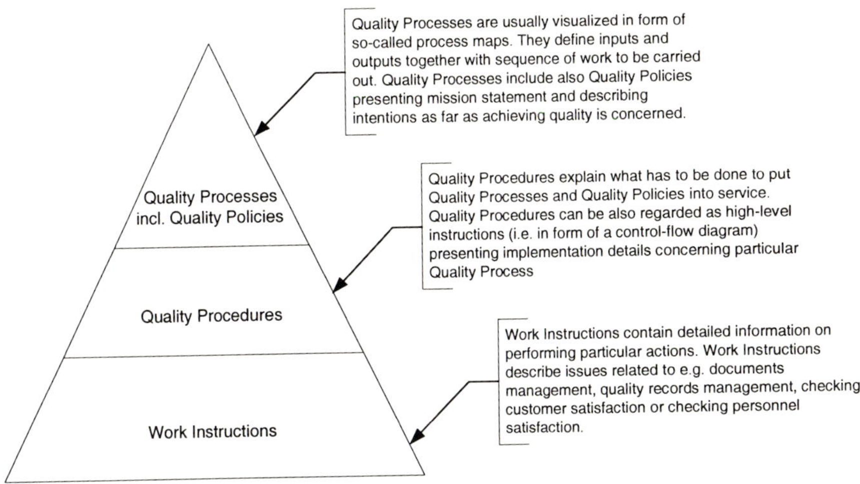

Fig. 1. Three-tier structure of Quality Manual.

annexes etc.). That makes the whole QMS documentation superfluously thick. Another drawback of Tricker's approach is form-orientation: Work Instructions focus on how to fill-in the forms used by the Quality Procedures. What we propose is to make Work Instructions shorter (some elements can be omitted, some, e.g. terminology, can be put together and placed in one section). Moreover, Work Instructions should describe practices specific for a given methodology of software development.

In our opinion, quality organization needs two things: general Quality Management System operating on a high abstraction level and a *Thesaurus* (knowledge database), which should materialize company's knowledge. In the thesaurus templates of e.g. Quality Plans, historical data concerning past projects etc can be deposited. This information will be indispensable during planning and improving software processes.

The clauses of ISO 9000:2000 can be split into two parts. One part describes the general Quality Management System (chapters 4, 5, and 6) while the other part specifies requirements for a methodology to be adopted by an ISO-9000 company (chapters 7 and 8 of ISO 9001:2000). In the remaining part of the paper we will focus on requirements imposed by chapters 7 and 8 of the ISO 9001:2000.

4 Product Realization: Modified XP Practices

To adopt XP practices to the needs of ISO-based product realization we suggest to use the XP Maturity Model (XPMM for short)[12]. That model resembles the SEI's Capability Maturity Model[14]. The XPMM's Key Process Areas associated with the maturity levels include: Customer Relationship Management and

Product Quality Assurance (Level 2), Pair Programming (Level 3), and Project performance (Level 4).

The XPMM maturity levels are useful as they allow to introduce XP to a company gradually (level by level). That provides flexibility to the software improvement process.

In the next section we describe XPMM's Key Process Areas together with modification, which are necessary to make the software development process compliant with ISO 9000:2000.

4.1 Customer Relationship Management

Customer satisfaction is of primary importance for both XP and ISO 9000. However, XP is much more specific about how to obtain that satisfaction. Here are the main XP practices directly influencing customer satisfaction:

CRM1: *User stories are used to describe requirements.* They are written on small pieces of paper using natural language. Usually the are very short and are just an introduction to the discussion between customer and the development team, so they do not have to be complete. User stories are not documented and maintained.

CRM2: *A development process is split into short releases (about 6-9 weeks) and each release is split into iterations (2-3 weeks).* Each iteration implements a set of user stories. When a release is finished the product is made available to the end users. This provides a fast feedback to the development team.

CRM3: *Planning game is used to create a release plan.* The user stories brought by the customer are evaluated by the developers. They estimate the effort required to implement each story and the technical risk. Knowing that, customer chooses the stories to be implemented in the current release.

CRM4: *A metaphor is chosen to facilitated communication with the customer.* The system is described in terms the customer is familiar with.

CRM5: *No functionality is added early.* The functionality to be implemented must be chosen by a customer representative, not by the development team.

The first problem, in the context of ISO 9000 is lack of written requirements (XP uses user stories instead of documented requirements). ISO 9001:2000 does not explicitly specify the need for existence of written requirements. However, the need for existence of written documentation follows from these clauses:

- *"(...)the organization shall determine (...) records needed to provide evidence that the (...) resulting product meets requirements."*([4], clause 7.1),
- *"The organization shall review the requirements related to the product (...) where the customer provides no documented statement requirement, the customer requirements shall be confirmed by the organization before acceptance."* ([4], clause 7.2.2),
- *"Inputs relating to product requirements shall be determined and records maintained. These inputs shall include functional and performance requirements (...). Requirements shall be complete, unambiguous and not in conflict with each other."*([4], clause 7.3.2).

The need for documented requirements is also emphasized by the CMM model ([14], KPA for Requirements Management, Ability 2). Thus, the problem arises how to introduce documented requirements to a lightweight methodology. To solve it one has to notice that XP was created to be lightweight to programmers and it puts extra work on the shoulders of other people, e.g. customer representative. We will follow that path. One of the roles in an XP team is the one of a tester, who implements test cases proposed by the customer. Our suggestion is to make the tester responsible for requirements documentation and management.

According to Beck, an XP tester is *"responsible for helping the customer choose and write functional tests"* and running them regularly [1]. Requirements and acceptance tests are on the same abstraction level, so tester seems to be the best person to take on the job of analyst responsible for requirements management.

4.2 Product Quality Assurance

Product quality assurance is addressed in ISO 9000 by clause 7.3.5 *Design and verification*: *"Verification shall be performed in accordance with planned agreements to ensure that the design and development outputs have met the design and development input requirements."*

XP implements this through the following practices:

PQA1: *Test-first coding.* It means that a programmer first writes a test, then starts coding. This helps to understand what we expect from a unit and removes "implementation bias" during testing.

PQA2: *All code must have unit tests.* That allows regression testing.

PQA3: *When a bug is found a test must be created.* That supports regression testing.

Moreover, XP supplements those practices with two others:

PQA4: *Continuous integration.* This should provide fast feedback on current system status.

PQA5: *Optimization is left till last.* Many optimizations require lots of effort and they are sources of potential bugs which are difficult to locate and fix. Thus, it is better for the product quality not to introduce and optimization if it is not necessary.

4.3 Pair Programming

Pair programming is specific to XP and it does not relate directly to any ISO 9000 clause. However, to implement pair programming in an efficient way, an open workspace lab is required. That is connected with clause 6.3 *Infrastructure* and 6.4 *Work environment.* Moreover, it is interesting to point out that ISO 9000 requires *conformity to product requirements*, not to *the defined processes.*

The pair programming practices are as follows:

PP1: *Code must be written to agreed standards.* This way it is much easier to understand and modify the code written by somebody else.
PP2: *All production code is pair programmed.* That is the main practice for XP projects.
PP3: *Only one pair integrates code at a time.* It is needed to ensure code consistency.
PP4: *Collective code ownership.* Everybody can change any piece of code if necessary.
PP5: *Use version management system.* It supports collective code ownership and continuous integration.

5 Monitoring and Measurement

In ISO 9001:2000 monitoring and measurement are described in very general terms. Extreme Programming gives hints how to implement ISO 9001 clauses related to this issue:

- *8.2.3 Monitoring and measurement of processes.* The aim is to *"demonstrate the ability of processes to achieve planned results"*. In the context of XP one should collect the following process metrics:
  - *Overtime* (day by day). XP assumes no overtime and one should know how far we are from the ideal process.
  - *Availability* of the customer representative to the development team to answer questions, resolve conflicts and create acceptance tests
  - *Project velocity* (i.e. the amount of time per week each team member can spend on his assignments). That data are used during planning.
  - *Integration log* to see how frequently new pieces of code are integrated with the system (XP suggests to have several integrations per day).
  - *Mode of production* for each piece of production code (pair or individual).
  - *Programming speed* (lines of code per hour, test cases per hour, acceptance tests per hour).
- *8.2.4 Monitoring and measurement of product.* In the context of software development the main measurement is test report showing the fraction of passed unit tests and acceptance tests (both unit tests and acceptance tests can comprise functional and performance tests).
- *8.3 Control of nonconforming product.* ISO 9001:2000 requires that *"when nonconforming product is concerned it shall be subject to re-verification to demonstrate conformity to the request"*. In XP before a new version of the system is checked-in to the baseline library all the unit tests should be passed.
- *8.5.2 Corrective action.* According to ISO 9001:2000 it is necessary to *"eliminate the cause of nonconformities in order to prevent recurrence."* To prevent recurrence of software defects XP requires to create test cases for each detected defect. Should the defect recur, the test cases will discover it before a new version of the system will be released.

The remaining clauses concerning monitoring and measurement (customer satisfaction, internal audit, analysis of data, continual improvement and preventive action are transparent with regard to XP.

6 Conclusions

The approach described in the paper is being implemented at the Software Development Studio (SDS for short). SDS is a software organization established at the Poznan University of Technology to allow students to get practical experience in software development and software process improvement [9]. Each year there are 11 projects developed for real customers. Each project involves 8 students: 4 from 3rd year, 2 from 4th year and 2 from 5th year of studies. Students play different roles in subsequent years, gaining experience in design and programming, project management and quality assurance. Each project lasts an academic year (9 months). The examples of the projects developed by students include: *Network Database System for Multiple-Choice Questions*, *Internet-Based Environment for Requirements Management*, *Internet-Based Traffic Analysis System.*

In the previous academic year the projects were split into two groups. One group developed software according to CMM Level 2, and the other applied pure XP. The XP projects suffered, among others, from: absence of precisely defined process, lack of adequate communication with customer, late delivery, and the most significant - maintenance problems. One of symptoms of that problems were the difficulties the 3rd year students had in writing their bachelor thesis [13].

This year we have used a modification of XP targeting at satisfying selected ISO 9000 clauses embracing requirements management and maintenance problems [10]. The aim of this experiment was to introduce ISO 9000 elements to XP approach.

In order to asses maturity of requirements engineering processes in SDS projects we used the Somerville-Sawyer model [17] based on a set of *good practices.* Sommerville and Sawyer have identified 66 practices and split them into 3 groups: basic, intermediate and advanced. Each practice can bring from 0 to 3 points, depending on how widely it is used by an organization. The highest maturity level is called *Defined.* Organizations at that level have more than 85 points in the basic practices and more than 40 points in the intermediate and advanced practices. The intermediate level is called *Repeatable* and organizations at this level have more than 55 points in basic practices. The lowest level is called *Initial* and an initial organization has fewer than 55 points in basic practices.

Our experiment shows that Extreme Programming modifications resulted in significant improvement from the Somerville-Sawyer point of view. Since XP addresses only 6 of the basic practices, 4 intermediate and 1 advanced practice, *"classical" XP* projects were assessed as Initial [10]. In contrast, this year projects developed according to the proposed methodology supported most of the 66 practices more or less directly and therefore were assessed as Repeatable. That can be considered as an important improvement indicator. However there is still a need for further research. These studies shall focus on implementing a Quality Management System combining ISO 9000 and XP.

References

1. Beck, K.: Extreme Programming: Embrace Change. Addison-Wesley, Boston (2000)
2. Beck, K., Fowler, M.: Planning Extreme Programming. Addison-Wesley, Boston (2001)
3. European Committee for Standardization: Quality Management Systems – Fundamentals and Vocabulary (ISO 9000:2000). European Committee for Standardization (2000)
4. European Committee for Standardization: Quality Management Systems – Requirements (ISO 9001:2000). European Committee for Standardization (2000)
5. European Committee for Standardization: Quality Management Systems - Guidelines for Performance Improvements (ISO 9004:2000). European Committee for Standardization (2000)
6. European Committee for Standardization: Software Process Assessment (ISO 15504:1998). European Committee for Standardization (1998)
7. European Information Technology Observatory: EITO2002 – 10th Edition 2002. European Information Technology Observatory, Brussels (2002)
8. Jeffries, R., Anderson, A., Hendrickson., C.: Extreme Programming Installed. Addison-Wesley, Boston (2001)
9. Nawrocki, J.: Towards Educating Leaders of Software Teams: A New Software Engineering Programme at PUT. Proceedring of SEES 98. Scientific Publishers OWN, Poznan (1998) 149–157
10. Nawrocki, J., Jasiñski, M., Walter, B., Wojciechowski, A.: Extreme Programming Modified: Embrace Requirement Engineering Practices. Proceedings of the 10th IEEE Joint International Requirements Engineering Conference. IEEE Press, Inc., Los Alamitos (2002) 303–310
11. Nawrocki, J., Wojciechowski, A.: Experimental Evaluation of Pair Programming. In: Maxwell, K., Oligny, S., Kusters, R., van Veenendaal, E. (eds.): Project Control: Satisfying the Customer. Proceedings of ESCOM 2001. Shaker Publishing (2001) 269–276
12. Nawrocki, J., Walter, B., Wojciechowski, A.: Toward Maturity Model for eXtreme Programming. Proceedings of the 27th EUROMICRO Conference, Los Alamitos. IEEE Computer Society (2001) 233–239
13. Nawrocki, J., Walter, B., Wojciechowski, A.: Comparison of CMM Level 2 and eXtreme Programming. Proceedings of the 7th European Conference on Software Quality, Helsinki, Finland. Lecture Notes in Computer Science 2349, Springer-Verlag (2002) 288–297
14. Paulk, M. C. et al.: The Capability Maturity Model: Guidelines for Improving the Software Process. Addison-Wesley, Reading MA (1994)
15. Seddon, J.: The Case Against ISO 9000. 2nd edn. Oak Tree Press, Dublin (2000)
16. Software Engineering Institute: Capability Maturity Model Integration. Version 1.1, Staged Representation. Carnegie Mellon University (2002)
17. Sommerville, I., Sawyer, P.: Requirements Engineering: A Good Practice Guide. John Wiley & Sons, Chichester (1997)
18. Tricker, R., Sherring-Lucas, B.: ISO 9000:2000 in Brief. Butterworth-Heinemann, Oxford (2001)

The Class Cohesion Using the Reference Graph G1 and G2

Wan-Kyoo Choi[1], Il-Yong Chung[1] Sung-Joo, Lee[1], and Hong-Sang Yoon[2]

[1] School of Computer Engineering, Chosun University, Kwangju, 501-759, Korea,
wkchoi@cafe.chosun.ac.kr, {iyc, sjlee}@mail.chosun.ac.kr

[2] Division of Computer, Electronics & Communiation Engineering, Kwangju University, 503-703, Kwangju, Korea,
hsyoon@kwangju.ac.kr

Abstract. Many measures have been proposed for measuring the cohesion of the class in the object-oriented paradigm. They, however, are inconsistent with the review of application because of members of the class that have no data interactions with other members. Some of the measures do not distinguish classes in terms of the interaction pattern. A solution to this problem is to exclude them from analysis and to consider the method invocation. However, this solution may be difficult to implement in the existing measures.Therefore, this paper introduces two reference graphs, referred to as G1 and G2. G1 and G2 easily exclude the members of the class that have no data interactions, and reflect the indirect reference by the method invocations. This paper shows that the existing measures be able to perform the measurement coinciding with the review of application by using G2.

1 Introduction

Software metrics have been proposed for the object-oriented paradigm to measure various attributes like software quality, programmer productivity, and program complexity. Among these, cohesion is an attribute, referred to the relatedness of module components. The more cohesive a module is, the easier it is to understand and maintain[1].

Many measures[2, 3, 4, 5, 6, 7, 8, 9] have been proposed for measuring cohesion in the object-oriented paradigm. By "type of connection", we can classify them into two categories[10]: Type1-measures that focus on counting pairs of methods that used or do not use common attributes, and Type2-measures that capture the extent to which individual methods use attributes or locally defined types.

A common problem of this measures, they are based on measuring cohesion in terms of only data interactions. That is, they do not take into account cases where there are no data interactions in the methods, but the data does belong together because it represents different attribute of an object[2].

The members of the class that have no data interaction are "access methods", "default constructors", "destructor", etc[10]. These members decrease the

M.H. Shafazand and A M. Tjoa (Eds.): EurAsia-ICT 2002, LNCS 2510, pp. 795–805, 2002.

class cohesion, because they form many pairs of methods that do not reference a common attribute in Type-1 measures and increase the number of possible references to attributes in Type2-measures. Constructors that reference all the attributes increase the class cohesion. However, in the review of application, the presence of these members not always makes the class be less cohesive or more cohesive. This causes the existing measures to be inconsistent with the review of application.

A solution to this problem is to exclude access methods, constructors and destructors or to count the invocation of an access method as reference to the attribute. But this solution may be difficult to implement in practice, because constructors and destructors are easily recognized but it is not always possible to recognize access methods automatically[10]. Also, when all of them are excluded from a class, attributes that are not referenced by any methods exist in a class.

For solving this difficulty, we introduce two reference graphs, referred to as G1 and G2, which can make the existing cohesion measures perform the measurement coinciding with the review of application. G1 and G2 exclude members(including access methods, constructors and destructors) that have no interactions with other members. They also regards the method invocations as the indirect references to attributes(e.g., instance variables). By using G1 and G2, we can perform the measurement considering not only the number of interactions but also their pattern, and particularly present the measurement coinciding with the review of application.

2 Background

By "type of connection" we refer to mechanisms that link elements within a class and thus make a class cohesive, and can classify the cohesion measures into two categories[10], namely Type1-measures and Type2-measures.

2.1 Type1-Measures

Type1-measures focus on counting pairs of methods that use or do not use common attributes.

Chidamber and Kemerer proposed a cohesion measure LCOM(Lack of COhesion in Methods) based on the number of disjoint sets of attributes that are used by the methods. LOCM is defined as follows[3]:

Consider a class C_1 with n methods $M_1, M_2, ..., M_n$. Let I_j=set of attributes used by the method M_j. There are n such sets $I_1, I_2, ..., I_n$. Let $P = \{(I_i, I_j)|I_i \cap I_j = \emptyset\}$ and $Q = \{(I_i, I_j)|Ii \cap I_j \neq \emptyset\}$. If all n sets $I_1, ..., I_n$ are $\emptyset$, then let $P = \emptyset$.

$$\begin{aligned} LCOM &= |P| - |Q|, if|P| > |Q| \\ &= \qquad 0, \ Otherwise \end{aligned} \tag{1}$$

LCOM is an inverse cohesion measure. A high value of LCOM indicates low cohesion and vice versa.

Hitz and Montazeri presented a graph theoretic version of LCOM proposed by Chidamber and Kemerer. They interpret LCOM as follows[4]:

Consider a class X with a set of attributes, I_X, and a set of methods of the class, M_X. Let $G_X(V, E)$ be a simple undirected graph representing the methods of the class, where $V = M_X$ and $E = \{(m, n) \in V \times V | (\exists i \in I_X : ((m \text{ accesses } i) \wedge (n \text{ accesses } i)) \vee (m \text{ calls } n) \vee (n \text{ calls } m))\}$. LCOM2(X) is then defined as the number of connected components of G_X, where $1 \leq LCOM2(X) \leq |M_X|$.

LCOM2 also is an inverse measure. LCOM2 consider the *access methods* by $(m \text{ calls } n) \vee (n \text{ calls } m)$. In the case where G_X consists of only one connected component, LCOM2=1. Hits and Montazeri define a measure for classes having LCOM=1 by taking into account the number of edges of the connected component:

$$C = 2|\frac{E - (n-1)}{(n-1)(n-2)}| \tag{2}$$

where C is a measure of the deviation of any given graph from the minimally cohesive case, E is the number of edges in the graph and n is the number of vertices(e.g., methods) in the graph.

Chen and Lu proposed a cohesion metric, referred to as LCOM3, which is based on the idea that operations(e.g., methods) with overlapping arguments tend to be related, thus making a class cohesive[5].

Let a class C have N operations, namely, $F(1), ..., F(N)$ and correspondingly there will be N sets of arguments, namely, $I(1), ..., I(N)$. Then,

$$LCOM3 = \frac{M}{N} \times 100\% \tag{3}$$

where M is the number of connected components in the graph with N vertices, namely the number of the disjoint sets in the graph.

LCOM3 also is an inverse measure. In the worst case where all the operations are disjoint, $M = N$ and LCOM3=100.

Bieman and Kang proposed two class cohesion measures to evaluate the relationship between class cohesion and private reuse in the system. They are based on the direct or indirect connectivity between a pair of methods. Two methods that use one or more common attributes are directly connected. Two methods that are connected through other directly connected methods are indirectly connected[6].

Let $NDC(C)$ be the number of directly connected methods in a class C. Let $NIC(C)$ is the number of indirectly connected methods in a class C. $NP(C) = N(N-1)/2$ is the maximum possible number of connections in a class. Then, TCC(Tight class cohesion) and LCC(Loose class cohesion) are defined as follows:

$$TCC(C) = NDC(C)/NP(C) \tag{4}$$

$$LCC(C) = (NDC(C) + NIC(C))/NP(C) \tag{5}$$

2.2 Type2-Measures

Type2-measures capture the extent to which individual methods use attributes or locally defined types.

Henderson-Sellers proposed the following measure, referred to as LCOM4[7]. Consider a set of methods $M_i (i = 1, ..., m)$ accessing a set of attributes $A_j (j = 1, ..., a)$. Let the number of methods that access each attribute be $\mu(A_j)$.

$$LCOM4 = \frac{\frac{1}{a} \sum_{j=1}^{a} \mu(A_j) - m}{1 - m} \tag{6}$$

LCOM4 yields 0, if each method of the class references every attribute of the class(called "perfect cohesion" by Henderson-Sellers). It yields 1, if each method of the class references only a single attribute. Values between 0 and 1 are to be interpreted as percentages of the perfect value.

Briand and Morasca proposed a set of cohesion measures for object-based systems(such Ada implementation). For the adaptation of the cohesion measures to object oriented systems, they see a class as a collection of data declarations and methods[8].

A data declaration a interacts with another data declaration b, if a change in a's declaration or use may cause the need for a change in b's declaration or use. We say there is a DD-interaction between declaration a and b. There is a DM-interaction between data declaration a and method m, if a DD-interacts with at least one data declaration of m.

Let $CI(c)$ be the set of all such DD- and DM-interactions, and $Max(c)$ be the set of all possible DD- and DM-interactions in the class interface. Measure RCI(ratio of cohesive interactions) is then defined as follows:

$$RCI = \frac{|CI(c)|}{|Max(c)|} \tag{7}$$

Park et al. proposed a class cohesion metric using the connection intensity between two methods[9].

Let I be a set of attributes defined within the same class. Let MP be a set of method-pairs, which are combination of all methods within a class. Then, CI(Connection Intensity of a method pairs) and CC(Class Cohesion) are defined as follow:

$$CI = \frac{|\{i \in I | m_1 \; and \; m_2 \; use \; i, (m_1, m_2) \in MP\}|}{|I|} \tag{8}$$

$$CC = \frac{\sum_{(m_i, m_j) \in MP} CI \, of \, (m_i, m_j)}{|MP|} \times 100 \tag{9}$$

3 Reference Graph G1 and G2

In object-oriented design, classes usually have "access method". An access method provides read or write access to an attribute of the class. Access methods typically reference only one attribute, namely the one they provide access to.

Access methods cause problems for Type1-measures and Type2-measures. That is, the presence of access methods artificially decreases the class cohesion. In Type1-measures, they form many pairs of methods that do not reference a common attribute can be formed. In Type2-measures, they artificially increase the number of possible references to attributes, while the number of actual references increase only by one[10, 11].

Constructors but default constructor typically reference all attributes and initialize them with no actual data interaction[2, 11]. This artificially increases the cohesion of the class. In Type1-measures, they generate many pairs of method that use a common attribute. In Type2-measures, they increase the number of actual references to attributes. Default constructor has none reference to attributes and therefore decreases the class cohesion. Destructors that have none reference to attributes cause the same result as the default constructors.

A solution to constructors and destructors is to exclude them from analysis. A solution to access methods is to count the invocation of an access method as reference to the attribute or to exclude them from analysis. However, this solution may be difficult to implement in practice because it is not always possible to recognize access methods automatically. Also, when all access methods are excluded from a class, attributes that are not referenced by any methods exist in a class[10].

For solving this difficulty, we introduce two reference graphs, referred as to G1 and G2. G1 regards the method invocations as the indirect reference to attributes. Instead of referencing an attribute directly, the methods(including access methods) may be used, which are not account for by this type of connection. In Type2-measures, the number of references to attributes is artificially decreased, and thus the class cohesion is decreased. This problem can be solved using G1. G2 exclude the members(including access methods, constructors and destructors) that have no data interaction between members from G1.

Let $V = \{v_1, v_2, ..., v_n\}$ a set of attributes and $M = \{m_1, m_2, ..., m_m\}$ a set of methods within a class C, where m_i is not constructor or deconstructor. If $m_i(m_i \in M)$ uses $v_i(v_i \in V)$, it is said that m_i directly references v_i, and it is denoted by $m_i \longrightarrow v_i$. If $m_i(m_i \in M)$ directly or indirectly calls $m_j(m_j \in M)$ and $m_j(m_j \in M)$ uses $v_i(v_i \in V)$, it is said that m_i indirectly references v_i, and it is denoted by $m_i \cdots > v_i$.

Definition 1: A reference graph $G1$ for a class C is a directed graph $G1 = (N, E)$, where $N = (V \cup M)$ and $E = \{(m, v) | m \longrightarrow v \; or \; m \cdots > v, \; m \in M, v \in V\}$.

Let I_{m_i} a set of attributes that is directly or indirectly referenced by a method $m_i(m_i \in M)$. Then $I = \bigcup_{i=1}^{|M|} I_{m_i}(m_i \in M)$, and $M^* = \{m_i | \, |I_{m_i}| \neq 1, m_i \in M\}$, and $I^* = \bigcup_{i=1}^{|M^*|} I_{m_i}(m_i \in M^*)$, and $V^* = (V - I) \cup I^*$.

Definition 2: A reference graph $G2$ for a class C is a directed graph $G2 = (N^*, E^*)$, where $N^* = (V^* \cup M^*)$ and $E^* = \{(m, v) | m \longrightarrow v \; or \; m \cdots > v, m \in M^*, v \in V^*\}$.

In Stack class of Fig.1, one class constructor is initialization function. It accesses all attributes in the class, and shares attributes with virtually all other methods. They therefore are excluded.

```
class Stack {
    int *array, top, size;
public:
    Stack(int s) {size = s;array = new int[size]; top = 0;}
    int IsEmpty() {return (top == 0);}
    int Size()  {return size;}
    int Vtop() {return array[top-1];}
    void Push(int item)
    {   if (top == size) printf("Empty");
        else array[top++] = item;
    }
    int Pop()
    {    if (IsEmpty()) {printf("Full"); return 0;}
         else  {--top; return array[top];
};
```

Fig. 1. Stack class

Fig.2 illustrates G1 for Stack class. In Fig.2, the rectangle and the oval indicate the method and the attribute, respectively. A link between rectangles indicates the method invocation. A link between a rectangle and an oval indicates that the method uses the attribute. A solid line means that a method directly uses an attribute, and a dot line means that a method indirectly uses an attribute. Fig.3 illustrates G2 for Stack class. Fig.3 excludes the methods(e.g., Size and IsEmpty) from Fig.2, which have no data interactions.

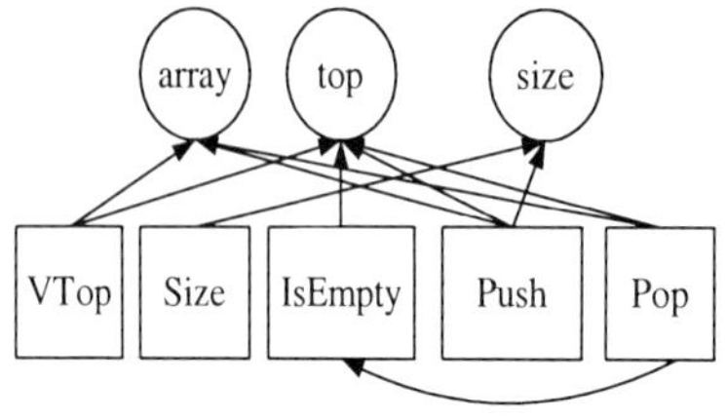

Fig. 2. G1 for Stack class

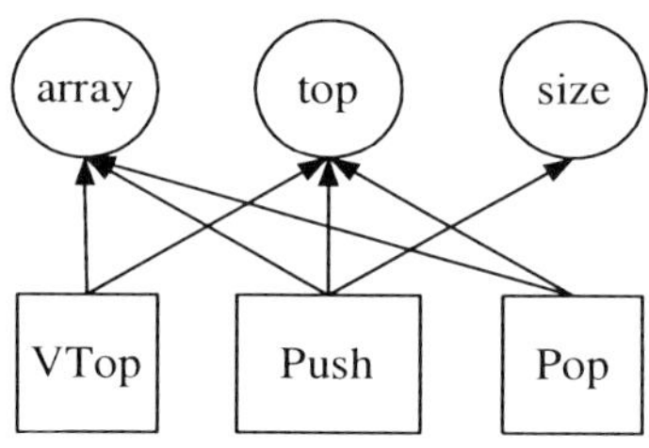

Fig. 3. G2 for Stack class

In Stack class of Fig.2,

V={array, top, size} M={VTop, Size, IsEmpty, Push, Pop}
I_{VTop}={array, top} I_{Size}={size}

$I_{IsEmpty}$={top} I_{Push}={array, top, size}
I_{Pop}={array, top} I={array, top, size}
M^*={VTop, Push, Pop} I^*={array, top, size}
V^*={array, top, size}.

4 Experiment and Results

4.1 Using the Reference Graph G2 in Type1-Measures

Fig.4 to Fig.8 illustrate G1s for Student class and Person class[12], List class[1], PB class and LIST class[2], respectively.

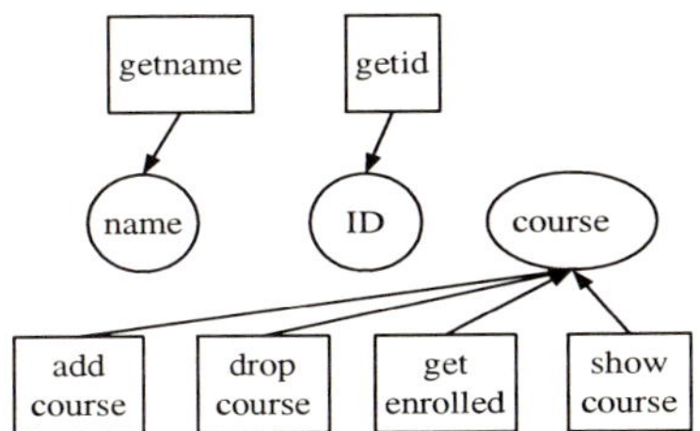

Fig. 4. G1 for Student class

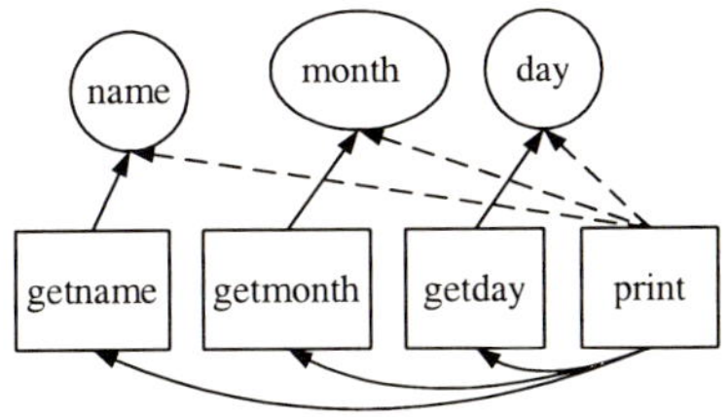

Fig. 5. G1 for Person class

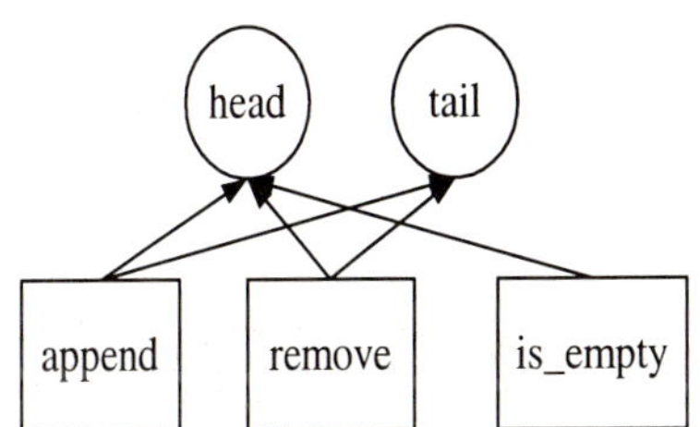

Fig. 6. G1 for List class

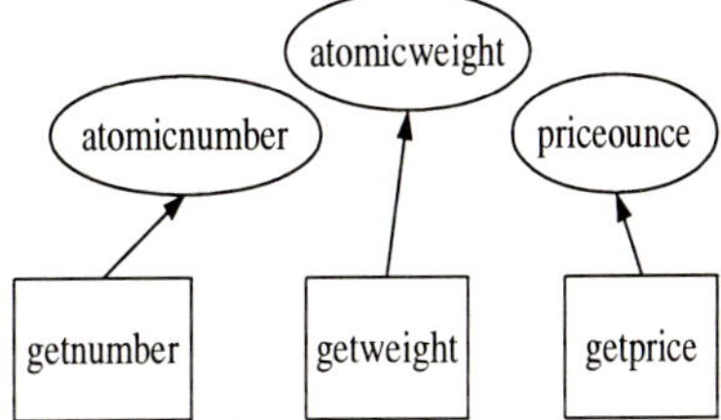

Fig. 7. G1 for PB class

In the review of application, the above classes are considerably cohesive classes that the attributes and methods belong together to but have no data interactions. LIST class should be a highly cohesive class as all the methods define the various operations on the abstract data type[2].

However, the existing Type1-measures indicate that all classes except List class have low cohesion or no cohesion, as like Table 1. This result is caused by the presence of the attributes and methods that have no data interactions between members within a class. That is, such members are *Size* and *IsEmpty* in Stack class, and all members in Student class, and *getname, getmonth* and *getday* in Person class, and *empty* in List class, and all members in PB class, and

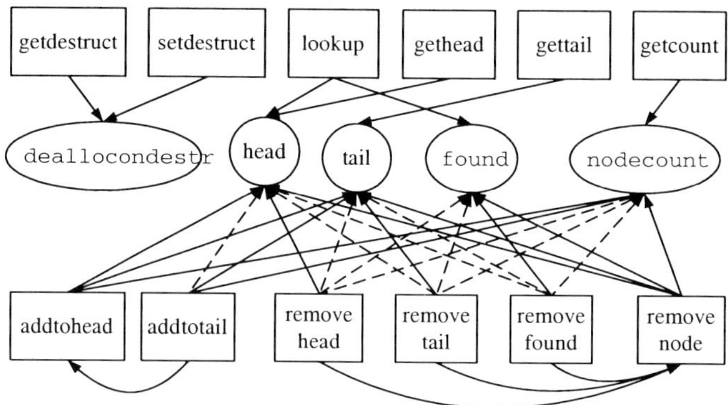

Fig. 8. G1 for LIST class

Table 1. The measured values of Type1-measures about six classes

Measures	Student	Person	List	PB	LIST	Stack
LCOM 3	3	6	0	3	16	0
LCOM2	3	1	1	3	2	1
C	0.1	0	0		0.23	0.5
LCOM3	50	100	33	100	17	20
TCC	0.4	0.5	0.66	0	0.37	0.7
LCC	0.4	0.5	1	0	0.69	1

getdestruct, setdestruct, gethead, gettail, getcount and *deallocondestr* in LIST class.

Particularly, PB class has only the three access methods and thus all of the existing measures indicate that it has no cohesion or low cohesion. In the review of application, however, PB class is clearly a cohesive class with methods and data belonging together[2]. If three access methods are replaced by one method, namely get(number, weight, price), all measures will be indicate that PB class is very cohesive. However, this is never preferred to more cohesive design.

This problem can be solved using G2. Fig.9 to Fig.11 illustrate G2s for Person class, List class and LIST class, respectively. In Student class and PB class, $M^* = V^* = \emptyset$, and thus G2s for them is not showed.

Table 2 shows the result of the measurement for the four classes except PB class and Student class. Table 2 indicates that the four classes are somewhat cohesive or very cohesive. Therefore, the existing measures also are able to perform the measurement coinciding with the review of application by using G2. However, the little modification of the existing Type1-measures is needed to measure such classes as PB class and Student class.

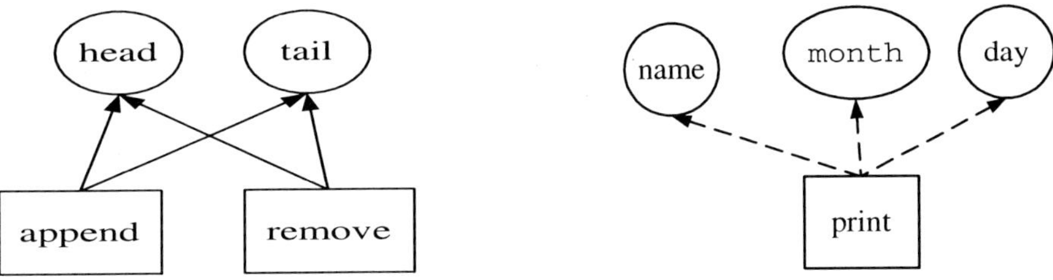

Fig. 9. G2 for Student class

Fig. 10. G2 for Person class

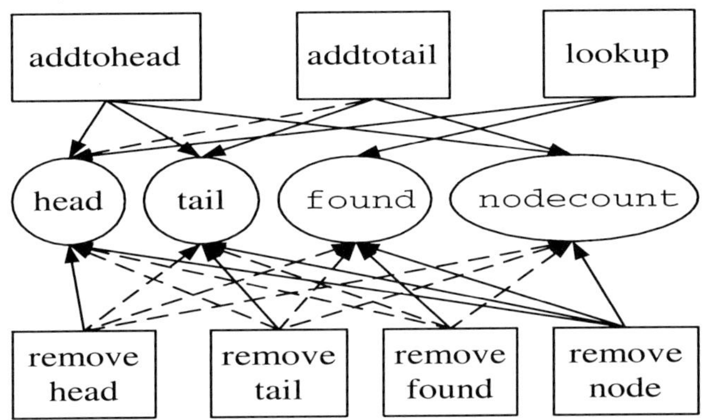

Fig. 11. G2 for List class

Table 2. The measured values of Type1-measures about four classes by using G2

Measures	Person	List	LIST	Stack
LCOM	0	0	0	0
LCOM2	1	1	1	1
C	0	0	0.46	1
LCOM3	100	50	14	33
TCC	1	1	0.61	1
LCC	1	1	1	1

4.2 Using the Reference Graph G2 in Type2-Measures

Table 3 shows the values to measure the six classes discussed in section 4.1 by using the existing Type2-measures.

According to LCOM4, RCI and CC, the six classes have low cohesion or no cohesion. As discussed in section 4.1, the six classes are considerably cohesive classes that the attributes and methods belong together to but have no data interactions. By using G2, the Type2-measures also are able to perform the measurement corresponding to the review of application, as like table 4.Table 4 indicates that the four classes are somewhat cohesive or very cohesive.

Table 3. The measured values of Type2-measures about six classes

Measures	Student	Person	List	PB	LIST	Stack
LCOM4	0.8	1	0.25	1	0.74	0.5
RCI	0.33	0.25	0.83	0.33	0.31	0.6
CC	0.13	0	0.66	0	0.09	0.23

Table 4. The measured values of Type2-measures about four classes by using G2

Measures	Person	List	LIST	Stack
LCOM4	0	0	0.16	0.33
RCI	1	1	0.85	0.77
CC	1	1	0.36	0.66

5 Conclusion

Generally, classes have the members that have no data interactions with other members, namely access methods, constructors, destructors, etc. In the existing cohesion measures, theses members artificially decrease or increase the cohesion of the classes. However, in the review of application, the presence of them not always makes classes more cohesive or less cohesive. A solution to this problem is to exclude them from analysis, but this solution is difficult to implement in practice by the existing representation for the class.

We, therefore, introduced new reference graphs for representing a class, referred to as G1 and G2. G1 and G2 could show the indirect reference by the method invocations and easily exclude the members of the class that have no data interactions. We showed that the existing Type1-measures and Type2-measures could present the enhanced measurement for the cohesion of the class by using G1 and G2. That is, they could present the results coinciding with the review of application.

References

1. L. M. Otto et al.: Developing Measures of Class Cohesion for Object-Oriented Software. 7th Annual Oregon Workshop on Software Metrics(1995).
2. Bindu Mehra: A Critique of Cohesion Measures in the Object-Oriented Paradigm. Masters Thesis, Department of Computer Science, Michigan Technological university, 1997.
3. S.R. Chidamber, C.H. Smith: A Metrics Suite for Object Oriented Design. IEEE Transactions on Software Engineering 20(1994), 636-638.
4. Martin Hitz, Behzad Montazeri: Measuring coupling and cohesion in object-oriented systems. Proc. of the International Symposium of Applied Corporate Computing(1995).
5. J.Y.Chen, J.F.lu: A new metric for object-oriented design. Information and Software Technology (1993), 232-240.

6. James M. Bieman, Byung-Kyoo Kang: Cohesion and reuse in an object-oriented paradigm. Proc. ACM Symposium on Software Reusability (1995), 259–262.
7. B. Henderson-Sellers, Software Metrics, Prentice Hall, Hemel Hempstaed, U.K., 1996.
8. Lionel C. Briand, S. Morasca, V.R. Basili: Defining and Validating High-Level Design Metrics. IEEE Transactions on Software Engineering 25 (1999), 722–743.
9. Sunghee Park et al.: Metrics Measuring Cohesion and Coupling in Object-Oriented Programs. Journal of Korean Information Science Society 25 (1998), 1779–1787.
10. Lionel C. Briand, John W. Daly, and Juergen Wuest: A Unified Framework for Cohesion Measurement. Proceedings of the 4th International Software Metrics Symposium (1997).
11. HeungSeok Chae, YongRae Kwon: A Cohesion Measure for classes in Object-Oriented Systems. Proceedings of the 5th International Symposium on Software Metrics (1998).
12. E. Horowitz, S. Sahni, D. Mehta: Fundamental of Data Structures in C++. Computer Science Press:New York, 1995.
13. E. Weyuker: Evaluating software complexity measure. IEEE Transaction Software Engineering 14(1988), 1357–1356.

Process-Oriented Interactive Simulation of Software Acquisition Projects

Tobias Häberlein[1] and Thomas Gantner[2]

[1] University of Ulm
Faculty of Computer Science
Department of Software Methodology
tobias.haeberlein@informatik.uni-ulm.de
[2] DaimlerChrysler
Research and Technology
thomas.gantner@daimlerchrysler.com

Abstract. The software acquisition process model GARP [3], developed at DaimlerChrysler, is a tool for process improvement and assessment. It comprises many, temporally, concurrently, and non-deterministically linked process steps and a huge amount of best practice information associated with them.
GARP is mainly designed for experts. We argue why it is essential in the process improvement environment to also have a facility which explains process dynamics and why an interactive simulation is suitable for this. The paper describes the design of an interactive simulation framework, which allows to map GARP (and with it any other similar process model) straightforwardly to a project simulation game.

1 Background and Overview

Software acquisition is concerned with the provision of an organisation's need for software. Its impact on business processes is increasing dramatically. Since more and more projects at DaimlerChrysler are concerned with software acquisition, the need for a generic acquisition process model emerges: A process model provides a means of reasoning about processes used to acquire and integrate software and additionally a defined process provides the basis for learning and improvement. Thus a systematic process-oriented approach for software acquisition helps to increase the benefits of investments in software products.

Particularly for us, the search for an acquisition process model which fulfils DaimlerChrysler's special needs is a major issue. The main requirements are:

- *Comprehensiveness.* Acquisition projects performed at DaimlerChrysler differ greatly among each other. The process model should be applicable to all of them and it should include all relevant company specific aspects.
- *(Temporally) Ordered Process Steps.* Our customers need *explicit* guidance and ordered process steps can be used as a work sequence.
- *Sufficient Detailing.* A main task is to give the customers advise on how exactly to do things better and provide them with specific hints.

M.H. Shafazand and A M. Tjoa (Eds.): EurAsia-ICT 2002, LNCS 2510, pp. 806–815, 2002.

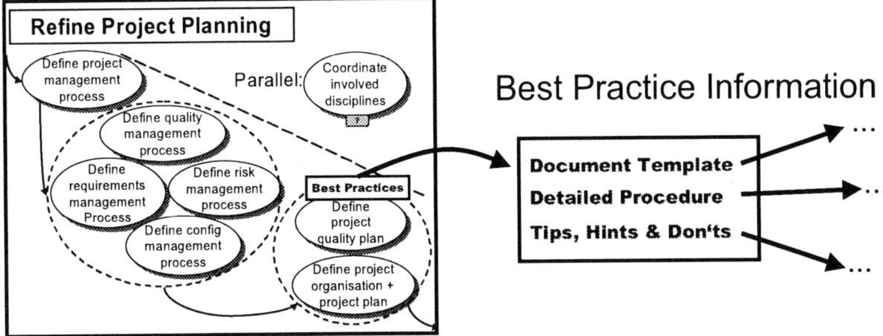

Fig. 1. Sample pages of GARP's process framework. One link of a process step to the corresponding best practice information is examplary shown.

As all these needs together are not fulfilled by any current acquisition process model, a DaimlerChrysler specific model was developed, called GARP, an abbreviation for "**G**eneric Software **A**cquisition **R**eference **P**rocess".

GARP is briefly described in chapter 2, since it forms the starting point for the design of the interactive simulation. Chapter 3 points out the importance of having a facility which explains the actual dynamics of an idealised acquisition project to novices and why an interactive simulation is a suitable facility for this. In chapter 4, past research in process simulation is reviewed and compared to our special requirements for such a system.

What should be the scope of such an interactive simulation, and how should it basically be designed? Chapter 5, which could be seen as the main contribution of this paper, tries to answer these questions. A process-oriented interactive simulation framework and its instantiation to a small part of GARP are presented. Chapter 6 describes techniques to embed best practice knowledge in the simulation framework and chapter 7 gives concluding remarks.

2 GARP: An Acquisition Process Model

Through integrating many existing software acquistion models and adapting them to our needs, a comprehensive acquisition process framework, called GARP, was developed which we want to describe briefly here (the reader is referred to a recent publication [3] for more detail).

GARP is made up of two parts: a process framework, which states *what* to do, and a best practice framework, which states *how* to perform the process steps. The process framework – the left part of figure 1 shows a sample page of it – is a graphical presentation of the process steps. Process steps are shown in small ovals, the temporal ordering is expressed through arrows and concurrent process steps are grouped in dashed ovals. The graphical presentation supports hyperlinks and some processes are linked to their respective subprocesses (which is not shown in figure 1).

GARP is enriched with best practices from popular software acquisition guides. To deal with the disparity in the kind of information provided by the best practice guides, the material is organised in three *information kinds*:

- *Document Template*: A proposal for the structure of the deliverable corresponding to the process step (usually in the form of a table of contents).
- *Detailed Procedure*: A "mini-process" describing the concrete implementation of the respective process step.
- *Tips, Hints and Don'ts*: The most common form of best practice information where suggestions or further information are given.

Since the graphical representation of GARP supports hyperlinks it is obvious to link processes to a corresponding set of best practices. This is shown in figure 1.

3 Promoting, Training, and Understanding Acquisition Processes

Process design should be followed by activities concerning its utilisation. The following points seem to be particularly important in this context:

1. Utilisation mainly depends on management compliance. As extensive documentation is seldom read thoroughly by busy managers we searched for representations to make the process model both understandable and attractive to decision makers. We searched for a sort of promotion.
2. For a project to run smoothly, *every* staff member should have a clear grasp of how the project dynamics should ideally look like. Many problems in complex projects stem from a lack of transparency of the project's process.
3. Due to the growing number of complex software acquisition projects at DaimlerChrysler, there is an increasing need for well-trained and experienced project leaders. Thus, another major issue is education and training.

Two things prevent GARP from being specially suited for promoting and understanding. Firstly, as the process model is comprehensive novices get easily lost in details. Secondly, since the processes are temporally ordered, it is neither fully capable of expressing the continuous dynamics of all the management processes, nor is it made for giving a grasp of how in detail the actual management of an acquisition process should work.

Providing a slim copy of GARP is not the solution and just another document on the desks of managers and project members. Ludewig [8] has expressed the idea to develop a teaching game where computer science students can play the role of a software project manager and this is – with a slightly different focus – an excellent means to tackle the above mentioned issues.

4 Past Research in Interactive Simulation and Our Needs

With the pioneer work of Abdel-Hamid and Madnick [1] software process simulation through system dynamics became popular. A technique called "Software

Process Flight Simulation" evolved. Such a flight simulator is basically an executable system dynamics model enhanced with a "cockpit", or interface, which contains the specific controls that reflect decision making in real projects. Software process flight simulators were used for prediction, for example for predicting the impact on productivity when moving an organisation to the highest CMM maturity levels [12], and for education and training [10]. Although this simulation methodology provides insight into the impact of important decisions, the "player" can hardly feel to be involved in the project's process. He or she can just handle some knobs and switches and then watch the outcome of the project.

Contrary to process flight simulation, the RENAISSANCE project [13] provides an approach for learning with a high degree of player involvement. The court of Urbino at around the 14th century is simulated through a multi-user role playing game over the Internet. Social life during renaissance was subject to plenty of subtle rules and users of the RENAISSANCE game should be enabled to experience this life as realistically as possible. A 3D-graphics rendering machine, local to each client, provides realistic impressions, and an extra authoring tool allows experts to describe and change the respective rules.

Interactive simulation approaches with a high player involvement have also been used in the software process world. However, there are surprisingly few approaches described in literature. The computer based training (CBT) module [11] developed at the IESE and the University of Kaiserslautern is based on a system dynamics model (calibrated with the help of Boehm's COCOMO model). A main feature of CBT's model is the player's ability to change parameters at any time during the simulation run and the possibility to apply for example code or design inspections – this is represented through special parameters in the system dynamics model.

The SESAM system [9], developed at the University of Stuttgart, is the result of a big research effort in game-like simulation of software projects. The system passed through several years of development. The basic idea of SESAM is to build up an understanding of project dynamics in a bottom-up manner i.e. just "small" things are simulated - there is *no* such thing like a process built into the system. SESAM is based on a static model described in an entity-relationship-like manner and a dynamic model which is basically a set of rules. The rules can both express sudden changes and continuous changes.

Our special needs. Like in SESAM, in an interactive simulation, as we want to understand it, everything but the project leader is simulated; the player of the simulation game takes its role.

But we need an interactive simulation methodology where a process step is first class citizen. This particularly means that a process description (like GARP), which expresses temporal ordering, concurrency and non-deterministic choice of process steps, should be mappable as directly as possible to an interactive simulation. We call this (desired) property *process-orientation*. Neither SESAM nor the CBT-module is process-oriented.

Furthermore, since both SESAM and the CBT module are tailored to the needs of a university environment (whereas our simulation is focused on an industrial setting), and neither of them is specially designed for software acquisition, it seemed reasonable to build a simulation framework tailored to our needs.

5 A Process-Oriented Interactive Simulation Framework

Since the process is never fixed – it is growing and changing as our experiences with it are growing – it must be easy to add or remove process steps from the simulation. This suggests every process step to be encapsulated in any kind of unit and a strong separation of the simulation's part containing the processes from a generative process engine – which we will also call *process interpreter* – that glues the processes together and builds the actual interactive simulation from them.

5.1 Process Description for Interactive Simulation

It has been pointed out in the past that process descriptions with an overemphasis on the modeling of tasks could be counterproductive since this limits human flexibility and leads to complex models. An alternative are entity process models [6] which focus on real-world objects (like requirements documents or test cases) which persist throughout the complete process.

Since it weakens complexity, we take a similar approach here which stresses the product-oriented aspect of software processes. Katayama's hierarchical and functional software process description method HFSP [7] proves to be useful in designing the simulation system: it is straightforward and due to its functional perspective quite near to a programming language. It emphasises the importance of the product-oriented aspect of software processes which should be the first thing to deal with.

Katayama treats process steps, in the first approximation, as mathematical funtions, denoted by $A(x_1, \ldots, x_n | y_1, \ldots, y_m)$, from input objects $x_1, \ldots, x_n$ to output objects $y_1, \ldots, y_m$. See figure 2 for a (simplified) part of GARP's acquisition processes in a HFSP-like notation. For reasons of brevity we provide a coarse model of an acquisition process. We don't model, for instance, the staff and all respective activities like hiring, motivating, training or firing staff members though it would be a simple refinement.

The textual ordering of an activity's subactivities is not significant. Temporal ordering, concurrency, and non-determinacy are expressed implicitly through object dependencies. The "tendering"-activity, for example, occurs prior to the "prepare-contract"-activity; the latter requires the "concrete supplier"-object which the "tendering"-activity provides. The subactivities of the "monitor-supplier"-activity may be activated in unspecified order or concurrently since there are no input/output - dependencies between them. The process is non-determinate since there are two definitions of the "procure-software"-activity.

An extension to the HFSP notation is that atomic actions, i.e. actions which are not further decomposable, are marked with either $\mathcal{A}$, for "active" or $\mathcal{P}$, for

```
types:
     concrete-supplier, contract, request-for-proposal,
     request-for-proposal-issued, sw-product

activities:
procure-software ( | sw-product) ⇒
     tendering ( | concrete-supplier)
𝒫    prepare-contract (concrete-supplier | conrete-supplier, contract, sw-product)
     monitor-supplier (concrete-supplier, contract, sw-product |
                            concrete-supplier, contract, sw-product)
𝒜    acquisition-acceptance (concrete-supplier, contract, sw-product | sw-product)

procure-software ( | sw-product) ⇒
𝒫    buy-COTS-product ( | sw-product)

monitor-supplier (concrete-supplier, contract, sw-product |
                     concrete-supplier, contract, sw-product) ⇒
𝒜    contract-change-mgmt (concrete-supplier, contract | concrete-supplier, contract)
𝒫    quality-control (concrete-supplier, sw-product | concrete-supplier, sw-product)
𝒫    risk-mgmt (concrete-supplier | concrete-supplier)

tendering ( | concrete-supplier) ⇒
𝒫    prepare-rfp ( | request-for-proposal)
𝒫    issue-rfp (request-for-proposal | request-for-proposal-issued)
𝒜    select-supplier (request-for-proposal-issued | concrete-supplier)
```

Fig. 2. The procure-software process, a simplified part of GARP, in an extended HFSP notation.

"passive". Passive actions can be activated solely by the player of the simulation. Active actions occur without the player's initiation – they can activate themselves which is why they are called "active" – and are actually *reactions* to external events. The "select-supplier"-action, for example, is active since the player must react to incoming offers. The simulation system handles active actions in a slightly different way.

A process description in extended HFSP notation is directly mappable to an interactive simulation.

5.2 Designing a Process-Oriented Interactive Simulation

Due to their high level programming paradigms, functional programming languages are a good means to rapidly prototype applications. In our setting *Scheme* [2] is specially suited since it has sophisticated GUI-building utilities (which is far from being a common feature in functional languages).

Process Steps as Scheme Functions. We can benefit from many advantages of functional languages [5] if we encapsulate a process step in a function of the programming language.

Figure 3 shows an extract of the simulation system's process part. The function *prepare-rfp* points out how activities can interact with the player: the function get-number opens a dialog, asks the player to enter the effort he wants to spend on the request for proposal and stores the provided value in *rfp-quality*.

```
  ⋮
(define prepare-rfp
  (lambda ()
    (let ([rfp-quality (get-number "How much effort do you want to
                                   spend on the rfp (in person hours)?")])
     ...
     (make-rfp ... rfp-quality ...))))

(define issue-rfp
  (lambda (request-for-proposal)
    ... (make-rfp-issued ...)))
  ⋮
```

Fig. 3. An extract of the simulation system's process part.

The *prepare-rfp*-function adjusts global parameters, e.g. resources spent, money earned, and many more (for reasons of brevity, however, this is not shown in figure 3). Here, GARP's best practice framework proves to be useful. It provides guidelines to decide which parameters should to used and how they should be changed if a certain action is activated. Finally, *prepare-rfp* returns a rfp-object.

These process step functions form the main input for the process interpreter.

The Process Interpreter. We call the program part responsible for building the simulation out of the list of process tuples the *process interpreter* due to the similarities to interpreters of program code.

The interpreter works up the process step functions and compiles them into an interactive simulation. Figure 4 shows what the process interpreter is doing examplary for one passive process step (the interpreter treats active process steps slightly different). The *object-base* contains objects (e.g. a supplier, a contract, a request for proposal) which are currently available to the player. First, the interpreter determines if the process step is *executable* by testing (arrow 1) if all input objects for the activity are available in the object base or, more formally, if the following condition holds (where p is the respective process step and $input_p$ the set of input objects of p): $\forall i \in input_p : \exists o \in object\text{-}base : \mathbf{type}(o) = i$. Only if this process step is executable, the interpreter creates a corresponding menu item. Finally the interpreter connects the menu item with the actual function (arrow 3) with its parameters instantiated to the respective objects in the *object-base* (arrow 2). After each execution of a process step by the player, the object base is updated (arrow 4) and the process interpreter passes through all the functions which represent process steps and adapts the menus accordingly as shown in figure 4.

Through this interpretation mechanism, temporal ordering, concurrency, and non-determinacy of process steps are properly expressed. The player "perceives" the temporal ordering through the dynamic change of the menus in the course of the simulation, the concurrency through the possibility to choose the execution

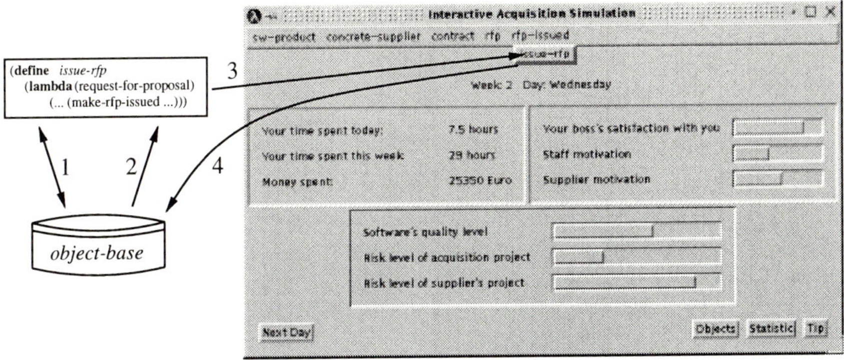

Fig. 4. The process interpreter's basic mode of operation

order of process steps, and the non-determinacy through the possibility to choose exactly one alternative.

6 Detailed Simulation Design Using Best Practice Information

So far, we described the simulation's "skeleton". The art, however, lies in filling it with live, i.e. in providing detailed implementations for the process step functions, as indicated by figure 3, and in determining the objects' attributes, the relationships between objects and the objects' dynamical behaviour. To move the "art", at least partially, to an engineering task we sketch how to systematically use best practice information (particularly information in GARP's best practice framework) for the detailed design.

As described in chapter 2, GARP divides best practices in information kinds. We will consider two information kinds here: *tips and hints* and *document templates.* Figure 5 shows examples of template and tip-and-hint best practices and how they influence and help in the detailed design of the simulation.

Document templates often influence the designer's decision about object attributes since templates provide a structure – and thus characteristics – of an object. The tip-and-hints – two examples of them are shown in figure 5 – can be quite diverse by their nature. One tip-and-hint indicates the risks which could be associated with the request for proposal. This influences the design of the subsequent process step function `select-supplier` because this is the point, where the risks can occur. A best practice which points out what types of contract exist, could influence the design of the `contract`-object's attributes and could lead to an extension of the `prepare-contract` process step function with a *choose feature* which lets the player select a contract type and/or an *information feature* which points the player at important information to educate him in software acquisition.

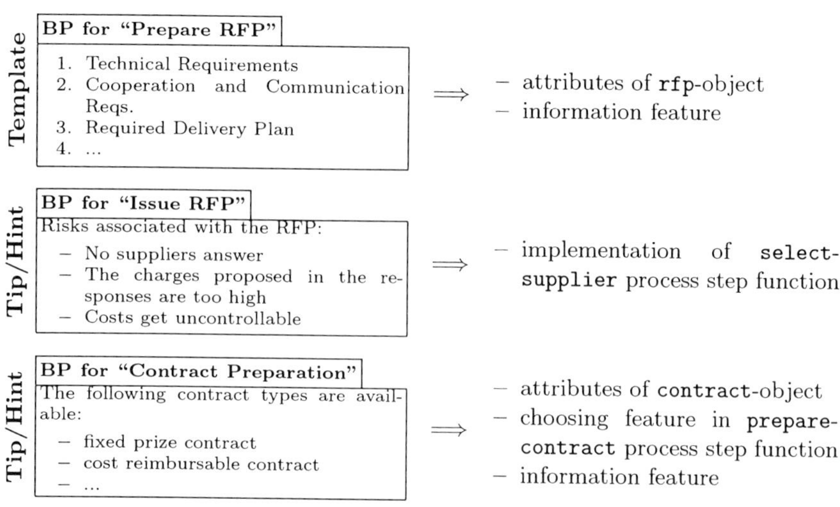

Fig. 5. Examples for using best practice information in building the interactive simulation

7 Discussion and Future Work

We presented an interactive simulation framework where process steps are first class citizens. Since our department's activity is mainly process improvement and our work is process-centered this is a vital feature to us. To the authors' knowledge, this approach differs from any other game-like simulation technique.

It could be worthwhile to broaden our perspective here and take the similarity to interactive storytelling into account – due to its process-orientation, our simulation technique also adheres to a "story"-plot. For interactive storytelling to be sensible, it should obey the so-called *story contract* [4] which states that

1. The author is responsible for the psychological integrity of the main characters.
2. The author is responsible for the sequencing and timing of plot events.
3. The audience must allow itself to be emotionally moved.

Replacing "author" with "simulator" and "audience" with "player", we obtain three rules of thumb for building sensible interactive simulations. The first rule stresses the importance to incorporate realistic dynamics into the simulation. The process interpreter cares for the second rule and the third rule pleads for building an attractive game.

Future Work. Since now, we do not utilise any system dynamics methods in our simulation. Searching for possibilities to incorporate system dynamics in process-oriented simulations would be a reasonable research effort.

A drawback of the presented simulation framework is that knowledge in writing Scheme functions is required in order to be able to build interactive simulations. Future extensions should provide facilites to build simulations without programming knowledge.

A multi-player interactive simulation with one player taking the role of the acquisition project leader and the other players taking the roles of managers of suppliers' software development teams could clarify the dynamics of acquisition processes even better. Playing against other persons emphasises the nature of software acquisition: each side wants to get out the most for themselves.

References

1. Tarek Abdel-Hamid and Stuart E. Madnick. *Software Project Dynamics: An Integrated Approach.* Prentice Hall, Eaglewood Cliffs, New Jersey, 1991.
2. R. Kent Dybvik. *The Scheme Programming Language.* Prentice Hall, 1996.
3. Thomas Gantner and Tobias Häberlein. GARP – the evolution of a software acquisition process model. In *Proceedings of the 7th European Conference on Software Quality 2002*, Lecture Notes in Computer Science 2349, pages 186–196. Springer-Verlag Heidelberg, 2002.
4. Andrew Glassner. Interactive storytelling: People, stories and games. In *Proceedings of the International Conference on Virtual Storytelling*, Lecture Notes in Computer Science 2197, pages 51–60. Springer-Verlag Heidelberg, 2001.
5. John Hughes. Why functional programming matters. *Computer Journal*, 32(2):98–107, 1989.
6. Watts S. Humphrey and Marc I. Kellner. Software process modeling: Principles of entity process models. In *Proceedings of the 11th International Conference on Software Engineering*, 1989.
7. Takuya Katayama. A hierarchical and functional software process description and its enaction. In *Proceedings of the 11th International Conference on Software Engineering*, pages 343–352, 1989.
8. Jochen Ludewig. Modelle der Software-Entwicklung – Abbilder oder Vorbilder? *GI Softwaretechnik-Trends*, 9(3):1–12, October 1989.
9. Jochen Ludewig and Marcus Deininger. Teaching software project management by simulation: The SESAM project. In Irish Quality Association, editor, *5th European Conference on Software Quality*, pages 417 – 426, Dublin, 1996.
10. Derek Merill and James Collofello. Improving software project management skills using a software project simulator. In *Frontiers in Education Conference*, 1997.
11. Dietmar Pfahl, Marco Klemm, and Günther Ruhe. Using system dynamics simulation models for software project management education and training. In *Proceedings of the 3rd Process Simulation Modeling Workshop*, London, UK, June 2000.
12. Howard A. Rubin, Margaret Johnson, and Ed Yourdon. Software process flight simulation. Dynamic modeling tools and metrics. *Information Systems Management*, summer 1995.
13. Massimo Zancanaro, Alessandro Cappelletti, Claudio Signorini, and Carlo Strapparara. An authoring tool for intelligent educational games. In *Proceedings of the International Conference on Virtual Storytelling*, Lecture Notes in Computer Science 2197, pages 61–68. Springer-Verlag Heidelberg, 2001.

Automatic Design Patterns Identification of C++ Programs

Félix Agustín Castro Espinoza, Gustavo Núñez Esquer, and Joel Suárez Cansino

Centro de Investigación en Tecnologías de Información y Sistemas
Universidad Autónoma del Estado de Hidalgo
Carretera Pachuca Tulancingo Km. 4.5 Ciudad Universitaria, Pachuca, Hidalgo
Tel. (+55-771) 71 7 20 00 Ext. 4884, 6738 y 6734
fax (+55-771) 71 7 21 09
México
{fcastro, gnunez, jsuarez}@ uaeh.reduaeh.mx

Abstract. A canonical model is formulated in this work for the representation of design patterns, in order to approach the problem of identification of design patterns in source code, and a system is developed as well, called DEPAIC++ (DEsign PAtterns Identification of C++ programs). Starting from the analysis of the structure of classes of a code written in C++ programming language, DEPAIC++ realizes a recognition to verify that the code analyzed is using or not design patterns. It is important to observe that this recognition is not necessary complete, since, in the source code, can be approximation to design patterns. This approach constitutes a support to the understanding process and, therefore, to the maintenance of source code with nonexistent or inappropriate documentation. The work also suggests that this approach allows to identify code reusability and code quality, which are inherited from the used design pattern.

1 Introduction

Along the last years, design patterns have gained much attention in the field of software engineering[4]. Many people consider them to be a promising approach for overcoming some fundamental problems concerning the design and reuse of object-oriented software[5].

Design patterns help you choose design alternatives that make a system reusable and avoid alternatives that comprise reusability. Design patterns can even improve the documentation and maintenance of existing systems by furnishing an explicit specification of class and object interactions and their underlying intent[4].

Design patterns could be used to design components in Component Based Development (CBD). One problem for software developers is to make sure that their own developments are reusable and that, in a given moment, they can be incorporated to a components base. A way to make it consists in verifying that the components were designed taking into account a design pattern, because this ensures their quality and also reusability.

M.H. Shafazand and A M. Tjoa (Eds.): EurAsia-ICT 2002, LNCS 2510, pp. 816–823, 2002.

Some characteristics that the components should fulfill to be reusable are the following: they should have, at least, generic and well designed conceptual architectures, closed to the modifications and open to the extensions (open-closed principle), in addition to be frequently used and taking care of minimizing problems of functional dependences.

In this case, we say that they have also been analyzed and designed of conformity to the characteristics previously mentioned. If the components are organized as design patterns, then this can greatly ensure their quality and also their reusability.

Automatic detection of design pattern instances is a useful aid for understanding and maintenance purposes of source code with nonexistent or inappropriate documentation; it is also a help to identify reusable code, such reusability is inherited of the used design pattern.

The paper is structured as follows: Section 2 provides a background information on design patterns. Section 3 introduces the focus taken to approach the problem previously mentioned. In section 4, the works related with this work are mentioned. Section 5 summarizes the results of the paper and discusses future research topics in DEPAIC++.

2 Design Patterns

Design Patterns are defined as a group of classes that work in collaboration in the solution of a problem that is presented recurrently under the same context, or of similar problems.

How is expressed in [4] a design pattern names, abstracts, and identifies the key aspects of a common design structure that make it useful for creating a reusable object oriented design. The design pattern identifies the participating classes and instances, their roles and collaborations, and the distribution of responsibilities. Each design pattern focuses on a particular object oriented design problem or issue. It describes when it applies, whether it can be applied in view of other design constraints, and the consequences and trade-offs of its use.

A design pattern has four essential elements:

1. *Name*, describe a design problem, its solution, and consequences in a word or two.
2. *Problem*, describes when to apply the pattern. It explains the problem and its context.
3. *Solution,* describes the elements that make up the design, their relationships, responsibilities, and collaborations.
4. *Consequences,* are the results and trade-offs of applying the pattern.

3 Automatic Design Patterns Identification

In order to approach the problem of identification of design patterns in source code, we propose that one form is to be based on its structure, that is to say, find in the source code relationships among classes that are common in the design patterns. Such relationships are: inheritance, aggregation, acquaintance, and creates[1].

In order to approach the problem before mentioned, a canonical model is formulated in this work for the representation of design patterns, and a system is developed as well, called DEPAIC++ (DEsign PAtterns Identification of C++ programs).

The operation of the canonical model is explained in section 3.1 and, later on, the architecture of DEPAIC++ will be described.

3.1 Canonical Model for the Representation of Design Patterns

The canonical model is useful to represent the structure of the design patterns in its canonical form. Likewise, the terminal symbols of this model are used to represent the structural relationships of the code source in their canonical form.

It is necessary to make clear that the rules of production of the model are not applied for the canonical form of the code, but rather the rules of the native language are used, in this case C++. Canonical model consists on a context free grammar, which is enunciated as follows:

Non terminal symbols:

<pattern>
<class>
<collection_def>
<def>
<comma>
<AGR>
<CRE>
<INHE>
<ACQ>
<end>
<id>

Terminal Symbols:

“,”
AGGREGATION
CREATES
INHERITS
ACQUAINTANCE
“ ” (blank)
ID (String)

Production rules:

<pattern> := <class><pattern>
<pattern> := <class>
<class> := <id><collection _def>
<class> := <id><INHE><id><collection _def>
<collection _def > := <def><collection _def>
<collection _def> := <def>

[1] These relationships are those that Erich Gamma mentions in his book that the design patterns can possess.

```
<def>          := <comma>
<def >         := <AGR><id>
<def >         := <ACQ><id>
<def >         := <CRE><id>
<def>          := <INHE><id>
<def>          := <end>
<comma>        := ,
<AGR>          := AGGREGATION
<CRE>          := CREATES
<INHE>         := INHERITS
<ACQ>          := ACQUAINTANCE
<end>          := " "
<id>           := ["a"-"z","A"-"Z","_"] (["a"-"z","A"-"Z","_","0"-"9"])*
```

Initial Symbol:
<pattern>

The user of canonical model should it uses the design pattern structure, the user should know the object-oriented paradigm, starting from this, he begins to apply the grammar of the model. To choose a class of the design pattern structure, the selection can be in a random way. Fig. 1 shows part of the derivation tree obtained of applying the canonical model to the design pattern structure.

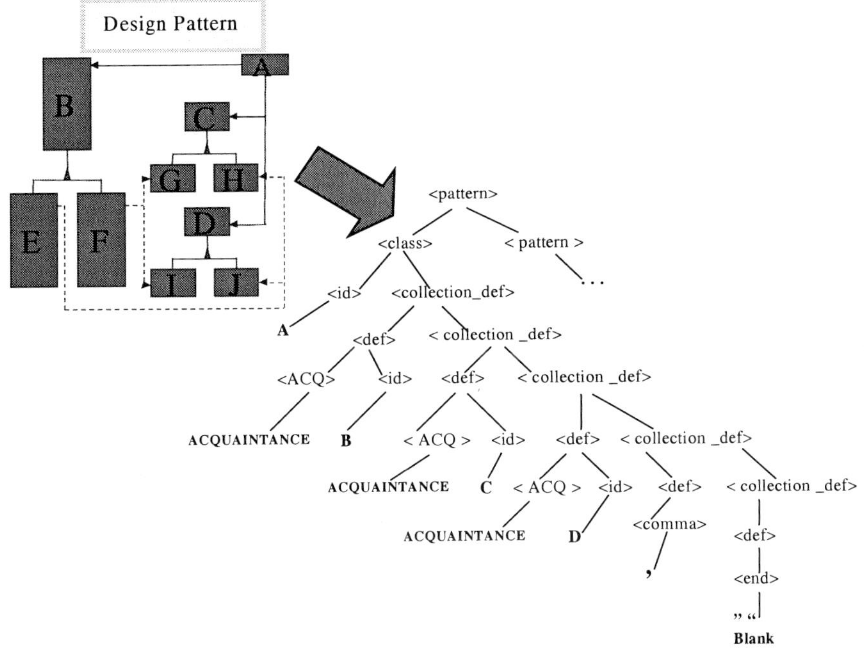

Fig. 1. Part of the derivation tree obtained of applying the canonical model.

3.2 Tool DEPAIC++

DEPAIC++ is composed by two modules that are: *Module for transformation of the C++ code to the canonical form* and *Module for recognition of design patterns*, which are also illustrated in Fig. 2. DEPAIC++ was implemented in Java programming language.

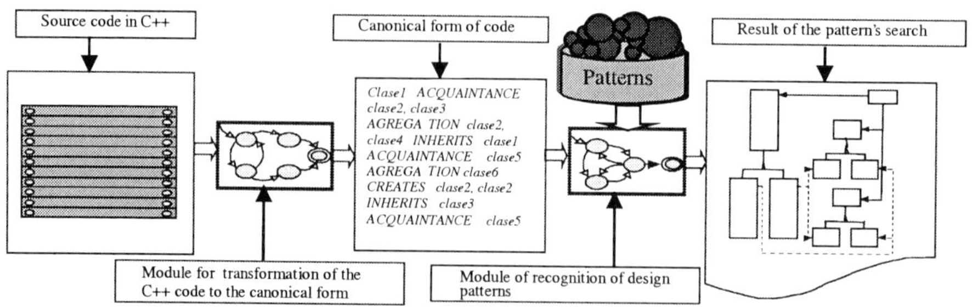

Fig. 2. General Structure of DEPAIC++.

3.2.1 Module for Transformation of the C++ Code to the Canonical Form

This module transforms the code written in C++ programming language to its corresponding canonical form. The canonical form is a string containing the names of all the classes and its relationships with other classes of the analyzed code. The relationships are separated by a comma.

In the analyzed code the module recognizes and generates the canonical form of the different types of relationships that can be in the design patterns: Inheritance, Aggregation, Acquaintance, and Creates.

It is important to mention that this module detects relationship cardinality. In addition to generate the canonical form of the analyzed code, it also provides a chart with all those relationships detected in the code related with C++ library.

Some considerations of this module are the following:

1. C++ code is only analyzed.
2. In DEPAIC++, errors are not verified in the analyzed code, therefore, the code should not have compilation and execution errors,
3. The analyzed code can be in one or several files.
4. There is no distinction in C++ implementation between aggregation and acquaintance. When a relationship has a pointer to instance, then it is considered as an acquaintance. If the relationship is implemented through a real instance, then is considered as an aggregation.

3.2.2 Module for Recognition of Design Patterns

This module analyzes the canonical form generated in the previous module and it detects, in this canonical form, if some design pattern was found by itself. The design patterns recognized in this module are: Abstract Factory, Composite, and Iterator, corresponding to each one of dimension of the classification proposed by Erich Gamma. Part of this module was generated using JavaCC[6]. It is important to mention that this module provides some design advice as: creation of superclass, abstract classes, etc. Furthermore, this module can show a window with the analyzed code.

The considerations for the construction of this module are the following:

1. We created the canonical forms of each one of the design patterns analyzed in this work, for this we use the Canonical Model for the Representation of Design Patterns, and the class diagram of each design pattern.
2. We created a parser with JavaCC. To formulate this, each canonical form and the design patterns behavior was analyzed. The types of classes or relationships belonging to design patterns, and which can repeat so that it continues being the same pattern.
3. For example, the AbstractFactory design pattern, can have some ConcreteFactories, and it continues being the same pattern. The restriction is that it inherits of AbstractFactories, and that creates Products.
4. Later on, this module match's both the canonical forms of design patterns and code and decide if the design pattern was found or not, and what missed to find it.

4 Related Work

Because realization of design patterns is an important field of research, many contributions on this topic have been made. We present only contributions, which are closely related to our work.

In the University of Maryland, Forrest Shull, , Walcélio Melo and Victor Basili, created "an Inductive Method for Discovering Design Patterns from Object-Oriented Software Systems" [7], they present an inductive method aimed at helping us discover design patterns in existing object-oriented software systems, which they called BACKDOOR (Backwards Architecting Concerned wit Knowledge Discovery of OO Relationships). It encompasses a set of procedures rigorously defined in order to be repeatable and usable by practitioners who are not acquainted with reverse architecting processes. The disadvantage of BACKDOOR is that the method is completely manual, therefore it requires a lot intervention of the user.

Jagdish Bansiya in his paper called "Automating Design-Pattern Identification" [8] present DP++, a tool that automates design pattern detection, identification, and classification in C++ programs. DP++ currently identifies most of the structural and a few of the behavioral patterns described in [4]. DP++ to pattern detection and identification focuses on the ability to identify key structural and functional relationships between classes and objects. The key structural relationships of interest for pattern identification are inheritance, aggregation, and uses. The disadvantage of DP++ is that it doesn't detect creational patterns.

Christian Krämer and Lutz Prechelt in their paper called "Design Recovery by Automated Search for Structural Design Patterns in Object Oriented Software" [9] present a system called Pat, where design information is extracted directly from C ++ header files and stored in a repository. The design patterns are expressed as Prolog rules and the design information is translated into facts. A single Prolog query is then used to search for all patterns. The disadvantage of PAT is that it only detects some structural patterns.

Kyle Brown in their thesis "Design Reverse-engineering and Automated Design Pattern Detection in Smalltalk" [10] proposed a tool called KT, KT detects the use of software design patterns in Smalltalk programs, this work examines the structure of design patterns, determines the nature of what makes a design pattern detectable by automated means, and outlines algorithms by which a small set of design patterns can be detected.

5 Conclusions and Further Work

In this paper, an approach to the Automatic Design Patterns Identification of C++ Programs is presented. We have shown that it is possible to detect design patterns from the analysis of the structure of classes of C++ code. The canonical model for the representation of design patterns presented here, provides an acceptable degree of formalization to the solution outlined in this work.

The developed tool, called DEPAIC++, uses the structural relationships between classes to identifying design patterns in C++ code. DEPAIC++ was tested with a group of programs that were designed using design patterns, and it detected them when should do it. Also, DEPAIC++ was tested with other programs that were not designed using design patterns and it found an approximation to each of them. Furthermore, in each case, DEPAIC++ found the characteristics missing to be a complete design pattern.

The approach presented here detects design patterns starting from a structural analysis of source code, but some design patterns implements different behavior in their solution. Therefore, to date we are adding a behavior analysis in source code, to detect with more accuracy design patterns.

To date, DEPAIC++ shows, as a string, the canonical form of design pattern and code, and we are working to show them in a UML class diagram, other extensions to DEPAIC++ include to analyze Java code.

References

1. Spitzer Tom: Component Architectures, DBMS, September (1997), pp. 56-66.
2. S. Novak Jr. Gordon: Software Reuse by Specialization of Generic Procedures through views. IEEE Transactions on Software Engineering, vol. 23, July (1997). Pp. 401–417.
3. Mili Hafedh, Mili Fatma, and Mili Ali: Reusing Software: Issues and Research Directions, IEEE Transactions on Software Engineering, Vol. 21, June (1995).
4. Gamma Erich, Helm Richard, Johnson Ralph, Vlissides John: Design Patterns Elements of Reusable Object-Oriented Software. Addison Wesley Professional Computing Series. (1995).
5. Schulz Benedikt, Genssler Thomas, Mohr Berthold and Zimmer Walter: On the Computer Aided Introduction of Design Patterns into Object-Oriented Systems" Proceedings of the TOOLS 27 Conference (Asia '98), IEEE Computer Society Press, (1998).

6. Sun Microsystems: Java Compiler Compiler. http://www.suntest.com/JavaCC/
7. Shull Forrest, Melo Walcélio, and Basili Victor: An Inductive Method for Discovering Design Patterns from Object-Oriented Software Systems. Technical Report, University of Maryland, computer Science Department, MD, 20742 USA. (1996).
8. Bansiya Jagdish: Automating Design-Pattern Identification. Dr. Dobb's Journal. http://www.ddj.com/ddj/1998/1998-06/lead/lead.htm. June 1998.
9. Krämer Christian, Prechelt Lutz: Design Recovery by Automated Search for Structural Design Patterns in Object-Oriented Software. Working Conference on Reverse Engineering, IEEE CS Press, Monterey CA, November 8–10, (1996).
10. Brown Kyle: Design Reverse-engineering and Automated Design Pattern Detection in Smalltalk. classic thesis. http://www2.ncsu.edu/eos/info/tasug/kbrown/thesis2.htm.

Specifying the Merging of Desires into Goals in the Context of Beliefs

Mehdi Dastani and Leendert van der Torre

The BOID Project
http://www.cs.vu.nl/~boid/
Utrecht University and Vrije Universiteit Amsterdam

Abstract. Rao and Georgeff's BDI_{CTL} logic is a popular specification and verification language for cognitive agent systems, in which desires and goals are unified into a single motivational attitude. In this paper desires and goals are distinguished to specify the merging of desires into goals, an important process in several agent systems such as Broersen *et al.*'s BOID system. We therefore introduce a BDGI_{CTL} logic. Moreover, we distinguish the merging of conflicting motivational attitudes such as desires into goals from the merging of conflicting informational attitudes such as knowledge bases or belief sources into beliefs, for which we use recent results in variants of belief revision known as semi-revision, fusion or merging. In particular, whereas belief merging is a generalization of revision, desire merging is a generalization of a kind of contraction known as (severe) withdrawal.

1 Introduction

Agent oriented software methodology [13] is the right abstraction level to analyze, design, and implement complex application domains. It is used for example for e-commerce applications and search bots that assist, actively support, and model the user. These application domains can be analyzed more intuitively and directly in terms of interacting agents or actors and high-level concepts such as beliefs, desires, goals, intentions, and plans. These agents are often called cognitive agents and the software systems are called cognitive agent systems.

Formal tools are necessary to guarantee the correctness of the analysis, design, and implementation. A popular specification and verification language for cognitive agent systems is Rao and Georgeff's BDI_{CTL} logic [10,12], the extension of so-called computational tree logic (CTL) with normal modalities for Beliefs, Desires and Intentions (BDI). CTL is used to specify traces of a cognitive agent through time while the belief, desire, and intention modalities are used to specify the epistemic states of the agent.

A drawback of the BDI_{CTL} formalism is that desires and goals have been unified into a single motivational attitude, such that the logic cannot be used to specify or verify how goals are generated. The crucial lack of expressive power is that BDI_{CTL} does not distinguish between motivational attitudes which can

M.H. Shafazand and A M. Tjoa (Eds.): EurAsia-ICT 2002, LNCS 2510, pp. 824–831, 2002.

conflict with beliefs (which we call desires or candidate goals) and motivational attitudes which cannot (which we call goals).

Our research question is how Rao and Georgeff's BDI_{CTL} has to be extended to specify or verify goal generation. In particular, we want to specify how *desires are merged into goals in the context of beliefs*. In our terminology, we are interested in the relation between beliefs, desires and goals, whereas Rao and Georgeff are interested in the relation between beliefs, goals and intentions. Our work is thus complementary to theirs.

The first step is the introduction of an extension of BDI_{CTL} called BDGI_{CTL}, that makes the distinction between desires and goals explicit. We assume that the following properties hold. They are called RG weak realism, because the formula represents a kind of realism proposed by Rao and Georgeff. These properties allow conflicts between beliefs and desires but not between beliefs and goals. For example, if you believe that it is not possible to be rich and to be a scientist at the same time, then you may have the desires to be both rich and a scientist, but you may not have the goals to be both.

$$\not\models D(p) \to \neg B(\neg p) \qquad \models G(p) \to \neg B(\neg p) \qquad \text{(RG weak realism)}$$

Moreover, desires are formalized by a non-normal modal logic [3,6], because for technical reasons explained later desires and goals are also distinguished by the following two logical properties, that express that goals are a logically closed set but desires are not.

$$\not\models (D(p) \land D(q)) \leftrightarrow D(p \land q) \qquad \models (G(p) \land G(q)) \leftrightarrow G(p \land q)$$

The second step is the use of recently introduced belief revision operators known as semi-revision, fusion or merging operators [1,4,5,8] to specify goal generation in cognitive agents. The idea is to take operators that merge belief sources into beliefs in the context of integrity constraints, and use them to merge desires into goals in the context of beliefs. Due to the implicit assumption that merged desires are considered as goals and used in a planning component, desires and goals should satisfy the following properties. They are called CL realism, because the formula represents a kind of realism proposed by Cohen and Levesque (see [10]). For example, you may believe that you cannot be both rich and a scientist, without having the desire or goal that this is so.

$$\not\models B(p) \to D(p) \qquad \not\models B(p) \to G(p) \qquad \text{(CL realism)}$$

These formulas should not be theorems of the logic, because it does not make sense for the agent to plan to achieve something which it already believes to be the case. Consequently, we distinguish desire merging from belief merging. In particular, whereas belief merging is a generalization of revision, desire merging is a generalization of a kind of contraction known as (severe) withdrawal [11].

The layout of this paper is as follows. In Section 2 we introduce BDGI_{CTL}. In Section 3 we consider belief and desire mergers.

2 Specifying the Merging of Desires into Goals

In this section we briefly repeat Rao and Georgeff's formalism [10] and we extend it with possibly conflicting desires. In our terminology desires in Rao and Georgeff's formalism are goals instead of desires, because the motivational attitudes in this formalism are not supposed to conflict with the beliefs. We therefore call their formalism BGI_{CTL} instead of BDI_{CTL}. In this paper we use an equivalent reformulation presented by Schild [12].

2.1 BGI_{CTL}

In contrast to Schild, we only consider the single agent case.

Definition 1. *[12, Def.4,6] The admissable formulae of BGI_{CTL} are categorized into two classes, state formulae and path formulae.*

S1 Each primitive proposition is a state formula.
S2 If α and β are state formulae, then so are $\alpha \wedge \beta$ and $\neg\alpha$.
S3 If α is a path formula, then $E\alpha$ and $A\alpha$ are state formulae.
S4 If α is a state formula, then $B(\alpha), G(\alpha), I(\alpha)$ are state formulae as well.
P If α and β are state formulae, then $X\alpha$ and $\alpha \cup \beta$ are path formulae.

The semantics of BGI_{CTL} involves two dimensions: an epistemic and a temporal dimension. The truth of a formula depends on both the epistemic world w and the temporal state s. A pair $\langle w, s\rangle$ is called a situation in which BGI_{CTL} formulae are evaluated. The relation between situations is traditionally called an accessibility relation (for beliefs) or a successor relation (for time).

Definition 2. *[12, Def.2,7] A Kripke structure $M = \langle W, \mathcal{R}_1, \ldots, \mathcal{R}_n, L\rangle$ is comprised of three components. The first component is an arbitrary non-empty set W containing all worlds relevant to M. The second component is a family of relations $\mathcal{R}_i \subseteq W \times W$. The remaining third component is an assignment function L. This function assigns a particular set of worlds to each primitive proposition. $L(p)$ contains all those worlds in which p holds.*

A Kripke structure $M = \langle \mathcal{S}, \mathcal{R}, \mathcal{B}, \mathcal{G}, \mathcal{I}, L\rangle$ forms a situation structure if each of the following three conditions is met.

1. *$\mathcal{S}$ is a set of situations.*
2. *$w = w'$ whenever $\langle w, s\rangle \mathcal{R} \langle w', s'\rangle$.*
3. *$s = s'$ whenever $\langle w, s\rangle \mathcal{B} \langle w', s'\rangle$ and similarly for $\mathcal{G}$ and $\mathcal{I}$.*

Schild [12, Section 3] does not present the semantic relation of CTL, but only the one of its extension CTL* (as well as the one of the μ-calculus). This extension is not considered in this paper. A speciality of both CTL and CTL* is that some formulae are not interpreted relative to a particular situation. These are the path formulae. What is relevant here are full paths. The reference to M is omitted whenever it is understood.

Definition 3. *A full path in* $M = \langle \mathcal{S}, \mathcal{R}, \mathcal{B}, \mathcal{G}, \mathcal{I}, L\rangle$ *is an infinite sequence* $\chi = \delta_0, \delta_1, \delta_2, \ldots$ *such that for every* $i \geq 0$, δ_i *is an element of* $\mathcal{S}$ *and* $\delta_i \mathcal{R} \delta_{i+1}$. *We say that a full path starts at* δ *iff* $\delta_0 = \delta$. *We use the following convention. If* $\chi = \delta_0, \delta_1, \delta_2, \ldots$ *is a full path in* M, *then* χ^i $(i \geq 0)$ *denotes* δ_i.

Let $M = \langle \mathcal{S}, \mathcal{R}, \mathcal{B}, \mathcal{G}, \mathcal{I}, L\rangle$ *be a situation structure,* δ *a situation, and* χ *a full path. The semantic relation* $\models$ *for* BGI_{CTL} *is then defined as follows:*

S1 $\delta \models p$ *iff* $\delta \in L(p)$.
S2 $\delta \models \alpha \wedge \beta$ *iff* $\delta \models \alpha$ *and* $\delta \models \beta$.
$\delta \models \neg\alpha$ *iff* $\delta \models \alpha$ *does not hold.*
S3 $\delta \models E\alpha$ *iff there is at least one full path* χ *in* M *starting at* δ *s.t.* $\chi \models \alpha$.
$\delta \models A\alpha$ *iff for every full path* χ *in* M *starting at* $\delta, \chi \models \alpha$.
S4 $\delta \models B(\alpha)$ *iff for every* $\delta' \in \mathcal{S}$ *such that* $\delta \mathcal{B} \delta', \delta' \models \alpha$.
$\delta \models G(\alpha)$ *iff for every* $\delta' \in \mathcal{S}$ *such that* $\delta \mathcal{G} \delta', \delta' \models \alpha$.
$\delta \models I(\alpha)$ *iff for every* $\delta' \in \mathcal{S}$ *such that* $\delta \mathcal{I} \delta', \delta' \models \alpha$.
P $\chi \models X\alpha$ *iff* $\chi^1 \models \alpha$.
$\chi \models \alpha \cup \beta$ *iff there is a* $i \geq 0$ *s.t.* $\chi^i \models \beta$ *and for all* $j (0 \leq j < i), \chi^j \models \alpha$.

2.2 BDGI$_{CTL}$

We extend BGI$_{CTL}$ with a new modal operator.

Definition 4. *The admissable formulae of* $BDGI_{CTL}$ *are defined by the formation rules of* BGI_{CTL} *together with the following rule.*

S4 If α *is a state formula, then* $D(\alpha)$ *is a state formula as well.*

For desires, we introduce minimal or non-normal modal logic [3,6]. The accessibility relation for desires is a binary relation between situations and sets of situations $\delta\mathcal{D}\{\delta_1, \ldots, \delta_n\}$.

Definition 5. *An extended Kripke structure* $M = \langle W, \mathcal{R}_1, \ldots, \mathcal{R}_n, L\rangle$ *is defined like a Kripke model, with the exception that we have either* $\mathcal{R}_i \subseteq W \times W$ *or* $\mathcal{R}_i \subseteq W \times 2^W$.

An extended Kripke structure $M = \langle \mathcal{S}, \mathcal{R}, \mathcal{B}, \mathcal{D}, \mathcal{G}, \mathcal{I}, L\rangle$ *forms an extended situation structure if each of the following three conditions is met.*

1. $\langle \mathcal{S}, \mathcal{R}, \mathcal{B}, \mathcal{G}, \mathcal{I}, L\rangle$ *is a situation structure,*
2. $\mathcal{D} \subseteq \mathcal{S} \times 2^{\mathcal{S}}$,
3. $s = s'_i$ *for* $i = 1, \ldots, n$ *whenever* $\langle w, s\rangle \mathcal{D}\{\langle w'_1, s'_1\rangle, \ldots, \langle w'_n, s'_n\rangle\}$

We choose the following definition of a semantic relation.[1]

Definition 6. *Let* $M = \langle \mathcal{S}, \mathcal{R}, \mathcal{B}, \mathcal{D}, \mathcal{G}, \mathcal{I}, L\rangle$ *be an extended situation structure and* δ *be a situation. The semantic relation* $\models$ *for* $BDGI_{CTL}$ *is the semantic relation* $\models$ *of* BGI_{CTL} *extended with:*

S4 $\delta \models D(\alpha)$ *iff* $\exists\{\delta_1, \ldots, \delta_n\}$ *such that* $\delta\mathcal{D}\{\delta_1, \ldots, \delta_n\}$, $\delta_i \models \alpha$ *for* $i = 1, \ldots, n$.

---

[1] Alternatively, we could have chosen an operator which is not closed under logical consequence, i.e. such that $\not\models D(p \wedge q) \rightarrow D(p)$. See [3,6] for details.

3 Logical Relations between Desires, Goals, and Beliefs

3.1 Belief Merging

In this section we briefly repeat Konieczny and Pino-Pérez operator Δ for merging belief bases with integrity constraints introduced in [4]. We first give some preliminary definitions.

Definition 7. *[4, Section 2] We consider a propositional language L over a finite alphabet P of propositional atoms. A belief base K is finite set of propositional formulae, which can be seen as the conjunction of the formulae of K. A belief set E is a multi-set of belief bases. We write $K \wedge K'$ for $K \cup K'$, $\bigwedge E$ for the conjunction of belief bases of E and $\sqcup$ for union of multi-sets. By abuse of language we write K for the multi-set $\{K\}$.*

A belief set E is consistent iff $\bigwedge E$ is consistent. Two belief sets E_1 and E_2 are equivalent, noted $E_1 \leftrightarrow E_2$, iff there exists a bijection f from $E_1 = \{K_1^1, \ldots, K_1^n\}$ to $E_2 = \{K_2^1, \ldots, K_2^n\}$ such that $f(K) \leftrightarrow K$.

Definition 8. *[4, Section 3] Let E be a belief set, IC a belief base (integrity constraints), and Δ an operator that assigns to each belief set E and belief base IC a belief base $\Delta_{IC}(E)$. Δ is a belief merging operator if and only if it satisfies the following properties:*

B0 $\Delta_{IC}(E) \vdash IC$
B1 *If IC is consistent, then $\Delta_{IC}(E)$ is consistent*
B2 *If $\bigwedge E \wedge IC$ is consistent, then $\Delta_{IC}(E) = \bigwedge E \wedge IC$*
B3 *If $E_1 \leftrightarrow E_2$ and $IC_1 \leftrightarrow IC_2$, then $\Delta_{IC_1}(E_1) \leftrightarrow \Delta_{IC_2}(E_2)$*
B4 *If $K \vdash IC$, $K' \vdash IC$, and $\Delta_{IC}(K \sqcup K') \wedge K$ is consistent, then $\Delta_{IC}(K \sqcup K') \wedge K'$ is consistent*
B5 $\Delta_{IC}(E_1) \wedge \Delta_{IC}(E_2) \vdash \Delta_{IC}(E_1 \sqcup E_2)$
B6 *If $\Delta_{IC}(E_1) \wedge \Delta_{IC}(E_2)$ is consistent, then $\Delta_{IC}(E_1 \sqcup E_2) = \Delta_{IC}(E_1) \wedge \Delta_{IC}(E_2)$*
B7 $\Delta_{IC_1}(E) \wedge IC_2 \vdash \Delta_{IC_1 \wedge IC_2}(E)$
B8 *If $\Delta_{IC_1}(E) \wedge IC_2$ is consistent, then $\Delta_{IC_1 \wedge IC_2}(E) \vdash \Delta_{IC_1}(E)$*

Most of these postulates are generalizations of belief revision postulates. (B0) states that the result of merging complies with the integrity constraints. (B1) ensures that, when the integrity constraints are consistent it is always possible to abstract a coherent piece of information from the belief set. (B2) says that, if possible, the result of the merging is simply the conjunction of the belief bases of the belief set with the integrity constraints, (B3) is the principle of irrelevance of syntax. The purely 'merging' postulates are (B4), (B5) and (B6). (B4) is what is called the fairness postulate. It ensures that when merging two belief bases, it cannot give full preference to one of them. (B5) and (B6) correspond to Pareto's conditions in social choice theory. Finally (B7) and (B8) state conditions on the conjunction of integrity constraints and make sure that 'closeness' is well-behaved. See [4,5] for further details and motivations.

3.2 Merging Operators Defined on Situation Structures

Intuitively, the operator $\mathcal{S}$ is the merging operator of an extended situation structure $M = \langle \mathcal{S}, \mathcal{R}, \mathcal{B}, \mathcal{D}, \mathcal{G}, \mathcal{I}, L \rangle$ iff for each situation δ, integrity constraints $IC = \{\alpha \mid \delta \models B\alpha\}$ and set $E = \{\{\alpha \mid \delta_i \models \alpha \text{ for } i = 1 \ldots n\} \mid \delta\mathcal{D}\{\delta_1, \ldots, \delta_n\}\}$ we have that $\Delta_{IC}(E) = \{\alpha \mid \delta \models G\alpha\}$.

This intuition explains our motivation to use non-normal modal logic. First, note that the multi-set E as defined above can contain several identical elements, because distinct situations can satisfy the same propositions. Second, suppose that $\mathcal{D}$ is a normal accessibility relation between worlds. In that case, we could only define a logically closed set of propositions like we defined IC and $\Delta_{IC}(E)$, not a set of conflicting ones.

However, this definition is still preliminary. One of the reasons is that postulate (B0) implies $B(p) \rightarrow G(p)$. As discussed in the introduction, this property called CL realism is considered counterintuitive [10].

3.3 Desire Merging Operators

The following definition defines merging for desires. Postulate B0 has been weakened, and the other postulates B1-B8 have been updated to incorporate this difference. We define desire bases and desire sets analogous to belief bases and belief sets.

Definition 9. *A desire base K is finite set of propositional formulae, which can be seen as the conjunction of the formulae of K. A desire set E is a multi-set of desire bases. We write $\bigwedge E$ for the conjunction of desire bases of E, etc.*

Definition 10. *Let E be a desire set, IC a belief base (integrity constraints), and ∇ an operator that assigns to each desire set E and belief base IC a desire base $\nabla_{IC}(E)$. ∇ is a desire merging operator if and only if it satisfies the following properties:*

D0 *If IC is inconsistent, then $\nabla_{IC}(E) = IC$*
D1 *If IC is consistent, then $\nabla_{IC}(E) \wedge IC$ is consistent*
D2a $\bigwedge E \vdash \nabla_{IC}(E)$
D2b *If $\bigwedge E \wedge IC$ is consistent, then $\nabla_{IC}(E) \vdash \bigwedge E$*
D3 *If $E_1 \leftrightarrow E_2$ and $IC_1 \leftrightarrow IC_2$, then $\nabla_{IC_1}(E_1) \leftrightarrow \nabla_{IC_2}(E_2)$*
D4 *If $K \vdash IC$, $K' \vdash IC$, and $\nabla_{IC}(K \sqcup K') \wedge K$ is consistent, then $\nabla_{IC}(K \sqcup K') \wedge K'$ is consistent*
D5 $\nabla_{IC}(E_1) \wedge \nabla_{IC}(E_2) \vdash \nabla_{IC}(E_1 \sqcup E_2)$
D6 *If $\nabla_{IC}(E_1) \wedge \nabla_{IC}(E_2) \wedge IC$ is consistent, then $\nabla_{IC}(E_1 \sqcup E_2) = \nabla_{IC}(E_1) \wedge \nabla_{IC}(E_2)$*
D7 *If $IC_1 \wedge IC_2$ is consistent, then $\nabla_{IC_1}(E) \vdash \nabla_{IC_1 \wedge IC_2}(E)$*
D8 *If $\nabla_{IC_1}(E) \wedge IC_1 \wedge IC_2$ is consistent, then $\nabla_{IC_1 \wedge IC_2}(E) \vdash \nabla_{IC_1}(E)$*

We choose for the borderline case in which the integrity constraints are inconsistent, that the output is a contradiction (D0). Moreover, we split the postulate D2 in two postulates D2a and D2b. The meaning of D1-D8 is analogous to the meaning of B1-B8. The differences are that it is no longer implied that IC follows from ∇ (D2, D7, D8), but IC is added to every the consistency check (D1, D6, D8). If a belief set consists of only one belief base K, then merging the belief set under integrity constraints IC is equivalent to revising K by IC, expressed by $K * IC$. Under similar conditions, merging a single desire base under integrity constraints IC is an operation known in the literature as 'severe withdrawal' [11] and expressed by $K - \neg IC$, where we write $\neg IC$ for $\{\neg \bigwedge IC\}$. This is basically a contraction operator that does not satisfy the recovery postulate, in our terminology $\nabla_{IC}(K) \wedge \neg IC \vdash K$.

The following proposition generalizes the well known Levi identity for merging operators and shows the close relation between belief merging and desire merging.

Theorem 1. *If ∇ is a desire merging operator, then we have that Δ defined by $\Delta_{IC}(E) = \nabla_{IC}(E) \wedge IC$ is a belief merging operator.*

4 Summary

Rao and Georgeff's BDI_{CTL} is one of the most popular specification or verification languages for cognitive agent systems. It describes the balance between beliefs, goals and intentions, but it does not describe where the beliefs and goals come from. However, in most agent systems these beliefs and goals are the result of merging respectively belief bases and candidate goals such as desires, obligations and norms. Logical analyses of the derivation of goals from desires using Reiter's default logic has been given in [2], a logical analysis of conflicts between desires has been given in [7], and the merging of candidate goals has been addressed in [5]. In this paper we have introduced merging in the specification language by introducing a new non-normal modal operator for desires. In a similar way we can build the extension of BDGI_{CTL} to $\text{KBDGI}_{\text{CTL}}$ with an additional operator K for belief sources, which have to be merged into beliefs. Another extension of BDI_{CTL} with belief revision operators can be found in [9].

As an example we have considered a recent merging operator proposed by Konieczny and Pino-Perez. We discussed one postulate of their logic, namely (B0), and we argued that this postulate is too strong, because it implies the Cohen-Levesque notion of realism expressed by $B(p) \rightarrow G(p)$. Moreover, we introduced a weakened version of their postulates. Whereas their operator is a generalization of revision, our operator is a generalization of (severe) withdrawal.

We believe that this paper is the beginning of a study of merging desires into goals in the context of beliefs. First, interesting properties have to be identified that can be expressed in the BDGI_{CTL} language. Second, the other belief merging postulates have to be considered, as well as other interesting merging operators proposals in the literature such as [1,8]. Third, the merging operator should not be fixed for the whole model but allowed to vary. Fourth, merging operators should not only be defined for finite sets of propositions, because even for a finite set of propositional atoms there are infinite sets of CTL formulas. Fifth, the language $BDGI_{CTL}$ must be extended such that it can distinguish – like the extended situation structure – the number of occurrences of a desire base. Sixth, the merging constraints should be axiomatized in an extension of $BDGI_{CTL}$. Seventh, to cover the analysis in [2], the merging operators have to be extended to conditionals. Eighth, as suggested by Richard Booth, sources should be represented explicitly in the formal framework such that we can distinguish between a belief and desire base from the same source, and from distinct sources.

Acknowledgements. Thanks to Richard Booth for pointing us to the relation between desire merging and (severe) withdrawal.

References

1. R. Booth. Social contraction and belief negotiation. In *Proceedings of the Eighth International Conference on Principles of Knowledge Representation and Reasoning (KR2002)*, pages 375–384. Morgan Kaufmann, 2002.
2. J. Broersen, M. Dastani, and L. van der Torre. Realistic desires. *Journal of Applied Non-Classical Logics*, To appear.
3. B. Chellas. *Modal Logic: An Introduction.* Cambridge University Press, 1980.
4. S. Konieczny and R. Pino Pérez. Merging with integrity constraints. In *Proceedings of the Fifth European Conference on Symbolic and Quantitative Approaches to Reasoning with Uncertainty (ECSQARU'99)*, volume 1638 of *LNCS*, pages 135–144, 1999. Extended version: S. Konieczny and R. Pino Pérez. Merging information under constraints: a qualitative framework. *Journal of Logic and Computation*, to appear.
5. S. Konieczny and R. Pino Pérez. On the frontier between arbitration and majority. In *Proceedings of the Eigth International Conference on Principles of Knowledge Representation and Reasoning (KR'02)*, 2002.
6. S. Kraus, K. Sycara, and A. Evenchik. Reaching agreements through argumentation; a logical model and implementation. *Articial Intelligence*, 104:1–69, 1998.
7. J. Lang, L. van der Torre, and E. Weydert. Two types of conflicts between desires (and how to resolve them). In *Proceedings of GTDT2001, AAAI Spring Symposium*, 2001.
8. T. Meyer. On the semantics of combination operations. *Journal of Applied Non-Classical Logics*, 11(1-2):59–84, 2001.
9. A. Rao and M. Georgeff. The semantics of intention maintenance for rational agents. In *Proceedings of the Fourteenth International Joint Conference on Artificial Intelligence (IJCAI-95)*, pages 704–710, 1995.
10. A.S. Rao and M.P. Georgeff. Decision procedures for BDI logics. *Journal of Logic and Computation*, 8(3):293–342, 1998.
11. H. Rott and M. Pagnucco. Severe withdrawal (and recovery). *Journal of Philosophical Logic*, 28:501–547, 1999.
12. K. Schild. On the relationship between BDI logics and standard logics of concurrency. *Autonomous Agents and Multi agent Systems*, 3(3):259–283, 2000.
13. Wooldridge M. J. and Jennings N. R. Software Engineering with Agents: Pitfalls and Pratfalls. *IEEE Internet Computing*, 3(3):20–27, May/June 1999.

The Illegal Copy Protection Using Hidden Agent [1]

Deok-Gyu Lee, Im-Yeong Lee, Jong-Keun Ahn, and Yong-Hae Kong

Division of Information Technology Engineering, SoonChunHyang University, #588, Eupnae-ri, Shinchang-myun, Asan-si, Choongchungnam-Do, 336-745, KOREA
{hbrhcdbr, imylee, jkahn00, yhkong}@sch.ac.kr
http://sec-cse.sch.ac.kr

Abstract. There have been researches into digital Watermarking technology or Fingerprinting vigorously to safeguard Protective rights for knowledge and poverty for digital contents. DRM(Digital Rights Management) is not only Protective rights for knowledge and poverty, but also management and systems that are necessary to put out, circulate and use for contents. This technology, DRM, encrypts contents to protect digital contents and they are sold users on. Sellers transmit contents with 'Usage Right' and a license including a key of encryption. The key of encryption decodes encoded files. The right of usage restricts users' application of contents. Even if digital contents that are applied the DRM are coped illegally and circulated, contents will be protected from that because a player of DRM checks existence of licenses and allows contents to be restored. However, this method might cause users to feel inconvenient since the users can only restore contents through the licenses offered by a player or a Smartcard. If radio as well as cable is used popularly in the future, there will be a lot of limits to use those kinds of players. In the method of that, the method using players need different players in order to work successfully in wired and wireless environment. In the case of using Smartcards, there might be a dangerous situation when the Smartcards disappeared. This paper proposes two kinds of ideas. One is protecting contents from illegal acts such as illegal copies when the contents are in the process of circulation. The other is the protocol that can give users convenience. Hidden Agents are used so that contents are protected from illegal copies and illegal use in the contents and cuts off those illegal acts. The Agent will be installed without any special setup. In addition, it can replace roles of Watermarking as a protection. Therefore, this paper shows the solution of illegal copies that happens frequently.

1 Introduction

To activate sales of digital contents in electronic commerce, 'protective rights for knowledge and property' should be studied at first. Digital contents can be copied and

[1] This work is supported in part by the Ministry of Information & Communication of Korea ("Support Project of University Information Technology Research Center" supervised by KIPA)

M.H. Shafazand and A M. Tjoa (Eds.): EurAsia-ICT 2002, LNCS 2510, pp. 832–841, 2002.

circulated unlike general off-line contents. Therefore, after a legal purchaser buys digital contents from a seller, the illegal redistribution of the contents should be considered to protect. There have been researches into digital Watermarking technology or Fingerprinting vigorously to safeguard 'Protective rights for knowledge and poverty' for digital contents. Many models of the DRM that is based on those technologies have been the proposed and used widely now. The DRM means those management and protection systems that are needed in publishing, circulating and using digital contents. And it means not only basic construct technology for building an integrated management system of contents, but also Infra technology of contents management such as DOI and INDECS, etc used all over the world. In a management session, another feature of DRM is a technology of application for protecting contents safely as a protection system. To protect digital contents safely, there are some practicable technologies such as protective technology for copyright of circulation and service control, and cryptogram technology for copyright, ownership, the right of using and digital Watermarking technology This paper will propose how to protect contents in the process of circulation and service for digital production among those technologies. This proposal will not only figure out illegal copies in circulation and service, but also protect copyright and the right of using. In existing proposed models, exclusive players, Smartcards or program installation has been used. In other words, the existing models should use a certain item to protect digital contents. It means that people are inconvenient to use that kind of models because of using the certain item. So this paper proposes the DRM model to prevent illegal copies without the above. Compared with the existing models like Smartcards and players, this proposed model could prevent illegal copies by using Agents included in contents. It never uses Smartcards and players.

2 Outline of Agents

Movable Agents are independent and autonomous so they can conduct many services by moving networks. This Figure 1 simply shows mobile Agents' motions of performance after moving from a local to a remote host. The Agent moves from host A to host B and collects the information that the Agent wants after approaching services and sources of the host B through defined interfaces. If finishing the job, it sends the information to the original server A. After acquiring the information an Agent wanted, the Agent moves other servers and do the same performance. The movable Agent is a processor that acts automatically for a user. If it starts to perform, it will leave the place that it was born at and collect the information that it wants by moving places through networks.

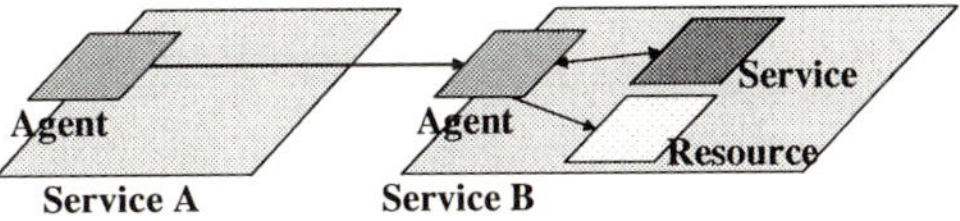

Fig. 1. Mobile Agents' motion of Performance

3 Elements of DRM and Analysis of Existing Model

3.1 Elements of DRM

A digital content goes through the step of formation, circulation, and consumption as writers' creature. In order to protect digital information, the function of DRM should be added to the each step above. In the preparatory stage of creation and flowing, Packagers are needed to encode and protect contents. In the stage of circulation and sales, Financial Clearinghouses and licensees that take charge of finance and the issue of license are needed.

In the stage of consumption, the DRM of Agents is needed to control restoration according to the encryption and the right of usage. The Figure 2 shows necessary the elements of DRM. The packager protects contents through the encryption. An encryption Algorithm is sent to a license clearinghouse in order to make a key needed in encrypting.

Packaged contents are sent to the purchasers who paid the bills in electronic commerce such as on-line shopping Malls, CD, E-mail (reference to step 2, 3 out of the Figure 2). The purchaser will get contents with license (reference to step 4 out of the Figure 2). The license has the information of the right that the purchasers can use contents legally and the cipher-keys that can decode encoded files. It means that the key decodes the cryptogram of contents and abstracts contents under the right of usage. The rights include the frequency of usages, the period of usages, the term of validity, and transfer to other tools, moving to other storing item.

Clearinghouses have been consisted of trade and finance, as well as a license clearinghouse. Financial Clearinghouses conduct necessary jobs of financial approval and the settlements of accounts followed by the approval in contents commerce including consumption, sales and circulation. License Clearinghouses, as it was mentioned before, might as well a general term as a server for issuing a license. It issues a license needed to decode a message of contents in code. Business models like IMPRIMATUR and MPEG have been using a technical term, 'Monitoring Authority'.

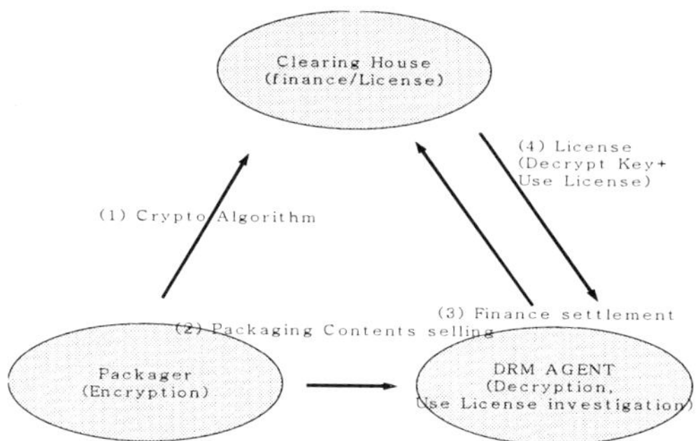

Fig. 2. Essential Elements of DRM

3.2 Analysis of Existing Models

DRM technology is necessary for digital contents to be circulated in a market and there is a conflict between the owners who has the copyright of contents and the Internet users who want to use contents on free. Therefore, there has been a slow development of that. Early DRM manufactures were aimed to make the Internet charged for, but they failed to do so. Now they have been developed many kinds of markets, for example, electronic mail, electronic documents, and circulation of software, etc.
Besides, there are two sales methods. One is an existing solution sales method. The other is the sales method that applies DRM, ASP models. Two kinds of methods support the total process of DRM application. In the side of working method, targets models have been introduced into a market. The target models are divided into license servers and Financial Clearinghouses that have a license and a bill system.
Korean DRM solution is consisted of three kinds of solution such as Intertrust, MS and Koreans' own technology.
Firstly, there are some companies, Pasu.com and Trust Tech, adopting Intertrust solution. Pasan.com has used the API supplied by Intertrust.co and provided a financial service by using Financial ClearingHouse and E-Book, A/V (Audio/ Video) contents solution accepting the complicated preservation algorithm. Trust Tech.co operates a license server by offering preservation of documents as well as E-mail solution to enterprises. It is necessary to set up exclusive client software on every media such as E-Book, Audio/Video, Image, etc.
Secondly, DRM Korea Co. is one of companies providing DRM as based on Microsoft. The company has offered contents solution to a small size of an A/V Entertainment market, charged a fee whenever it authenticates licensees for encoded contents. However, it could not prepare the Financial Clearinghouse linked with licensees. This solution uses the Window Media Player of Microsoft as client software. It does not need another S/W for A/V contents but it should have exclusive client software for E-Book.
Thirdly, domestic companies of DRM solution such as Markany Co. and Dreamintech Co. have supported DRM solution on the basis of Watermarking and Public key infrastructure as cipher special enterprises.

4 Proposal Scheme

This paper prevents illegal copies by using a Hidden Agent and encourages existing systems to figure out problems of inserting Watermarking and payment for early contents.

4.1 Terms Desired of the Proposal Scheme

- Hidden Agents should satisfy the terms desired as fallow.
- The proposed a Hidden Agent should be in the contents.

The Hidden Agent that exists in the contents cannot be removed by optional request of users. If the Hidden Agent is deleted, all contents will be deleted together.

- Hidden Agents will fulfil after offering contents.
  The Hidden Agent that exists in the contents is provided with contents.
- As soon as the contents practice on user's computer, Agents load.
- Hidden Agents will be loaded in booting.
  After an operating system starts, Hidden Agents will be loaded to prevent illegal copies of contents.
- Hidden Agents include a factor.
  To prevent illegal copies of contents, a factor should have its ID and value of key. Also, creation argument can keep away piracy by creation argument and offer argument as argument that Hidden Agent can create directly.
- T that exist on Hidden Agent interior changes at COPY instruction.
  Hidden Agent T that is inside factor is Time-stamp at early contents last month. Inside factor T again when copy is achieved T by Time of computer capable of copy again update do. Therefore, copy for the contents can not be achieved again.

4.2 Illegal Copies of Contents

- In the case of copying contents without the authority of usage or in the case of copying contents without acquiring the authority of the usage.
- The contents are obtained illegally on the Internet.

4.3 Whole System Model

The follow schematizes the whole system of DRM.

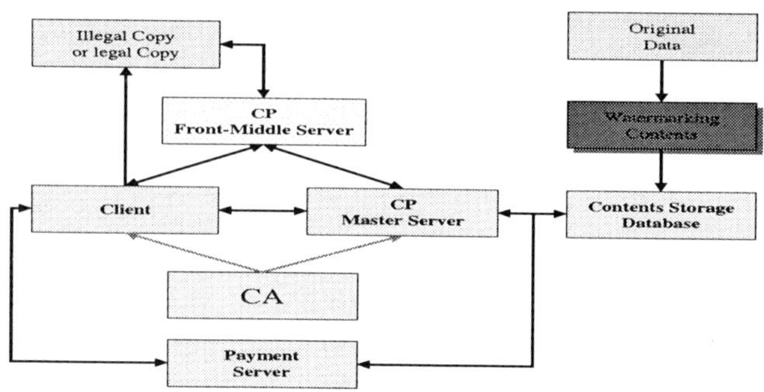

Fig. 3. Whole System Model

4.4 Constituents

Explanation of constituent items in this system is followed by next.

- **A User:** wants to purchase contents has a bill and the right of usage. With CP Master Server offer a key for need in contents
- **CP Master Server:** cover user's registration and have the right of ownership. Also, it creates a key for offering contents as a user does so.
- **CP Front-Middle Server:** communicate with Hidden Agents offered within contents to prevent illegal copies. This item receives user's information from CP Master Servers. A user is authorized by Hidden Agents can copy on the basis of user's information.
- **Payment Server:** is located between users and CP Master Servers as an item for payments.
- **Contents Database:** is offered contents undergone Watermarking by a writer. In case that a writer is a CP, the CP will get copyright. On the other hand, just in cast of the existing a writer, the Contents Database will get copyright.
- **CA (Certificate Authentication):** is constituted to use signature and put to practical use with payment system and Contents Database afterward.

4.5 System Parameters

This paper explains exchange of keys for offering contents. Hidden Agents need a system coefficient.

· **ID :** User Identify · **S :** Contents class to buy
· **M :** Payment information for Contents
· $\mathbf{M_{Key}}$ **:** Master key for communication and certification that CP Master Server creates
· $\mathbf{A_T}$ **:** Agent
· $\mathbf{A_{Key}}$ **:** Key price for certification, AKey = $1/2H(ID||M_{Key}||S||M)$ = High position 1/2 bit
· $\mathbf{E_{Key}}$ **:** Key price for password, EKey = $1/2H(ID||M_{Key}||S||M)$ = Low position 1/2 bit
· **H :** Hash Function · **D :** Agent value, $D = A_{Key}(E_{Key}(T))$
· $\mathbf{PK_{User}}$, $\mathbf{SK_{User}}$ **:** User's public key and private key encryption and decryption
· $\mathbf{PK_{CM}}$, $\mathbf{SK_{CM}}$ **:** CP Master Server's public key and private key encryption and decryption
· $\mathbf{PK_{CMS}}$, $\mathbf{SK_{CMS}}$ **:** CP Front-Middle Server's public key and private key encryption and decryption
· **T :** Of in transmission Time-Stamp · **T' :** Time-Stamp at COPY
· $\mathbf{C_{Info}}$ **:** Contents Information

4.6 Explanation for Each Stage of Protocols

In this paragraph, this paper gives an explanation for each stage of protocols.

There are three steps totally, 'contents offering stage', 'contents payment stage', and 'verification stage of contents' through illegal copies. The rest except 'contents payment stage' will follow the existing system for the matters of payment. You can see the detail explanation for each stage.

4.6.1 Contents Offering Stage

The next paragraph explains exchanging keys between users, CP master Servers and CP Front Middle servers, as well as the process of providing user's information.

Phase 1. User registers and encrypts user ID and a sort (S) of contents that the user wants, and payment information and transmits to CP Master server.

- $PK_{CM}(SK_{User}(ID)||ID||S||M)$

Phase 2. Create key that need Master key and encryption using information that offer from user.

- M_{Key} : Master Key
- $A_{Key} = 1/2H(ID||M_{Key}||S||M)$
- $E_{Key} = 1/2H(ID||M_{Key}||S||M)$

Phase 3. Sends Master Key and the factors created in the master servers to Front-Middle servers.

- $PK_{CMS}(SK(M_{Key})||ID||S||M)$

Phase 4. Master servers send the Hidden Agent that is suitable for users after inserting in the contents. This time, COPY city T cost changes that T has copied point of time.

- $PK_{User}(C_{A(CInfo||AKey||EKey||AKey(EKey(T))||T)})$

Phase 5. Agent is stored on user's computer.

- $A(C_{Info}||A_{Key}||E_{Key}||A_{Key}(E_{Key}(T)||T')$

Phase 6. Transmit Time-Stamp cost that is passed to user by CP Front-Middle Server.

- $PK_{CMS}(SK_{CM}(T))$

Phase 7. User sends is Finish Message.

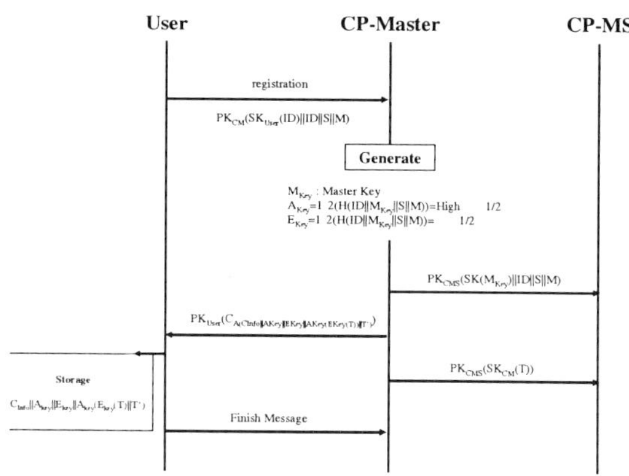

Fig. 4. Contents offering Stage

4.6.2 Verification Stage of Contents Illegally Copied

In case that a user wants to copy contents illegally, what kind of reaction from Hidden Agents is explained as follow:

As it was explained previously, Hidden Agents conduct after supplying contents. If a user commands orders like 'COPY' and 'MOVE' on an operating system, a Hidden Agent will start to work by using the keys received from a server. The Agent encrypts ID, S, M, L, R and sends them to Front-Middle Servers. The Front-Middle sever will work only with a Hidden Agent.

This time, T value in Hidden Agent interior means early T value. Also, Front-Middle Sever works with Hidden Agent.

If the Agent cannot connect with a server, the right of copy won't be provided for a user.

Phase 1. If the order of copy acts on user's computer, a Hidden Agent will not only run automatically, but also encode S, M, and T(Hidden Agent inside factor). Then the Agent will send them to a Front-Middle server.

· $PK_{CMS}(SK_{User}(ID)||A_{AKey(EKey(T'))}||S||M) = PK_{CMS}(SK_{User}(ID)||D||S||M)$

Phase 2. The Front-Middle sever will calculate 'T`' by using 'T', compare them with its database. It allows a user the right of copy.

· $A_{Key}(A_{Key}(E_{Key}(T')))$
$= E_{Key}(E_{Key}(T'))$
$= T'$

· Compare T=T'

Phase 3. Transmit competence about COPY to user.

· $PK_{User}(A_{(AKey(EKey(Yes\ or\ No)))})$

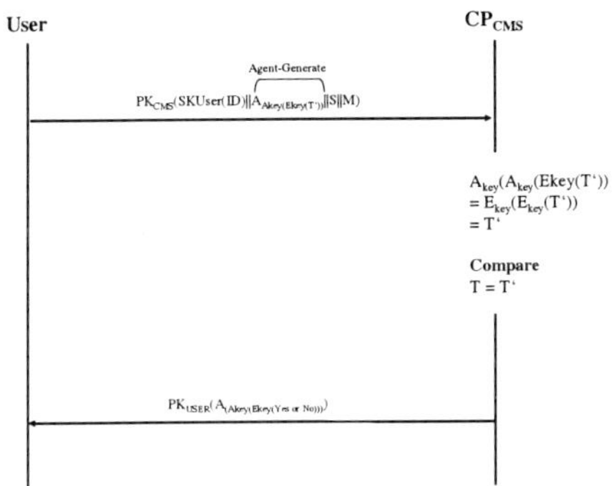

Fig. 5. Verification stage of contents illegally copied

5 Inquiry into Proposed System

This paper proposes how to prevent illegal copies by using a Hidden Agent. Through inserting a Hidden Agent in the contents, a user has the limitation of the right of usage of contents. A user can receive the right of ownership only from CP and use contents with copyright. Even if a user copies contents illegally, the user cannot get the value of 'R'. It means that Hidden Agents and Front-Middle severs cannot calculate 'D', 'L' so the illegal copy won't be verified. When a user copies all contents illegally, the value of 'R' will be changed to distinguish itself from original value. Through this process, illegal copies will be prevented.

Even if a user tries to copy contents with a legal way, the user cannot copy it successfully since Hidden Agents have the right of copy for a Front-Middle server.

Another feature of preventing illegal copies is that when a Hidden Agent doesn't have the right of a Front-Middle server, a user cannot delete illegal copy because it doesn't allow contents to be copied. If some files circulated by illegal copies, a Hidden Agent locating in the contents will check its T and verify responsibility for the fulfillment of contents.

5.1 Analysis and Compare with Existing Method

In this section, a proposed method will be compared and analyzed with an existing method.

The Table 1 shows the result after comparing and analyzing a proposed method with existing one. Some existing systems can prevent illegal copies of MP3, but there is a bad situation that a devil user can distribute data and keys of MP3. In such case, it is hard to prevent those illegal copies. However, this proposed system could prevent even those damages when such illegal distribution happens because it has keys of a Hidden Agent and CP Front-Middle severs. In addition, since there is the value of 'R' that is created in the Hidden Agent, illegal distribution of MP3 data can be prevented.

Table 1. Analysis of Proposal Scheme

	Contents illegal distribution protect	Contents eaves-dropping	On/Off-Line Application	Contents robust-ness	Indepen-dence player
Pasu.com	×	O	×	×	O
TrustTech	O	O	△	×	×
DRM Korea	O	×	△	O	×
DreamIntech	×	O	×	×	O
MarkAny	×	×	×	O	O
Proposal Scheme	O	O	O	O	×

Existing systems such as Pasu.com, Dreamtech, and Markany have used independent players to prevent that kind of situation, but it seems to be difficult for existing systems to expand into wireless environment and to be impossible for original contents to be potent because of frequent ciphers. Besides the proposed method has convenient feature, comparing with existing systems. The great feature is that the proposed Agent just restricts the right of copy however the old systems should pass all time-consuming processes of Authentication for each the content and then it can play the contents.

After a user purchases contents at first, he/she has to get authentication every time. This process makes the user feel irritated. However, the proposed method only has some limitation to use some orders like 'COPY' and 'MOVE'. It means that a user can use contents as the same as regular contents.

6 Conclusion

There have been many researches into DRM. Protection of contents in distributing and managing is the most important part of processes. An existing method using an exclusive player and a Smartcard has some obstacles that a user feels inconvenient because of frequent authentication for getting contents. This proposed method has made effort to solve the problems.

This paper proposes the DRM models for preventing illegal copies by using a Hidden Agent. Users can use an existing system without alteration and even they don't know the existence of a Hidden Agent. Moreover, this method can be used wired and wireless environment since it is different from an existing system, especially an exclusive player. A user doesn't have to make effort to install a Hidden Agent in the contents. Through a Hidden Agent, illegal copies will be prevented and people easily can approach whole DRM models. To improve the proposed method more efficient and convenient, this paper suggests homework like how to connect the right of ownership of original contents with payments, and how to supply anonymous users' contents. Those DRM technologies will encourage the changes of current software circulation system in which people can purchase products like CD off-line, as well as on-line sales such as digital contents for entertainment and amusement.

References

1. Wagner N. R, 1983. "Fingerprinting", IEEE Symposium on Security and Privacy
2. Lee K. H. 2001 "A Contents Protection Technology based on Mobile Agent," Korea Multimedia Society review. pp 164–171
3. Vigna A., 1998, "Cryptographic traces for Mobile Agents," In: G. Vigna (Ed.), Mobile Agents and Security, Springer-Verlag, Lecture Note in Computer Science 1419, pp 99–113
4. Hohl F., 1998, "Time Limited Blackbox Security: Protecting Mobile Agents from Malicious Hosts", In: G. Vigna(Ed.), Mobile Agents and Security, Springer-Verlag, Lecture Note in Computer Science 1419, pp 137–153
5. Sander T. & Tschudin, 1997, "Toward Mobile Cryptography", International Computer Security Institute (ICSI), TR-97-049
6. http://www.intertrust.com
7. http://www.uspto.gov
8. http://www.markany.com
9. http://www.dreamintech.co.kr

Mobile Agent-Based Misuse Intrusion Detection Rule Propagation Model for Distributed System

Tae-Kyung Kim, Dong-Young Lee, and T.M. Chung

Real-Time Systems Laboratory,
School of Information and Communication Engineering,
SungKyunKwan University,
Chunchun-dong 300, Jangan-gu, Suwon, Kyunggi-do,
Republic of Korea
{tkkim,dylee,tmchung}@rtlab.skku.ac.kr

Abstract. This paper describes the rule propagation model for the misuse detection methods using mobile agents. Approaches to detecting intrusions can be broadly classified into two categories: Anomaly Detection and Misuse Detection. Misuse detection is best suited for reliably detecting known use patterns. Misuse detection systems can detect many or all known attack patterns, but they are of little use for as yet unknown attack methods [1]. Therefore, the introduction of mobile agents to provide computational security by constantly moving around the Internet and propagating rules is presented as a solution to misuse detection. This work presents a method of use of mobile agent mechanisms to add mobility features to the process of rule propagation. This approach presents significant advantages in terms of spreading rules rapidly, increasing scalability and providing fault tolerance.

1 Introduction

Significant progress has been made in the improvement of computer system security. On the other hand, attempted attacks and successful invasions involving an increasing number of computers have become frequent. Thus, security has become a key word for most companies worldwide. Intrusion detection is defined [2] as, "The problem of identifying individuals who are using a computer system without authorization and those who have legitimate access to the system but are abusing their privileges." Intrusion-detection systems aim at detecting attacks against computer systems and networks. There are two complementary trends in intrusion detection: misuse detection [3][4] uses knowledge accumulated about attacks and looks for evidence of the exploitation of these attacks, and anomaly detection [3] which builds a reference model of the usual behavior of the information system being monitored and looks for deviations from the observed usage. And many of intrusion detection systems use misuse detection method. According to the data of information security 21c [5], 7% of intrusion detection systems use anomaly detection, 43% use misuse detection, 17% use anomaly and misuse detection at the same time, and 33% use other methods.

The drawbacks of the misuse detection approaches are that they have the difficulty of gathering the required information on the known attacks and keeping it up to date

M.H. Shafazand and A M. Tjoa (Eds.): EurAsia-ICT 2002, LNCS 2510, pp. 842–849, 2002.

with new vulnerabilities and environments. In developing misuse intrusion detection, we propose a mobile agent-based rule propagation method. This method rapidly spreads the rules, which have information about new forms of attacks, to other's misuse detection systems.

Prior to designing a mobile agent-based rule propagation approach, we investigated patterns of the intrusion detection rules. The intrusion detection rules can be divided into two systems: A forward-chaining rule-based system and a backward-chaining rule-based system [6]. A forward-chaining rule-based system is data-driven: each fact asserted may satisfy the conditions under which new facts or conclusions are derived. Alternatively, backward-chaining rule-based systems employ the reverse strategy; starting from a proposed hypothesis they proceed to collect supportive evidence. Our proposed system uses the forward-chaining rule-based system. The person who finds the intrusion first creates an intrusion detection rule and the rule is propagated to other intrusion detection systems that are not intruded upon by the above intrusion using a mobile agent.

This paper mainly describes dynamic agent-based rule propagation for improving misuse detection efficiency. In section 2, some related works and a mobile agent description are shown. In section 3, the design of mobile agent-based misuse intrusion detection rule propagation model is clarified. Section 4, summarizes this paper.

2 Related Works

In this section, we introduce the overview of mobile agent and describe briefly the characteristics of the typical agent-based intrusion detection systems (IDSs) – EMERALD, AAFID and IA-NSM.

2.1 Overview of Mobile Agent

A mobile agent is a kind of independent program, which can migrate from one node to another node in a distributed network by itself. Compared to traditional distributed techniques such as the Client-Server model and Code On Demand model, a mobile agent has advantages as follows:

- ♦ can make proper use of existing resources so as to fulfill user's assignment
- ♦ can debase network traffic
- ♦ can balance network load
- ♦ support fault-tolerance
- ♦ support mobile user
- ♦ support customized services

A mobile agent has its own lifecycle including the states of create, halting, executing, service searching, arriving new host, migrating, returning to the original host and terminating. We use this agent technology in misuse intrusion detection system to spread the rule, which contains the intrusion information. A mobile agent consists of three parts: resource, function and rule information. Figure 1 shows the constitution of mobile agent [17].

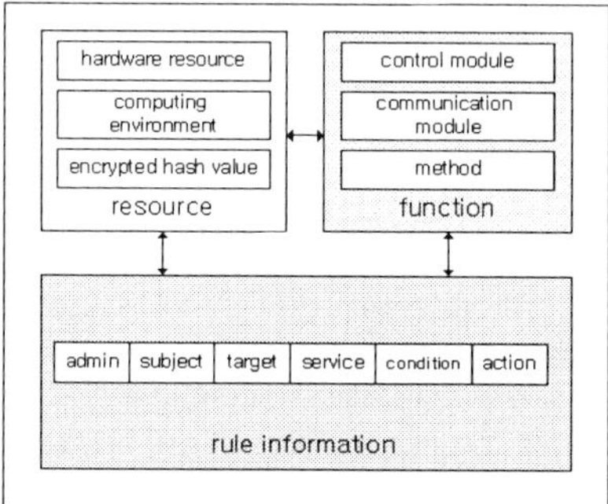

Fig. 1. Constitution of mobile Agent

The resource section contains the hardware resource, computing environment and encrypted hash value. Computing environment means the serialized state of the agent encoded in the method part. The encrypted hash value is used to check the integrity of the agent. The function part includes control module, communication module and method. The control module controls all functions of the mobile agent including authentication and authorization. The communication module provides a secure communication channel between the agent and IDS. Method module includes a concrete computing code and program [17]. Rule information in the agent is composed of six elements. The elements are administrator, subject, target, service, condition and action information. Admin indicates the operator who makes the intrusion detection rule. Subject describes the object which needs the security service. Target shows the object that needs to be protected from attacks. Service shows which service is controlled by the intrusion detection rule. Condition shows the condition that must be satisfied by an event. Action describes the action that is performed if the condition is satisfied. Mobile agents offer a new paradigm for distributed system environments, but there are two kinds of security issues specific to a mobile agent [17]: malicious agent – protection of the host against agent and protection of other agents, and malicious host – protection of the agent from the host and protection of the network.

We have suggested a security model that provides safety from a malicious agent and a malicious host. Encrypted hash values guarantee the integrity of the mobile agent and protect unauthorized modification of the mobile agent. A trusted third party authenticates a mobile agent. Thus this system is protected from the activities of malicious agents and malicious hosts.

2.2 The Characteristics of the Typical Agent-Based IDSs

Agent technology has been academically applied in a variety of fields, particularly in artificial intelligence, distributed systems, software engineering and electronic commerce. Generally, an agent can be defined as a software program capable of executing a complex task on behalf of a user [7]. The use of autonomous agents have been proposed by some other authors as a form of constructing non-monolithic intrusion detection systems [8, 9, 10]. The capacity of some autonomous agents to maintain specific information of their application domains, in this case security, confers great flexibility on these agents and hence, on the entire system. The characteristics of typical agent-based IDS is follows:

♦ EMERALD

The SRI(Stanford Research Institute) EMERALD(Event Monitoring Enabling Response to Autonomous Live Disturbance) project addresses the problems of network intrusion via TCP/IP data streams. Network surveillance monitors observe local area network traffic and submit analysis reports to an enterprise monitor, which correlates the reports. EMERALD appears to concentrate the intelligence in a central system and does not incorporate any agent technology [12][13][14].

♦ AAFID

The Autonomous Agents For Intrusion Detection (AAFID) Project is based on independent entities called autonomous agents that perform distributed data collection and analysis. Centralized analysis is done on a per-host and per-network basis by higher-level entities called transceivers and monitors. The architecture allows for computation to be performed (and thus, for intrusion detection to happen) at the point where enough information is available. This can be at the agent, transceiver or monitor level [15].

♦ IA-NSM

The Intelligent Agents for Network Security Management (IA-NSM) Project for Intrusion Detection using intelligent agent technology provides a flexible integration of a multi-agent system in a classical networked environment to enhance its protection level against inherent attacks [16].

As previously mentioned, a mobile agent technology is a growing area of research and new application development in telecommunications. Until now, there was no mobile agent-based rule propagation system in Intrusion Detection Systems. In this paper, the mobile agents are used as carrier and especially negotiate with other intrusion detection systems about intrusion detection rules. And also, we present the mobile agent-based detection rules propagation model for the misuse IDS in distributed network environments.

3 Misuse Intrusion Detection Rule Propagation System Design

This section describes the design details of the mobile agent-based rule propagation system and clarifies the objectives in designing a mobile agent-based rule propagation system to solve the problems of misuse detection. We set up the following goals:

- ♦ Propagate rules in dynamic and distributed environments
- ♦ Security of the mobile agent-based rule propagation system
- ♦ Authentication of the mobile agent
- ♦ Negotiation of the mobile agent and intrusion detection system

When the administrator managing the intrusion detection system finds new attacks, he enters information about the new attacks into the User Interface. Then the information is transmitted to the Rule Executor. The Rule Executor decides whether this rule is applied locally or globally. If the rule is specified locally, the rule is added to the Knowledge-base. Otherwise the Rule Executor generates a rule about new attacks using the Knowledge-base and sends the rule to the Agent Generator. The Agent Generator generates a mobile agent, which contains rule information. The

mobile agent migrates to the other IDSs through the Internet, and moves dynamically in distributed environments to propagate the intrusion detection rule.
Architecture of the mobile agent generation is shown in Fig. 2.

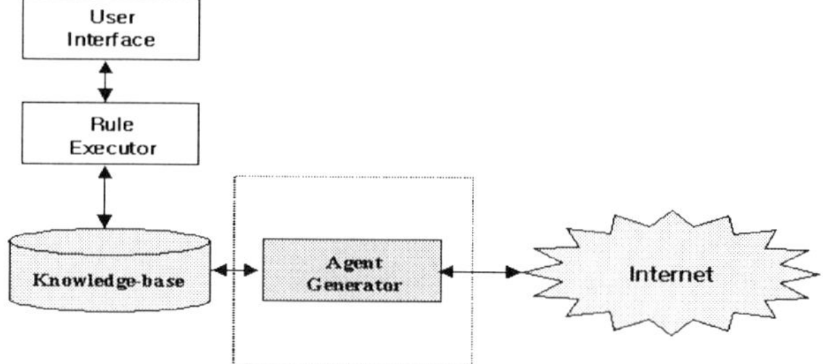

Fig. 2. Architecture of the mobile agent generation

The message types defined are summarized in the following table:

Table 1. Message types between a mobile agent and IDS

Message	Description
Request	request for information
Reply	reply that answers a request
Rule	rule up and download between mobile agent and IDS
Ack	Acknowledges a rule message
Status	informs status of mobile agent and IDS

The architecture of a mobile agent-based rule propagation system (MARS) is shown in Fig. 3.

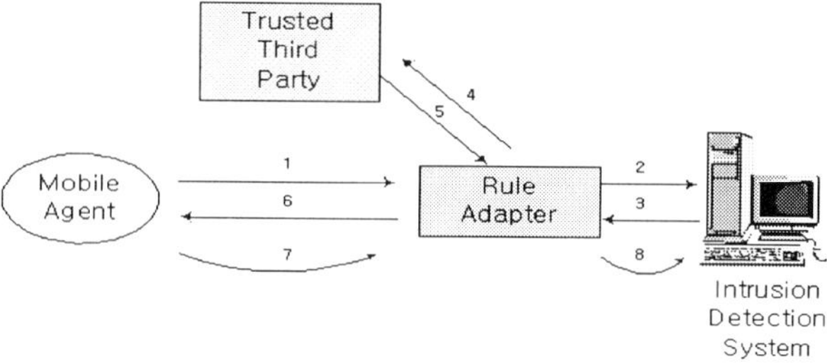

Fig. 3. Architecture of a mobile agent-based rule propagation system (MARS)

The Rule adapter is a function of software that communicates with mobile agents and makes decisions as to whether a rule is necessary to the intrusion detection system. A trusted third party is an authentication server that authenticates the mobile agent. When a mobile agent arrives at the rule adapter (1), The Rule adapter checks the policy conflict (2). If there is no conflict between the rules of the mobile agent and the rules of the intrusion detection system (3), then the rule adapter authenticate the mobile agent using the trusted thirty party (4,5,6). And the rule adapter checks the integrity of the rule using the hash value in the mobile agent (7). Finally, the rule having the information about the new attack is added to the rules of the intrusion detection system. When there is conflict between the rules of the agent and the intrusion detection rules, the rule adapter informs the mobile agent of the conflict. And the mobile agent returns to the original host to notify the conflict information. Therefore the rule adapter serves as a mediator between the mobile agent and the IDS to offer the capability of negotiation. Fig. 4 depicts the difference between a conventional Intrusion Detection System and MARS.

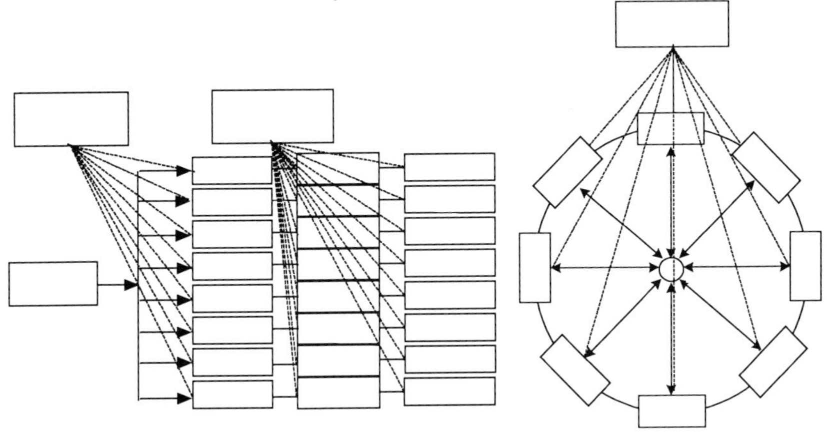

Fig. 4. The comparison of the performance of a conventional IDS and MARS

We listed the workload data of the above two systems like this.

- ♦ i : transmission workload between trusted third party and ids
- ♦ j : transmission workload between one ids and another ids
- ♦ k : transmission workload between mobile agent and ids
- ♦ l : transmission workload between administrator and ids
- ♦ m : the workload of analysis for intrusion detection rule
- ♦ n : the workload of rule creation for intrusion detection system
- ♦ p : authentication and authorization of mobile agent, packet and ids
- ♦ q : transmission workload of packet from the person who finds the intrusion to ids

Table 2. The workload data of conventional and MARS IDS

Action	Conventional IDS	MARS IDS
Transmission packet	Nq	Q
authentication & authorization	N(i+p)	2N(i+p)
Information analysis	N(l+m)	l+m
rule creation	N(l+n)	l+n
rule propagation		k+(N-1)j

N : Number of IDS

So, the total workload data of a conventional IDS is N(i+p+q+2l+m+n). As the number of IDS increase, the total workload increases (i+p+q+2l+m+n) times. The total workload data of MARS IDS is q+2N(i+p)+2l+m+n+k+(N-1)j. And q+2l+m+n+k is fixed. So as the number of IDS increases, the total workload data increases (2i+2p+j) times. The value of authentication & authorization for the MARS IDS is 2N(i+p) because of mutual authentication & authorization between the IDS and the mobile agent. Fig. 5 depicts the difference between the two systems with respect to total workload data and number of IDS.

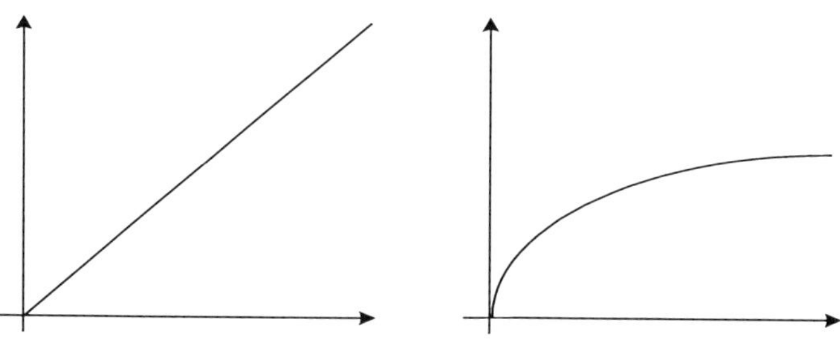

Fig. 5. Sum of workload data

As the number of IDS increases, the MARS IDS is more efficient than the conventional IDS. And mutual authentication between the IDS and the mobile agent guarantees the security of the rules.

4 Conclusion and Future Works

Misuse-based intrusion detection method is used more than any other intrusion detection methods. But this method has the difficulty of protecting a system from unknown attacks. We have presented a mobile agent-based rule propagation model, MARS, that solves this problem. The mobile agent can move around the IDSs and quickly propagate the rules. And negotiation between the mobile agent and IDS makes it possible to select, reject, upload and download the rules. And the comparison of the workload of conventional IDS with MARS shows that the cost of MARS is less than that of conventional IDS. We have also proposed a constitution of a mobile agent and a rule adapter to solve the security issues of mobile agent.

In future work we plan to improve and extend MARS in a variety of areas, including enhancing the ability of the rule adapter to alert or cooperate with other security systems such as Firewall and ESM (Enterprise Security Management), and enhancing the mobile agent to efficiently communicate with other security systems in analyzing, complementing, reinforcing, verifying, adjusting monitoring and responding.

References

1. R. G. Bace. Intrusion Detection, Macmillan Technical Publishing 2000.
2. B. Mukherjee, T. L. Heberlein and K. N. Levitt. Network Intrusion Detection. IEEE Network, May/June 1994.
3. R. Jagannathan, T. Lunt, D. Anderson, C. Dodd, F. Gilham, C. Jalali, H. Javitz, P. Neumann, A. Tamaru, and A. Valdes. System Design Document: Next-Generation Intrusion Detection Expert System (NIDES). Technical Report A007/A008/A009/ A011/A012/A014, SRI International, March 1993.
4. S. Kumar and E. Spafford. "A Pattern Matching Model for Misuse Intrusion Detection," Proceedings of the Seventeenth National Computer Security Conference, Oct. 1994.
5. Information Security 21c, the history and kinds of intrusion detection system, http://www.securityinformation.com, July 2001.
6. U. Lindqvist and P. A. Porras. Detecting computer and network misuse through the Production-Based Expert System Toolset (PBEST). In Proceedings of the 1999 Symposium on Security and Privacy, Oakland, California, May 1999.
7. H. S. Nwana. Software Agents: an Overview. Knowledge Engineering Review, 1996.
8. M. Crosbie and G. H. Spafford. Defending a Computer System using Autonomous Agents. Technical Report No. 95-022, Dept. of Comp. Sciences, Purdue University, March 1996.
9. M. Crosbie, and E. H. Spafford. "Active Defense of a Computer System using Autonomous Agents", Technical Report CSD-TR-95-008, Department of Computer Sciences, Purdue University, 1995.
10. Balasubramaniyan, Jai, J. O. Garcia-Fernandez, E. H. Spafford, and D. Zamboni. An Architecture for Intrusion Detection using Autonomous Agents. Department of Computer Sciences, Purdue University; Coast TR 98-05; 1998.
11. S. Stolfo, A. Prodromidis, S. Tselepsis, W. Lee, D. Fan and P. Chan. JAM: Java Agents for Metalearning over Distributed Databases. In Prod. Third Intl. Conf. Knowledge Discovery and Data Mining, 1997.
12. G. G. Helmer, J. S. K. Wong, V. Honavar, and L. Miller. Intelligent agents for intrusion detection. In Proceedings, IEEE Information Technology Conference, pages 121--124, Syracuse, NY, September 1998.
13. A. Porras and P. G. Neumann. EMERALD: Event Monitoring Enabling Responses to Anomalous Live Disturbances. In Proceedings of the National Information Systems Security Conference, Oct 1997.
14. A. Porras and A. Valdes. "Live Traffic Analysis of TCP/IP Gateways," in Networks and Distributed Systems Security Symposium, March 1998.
15. B. Jai, J. O. Garcia-Fernandez, E. H. Spafford, and D. Zamboni. An Architecture for Intrusion Detection using Autonomous Agents. Department of Computer Sciences, Purdue University; Coast TR 98-05; 1998.
16. K. Boudaoud, H. Labiod, R. Boutaba, Z. Guessoum. Network security management with intelligent agents. Network Operations and Management Symposium, 2000. NOMS 2000.
17. L. Qi, L. Yu. "Mobile agent-based security model for distributed system", Systems, Man, and Cybernetics, 2001 IEEE International Conference, 2001.

H-Colorings of Large Degree Graphs*

Josep Díaz[1], Jaroslav Nešetřil[2], Maria Serna[1], and Dimitrios M. Thilikos[1]

[1] Departament de Llenguatges i Sistemes Informàtics, Universitat Politècnica de Catalunya, Campus Nord – Mòdul C5, Desp. 211b, c/Jordi Girona Salgado, 1-3. E-08034, Barcelona, Spain {diaz,mjserna,sedthilk}@lsi.upc.es

[2] DIMATIA-ITI and Department of Applied Mathematics, Charles University, Malostranské nám. 25, 118 00 Prague, Czech Republic nesetril@kam.ms.mff.cuni.cz

Abstract. We consider the H-coloring problem on graphs with vertices of large degree. We prove that for H an odd cycle, the problem belongs to P. We also study the phase transition of the problem, for an infinite family of graphs of a given chromatic number, i.e. the threshold density value for which the problem changes from P to NP-complete. We extend the result for the case that the input graph has a logarithmic size of small degree vertices. As a corollary, we get a new result on the chromatic number; a new family of graphs, for which computing the chromatic number can be done in polynomial time.

1 Introduction

Given a graph G, with vertex set $V(G)$ and edge set $E(G)$, a *k-coloring* of G is a mapping of V to $\{1, \ldots, k\}$ such that no two vertices on the same edge receive the same color. Given graphs G and H, an *homomorphism* of G to H is an edge preserving mapping of $V(G)$ to $V(H)$.

For any fixed graph H, the *H-coloring problem* consists in deciding whether there is a homomorphism of a given input graph G into H.

As usual, let K_k denote a complete graph on k vertices. Then, the problem of deciding if there is a K_k -coloring of G is the problem of deciding if G is k-colorable. Thus the H-coloring problem naturally generalizes decision problems related to the chromatic number. Further examples of H-coloring problems include *circular chromatic number* [7], *T-colorings* and problems related to *channel assignments problems* [6].

For general graphs, the complexity of the H-coloring problem is well known, the problem is in P if H is bipartite, otherwise it is NP-complete [5]. In this paper, we are interested in studying the complexity of the H-coloring for certain

* This research was supported by the IST Program of the EU under contract number IST-99-14186 (ALCOM-FT) and the Spanish CICYT project TIC2000-1970-CE. The second author was also supported by the Czech Ministry of Education project LN00A056 and the forth author was further supported by the Ministry of Education and Culture of Spain (Resolución 31/7/00 – BOE 16/8/00).

M.H. Shafazand and A M. Tjoa (Eds.): EurAsia-ICT 2002, LNCS 2510, pp. 850–857, 2002.

types of dense input graphs. Given a constant $0 \le \alpha < 1$, a graph G is said to be *α-dense* if $\delta(G) \ge \alpha|G(V)|$, where $\delta(G)$ denotes the *minimum vertex degree* of G.

For any fixed graph H and a constant $0 \le \alpha < 1$, the *(H, α)-coloring problem* consists in: Given an *α-dense* graph G, determine whether there is an H-coloring of G. Notice that α is a parameter of the problem.

From the previous definition, it follows that for any two constants α, β such that $0 \le \alpha \le \beta < 1$, we get:

- If the (H, α)-coloring problem is in P, then the (H, β)-coloring problem is in P,
- if the (H, β)-coloring problem is NP-complete , then the (H, α)-coloring problem is NP-complete,

where the definitions of the complexity classes P and NP-complete are the well known ones. These properties motivates the following definition:

Definition 1. *Given a graph H the* complexity threshold *$c(H)$ is defined as the* $\inf\{\alpha \mid (H, \alpha)\text{-coloring} \in P\}$.

One immediate consequence of the previous definition is the fact that $c(H) = 0$ if and only if for every $0 < \alpha < 1$, the (H, α)- coloring problem is in P. This is the case when H is a bipartite graph, as in this case even the $(H, 0)$-coloring problem is in P.

For every graph H, the threshold $c(H)$ exists. A question of interest is whether $c(H)$ can be defined alternatively as

$$\sup\{\alpha \mid (H, \alpha)\text{-coloring} \in \text{NP-complete}\}.$$

A second question is whether there are graphs H for which

$$c(H) = \min\{\alpha \mid (H, \alpha)\text{-coloring} \in \text{P}\}.$$

All non-bipartite examples for which we can compute the exact threshold indicate that $c(H)$ is never attained. A positive answer to the first question follows from an affirmative solution to the following *dichotomy conjecture*:

> For every H and every α, the *(H, α)-coloring problem* is either NP-complete or P.

A partial affirmative answer to the previous conjecture was given by Edwards, for a particular case. He considered the graph family $\{K_k\}_{k \ge 3}$, where for each k, $c(K_k) = (k-3)/(k-2)$ (see Theorem 2.5 of [3]).

In Sections 2 and 3, we extend Edwards' results. We give some examples of particular families of graph for which the Dichotomy conjecture is true. We also determine the complexity threshold, for infinitely many graphs with a given chromatic number. In Section 4, we relax the notion of density, by allowing a small number of low degree vertices. It is interesting to notice taht we obtain the same results, as in the density case.

We use the standard notation from graph theory. Given two graphs G and G' the graph $G \oplus G'$ is formed by taking one copy of G and G', and join by edges all vertices of G with all vertices of G'. Given G and a $v \in V(G)$, $\mathcal{N}[v]$ denotes the set of neighbors of v. Given G and a subset $S \subseteq V(G)$, the *induced subgraph* $G[S]$ has as vertex set S and edge set $\{(u,v)|u,v \in S \wedge (u,v) \in E(G)\}$. Given a graph G and a subgraph G' of G, we say that $v \in V(G)$ is *completely joined* to G' if for all $u \in V(G')$, we have $(u,v) \in E(G)$. The *chromatic number* $\chi(G)$ is the minimum number of colors needed to color G. The *clique number* $\omega(G)$ is the maximum k such that K_k is a subgraph of G. A cycle on $2k+1$-vertices is denoted by C_{2k+1}, and $\overline{C}_{2k+1}$ is the complement graph of C_{2k+1}.

Through the paper, we shall work under the plausible hypothesis that $\mathsf{P} \neq \mathsf{NP}$.

2 Exact Thresholds

We begin the section, stating a technical lemma due to Edwards, which will be needed later.

Lemma 1. *[3] For any integer $k \geq 3$, let α be a fixed rational such that $(k-3)/(k-2) < \alpha < 1$, and let G be an α-dense graph, with $|V(G)| = n$. Then, there exits a $U \subseteq V(G)$, and a set $\mathcal{T}$ of $(k-2)$-cliques in G (not necessarily disjoint), such that:*

- *$G[U]$ is the union of all cliques in $\mathcal{T}$.*
- *$|\mathcal{T}| = O(\log n)$.*
- *Every vertex in $V(G) \setminus U$ is completely joined to at least one clique in $\mathcal{T}$.*

Furthermore, U can be computed in polynomial time.

Using the previous lemma, we prove the main result in this section, we characterize an infinite family $\{H_i\}$ of graphs, for which $c(H_i) \leq (k-3)/(k-2)$.

Theorem 1. *For any fixed integer $k \geq 3$, assume that H satisfy $\chi(H) = k$, and that any subgraph of H isomorphic to K_{k-2} is contained in at most two subgraphs of H isomorphic to K_{k-1}. Then for $\frac{k-3}{k-2} \leq \alpha \leq 1$, the (H,α)-coloring problem belongs to P.*

Proof. Let G be an α-dense instance to the (H,α)-coloring problem, and let U be as in the statement of the previous lemma. Consider any fixed homomorphism f of $G[U]$ to H. By lemma 1, any given $v \in V(G) \setminus U$ is completely joined with at least one $(k-2)$-clique. Therefore, by the hypothesis on H, if $F(v)$ is the set of possible vertices of H that can extend f to the whole $U \cup \{v\}$, then $|F(v)| \leq 2$.

First, we prove that in polynomial time, we can decide if f can be extended to the whole G. Let $V(G) \setminus U = \{v_1, \ldots, v_m\}$, and $V(H) = \{1, \ldots, r\}$. Define the instance of the 2-SAT problem as follows, the set of variables is $x_{i,j}$, $1 \leq i \leq m, 1 \leq j \leq r$ each of them indicating whether $f(i) = j$, and the clauses are:

1. $\{x_{i,j} | j \in F(i)\}$, for $1 \le i \le m$,
2. $\{\overline{x}_{i,j}, \overline{x}_{i,l}\}$ for $1 \le i \le m$, $1 \le j < l \le r$,
3. $\{\overline{x}_{i,j}, \overline{x}_{h,l}\}$ if $(v_i, v_h) \in E(G)$, and $(i,l) \notin E(H)$.

It is easy to see that the homomorphism f on $G[U]$ can be extended to G if and only if this instance of 2-SAT has a satisfying assignment. As 2-SAT belongs to the class P, then the problem of deciding the extension of f to G is also in P. Hence, to determine if there exists an H-coloring f of G, we try all possible homomorphisms of $G[U]$ to H, until obtaining a valid f (otherwise, the answer is NO). As the total number of possibilities for the homomorphism f is $r^{|U|} \le r^{O(\log n)}$, then exhaustive search together with the 2-SAT algorithm can decide, in polynomial time, if G is H-colorable.

As a corollary we obtain that the exact threshold for odd cycles is 0.

Corollary 1. *For every* $0 < \alpha < 1$ *and for every* $k \ge 1$*, the* (C_{2k+1}, α)*-coloring problem is in the class* P*.*

Note that also for even cycles and all bipartite graphs H, the threshold is 0, and these are the only known case where the threshold is attained.

Another result we can obtain from the previous theorem gives us an exact threshold for infinitely many graphs of a given chromatic number:

Corollary 2. $c(C_{2k+1} \oplus K_{k-3}) = (k-3)/(k-2)$ *for every* $k \ge 3$.

Proof. By the the previous theorem, for $\alpha(k-3)/(k-2)$ the $(C_{2k+1} \oplus K_{k-3}, \alpha)$-coloring is in P. To finish the proof that $(k-3)/(k-2)$ is a phase transition for the $(C_{2k+1} \oplus K_{k-3}, \alpha)$-coloring, we need to prove completeness. Let us consider the following reduction:
For any fixed constant k, given a graph G construct a graph G' in the following way. $V(G')$ consist of $k-2$ copies of the vertices $V(G)$. To form $E(G')$, join by an edge all vertices in different copies, also add the edges $E(G)$ to one of the copies. Then, G is C_{2k+1}-colorable if and only if G' is $C_{2k+1} \oplus K_{k-3}$-colorable. Moreover, $|V(G')| = (k-2)|V(G)|$, and every vertex in $V(G')$ has degree at least $\frac{k-3}{k-2}|V(G')|$. Therefore, G' is $\frac{k-3}{k-2}$-dense.

3 Threshold Bounds

We have seen that there are graphs for which their threshold is 0, e.g. all cycles. Next, we prove that the threshold of a graph can never be 1,

Proposition 1. *For every graph* H, $c(H) < 1$.

Proof. Let H be a fixed graph, and let $\omega = \omega(H)$. It follows from the classical result of Turán (see e.g. [2], page 108), that a graph with more than $|G(V)|^2/2(1-1/\omega)$ edges must contain a $K_{\omega+1}$, and therefore, the graph can not be homomorphic to H. It follows that for any graph H, $c(H) \le 1 - 1/\omega < 1$.

In the following result we present a family $\{H_k\}_{k>1}$, for which we can compute sharp bounds for the complexity threshold of the (H,α)-coloring problem,

Proposition 2. *For all $k > 1$, the $(K_3 \oplus \overline{C}_{2k+1}, \alpha)$-coloring problem is:*

- *NP-complete, if $0 \leq \alpha \leq 1 - \frac{3}{3k+1}$,*
- *P, if $1 > \alpha > 1 - \frac{k}{k+1}$.*

Proof. Let us consider the following reduction:
For any fixed constant k, given a graph $G = (V,E)$ such that the number of edges is a multiple of k, construct a graph $G' = (V',E')$ in the following way: Take a $\overline{C}_{2k+1}$ and replace each vertex of it by $|V(G)|/k$ independent vertices. Denote this graph by $\overline{F}_{2k+1}$, then $G' = \overline{F}_{2k+1} \oplus G$. Notice $|V'| = (\frac{2k+1}{k}+1)|V|$. Then G is 3-colorable if and only if G' is $K_3 \oplus \overline{C}_{2k+1}$-colorable. Furthermore, every vertex in G' has degree at least $(\frac{2k-2}{k}+1)|V|$, which implies that α for G is less or equal to $\frac{((2k-2)/k)+1}{((2k+1)/k)+1}$.

To prove the P part, recall that for any $k > 1$ $\omega(\overline{C}_{2k+1}) = k$. Furthermore, any $(k-1)$-clique in $\overline{C}_{2k+1}$ is contained in at most two cliques of size k. Therefore, any $(k+2)$-clique in $K_3 \oplus \overline{C}_{2k+1}$ is contained in at most two cliques of size $k+3$. Using Theorem 1, we get the second part of the statement.

The second statement of Proposition 2 gives us better bounds that the one produced by Proposition 1, as $\omega(K_3 \oplus \overline{C}_{2k+1}) = k+3$, so the upper bound produced by Proposition 1 is $1 - \frac{1}{k+3}$, which is larger than $1 - \frac{k}{k+1}$.

It is still open to decide if for values $1 - \frac{3}{3k+1} < \alpha \leq 1 - \frac{1}{k+1}$, the problem is in P or NP-complete, but notice that for $k = 6$, both values are above 0.8, and differ in 0.1. This means that, for values of $k \geq 6$, the decision problem is NP-complete for most of the dense graphs. The value of the bounds seems to indicate that, the complexity threshold for the given graphs coincides with the threshold for K_{k+4}.

The next result gives a necessary condition to guarantee a complexity threshold of at least 1/2.

Proposition 3. *For any graph H, such that for some $x \in V(H)$ the graph $H[\mathcal{N}[x]]$ has $\chi(H[\mathcal{N}[x]]) > 2$, the $(H, \frac{1}{2})$-coloring problem is NP-complete.*

Proof. Let H be a graph as in the hypothesis of the Proposition. For every $x \in V(H)$, let $H_x = H[\mathcal{N}[x]]$. Consider the graph H' obtained as the disjoint union of the graphs H_x such that $\chi(H_x) > 2$. As each H_x is not bipartite, then H' is not bipartite and hence, the H'-coloring problem is NP-complete.

Given a connected graph G', define G in the following way; $V(G) = V(G') \times \{0,1\}$, and the edges

$$E(G) = \{((v,0),(u,0)) | (v,u) \in E(G')\} \cup \{((v,0),(u,1)) | \forall u,v \in V(G')\}.$$

If f is an H'-coloring of G', there is a $x \in H$ such that f is a H_x-coloring of G'. The mapping $g : G \to H$ defined as $g(v,0) = f(v)$ and $g(v,1) = x$ is an

H-coloring of G. On the other hand, if g is an H-coloring of G, then for any $v \in V(G')$, the mapping $f(v) = g(v, 0)$ is an H'-coloring of G'. So the above construction is a polynomial-time reduction from the H'-coloring problem to the $(H, \frac{1}{2})$-coloring problem.

These investigations lead to the following interesting problem:

Characterize graphs H with $c(H) = 0$.

Presently we do not know any 3-chromatic graph H for which the (H, α)-coloring problem is NP-complete for some $\alpha > 0$. A candidate for such a graph H is the particular subdivision of K_4 where we subdivide each edge of a triangle by 2 points. The resulting graph has 10 vertices and it is not C_5-colorable. By the general theorem in [5], H-coloring is NP-complete, and in this particular case it may be seen easily by the reduction from 3NAESAT given in Figure 1. However this reduction and the general reduction in [5] produce graphs with constant minimum degree and thus it does not yield NP-completeness of the (H, α)-coloring problem for any $\alpha > 0$. Perhaps $c(H) = 0$ holds for any 3-chromatic graph H.

4 Almost α-Dense Graphs

In this section, we consider the complexity of the H-coloring problem, for input graphs with vertices of high degree, except for a subset of logarithmic size.

We say that a graph G is *almost α-dense* if there exists a partition $\{V_1, V_2\}$ of $V(G)$, with $|V_1| = O(\log |V(G)|)$, and such that for every $v \in V_2$, the degree $d_G(v) \geq \alpha |V(G)|$.

We will consider now the H-coloring problem on almost dense graphs: For any fixed graph H, and a constant $0 \leq \alpha < 1$, the *almost-(H, α)-coloring problem* consists in: Given an almost–α-dense graph G, decide whether there is an homomorphism of G into H. In the same way as it has been done in Definition 1, given a graph H, define its *complexity threshold* $\tilde{c}(H)$ as the $\inf\{\alpha \mid \text{almost}(H, \alpha)\text{-coloring} \in \mathsf{P}\}$.

Theorem 2. *Let $k \geq 3$, and $0 < \epsilon < 1$, assume that the graph H satisfy $\chi(H) = k$, and let any subgraph of H isomorphic to K_{k-2}, be contained in at most two subgraphs of H isomorphic to K_{k-1}. Then, for any $\alpha > (k-3)/(k-2) + \epsilon$, the almost (H, α)-coloring problem belongs to* P.

Proof. Let G be an almost α-dense graph, and let $S = \{u \in V(G) \mid d(u) \geq \alpha n\}$, and let $R = V(G) - U$. We claim that for $n \geq n_0 = n_0(\epsilon)$ the graph $G' = G[U]$ is $\alpha - \epsilon$-dense. This follows from the fact that assuming that $|R| = r$, for any $u \in S$ we have $d_{G'}(u) \geq \alpha n - p$, given ϵ for n big açenough it holds that $\alpha n - p \geq (\alpha - \epsilon)(n - p)$ as $p = O(\log n)$. Now, using Lemma 1 we can construct a decomposition (U, P) of S. Recall that $\mathcal{T}$ is a set of $(k-2)$-cliques in G' with vertex set U. Let f be a homomorphism of $G[U \cup R]$ to H. For any valid H-coloring f of $G[U \cup R]$, and each $v \in P$, there exists a set $C(v)$ of possible

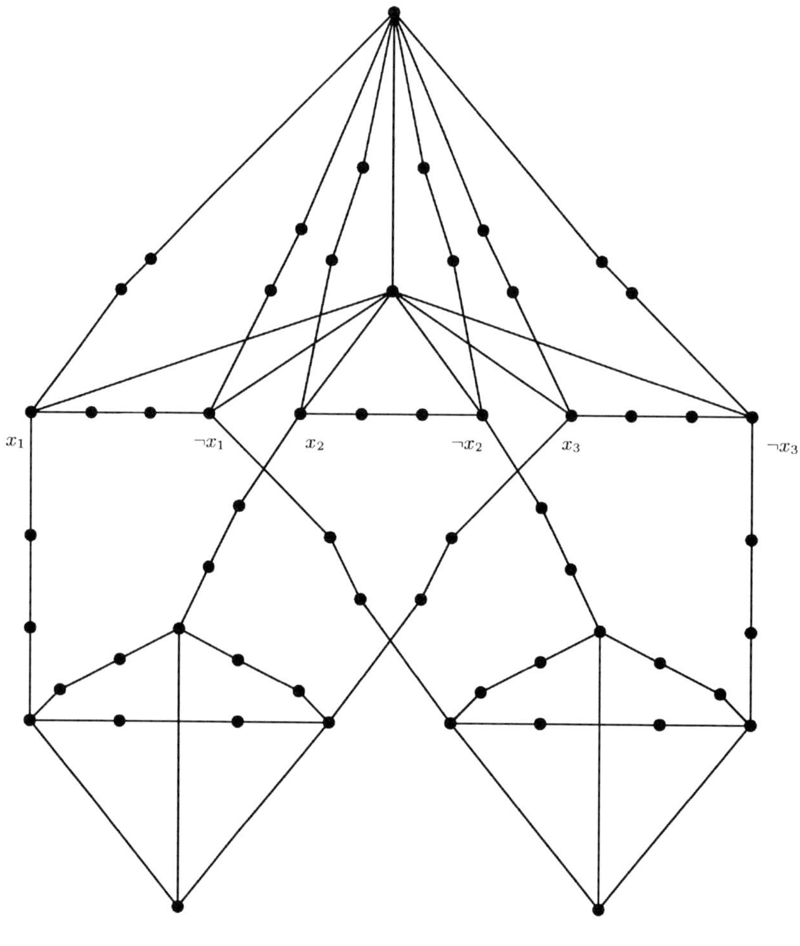

Fig. 1. A sketch of the reduction.

extensions of f to the whole G. then $|C(v)| \leq 2$ as, from Lemma 1 each vertex in P is completely joined to one or more of the cliques in $\mathcal{T}$, which have already been assigned $k-2$ *colors* in H. In the same manner as in the proof of Theorem 1, we can define a polynomial reduction between the problem of extending a H-coloring of $G[U \cup L]$ to a H-coloring of G, and 2-SAT, so it is possible to decide whether an H-coloring of $G[U \cup L]$, can be extended to a H-coloring of G. Since $|U \cup L| = O(\log n)$, exhaustive search plus the reduction yield a polynomial time algorithm to decide the almost (H, α)-colorability of the given G.

As Corollary to the previous theorem, we state the particular case, when $H = K_k$.

Corollary 3. *For any* $\alpha > (k-3)/(k-2)$ *the problem of deciding whether an almost* α*-dense graph* G *is* k*-colorable, is polynomially solvable.*

Corollary 4. *For every* $0 < \alpha < 1$ *and for every* $k \geq 1$*, the almost* (C_{2k+1}, α)*-coloring problem is in the class P.*

Corollary 5. *For any* $\alpha > (k-3)/(k-2)$*, the almost* $(C_{2k+1} \oplus K_{k-3}, \alpha)$*-coloring problem is in the class P*

Notice that the previous corollaries, state that for graphs such that we can compute the phase transition for density, the same phase transition holds for almost density. However, the time bounds might be too large, as the minimum value of n could be high, as ϵ approaches zero.

Acknowledgement. We thank K. Edwards for providing the authors with a copy of his work in a record time and to J. Kratochvíl and three anonymous referees, for comments which improved the quality of this paper.

References

1. R. J. Anderson and E. W. Mayr. Approximating P-complete problems. Technical report, Stanford University, 1986.
2. B. Bollobás: Modern Graph Theory, Springer 1998.
3. K. Edwards. The complexity of colouring problems on dense graphs. *Theoretical Computer Science*, 43:337–343, 1986.
4. P. Erdös. On the structure of linear graphs. *Israel Journal of Mathematics*, 1:156–160, 1963.
5. P. Hell and J. Nešetřil. On the complexity of H-coloring. *Journal of Combinatorial Theory, series B*, 48:92–110, 1990.
6. F. S. Roberts: T-colorings of Graphs Recent Results and Open Problems, Discrete Math. 93(1991),229-245
7. X. Zhu. Circular chromatic number: A survey. In: Combinatorics, Graph Theory, Algorithms and Applications - DIMATIA Survey Collection (M.Fiedler, J. Kratochvíl, J. Nešetřil, eds.), North Holland (2000)

Hyper-Star Graph: A New Interconnection Network Improving the Network Cost of the Hypercube

Hyeong-Ok Lee[1], Jong-Seok Kim[2], Eunseuk Oh[3], and Hyeong-Seok Lim[4]

[1] Department of Computer Education, Sunchon National University, 315 Maegok-dong, Sunchon, Chonnam 540-742, Korea
oklee@sunchon.ac.kr

[2] Department of Computer Science, Sunchon National University, 315 Maegok-dong, Sunchon, Chonnam 540-742, Korea
rockhee@sunchon.ac.kr

[3] Department of Computer Science, Texas A&M University, College Station, TX 77843-3112, USA
eunseuko@cs.tamu.edu

[4] Department of Computer Science, Chonnam National University, 300 Yongbong-dong, Buk-gu, Kwangju 500-757, Korea
hslim@chonnam.chonnam.ac.kr

Abstract. In this paper, we introduce the *hyper-star graph* $HS(n,k)$ as a new interconnection network, and discuss its properties such as fault-tolerance, scalability, isomorphism, routing algorithm, and diameter. A hyper-star graph has merits when degree $\times$ diameter is used as a desirable quality measure of an interconnection network because it has a small degree and diameter. We also introduce a variation of $HS(2k,k)$, *folded hyper-star graphs* $FHS(2k,k)$ to further improve the cost degree $\times$ diameter of a hypercube: when $FHS(2k,k)$ and an n-dimensional hypercube have the same number of nodes, degree $\times$ diameter of $FHS(2k,k)$ is less than $(k+1)(\lceil logK \rceil + 1)$ whereas a hypercube is n^2, where $K = \binom{2k}{k}$. It shows that $FHS(2k,k)$ is superior to a hypercube and its variations in terms of the cost, degree $\times$ diameter.

1 Introduction

An interconnection network can be represented as an undirected graph where a processor is represented as a node, and a communication channel between processors as an edge between corresponding nodes. Tree, mesh, hypercube, and star graph are popular interconnection networks. Measures of the desirable properties for interconnection networks include degree, connectivity, scalability, diameter, fault tolerance, and symmetry [3]. There is a trade-off between degree, related to hardware cost, and diameter, related to transmission time of messages. Thus, the network cost, defined by degree $\times$ diameter, is introduced as a measure for interconnection networks [4,5,7].

M.H. Shafazand and A M. Tjoa (Eds.): EurAsia-ICT 2002, LNCS 2510, pp. 858–865, 2002.

An n-dimensional hypercube Q_n consists of 2^n nodes and $n2^{n-1}$ edges. The degree and diameter of Q_n are n, and the network cost is n^2. The attractive properties of a hypercube is node and edge symmetry, a simple recursive structure, and an efficient embedding into other interconnection networks such as ring, tree, pyramid, and mesh. However, as the dimension of a hypercube increases, the degree of the hypercube also increases [5]. To improve such a shortcoming of hypercube, its variations such as butterfly, cube-connected-cycle(CCC), de Brujin, and shuffle exchange were introduced. Comparing degree, the hypercube has a rather large diameter and average distance between nodes because the hypercube does not utilize edges efficiently. Multiply-twisted cube [4], twisted n-cube [1], folded hypercube(FH) [6], and hierarchical cubic network(HCN) [8] were introduced to better resolve this shortcoming.

An n-dimensional star graph S_n consists of $n!$ nodes and $\frac{(n-1)n!}{2}$ edges. S_n has node and edge symmetry, a smaller degree and diameter than the hypercube [2]. However, embedding other networks into S_n is rather complicated and many nodes are not equally loaded. Expanding S_n to S_{n+1} is also impractical because a large number of nodes must be added. In this paper, we introduce a new interconnection network to improve the shortcomings of hypercube and star graphs, and demonstrate its desirable properties as an interconnection network. The suggested network has attractive properties such as better scalability, a simple routing algorithm, maximum fault-tolerance, and lower network cost than the hypercube and its variations. We also introduce the *folded hyper-star graph* to further improve the network cost.

2 Topological Properties of Hyper-Star Graph

Let $HS(n,k) = (V(HS(n,k)), E(HS(n,k)))$ be a *hyper-star graph*, where a node is represented by the string of n bits $s_1 s_2 \cdots s_i \cdots s_n, s_i \in \{0,1\}$, and $k(n > k)$ bits are "1". Two nodes $u = s_1 s_2 \cdots s_i \cdots s_n$ and $v = s_i s_2 \cdots s_1 \cdots s_n$ are connected by an edge $(u,v) \in E(HS(n,k))$ if and only if s_i is a complement of s_1, and the bit string of v is obtained by exchanging s_1 and s_i, $2 \leq i \leq n$, in the bit string of u. That is, $HS(n,k)$ is defined as

$$
\begin{aligned}
V(HS(n,k)) &\stackrel{def}{=} s_1 s_2 \cdots s_i \cdots s_n, s_i \in \{0,1\}, \text{ where } |s_i = "1"| = k, \\
E(HS(n,k)) &\stackrel{def}{=} (u,v), \text{ where } u = s_1 s_2 \cdots s_i \cdots s_n, v = s_i s_2 \cdots s_1 \cdots s_n, \\
&\qquad s_1 = \bar{s_i}
\end{aligned}
$$

From the above definition, since nodes of $HS(n,k)$ can be represented by k-combinations of the set of n bit strings such that the number of "1" is k and the number of "0" is $n-k$, the number of nodes in $HS(n,k)$, $|V(HS(n,k))|$ is $\binom{n}{k} = \frac{n!}{k!(n-k)!}$. The degree of $HS(n,k)$ is $n-k$ if the left-most symbol in a node is "1" and k if the left-most symbol is "0". Thus, a hyper-star graph $HS(n,k)$ is a regular interconnection network when the length of a bit string is $2n$ and $k = n$, otherwise, it is irregular. Nodes of the hyper-star graph $HS(n,k)$ represented by a bit string $s_1 s_2 \cdots s_i \cdots s_n$ can be divided by nodes whose symbol s_1 is "0" or

"1". Thus, $HS(n,k)$ is a bipartite graph and has only an even length of cycles if it contains cycles. An edge connecting two nodes $u = s_1 s_2 \cdots s_i \cdots s_n$ and $v = s_i s_2 \cdots s_1 \cdots s_n$ is called $i - dimensionaledge$, where $s_1 = \bar{s_i}$.

2.1 Scalability

A hyper-star graph $HS(n,k)$ can be constructed as a regular network from irregular networks, $HS(n-1,k-1)$ and $HS(n-1,k), (n > 2, k = \frac{n}{2})$ by adding edges. Its construction is as follows. Suppose $s_1 s_2 ... s_{n-1}$ to be a bit string of a node in the hyper-star graph $HS(n-1,k-1)$. Then the number of nodes whose symbol s_1 is "1" is $\binom{n-2}{k-2}$ and their degrees are $n-k$. In the case s_1 ="0", the number of such nodes is $\binom{n-2}{k-1}$ and degrees are $k-1$. Similarly, suppose $s'_1 s'_2 \cdots s'_{n-1}$ to be a bit string of a node in $HS(n-1,k)$, then the number of nodes whose symbol s'_1 is "1" is $\binom{n-2}{k-1}$ and their degrees are $n-k-1$. In the case s'_1="0", the number of such nodes is $\binom{n-2}{k}$ and degrees are k. For any two nodes $u = s_1 \cdots s_{n-1}$ in the $HS(n-1,k-1)$ and $v = s'_1 \cdots s'_{n-1}$ in the $HS(n-1,k)$, u and v are connected if $s_1 = \bar{s'_1}$ and $s_i = s'_i, 2 \le i \le n-1$, and new bits s_n and s'_n are added to u and v, respectively.

Further, for the n bit string of the $HS(n,k), s_n$="1" and s'_n="0" since the number of symbol "1" is k, a new added bit to $HS(n-1,k-1)$ must be "1", and a new added bit to $HS(n-1,k)$ must be "0". The number of nodes whose symbol s_1 is "0" in $HS(n-1,k-1)$ and the number of nodes whose symbol s'_1 is "1" in $HS(n-1,k)$ are $\binom{n-2}{k-1}$. Thus, the degree of connected nodes between $HS(n-1,k-1)$ and the $HS(n-1,k)$ is k. Therefore, the graph generated by connecting $HS(n-1,k-1)$ and $HS(n-1,k)$ is a regular graph of degree k and the number of nodes $\binom{n-1}{k-1} + \binom{n-1}{k} = \binom{n}{k}$, which results in $HS(n,k)$. Fig. 1 shows the $HS(6,3)$ generated by connecting $HS(5,2)$ and $HS(5,3)$.

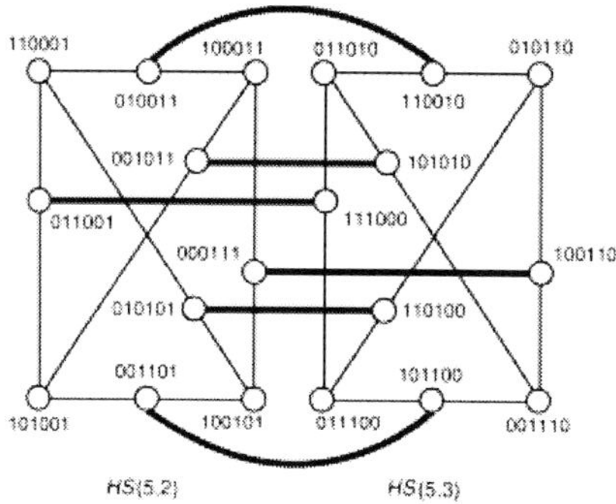

Fig. 1. The $HS(6,3)$ generated by connecting $HS(5,2)$ and $HS(5,3)$

Theorem 1. *A hyper-star graph $HS(n,k)$ is isomorphic to a hyper-star graph $HS(n,n-k)$.*

Since $HS(n,k)$ is isomorphic to $HS(n,n-k)$, throughout the paper, we assume that k $\leq \frac{n}{2}$ unless explicitly mansion.

2.2 Connectivity

In interconnection networks, node connectivity or edge connectivity is an important measurement to evaluate the network as still functional. For a graph G, let its node connectivity, edge connectivity, and degree be $\kappa(G)$,$\lambda(G)$, and $\delta(G)$, respectively. Then, it has been known that $\kappa(G) \leq \lambda(G) \leq \delta(G)$ [1, 2]. Since the degree of the $HS(n,k)$ is decided by k, we show that the hyper-star graph $HS(2n,n)$ is maximally fault tolerant.

Theorem 2. *For a hyper-star graph* $HS(2n,n), k(HS(2n,n)) = \lambda(HS(2n,n)) = n$.

Proof. Let P be a set of $n-1$ nodes that will be removed from the $HS(2n,n)$.
Case 1 The first bit string s_1 of all nodes($\in P$) is "0"

If there is a node u in $HS(2n,n)$ that has all $n-1$ neighbors in P , there exists at least one non-faulty neighbor of $u$whose s_1 is "0". If some neighbors of u are not in P, it is easy to see that $HS(2n,n)$ is connected. Similarly, we can show the case where the first bit string s_1 of all nodes $\in P$ is "1".
Case 2 The first bit string s 1 of some nodes($\in P$) is "1"

The degrees of all nodes in $HS(2n,n)$ are n and n nodes incident on a node u are connected to u if their first bit of bit strings are the complement of the first bit of u. Thus, if $n-1$ nodes in P are consisting of nodes having their first bit as "1" and nodes having their first bit as "0", $HS(2n,n)$ is always connected. From the above discussion, we can see that the $HS(2n,n)$ is connected even if we remove at most $n-1$ nodes from it. It shows that $k(HS(2n,n)) \geq n$. Also, since $HS(2n,n)$ is a regular graph of degree $n, k(HS(2n,n)) \leq n$. Therefore, $k(HS(2n,n)) = n$. Similarly, we can show $\lambda(HS(2n,n)) = n$ □

3 Routing Algorithm and Diameter

In this section, we present a routing algorithm and derive the diameter of the $HS(2n,n)$.

3.1 Routing Algorithm

Let a node S be the source node and a node D be the destination node in a hyper-star graph $HS(2n,n)$, then the shortest path between S and D can be regarded as the process of changing the bit string of S to that of D. A routing scheme to construct the shortest path is as follows.

For two nodes $S = s_1 s_2 \cdots s_n$ and $D = d_1 d_2 \cdots d_n$, denote by $R = r_1 r_2 \cdots r_n$ the bit string obtained by applying *Exclusive-OR* operation between S and D. We use $\oplus$ to represent the Exclusive-OR operator. Each i-dimensional edge, which leads to two identical ith bits from bits where r_i ="1", belongs to a shortest path. A path is continuously extended, provided there is a bit "1" in R. Fig. 2 shows the algorithm that constructs a shortest path between any two nodes. Let P_s be the set of nodes on a shortest path from S to D. Initially, $P_s = \{S\}$.

```
Shortest_path_1(S, D, P_S)
begin
    if(S = D) then
        return P_s;
    obtain R = r_1 r_2 ··· r_i ··· r_n , where r_i = s_i ⊕ d_i, 1 ≤ i ≤ n;
    let p = {i | r_i = 1, 2 ≤ i ≤ n} and q = {j | (S, S') ∈ E(HS(n, k)), 2 ≤ j ≤ n},
        where jth bits in S and S' complement;
    if (| p |> 0) then
       find a node S' such that (S, S') is an i-dimensional edge, where
            i = min{p ∩ q};
        P_s = P_s ∪ {S'};
        S = S';
    call Shortest_path_1(S, D, P_S);
end
```

Fig. 2. The shortest path routing algorithm of the $HS(n, k)$

In each iteration satisfying the condition r_i ="1", one bit from the source node S is fixed to be identical to its corresponding bit in the destination node D, and the number of different bits between S and D is reduced exactly by 1. Thus, it is easy to see that the algorithm **Shortest_path_1** is optimal in terms of the length of a path.

3.2 Diameter

Let $dia(HS(n, k))$ be the diameter of a hyper-star graph $HS(n, k)$. Then, $dia(HS(n, k))$ is given in the following theorem.

Theorem 3. *The diameter of a hyper-star graph $HS(n, k)$ is $n - 1$.*

Proof. The diameter of a hyper-star graph $dia(HS(n, k))$ is the same as the maximum number of indices i, $2 \le i \le n$, such that r_i is "1" in the shortest path routing algorithm. Based on the condition of κ, we can have different diameters as follows.

$$dia(HS(n,k)) = \begin{cases} 2 & \text{if } k = 1 \\ n-1 & \text{if } k = \frac{n}{2} \\ 2k & \text{if } k < \frac{n}{2} \end{cases}$$

All hyper-star graph $HS(n, k)$ whose k is 1 have star structure. That is, all other nodes are incident on one node. Thus, the diameter of $HS(n, k)$ is 2. Suppose for two nodes u and v in a regular hyper-star graph such that $k = \frac{n}{2}$, we apply Exclusive-OR operations on their bit stings. Then, the maximum number of bits such that r_i is "1" is n. However, after going through the other $n-1$ nodes from u, bit strings of u and v become identical. Thus, the diameter of $HS(n, \frac{n}{2})$ is $n-1$. For an irregular hyper-star graph such that $k < \frac{n}{2}$, the maximum number of indices i when r_i is "1" is $2k$. Since we assume that $k < \frac{n}{2}$, it is less than diameter of $n-1$ obtained by Case 2. From the above discussion, the diameter of a hyper-star graph $HS(n, k)$ is $n-1$. □

The above theorem shows that Hyper-Star graphs improve diameter of hypercube networks. We will try to further improve it in the following section.

4 Folded Hyper-Star Graph

In this section, we introduce a *folded hyper-star graph* constructed by adding an edge to each node of a hyper-star graph to improve diameter. We first provide the definition of a folded hyper-star graph $FHS(2n, n)$ and analyze its properties. Since a hyper-star graph $HS(n, k)$ has different degrees depending on k, to simplify the explanation, we consider a regular hyper-star graph with $n = 2k$ and provide the method to add edges to it. After that, we introduce a simple and effcient routing algorithm in $FHS(2n, n)$.

4.1 Definition and Topological Properties

A folded hyper-star graph $FHS(2n, n)$ is a graph where edges are added to connect any two nodes whose bit strings are complements in the hyper-star graph $HS(2n, n)$. That is, a folded hyper-star graph $FHS(2n, n)$ is defined as

$$\begin{aligned} V(FHS(2n,n)) &= V(HS(2n,n)) \\ E(FHS(2n,n)) &= E(HS(2n,n)) \cup e \text{ ,where } e = \{(u,v) \mid u = s_1 s_2 \cdots s_{2n}, \\ &\qquad v = \bar{s}_1 \bar{s}_2 \ldots \bar{s}_{2n}\} \end{aligned}$$

An edge in the set of edges e is called a $c-edge$. For two nodes u and v, denote a c-edge connecting them by $(u, v)_c$. Fig. 3 shows a folded hyper-star graph $FHS(4, 2)$.

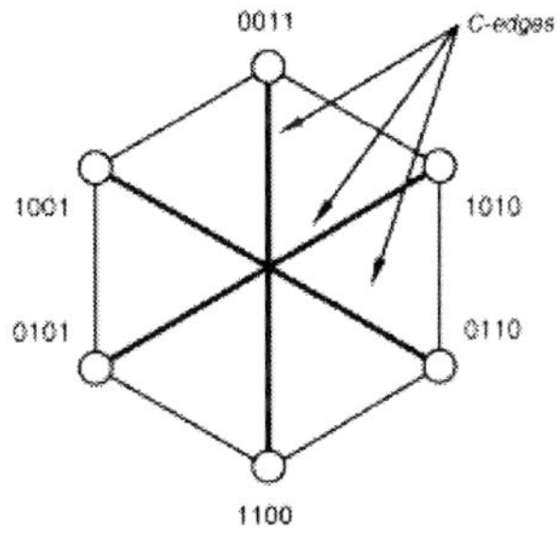

Fig. 2. $FHS(4, 2)$ graph

We can easily see that the number of nodes in $FHS(2n, n)$ is the same as that of $FS(2n, n)$, while the degree of $FHS(2n, n)$ is $n+1$. Since new edges are added to $FHS(2n, n)$ if bit strings of any two nodes in $HS(2n, n)$ complement, $FHS(2n, n)$ is a bipartite graph. The diameter of a hyper-star graph $HS(2n, n)$ is $2n - 1$ if bit strings of two nodes complement. On the other hand, such nodes are connected by an edge in $FHS(2n, n)$. Thus, it is easy to see that the diameter of $FHS(2n, n)$ is n. Since the network cost is defined as degree $\times$ diameter, the network cost of a folded hyper-star graph $FHS(2n, n)$ is n^2+n. Thus, the folded Hyper-Star graph is superior to the hypercube and its variations with the same number of nodes in terms of the network cost. We present comparisons between $FHS(2n, n)$ and other interconnection networks below. Compared models are the hypercube and its other variations such as folded hypercube [6], multiply-twisted cube [4], and hierarchical cubic networks(HCN) [8]. Basic parameters such as the size, degree, diameter and the network cost of folded hyper-star graph, hypercube and its other variations are shown in Table 1.

Table 1. Network cost of the hypercube and its variations

Network Model	Size	Degree	Diameter	Network Cost
Hypercube	2^n	n	n	n^2
Folded Hypercube	2^n	$n+1$	$\lceil \frac{n}{2} \rceil$	$\approx 0.5n^2$
Multiply-Twisted Cube	2^n	n	$\lceil \frac{n+1}{2} \rceil$	$\approx 0.5n^2$
$HCN(\frac{n}{2}, \frac{n}{2})$	2^n	$\frac{n}{2}+1$	$\frac{n}{2} + \lceil \frac{n+2}{6} \rceil + 1$	$\approx 0.33n^2$
Folded $HS(2k, k)$	$\binom{2k}{k}$	$k+1$	k	k^2+k

4.2 Routing Algorithm

Let $R = r_1 r_2 ... r_{2n}$ be a bit string obtained by applying Exclusive OR operations between the source node S and the destination node D. Also, let M be the set of nodes whose number of "1" is greater than n. When a node belongs to the set M , the routing scheme for a folded hyper-star graph $FHS(2n, n)$ chooses a node first such as to utilize c-edges efficiently. The set of nodes P_s on a shortest path from S to D is constructed by the algorithm called **Shortest_path_2**. Initially, P_s = {S}.

5 Conclusion

A hyper-star graph $HS(n, k)$ as a new interconnection network was introduced to improve the interconnection cost, defined by degree $\times$ diameter, of hypercube. Its topological properties and routing scheme were discussed to demonstrate its superiority. Its variation folded hyper-star graph $FHS(n, k)$ was also introduced

```
Shortest_path_2(S, D, P_s)
begin
    k=0;
    if (S = D) then
        return P_s;
    obtain R = r_1r_2...r_i...r_2n, where r_i = s_i ⊕ d_i, 1 ≤ i ≤ 2n;
    let p = {i|r_i = 1, 2 ≤ i ≤ 2n}, q_1 = (S, S')_c, and q_2 = {j|(S, S")
      ∈ E(FHS(2n, n)) − q_1}, where jth bits in S and S" complement;
    for i = 2 to 2n do
        if (r_i = "1") then k = k + 1;
        i = i + 1;
    if (k > n) then
        P_s = P_s ∪ {S'};
        S = S';
    else
        if (|p| > 0) then
          find a node S" such that (S, S") is an j-dimensional edge, where
            j = min{p ∩ q_2};
          P_s = P_s ∪ {S"};
          S = S";
    call Shortest_path_2(S, D, P_s)
end
```

Fig. 4. The shortest path routing algorithm of the $FHS(2n, n)$

to further improve the network cost. When $FHS(2k, k)$ and the n-dimensional hypercube have the same number of nodes, the network cost, degree × diameter of $FHS(2k, k)$, is less than $(k+1)(\lceil logK \rceil + 1)$ and the hypercube is n^2, where $K = \binom{2k}{k}$. It shows that $FHS(2k, k)$ utilizes edges more efficiently than the hypercube and its variations.

References

1. Abraham, S.: The Twisted Cube Topology for Multiprocessor: A Study in Network Asymmetry. J. Parallel and Distributed Computing, Vol.13, (1991) 104–110
2. Akers, S.B., Harel, D., Krishnamurthy, B.: The Star Graph: An Attractive Alternative to The n-Cube. Proc. Int'l Conf. on Parallel Processing, Aug, (1987) 393–400
3. Akers, S.B., Krishnamurthy,B.: A Group-Theoretic Model for Symmetric Interconnection Network. IEEE Trans. Comput., Vol.38, No.4, (1989) 555–565
4. Efe, K.: A Variation on the Hypercube with Lower Diameter. IEEE Trans. Comput., Vol. 40, No. 11, (1991) 1312–1316
5. Harary, F., Hayes, J.P., Wu, H.-J.: A Survey of the Theory of Hypercube Graphs. Comput. Math. Appl., Vol.15, (1988) 277–289
6. Lai, C.-N., Chen, G.-H., Duh, D.-R.: Constructing One-to-Many Disjoint Paths in Folded Hypercubes. IEEE Trans. Comput., Vol. 51, No. 1, (2002) 33–45
7. Parhami, B., Kwai, D.-M.: A Unified Formulation of Honeycomb and Diamond Networks. IEEE Trans. Parallel and Distributed Systems, Vol. 12, No. 1, Jan. (2001) 74–80
8. Yun, S.-K., Park, K.-H.: Comments on 'Hierarchical Cubic Networks'. IEEE Trans. Parallel and Distributed Systems, Vol. 9, No. 4, (1998) 410–414

Sequential Consistency as Lazy Linearizability

Michel Raynal

IRISA – Campus de Beaulieu, 35042 Rennes Cedex, France

Abstract. This paper shows that actually sequential consistency is a form of "lazy" atomic consistency. More precisely, it proposes a new particularly simple sequential consistency protocol that orders the conflicting operations on each object separately, and appropriately invalidates object copies to prevent consistency violation. When compared to invalidation-based protocols that ensure atomic consistency (such as Li-Hudak's protocol), the proposed protocol can be seen as using lazy invalidation. Hence, in addition to a new consistency protocol, the paper provides a new insight into the concepts and mechanisms that underlie consistency protocols: while atomic consistency is based on physical time and requires eager invalidation, sequential consistency is based on logical time and needs only lazy invalidation.

1 Introduction

The definition of a consistency criterion is crucial for the correctness of a multiprocess program. Basically, a consistency criterion defines which value has to be returned when a read operation on a shared object is invoked by a process. This paper focuses on *sequential consistency* [3]. This criterion lies between atomic consistency and causal consistency. Informally it states that a multiprocess program executes correctly if its results could have been produced by executing that program on a single processor system. This means that an execution is correct if we can totally order its operations in such a way that (1) the order of operations in each process is preserved, and (2) each read gets the last previously written value, "last" referring here to the total order. The difference between atomic consistency and sequential consistency lies in the meaning of the word "last". This word refers to real-time when we consider atomic consistency, while it refers to a logical time notion when we consider sequential consistency (namely the logical time defined by the total order).

It has been shown that determining whether a given execution is sequentially consistent is an NP-complete problem [7]. This has an important consequence as it rules out the possibility of designing efficient sequential consistency protocols (i.e., protocols that provide sequentially consistent executions and just these). This means that, in order to be able to design efficient sequential consistency protocols, we do have to impose additional constraints on executions. We have introduced such an approach and investigated such additional constraints in a previous work [5] where several additional constraints have been proposed. One of these constraints (that we called *OO*-constraint) is the following. Let

M.H. Shafazand and A M. Tjoa (Eds.): EurAsia-ICT 2002, LNCS 2510, pp. 866–873, 2002.

two operations *conflict* if both are on the same object and one of them is a write. Let us say that an execution satisfies the *OO-constraint* if any pair of conflicting operations are ordered. It is shown in [5] that an *OO*-constrained execution is sequentially consistent if its read operations are legal (i.e., do not provide overwritten values). This approach shows that a sequential consistency protocol can be obtained by combining two mechanisms: one providing the *OO*-constraint and the other providing read legality. This paper proposes such a protocol building a sequentially consistent shared memory abstraction on top of an asynchronous message-passing distributed system. So, the protocol ensures that, for each object separately, conflicting operations are ordered. Moreover, object copies are duplicated and an invalidation mechanism guarantees that, if a copy is present in a node, it can be legally read.

In a very interesting way, it appears that this protocol is very close to Li-Hudak's protocol implementing atomic consistency [4]. Actually, both protocols are based on the *OO*-constraint (although this constraint is not explicitly identified in [4]), and mainly differ in the time moment at which object copies are invalidated. Li-Hudak's protocol is based on an eager invalidation, while the proposed protocol is based on a lazy invalidation. More precisely, the Li-Hudak's atomic consistency protocol invalidates all the copies of an object x each time x is written. This brute force invalidation strategy ensures that no read operation can get an overwritten value. Differently, the proposed sequential consistency protocol does not invalidate the copies of x at the time x is written. For any object x, old values can coexist with a new value as long as the reading of an old value does not violate the legality of read operations. In that sense, the invalidation mechanism is lazy. So, the proposed protocol shows that sequential consistency can be obtained from an atomic consistency protocol by appropriately delaying the invalidation of old values. Intuitively, this is in perfect agreement with the fact that the definition of atomic consistency involves real-time, while the definition of sequential consistency involves only logical time. (Multi-object operations and fault-tolerance are addressed in [6].)

2 The Consistent Shared Memory Abstraction

A shared memory system is composed of a finite set of sequential processes $p_1, \dots, p_n$ that interact via a finite set X of shared objects. Each object $x \in X$ can be accessed by read and write operations. A write into an object defines a new value for the object; a read allows to obtain a value of the object. A write of value v into object x by process p_i is denoted $w_i(x)v$; similarly a read of x by process p_j is denoted $r_j(x)v$ where v is the value returned by the read operation; *op* will denote either r (read) or w (write). For simplicity, we assume all values written into an object x are distinct. Moreover, the parameters of an operation are omitted when they are not important. Each object has an initial value (it is assumed that this value has been assigned by an initial fictitious write operation).

Histories are introduced to model the execution of shared memory parallel programs. The *local history* (or local computation) $\widehat{h}_i$ of p_i is the sequence of operations issued by p_i. If $op1$ and $op2$ are issued by p_i and $op1$ is issued first, then we say "$op1$ precedes $op2$ in p_i's process-order", which is noted $op1 \rightarrow_i op2$. Let h_i denote the set of operations executed by p_i; the local history $\widehat{h}_i$ is the total order $(h_i, \rightarrow_i)$.

Definition 1. *An* execution history *(or simply history, or computation)* $\widehat{H}$ *of a shared memory system is a partial order* $\widehat{H} = (H, \rightarrow_H)$ *where* $H = \bigcup_i h_i$*, and* $op1 \rightarrow_H op2$ *is defined as follows:* i) $\exists\ p_i\ : op1 \rightarrow_i op2$ *(in that case,* $\rightarrow_H$ *is called* process-order *relation), or* ii) $op1 = w_i(x)v$ *and* $op2 = r_j(x)v$ *(in that case* $\rightarrow_H$ *is called* read-from *relation), or* iii) $\exists op3\ : op1 \rightarrow_H op3$ *and* $op3 \rightarrow_H op2$.

The legality concept is the key notion on which are based definitions of shared memory consistency criteria. From an operational point of view, it states that, in a legal history, no read operation can get an overwritten value.

Definition 2. *A read operation* $r(x)v$ *is* legal *if: (i)* $\exists\ w(x)v\ : w(x)v \rightarrow_H r(x)v$ *and (ii)* $\nexists\ op(x)u\ : (u \neq v) \wedge (w(x)v \rightarrow_H op(x)u \rightarrow_H r(x)v)$*. A history* $\widehat{H}$ *is* legal *if all its read operations are legal.*

Sequential consistency has been proposed by Lamport in 1979 to define a correctness criterion for multiprocessor shared memory systems [3]. A system is sequentially consistent with respect to a multiprocess program, if "*the result of any execution is the same as if (1) the operations of all the processors were executed in some sequential order, and (2) the operations of each individual processor appear in this sequence in the order specified by its program*".

This informal definition states that the execution of a program is sequentially consistent if it could have been produced by executing this program on a single processor system. Formally, a history $\widehat{H} = (H, \rightarrow_H)$ is *sequentially consistent* if it has a legal linear extension [6] (i.e., its operations can be totally ordered while respecting legality and process order).

Atomic consistency considers that operations take time and consequently its definition is based on the real-time occurrence order of operations [2]. Let $\prec_{rt}$ be a real-time precedence relation on operations defined as follows: $op_1 \prec_{rt} op_2$ if op_1 was terminated before op_2 started. Let us notice that $\prec_{rt}$ is a partial order relation as two operations whose executions overlap in real-time are not ordered. A history $\widehat{H} = (H, \rightarrow_H)$ is *atomically consistent* if it has a legal linear extension that includes $\prec_{rt}$. This means that, to be atomically consistent, $\widehat{H}$ must have a legal linear extension $\widehat{S} = (H, \rightarrow_S)$ such that $\forall op_1, op_2 :\ (op_1 \prec_{rt} op_2) \Rightarrow (op_1 \rightarrow_S op_2)$. The linear extension $\widehat{S}$ has to keep real-time order. It is easy to see why invalidation-based atomic consistency protocols use an eager invalidation strategy: this ensures that the real-time occurrence order on operations cannot be ignored, a read always getting the last value (with respect to real-time).

A property P of a concurrent system is *local* if the system as a whole satisfies P whenever each individual object satisfies P. The following theorem is a main result of [2]. It states that linearizability is a local property. let $\widehat{H}|x$ ($\widehat{H}$ at x) be the projection of $\widehat{H}$ on x (i.e., $\widehat{H}|x$ includes only the operations that involve x).

Theorem 1. [2] *$\widehat{H}$ is linearizable (atomically consistent) iff, for each object x, $\widehat{H}|x$ is linearizable (atomically consistent).*

This theorem is important as it states that a concurrent system can be designed in a modular way: a linearizable object can be implemented independently of the other objects. Unfortunately, sequential consistency is not a local property [2]. As we will see in Section 3, the objects have to cooperate to guarantee sequential consistency. Hence, there is a tradeoff between the *locality* of a consistency criterion and the *timeliness* (eager *vs* lazy) of the invalidation strategy used in the protocol that implements it.

The Constraint-Based Approach. To address sequential consistency, we have developed in [5] a method that consists in imposing additional constraints on histories in order to be able to design efficient sequential consistency protocols. Let two operations *conflict* if both are on the same objet x and one of them is a write. The WW-constraint and the OO-constraint have been introduced in [5]. A history $\widehat{H}$ satisfies the *WW-constraint* if all its write operations are totally ordered. A history $\widehat{H}$ satisfies the *OO-constraint* if any pair of conflicting operations are ordered (so, when $\widehat{H}$ satisfies the OO-constraint, the operations on each object $x \in X$ follow the reader/writer synchronization discipline.) Among the theorems proved in [5], the rest of the paper is interested in the following:

Theorem 2. [5] *Let $\widehat{H} = (H, \rightarrow_H)$ be a history that satisfies the OO-constraint (resp. the WW-constraint). $\widehat{H}$ is sequentially consistent if and only if it is legal.*

Several protocols providing a sequentially consistent shared memory abstraction on top of an asynchronous message passing distributed system have been proposed (e.g., [1,5]). Although they do not explicitly identify the WW-constraint, a lot of them are implicitly based on it. Due to space limitation they are not discussed here (see [6]). To our knowledge, to date no sequential consistency protocol fully based on the OO-constraint has been proposed.

3 An *OO*-Constraint-Based SC Protocol

This section presents an OO-constraint-based protocol that implements sequential consistency on top of a distributed system. By a slight abuse of language, we say "that value of x is legal" when, assuming a read operation returning it, that read operation is legal.

3.1 Underlying System

The concurrent program is made up n processes $p_1, \ldots, p_n$ sharing m objects (denoted $x, y, \ldots$). The underlying system is made up of $n+m$ sites, divided into n process-sites and m object-sites (hence, without ambiguity, p_i denotes both a process and the associated site; M_x denotes the site hosting and managing x). The sites communicate through reliable channels by sending and receiving messages. There are assumptions neither on the processing speed of the sites, nor on message transfer delays. Hence, the underlying distributed system is asynchronous.

To be as modular as possible, and in the spirit of clients/servers architectures, the proposed solution allows a process and an object to communicate, but no two processes -nor two objects- are allowed to send messages to each other. This constraint makes easier the addition of new processes or objects into the system.

3.2 *OO*-Constraint and Legality

The *OO*-constraint is ensured by having a manager M_x for each object x. This makes the protocol really distributed: there is no centralized control. Basically, the manager M_x orders the write operations on the object x it is associated with. As in other protocols (e.g., [4]), in order to get an efficient protocol, object values are cached and the last writer of an object x is considered as its *owner* until the value it wrote is overwritten or read by another process. The combination of value caching and object ownership allows processes to read cached values and to update the objects they own for free (i.e., without communicating with the corresponding managers). According to the read/write access pattern, this can be very efficient.

The read operations are then appropriately scheduled in order (1) to ensure the read/write and write/read synchronization required to satisfy the *OO*-constraint, and (2) to get legal read operations. This is the main part of the protocol. It is done in the *check_legality* procedure executed by a process p_i each time it questions a manager M_x.

3.3 Control Variables

Local variables of a process. As indicated previously, the protocol is based on cached copies and an invalidation strategy. A process p_i manages the following local variables:

- $present_i[x]$ is a boolean variable that is true when p_i has a legal copy of x. This means that p_i can then locally read its cached copy of x.
- $C_i[x]$ is a cache containing a copy of x. Its value is meaningful only when $present_i[x]$ is true.
- $owner_i[x]$ is a boolean variable that is true if p_i is the last writer of x. Let us note that we have $owner_i[x] \Rightarrow present_i[x]$.

Local variables of an object manager. The site M_x associated with the object x manages the following local variables:

- C_x contains the last value of x (as known by M_x).
- $owner_x$ contains a process identity or $\perp$. It has the following meaning. $owner_x = i$ means that no process but p_i has the last value of x (it is possible that M_x does not know this value). Differently, if $owner_x = \perp$, M_x has the last copy of x. The owner of an object x (if any) is the only process that can modify it.

– $hlw_x[1..n]$ is a boolean array, such that $hlw_x[i]$ is true iff p_i has the last value of x (hlw_x stands for "*h*old *l*ast *w*rite of x").

System initial state. Initially, only M_x knows the initial value of x which is kept in C_x. So, we have the following initial values: $\forall p_i$: $\forall x$: $owner_i[x] = false$, $present_i[x] = false$, $C_i[x]$= undefined, $owner_x = \bot$, $hlw_x[i] = false$.

upon reception of *write_req* (v) **from** p_i:
(1) **if** $(owner_x \neq \bot)$ **then** send *downgrade_req* (x) to p_{owner_x};
(2) wait *downgrade_ack* $(-)$ from p_{owner_x} **endif**;
(3) $C_x \leftarrow v$; $owner_x \leftarrow i$;
(4) $hlw_x[i] \leftarrow true$; $\forall j \neq i$: $hlw_x[j] \leftarrow false$;
(5) send *write_ack* () to p_i

upon reception of *read_req* () **from** p_i:
(6) **if** $(owner_x \neq \bot)$ **then** send *downgrade_req* (x) to p_{owner_x};
(7) wait *downgrade_ack* (v) from p_{owner_x};
(8) $C_x \leftarrow v$; $owner_x \leftarrow \bot$ **endif**;
(9) $hlw_x[i] \leftarrow true$;
(10) send *read_ack* (C_x) **to** p_i

upon reception of *check_req* () **from** p_i:
(11) **let** $d = \big(hlw_x[i] \wedge$ (no *write_req* is currently processed) $\big)$;
(12) send *check_ack* (d) **to** p_i

Fig. 1. Behavior of an Object Manager M_x

3.4 Behavior of an Object Manager

The behavior of the manager M_x of the object x is depicted in Figure 1. It is made up of three statements that describes what M_x does when it receives a message. The *write_req* and *read_req* messages are processed sequentially[1]. The *check_req* messages are processed as soon as they arrive (let us notice they do not modify M_x's context).

– M_x receives *write_req* (v) from p_i. In that case, p_i has issued a $w(x)v$ operation. If x is currently owned by some process, M_x first downgrades this previous owner (lines 1-2). Then, M_x sets C_x and $owner_x$ to their new values (line 3), updates the boolean vector hlw_x accordingly (line 4), and sends back to p_i an ack message indicating it has taken its write operation into account (line 5).
– M_x receives *read_req* (v) from p_i. In that case, p_i has issued a $r(x)$ operation, and has not a legal copy of x. If x is currently owned by some process, M_x

[1] Assuming no message is indefinitely delayed, it is possible for M_x to reorder waiting messages if there is a need to favor some processes or some operations.

```
operation w_i(x)v:
(1)  if (¬owner_i[x]) then send write_req (v) to M_x;
(2)                        wait write_ack () from M_x;
(3)                        owner_i[x] ← true; present_i[x] ← true;
(4)                        check_legality (x) endif;
(5)  C_i[x] ← v;
(6)  return ()

operation r_i(x):
(7)  if (¬present_i[x]) then send read_req () to M_x;
(8)                          wait read_ack (v) from M_x;
(9)                          present_i[x] ← true; C_i[x] ← v;
(10)                         check_legality (x) endif;
(11) return (C_i[x])

procedure check_legality (x):
(12) ∀y such that (present_i[y] ∧ ¬owner_i[y] ∧ y ≠ x)
(13)          do send check_req () to M_y;
(14)             wait check_ack (d) from M_y;
(15)             present_i[y] ← d enddo

upon reception of downgrade_req (x):
(16) owner_i[x] ← false;
(17) send downgrade_ack (C_i[x]) to M_x
```

Fig. 2. Behavior of a Process p_i

first gets the last value of x from this process and downgrades this previous owner (lines 6-8). Then, it sends the current value of x to p_i (line 10) and updates $hlw_x[i]$ accordingly (line 9).

- M_x receives *check_req* () from p_i. In that case, p_i has got a new value for some object and queries M_x to know if it still has the last value of x. As we will see, this inquiry is necessary for p_i to ensure its locally cached object values are legal. M_x answers *true* if p_i has the last value of x, otherwise it returns *false*.

3.5 Behavior of a Process

The behavior of a process p_i is described in Figure 2. It is made of three statements that are executed atomically (the procedure *check_legality*(x) invoked at line 4 and at line 10 is considered as belonging to the invoking statement). The return statement terminates each write (line 6) and read operation (line 11).

- p_i invokes $w_i(x)v$. If p_i is the current owner of x, it was the last process that updated x. In that case, it simply updates its current cached copy of x (line 5). This value will be sent to M_x by p_i when it will leave its ownership of x (lines 16-17).

If p_i is not the owner of x, it first becomes the current owner (line 3) by communicating with M_x (lines 1-2). As we have seen in Figure 1, this entails the downgrading of the previous owner if any. Then, p_i executes the *check_legality*(x) procedure. This procedure is explained in a next item. It is the core of the protocol as far as the *OO*-constraint and the legality of read operations are concerned.

- p_i invokes $r_i(x)$. If the local copy of x does not guarantee a legal read (*present*$_i[x]$ is false), p_i gets the last copy from M_x (lines 7-8), updates accordingly its context (line 9) and executes the *check_legality*(x) procedure (see next item).
- The *check_legality*(x) procedure. As we just claimed, this procedure is at the core of the protocol. It is invoked each time p_i questions M_x. This occurs when its context is modified because (1) it becomes the new owner of x (lines 1-4), or (2) it reads the value of x from M_x (lines 7-10). The aim of the *check_legality*(x) procedure is to guarantee that all the values defined as present in p_i's local cache will provide legal read operations. To attain this goal, for each object y not currently owned by p_i but whose local readings by p_i were previously legal, p_i questions M_y to know if its current copy is still legal (line 13). M_y answers p_i if it has the last value of x (line 14), and accordingly p_i updates *present*$_i[x]$. The *check_legality* procedure acts as a reset mechanism that invalidates object copies that cannot be guaranteed to provide legal read operations.
- p_i receives *downgrade_req*(x). In that case, p_i is the current owner of x. It leaves it ownership on x (line 16) and sends the last value of x to M_x (line 17). Note that the current value of x is still present at p_i.

Theorem 3. *The protocol described in Figures 1 and 2 implements sequential consistency.* (Proof in [6]).

References

1. Attiya H. and Welch J.L., Sequential Consistency versus Linearizability. *ACM Transactions on Computer Systems*, 12(2):91-122, 1994.
2. Herlihy M.P. and Wing J.L., Linearizability: a Correctness Condition for Concurrent Objects. *ACM TOPLAS*, 12(3):463-492, 1990.
3. Lamport L., How to Make a Multiprocessor Computer that Correctly Executes Multiprocess Programs. *IEEE Transactions on Computers*, C28(9):690-691, 1979.
4. Li K. and Hudak P., Memory Coherence in Shared Virtual Memory Systems. *ACM Transactions on Computer Systems*, 7(4):321-359, 1989.
5. Mizuno M., Raynal M. and Zhou J.Z., Sequential Consistency in Distributed Systems. Springer Verlag LNCS #938, pp. 224-241, Dagsthul (Germany), 1994.
6. Raynal M., Sequential Consistency as Lazy Linearizabilty. *Research Report # 1437*, IRISA, Université de Rennes 1 (France), 2002, 13 pages. http://www.irisa.fr/bibli/publi/pi/2002/1437/1437.html.
7. Taylor R.N., Complexity of Analyzing the Synchronization Structure of Concurrent Programs. *Acta Informatica*, 19:57-84, 1983.

Embedding Full Ternary Trees into Recursive Circulants*

Cheol Kim[1], Jung Choi[2], and Hyeong-Seok Lim[2]

[1]Division of Computer and Digital Multimedia, Dongshin University,
252 Daeho-dong, Naju, Chonnam 520-714, Republic of Korea
ckim@blue.dongshinu.ac.kr
[2]Department of Computer Science, Chonnam National University,
300 Yongbong-dong Buk-gu, Kwangju 500-757, Republic of Korea
{jung,hslim}@chonnam.chonnam.ac.kr

Abstract. Recursive circulant $G(N,d)$ is a circulant graph with N vertices and jumps of power of d. In this paper we show that k level full ternary trees whose number of vertices are not more than N can be embedded into recursive circulants $G(N,2)$ and $G(N,3)$ with dilation 2.

1 Introduction

Trees play an important role as data structures and as the computational graphs for solving strategies and algorithms underlying divide-and-conquer problem. This fact derives an important consideration for a network topology whether there exists good mappings from various kinds of trees to the topology. The mapping of a data structure or an interconnection structure into another has been studied as graph embedding. There have been many papers on embedding trees into other interconnection networks[1,3,5,6,8,10]. In some applications such as n-degree polynomial multiplication problems, problems can be divided into 3 subproblems. If we use ternary tree structure to solve the n-degree polynomial multiplication problems, it needs $n^{1.53}$ multiplications while the conventional algorithm needs n^2 multiplications[15].

In this paper, we consider the problem of mapping full ternary trees into the interconnection networks called recursive circulants. Recursive circulant graphs were introduced in 1994 by Park and Chwa[12] as new topologies for the interconnection networks. Recursive circulants $G(N,d)$, where $d \geq 2$, has N vertices labeled by integers between 0 and N-1. Two vertices v, w are joined by an edge if and only if there exists i, $0 \leq i \leq \lceil \log_d N \rceil - 1$, such that $v + d^i \equiv w \pmod N$. That is, $G(N,d)$ is a circulant graph with N vertices and jumps of all powers of d less than N. Examples of $G(8,2)$ and $G(8,3)$ are shown in Fig. 1.

* This work was supported by grant No. R05-2002-000-01443-0 from the Basic Research Program of the Korea Science & Engineering Foundation

M.H. Shafazand and A M. Tjoa (Eds.): EurAsia-ICT 2002, LNCS 2510, pp. 874–882, 2002.

Recursive circulants are a class of circulant graphs. Circulants graphs, also known as star-polygons[4], are a generalization of the graphs constructed by Harary [7] for achieving maximum connectivity. Every circulant graph is a vertex transitive graph and a Cayley graph. Also, circulant graphs are a class of generalized chordal ring graphs [2], which is an important topology for the interconnection networks.

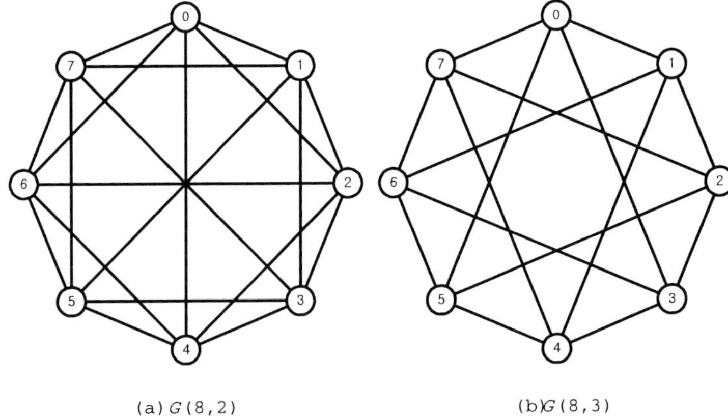

Fig. 1. Examples of recursive circulants $G(8,2)$ and $G(8,3)$

Recursive circulants $G(2^m,2)$ and $G(2^m,3)$ are the major concern in this paper. $G(2^m,2)$, also known as a barrel shifter or PM2I, is a regular graph with degree $2m$-1. $G(2^m,2)$ has both $G(2^m,4)$ and the m-dimensional hypercube Q_m as subgraphs. Network metrics for recursive circulants, such as connectivity, diameter, average distance, and visit ratio, are analyzed in [9,11,14].

The paper is organized as follows. In section 2, terminology and definitions are given. In section 3, we present the embedding of full ternary trees into recursive circulants $G(N,2)$ and $G(N,3)$. In section 4, a summary of this paper and further remarks are given.

2 Preliminaries

The *embedding* of a guest graph $G = (V_G, E_G)$ into a host graph $H = (V_H, E_H)$ is defined by a mapping $\phi : G \rightarrow H$ consisting of two mapping $\phi_V : V_G \rightarrow V_H$ and $\phi_E : E \rightarrow \rho(H)$. Here, $\rho(G)$ denotes the set of paths in a graph $G = (V, E)$. The mapping ϕ_E maps each edge $(v, w) \in E_G$ to a path $p \in \rho(H)$ connecting $\phi_V(v)$ and $\phi_V(w)$. We call an embedding *one-to-one* if the mapping ϕ_V is 1-1.

One of the most important factors to measure cost of an embedding is dilation. If $(v, w) \in E$, then the distance between $\phi_V(v)$ and $\phi_V(w)$ in H is called the *dilation* of the edge (v, w). The maximal dilation over all edges of G is called the dilation of the embedding ϕ.

The number of vertices of a guest graph, which are mapped onto a vertex v in the host graph, is called the *load* of the vertex v. The load of embedding ϕ is the maximal load of a vertex in the host graph. The *expansion* of the embedding ϕ is the ratio $|V_H|/|V_G|$.

The *congestion* of an edge $e' \in E_H$ is the number of paths in $\{\phi_E(e) \mid e \in E_G\}$, containing e'. The *edge-congestion* is the maximal congestion over all edges in H. The *congestion* of an vertex $v \in V_H$ is the number of paths in $\{\phi_E(e) \mid e \in E_G\}$, containing v. The *node-congestion* is the maximal congestion over all vertices in H.

The results on embedding fibonacci trees, full quaternary trees, and full binary trees into recursive circulants are given in [10]. Embeddings among recursive circulants and hypercubes are presented in [7]. An m-dimension hypercube Q_m is subgraph of $G(2^m,2)$, so Q_m can be embedded in $G(2^m,2)$ with dilation 1 and congestion 1. To the contrary, $G(2^m,2)$ can be embedded in Q_m with dilation 2 and congestion 4.

As an approach to embedding problems in recursive circulants $G(N,2)$ and $G(N,3)$, we use a graph labeling problem called d-edge labeling. A *labeling* on a graph $G(V,E)$ is a one-to-one mapping of the vertices V into distinct integers $\{1,2,\ldots,|V|\}$. Each edge of the graph has an edge label induced by the labeling. The edge label of an edge is the absolute difference between the labels of end-vertices of the edge. A d-edge labeling on a graph H is defined to be a labeling such that the set of edge labels is a subset of $\{d^0, d^1, d^2, \ldots\}$ for some integer d, that is , each edge label should be a power of d. Here we are interested in the case where $d = 2$ or 3. From the definition of d-edge labeling and recursive circulants $G(N, d)$, it is easily implied that if a graph H has a d-edge labeling then H is a subgraph of $G(N,d)$ with $N \geq |V(H)|$, that is, H can be embedded into $G(N,d)$ with dilation 1 and congestion 1.

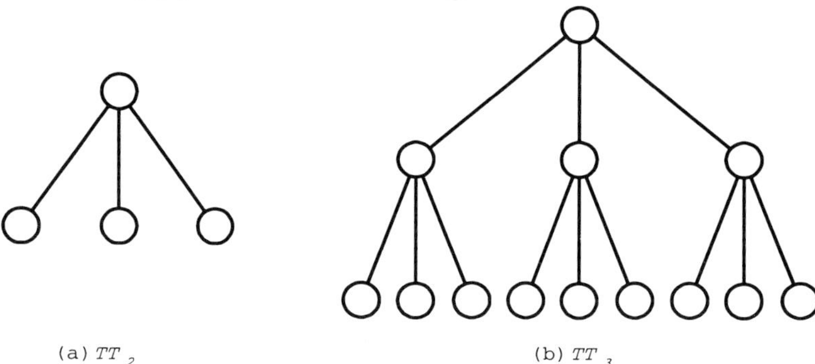

Fig. 2. Examples of full ternary trees

3 Embedding Full Ternary Trees into Recursive Circulants

We denote by TT_k the full ternary tree with height k. The root vertex of TT_k has three subtrees TT_{k-1}, left-subtree, center-subtree, and right-subtree. TT_k has $(3^k - 1) / 2$ vertices and 3^{k-1} leaf vertices. Examples of full ternary trees are shown in Fig. 2.

Here, we introduce a simple transformation to find an efficient labeling on TT_k. Our labeling scheme is performed on a transformed TT_k. The definition of the transformation and the properties of the transformed trees are as follows.

Definition 1. Detoured ternary tree, DT_k is a transformed tree of TT_k by the transformation as follows.

(1) For each internal vertex, place an edge between the center child and right child.

(2) For each internal vertex, cut the edge connecting the right child.

Let $L_{k,1}$ be the set of leaves of DT_k. The number of vertices in $L_{k,1}$ is $2 \cdot 3^{k-2}$. Let $L_{k,2}$ be the set of leaves of tree obtained by deleting the vertices in $L_{k,1}$ from DT_k. The number of vertices in $L_{k,2}$ is 3^{k-2}. DT_{k-1} is obtained by deleting the vertices in $L_{k,1}$ and $L_{k,2}$ from DT_k. Each vertex in $L_{k,2}$ is adjacent to vertices in $L_{k-1,1} \cup L_{k-1,2}$, and each vertex of $L_{k,1}$ is adjacent to vertices in $L_{k-1,1} \cup L_{k-1,2} \cup L_{k,2}$. Examples of 2-edge labelings and 3-edge labelings on ternary DT_3 are shown in Fig. 3(*c*) and Fig. 4(*c*).

Lemma 1. TT_k can be embedded into DT_k with dilation 2, expansion 1, and congestion 2.

Proof. In our transformation, for each internal vertex, the edge connecting the right child in TT_k is replaced to a path with length 2 in DT_k, to the right child via the center child. Edges incident to the left child and to the center child in TT_k are not changed in DT_k. The edge joining each internal vertex and its center child in TT_k is shared by two paths, from the internal vertex to the center child and to the right child via the center child in DT_k. The path to the right child is a mapping of an edge, from each internal vertex to the right child in TT_k. Thus, the proof is given.

DT_k has $(3^k - 1) / 2$ vertices, so vertex labels 1, 2, …, $(3^k - 1) / 2$ are used. Our labeling scheme on DT_k satisfies the following properties.

(1) The label of the root is 1.

(2) The labels on the vertices in $L_{k,2}$ are consecutive integers, $(3^{k-1} - 1) / 2 + 1$, $(3^{k-1} - 1) / 2 + 2$, …, $(3^{k-1} - 1) / 2 + 3^{k-2}$, and the labels on the vertices in $L_{k,1}$ are consecutive integers, $(3^{k-1} - 1) / 2 + 3^{k-2} + 1$, $(3^{k-1} - 1) / 2 + 3^{k-2} + 2$, …, $(3^k - 1) / 2$.

(3) The labeling of the tree obtained by removing the vertices of $L_{k,1}$ and $L_{k,2}$ from DT_k is the 2- or 3-edge labeling of DT_{k-1}.

In our labeling scheme, the 2-edge labeling on DT_{k+1} is obtained by joining vertices to the 2-edge labeled DT_k and by assigning consecutive integers to the vertices so that each edge label of the attached edge is a power of 2. The labeling scheme has two steps. First, attach a vertex to each vertex belonging to $L_{k,1} \cup L_{k,2}$ and assign consecutive integers to the attached vertices so that each edge label of the attached edge is a power of 2. Then, the set of attached vertices is $L_{k+1,2}$. Next, attach a vertex to each vertex in $L_{k,1} \cup L_{k,2} \cup L_{k+1,2}$ and assign consecutive integers to the attached vertices, $L_{k+1,1}$, so that each edge label of the attached edge is a power of 2. An example of the 2-edge labeling scheme is shown in Fig. 3.

The 3-edge labeling on DT_{k+1} is similar to the 2-edge labeling on DT_{k+1} except that the difference of each matched pair is a power of 3. An example of the 3-edge labeling scheme is shown in Fig. 4.

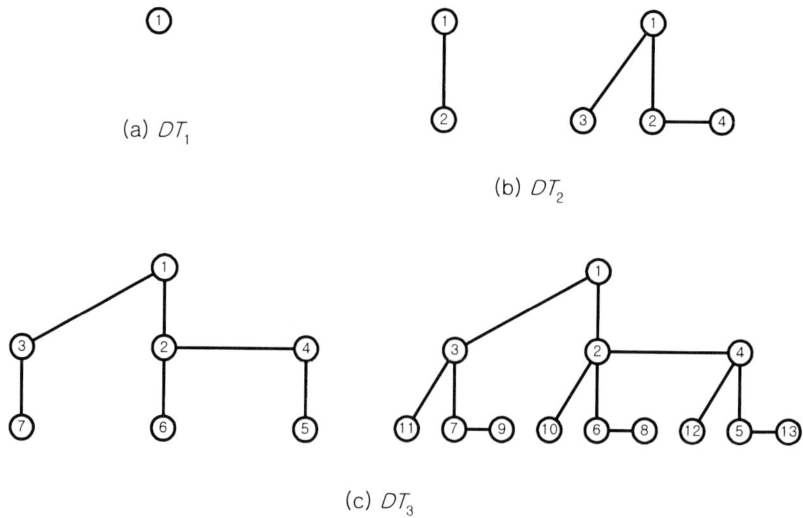

Fig. 3. 2-edge labelings on DT_1, DT_2, and DT_3

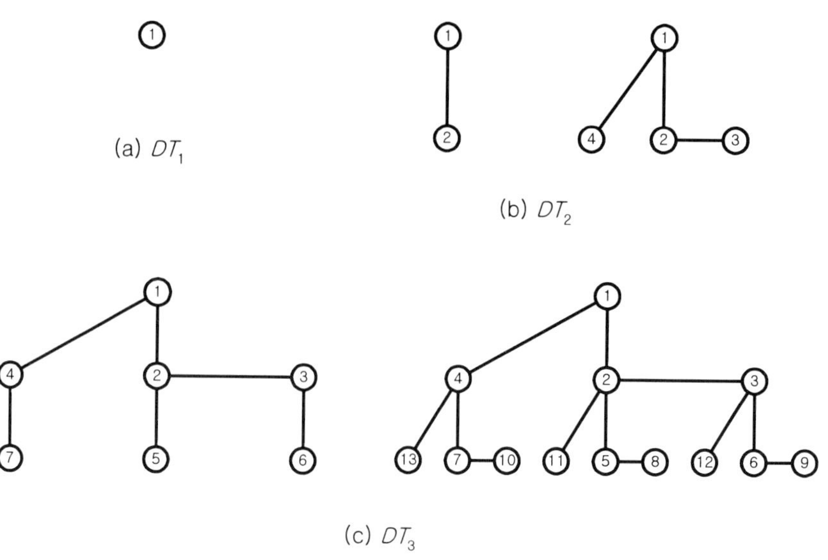

Fig. 4. 3-edge labelings on DT_1, DT_2, and DT_3

Lemma 2[10]. For each integer from i to $i + k$, there is a matching of distinct integers from $i + k + 1$ to $i + 2k + 1$ such that difference of each matched pair is a power of 2.

Theorem 1. Every DT_k has a 2-edge labeling

Proof. The proof will be given by induction on k. For $k \leq 2$, the theorem holds from the labeling given in Fig. 3(a),(b). Also, the labelings satisfy the previous labeling properties. Assume that for any integer k-1, there exists a 2-edge labeling on DT_{k-1} satisfying the labeling properties. By the assumption, the vertices in $L_{k-1,1} \cup L_{k-1,2}$, are labeled with 3^{k-2} consecutive integers, $\{(3^{k-2} - 1) / 2 + 1, \ldots, (3^{k-1} - 1) / 2\}$. Attach a vertex to each vertex in $L_{k-1,1} \cup L_{k-1,2}$. Then, the set of attached vertices is $L_{k,2}$. By Lemma 2, we can assign 3^{k-2} consecutive integers, $\{(3^{k-1} - 1) / 2 + 1, \ldots, (3^{k-1} - 1) / 2 + 3^{k-2}\}$, to the vertices in $L_{k,2}$ so that each edge label of the attached edge is a power of 2. Next, attach a vertex to each vertex belonging to $L_{k-1,1} \cup L_{k-1,2} \cup L_{k,2}$. The set of attached vertices is $L_{k,1}$. Again, by Lemma 2, we can assign $2 \cdot 3^{k-2}$ consecutive integers, $\{(3^{k-1} - 1) / 2 + 3^{k-2} + 1, \ldots, (3^k - 1) / 2\}$ to the vertices of $L_{k,1}$ so that each edge label of the attached edge is a power of 2. The labeling on the level k satisfying the constraint can be obtained by the method we describe here. Thus, the theorem holds.

Corollary 1. DT_k with vertices 2^m or less can be embedded in $G(2^m,2)$ with dilation 1 and congestion 1.

Corollary 2. TT_k with vertices 2^m or less can be embedded in $G(2^m,2)$ with dilation 2 and congestion 2.

There exists dilation 2 and congestion 2 embedding between TT_k and DT_k. A 2-edge labeled DT_k is a subgraph of recursive circulants $G(N,2)$ and can be embedded into $G(2^m,2)$ with dilation 1. By applying two embeddings consecutively, TT_k with 2^m vertices or less can be embedded in $G(2^m,2)$ with dilation 2. Fig. 3(c) is an example of 2-edge labeling of DT_3 and Fig. 6(a) is an example of its embedding in $G(2^4,2)$.

Lemma 3. For integers $n = 3^k$ and $n = 2 \cdot 3^k$, and each integer from $a + 1$ to $a + n$, there is a matching of distinct integers from $a + n + 1$ to $a + 2n$ such that the difference of each matched pair is a power of 3.

Proof. For $n = 3^k$, integers $a + 1, a + 2, a + 3, \ldots, a + n$ are matched to $a + n + 1, a + n + 2, a + n + 3, \ldots, a + 2n$, respectively. The difference of each pair of two integers is 3^k. For $n = 2 \cdot 3^k$, first, integers $a + 1, a + 2, a + 3, \ldots, a + 3^k$ are matched to $a + 3 \cdot 3^k + 1, a + 3 \cdot 3^k + 2, a + 3 \cdot 3^k + 3, \ldots, a + 4 \cdot 3^k$, respectively. The difference of each pair of two integers is 3^{k+1}. Next, integers $a + 3^k + 1, a + 3^k + 2, a + 3^k + 3, \ldots, a + 2 \cdot 3^k$ are matched to $a + 2 \cdot 3^k + 1, a + 2 \cdot 3^k + 2, a + 2 \cdot 3^k + 3, \ldots, a + 3 \cdot 3^k$, respectively. The difference of each pair of two integers is 3^k.

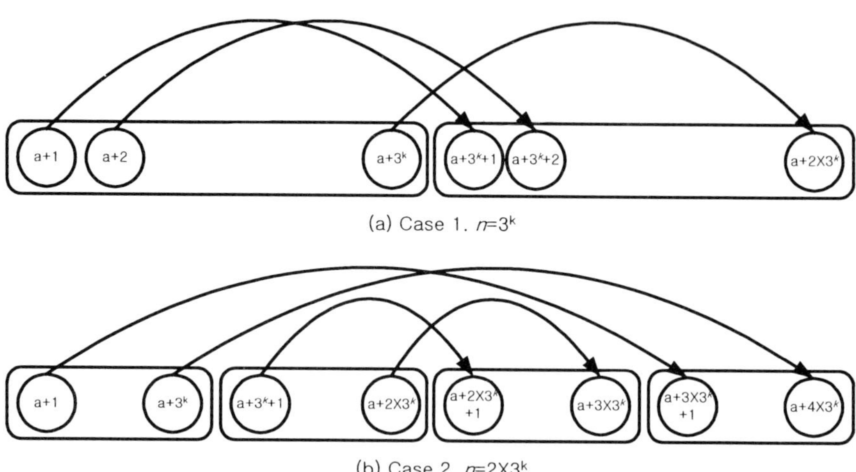

Fig. 5. Matching pairs difference of power of 3

Theorem 2. Every DT_k has a 3-edge labeling

Proof. The proof will be given by induction on k. For $k \leq 2$, the theorem holds from the labeling given in Fig. 4(a),(b). Also, the labelings satisfy the previous labeling properties. Assume that for any integer k-1, there exists a 3-edge labeling on $DT_{k\text{-}1}$ satisfying the labeling properties. By the assumption, the vertices in the set, $L_{k\text{-}1,1} \cup L_{k\text{-}1,2}$, are labeled with $3^{k\text{-}2}$ consecutive integers, $\{(3^{k\text{-}2} - 1) / 2 + 1, \ldots, (3^{k\text{-}1} - 1) / 2\}$. Attach a vertex to each vertex belonging to $L_{k\text{-}1,1} \cup L_{k\text{-}1,2}$. Then, the set of attached vertices is $L_{k,2}$. By Lemma 3, we can assign $3^{k\text{-}2}$ consecutive integers, $\{(3^{k\text{-}1}\text{-}1)/2 + 1, \ldots, (3^{k\text{-}1} - 1) / 2 + 3^{k\text{-}2}\}$, to the vertices of $L_{k,2}$ so that each edge label of the attached edge is a power of 3. Next, attach a vertex to each vertex belonging to $L_{k\text{-}1,1} \cup L_{k\text{-}1,2} \cup L_{k,2}$. Then the set of attached vertices is $L_{k,1}$. Again, by Lemma 3, we can assign $2{\cdot}3^{k\text{-}2}$ consecutive integers, $\{(3^{k\text{-}1} - 1) / 2 + 3^{k\text{-}2} + 1, \ldots, (3^k - 1) / 2\}$ to the vertices of $L_{k,1}$ so that each edge label of the attached edge is a power of 3. The labeling on level k satisfying the constraint can be obtained by the method we describe here. Thus, the theorem holds.

Corollary 3. DT_k with vertices 2^m or less can be embedded in $G(2^m,3)$ with dilation 1 and congestion 1.

Corollary 4. TT_k with vertices 2^m or less can be embedded in $G(2^m,3)$ with dilation 2 and congestion 2.

By applying two embeddings consecutively, TT_k into DT_k with dilation 2 and congestion 2 embedding, DT_k into $G(2^m,3)$ with dilation 1 and congestion 1 embedding, TT_k with 2^m vertices or less can be embedded in $G(2^m,3)$ with dilation 2.

Fig. 4(c) is an example of 3-edge labeling of DT_3 and Fig. 6(b) is an example of its embedding in $G(2^4,3)$.

Corollary 5. TT_k with vertices 2^m or less can be embedded in Q_m with dilation 4.

By composing two embeddings, embedding a ternary tree into recursive circulants with dilation 2, and embedding recursive circulants into hypercubes with dilation 2, embedding a ternary tree with vertices 2^m or less into hypercube with dilation 4 is obtained.

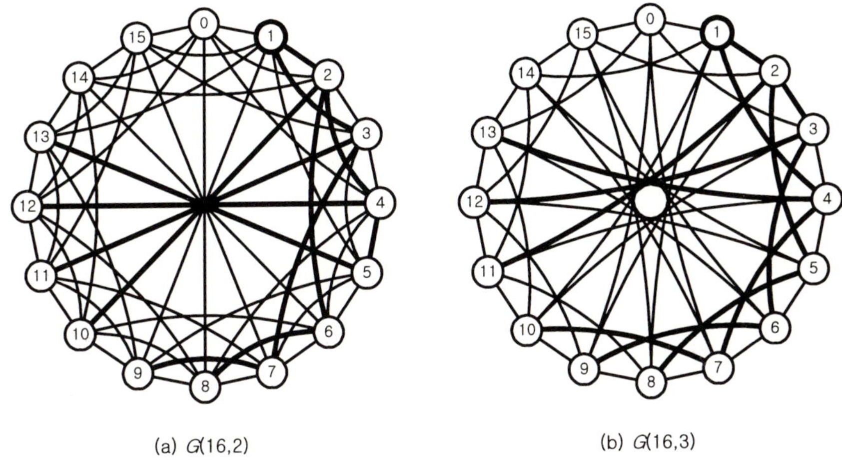

Fig. 6. Embeddings TT_3 into recursive circulants

4 Conclusion

In this paper we have investigated the embedding of full ternary trees into recursive circulants. We used d-edge labeling scheme to solve the embedding problems. We introduced a new type of tree DT_k induced from a transformation on full ternary tree TT_k, and presented a 2- and 3-edge labeling scheme on DT_k. Using these results and relationship between DT_k and TT_k, we embedded TT_k in recursive circulants $G(2^m,2)$ and $G(2^m,3)$ with dilation 2, congestion 2 and the smallest expansion. By the fact that $G(2^m,2)$ can be embedded in Q_m with dilation 2, we embedded TT_k in hypercube Q_m with dilation 4.

2-edge labeling problems on other trees and graphs, such as full k-ary trees, arbitrary trees, and meshes remain open.

References

1. N. Bagherzandeh, M. Dowd, N. Nassif, Embedding an Arbitrary Binary Tree into the Star Graph, IEEE Transactions on Computers, Vol. 45, No. 4, pp. 475-481, 1996.
2. J. C. Bermond, C. Delorme, J. J. Quisquater, Strategies for interconnection networks: some methods for graph theory, J. Parallel Distrib. Computer Vol. 3 pp. 433-449, 1986.
3. S. Bezrukov, Embedding complete trees into the Hypercube, Discrete Applied Mathematics, Vol. 100, No. 2-3, pp. 101-119, 2001.
4. F. T. Boesch and A. P. Felzer, A general class of invulnerable graphs, Networks Vol. 11, pp. 261-283, 1972.
5. W.-K. Chen and M. F. M. Stallmann, On embedding binary trees into hypercubes, Journal of Parallel and Distributed Computing, Vol. 24, No. 2, pp. 132-138, 1995.
6. A. K. Gupta and H. Wang, On Embedding Ternary Trees into Boolean Hypercubes, Proceeding of 4th IEEE Symposium on Parallel and Distributed Processing, pp. 230-235, 1992.
7. F. Harray, The maximum connectivity of a graph, Proc. Natl. Acad. Sci. USA Vol. 48, pp. 1142-1146, 1962.
8. V. Heun and E. W. Mayr, A new efficient algorithm for embedding an arbitrary binary tree into its optimal hypercube, Journal of Algorithms, Vol. 20, pp. 375-399, 1996.
9. H.-C. Kim, S.-B. Kim, K.-Y. Chwa, Fault diameter of recursive circulants. Journal of KISS. Vol. 21, No. 2, pp. 663-665, 1994.
10. H.-S. Lim, J.-H. Park, K.-Y. Chwa, Embedding trees in recursive circulants, Discrete Applied Mathematics, Vol. 69, pp. 83-99, 1996.
11. C. Micheneau, Disjoint Hamiltonian cycles in recursive circulant graphs, Information Processing Letters Vol. 61, pp. 259-264, 1997.
12. J.-H. Park and K.-Y. Chwa, Recursive circulants: a new topology for multicomputer networks, IEEE International Symposium on Parallel Architectures, Algorithms and Networks, Kanazawa, Japan, Dec. 1994.
13. J.-H. Park and K.-Y. Chwa, Recursive circulants and their embeddings among hypercubes. Journal of KISS. Vol. 22, pp. 1736-1745, 1995.
14. J.-H. Park, Hamiltonian decomposition of recursive circulants, International Symposium on Algorithms and Computation ISAAC'98 (LNCS #1533), Taejon, Korea, pp. 297-306, Dec. 1999.
15. R. Sedgewick, Algorithms in C, Addison-Wesley Publishing Company, Inc., 1990.

A Handoff Priority Scheme for TDMA/FDMA-Based Cellular Networks

Kil-Woong Jang, Sun-Woo Lee, and Ki-Jun Han

Dept. of Computer Engineering, Kyungpook National Unversity, Daegu, Korea
{jangkw, sunwlee}@netopia.knu.ac.kr and {kjhan}@bh.knu.ac.kr

Abstract. In this paper, we propose a new handoff scheme, which is designed to efficiently carry out handoffs in TDMA/FDMA-based cellular networks. The new handoff scheme is based on conventional handoff methods, which include the channel reservation, carrying and sub-rating methods to assign channels for handoffs. We evaluated the proposed handoff scheme using Markov chain with a two-cell model. The analytical results indicate that the proposed handoff scheme may offer better performance than the conventional handoff schemes in terms of the handoff blocking probability.

1 Introduction

Handoff is one of the most important techniques in cellular networks. Generally, a handoff is a mechanism that transfers a call from one cell to another while the mobile station moves through the coverage region of the cellular networks. A handoff call may be blocked if there is no free channel for the next cell. Blocking a handoff call is less desirable than blocking a new call. Therefore, several handoff schemes have been developed to prioritize handoff calls [1]. Now, two popular schemes are briefly described below.

The channel reservation scheme (CRS) offers a method of achieving successfully handoffs by reserving a number of channels exclusively for handoff calls [2]. In each cell, the channels are divided into two subset groups that are reserved for new calls and handoff calls. Some channels are allocated for new and handoff calls, while others are only allocated for handoff calls. In particular, in each cell a threshold is set. Thus, if the number of channels currently being used in the cell is below that threshold, both new and handoff calls are accepted. However, if the number of channels being used exceeds this threshold, incoming new calls are blocked and only handoff calls are admitted.

The channel carrying scheme (CCS) allows a mobile terminal to carry its current channels into the next cell without blocking the handoff request. Due to channel movement, however, it shortens the reuse distance and may violate the minimum reuse distance requirement. Here, we let δ be the minimum reuse distance. This scheme uses a (δ+1)-channel assignment scheme to ensure the minimum reuse distance requirement is not violated [1]. In addition, in order to ensure the same channels do not get closer than δ, this scheme restricts channel movement in the following way: if the configuration of a cellular system is

M.H. Shafazand and A M. Tjoa (Eds.): EurAsia-ICT 2002, LNCS 2510, pp. 883–890, 2002.

assumed to be hexagonal, channels in a cell divide into six groups. Channels of each group can then be carried into a particular cell. Detailed procedures of the channel carrying scheme can be found in [1].

In this paper, we propose a new handoff priority scheme that improves the performance of the existing approaches discussed thus far.

2 Proposed Handoff Scheme

In the proposed handoff scheme (PHS), we first divide channels in a cell into two groups, as shown in Fig. 1. The two groups are the normal channel group and the handoff channel group. Normal channels (H_{NO}) are assigned for both new and handoff calls. Conversely, handoff channels (H_{HO}) are exclusively allocated to handoff calls.

The way handoff channels are allocated depends on whether the next cell has an idle channel for the incoming handoff call. If the next cell does not have an idle channel for the incoming handoff call, the handoff call is assigned an outgoing carrying channel (H_{CO}) of an idle carrying channel (H_C) in the current cell. That is, if there is no available channel in the next cell during handoff, the mobile station is allowed to carry an idle carrying channel into the next cell. In order to avoid co-channel interference, due to channel mobility, we use a channel locking policy [2], where channels with the same channel set cannot be used when the channels in other cells are being used. However, in PHS, we modify the channel locking policy as follows. We first divide this policy into two classes: full-channel locking and half-channel locking. Full-channel locking is used when an incoming handoff call occupies an incoming carrying channel (H_{CI}) or H_C at the current cell, while half-channel locking is used when an outgoing handoff call with a H_{CO} is carried into the next cell.

The PHS exploits the channel sub-rating method proposed for cellular systems, which enhances system performance in terms of new call and handoff blocking probabilities. This method allows an occupied full-rate channel by temporarily dividing into two half-rate channels. In other words, one half-rate channel serves the existing call, while the other serves a handoff call. This method

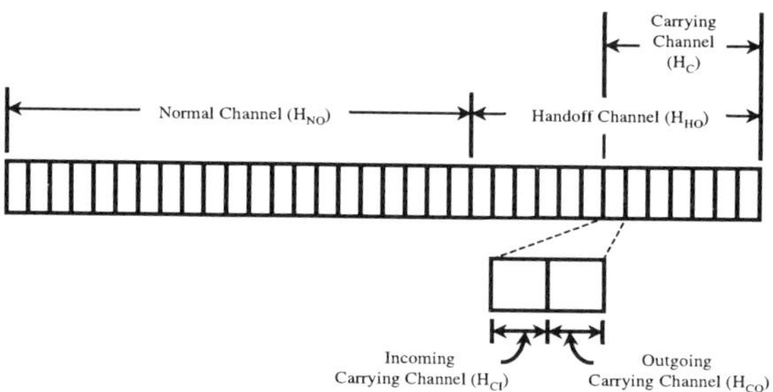

Fig. 1. Allocation of channels in the PHS

exploits the ability of mobile station of voice services to tolerate the use of half-rate vocoders with some degradation in quality of service (QoS) [3].

To describe the PHS, we focus our attention on two particular cells, which we call the current cell C_A and the next cell C_B. The handoff procedure is attempted as follows: when a handoff call request is received from C_B, the normal channel is first scanned into C_A and assigned if a free normal channel is available. Otherwise, the handoff channel is then scanned to find a free handoff channel in C_A. The handoff call is accepted into C_A if a free handoff channel is available.

However, if there is no available handoff channel in C_A, the mobile station requesting the handoff checks for the possibility of being able to carry its currently occupying channel to C_B. If the mobile station has been using a normal channel in C_B and there is an idle H_{CO} in C_B at the same time, it then exchanges its current normal channel with the handoff channel and carries it to C_A. In this case, C_A requests half-channel locking to cells with in the same channel set. If this channel in the cells with in the same channel set is currently being used, the station using this H_{CO} is changed to a H_{CI}.

On the other hand, if the mobile station has been using a handoff channel in C_B and there is an idle H_{CO} in this cell, then it exchanges its current channel with the H_{CO} and carries it to C_A. In this case, C_A also requests half-channel locking to the cells that have the same channel set. In addition, if this channel is currently being used in the cells that have the same channel set, the station using the H_{CO} is changed to a H_{CI}.

If there is no available H_{CO} in C_B, the H_C is divided into two subchannels: H_{CI} and H_{CO}. Thus, the H_{CI} is assigned to the original mobile station while it exchanges its current channel with the H_{CO} and carries it to C_A. Additionally, if the H_C in the cells is currently being used with in the same channel set, the H_{CI} serves the ongoing call. Finally, if the subscriber fails to find a channel, then the handoff attempt is terminated.

When a call using a normal channel in C_A is terminated, the normal channel is replaced by the carried channel if there is a carried channel being used by other mobile stations in C_A. Following this, the released handoff channel is returned to the cell that originally assigned it. If the released channel sets full-channel locking, the mobile station with the half-rate channel occupies the full-rate channel with the original bandwidth.

3 Performance Evaluation

In this section, we present a Markov model to obtain the blocking probability of the new and handoff calls, designated by B_N and B_H, respectively. These are the most important QoS factors in which we are interested.

It is impossible to analyze of the entire planar system. In general, performance analysis is done using one cell [6]. However, a one-cell model does not accurately capture the features of handoff schemes. To reduce the limitations of a one-cell model, we consider a two-cell model [1]. To develop the Markov chain, we made some assumptions:

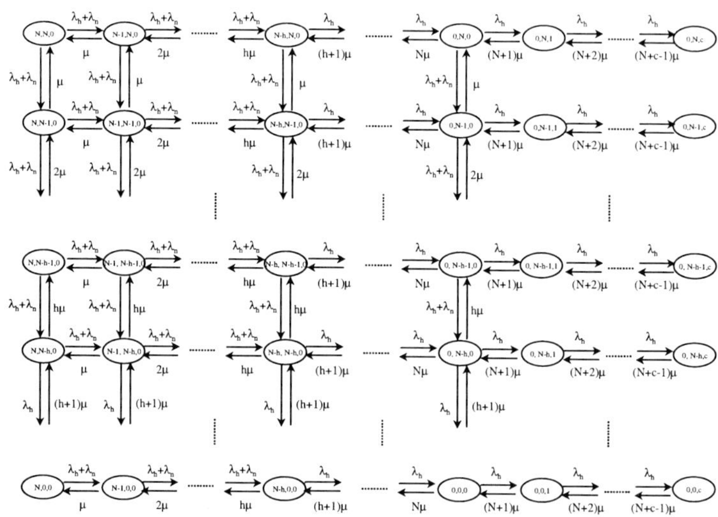

Fig. 2. State transition diagram of the Markov chain for the PHS

1. Traffic is symmetrically distributed over both cells.
2. New calls arrive according to the Poisson process at a rate λ_n.
3. Handoff calls arrive according to the Poisson process at a rate λ_h.
4. Call duration time is exponentially distributed with a mean $1/\mu$.

In addition, we define notations to describe the Markov chain. Here, we assume that the values of H_{NO} and H_C in a cell are h and c, respectively.

1. $\psi_A \in \{0, ..., N\}$ and $\psi_B \in \{0, ..., N\}$ are the number of available channels in cell C_A and cell C_B, respectively.
2. $\psi_C \in \{0, ..., c\}$ is the number of carried channels in a cell.
3. (ψ_A, ψ_B, ψ_C) is the state of the Markov chain.
4. $P(i, j, k)$ is the steady state probability of the state $\{\psi_A{=}i,\ \psi_B{=}j,\ \psi_C{=}k\}$.

We develop a Markov chain of the PHS, as shown in Fig. 2. We now show that it is easy to obtain a closed-form solution for this Markov chain. We note that

$$P(i,0,0) = \frac{\lambda_n + \lambda_h}{(N-i)\mu} P(i+1,0,0), \quad (N-h \le i < N) \tag{1}$$

Using this iterative equation (1), we can obtain

$$P(i,0,0) = \frac{1}{(N-i)!}\rho_1^{(N-i)} P(N,0,0) = S_1 P(N,0,0)$$
$$where \quad \rho_1 = \frac{\lambda_n + \lambda_h}{\mu}, \quad S_1 = \frac{1}{(N-i)!}\rho_1^{(N-i)} \tag{2}$$

We note that

$$P(i,0,0) = \frac{\lambda_h}{(N-i)\mu} P(i+1,0,0), \quad (0 \le i < N-h) \tag{3}$$

Using this iterative equation (3), we can obtain

$$P(i,0,0) = \frac{h!}{(N-i)!} \rho_2^{(N-i-h)} P(N-h,0,0) = S_2 P(N,0,0)$$
$$where \quad \rho_2 = \frac{\lambda_h}{\mu}, \quad S_2 = \frac{h!}{(N-i)!} \rho_2^{(N-i-h)} S_1 \tag{4}$$

We note that

$$P(i,j,0) = \frac{(N-j)\mu}{\lambda_h} P(i,j-1,0), \quad (0 \le i \le N) \cup (0 < j < N-h) \tag{5}$$

Using this iterative equation (5), we can obtain

$$P(i,j,0) = \frac{(N-j)!}{h!} \rho_3^{(N-j-h)} P(i,0,0) = S_3 P(i,0,0)$$
$$where \quad \rho_3 = \frac{\mu}{\lambda_h}, \quad S_3 = \frac{(N-j)!}{h!} \rho_3^{(N-j-h)} \tag{6}$$

We note that

$$P(i,j,0) = \frac{(N-j)\mu}{\lambda_n + \lambda_h} P(i,j-1,0), \quad (0 \le i \le N) \cup (N-h \le j \le N) \tag{7}$$

Using this iterative equation (7), we can obtain

$$P(i,j,0) = (N-j)! \rho_4^{(N-j)} P(i,N-h,0) = S_4 P(i,0,0)$$
$$where \quad \rho_4 = \frac{\mu}{\lambda_n + \lambda_h}, \quad S_4 = (N-j)! \rho_4^{(N-j)} S_3 \tag{8}$$

We note that

$$P(i,j,k) = \frac{\lambda_h}{(N+k)\mu} P(i,j,k-1), \ (0 \le i \le N) \cup (0 \le j \le N) \cup (0 < k \le c) \tag{9}$$

Using this iterative equation (9), we can obtain

$$P(i,j,k) = \frac{N!k!}{(N+k)!} \rho_5^k P(i,j,0) = S_5 P(i,j,0)$$
$$where \quad \rho_5 = \frac{\lambda_h}{\mu}, \quad S_5 = \frac{N!k!}{(N+k)!} \rho_5^k \tag{10}$$

Due to the chain regularities, all the values $P(i,j,k)$ are expressed as functions of the value $P(N,0,0)$ using the above equations. The value of $P(N,0,0)$

is finally determined by imposing the normalization condition, which simplifies as follows:

$$1 = \sum_{i=0}^{N} \sum_{j=0}^{N} \sum_{k=0}^{c} P(i,j,k) \tag{11}$$

Therefore, we can obtain

$$P(N,0,0) = \left(\left(\sum_{i=0}^{N-h-1} S_2 + \sum_{i=N-h}^{N} S_1 \right) \left(\sum_{j=0}^{N-h-1} S_3 + \sum_{j=N-h}^{N} S_4 \right) \sum_{k=0}^{c} S_5 \right)^{-1} \tag{12}$$

The handoff rate λ_h in the two-cell model is actually dependent of new call rates. Over all the states, λ_h is given by

$$\lambda_h = \nu \lambda_n \tag{13}$$

We can iteratively solve the Markov chain in Fig. 2 because $P(i,j,k)$ depends on λ_h. Having determined $P(i,j,k)$, we can then obtain B_N and B_H by summing the appropriate states, which are defined as the probability that all channels are active. Therefore, B_N and B_H are given by

$$B_N = \sum_{i=0}^{N-h} \sum_{j=0}^{N} \sum_{k=0}^{c} P(i,j,k) \tag{14}$$

$$B_H = \sum_{j=0}^{N} P(0,j,c) \tag{15}$$

Similarly, we develop the Markov chains of the CRS and the CCS under the assumptions previously defined. In CRS, since no channel mobility is allowed in this scheme, the factor (ψ_A, ψ_B) suffices to characterize the state of the two-cell model. B_N and B_H for the channel reservation scheme can be obtained by

$$B_N = \sum_{i=0}^{N-h} \sum_{j=0}^{N} P(i,j) \tag{16}$$

$$B_H = \sum_{j=0}^{N} P(0,j) \tag{17}$$

In CCS, the channel reuse distance is given $\delta + 1$ to ensure the minimum reuse distance requirement. Therefore, the total number of channels in a cell is fewer than the other two schemes. We denote that the N' is the total number of channels in a cell for the CCS. As in PHS, the factor (ψ_A, ψ_B, ψ_C) characterizes the state of the two-cell model for the CCS. Therefore, P_N and P_H for the CCS can be obtained by

$$B_N = \sum_{k=0}^{N'/6} \sum_{j=0}^{N'-k} P(0,j,k) \tag{18}$$

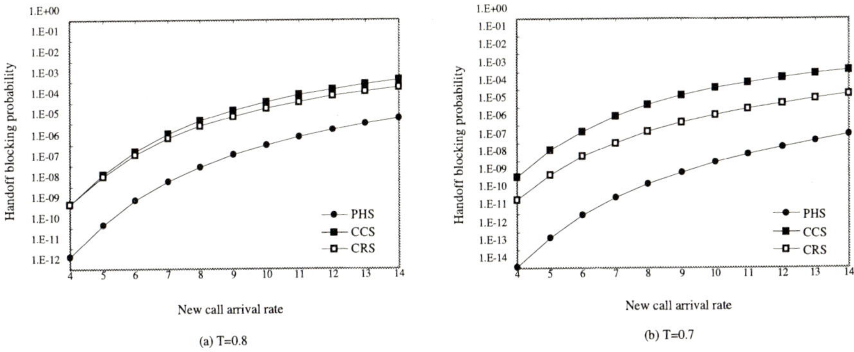

Fig. 3. Handoff blocking probability

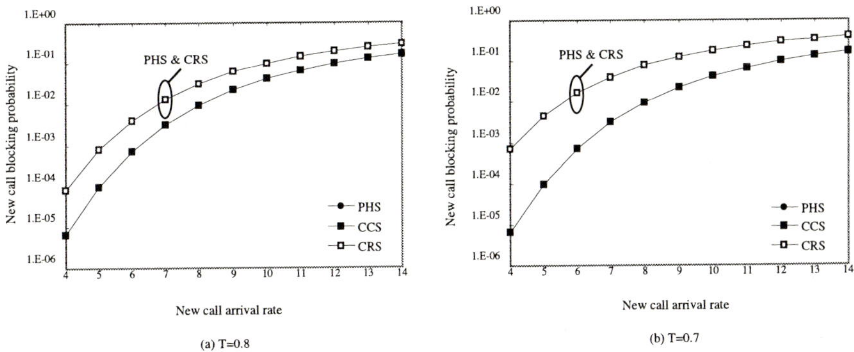

Fig. 4. New call blocking probability

$$B_H = \sum_{j=0}^{N'-N'/6} P(0, j, N'/6) + \sum_{k=0}^{N'-N'/6-1} P(0, N' - k, k) \tag{19}$$

We computed the Markov chain using several parameters. Two different values, 0.8 and 0.7, were used for h/N, which is hereafter denoted by T. In addition, we set N=20 for the PHS and the CRS, while N'=18 for the CCS. The parameters for the Markov chain, μ, ν and c, were assigned 1, 0.3 and $(N - h)/2$, respectively.

Fig. 3 shows the handoff blocking probability at various traffic loads. In this figure, we can clearly see that our scheme performs much better than the CRS and CCS under all traffic loads. The PHS offers a lower handoff blocking probability than the the CCS. This is because the number of occupied channels in a cell is greater than the CCS, whose minimum reuse distance is $\delta + 1$ instead of δ. In addition, the PHS may offer a lower handoff blocking probability than the CRS due to channel mobility. Fig. 4 shows the new call blocking probability for the three handoff schemes. In this figure, we can see that the new call blocking probability of the PHS is very close to the CRS. However, our scheme offers a higher new call blocking probability than the CCS. This is because we reserve channels for handoffs. Therefore, the number of channels for new calls in the PHS is smaller than the CCS. From Figs. 3 and 4, we can see that handoff calls have more access opportunities than new calls because the PHS reserves and carries the channel for handoff calls.

4 Conclusions

In this paper, we presented a new handoff priority scheme for efficient handoffs in TDMA/FDMA-based cellular networks. The proposed handoff scheme is based on conventional handoff methods, which include the channel reservation, carrying and sub-rating methods to assign channels for handoffs. First, we reserve a number of channels only for handoffs. Second, if there is no available channel in the next cell during handoff, the mobile station is allowed to carry a movable channel into the next cell. Finally, in order to avoid co-channel interference due to channel mobility, we adopt the sub-rating method for its carried channel.

We evaluated the proposed handoff scheme using Markov analysis. The results indicate that the proposed handoff scheme may offer a lower handoff blocking probability than the channel reservation and channel carrying schemes under all traffic loads.

References

1. 1. J. Li, Ness B. Shroff and E. K. P. Chong.: Channel Carrying: A Novel Handoff Scheme for Mobile Cellular Networks. IEEE/ACM Trans. on Networking, vol. 7, no. 1, Feb. (1999) 38-50
2. R. Ramjee, R. Nagarajan and D. Towsley.: On Optimal call admission control in cellular networks. Proc. IEEE INFOCOM, San Francisco, CA, (1996) 35-42
3. W. Zhuang, B. Bensaou and K. C. Chua.: Adaptive quality of service handoff priority scheme for mobile multimedia networks. IEEE Trans. Veh. Technol., vol. 49, No. 2, Mar. (2000) 494-505
4. L. J. Cimini, G. J. Foschini, C. L. I and Z. Miljanic.: Call blocking performance of distributed algorithms for dynamic channel assignment. IEEE Trans. Commun., vol. 42, no. 8, Aug. (1994) 2600-2607
5. M.H. Chiu and M. A. Bassiouni.: Predictive Schemes for Handoff Prioritization in Cellular Networks Based on Mobile Positioning. IEEE J. Select. Areas Commun., vol. 18, no. 3, Mar. (2000) 510-522
6. W. S. Jeon and D. G. Jeong.: Comparison of Time Slot Allocation Strategies for CDMA/TDD Systems. IEEE J. Select. Areas Commun., vol. 18, no. 7, July (2000) 1271-1278
7. S. Choi and G. Shin.: Predictive and adaptive bandwidth reservation for hand-offs in QoS-sensitive cellular networks. Proc. ACM SIGCOMM'98, (1998) 155-166

On Delay Times in a Bluetooth Piconet: The Impact of Different Scheduling Policies

Jelena Mišić and Vojislav B. Mišić

The Hong Kong University of Science and Technology
Clear Water Bay, Kowloon
Hong Kong SAR, China.

Abstract. The performance of a single Bluetooth piconet is analyzed using the theory of M/G/1 queues with vacations. Analytical results for probability distributions of packet access time and service cycle time are derived. Two scheduling policies, limited and exhaustive service, are considered. In general, exhaustive scheduling was found to perform better than limited service. Results were confirmed through simulations.

Keywords: *Bluetooth, Bluetooth piconet, queueing analysis, queues with vacations, limited service scheduling, exhaustive service scheduling*

1 Introduction

Among modern wireless LAN technologies, Bluetooth holds great promise – not only as a simple cable replacement solution, but also as a feasible solution for short range ad hoc networking [2, 11]. A number of distinctive features have significant impact on the performance of Bluetooth networks; yet performance analyses of Bluetooth networks, esp. at the media access control (MAC) level, are still scarce, and a number of issues have yet to be resolved. Among the most prominent of these is the actual scheduling policy, which is not prescribed by the current Bluetooth specification [2].

The performance of different scheduling algorithms has been analyzed to some extent; a brief overview of those results is given in section 2. Most of these were based on discrete event simulation, although an analytical approach based on the theory of queues with vacations is also feasible [13]. In this paper, we present a queueing analysis of the Bluetooth bidirectional polling system within a single piconet, consisting of $m \leq 8$ members: one master device and $m - 1$ slave devices, as shown in Fig. 1.

We limit our analysis to two scheduling policies, neither of which requires that the piconet master knows the status of the slaves' queues, and therefore does not incur additional signaling overhead between the master and the slaves. In *limited service scheduling*, piconet master polls each slave once, and then moves on to the next slave, regardless of whether an actual data transmission has taken place or not. In *exhaustive service scheduling*, the master polls the same slave as long as there are packets to be transmitted to and/or from that slave, and moves on only when both queues are empty; this condition is detected when the slave responds with a NULL packet to a

M.H. Shafazand and A M. Tjoa (Eds.): EurAsia-ICT 2002, LNCS 2510, pp. 891–898, 2002.

POLL packet sent by the master. Note that, regardless of the service discipline, we assume that the selection of next slave to be polled follows the round-robin principle.

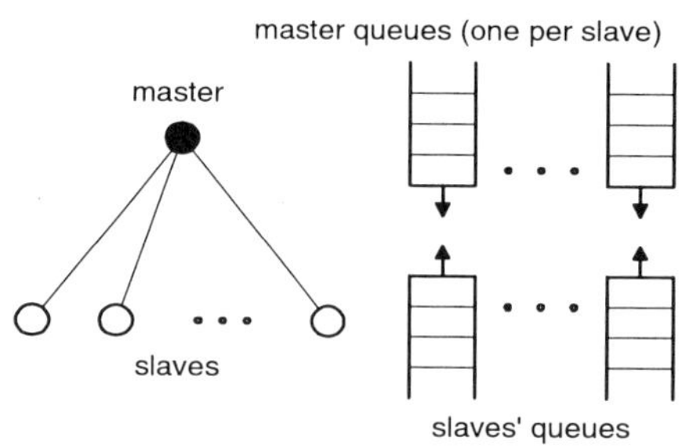

Fig. 1. A single piconet and its queueing model.

The rest of the paper is organized as follow: after an overview of related work in Sec. 2, we discuss the analytical approach to finding the delays in case of limited service and exhaustive service scheduling, in Secs. 3 and 4, respectively. Then we discuss the performance of these two policies, through analytical solutions for delay variables, as well as simulation results. Finally, we conclude the paper and give some pointers for future research.

2 Related Work

Scheduling policy is one among several important issues not yet prescribed by the Bluetooth specification [2]. So far, most authors have analyzed just the simplest scheduling policies, limited service and exhaustive service, or variations thereof. Among them, Johansson *et al.* discuss the performance of different scheduling algorithms using discrete event simulation [6], and conclude that limited service scheduling offers lower waiting time than the exhaustive service one, esp. under medium to heavy traffic load. They also argue strongly in favor of using longer packets (i.e., three- and five-slot ones) because of their greater efficiency, compared to the single-slot packets.

Kalia *et al.* examine the master-driven TDD protocol used in Bluetooth and propose several scheduling and segmentation-reassembly (SAR) policies at the MAC level [7]. They have performed simulations of a simple Bluetooth piconet in order to measure throughput, and have shown that more sophisticated scheduling policies, such as exhaustive service scheduling, can improve performance compared to the simple limited service round robin scheduling. In a subsequent paper [8], more information on the performance of different scheduling policies is provided, together with some guidelines—albeit without too much elaboration—about the preferred distribution of packet lengths. However, their work relies on the assumption that the master is aware of the state of slaves' queues, so that optimum scheduling decisions can be made. Such knowledge cannot be readily obtained in Bluetooth, where the master always addresses one slave only, hence the information about the current state of other slaves is simply not available. Furthermore, the structure of Bluetooth packet headers makes it quite difficult to signal even the state of the polled slave's queue, and such information can only be exchanged at the expense of the payload space within a packet [2]. Still, such an assumption might be justifiable from the viewpoint of higher layers of the protocol stack, which by default incur some administrative overhead.

Capone *et al.* compare the performance of limited service (called 'pure round robin') and exhaustive service scheduling, as well as that of some variations thereof, through discrete event simulation [3]. They found that the exhaustive service scheduling offers best performance, except at very high traffic loads where the limited service policy provides lower overall delays. However, the focus of their work is the impact of scheduling policies on the performance of higher-level protocol layers, including TCP/IP, rather than the MAC layer. Hence, their traffic model includes transfer of sequences of packets, patterned to closely resemble file transfers, rather than individual packets.

Other authors have also investigated the scheduling problem from the viewpoint of higher layers of the protocol stack, by simulating the performance of TCP/IP traffic transmitted using the Bluetooth MAC [4, 9]. The performance of constant-bit rate traffic, such as voice, that uses the UDP protocol under the same circumstances is also analyzed through simulations.

It is interesting to note that most of the aforementioned work is based exclusively on simulations, although an alternative approach based on theory of queueing systems with vacations [13] – adapted, of course, to the characteristics of Bluetooth networks – is also possible. Such adaptations are necessary because previous work on analysis of service policies that include bidirectional master-slave traffic cannot be directly applied to Bluetooth: for example, some authors analyze unidirectional traffic only [12], while others use a combination of exhaustive and limited service for downlink and uplink traffic, respectively [10]. In particular, no paper that we know of deals explicitly with the master-driven TDD communication mechanism as used in Bluetooth. Still, the queueing theoretic analysis turns out to be feasible, and the solutions can be obtained in closed form, as will be seen in subsequent discussion.

3 Performance of Limited Service Scheduling

The operation of a piconet may be characterized with two main performance indicators: access delay W_a, the time a data packet has to wait in the uplink queue of the source device before it is serviced, and end-to-end delay W_e, which stands for the time a packet has to spend in 'transit', i.e., from the moment it enters the uplink queue at the source device, to the time it arrives at its destination device. Our analysis will focus on those two delays and their dependency on the chosen scheduling policy.

We assume that each slave generates bursts (batches) of packets that follow a Poisson distribution with arrival rate λ. All packets within a burst have the same destination node, and the distribution of destination nodes is uniform. The master device itself does not generate any traffic – it just routes packets between their corresponding source and destination devices.

The probability distribution of length of the packet burst may be described with the probability generating function (PGF) $G_b(x) = \sum_{k=0}^{\infty} b_k x^k$, where b_k is the probability that the burst will contain exactly k packets. Mean value of the burst length is $\overline{B} = G_b'(1)$, while its second factorial moment is defined as $\overline{B^{(2)}} = E[B(B-1)] = G_b''(1)$. The equivalent Laplace-Stieltjes transform (LST) of the probability distribution may be obtained by substituting the variable x with e^{-s} [13]; for example, the LST of the packet burst length PDF is $G_b^*(s) = \sum_{k=0}^{\infty} b_k e^{-ks}$.

The probabilities of packets being one, three, and five slots long are p_1, p_3 and $p_5 = 1 - p_1 - p_3$, respectively. The corresponding probability generating function (PGF) is $G_p(x) = p_1 x + p_3 x^3 + p_5 x^5$. First and second moments of the packet length distribution are equal to $\overline{L} = G_p'(1)$ and $\overline{L^2} = G_p''(1) + G_p'(1)$, respectively, and its LST is $G_p^*(s) = p_1 e^{-s} + p_3 e^{-3s} + p_5 e^{-5s}$.

Piconet service cycle time. Let us define the service cycle as the time in which the piconet master services all of its slaves once. Since our model is symmetrical with respect to slave devices, it suffices to consider just one master-slave channel which may be modeled as a pair of queues, as shown in Fig. 1. Under the assumptions outlined above, the burst arrival rates for the slave (uplink) and master (downlink) queue will be $\lambda_u = \lambda$ and $\lambda_d = \lambda$, respectively. If X_c denotes the length of the piconet cycle time, the probabilities that the channel queues are not empty will be $P_d = \lambda_d \overline{B}\, \overline{X_c}$ and $P_u = \lambda_u \overline{B}\, \overline{X_c}$ for master and slave queues, respectively. The durations of the downlink and uplink communications may be described with the following PGFs:

$$G_d(x) = (P_d p_1 + (1 - P_d))x + P_d p_3 x^3 + P_d p_5 x^5 \tag{1}$$

$$G_u(x) = (P_u p_1 + (1 - P_u))x + P_u p_3 x^3 + P_u p_5 x^5 \tag{2}$$

while the PGF for the duration of the cycle time will be

$$G_{X_c}(x) = (G_d(x) G_u(x))^{(m-1)} \tag{3}$$

The mean value of the service cycle time may be obtained as $\overline{X_c} = G_{X_c}'(1)$, and its second moment as $\overline{X_c^2} = G_{X_c}''(1) + G_{X_c}'(1)$.

Server vacation time. In queueing theory, a vacation may happen in the case when multiple queues are serviced by a single server [13]. When the server finds an empty client queue, it goes on to service other clients, i.e., takes a vacation until the next visit to this client queue. The vacation time is the time while the server is busy servicing other client queues (of which there are $m - 2$). In a Bluetooth piconet, a server vacation starts when the master polls a slave and finds its uplink queue to be empty (slave has replied with a NULL packet), and decides to move on to next slave. (This does not apply to the downlink queue, as the TDD communications in Bluetooth piconets are, by definition, master-driven [2].) The duration of the vacation period V_l (where the index l stands for limited service scheduling) may be described with the following PGF:

$$G_{V_l}(x) = x G_d(x) \left(G_d(x) G_u(x)\right)^{m-2} \tag{4}$$

and its first and second moments are $\overline{V_l} = G_{V_l}'(1)$ and $\overline{V_l^2} = G_{V_l}''(1) + G_{V_l}'(1)$, respectively.

Access and end-to-end delay calculation. We are now able to calculate the LST for distribution of the access delay at the slave queue as [13]:

$$W_a^*(s) = \frac{1 - V_l^*(s)}{s\overline{V_l}} \cdot \frac{s(1 - \lambda \overline{B}\, \overline{X_c})}{s - \lambda + \lambda G_b(X_c^*(s))} \cdot \frac{1 - G_b(X_c^*(s))}{\overline{B}\,(1 - X_c^*(s))} \tag{5}$$

where $V_l^*(s)$ and $X_c^*(s)$ denote the LST of the probability distributions of vacation time and cycle time, respectively. The average access delay at the slave is then calculated as

$\overline{W_a} = -W_a^{*'}(0)$, or:

$$\overline{W_a} = \frac{\lambda \overline{B}\ \overline{X_c^2}}{2(1 - \lambda \overline{B}\ \overline{X_c})} + \frac{\overline{B^{(2)}}\ \overline{X_c}}{2\overline{B}(1 - \lambda \overline{B}\ \overline{X_c})} + \frac{\overline{V_l^2}}{2\overline{V_l}} \tag{6}$$

The queueing delay at the piconet master may be described with an expression similar to eqn. 5. The LST of the end-to-end delay is $W_e^*(s) = W_a^*(s)W_m^*(s)$ and, consequently, the mean value of the end-to-end delay is $\overline{W_e} = \overline{W_a} + \overline{W_m}$. However, the burstiness of the traffic in the downlink (master) queue will differ from that of the traffic in the slave (uplink) queue. This difference is due to the fact that bursts from different source slaves with the same destination will become interleaved in the same downlink queue, which will in turn lead to an equivalent decrease in burst length. Exact analysis of this phenomenon is fairly involved, esp. in the context of Bluetooth transmission mechanisms, and we will only present an approximate model for the decrease of the burst length $\overline{B} = 1/p_B$, where p_B is the parameter of geometric distribution [5]. The probability that the two bursts will not overlap in time (and, consequently, that they will not be interleaved in the same downlink queue) will be $1 - 1/(m-2)^2$. Therefore, given the parameter p_B of the geometric distribution of burst length at the source slave, the equivalent parameter p_{Bm} of the packet burst at the master (downlink) queue will be

$$p_{Bm} = p_B\left(1 - \frac{1}{(m-2)^2}\right) + \frac{1}{(m-2)^2} \tag{7}$$

The new equivalent mean burst length will be $\overline{B_m} = 1/p_{Bm}$. In order to maintain the same server utilization under decreased burst length, the burst arrival rate has to be scaled so that $\lambda_{dm}\overline{B_m} = \lambda_d\overline{B}$. With these changes, the mean queueing delay in master downlink queues will become

$$\overline{W_m} = \frac{\lambda_{dm}\overline{B_m}\ \overline{X_{c1}^2}}{2(1 - \lambda_{dm}\overline{B_m}\ \overline{X_{c1}})} + \frac{\overline{B_m^{(2)}}\ \overline{X_{c1}}}{2\overline{B_m}(1 - \lambda_{dm}\overline{B_m}\ \overline{X_{c1}})} + \frac{\overline{V_1^2}}{2\overline{V_1}} \tag{8}$$

4 Performance Analysis of the Piconet under the Exhaustive Service Policy

In case of exhaustive service scheduling policy, a number of packet pairs (downlink followed by an uplink) may be exchanged between the master and a single slave during a single visit to that slave. The actual number is equal to the number of packets in the downlink or the corresponding uplink queue, whichever is larger, plus one; this exchange is required for coordination at the beginning or at the end of the exchange. Let X_{ms} denote the time to serve a single channel, i.e., the time to empty both channel queues for one particular slave. Its PGF is

$$X_{ms}(x) = e^{(\lambda_u+\lambda_d)X_c(x)(G_b(G_p(x))-1)} \cdot e^{|\lambda_u-\lambda_d|X_c(x)(G_b(x)-1)} \cdot x^2 \tag{9}$$

Let X_c denote the piconet service cycle time; its PGF is $X_c(x) = X_{ms}^{m-1}(x)$. Then, the PGF of the server vacation time is

$$G_{V_e}(x) = x^2 X_{ms}^{m-2}(x) e^{\lambda_d X_c(x)(G_b(xG_p(x))-1)} \tag{10}$$

and its first and second moments are $\overline{V_e} = G'_{V_e}(1)$ and $\overline{V_e^2} = G''_{V_e}(1) + G'_{V_e}(1)$, respectively.

The LST for the access delay probability distribution is:

$$W_a^*(s) = \frac{1 - V_e^*(s)}{s\overline{V_e}} \cdot \frac{1 - G_b(G_p^*(s))}{\overline{B}\,(1 - G_p^*(s))} \cdot \frac{s(1 - 2(m-1)\lambda\overline{B}\,\overline{L})}{s - 2(m-1)\lambda + 2(m-1)\lambda G_b(G_p^*(s))} \tag{11}$$

which translates into the mean access delay of the form

$$\overline{W_a} = \frac{2(m-1)\lambda\overline{B}\,\overline{L^2}}{2(1 - 2(m-1)\lambda\,\overline{B}\,\overline{L})} + \frac{\overline{B^{(2)}}\,\overline{L}}{2\overline{B}(1 - 2(m-1)\lambda\overline{B}\,\overline{L})} + \frac{\overline{V_e^2}}{2\overline{V_e}} \tag{12}$$

The LST of delay at the master queue, as well as of the end-to-end delay, have the same form as in the previous case, although, of course, the individual component times differ.

5 Performance Comparison

In order to validate our theoretical analysis, we have designed a Bluetooth piconet simulator using the Artifex Petri Net simulation engine [1]. We considered a piconet with a maximum size of $m = 8$, with Poisson-distributed burst arrivals and geometrically distributed packet burst size. Mean packet length was $\overline{L} = 3$, assuming that $p_1 = p_2 = p_3 = 1/3$. Mean access delay times as functions of burst arrival rate λ and mean burst size $\overline{B}$ are shown in Fig. 2, while the corresponding dependencies of mean end-to-end delay times are shown in Fig. 3. In both figures, the diagram on the left corresponds to limited service scheduling, while the one on the right corresponds to exhaustive service. We have also plotted the ratios of access delay and end-to-end delay times as functions of mean burst size, Fig. 4(a), and burst arrival rate, Fig. 4(b).

The exhaustive service scheduling policy clearly performs better in a wide range of burst arrival rates and mean burst sizes; the difference decreases only at very high packet arrival rates and/or long burst sizes (which of little practical value as the overall delays will be unacceptably long anyway). The difference is larger with access delay than with end-to-end delay; yet even in the latter case, the exhaustive service scheduling outperforms limited service scheduling by 20 to 25% on the average. In particular, the exhaustive service appears to be more capable of accommodating bursty traffic, as witnessed by the much flatter surface shown in Fig. 2(b). It may be interesting to note that the limited service scheduling performs slightly better for non-bursty traffic (i.e., Poisson packet arrivals with $B = 1$). This is probably due to the fact that, under exhaustive service, the presence of a packet in a queue will cause the master to remain with the slave, and the probability of the next packet arriving within a single time frame is usually low. (Under limited service scheduling, the master polls each slave only once, regardless of the presence or absence of a data packet in any of the queues.)

Our results confirm the conclusions of other authors [3, 4, 8]. Limited service has been found to perform marginally better at very high traffic loads in [3]; our results do indeed show that the ratios of delay times increase towards one at high arrival rates.

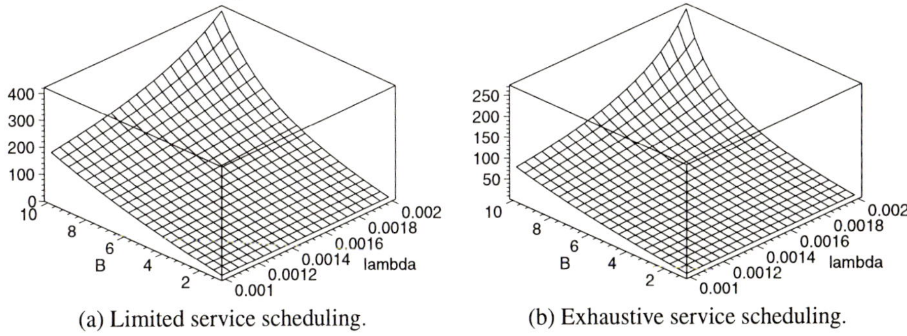

(a) Limited service scheduling. (b) Exhaustive service scheduling.

Fig. 2. Analytical solutions for mean access delay as a function of packet burst arrival rate and mean burst size.

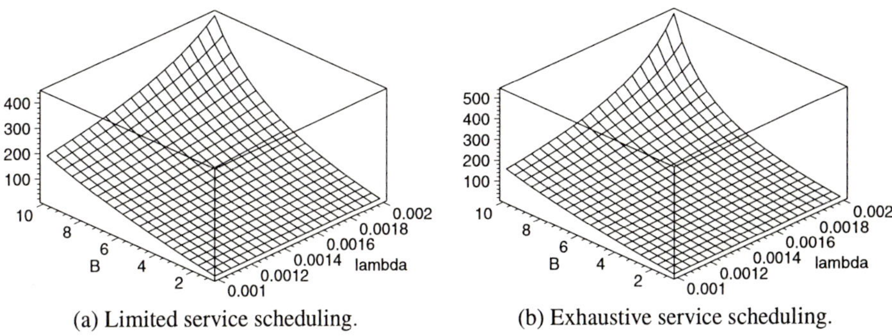

(a) Limited service scheduling. (b) Exhaustive service scheduling.

Fig. 3. Analytical solutions for end-to-end delay as a function of packet burst arrival rate and mean burst size.

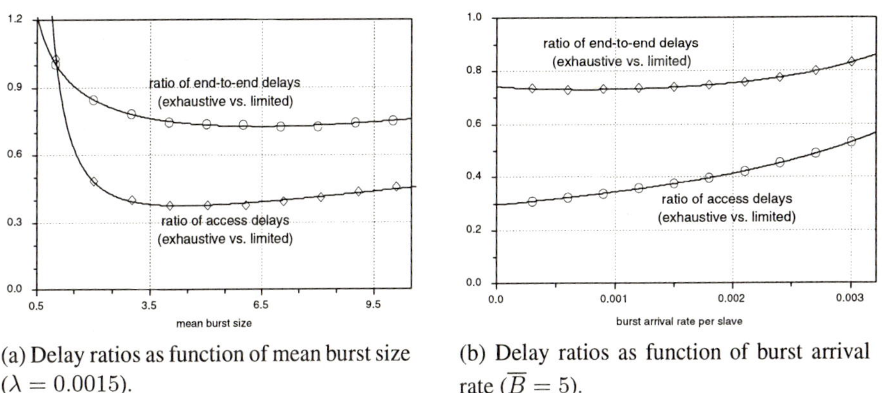

(a) Delay ratios as function of mean burst size ($\lambda = 0.0015$).

(b) Delay ratios as function of burst arrival rate ($\overline{B} = 5$).

Fig. 4. Comparing the performance of limited service and exhaustive service scheduling: simulation results.

6 Concluding Remarks

In summary, we may conclude that exhaustive scheduling offers better performance than limited service scheduling in Bluetooth piconets. An improved scheduling policy that allows a fixed number of packets to be transmitted to and from a slave during a single visit, was found to offer much better performance than limited service, while at the same time providing a measure of flexibility that can be beneficial in the presence of QoS constraints. Moreover, we have developed a queueing theoretical analysis framework that is applicable to any scheduling policy. Promising directions for further research appear to be analysis of more complex Bluetooth networks (scatternets), as well as the investigation of other scheduling policies.

References

1. *Artifex v.4.2.1.* Artis Software Inc., San Jose, CA, 2001.
2. Bluetooth SIG. *Specification of the Bluetooth System.* Version 1.1, Feb. 2001.
3. A. Capone, M. Gerla, and R. Kapoor. Efficient polling schemes for Bluetooth picocells. In *Proceedings of IEEE International Conference on Communications*, volume 7, pages 1990–1994, Helsinki, Finland, June 2001.
4. A. Das, A. Ghose, A. Razdan, H. Saran, and R. Shorey. Enhancing performance of asynchronous data traffic over the Bluetooth wireless ad-hoc network. In *Proceedings Twentieth Annual Joint Conference of the IEEE Computer and Communications Societies IEEE INFOCOM 2001.*, volume 1, pages 591–600, Anchorage, AK, Apr. 2001.
5. G. R. Grimmett and D. R. Stirzaker. *Probability and Random Processes.* Oxford University Press, Oxford, UK, 2nd edition, 1992.
6. N. Johansson, U. Körner, and P. Johansson. Performance evaluation of scheduling algorithms for Bluetooth. In D. H. K. Tsang and P. J. Kuhn, editors, *Proceedings of BC'99 IFIP TC 6 Fifth International Conference on Broadband Communications*, pages 139–150, Hong Kong, Nov. 1999.
7. M. Kalia, D. Bansal, and R. Shorey. MAC scheduling and SAR policies for Bluetooth: A master driven TDD pico-cellular wireless system. In *Proceedings Sixth IEEE International Workshop on Mobile Multimedia Communications (MOMUC'99)*, pages 384–388, San Diego, CA, Nov. 1999.
8. M. Kalia, D. Bansal, and R. Shorey. Data scheduling and SAR for Bluetooth MAC. In *Proceedings VTC2000-Spring IEEE 51st Vehicular Technology Conference*, volume 2, pages 716–720, Tokyo, Japan, May 2000.
9. A. Kumar, L. Ramachandran, and R. Shorey. Performance of network formation and scheduling algorithms in the Bluetooth wireless ad-hoc network. *Journal of High Speed Networks*, 10:59–76, Oct. 2001.
10. D. R. Manfield. Analysis of a priority polling system for two-way traffic. *IEEE Transactions on Communications*, 33(9):1001–1006, Sept. 1985.
11. B. A. Miller and C. Bisdikian. *Bluetooth Revealed: The Insider's Guide to an Open Specification for Global Wireless Communications.* Prentice-Hall, Upper Saddle River, NJ, 2000.
12. H. Takagi. Queuing analysis of polling models. *ACM Computing Surveys*, 20(1):5–28, Mar. 1988.
13. H. Takagi. *Queueing Analysis*, volume 1: Vacation and Priority Systems, Part 1. North-Holland, Amsterdam, The Netherlands, 1991.

Intelligent Paging Strategy in 3G Personal Communication Systems*

I. Saha Misra[1], S. Karmakar[1], M.S. Mahapatra[1], P.S. Bhattacharjee[2], Debasish Saha[3], and A. Mukhertjee**[4]

[1] Depart. of Electronics & Telecomm. Engineering, Jadavpur University, Kolkata 700 032, India., itimisra@cal.vsnl.net.in

[2] Telephone Bhawan, 34 B. B. D. Bag, Kolkata 700 001, India

[3] MIS and Computer Science, IIMC, Joka, Kolkata 700 104, India, ds@iimcal.ac.in

[4] PricewaterhouseCoopers Ltd., Salt Lake, Kolkata 700091, India, amitava.mukherjee@in.pwcglobal.com

Abstract. In third generation personal communication system a significant amount of signaling load will have to be carried by the finite capacity radio bandwidth owing to the diverse functions of the huge number of the mobile terminals. Here, we propose and analyze an intelligent paging method that reduces considerable amount of paging signaling load and cost of locating an MT subject to a predefined delay constraint. The price to be paid for this saving of radio bandwidth is the extra processing power. An analytical model is developed so as to describe the performance for different CMR. The performance of our proposed scheme is compared with [1] and it is shown that the method operates well under different delay bounds.

1 Introduction

IN personal cellular communication network (PCN), unlike simple static networks like telephone, as the user changes his position from one location to another, the network must keep track of the present location of the user. The MT has to track following two basic processes: location registration and terminal paging. Location registration is associated with the update of the current position of the mobile terminal. The location information of each MT is generally kept on a database called HLR (Home Location Register). Whenever an MT does a location update, the corresponding database is modified accordingly. Terminal paging is needed to find an MT whenever a call comes to that MT, the cells in that location area is paged according to some selected algorithm.

In this paper we focus on the paging problem under different delay bounds. The two main factors of paging are delay and cost. Delay factor comes for QoS requirement for multimedia services and cost of paging is related to the efficiency of bandwidth utilization and should be minimized under delay bounds

* This work is supported by UGC and AICTE, India

** Author of correspondence

M.H. Shafazand and A M. Tjoa (Eds.): EurAsia-ICT 2002, LNCS 2510, pp. 899–906, 2002.

[2–4]. If the MT is situated nearby the center cell (the cell where last location registration was done) then many unwanted cells are to be paged for blanket paging increasing the cost of paging. So for searching the MT, a well-defined cost effective paging strategy has to be chosen. Many paging schemes have been introduced [2–6] to improve the efficiency of bandwidth utilization. The paging cost is depended on the location update procedure and location. In this paper, we propose a new paging scheme that introduces both the randomness and directional probability (DP) of the MT. It considers good aspect of both SP and SDF paging. We have divided each hexagonal location area into six sectors. Each sector is given a probability according to the direction of entry of the MT. This paper is organized as follows. Section 2 describes the system model. In section 3 analytical model is given. Section 4 deals with performance evaluation, with the results and discussions. Conclusion is drawn in section 5.

2 System Model

For paging we have introduced a new paging strategy namely Sectored Sequential Paging or in short SSP. In this scheme we have divided the LA into different sub-areas. Now, according to the direction of entry of an MT into a location area, it is again divided into six sectors. Figure1 shows how a hexagonal cell is divided. It also shows the probable direction of entry point and exit point within the cell. It is assumed that the edge at which the MT enters has the least probability to exit and the edge opposite to it has the highest probability of being the exit point i.e., each sector has some weighted probability according the direction of entrance.

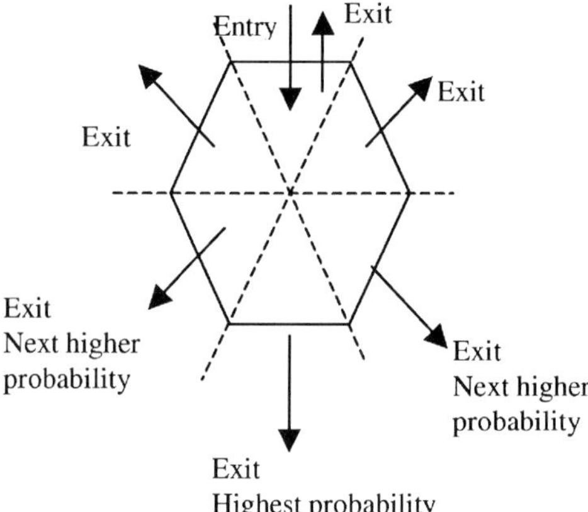

Fig. 1. Sectorization of a hexagonal cell showing directional movement of MT.

Now, when a call comes to a certain MT, the sub-area with the highest occupancy probability (according to the RWM) that falls in the sector having the highest probability is paged first. Next the same sub-area of the sector having the next higher probability is paged in the same fashion. When a sub-area is paged completely, paging proceeds to the sectored sub-area with the next highest probability. In this manner all the sub-areas in the six sectors are paged until the MT is found.

When dividing the LA into sectors fraction of a cell may fall under one sector and the remaining on another sector as shown in Figure 2. This causes the problem of fractional cell paging in analytical model. In practice a fraction of a cell cannot be paged. It should be fully paged or not to be paged at a time. So, when a half-cell is paged first, it is excluded in the second sector. Also an entire sector need not be paged in a single polling cycle. We may divide a sector into a number of sub-sectors as in SDF [1]. This has an advantage that the number of cells paged in a single polling cycle gets reduced thereby making it feasible to handle blocking and also reducing cost.

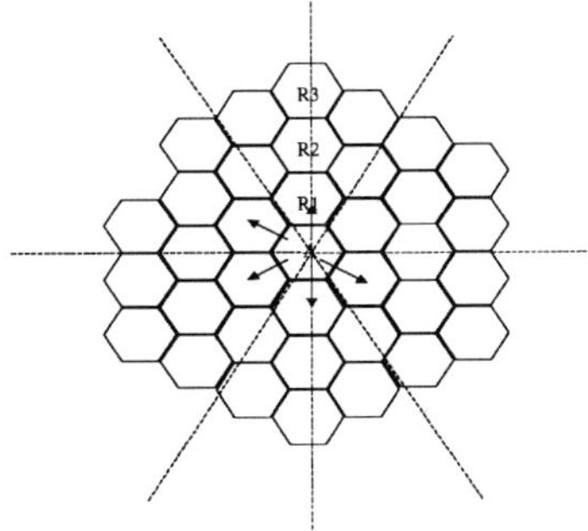

Fig. 2. Sharing of cells within the sectors.

3 Analytical Model

3.1 RWM Model

We have assumed that,

i. All the cells are of same size and the cells are of hexagonal shape
ii. The cell residence time of an MT follows Gamma distribution with mean $1/\lambda$ m

We denote the pdf for cell residence time is $f_m(t)$, which has Laplace-Stieltjes transform $f_m{}^*(s)$. For the ease of analytical modeling, LA is divided into rings surrounding the center cell. Number of cells denoted in ith ring is given by g(i) = 6i [1]. For movement threshold value d, the movement-based location update scheme guarantees the MT is within a maximum distance (d-1) starting from the center cell. This is the residing area of the MT.

If the call arrival to the MT follows Poisson distribution with rate $\lambda_c, \alpha(K)$ being the probability that there are K boundary crossings between two call arrivals is calculated as given in [7]. We assume MT moves to each of its neighboring cell from its residing cell with equal probability 1/6. The 2-D hexagonal RWM can be converted into a simple 1-D hexagonal model with one absorbing and one reflecting state and P_K is formed from KxK transition matrix of 1-D hexagonal RWM. Hence the probability β (k, K) for the hexagonal random walk model is calculated from [1].

3.2 Model for Sectored Sequential Paging

Pseudo RWM

For SSP, LA is divided into six sectors as shown in Fig.2. We consider a probability distribution function for selecting a sector with the following equation

$$\mu(r1, r2) = \frac{1}{r1} + \frac{1}{r2} \times \cos(\theta/2) \tag{1}$$

Where, the first term is for randomness and the second term is for direction based movement of the MT. θ is the angle of deviation from the angle at which the MT has entered the center cell. Depending on the mobility pattern of the MT proper weight has to be given to the first and second term of the above equation. Indeed the movement of MT is not as random as RWM. So concept of introducing directional probability comes into play. As the sum of all probability of being the MT in all sectors is equal to one i.e.

$$\sum_{i=1}^{6} \mu_{\mathrm{i}}(\mathrm{r}1, r2) = 1 \tag{2}$$

$$r2 = \frac{6 \times r1}{r1 - 2.9} \tag{3}$$

By varying one of r1 or r2 we can fix the value of μ_{i}.

To get the effective number of cells to be paged to find an MT, we use the novel scheme of multiplying the cells being paged with a function $\chi(\mu)$.

For increasing value of μ, χ decreases. The example of such functions are $[1 - \mu]$, $[1 - \exp\,(\mu)]$ or $[\exp\,(-\mu)]$. The function should be chosen according to LA geography and movement of MT. If the area topology forces the MT more to follow straight-line direction to some extent, we choose functions such as $[\exp\,(-\mu)]$ which varies rapidly with change in μ. If such is not the case then linear function like $[1\ -\mu]$ can be chosen.

If the cluster of cells being paged simultaneously is referred as N (i, j), i = 1,2 … 6, j=1,2 … n, n = number of sub-area then the effective number of cells being paged is $\sum$ N(i,j)$^*\chi(\mu)$ over the entire LA. Let us call $\mathrm{w_k}$ (i, j) being the total effective number of cells being paged simultaneously in k-th paging delay in i-th sector and j-th sub-area.

The average paging cost for detection of an MT per call is

$$C\nu = \sum_{i=1}^{6} \sum_{j=0}^{l-1} \frac{\rho(j)w(i,j)}{6} x\,\mathrm{V} \tag{4}$$

Where, V is the paging cost /cell. Here ρ (j) is replaced by ρ (j)/6 because each sub-area is divided into six equal sectors for which occupancy probability in each sectored sub-area has become 1/6 of the original occupancy probability. According to [1] if π_i be the probability that the MT is located in any of the ring-i at the time of call arrival to an MT, then,

$$\pi_i = \sum_{k=0}^{\infty} \alpha(K)\beta(i, k\mathrm{mod}d) \tag{5}$$

and $\rho(j)$ is the probability that MT is in the j-th sub-area

$$\rho(j) = \sum_{r_i \in A_j} \pi_i\,. \tag{6}$$

Normal Distribution Model

The directive pseudo random walk model was modeled by the two variables r1 and r2. The Normal distribution with mean =0 has the form:

$$f(x) = \frac{1}{\sqrt{2\pi\sigma}} \exp(-\frac{x^2}{2\sigma^2}) \tag{7}$$

By varying the variance s we can change the shape of the distribution. The model that is being discussed here takes as input two values:

- The probability of occupancy at the 0 th sector (Δ)
- The variance of the Normal distribution curve (σ)

Let us consider that at $x = x', \Delta = f(x)$, we divide area between x' and 3σ in the positive X direction into three parts. It is repeated for the negative X side also. The area under the curve in each of the three areas going from center to outward is the probability of sector 1, 2, 3 respectively. The probability is assigned to μ_i according to the sector number and the process that is described in the previous section repeated to find the paging delay.

4 Performance Evaluation

The paging cost is determined both for SDF strategy and for SSP strategy considering directional probability. The cost is determined under different polling delay for various calls to mobility ratio and is compared for both the pseudo random and normal distribution SSP paging strategy.

In Table 1, the variation of total cost with respect to movement threshold value d, for different values of delay (η) call to mobility ratio (CMR) are given. In each table paging cost is compared both for SDF (shortest distance first [1]) and SSP (sector sequential paging), our proposed scheme. Let us first consider the variation of CMR for delay (η) = 5 and other parameters as written. From Table 1, it is seen that paging cost can be reduced for an optimum value of movement threshold value d. The most prominent things that are observed from this table are that cost/call is less for SSP than SDF for every CMR ratio. This effect is very much prominent for low value of CMR. This is because as we consider the directional probability of MT in SSP, for high mobility, cost is approximately determined by paging cost. So in SSP as the paging strategy is based on directional probability of MT, less number of polling is required to find an MT i.e redundant polling is avoided.

Table 1. Comparison of paging Cost per Call for SDF and SSP Polling delay $\eta = 5, \gamma = 1, \sigma$ (variance) $= 1.0, \Delta = .8$ (first sector probability)

d	Paging cost /Call CMR=0.1		Paging cost /Call CMR=1		Paging cost /Call CMR=10	
	SDF	SSP	SDF	SSP	SDF	SSP
1	0.000	0.000	0.000	0.000	0.000	0.000
2	00.00	0.000	00.00	0.000	0.000	0.000
3	0.354	0.000	0.403	0.000	0.226	0.000
4	10.913	0.000	6.202	0.000	0.841	0.000
5	15.354	0.791	7.957	0.922	0.854	0.359
6	19.318	1.868	7.667	2.265	1.508	0.722
7	25.499	2.223	9.537	3.135	1.083	1.339
8	29.489	3.236	12.531	4.851	2.075	1.969
9	33.728	3.726	13.208	5.902	2.103	2.304
10	34.374	16.118	11.805	20.513	2.796	6.027

Table 2. Comparison of paging Cost Per Call for SDF and SSP for different r2 $\eta = 5, \gamma = 1$ and CMR =1.0

d	PAGING COST/CALL R2=5		Paging Cost /Call R2=10 r2=15		Paging Cost /Call R2=25 r2=50	
	SDF	SSP	SSP	SSP	SSP	SSP
1	0.000	0.000	0.000	0.000	0.000	0.000
2	0.000	0.000	0.000	0.000	0.000	0.000
3	0.360	0.000	0.000	0.000	0.000	0.000
4	6.201	0.000	0.000	0.000	0.000	0.000
5	8.075	0.963	0.969	0.971	0.972	0.974
6	7.368	2.320	2.331	2.334	2.337	2.339
7	9.904	3.162	3.170	3.173	3.175	3.177
8	12.520	4.898	4.910	4.914	4.917	4.914
9	13.164	5.949	5.961	5.965	5.968	5.971
10	11.507	20.686	20.73	20.745	20.756	20.765

Table 3. Comparison of paging Cost Per Call for SDF and SSP for different $\Delta : \eta = 5, \gamma = 1, \sigma = 1.0$ and CMR =1.0

d	PAGING COST/CALL Δ=0.8		Paging Cost /Call Δ=0.7 Δ=0.6		Paging Cost /Call Δ=0.5 Δ=0.3	
	SDF	SSP	SSP	SSP	SSP	SSP
1	0.000	0.000	0.000	0.000	0.000	0.000
2	0.000	0.000	0.000	0.000	0.000	0.000
3	0.403	0.000	0.000	0.000	0.000	0.000
4	6.202	0.000	0.000	0.000	0.000	0.000
5	7.957	0.922	0.929	0.936	0.943	0.958
6	7.667	2.265	2.274	2.283	2.292	2.310
7	9.537	3.135	3.139	3.142	3.146	3.153
8	12.530	4.851	4.858	4.864	4.871	4.884
9	13.208	5.902	5.909	5.915	5.932	5.936
10	11.805	20.513	20.539	20.560	20.587	20.367

On the other hand when CMR is high, cost between two call arrivals is mainly dependent on location update cost rather than paging cost. Effect of paging is less prominent in SDF, but significant cost reduction is achieved by SSP. Moreover for optimum value of d, cost effective ness is high in SSP. As d crosses its optimum value, difference of cost reduction between SSP and SDF is in decreasing order. This is because of the fact that as number of cluster of cells to be paged simultaneously is gone up to maintain the same paging delay. Also the cost of unsuccessful attempt is high affecting paging cost/call.

We have studied the variation of r2 for pseudo random model and variation of first sector probability, Δ, for normal distribution case on the cost and is given in Table 2 and Table 3. Changing r2 from high to low value means increase of randomness. The number of cell in each sector is equal and the directional probability μ is multiplied with the number of cells in each sectored sub area. When in one-sector sub area μ increases, it decreases in other areas. This ultimately balances the changing effect of r2 for small location area (LA). But for large LA effect of r2 comes into play this valid for Δ also. In Table 2 to Table 3, cost of paging in SDF remains same for different r2 and Δ. So only SSP cost variation are given for different r2 and Δ. Comparison of Table 2 and 3 proves the validity of the simple mathematical model we have proposed for SSP.

5 Conclusion

In this paper the authors have tried to reduce the paging cost per call using an intelligent paging scheme known as sector sequential paging. The main drawback of RWM is that direction of the mobile users is not taken into consideration. In general, a mobile user travels with a specific destination in mind and location of the users can be correlated with its movement direction. In SSP both the randomness and directional probability of the mobile users are considered. The paging cost per call is compared with the SDF paging strategy as given in [1]. It is seen that SSP system reduces the paging cost considerably for different polling delay and thus saving the radio resource bandwidth.

References

[1] Akyildiz, I.F., Ho, J.S.M., and Lin, Y.B., "Movement-Based Location Update and Selective paging for PCS Networks", IEEE Trans. On Networking, Vol. 4, Aug. 1996, pp. 629–638.

[2] Rezaiifar, R., and Makowski, A.M., "From optimal search theory to sequential paging in cellular networks", IEEE J. on Selected Areas in Communication, Vol. 15, Sep. 1997, pp. 1253–1264.

[3] Akyildiz, I.F., McNair, J., Ho, J., Uzunalioglu, H., and Wang, W., "Mobility management in next-generation wireless systems," Proc. IEEE, vol. 87, pp. 1347–1384, Aug. 1999.

[4] Madhow, U., Honig, M., and Steiglitz, K., "Optimization of wireless resources for personal communications mobility tracking," IEEE/ACM Trans. Networking, vol. 3, pp. 698–707, Dec. 1995.

[5] Ho, J.S.M., and Akyildiz, I., "Mobile User Location Update and Paging Under Delay Constraints", ACM-Baltzer Journal of Wireless Networks, 1(4):413–425, December 1995.

[6] Rose, C., "Minimizing the average cost of paging and registration: A timer-based method," ACM/Baltzer J. Wireless Networks, vol. 2, no. 2, pp. 109–116, June 1996.

[7] Lin, Y. B., "Reducing location update cost in a PCS network," IEEE/ACM Trans. On Networking, Vol. 5, Feb. 1997, pp. 25–33.

Experience from Mobile Application Service Framework in WIP

Shinyoung Lim[1] and Youjin Song[2]

[1]ETRI, Gajong-dong, Yusong-gu, Taejon, 305-350, Korea
sylim@econos.etri.re.kr
[2]Department of Information Industrial Systems, Dongguk University, Kyoungju, Korea
song@mail.dongguk.ac.kr

Abstract. As there is a rapid growth of service market and technologies of mobile applications, many kinds of platforms, service, business models, devices, and requirements for mobile business service are under consideration. In case of Korea, it has been experienced for years in applying mobile application service to the real world. In technical domain, one of the hot issues in development of application service is different development environment of each wireless carrier companies. This issue causes a problem to the application service providing companies in developing the same service to fit each different platform for their customers, the wireless carrier companies. A forum called Korea Wireless Internet Standardization (KWIS) was established for solving technical problems in developing the mobile application service. In April, 2002, the forum annonuced the recommended technical specifications of 'Wireless Internet Platform (WIP)' as one of the draft domestic de facto standards in Korea. In this paper, the common application service framework of 'Wireless Internet Platform' is presented. The proposed framework, called 'Mobile Application Service (MAS)' framework, is composed of Application Adaptation Layer (AAL) and Handset Adaptation Layer (HAL). In this paper, the AAL, the application component of the MAS framework, is discussed. The requirements analysis of content provider, handset vendor, wireless carrier, and mobile user for mobile service application are discussed in this paper.

1 Introduction

The broad definition of the wireless Internet is the Internet that is able to provide users with seamless connections to their mobile computing devices, the transparent allocation, handover, and management of mobile IP address, and realtime multimedia contents service. One of the definite application areas of the wireless Internet is mobile commerce. It also covers mobile game, mobile contents, and telematics service. The wireless Internet is consisted of two major technologies: the World Wide

M.H. Shafazand and A M. Tjoa (Eds.): EurAsia-ICT 2002, LNCS 2510, pp. 907–915, 2002.

Web (WWW) and wireless communications. The mobile commerce service will bring the rapid growth in opportunity of business service. It will enable us to do business with the mobile computing devices at any time and any where.

In the mobile commerce service environments, two of the primary requirements of customers will be the security and compatibility of their own enterprise and business information [6]. For example, when a customer pays for presents by his/her credit card, the customer's credit card information should be encrypted between customer's mobile device and the credit card company's service system for preventing someone from looking into it or changing it. In the case of busness or enterprise assets in the form of digital, it is required to be protected in the wireless Internet environments [5]. And the other requirement of compatibility also needs to be considered seriously. The contents and format of digitalized business and enterprise documents need to be compatible to any kind of computing device, Internet, and applications. It means the converting module of document is required for both wired Internet and wireless Internet. However, it is hard to satisfy the requirements when there are many wireless carrier with the different technical specifications. In case of Korea, the current status of mobile communication service market is divided by 3 major wireless carriers such as SK Telecom, KTF, and LG TeleCom. At present, the estimated number of domestic subscribers of mobile communication service is 47 millions, and among them the number of wireless Internet subscribers is 35 millions. The estimated share of the mobile communication service is SK Telecom has 50%, KTF has 35%, and LG TeleCom has 15%.

In this paper, the application component of the Mobile Application Service(MAS) framework is discussed. The merits of the MAS framework are common API for diverse types of multimedia contents, more effective in function and performance than conventional platforms, and standards for handset layer porting. As the WIP is recommended as a domestic standard for mobile application service in Korea, the most beneficial parts will be content provider and mobile user [7]. The application areas of the proposed framework are discussed briefly in the fields of LBS (Location Based Service), mobile payment, mobile CRM (Customer Relationship Management), and mobile ASP (Applications Service Provider).

2 Overview of the Wireless Internet Platform (WIP)

In this paper, the discussion of the run time environments of the wireless communication service will not be included because the recommended standards just define the requirement of the run time environment and it remains to the implementors for accepting the recommended standards. However, the run time environments of current wireless communication service platforms are interpreter and binary down load methods. In case of interpreter method, it is run by Virtuam Machine (VM) such as Java VM, KVM (Kilo VM), GVM, XVM, and SK-VM. And in case of binary down load method as known as native method is run by MAP and Brew. And the programming languages are Java in Java VM, and SK-VM, and C (C++) in GVM, MAP, Stinger, Brew, and Window CE. Therefore, the current run time environments and the programming languages of the mobile communication platforms give implementors heavy burden in providing different types of contents and software for running service. And the discussion of current wireless Internet protocols such as Wireless Applications Protocol (WAP) and i-mode are not included in this paper because these will be independent from the WIP as well as the WAS framework. Being the problems are discussed in KWIS forum, which is established early in 2001 for solving technical problems in developing the mobile application programs in Korea, the proposed framework is to be run on any of the current platforms and wireless communication protocols.

2.1 The Requirement Analysis

The kinds of the wireless Internet entities are content provider, handset vendor, wireless carrier, and mobile user. Before the requirement analysis, it is necessary to consider the common security requirements in the wireless Internet service as follows in Table 1.

Table 1. The security requirements of the wireless Internet service

Security Issues	Requirements
Confidentiality	Guarantees that communications and transactions are private and confidential.
Authentication	Knowing whom you are communicating with.
Integrity	Verifies the information being communicated is correct and has not been modified.
Non-repudiation	Guarantees that agreements can be legally enforced.

The following summary is the requirement analysis of each party in the wireless Internet service as shown in Table 2.

The overall functional requirements for the wireless Internet service are able to summarized as 5 categories as follows; bring together as the standard API specifications for diverse type of digital contents, more efficient in performance and function than conventional platforms, the standard handset's porting layer, the standard SDK(Software Development Kit) for the standard platform, and the authentication procedure for maintaining the compatibility of standard platform. And the performance requirements of the wireless Internet service are the more efficient run time environment than any other platforms for diverse multimedia contents service.

The forum defined the 8 characteristics for the platform from the requirement analysis as follows; fast loading and execution, download applications in the form of machine code, support C/C++ and Java programming languages, API download and upgrade by way of the wireless Internet, providing security service of access control, support multi-applications, memory compression management, and control of diverse peripherals. The underlying principles of WIP is originated from the specifications

Table 2. Requirement analysis of the wireless Internet service

Entity	Requirements	Relationships
Content Provider	More market base per SKU(Stock Keeping Unit, SKU means here the number of software programs by platforms per content) Enhanced SDK(Software Development Kit) Standard API Enhanced performance Simple Authentication of outcomes	Wireless Carrier Handset Vendor
Handset Vendor	More market base per engineering resource Efficient engineering resource management Simplified porting Scalability	Wireless Carrier Content Provider
Wireless Carrier	Easy provision of data service to mobile user Stable in information security Diverse multimedia service Establish & Support Developer's Community	Mobile User Handset Vendor Content Provider
Mobile User	Access any kind of mobile portal service High quality and diverse content service Reasonable price	Wireless Carrier Content Provider

are to be satisfied these 8 characteristics as well as the functional and performance requirements of the wireless Internet service.

2.2 The Architecture of the Wireless Internet Platform (WIP)

Extracting the core specifications from the full specifications of the Wireless Internet Platform (WIP), the functions of handset hardware, native system software, hanset adaption module, run time engine, APIs, and application programs are the areas of the core functional specifications of WIP. Actually, in the WIP specifications, only the handset adaption and APIs are included and the other parts of functions of the wireless Internet platform are considered as the requirements to the handset vendors whether they accept it or not. For example, the run time engine part is required as the mode of down load of binary code for its maximum performance.

The core functions of the WIP are the handset adaption and APIs which are called 'Handset Adaptation Layer(HAL)' and 'Application Adaptation Layer (AAL)', respectively. The HAL defines the specifications for supporting the hardware independent on platform porting as an abstract layer. And the AAL defines the specifications for application programming interface (API) of the wireless Internet platform. The AAL supports the C/C++ and Java programming languages.

2.3 The Service Profile of the Wireless Internet Platform (WIP)

Figure 1 depicts the WIP layers. The WIP cosists of three layers of the common wireless Internet API layer, run time engine, and handset adaptation layer. The run time engine defines requirements to the implementators. The native system software defines the operating system of handset device.

The HAL defines functions of starting the platform and transferring the events from the upper layer of HAL to the lower layer of HAL.

The HAL API covers calls, handset device, network, serial communication, short message service, sound, time, code conversion, file system, vocoder, input method, font, frame buffer, and virtual key. The AAL defines functions of adaptive functions for kernel, graphics, database, file system, network, media player, serial port, phone for call and Short Message Service(SMS), and User Interface(UI) components.

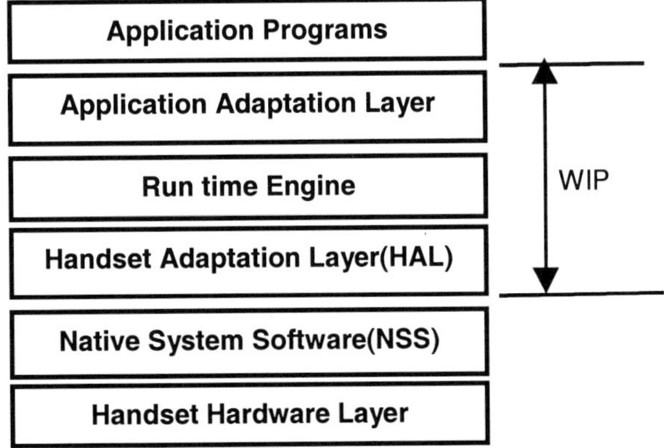

Fig. 1. Architecture of Wireless Internet Platform(WIP)

3 Proposed Framework

In the figure 2, the the proposed MAS framework is presented. The proposed framework is able to be defined as the set of API for wireless communications for the mobile applications. The MAS framework is composed of basic API, API manager, application manager, and extended API Set. The basic API is the set of basic API for the platform used by application programmers. At present, only the basic API is defined and the other specifications of MAS are under reviewing by the KWIS form.

3.1 Design Problems

One of the major technical problems between content providers and wireless carriers is as follows; content providers are required to develop each different content formats and software for different running environments of the wireless carrier companies. It reacts as a resistance to activation and growth in the mobile communication service market. It also causes handset vendors to consume engineering resources excessively to each different handsets.

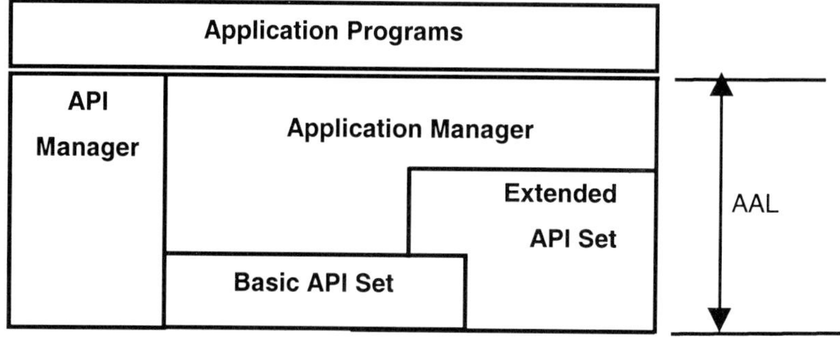

Fig. 2. Mobile Application Service(MAS) Framework

However, for its unique characteristics of mobile application service, the mobile application service providers try to find the solution for the interoperability, compatibility and security with the (M X N) problem of each different wireless carriers' technical specifications such as different programming language, program run time environment, and handset specifications.

As discussed the requirements and 8 characteristics for the platform in the chapter 2 in this paper, table 3 shows the provision of 8 principles of mobile application service with the KWIS WIP. The 8 princiles of mobile application service is originated from the proposed 8 characteristics of WIP.

Table 3. Comparisons of principles of wireless Internet service

Principles	Characteristics	Proposed Framework
Machine code of application programs	download applications in the form of machine code	Avaiable of execution of the downloaded application program in the form of machine code
Execution of multi-applications	support multi-applications	Provide a criteria of process priority and execution of multi programs independently
Programming Languages	support C/C++ and Java programming languages	Provide both Java and C languages
Platform Security	providing security service of access control	Provide access control to the API and directory at the security level: public level, content provider level, and system level
API Add/Update	API download and upgrade by way of the wireless Internet	Provide version management for API update, delete, renewal, and up-and-down load by wireless Internet
Memory Management	memory compression management	Provide HEAP memory management of automatic memory release, memory compaction, Java garbage collection, Java stack, and shared memory
Application Program Management	fast loading and execution	Provide application program lifecycle management and application program download
Multi-lingual Service	control of diverse peripherals	Provide Uni Code and Extended Uni Code(EUD)

3.2 Basic API Set

The core part of the proposed MAS framework is the Basic API Set. There are 11 elements of the Basic API Set defined for Java application programmers. The elements are for core system, high level io, utilities, low level io, system, additional ip, graphics, database, UI component, handset, and media. The detailed definition of the basic API set are shown in Table 4.

Table 4. Definition of the Basic API Set

Category	Definition
Core System (java.lang)	Support the same or part of J2SE 'java.lang'
High Level IO (java.io)	Support the same or part of J2SE 'java.io'
Utilities (java.util)	Support the same or part of J2SE 'java.util'
Low Level IO (org.kwis.msf.io)	Support the same or part of J2SE 'java.net'
System (org.kwis.msf.core)	Provide shared memory API of loading, execution, and multi-application program for supporting multi-application program and dynamic library
Additional IO (org.kwis.msp.io)	Support the same or part of J2SE file APIs. Support API for PPP connection
Graphics (org.kwis.msp.lcdui)	Composed of diverse drawing API for screen or off screen frame buffer Support API for encoding and decoding of diverse image formats(bmp, png, gif, agif)
Database (org.kwis.msp.db)	Composed of API for storing, retrieving, searching, and managing data by record
UI Component (org.kwis.msp.lwc)	Composed of API for diverse high quality graphic interface to graphic package
Hansdet (org.kwis.msp.handset)	Composed of API for specific information, back light, and Call
Media (org.kwis.msp.media)	Composed of API for media processing functions of sound or moving pictures, tone recovery, sound recording, volume control

The C version of the basic API set is also defined in MAS framework. However, the figure 3 shows a class diagram of some parts of the set in the table 4 such as io, database, and media.

3.3 API Manager, Application Manager, and Extended API Set

The API manager performs the tasks of adding and renewal of elements in basic API Set and extended API Set. And the API manager is also able to download the Application Manager.

The Application Manager is responsible for the overall application management of read, download, implement, execution, deletion, and security.

And the Extended API Set is a set of API for draft standard API. These three parts of the MAS framework are now under reviewing by KWIS form.

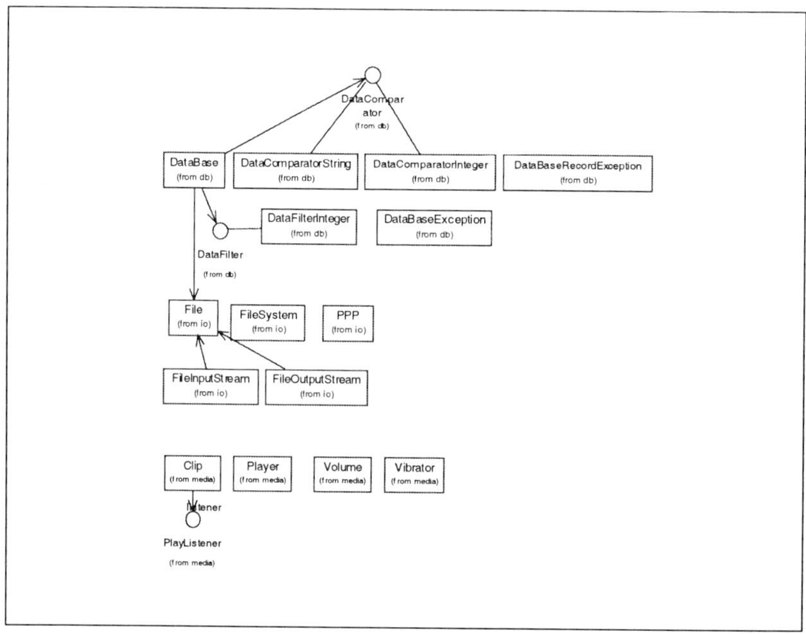

Fig. 3. Java Class Diagram for io, database, and media in the Basic API Set

4 Conclusions

In this paper, the Mobile Application Service(MAS) framework is proposed. The current status, problems, issues, and principles in moble communication service as well as the requirement analysis, the recommended domestic standard of the WIP(Wireless Internet Platform), and its some part of the profile are discussed in this paper. From the discussion, the motivation and necessity of proposing the Mobile Application Service(MAS) framework is derived. Althogh the MIP and the MAS framework are the recommended domestic standards in Korea at present, it will be able to refer and consider as one of the proposed draft international standards in mobile communication service. The merits of the framework are common API for diverse types of multimedia contents, more effective in function and performance than conventional platforms do, and standards for handset layer porting. As the framework is recommended as a domestic standard for mobile application service in Korea, the most beneficial parts will be content provider and mobile user. There are four major application areas of the MAS framework. They are LBS(Location Based Service), mobile payment, mobile CRM(Customer Relationship Management), and mobile ASP(Application Service Provider) [1, 2, 3, 4]. Of course, there will be many applications of the MAS framework in the form of business models.

For further study of this area, the verification and validation of the framework model by design and implementing some pilot-scale service system is necessary. It is also required to develop evaluation and authentication methodologies with the assistance of toolkits for the granuality of the QoS of the pilot-scale service system.

References

[1] www.baltimore.com, Baltimore telepathy-Making Mobile Commerce Secure, 2000.

[2] http://www.mobilemcommerce.com

[3] J Davison etc., Mobile E-commerce, Ovum, 2000.

[4] Katrina bond, Danny Williams, Mobile Ecommerce, Analysis, Analysis Publication.

[5] G. Horn and B. Preneel, "Authentication and Payment in Future Mobile Systems", Computer Security ESORICS '98, Lecture Notes in Computer Science, 1485, pp. 277–293, 1998.

[6] Vesna Hassler, "Security fundamentals for e-commerce", 2001.

[7] TTA, "Wireless Internet Platform(WIP) Specifications(in Korean version)", 2002.

An Efficient Approach to Improve TCP Performance over Wireless Networks

Satoshi Utsumi[1], Salahuddin M.S. Zabir[2], and Norio Shiratori[2]

[1] Graduate School of Information Sciences, Tohoku University, Japan
u-satoshi@shiratori.riec.tohoku.ac.jp

[2] Resarch Institute of Electrical Communication, Tohoku University, Japan
{szabir, norio}@shiratori.riec.tohoku.ac.jp

Abstract. We propose a new end-to-end flow control scheme we named as "TCP Identification & Revivable Window (TCP-I&RW)" to improve TCP performance in wireless networks supporting link level retransmissions for link errors. The TCP sender in TCP-I&RW places an identification tag for every data segment it sends. Like conventional TCP protocols, if a segment loss is detected, it first infers a congestion, lowers the sending rate and retransmits with a different identification tag. Sender figures out the actual cause of segment loss depending on identification response field in the acknowledgement (ACK) corresponding to the retransmitted data. If the segment loss is not due to congestion, the TCP sender revives its transmission rate to the value prior to the retransmission. As such, erroneous detection of congestion is avoided. This ensures an improved throughput for TCP over wireless links. Experiments show our proposed new scheme can achieve better performance than existing well established schemes in the cellular network and also in the satellite network.

1 Introduction

In recent years various wireless links have been incorporated into the Internet. Therefore, studies about TCP in wireless networks deserve special attention [1] [2] [3].

Wireless links are generally prone to a higher link error rate than their wired counterparts. Since conventional TCP protocols are designed to be used in wired networks with low link error rates, their designs do not take link errors into account. That is, they assume segment losses occur solely due to congestion in networks [4]. Whenever a segment loss occurs, any conventional TCP infers a congestion and the TCP sender reduces its sending rate as a remedial measure. Therefore, when such TCP protocols are used over wireless networks, the TCP senders interpret all the losses to be originating from congestion and frequently lower their transmission rates unnecessarily. As such, deployment of conventional TCP protocols over wireless links results in a decreased throughput [1] [3].

In this paper, we propose a new variant of TCP protocol to solve the problem. When a segment loss occurs, our protocol first behaves like other TCP

M.H. Shafazand and A M. Tjoa (Eds.): EurAsia-ICT 2002, LNCS 2510, pp. 916–925, 2002.

protocols by lowering the TCP sender's transmission rate. It then figures out the actual cause of each segment loss using some tricky identification tags in the corresponding ACKnowledgement. If the tags suggest that the loss is due to link error, it revives the transmission rate prior to the detected loss. Our proposed protocol thus alleviates unnecssary decrease in TCP transmission rate and corresponding throughput degradation. We name this new TCP protocol as "TCP Identification & Revivable Window (TCP-I&RW)". TCP-I&RW requires wireless links facilitate retransmissions on data link layer (hereinafter, mentioned as link level retransmissions) to recover from link errors and modification of the TCP sender. TCP receiver is kept unchanged if it supports commonly implemented timestamp option. Our scheme does not pose implementation problems like the other well established existing wireless TCP solutions to be outlined in section 3. In addition, experiments show this scheme yields higher performance than existing ones in the cellular network and in the satellite network.

We organize the rest of this paper as follows. Section 2 outlines TCP characteristics over wireless links. In section 3 we describe well established existing wireless TCP solutions. We present our proposed TCP-I&RW in section 4. In section 5 we evaluate the performance implications of our scheme. Finally we conclude in section 6.

2 TCP Issue over Wireless Links

The conventional TCP senders detect segment losses in either of the following ways[5].

1. Retransmission Timeout: The TCP sender decides that the segment was lost in the network, if it does not receive the corresponding ACK (Acknowledgement) within a certain time RTO (Retransmission Timeout) after the TCP sender had sent the segment. RTO is determined based on RTT(Round Trip Time) of the connection.
2. Fast Retransmit: When the TCP receiver receives any segment which is not the expected next in-sequence, it sends a duplicate ACK (dupack) containing the same acknowledgement number as the previous ACK. The TCP sender decides that the segment was lost in network, if it receives three dupacks consecutively.

Conventional TCP protocols can perform well in wired networks, where the segment losses due to link errors can be ignored [4]. But conventional TCP senders shrink the congestion window unnecessarily for the segment losses due to link errors in wireless networks, which is prone to a high link error rate. As such, conventional TCP protocols cannot perform well in wireless networks.

Link level retransmissions can be used to hide the influence of link error on wireless links from the end hosts. But link level retransmissions alone cannot always succeed in this effort for TCP because of the following two reasons.

1. If it takes too much time to retransmit on wireless links, timeout of TCP may occur [6].

2. When link level retransmissions make the order of the segment arrivals at the TCP receiver different from the order of the segment transmissions at the TCP sender, the TCP receiver sends dupacks [7].

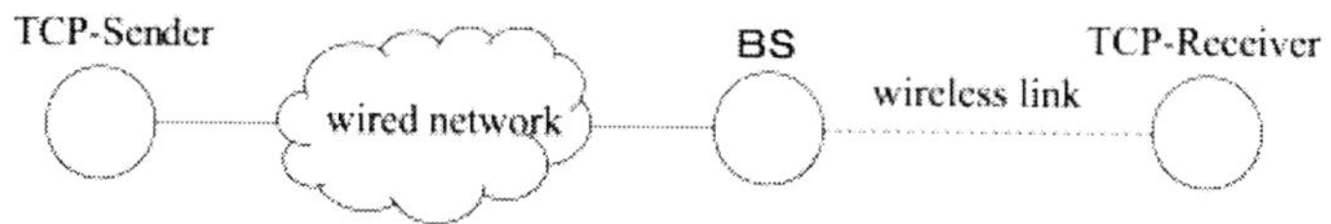

Fig. 1. System Model

3 Related Works

Snoop scheme [1] and Delayed Duplicate ACKs scheme [2] are the most prominent existing solutions to improve TCP performance in wireless networks using retransmissions on wireless links. We provide an outline of these schemes using the network model as in Fig. 1. In this system, there is a base station BS between one TCP sender and one TCP receiver. The TCP sender is connected to BS by the wired network and the TCP receiver is connected to BS by the wireless link. The details are described in section 5.

3.1 Snoop Scheme

Snoop is a TCP-aware retransmission scheme on the wireless link. Snoop is implemented at the base station BS, where it observes TCP headers of segments passing through BS. Snoop behaves as follows.

1. Snoop buffers TCP data segments sent by the TCP sender until the corresponding ACKs pass through BS.
2. If Snoop receives a dupack from the TCP receiver and the corresponding lost segment is in the buffer, it drops the dupack and retransmits the corresponding segment.
3. If Snoop has not received the ACK after sending a TCP data segment until timeout (determined by Snoop) occurs, then it retransmits the data segment.

Snoop has the following problems on the implementation.

1. Per-flow management is needed on BS. So a large amount of resources on BS is consumed.
2. If IP encryption is used, this scheme is not usable.

Also TCP performance is affected owing to the following reasons.

1. Link error detections based on TCP-dupacks is not efficient.
2. If the round trip time between BS and TCP receiver are large, timeout may occur.

3.2 Delayed Duplicate ACKs Scheme

Delayed Duplicate ACKs(DDA) scheme uses (TCP-unaware) link level retransmissions on the wireless link. If the order of the segment arrival at the TCP receiver is different from the order of the segment transmission at the TCP sender, then the TCP receiver delays to generate the third dupack (and later dupacks) for a certain optimal time *d*. If the TCP receiver receives the segment retransmitted by link level retransmissions during the time *d*, then the TCP receiver doesn't send the third dupack.

DDA scheme has problems on the implementation. It is difficult to estimate the optimal time *d* both statically and dynamically.

Also TCP performance is affected owing to the following reasons.

1. If the time *d* is too large, throughput may decrease due to timeout of TCP.
2. If segment losses occur due to congestion, the error recovery is unnecessarily delayed for the time *d*. As such, the benefits of using dupack vanish.

4 Our Scheme

In this section, we propose a new scheme to avoid TCP performance degradation in the wireless network. This scheme has the following features.

1. Wireless links have (TCP-unaware) link level retransmissions.
2. This scheme needs to modify only the TCP sender. It is not needed to modify the TCP receiver that implements the TCP timestamp option [8].

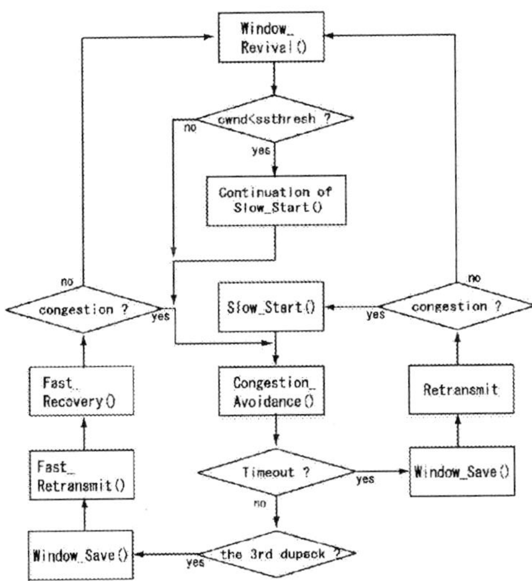

Fig. 2. Flowchart of TCP-I&RW

4.1 New End-to-End Flow Control: TCP Identification & Revivable Window

In our scheme, the congestion window shrinks upon timeout or receiving three dupacks. However, the corresponding parameters (that is, **cwnd** and **ssthresh**) are saved for a certain time. If the link level retransmissions on the wireless link recovers segment losses due to link errors, our proposed new TCP sender revives the congestion window to its pre-congestion state by restoring the saved parameter values. The details of TCP-I&RW sender's functions are as follows.

1. The TCP-I&RW sender places an identification distinguishing the data segment itself at the timestamp field on the TCP header of each data segment.
2. The TCP-I&RW sender memorizes the parameters about the congestion window at the time when it retransmits the data just after the detection of segment loss.
3. The TCP-I&RW sender may restore the parameters about the congestion window standing on the identification response, which contains the same value as the identification on the corresponding data segment, placed at the timestamp echo response field on the TCP header of the ACK corresponding to the retransmitted data. This restoration is called as "the window revival" in this paper.

Upon receiving the ACK corresponding to the retransmitted data, the TCP-I&RW sender perceives the reason of the segment loss by the identification response as follows.

1. If the identification response on the ACK contains the same identification value as the originally transmitted (i.e. not retransmitted) data segment, the TCP-I&RW sender perceives that the segment loss is due to link error and that the segment is recovered by the link level retransmission. That is, the segment loss is not due to congestion.
2. If the identification response on the ACK contains the same identification value as the retransmitted data segment, the TCP-I&W sender perceives that the segment loss may be due to congestion.

If the TCP-I&RW sender decides that the segment loss is not due to congestion (i.e. due to link error) upon receiving the ACK corresponding to the retransmitted data, it returns the parameters to the condition before the segment loss detection. We call this end-to-end flow control mechanism as TCP Identification & Revivable Window (TCP-I&RW). TCP-I&RW operates on the following algorithms: *Slow_Start()*, *Congestion_Avoidance()*, *Fast_Retransmit()*, *Fast_Recovery()*, *Window_Save()* and *Window_Revival()*. First four algorithms have been already implemented in TCP-Reno [4], later two algorithms are proposed in this paper. Fig. 2 shows the flowchart of TCP-I&RW.

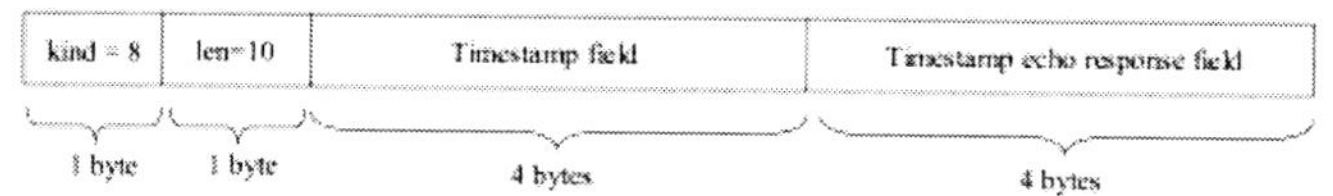

Fig. 3. TCP timestamp option field

4.2 Identification

The TCP-I&RW sender of TCP-I&RW places a new identification tag distinguishing the entire data segment itself at the timestamp field on the TCP header

Fig. 4. Identification

of each data segment sent by the TCP sender. The TCP receiver with the TCP timestamp function copies the identification to the timestamp echo response field on the TCP header of the ACK corresponding to the data segment. Fig. 3 shows the form of the timestamp option field on the TCP header.

The identification is composed of the data identification (DID) and the segment identification (SID) as shown in Fig. 4. The value of DID is unique for each data. The value of SID does not overlap among the segments with the same data.

Since the identification is implemented using the TCP timestamp fields, the value of the identification should support the PAWS (Protection Against Wrapped Sequence number) algorithm [8], which is implemented by the TCP timestamp option. Otherwise, the TCP receiver may drop valid data segments. To support PAWS, we propose the following methods to decide the value of the identifications as follows.

1. DID is put before SID as more significant bits on the TCP timestamp field (as Fig. 4).
2. The TCP sender makes the value of DID of the later data larger.

If these rules are not followed, TCP receiver implementing PAWS may drop the data segment.

If the identification code on the un-received data segment is predictable, *aggressive TCP receivers* may manipulate the identification response on the ACK to occupy an unfairly high share of the link bandwidth. The TCP sender therefore should decide SID randomly (ideally, using a random number generator).

4.3 Window Revival

When the segment loss is detected by timeout or three dupacks, TCP-I&RW behaves the same as the conventional TCP protocols. That is, since the TCP sender cannot judge the reason of the segment loss (i.e. congestion or link error) just after the segment loss is detected, the TCP sender shrinks the congestion window moderately. In TCP-I&RW the TCP sender saves the parameters about the congestion window (that is, **cwnd** and **ssthresh**) just before shrinking the congestion window (*Window_Save()*). If the TCP sender perceives that the segment loss is not due to congestion by the identification response on the ACK corresponding to the retransmitted data, it returns the parameters about the congestion window to their values just before shrinking the congestion window (*Window_Revival()*).

5 Evaluation

In this section, we evaluate TCP-I&RW scheme (that is, TCP-I&RW with link level retransmissions) and other related flow control schemes using the ns-2 simulator [9]. We consider cellular networks and satellite networks for the evaluation. Our performance metrics are as follows.

1. Throughput (for cellular networks).
2. Bandwidth utilization (for satellite networks).

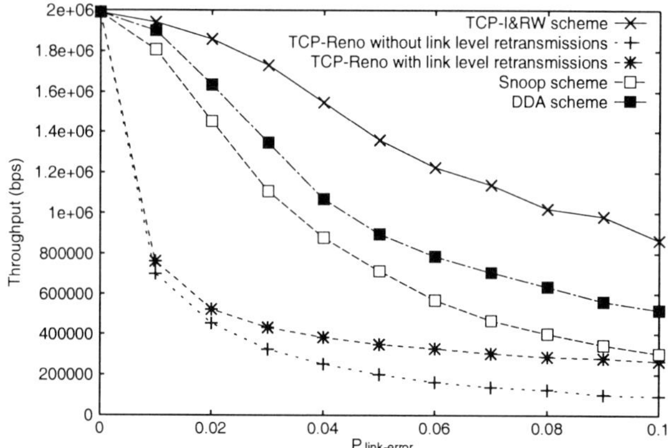

Fig. 5. TCP throughput in the cellular network ($P_{congestion} = 0$)

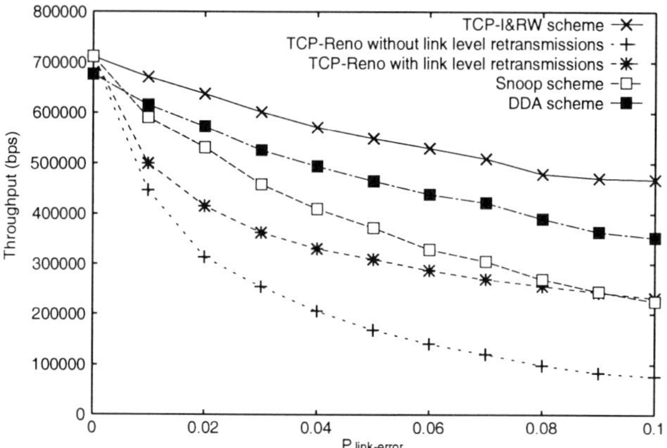

Fig. 6. TCP throughput in the cellular network ($P_{congestion} = 0.01$)

The other flow control schemes are (1) TCP-Reno without link level retransmissions, (2) TCP-Reno with link level retransmissions, (3) Snoop scheme (TCP-Reno based) and (4) DDA scheme (TCP-Reno based).

The TCP sender is assumed to be performing a bulk data transfer. Each TCP data segment contains 1000 bytes, while each TCP ACK contains 40 bytes. We use NAK-based selective repeat [10] as link level retransmissions on the wireless link. Each link level NAK contains 16 bytes. The wireless link has a priority

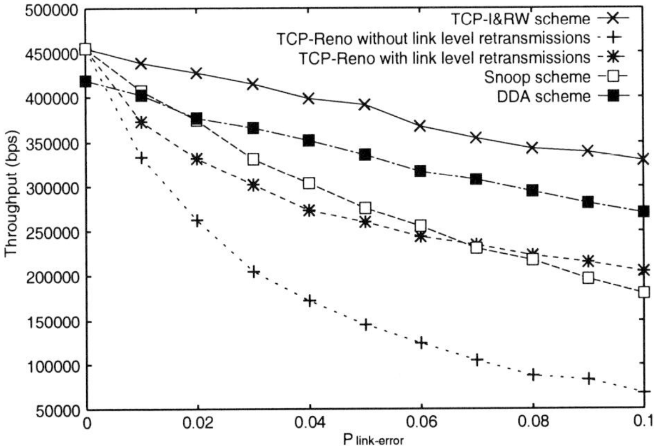

Fig. 7. TCP throughput in the cellular network ($P_{congestion} = 0.02$)

scheduling queue holding at most 50 segments. That is, each TCP data segment retransmitted by the wireless link and each link level NAK has priority to send before other segments in the queue. TCP ACKs and link level NAKs are assumed not to be lost due to link error, as these are sufficiently small. The time d of DDA scheme is statically set twice the propagation delay (i.e. round trip time) on the wireless link.

5.1 Cellular Networks

We measure throughput of a flow on the wireless link which is available exclusively for the flow. We evaluate on the system as the same as shown in Fig. 1. Generally a cellular network has this form. Segment losses due to congestion in the wired network are assumed. The propagation delay on the path in the wired network is set 10ms (both up and down). The bandwidth and the propagation delay on the wireless link are set 2Mbps and 50ms respectively (both up and down). We measure the average throughput of a flow which is generated during 120 seconds (about 1000 round trip times) by the TCP sender in this system. The data segments get lost due to link errors with a probability $P_{link-error}$ varied from 0% to 10% at 1% intervals there. We observe cases when the data segments are lost due to congestion In the wired network with a probability $P_{congestion}$ 0%, 1% and 2% respectively.

Fig. 5, 6 and 7 show the results when the segment loss rates due to congestion in the wired network are 0%, 1% and 2% respectively. We can observe TCP-I&RW scheme realizes the best throughput independent of the segment loss rates due to congestion and due to link error.

We consider and compare the throughput of TCP-I&RW scheme and DDA scheme. TCP-I&RW scheme realize higher throughput than DDA scheme at all combination of $P_{congestion}$ and $P_{link-error}$.

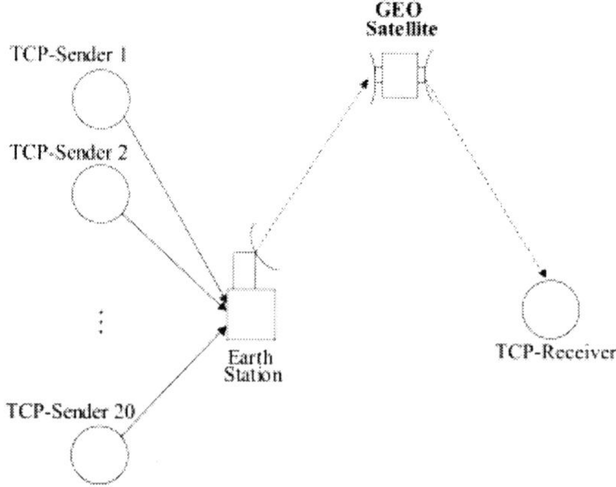

Fig. 8. Simulation scenario for the satellite network

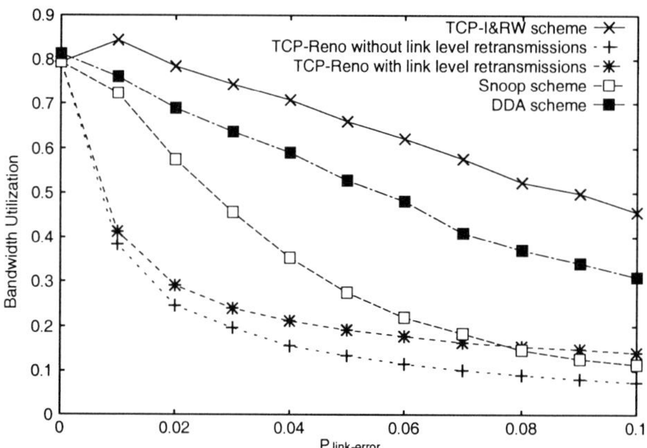

Fig. 9. Bandwidth utilization on the satellite link

5.2 Satellite Networks

We measure bandwidth utilization on the wireless link which is shared by many flows. We evaluate on the system shown in Fig. 8. In this system, there are an earth station between 20 TCP senders and one TCP receiver. The TCP senders are connected to the earth station by wired links. The earth station and the TCP receiver form the wireless link with the satellite. We assume this satellite is a Geosynchronous Earth Orbit (GEO) Satellite. The bandwidth and the propagation delay of the wired link are 50Mbps and 10ms (both up and down). The bandwidth and the propagation delay of the satellite link are 10Mbps and 225ms respectively (both up and down). We measure the average throughput of a flow which is generated during 570 seconds (about 1000 round trip times) by the TCP sender. We measure bandwidth utilization in this system where the data segments get lost due to link error with a probability $P_{link-error}$ varied from 0% to 10% at 1% intervals.

Bandwidth utilization is defined as the ratio of the total throughput of the all flow sharing the link to the link bandwidth. Simulation results obtained for bandwidth utilization on the satellite link is shown in Fig. 9. This shows that TCP-I&RW scheme can realize the largest bandwidth utilization except when $P_{link-error}$ is 0%.

6 Conclusion

Snoop scheme and DDA scheme are the representative ones to improve TCP performance over wireless links prone to link errors. But both Snoop and DDA schemes have problems on the implementation. TCP-I&RW scheme proposed in this paper improves TCP performance over wireless links without these problems. Experiments show that TCP-I&RW scheme can get higher throughput than the existing schemes in the cellular network. Experiments also show that the TCP-I&RW scheme can get higher bandwidth utilization than the existing schemes in the satellite networks.

References

1. H. Balakrishnan, V. Padmanabhan, S. Seshan, and R. Katz, "A comparison of mechanisms for improving TCP performance over wireless links," in IEEE/ACM Transmissions on Networking, December 1997, pp. 756-769.
2. N. Vaidya, M. Mehta, C. Perkins, and G. Montenegro, "Delayed Duplicate Acknowledgements: A TCP-Unaware Approach to Improve Performance of TCP over Wireless," Technical Report 99-003, Computer Science Dept., Texas A&M University, February 1999. Also, to appear in Journal of Wireless Communications and Wireless Computing (Special Issue on Reliable Transport Protocols for Mobile Computing).
3. A. Bakre and B. Badrinath, "I-TCP: Indirect TCP for mobile hosts," in Proc. 15th International Conf. On Distributed Computing Systems (ICDCS), May 1995.
4. V. Jacobson, "Congestion avoidance and control," in Proc. ACM SIGCOMM, August 1988.
5. W.R. Stevens, TCP/IP Illustrated. Addison-Wesley, 1994.
6. A. DeSunibem, N.Chuah, and O.Yue, "Throughput performance of transport-layer protocols over wireless LANs," in Proc. Globecom '93, December 1993.
7. H. Balakrishnan, S.Seshan, and R.Katz, "Improving reliable transport and hand-off performance in cellular wireless networks," ACM Wireless Networks, vol. 1, December 1995, pp. 469-482.
8. V. Jacobson, R. Braden, and D. Borman, "TCP Extensions for High Performance ", IETF RFC 1323 , May 1992.
9. K. Fall and K. Varadhan, "ns Notes and documentation," tech. Rep., the VINT Project (UC Berkeley, LBL, USC/ISI, and Xerox PARC, April 2002.
10. "Data service options for wideband spread spectrum systems: Radio link protocol," V&V version, TIA/EIA/IS-707-A.2, PN-4145.2, July 1998.

Extended Hexagonal Constellations as a Means of Multicarrier PAPR Reduction

Ali Pezeshk[1] and Babak H. Khalaj[2]

[1, 2] Department of Electrical Engineering, Sharif University of Technology
P. O. Box 11365-8639 Tehran, Iran,
[1] apezeshk@mehr.sharif.edu,
[2] khalaj@sharif.edu

Abstract. An efficient, easy to implement method for OFDM and DMT Peak to Average Power Ratio (PAPR) reduction is proposed based on constellation extension. A PAPR reduction of more than 5 dB at clip rate of 10^{-5}, is achieved with no rate loss and improved noise margin with low complexity. Various versions of the proposed method have been studied and compared in terms of PAPR reduction, complexity and average power, with each other and earlier methods.

1 Introduction

The Orthogonal Frequency Division Multiplexing (OFDM) and Discrete Multitone (DMT) have proven to be economical and easy to implement multicarrier modulation techniques. The high resistance of OFDM to channel fading has made it a favorable candidate for wireless access.

One of the major drawbacks of OFDM and DMT is their high Peak to Average Power Ratio (PAPR) which obliges the designer to leave high back offs for amplifiers and use higher resolution analog-to-digital-converters to prevent the signal from being clipped or carrier intermodulation to occur.

Various PAPR reduction methods such as optimization of Partial Transmit Sequences (PTS) [1,2], phase optimization through Selected Mapping (SLM) [3, 4], tone reservation [5,6] and tone injection [7,8] have been proposed. In addition, coding techniques such as use of Golay complementary sequences and Reed-Muller codes [9] and other schemes based on scrambling [10] or interleaving [11] are proven effective in reduction of PAPR.

It should be noted that most of the methods are based on the same idea of selecting the time domain signal to be transmitted from a set of different representations with the constraint of minimization of PAPR. This approach, in many cases requires some extra information about the way of selection to be transmitted to the receiver. This overhead needs to be well protected against possible errors introduced by channel since it usually contributes to the way a number or all of the symbols in a frame should be decoded. In some other cases some redundancy should be provided in the transmitted frame. Therefore, in general a trade-off between the rate reduction and the amount of PAPR reduction has to be made or schemes introducing no overhead shall be utilized.

M.H. Shafazand and A M. Tjoa (Eds.): EurAsia-ICT 2002, LNCS 2510, pp. 926–936, 2002.

In this paper, a new method to reduce PAPR, based on extension of hexagonal constellations is proposed, resulting in no rate reduction while providing a substantial amount of PAPR reduction. The scheme is shown to be of low complexity and increase the noise margin.

2 OFDM Transmission

A data packet in OFDM/DMT transmission, called a frame, is a complex vector $\vec{X} = [X_0, X_1, \cdots, X_{N-1}]$ consisting of concatenation of data symbols to be transmitted on subchannels and constraining X_0 and $X_{N/2}$ to be real and $X_k = X^*_{N-k-1}, 1 \leq k \leq \frac{N}{2} - 1$, for the transmitted signal to be real valued.

From the above frame structure, the time domain signal which is transmitted is obtained by reconstructing the continuous signal from its samples x_n, where we have:

$$\vec{x} = [x_0, x_1, \cdots, x_{N-1}] = \mathcal{IFFT}\left\{\vec{X}\right\} \tag{1}$$

so,

$$x_n = \frac{1}{\sqrt{N}} \sum_{k=1}^{N-1} X_k e^{j2\pi kn/N} \tag{2}$$

The PAPR of an OFDM/DMT frame is defined as[1]:

$$PAPR = \frac{\max_t \left(x^2(t)\right)}{\mathcal{E}\left\{\frac{1}{T}\int_T x^2(t)dt\right\}}, \tag{3}$$

where T is the duration of the time domain frame signal and $\mathcal{E}\{\cdot\}$ denotes the statistical expectation.

Since we have the time domain samples, x_k, it would be easier to compute PAPR from the discrete time samples. As denoted by [12,13], the peak value of the continuous time domain signal would not necessarily be equal to the maximum of time domain sample. Oversampling would then be necessary to get accurate results.

The constellations used on subchannels of OFDM/DMT, are Square QAM (SQAM). As a first try to reduce the peak values, one might think of reducing $|X_k|$ for $0 \leq k < N$ while retaining the symbol error rate. This would lead to the famous sphere packing problem [14]. The problem would then be to pack M (number of constellation symbols) circles of unit radius, in the smallest circle possible while no two unit circles collide.

There are three major problems with this approach:

1. The average power would also be reduced this leading to a lower PAPR reduction level.
2. The best circle-in-circle-packings for large number of circles are not known yet [15,16]. Even computer optimizations for more than about 200 circles are not available [17].

[1] Another measure is crest factor which is the square root of PAPR [2]

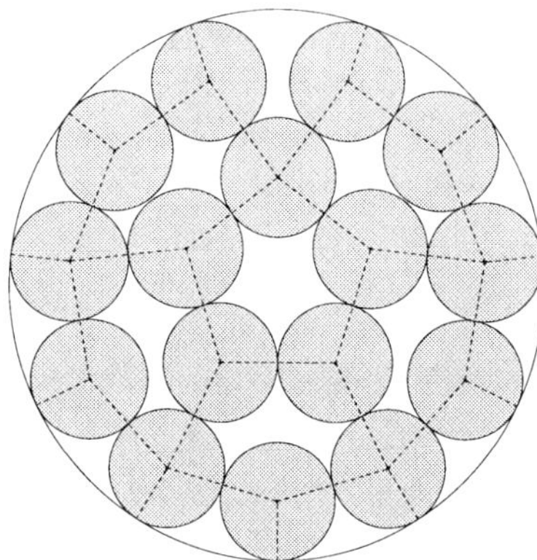

Fig. 1. Best packing of 16 circles of unit radius in a circle [18]

3. The Voronoi cells do not necessarily have regular and/or equal shapes, so the distance of the received point to all symbols shall be computed requiring a more complex receiver. In addition, the same receiver could not be used for different subchannels. (Fig. 1)

The proposed scheme, is based on a simpler form of the aforementioned idea, that also extends the basic ideas in PTS [1] and tone injection [7,8] to a new framework.

3 Proposed Method

The densest lattice in the plane is the hexagonal lattice, A_2 (cf. [19]). Note should be taken that this is true for the limiting case of covering whole plane, whereas the sublattices of the hexagonal lattice with M points do not necessarily give the densest packing in a predefined shape (subject to our average power constraint a circle) as can be seen from Fig. 1.

The hexagonal constellations while being denser than the square constellations, are much easier to implement than the optimum packings of circles-in-circles, since they are subsets of a single lattice. This allows the use of the same receiver for all subchannels in a DMT system, where different subchannels have distinct bit-loadings. Another superiority of the hexagonal constellations compared to the optimum packings is the equality of the decision regions (Voronoi cells), which makes the symbol detection much simpler.

It is widely believed that the hexagonal constellation symbols are hard to decode, and lower complexity schemes such as [24], may not be used for OFDM and/or DMT to decode the received symbols. However, the ingenious simple approach by C. Fu [25], originally intended to be used in computer games, can be shown to fit the needs of multicarrier systems the best. The method is based on the fact that the isometric projection of a lattice of cubes with centers lying on the plane $x + y + z = 0$, is the hexagonal lattice. So, the problem reduces to the trivial case of finding the cube, to which the received point belongs, which has a complexity of $O(1)$.

Constellation Extension

As the simulation results show, there would not be a significant improvement in PAPR if hexagonal constellations are simply used instead of square constellations. But, since there are more points on this lattice in a circle of given radius than points on the square lattice $(\mathbb{Z}^2)$ in a circle of same radius, assuming same distance between lattice points, more symbols can be assigned to have two or more representations from the points inside the circle on the hexagonal lattice. By selecting the right representation, the proposed method is able to reduce the PAPR further. Consequently, the constellation can be *extended* to different total number of points where some of the symbols are assigned to have two representations, while the rest are assigned a single representation. (Fig. 2)

We may construct A_2 in a layer by layer approach, where each layer consists of the points on the edges of a hexagon whose vertices are at distance $D \in \mathbb{N}$ from the origin (assuming unit distance between lattice points). So, the points on inner layers would have less power than those on outer layers. This suggests assigning a second representation to the symbols on the outer layers, since they would have a higher contribution in the instantaneous power. In this way, there would be a lower increase in average power.

As can be seen in Fig. 2, the inner 7 symbols have a single representative point assigned to them while the outer 9 symbols have two.

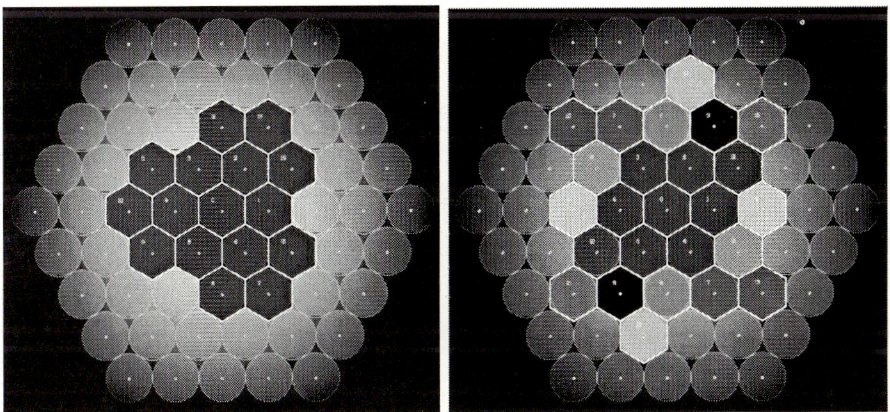

Fig. 2. *Left* – A normal 16 symbol Hexagonal QAM (HQAM) and its associated Voronoi cells, *right* – An Extended-Hexagonal-QAM (EHQAM) constellation for the same 16 symbols. The inner 7 symbols have a single representation, while the 9 outer symbols (designated as regional pairs of the same color), each have two representations.

Now, for those symbols in a frame, where we have a choice for selecting between the two representations, the representation which results in a lower PAPR will be selected. Since the PTS with phase shifts from the set $\{0, \pi\}$ has been well studied, we have assigned the symbol representations to be only π radians out of phase with each other.

Also in order to be able to use the same receiver for all subchannels in DMT, where loading of subchannels are not equal, we haven't used the so called optimum hexagonal constellations [20]. It can be shown that the difference between the constellations we have used and the optimum constellation vanishes as the number of symbols increases.

We define the constellation extension factor κ as:

$$\kappa \triangleq \frac{\text{The number of points on the extended constellation}}{\text{The number of points on the original constellation}} \tag{4}$$

The idea of extension has also been proposed by [7,8] in a different manner under the name generalized constellations. In this method, the original constellation is repeated around itself, so each symbol would have 9 (or more) representations resulting in an extension factor of $\kappa = 9$ or higher. This means that for a DMT transceiver, over 3 more bits could have been transmitted in each subchannel where the symbol representations are chosen to be on the extension points. (Note that the increase in average power is assumed as tolerable when reducing PAPR.)

Our proposed scheme, suggests keeping $\kappa < 2$, so the number of excess bits which could have been sent on each subchannel becomes a fraction of 1, since it is more complex to realize fractional loading increments, true no rate loss would thus be obtained.

It should be noted that the number of symbols out of the total M symbols, having a second representation when $\kappa < 2$ could easily be proven to be $(\kappa - 1)\, M$.

Optimization Algorithm

The optimization phase, i.e. selection of the representation of symbols in a frame, is based on a low complexity hill-climbing method by [21] for PTS which gives a suboptimal solution. We have slightly modified the method to fit to our requirements and improved it as discussed below.

A flip vector $\vec{F} = \left[F_0, F_1, \ldots F_{(N/2)-1}\right] \in \mathbb{Z}_2^{N/2}$ is defined where:

$$F_i = \begin{cases} 0, \text{ if the } 1^{\text{st}} \text{ representation on } i^{\text{th}} \text{ subchannel is selected} \\ 1, \text{ if the } 2^{\text{nd}} \text{ representation on } i^{\text{th}} \text{ subchannel is selected} \end{cases} \tag{5}$$

1. Choose the first representation for all symbols $\left(\text{set } F_i = 0, 0 \le i \le \frac{N}{2} - 1\right)$
2. Start from the first subchannel:
   a) If the symbol has a second representation, flip that subchannel, i.e. use the second representation. If PAPR of the resulting frame is reduced, keep this setting, (set $F_i = 1$, for the symbol being flipped), otherwise return the symbol to its original representation.

   b) If the symbol is currently flipped, $F_i = 1$, check if flipping it back results in a lower PAPR and if true set $F_i = 0$.
   c) Consider next subchannel, increment i, and go to 2.a.
3. Re-iterate through step 2.a, a number of times to refine the result.

Note that the flips in steps 2.a and 2.b do not require an $\mathcal{IFFT}$ for each step since only one subchannel is involved. Let X'_l be the first representation of the symbol to be transmitted on the lth subchannel, and X''_l be its second representation. The new time frame may be computed as follows, knowing that $X'_l = -X''_l$ and substituting for X''_l in (2) and using conjugate symmetry:

$$\begin{aligned}\breve{x}_n &= x_n - \frac{2}{\sqrt{N}} X'_l e^{j2\pi ln/N} - \frac{2}{\sqrt{N}} \left(X'_l\right)^* e^{j2\pi(N-l+1)n/N} \\ &= x_n + (-1)^{n+1} \frac{4}{\sqrt{N}} |X'_l| \cos\left(\frac{2\pi nl}{N} + \angle X'_l\right) \end{aligned} \quad (6)$$

where, $\breve{x}_n$ is the nth time domain sample of the frame resulting from flipping its lth symbol representation.

4 Simulation Results

OFDM with 64 complex subchannels with 64-symbol constellations have been used for simulation with an oversampling factor of 4 as suggested analytically by [22] and experimentally by [23]. The two real channels have been silenced for simplicity. Three different constellation types have been studied:

1. Square QAM (SQAM)
2. Hexagonal QAM (HQAM)
3. Extended Hexagonal QAM (EHQAM)

where, in the case of EHQAM, three different constellation extension factors, $\kappa \in \left\{\frac{91}{64}, \frac{109}{64}, \frac{121}{64}\right\}$ have been simulated and studied.

Step 3 of the optimization algorithm (the refinement step) is performed for 4 different number of times for each of the extension factors: not performed (normal optimization), 1, 2, and 3 times refinement. The assignment of constellation points to symbols has been performed by moving along a helical path as shown in Fig. 3, assigning the points along the path in succession to symbols 0 to 64, respectively. Note should be taken that since the symbols are assumed to be equiprobable, this would not affect the results of the simulation when running for large number of instances. Any permutation on the assignment of constellation points to symbols results in general to a permutation on the PAPR values of possible frames.

Fig. 4, top, middle, and bottom compare the complementary cumulative distribution functions (CCDFs) and the probability distribution functions (PDFs) for PAPR of 150 000 frames for the three constellation types with different refinement levels for EHQAM with $\kappa = \frac{121}{64}$, $\kappa = \frac{109}{64}$ and $\kappa = \frac{91}{64}$, respectively. The CCDF curves could be interpreted as the probability of clip.

Table 1. Comparison of mean and standard deviation of PAPR and average frame power (AVP) for SQAM, HQAM and EHQAM.

	SQAM	HQAM	EHQAM ($\kappa = 121/64$)		EHQAM ($\kappa = 109/64$)		EHQAM ($\kappa = 91/64$)	
			Normal Optimization	1x refined	Normal Optimization	1x refined	Normal Optimization	1x refined
Mean PAPR	9.7347	9.7323	7.8279	7.2348	7.8269	7.3427	7.8555	7.5649
STD PAPR	1.0193	1.0075	0.5768	0.4966	0.5824	0.5053	0.5437	0.4891
Mean AVP	31.2025	30.4660	32.9628	32.9628	32.1832	32.1832	31.1410	31.1410
STD AVP	0.3397	0.3217	0.3473	0.3473	0.3841	0.3841	0.3792	0.3792

As can be seen, the refinement procedure virtually has no effect after one time refinement. The curves of CCDF of PAPR for 2 and 3 times refinements are not distinguishable for clip rates of less than 1%. This suggests that two passes on the frame symbols is essentially sufficient. In all three cases (three extension factors) the PAPR is reduced by about 5 dB at clip rates of about 10^{-5}.

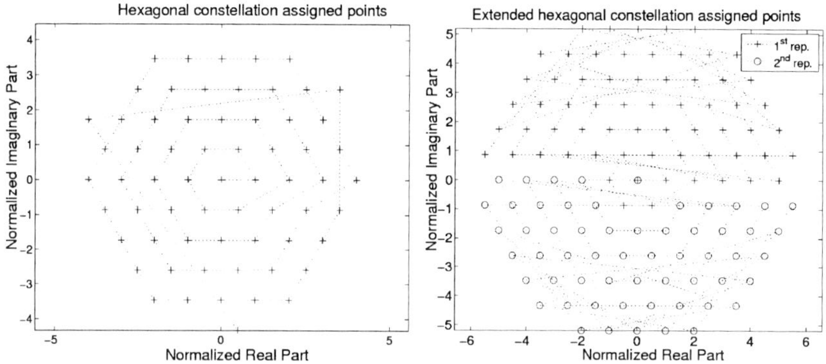

Fig. 3. Constellation symbol assignment: *left* – Normal 64-HQAM, *right* – 64-EHQAM with $\kappa = \frac{109}{64}$, the 1st and 2nd representations are pointed out explicitly.

Table 1 summarizes the standard deviation and mean value of PAPR and the average frame power (AVP) for some different cases studied[2]. It can be seen that for EHQAM with extension factor $\kappa = \frac{91}{64}$, the AVP has reduced by about 0.1 dB while retaining its PAPR reduction characteristics.

[2] The values for average frame power shown in the table are different from actual values by an additive constant, since non-normalized $\mathcal{IFFT}$ is used.

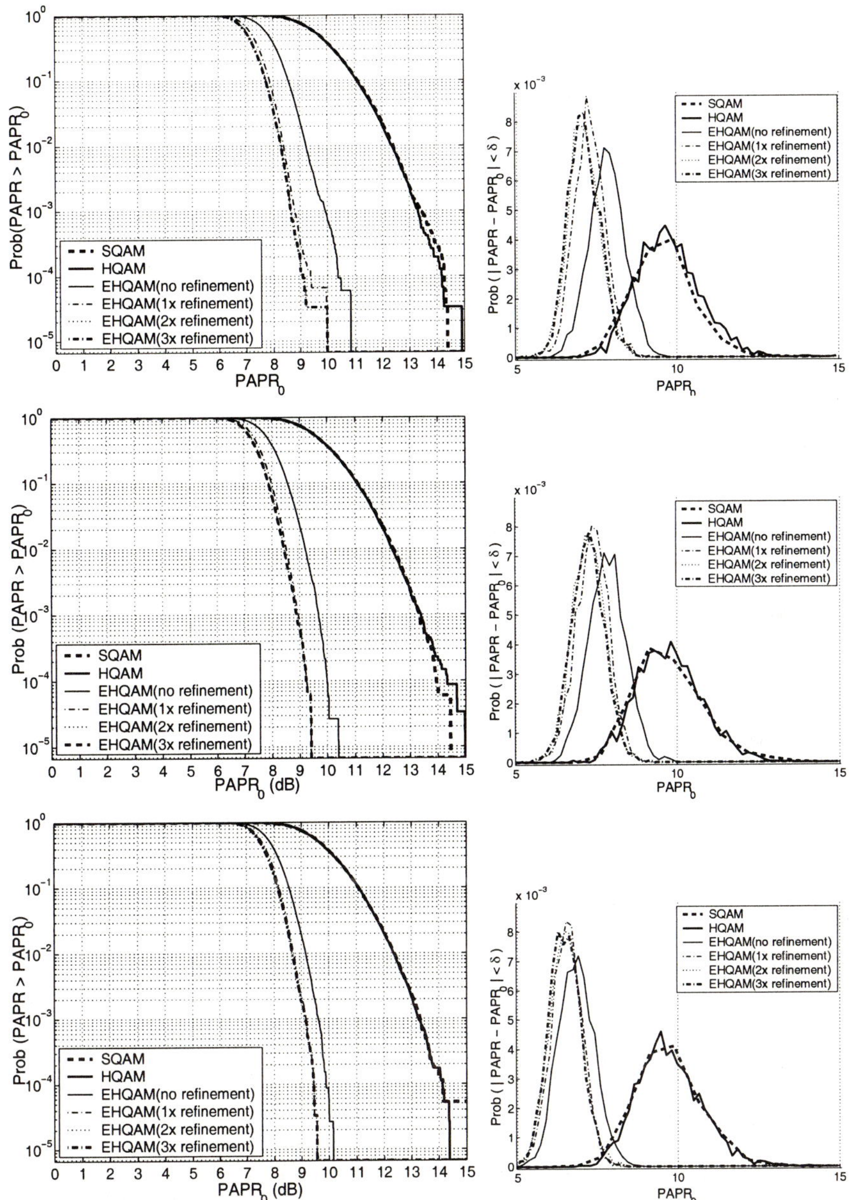

Fig. 4. *Left* – Plots of CCDF of PAPR, *right* – Plots of PDF of PAPR, for *top* – $\kappa = \frac{121}{64}$, *middle* – $\kappa = \frac{109}{64}$, *bottom* – $\kappa = \frac{109}{64}$, for SQAM, HQAM, EHQAM: Normal optimization and 1–3x refined. (The plots for PDFs are downsampled and smoothed for the sake of clarity.)

Fig. 5, left, compares the PDFs of AVP for the three studied extension factors. The values of AVP for different refinement levels in the optimization phase, is the same, as can be seen in the PDFs and also in Table 1. This is quite reasonable, since for symbols with two representations, X_l' and X_l'', we have $|X_l'|^2 = |X_l''|^2$, and by Parseval's theorem, the AVP of the frames with any number of flips would be the same.

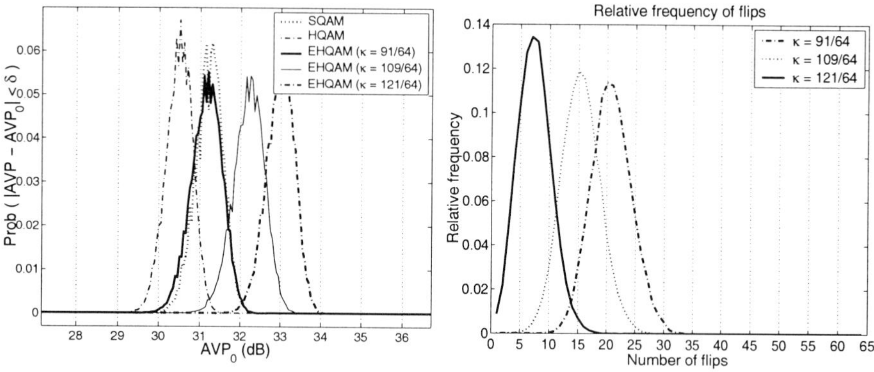

Fig. 5. *Left* – PDFs of average frame power (AVP) for SQAM, HQAM and EHQAM, *right* – Relative frequency of flips in optimized (3x refined) EHQAM, for extension factors $\kappa = \frac{91}{64}$, $\kappa = \frac{109}{64}$ and $\kappa = \frac{121}{64}$.

As a measure of complexity, the number of flips performed in the frame, were studied. The results are summarized as relative frequency plots in Fig. 5, right. It can be seen that for EHQAM with extension factor $\kappa = \frac{91}{64}$, the number of flips is substantially smaller than the other two cases. This was also presumable, since merely about 42% of the symbols have two representations while for $\kappa = \frac{109}{64}$ and $\kappa = \frac{121}{64}$, 70% and 89% of the symbols have a second representation, respectively.

5 Conclusion

A method for PAPR reduction of OFDM and DMT based on proper extension of hexagonal constellations is introduced. The scheme is proved to be effective in several aspects namely a reduction in PAPR by more than 5 dB at clip rate of 10^{-5} is possible with decreased average power (which means increased noise margin), at low complexity. Since no overhead information on how to decode a number of subchannels is required to be transmitted the symbol error rate would not increase. As deducible from simulations results, higher constellation extension factors result in higher PAPR reduction at the expense of higher complexity and increase of average power. It can also be observed from the results that use of normal hexagonal constellations, neither increases, nor substantially decreases the PAPR, while reducing the average power by about 0.8 dB.

As compared to PTS, both the proposed method and PTS are actually of exponential complexity, rendering them useless as the number of channels increases. In case of the sub-optimal optimization schemes, the proposed method has the superiority of requiring a single oversampled $\mathcal{IFFT}$ whereas the PTS method requires an oversampled $\mathcal{IFFT}$ for each decision. On the other hand, the number of symbols with two representations in each frame is presumed to be dependent on the number of subchannels, while for the PTS approach the number of partial transmit sequences seems to be taken as a constant not depending on the number of subchannels. The amount of PAPR reduction by the proposed method and PTS with two phases seems to be comparable. The proposed method is also superior to many other schemes in that no rate reduction is imposed on the system.

Acknowledgment. The authors want to thank Dr. Pedram Safari of Sharif University, Department of Mathematics, for guiding discussions and helpful comments on sphere packings, Prof. Frank R. Kschischang of University of Toronto, Department of Electrical and Computer Engineering, for kindly supplying us with [7] and Dr. Masoud Sharif of California Institute of Technology, Department of Electrical Engineering, for reviewing the first drafts of the paper and profound comments.

References

1. Müller, H. and Huber, J. B.: OFDM with Reduced Peak-to-Average Power Ratio by Optimum Combination of Partial Transmit Sequences. Electronics Letters, **33**, No. 5, (1997), 368–369
2. Friese, M.: Multitone Signals with Low Crest Factor. IEEE Trans. Communications, **45**, No. 10, (1997), 1338–1344
3. Bäuml, R., Fischer, R., Huber, J.: Reducing the peak-to-average power ratio of multicarrier modulation by selected mapping. Electronic Letters, **32**, No. 22, (1996), 2056–2057
4. Mestdagh, D. J. G., Spruyt, P. M. P.: Method to Reduce the Probability of Clipping in DMT-based Transceivers. IEEE Trans. Communications, **44**, (1996), 1234–1238
5. Gatherer, A., Polley, M.: Controlling Clipping Probability in DMT Transmission. Asilomar Conference on Signals, Systems, and Computers, (1998), 578–584
6. Tellado, J, Cioffi, J. M.: PAR Reduction in Multicarrier Transmission Systems. Delayed Contribution ITUT 4/15, D.150 (WP 1/15), Geneva, (1998)
7. Kschischang, F. R., Narula, A., Eyuboglu, V.: A new approach to PAR control in DMT systems. UAWG contribution, No. TG/98-127, (1998), 1–5.
8. Tellado, J., Cioffi, J. M.: PAR reduction with minimal or zero bandwidth loss and low complexity. Contribution T1E1.4, (1998), 98–173
9. Davis, J. A., Jedwab, J.: Peak-to-Mean power Control in OFDM, Golay Complementary Sequences and Reed-Muller Codes. IEEE Trans. Information Theory, **45**, (1999), 2397–2417
10. Van Eetvelt, P., Wade, G., Tomlinson, M.: Peak to average power reduction for OFDM schemes by selective scrambling. Electronic Letters, **32**, (1996), 1963–1964

11. Jayalath, A. D. S., Tellambura, C.: Reducing the peak-to-average power ratio of orthogonal frequency division multiplexing signal through bit or symbol interleaving. Electronic Letters, **36**, (2000), 1161–1163
12. Tellambura, C.: Use of m-sequences for OFDM peak-to-average power ratio reduction. Electronic Letters, **33**, (1997), 1300–1301
13. Wulich, D.: Comments on the Peak Factor of Sampled and Continuous Signals. IEEE Communications Letters, **4**, No. 7, (2000), 213–214.
14. Cioffi, J. M.: Advanced communications 1. EE 379A course reader, Stanford University, Stanford, CA., (2001)
15. Weisstein, Eric W.: Circle packing. World of mathematics, `http://mathworld.wolfram.com/CirclePacking.html`
16. Peterson, Ivars: Pennies in a tray. The Mathematical Association of America, MAA online, Nov. 25, (1996), `http://www.maa.org/mathland/mathland_11_25.html`
17. Specht, E.: The best known packings of equal circles in the unit circle. Nov. 30, (2000), `http://hydra.nat.uni-magdeburg.de/packing/cci/cci.html`
18. Specht, E.: 16 circles in the unit circle. `http://hydra.nat.uni-magdeburg.de/packing/cci/cci16.html`
19. Conway, J. H., Sloane, N. J. A.: Sphere packings, lattices and groups. Springer Verlag, 2nd edition, (1993)
20. Murphy, D.: High-order optimum hexagonal constellations. Proceedings of the IEEE Personal Indoor and Mobile Communications Conference, London, UK, (2000), 143–146.
21. Cimini Jr., L. J., Sollenberger, N.R.: Peak-to-Average Power Ratio Reduction of an OFDM Signal using Partial Transmit Sequences. IEEE Communications Letters, **4**, No. 3, (2000), 86–88
22. Sharif, Massud, Khalaj, Babak H.: Peak to mean envelope power ratio of oversampled OFDM signals: An analytical approach. Proc. IEEE International Conference on Communications, Helsinki, Finland, **5**, (2001), 1476–1480
23. Tellado-Mourelo, J.: Peak to average power reduction for multicarrier modulation. Ph.D. dissertation, Stanford University, Stanford, CA., (1999)
24. Mow, W. H.: Fast decoding of the hexagonal lattice with applications to power efficient multi-level modulation systems. Proc. IEEE ICCS/ISITA 92, Singapore, (1992), 370–373
25. Fu, Charles: Hexagonal coordinates. On news group `rec.games.progrmammer`, available at `http://www-cs-students.stanford.edu/~amitp/Articles/Hexagon2.html`
26. Müller, Stefan H., Huber, J. B.: A Comparison of Peak Power Reduction Schemes for OFDM. Proc. IEEE, Globecom 97, (1997), 1–5
27. Li, Xiaodong, Cimini, Jr., L. J.: Effects of clipping and filtering on the performance of OFDM. IEEE Communications Letters, **2**, No. 5, (1998), 131–133

Adaptive Application-Centric Management in Meta-computing Environments

Yoonhee Kim[1] and Sung-Yong Park[2]

[1] Dept. of Computer Science, Sookmyung Women's University, Seoul, Korea
yulan@sookmyung.ac.kr
[2] Dept. of Computer Science, Sogang University, Seoul, Korea
parksy@ccs.sogang.ac.kr

Abstract. An explosive growth in the number and diversity of resources in meta-computing environment has caused the high degree of difficulty of management of large-scale applications due to factors such as platform oriented management architectures, lack of coordination and compatibility among heterogeneous network management systems, and the dynamic characteristics of networks and application requirements. Even though previous researches in management solved many problems related to resource management including network management, they are not enough to provide integrated management services for an application point of view. To achieve the services, the matter of heterogeneity and dynamic adaptability in applications and resources should be addressed. This paper presents APAM (Adaptive and Proactive Application Management System), an integrated management model that provides a slim and adaptive management services based on the characteristics of an application and its adaptive fault management implementation.

1 Introduction

The emerging high speed networks and the advances in computing technology are important driving forces to merge the communications and computing technologies that has resulted in an explosive growth in network complexity, size and networked applications. The management and control of large-scale distributed computing systems and their applications is a challenging research problem due to the huge amount of data that needs to be collected and coordinated, the heterogeneity and the independence of resources required by these services (e.g., database servers, domain name servers, various middleware). They also run under different organizations and administration policies.

Most of current network management systems have focused on the monitoring and management of physical network components (e.g., servers, switches, routers, links) and have not adequately addressed the application service management issues and needs. Moreover, since the integration of software development and management has not been well understood and investigated [1,2], the current practice of application development does not consider the application management issues. Even if they do consider the management issues, it will be a force-fitting approach that is applied at the end of the development cycle. As a result, building an integrated network

M.H. Shafazand and A M. Tjoa (Eds.): EurAsia-ICT 2002, LNCS 2510, pp. 937–944, 2002.

management system to efficiently and adaptively manage both the computing systems and their application services is very difficult.

In this paper we present the architecture of an integrated application-centric network management system for meta-computing environment, APAMS (Adaptive and Proactive Application Management System), and propose an integrated application development and management framework to achieve adaptive and proactive application service management. Our approach is to identify and incorporate the management functions during all phases of an application life cycle -- specification, development, deployment, operations, and maintenance. This leads to the development of efficient data structures (e.g., sensors) that accurately describe the current state of the application at runtime and effector data structures (e.g., actuators) that allow control and management systems to efficiently stop, resume, and migrate the execution of application services. As an initial study of application service management, we assume that an application is already composed of multiple modules (i.e. tasks). We also assume that their requirements and management strategies are assigned accordingly. Based on the assumption, APAMS has to provide the services for the configuration and deployment of any application service in addition to the interaction with the network management system.

Recently, the use of mobile agent systems in network management tasks has been an active research topic [10]. One of the pioneering works has been the Management by Delegation (MbD) approach [3]. This approach dynamically distributes the management computations from the NMS to the elastic servers (MbD-servers) at the devices where the managed resources are located. The mobile code infrastructure closely interacts with that of the network management architecture. In active network research area, NetScript 9 uses SNMP to manage IP router in active nodes, accessing routing information. [4,5,8,9] proposed various active network approaches to control network with mobile agents and [6,7] applies mobile agents to build applications within telecommunication network management. These solutions have limitations mainly due to the difficulties associated with accessing all the resources from the mobile agents. On the other hand, our approach is unique in that we mainly focus on developing management techniques that are proactive, open, and general so they can be applied to managing the properties or functionalities (performance, fault, security, etc.) of any application services and networks. Furthermore, our approach aims at integrating the application development process with management in order to build efficient, robust data structures that will enable us to dynamically adapt and change the application execution environments at runtime.

This paper is organized as follows. Section 2 describes the architecture of APAMS. Section 3 presents the preliminary results of the application fault management implementation with two different control scenarios. Section 4 concludes this paper.

2 Overview of APAMS Architecture

The APAMS consists of Application Management Editor (AME) and Management Computing System (MCS). The AME module provides the application developers with all the tools required to specify the appropriate control and management schemes to maintain any quality of service requirements or application attributes/ functionalities (e.g., performance, fault, security, etc.). The MCS provides the core active management services to enable the efficient, proactive management of a wide range of network applications. The MCS creates Application Execution Environment (AEE) at runtime and deploys the application and its services over the network automatically. The services offered by the MCS are implemented using mobile agents and they can be selected dynamically by invoking corresponding mobile agent template for the service implementation. A general agent based and adaptive methodology achieves automated deployment and configuration of application and proactive management of application faults, performance, security, etc. The architecture of APAMS is shown in Fig. 1. The main APAMS modules include Application Management Editor (AME), Management Computing System (MCS), Application Delegated Manager (ADM), Application Service Template (AST), and Network Information Service (NIS).

- The Application Management Editor (AME): The Application Management Editor is a graphical user interface for developing an application with pre-developed tasks and specifying management requirements to configure and manage the execution of the application. It provides application developers with the services required for specifying and characterizing the application requirements in terms of performance, fault, security, and also specifying the appropriate management schemes to maintain the application requirements.
- Management Computing System (MCS): Once the application management requirements are defined using the AME, the next step is to utilize the MCS services to build the appropriate application execution environment that can dynamically control the allocated resources to maintain the application requirements during the application execution. MCS defines strategies required to analyze the dependencies among components, deploy them automatically, set up the application execution environment and then run the application.
- Application Delegated Manager (ADM): The MCS assigns one ADM to manage one or several application attributes (performance, fault, security, etc.). We can view ADM as the operating system of the application service to be managed in the system. For each application task (component) (we are using task and component interchangeably here), the ADM launches an appropriate Task Agent (TA) to monitor and manage the application task execution. The TA monitors the task execution using appropriate task sensors and intervenes using the task actuator whenever the task execution on the assigned machine cannot meet its requirements. In this case the task actuator can suspend, save task execution state, or migrate the task execution to another remote machine. The appropriate management scheme can be selected at runtime depending on the fault management strategy specified by the user.
- Application Service Template (AST): The application requirements and its execution environment are stored in a template repository. The template can be viewed as the blueprint of the application execution environment. The MCS searches the Template Repository for the appropriate template that can meet all the requirements of a given application service
- Network Information Service (NIS): The NIS is responsible to collect management information not only about the network devices, but also information related to computer processes, file systems, user access information and patterns, and protocols. The NIS also performs tasks to manage the network devices, protocol functions, computer processes and file systems. The common interface of NIS creates uniform data format from data in various formats given by diverse

protocols or platform-oriented management systems. The data is forwarded to MCS, which is responsible for offering appropriate data to management components.

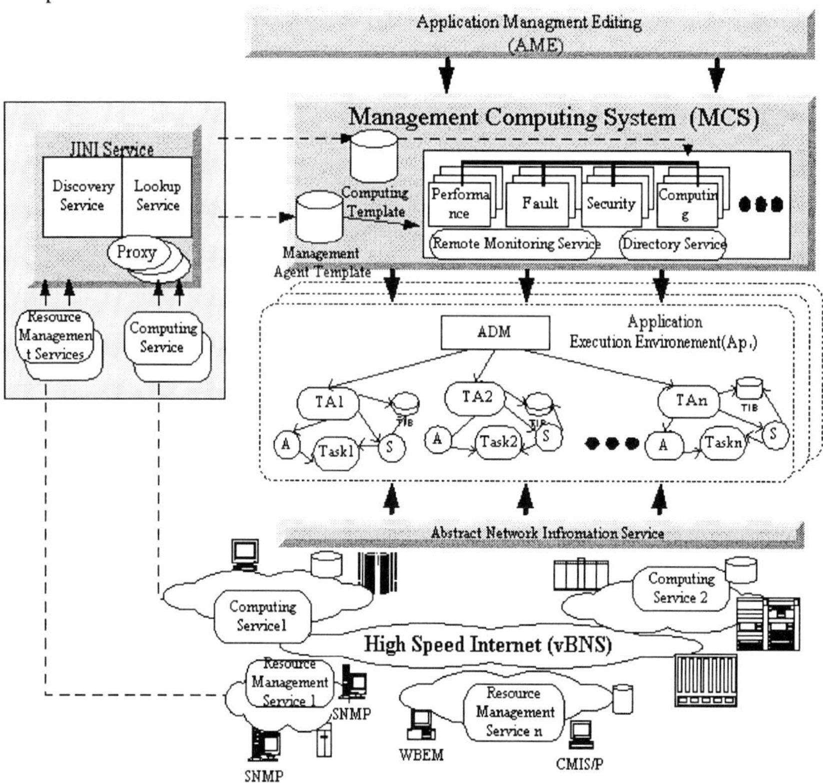

Fig. 1. APAMS Architecture

The main functions offered by the AME are controlling the application editor workplace and storing the application management requirements in application service template repository. The AME provides menu-driven task libraries that are grouped in terms of their functionalities. Users can develop applications with selected tasks and create relationship between the tasks. Each task creates a management specification window, which includes management requirement fields such as dependencies among tasks, monitoring parameters, and fault tolerance strategy.

The dependency analysis service, which is one of the MCS services, queries the template repository to obtain the information about the task dependencies. The dependency attributes denote which software components must be present in the system before the application tasks can be successfully installed. With the help of the dependency analysis, the automatic application deployment service creates the deployment plan for each application task. The service first locates available candidate resources for the application tasks. This work is performed by the task scheduling algorithm in the deployment service. Based on the resource assignment of the application tasks, the appropriate executables and their locations are identified.

When a new application is created, an AST is created in the template repository. The AST provides a uniform format for representing application-specific information. The AST includes two kinds of information structures: invariant and variant. Invariant application information is common to all applications, while variant information is specific to an application based on its characteristics. The representation of application templates is realized in an application template class, which defines the standard format that should be followed by every application creation. With the concept of object-oriented design, an application template object is instantiated from a template class.

Our methodology to achieve proactive management of any functionality or property of an application service is based on three procedures: Change Detection, Change Verification and Analysis, and Adaptation procedures. The Change Detection procedure of an application service is responsible for detecting the conditions in which the monitored property (fault, performance, security) of an application service deviates from the acceptable behavior or operation; the application performance degrades severely due to traffic conditions, or due to software or hardware failures, or application security has been compromised. The Change Verification and Analysis procedure is entered to make sure that the detected change is real and is not caused by false alarms. Once the change event is verified and its type is identified, the Adaptation procedure is invoked to execute the appropriate adaptation scheme. The APAMS management services are targeted to managing both the applications as well as their resources (hardware and/or software systems). In our approach we use the word "*change*" so that it can be replaced by performance, fault, security or any other attributes. In this paper we focus on the fault management of application services. However, the same approach can be applied to managing other attributes. Furthermore, the APAMS management services are targeted to managing both the applications as well as their resources (hardware and/or software systems).

In what follows, we discuss in further detail our implementation approach by using a matrix multiplication application and then how to manage the application faults using the APAMS management services.

3 Adaptive Fault Management

The main goal of the application fault management is to efficiently recover from hardware/software failures of the system resources. Redundancy is an important technique to detect and recover from component failures in the system. The redundancy can be in the form of hardware, software, or time [11]. As the system increases its complexity, more sophisticated techniques are needed to manage those redundancies. In addition, the fault management scheme must be flexible and adaptive. In SCOP [12], a design methodology is proposed to introduce support techniques to reduce the resource cost of fault-tolerant software, both in space and time, by providing designers with a flexible redundancy architecture in which dependability and efficiency can be adjusted dynamically at run time. In another work [12,13], the use of mobile agents to support adaptive fault tolerance is implemented. In our adaptive application fault-tolerance approach, we use mobile agents to efficiently manage the redundancy.

We use redundancy techniques and task migration to implement the control functions required to dynamically manage application fault. In this paper, we evaluate three techniques to manage the application fault: active redundancy, passive redundancy and migration. Each technique is implemented as an agent template as shown in Fig. 2.

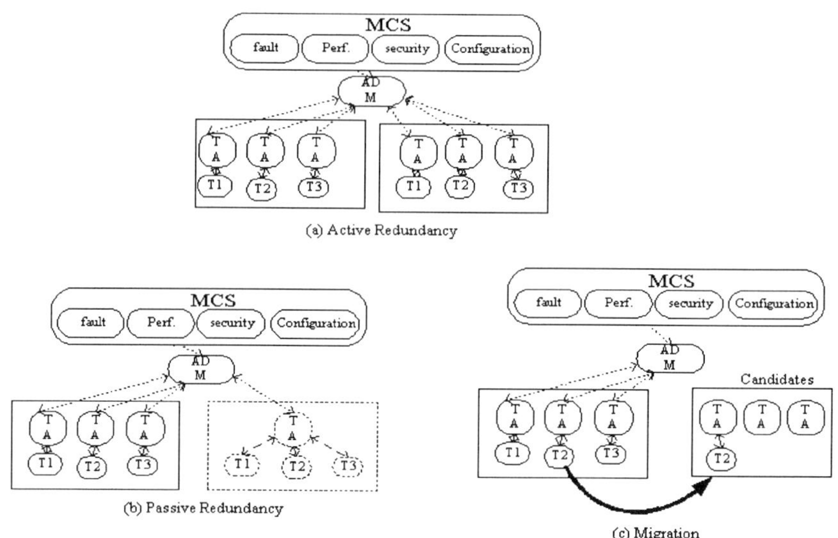

Fig. 2. Management schemes

The active redundancy scheme duplicates the execution of the application on two machines (see Fig. 2 (a)). In this scheme, the task agent picks up the results from the faster machine that completes the task execution first. When a fault occurs, the secondary execution is automatically chosen without any adaptation plan. The passive redundancy assigns each task to a primary machine that will run the task and another machine to be used as a backup whenever faults are detected on the assigned machine (see Fig. 2 (b)). The backup machine is kept-up-to-date in order to be ready to resume the task execution from the last updated checkpoint. The main advantage of this approach is that it needs less resources than the active redundancy approach. In this scheme, one backup machine can be used as a backup machine to several tasks. The third approach does not introduce redundancy and recovers faults by task migration (see Fig. 2 (c)). The overhead of task migration is relatively higher than redundancy schemes.

In addition to reducing the time for fault detection, active redundancy technique simplifies the communication between task agents. Fig. 2 (a) shows the overhead incurred by applying this redundancy scheme to adaptively manage the faults of three applications with three tasks each. In the small application case (execution time is around 60s), the overhead incurred in using our scheme to detect and recover from one task failure, two task failures, and three task failures are 0.10%, 0.18%, and 0.22%, respectively. For medium and large applications, the overhead in managing one, two or three task failures is very small (less than 0.02%).

In the scenario using passive redundancy in managing the application faults, we assign the task to two machines: one is designated as the primary machine while the second machine is designated as the backup
machine. Fig. 2 (b) shows the overhead incurred in applying this redundancy technique to manage the faults of three applications. For a small application with three tasks, the overhead incurred to manage one task failure, two task failures, and three task failures are 0.18%, 0.26%, and 0.42%. For a medium to large size application, the overhead to manage one, two or three task failures is very small.

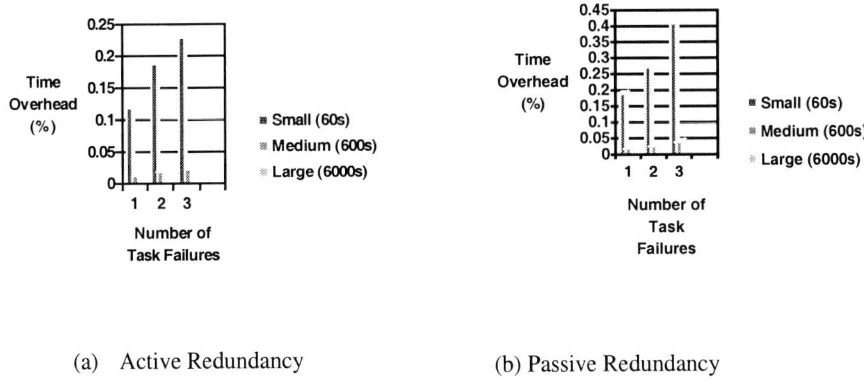

(a) Active Redundancy (b) Passive Redundancy

Fig. 3. Overheads of Redundancy Techniques

It is clear from the experimental results that our approach is very efficient, especially, for large parallel/distributed applications. Furthermore, the use of mobile agents and agent templates, we can dynamically select the appropriate redundancy technique at runtime depending on the system load and number of available resources

4 Conclusion

In this paper, we presented our approach to provide adaptive application-centric management and to implement an Adaptive and Proactive Application Management System (APAMS). As APAMS builds adaptive management environment (i.e. AEE) based on the management requirements of an application, management had been performed in timely manner and low overheads. The experimental results of the APAMS management services to manage the performance and fault tolerance execution of three applications of different sizes demonstrate that our agent-based approach can lead to low overhead in fault management. We are currently implementing additional services to balance the load across the network resources and maintain the system and application security requirements to provide efficient management for meta-computing environment.

References

1. G. Kar, A. Keller, S. Calo, "Managing Application Services over Service Provider Networks: Architecture and Dependency Analysis," 2000 IEEE/IFIP Network Operations and Management Symposium, Hawaii, April, 2000
2. V. Machiraju, M. Dekhil, K. Wurster, P. Garg, M. Griss and J. Holland, "Toward Generic Application Auto-discovery, 2000 IEEE/IFIP Network Operations and Management Symposium, Hawaii, April, 2000
3. G. Goldszmidt and Y. Yemini, "Distributed Management by Delegation", in 15th international Conference on Distbuted Computing, June 1995.
4. Danny Raz, Y. Shavitt, "An active Network Approach to Efficient Network Management," DIMACS Technical Report 99-25.
5. J. Nicklisch, J. Quittek, A. Kind, S. Arao, "INCA:An Agent-based Network Control Architecutre," international workshop on agents in telecommunication application-IATA'98, AgentWorld'98, Paris, France.
6. Magedanz, Eckardt, "Mobile Software Agents: A new Paradigm for Telecommunications Management", IEEE/IFIP Network Operations and Management Symposium (NOMS) Kyoto, Japan, April, 1996.
7. Krause and Magadantz, Mobile Service Agents Enabling "Intelligence on Demand" in Telecommunications", Proceedings of IEEE GLOBECOM'96 London, UK, Nov. 1996.
8. Joseph Kiniry, D. Simmerman "A Hands-on look at java mobile agents", IEEE Internet Computing 1(4):21–30, July/August 1997.
9. Akhil, Sahai, Christine Morin, "Towards Distributed and Dynamic Network Management," Proceedig of the IEEE/IFIP Network Operations and Network Symposium (NOMS), Feb, 98 New Orleans, USA.
10. Y. Yemini, and S. de Silva, "Toward programmable networks, In IFIP/IEEE Intl. Workshop on Distributed Systems: Operations and Management, October 1996.
11. Avizienis. "Fault-Tolerant Systems." IEEE Transactions on Computers, C-25(12):1304–1312, December 1976.
12. J. Xu, A. Bondavalli, F. D. Giandomenico. "Dynamic Adjustment of Dependability and Efficiency in Fault-Tolerant Software", in "Predictably Dependable Computing Systems", B. Randell, J. C. Laprie, H. Kopetz and B. Littlewood Ed., Springer-Verlag, 1995, pp.155–172.
13. S. Bagchi, K. Whisnant, Z. Kalbarcyzk, R.K. Iyer. "Chameleon: Adaptive Fault Tolerance Using Reliable, Mobile Agents", The 27th Fault Tolerance Computer Symposium, Munich, Germany, June 23-25 1998[1]

[1] The publication of this paper was supported in part by Institute for Applied Science and Technology of sogang University.

The Weakest Failure Detector for Solving Election Problems in Asynchronous Distributed Systems

Sung-Hoon Park

Dept. of Computer Science, NamSeoul University, Chung-Nam 330-800, Korea
spark@nsu.ac.kr

Abstract. This paper is about the weakest failure detector to solve the Election problem in asynchronous distributed systems. We first discuss the relationship between the *Election* problem and the *Consensus* problem in asynchronous distributed systems with unreliable failure detectors. Chandra and Toueg have stated that Consensus is solvable in asynchronous systems with unreliable failure detectors. But, in contrast to the Consensus problem, the *Election* problem is impossible to solve with unreliable failure detectors even with a single crash failure. More precisely, the weakest failure detector that is needed to solve this problem is a *Perfect Failure Detector*, which is strictly stronger than the weakest failure detector that is needed to solve Consensus.

1 Introduction

To elect a *Leader* (or Coordinator) in a distributed system, an *agreement* problem must be solved among a set of participating processes. This problem, called the *Election* problem, requires the participants to agree on only one leader in the system [1]. The problem has been widely studied in the research community [2,3,4]. One reason for this wide interest is that many distributed protocols need an election protocol.

The Election problem is described as follows. At any time, there is at most one process that considers itself a *leader* and all other processes consider it as to be their only leader. If there is no leader, a leader is eventually elected.

Consensus and Election are similar problems in that they are both agreement problems. The so-called FLP impossibility result, which states that it is impossible to solve any non-trivial agreement in an asynchronous system even with a single crash failure, applies to both problems [5]. The starting point of this paper is the fundamental result of Chandra and Toueg [6], which states that Consensus is solvable in asynchronous systems with unreliable failure detectors.

An interesting question is then whether the Election problem can also be solved in asynchronous systems with unreliable failure detectors. The answer to this question is "No". This means that the Election problem is harder than the Consensus problem. However, in contrast to initial intuition, the reason Election is harder than Consensus is not upon its *Liveness* condition. The difficulty in solving Election is actually upon its *Safety* condition (all the processes connected the system never disagree on the leader). This condition requires precise knowledge about failures which unreliable

M.H. Shafazand and A M. Tjoa (Eds.): EurAsia-ICT 2002, LNCS 2510, pp. 945–952, 2002.

failure detectors cannot provide. More precisely, the weakest failure detector that is needed to solve this problem is a *Perfect Failure Detector*, which is strictly stronger than the weakest failure detector that is needed to solve Consensus.

The rest of the paper is organized as follows. Section 2 describes motivations and the related works. In Section 3 we describe our system model. In Section 4 and 5 we show that the weakest failure detector for solving Election is a *Perfect failure Detector*. Finally, Section 6 summarizes the main contributions of this paper and discusses related and future work.

2 Motivations and Related Works

In recent years, several paradigms have been identified to simplify the design of fault-tolerant distributed applications in a conventional static system. Election is among the most noticeable, particularly since it is closely related to group communication, which (among other uses) provides a powerful basis for implementing active replications.

It was shown in [5] that the Consensus problem cannot be solved in an asynchronous system if even a single crash failure can occur. The intuition behind this widely cited result is that in an asynchronous system, it is impossible for a process to distinguish between another process that has crashed and one that is merely very slow. The consequences of this result have been enormous, because most real distributed systems today can be characterized as asynchronous, and Consensus is an important problem to be solved if the system is to tolerate failures.

As a result, the Consensus problem has frequently been used as a yardstick of computability in asynchronous fault-tolerant distributed systems. That means that if any problem is harder than Consensus, it also cannot be solved in asynchronous systems.

The asynchronous model of computation is especially popular in practice because unpredictable workloads are sources of asynchrony in many real systems. Therefore rendering any synchrony assumption is valid only probabilistically. Thus, the impossibility of achieving Consensus reveals a serious limitation of this model for fault-tolerant applications such as the Election problem. Because Consensus is such a fundamental problem, researchers have investigated various ways of circumventing the impossibility.

Actually, the main difficulty in solving such a problem in presence of process crashes lies in the detection of crashes. As a way of getting around the impossibility of Consensus, Chandra and Toug extended the asynchronous model of computation with unreliable *failure detectors* and showed that the Consensus problem is solvable even with unreliable failure detectors [8].

3 Model and Definitions

Our model of asynchronous computation with failure detection is the one described in [8]. In the following, we only recall some informal definitions and results that are needed in this paper.

3.1 Processes

We consider a distributed system composed of a finite set of processes $\Omega=\{p_1,p_2,..,p_n\}$ completely connected through a set of channels. Communication is by message passing, *asynchronous* and *reliable*. Processes fail by crashing; Byzantine failures are not considered.

Asynchrony means that there is no bound on communication delays or process relative speeds. A reliable channel ensures that a message, sent by a process p_i to a process p_j, is eventually received by p_j if p_i and p_j are correct (i.e. do not crash).

To simplify the presentation of the model, it is convenient to assume the existence of a discrete global clock. This is merely a fictional device inaccessible to processes. The range of clock ticks is the set of natural numbers. A history of a process $p_i \in \Omega$ is a sequence of events $h_i = e_i^0 \cdot e_i^1 \cdot e_i^2 \cdots e_i^k$, where e_i^k denotes an event of process p_i occurred at time k. Histories of correct processes are infinite. If not infinite, the process history of p_i terminates with the event crash_i^k (process p_i crashes at time k). Processes can fail at any time, and we use f to denote the number of processes that may crash. We consider systems where at least one process correct (i.e. $f < |\Omega|$).

A failure detector is a distributed oracle which gives hints on failed processes. We consider algorithms that use failure detectors. An algorithm defines a set of runs, and a run of algorithm A using a failure detector D is a tuple $R = < F, H, I, S, T>$: I is an initial configuration of A; S is an infinite sequence of events of A (made of process histories); T is a list of increasing time values indicating when each event in S occured; F is a failure pattern that denotes the set $F(t)$ of processes that have crashed at any time t; H is a failure detector history, which gives each process p and at any time t, a (possibly false) view $H(p,t)$ of the failure pattern: $H(p,t)$ denotes a set of processes, and $q \in H(p,t)$ means that process p *suspects* process q at time t.

3.2 Failure Detector Classes

Failure detectors are distributed *oracles* related to the detection of failures. A failure detector of a given class is a device that gives hints on a set of processes that it suspects to have crashed. Failure detectors are abstractly characterized by *completeness* and *accuracy* properties [8]. Completeness characterizes the degree to which crashed processes are permanently suspected by correct processes. Accuracy restricts the false suspicions that a process can make.

Two completeness properties have been identified. *Strong Completeness*, i.e. there is a time after which every process that crashes is permanently suspected by every correct process, and *Weak Completeness*, i.e. there is a time after which every process that crashes is permanently suspected by some correct process.

Four accuracy properties have been identified. *Strong Accuracy*, i.e. no process is never suspected before it crashes. *Weak Accuracy*, i.e. some correct process is never suspected. *Eventual Strong Accuracy* ($\Diamond$Strong), i.e. there is a time after which correct processes are not suspected by any correct process; and *Eventual Weak Accuracy* ($\Diamond$Weak), i.e. there is a time after which some correct process is never suspected by any correct process. A failure detector class is a set of failure detectors characterized by the same completeness and the same accuracy properties (Fig. 1).

For example, the failure detector class P, called *Perfect Failure Detector*, is the set of failure detectors characterized by Strong Completeness and Strong Accuracy. Failure detectors characterized by Strong Accuracy are reliable: no false suspicions are made. Otherwise, they are unreliable.

Completeness	Accuracy			
	Strong	Weak	◇Strong	◇Weak
Strong	P	S	$\diamond P$	$\diamond S$
Weak	Q	W	$\diamond Q$	$\diamond W$

Fig. 1. Failure detector classes. A failure detector class is a set of failure detectors characterized by the same completeness and the same accuracy properties.

For example, failure detectors of S, called Strong Failure Detector, are *unreliable*, whereas the failure detectors of P are *reliable*.

3.3 Reducibility and Transformation

The notation of *problem reduction* first has been introduced in the problem complexity theory [10], and in the formal language theory [9]. It has been also used in the distributed computing [11,12]. We consider the following definition of problem reduction. An algorithm A *solves* a problem B if every run of A satisfies the specification of B. A problem B is said to be *solvable with* a class C if there is an algorithm which solves B using any failure detector of C. A problem B_1 is said to be reducible to a problem B_2 with class C, if any algorithm that solves B_2 with C can be transformed to solve B_1 with C. If B_1 is not reducible to B_2, we say that B_1 is *harder than* B_2.

A failure detector class C_1 is said to be *stronger than* a class C_2, (written $C_1 \geq C_2$), if there is an algorithm which, using any failure detector of C_1, can emulate a failure detector of C_2. Hence if C_1 is stronger than C_2 and a problem B is solvable with C_2, then B is solvable with C_1. The following relations are obvious: $P \geq Q$, $P \geq S$, $\diamond P \geq \diamond Q$, $\diamond P \geq \diamond S$, $S \geq W$, $\diamond S \geq \diamond W$, $Q \geq W$, and $\diamond Q \geq \diamond W$. As it has been shown that any failure detector with *Weak Completeness* can be transformed into a failure detector with *Strong Completeness* [8], we also have the following relations: $Q \geq P$, $\diamond Q \geq \diamond P$, $W \geq S$ and $\diamond W \geq \diamond S$. Classes S and ◇P are incomparable.

3.4 The Election Problem

The Election problem is described as follows: At any time, at most one process considers itself the leader, and at any time, if there is no leader, a leader is eventually elected. More formally, the Election Problem is specified by the following two properties:

- *Safety*: All processes connected the system never disagree on a *leader*.

- *Liveness*: All processes should eventually progress to be in a state in which all processes connected to the system agree to the *only one* leader.

3.5 The Consensus Problem

In the *Consensus* problem (or simply Consensus), every participant *proposes* an input value, and correct participant must eventually *decide* on some common output value [7,13]. Consensus is specified by the following conditions.

- *Agreement*: no two correct participant decide different values;
- *Uniform-Validity*: if a participant decides v, then v must have been proposed by some participant;
- *Termination*: every correct participant eventually decide.

Chandra and Toueg have stated the following two fundamental results [6]:

1. If $f < |\Omega|$, Consensus is solvable with either S or W.
2. If $f < \lceil |\Omega|/2 \rceil$, Consensus is solvable with either $\Diamond S$ or $\Diamond W$.

4 Impossibility of Solving the Election Problem with Unreliable Failure Detectors

In this section, we show that the Election problem is not solvable in asynchronous systems with unreliable failure detectors. This impossibility result holds even with the assumption that at most one process may crash. Though a Strong Failure Detector is sufficient to solve Consensus, it is not sufficient to solve Election. More precisely, we show that if $f > 0$, Election can not be solved with either $\Diamond P$ or S.

Theorem 1. *If $f > 0$, Election can not be solved with either $\Diamond P$ or S.*

PROOF (by contradiction). Consider a failure detector D of $\Diamond P$ (respectively of S). We assume for a contradiction that there exists a deterministic election protocol E that can be combined with a failure detector D such that $E + D$ is also an election protocol. Consider an algorithm A combined with $E + D$ which solves Election and a run $R = < F, H_D, I, S, T >$ of A. We assume that only two processes P_i and P_j are correct. Consider that P_i is a leader at time (R, t_0). At time (R, t_1) where $t_1 > t_0$, the process P_j sends a message to confirm whether the leader is alive. At time (R, t_2) where $t_2 > t_1$, the process P_i sends a reply message to the process P_j. But at time (R, t_3) where $t_3 > t_1$, P_j falsely suspects other process P_i by the Weak accuracy property of the unreliable failure detector D in some run. At time (R, t_4) where $t_4 > t_3$, P_j considers itself a leader by delaying the receipt of the reply message sent by P_i until t_5, where $t_5 > t_4$. Thus at time (R, t_5) both P_i and P_j consider themselves the leader, violating the assumption that A is an election protocol. But after time t_5, all the processes except P_i and P_j are suspected. Hence there is a time after which every process that crashes is permanently suspected by every correct process. So H_D satisfies Strong Completeness.

Consider Accuracy.

- If D is of class $\diamond P$, H_D satisfies *Eventual Strong Accuracy*, i.e. there is a time after which correct participants are never suspected by any correct participant. As P_i and P_j are never suspected after time t_5 in H_D, then H_D satisfies *Eventual Strong Accuracy*.
- If D is of class S, H_D satisfies *Weak Accuracy*, i.e. some correct participant is never suspected in H_D. As P_j is never suspected by the correct process P_i, H_D satisfies Weak accuracy.
- This is a contradiction.

By the relation between failure detector classes, we have the following Corollary.

Corollary 1. *If $f > 0$, Election is not solvable with either $\diamond Q$, $\diamond S$, W* or *$\diamond W$.*

5 The Weakest Failure Detector for Solving the Election Problem

In the previous section, we showed that the Election problem is not solvable in asynchronous systems with unreliable failure detectors. Then, what is the weakest failure detector that is needed to solve this problem in asynchronous distributed systems? In this section, as the answer to this question, we show that a *Perfect Failure Detector* is the weakest failure detector for solving Election.

Theorem 2. *If $f > 0$, a failure detector of class Q is sufficient to solve Election.*

PROOF: The Election problem can be solved using an election protocol E combined with the following algorithm A and a failure detector D belonging to class Q:

1. Each process has a unique ID number that is known by all processes a *priority*.
2. The leader is initially the process with the lowest ID number.
3. If a process detects a failure, it broadcasts this information to all other processes. Upon receiving such a message, the receiver detects the failure.
4. When a process detects the failure of all processes with lower ID numbers, then that process becomes the leader.

The proof that the protocol E satisfies Election is as followings.

- *Safety* (proof by contradiction). The election protocol E starts only when the current leader has failed by the strong Accuracy property of D which belongs to the class Q (line 3). Assume that the current leader has crashed at time (R, t_1) in some run R of the protocol E. We assume for the contradiction that two processes, P_i and P_j ($P_i \neq P_j$) are elected as leaders at time (R, t_5), where $t_5 > t_1$. To be elected as a leader, they must have detected the failure of all processes with lower ID numbers (line 4). That means that P_i and P_j have detected the failure of all processes with lower ID numbers at a time (R, t_3) and at a time (R, t_4) respectively, where $t_1 < t_3$, $t_4 < t_5$. As two processes are different each other ($P_i \neq P_j$), at least one of them has a lower ID number (line 1). Thus, one process falsely suspected the other. But it is contradiction to the assumption that a failure detector D of class Q has the strong Accuracy property, i.e. no process is never suspected before it crashes.

- *Liveness*. In case of the current leader's failure at time (R, t_1) in some run of E, some correct process eventually detects the leader's crash by the *Weak Completeness* property of the failure detector D of class Q (i.e. there is a time after which every process that crashes is permanently suspected by some correct process) and broadcasts this information at time (R, t_2) where $t_2 > t_1$ (line 3). By the reliable channel assumption, every correct process eventually receives the information and starts to detect the failures of all processes with lower ID (line 3). Because the process detected a failure broadcasts this information to all other processes (line 4), every correct process eventually suspects every failed process. Thus at time (R, t_3) where $t_3 > t_2$, at least one process eventually has completed the detection of all failed processes with lower ID. That process becomes a leader at time (R, t_4) where $t_4 > t_3$ (line 4).

Theorems 1, 2 and Corollary 1 together show that a failure detector of class Q that satisfies Weak Completeness and Strong Accuracy is the weakest failure detector sufficient to solve Election. However, the failure detectors belonging to class Q are strong enough to implement a Perfect Failure Detector, as shown in [8]. Hence, we have the following theorem.

Theorem 3. *A weakest failure detector to solve Election is the Perfect Failure Detector.*

PROOF: As we mentioned it above, Theorems 1, 2 and Corollary 1 together show that a failure detector belonging to the class Q is the weakest failure detector sufficient to solve Election. It is shown in [8] that a failure detector of class Q satisfying Strong Accuracy and Weak Completeness can be used to implement a Perfect Failure Detector P. Therefore a Perfect Failure Detector is the weakest failure detector that is sufficient to solve Election. This theorem follows from Theorems 1, 2 and Corollary 1.

6 Concluding Remarks

The importance of this paper is in extending the applicability field of the results, which Chandra and Toueg have studied on solving problems, into the Election problem in asynchronous system (with crash failures and reliable channels) augmented with unreliable failure detectors.

More specifically, what is the weakest failure detector for solving the Election problem in the asynchronous system? As an answer to this question, we showed that Perfect failure Detector P is the weakest failure detector to solve the Election problem in asynchronous systems. Though S or W are sufficient to solve Consensus, we showed that they are not sufficient to solve Election. Therefore the Election problem is strictly harder than the Consensus problem even when assuming a single crash.

Determining that a problem Pb_1 is harder than a problem Pb_2 has a very important practical consequence, namely, the cost of solving Pb_1 cannot be less than that of solving Pb_2. That means that the cost of solving Election cannot be less than that of solving Consensus.

The applicability of these results to problems other than Consensus has been discussed in [6,13,14,15]. To our knowledge, it is however the first time that Election problems are discussed in asynchronous systems with unreliable failure detectors.

References

1. G. LeLann: Distributed Systems–towards a Formal Approach. Information Processing 77, B. Gilchrist, Ed. North–Holland, 1977
2. H. Garcia-Molina: Elections in a Distributed Computing System. IEEE Transactions on Computers, C-31 (1982) 49–59
3. H. Abu-Amara and J. Lokre: Election in Asynchronous Complete Networks with Intermittent Link Failures. IEEE Transactions on Computers, 43 (1994) 778–788
4. G. Singh: Leader Election in the Presence of Link Failures. IEEE Transactions on Parallel and Distributed Systems, 7 (1996) 231–236
5. M. Fischer, N. Lynch, and M. Paterson: Impossibility of Distributed Consensus with One Faulty Process. Journal of ACM, (32) 1985 374–382
6. T. Chandra and S.Toueg: Unreliable Failure Detectors for Reliable Distributed Systems. Journal of ACM, 43 (1996) 225–267
7. D. Dolev and R Strong: A Simple Model For Agreement in Distributed Systems. In: B. Simons and A. Spector (eds.): Fault-Tolerant Distributed Computing. Lecture Notes in Computer Science, Vol.448. Springer-Verlag, Berlin Heidelberg New York (1987) 42–50
8. T. Chandra, V. Hadzilacos and S. Toueg: The Weakest Failure Detector for Solving Consensus. Journal of ACM, 43 (1996) 685–722
9. J. E. Hopcroft and J. D. Ullman: Introduction to Automata Theory, Languages and Computation. Addison Wesley, Reading, Mass., 1979
10. Garey M.R. and Johnson D.S: Computers and Intractability: A Guide to the Theory of NP-Completeness. Freeman W.H & Co, New York, 1979
11. Eddy Fromentin, Michel RAY and Frederic TRONEL: On Classes of Problems in Asynchronous Distributed Systems. In Proceedings of Distributed Computing Conference. IEEE, June 1999
12. Hadzilacos V. and Toueg S: Reliable Broadcast and Related Problems. Distributed Systems (Second Edition), ACM Press, New York, pp.97–145, 1993
13. V. Hadzilacos, "On the Relationship between the Atomic Commitment and Consensus Problems," In Fault-Tolerant Distributed Computing, pp. 201–208. B. Simons and A. spector ed, Springer Verlag (LNCS 448), 1987
14. Schiper and A. Sandoz: Primary Partition: Virtually-Synchronous Communication harder than Consensus. In Proceedings of the 8th Workshop on Distributed Algorithms, 1994
15. R. Guerraoui and A. Schiper: Transaction model vs. Virtual Synchrony model: bridging the gap. In: K. Birman, F. Mattern and A. Schiper (eds.): Distributed Systems: From Theory to Practice. Lecture Notes in Computer Science, Vol. 938. Springer- Verlag, Berlin Heidelberg New York (1995) 121–132

From Lens to Flow Structure

David Fauthoux and Jean-Paul Bahsoun

Institut de Recherche en Informatique de Toulouse
31400, Toulouse, France
{fauthoux, bahsoun}@irit.fr

Abstract. More than a basic aggregation of objects, a program can be modelled as a semantically rich structure of micro-components. In this paper, we study program construction, our starting point being the object-oriented framework requirements. We put under the lights the component form which matches our reuse goals - the "lens" - and we show that it is possible to identify clearly the program structure as a set of "semantic flows". For example, in a graphical user interface, a drawing flow manages and orders the displayed components.
This program model enables complex, rich and free structuring. Prevented from unsound component adaptation or composition, the programmer is free to define his own structural flows.

1 Introduction

To introduce our model, we explore in this part the requirements that a framework may fulfill. We consider as "framework" a set of components, where the components can be combined, and therefore where they are usually complementary.

Building a piece of software from such a framework consists in reusing the framework components, to adapt them to achieve the software goal, and last but not least to compose them.

As before, the goals while building a framework can be divided into two parts:

- *Extensibility* : The components must be written in order to be adapted or extended.
- *Composition* : The components must be written in order to be composed [14].

Extensibility gathers the following skills:

- easy adapting: neither complex programming protocol nor excessive strictness from the developer.
- rich adapting: to let the developer free to adapt the component as he needs.
- sound adapting: to forbid as soon as possible to the developer, if possible at compile time, to use a component while he is breaking its encapsulation or corrupting its properties, [16].
- targeted maintenance: fixing, improving or upgrading one component must not interfere with the other ones, and must not lead to modify any other component.

M.H. Shafazand and A M. Tjoa (Eds.): EurAsia-ICT 2002, LNCS 2510, pp. 953–964, 2002.

Composition gathers the following skills:

- sound composition: to forbid to the developer to compose components that are not planned to, [14].
- maximal composition: to let the developer free to combine all the components that do not forbid it.
- late composition [4, 11]: components must be cut up enough to let the developer free to compose wanted and only wanted properties.

In order to demonstrate that all these skills can be fulfill, we will use a key concept: the interface. considering an interface as a named set of one or more methods signatures.

Within the object-oriented late-binding context, and used to separate implementation from specification, this concept enables the dynamic genericity (a component can be linked to a set of other ones, each one matching a specified interface), the sound composition (a component is designed to be composed if it matches a required interface) and the shielding (referencing the interface of an object instead of its class restricts access to specific methods).

2 Proposing an Extremely Fine Grained Model

Inspired by [10, §2.1] and [7, §2], we think that we can reach interesting goals for the framework construction by choosing the opposite way of heavy objects making. Our observations:

- it is difficult to adapt or modify the heavy components coming from a framework;
- it is difficult to fix or maintain a heavy component due to its complex managing;
- a heavy object contains a lot of properties. The developer cannot use, adapt or aggregate partially some specific object properties.

We propose here an extremely fine grained model:

- only one property per class;
- on the contrary to [8]'s point of view, adding property (when extending) will *not* amounts to adding methods.

2.1 Introducing the "Lens" Model

Trying to agree with the "essential ingredients of any extensible system", from [15], we introduce here the function-class concept, or "lens" concept.

An interface, or a "lightweight interface", declares one and only one property. All classes implementing such a "mother interface" ensure matching the interface specification. Such a class is a semantic refinement of the interface [17, 14]. Therefore, we can consider that the interface declares, along to its property, a set of classes, each one implementing the property or an extension of this property.

This set of small components ensures that each one can be composed with each other.

This fine grained model enables the framework user to compose, without any restriction, any component with any other one. Actually, consider a component that matches the mother interface and another one implementing an extension for it: the latter matches also the mother interface. And then, after being composed with a component, it can be used to be composed again with another one.

Any aggregation is possible within the components set, according to the success story: "a wrapper can be wrapped again" [4, §5]. The user can combine all the extensions in the set of components (declared by the mother interface). Extending or modifying a property comes to aggregate a wrapping class.

Sound composition is achieved because each component enabling itself to be wrapped declares implementing the mother interface, so it foresees to be wrapped. Wrapping such a component with any other component (declared to be able to) is then correct.

In the following diagram, `I` symbolizes the mother interface, while `C1`, `C2`... symbolizes the semantic refinements. The vertical arrows shows the adaptation phase (adapting or extending the mother property by defining a new refinement). The horizontal dotted box surrounds the classes that can be composed, within the composition phase.

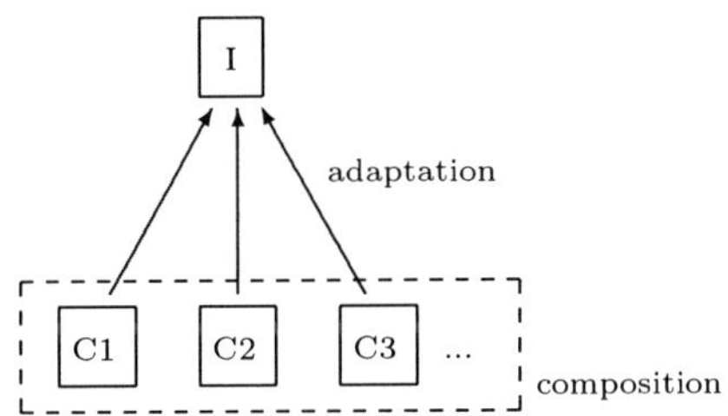

Example Let a `Drawer` interface that defines the drawability property. This interface specifies a drawing method.

Let an `ImageDrawer` class that can draw a house.

Let a `RotateDrawer` class that rotates drawing of 45 degrees.

Let an `AlphaDrawer` class that semi-transparently renders drawing. These three classes implements `Drawer` specification.

If we want to draw the house with a 45 degrees rotation, we combine a `RotateDrawer` with an `ImageDrawer` (objects, instances of both classes), and then we call the drawing method.

If we want to draw semi-transparently the house, we combine an `AlphaDrawer` with an `ImageDrawer`.

If we want to semi-transparently draw the house with a 45 degrees rotation, we combine all the three classes.

We can decide to compose the properties at any time: maximal and late composition.

We can *only* graft a rotation on a class that can be drawn (e.g. that matches and then ensures the `Drawer` interface) and we *know* that the rotation can only be grafted on a `Drawer`: sound and guided composition.

If we decide to improve the rotation algorithm, we modify only the`RotateDrawer` class: easy adapting and targeted maintenance.

We can add a new refinement to the set declared by the Drawer interface. For example, we can create the `ScaleDrawer` class that scales the drawing. Without modifying or improving any other class, this `ScaleDrawer` class will be combinable with any other class of the set: rich and easy adapting.

Thus, with this model, the programmer manipulates little and easy to understand objects. He gains reuse requirements and he is protected from reuse faults at compile time. Plus, handling this micro-components model does not require any language extension (from the common object-oriented languages). The programmer is free to use it without any supplementary lore (like semantics of new keywords) or tool (like a pre-compiler).

2.2 Writing Our Model with Java

Our challenge is now to implement our model without adding any new keyword in the language. We want our model simple to write and directly usable[1]. An interface will define a property and a set of classes refining this property. Two roles for the interface: to show to the external part what is the accessible method (an interface is a window) and to express the sound wrapping (a class will wrap other objects by defining which interface it wraps: which interface these objects implement).

Code in Java.[2]

```
public interface Drawer {
    public void draw(Graphics g);
}

public class ImageDrawer implements Drawer {
    public ImageDrawer() {
    }
    public void draw(Graphics g) {
        // draw a house on g
    }
}
public class RotateDrawer implements Drawer {
    private Drawer wrappee;
    public RotateDrawer(Drawer wrappee) {
        this.wrappee = wrappee;
    }
```

[1] It is important to note that our model meets the implementing way of the Java streams (cf `java.io` in Sun API). See Appendix 2.

[2] It is possible to define a coding scheme to write a mother interface and a refinement of the property. See Appendix 1.

```
    public void draw(Graphics g) {
        // rotate g of 45 degrees
        wrappee.draw(g);
    }
}
```

2.3 Diagram

We present here a diagram[3] to express easily what a wrapper can wrap and how it can be wrapped.

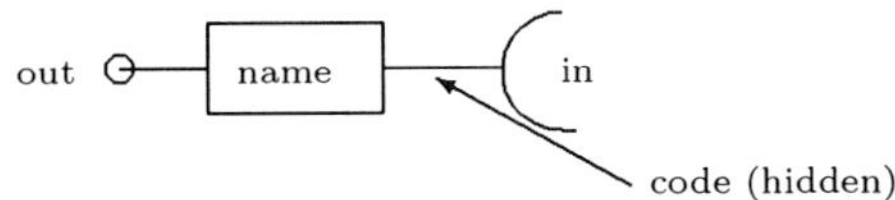

Out plug: the wrapper interface, what is public to the external part.
In plug: on which type of object the wrapper is grafted.

Example

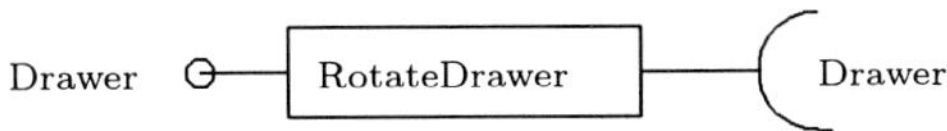

When we have a group of components with their diagrams, it is easy to create a global diagram that expresses the wanted composition. Combining diagrams leads to completely express the composition aggregation.

Example

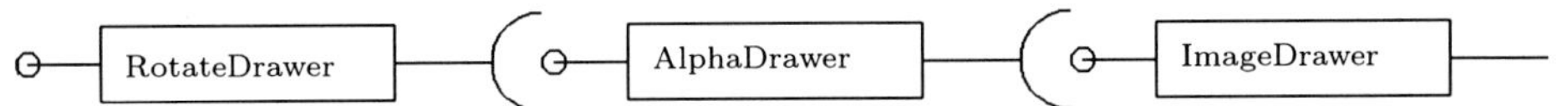

We can see in the last example that a leaf (a wrapper without any wrappee, or better, a class that directly defines a kind of refinement of the property) has a simplified diagram. It is not necessary for such a component to have an *in* plug (see `ImageDrawer`).

The diagrammatic notation follows our goals because it expresses sound composition: what is forced (where the wrapper must be grafted) and what is possible (the wrapper can be grafted on every component that matches its *out* plug).

The diagrams express the independence from a component to other ones: the components are linked only by *in/out* plugs, their codes are hidden and independent.

The diagrams express the composition easiness and richness: writing a global composition diagram from some component diagrams is fast and expressive.

[3] Looking like an electric plug, this diagram is inspired by (1) Oscar Nierstrasz [14] component diagram with required input and output plugs and (2) the pictorial view of mixins by Davide Ancona and Elena Zucca [2].

2.4 Why the "Lens" Term

The lens notion contains our presented idea: "one property per class".

In a set of classes defined by a mother interface, the property does not change. The classes are only semantic refinements and are not able to add more methods than there are in the interface. Thus, we can imagine to place an eye in front of the composition chain and to see to the end of the chain. The lenses placed in the chain do not change the property but only refine it, as a real lens that continues to enable the view sense but changes or corrects the focus, aperture, brightness or accuracy. As a real lens, it is impossible to know its internal properties. It is only possible to know its effect on the view sense.

The "lens" notion expresses the parallel between what is possible to see (to know) and what is hidden (internal).

Finally, a lens can be combined with another one to apply the properties of both lenses. Thus, the "lens" notion expresses our composition model.

3 Flows in the Program Structure

In this section, we aim at showing that a program structure becomes explicit when it is built on our model.

3.1 Tree Composition

As Wolfgang Weck [17], we think that linear composition is not enough. Enabling multi-wrappees for one wrapper, or multi-lensees for one lens, leads to enable the programmer to build tree composition structures. We think that our model sufficiently prevents programming faults by binding the component code in its hidden part and by forcing the component to expose its *in* and *out* plugs. Thus, composition is greatly free and expressive.

Within our Java implementation, a lens with multiple lensees will have such a diagram and such a constructor:

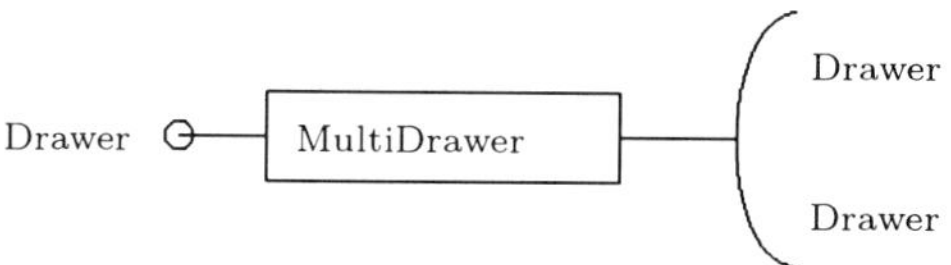


```
public ExampleLens(LensInterface lens1, LensInterface lens2, etc...)
```

3.2 A More Complete Example

Let us change the `ImageDrawer` class to be able to be specialized with a chosen image. We can now instantiate `ImageDrawer` into two different objects: the first to draw a house, the second to draw a cloud. Let us add a new functionality to our framework: ability to move drawing (`Mover`), for example by translating the drawing zone (`Graphics`).

We aim at building a program that draws a semi-transparent and moving cloud in front of a house. We get the following diagram:

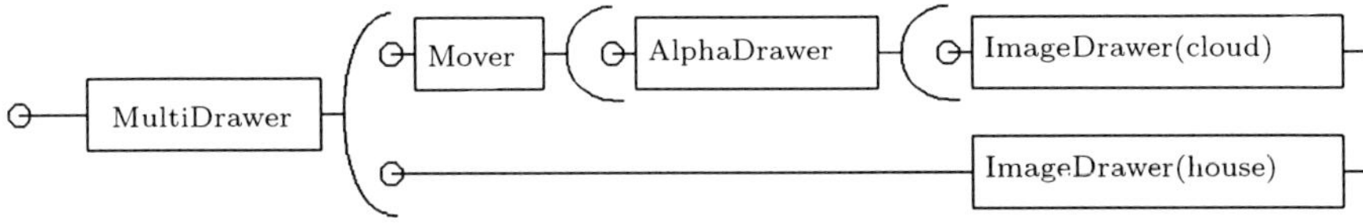

ritten in Java, we have the following lens chain to draw the cloud:

```
Drawer cloudDrawer = new ImageDrawer(cloud); //the alpha component wraps the cloud drawer:
Drawer transparentCloudDrawer = new AlphaDrawer(cloudDrawer);
Drawer cloudChain = new Mover(transparentCloudDrawer);
```

Same definition within one line:

```
Drawer cloudChain = new Mover(new AlphaDrawer(new ImageDrawer(cloud)));
```

Finally, we have the following program structure:

```
Drawer root = new MultiDrawer(cloudChain, new ImageDrawer(house));
```

3.3 Structure Is Function

As opposed to common program building, in our model, the program behavior is not managed by some internal code. A global program algorithm is no more implemented by writing code in a large method in any object. Such a code is too hard to write, too hazardous to modify and quite impossible to maintain. In our model, the program function is no more achieved with internal tricky code but is defined by its structure. Combining fine grain objects in the composition phase is the way to make the program behavior.

We have a global semantic value. For example, the draw property in the `Drawer` interface. We have semantic refinements (the classes that implements interface). And in the composition phase, the programmer composes these semantic bricks. Thus, it is no more at object-writing phase that the programmer defines the program semantics but at structure-building phase. The building notion is tied to the composition notion. The program structure is its semantics; its semantics is a composition of fine grain objects semantics.

Moreover, each lens brick is a semantic refinement acting on a kind of brick. A lens acts as a function on another lens. Therefore, structuring a program is choosing how and when the functional lenses will be applied on the other ones. Thus, in our model, the program structure defines its function.

With the previous example, we can give an intuitional script to explicit the program function:

```
f = multi(move(alpha(cloud)), house)
```

Our model enables oneself to identify clearly and at correct time (according to our "late composition" goal) the program semantics. A program is written as a function. The internal codes define the semantics of the fine grain components. And the program publicizes its structure, the latter containing a strong and explicit semantics.

In the following, we will see that we can continue on this way to build more complex programs that combine different kind of functions.

3.4 Let Us Identify "Flows"

Based on the previous separation of the global semantics and the local semantics, we introduce here the "flow" notion.

Analyzing the use of the functional tree by a program external part, we can separate the building of the lenses chain (when the developer chooses which property refinements to apply) from its using (when the developer uses the defined property).

In our example, the external part calls the`draw` method of the tree root component. The external part accesses only the global property. The semantic refinements of the lenses do not matter any more. The building of the lenses chain is independent from its using.

Moreover in our example, a developer builds the`Drawer` lenses chain to paint a cloud on a house, and he publicizes it. Then, another developer can use it as a global `Drawer`. This last developer knows that the chain paints something correct but he does not need to know how.

The program external part uses a brick pack with a global reference to it. It needs only a reference to the tree root.
Example:

```
Drawer root = ...;
root.draw(g);
```

We can see here that the reference will be declared (according to our choices, see previous section) with an interface. A reference hence points a lenses tree that defines the property implementation, but the programmer will inevitably use the global method (he is not allowed to use the specific methods of the class).

Thus, in such a model, and better than in its common using, the reference has a semantic value:

- a global semantic value declared by the interface: the property.
- an internal semantic value defined by the lenses tree.

We will use now the "flow" term when a reference will point a tree of lenses composition. Such a "flow" term matches with the diagram, which is entered by the root lens, and travelled trough the internal code of the lenses, from a lens to the next[4].

[4] Initially, this term comes from Data Flow hardware architectures, because in both cases, a flow is independent from the remainder of the program.

4 Related Work

Mixins. The current research about mixins [1, 8, 5, 7, 3] is oriented, as inheritance questions, to class/object design. Our research is axed to software and framework building.

Mixins are presented as a new object-oriented technique. Thus, many new keywords come to complete or replace other keywords of the common object-oriented languages. We think that the mixins merge two hidden axes: object creation and component composition. On the opposite way, we try to clearly separate and analyse these two axes.

Design Patterns. The Design Patterns [9] point of view aims at identifying the class structures that are usually used, and their interactions. It aims at linking class structures to such a situation, to solve such a problem, or to achieve such a goal. Then, patterns are written, and developers try to match these patterns when they face the associated problems.

The Design Patterns comes to classify software building techniques. It is an experimental orientation. We think this orientation hides that a program gets a semantics beyond the semantics of its contained classes. Our model proposes to clearly explicit this semantics as a component structure. We think that only one pattern is fundamental: a restrictive form of the "wrapper", also known as "Decorator" or "Adapter". The differences between these three previous patterns are no longer useful in our restricted model.

Generic Wrappers. As for the mixins point of view, the Generic Wrappers [4] research proposes a new or improved object-oriented technique. Thus, new keywords are added to Java and the language is reformed to match the Generic Wrapper model (for example, the self notion is extended to enable the programmer to call the mother wrapper). Therefore, any developer that wants to use the Generic Wrapper model needs to learn its differences from well-known object-oriented models and needs a tool to compile his programs with this model. On the contrary, our model does not need any new keyword or special tool.

In Generic Wrappers, the code inheritance problems are reported to the wrapping model [4, §5-6]. That is why our model searches to be emancipated from the code inheritance and solely collects the reuse goal of this notion.

5 Conclusion

By this conclusion, we group the used terms in this paper. We have defined a fine grained model where the components are micro-objects that refine a property. This property is declared by a mother lightweight interface (all the refining components implement this interface) and the components can be composed each other. They are so called "lens"; the component on which a lens is wrapped is called a "lensee" (according to the "wrappee" term in Generic Wrappers).

We have replaced the rough reference notion in object-oriented language by the "flow" notion. When lenses are composed, they form a tree, defining the implementation of the property. This implementation is hidden to the remainder of the program, but the property gets a global semantic value (declared by the mother interface). As in common object-oriented languages and without complicating the self notion, the tree is travelled through the lenses to apply the property at runtime. A flow is created at the aggregation time, clearly separated from runtime. The aggregation phase enables the programmer to build complex but clear and expressive program structures.

We have implemented a numerous classes framework[5], matching our lens model, which covers graphics, files input/output, keyboard and mouse managing, sound playing, symbolic graphes programming and time modelling.

The proposed implementations show different kind of flows, each flow using many specific lenses. We have then defined some lenses which join different flows, acting as nodes. For example, a drawing flow and a time flow intersect to manage animations.

In our future work, we want to fill the formalization gap. According to the easy crossing from the diagram to the code and from the code to the diagram, we think a mathematical formalization for the program structure may be accessible without a heavyweight work. We want to meet the work about "behavioral interface contracts" [6] and "contracts object composition" [16].

We want to separate the core language (which defines the internal code of the components) from the structure language (which defines composition of the components). The latter can be viewed as a meta-language which enables to write the program function.

References

1. D. Ancona, G. Lagorio, and E. Zucca. Jam - a smooth extension of java with mixins. In *ECOOP*, pages 154–178, 2000.
2. D. Ancona and E. Zucca. A theory of mixin modules: Basic and derived operators. *Mathematical Structures in Computer Science*, 8(4):401–446, 1998.
3. D. Ancona and E. Zucca. A theory of mixin modules: Algebraic laws and reduction semantics, 1999.
4. M. Bchi and W. Weck. Generic wrappers. In *ECOOP '2000 — Object-Oriented Programming 14th European Conference, Sophia Antipolis and Cannes, France*, pages 201–225. Springer-Verlag, 2000.
5. G. Bracha and W. Cook. Mixin-based inheritance. In N. Meyrowitz, editor, *Proceedings of the Conference on Object-Oriented Programming: Systems, Languages, and Applications / Proceedings of the European Conference on Object-Oriented Programming*, pages 303–311, Ottawa, Canada, 1990. ACM Press.
6. R. B. Findler and M. Felleisen. Behavioral interface contracts for java.

[5] available from: `http://www.irit.fr/ACTIVITES/LILaC/Pers/Fauthoux/`

7. R. B. Findler and M. Flatt. Modular object-oriented programming with units and mixins. In *Proceedings of the ACM SIGPLAN International Conference on Functional Programming (ICFP '98)*, volume 34(1), pages 94–104, 1999.
8. M. Flatt, S. Krishnamurthi, and M. Felleisen. Classes and mixins. In *Conference Record of POPL 98: The 25TH ACM SIGPLAN-SIGACT Symposium on Principles of Programming Languages, San Diego, California*, pages 171–183, New York, NY, 1998.
9. E. Gamma, R. Helm, R. Johnson, and J. Vlissides. *Design Patterns: Elements of Reusable Object-Oriented Software.* Addison Wesley, Massachusetts, 1994.
10. G. Kniesel. Type-safe delegation for run-time component adaptation. In R. Guerraoui, editor, *ECOOP '99 — Object-Oriented Programming 13th European Conference, Lisbon Portugal*, volume 1628, pages 351–366. Springer-Verlag, New York, NY, 1999.
11. B. Kucuk, M. Alpdemir, and R. Zobel. Customizable adapters for blackbox components, 1998.
12. S. Microsystems. java.sun.com, 2002.
13. L. Mikhajlov and E. Sekerinski. A study of the fragile base class problem. In E. Jul, editor, *ECOOP '98 — Object-Oriented Programming, 12th European Conference , Brussels, Proceedings*, volume 1445, pages 355–382. Springer-Verlag, 1998.
14. O. Nierstrasz and T. D. Meijler. Research directions in software composition. *ACM Computing Surveys*, 27(2):262–264, 1995.
15. C. Szyperski. Independently extensible systems – software engineering potential and challenges. In *Proceedings of the 19th Australian Computer Science Conference*, Melbourne, Australia, 1996.
16. W. Weck. Inheritance using contracts object composition. In *ECOOP Workshops*, pages 384–388, 1997.
17. W. Weck and C. Szyperski. Do we need inheritance?

Appendix 1: Lens Coding Scheme in Java

We propose in this section an intuitive way to normalize the lens coding. This proposal is as short as possible to be understood "at a glance".

To define a mother interface:

```
public interface MotherInterfaceName {
    public ... // method that declares property
}
```

To define a refinement of this property that is wrapping another object of the set (defined by the mother interface):

```
public class RefinementName implements MotherInterfaceName {

    private MotherInterfaceName wrappee;

    public RefinementName(MotherInterfaceName wrappee) {
        this.wrappee = wrappee;
    }
    public ... { // method declared in interface
        // algorithm to apply refinement on wrappee
    }
}
```

Appendix 2: Java Streams API

In Sun java.io v1.3 API [12], an `InputStream` defines the property (being able to read bytes) and many classes refine this property. For example, the `BufferedInputStream` class adds a buffering filter while reading bytes from a stream. The `BufferedInputStream` class is a wrapper and its wrappee is forced to be an `InputStream`, but is whatever `InputStream` the programmer chooses.

Thus, the Java streams can be linearly composed. The composition form expresses and directly implements the chosen algorithm to read bytes. The user chooses this algorithm as late as possible: at composition time.

A difference with our model: Java developers merge the mother interface notion and the basic implementation of the property in one class (`InputStream`), and then force the programmer to inherit it in order to create a refinement. Therefore, the programmer faces the problems of the code inheritance [17, 13]. He does not know which methods he can extend without danger, he does not know how `InputStream` methods interact and he is not protected from a possible modification of the `InputStream` class.

Finally, Java developers create a new semantic interpretation of a data stream by adding methods when they extend the `InputStream` class. For example, the `DataInputStream` class contains a `readChar` method that interprets a byte stream as a character stream and returns the casting of two bytes in one char. Within this programming way, the `read` method of the mother class should become unavailable. In our model, such an adaptation is out of the bounds of the mother property, and a new property has to be defined.

ADML: A Language for Automatic Generation of Migration Plans

Jennifer Pérez, José A. Carsí, and Isidro Ramos

Department of Information Systems and Computation
Valencia University of Technology
Camino de Vera s/n
E-46071 Valencia – Spain
{jeperez|pcarsi|iramos}@dsic.upv.es

Abstract. This paper presents a solution to the data evolution problem of information systems. This solution follows an automatic approach that reduces the number of people and the time invested in the software maintenance process. It uses the information of UML-like OASIS OO conceptual schemas, representing system evolution, in order to automatically generate a data migration plan. This work defines a specification language (ADML) for migration plans which are automatically generated using patterns. The plan execution migrates data from the database of the initial conceptual schema to the database of the evolved conceptual schema. This solution has been implemented in a migration tool (ADAM).

Keywords: Data migration plan, migration language, patterns, migration expressions, automatic generation.

1 Introduction

Information systems have a dynamic nature and they frequently undergo changes. These changes are due to volatile business rules or to external factors. For this reason, system management tools must be able to deal with them in an efficient and complete way.

Existing model compilers generate the database schemas and the application code from its conceptual schemas. The application generation can be complete, like Oblog Case ([Ser94]), OO-Method/CASE ([Pas97]), or partial, like Rational Rose ([Rat02]), System Architect ([Sys02]), Together ([Tog02]) and others. However, these modeling tools do not deal with the data evolution problem. In other words, they do not take the evolution of the data stored in databases into account. In this paper a solution to the data evolution problem is presented.

Nowadays, our solution is used together with OO-Method/CASE, a model compiler generating automatically applications with three layer architecture (presentation, business logic and persistence). A formal language OASIS

M.H. Shafazand and A M. Tjoa (Eds.): EurAsia-ICT 2002, LNCS 2510, pp. 965–974, 2002.

([Pas95]) based in dynamic logic is used as design language allowing the automatic compilation of UML-like models.

Our starting point to implement data evolution is the OASIS ([Pas95]) conceptual schemas that represent system evolution (the old and the new schemas). From the new schema, the OO-Method/CASE compiler generates a new code and a new empty database. Therefore, the information system data remaining in the old database must be correctly transferred to the new one. We propose a solution for the migration of data, consisting of three steps:

1.- Old and new conceptual schemas are compared using different comparison criteria ([Sil02]) to automatically obtain the differences between them.

2.- Changes on the data are obtained from these differences and are specified as migration expressions using a migration language (ADML). These expressions constitute a data migration plan ([Per01-b]).

3.- The execution of the plan migrates data from an old database to a new database ([Ana01]).

This approach has been implemented in a migration tool (ADAM) developed by the Department of Information Systems and Computation of the Valencia University of Technology in collaboration with the industrial partner CONSOFT. We have obtained very good results applying this solution. The time invested and the number of people involved in data evolution process have decreased up to 80%.

ADML captures all changes undergone in the schema and its related updates on the persistence layer. It is accurate and complete. This language is not ambiguous and has a high abstraction level. ADML and different migration expression patterns ([Per01-a]) are used to generate a data migration plan based on the OASIS object-oriented model.

The structure of the paper is the following: first, the process of generating expressions is explained and then, the ADML language and patterns used to generate a data migration plan are introduced. In section 3, the article presents a brief summary of related work. Finally, future work and conclusions are presented in section 4.

2 A Data Migration Language (ADML) and the Patterns Used to Capture Migration Features

A data migration plan is a set of data changes that are specified using a migration language. Its execution allows us to transfer data from an old database to a new database. A data migration plan must include all the necessary changes in order to perform a correct migration and in the right order. It is structured as the composition of different elements: modules, changes and migration expressions. The highest abstraction level of this composition is constituted by migration modules whereas migration expressions are at the lowest level. The definition of these elements is as follows:

- *Migration expressions:* Migration expressions are those expressions that can be specified in a data migration plan. Each type of migration expression has a different semantic and follows different syntactic restrictions.
- *Changes:* A change is the set of migration expressions that specify the updates undergone and the filters applied on the old element's[1] instances.
- *Migration modules :* A migration module has a transactional behavior when it is executed by ADAM tool. The composition of modules using aggregations forms a data migration plan. Finally, it is important to note that there is one migration module for each class, aggregation and specialization of the new conceptual schema.

ADML is declarative and is used to specify all expressions that make up a data migration plan. Object-oriented conceptual schema elements are the data that are managed by the migration language. This language follows the object-oriented model. ADML makes use of path expressions to specify:

- The changes undergone in a conceptual schema element.
- The data belonging to the old conceptual schema.
- The filters that will be applied on the data, if necessary.

In the paper, the well known example order-customer-products in Figures 1 and 2 is used. We take a part of the complete example and we focus on the language and the generation of the migration expressions of this plan. The following OASIS schemas capture the structure of the order management of an information system. The new conceptual schema (figure 2) improves the previous one (figure 1). This improvement consists in a classification of the customers, a better management of the orders and some corrections of model errors. The example is the following:

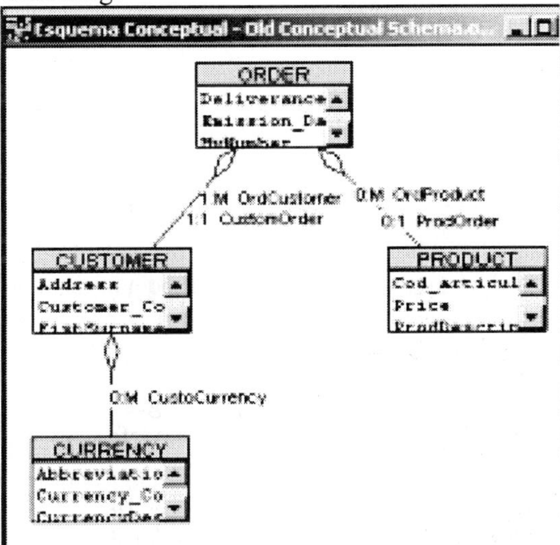

Fig. 1. OASIS Old Conceptual Schema of the IS Orders Management

[1] The conceptual schema elements that has been considered for the generation of a data migration plan are: classes, attributes, aggregations, associations and inheritances.

It must be taken into account that these schemas are modeled using OASIS. However, if we model them using UML, their appearance would be different. This is due to the fact that in OASIS the association is considered as a kind of aggregation and thus, the referential aggregation and the inclusive aggregation are represented by a lozenge. This is the reason why UML associations are modeled like OASIS referential aggregations. For example: The relationship between the classes *Customer* and *Currency* is an association represented with a OASIS referential aggregation.

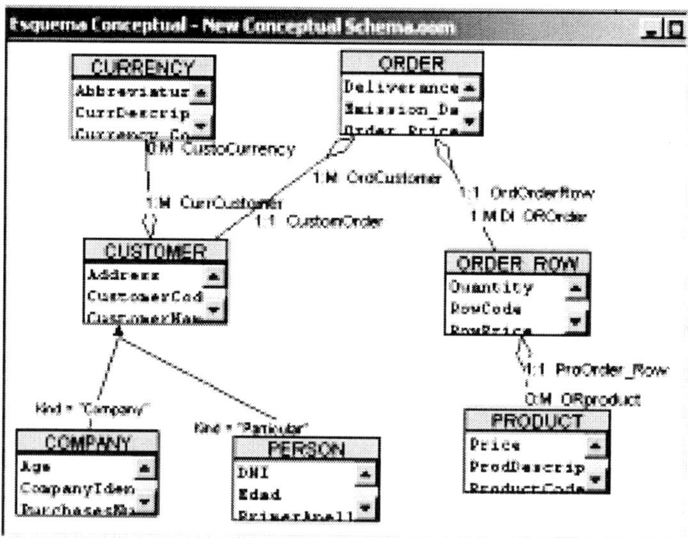

Fig. 2. OASIS New Conceptual Schema of the IS Orders Management

2.1 A Data Migration Language (ADML)

The ADML specifies of the data evolution by means of migration expressions. Its main advantage is the independence from any DBMS due to its high abstraction level. The high abstraction level allows us to express a migration plan in an easy and user friendly way and it does not need to take into account implementation details. In addition, the migration language can specify five kinds of expressions allowing us to define the migration module of each kind of conceptual schema element. These migration expressions are the following:

- *Origin:* The origin specifies the data source for an element of the new conceptual schema. The data source is the old element's contents that the analyst wants to migrate to the new one. The origin must be specified for all the element migrations of the plan. For example, the origin of the *Quantity* attribute of *Order_Row* class is the *Number* attribute of *Order* class. The origin of *Quantity* is specified by the migration tool as follows:

  ```
  Origin: OldCS.Order.Number
  ```

- *Filter*: The filter selects the population that will be migrated from an old element to another new element. The filter can restrict the population of a class or an aggregation. The filter is a boolean condition that will be evaluated at execution.

For example, if the analyst of this system is only interested in the orders made after 1999, then this restriction can be specified by means of a filter in the *Order* class and in the *Order_ Row* class. The *Order_ Row* filter is specified as follows:

```
Filter: OldCS.Order.Issue_date > 31/12/1999
  Origin: OldCS.Order
```

- *Sorting*: The migration order of the instances is represented in the sorting section. This expression specifies the order using the class attributes and gives its direction (ascendant or descendant). For example, if the analyst wants to migrate the orders sorted by *emission_date* and *customer* in ascending order, the *Order_ Row* order is specified in the migration tool as follows:

```
Filter: OldCS.Order.Issue _date > 31/12/1999
 Origin: OldCS.Order
 Sorting: OldCS.Order.Issue _date ASC,
          OldCS.Order.CustomOrder.Cod_customer ASC
```

- *Transformation function*: The transformation function is the migration expression that specifies the changes that the data will undergo during the process of migration. When the data need more than one change, the transformation function is the composition of changes. This expression transforms the old data to be compliant with the new element definition. That is, the transformation function execution makes the data evolution of one element. This migration expression can only be specified when migrating attributes, specializations and aggregations. The transformation function always is executed. However, if it depends on a condition, it will only be executed when the condition is true. For example, the migration tool applies a type conversion on the *Number* attribute to migrate its data to the *Quantity* attribute. The transformation function of the *Quantity* attribute is specified by the migration tool as follows:

```
Origin: OldCS.Order.Number
  Transformation function: RealToNatural(OldCS.Order.Number)
```

- *Condition*: The condition is the migration expression that determines which transformation function will be executed. This migration expression can only be specified for attributes. Depending on the case, one attribute can have more than one possible transformation function associated to it. Each one is represented by means of a pair (condition, transformation function). The value of the condition determines which transformation function will be executed. For example, the migration tool applies a type conversion on the *Number* attribute to migrate its data to the *Quantity* attribute. This conversion will only be executed if the *Number* has decimals. The transformation function of the *Quantity* attribute is specified in the migration tool as follows:

```
Origin: Order. Number
     Condition: HasDecimals(OldCS.Order. Number) = true
     Transformation function: RealToNatural(OldCS.Order.Number)
     Condition: HasDecimals (OldCS.Order. Number) = false
     Transformation function: OldCS.Order.Number
```

ADML specifies conceptual schema elements and relationships between them. The navigation through elements gives flexibility and expressivity to construct migration expressions. For example, one can express a filter of the *Customer* class by means of the *Emission_Date* attribute which belongs to the *Order*class that is related to the *Customer* class (`EXIST(CUSTOMER.OrdCustomer.Emission_Date > 31/12/1999)`).

For the moment, all elements that can be specified in a migration plan must belong to the old conceptual schema. This is a limitation of the first version of ADML language that we are currently solving. Migration languages must allow us to specify new conceptual schema elements, despite the fact that this flexibility implies a lot of complexity in the migration tasks. A good example of this increase in complexity is the setting of the *migration order* ([Aba01]). The inclusion of new elements in a data migration plan is more complex than specifying only old elements, because a new element must be migrated before using it in a migration expression.

ADML uses simple and iterative control structures. These simple structures analyze different migration cases of an element. The iterative structures can treat each one of the instances of an element. The migration language is more flexible thanks to these structures and is able to specify the same expression in different ways. For example, one may specify a condition and a transformation function using different expressions or specify a transformation function including an "if statement" with the condition and the transformation function :

```
Transformation function: if HasDecimals(OldCS.Order. Number) = true then
                                   RealToNatural(OldCS.rder.Number)
                         Else
                                   OldCS.Order.Number
                         End if
```

The last feature of the grammar are operators. It has different operators (arithmetic, boolean and aggregate functions) to express actions that the analyst might need. For this reason, it was necessary to build new operators to perform specific actions that were not possible with the standard operators. For example: FIRST(data source, condition, number of elements). This operator is useful when the cardinality of a relationship has been changed, because it allows to migrate only the highest cardinality of elements that satisfy the condition.

2.2 Generation of Migration Expressions

ADAM automatically generates the first version of a data migration plan. This version could be the final one; however, it can be updated by the analysts to suit their purpose.

The automatic generation of a data migration plan implies generating its structure and its contents. The structure generation creates a migration module for each new element and includes them in the data migration plan following the migration order. The content generation consists in automatically constructing all the migration expressions of a complete migration plan. This paper is focused on the generation of the contents using ADML and a set of patterns ([Alexander79] y [Gamma94]). To do this, two data sources are necessary:

- *Mappings*: Mappings specify the data sources of the new conceptual schema elements ([Sil02]). This information is generated previously by ADAM tool. The mappings are produced by comparing the old element and the new one in order to identify the differences between them.

– *The conceptual schemas:* Conceptual schemas contain all the properties and the types of their elements. The migration tool needs this information in order to determine which properties have been changed between an old element and a new element in the previous mapping. The properties of the elements must also be known in order to determine the pattern that the migration tool must apply.

There is a pattern for each one of the element properties that can be changed by schema evolution and for all the combinations of the patterns. This set of patterns is included in a pattern catalog [Per-a] (P-01, P-02). Each pattern has a set of migration expressions which allows us to specify the correct transformation of data.

The generation of the migration expressions for a new element consists of determining which old element is related to it in the mapping and consulting their different properties. Later, ADAM applies the specific element pattern that specifies the generation for the updated properties and generates the migration expressions. Finally, the tool associates these expressions to the new element.

When the data migration plan is executed, the migration expressions of an element which are generated will be evaluated in its instances and these will be migrated to the new database. These instances will satisfy the properties of the new database.

Each pattern is composed by several sections that explain different qualities of the pattern. An example of migration expression pattern arises when the "name", the "data type" and the "not null value" properties of an attribute change. The solution and the example sections of this pattern (**P-12**) are as follows:

Solution

The solution presents the generic migration expression that specifies the attribute changes of "name", "data type" and "not null value" properties. In this case, as in the **P-04** and **P-08** patterns, it is necessary to do a type conversion in the transformation function as follows:

```
old_data_typeTOnew_data_type (old_attribute)
```

This pattern is a composition of the name, data type and "not null value" property patterns (**P-03, P-04** and **P-05** or **P-08** and **P-10** patterns). The migrations expressions that express these changes are the following:

`Filter:` <IDENT_clase> . <IDENT_attr> <> NULL

`Transformation Function:` <generic_func> '(' <IDENT_clase> . <IDENT_attr> ')'

Example

The origin of *Quantity* attribute of *Order_ Row* class is the *Number* attribute of *Order* class. The *Number* attribute of the old conceptual schema is called *Quantity* in the conceptual schema. In addition, *Number* allowed null values and had float type, and now *Quantity* does not allow null values and has natural type in order to correct a model error of the old schema.

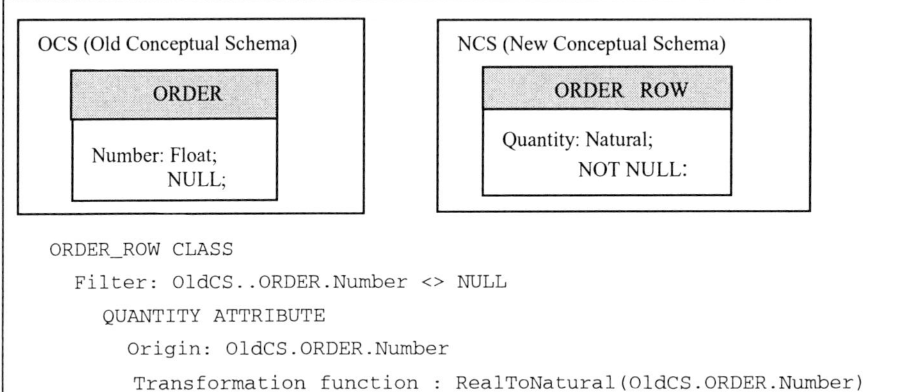

Fig. 2 Solution and Example sections of the pattern P-12: Transformation function for the change of name, data type and not null value properties

After applying the necessary patterns to automatically generate the migration module of the *Row_Order* class and specifying the sentences that the analyst desires, the migration tool obtains the following code:

```
ORDER_ROW CLASS
  Origin: OldCSOrder
Sorting: OldCS.Order.Emission_date ASC,
         OldCSOrder.CustomOrder.Cod_customer ASC
   Filter:( OldCSOrder.Emission_date > 31/12/1999) AND
           (OldCSOrder.Number <> NULL)
      ROW_CODE ATTRIBUTE
        Origin:λ
        Transformation function: 1
      END ROW_CODE ATTRIBUTE
      QUANTITY ATTRIBUTE
        Origin: OldCSOrder.Number
        Condition: Decimal(OldCSOrder. Number) <> 0
        Transformation function: RealToNatural(OldCSOrder.Number)
        Condition: Decimal(OldCSOrder. Number) = 0
        Transformation function: OldCSOrder.Number
      END QUANTITY ATTRIBUTE
      ROW_PRICE ATTRIBUTE
        Origin: OldCSOrder.Order_price
        Transformation function: OldCSOrder.Order_price
        END ROW_PRICE ATTRIBUTE
END ORDER_ROW CLASS
```

3 Related Work

Nowadays, data evolution support is given by DBMS. Relational DBMS such as ORACLE ([Ora02]) or SQL Server ([Sql02]) and object-oriented DBMS like O2 ([Ard02]), Poet ([Poe02]) or Versant ([Ver02]).

Several DBMS allow for data migration using their ETL (Extract, Transforma & Load) tools. This migration can be done by means of SQL statements or user defined scripts which can be executed on the database. However, these tools do not provide automatic support to generate these statements and scripts as ADAM tool does. For this reason, DB administrators must write migration code by hand. ETL tools offer a friendly interface to migrate data from an old database to a new database. However, they do not solve two important problems at software maintenance stage: the high number of people required for the migration process and the high temporal cost.

The closest point of view to our approach is the TESS tool presented by Barbara Staund Lerner in her paper [Sta00]. These approaches have two things in common, both processes are automatic and both tools are based on schema evolution. TESS processes an intermediate language generated from a relational schema code. This is an important difference from our approach, because we analyze the OO conceptual schemas and we do not have to translate to an intermediate language. The OO conceptual schemas give us a higher abstraction level and save us from the translation processes. It also implies that we do not have to take implementation details into account and so our process is more agile and has a higher abstraction level than the TESS process.

4 Conclusion and Future Work

The paper presents a solution to the data evolution problem of information systems. It explains the importance of this evolution feature and the need to support data changes in information systems.

The presented solution uses conceptual schemas, which represent system evolution, to generate a data migration plan and evolve data. The automatic generation gives us a first version of a data migration plan. The contents of a data migration plan are defined by a set of patterns and they are composed of migration expressions that are specified using our migration language (ADML).

The high abstraction level of ADML allows us to be independent from the underlying DBMS. In order to use a given DBMS, it is only necessary to construct its specific "driver". We do not take implementation details into account at the comparison and plan generation processes. The high automation of this solution also reduces the time invested in the maintenance stage and the number of employees involve in the data evolution task up to 80 % in our own experiences.

Finally, it is important to note that it is possible to improve the automatic migration process. Therefore, there are some aspects that we must solve in the future: the migration language needs to be able to specify migration expressions that contain elements of the new conceptual schema and the migration process should take into account the dynamics of the system (States Transition Diagrams (STD)), etc.

References

[Aba01] Abad, S, Carsí, J.A, Ramos, I, *How to obtain a data migration order for the elements of an OO conceptual schema*, Workshop on Evolution in the VI conference on Software Engineering and Databases , Almagro, Ciudad Real, Spain, November 2001 (in Spanish)

[Ana01] Anaya, V. *Generation of transformation modules to migrate data between databases using a data migration plan*, Msc Project, Computer Science Faculty, Valencia University of Technology, September 2001 (in Spanish)

[Alexander79] Christopher Alexander. *The Timeless Way of Building*. Oxford University Press. 1979.

[Ard] Ardent Software, *O2*, http://www.ardent.com/

[Gamma94] Eric Gamma, Richard Helm, Ralph Johnson, John Vlissides. *Design Patterns: Elements of Reusable Object-Oriented Software*. Addyson-Wesley.1994.

[Ora] Oracle Corporation, *Oracle*, http://www.oracle.com

[Pas95] Pastor O. Et al., *OASIS versión 2 (2.2): A Class-Definition Language to Model Information Systems using an Object-Oriented Approach*, SPUPV-95.788, Universidad Politécnica de Valencia, España, 1995.

[Pas97] Pastor O. Et al, *OO-METHOD: A Software Production Environment Combining Conventional and Formal Methods*, Procc. of 9th International Conference, CaiSE97, Barcelona, 1997.

[Per01-a] Pérez, J. *Generation of a data migration plan for OASIS conceptual schemas*, Msc Project, Computer Science Faculty, Valencia University of Technology, September 2001 (in Spanish)

[Per01-b] Pérez J., Carsí J.A, Ramos I., Anaya V., Silva J.F., *Data migration of conceptual schemas*, Workshop on Evolution in the VI conference on Software Engineering and Databases , Almagro, Ciudad Real, Spain, November 2001 (in Spanish)

[Poe] POET Software, *POET*, http://www.poet.com

[Rat] Rational Software, *Rational Rose*, http://www.rational.com/products/rose/

[Ser94] Sernadas A., Costa J.F., Sernadas C., "Object Specifications Through Diagrams: OBLOG Approach" INESC Lisbon 1994

[Sil02] Silva, J.F.., Carsí, J.A., Ramos, I., *Theoric analyze of the criteria of OO conceptual schemas comparison,* Ingeniería Informática Magazine, ISSN:0717-4197, Enero, http://www.inf.udec.cl/revista/edicion7/jsilva.htm

[Sys] System Architec, http://www.popkin.com/products/sa2001/systemarchitect.htm

[Sql] Microsoft, *SQL Server*, http://www.microsoft.com/sql

[Sta00] Staund Lerner, B., *A Model for Compound Type Changes Encountered in Schema Evolution*, ACM Transactions on Database Systems (TODS) Marzo 2000, Volumen 25 número 1

[Tog] TogetherSoft Corporation http://www.togethersoft.com/

[Ver] Versant Object Technology, *Versant*, http://www.versant.com/

Considerations for Using Domain-Specific Modeling in the Analysis Phase of Software Development Process

Kalle Korhonen

University of Jyväskylä
kaosko@cc.jyu.fi

Abstract. Recent studies claim that domain-specific modeling may highly increase development productivity in specific well-defined domains. Domain-specific modeling is most often used only in the design phase of software development process, while general-purpose modeling techniques are used in the other phases. Integrating general-purpose modeling techniques with domain-specific modeling might be problematic, which is why in this paper we consider if it is feasible to extend domain-specific modeling to cover analysis phase activities in addition to design phase activities. Essentially, we discuss the different activities in analysis phase and consider whether or not domain-specific techniques can be used in them and is it possible to gain any benefits in doing so.

1 Introduction

In this paper, we theoretically evaluate whether or not domain-specific modeling (DSM) techniques can be used in some of the activities in analysis phase and what are the possible benefits for doing that. Some studies (e.g. [1, 2]) claim exceptional high increases in development productivity using DSM approaches. Naturally, we want to study if it is possible to gain any additional benefits by utilizing DSM techniques sooner in the development process than in physical design and implementation phases. We will first discuss what domain-specificity means in software development and what kind of DSM techniques there are. Next, we consider different analysis phase activities and how DSM techniques could be used in them. Finally, we evaluate theoretically the possible benefits that DSM could offer for analysis phase activities.

When we speak about software domains we usually refer to some business domain, like mobile phone or ERP domains. However, domains can also be differentiated in other ways, e.g. based on their visibility in architecture (user-interface, middleware domains) or dependency of run-time environment (operating systems like Windows etc.). Domain area is defined in [3] as "*an area of knowledge or activity characterized by a set of concepts and terminology understood by practitioners in that area*". Thus, we can argue that any software is always related to some domain even if it is not related to a business domain. Reciprocally, business domains are usually relevant only for entire software systems whereas e.g. general-purpose user-interface components, such as Java Beans, are most often business domain independent.

M.H. Shafazand and A M. Tjoa (Eds.): EurAsia-ICT 2002, LNCS 2510, pp. 975–981, 2002.

Even though we can say that software is always related to some domain it does not automatically mean that using domain-specific techniques help developing it. To justify the use of domain-specific modeling techniques the domain should be so specific that general-purpose modeling techniques are not well suited in modeling it. Even if a domain is somewhat different to other common software domains, but very small or simple, i.e. the domain model contains only a few domain-specific elements, it does not justify the use of any DSM techniques. A domain should also be stable and well defined that domain-specific modeling techniques could offer substantial advantage over using general-purpose methods. Compared to general-purpose methods like UML, the benefits of a DSL–based development approach should increase when a domain becomes more stable and more specific, as shown in Figure 1. Making any domain-specific technique is costly [4] [5], thus it does not make sense to spend resources on making them if they are not used more than once, even if the actual development would be faster with utilizing them.

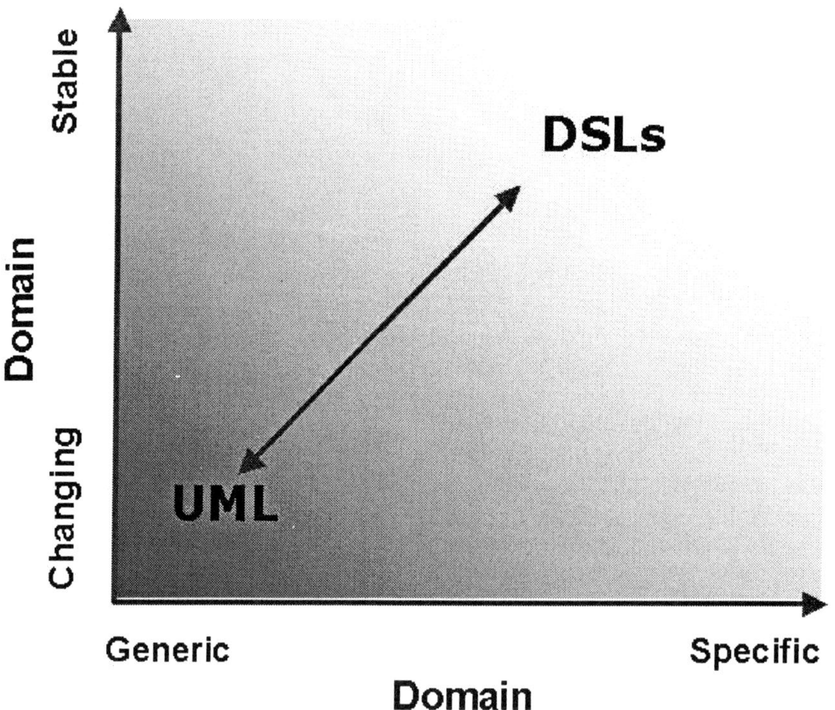

Fig. 1. Applicability of DSLs versus UML [6]

Generally, *domain-specificity means that some technique or subject is applicable for limited domain only*. However, the level of a method's domain-specificity can vary greatly. For example, we could argue that even UML [7] is a domain-specific language because it is more suitable for some domain areas than others. Still, UML is clearly much less domain-specific than for example VHDL[1]. Some domain-specific modeling techniques are often used in ordinary software development process, even if the method used is considered to be general-purpose method. For example, most common and easy to implement domain-specific modeling technique is to use domain-specific notation, e.g. some symbols to describe some objects in that domain or to extend UML to better suit the needs of a particular domain or of a software development company. Some domain-specific notations may even be developed unintentionally by itself in time, when developers replace frequently occurring formations of symbols with one easier-to-use symbol [9]. An example of more extensive use of domain-specific techniques would be to create a complete modeling language for one domain only. Then complete use of domain-specific techniques would mean developing a full domain-specific language (DSL) that substitutes the use of general-purpose language such as Java or C++. While there are various example of successful implementations of text-based DSLs [10] [1], developing (and to some degree, also using) such languages require advanced development skills and lot of knowledge about implementing programming languages. Simply by considering implementations DSLs in use today that we know of, it seems that DSLs tend to work better at domains with low level of abstraction such as closely hardware-related software domains.

The goal of this study is to consider if DSM techniques can be beneficially used in the analysis phase of software development process. This goal is derived from the needs of developers of Nokia Mobile Phones' domain-specific method (DSM) Plato, which is a subject of case studies in the RAMSES project [11] that this study is part of. Plato is a DSL for developing mobile phone software and it will be used as an example of domain-specific method throughout this study.

Most reported studies of DSLs propose that design DSLs are more relevant than analysis DSLs (e.g. [12], [1]). However, the transition from analysis phase carried out using general-purpose method to domain-specific design method can be problematic [12] if there are no appropriate mappings from one phase to the other and if the earlier phase does not properly support the objectives of the latter phase. Thus, it is worth considering if we can extend domain-specific techniques to be used also in analysis phase. In order to do that we must clarify what is meant with analysis phase and what are the goals and the activities in that phase. While we concentrate on single phase in a typical software development process, we do not study the suitability of different process models in particular domain or try to develop a domain-specific process.

---

[1] VHDL is an acronym that stands for VHSIC Hardware Description Language. VHSIC is another acronym that stands for Very High Speed Integrated Circuits. VHDL can be used for documentation, verification, and synthesis of large digital designs [8].

2 Characteristics of Analysis Phase

The analysis phase is often described as defining the "what" of the software whereas design defines the "how" of the system [3]. While the transition from one to the other was often problematic in traditional structured methods, one claim of object-oriented methods was that they offered a seamless transition from analysis to design [13]. According to this easy-transition view, the design model progressively expands the analysis model using analysis objects as they are whilst introducing and adding other design objects. Thus, it would seem logical that if domain-specific notation is used while making of design models, the same or properly extended (or reduced for simplicity) set of notation should be used in analysis. However, [13] raises the question of how design objects can be both abstractions of something in a problem domain, as analysis objects are commonly defined, and objects in a solution space. The use of the same notation can conceal the fact that the objects in analysis and design represent inherently different things. Although the problem was already noticed in early 90's [14], still methods continue to use analysis objects as a central part of their design class models, as proposed in methods such as RUP [3] and OMT++ [15]. Yet those methods are popular and even successful [16]. The paradox can be explained by accepting that the object-oriented analysis already considers design objects, as [13] points out. [3] actually acknowledges that RUP's object-oriented analysis is effectively a first cut at design.

Developers of OPEN [16] see the analysis to design problem differently. They claim that when human brains analyze the problem it rapidly leads to thoughts of likely solutions, which by definition, is the beginning of design. They suggest that analysis and design should be merged at the macro level, even though they still happen at the micro level. The tasks will certainly occur as a part of the human thought process but they do so on a time scale of minutes or seconds and not on the time scale of project management. However, while not acknowledging the distinction between analysis and design OPEN emphasizes a difference between logical and physical design. Logical design relates to analysis and logical models are created from it, thus it is called modeling. Physical design refers to implementation, containing detailed design and actual programming.

Accepting OPEN's point of view on characteristics of analysis, we can differentiate two types of analysis activities. The first type is classic answer to "what" –problem or analysis of real-world business objects. This type is often called *domain modeling*. It is actually a subset of business modeling [3], which, in turn, is the result of analysis done for business-process re-engineering. It must be stressed that while *domain modeling* and *domain-specific modeling* are similar terms, they address different issues. Domain modeling is clearly an analysis phase activity whereas domain-specific modeling considers design issues in a defined domain. The second type of analysis activity is typical object-oriented analysis using class symbols; really more of a "first-cut" to design than actual analysis. Thus, we call it *logical design* [16] to differentiate it from pure analysis. Now the research question that needs to be answered is whether or not we can use domain-specific modeling techniques in either type of these activities.

There are several entire methods for business process re-engineering such as [17]. The level of formality in domain models is often lower than in design models, hence they are often filled with textboxes, arrows and symbols that have specific meaning in the context. Because the diagrams describe real-world functions and objects the symbols tend to be easy-to-read, such as icons of computers or stickmen, which eases the readability of the notation-rich diagrams [3]. Thus, at least one of the characteristics of domain-specific modeling techniques, using specific symbols for that domain, is fulfilled. However, it is questionable if those symbols taken into use in one domain could be successfully used in other domain in the same meaning. Theoretically, there could be some benefits in using domain-specific notation to model two large domains that are alike but so different from typical business domains that it would be inconvenient to model them with a general-purpose notation. In practice, lower formality requirements for domain modeling probably diminishes any actual benefits. More importantly, it would also be illogical to use domain-specific techniques when making the domain model, as in essence we are trying to define what the domain is. Indeed, if there were two domains we could model with a domain-specific domain modeling technique, the models and thus also the domains should be the same.

3 Domain-Specific Modeling in Logical Design

Considering the above presented points, it would seem unlikely that domain-specific modeling techniques would bring any substantive benefits into domain modeling or business re-engineering. However, it could facilitate our latter type of analysis, logical design. While this type is often called analysis, as is the case in Rational's Unified Process (RUP) [3], it is intended to help software developers to jump into hard-core design or *physical design* [16] phase. Its aim is to create general structure of the system, often achieved through architecture modeling such as Kruchten's 4+1 [18], including logical design using analysis class diagrams among other things.

In current object-oriented methods such as RUP [3] class diagrams are used both in logical design and in physical design phase. It is argued that allowing developer to use similar objects in both phases offers seamless transition from one phase to the other, which has been one of the means to market object-oriented methods over traditional structural methods [13]. Therefore, extending domain-specific physical design to logical design could be beneficial as well. However, as Kaindl states, the approach that the same model is used for different purposes, can also be problematic. Often the developers do not make proper logical design because the model does not prevent them to focus on detailed design straight away, while the goal in making a logical design should be to help the developer to plan ahead what should be designed later in physical design.

Let us consider the possible benefits in using DSM techniques in logical design with an example. We already mentioned Nokia's domain-specific method Plato, which we will now shortly describe. In Plato, the physical design is done in a powerful visual model-based programming environment that generates executable code straight from the models. Developers develop the software working in a graphical environment and by connecting symbols visually together and filling in their custom attribute values. Each one of those symbols represents some abstract elements in the mobile phone domain. The symbols have complex mapping to code blocks, but the implementation is hidden from a developer. The graphical models that are composed are custom state diagrams, from which the application code is automatically generated. Advanced method engineering group has made Plato method, offering the playing blocks for the application developers. While developing the kind of an environment has required a lot of resources, the software development with it is much faster compared to ordinary software development [19].

One of the major problems hampering the Plato development process is the difficulty to move from analysis phase (logical design) made with in-house developed, a general-purpose, UML-based method called OMT++ to physical design phase in Plato environment, as pointed out in [12]. The problem is that the logical design phase does not support problem solving in the physical design phase. The logical design is executed with traditional o-o techniques, but the Plato's physical design is not object-oriented as such, but state-based. Thus extending domain-specificity to logical design would seem to be reasonable and beneficial. In Plato's case, logical design should help developers design what should be modeled in which diagrams and what is the relations between the design diagrams. Ideally, the method would allow a developer to explicitely state what kind of relationships there are between the numerous diagrams and jumping quickly between logical and physical design.

4 Summary

Since logical design is a design activity, while often count as part of the analysis phase, we would expect that it is possible to use domain-specific techniques in it and that there are possible benefits in doing so. In theory, using domain-specific techniques for logical design should result in seamless transition from logical to physical design and better support for designing tasks to be done in the physical design. Our intention is to empirically prove that in a sample DSL. Our plan is to create a domain-specific analysis phase for our already existing design time experimental DSL [20] and compare analysis phase using DSM techniques to analysis phase using general-purpose method in the sample DSL.

To summarize, we have discussed in this paper what domain-specific modeling is, what are the activities in the analysis phases of a software development process and possibilities to extend DSM techniques to be used in some of those activities. We have separated two types of tasks called analysis: domain analysis, which purpose is to define a domain and analysis of the problem that is actually a logical design. We believe that for domain modeling, domain-specific techniques have little or no use at all. However, domain-specific techniques seem to be well suited for logical design if also physical design is executed using domain-specific techniques. We claim that the purpose of logical design is to help developer to divide the problem in smaller and allowing him to plan ahead for the physical design.

This paper does not completely answer whether or not domain-specific techniques can be used in analysis phase of software development. Rather, we suggest that it may well be beneficial for some analysis activities. Hence, we point out an important topic in this area: development of comprehensive domain-specific methods considering analysis phase activities among other things and an empirical study of its usefulness. This topic is left for future studies.

References

1. Kieburtz, R., et al., *A Software Engineering Experiment in Software Component Generation*. 1996, Oregon Graduate Institute of Science & Technology: Portland, OR.
2. MetaCase Consulting, *MetaEdit+ Revolutionized the Way Nokia Develops Mobile Phone Software*. 1999, MetaCase Consulting, Inc.
3. Jacobson, I., G. Booch, and J. Rumbaugh, *The Unified Software Development Process*. 1999: Addison-Wesley.
4. Kieburtz, R.B., *Defining and Implementing Closed, Domain-Specific Languages*. 2000, Oregon Graduate Institute of Science & Technology: Beaverton, Oregon USA.
5. Korhonen, K. *Motivation and Hypothesis for Comparison between Component Frameworks and DSL Paradigms*. in *OOPSLA Workshop on Domain-Specific Visual Languages (DSVL'01)*. 2001. Tampa, FL, USA.
6. Korhonen, K.e., *Comparison between UML and DSVLs. A presentation given in OOPSLA'01 DSVL Workshop*. 2001: Tamba, FL, USA.
7. Rumbaugh, J., I. Jacobson, and G. Booch, *The Unified Modeling Language Reference Manual*. 1998: Addison-Wesley.
8. IEEE Std 1076-1993, *IEEE Standard VHDL Language Reference Manual*. 1994, IEEE.
9. D'Souza, D.F. and A.C. Wills, *Objects, Components and Frameworks with UML: The Catalysis Approach*. 1999: Addison-Wesley.
10. Hudak, P., *Building Domain-Specific Embedded Languages*. Computing Surveys, 1996. **28A**(4).
11. Pohjonen, R. and J.-P. Tolvanen, *Reuse in Advanced Method Support Environments (RAMSES): Project Plan*. 1999, Information Technology Research Institute: Jyväskylä.
12. Korhonen, K., *Build Your Own Lego: Components, Architecture and Processes in Component-Based Development*, in *Department of Computer Science and Information Systems*. 2000, University of Jyväskylä: Jyväskylä, Finland.
13. Kaindl, H., *Difficulties in the Transition from OO Analysis to Design*. IEEE Software, 1999(September/october): p. 94–102.
14. Hoydalsvik, G. and G. Sindre. *On the Purpose of Object-Oriented Analysis*. in *Conference on Object-Oriented Systems, Languages, and Applications*. 1993: ACM SIGPLAN Notices.
15. Jaaksi, A., et al., *Tried & True Object Development: Industry-Proven Approaches with UML*. 1999: Cambridge University Press.
16. Graham, I., B. Henderson-Sellers, and H. Younessi, *The OPEN Software Process Specification*. 1997: Addison-Wesley.
17. Nomura, V. and K. O'Connor, *Knowledge Value Added and Activity Based Costing*. 2000.
18. Kruchten, P.B., *The 4+1 View Model of Architecture*. IEEE Software, 1995. **12**(6): p. 42–50.
19. MetaCase Consulting, *Domain-Specific Modelling: 10 Times Faster than UML*. 2001, MetaCase Consulting, Inc.: Jyväskylä.
20. Korhonen, K. *An empirical study on an experimental DSL for product-line development*. in *The 6th World Multiconference on Systemics, Cybernetics and Informatics*. 2002. Orlando, FL, US.

Organizations and Normative Agents

Mehdi Dastani, Virginia Dignum, and Frank Dignum

Institute of Information and Computing Sciences
Utrecht University
The Netherlands
{mehdi,virginia,dignum}@cs.uu.nl

Abstract. The overall behavior of multiagent systems can be controlled by designing agent organizations and allowable interactions between autonomous and goal directed agents such as 3APL agents. The rules of behavior for individual agents are described using concepts from organization theories such as roles and norms. In this paper, a framework for agent organization is discussed and its related concepts such as norms and roles are formalized. It is argued that organizational roles and norms influence and determine the goals of individual agents whenever they are involved in the corresponding organization. Some formal properties of the relation between individual agents and their organizations is presented.

1 Introduction

Software agents and in specific Multi-agent systems [5,6] are one of the most promising areas in the field of computer science. Software agents get some knowledge about the world in which they operate, such that they can solve most of the minor problems they encounter in operation by themselves, without intervention of the user. This has a large advantage over traditional systems for which the environment of the system had to be completely predictable or otherwise the system would not function correctly. This might seem trivial, but if I am a user that only know that there exists a handy program for calculating the contents of a cilinder, but do not know whether I should give it the radius or diameter of the cilinder as argument I cannot use the program. Especially in open environments such as the Internet or Intranets agents will be able to react more flexible and cooperatively with other systems than traditional software.

However, the autonomy of the agents also has a downside. If one creates a system with a number of autonomous agents it becomes unpredictable what the outcome of their interactions will be. This so-called emerging behavior can be interesting in settings where the multi-agent system is used to simulate a group of people and one tries to find out which factors influence the overall behavior of the system. E.g. some studies have been done in which groups of selfish agents are compared with cooperative agents. (In general the system with cooperative agents produces better results for the individual agents).

However, in settings where the multi-agent system is used to implement a system with a specific goal one does not want this emergent behavior to diverge

M.H. Shafazand and A M. Tjoa (Eds.): EurAsia-ICT 2002, LNCS 2510, pp. 982–989, 2002.

from the overall goal of the system. E.g. if the system is designed to get up-to-date information about the indexes of the stock exchanges in London, Frankfurt and new York, one does not want the system to evolve into a system that gives up-to-date information about bonds in Tokyo and Sydney (even though it still gives financial information). In order to limit the autonomy of the agents in these situations and ensure a certain behavior of the overall system we have to design agent organizations and allowable interactions between the agents in these organizations. The rules of behavior for the agents within the organization are described using "norms". Of course it is important that we also define how the agents use these norms to govern their behavior as this determines the interaction between the individual agents and the multi agent system. In this paper we will explore this interface between the individual agent with its autonomy and the society with its norms.

In the next section we will describe multi-agent systems from the organizational point of view and indicate which choices should be made and what concepts are used to describe this view. In section 3 a model to design organizational multi-agents systems is described. In section 4 we will introduce 3APL, an agent programming language that is used to specify the behavior of certain types of individual agents. In section 5 we will describe the mapping of the social aspects of the agent organization into the description of the behavior of the individual 3APL agents. An example of such a mapping is given in section 6.

2 Agents Societies

The design of multi-agent systems must consider organizational characteristics such as stability over time, some level of predictability, and commitment to aims and strategies. The development of multi-agent systems calls for models, languages and methodologies to represent communication, interaction, roles and other concepts that characterize multi-agent systems. Such modelling primitives are usually not provided by (single) agent languages. Furthermore, traditional multi-agent models and architectures often assume an individualistic perspective in which agents are taken as autonomous entities pursuing their own individual goals based on their own beliefs and capabilities. In this perspective global behavior emerges from individual interactions and cannot easily be managed or specified externally.

Agent societies are an effective platform to model organizations because they provide mechanisms to allow organizations to advertise their capabilities, negotiate their terms, exchange rich information, and synchronize processes and workflow at a high-level of abstraction. From an organizational perspective, it is the society goals that determine agent roles and interaction norms. In agent societies, individual agents are therefore seen as actors that perform role(s) described by the society design; they interact with each others to accomplish their goals by playing their roles. However, the agent's own capabilities will determine the specific way an agent enacts its role(s). Therefore, frameworks for agent societies must combine models that describe the structure and characteristics of an organization with models that specify the interests and capabilities of the involved individuals.

In [2], we proposed a framework for agent societies that incorporates organizational and individual perspectives as described above. The framework consists of three interrelated models each describing different aspects of the society: organizational model, agent model and social model. We assume that individual agents are designed independently from the organization. The social model provides a dynamic link between agents and organization. Organizational model describes the desired or intended behavior and the overall structure of the society. This model does not include or refers to agents, but only to roles, which are described in terms of externally perceived actions and behavior. Agent model is used to describe the agents that will participate in the society, in terms of their capabilities, goals and interaction patterns. Social model specifies the relation between organizational roles and specific agents. In this model, the organizational model is populated by agents that fulfil the designed roles and interact according to the defined rules.

3 Organizational Model

The organizational model specifies the structure of an agent society according to the requirements of the organization itself. In its most simplified form, a society can be defined in terms of its objectives (goals) and the norms that regulate interaction in the society. The goals of a society are specified in terms of roles that correspond to the different stakeholders in the domain. That is, an overall goal of a society is represented as an hierarchy of subgoals that correspond to the goals of the different roles. A role is defined as a triple $< G, N, R >$, where G, N, and R stand respectively for the goals, the norms, and the interaction rules that are associated with the role. Furthermore, the organizational model is split into two parts: facilitation and operation. The facilitation layer provides the backbone of the society and consists of institutional agent roles, which are designed to enforce the social behavior of agents in the society and assure the global activity of the society. The operational layer models the overall objectives and intended action of the society and consists of domain related roles.

For example, consider the case of a trading society. The overall goal of the society is to generate transactions. This goal can be split into a facilitation component that aims at the regulation of those transactions, and an operational part where transactions are generated. Typical domain related and operational stakeholders in such a society are sellers and buyers, which exhibit autonomous behavior in the society. These will be specified as roles in the operational component of the model. Finally the activity of the facilitation layer can be described in terms of a registrar role that regulates the participants and a market master role that regulates the transactions and supports the matching between sellers and buyers.

The roles at the lowest level can be described as follows. The (S)eller represents an entity that wants to exchange its goods for money; the (B)uyer rep-

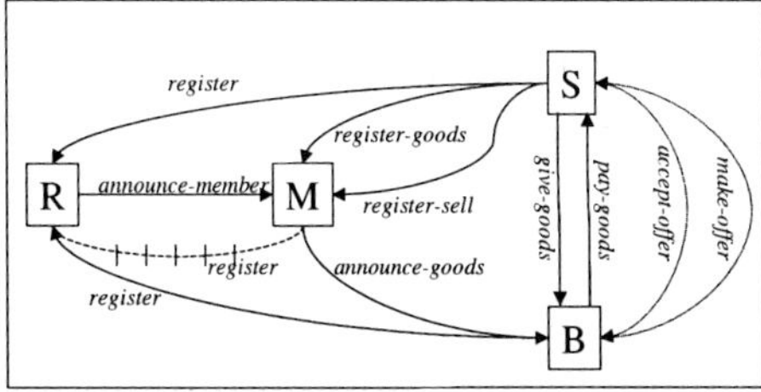

Fig. 1. The role interactions in the organization.

resents an entity that wants to exchange its money for goods; the (M)arket master takes care of introducing potential sellers to potential buyers; and the (R)egistrar keeps track of who are the sellers and buyers at each moment. The interaction between these roles are depicted in figure 1. The roles can be specified by considering the norms that regulate these roles in terms of its obligations (O), permissions (P) and prohibitions (F). In this example, the role *Registrar* has the obligation to register buyers and sellers and to announce new buyers and sellers to the Market Master, the role *Market Master* has the obligation to announce goods for sale to Buyers and is forbidden to register as buyer or as seller. The role *Buyer* is obliged to register as buyer and to pay for the bought goods, and is permitted to make offers to sellers. Finally, the role *Seller* is obliged to register as seller, to register goods to sell with Market master, to supply sold goods to buyers, and to indicate the sold goods to the Marker master. The way actions are performed can be described in interaction protocols, e.g. *pay goods* can be implemented as sending a credit card number.

4 Agent Model

In this section, we consider a particular model for social agents called 3APL. These agents are cognitive agents that have beliefs and goals as mental attitudes and can revise or modify their goals by means of the so-called practical reasoning rules. Moreover, they are assumed to be capable of performing certain basic actions such as sensing and communication actions as well as mental updates. The specification of 3APL is introduced in [4]. A 3APL agent can be specified in terms of its beliefs, goals, and goal revision rules. These rules are called practical reasoning rules.

Definition 1 (Beliefs). *Given a set of domain variables VAR and functions (constants are zero-place functions), the set of domain terms is defined as usual. Let $t_1, \ldots, t_n$ be terms referring to domain elements and Pred be a set of domain predicates, then the set of programming constructs for belief formula, BF, is defined as follows: If $p \in Pred$, then $p(t_1, \ldots, t_n) \in BF$, and If $\varphi, \psi \in BF$, then $\neg\varphi, \varphi \wedge \psi \in BF$. All variables in the BF formula are universally quantified with maximum scope.*

Agents are assumed to be able to perform a set of basic actions. These actions, which are also called basic capabilities, are parameterized actions and are defined in terms of pre- and post-conditions.

Definition 2. *Let t_i be a term denoting a domain element, A_i be an action symbol and $\phi, \psi \in BF$. Then, $\{\phi\}A_i(t_1, \ldots, t_n)\{\psi\}$ is a parameterized basic action with ϕ and ψ as pre- and post-conditions, respectively. The set of basic actions is denoted by Bact.*

Goals in 3APL can be best understood as partial plans which should be performed by the agents. The goals are based on action operators, primitive actions, or other plans. We also assume a set of goal variables, $GVAR$ ($GVAR \cap VAR = \emptyset$) that can be used to refer to unidentified goals.

Definition 3 (Goals). *Let Bact be a set of basic actions, BF be a set of belief sentences, $\varphi \in BF$, and $\pi_1, \pi_2 \in G$. Then, the set of 3APL goals (G) are defined as follows: $GVAR, Bact, BF \subseteq G$, and $(\varphi?)$, $(\pi_1;\pi_2)$, $(\pi_1+\pi_2)$, $(\pi_1||\pi_2)$ $\in G$.*

Practical reasoning rules are at the heart of the functioning of 3APL agents. For each goal of the agent, that is not a basic action, there should be a practical reasoning rule indicating how the goal can be achieved. However, the practical reasoning rules can also be applied to revise agent's goals that are not achievable, basic actions that are blocked, to optimize agent's goals, to generate some sort of reactive behavior, or to define achievement goals (i.e. procedures).

Definition 4 (Practical Reasoning Rules). *Let $\pi_h, \pi_b \in G$ and $\varphi \in BF$, then a practical reasoning rule is defined as follows: $\pi_h \leftarrow \varphi \mid \pi_b$. This rule can be read as follows: if the agent's goal is π_h and the agent believes φ, then π_h can be replaced by π_b.*

The 3APL agent functions as follows: at each cycle the agent checks what its current goal is and tries to execute the goal if it is a basic action or find a practical reasoning rule whose head can be unified with the current goal and whose guard is true. If the agent succeeds in finding such a rule, it rewrites the current goal to the body of the rule (given the unification of the variables in the goal and the rule). This process continues until the agent has no more goals left or no rule is applicable. Given the definition of beliefs, goals and practical reasoning rules, a 3APL agent can be specified as follows:

Definition 5. *A 3APL agent is a triple $< \Pi, \sigma, \Gamma >$, where Π is a set of goals, σ is a set of belief formula, and Γ is a set of practical reasoning rules.*

For example, let $\{\neg p(a)\}$ $A()$ $\{p(a)\}$ and $\{p(a)\}$ $B()$ $\{\neg p(a)\}$ be two basic actions. Then, $< \ \Pi = \{A(); B()\}\ ,\ \sigma = \{q(b)\}\ ,\ \Gamma = \{A(); X \leftarrow p(a) \mid X; A()\}\ >$ is an agent which has a goal to do first $A()$ and then $B()$, believes $p(a)$, and has a goal revision rule which states that whenever it has to do $A()$ and after that something else, but believes $p(a)$ (i.e. the precondition of $A()$ is not satisfied), then it delays the execution of $A()$ and does X first.

5 Social Model

A trivial characterization of the social model for goal directed agents such as 3APL is to relate the goals and interaction rules that are associated with the organizational role to the goals and practical reasoning rules of role playing 3APL agents. Although this characterization seems to be intuitive, it does not characterize the interaction between role-related goals and practical reasoning rules with the individual goals and practical reasoning rules of the agent itself. Moreover, this characterization does not explain the relation between norms that are associated with the organizational roles and the model of 3APL agents.

In order to characterize the relation between role-related and individual goals and practical reasoning rules, the specification of agents can be extended to include, besides goalbase Π and practical reasoning rules Γ, two new components called normative goalbase and social interaction rules, respectively. The first component contains role-related goals and the second component contains role-related interaction rules.

Definition 6. *A social 3APL agent is a 5-tuple* $< \Pi, \sigma, \Gamma, \Delta, \Upsilon >$ *where* Π *is the set of agent's goals,* σ *its beliefs,* Γ *its practical reasoning rules,* Δ *its normative goals, and* Υ *its social interaction rules.*

The reason to distinguish between individual and role-related goals and rules is to indicate goals and rules that the agent should pursue when the agent plays the role. The social model, i.e. the relation between individual and role-related goals and practical reasoning rules, can be characterized by the following definition.

Definition 7. *Let* G_ρ *and* R_ρ *be goals and interaction rules associated with the organizational role* ρ. *A 3APL agent* $\alpha =< \Pi, \sigma, \Gamma, \Delta, \Upsilon >$ *plays the organizational role* ρ *if it satisfies the following property:* $G_\rho \subseteq \Delta \ \wedge \ R_\rho \subseteq \Upsilon$.

Moreover, we allow agents to aim at achieving their individual goals, i.e. the goals that are not associated with its role but are compatible with it. We assume that Δ and Υ contain compatible goals and interaction rules.

Definition 8. *Let* ϕ, ψ, Φ, *and* Ψ *be respectively a goal, a rule, a set of goals and a set of rules. Let also* $compatible_g(\phi, \Phi)$ *indicate that* ϕ *is compatible with* Φ *and* $compatible_r(\psi, \Psi)$ *indicate that* ψ *is compatible with* Ψ. *A 3APL agent* $\alpha =< \Pi, \sigma, \Gamma, \Delta, \Upsilon >$, *which plays the organizational role* ρ, *satisfies the following property:* $\forall \pi \in \Pi(compatible_g(\pi, \Delta) \leftrightarrow \pi \in \Delta) \ \wedge \ \forall \gamma \in \Gamma(compatible_r(\gamma, \Upsilon) \leftrightarrow \gamma \in \Upsilon)$.

We can now demand that agents should aim at achieving their normative goals and apply their social interaction rules when they are playing the corresponding role. This demand does not mean that the agents cannot aim at achieving their non-normative goals since definition 8 allows agents to import some of their non-normative goals to their normative goalbase.

Proposition 1. *In an organization, agents aim at achieving their non-normative goals if the goals are compatible with their organizational roles.*

This proposition follows immediately from definitions 7 and 8.

The relations between norms and goals and interaction rules are discussed in [1]. According to this study, norms are related to goals in two ways. First, one may consider norms as a source of goals or, in other words, goals are generated by norms. Thus, when for example an agent participates in a social setting, the corresponding norms determine the goals of the agent. This type of goals are called normative goals. Second, norms may be considered as a filter or selection mechanism on agent's goals, or in other words, goals that are incompatible with norms are filtered out. Thus, when an agent plays a role, the agent should aim at achieving only those goals that are compatible with the norms associated with the role.

The goals generation view of norms can be expressed in 3APL by means of reactive practical reasoning rules. Note that the head of reactive rules are assumed to be empty. Such a rule has the form $\leftarrow \phi \mid \pi$, where the guard ϕ indicates the belief condition of the agent which needs to hold to generate the goal π; this condition should indicate the situation in which the agent can play a role. Note that reactive rules do not revise any existing goal, but rather introduce or generate new goals. The generated goals from these reactive rules will be included in the normative goalbase component of the agent.

Definition 9. *Let* $\alpha =< \Pi, \sigma, \Gamma, \Delta, \Upsilon >$ *be an agent that plays the role* $\rho =< G, N, R >$ *and* I *be the set of reactive rules that corresponds with norms in* N. *The agent* α *plays the role* ρ *if* $I \subseteq \Upsilon$.

The second view of norms, the goal selection view, can be implemented in 3APL in various ways. One way to do this is to require that agent's non-normative goals can be imported to the agent's normative goalbase only if they are compatible with the norms imposed by the social setting. This means that this aspect of norms specifies the notion of compatibility relation used in definition 8, which can be defined in various ways. For example, two goals π and π' can be defined as compatible if they are identical, if they have a subgoal relationship, i.e. $\pi = \pi_1 \cdots \pi' \cdots \pi_n$, or if one is an instantiation of the other. We will not explore this aspect of norms in more detail.

6 Example Revisited

The relevant parts of a 3APL agent that plays the registrar role, as explained in the example of section 3 can be formulated as follows:

$\Delta = \{register()\}$

$\Upsilon = \{register() \leftarrow \top \mid$

$(register\_as\_seller(Seller, Prod\_id) \; ; \; announce\_member(Seller) +$

$register\_as\_buyer(Buyer, Prod) \; ; \; announce\_member(Buyer));$

$register()\}.$

The goal *register*() is assumed to be the role-related goal of the registrar. The rule in Υ indicates that the registrar should either register a seller and announce him to the master or do it for a buyer. This activity continues by keeping the goal *register*() at the end of this rule (recursive call). In this example,

$received(\alpha, \beta, F, \phi)$ is a belief formula indicating that agent β (receiver) has received message ϕ from agent α (sender); F indicates the modality of the message (also called speech act [3]) such as *request*, *inform*, or *agree*. Finally, the goals, such as *register_as_seller*, are basic actions defined as follows:

1) $\{received(Seller, self, request, register\_as\_seller(Seller, Prod\_id))\}$
$register\_as\_seller(Seller, Prod\_id)$
$\{seller\_agent(Seller, Prod\_id)\}$

2) $\{received(Buyer, self, request, register\_as\_buyer(Buyer, Prod))\}$
$register\_as\_buyer(Buyer, Prod)$
$\{buyer\_agent(Buyer, Prod)\}$

3) $\{\neg announced\_to\_master(Agent), seller\_agent(Agent, P) \vee buyer\_agent(Agent, P)\}$
$announce\_member(Agent)$
$\{announced\_to\_master(Agent)\}$

7 Conclusion

In order to profit from the flexibility of autonomous agents while at the same time ensuring the robustness of the complete system, one needs to define an organizational structure for the multi-agent system. This organizational structure defines the goal of the overall system and indicates the behavioral boundaries for the individual agents within the system. Because agents come with their own goals and capabilities, we need to specify how the agent fits into a certain role it takes in the multi-agent organization. In order to implement this fitting between agents and roles in an organization we added a social component to the agent architecture. This component takes care that the goals of the agent fit within those for the role it plays and that the agent fulfills all norms belonging to the particular role. We have shown how this can be implemented in systems for which the agents are specified in 3APL which is a typical goal-directed agent language. Due to space limitations we could not explore all possible relations between the agents' goals and capabilities and those required by the role the agent plays. We hope to fully explore this part in subsequent research.

References

1. C. Castelfranchi, F. Dignum, C. Jonker, and J. Treur. Deliberate normative agents: Principles and architecture. In *Proceedings of ATAL-99*, Orlando, FL, 1999.
2. V. Dignum, J.J. Meyer, and H. Weigand. Towards an organisational model for agent societies using contracts. In *Proceedings of AAMAS'02, To appear*, Bologna, Italy, 2002.
3. FIPA. http://www.fipa.org/.
4. K. Hindriks, F. de Boer, W. van der Hoek, and J. J. Meyer. Agent programming in 3APL. *Autonomous Agents and Multi-Agent Systems*, 2(4):357–401, 1999.
5. G. Weiss, editor. *Multiagent systems.* MIT, 1999.
6. M. Wooldridge and N. R. Jennings. Intelligent agents: Theory and practice. *Knowledge Engineering Review*, 10(2), 1995.

A Framework for Agent-Based Software Development

Behrouz Homayoun Far

Faculty of Engineering, University of Calgary,
2500 University Drive N.W.
Calgary, T2N 1N4, Alberta, Canada
far@acm.org
http://www.enel.ucalgary.ca/People/far

Abstract. In software engineering community there is an increasing effort of design and development of multi-agent systems (MAS). However, agent system development is currently dominated by informal guidelines, heuristics and inspirations rather than formal principles and well-defined engineering techniques. In this paper we present a framework for agent-based software development called Agent-SE. We present methods to generate organizational information for cooperative and coordinative agents.

1 Introduction

Nowadays, an increasing number of software projects are being revised and restructured in terms of multi-agent systems (MAS). Software agents are considered as a new experimental embodiment of computer programs and are being advocated as a next generation model for engineering complex, heterogeneous, scalable, open, distributed and networked systems. However, agent system development is currently dominated by informal guidelines, heuristics and inspirations rather than formal principles and well-defined engineering techniques. There are some ongoing initiatives by the Foundation for Intelligent Physical Agents (FIPA) (http://www.fipa.org) and some other institutions to produce guidelines and standards for heterogeneous, interacting agents and agent-based systems. However, such initiatives fall short to address the design, development, issues in MAS. In this paper we present a framework for agent-based software development called Agent-SE. We present methods to generate organizational information for cooperative and coordinative agents.

The structure of this paper is as follows. In Section 2 the knowledgeability of the agent based systems is discussed. In Section 3 the Agent-SE approach is introduced. In Section 4 a method to derive organizational knowledge is presented. Finally, a conclusion is given in Section 5.

M.H. Shafazand and A M. Tjoa (Eds.): EurAsia-ICT 2002, LNCS 2510, pp. 990–997, 2002.

2 Knowledgeability in MAS

Traditional software engineering can handle data and information. Data is defined as a sequence of quantified or quantifiable symbols. Information is about taking data and putting it into a meaningful pattern. Knowledge is the ability to use that information. Knowledgeability is an integral part of MAS paradigm. Agents in order to interact and work proactively must be knowledgeable in their area of expertise. Knowledgeability can be defined in terms of cognitive capabilities and interactions. The following subsections elaborate this.

2.1 Cognitive Capabilities

Three main capabilities of agents in MAS are representing, using and sharing the knowledge. Symbol structure (SS) is used to model individual agent's knowledge structure (see Section 4.2). Ability to use the knowledge can be realized by having a knowledge-base in the form of SS and mechanisms for problem solving using the knowledge base. Finally, ability to share the knowledge depends on ontologies for the domain and task. Mechanisms for using and sharing knowledge are presented in [5].

2.2 Interaction in MAS

Basic agents' interactions are cooperation, coordination and competition.

1. *Cooperation:* Cooperation is revealing the agent's goal and the knowledge behind it to the other party. In cooperation both agents have a common goal.
2. *Coordination:* Coordination is revealing the agent's goals and the knowledge behind it to the other party. In coordination, the agents have separate goals.
3. *Loose Competition:* Loose competition is revealing only the agent's goal but masking the knowledge behind it to the other party.
4. *Strict Competition:* Strict competition is neither revealing the agent's goal nor the knowledge behind it to the other party.

Figure 1 shows decision making mechanism based on agents' interaction. Conventionally, it is believed that for a pair of agents to interact, each should maintain a model of the other agent, as well as a probabilistic model of future interactions [4]. This is unnecessary when using SS representation in cooperation and coordination cases [5].

3 The Agent-SE Approach

In this section we propose a method for multiagent system design, called Agent-SE. The Agent-SE design steps are as follows:

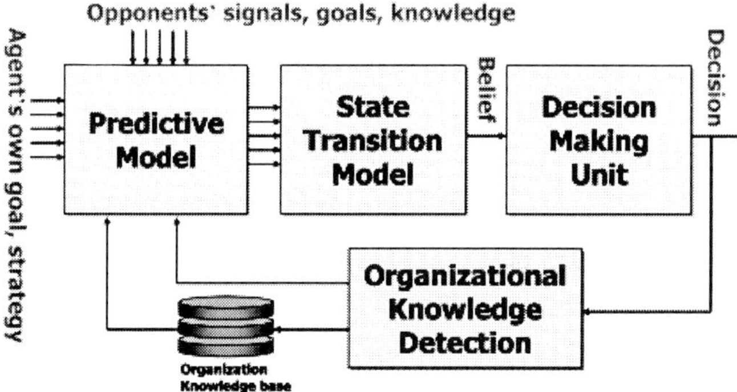

Fig. 1. Decision making mechanism

1. Design use-cases that show the interactions of the MAS with the external world.
2. Devise a candidate set of agents that can realise the use-cases based on functions, inputs and outputs.
3. Design the task ontology of the problem.
4. Build an abstraction model and add interactions and signal level organizational relationships using the task ontology.
5. Design each agent and its internal intra-actions using conventional SE techniques (preferably, object-oriented design with UML, etc.) and a predefined agent model, if necessary.
6. Based on the domain ontology, design each agent's knowledge-base using Symbol Structure. (See Section 4).
7. Derive and record symbol level organizational properties based on interactions of pairs of cooperative or coordinative agents. (See Section 4).

Organizational formation, maintenance and updating are typical of the dynamic nature of groupings in complex systems. Agent-based systems require computational mechanisms for dynamic formation, maintenance and disbanding of organizations. One such mechanism is presented in Section 4.

4 Organization

Organization is a goal directed coalition of software agents in which the agents are engaged in one or more tasks. Control, knowledge and capabilities are distributed among the agents.

The already proposed organizational models for multiagent systems have certain drawbacks. First, they cannot explain the organizational knowledge in terms of its comprising agents without reference to any other intermediary concepts. Second, they cannot provide frameworks for comparing and evaluating different organizations. Third, the organizational knowledge base cannot be updated dynamically, accounting for different configuration of the participant agents. Finally, they cannot explain the need for services of a certain agent in an organization. All of these factors are necessary in organization design and are addressed in our research.

4.1 Assumptions

Intelligence of Pair (IoP): The already proposed theories have implicitly assumed that Organizational Intelligence (OI) exists and implemented using a meta-agent (e.g., directory and ontology service agents) [1]. However there are certain difficulties in both logical formulation and actual implementation of such theories. This is mainly due to ignoring the dynamic interactions among the agents when devising the components of OI.

A point in our research is that in a purposeful (i.e., not random) organization, OI is a property of interaction among agents and can only be ascribed to at least a pair of agents. We call this Intelligence of Pair (IoP) assumption.

History of Patterns (HoP): In biological coalitions, participants may have a kind of role or function (during interaction with the other participants), if they show some persistence in their profile of actions over time. The same could be devised for artificial coalitions. As a matter of fact, it is not difficult to find organizations that display non-random persistent and repeated patterns of actions [1].

Agents act and perform in a physical world. Their past experiences can be recorded and explained in terms of their histories, that is, their profile of actions and states that they go through. Intuitively, histories can display certain patterns. A basic feature of state representation is that it assigns a certain characteristic to its reference agent. Therefore it is possible to define OI patterns with reference to agents' history.

Another point is that OI patterns emerge from discovering a persisted state or an ordered pattern in the agent's profile of actions. We call this History of Patterns (HoP) assumption. IoP and HoP assumptions account for dynamic interactions and a computation method based on this assumption is proposed below.

4.2 Modeling

Symbol structure (SS) is used to model individual agent's knowledge structure. SS is a finite connected multi-layer bipartite graph. There are two kinds of nodes in each layer of SS: concepts (c) and relations (r). One source of difficulty when processing concepts, is distinguishing a concept at various levels of abstraction, as well as differentiating between generic concepts and their instances. Function *type* is defined to ease such differentiation. The function type maps concepts and relations onto a set T. The elements of T are called type labels. Type hierarchy provides a means of evaluating a concept at various levels. The type hierarchy is a partial ordering defined over the set of type labels, T.

We can mention that SS is semantically richer than semantic networks because both the concepts and relations are augmented with types. Furthermore, it explicitly defines mechanisms for combining and using the private knowledge together with the agents' profile of actions in the MAS environment.

4.3 Reasoning Rules

Join rule: Join rule merges identical concepts. If a concept c in symbol structure u is identical to a concept d in symbol structure v, then let w be the symbol structure obtained by deleting d and linking to c all arcs of relations that had been linked to d.

Simplification rule: Redundant relations of the same type linked to the same concept in the same order can be reduced by deletion all but one.

Generalization/Specialization rule: For two arbitrary levels p and q of any SS, if p is identical to q except that some type labels of the nodes of q are restricted to subtypes of the same nodes in p, then p is a specialization of q, and q is called a generalization of p.

4.4 Computational OI

Here we propose a method for generating organizational concepts based on the IoP and HoP assumptions and definitions of interaction. In case of *cooperation* and *coordination* the agent's private knowledge is exposed to the other party. In this method, first, a pair of agents are selected and by using join, simplification, generalization and specialization rules their SS are merged (see Figure 2). There are finite sets of *actions* and *states* associated with each agent. Each agent's state transition is represented by a non-deterministic finite state automaton (NFSA).

Both *actions* and *states* take concepts and relations as their attributes. For each agent, the NFSA state transition model has an initial state *idle* and final state *goal*. A sequence of actions that convert the *idle* state to *goal* is a plan of actions towards a goal. Theoretically, all such sequences form a *regular language (RL)* whose elements are generated by a *regular grammar (RG)* equivalent to the NFSA.

As both actions and states take concepts and relations of SS as their attributes, in the case of cooperation and coordination, a joint sequence of actions can be generated by matching the actions whose attributes are concepts or relations belonging to the same type hierarchy. Such sequences form the joint plans of actions handled

partially by either of the agents pair. The proposed algorithm is as follows:

1. Select an agent pair, Agent (G_1) and (G_2).
2. Merge their SS using join, simplification, generalization and specialization rules.
3. Select mode of operation based on external information.

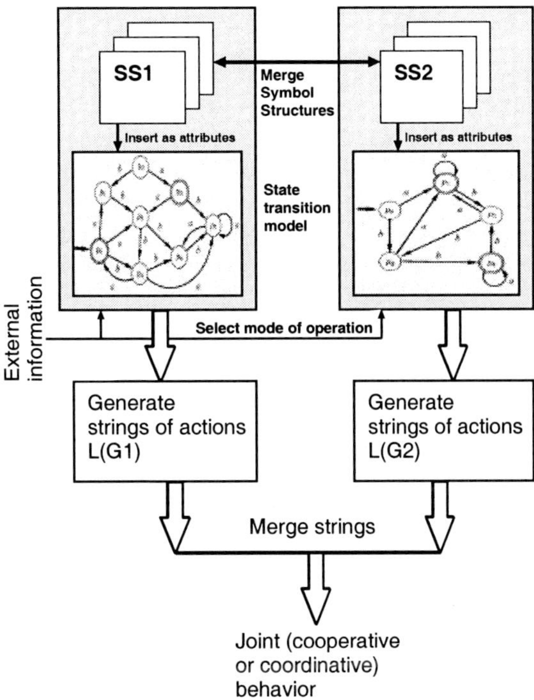

Fig. 2. Deriving organizational knowledge

4. Select the NFSA model and generate sequences of actions for the selected mode.
5. Compare two such sequences $\omega_i \in \mathcal{L}_{G_1}$ and $\omega_j \in \mathcal{L}_{G_2}$ where $\mathcal{L}_{G_1}$ and $\mathcal{L}_{G_2}$ are regular languages of agent (G_1) and (G_2), respectively.
6. For a common action $a \in \omega_i$ and $a \in \omega_j$ check the type of attributes of a. If the attributes belong to the same type hierarchy, merge the sequences from that point on after adjusting the types.
7. Record such joint sequences and check for possible repetition and/or persistence patterns in the future course of actions.
8. When repetition and/or persistence becomes present, add such strings to the *organizational knowledge base.*

Examples of our *electronic commerce* project [2] are presented below. In the *electronic commerce* project seven types of agents, *customer*, *dealer*, *manufacturer*, *delivery*, *banker*, *search* and *catalog* agents work together and/or compete to do business on the Web [2]. By default, the *dealer* agent only knows about the goods to be sold, the *delivery* agent knows about transporting goods, *banker* agent has information on customers and their credit and/or cash accounts and *customer* agent is a personal assistant agent for a human user and has information on the user's preferences, etc.

4.5 Example: Cooperation

Let's consider a case of a *dealer* agent and a *delivery* agent cooperating to sell and deliver an article of commerce, such as a PC to a human user. A portion of SS for *dealer* and a *delivery* agents is shown in the upper portion of Figure 3. It is visible that the concept `Goods` is a super-type for `PC` for both agents. However, based on the agents' roles, the concepts may have different data associated with them. For example, `PC` in *dealer* agent's SS may be associated with CPU, memory, etc. However, for *delivery* agent the same concept may have weight and size as its attributes.

The lower part of Figure 3 depicts an example of the sequence of actions for the *dealer* and *delivery* agents. It is shown that the action `transport`(*Goods*,...) for the *delivery* and `transport`(*PC*,...) for the *dealer* agents have attributes belonging to the same type hierarchy. Therefore, the strings can be merged by adjusting the type and changing `transport`(*Goods*,...) to `transport`(*PC*,...) for the *delivery* agent and let the plan be executed by assigning this action to the *delivery* agent.

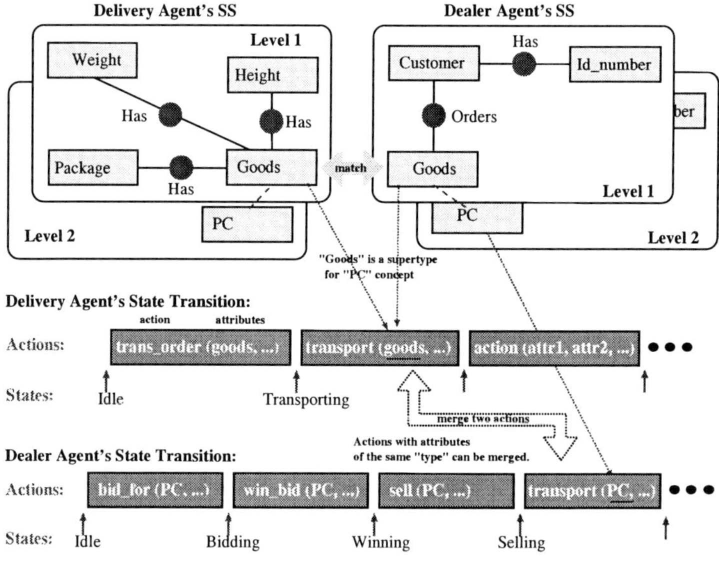

Fig. 3. A portion of SS for dealer agent

4.6 Example: Coordination

Let's consider a case of a *dealer* agent and a *banker* agent coordinating their operations in order to exchange information related to a particular human user. Apparently the *dealer* and *banker* agents have separate goals but they coordinate their activities to help each other.

Let's assume that the *dealer* agent already receives a purchase order from a human user through his/her *customer* agent. A portion of the *dealer* agent's SS is shown in the upper right part of Figure 3.

The *dealer* agent interacts with the *banker* agent by sharing its SS with that of the *banker* agent. By merging the two SS, the `customer` concept is common between the two SS and in this way the *dealer* agent can verify that the `customer` has an `Account` and from the *banker* agent's SS verify that there are either `Cash` or `Credit` accounts available. Also it can verify that the user may also use a `Cash Card`. Therefore, the *dealer* agent can contact the user to get data for `Cash` or `Credit` the accounts. It can further verify genuineness of the data supplied by the user by consulting the *banker* agent again. In this simple example, knowledge sharing is used to enable the *dealer* agent to successfully proceed with the selling task in spite of possessing only a limited amount of knowledge about the user without implementing a redundant customer database within the *dealer* agent.

5 Conclusions

In this paper, we proposed a method for MAS design. We show that multiagent system design can be achieved at the expense of additional design steps including, design of domain ontology and agents' internal knowledge structure. Actually, designing domain ontology is not a totally new task in software engineering practice. It has been carried out informally in almost all of the contemporary software engineering techniques in the form of data definitions, glossary or non-functional specification. Applications using the framework and techniques described in this paper, such as a multi-agent system for electronic commerce [2,3] have already been developed.

References

1. K.M. Carley and L. Gasser, "Computational Organization Theory," in Multiagent Systems: A Modern Approach to Distributed Artificial Intelligence, ed. G. Weiss, G., pp. 299–330, MIT Press, 1999.
2. Far, B.H., et al., "An Integrated Reasoning and Learning Environment for WWW Based Software Agents for Electronic Commerce," IEICE Trans. Inf. and Syst., vol. E81-D, no. 12, pp. 1374–1386, 1998.
3. Far, B.H., et al., "Formalization of Organizational Intelligence for Multiagent System Design," IEICE Trans. Inf. and Syst., vol. E83-D, no. 4, pp. 599–607, 2000.
4. Huhns, M.N. and Stephens, L.A., "Multiagent Systems and Societies of Agents," in Multiagent Systems: A Modern Approach to Distributed Artificial Intelligence, ed. G. Weiss, pp. 79–120, MIT Press, 1999.
5. Onjo, H., and Far, B.H., "A Unified View of Heterogeneous Agents' Interaction," Transactions of the IEICE, vol. E84-D, no. 8, pp. 945–956, 2001.

Application of Agent Technologies in Extended Enterprise Production Planning

V. Mařík, M. Pěchouček, J. Vokřínek, and A. Říha

Gerstner Laboratory, Czech Technical University in Prague
{marik, pechouc, vokrinek, riha}@labe.felk.cvut.cz

Abstract. By reusing and extending the classical multi-agent production planning infrastructure, the extra-enterprise technology helps to organize external business relations of an enterprise and optimize their supply chains. It is obvious that small and medium enterprises that operate in project-driven manufacturing need to share their available resources in the most optimal way. Apart from technological infrastructure that is needed to achieve interoperability across the extended enterprise, this paper emphasis the role of shared ontologies and semantic aspects of the inter-agent communication.

1 Introduction

ExPlanTech [1] is a production-planning tool based on multi-agent technology. ExPlanTech was originally designed for the intra-enterprise production planning in the project oriented manufacturing environment. ExPlanTech integrates existing software systems that administer the production processes within the enterprise. An agentification technique is used to convert the subsystems into the form of agents. The structure of the multi-agent system mirrors the structure of the enterprise. ExPlanTech operates with the production data provided directly by the enterprise information system and elaborates production plan upon it.

There were several reasons that led us towards extension of the ExPlanTech over the boundaries of one enterprise. The main reason was to improve production planning, increase accuracy of the planning processes with taking into account the status of suppliers or possible collaborators. Even though the production planning has very small variability within the single factory, there are possibilities to subcontract various cooperating enterprises, which may overtake responsibility for a part of an order. The optimal production plan should balance the available internal or external resources. The efficient supply chain and service provision management can be handled using the agent-based approach on the *intra-enterprise* and *extra-enterprise* levels.

The problem of managing external resources should be solved by a special set of extra-enterprise agents. Remote access to the planning process is also very important feature of the system. Extra-enterprise agents can be designed for accessing the system remotely and providing users (managers, customers) by appropriate information.

First, we discuss the extra enterprise agents that can be used either in a passive or in an active way in remote interaction with the multi-agent production planning tool,

M.H. Shafazand and A M. Tjoa (Eds.): EurAsia-ICT 2002, LNCS 2510, pp. 998–1007, 2002.

or with its results, respectively. Moreover, we introduce a new type of agent that allows mutual interconnection among enterprises and fully supports the supply chain management. The implementation aspects are discussed in the paper as well.

2 ExPlanTech Architecture

2.1 Intra-Enterprise Agents

The architecture of the intra-enterprise ExPlanTech community [1] consists of a Configurator Agent, several Workshop (Scheduler) Agents, and a Database Agent.

- The **Configurator Agent** (CA) cooperates with the community of Scheduler Agents and provides them with production data about orders. The CA is responsible for synchronizing the entire plan. The CA decides which SA is the best one for cooperating on a particular task. This decision is based on the several parameters of the SAs (load, capabilities, price, priority or restrictions, etc.). This, for example, allows to use outsourcing of the part of the task to the partner enterprise when internal departments are overloaded.
- Each **Scheduler Agent** (SA) represents a particular workshop/department within the factory. The Scheduler Agents create schedules according to the order/task specification received from the CA with respect to the workshop capacity, order priority, etc.
- **Database Agent** (DBA) manages the data concerning tasks and plans, stores the data in a local database and communicates with the ERP system in the enterprise. DBA is able to make a full backup of the system and store it in the database. It is useful when system is stopped or crashes.

2.2 Extra-Enterprise Agents

While the community of intra-enterprise agents makes together an operational tool for planning of manufacturing processes within the enterprise, it is often required to inspect the orders and availability of resources from outside of the company. The production managers also want to change existing plans or accept new orders remotely. For this purposes, we have investigated and developed a community of Extra-Enterprise (EE) agents (see Fig. 1) that are categorized into Passive Extra-Enterprise (PEE) and Active Extra-Enterprise (AEE) agents.

Using modern information/communication technology (e.g. ordinary web browser, mobile PDA devices, Wireless Application Protocol (WAP), etc.) the Extra-Enterprise agents have been designed to be used either by off-site (travelling) production managers or by customers. EE access technology is developed not only for viewing the production related information, but also, more importantly, for initiating/updating the production planning process by remote users. Such an agent can either be executed within the enterprise multi-agent system and managed via the Internet, or it can be run from outside of the enterprise and connected with the intra-enterprise community. Only authenticated agents/human users can access and manage the intra-enterprise production planning data.

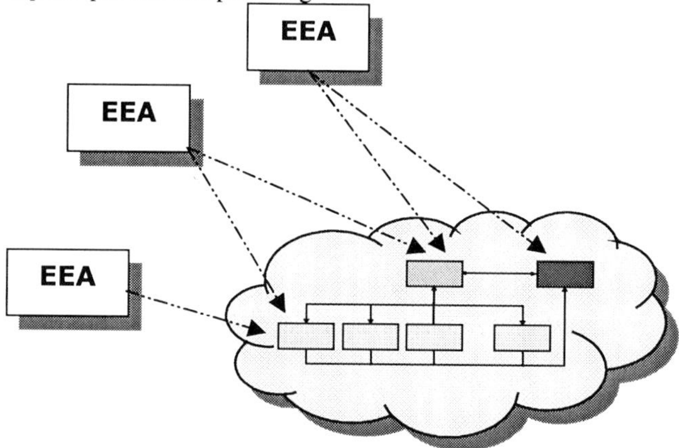

Fig. 1. Extra-Enterprise agents connected to the intra-enterprise multi-agent production planning system.

The Passive Extra-Enterprise agent may passively access the production planning data from outside of the company. It allows the user to monitor:

- *orders* – user (typically a customer or a factory manager) watches the production plan for single orders. Order specification and the relevant plan can be inspected. The user can obtain detailed data on the orders and the progress of production.
- *resources* – allows the user (a factory manager) to check single workshops/departments, task allocation and available capacities. Shall be noted that the user retains his full control of the production plans (including monitoring) via the standard GUIs of the agents.

The data are not stored in the database but it is gathered on-line directly from the agent community, according to user access rights.

The Active Extra-Enterprise agent may trigger the entire course of production planning. AEE agent negotiates with the responsible agent in the system in order to specify the production requirements, the deadline and available budget. However, if its access rights permit, it can also subcontract individual agents for only parts of the products. AEE is used for remote planning of new orders in two possible ways:

- *Permanent planning* – the result of this type of planning is a plan with permanent changes in a global production plan and given resources allocation.
- *Competitive planning* – it is designed for evaluating suitability of acceptance of a new order with no final impact on the plan and agents' resources (temporary planning). This type of planning is useful for comparing alternative plans for several orders. The user finds out differences between obtained plans and should make decision which order should be accepted.

2.3 Extended Enterprise

The Extra-Enterprise agents offer an ideal mean for turning the intra-enterprise production planning multi-agent system into a planning infrastructure within an extended enterprise [4,5,6,7]. Within the collaborative cluster of enterprises it is important

i) to advertise available resources to other enterprises and to explore capabilities of partners willing to cooperate (outsourcing, capacity sharing) and

ii) to interconnect several enterprises together, so that resources within the enterprises may be freely shared.

Enterprise-to-Enterprise agent – The main role of the E2E is to link particular enterprises into an extended enterprise. Each enterprise may be extended over its boundaries by the E2E that can represent either one physical workshop or can be viewed as a wrapper for the entire enterprise (see Fig. 2). E2E acts on behalf of the cooperating enterprise and allows the intra-enterprise agents to subcontract it for a specific part of supply. A partner enterprise (organization) could advertise the status of services and resources, which are currently available. This shared (public) knowledge should be prepared in a standard form with respect to the agreed ontology.

The planning system uses external resources represented by this E2E agent in the same way like it uses the internal resources represented by a set of Scheduler Agents. The E2E agent acts as the Scheduler Agent on the one side and the Active Extra-Enterprise agent on the other side (see Fig. 3). Functionality of the E2E agent is fully transparent for both enterprises. E2E agent can also translate messages to appropriate format, which is understandable in the enterprises.

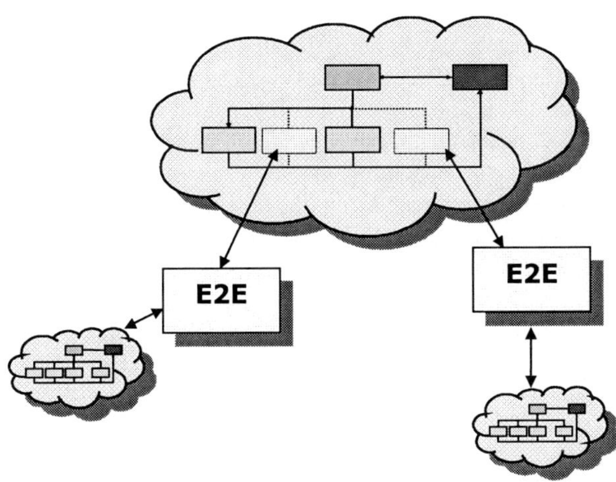

Fig. 2. Enterprise-to-Enterprise agents representing cooperating enterprise as a Scheduler Agent

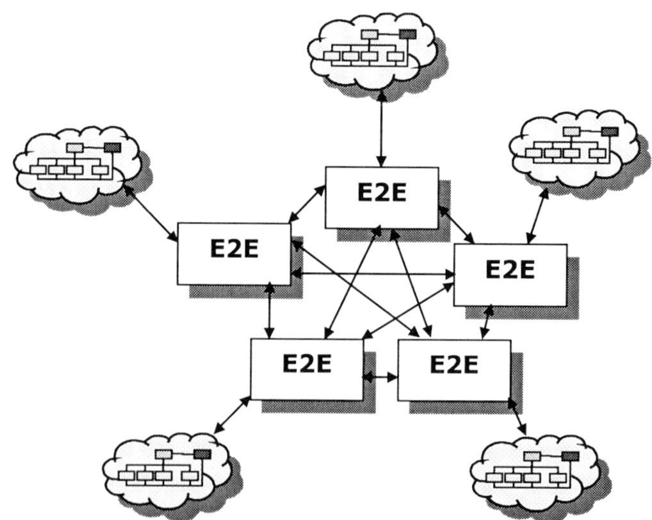

Fig. 3. Several enterprises connected together using Enterprise-to-Enterprise agents.

3 Interaction among the Agents

The ExPlanTech multi-agent system has been implemented in Java. In order to meet full FIPA compliancy it has been build upon the JADE (Java Agent Development Environment). The ExPlanTech agents are extensions of the JADE agent class. JASE implementation offers the agent interaction infrastructure that includes AMS (agent management system), DF (directory facilitator) and ACC (agent communication channel) for interaction with agents being located on different platforms. The agents in the ExPlanTech exchange FIPA-ACL messages with a rather limited vocabulary of performatives (speech acts). The context sensitive action specification is always a part of the content of the message that is encoded in the XML (eXtensible Markup Language). Examples of these actions include DeletePlan, MoveTask, MakePlan, CapacityRequest and others. Apart from the agent management related protocols (registration, deregistration, etc.), ExPlanTech uses primarily the FIPA-request protocol. Such a conversation is initiated by a message with the 'request' performative:

```
(request
    :sender operator@mas0:1099/JADE
    :receiver workshop@mas0:1099/JADE
    :language XML
    :ontology plan-task-ontology
    :protocol fipa-request
    :content
       (<?xml version="1.0" encoding="UTF-8"?>
       <plan-task>
              <task-id>Z_211-0242</task-id>
              <time>10</time>
              <day-load>8</day-load>
```

```
            <priority>1</priority>
            <start-day>20011026T00</start-day>
            <dead-line>20011120T00</dead-line>
            <strategy>E</strategy>
        </plan-task >)
)
```

The `operator@mas0:1099/JADE` agent sent this message to the `workshop@mas0:1099/JADE` agent requesting it to perform an action `<plan-task>` with parameters specifying the properties of the task to be planned (such as id, deadline, strategy, ...). This message includes information about the used content language, conversation protocol and ontology as well.

The receiver replies with the message with an `'agree'` speech act. Here the agent `workshop@mas0:1099/JADE` informs the agent `operator@mas0:1099/JADE` about its agreement with the suggested request. This message is sent as soon as possible and it follows only the agent's decision to accept the request.

After this the `workshop@mas0:1099/JADE` agent starts to work on planning the requested task. Upon achieving this request it sends a message with an `'inform'` speech act. In this message the agent `workshop@mas0:1099/JADE` informs the agent `operator@mas0:1099/JADE` about having managed to plan the requested task. In the content of the message there are embedded properties of the resulting plan.Apart from this classical protocol, the ExPlanTech strongly relies on the subscribe-inform communication mechanism, for which there is no specific conversation protocol defined in FIPA. In such a case a sender subscribes a receiver for a certain type of information. The receiver replies with an appropriate `'inform'` message. Upon each update of the truthfulness of its previous reply, the receiver sends an updated `'inform'` message to the sender. See bellow for en example of such a conversation mechanism:

```
(subscribe
    :sender workshop@mas0:1099/JADE
    :receiver DBA@mas0:1099/JADE
    :language XML
    :ontology capacity-ontology
    :content
       (<?xml version="1.0" encoding="UTF-8"?>
       <change-capacity/>
       )
)
```

The `workshop@mas0:1099/JADE` agent subscribes the `DBA@mas0:1099/JADE` agent for information about new capacities and capacity changes that are propagated from the ERP system (such as extra shifts or holidays) automatically. This happens at the beginning of the community lifecycle. Upon such an event (change of the capacity of the workshop) the `DBA@mas0:1099/JADE` agent automatically sends an `'inform'` message to the `workshop@mas0:1099/JADE` agent:

```
(inform
    :sender DBA@mas0:1099/JADE
    :receiver workshop@mas0:1099/JADE
    :language XML
    :ontology capacity-ontology
```

```
    :content
       (<?xml version="1.0" encoding="UTF-8"?>
<request-change-capacity-message>
        <day-capacity-list>
        <day-capacity>
            <day>20011120T000000000</day>
            <capacity>10</capacity>
        </day-capacity>
    </day-capacity-list>
</request-change-capacity-message>)
  )
```

There are several ontologies in the system which specify semantics of the inter-agent communication (e.g. TaskOntology, PlanOntology, ChangePlanTaskOntology, TaskMoveOntology etc.). Some of them are simple, some of them are complex but the structure and the way of how they are defined complies with the FIPA specifications.

4 Extra-Enterprise Access

The concept of Extra-Enterprise agents and the extended enterprise was implemented and integrated in the web server. Extra-Enterprise agents are created on the web server side and connected to the JADE [2] multi-agent platform using JAVA servlet technology (see Fig. 4) Dynamic HTML pages are based on the information downloaded from the agents using the FIPA standardized communication [3]. Users are logged to the web portal according to their access rights (see below). Servlets running on the server are directly connected to the extra-enterprise agents. Extra-Enterprise agents (both Passive and Active) communicate with the rest of the community and pass the information back to the servlets. Users can inspect in the web pages only information on their own tasks and workshops.

There may be more web servers, each hosting different EEAs. Each web server is a member of a different JADE platform and connects to the rest of the community by the HTTP protocol means, while EEAs on a single server communicate via RME.

Besides the above described functionality of the EE agents, such as passive inspection of tasks, plans, workshops, their loads and schedules or specification of new orders, and active planning (permanently or competitively), the users can define and create own collaboration models (so called virtual workshops). This approach is particularly suitable for a potential supplier, who wishes to join the respective supply chain but does operate neither intra-enterprise production planning nor the extended enterprise multi-agent system. By means of a virtual workshop, the supplier can submit specific information about the type, amount, costs and delivery dates of particular pieces of a supply that it offers to the supply chain. Other supply chain members can incorporate this model into their planning strategies. Virtual workshops are simple implementations of the E2E agent. These virtual workshops are created on the server and connected to the community. It is also possible to construct a virtual workshop at the user-side enterprise.

The Extra-Enterprise agents providing access to the intra-enterprise community share the defined set of ontologies with the Intra-Enterprise agents. Semantically the agents refer to the modelled factory and thus there in no need to worry about integrating different ontologies.

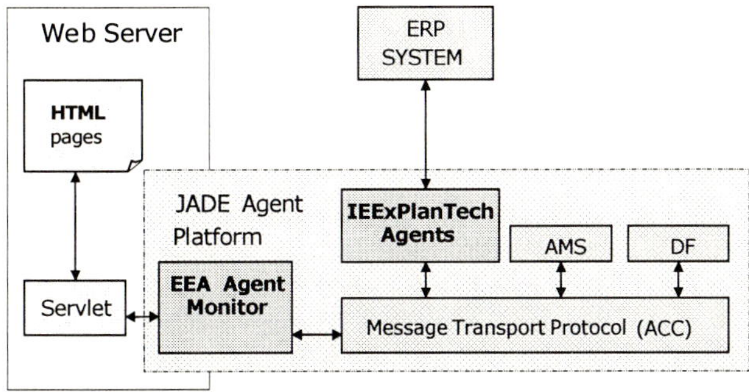

Fig. 4. Extra enterprise technology integration

5 Enterprise-to-Enterprise Connection

E2E is a key agent for linking enterprises together. E2E agent serves as a gate to the system (see above). The community of E2E agents negotiates about service and resources that the particular intra-enterprise communities require and/or provide. Negotiated service requests are propagated to the IE agents which update their plans accordingly. The negotiated service offers are sent to the other IE agents which incorporate the agreed subcontracting into their plans. This process is fully transparent and doesn't require any changes in functionality of the other agents than E2E.

There are several ways how the E2E agents can be connected to different enterprises:

- *Unique E2E on each platform* (system/enterprise) – Each enterprise owns one E2E agent that is connected to other enterprises. This agent maintains all the restrictions and knowledge for cooperation with other E2E agents.
- *Unique E2E for each connection* – Each E2E is dedicated for cooperation with only one E2E. This kind of agent maintains information on cooperation with the only one enterprise. Therefore, we need to run more then one E2E on each platform.
- *Semi-unique E2E* – This is a combination of the above two cases. There are some E2E agents on each platform and each of them can be connected to the several enterprises (more exactly, to E2E agents on different platforms/enterprises). But here is always restriction to a unique connection: two enterprises can be connected together via exactly one pair of E2E agents (each in one enterprise).

The is specific information required for establishing the cooperation: the location of the cooperating enterprise (e.g. IP address, port), authentication and restrictions. Restrictions specify primarily resources that will be available for the cooperating enterprise. In the case of connecting trusted known enterprises (a standard factory supplier or customer), there is no problem with localization and authentication of these enterprises. But there are open problems how to identify new cooperating eneterprises and how to authenticate them. The framework for dynamic enterprise reconfiguration is still an open problem.

So far, we have been relying on a rather strong assumption that the E2E agents connect enterprises each operating the ExPlanTech system for their resource allocation and production planning. This is what we can hardy rely upon when dynamically creating and reconfiguring a virtual enterprise organising the supply chain. Operational functionality of the E2E agents community does not depend on internal intra-enterprise architecture while it strongly depends on a shared ontology.

An ideal thing would be to impose a widely accepted product and manufacturing based ontology to be shared within the E2E agents. There were many attempts to come to a mutual agreement on such ontology. In manufacturing industry there is STEP (Standard for the Exchange of Product Model Data) – a comprehensive ISO standard (ISO 10303) which is used for representation and exchange of engineering product data and which specifies the EXPRESS language for product data representation in any kind of industry. Similar initiative rather focused on the agent and system interoperability aspects has been followed within FIPA in a Product Design and Manufacturing (PDM) working group, which specializes in (i) supply chain management, (ii) planning and scheduling, and (iii) holonic control devices. Must be noted that despite a considerable effort and attention no significant results have been achieved yet.

E2E agents can also speak semantically different languages and use different ontologies. In such a case it is inevitable to implement an automated ontology translation mechanism that would allow agents to interact in the same semantical framework while using dissimilar ontologies for running the conversation. Such a translation can be carried out by the E2E agents autonomously. When an ontological disagreement is identified, the agents analyze each own ontology so that the least general ontological term upon which they share the semantic meaning is detected. By negotiating about properties and constraints of the derived, more specific terms, they can construct mapping between different ontological terms. However the E2E agents need to share at least the meta-ontology for doing the ontological transformation.

Alternatively, the ontology transformation can be carried out by a specialized agent – an ontology-agent. It can store a library of different ontologies and provide the respective agents with a translation when needed. It can also help in solving a conflict by analyzing two partially known ontologies.

6 Conclusions

In this paper we have illustrated strengths of the multi-agent approach to intra-enterprise project-driven production planning and potentials of extending the general technology to the extra-enterprise level. The ExPlanTech multi-agent community has been originally designed so that each agent represents/models a particular manufacturing unit within the respective factory (e.g. department, workshop, CNC machine, warehouse). Agents' interaction simulates possible variations of the course of manufacturing. Such a model facilitates interaction-based production planning and flexible, intelligent replanning once a production unit fails to operate, gets overloaded or simply a high priority order is in conflict with the already planned order.

There is a need to extend enterprises over their boundaries and to interconnect existing planning tools so that resource can be shared and enterprise planning can be improved. As a result of having available real-time, up-to-date information about collaborators' available resources, such a community of interconnected enterprises (extended enterprise) is expected to act more efficiently and flexibly. This configuration of the infrastructure should make the on-demand-product manufacturing shorter and meeting customer specific demands more accurate.

One of the weakest points in the extended enterprise approach comes from different semantic meanings of communication between different members of a *virtual enterprise*. There is a strong demand for a shared, more general manufacturing and business process ontology.

Acknowledgments. This research work has been supported by the European Commission contract No. IST-1999-20171 (ExPlanTech) and by the Ministry of Education, Youth and Sports of the Czech Republic within the frame of the project No. MSM212300013.

References

1. Říha, A., Pěchouček, M., Vokřínek, J., and Mařík, V.: ExPlanTech: Exploitation of Agent-based Technology in Production Planning. In: V. Marik, O. Stepankova, H. Krautwurmova, M. Luck (Eds.): Multi-Agent Systems and Application II, LNAI 2322, Springer Verlag, Heidelberg, 2002
2. Bellifemine, F., Poggi, A., and Rimassa, G.,: Developing Multi-agent Systems with JADE. In: *Seventh International Workshop on Agent Theories, Architectures, and Languages (ATAL-2000)*, Boston, MA, 2000
3. FIPA: Agent Management. *In http://www.fipa.org*, Geneva, Switzerland, 1998
4. Fischer, K., Müller, J. P., Heimig, I., and Scheer, A.W.: Intelligent Agents in Virtual Enterprises. In *Proceedings of the First International Conference on Practical Applications of Intelligent Agents and Multiagents (PAAM'96)*, London, 1996
5. Camarinha-Matos, L.M., Afsarmanesh, H., Garita, C., and Lima, C.: Towards an Architecture for Virtual Enterprises. *Journal of Intelligent Manufacturing*, Vol. 9, No. 2, April 1998
6. Camarinha-Matos, L.M., Afsarmanesh, H., and Mařík,V. (Eds.): *Intelligent Systems for Manufacturing – Multi-Agent Systems and Virtual Organizations*. Kluwer Academic Publishers, Massatchusetts, 1998
7. Camarinha-Matos, L.M., and Afsarmanesh, H.: *Infrastructures for Virtual Enterprises – Networking Industrial Enterprises*. Kluwer Academic Publishers, 1999

Zamin*: An Artificial Ecosystem

Ramin Halavati and Saeed Bagheri Shouraki

Computer Eng. Department,
Sharif University of Technology,
Tehran, Iran
{halavati,sbagheri}@ce.sharif.edu

Abstract. One of the major goals of Artificial Life is to know 'life-as-it-could- be'. Several platforms for ALife simulations are presented but not all are suitable for every cognitive study in ALife. Platforms designed for studying the origin of life are based on a world of simple rules and simple objects, capable of producing complex organisms. But such worlds are not suitable for cognitive studies focused on complex behaviors, because they are too slow for emergence and simulation of such behaviors. A platform called Zamin is presented, which is believed to be fast, expandable, realistic, and suitable for cognitive simulations aimed to study the emergence of complex behaviors.

1 Introduction

One of the major goals of ALife is to know more about life. As Langton states in [7], we can study 'life-as-we-know-it' in biology but 'life-as-it-could-be' in ALife. Several different models for ALife have been proposed such as:

- Cellular Automata based models in which a set of cells and the rules specifying their interactions develop some form of self organization such as [10] and [8].
- Artificial Chemistry models in which the system includes some artificial molecules and rules specifying their interactions such as [3]
- Artificial life models whose subject is a computer program which is evolved in a specific environment such as Tierra [12]
- Artificial Life models which simulate an cosystem such as ERL [1], ECHO [4], [5] and LEE [9].

The main goal in the three first model categories is usually to study the evolution of self organization and organisms with the ability of self-replication. Thus, all such systems include simple, primitive rules to show how a self-organized creature can be evolved inside a model with simple primary rules and limits.

Apart from the fact that these models satisfy what they are designed for, i.e. to develop self-replicative organisms, having the simplest possible rules and minimum possible limits, they construct some sort of open-ended evolution [12].

* Zamin stands for 'Zoological Agents for Modification and Improvement of Neocreatures'. Zamin also means 'Earth' in Persian.

M.H. Shafazand and A M. Tjoa (Eds.): EurAsia-ICT 2002, LNCS 2510, pp. 1008–1016, 2002.

But due to their low-level rules, they have the negative side effect of a very long time necessary for the evolution of creatures with complex behavior. This is because any complex behavior in such systems must be comprised of many small behaviors and must execute many of the system's simple rules. (For example, a creature in an artificial chemistry model may need to execute several thousand of the system's general propose rules to transmit a meaningful message to another creature whilst an organism in a system with a primitive communication command needs to execute just one.) Therefore, not only the evolution of an organism able to command such a long sequence of actions takes a long time, but also each execution of such long-sequence commands would consume so much CPU time of the computer running them and hence, the simulation would become too slow.

On the other hand,systems such as ECHO, ERL and LEE simulate high-level creatures which exhibit several complex behaviors such as reproduction, attacking, defending, trading, hiding, etc. These systems may not show the emergence of low level phenomena and are not as open-ended as the previous group, but due to the previously stated problems of those models, they are the better choice for a cognitive researcher in ALife whose subject of study is advanced behaviors such as group working, communication, etc.

The target of project Zamin is to provide a suitable platform for cognitive ALife so that it would be both fast (where the evolution of complex behaviors can be seen in acceptable time) and realistic (so that the emergence of behaviors already seen in real organisms, can be traced in this system).

2 The Base Model

As stated in previous section, the design of Zamin project could be either based on a primitive model and at-most, accompany it with some sort of coevolution of the system rules to resolve the previously stated speed problems, (in this case, the system rules get more complicated while the organisms evolve and get more complex) or start from a high-level structure and a bigger assumption about the beginning of the artificial world.

To go on with the first choice, there must be some sort of (most preferably automatic) tools to identify the behaviors already evolved in the system, and replace them with fast, primitive system commands. For example, whenever organisms with self-replication phenomenon have emerged, the system must automatically replace them with organisms using a simple system command for replication, instead of operating long sequences of actions, formerly required to reproduce. This task has two serious problems: First, the design of behavior identifier tool, aimed to discover the very complicated patterns of organisms' behaviors, and second, the replacement tool, which must apply the command replacement, so that it would not alter the organisms' external behaviors while changing the internal commands.

Instead of the above hard-to-achieve solution, as the project's target is to study the evolution of complex behavior and not the organism's structure, a high level structure can be assumed from the beginning and just use the evolution to manage the operation of this structure. For example, being able to reproduce, eat, move around, communicate, etc.can be assumed from the beginning, but the control parameters of these functions can be left to organisms' genetic code. The advantages of this approach against the former are:

- The experiment's time is not wasted by reaching to phenomena whose possibility of emergence by evolution is previously proved.
- The system is speeded up by assuming high level command set so that the organisms need not follow up long sequences of commands to show complex behaviors.
- The organisms become highly expressible,easier to understand,and can be more similar to the natural organisms we currently know.

Among existing high-level alife platforms, ERL, ECHO and LEE seemed to be the most promising ones. As ECHO and LEE assume a much more abstract world than ERL and one of project's targets is to create a platform suitable for evolution of organisms similar to real ones, ERL was chosen as the base platform.

2.1 ERL

The *'Evolutionary Reinforcement Learning'* model, presented in [1], consists of a simple lattice which is the world of four different types of creatures. The first type, which are called the agents, are the main subject of study. An agent can move to any of the four adjacent cells of the cell it is currently in, it can attack to and weaken another organism in an adjacent cell, eat a dead body or a plant in its own cell and get stronger, climb the trees that exist in the lattice and hide above them and finally, reproduce asexually if it is healthier than a certain limit. The second type are the predators, whose capabilities are almost the same as agents, except for they can not eat plants, they don't reproduce (they are created in certain time intervals) and their behavior is hard-coded. The third objects are the trees which are just used by the agents to hide on and the forth are the plants which are the green energy resource of the agents.

Each agent has an internal energy level which is gradually exploited during the life time of the agent and must be kept above a certain limit to prevent death. This is accomplished by eating the dead bodies of the other agents and the plants existing in the lattice. Once an agent's energy gets above a certain limit, it can choose to reproduce by cloning its genome. Each agent possesses two feed forward neural networks as its brain. The weights of these networks are initiated according to the genetic code which the agent receives from its parent. The first network is used to produce the behavior and is called Action Network. The second network is to generate a measure of how good the agent's current state is and is called Evaluation Network. The result of the evaluation network is used to adjust the weights of the action network. The life cycle of an agent is as follows:

a. Each agent is born inside a cell of the lattice with some specific energy level. Its brain networks weights are initialized based of the genetic code received from its parent.
b. The contents of the cells in a 4 cell radius on the 4 basic directions of agent's location are given as input to its action-network. Thus, the action net has 16 inputs, each stating what type of object (Agent/Predator/Tree/Plant) is in each of the 16 cells and how much its energy level is, if it is one of the first two types.
c. The action network chooses to reproduce or to move to one of the 4 adjacent cells. If the choice is to reproduce and agent's energy level is above a certain limit, the organism's genome is cloned and a new organism is created. But if a movement choice is selected, the organism moves to the selected cell and an action appropriate to the content of that cell is applied (e.g. if it has chosen to move into a cell with a tree,it will climb it, or if it has chosen to move into a cell with a predator, it will attack the predator, etc.).
d. The same input as the one given to action-network is given to evaluation network and its result is used as the measure for a back-propagation algorithm which adjusts the weights of the action network.
e. If the agent's energy level is above zero, the process proceeds with step *b* after steps *b* to *e* are applied for all other agents, otherwise, it is killed and its body remains as food for the remaining agents and predators.

2.2 Weaknesses of ERL for Project Zamin

a. The agents' external sensors are too simple and they can only understand the type and a very simple qualitative factor about the organisms around them. This very simple input pattern prevents the organisms from making decisions based on reach information about outside world.
b. The agents have no internal-sensors and can't make decisions due to their own internal state while real organisms make many of their decisions that way.
c. The agents have no memory except their network weights. Thus,they can't have internal state and in result, they can not plan for long time behavior. In this case, a behavior needing a sequence of actions must be redecided at each step and therefore it is hard to evolve.
d. All agents have the same body. This lack of variety would probably lead to evolution of organisms which all have the same behaviors. But it is known that competitions have a major effect on the evolution speed [2]. This problem is resolved in ERL using the predators. But as the predators are hard-coded and can not enter the arm race,they can not be dynamic forces for evolution.
e. The agents' behavioral choices are too few. Due to this and the lack of states, the evolution of complex behaviors is not so much promising.

3 The Zamin Model

As stated in the previous section, ERL is a good start point to design a platform suitable for the emergence of complex behaviors, but it still has some flaws. This section deals with some ways to work out these weaknesses. The first part explains two important points that must be considered throughout the design of such systems and the second part deals with the add ups to ERL.

3.1 Design Considerations

System Speed. As such models have their root in evolution and interaction of the artificial creatures, they need to simulate thousands of creatures in the artificial environment at the same time and also pass several thousands generations, searching through evolutionary time-space. Thus,the system must be fast enough so that each experiment can be completed in a reasonable time. It is more necessary to pay attention to this fact when we are now repeatedly warned about getting closer and closer to our technological boundaries in making faster and smaller computers.

The size of Genetic Space. As ALife models like ECHO, ERL, LEE and Zamin start the evolution of their complex organisms with random initial genomes to avoid biasing, the size and complexity of genetic space becomes an important factor for the success of the model: The more complex the organisms are, the bigger the genome must be. Thus,as the genetic space grows exponentially with the increment of genomes size and complexity, the chance of starting from random points in this space and reaching to successful organism decreases.

As the conclusion of the above paragraph, an Alife simulation of this family must:

1. Try to keep the genetic space as small and simple as possible.
2. When ever possible, start the evolution with simpler structures and transform the organisms (and/or the environment) to more complex ones when the simple ones have show successful results. For example, a model like ours can start with a simple pattern-memorize-and-remember brain at the beginning of the simulation (a brain with fewer and simpler genes), and replace it by a neural network with the same initial functionality when the organisms have found successful behaviors.

3.2 Differences between Zamin and ERL

External Sensors. To overcome the problem stated in 2.2.a, there are two ways to extend the organisms external sensors. The first choice is to add the same sensory inputs as that of the real organisms: 3 separate sensors namely seeing,hearing and smelling and the body which is sensed by any of these three sensors is created by the cooperation of the entire genome of the creature being sensed. This can be modeled by having complex formulas,which map the entire (or some parts) of the visited organism's genome to the sensory input of the visitor. Although this modeling seems quite similar to real organisms' input preceptors, it has a major problem for Alife modeling. As the inputs are complex patterns created from the entire genome of the organisms, they need to be decoded in the brain, to be understood which in turn needs a complex brain and input processing system. Considering the point stated in 3.1.2, it is clear that this method does not have a good chance for being a successful starting point.[1]

Another choice is to use *Sensor Processing* which is now well accepted in control systems (such as [6]): Instead of forcing the controller make all decisions, have more complex sensors which process what they sense and give the controller richer data. Likewise, as the subject of this project is to investigate the evolution of complex behavior and not the structure, a preprocessing unit can be assumed for the organisms' brain, some thing whose possibility of evolution is well proved (e.g. [13]). In this modeling,the advanced preprocessing unit gives qualitative information about every phenomenon of the visited organism to the visitor (such as how far it is, how strong it is, is it looking at this direction,how old is it, is it a carnivore, etc.) The genome's role for this unit is just to specify which of these inputs are given to each part of the brain and what the precision of their datum is.

The advantages of this method over the first are:

- As the input data are meaningful,organisms need a much simpler brain.
- As the inputs are meaningful to us, further analysis of their brain is easier.

Internal Sensors. To overcome the weakness stated in 2.2.b, several internal sensors were added to organisms to help them make better decisions such as the organism's energy-level which is a measure of its health,the net change of energy level due to the last performed action and several flags indicating what the organism has done recently.

Memory. As stated in 2.2.c, the previous model assumed no memory for its agents. A gene is added to Zamin's organisms, stating how many bits of memory each organism has. These memory bits can be used as input for action and evaluation networks and are set by the action network (the action-network has some extra outputs, not used in action selection, which are fed into memory).

Physical Differences. To make different creatures (2.2.d), some genes specifying the physical capabilities of the organisms are added such as a measure of how good they can digest herbs or flesh, how fast they use their energy and how fast they can move.

[1] We have made a simulation of this method to verify the guess. See section 4 for results.

Behaviors. To partly overcome the 2.2.e problem (this problem must be more worked out in further expansions of the system), the action selection system is changed so that the organism is no more forced to choose an action regarding the cell it enters. Therefore, the organism must implicitly choose to eat a plant, if there is one in its cell or bite another organism,if it faces one.

Genetic Structure. *Mutation Controllers:* In real organisms, the mutation frequency is not the same for all genes and some genes are much more stable trough time. This stability is sometimes used as a measure of importance of the gene for the life of the creature (see [2]). Each group of genes in the genome of our organisms has a mutation- controller factor which states the probability of mutation for this group,in compare with the other genes. These mutation-controllers are subject of evolution as well and therefore,change through time.

Processing Unit. The last difference of our organisms with the previous model agents, which took place due to items 3.1.1 and 3.1.2 was about their brain. The previous model used two neural networks with backward propagation learning algorithm. This approach is both too slow and too complicated to start with a blind selection of genome.[2] Thus we chose a much simpler method using case based learning. The advantages of the case based reasoning over neural networks in this project are as follows:

- Learning and remembering is much faster in this model.
- This structure can learn simple patterns very fast (as it just stores them) and this is too much worthy for organisms with short life-times (what that is expected at least for the beginning of evolution).
- This structure needs a match smaller genetic code in compare with neural networks.
- This structure is easily upgradable to a Fuzzy controller and Fuzzy controllers have shown a great success in machine learning in recent years.

4 Implementation Results

We made two implementations of the above model: One using the complex external sensors and neural network brain to verify the assertions made in 3.2.1 and 3.2.7 and the second with the advanced input preprocessor and case based reasoning. Table 1 summarizes their comparison.

As a sample of Zamin results, figure 1-A shows the gradual change of the gene which shows the ability of digesting herbs or flesh in two tribes, one evolving to become carnivore and the other herbivore. As it is seen in both tribes the gene has converged to a limiting value which maximizes the organism's benefit. As another sample, in figure 1-B,we can see the total number of plants ate in each time units in two carnivore and herbivore tribes.

[2] We have also made a simulation with neural networks. See brief results in section 4.

Table 1. Brief comparison of the two implementations.

	Neural Net + Complex Sensors	Case Based Learning + Advanced Preprocessors
Speed		**60% faster**
Average Organisms Age	**25 system time unit**	**170 system time unit**
Maximum Organism Age	**50 system time unit**	**1800 system time unit**
Gene change	**Chaotic / None**	**Smooth and Purposeful**

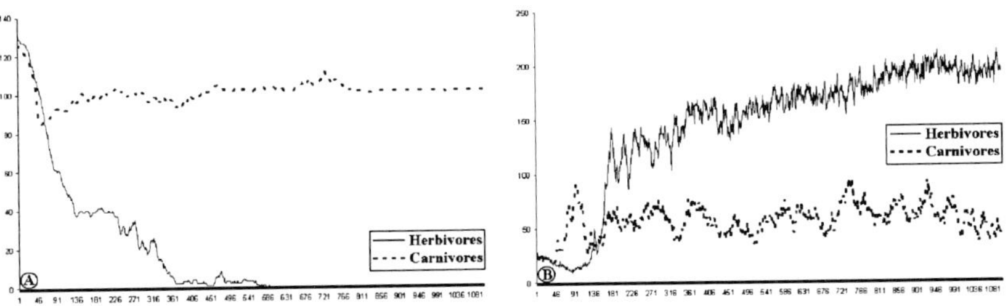

Fig. 1. The evolution of carnivore and herbivore tribes.

5 Conclusion and Further Works

The size of the organisms' genetic state-space and processing speed of the computer running the simulation are two very important factors in success of an artificial life model designed to study the emergence of advanced behaviors such as communication, group working, knowledge acquisition and using tools. These two limitations make a serious distinction in design of systems aim d to study such behaviors from systems designed to study emergence of life and self-organization. The former needs simplicity but the latter, speed.

Zamin is an artificial ecosystem based on the ERL model proposed in [1]. ERL agents had very limited input data and their brain model consisted of two neural networks with back-propagation learning algorithm which results in slow speed in simulations with a large population and many generations. Zamin is designed to have organisms with much more input data and faster brain besides keeping the genetic state-space small enough, so that a random initiation of the world would have the chance of successful life for its inhabitants.

Zamin simulations show that it has meaningful and rational behavior, so it is realistic; it reaches to organisms with long life time and almost complicated living patterns, so its successful in keeping the genetic-space appropriate for a random unbiased start; its simulations reach to acceptable results in some hours on regular PCs, so it is fast; and it is quit expandable, appropriate for cognitive studies. Wc are now increasing Zamin's environmental complexity and organisms' action choices to increase their behavioral complexity and replacing their brains with a three layer fuzzy rule set to investigate the emergence of meaning and communication.

Acknowledgements. We must thank Dr. Mohammad Reza Kangavari, Shervin Vakili and Yasamin Mokri for their helpful ideas and support through out the project.

References

1. D.H. Ackley, M. Littman. (1992) "Interactions Between Learning and Evolution." In C. G. Langton, C. Taylor, J.D. Farmer, and S. Rasmussen, editors, Artificial Life II, Volume X of SFI Studies in the Sciences of Complexity, pages 487–507, Addison-Wesley, Redwood City, CA, 1992.
2. R. Dawkins (1997) "The Blind Watchmaker", W W. Norton & Company Inc.
3. P. Dittrich, J. Ziegler, W. Banzhaf. (2000) "Artificial Chemistries, A Review" http://ls11-www.cs.uni-dortmund.de
4. S. Forrest, P.T. Hraber, T. Jones. 1996 "The Ecology of ECHO"
5. J.H. Holland. (1975) "Adaptation in Natural and Artificial Systems" MIT Press. Cambridge, MA 1992. Second Edition (First Edition, 1975).
6. T.Kanade, R.T. Collins, A.J. Lipton, P. Burt & R. Wixson (1998) "Advances in Cooperative Multi-Sensor Video Surveillance"
7. C.G. Langton. (1992) "Preface." In C.G. Langton, C. Taylor, J.D. Farmer, and S. Rasmussen, editors, Artificial Life II, Volume X of SFI Studies in the Sciences of Complexity, pages xiii-xviii, Addison-Wesley, Redwood City, CA, 1992.
8. C.G. Langton. (1992) "Life at the Edge of Chaos." In C.G. Langton, C. Taylor, J.D. Farmer, and S. Rasmussen, editors, Artificial Life II, Volume X of SFI Studies in the Sciences of Complexity, pages 41–91, Addison-Wesley, Redwood City, CA, 1992.
9. F. Menczer, R.K. Belew. (1993) "Latent Energy Environments: A Tool for Artificial Life Simulations".
10. J. von Neumann. (1966) "Theory of Self-Reproducing Automata". University of Illinois Press, Illinois, 1966. Edited and completed by A.W. Burks.
11. S. Rose. (1976) "The Conscious Brain", Random House Incorporated.
12. T.S. Ray. (1992) "An approach to the Synthesis of Life." In C.G. Langton, C. Taylor, J. D. Farmer, and S. Rasmussen, editors, Artificial Life II, Volume X of SFI Studies in the Sciences of Complexity, pages 371–401, Addison-Wesley, Redwood City, CA, 1992.
13. C. Slocum, D.C. Downey, R.D. Beer. (2000) "Further Experiments in the Evolution of Minimally Cognitive Behavior"

Author Index

Lecture Notes in Computer Science

For information about Vols. 1–1234

please contact your bookseller or Springer-Verlag

Vol. 2344: I. Chong (Ed.), Information Networking. Proceedings, Part II, 2002. XX, 825 pages. 2002.

Vol. 2426: J.-M. Bruel, Z. Bellahsène (Eds.), Advances in Object-Oriented Information Systems. Proceedings, 2002. IX, 314 pages. 2002.

Vol. 2427: M. Hannebauer, Autonomous Dynamic Reconfiguration in Multi-Agent Systems. XXI, 284 pages. 2002. (Subseries LNAI).

Vol. 2428: H. Hermanns, Interactive Markov Chains. XII, 217 pages. 2002.

Vol. 2499: P. Druschel, F. Kaashoek, A. Rowstron (Eds.), Peer-to-Peer Systems. Proceedings, 2002. IX, 339 pages. 2002.

Vol. 2430: T. Elomaa, H. Mannila, H. Toivonen (Eds.), Machine Learning: ECML 2002. Proceedings, 2002. XIII, 532 pages. 2002. (Subseries LNAI).

Vol. 2431: T. Elomaa, H. Mannila, H. Toivonen (Eds.), Principles of Data Mining and Knowledge Discovery. Proceedings, 2002. XIV, 514 pages. 2002. (Subseries LNAI).

Vol. 2432: R. Bergmann, Experience Management. XXI, 393 pages. 2002. (Subseries LNAI).

Vol. 2433: A.H. Chan, V. Gligor (Eds.), Information Security. Proceedings, 2002. XII, 502 pages. 2002.

Vol. 2434: S. Anderson, S. Bologna, M. Felici (Eds.), Computer Safety, Reliability and Security. Proceedings, 2002. XX, 347 pages. 2002.

Vol. 2435: Y. Manolopoulos, P. Návrat (Eds.), Advances in Databases and Information Systems. Proceedings, 2002. XIII, 415 pages. 2002.

Vol. 2436: J. Fong, C.T. Cheung, H.V. Leong, Q. Li (Eds.), Advances in Web-Based Learning. Proceedings, 2002. XIII, 434 pages. 2002.

Vol. 2437: G. Davida, Y. Frankel, O. Rees (Eds.), Infrastructure Security. Proceedings, 2002. XI, 339 pages. 2002.

Vol. 2438: M. Glesner, P. Zipf, M. Renovell (Eds.), Field-Programmable Logic and Applications. Proceedings, 2002. XXII, 1187 pages. 2002.

Vol. 2439: J.J. Merelo Guervós, P. Adamidis, H.-G. Beyer, J.-L. Fernández-Villacañas, H.-P. Schwefel (Eds.), Parallel Problem Solving from Nature – PPSN VII. Proceedings, 2002. XXII, 947 pages. 2002.

Vol. 2440: J.M. Haake, J.A. Pino (Eds.), Groupware: Design, Implementation and Use. Proceedings, 2002. XII, 285 pages. 2002.

Vol. 2441: Z. Hu, M. Rodríguez-Artalejo (Eds.), Functional and Logic Programming. Proceedings, 2002. X, 305 pages. 2002.

Vol. 2442: M. Yung (Ed.), Advances in Cryptology – CRYPTO 2002. Proceedings, 2002. XIV, 627 pages. 2002.

Vol. 2443: D. Scott (Ed.), Artificial Intelligence: Methodology, Systems, and Applications. Proceedings, 2002. X, 279 pages. 2002. (Subseries LNAI).

Vol. 2444: A. Buchmann, F. Casati, L. Fiege, M.-C. Hsu, M.-C. Shan (Eds.), Technologies for E-Services. Proceedings, 2002. X, 171 pages. 2002.

Vol. 2445: C. Anagnostopoulou, M. Ferrand, A. Smaill (Eds.), Music and Artificial Intelligence. Proceedings, 2002. VIII, 207 pages. 2002. (Subseries LNAI).

Vol. 2446: M. Klusch, S. Ossowski, O. Shehory (Eds.), Cooperative Information Agents VI. Proceedings, 2002. XI, 321 pages. 2002. (Subseries LNAI).

Vol. 2447: D.J. Hand, N.M. Adams, R.J. Bolton (Eds.), Pattern Detection and Discovery. Proceedings, 2002. XII, 227 pages. 2002. (Subseries LNAI).

Vol. 2448: P. Sojka, I. Kopecÿek, K. Pala (Eds.), Text, Speech and Dialogue. Proceedings, 2002. XII, 481 pages. 2002. (Subseries LNAI).

Vol. 2449: L. Van Gool (Ed.), Pattern Recognotion. Proceedings, 2002. XVI, 628 pages. 2002.

Vol. 2451: B. Hochet, A.J. Acosta, M.J. Bellido (Eds.), Integrated Circuit Design. Proceedings, 2002. XVI, 496 pages. 2002.

Vol. 2452: R. Guigó, D. Gusfield (Eds.), Algorithms in Bioinformatics. Proceedings, 2002. X, 554 pages. 2002.

Vol. 2453: A. Hameurlain, R. Cicchetti, R. Traunmüller (Eds.), Database and Expert Systems Applications. Proceedings, 2002. XVIII, 951 pages. 2002.

Vol. 2454: Y. Kambayashi, W. Winiwarter, M. Arikawa (Eds.), Data Warehousing and Knowledge Discovery. Proceedings, 2002. XIII, 339 pages. 2002.

Vol. 2455: K. Bauknecht, A M. Tjoa, G. Quirchmayr (Eds.), E-Commerce and Web Technologies. Proceedings, 2002. XIV, 414 pages. 2002.

Vol. 2456: R. Traunmüller, K. Lenk (Eds.), Electronic Government. Proceedings, 2002. XIII, 486 pages. 2002.

Vol. 2457: T. Yakhno (Ed.), Advances in Information Systems. Proceedings, 2002. XII, 436 pages. 2002.

Vol. 2458: M. Agosti, C. Thanos (Eds.), Research and Advanced Technology for Digital Libraries. Proceedings, 2002. XVI, 664 pages. 2002.

Vol. 2459: M.C. Calzarossa, S. Tucci (Eds.), Performance Evaluation of Complex Systems: Techniques and Tools. Proceedings, 2002. VIII, 501 pages. 2002.

Vol. 2460: J.-M. Jézéquel, H. Hussmann, S. Cook (Eds.), ÇUMLÈ 2002 – The Unified Modeling Language. Proceedings, 2002. XII, 449 pages. 2002.

Vol. 2461: R. Möhring, R. Raman (Eds.), Algorithms – ESA 2002. Proceedings, 2002. XIV, 917 pages. 2002.

Vol. 2462: K. Jansen, S. Leonardi, V. Vazirani (Eds.), Approximation Algorithms for Combinatorial Optimization. Proceedings, 2002. VIII, 271 pages. 2002.

Vol. 2463: M. Dorigo, G. Di Caro, M. Sampels (Eds.), Ant Algorithms. Proceedings, 2002. XIII, 305 pages. 2002.

Vol. 2464: M. O'Neill, R.F.E. Sutcliffe, C. Ryan, M. Eaton, N. Griffith (Eds.), Artificial Intelligence and Cognitive Science. Proceedings, 2002. XI, 247 pages. 2002. (Subseries LNAI).

Vol. 2465: H. Arisawa, Y. Kambayashi, V. Kumar, H.C. Mayr, I. Hunt (Eds.), Conceptual Modeling for New Information Systems Technologies. Proceedings, 2001. XVII, 500 pages. 2002.

Vol. 2467: B. Christianson, B. Crispo, J.A. Malcolm, M. Roe (Eds.), Security Protocols. Proceedings, 2001. IX, 241 pages. 2002.

Vol. 2469: W. Damm, E.-R. Olderog (Eds.), Formal Techniques in Real-Time and Fault-Tolerant Systems. Proceedings, 2002. X, 455 pages. 2002.

Vol. 2470: P. Van Hentenryck (Ed.), Principles and Practice of Constraint Programming – CP 2002. Proceedings, 2002. XVI, 794 pages. 2002.

Vol. 2471: J. Bradfield (Ed.), Computer Science Logic. Proceedings, 2002. XII, 613 pages. 2002.

Vol. 2473: A. Gomez-Perez, V.R. Benjamins, Knowledge Engineering and Knowledge Management. Proceedings, 2002. XI, 402 pages. 2002. (Subseries LNAI).

Vol. 2474: D. Kranzlmüller, P. Kacsuk, J. Dongarra, J. Volkert (Eds.), Recent Advances in Parallel Virtual Machine and Message Passing Interface. Proceedings, 2002. XVI, 462 pages. 2002.

Vol. 2475: J.J. Alpigini, J.F. Peters, A. Skowron, N. Zhong (Eds.), Rough Sets and Current Trends in Computing. Proceedings, 2002. XV, 640 pages. 2002. (Subseries LNAI).

Vol. 2476: A.H.F. Laender, A.L. Oliveira (Eds.), String Processing and Information Retrieval. Proceedings, 2002. XI, 337 pages. 2002.

Vol. 2477: M.V. Hermenegildo, G. Puebla (Eds.), Static Analysis. Proceedings, 2002. XI, 527 pages. 2002.

Vol. 2478: M.J. Egenhofer, D.M. Mark (Eds.), Geographic Information Science. Proceedings, 2002. X, 363 pages. 2002.

Vol. 2479: M. Jarke, J. Koehler, G. Lakemeyer (Eds.), KI 2002: Advances in Artificial Intelligence. Proceedings, 2002. XIII, 327 pages. (Subseries LNAI).

Vol. 2480: Y. Han, S. Tai, D. Wikarski (Eds.), Engineering and Deployment of Cooperative Information Systems. Proceedings, 2002. XIII, 564 pages. 2002.

Vol. 2483: J.D.P. Rolim, S. Vadhan (Eds.), Randomization and Approximation Techniques in Computer Science. Proceedings, 2002. VIII, 275 pages. 2002.

Vol. 2484: P. Adriaans, H. Fernau, M. van Zaanen (Eds.), Grammatical Inference: Algorithms and Applications. Proceedings, 2002. IX, 315 pages. 2002. (Subseries LNAI).

Vol. 2486: M. Marinaro, R. Tagliaferri (Eds.), Neural Nets. Proceedings, 2002. IX, 253 pages. 2002.

Vol. 2487: D. Batory, C. Consel, W. Taha (Eds.), Generative Programming and Component Engineering. Proceedings, 2002. VIII, 335 pages. 2002.

Vol. 2488: T. Dohi, R. Kikinis (Eds), Medical Image Computing and Computer-Assisted Intervention – MICCAI 2002. Proceedings, Part I. XXIX, 807 pages. 2002.

Vol. 2489: T. Dohi, R. Kikinis (Eds), Medical Image Computing and Computer-Assisted Intervention – MICCAI 2002. Proceedings, Part II. XXIX, 693 pages. 2002.

Vol. 2491: A. Sangiovanni-Vincentelli, J. Sifakis (Eds.), Embedded Software. Proceedings, 2002. IX, 423 pages. 2002.

Vol. 2493: S. Bandini, B. Chopard, M. Tomassini (Eds.), Cellular Automata. Proceedings, 2002. XI, 369 pages. 2002.

Vol. 2495: C. George, H. Miao (Eds.), Formal Methods and Software Engineering. Proceedings, 2002. XI, 626 pages. 2002.

Vol. 2496: K.C. Almeroth, M. Hasan (Eds.), Management of Multimedia in the Internet. Proceedings, 2002. XI, 355 pages. 2002.

Vol. 2498: G. Borriello, L.E. Holmquist (Eds.), UbiComp 2002: Ubiquitous Computing. Proceedings, 2002. XV, 380 pages. 2002.

Vol. 2499: S.D. Richardson (Ed.), Machine Translation: From Research to Real Users. Proceedings, 2002. XXI, 254 pages. 2002. (Subseries LNAI).

Vol. 2502: D. Gollmann, G. Karjoth, M. Waidner (Eds.), Computer Security – ESORICS 2002. Proceedings, 2002. X, 281 pages. 2002.

Vol. 2503: S. Spaccapietra, S.T. March, Y. Kambayashi (Eds.), Conceptual Modeling – ER 2002. Proceedings, 2002. XX, 480 pages. 2002.

Vol. 2504: M.T. Escrig, F. Toledo, E. Golobardes (Eds.), Topics in Artificial Intelligence. Proceedings 2002. XI, 432 pages. 2002. (Subseries LNAI).

Vol. 2505: A. Corradini, H. Ehrig, H.-J. Kreowski, G. Rozenberg (Eds.), Graph Transformations. Proceedings, 2002. IX, 459 pages. 2002.

Vol. 2508: D. Malkhi (Ed.), Distributed Computing. Proceedings, 2002. X, 371 pages. 2002.

Vol. 2509: C.S. Calude, M.J. Dinneen, F. Peper (Eds.), Unconventional Models in Computation. Proceedings, 2002. VIII, 331 pages. 2002.

Vol. 2510: H. Shafazand, A Min Tjoa (Eds.), EurAsia-ICT 2002: Information and Communication Technology. Proceedings, 2002. XXIII, 1020 pages. 2002.

Vol. 2511: B. Stiller, M. Smirnow, M. Karsten, P. Reichl (Eds.), From QoS Provisioning to QoS Charging. Proceedings, 2002. XIV, 348 pages. 2002.

Vol. 2514: M. Baaz, A. Voronkov (Eds.), Logic for Programming, Artificial Intelligence, and Reasoning. Proceedings 2002. XIII, 465 pages. 2002. (Subseries LNAI).

Vol. 2516: A. Wespi, G. Vigna, L. Deri (Eds.), Recent Advances in Intrusion Detection. Proceedings, 2002. X, 327 pages. 2002.

Vol. 2521: A. Karmouch, T. Magedanz, J. Delgado (Eds.), Mobile Agents for Telecommunication Applications. Proceedings 2002. XII, 317 pages. 2002.

Vol. 2526: A. Colosimo, A. Giuliani, P. Sirabella (Eds.), Medical Data Analysis. Proceedings 2002. IX, 222 pages. 2002.